Introduction to Ceramics

Second Edition

W. D. Kingery

PROFESSOR OF CERAMICS
MASSACHUSETTS INSTITUTE OF TECHNOLOGY

H. K. Bowen

ASSOCIATE PROFESSOR OF CERAMICS
MASSACHUSETTS INSTITUTE OF TECHNOLOGY

D. R. Uhlmann

PROFESSOR OF CERAMICS AND POLYMERS
MASSACHUSETTS INSTITUTE OF TECHNOLOGY

A Wiley–Interscience Publication

JOHN WILEY & SONS

New York • Chichester • Brisbane • Toronto • Singapore

Library of Congress Cataloging in Publication Data:

Kingery, W. D.
Introduction to ceramics.

(Wiley series on the science and technology of materials)
"A Wiley–Interscience publication."
Includes bibliographical references and index.
1. Ceramics. I. Bowen, Harvey Kent, joint author. II. Uhlmann, Donald Robert, joint author. III. Title.

TP807.K52 1975 666 75-22248
ISBN 0-471-47860-1

Printed and bound by Courier Companies, Inc.

22 23 24 25 26 27 28 29 30

Preface to the Second Edition

During the fifteen years which have passed since the first edition was published, the approach described has been widely accepted and practiced. However, the advances made in understanding and controlling and developing new ceramic processes and products have required substantial modifications in the text and the introduction of a considerable amount of new material.

In particular, new and deeper understanding of the structure of noncrystalline solids and the characteristics of structural imperfections, new insight into the nature of surfaces and interfaces, recognition of spinodal decomposition as a viable alternative to classical nucleation, recognition of the widespread occurrence of phase separation, development of glass-ceramics, clearer understanding of some of the nuances of sintering phenomena, development of scanning electron microscope and transmission electron microscope techniques for the observation of microstructure, a better understanding of fracture and thermal stresses, and a myriad of developments relative to electrical, dielectric, and magnetic ceramics have been included. The breadth and importance of these advances has made a single author text beyond any individual's competence.

The necessary expansion of material related to physical ceramics, and the recent availability of excellent texts aimed at processing and manufacturing methods [F. N. Norton, *Fine Ceramics*, McGraw-Hill, New York (1970); F. H. Norton, *Refractories*, McGraw-Hill, New York (1961); F. H. Norton, *Elements of Ceramics*, second ed., Addison Wesley Publ. Co. (1974); F. V. Tooley, ed., *Handbook of Glass Manufacture*, 2 Vols., Ogden Publ. Co. (1961); A. Davidson, ed., *Handbook of Precision

Engineering, Vol. 3, Fabrication of Non-Metals, McGraw-Hill Publ. Co. (1971); *Fabrication Science*, Proc. Brit. Ceram. Soc., No. 3 (1965); *Fabrication Science, 2*, Proc. Brit. Ceram. Soc., No. 12 (1969); Institute of Ceramics Textbook Series: W. E. Worrall, *1: Raw Materials*; F. Moore, *2: Rheology of Ceramic Systems*; R. W. Ford, *3: Drying*; W. F. Ford, *4: The Effect of Heat on Ceramics*, Maclaren & Sons, London (1964–1967), *Modern Glass Practice*, S. R. Scholes, rev. C. H. Green, Cahners (1974)] has led us to eliminate most of the first edition's treatment of these subjects. We regret that there is still not available a single comprehensive text on ceramic fabrication methods.

While we believe that structure on the atomic level and on the level of simple assemblages of phases has developed to a point where lack of clarity must be ascribed to the authors, there remain areas of great interest and concern that have not seen the development of appropriate and useful paradigms. One of these, perhaps the most important, is related to the interaction of lattice imperfections and impurities with dislocations, surfaces and grain boundaries in oxide systems. Another is related to ordering, clustering, and the stability of ceramic solid solutions and glasses. A third is methods of characterizing and dealing with the more complex structures found for multi-phase multi-component systems not effectively evaluated in terms of simple models. Many other areas at which the frontier is open are noted in the text. It is our hope that this book will be of some help, not only in applying present knowledge, but also in encouraging the further extension of our present understanding.

Finally, the senior author, Dr. Kingery, would like to acknowledge the long term support of Ceramic Science research at M.I.T. by the Division of Physical Research of the U. S. Atomic Energy Commission, now the Energy Research and Development Agency. Without that support, this book, and its influence on ceramic science, would not have developed.

We also gratefully acknowledge the help of our many colleagues, especially, R. L. Coble, I. B. Cutler, B. J. Wuensch, A. M. Alper, and R. M. Cannon.

W. D. KINGERY
H. K. BOWEN
D. R. UHLMANN

Cambridge, Massachusetts
June 1975

Contents

III DEVELOPMENT OF MICROSTRUCTURE IN CERAMICS 265

7 Ceramic Phase-Equilibrium Diagrams 269

8 Phase Transformation, Glass Formation, and Glass-Ceramics 320

9 Reactions with and between Solids 381

Introduction to Ceramics

part I

INTRODUCTION

This book is primarily concerned with understanding the development, use, and control of the properties of ceramics from the point of view of what has become known as *physical ceramics.*

Until a decade or so ago, ceramics was in large part an empirical art. Users of ceramics procured their material from one supplier and one particular plant of a supplier in order to maintain uniformity (some still do). Ceramic producers were reluctant to change any detail of their processing and manufacturing (some still are). The reason was that the complex systems being used were not sufficiently well known to allow the effects of changes to be predicted or understood, and to a considerable extent this remains true. However, the fractional part of undirected empiricism in ceramic technology has greatly diminished.

Analysis of ceramics shows them to be a mixture of crystalline phases and glasses, each of many different compositions, usually combined with porosity, in a wide variety of proportions and arrangements. Experience has shown that focusing our attention on the structure of this assemblage in the broadest sense, from the viewpoint of both the origin of the structure and its influence on properties, is a powerful and effective approach. This concentration on the origin of structure and its influence on properties is the central concept of physical ceramics.

To be fully fruitful, structure must be understood in its most comprehensive sense. On one hand, we are concerned with atomic structure—the energy levels in atoms and ions that are so important in understanding the formation of compounds, the colors in glazes, the optical properties of lasers, electrical conductivity, magnetic effects, and a host of other characteristics of useful ceramics. Equally important is the way in which atoms or ions are arranged in crystalline solids and in noncrystalline glasses, from the point of view of not only lattices and ideal structures but also the randomness or ordering of the atoms and lattice defects such as vacant sites, interstitial atoms, and solid solutions. Properties such as heat conduction, optical properties, diffusion, mechanical deformation, cleavage, and dielectric and magnetic properties are influenced by these considerations. Departures from crystalline perfection at line imperfections called dislocations and at surfaces and interfaces also have a critical influence on many, perhaps most, properties of real systems.

At another level, the arrangement of phases—crystalline, glass, porosity—is often controlling, as is the nature of the boundaries between phases. A tenth of a percent porosity changes a ceramic from transparent to translucent. A change in pore morphology changes a ceramic from gastight to permeable. A decrease in grain size may change a ceramic from weak and friable to strong and tough. One refractory with crystal–crystal bonds and a large glass content withstands deformation at high temperature; another with a much smaller glass content penetrating between grains deforms readily. Changing the arrangement of phases can change an insulator into a conductor, and vice versa. The separation of a glass into two phases by appropriate heat treatment can dramatically alter many of its properties and increase or decrease its usefulness.

These observations are of academic interest, but even more they provide a key to the successful preparation and use of real ceramics. This approach provides us with the basis for understanding the source, composition, and arrangement of the phases that make up the final product; in addition, it provides the basis for understanding the resultant properties of a mixture of two or more phases. Such understanding must ultimately provide the basis for effective control and use; it is not only more satisfying intellectually but also more useful practically than the alternative—trying to learn by rote the characteristics of thousands of different materials. A further advantage of the method of this book is that it provides a single basis for understanding the preparation, properties, and uses of both new ceramics and traditional compositions.

1

Ceramic Processes and Products

We define ceramics as the art and science of making and using solid articles which have as their essential component, and are composed in large part of, inorganic nonmetallic materials. This definition includes not only materials such as pottery, porcelain, refractories, structural clay products, abrasives, porcelain enamels, cements, and glass but also nonmetallic magnetic materials, ferroelectrics, manufactured single crystals, glass-ceramics, and a variety of other products which were not in existence until a few years ago and many which do not exist today.

Our definition is broader than the art and science of making and using solid articles formed by the action of heat on earthy raw materials, an extension of the Greek word *keramos*, and is much broader than a common dictionary definition such as "pottery" or "earthenware." Modern developments in methods of fabrication, the use of materials to close specifications, and their new and unique properties make traditional definitions too restrictive for our purposes. The origination of novel ceramic materials and new methods of manufacture requires us to take a fundamental approach to the art and science and a broad view of the field.

1.1 The Ceramic Industry

The ceramic industry is one of the large industries of the United States, with an annual production of nearly $20 billion in 1974.

One important characteristic of the ceramic industry is that it is basic to the successful operation of many other industries. For example, refractories are a basic component of the metallurgical industry. Abrasives are essential to the machine-tool and automobile industries. Glass products are essential to the automobile industry as well as to the architectural, electronic, and electrical industries. Uranium oxide fuels are essential to the nuclear-power industry. Cements are essential to the architectural and building industry. Various special electrical and magnetic ceramics are essential to the development of computers and many other electronic devices. As a matter of fact, almost every industrial production line, office, and home is dependent on ceramic materials. Newly designed devices incorporate ceramic materials because of their useful chemical, electrical, mechanical, thermal, and structural properties.

Perhaps even more important than being useful or necessary of themselves are those situations in which the feasibility or effectiveness of a large system depends critically on its ceramic components. For example, building bricks are a useful ceramic product and an important part of the ceramic industry. An understanding of their composition and structure has led to improvements in properties and usefulness. These have led to comparable increases in the value of the ultimate product (buildings). However, the magnetic cores used in the memory system of a large electronic computer are different in that they carry out an essential function critical for the entire design. The efficiency of a large, precise, and expensive system is essentially determined by the functioning of ceramic magnets. Improvements in their properties are important for the operation of the entire computer. Consequently, any understanding and improvement of properties is tremendously magnified in value, owing to the complexity and high cost of the final assembly. There are many similar examples in which a ceramic component determines in large part the functioning or even the feasibility of an entire system. This leverage in the importance of ceramic materials has in many cases led to intensive research toward a better understanding of properties, often out of all proportion to their dollar value.

That is, ceramics are important, first, because they comprise a large and basic industry and, second, because their properties are critical for many applications.

1.2 Ceramic Processes

A major characteristic of ceramics familiar to everyone is that they are brittle and fracture with little or no deformation. This behavior stands in contrast to metals, which yield and deform. As a result, ceramics cannot be formed into shape by the normal deformation processes used for metals. Two basic processes have been developed for shaping ceramics. One is to use fine ceramic particles mixed with a liquid or binder or lubricant or pore spaces, a combination that has rheological properties (classically the plasticity of a clay–water mixture) which permit shaping. Then by heat treatment the fine particles are agglomerated into a cohesive, useful product. The essentials of this procedure are first to find or prepare fine particles, shape them, and then stick them back together by heating. The second basic process is to melt the material to form a liquid and then shape it during cooling and solidification; this is most widely practiced in forming glasses. For completeness, we should also mention forming shapes in a mold or by dipping a form with a slurry containing a ceramic binder such as portland cement or ethyl silicate.

Raw Materials. The types of minerals found in nature are controlled mainly by the abundance of the elements and their geochemical characteristics. Since oxygen, silicon, and aluminum together account for 90% of the elements in the earth's crust, as shown in Fig. 1.1, it is not surprising that the dominant minerals are silicates and aluminum silicates. These, together with other mineral compounds of oxygen, constitute the great bulk of naturally occurring ceramic raw materials.

The mineral raw materials used in the ceramic industry are mainly

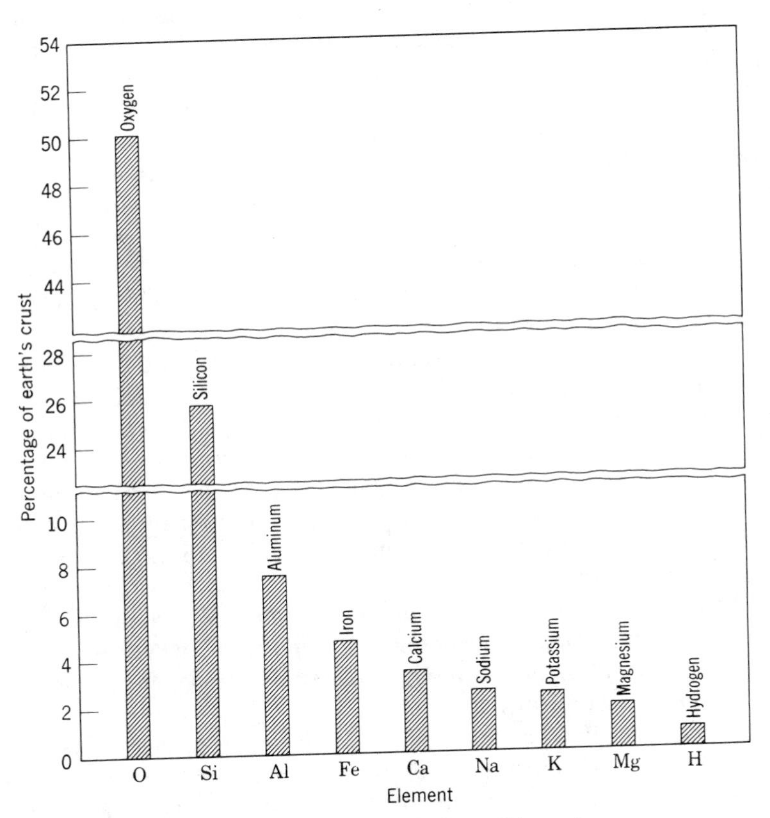

Fig. 1.1 Abundance of common elements in the earth's crust.

inorganic nonmetallic crystalline solids formed by complex geologic processes. Their ceramic properties are largely determined by the crystal structure and the chemical composition of their essential constituents and the nature and amounts of accessory minerals present. The mineralogic characteristics of such materials and therefore their ceramic properties are subject to wide variation among different occurrences or even within the same occurrence, depending on the geological environment in which the mineral deposit was formed as well as the physical and chemical modifications that have taken place during subsequent geological history.

Since silicate and aluminum silicate materials are widely distributed, they are also inexpensive and thus provide the backbone of high-tonnage products of the ceramic industry and determine to a considerable extent its form. Low-grade clays are available almost everywhere; as a result, the manufacture of building brick and tile not requiring exceptional properties is a localized industry for which extensive beneficiation of the raw material is not appropriate. In contrast, for fine ceramics requiring the use of better-controlled raw materials, the raw materials are normally beneficiated by mechanical concentration, froth floatation, and other relatively inexpensive processes. For materials in which the value added during manufacture is high, such as magnetic ceramics, nuclear-fuel materials, electronic ceramics, and specialized refractories, chemical purification and even chemical preparation of raw materials may be necessary and appropriate.

The raw materials of widest application are the clay minerals—fine-particle hydrous aluminum silicates which develop plasticity when mixed with water. They vary over wide limits in chemical, mineralogical, and physical characteristics, but a common characteristic is their crystalline layer structure, consisting of electrically neutral aluminosilicate layers, which leads to a fine particle size and platelike morphology and allows the particles to move readily over one another, giving rise to physical properties such as softness, soapy feel, and easy cleavage. Clays perform two important functions in ceramic bodies. First, their characteristic plasticity is basic to many of the forming processes commonly used; the ability of clay–water compositions to be formed and to maintain their shape and strength during drying and firing is unique. Second, they fuse over a temperature range, depending on composition, in such a way as to become dense and strong without losing their shape at temperatures which can be economically attained.

The most common clay minerals and those of primary interest to ceramists, since they are the major component of high-grade clays, are based on the kaolinite structure, $Al_2(Si_2O_5)(OH)_4$. Other compositions often encountered are shown in Table 1.1.

Table 1.1. Ideal Chemical Formulas of the Clay Minerals

Kaolinite	$Al_2(Si_2O_5)(OH)_4$
Halloysite	$Al_2(Si_2O_5)(OH)_4 \cdot 2H_2O$
Pyrophyllite	$Al_2(Si_2O_5)_2(OH)_2$
Montmorillonite	$\left(Al_{1.67}\begin{matrix}Na_{0.33}\\Mg_{0.33}\end{matrix}\right)(Si_2O_5)_2(OH)_2$
Mica	$Al_2K(Si_{1.5}Al_{0.5}O_5)_2(OH)_2$
Illite	$Al_{2-x}Mg_xK_{1-x-y}(Si_{1.5-y}Al_{0.5+y}O_5)_2(OH)_2$

A related material is talc, a hydrous magnesium silicate with a layer structure similar to the clay minerals and having the ideal formula $Mg_3(Si_2O_5)_2(OH)_2$. Talc is an important raw material for the manufacture of electrical and electronic components and for making tile. Asbestos minerals are a group of hydrous magnesium silicates which have a fibrous structure. The principal variety is chrysotile, $Mg_3Si_2O_5(OH)_4$.

In addition to the hydrous silicates already discussed, anhydrous silica and silicate materials are basic raw materials for much of the ceramic industry. SiO_2 is a major ingredient in glass, glazes, enamels, refractories, abrasives, and whiteware compositions. It is widely used because it is inexpensive, hard, chemically stable, and relatively infusable and has the ability to form glasses. There is a variety of mineral forms in which silica occurs, but by far the most important as a raw material is quartz. It is used as quartzite rock, as quartz sand, and as finely ground *potter's flint*. The major source of this material is sandstone, which consists of lightly bonded quartz grains. A denser quartzite, *gannister*, is used for refractory brick. Quartz is also used in the form of large, nearly perfect crystals, but these have been mostly supplanted by synthetic crystals, manufactured by a hydrothermal process.

Together with quartz, which serves as a refractory backbone constituent, and clay, which provides plasticity, traditional triaxial porcelains (originally invented in China) include feldspar, an anhydrous aluminosilicate containing K^+, Na^+, or Ca^{2+} as a flux which aids in the formation of a glass phase. The major materials of commercial interest are potash feldspar (microcline or orthoclase), $K(AlSi_3)O_8$, soda feldspar (albite), $Na(AlSi_3)O_8$, and lime feldspar (anorthite), $Ca(Al_2Si_2)O_8$. Other related materials sometimes used are nepheline syenite, a quartzfree igneous rock composed of nephelite, $Na_2(Al_2Si_2)O_8$, albite, and microcline; also wollastonite, $CaSiO_3$. One group of silicate minerals, the sillimanite group, having the composition Al_2SiO_5, is used for the manufacture of refractories.

Most of the naturally occurring nonsilicate materials are used primarily as refractories. Aluminum oxide is mostly prepared from the mineral bauxite by the Bayer process, which involves the selective leaching of the alumina by caustic soda, followed by the precipitation of aluminum hydroxide. Some bauxite is used directly in the electric-furnace production of alumina, but most is first purified. Magnesium oxide is produced both from natural magnesite, $MgCO_3$, and from magnesium hydroxide, $Mg(OH)_2$, obtained from seawater or brines. Dolomite, a solid solution of calcium and magnesium carbonates with the formula $CaMg(CO_3)_2$, is used to make basic brick for use in the steel industry. Another refractory widely used for metallurgical purposes is chrome ore, which consists primarily of a complex solid solution of spinels, $(Mg,Fe)(Al,Cr)_2O_4$, which make up most of the material; the remainder consists of various magnesium silicates.

Other mineral-based materials which are widely used include soda ash, $NaCO_3$, mostly manufactured from sodium chloride; borate materials including kernite, $Na_2B_4O_7 \cdot 4H_2O$, and borax, $Na_2B_4O_7 \cdot 10H_2O$, used as fluxing agents; fluorspar, CaF_2, used as a powerful flux for some glazes and glasses; and phosphate materials mostly derived from apatite, $Ca_5(OH,F)(PO_4)_3$.

Although most traditional ceramic formulations are based on the use of natural mineral materials which are inexpensive and readily available, an increasing fraction of specialized ceramic ware depends on the availability of chemically processed materials which may or may not start directly from mined products and in which the particle-size characteristics and chemical purity are closely controlled. Silicon carbide for abrasives is manufactured by electrically heating mixtures of sand and coke to a temperature of about 2200°C where they react to form SiC and carbon monoxide. Already mentioned, seawater magnesia, Bayer alumina, and soda ash are widely used chemical products. In the manufacture of barium titanate capacitors, chemically purified titania and barium carbonate are used as raw materials. A wide range of magnetic ceramics is manufactured from chemically precipitated iron oxide. Nuclear-fuel elements are manufactured from chemically prepared UO_2. Single crystals of sapphire and ruby and also porefree polycrystalline aluminum oxide are prepared from aluminum oxide made by precipitating and carefully calcining alum in order to maintain good control of both chemistry and particle size. Special techniques of material preparation such as freeze-drying droplets of solution to form homogeneous particles of small size and high purity are receiving increasing attention, as is the vapor deposition of thin-film materials in a carefully controlled chemical and physical form. In general, raw-material preparation is clearly headed toward the increasing use of mechanical, physical, and chemical purification and upgrading of raw materials together with special control of particle size and particle-size distribution and away from the sole reliance on materials in the form found in nature.

Forming and Firing. The most critical factors affecting forming and firing processes are the raw materials and their preparation. We have to be concerned with both the particle size and the particle-size distribution of the raw materials. Typical clay materials have a particle-size distribution which ranges from 0.1 to 50 microns for the individual particles. For the preparation of porcelain compositions the flint and feldspar constituents have a substantially larger particle size ranging between 10 and 200 microns. The fine-particle constituents, which for special ceramics may be less than 1 micron, are essential for the forming process, since colloidal suspensions, plastic mixes with a liquid-phase binder, and dry pressing all depend on very small particles flowing over one another or remaining in a stable suspension. For suspensions, the settling tendency is directly proportional to the density and particle size. For plastic forming the coherence of the mass and its yield point are determined by the capillarity of the liquid between particles; this force is inversely proportional to the particle size. However, if all the material were of a uniformly fine particle size, it would not be feasible to form a high concentration of solids. Mixing in a coarser material allows the fines to fill the interstices between the coarse particles such that a maximum particle-packing density is achieved at a ratio of about 70% coarse and 30% fine material when two particle sizes are used. In addition, during the drying process, shrinkage results from the removal of water films between particles. Since the number of films increases as the particle size decreases, bodies prepared with a liquid binder and all fine-particle materials have a high shrinkage during drying and the resultant problems of warping and distortion.

In addition to a desired particle size and particle-size distribution, intimate mixing of material is necessary for uniformity of properties within a body and for the reaction of individual constituents during the firing process. For preparing slurries or a fine-grain plastic mass, it is the usual practice to use wet mixing, with the raw materials placed together in ball mills or a blunger. Shearing stresses developed in the mixing process improve the properties of a plastic mix and ensure the uniform distribution of the fine-grain constituent. For dewatering the wet-milled mix, either a filter press may be used, or more commonly spray-drying, in which droplets of the slurry are dried with a countercurrent of warm air to maintain their uniform composition during drying. The resulting aggregates, normally 1 mm or so in size, flow and deform readily in subsequent forming.

Since the firing process also depends on the capillary forces resulting from surface energy to consolidate and densify the material and since these forces are inversely proportional to particle size, a substantial percentage of fine-particle material is necessary for successful firing. The clay minerals are unique in that their fine particle size provides both the capability for plastic forming and also sufficiently large capillary forces for successful firing. Other raw materials have to be prepared by chemical precipitation or by milling into the micron particle range for equivalent results to be obtained.

Perhaps the simplest method of compacting a ceramic shape consists of forming a dry or slightly damp powder, usually with an organic binder, in a metal die at sufficiently high pressures to form a dense, strong piece. This method is used extensively for refractories, tiles, special electrical and magnetic ceramics, spark-plug insulators and other technical ceramics, nuclear-fuel pellets, and a variety of products for which large numbers of simple shapes are required. It is relatively inexpensive and can form shapes to close tolerances. Pressures in the range of 3000 to 30,000 psi are commonly used, the higher pressures for the harder materials such as pure oxides and carbides. Automatic dry pressing at high rates of speed has been developed to a high state of effectiveness. One limitation is that for a shape with a high length-to-diameter ratio the frictional forces of the powder, particularly against the die wall, lead to pressure gradients and a resulting variation of density within the piece. During firing these density variations are eliminated by material flow during sintering; it necessarily follows that there is a variation in shrinkage and a loss of the original tolerances. One modification of the dry-pressing method which leads to a more uniform density is to enclose the sample in a rubber mold inserted in a hydrostatic chamber to make pieces by hydrostatic molding, in which the pressure is more uniformly applied. Variations in sample density and shrinkage are less objectionable. This method is widely used for the manufacture of spark-plug insulators and for special electrical components in which a high degree of uniformity and high level of product quality are required.

A quite different method of forming is to extrude a stiff plastic mix through a die orifice, a method commonly used for brick, sewer pipe, hollow tile, technical ceramics, electrical insulators, and other materials having an axis normal to a fixed cross section. The most widely practiced method is to use a vacuum auger to eliminate air bubbles, thoroughly mix the body with 12 to 20% water, and force it through a hardened steel or carbide die. Hydraulic piston extruders are also widely used.

The earliest method of forming clay ware, one still widely used, is to add enough water so that the ware can readily be formed at low pressures.

This may be done under hand pressure such as building ware with coils, free-forming ware, or hand throwing on a potter's wheel. The process can be mechanized by soft-plastic pressing between porous plaster molds and also by automatic *jiggering*, which consists of placing a lump of soft plastic clay on the surface of a plaster-of-paris mold and rotating it at about 400 rpm while pulling a profile tool down on the surface to spread the clay and form the upper surface.

When a larger amount of water is added, the clay remains sticky plastic until a substantial amount has been added. Under a microscope it is seen that individual clay particles are gathered in aggregates or flocs. However, if a small quantity of sodium silicate is added to the system, there is a remarkable change, with a substantial increase in fluidity resulting from the individual particles being separated or *deflocculated*. With proper controls a fluid suspension can be formed with as little as 20% liquid, and a small change in the liquid content markedly affects the fluidity. When a suspension such as this is cast into a porous plaster-of-paris mold, the mold sucks liquid from the contact area, and a hard layer is built on the surface. This process can be continued until the entire interior of the mold is filled (solid casting) or the mold can be inverted and the excess liquid poured out after a suitable wall thickness is built up (drain casting).

In each of the processes which require the addition of some water content, the drying step in which the liquid is removed must be carefully controlled for satisfactory results, more so for the methods using a higher liquid content. During drying, the initial drying rate is independent of the water content, since in this period there is a continuous film of water at the surface. As the liquid evaporates, the particles become pressed more closely together and shrinkage occurs until they are in contact in a solid structure free from water film. During the shrinkage period, stresses, warping, and possibly cracks may develop because of local variations in the liquid content; during this period rates must be carefully controlled. Once the particles are in contact, drying can be continued at a more rapid rate without difficulty. For the dry-pressing or hydrostatic molding process, the difficulties associated with drying are avoided, an advantage for these methods.

After drying, ceramic ware is normally fired to temperatures ranging from 700 to 1800°C, depending on the composition and properties desired. Ware which is to be glazed or decorated may be fired in different ways. The most common procedure is to fire the ware without a glaze to a sufficiently high temperature to mature the body; then a glaze is applied and fired at a low temperature. Another method is to fire the ware initially to a low temperature, a bisque fire; then apply the glaze and mature the body and glaze together at a higher temperature. A third method is to apply the glaze to the unfired ware and heat them together in a one-fire process.

During the firing process, either a viscous liquid or sufficient atomic mobility in the solid is developed to permit chemical reactions, grain growth, and sintering; the last consists of allowing the forces of surface tension to consolidate the ware and reduce the porosity. The volume shrinkage which occurs is just equal to the porosity decrease and varies from a few to 30 or 40 vol%, depending on the forming process and the ultimate density of the fired ware. For some special applications, complete density and freedom from all porosity are required, but for other applications some residual porosity is desirable. If shrinkage proceeds at an uneven rate during firing or if part of the ware is restrained from shrinking by friction with the material on which it is set, stresses, warping, and cracking can develop. Consequently, care is required in setting the ware to avoid friction. The rate of temperature rise and the temperature uniformity must be controlled to avoid variations in porosity and shrinkage. The nature of the processes taking place is discussed in detail in Chapters 11 and 12.

Several different types of kilns are used for firing ware. The simplest is a skove kiln in which a benchwork of brick is set up inside a surface coating with combustion chambers under the material to be fired. Chamber kilns of either the up-draft or down-draft type are widely used for batch firing in which temperature control and uniformity need not be too precise. In order to achieve uniform temperatures and maximum use of fuel, chamber kilns in which the air for combustion is preheated by the cooling ware in an adjacent chamber, the method used in ancient China, is employed. The general availability of more precise temperature controls for gas, oil, and electric heating and the demands for ware uniformity have led to the increased use of tunnel kilns in which a temperature profile is maintained constant and the ware is pushed through the kiln to provide a precise firing schedule under conditions such that effective control can be obtained.

Melting and Solidification. For most ceramic materials the high volume change occurring during solidification, the low thermal conductivity, and the brittle nature of the solid phase have made melting and solidification processes comparable with metal casting and foundry practice inappropriate. Recently, techniques have been developed for unidirectional solidification in which many of these difficulties can be substantially avoided. This process has mainly been applied to forming controlled structures of metal alloys which are particularly attractive for applications such as turbine blades for high-temperature gas turbines. So far as we are aware, there is no large scale manufacture of ceramics in this way, but we anticipate that the development of techniques for the unidirectional solidification of ceramics will be an area of active research during the next decade.

Another case in which these limitations do not apply is that of glass-forming materials in which the viscosity increases over a broad temperature range so that there is no sharp volume discontinuity during solidification and the forming processes can be adjusted to the fluidity of the glass. Glass products are formed in a high-temperature viscous state by five general methods: (1) blowing, (2) pressing, (3) drawing, (4) rolling, and (5) casting. The ability to use these processes depends to a large extent on the viscous flow characteristics of the glass and its dependence on temperature. Often surface chilling permits the formation of a stable shape while the interior remains sufficiently fluid to avoid the buildup of dangerous stresses. Stresses generated during cooling are relieved by annealing at temperatures at which the force of gravity is insufficient to cause deformation. This is usually done in an annealing oven or lehr which, for many silicate glasses, operates at temperatures in the range of 400 to 500°C.

The characteristics most impressive about commercial glass-forming operations are the rapidity of forming and the wide extent of automation. Indeed, this development is typical of the way in which technical progress affects an industry. Before the advent of glass-forming machinery, a major part of the container industry was based on ceramic stoneware. Large numbers of relatively small stoneware potters existed solely for the manufacture of containers. The development of automatic glass-forming machinery allowing the rapid and effective production of containers on a continuous basis has eliminated stoneware containers from common use.

Special Processes. In addition to the broadly applicable and widely used processes discussed thus far, there is a variety of special processes which augment, modify, extend, or replace these forming methods. These include the application of glazes, enamels, and coatings, hot-pressing materials with the combined application of pressure and temperature, methods of joining metals to ceramics, glass crystallization, finishing and machining operations, preparation of single crystals, and vapor-deposition processes.

Much ceramic ware is coated with a glaze, and porcelain enamels are commonly applied on a base of sheet steel or cast iron as well as for special jewelry applications. Glazes and enamels are normally prepared in a wet process by milling together the ingredients and then applying the coating by brushing, spraying, or dipping. For continuous operation, spray coating is most frequently used, but for some applications more satisfactory coverage can be obtained by dipping or painting. For porcelain enamels on cast iron, large castings heated in a furnace are coated with a dry enamel powder which must be distributed uniformly over the surface, where it fuses and sticks. In addition to these widely used processes, special coatings for technical ware have been applied by flame spraying to obtain a refractory dense layer; vacuum-deposited coatings have been formed by evaporation or cathodic sputtering; coatings have been applied by chemical vapor deposition; electrophoretic deposition has been applied; and other specialized techniques have had some limited applications.

To obtain a high density together with fine particle size, particularly for materials such as carbides and borides, the combination of pressure with high temperature is an effective technique mostly used for small samples of a simple configuration. At lower temperatures, glass-bonded mica is formed in this way for use as an inexpensive insulation. One of the main advantages of the hot-pressing method is that material preparation is less critical than for the sintering processes, which require a high degree of material uniformity for successful applications of the highest-quality products. The main difficulties with hot-pressing techniques are applying the method to large shapes and the time required for heating the mold and sample, which makes the method slow and expensive.

For many applications, joining processes are necessary to form fabricated units. In manufacturing teacups, for example, the handle is normally molded separately, dipped in a slip, and stuck on the body of the cup. Sanitary fixtures of complex design are similarly built up from separately formed parts. For many electronic applications requiring pressure-tight seals, it is necessary to form a bond between metals and ceramics. For glass-metal seals, the main problem is matching the expansion coefficient of the glass to that of the metal and designing the seal so that large stresses do not develop in use; special metal alloys and sealing glasses have been designed for this purpose. For crystalline ceramics, the most widely applied method has been to use a molybdenum-manganese layer which, when fired under partially oxidizing conditions, forms an oxide that reacts with the ceramic to give an adhesive bonding layer. In some cases, reactive metal brazes containing titanium or zirconium have been used.

One of the most important developments in ceramic forming has been to use a composition which can be formed as a glass and then transformed subsequent to forming into a product containing crystals of controlled size and amount. Classic examples of this are the striking gold-ruby glasses, in which the color results from the formation of colloidal gold particles. During rapid initial cooling, nucleation of the metal particles occurs; subsequent reheating into the growth region develops proper crystallite sizes for the colloidal ruby color. In the past 10 years there has been extensive development of glasses in which the volume of crystals formed is much larger than the volume of the residual glass. By controlled nucleation and growth, glass-ceramics are made in which the advantage of automatic glass-forming processes is combined with some of the desirable properties of a highly crystalline body.

For most forming operations, some degree of finishing or machining is required which may range from fettling the mold lines from a slip-cast shape to diamond-grinding the final contour of a hard ceramic. For hard materials such as aluminum oxide, as much machining as feasible is done in the unfired state or the presintered state, with final finishing only done on the hard, dense ceramic where required.

A number of processes have been developed for the formation of ceramics directly from the vapor phase. Silica is formed by the oxidation of silicon tetrachloride. Boron and silicon carbide fibers are made by introducing a volatile chloride with a reducing agent into a hot zone, where deposition occurs on a fine tungsten filament. Pyrolytic graphite is prepared by the high-temperature deposition of graphite layers on a substrate surface by the pyrolytic decomposition of a carbon-containing gas. Many carbides, nitrides, and oxides have been formed by similar processes. For electronic applications, the development of single-crystal films by these techniques appears to have many potential applications.

Thin-wafer substrates are formed by several techniques, mostly from alumina. A widely used development is the technique in which a fluid body is prepared with an organic binder and uniformly spread on a moving nonporous belt by a doctor blade to form thin, tough films which can subsequently be cut to shape; holes can be introduced in a high-speed punch press.

There is an increasing number of applications in which it is necessary or desirable to have single-crystal ceramics because of special optical, electrical, magnetic, or strength requirements. The most widespread method of forming these is the Czochralski process, in which the crystal is slowly pulled from a molten melt, a process used for aluminum oxide, ruby, garnet, and other materials. In the Verneuil process a liquid cap is maintained on a growing boule by the constant-rate addition of powdered material at the liquid surface. For magnetic and optical applications thin single-crystal films are desirable which have been prepared by epitaxial growth from the vapor phase. Hydrothermal growth from solution is widely used for the preparation of quartz crystals, largely replacing the use of natural mineral crystals for device applications.

1.3 Ceramic Products

The diversity of ceramic products, which range from microscopic single-crystal whiskers, tiny magnets, and substrate chips to multiton refractory furnace blocks, from single-phase closely controlled compositions to multiphase multicomponent brick, and from porefree transparent crystals and glasses to lightweight insulating foams is such that no simple classification is appropriate. From the point of view of historical development and tonnage produced, it is convenient to consider the mineral-raw-material products, mostly silicates, separately from newer nonsilicate formulations.

Traditional Ceramics. We can define traditional ceramics as those comprising the silicate industries—primarily clay products, cement, and silicate glasses.

The art of making pottery by forming and burning clay has been practiced from the earliest civilizations. Indeed, the examination of pottery fragments has been one of the best tools of the archeologist. Burnt clayware has been found dating from about 6500 B.C. and was well developed as a commercial product by about 4000 B.C.

Similarly, the manufacture of silicate glasses is an ancient art. Naturally occurring glasses (obsidian) were used during the Stone Age, and there was a stable industry in Egypt by about 1500 B.C.

In contrast, the manufacture of portland cement has only been practiced for about 100 years. The Romans combined burned lime with volcanic ash to make a natural hydraulic cement; the art seems then to have disappeared, but the hydraulic properties of lightly burned clayey limes were rediscovered in England about 1750, and in the next 100 years the manufacturing process, essentially the same as that used now, was developed.

By far the largest segment of the silicate ceramic industry is the manufacture of various glass products. These are manufactured mostly as sodium-calcium-silicate glasses. The next largest segment of the ceramic industry is lime and cement products. In this category the largest group of materials is hydraulic cements such as those used for building construction. A much more diverse group of products is included in the classification of whitewares. This group includes pottery, porcelain, and similar fine-grained porcelainlike compositions which comprise a wide variety of specific products and uses. The next classification of traditional ceramics is porcelain enamels, which are mainly silicate glasslike coatings on metals. Another distinct group is the structural clay products, which consist mainly of brick and tile but include a variety of similar products such as sewer pipe. A particularly important group of the traditional ceramics industry is refractories. About 40% of the refractory industry consists of fired-clay products, and another 40% consists of heavy nonclay refractories such as magnesite, chromite, and similar compositions. In addition there is a sizable demand for various special refractory compositions. The abrasives industry produce mainly silicon carbide and aluminum oxide abrasives. Finally, a segment of the ceramic industry which does not produce ceramic products as such is concerned with the mineral preparation of ceramic and related raw materials.

Most of these traditional ceramics could be adequately defined as the silicate industries, which indeed was the description originally proposed for the American Ceramic Society in 1899. The silicate industries still compose by far the largest part of the whole ceramic industry, and from this point of view they can be considered the backbone of the field.

New Ceramics. In spite of its antiquity, the ceramic industry is not stagnant. Although traditional ceramics, or silicate ceramics, account for the large bulk of material produced, both in tonnage and in dollar volume, a variety of new ceramics has been developed in the last 20 years. These are of particular interest because they have either unique or outstanding properties. Either they have been developed in order to fulfill a particular need in greater temperature resistance, superior mechanical properties, special electrical properties, and greater chemical resistivity, or they have been discovered more or less accidentally and have become an important part of the industry. In order to indicate the active state of development, it may be helpful to describe briefly a few of these new ceramics.

Pure oxide ceramics have been developed to a high state of uniformity and with outstanding properties for use as special electrical and refractory components. The oxides most often used are alumina (Al_2O_3), zirconia (ZrO_2), thoria (ThO_2), beryllia (BeO), magnesia (MgO), spinel ($MgAl_2O_4$), and forsterite (Mg_2SiO_4).

Nuclear fuels based on uranium dioxide (UO_2) are widely used. This material has the unique ability to maintain its good properties after long use as a fuel material in nuclear reactors.

Electrooptic ceramics such as lithium niobate ($LiNbO_3$) and lanthanum-modified lead zirconate titanate (PLZT) provide a medium by which electrical information can be transformed to optical information or by which optical functions can be performed on command of an electrical signal.

Magnetic ceramics with a variety of compositions and uses have been developed. They form the basis of magnetic memory units in large computers. Their unique electrical properties are particularly useful in high-frequency microwave electronic applications.

Single crystals of a variety of materials are now being manufactured, either to replace natural crystals which are unavailable or for their own unique properties. Ruby and garnet laser crystals and sapphire tubes and substrates are grown from a melt; large quartz crystals are grown by a hydrothermal process.

Ceramic nitrides with unusually good properties for special applications have been developed. These include aluminum nitride, a laboratory refractory for melting aluminum; silicon nitrides and SiAlON, commercially important new refractories and potential gas turbine components; and boron nitride, which is useful as a refractory.

Enamels for aluminum have been developed and have become an important part of the architectural industry.

Metal-ceramic composites have been developed and are now an important part of the machine-tool industry and have important uses as refractories. The most important members of this group are various carbides bonded with metals and mixtures of a chromium alloy with aluminum oxide.

Ceramic carbides with unique properties have been developed. Silicon carbide and boron carbide in particular are important as abrasive materials.

Ceramic borides have been developed which have unique properties of high-temperature strength and oxidation resistance.

Ferroelectric ceramics such as barium titanate have been developed which have extremely high dielectric constants and are particularly important as electronic components.

Nonsilicate glasses have been developed and are particularly useful for infrared transmission, special optical properties, and semiconducting devices.

Molecular sieves which are similar to, but are more controlled than, natural zeolite compositions are being made with controlled structures so that the lattice spacing, which is quite large in these compounds, can be used as a means of separating compounds of different molecular sizes.

Glass-ceramics are a whole new family of materials based on fabricating ceramics by forming as a glass and then nucleating and crystallizing to form a highly crystalline ceramic material. Since the original introduction of Pyroceram by the Corning Glass Works the concept has been extended to dozens of compositions and applications.

Porefree polycrystalline oxides have been made based on alumina, yttria, spinel, magnesia, ferrites, and other compositions.

Literally dozens of other new ceramic materials unknown 10 or 20 years ago are now being manufactured and used. From this point of view the ceramic industry is one of our most rapidly changing industries, with new products having new and useful properties constantly being developed. These ceramics are being developed because there is a real need for new materials to transform presently available designs into practical, serviceable products. By far the major hindrance to the development of many new technologically feasible structures and systems is the lack of satisfactory materials. New ceramics are constantly filling this need.

New Uses for Ceramics. In the same way that the demand for new and better properties has led to the development of new materials, the availability of new materials had led to new uses based on their unique properties. This cycle of new ceramics–new uses–new ceramics has accelerated with the attainment of a better understanding of ceramics and their properties.

One example of the development of new uses for ceramics has occurred in the field of magnetic ceramic materials. These materials have hysteresis loops which are typical for ferromagnetic materials. Some have very nearly the square loop that is most desirable for electronic computer memory circuits. This new use for ceramics has led to extensive studies and development of materials and processes.

Another example is the development of nuclear power, which requires uranium-containing fuels having large fractions of uranium (or sometimes thorium), stability against corrosion, and the ability to withstand the fissioning of a large part of the uranium atoms without deterioration. For many applications UO_2 is an outstanding material for this fuel. Urania ceramics have become an important part of reactor technology.

In rocketry and missile development two critical parts which must withstand extreme temperatures and have good erosion resistance are the nose cone and the rocket throat. Ceramic materials are used for both.

For machining metals at high speeds it has long been known that oxide ceramics are superior in many respects as cutting tools. However, their relatively low and irregular strength makes their regular use impossible. The development of alumina ceramics with high and uniform strength levels has made them practicable for machining metals and has opened up a new field for ceramics.

In 1946 it was discovered that barium titanate had a dielectric constant 100 times larger than that of other insulators. A whole new group of these ferroelectric materials has since been discovered. They allow the manufacture of capacitors which are smaller in size but have a larger capacity than other constructions, thus improving electronic circuitry and developing a new use for ceramic materials.

In jet aircraft and other applications metal parts have had to be formed from expensive, and in wartime unobtainable, alloys to withstand the moderately high temperatures encountered. When a protective ceramic coating is applied, the temperature limit is increased, and either higher temperatures can be reached or less expensive and less critical alloys can be substituted.

Many further applications of ceramics which did not even exist a few years ago can be cited, and we may expect new uses to develop that we cannot now anticipate.

Suggested Reading

1. F. H. Norton, *Elements of Ceramics*, 2d ed., Addison Wesley Publishing Company, Inc., Reading, Mass., 1974.
2. F. H. Norton, *Fine Ceramics*, McGraw-Hill Book Company, New York, 1970.
3. F. H. Norton, *Refractories*, 4th ed., McGraw-Hill Book Company, New York, 1968.
4. Institute of Ceramics Textbook Series:
   (*a*) W. E. Worrall, *Raw Materials*, Maclaren & Sons, Ltd., London, 1964.
   (*b*) F. Moore, *Rheology of Ceramic Systems*, Maclaren & Sons, Ltd., London, 1965.
   (*c*) R. W. Ford, *Drying*, Maclaren & Sons, Ltd., London, 1964.
   (*d*) W. F. Ford, *The Effect of Heat on Ceramics*, Maclaren & Sons, Ltd., London, 1967.
5. "Fabrication Science," *Proc. Brit. Ceram. Soc.*, No. 3 (September, 1965).
6. "Fabrication Science: 2," *Proc. Brit. Ceram. Soc.*, No. 12 (March, 1969).
7. J. E. Burke, Ed., *Progress in Ceramic Science*, Vols. 1–4, Pergamon Press, Inc., New York, 1962–1966.
8. W. D. Kingery, Ed., *Ceramic Fabrication Processes*, John Wiley & Sons, Inc., New York, 1958.
9. F. V. Tooley, Ed., *Handbook of Glass Manufacture*, 2 Vols., Ogden Publishing Company, New York, 1961.
10. A. Davidson, Ed., *Fabrication of Non-metals: Handbook of Precision Engineering*, Vol. 3, McGraw-Hill Book Company, New York, 1971.

part II

CHARACTERISTICS OF CERAMIC SOLIDS

The ceramic materials with which we are concerned may be single crystals, wholly vitreous, or mixtures of two or more crystalline or vitreous phases. Pore spaces are also a principal phase in most ceramic materials. As the basis for understanding the properties of real ceramics, it is essential to have an understanding of the properties of single crystals and noncrystalline solids. In Part II we consider the properties of ceramic solids as a single phase without regard to their source or the effects of combining with other materials.

In Chapter 2 we consider the structure of crystalline ceramics. The nature of the atomic arrangements, the forces between atoms, and the location of atoms in a crystalline lattice are important parameters basic to the properties of the crystal. In Chapter 3 we consider noncrystalline solids. The atomic structure of these materials is quite different from that of crystals, and many of their properties are intimately related to the noncrystalline nature of the atomic arrangement.

Both crystalline and noncrystalline materials depart from ideal structures in many respects. Some of their properties are strongly dependent on the nature of departures from perfect crystallinity or perfect randomness. Consequently, in Chapter 4 we consider structural imperfections, their source, and properties. In Chapter 5 we consider the surfaces and interfaces as a separate characteristic property. Finally, in Chapter 6 we consider the question of atomic mobility; this is important to many properties and is intimately related to the structure of solids. An understanding of these five chapters is essential to understanding the properties of more complex ceramics.

In describing the characteristics of ceramic solids, two different points of view have been useful. One is to consider them from an atomistic point of view, defining as closely as possible the location of atoms relative to one another, the interaction between atoms, the motion of atoms relative to one another, and the influence of changed conditions, such as increased temperature, on atomic behavior. This point of view leads to an understanding of structure and an insight into atomic interaction that is essential for developing models and generalizations about the complex phenomena we wish to understand. It is an approach that first became practical about 60 years ago with the discovery of X-ray diffraction by crystals and has continued to depend strongly on observations of the interaction between radiation and matter.

A second and equally useful viewpoint is to consider the macroscopic properties of matter independent of conjectures about the details of atomic characteristics and interaction. This, the thermodynamic approach, depends on the observation that the state of matter at equilibrium, whether gaseous, liquid or crystalline, is determined by the thermodynamic variables which describe the system (temperature, volume, pressure, composition). The interrelationship of these variables to the state of the system has been formally developed in the principles of thermodynamics, which are based on three fundamental laws. The first law requires that the internal energy E of a system be conserved. The second law introduces another function, the entropy S, a measure of randomness which determines the direction of all spontaneous processes: the entropy of the world tends toward a maximum. Thus the change in the entropy of the system and the surroundings during any process is always toward greater randomness:

$$dS_{\text{system}} + dS_{\text{surroundings}} \geq 0 \tag{1}$$

At equilibrium, the entropy change is zero, and therefore this equation serves as a definition of thermodynamic equilibrium. The third law sets the zero-point entropy of matter at the absolute zero of temperature: the entropy of a perfect crystal at 0°K is zero.

From these three fundamental laws and from the definition of internal energy and entropy other useful state functions are defined: the enthalpy, or heat content H, the Gibbs free energy G, and the Helmholtz free energy F. The Gibbs free energy ($G = E + PV - TS = H - TS$) is the state function most commonly used to describe the equilibrium state of the system. For example, at equilibrium (0°C and 1 atm pressure), ice and water can coexist, and the free energy of water is equal to that of ice. From experience, we know that there is an enthalpy change, the heat of fusion, and also an entropy change associated with this equilibrium reaction:

$$\Delta G = 0 = \Delta H - T_e \, \Delta S \tag{2}$$

$$\Delta S_{\text{ice to water}} = \frac{\Delta H_{\text{fusion}}}{T_e(273°\text{K})} \tag{3}$$

When we deal with phases of variable-composition (gaseous, liquid or solid solutions), the Gibbs free energy is not only a function of temperature and pressure but also of composition. If X_i is the mole fraction of the ith component,

$$dG = -S\,dT + V\,dP + \sum_i \mu_i \, dX_i \tag{4}$$

where we define the chemical potential μ_i as the change in the free energy of the system with respect to a change in the concentration of the ith component at constant temperature and pressure:

$$\mu_i = \left(\frac{\partial G}{\partial X_i}\right)_{T, P, X_{j \neq i}} \tag{5}$$

If the system is at equilibrium, each of the terms in Eq. 4 must be independent of time and of position in the system, that is, uniform temperature (thermal equilibrium), uniform pressure (mechanical equilibrium), and uniform chemical potential of each component (chemical equilibrium). In a multiphase system, this means that the chemical potential of a particular component must be the same in each phase.

As we proceed, it will become increasingly clear that the energy-matter relationships both on an atomistic scale and on a scale of macroscopic assemblages are areas of knowledge directly pertinent to ceramics. Many excellent texts are available; particularly recommended C. Kittel, *Introduction to Solid State Physics*, and R. A. Swalin, *Thermodynamics of Solids*.

2

Structure of Crystals

In this chapter we examine the structure of crystalline solids, solids characterized by an orderly periodic array of atoms. The three states of matter—gaseous, liquid, solid—can be represented as in Fig. 2.1. In the gaseous state, atoms or molecules are widely scattered and are in rapid motion. The large average separation between atoms and nearly elastic interactions allow the application of the well-known ideal gas laws as a good approximation at low and moderate pressures. In contrast, the liquid and solid states are characterized by the close association of atoms, which to a first approximation can be regarded as spherical balls in contact with springs between them representing interatomic forces. In liquids there is sufficient thermal energy to keep the atoms in random motion, and there is no long-range order. In crystals, the attractive forces of interatomic bonding overcome the disaggregating thermal effects, and an ordered arrangement of atoms occurs. (In glasses, considered in Chapter 3, a disordered arrangement persists even at low temperatures.) This chapter is concerned with the structure of the orderly periodic atomic arrangements in crystals. What we consider here are ideal crystal structures. Later, in Chapters 4 and 5, we consider some of the important departures from ideality.

In order to understand the nature and formation of crystal structures, it is essential to have some understanding of atomic structure. We present some results of quantum theory relating to atomic structure in the first section. Some additional aspects of quantum theory are brought in later as needed (particularly in connection with electrical and magnetic properties). However, we strongly urge students who have not done so to learn as much as possible about modern atomic physics as a basis for a better understanding of ceramics.

2.1 Atomic Structure

The basis for our present understanding of the structure of the atom lies in the development of quantum theory and wave mechanics. By about

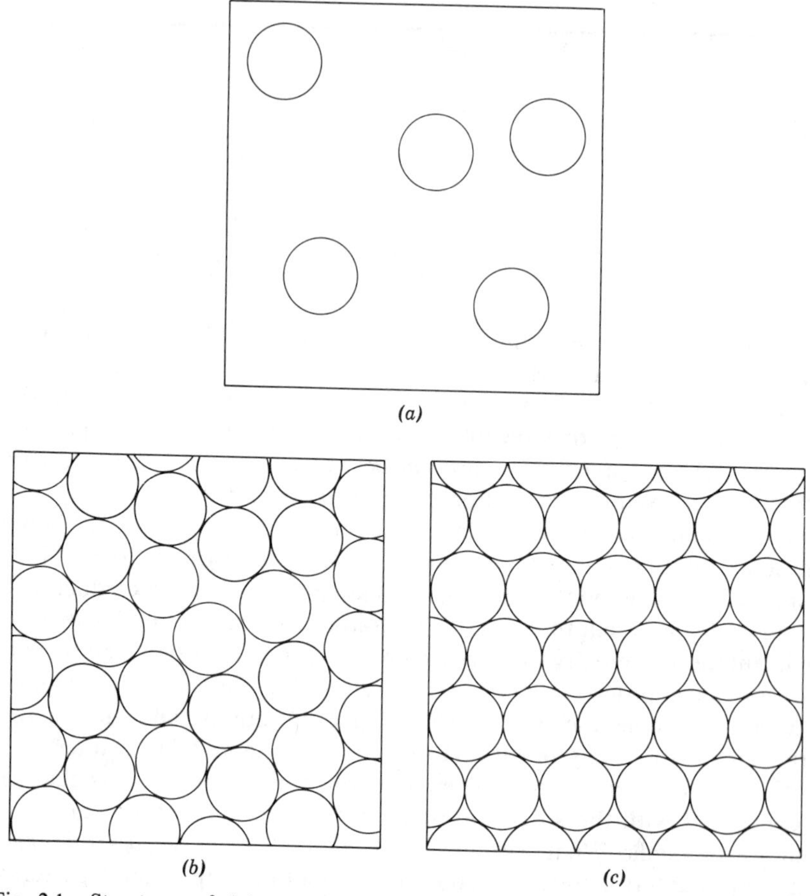

Fig. 2.1. Structures of (*a*) gas with widely separated molecules, (*b*) liquid with no long-range order, and (*c*) crystal with atoms or molecules having an ordered pattern.

1900, extensive spectroscopic data for the series of spectral lines emitted by various atoms, the frequency dependence of thermal radiation, and the characteristics of photoelectric emission could not be satisfactorily explained on the basis of classical continuum physics. Planck (1900) successfully explained thermal radiation by assuming that it is emitted discontinuously in *energy quanta* or *photons* having an energy $h\nu$, where ν is the frequency and $h = 6.623 \times 10^{-34}$ J-sec is a universal constant. Einstein (1905) used this same idea to explain photoemission. About 10 years later Bohr (1913) suggested an atomic model in which electrons can move only in certain stable orbits (without radiation) and postulated that transitions between these stable energy states produce spectral lines by emission or absorption of light quanta. This concept leads to a satisfactory explanation of observed series of spectral lines.

The Bohr Atom. In the Bohr atom, Fig. 2.2, quantum theory requires that the angular momentum of an electron be an integral multiple of $h/2\pi$. The integral number by which $h/2\pi$ is multiplied is called the *principal quantum number n*. As n increases, the energy of the electron increases and it is farther from the positively charged nucleus. In addition to the principal quantum number, electrons are characterized by secondary integral quantum numbers: l corresponding to a measure of eccentricity of the orbit varies from 0 to $n-1$, called s $(l=0)$, p $(l=1)$, d $(l=2)$, f $(l=3)$ orbitals; m corresponding to a measure of ellipse orientation takes integral values from $-l$ to $+l$; s corresponding to the direction of electron spin is either positive or negative. As the values of n and l increase, the energy of their electron orbits also increases in general.

A further restriction on atom structure is the *Pauli exclusion principle* that no two electrons can have all quantum numbers the same in any one atom. As the number of electrons in an atom increases, added electrons fill orbits of higher energy states characterized by larger principal quantum numbers. The number of electrons that can be accommodated in successive orbitals in accordance with the Pauli exclusion principle determines the periodic classification of the elements.

Electron configurations are characterized by the principal quantum number (1, 2, 3, . . .) and the orbital quantum number (s, p, d, f) together with the number of electrons that can be accommodated at each energy level in accordance with the Pauli exclusion principle (up to 2 electrons for s orbitals, 6 for p orbitals, 10 for d orbitals, and 14 for f orbitals). The resulting electron configurations in a periodic table of the elements are given in Table 2.1.

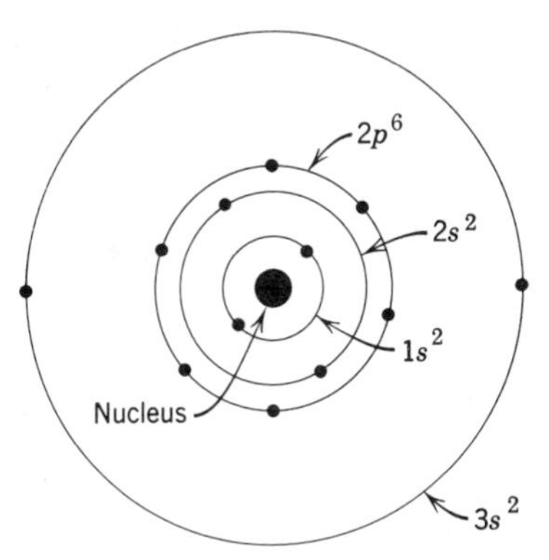

Fig. 2.2. Structure of the Bohr atom (magnesium).

Table 2.1. Periodic Classification

Group I	II	III	IV
1 H $1s$			
3 Li $2s$	4 Be $2s^2$	5 B $2s^22p$	6 C $2s^22p^2$
11 Na $2p^63s$	12 Mg $2p^63s^2$	13 Al $3s^23p$	14 Si $3s^23p^2$

Transition Elements

19 K $3p^64s$	20 Ca $3p^64s^2$	21 Sc $3d4s^2$	22 Ti $3d^24s^2$	23 V $3d^34s^2$	24 Cr $3d^44s^2$	25 Mn $3d^54s^2$	26 Fe $3d^64s^2$	27 Co $3d^74s^2$
37 Rb $4p^65s$	38 Sr $4p^55s^2$	39 Y $4d5s^2$	40 Zr $4d^25s^2$	41 Nb $4d^45s$	42 Mo $4d^55s$	43 Tc $4d^65s$	44 Ru $4d^75s$	45 Rh $4d^85s$

Rare Earths

55 Cs $5p^66s$	56 Ba $5p^66s^2$	57 La $5p^65d6s^2$	58 Ce $4f^26s^2$	59 Pr $4f^36s^2$	60 Nd $4f^46s^2$	61 Pm $4f^56s^2$	62 Sm $4f^66s^2$	63 Eu $4f^76s^2$	64 Gd $4f^75ds^2$	65 Tb $4f^85d6s^2$	66 Dy $4f^{10}6s^2$	67 Ho $4f^{11}6s^2$	68 Er $4f^{12}6s^2$	69 Tm $4f^{13}6s^2$
87 Fr $6p^27s$	88 Ra $6p^27s^2$	89 Ac $6d7s^2$	90 Th $6d^27s^2$	91 Pa $5f^26d7s^2$	92 U $5f^36d7s^2$	93 Np $5f^57s^2$	94 Pu $5f^67s^2$	95 Am $5f^77s^2$	96 Cm $5f^76d7s^2$	97 Bk $5f^86d7s^2$	98 Cf $5f^96d7s^2$			

70 Yb $4f^{14}6s^2$

of the Elements

V	VI	VII	0
			2 He $1s^2$
7 N $2s^22p^3$	8 O $2s^22p^4$	9 F $2s^22p^5$	10 Ne $2s^22p^6$
15 P $3s^23p^3$	16 S $3s^23p^4$	17 Cl $3s^23p^5$	18 A $3s^23p^6$

28 Ni $3d^84s^2$	29 Cu $3d^{10}4s$	30 Zn $3d^{10}4s^2$	31 Ga $4s^24p$	32 Ge $4s^24p^2$	33 As $4s^24p^3$	34 Se $4s^24p^4$	35 Br $4s^24p^5$	36 Kr $4s^24p^6$
46 Pd $4d^{10}$	47 Ag $4d^{10}5s$	48 Cd $4d^{10}5s^2$	49 In $5s^25p$	50 Sn $5s^25p^2$	51 Sb $5s^25p^3$	52 Te $5s^25p^4$	53 I $5s^25p^5$	54 Xe $5s^25p^6$

71 Lu $4f^{14}5d6s^2$	72 Hf $5d^26s^2$	73 Ta $5d^36s^2$	74 W $5d^46s^2$	75 Re $5d^56s^2$	76 Os $5d^66s^2$	77 Ir $5d^9$	78 Pt $5d^96s$	79 Au $5d^{10}6s$	80 Hg $5d^{10}6s^2$	81 Tl $6s^26p$	82 Pb $6s^26p^2$	83 Bi $6s^26p^3$	84 Po $6s^26p^4$	85 At $6s^26p^5$	86 Rn $6s^26p^6$

Electron Orbits. Although the Bohr model of the atom was successful in quantitatively explaining many spectral data, the stabilization of certain electron orbits and the fine structure of spectral lines remained unexplained. De Broglie (1924) postulated that the dualism of observed light phenomena, which can be discussed either as wave phenomena or from the standpoint of the energy and momentum of photons, is quite general. According to the Planck and de Broglie equations,

$$\text{Energy:} \qquad E = h\nu$$

$$\text{Momentum:} \qquad mv = \frac{h}{\lambda} \tag{2.1}$$

where m is mass, v velocity, and λ wavelength, the motion of any particle is correlated to a wave phenomena of fixed frequency and wavelength. These relationships have been experimentally confirmed by X-ray, electron, and neutron diffraction. For stable electron orbits it is necessary to avoid destructive interference. A standing wave results when the orbit circumference corresponds to an integral number of wavelengths (Fig. 2.3).

Limitations fixed on the wave motion by the de Broglie equations, the particle mass, and energy are incorporated in the *Schrödinger wave equation*, which for an electron is

$$\frac{h^2}{8\pi^2 m}\left(\frac{\partial^2\psi}{\partial x^2} + \frac{\partial^2\psi}{\partial y^2} + \frac{\partial^2\psi}{\partial z^2}\right) - P\psi = \frac{h}{2\pi i}\frac{\partial\psi}{\partial t} \tag{2.2}$$

where P is the particle potential energy and $i = \sqrt{-1}$. Solutions of this equation give the pattern of the wave function ψ in space. The square of

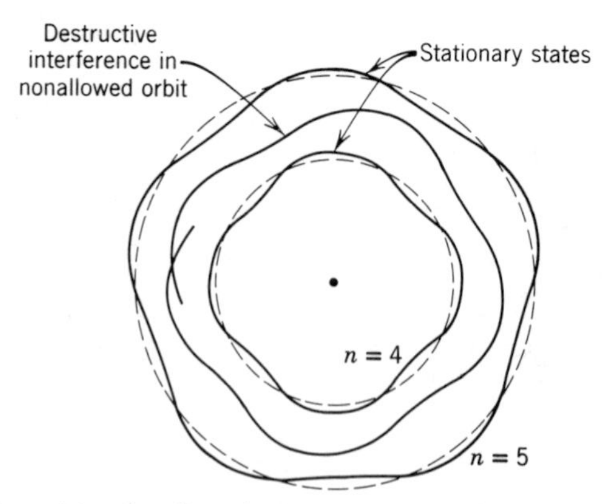

Fig. 2.3. Stationary states in allowed electron orbits and destructive interference in nonallowed orbit. From A. R. von Hippel, *Dielectrics and Waves*, John Wiley & Sons, New York, 1954.

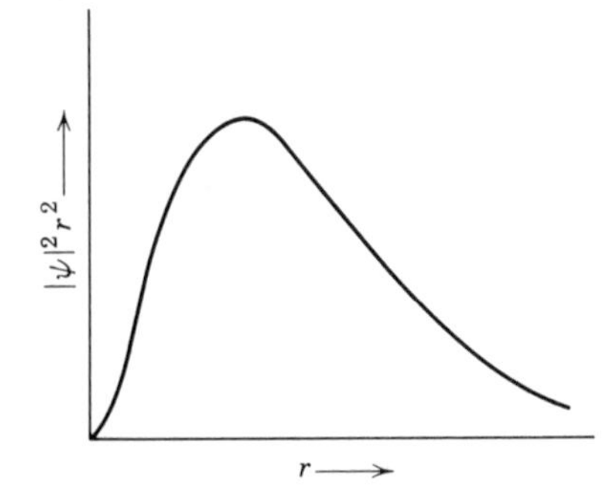

Fig. 2.4. Probability of finding the electron at a distance r from the nucleus for the $1s$ electron in the hydrogen atom.

its absolute value $|\psi|^2$ represents the probability of finding the electron in the enclosed volume element dv. For a number of relatively simple cases the distribution of electrons in space has been demonstrated. In its representation as a standing wave it must be viewed as smeared out over a probability pattern.

The simplest atom is hydrogen, which has a nucleus composed of one proton and, in the ground state, one $1s$ electron. This electron has spherical symmetry with a maximum probability distribution at a radial distance of about 0.5 Å (Fig. 2.4), which corresponds closely with the radius of the first Bohr orbit. For higher atomic numbers, the $1s$ electron distribution is similar except that the higher nuclear charge Ze makes them more tightly bound and closer to the nucleus. The $2s$ electrons also have spherical symmetry but are higher energy states and are farther from the central core of the positive nucleus and $1s$ electrons. In lithium, for example, the average radius of the $2s$ electrons is about 3 Å, whereas the average radius of the core is only about 0.5 Å. In contrast, the p orbitals are dumbbell-shaped (Fig. 2.5) with the three orbitals extending along orthogonal axes.

The fact that all but the outer few electrons form with the nucleus a compact stable core means that the few highest-energy electrons determine in large extent many properties of the elements. This can be seen from the periodic arrangement in Table 2.1.

The group 0 elements (He, Ne, Ar, Kr, Xe, and Rn) are characterized by a completed outer shell of electrons (the rare gas configuration). In helium, for example, the $n = 1$ shell is completely filled. Because of the increased nuclear charge it is much more difficult to remove an electron (energy required = 24.6 eV compared with 13.6 eV for hydrogen*) than is

*A unit of energy frequently used in discussing properties of atoms and molecules is the electron volt. This is an energy unit equal to the energy of an electron accelerated through a potential of 1 volt. As an energy unit is equal to 1.6×10^{-19} joule, since the charge on an electron is equal to 1.6×10^{-19} coulomb and $1\text{ eV} = (1\text{ volt})\ (1.6 \times 10^{-19}\text{ coulomb}) = 1.6 \times 10^{-19}$ joule. One electron volt per molecule equals 23.05 kcal/mole.

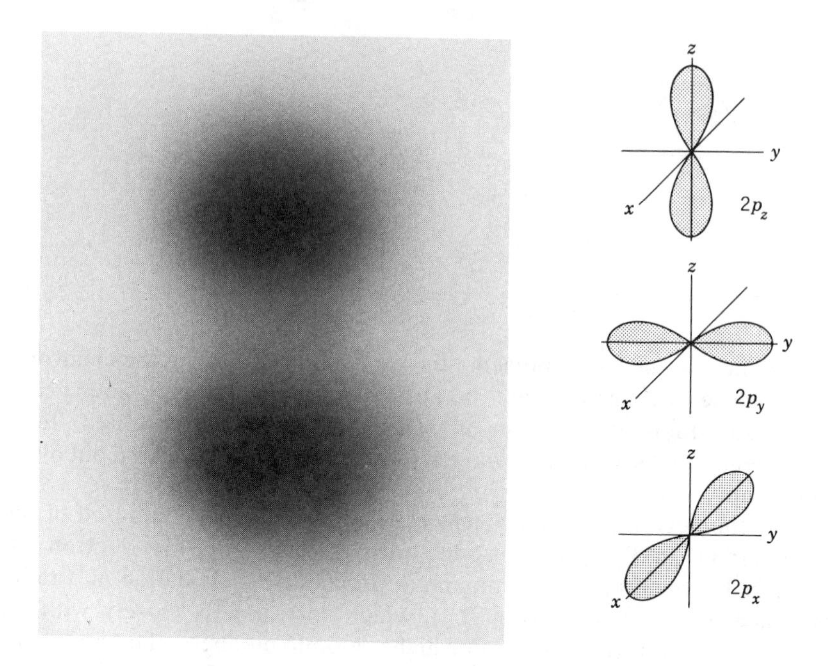

Fig. 2.5. Probability density contours for the dumbbell-shaped p orbitals.

the case for hydrogen (see Table 2.2 for the ionization energies of the elements). Since there are no vacant sites in the $n = 1$ shell, adding an electron would put it in the $2s$ orbital far from a neutral core—not a stable configuration. Consequently, helium is one of the most inert elements. Similar considerations apply to the other rare gases.

The group I elements are characterized by an outer s^1 orbital such as that illustrated in Fig. 2.6. In lithium ($1s^2, 2s^1$) the outer electron is at an average radius of about 3 Å and can be easily removed from the inner core of the nucleus and $1s^2$ electrons (ionization potential = 5.39 eV) to form the Li^+ ion. The ease of ionization makes lithium highly reactive and *electropositive* in chemical reactions. Removal of a second electron requires a much higher energy so that lithium is always monovalent, as are other group I elements.

In group II elements there is an outer s^2 shell from which two electrons are lost with an approximately equal expenditure of energy. These elements are electropositive and divalent. Similarly in the group III and IV elements there are three and four *outer* electrons; these elements are less electropositive with typical valencies of +3 and +4. The group V elements are characterized by an outer configuration of s^2 plus three

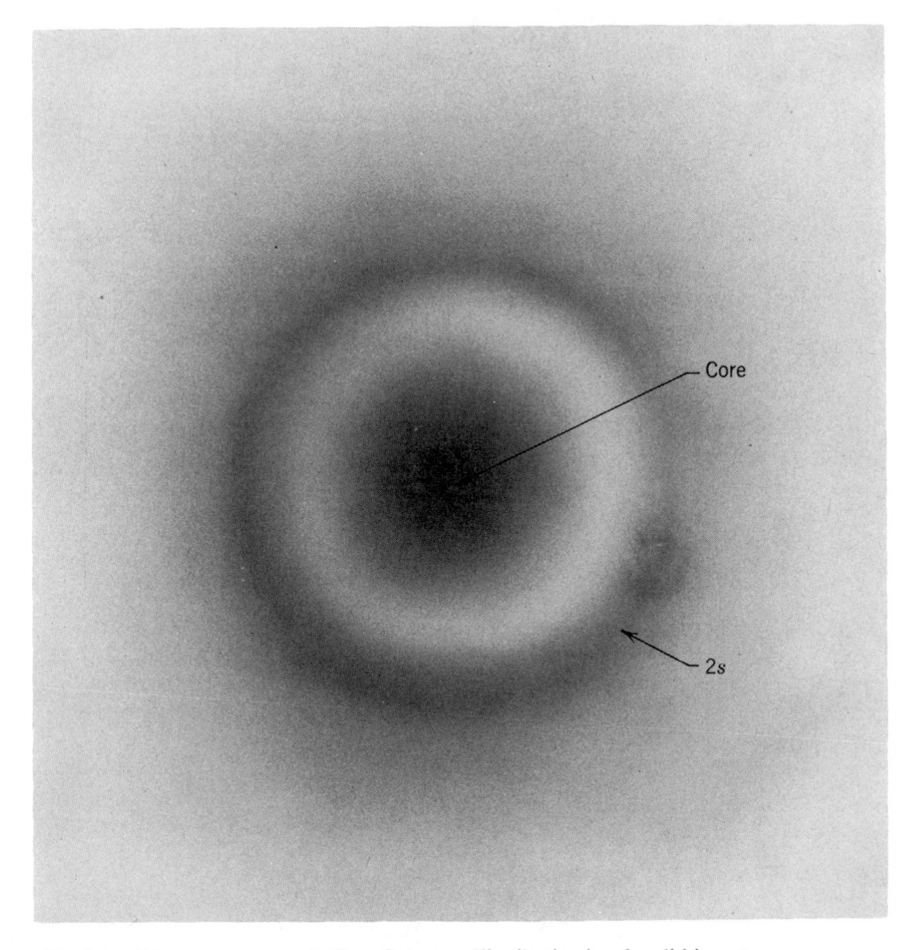

Fig. 2.6. Schematic representation of electron distribution in a free lithium atom.

other outer electrons (p^3 or d^3) and typically exhibit either a $+3$ or $+5$ valence. In some cases nitrogen and phosphorus gain additional electrons to fill completely the p orbital to form negative ions.

The formation of negative ions is characteristic of the group VII elements, which contain five electrons in the outer p orbital. The addition of one electron forms a stable F^- ion, for example. The binding energy for this additional electron in fluorine is 4.2 eV, called the *electron affinity*. This binding energy arises because in the $2p$ orbital the additional electron is not completely screened from the nucleus by other electrons,

Table 2.2. Ionization Energies of the Elements[a]

		Reaction[b]						Reaction[b]			
Z	Element	I	II	III	IV	Z	Element	I	II	III	IV
1	H	13.595				39	Y	6.38	12.23	20.5	
2	He	24.581	54.403			40	Zr	6.84	13.13	22.98	34.33
3	Li	5.390	75.619	122.419		41	Nb	6.88	14.32	25.04	38.3
4	Be	9.320	18.206	153.850	217.657	42	Mo	7.10	16.15	27.13	46.4
5	B	8.296	25.149	37.920	259.298	43	Tc	7.28	15.26		
6	C	11.256	24.376	47.871	64.476	44	Ru	7.364	16.76	28.46	
7	N	14.53	29.593	47.426	77.450	45	Rh	7.46	18.07	31.05	
8	O	13.614	35.108	54.886	77.394	46	Pd	8.33	19.42	32.92	
9	F	17.418	34.98	62.646	87.14	47	Ag	7.574	21.48	34.82	
10	Ne	21.559	41.07	63.5	97.02	48	Cd	8.991	16.904	37.47	
11	Na	5.138	47.29	71.65	98.88	49	In	5.785	18.86	28.03	54.4
12	Mg	7.644	15.031	80.12	109.29	50	Sn	7.342	14.628	30.49	40.72
13	Al	5.984	18.823	28.44	119.96	51	Sb	8.639	16.5	25.3	44.1
14	Si	8.149	16.34	33.46	45.13	52	Te	9.01	18.6	31	38
15	P	10.484	19.72	30.156	51.354	53	I	10.454	19.09		
16	S	10.357	23.4	35.0	47.29	54	Xe	12.127	21.2	32.1	
17	Cl	13.01	23.80	39.90	53.5	55	Cs	3.893	25.1		
18	Ar	15.755	27.62	40.90	59.79	56	Ba	5.210	10.001		
19	K	4.339	31.81	46	60.90	57	La	5.61	11.43	19.17	
20	Ca	6.111	11.868	51.21	67	72	Hf	7	14.9		

21	Sc	6.54	12.80	24.75	73.9	73	Ta	7.88	16.2		
22	Ti	6.82	13.57	27.47	43.24	74	W	7.98	17.7		
23	V	6.74	14.65	29.31	48	75	Re	7.87	16.6		
24	Cr	6.764	16.49	30.95	50	76	Os	8.7	17		
25	Mn	7.432	15.636	33.69		77	Ir	9			
26	Fe	7.87	16.18	30.643		78	Pt	9.0	18.56		
27	Co	7.86	17.05	33.49		79	Au	9.22	20.5		
28	Ni	7.633	18.15	35.16		80	Hg	10.43	18.751	34.2	
29	Cu	7.724	20.29	36.83		81	Tl	6.106	20.42	29.8	50.7
30	Zn	9.391	17.96	39.70		82	Pb	7.415	15.028	31.93	42.31
31	Ga	6.00	20.51	30.70	64.2	83	Bi	7.287	16.68	25.56	45.3
32	Ge	7.88	15.93	34.21	45.7	84	Po	8.43			
33	As	9.81	18.63	28.34	50.1	85	At				
34	Se	9.75	21.5	32	43	86	Rn	10.746			
35	Br	11.84	21.6	35.9	47.3	87	Fr				
36	Kr	13.996	24.56	36.9		88	Ra	5.277	10.144		
37	Rb	4.176	27.5	40		89	Ac	6.9	12.1	20?	
38	Sr	5.692	11.027	. . .	57						

[a] Values in electron volts. The values are obtained from the ionization potentials in Charlotte E. Moore, *Atomic Energy Levels as Derived from the Analyses of Optical Spectra* (Circular of the National Bureau of Standards 467, Government Printing Office, Washington, D.C., 1949–1958, vol. III). Multiply by 23.053 to convert from electron volts to kilocalories per mole.

[b] For reactions:

I. $m^{\circ}(g) = m^{+}(g) + e^{-}$.

II. $m^{+}(g) = m^{2+}(g) + e^{-}$.

III. $m^{2+}(g) = m^{3+}(g) + e^{-}$.

IV. $m^{3+}(g) = m^{4+}(g) + e^{-}$.

and the nuclear attractive force predominates over the repulsion forces of its companion electrons. In contrast, a second electron, which must enter the $3s$ orbital, is not stable; this electron finds an electrostatic repulsion force from the negative F^- core. In much the same way, an electron affinity occurs for the group VI elements, which tend to form divalent negative ions.

As the atomic number and number of electrons increase, the relative stability of energy levels of different orbitals becomes nearly the same. Orbitals fill in the order $1s$, $2s$, $2p$, $3s$, and $3p$, but then the $4s$ orbital becomes more stable and fills before the $3d$. However, they are nearly at the same energy level, and chromium has a $3d^5 4s$ configuration, in which both are incomplete. Elements with an incomplete d shell are called the *transition elements.* They have similar chemical properties, since the filling of the inner $3d$ shell has little effect on the ionization potential and properties of the $4s$ electrons. They also characteristically form colored ions and have special magnetic properties as a result of their electronic structure. Other series of transition elements occur with incomplete $4d$ and $5d$ shells. A similar and even more pronounced effect occurs for the *rare earth elements* in which the inner $4f$ shell is incompletely filled.

2.2 Interatomic Bonds

The principal forces that result in the formation of stable inorganic crystals are the electrostatic attractions between oppositely charged ions (as in KCl) and the stability of a configuration in which an electron pair is shared between two atoms (as in H_2, CH_4).

Ionic Bonds. The nature of ionic bonding can be illustrated by the formation of a KCl pair. When a neutral potassium atom is ionized to form K^+, there is an expenditure of 4.34 eV, the ionization energy. When a neutral chlorine atom adds an electron to form Cl^-, there is an energy gain of 3.82 eV, the electron affinity. That is, ionizing both requires a net expenditure of 0.52 eV (Fig. 2.7). As the positive and negative ions approach, there is a coulomb energy of attraction, $E = -e^2/4\pi\epsilon_0 R$ joules, where e is the charge on an electron and ϵ_0 is the permittivity of free space. The molecule becomes more stable as the ions approach. However, when the closed electron shells of the ions begin to overlap, a strong repulsive force arises. This repulsion force is due to the Pauli exclusion principle, which allows only one electron per quantum state. Overlapping of the closed shells requires that electrons go to higher energy states. In addition, the wave functions of the ions are distorted as the ions approach, so that the energy of each quantum state continuously increases as the separation decreases. This repulsion energy rises rapidly

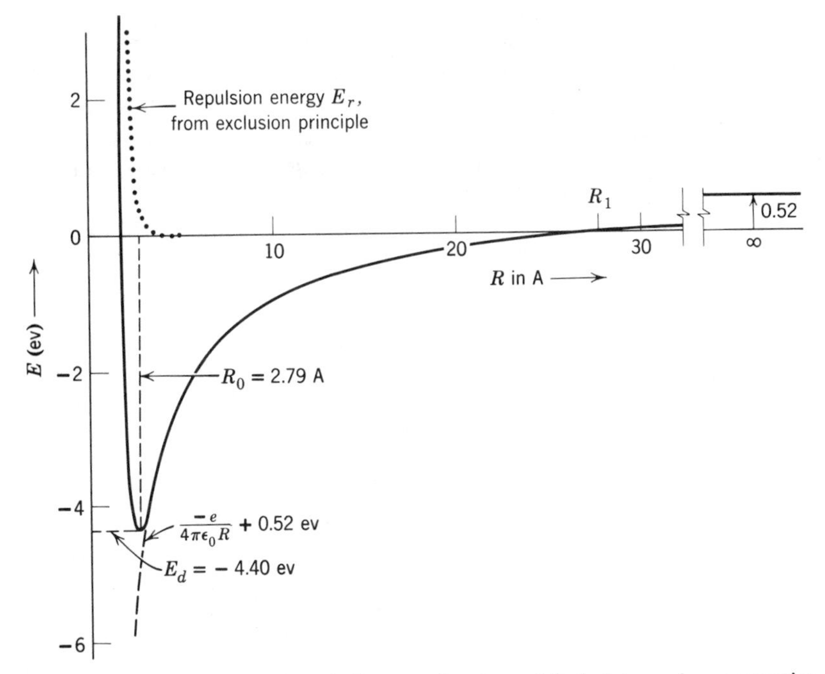

Fig. 2.7. Total energy of K^+ and Cl^- as a function of their internuclear separation R (reference 2).

when interpenetration of electron shells begins but makes little contribution at large ion separations. The assumption that this energy term varies as $1/R^n$, where n is a number typically of the order of 10, results in a satisfactory description of this behavior. The total energy of the KCl pair is

$$E = -\frac{e^2}{4\pi\epsilon_0 R} + \frac{B}{R^n} + 0.52 \text{ eV} \tag{2.3}$$

The empirical constant B and the exponent n may be evaluated from physical properties, as will be seen shortly. The combined effect of a decreasing energy term from the coulombic attraction and an increasing energy term from the repulsion force leads to an energy minimum (Fig. 2.7). This occurs at a configuration in which the net energy of formation of the KCl pair from the isolated atoms is about −4.4 eV.

The alkali halide compounds are largely ionic, as are compounds of group II and group VI elements. Most other inorganic compounds have a partly ionic–partly covalent character.

Covalent Bonds. The situation which leads to the formation of a stable hydrogen molecule H_2 is quite different from that considered for KCl. Here we consider the approach of two hydrogen atoms, each with one 1*s* electron. The potential energy of an electron is zero when it is far from the proton and a minimum at each proton. Along the line between protons the potential energy of the electron increases, but it always remains lower than that of a free electron (Fig. 2.8*a*). As the nuclei approach, there is a

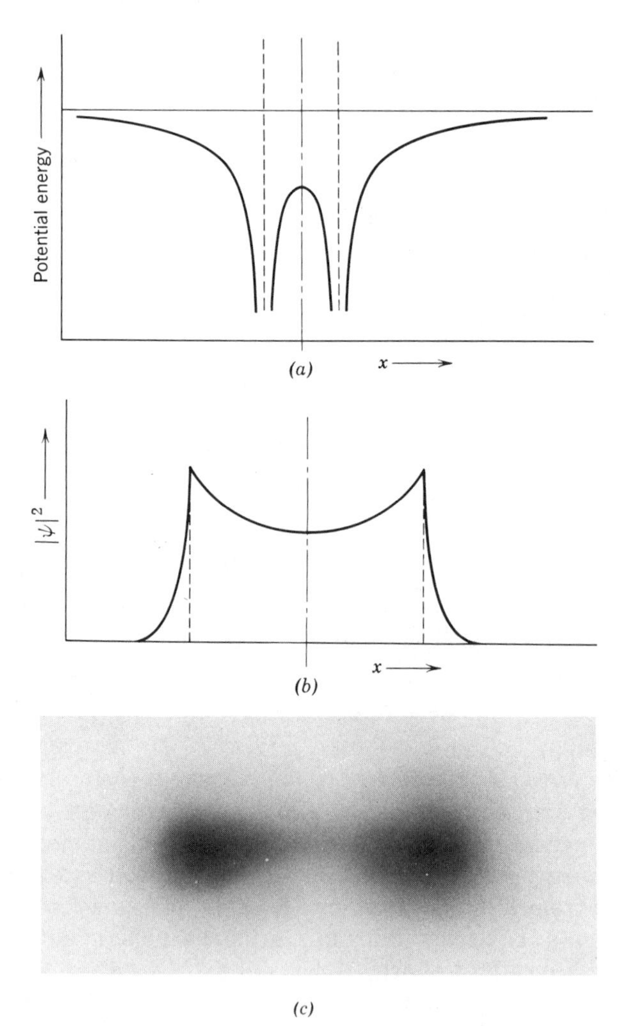

Fig. 2.8. (*a*) Potential energy and (*b*) and (*c*) electron density along a line between protons in the hydrogen molecule.

greater probability of finding an electron along a line between the protons, and a dumbbell distribution is found to be the most stable. The energy gained from the concentration of the electrons between protons increases as the protons get closer together. However, the repulsive force also increases, leading to an energy minimum similar in general shape to that of Fig. 2.7. This electron distribution, or wave function, which makes the total energy a minimum is the stable one for the system. A pair of electrons forms a stable bond, since only two electrons can be put into the wave function of lowest energy (the exclusion principle). A third electron would have to go into a quantum state of higher energy, and the resulting system would be unstable.

Covalent bonds are particularly common in organic compounds. Carbon, which has four valence electrons, forms four electron pair bonds which are tetrahedrally oriented in four equivalent sp^3 orbitals, each of which is similar in electron distribution to the contour map illustrated in Fig. 2.8*c*. This strong directional nature of covalent bonds is distinctive.

Van der Waals Bonds. An additional bonding force is the weak electrostatic forces between atoms or molecules known as the van der Waals, or dispersion, forces. For any atom or molecule there is a fluctuating dipole moment which varies with the instantaneous positions of electrons. The field associated with this moment induces a moment in neighboring atoms, and the interaction of induced and original moments leads to an attractive force. The bonding energies in this case are weak (about 0.1 eV) but of major importance for rare gases and between molecules for which other forces are absent.

Metallic Bond. The cohesive force between metal atoms arises from quantum-mechanical effects among an assemblage of atoms. This type of bond is discussed in the following section on bonding in solids.

Intermediate Bond Types. Although the structure of KCl can be regarded as almost completely ionic and that of H_2 as completely covalent, there are many intermediate types in which a bond may be characterized by an ionic electron configuration associated with an increased electron concentration along the line between atom centers.

Pauling has derived a semiempirical method of estimating bond type on the basis of an *electronegativity* scale. The electronegativity value is a measure of an atom's ability to attract electrons and is roughly proportional to the sum of the electron affinity (energy to add an electron) and ionization potential (energy to remove an electron). The electronegativity scale of the elements is shown in Fig. 2.9. Compounds between atoms with a large difference in electronegativity are largely ionic, as shown in Fig. 2.10. Compounds in which atoms have about the same electronegativity are largely covalent.

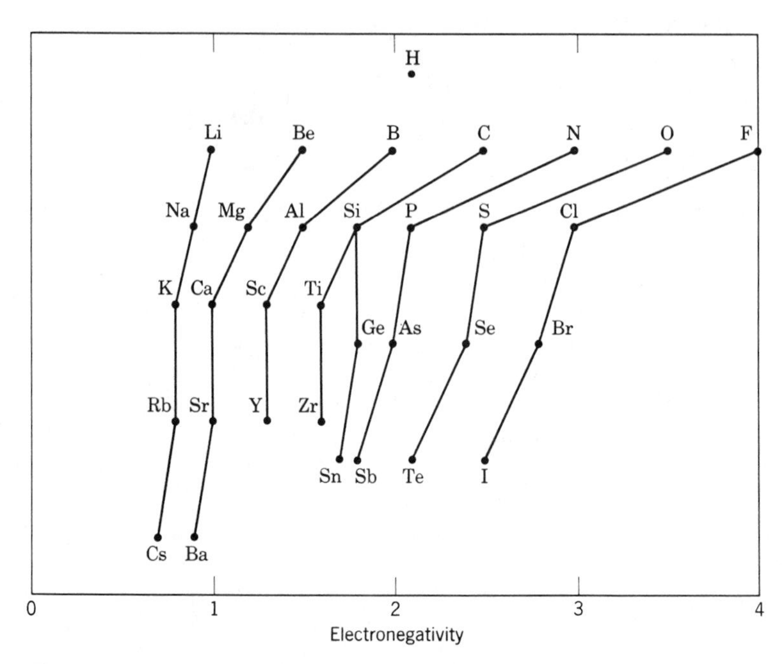

Fig. 2.9. Electronegativity scale of the elements (reference 3).

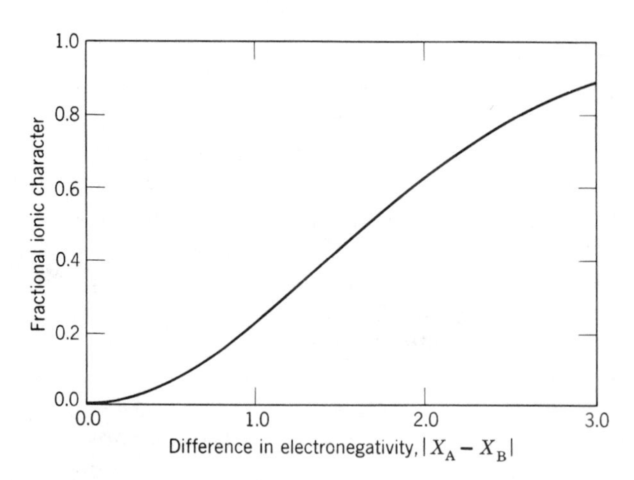

Fig. 2.10. Fraction of ionic character of bond A—B related to the difference in electronegativity $X_A - X_B$ of the atoms (reference 3).

2.3 Atomic Bonding in Solids

The forces between atoms in solids are similar to those already discussed, with the added factor that complex units fit together in crystalline solids with a periodicity that minimizes electrostatic repulsive forces and allows solids to have bonds which match at energetically favorable angles and spacings. It is useful to consider bonding in solids in classes based on the major contribution to bond development. As for molecules, however, intermediate cases are common. The major characteristic determining bond energy and bond type is the distribution of electrons around the atoms and molecules. We can generally class solids as having ionic, covalent, molecular, metallic, or hydrogen bond structures.

Ionic Crystals. In ionic crystals the distribution of electrons between ions is the same as for the single ionic bond discussed previously. In a crystal, however, each positive ion is surrounded by several negative ions, and each negative ion is surrounded by several positive ions. In the sodium chloride structure (Fig. 2.11), for example, each ion is surrounded by six of the opposite charge. The energy of the assemblage varies with interionic separation in much the same way as in Fig. 2.7.

The energy of one ion of charge $Z_i e$ in a crystal such as NaCl may be obtained by summing its interaction, as given by Eq. 2.3, with the other j ions in the crystal:

$$E_i = \sum_j \left(\frac{Z_i e Z_j e}{4\pi\epsilon_0 R_{ij}} + \frac{B_{ij}}{R_{ij}^{\,n}} \right) \tag{2.4}$$

where R_{ij} is the distance between the ion under consideration and its jth neighbor of charge $Z_j e$. Subscripts have been added to the empirical constant B to take into account that its value may be different for interactions between the different species of ions. For simplicity we have neglected adding in the constant difference between the ionization potential and electron affinity (this, in effect, defines zero energy as when the set of ions rather than neutral atoms is at infinite separation). The total energy of the crystal may be obtained by adding the contributions (Eq. 2.4) of every ion in the crystal, but the result must be multiplied by 1/2: the interaction of an ion pair ij represents the same contribution as ji, and simply summing Eq. 2.4 over the entire crystal would include each interaction twice.

We would expect the energy of each ion in the NaCl structure to be the same, so that the summation of Eq. 2.4 over the $2N$ ions or N "molecules" of NaCl may be accomplished by the multiplication of Eq.

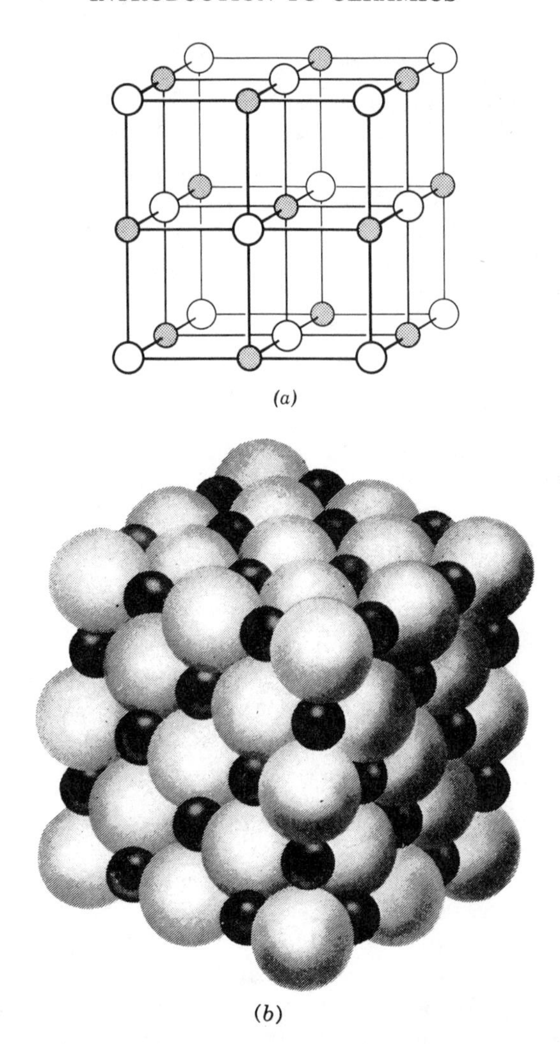

(a)

(b)

Fig. 2.11. Crystal structure of sodium chloride.

2.4 by $2N \times 1/2$:

$$E = \frac{1}{2}\sum_i E_i = \frac{1}{2}(2NE_i) = N\sum_j \left(\frac{Z_i Z_j e^2}{4\pi\epsilon_0 R_{ij}} + \frac{B_{ij}}{R_{ij}^n}\right) \tag{2.5}$$

The nature of the summation depends on the ion separation as well as the atomic arrangement. If we let $R_{ij} = R_0 x_{ij}$, where R_0 is some characteristic separation (usually taken as the interionic separation), then

$$E = N\left(-\frac{|Z_1||Z_2|e^2}{4\pi\epsilon_0 R_0}\alpha + \frac{C}{R_0{}^n}\right) \tag{2.6}$$

where

$$\alpha = \sum_i \frac{-(Z_i/|Z_i|)(Z_j/|Z_j|)}{x_{ij}} \tag{2.7}$$

and

$$C = \sum_i \frac{B_{ij}}{x_{ij}{}^n} \tag{2.8}$$

The quantity α is called the *Madelung constant.* From the way in which it is defined, its value depends only on the geometry of the structure and may be evaluated once and for all for a particular structure type. For the NaCl structure type, $\alpha = 1.748$; for the CsCl structure, 1.763; for the zinc blende structure, 1.638; and for wurtzite, 1.641. Physically, the Madelung constant represents the coulomb energy of an ion pair in a crystal relative to the coulomb energy of an isolated ion pair; α is larger than unity but not very much so. It may also be noted that the Madelung constant for the NaCl structure type (six nearest neighbors) is less than 1% smaller than that for the CsCl structure type (eight nearest neighbors). The Madelung constants for the wurtzite and zinc blende structures (4 nearest neighbors), which differ only in second-nearest-neighbor arrangements are even more similar. The coulomb energies of different arrangements of ions in a crystal may therefore be seen to differ only by relatively minor amounts.

The series which provides the value of C in Eq. 2.8 might be expected to converge rapidly because the repulsive interaction between ions is short-range. Unfortunately, it depends not only on structure type but also on the particular chemical compound in question, since B_{ij} is different for different species of ions. The value of C, however, may be evaluated by noting that the energy of the crystal is a minimum when the ions are separated by R_0. Differentiating Eq. 2.6 with respect to R_0, setting the result equal to zero, and solving for C provides

$$C = \frac{\alpha|Z_i||Z_j|e^2}{4\pi\epsilon_0 n} R_0{}^{n-1} \tag{2.9}$$

so that Eq. 2.6 may be written

$$E = -\frac{N\alpha|Z_i||Z_j|e^2}{4\pi\epsilon_0}\frac{1}{R_0}\left(1 - \frac{1}{n}\right) \tag{2.10}$$

The value of n may, in turn, be calculated from measurement of the compressibility of the crystal. It usually has a value of the order of 10, so that the repulsive interaction between ions increases the total energy of the crystal by only 10% or so of the coulomb energy.

Ionic crystals are characterized by strong infrared absorption, transparency in the visible wavelengths, and low electrical conductivity at low temperatures but good ionic conductivity at high temperatures. Compounds of metal ions with group VII anions are strongly ionic (NaCl, LiF, etc.). Compounds of metals with oxygen ions are largely ionic (MgO, Al_2O_3, ZrO_2, etc.). Compounds with the higher-atomic-weight elements of group VI (S, Se, Te), which have lower electronegativity (see Figs. 2.9 and 2.10), are increasingly less ionic in character. The strength of ionic bonds increases as the valence increases (Eq. 2.6). The electron distribution in ions is nearly spherical, and the interatomic bond, since it arises from coulombic forces, is nondirectional in nature. The stable structure assumed by an ionic compound thus tends to be one in which an ion obtains the maximum number of neighbors (or *coordination number*) of opposite charge. Such structures therefore depend on obtaining maximum packing density of the ions.

Covalent Crystals. Each single bond in a covalent crystal is similar to the bond between hydrogen atoms discussed in the previous section. A pair of electrons is concentrated in the space between the atoms. Covalent crystals form when a repetitious structure can be built up consistent with the strong directional nature of the covalent bond. For example, carbon forms four tetrahedral bonds. In methane $\cdot CH_4$ these are used up in forming the molecule so that no electrons are available for forming additional covalent bonds and no covalent crystal can be built up. In contrast, carbon itself forms a covalent crystal, diamond, with bonds arrayed periodically. In the diamond structure each carbon atom is surrounded by four other carbon atoms (Fig. 2.12). This structure, with

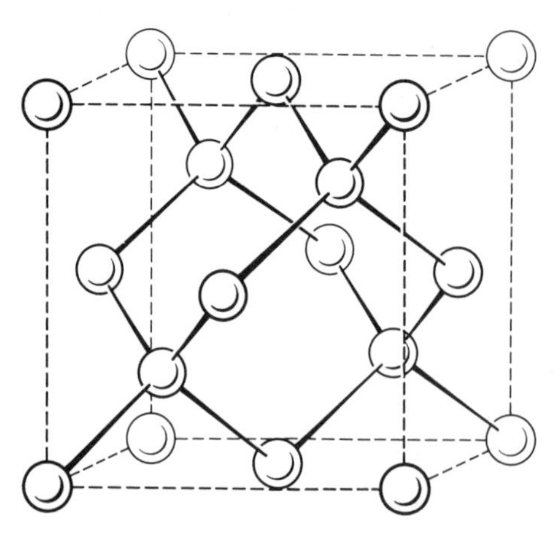

Fig. 2.12. Crystal structure of diamond.

tetrahedral (fourfold) coordination, does not allow dense packing of the atoms in space to get the maximum possible number of bonds, but the open structure is required by the directed nature of the bonds.

Covalent crystals, such as diamond and silicon carbide, have high hardness, high melting points, and (when specimens are pure) low electrical conductivities at low temperatures. Covalent crystals are formed between atoms of similar electronegativity which are not close in electronic structure to the inert gas configuration (i.e., C, Ge, Si, Te, etc.). In addition to purely covalent crystals, most other crystals also have a significant contribution of covalent bond nature, as illustrated by Fig. 2.10. Although the empirical curve there may be taken as a guide, it is difficult to resolve intermediate cases with much confidence.

Molecular Crystals. Organic molecules, such as methane, and inert gas atoms are bound together in the solid phase by means of weak van der Waals forces. Consequently, these crystals are weak, easily compressible, and have low melting and boiling points. Although these forces occur in all crystals, they are only important when other forces are absent. One place in ceramics in which they may come into play is in the bonding together of silicate sheet structures in clays.

Hydrogen Bond Crystals. A special, but common, bond in inorganic crystals is due to a hydrogen ion forming a rather strong bond between two anions. The hydrogen bond is largely ionic and is formed only with highly electronegative anions: O^{2-} or F^{-}. The proton can be viewed as resonating between the positions O–H—O and O—H–O. The resultant bond is important in the structure of water, ice, and many compounds containing hydrogen and oxygen, such as hydrated salts. It is responsible for the polymerization of HF and some organic acids and in the formation of a number of inorganic polymers of importance to inorganic adhesives and cements.

Metal Crystals. A prominent characteristic of metals is their high electrical conductivity, which implies that a high concentration of charged carriers (electrons, able to move freely). These electrons are called *conduction* electrons. As a first crude approximation, metals may be regarded as an array of positive ions immersed in a uniform *electron cloud*, and this is not too far from the truth for the alkali metal crystals; in these the bonding energy is much less than for the ionic alkali halides, for example. In the transition metals the inner electronic orbitals contribute to electron concentration (electron pair bonds) along lines between atom centers, and stronger bonding results.

The characteristic electron mobility of metals can best be understood by considering the changes that occur in electronic energy states as a number of atoms come together to form a crystal. Bringing atoms

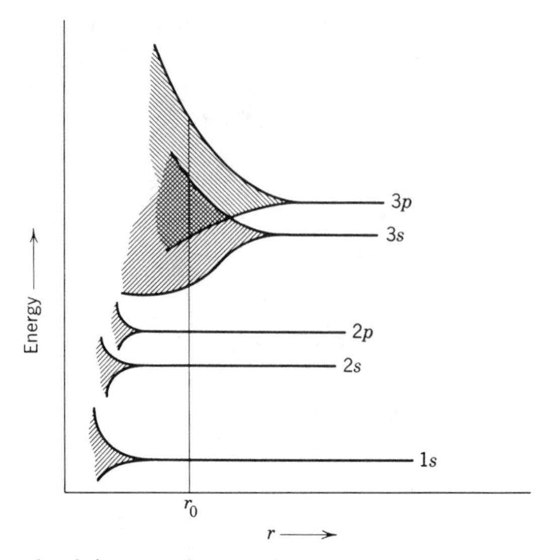

Fig. 2.13. Energy levels in magnesium broaden and become bands as the atoms are brought closer together (schematic).

together leaves the total number of quantum states with a given quantum number unchanged, but as atoms are brought together, interaction between orbitals is increasing the number of electrons with the same quantum number. Energy levels broaden and become *allowed bands,* in which the spacing between individual electron energy levels is so close that they can be considered continuous bands of allowed energy (Fig. 2.13). In metals the higher-energy allowed bands, or permitted energy levels, overlap and are incompletely filled with electrons. This allows relatively free movement of the electrons from atom to atom without the large energies which are required for dielectrics, in which electrons must be raised in energy to a new band level before conduction is possible.

2.4 Crystal Structures

Crystals are composed of periodic arrays of atoms or molecules, and an understanding of crystal properties can be very rapidly developed if we know the ways in which periodicity is obtained. The stablest crystal structures are those that have the densest packing of atoms consistent with other requirements, such as the number of bonds per atom, atom sizes, and bond directions. As a basis for further discussion, it is essential to have a clear picture of how spherical atoms can be stacked together. It

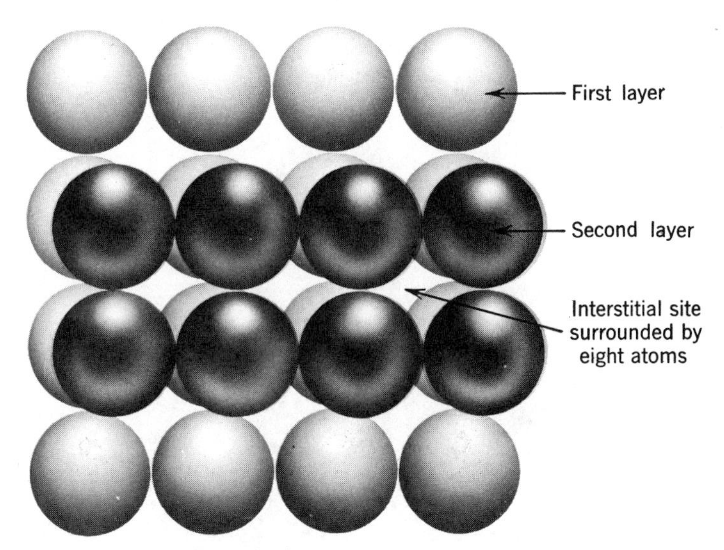

Fig. 2.14. Simple cubic packing of spheres.

is best to do actual experiments with spheres such as ping pong balls, cork balls, and other models which allow study in three dimensions.

Simple Cubic Structure. One way in which spheres can be packed together is in a simple cubic array (Fig. 2.14). Each sphere has four adjacent spheres in the plane of the paper, one above, and one below for a total of six nearest neighbors. In addition, there are interstices surrounded by eight spheres. These interstices are also in a cubic array, with one hole for each sphere. This kind of packing is not very dense, having a total of 48% void space.

Close-Packed Cubic Structure. Another arrangement of spheres has cubic layers with the second layer placed above the spaces in the bottom layer, as illustrated in Fig. 2.15*a*. When a third layer is put on above the first, we have the basis for a dense-packed structure in which each sphere has twelve nearest neighbors, four in the plane of the paper, four above, and four below. This kind of packing is more dense than the simple cubic structure; it has a void volume of only 26%. The same structure can be built up from hexagonal layers of spheres having six closest neighbors in the plane of the paper, three above, and three below to give a total of twelve, as shown in Fig. 2.15*b*. Other views of this arrangement which show the cubic symmetry are given in Fig. 2.15*c* and *d*. The simplest unit cell which gives this structure when periodically repeated is the face-centered cubic one illustrated in Fig. 2.15*f*.

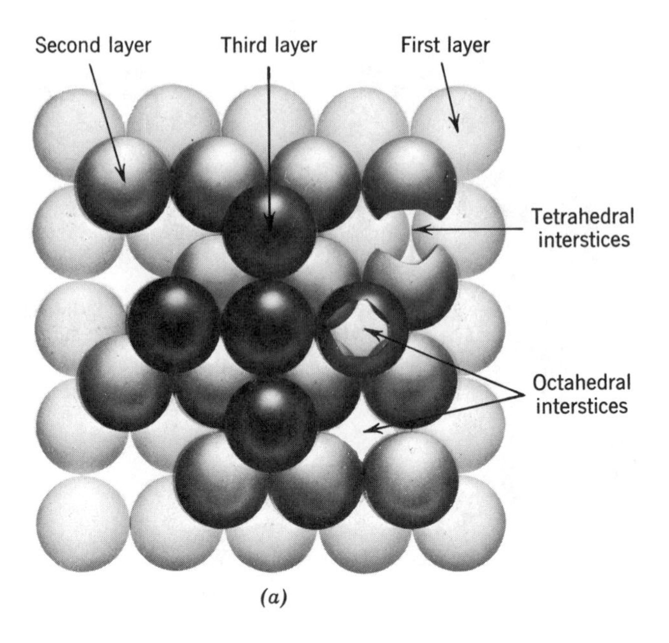

(a)

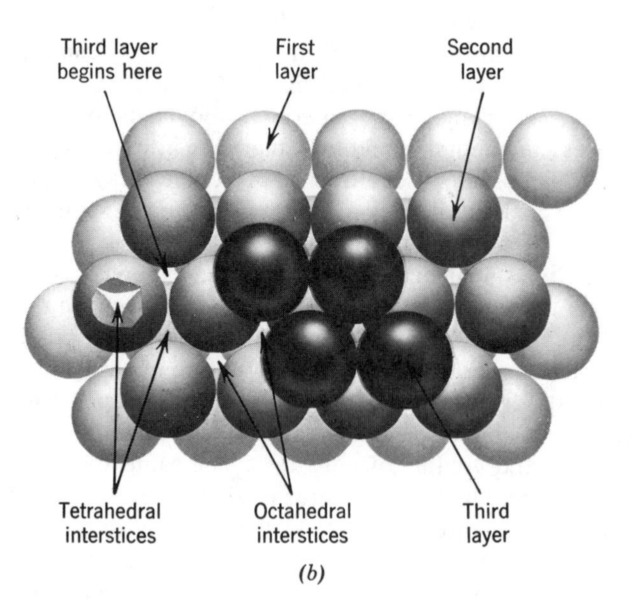

(b)

Fig. 2.15. Various aspects of face-centered cubic packing of spheres. See text for discussion.

(c)

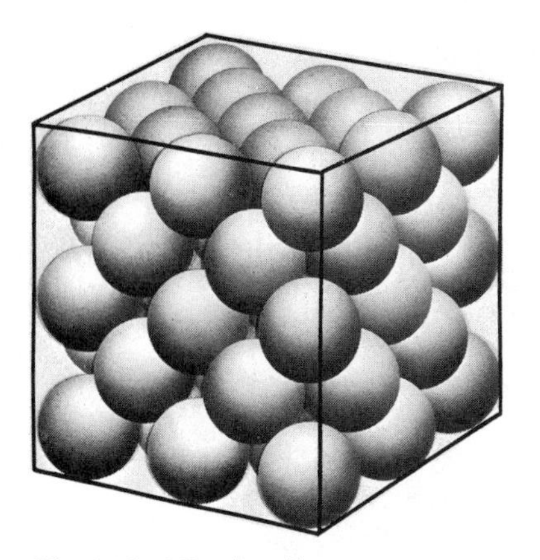

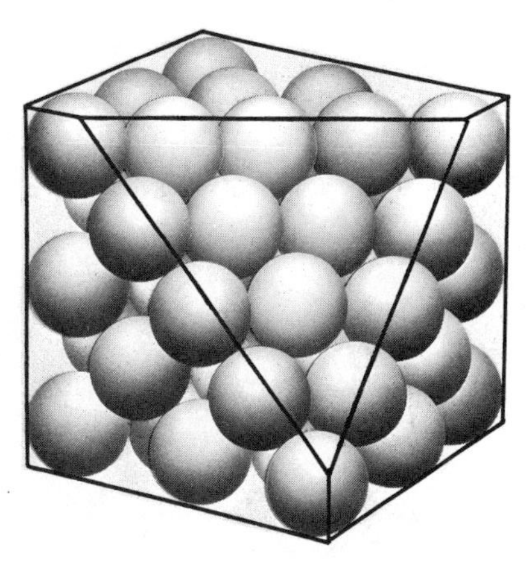

(d)

Fig. 2.15 (*Continued*)

In contrast to the simple cubic packing there are two kinds of interstices in the face-centered cubic array. There are *octahedral* holes surrounded by six atoms and *tetrahedral* holes surrounded by four atoms (Fig. 2.15*e*). In each unit cell containing a total of four atoms there are four octahedral interstices and eight tetrahedral interstices arranged with cubic symmetry, as shown in Fig. 2.15*f*. This is difficult to visualize easily,

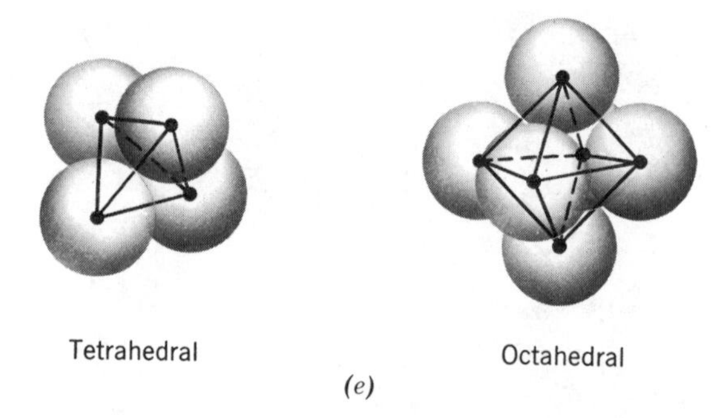

(e)

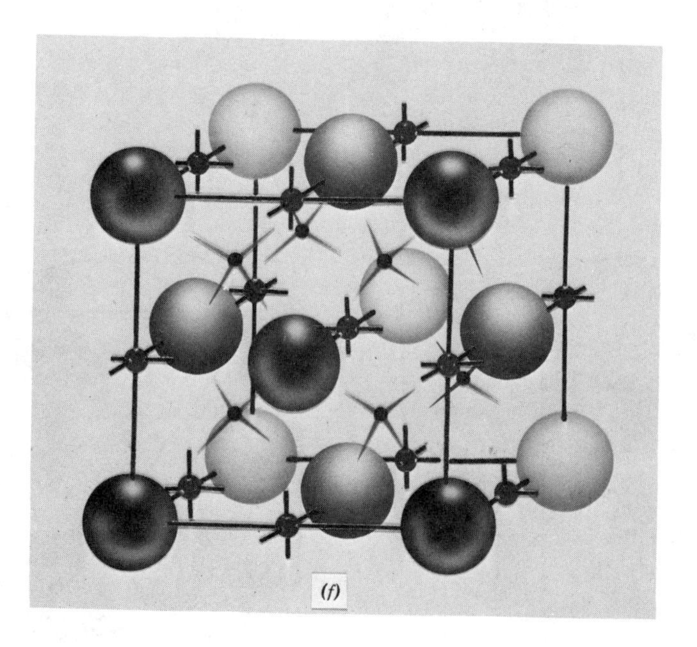

(f)

Fig. 2.15 (*Continued*)

but Fig. 2.15*a*, *b*, and *f* should be compared so that the nature and distribution of interstitial sites is clear.

Close-Packed Hexagonal Structure. In the close-packed cubic structure, the plane of densest atomic packing is a plane in which each atom is surrounded by six others in hexagonal symmetry, as shown in the cutaway view of Fig. 2.15*d*. If we start with a layer of atoms closely packed and then add a second layer, we can go on to add a third layer in two ways. If the third layer is not directly above previous layers, we end up with a face-centered cubic lattice (Fig. 2.15*f*). However, if the third layer is directly above the first (Fig. 2.16*a*), the packing is equally dense but has a different structure, called hexagonal close packing. From a side view the face-centered cubic lattice corresponds to stacking layers of displacement *a*, *b*, *c*, *a*, *b*, *c*, whereas the hexagonal close-packing structure corresponds to stacking of layers *a*, *b*, *a*, *b* (Fig. 2.16*b*). Although the two packing arrangements have the same density, there is a difference in the arrangement of the atoms and interstices. It is very instructive to work out the structural features of the hexagonal close-packed lattice, such as those that are illustrated in Fig. 2.15, for the face-centered cubic structure.

Space Lattices. As implied previously, only certain geometric forms can be repeated periodically to fill space. By systematically considering the various symmetry operations needed to develop a periodic structure that fills space, it can be shown that there are 32 permissible arrangements of points around a central point. These require 14 different *Bravais* or *space lattices*, as illustrated in Fig. 2.17. The conventional unit cells derived from these space lattices are described in terms of unit-cell axes and angles (Fig. 2.18). The lattices are grouped into six systems—triclinic, monoclinic, orthorhombic, tetragonal, hexagonal and cubic—in order of increasing symmetry.

Geometrical features in a lattice, such as directions and planes, are most conveniently described relative to the unit-cell edges. Directions are specified with the three indices which give the multiples of the cell edges necessary as components to achieve a given bearing. A negative component is indicated by a bar over the index. The three indices are enclosed in square brackets to distinguish a direction from other geometrical features such as points or planes. Several directions are indicated in Fig. 2.19. In symmetrical lattices several different directions are equivalent. A whole set of equivalent directions is indicated by the symbol ⟨ ⟩ about the indices of one representative direction. For example, ⟨100⟩ in a cubic crystal stands for the set of six equivalent directions along the cell edges: [100], [010], [001], $[\bar{1}00]$, $[0\bar{1}0]$, $[00\bar{1}]$.

Crystallographic planes are defined in terms of their intercepts on the cell edges. The intercepts themselves are not used, since this would necessitate use of the symbol ∞ if a plane happened to be parallel to one of the cell edges. Instead, the integers used, called *Miller indices*, are the reciprocals of the intercepts multiplied by the factor necessary to convert them to integers. Indices of planes are placed within parentheses to distinguish them from directions. For the plane in Fig. 2.19*b*, for example, the intercepts are 1, ∞, ∞. Their reciprocals are 1, 0, 0, and the Miller indices assigned are (100). In Fig. 2.19*d* the intercepts are ∞, 2, 4, their

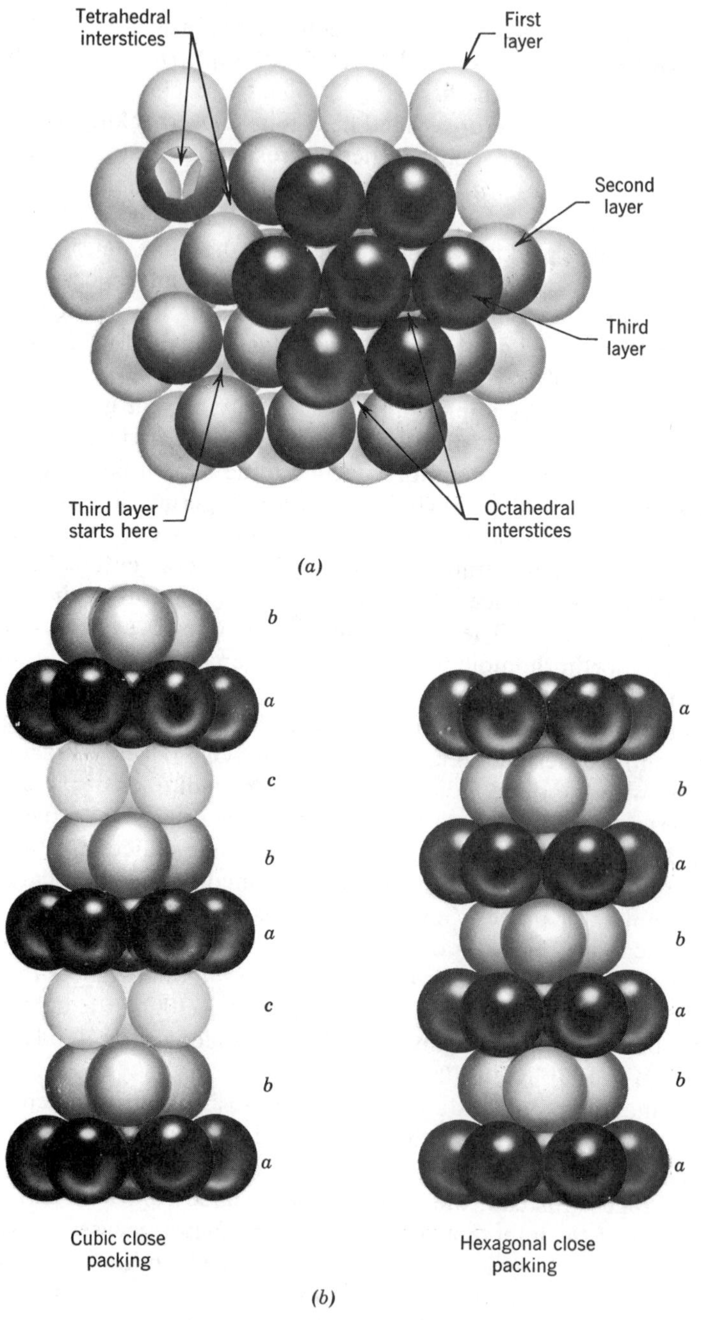

Fig. 2.16. Development of hexagonal close packing.

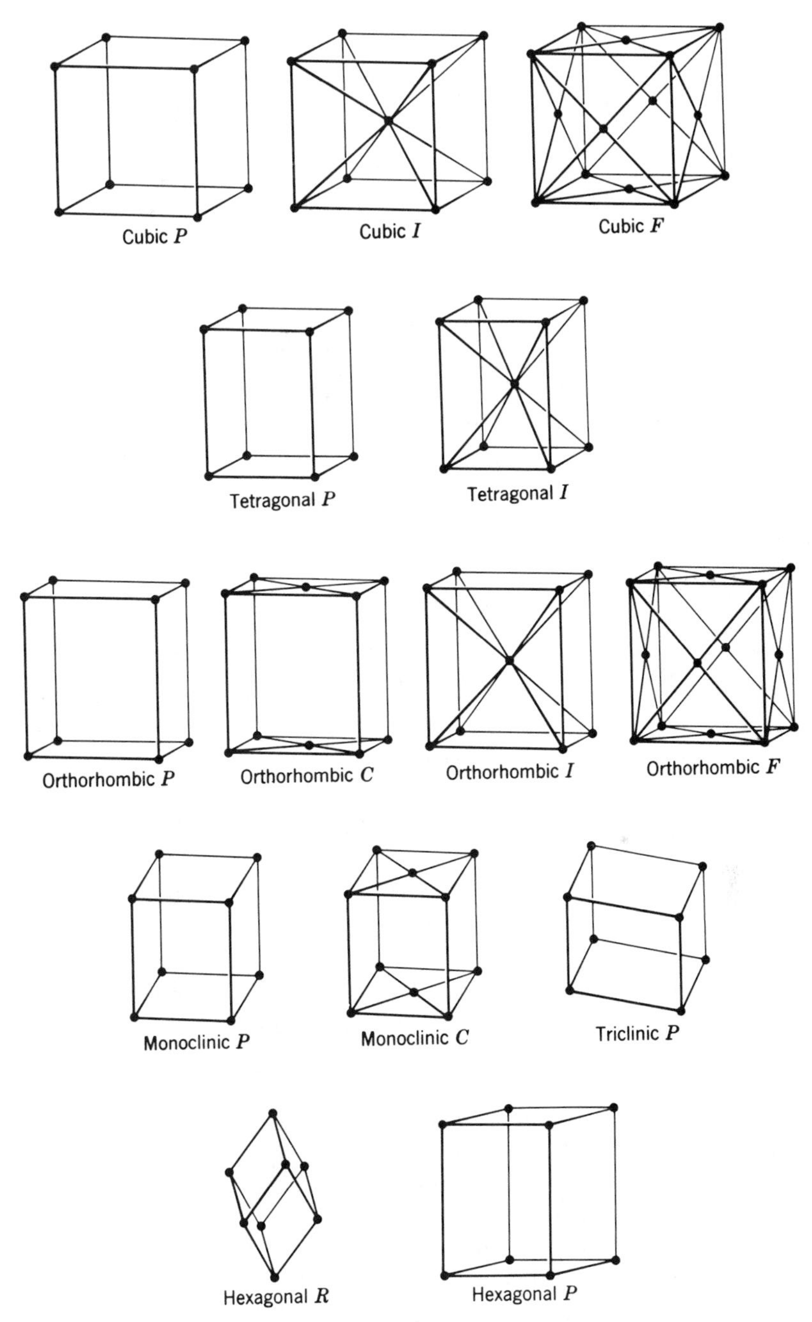

Fig. 2.17. Fourteen Bravais or space lattices.

System	Number of Lattices in System	Nature of Unit-Cell Axes and Angles	Lengths and Angles to Be Specified
Triclinic	1	$a \neq b \neq c$ $\alpha \neq \beta \neq \gamma$	a, b, c α, β, γ
Monoclinic	2	$a \neq b \neq c$ $\alpha = \gamma = 90^\circ \neq \beta$	a, b, c β
Orthorhombic	4	$a \neq b \neq c$ $\alpha = \beta = \gamma = 90^\circ$	a, b, c
Tetragonal	2	$a = b \neq c$ $\alpha = \beta = \gamma = 90^\circ$	a, c
Hexagonal	2	$a = b \neq c$ $\alpha = \beta = 90^\circ$ $\gamma = 120^\circ$	a, c
Cubic	3	$a = b = c$ $\alpha = \beta = \gamma = 90^\circ$	a

(*a*)

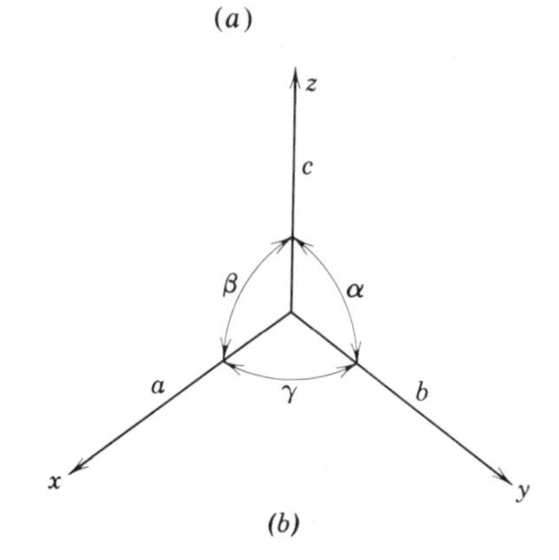

(*b*)

Fig. 2.18. (*a*) The six systems containing fourteen Bravais or space lattices and conventional unit cells of crystals. (*b*) The lengths and angles to be specified.

reciprocals 0, 1/2, 1/4, and the Miller indices (0 2 1). An entire set of equivalent planes is denoted by braces about the Miller indices of one representative plane. Thus {100} in a cubic crystal represents the set of six cube faces (100), (010), (001), ($\bar{1}$00), (0$\bar{1}$0), and (00$\bar{1}$). In cubic crystals the direction [*hkl*] is always perpendicular to the plane having the same indices. This is not generally true in any other crystal system.

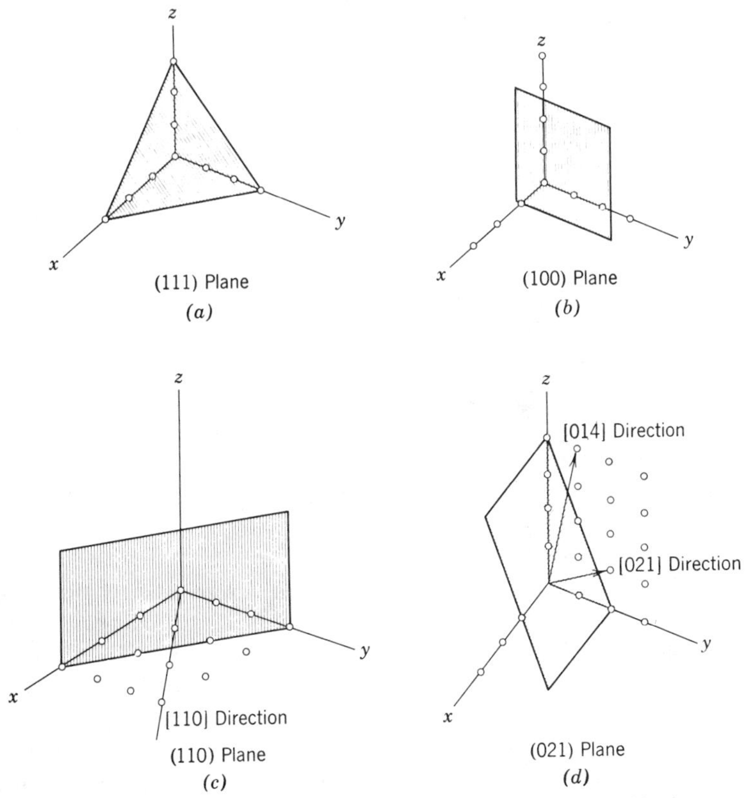

Fig. 2.19. Miller indices of selected planes and directions in a crystal lattice.

In the examples for sets of equivalent planes or directions given above it may be seen that the indices of all members of the set are related through a permutation of their order. This occurs because the symmetry operations which relate equivalent features also transform the cell edges into one another. This is not the case in the hexagonal system in which, for example, the six faces of a hexagonal prism may have indices (110), ($\bar{1}$20), ($\bar{2}$10), ($\bar{1}\bar{1}$0), (1$\bar{2}$0), and (2$\bar{1}$0), which bear no obvious relationship to one another. This situation may be remedied by defining a fourth redundant axis opposite in direction to the vector sum of *a* and *b*. The Miller index of this axis turns out to be minus the sum of the first two. The Miller indices (*hkl*) for a hexagonal crystal are therefore expanded and written $h, k, -(h+k), l$. Some writers prefer to omit the redundant index, and planes for hexagonal crystals are sometimes expressed $(hk \cdot l)$. The reader should verify that, on inclusion of the fourth index, all six faces of the hexagonal prism given above are related by permutation of position and sign of the same integers.

2.5 Grouping of Ions and Pauling's Rules

In crystals having a large measure of ionic bond character (halides, oxides, and silicates generally) the structure is in large part determined on the basis of how positive and negative ions can be packed to maximize electrostatic attractive forces and minimize electrostatic repulsion. The stable array of ions in a crystal structure is the one of lowest energy, but the difference in energy among alternative arrays is often very slight. Certain generalizations have been made, however, which successfully interpret the majority of ionic crystal structures which are known. These generalizations have been compactly expressed in a set of five statements known as *Pauling's rules.*

Pauling's first rule states that a coordination polyhedron of anions is formed about each cation in the structure. The cation–anion distance is determined by the sum of their radii. The coordination number (i.e., the number of anions surrounding the cation), is determined by the ratio of the radii of the two ions. The notion that a "radius" may be ascribed to an ion, regardless of the nature of the other ion to which it is bonded, is strictly empirical. Its justification is the fact that self-consistent sets of radii may be devised which successfully predict the interionic separations in crystals to within a few percent. The reason why the radius ratio of two species of ions influences the coordination number is apparent from Fig. 2.20. A central cation of given size cannot remain in contact with all surrounding anions if the radius of the anion is larger than a certain critical value. A given coordination number is thus stable only when the ratio of cation to anion radius is greater than some critical value. These limits are given in Fig. 2.21. In a crystal structure the anion is also surrounded by a coordination polyhedron of cations. Critical radius ratios also govern the coordination of cations about anions. Since anions are

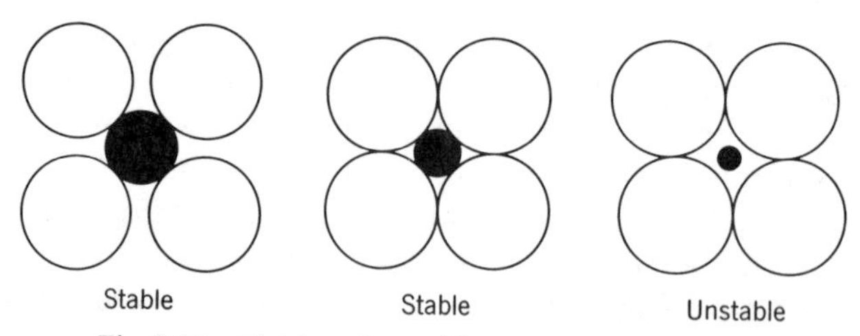

Fig. 2.20. Stable and unstable coordination configurations.

Coordination Number	Disposition of Ions about Central Ion	Range of $\frac{\text{Cation Radius}}{\text{Anion Radius}}$ Ratio
8	Corners of cube	≥ 0.732
6	Corners of octahedron	≥ 0.414
4	Corners of tetrahedron	≥ 0.225
3	Corners of triangle	≥ 0.155
2	Linear	≥ 0

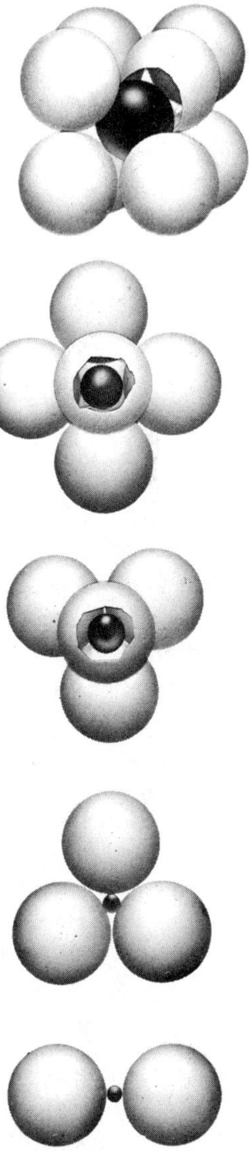

Fig. 2.21. Critical radius ratios for various coordination numbers. The most stable structure is usually the one with the maximum coordination number allowed by the radius ratio.

Table 2.3. Ionic Crystal Radii
(Coordination Number = 6)

Ag^{1+}	Al^{3+}	As^{5+}	Au^{1+}	B^{3+}	Ba^{2+}	Be^{2+}	Bi^{5+}	Br^{1-}	C^{4+}	Ca^{2+}	Cd^{2+}	Ce^{4+}
1.15	0.53	0.50	1.37	0.23	1.36	0.35	0.74	1.96	0.16	1.00	0.95	0.80
Cl^{1-}	Co^{2+}	Co^{3+}	Cr^{2+}	Cr^{3+}	Cr^{4+}	Cs^{1+}	Cu^{1+}	Cu^{2+}	Dy^{3+}	Er^{3+}	Eu^{3+}	F^{1-}
1.81	0.74	0.61	0.73	0.62	0.55	1.70	0.96	0.73	0.91	0.88	0.95	1.33
Fe^{2+}	Fe^{3+}	Ga^{3+}	Gd^{3+}	Ge^{4+}	Hf^{4+}	Hg^{2+}	Ho^{3+}	I^{1-}	In^{3+}	K^{1+}	La^{3+}	Li^{1+}
0.77	0.65	0.62	0.94	0.54	0.71	1.02	0.89	2.20	0.79	1.38	1.06	0.74
Mg^{2+}	Mn^{2+}	Mn^{4+}	Mo^{3+}	Mo^{4+}	Na^{1+}	Nb^{5+}	Nd^{3+}	Ni^{2+}	O^{2-}	P^{5+}	Pb^{2+}	Pb^{4+}
0.72	0.67	0.54	0.67	0.65	1.02	0.64	1.00	0.69	1.40	0.35	1.18	0.78
Rb^{1+}	S^{2-}	S^{6+}	Sb^{5+}	Sc^{3+}	Se^{2-}	Se^{6+}	Si^{4+}	Sm^{2+}	Sn^{2+}	Sn^{+4}	Sr^{2+}	Ta^{5+}
1.49	1.84	0.30	0.61	0.73	1.98	0.42	0.40	0.96	0.93	0.69	1.16	0.64
Te^{2-}	Te^{6+}	Th^{4+}	Ti^{2+}	Ti^{4+}	Tl^{1+}	Tl^{3+}	U^{4+}	U^{5+}	V^{2+}	V^{5+}	W^{4+}	W^{6+}
2.21	0.56	1.00	0.86	0.61	1.50	0.88	0.97	0.76	0.79	0.54	0.65	0.58
Y^{3+}	Yb^{3+}	Zn^{2+}	Zr^{4+}									
0.89	0.86	0.75	0.72									

Source. R. D. Shannon and C. T. Prewitt, *Acta Cryst.*, **B25**, 925 (1969).

generally larger than cations, as shown in Tables 2.3 and 2.4, the critical radius ratio for a structure is almost always determined by the coordination of anions about the cations. This is why Pauling's first rule emphasizes the cation coordination polyhedron. For a given pair of ions, the radius ratio places an upper limit on the coordination number of the cation. In general, geometry would permit the structure to form with any

Table 2.4. Ionic Crystal Radii
(Coordination Number = 4)

Ag^{1+}	Al^{3+}	As^{5+}	B^{3+}	Be^{2+}	C^{4+}
1.02	0.39	0.34	0.12	0.27	0.15
Cd^{2+}	Cr^{4+}	Cu^{2+}	F^{1-}	Fe^{2+}	Fe^{3+}
0.84	0.44	0.63	1.31	0.63	0.49
Ga^{3+}	Ge^{4+}	Hg^{2+}	Li^{1+}	Mg^{2+}	N^{5+}
0.47	0.40	0.96	0.59	0.49	0.13
Na^{1+}	Nb^{5+}	O^{2-}	P^{5+}	Pb^{2+}	S^{6+}
0.99	0.32	1.38	0.33	0.94	0.12
Se^{6+}	Si^{4+}	V^{5+}	W^{6+}	Zn^{2+}	
0.29	0.26	0.36	0.41	0.60	

one of a number of smaller coordination numbers. The stablest structure, however, always has the maximum permissible coordination number, since the electrostatic energy of an array is obviously decreased as progressively larger numbers of oppositely charged ions are brought into contact. The critical ratios presented in Fig. 2.21 are useful but are not always followed. The reason for this is that the packing considerations have considered the ions to be rigid spheres. A coordination number larger than that permitted by the radius ratio would be assumed if the electrostatic energy gained by increasing the coordination number exceeded any energy expended in deforming the surrounding ions. This consideration becomes especially important when the central cation has high charge or when the surrounding anions have a high atomic number and are large and easily deformed. Similarly, contributions of directional covalent bonding have an effect. Some experimentally observed coordination numbers are compared with predicted values in Table 2.5.

The first rule focuses attention on the cation coordination polyhedron as the basic building block of an ionic structure. In a stable structure such units are arranged in a three-dimensional array to optimize second-

Table 2.5. Coordination Number and Bond Strength of Various Cations with Oxygen

Ion	Radius (CN = 6)	Predicted Coordination Number	Observed Coordination Number	Strength of Electrostatic Bond
B^{3+}	0.16	3	3, 4	1 or 3/4
Be^{2+}	0.25	4	4	1/2
Li^{+}	0.53	6	4	1/4
Si^{4+}	0.29	4	4, 6	1
Al^{3+}	0.38	4	4, 5, 6	3/4 or 1/2
Ge^{4+}	0.39	4	4, 6	1 or 2/3
Mg^{2+}	0.51	6	6	1/3
Na^{+}	0.73	6	4, 6, 8	1/6
Ti^{4+}	0.44	6	6	2/3
Sc^{3+}	0.52	6	6	1/2
Ar^{4+}	0.51	6	6, 8	2/3 or 1/2
Ca^{2+}	0.71	6, 8	6, 7, 8, 9	1/4
Ce^{4+}	0.57	6	8	1/2
K^{+}	0.99	8, 12	6, 7, 8, 9, 10, 12	1/9
Cs^{+}	1.21	12	12	1/12

Source. Reference 3.

nearest-neighbor interactions. A stable structure must be electrically neutral not only on a macroscopic scale but also at the atomic level. Pauling's *second rule* describes a basis for evaluating local electrical neutrality. We define the *strength* of an ionic bond donated from a cation to an anion as the formal charge on the cation divided by its coordination number. For example, silicon, with valence 4 and tetrahedral coordination, has bond strength $4/4 = 1$; Al^{3+} with octahedral coordination has bond strength $3/6 = 1/2$. (The same considerations are applied regardless of whether all coordinating anions are the same chemical species; the bond strength of Al^{3+} is 1/2 in both the structure of Al_2O_3, Fig. 14.28, where the six anion neighbors are O^{2-}, and in that of kaolinite, Fig. 2.35, where the anions surrounding Al^{3+} are $4OH^-$ and $2O^{2-}$.) The second rule states that in a stable structure the total strength of the bonds reaching an anion from all surrounding cations should be equal to the charge of the anion. For example, in the Si_2O_7 unit, Fig. 2.22*a*, two bonds of strength 1 reach the shared oxygen ion from the surrounding silicon ions; the sum of the bonds is thus 2, the valence of the oxygen ion. (Note that this implies that, in a silicate based on Si_2O_7 units, no additional cation may be bonded to this shared oxygen.) Similarly in the structure of spinel $MgAl_2O_4$, Fig. 2.25, each O^{2-} is surrounded by one Mg^{2+} which donates a bond of strength 2/4 and three Al^{3+} which donate three bonds of strength 3/6.

Pauling's *third rule* further concerns the linkage of the cation coordination polyhedra. In a stable structure the corners, rather than the edges and especially the faces, of the coordination polyhedra tend to be shared. If an edge is shared, it tends to be shortened. The basis of this rule is again

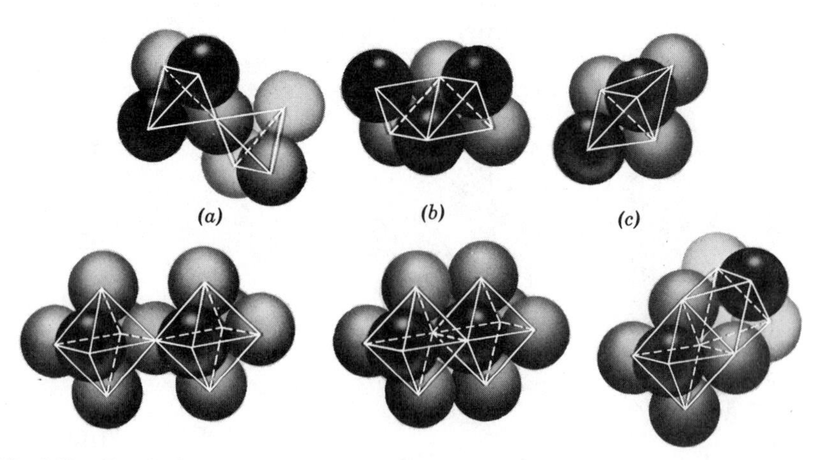

Fig. 2.22. Tetrahedra and octahedra linked by sharing (*a*) corner, (*b*) edge, and (*c*) face (reference 3).

geometrical. The separation of the cations within the polyhedron decreases as the polyhedra successively share corners, edges, and faces and the repulsive interaction between cations accordingly increases. Pauling's *fourth rule* states that polyhedra formed about cations of low coordination number and high charge tend especially to be linked by corner sharing. That this is true may be appreciated by recognizing that the repulsive interaction between a pair of cations increases as the square of their charge and that the separation of cations within a coordination polyhedron decreases as the coordination number becomes smaller. A *fifth rule* states that the number of different constituents in a structure tends to be small. This follows from the difficulty encountered in efficiently packing into a single structure ions and coordination polyhedra of different sizes.

2.6 Oxide Structures

Most of the simple metal oxide structures can be built up on the basis of nearly close-packed oxygen ions, with cations placed in available interstices; this similarity is illustrated for a large number of structures in Table 2.6 and is emphasized in the discussion of common structures.

Rock Salt Structure. Many halides and oxides crystallize in the cubic rock salt structure which has already been illustrated in Fig. 2.11. In this structure the large anions are arranged in cubic close packing and all the octahedral interstitial positions are filled with cations. Oxides having this structure are MgO, CaO, SrO, BaO, CdO, MnO, FeO, CoO, and NiO. The coordination number is 6 for both cation and anion. For stability the radius ratio should be between 0.732 and 0.414, and the anion and cation valences must be the same. All the alkali halides except CsCl, CsBr, and CsI crystallize with this structure, as do the alkaline earth sulfides.

Wurtzite Structure. In beryllium oxide the radius ratio is 0.25, requiring tetrahedral coordination of four oxygen about each beryllium ion. The bond strength is then equal to one-half so that each oxygen must be coordinated with four cations. These requirements can be met with hexagonal packing of the large oxygen ions, with half the tetrahedral interstices filled with beryllium ions so as to achieve maximum cation separation (Fig. 2.23). This structure also occurs for wurtzite, ZnS, and is commonly known as the wurtzite structure.

Zinc Blende Structure. Another structure having tetrahedral coordination is the zinc blende structure illustrated in Fig. 2.24. This structure is based on cubic close packing of the anions. A BeO polymorph with this structure has been observed at high temperatures.

Table 2.6. Simple Ionic Structures Grouped According to Anion Packing

Anion Packing	Coordination Number of M and O	Sites by Cations	Structure Name	Examples
Cubic close-packed	6:6 MO	All oct.	Rock salt	NaCl, KCl, LiF, KBr, MgO, CaO, SrO, BaO, CdO, VO, MnO, FeO, CoO, NiO
Cubic close-packed	4:4 MO	1/2 tet.	Zinc blende	ZnS, BeO, SiC
Cubic close-packed	4:8 M_2O	All tet.	Antifluorite	Li_2O, Na_2O, K_2O, Rb_2O, sulfides
Distorted cubic close-packed	6:3 MO_2	1/2 oct.	Rutile	TiO_2, GeO_2, SnO_2, PbO_2, VO_2, NbO_2, TeO_2, MnO_2, RuO_2, OsO_2, IrO_2
Cubic close-packed	12:6:6 ABO_3	1/4 oct. (B)	Perovskite	$CoTiO_3$, $SrTiO_3$, $SrSnO_3$, $SrZrO_3$, $SrHfO_3$, $BaTiO_3$
Cubic close-packed	4:6:4 AB_2O_4	1/8 tet. (A) 1/2 oct. (B)	Spinel	$FeAl_2O_4$, $ZnAl_2O_4$, $MgAl_2O_4$
Cubic close-packed	4:6:4 $B(AB)O_4$	1/8 tet. (B) 1/2 oct. (A, B)	Spinel (inverse)	$FeMgFeO_4$, $MgTiMgO_4$
Hexagonal close-packed	4:4 MO	1/2 tet.	Wurtzite	ZnS, ZnO, SiC
Hexagonal close-packed	6:6 MO	All oct.	Nickel arsenide	NiAs, FeS, FeSe, CoSe
Hexagonal close-packed	6:4 M_2O_3	2/3 oct.	Corundum	Al_2O_3, Fe_2O_3, Cr_2O_3, Ti_2O_3, V_2O_3, Ga_2O_3, Rh_2O_3
Hexagonal close-packed	6:6:4 ABO_3	2/3 oct. (A, B)	Ilmenite	$FeTiO_3$, $NiTiO_3$, $CoTiO_3$
Hexagonal close-packed	6:4:4 A_2BO_4	1/2 oct. (A) 1/8 tet. (B)	Olivine	Mg_2SiO_4, Fe_2SiO_4
Simple cubic	8:8 MO	All cubic	CsCl	CsCl, CsBr, CsI
Simple cubic	8:4 MO_2	1/2 cubic	Fluorite	ThO_2, CeO_2, PrO_2, UO_2, ZrO_2, HfO_2, NpO_2, PuO_2, AmO_2
Connected tetrahedra	4:2 MO_2	...	Silica types	SiO_2, GeO_2

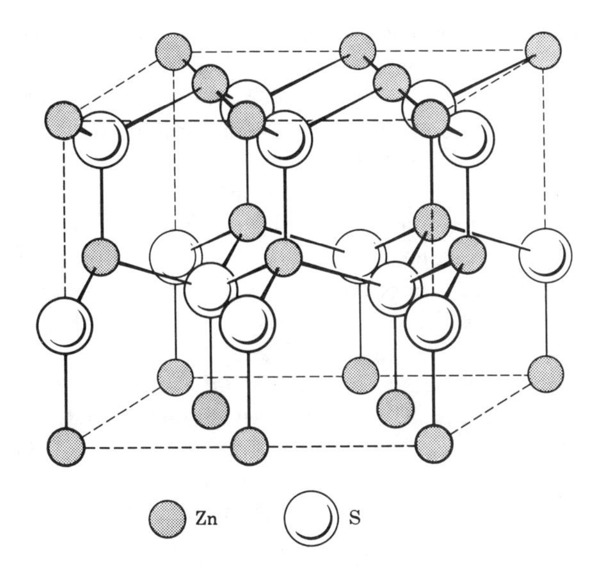

Fig. 2.23. Wurtzite (ZnS) structure (also BeO, and oxygen positions in H_2O).

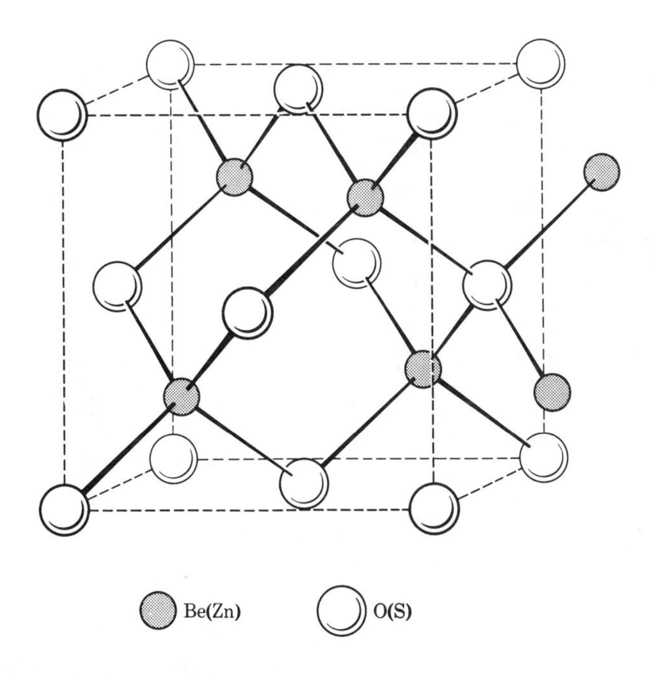

Fig. 2.24. Zinc blende (ZnS) structure.

Spinel Structure. A number of oxides of the general formula AB_2O_4, such as magnesium aluminate (spinel), $MgAl_2O_4$, have a cubic structure which can be viewed as a combination of the rock salt and zinc blende structures. The oxygen ions are in face-centered cubic close packing. As shown in Fig. 2.15*f*, for a subcell of this structure there are four atoms, four octahedral interstices, and eight tetrahedral interstices. This makes a total of twelve interstices to be filled by three cations, one divalent and two trivalent. In each elementary cell two octahedral sites are filled and one tetrahedral. Eight of these elementary cells are arranged so as to form a unit cell containing 32 oxygen ions, 16 octahedral cations, and 8 tetrahedral cations, as illustrated in Fig. 2.25.

Two types of spinel occur. In the *normal* spinel the A^{2+} ions are on tetrahedral sites and the B^{3+} ions are on octahedral sites. (This is the structure observed for $ZnFe_2O_4$, $CdFe_2O_4$, $MgAl_2O_4$, $FeAl_2O_4$, $CoAl_2O_4$, $NiAl_2O_4$, $MnAl_2O_4$, and $ZnAl_2O_4$.) In the *inverse* spinels, the A^{2+} ions and half the B^{3+} ions are on octahedral sites; the other half of the B^{3+} are on

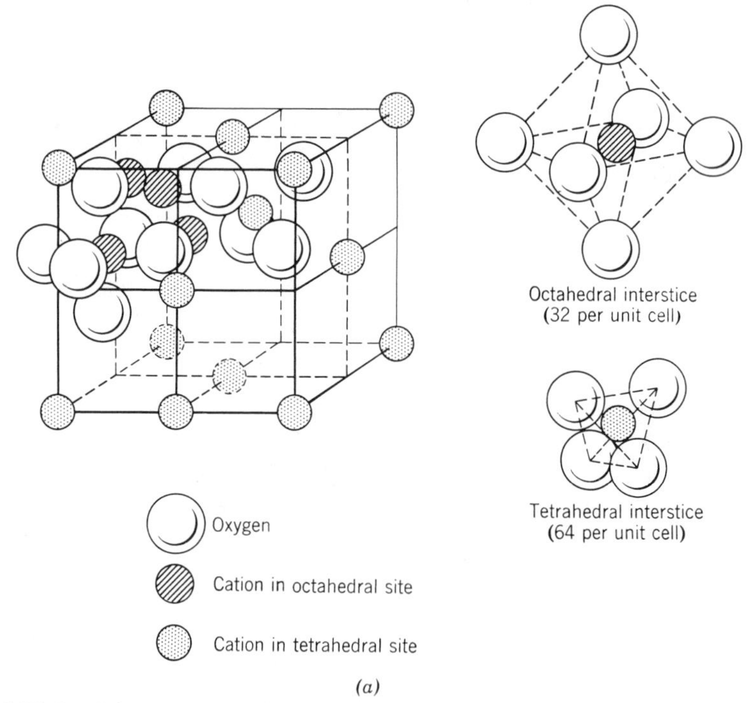

Fig. 2.25(*a*). Spinel structure. From A. R. von Hippel, *Dielectrics and Waves*, John Wiley & Sons, New York, 1954.

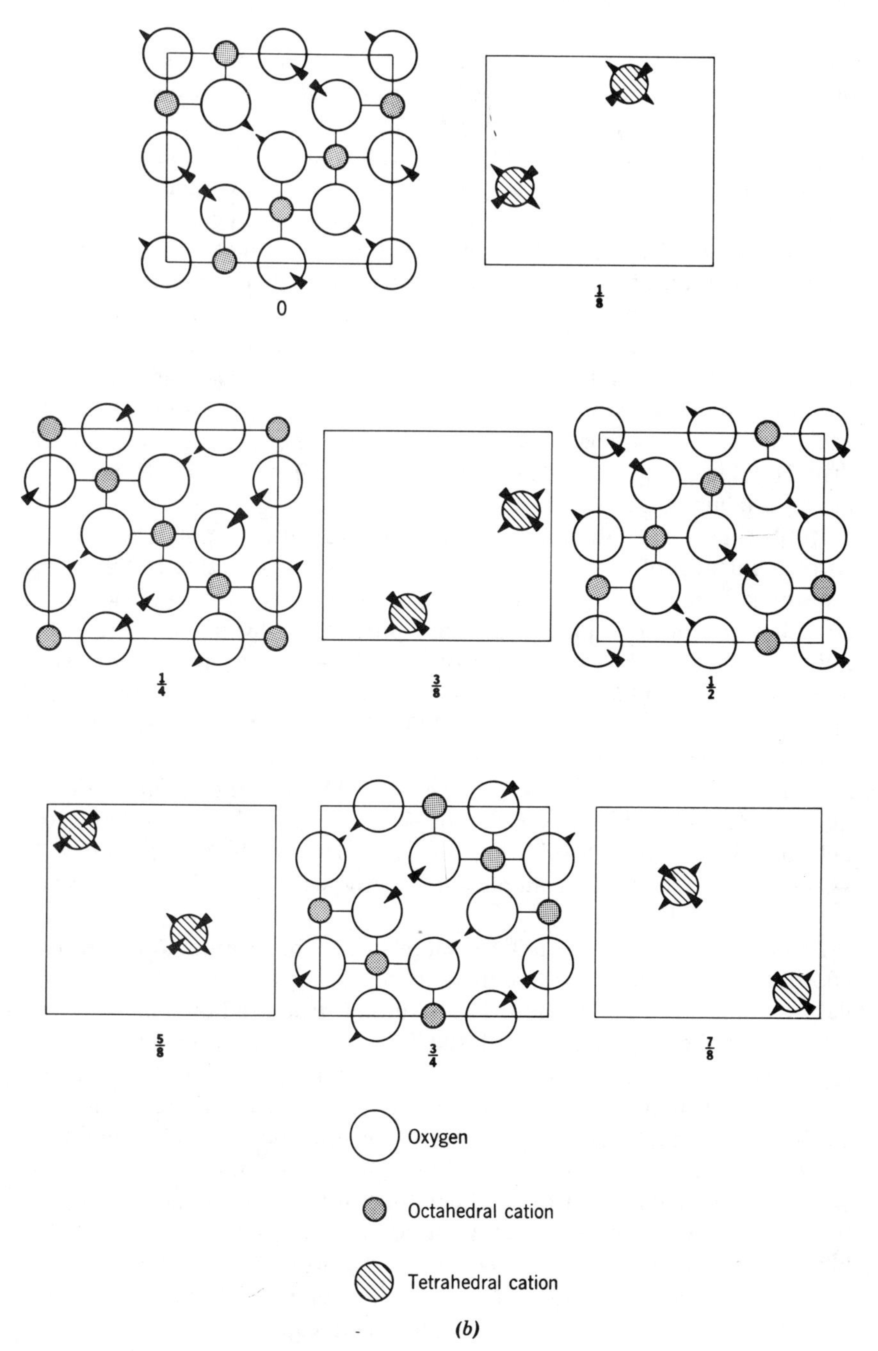

Fig. 2.25(*b*). Layers of atoms parallel to (001).

tetrahedral sites, $B(AB)O_4$. This is the more common structure and is observed for $FeMgFeO_4$, $FeTiFeO_4$, Fe_3O_4, $ZnSnZnO_4$, $FeNiFeO_4$, and many other ferrites of importance for their magnetic properties.

Corundum Structure. In Al_2O_3, the preferred coordination number for aluminum is 6 so that with a valence of 3 there is a bond strength of one-half; this requires four Al^{3+} adjacent to each O^{2-}. This is achieved by nearly hexagonal close packing of the oxygen ions, with aluminum ions filling two-thirds of the octahedral sites. Subsequent similar layers are built up such that maximum spacing of the Al^{3+} ions is achieved.

The basic similarity between these oxide structures can best be seen by taking a cut parallel to the plane of closest packing—basal plane in hexagonal close packing, (111) in cubic close packing. In MgO and Al_2O_3 each cation is in an octahedral site. In BeO the cations are regularly distributed in tetrahedral sites, and in spinel there are two kinds of layers, giving a combination of these.

Rutile Structure. In rutile, TiO_2, the coordination number for Ti is 6 with a valence of +4, leading to a bond strength of two-thirds and requiring threefold coordination of Ti^{4+} around each oxygen ion. The structure is more complex than those previously discussed. Cations fill only half the available octahedral sites, and the closer packing of oxygen ions around the filled cation sites leads to the distortion of the nearly close-packed anion lattice. GeO_2, PbO_2, SnO_2, MnO_2, and several other oxides crystallize in this structure.

Cesium Chloride Structure. In cesium chloride, the radius ratio requires eightfold coordination. Since the bond strength is one-eighth, the chlorine is also in eightfold coordination. This leads to a structure in which the Cl^- ions are in a simple cubic array, with all the interstices filled with Cs^+ ions (Fig. 2.26).

Fluorite Structure. In ThO_2 the large size of the thorium ion requires a coordination number of 8, leading to a bond strength of one-half and four valency bonds to each oxygen. The resulting structure has a simple cubic packing for the oxygen ions with the Th^{4+} in half the available sites with eightfold coordination. This is similar to the cesium chloride structure, but only half of the cation sites are filled; it is the structure of fluorite, CaF_2, for which it is named. As seen in Fig. 2.27, the unit cell is based on the face-centered cubic packing of the cations. A notable feature is the large void in the center of the unit cell (one of the unoccupied positions in the simple cubic fluorine array). In addition to ThO_2, both TeO_2 and UO_2 have this structure, and ZrO_2 has a distorted (monoclinic) fluorite structure. The large number of vacant sites allows UO_2 to be used as a unique nuclear fuel in which fission products cause little difficulty; they are accommodated on the vacant lattice positions.

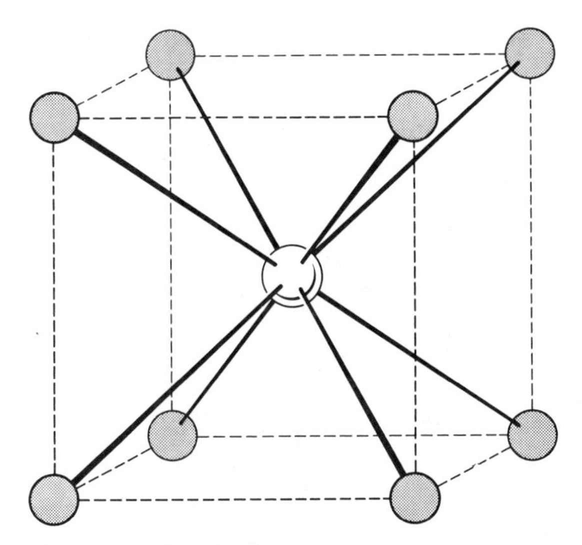

Fig. 2.26. Cesium chloride structure.

Antifluorite Structure. An oxide structure consisting of a cubic close-packed array of oxygen atoms with cations arranged in the tetrahedral sites has cations and anions just reversed from the normal fluorite lattice (Fig. 2.27). This structure is observed for Li_2O, Na_2O, and K_2O.

Perovskite Structure. Previous oxide structures have been based on close packing of anions. A somewhat different structure occurs where large cations are present which can form a close-packed structure along with the oxygen ions. This is the case for perovskite, $CaTiO_3$, in which the Ca^{2+} and O^{2-} ions combine to form a close-packed cubic structure with the smaller, more highly charged Ti^{4+} ions in octahedral interstices. The structure is illustrated in Fig. 2.28. Each O^{2-} is surrounded by four Ca^{2+} and eight O^{2-}; each Ca^{2+} is surrounded by twelve O^{2-}. In the center of the face-centered cubic unit cell the small, highly charged Ti^{4+} is octahedrally coordinated to six O^{2-}.

We can apply Pauling's rules for bond strength and coordination number. The strength of the Ti–O bond is two-thirds; each Ca–O bond is one-sixth. Each oxygen is coordinated with two Ti^{4+} and four Ca^{2+} for a total bond strength of $4/3 + 4/6 = 2$, which is equal to the oxygen valency.

The perovskite structure is observed for $CaTiO_3$, $BaTiO_3$, $SrTiO_3$, $SrSnO_3$, $CaZrO_3$, $SrZrO_3$, $KNBO_3$, $NaNBO_3$, $LaAlO_3$, $YAlO_3$, and $KMgF_3$, among others. Similar structures (close-packed large cations and anions along with smaller cations and interstitial sites) occur for other compositions such as $K_2SiF_6(KSi_{1/2}F_3)$.

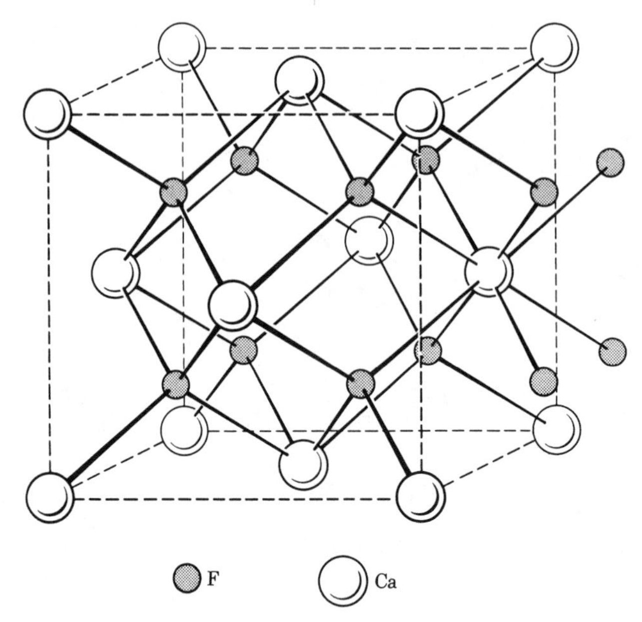

Fig. 2.27. Fluorite structure.

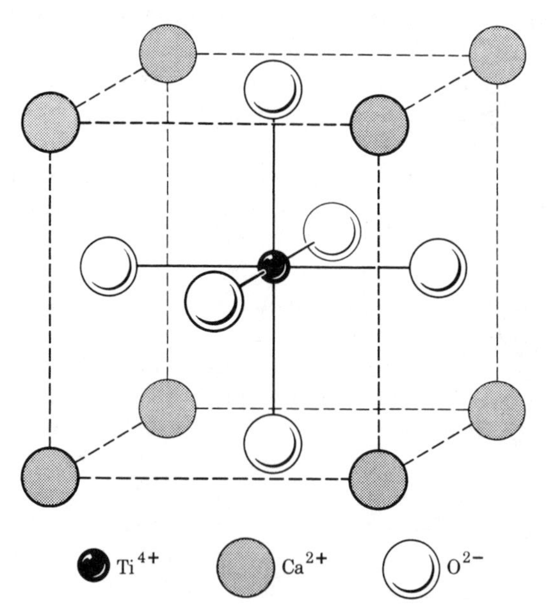

Fig. 2.28. Perovskite structure (idealized).

Ilmenite Structure. Ilmenite, $FeTiO_3$, is a derivative of the Al_2O_3 or Fe_2O_3 structure. Half the cation sites are occupied by Fe^{2+} and half by Ti^{4+}; alternate layers of cations are all Fe^{2+} and all Ti^{4+}. This structure is found for $MgTiO_3$, $NiTiO_3$, $CoTiO_3$, and $MnTiO_3$. $LiNbO_3$ is a different variety of derivative structure in which each layer of cations contains an ordered arrangement of Li and Nb.

Derivative Structures. In comparing crystal structures, one frequently is struck by the similarity of a complex structure to some simpler atomic arrangement. The symmetry and regularity of some simple structure is often perturbed to produce a more complex array of atoms. Possible mechanisms include the ordered substitution of several different species of atoms, the ordered omission of atoms, the addition of an atom to an unoccupied site (*stuffing*), and the distortion of the atomic array. Several or all of these mechanisms may be present in a single structure. Buerger* calls such structures *derivative structures.* A *superstructure* is a special type of derivative structure in which the perturbation cause the unit cell of the derivative to be larger than that of the basic structure.

The mechanisms of derivative structure closely resemble those involved in solid solution. It is important to note, however, that the concept concerns only a geometric relation between two structure types and is not meant to imply anything concerning the genesis of the derivative. For example, the structure of zinc blende, ZnS, may be viewed as a derivative of the diamond structure in which Zn and S replace the C atoms in the diamond structure. This does not imply that zinc blende crystals form when equal parts of Zn and S are dissolved in diamond. Chalcopyrite, $CuFeS_2$, is an example of a derivative of zinc blende which forms a superstructure. Layers of ordered Cu and Fe are substituted for Zn in such a fashion that the resulting structure is tetragonal and has a lattice constant c double that of the zinc blende cell. Other examples of a substitutional derivative structure which have been discussed above are ilmenite, $FeTiO_3$, and $LiNbO_3$, which are based on the structure of Al_2O_3.

Derivative structures and superstructures involving the ordered omission of atoms frequently occur in nonstoichiometric materials with high vacancy concentrations. The phases Cr_2S_3, Cr_3S_4, Cr_5S_6, Cr_7S_8, which occur in the system Cr-S, are all derivatives of the NiAs structure type; their compositions reflect different numbers and ordering schemes of vacancies. Stoichiometric CrS is a distorted monoclinic derivative of NiAs. A number of silicates are stuffed derivatives of the network structures found in the high-temperature forms of silica; Al^{3+} replaces part of Si^{4+}, and other atoms are stuffed into interstices to maintain charge

J. Chem. Phys., **15**, 1 (1947).

balance. The stuffed atom stabilizes a network which, in pure silica, would collapse to a less open framework at lower temperatures. High carnegieite, $NaAlSiO_4$, kalsilite, $KAlSiO_5$, and high eucryptite, $LiAlSiO_4$, are stuffed derivatives of the high-temperature forms of the structures of the cristobalite, tridymite, and quartz forms of silica, respectively. Stuffed derivatives of silica are frequently the crystalline phase which forms when a silicate glass devitrifies.

Common Features of Oxide Structures. Without question, the most striking feature of the oxide structures discussed thus far is the close relationship to and dependence on close-packed oxygen arrays. Viewed on this basis, similarities between structures otherwise difficult to discern are striking. This makes it essential that students obtain a good grasp of these packing systems. In particular, the cubic close-packed structure and the distribution of octahedral and tetrahedral interstices should be thoroughly familiar to the student.

2.7 Silicate Structures

Atomic arrangements in hundreds of silicates having complex chemical compositions have in their basic structures a beautiful simplicity and order. At the same time the details of many of the silicate structures are complex and difficult to illustrate without three-dimensional models, and we will not attempt to give precise structure information. (Reference 9 is recommended.)

The radius ratio for Si–O is 0.29, corresponding to tetrahedral coordination, and four oxygen ions are almost invariably arrayed around a central silicon. With a bond strength of 1, oxygen ions may be coordinated with only two silicon atoms in silica; this low coordination number makes close-packed structures impossible for SiO_2, and in general silicates have more open structures than those discussed previously. The SiO_4 tetrahedra can be linked in compounds such that corners are shared (Fig. 2.22) in several ways. Some of these are illustrated in Fig. 2.29. There are four general types. In orthosilicates, SiO_4^{4-}, tetrahedra are independent of one another; in pyrosilicates, $Si_2O_7^{6-}$, ions are composed of two tetrahedra with one corner shared; in metasilicates, SiO_3^{2-}–$(SiO_3)_n^{2n-}$, two corners are shared to form a variety of ring or chain structures; in layer structures, $(Si_2O_5)_n^{2n-}$, layers are made up of tetrahedra with three shared corners; in the various forms of silica, SiO_2, four corners are shared.

Silica. Crystalline silica, SiO_2, exists in several different polymorphic forms corresponding to different ways of combining tetrahedral groups with all corners shared. Three basic structures—quartz, tridymite, cristobalite—each exist in two or three modifications. The most stable forms are low quartz, below 573°C; high quartz, 573 to 867°C; high

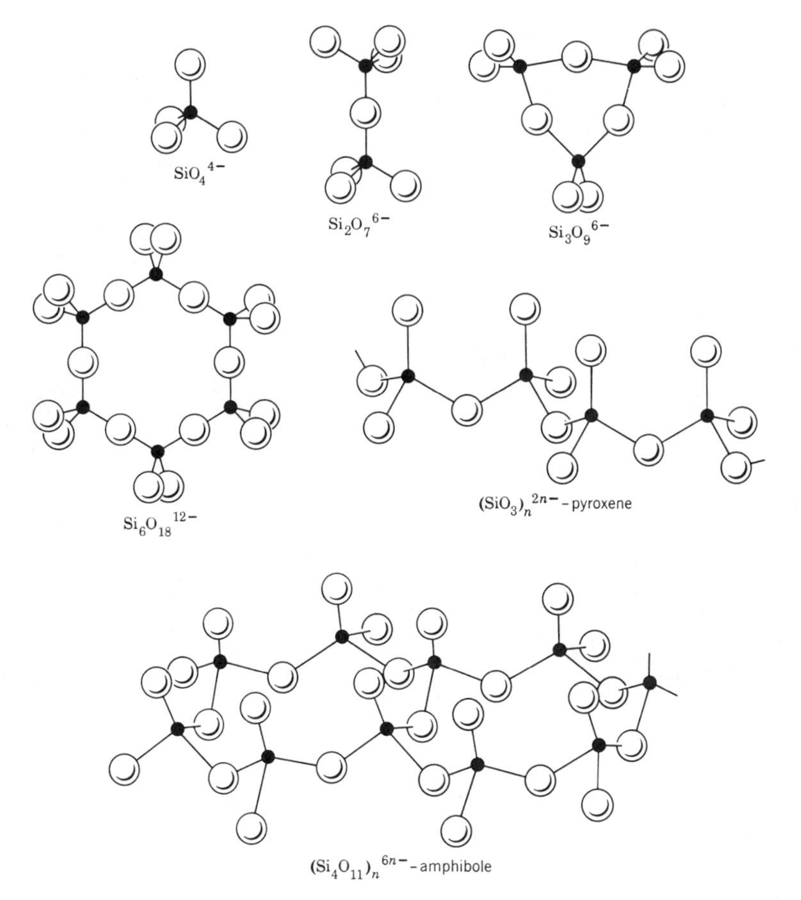

Fig. 2.29. Some silicate ions and chain structures. Planar configurations are considered in the next section.

tridymite, 867 to 1470°C; high cristobalite, 1470 to 1710°C; and liquid, above 1710°C. The low-temperature modifications are distorted *derivative* structures of the basic high-temperature forms. (A derivative structure in this sense is one that can be derived from a basic structure of greater symmetry by distorting the structure in space rather than substituting different chemical species.) We confine our attention to the basic high-temperature forms.

High quartz has a structure which can be viewed as composed of connected chains of silica tetrahedra, as illustrated in Fig. 2.30. Compared to the close-packed structures discussed in the last section, this is a relatively open structure; for example, the density of quartz is 2.65 g/cm^3, compared with 3.59 for MgO and 3.96 for Al_2O_3. However, quartz has a

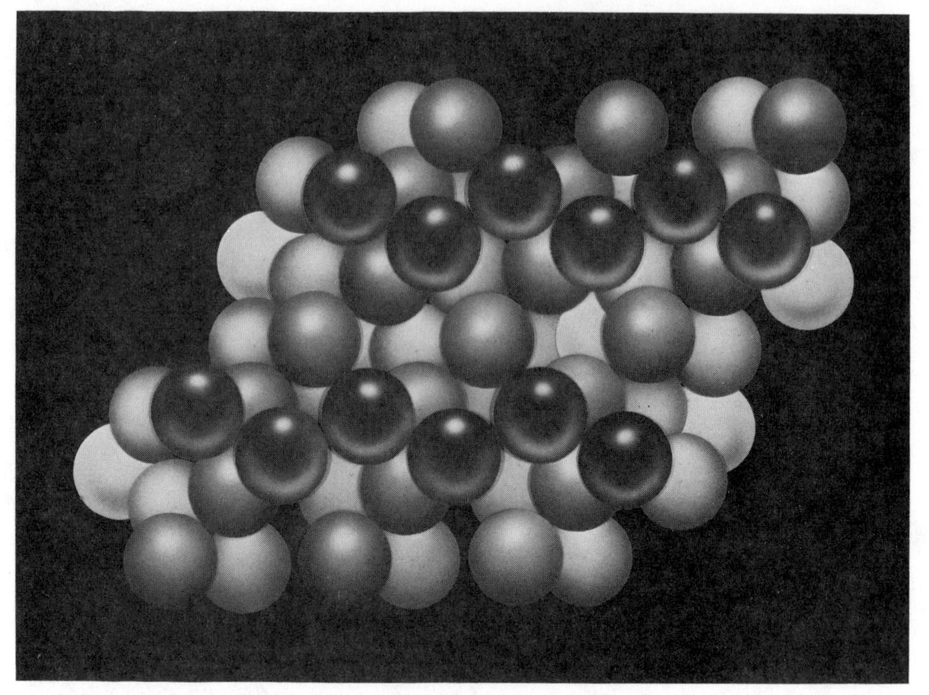

Fig. 2.30. Structure of high quartz, looking down on basal plane.

greater density and closer packing than either of the high-temperature forms, tridymite ($\rho = 2.26$) and cristobalite ($\rho = 2.32$), illustrated in Fig. 2.31.

Orthosilicates. This group includes the olivine minerals (forsterite, Mg_2SiO_4, and solid solutions with Fe_2SiO_4), the garnets, zircon, and the aluminosilicates—kyanite, sillimanite, andalusite, and mullite. The structure of forsterite, Mg_2SiO_4, is similar to that found for chrysoberyl, Al_2BeO_4. The oxygen ions are nearly in a hexagonal close-packed structure with Mg^{2+} in octahedral and Si^{4+} in tetrahedral sites. (From a coordination point of view this assembly can also be considered an array of SiO_4^{4-} tetrahedra with Mg^{2+} ions in the octahedral holes.) Each oxygen ion is coordinated with one Si^{4+} and three Mg^{2+} or with two Si^{4+}.

The structure of kyanite, Al_2SiO_5, consists of nearly cubic close-packed oxygen ions with Si^{4+} in tetrahedral and Al^{3+} in octahedral sites. However, the polymorphic forms andalusite and sillimanite have much more open structures, with SiO_4 tetrahedra coordinated with AlO_6 octahedra. Mullite, $Al_6Si_2O_{13}$, a common constituent of fired clay products, has a structure similar to that of sillimanite (compare $Al_{16}Si_8O_{40}$ and $Al_{18}Si_6O_{39}$).

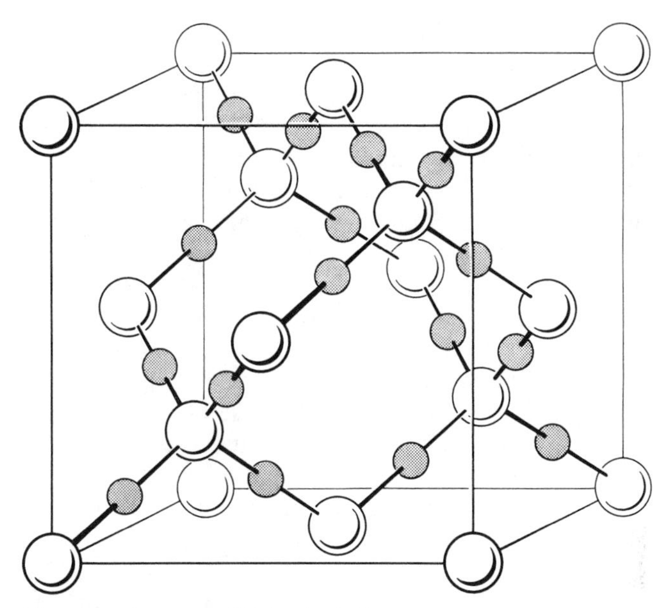

Fig. 2.31. Structure of high cristobalite.

Pyrosilicates. Crystalline silicates containing $Si_2O_7^{6-}$ ions are rare.

Metasilicates. Silicates containing $(SiO_3)_n^{2n-}$ ions are of two types—cyclic or chain arrangements of the silica tetrahedra. Some of the discrete cyclic ions observed are the $Si_3O_9^{6-}$ (such as in wollastonite, $CaSiO_3$) and $Si_6O_{18}^{12-}$ (in beryl, $Be_3Al_2Si_6O_{18}$) ions. Minerals with chain structures comprise a large group. Those with compositions corresponding to single chain, $(SiO_3)_n^{2n-}$, are the pyroxenes, and those with double chains, $(Si_4O_{11})_n^{6n-}$, the amphiboles. The silicate chain structures are built up as shown in Fig. 2.29. The pyroxenes include enstatite, $MgSiO_3$; diopside, $MgCa(SiO_3)_2$; spodumene, $LiAl(SiO_3)_2$; and jadeite. The amphiboles include tremolite, $(OH)_2Ca_2Mg_5(Si_4O_{11})_2$, in which isomorphic substitution is widespread. The asbestos minerals are amphiboles.

Framework Structures. Many important silicate structures are based on an infinite three-dimensional silica framework. Among these are the feldspars and the zeolites. The feldspars are characterized by a framework formed with Al^{3+} replacing some of the Si^{4+} to make a framework with a net negative charge that is balanced by large ions in interstitial positions, that is, albite, $NaAlSi_3O_8$; anorthite, $CaAl_2Si_2O_8$; orthoclase, $KAlSi_3O_8$; celsian, $BaAl_2Si_2O_8$; and the like. The network structures are similar in nature to the cristobalite structure illustrated in Fig. 2.31, with the alkali or alkaline earth ions fitting into interstices. Only the large positive ions form feldspars; smaller ones that enjoy octahedral coordination form chains or layer silicates.

Much more open alumina-silica frameworks occur in the zeolites and ultramarines. In these compounds the framework is sufficiently open for there to be relatively large channels in the structure. The alkali and alkaline earth ions present can be exchanged in aqueous solutions, leading to their use as water softeners. In addition, these channels can be used as molecular sieves for filtering mixtures on the basis of molecular size. The size of the channels in the network depends on the composition.

Derivative Structures. Defining a derivative structure as one derived from a simpler basic structure, there are many of these which are closely related to the structures of silica. One way that this can occur is by distorting the basic structure. This is the case of quartz, tridymite, and cristobalite, all of which have low-temperature forms that are distorted from the more symmetrical high-temperature forms. This distortion occurs by the shifting of ions, schematically illustrated in Fig. 2.32.

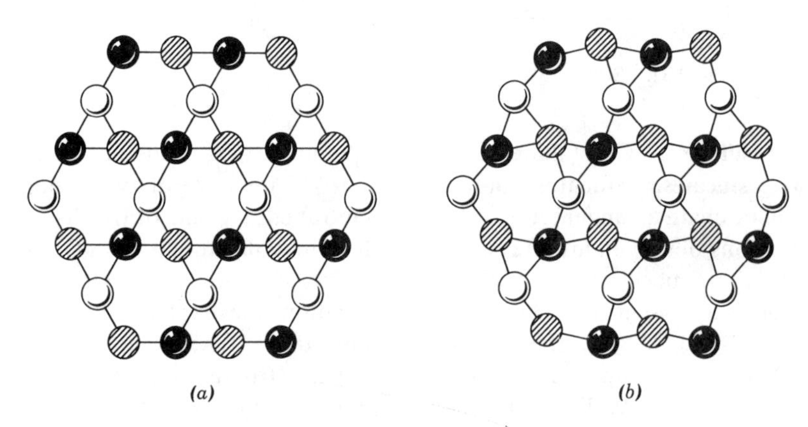

Fig. 2.32. Schematic illustration of relationship between (*a*) high-temperature and (*b*) low-temperature forms of quartz.

Another way of forming derivative structures is by substituting different chemical species. When this is accompanied by a change in valence, additional ions must be substituted. This leads to a wide variety of stuffed silica structures* in which Al^{3+} replace Si^{4+}, and other atoms are stuffed into interstices in the structure in order to maintain a charge balance. The interstices in the quartz structure are relatively small, suitable only for Li^{+} or Be^{2+} ions. Eucryptite, $LiAlSiO_4$, is a stuffed derivative of quartz.

*M. J. Buerger, *Am. Miner.*, **39**, 600 (1954).

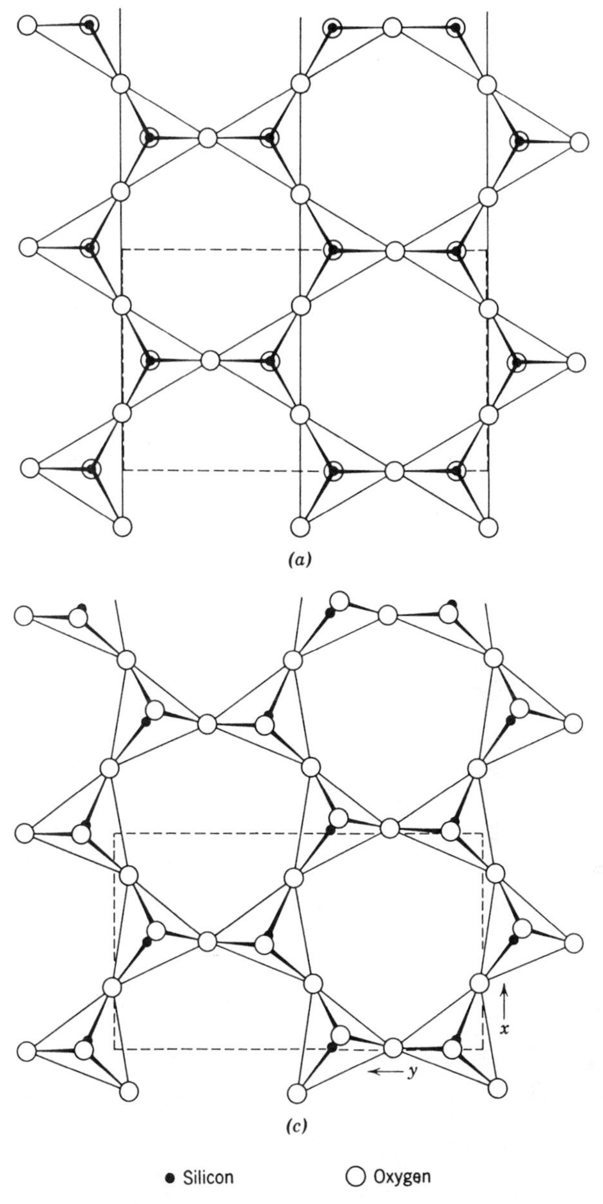

Fig. 2.33. Atomic arrangements of Si_2O_5 and $AlO(OH)_2$ layers. Patterns (*a*) and (*b*) are idealized, and (*c*) and (*d*) are the distorted arrangements found to occur in kaolinite and dickite by R. E. Newnham and G. W. Brindley, *Acta Cryst.*, **9**, 759 (1956); **10**, 89 (1957). From G. W. Brindley, in *Ceramic Fabrication Processes*, W. D. Kingery, Ed., Technology Press, Cambridge, Mass., and John Wiley & Sons, New York, 1958.

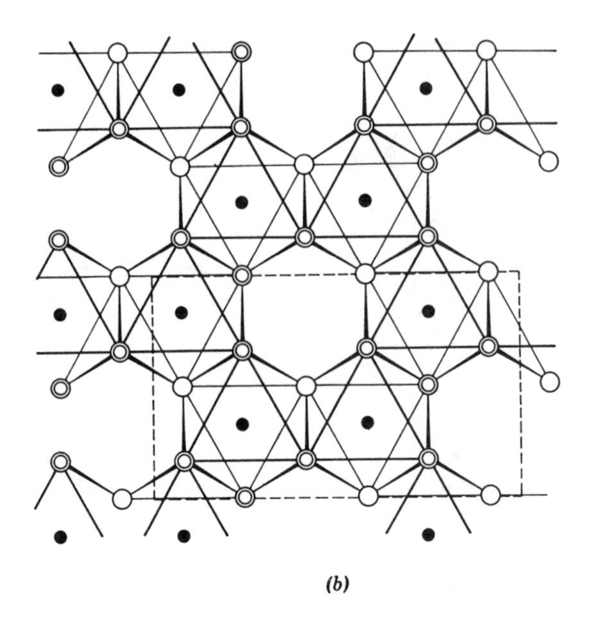

(b)

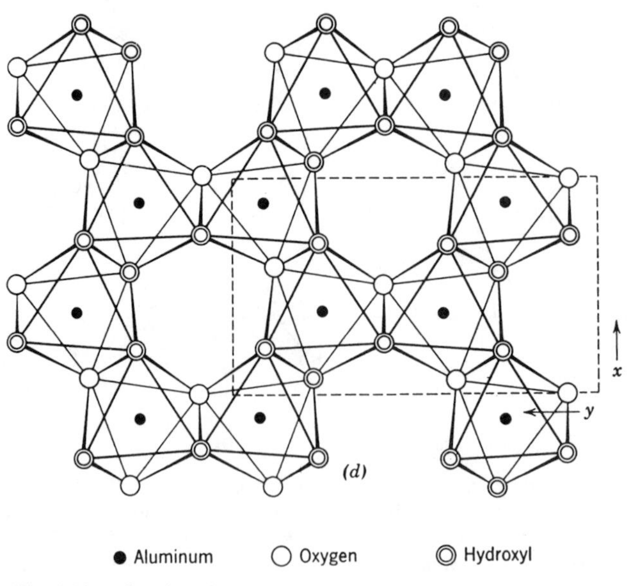

Fig. 2.33 (*Continued*)

The interstices in tridymite and cristobalite are larger, and there are many structures derived from these basic ones. Stuffed derivatives of tridymite are the most common and include nepheline, $KNa_3Al_4Si_4O_{16}$, several forms of $KAlSiO_4$, and many others. Stuffed derivatives of cristobalite include carnegeite, $NaAlSiO_4$.

2.8 The Clay Minerals

The clay minerals consist of hydrated aluminum silicates that are fine-grained and usually have a platy habit. The nature and properties of the clay minerals are determined in large extent by their structures which have been described in detail for many of them (reference 6).

The crystal structures of the common clay minerals are based on combinations of an $(Si_2O_5)_n$ layer of SiO_4 tetrahedra joined at the corners with an $AlO(OH)_2$ layer of alumina octahedra. These layer structures are illustrated in Fig. 2.33. If the oxygen ions projecting down from the Si_2O_5 plane are built into the $AlO(OH)_2$ plane, the layers can be combined to give the composition $Al_2(Si_2O_5)(OH)_4$, which is the most common clay mineral kaolinite. The resultant structure is illustrated in Fig. 2.34. The other basic clay mineral structure is that of the montmorillonite clays and is typified by pyrophyllite, $Al_2(Si_2O_5)_2(OH)_2$. In this structure there are Si_2O_5 sheets both above and below a central $AlO(OH)$ layer.

Different clay minerals are built up from different layer combinations and with different cations. Isomorphous substitution of cations is common, with Al^{3+} and sometimes Fe^{3+} substituting for some of the Si^{4+} ions in the tetrahedral network and Al^{3+}, Mg^{2+}, Fe^{2+}, and others substituting for

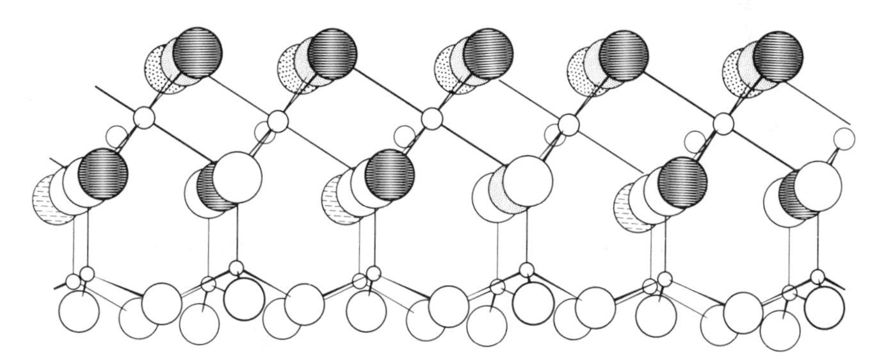

Fig. 2.34. Perspective drawing of kaolinite with Si—O tetrahedrons on the bottom half of the layer and Al—O, OH octahedrons on the top half. From G. W. Brinkley, in *Ceramic Fabrication Processes*, W. D. Kingery, Ed., Technology Press, Cambridge, Mass., and John Wiley & Sons, New York, 1958.

one another in the octahedral network. These isomorphous substitutions lead to a net negative charge on the structure. This negative charge is balanced in the mica structure by potassium ions which occupy positions between the large open cavities in the Si_2O_5 sheets. Occasional substitutions which lead to a negative charge are balanced in the clay minerals by loosely held positive ions which fit on the surface of the particles or between the layers. These ions are more or less readily exchanged and are responsible for the observed *base exchange capacity.* For example, a

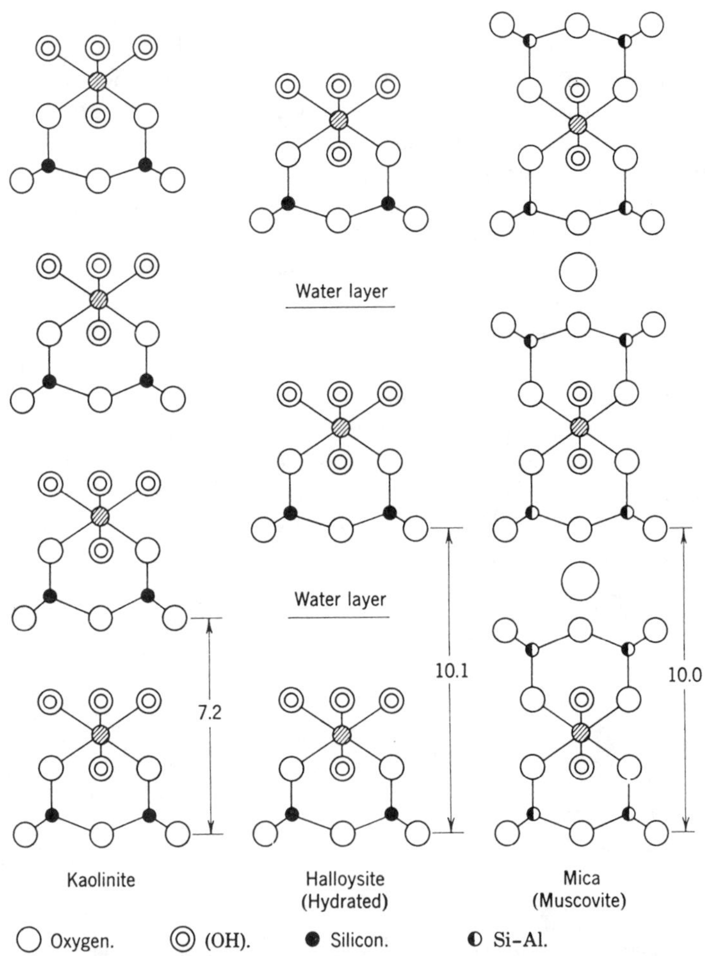

Fig. 2.35. Layer structure of clays and similar materials. From G. W. Brindley, in *Ceramic Fabrication Processes*, W. D. Kingery, Ed., Technology Press, Cambridge, Mass., and John Wiley & Sons, New York, 1958.

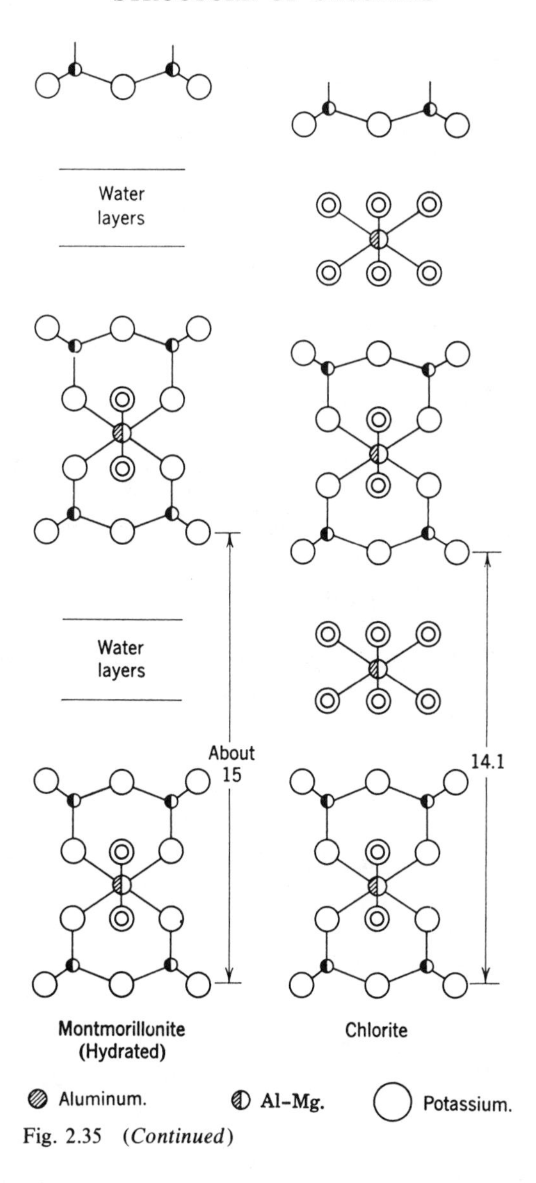

Fig. 2.35 (*Continued*)

natural clay containing absorbed Ca^{2+} can react with sodium silicate to form insoluble calcium silicate and the sodium clay:

$$\text{Clay} - \text{Ca}^{2+} + \text{Na}_2\text{SiO}_3 = \text{Clay}\begin{matrix} \diagup \text{Na}^+ \\ \diagdown \text{Na}^+ \end{matrix} + \text{CaSiO}_3 \tag{2.11}$$

These reactions are particularly important in determining the properties of aqueous suspensions of the clay minerals.

Several of the layer structures observed for clay minerals are illustrated in Fig. 2.35.

2.9 Other Structures

The structures of most other crystals which are of importance in ceramics are rather closely related to those already described for the oxides or the coordination structures of the silicates. A few general groups are of interest.

Gibbsite. The gibbsite structure, $Al(OH)_3$, is one in which each Al^{3+} is surrounded by six OH^- in a layer structure. A similar structure is observed for brucite, $Mg(OH)_2$, except that in this crystal all the octahedral sites are filled.

Graphite. Graphite (Fig. 2.36) has a layer structure in which the carbon atoms in the basal plane are held together by strongly directed covalent bonds in a hexagonal array. In contrast, the bonds between layers are weak van der Waals forces so that the structure has very strong directional properties. For example, the linear thermal expansion coefficient in the plane of the layers is about $1 \times 10^{-6}/°C$, whereas in the direction normal to the layers it is $27 \times 10^{-6}/°C$. Boron nitride, BN, has a similar structure.

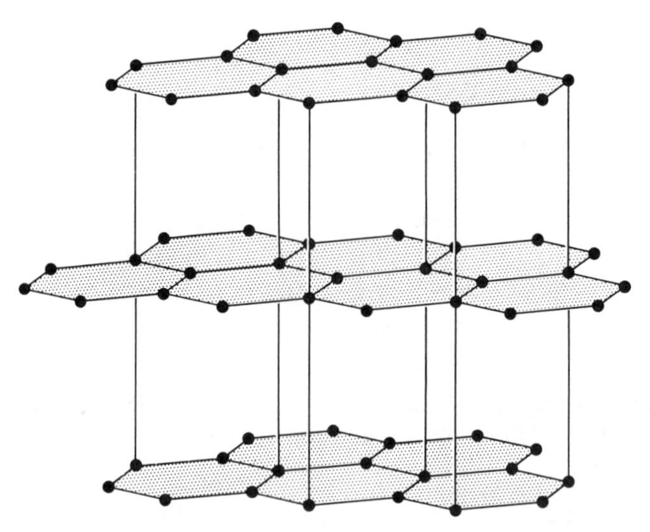

Fig. 2.36. Graphite structure.

Carbides. Carbide structures are fixed mainly by the small size of the carbon atom, which can readily fit into interstitial positions. Consequently, most of the transition metal carbides tend to have close-packed metal atoms with carbon atoms in the interstices. The metal-carbon bonding in these structures is intermediate between covalent and metallic. Compounds of carbon with atoms of similar electronegativity, as in SiC, are completely covalent. One common form of SiC has a structure similar to wurtzite (Fig. 2.24).

Nitrides. Nitride structures are similar to carbides; the metal-nitrogen bonding is usually less metallic in nature than the metal-carbon bonds.

2.10 Polymorphism

Polymorphs are different crystalline modifications of the same chemical substance, and the word polymorphism is used to describe the general relations among the several phases of the same substance without regard to the number of phases being considered. The crystallographic aspects of polymorphism have been considered in some detail by Buerger (reference 5). Many materials exist in more than one crystallographic form. For zirconia, ZrO_2, the stable form at room temperature is monoclinic, but there is a transition at about 1000°C to a tetragonal form. This transition is accompanied by a large volume change, and a disruption of ceramic bodies made with pure zirconia results. Although hexagonal α-alumina is the thermodynamically stable form of Al_2O_3 at all temperatures, a cubic form, γ-alumina, can also be formed under some circumstances. Many other materials important to ceramics exist in different polymorphic forms (C, BN, SiO_2, TiO_2, As_2O_3, ZnS, FeS_2, $CaTiO_3$, Al_2SiO_5, etc.). A ceramic material which is particularly rich in polymorphic forms is silica.

Polytypism. Polytypism is used to denote a special type of polymorphism in which the different structures assumed by a compound differ only in the order in which a two-dimensional layer is stacked. The wurtzite and sphalerite forms of ZnS, for example, are polytypes, since they differ only in the order in which sheets of tetrahedra are arranged. Other polytypes are known for ZnS. The effect is common in layer structures (e.g., MoS_2, CdI_2, graphite, and layer silicates such as the clay minerals). Silicon carbide, a ceramic material of considerable importance, holds the honor of being the material which displays the richest collection of polytypic forms. The basic unit of these structures is the tetrahedral layer as in ZnS. At least 74 distinct stacking sequences have been found in crystals of SiC, some of which require lattice constants of up to 1500 Å to define the distance over which the stacking sequence repeats.

Thermodynamic Relations. Which of a group of polymorphs is stable over a particular range of temperatures is governed by their free energies. The polymorphic form that has the lowest free energy is the most stable, and others tend to transform into it. The free energy of each phase is given by the relation

$$G = E + PV - TS \qquad (2.12)$$

where E is the internal energy, largely determined by the structure energy, P is the pressure, V is the volume, T is the absolute temperature, and S is the entropy of a particular crystalline form. The PV product is small and does not change much with temperature or transformations, and we can neglect it in this discussion. At the absolute zero the free energy is fixed by the internal energy. However, as the temperature increases, the TS term becomes increasingly important. At a sufficiently high temperature some other polymorphic form with a larger entropy may achieve a lower free energy in spite of its larger internal energy. Thermodynamic relationships between polymorphic forms indicating regions of stability are illustrated in Fig. 2.37.

The differences $E_2 - E_1$ and $E_3 - E_2$ are the constant volume heats of transformation and are always positive in transforming from a low-temperature form to a high-temperature form. In addition, it can be shown that the entropy of the high-temperature form must be larger than the entropy of the low-temperature form. The increased structure energy and

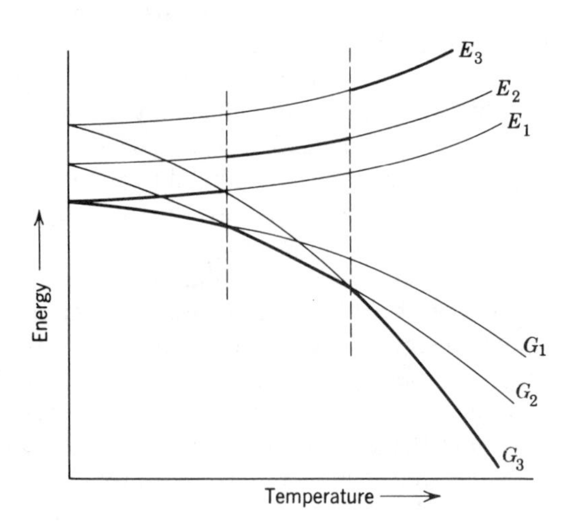

Fig. 2.37. Relationship between internal energy E and free energy G of polymorphic forms; $E_3 > E_2 > E_1$ and $S_3 > S_2 > S_1$.

increased entropy of the high-temperature forms go hand in hand and correspond to more open structures.

Structural Characteristics of Transformations. Polymorphic transformations can be classed in two general types, depending on the kind of changes occurring in the crystal. They can also be classified in two general types with regard to the speed of transformation. *High-low* transformations correspond to a change in the secondary coordination by a distortion of the structure without breaking bonds or changing the numbers of nearest neighbors. This can occur by merely displacing atoms from their previous positions. These transformations are rapid and occur at a definite temperature. In contrast, some other transformations involve a substantial change in the secondary coordination, requiring that bonds be broken and a new structure reconstructed. To break and reform bonds requires greater energy, and consequently these transformations occur but slowly. Frequently high-temperature forms can be cooled to room temperature in a metastable state without the polymorphic transformation occurring at all.

Displacive Transformation. Structurally, the least drastic kind of transformation is one that requires no change in the first coordination of the atoms. The energy change is accounted for by a change in the secondary coordination. If we start with a highly symmetric structure, as illustrated in Fig. 2.38*a*, it can be transformed into the forms shown in Fig. 2.38*b* or *c* merely by distorting the structure without breaking any bonds or changing the basic structure. The distorted form is a derivative structure of the starting material. This kind of transformation is called a displacive transformation. It characteristically occurs rapidly and is sometimes referred to as a high-low transformation.

If we consider the energy relationships among the structures illustrated in Fig. 2.38, calling 2.38*a* the open form, it is apparent that on a displacive transformation to the collapsed form, Fig. 2.38*b*, the structural energy of the system is lowered, because the distance between secondary coordination circles is decreased. Consequently, the distorted form is a low-temperature form of lower structural energy.

There are a number of characteristics of the polymorphic forms related to displacive transformations in silicates. Some of these are that (1) the high-temperature form is always the open form; (2) the high-temperature form has the larger specific volume; (3) the high-temperature form has the larger heat capacity and a higher entropy; (4) the high-temperature form has the higher symmetry—in fact the low-temperature form is a derivative structure of the high-temperature form; and (5) as a consequence of having an initial and a reverse collapsed form, transformations to lower temperature forms commonly result in twins.

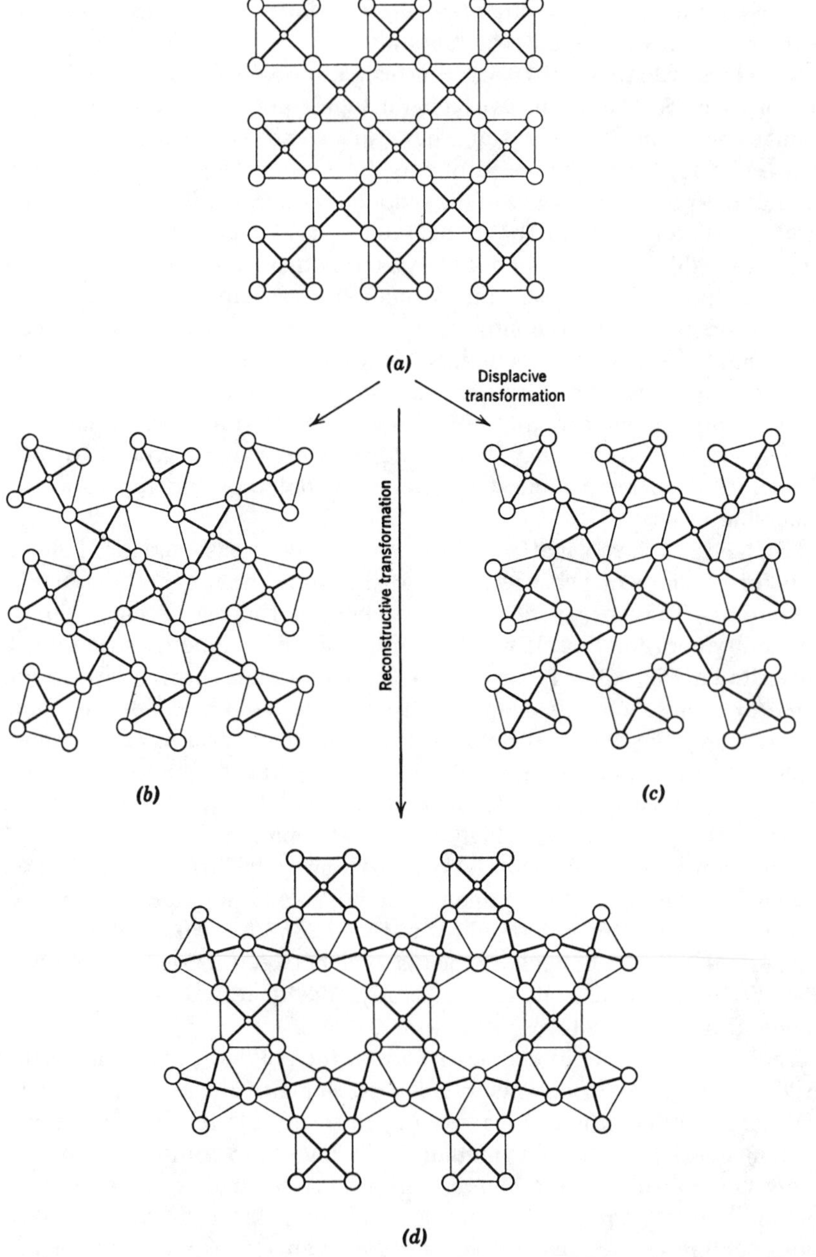

Fig. 2.38. Open form of structure *a* showing displacive transformations into collapsed forms *b* and *c* and reconstructive transformation into basically different form *d* (reference 5).

A particular type of displacive transformation which has been extensively studied in metals is called the *martensite* transformation. Because it is diffusionless and requires only a shear of the parent structure to obtain the new phase, the rate of transformation at any temperature occurs nearly instantly. It is well documented for the austenite (cubic)–martensite (tetragonal) transformation in steels, and in ceramics by the cubic $BaTiO_3$–tetragonal (ferroelectric) $BaTiO_3$ transition and the tetragonal–monoclinic inversion in ZrO_2. In Fig. 2.39, the zirconia inversion is shown. There are two internally twinned martensitic lathes in a twinned matrix (Fig. 2.39*a*). No thermally activated diffusion is required, and the shear at the coherent interface between the monoclinic and tetragonal phases is thought to be accommodated by a series of dislocations. The large hysteresis in the transformation temperature is attributed to the large difference in specific volume of the phases. Figure 2.39*b* shows another type of martensite plate viewed edge-on; in this case neither the matrix nor the plates are twinned.

Reconstructive Transformation. Another way of changing the secondary coordination is by completely altering the structural relationships, as in the transformation from Fig. 2.38*a* to *d*. Here the change in structure cannot be arrived at simply by displacing atoms, but interatomic bonds must be broken. The energy required for this breaking up of the structure is recovered when the new structure is formed. In contrast, there is no activation energy barrier to a displacive transformation. Consequently, transformations of this structural type are frequently sluggish. High-temperature forms can often be cooled below the transformation temperature without reverting to the thermodynamic stable form. This kind of transformation has been called a reconstructive transformation.

A reconstructive transformation can take place in several ways. One way is the nucleation of a new phase and growth in the solid state. In addition, if there is an appreciable vapor pressure, the unstable modification can vaporize and condense as the more stable lower-vapor-pressure form. Similarly, many transformations are speeded up by the presence of a liquid in which the greater solubility of the unstable form allows it to go into solution and then precipitate as the more stable form. (This method is used in the manufacture of refractory silica brick in which the addition of a small percentage of lime acts as a flux at the firing temperature, dissolving the quartz and precipitating silica as tridymite. Tridymite is desirable because the high-low inversion involves a much smaller volume change than the high-low quartz transformation.) Reconstructive transformations can also be speeded by the addition of mechanical energy. As is clear from the structural changes, reconstructive transformations require a high activation energy and frequently do not take place at all

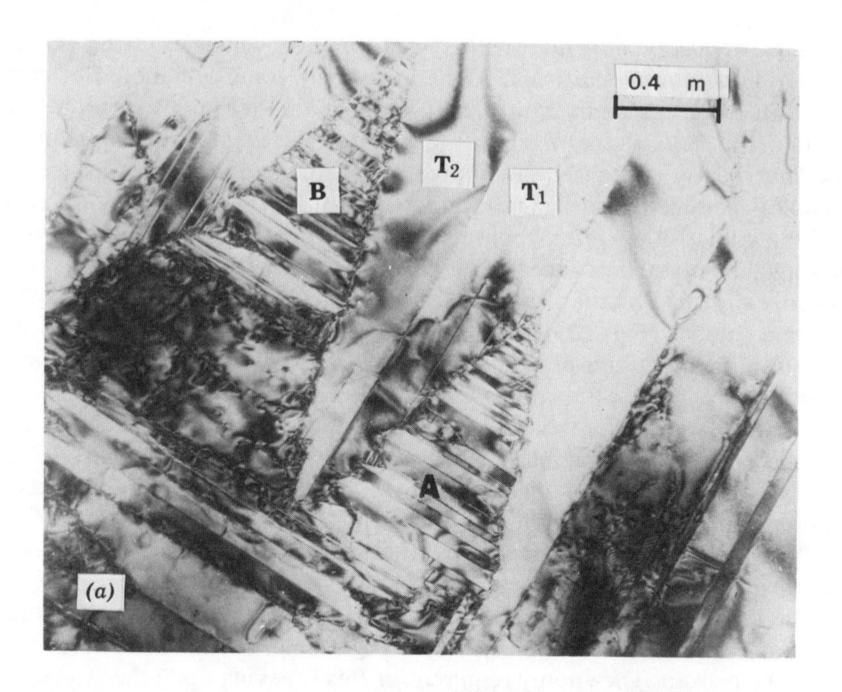

Fig. 2.39. Martensitic transformation in ZrO_2. (*a*) Monoclinic plates (A and B) with fine twins within the plates. The regions between the plates are also twinned (T_1 and T_2). (*b*) Edge-on view of monoclinic plates (which are not twinned) in a tetragonal matrix. From Bansal and Heuer, *Acta Met.*, **20**, 1281 (1972).

unless aided by the presence of a solvent, mechanical work, or other methods of circumventing the activation energy barrier.

Silica. Of particular interest to silicate technology are the transformations observed for silica (Table 2.7). The stable form at room temperature is low quartz, which transforms into high quartz with a displacive transformation at 573°C. Quartz transforms only slowly into the stable tridymite form at 867°C. Indeed, there is some evidence that tridymite cannot be formed at all from quartz without the presence of other impurities. Tridymite remains the stable form until 1470°C, when it transforms into cristobalite, another reconstructive transformation. Both cristobalite and tridymite have displacive transformations on cooling from high temperatures. High cristobalite transforms by distorting its structure into low cristobalite at 200 to 270°C. High tridymite transforms into middle tridymite at 160°C, and this transforms into low tridymite at 105°C. These transformations, which involve a total of seven different polymorphic forms of silica, involve three basic structures. The transformations among the basic structures are sluggish reconstructive transformations which, if they occur at all, take place only slowly and require the addition of materials which act as a solvent in order to occur in reasonable times. In contrast, the displacive transformations between the high- and low-temperature forms of each basic structure occur rapidly and cannot be restrained from taking place. This is particularly important for the high-low quartz transformation, which involves a substantial volume change which can lead to the fracture of bodies containing large amounts of quartz and frequently results in the fracture of quartz grains in ceramic bodies with a consequent reduction in the strength.

Table 2.7. Polymorphic Forms of Silica

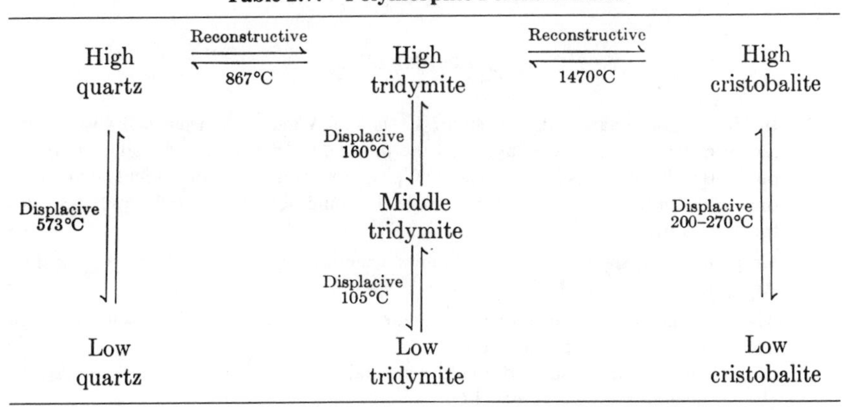

Suggested Reading

1. C. Kittel, *Introduction to Solid State Physics*, 4th ed., John Wiley & Sons, Inc., New York, 1971.
2. R. Sproull, *Modern Physics: A Textbook for Engineers*, John Wiley & Sons, Inc., New York, 1956.
3. L. Pauling, *Nature of the Chemical Bond*, 3d ed., Cornell University Press, Ithaca, N.Y., 1960.
4. A. F. Wells, *Structural Inorganic Chemistry*, 3d ed., Clarendon Press, Oxford, 1962.
5. M. J. Buerger, "Crystallographic Aspects of Phase Transformations," *Phase Transformations in Solids*, R. Smoluchowski, J. E. Mayer, and W. A. Weyl, Eds., John Wiley & Sons, Inc., New York, 1951, p. 183.
6. G. W. Brindley, Ed., *X-Ray Identification and Crystal Structures of the Clay Minerals*, Mineralogical Society, London, 1951.
7. R. C. Evans, *An Introduction to Crystal Chemistry*, 2d ed., Cambridge University Press, London, 1964.
8. R. W. G. Wyckoff, *Crystal Structures*, Vols. 1–4, Interscience Publications, New York, 1948–1953.
9. L. Bragg, G. F. Claringbull, and W. H. Taylor, *Crystal Structure of Minerals*, Cornell University Press, Ithaca, N.Y., 1965.
10. Ajit R. Verma and P. Krishna, "Polymorphism and Polytypism in Crystals," John Wiley & Sons, Inc., New York, 1966.
11. W. B. Pearson "Handbook of Lattice Spacings and Structures of Metals and Alloys," Pergamon Press, Oxford, 1958; Vol. II, 1967.
12. *Structure Reports*, Published for the International Union of Crystallography, Oosthoek Publishing Company, Utrecht. Each volume contains critical summation of all crystal structures determined in a given year (presently complete through Vol. 29, 1967).

Problems

2.1. In sixfold coordination the radius of K^+ is 1.38 Å and of oxygen is 1.40 Å. Some recent measurements have suggested the *possibility* that the K – O radius sum in a potassium silicate glass is 2.4 Å. Can you propose a conjecture that might account for this? Explain how your conjecture would or would not comply with Pauling's rules for ionic structures.

2.2. Graphite, mica, and kaolinite have similar structures. Explain the differences in their structures and resultant differences in properties.

2.3. The structure of lithium oxide has anions in cubic close packing with Li^+ ions occupying all tetrahedral positions.
  (*a*) Compute the value of the lattice constant.
  (*b*) Calculate the density of Li_2O.

(*c*) What is the maximum radius of a cation which can be accommodated in the vacant interstice of the anion array?
(*d*) Calculate the density of the 0.01 mole% SrO solid solution in Li_2O.

2.4. Sketch the atomic plan of the (110) and the (111) planes of MgO. Show direction of closest packing. Point out tetrahedral and octahedral sites.

2.5. (*a*) Starting with cubic close-packing of oxygen ions, sketch the type and position of interstices suitable as cation sites. What is the ratio of octahedral sites to oxygen ions? What is the ratio of tetrahedral sites to oxygen ions?
(*b*) Explain on the basis of bond strength and Pauling's rules what valency ions are required to have stable structures in which:
1. All octahedral sites are filled.
2. All tetrahedral sites are filled.
3. Half the octahedral sites are filled.
4. Half the tetrahedral sites are filled.

Give an example of each.

2.6. Very briefly explain the meaning of the following terms:
Isomorphism
Polymorphism
Polytypism
Antistructure (e.g., "anti"-fluorite structure)
Inverse structure (e.g., "inverse" spinel structure)

2.7. The atomic weight of Si and Al are very similar (28.09 and 26.98 respectively), yet the densities of SiO_2 and Al_2O_3 are quite different (2.65 and 3.96 respectively). Explain this difference in terms of crystal structure and Pauling's rules.

2.8. Barium titanate, $BaTiO_3$, an important ferroelectric ceramic, crystallizes with the perovskite structure type.
(*a*) What is the lattice type?
(*b*) What are the coordination numbers of the ions in this structure?
(*c*) Does the structure obey Pauling's rules? Discuss fully.

2.9. Sketch the crystal structure of pyrophyllite, talc, and montmorillonite, illustrating any differences. Explain how these differences are related to differences in cation exchange capacity.

2.10. (*a*) Calculate the lower limiting cation–anion radius ratio for threefold coordination.
(*b*) For B^{+3} ($r_c = 0.20$ Å) and O^{2-} ($r_a = 1.40$ Å), what coordination number would be predicted?
(*c*) Triangular coordination is normally observed experimentally. Explain.

2.11. Asbestos minerals such as tremolite $(OH)_2Ca_2Mg_5(Si_4O_{11})_2$ have a fibrous habit; talc $(OH)_2Mg_3(Si_2O_5)_2$ has a platy habit. Explain this difference in terms of O/Si ratio and bonding between silica tetrahedra.

2.12. Give explanations for the following observations:
(*a*) Many ceramics are layer silicates with structures which contain a layer of octahedrally coordinated Al and a layer of tetrahedrally coordinated Si. In such structures Al often substitutes for Si, but Si never substitutes for Al (ionic radii are 0.41, 0.50, and 1.40 Å for Si^{4+}, Al^{3+}, and O^{2-}, respectively).
(*b*) Many oxides are based on cubic close-packed arrays of anions. Relatively few structures are based on hexagonal close packing in spite of the fact that the densities of both arrays are equal.

(*c*) Silicates have structures in which SiO_4^{4-} tetrahedra share vertices to form chains, rings, sheets, etc. In phosphates (PO_4^{3-}) and sulfates (SO_4^{2-}) similar tetrahedra are found, but they are always isolated. Yet $AlPO_4$ has a structure which corresponds to that of quartz, SiO_2.

(*d*) The alkaline earth oxides, MgO, SrO, BaO, all have the rock salt structure. The hardness and melting points of the compounds decreases in the order given.

(*e*) MgO (rock salt structure) and Li_2O (antifluorite structure) both are based on cubic close-packed oxygen, with cations occupying interstices in the array. Yet predominant point defect in MgO is of the Schottky type; that in Li_2O is of the Frenkel type.

2.13. The garnets $Mg_3Al_2(SiO_4)_3$ and $Fe_3Al_2(SiO_4)_3$ are isomorphous in a manner similar to Fe_2SiO_4 and Mg_2SiO_4. They are not isomorphous with $Ca_3Al_2(SiO_4)_3$, and neither is Mg_2SiO_4 or Fe_2SiO_4 isomorphous with Ca_2SiO_4. Give an explanation for this based on ionic size and coordination numbers. On the basis of your theory predict a mineral not mentioned in this question that will be isomorphous with Mg_2SiO_4, one for Ca_2SiO_4, one for $Mg_3Al_2(SiO_4)_3$, and one for $Ca_3Al_2(SiO_4)_3$.

2.14. A certain engineer was asked to identify some plate-shaped crystals that had crystallized from a glass melt. The X-ray diffraction pattern showed they were a single phase (only one kind of crystal structure), but the chemical analysis indicated a complicated formula $KF \cdot AlF \cdot BaO \cdot MgO \cdot Al_2O_3 \cdot 5MgSiO_3$ on an empirical basis. If he called you in as a consultant, would you be able to show him this is related to muscovite (potassium mica) and to the talc or pyrophillite crystals? Show what substitutions in talc or pyrophillite that have been made to produce this crystal.

3

Structure of Glasses

Even though a majority of natural and manufactured solids are crystalline in nature, as discussed in Chapter 2, materials which are not crystalline are of great importance for both traditional and newly developed ceramics. One important class is the liquid silicates, the properties of which are an essential part of the ceramist's knowledge in the formulation of glasses, glazes, and enamels. Solid glasses, of which the silicates are the technologically most important group, usually have a more complex structure than the liquids from which they are derived, and recent studies indicate a complexity which is still not well understood, although the broad structural characteristics seem reasonably clear. A more newly developed class of materials is thin films deposited as noncrystalline solids from the vapor phase, about which even less is known as to structural details. In each of these classes of ceramic materials the short-range order is preserved in the immediate vicinity of any selected atom, that is, the first coordination ring; the longer-range order characteristic of the ideal crystal is dissipated in a way characterized by diversity among different systems and by difficulty in precise description.

We focus our attention on glasses, which are by far the most important group of inorganic noncrystalline solids. The structure of glasses may be considered on three scales: (1) the scale of 2 to 10 Å, or that of local atomic arrangements; (2) the scale of 30 to a few thousand angstroms, or that of submicrostructure; and (3) the scale of microns to millimeters or more, or that of microstructure and macrostructure. In this chapter we consider the atomic structure and the submicrostructure of glasses; the consideration of microstructural features is deferred to Chapter 11.

3.1 Glass Formation

Glasses are usually formed by solidification from the melt. The structure of glasses can be clearly distinguished from that of liquids, since glass structure is effectively independent of temperature. This can best be seen by a plot of the specific volume of the crystal, liquid, and glass as a function of temperature (Fig. 3.1). On cooling the liquid, there is a discontinuous change in volume at the melting point if the liquid crystallizes. However, if no crystallization occurs, the volume of the liquid decreases at about the same rate as above the melting point until there is a decrease in the expansion coefficient at a range of temperature called the glass transformation range. Below this temperature range the glass

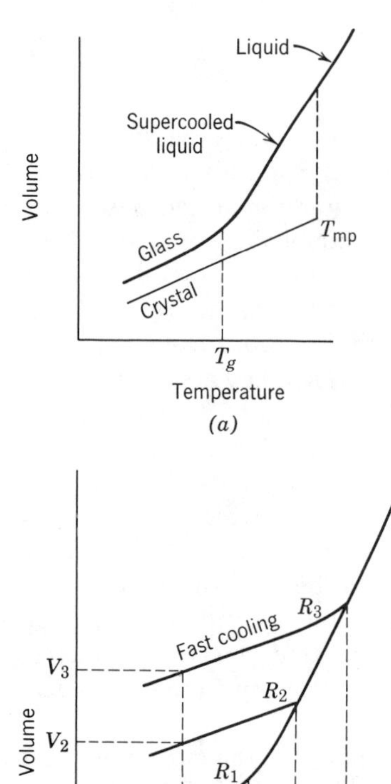

Fig. 3.1. Schematic specific volume-temperature relations. (*a*) Relations for liquid, glass, and crystal; (*b*) glasses formed at different cooling rates $R_1 < R_2 < R_3$.

structure does not relax at the cooling rate used. The expansion coefficient for the glassy state is usually about the same as that for the crystalline solid. If slower cooling rates are used so that the time available for the structure to relax is increased, the supercooled liquid persists to a lower temperature, and a higher-density glass results. Similarly, by heating the glassy material in the annealing range, in which slow relaxation can occur, the glass structure in time approaches an equilibrium density corresponding to the supercooled liquid at this temperature.

A concept useful in discussing the properties of glasses is the glass transition temperature T_g, which corresponds to the temperature of the intersection between the curve for the glassy state and that for the supercooled liquid (Fig. 3.1). Different cooling rates, corresponding to different relaxation times, give rise to a different configuration in the glassy state equivalent to different points along the curve for the supercooled liquid. In the transition range the time for structural rearrangements is similar in magnitude to that of experimental observations. Consequently the configuration of the glass in this temperature range changes slowly with time toward the equilibrium structure. At somewhat higher temperatures the structure corresponding to equilibrium at any temperature is achieved very rapidly. At substantially lower temperatures the configuration of the glass remains sensibly stable over long periods of time.

In discussing the structural characteristics of glasses, reference is often made to the structure of a particular glassy material. It should be noted, however, that any determination of glass structure is only meaningful within limits seen from the volume-temperature relations shown in Fig. 3.1. As the liquid is cooled from a high temperature without crystallizing, a region of temperature is reached in which a bend appears in the volume-temperature relation. In this region, the viscosity of the material has increased to a sufficiently high value, typically about 10^{12} to 10^{13} P, so that the sample exhibits solidlike behavior. As shown in Fig. 3.1*b*, the glass transition temperature increases with increasing cooling rate, as do the specific volumes of the glasses which are formed. In the case shown, the specific volume of the glass at temperature T_0 can be V_1 or V_2 or V_3, depending on which of the three cooling rates was used in forming the glass. The maximum difference in specific volume obtainable with variations in the cooling rate is typically in the range of a few percent; only within this range can one speak of the structure of a glass without carefully specifying its mode of formation.

Noncrystalline solids can be formed in other ways besides cooling from the liquid state, and their structure may differ significantly from glasses formed by the cooling of liquids. Among these alternative methods, the most widely used and most effective method for materials which are difficult to form as noncrystalline solids is condensation from the vapor onto a cold substrate. When a vapor stream formed by electron-beam evaporation, sputtering, or thermal evaporation impinges on the cold substrate, thermal energy is extracted from the atoms before they can migrate to their lowest free-energy configuration (the crystalline state).

Another method of forming glasses is by electrodeposition; Ta_2O_5, Ge, and certain Ni–P alloys are among the materials which have been prepared in this way. Noncrystalline solids can also be formed by chemical reactions. Silica gel, for example, can be manufactured from ethyl silicate by the reaction

$$Si(O\,Eth)_4 \xrightarrow[\text{catalyst}]{H_2O} Si(OH)_4 \xrightarrow{-H_2O} SiO_2 \tag{3.1}$$

In this reaction the SiO_2 resulting from the condensation of the silicic acid is noncrystalline. A similar silica gel can be formed by the reaction of sodium silicate with acid. These reactions are particularly effective in the case of hydrogen-bonded structures in aqueous media. For example, the reaction

$$Al_2O_3 + 6H_3PO_4 \rightleftarrows 2Al(H_2PO_4)_3 + 3H_2O \tag{3.2}$$

forms a noncrystalline gel in which hydrogen bonding predominates. Like silica gel it makes a good inorganic cement.

On the scale of atomic structure, the distinguishing structural characteristic of glasses, like the liquids from which many are derived, is the absence of atomic periodicity or long-range order. Such a lack of periodicity does not, however, imply the absence of short-range order, on a scale of a few angstroms. The short-range order which characterizes a particular glass or liquid may be described in terms of an atom-centered coordinate system and is frequently represented in terms of radial distribution functions.

The radial distribution function $\rho(R)$ is defined as the atom density in a spherical shell of radius R from the center of a selected atom in the liquid or glass. The radial distribution function for a Se glass, determined from X-ray diffraction studies, is shown in Fig. 3.2. As shown there, modulations in the radial density of atoms are observed for interatomic separations of the order of a few angstroms; for large distances the observed atom density approaches the average value ρ_0. The approach of the actual radial-density function to the average atom density at large distances reflects the absence of structure on such a scale. Hence, a precise description can be given to the scale on which short-range order is observed, that is, the scale on which significant modulations are seen in the radial-density function, the scale of a few angstroms.

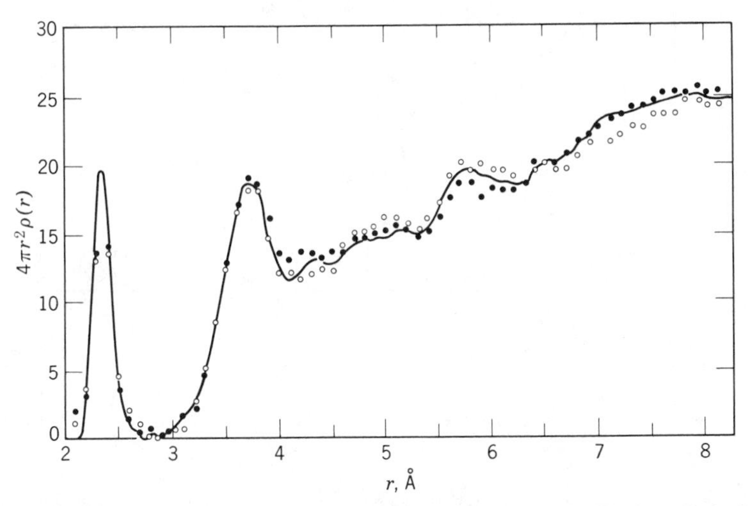

Fig. 3.2. Radial distribution function for glassy selenium. From R. Kaplow, T. A. Rowe, and B. L. Averbach, *Phys. Rev.*, **168**, 1068 (1968).

3.2 Models of Glass Structure

A number of models have been suggested to describe the structure of glasses.

Crystallite Model. X-ray diffraction patterns from glasses generally exhibit broad peaks centered in the range in which strong peaks are also seen in the diffraction patterns of the corresponding crystals. This is shown in Fig. 3.3 for the case of SiO_2. Such observations led to the suggestion that glasses are composed of assemblages of very small crystals, termed crystallite, with the observed breadth of the glass diffraction pattern resulting from particle-size broadening. It is well established that measurable broadening of X-ray diffraction peaks occurs for particle sizes or grain sizes smaller than about 0.1 micron. The broadening increases linearly with decreasing particle size. This model was applied to both single-component and multicomponent glasses (in the latter case, the structure was viewed as composed of crystallites of compositions corresponding to compounds in the particular system), but the model is not today supported in its original form, for reasons discussed in the next section.

Random-Network Model. According to this model, glasses are viewed as three-dimensional networks or arrays, lacking symmetry and periodicity, in which no unit of the structure is repeated at regular intervals. In the case of oxide glasses, these networks are composed of oxygen polyhedra.

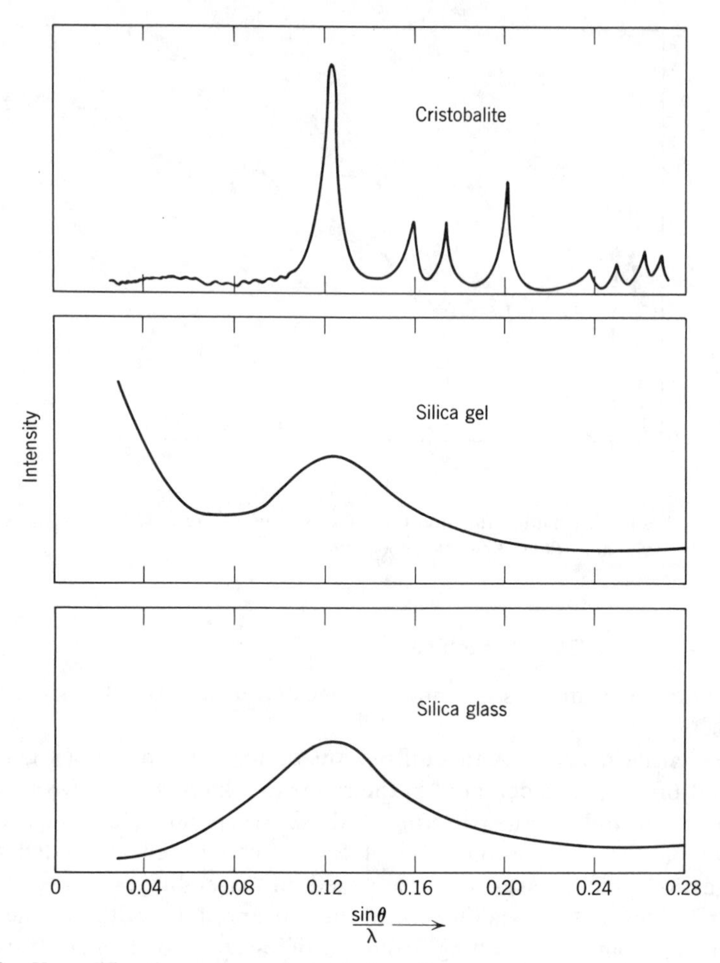

Fig. 3.3. X-ray diffraction patterns of cristobalite, silica gel, and vitreous silica. From B. E. Warren and J. Biscal, *J. Am. Ceram. Soc.*, **21**, 49 (1938).

Adopting the hypothesis that a glass should have an energy content similar to that of the corresponding crystal, W. H. Zachariasen* considered the conditions for constructing a random network such as shown in Fig. 3.4 and suggested four rules for the formation of an oxide glass:

1. Each oxygen ion should be linked to not more than two cations.
2. The coordination number of oxygen ions about the central cation must be small, 4 or less.

J. Am. Chem. Soc., **54**, 3841 (1932).

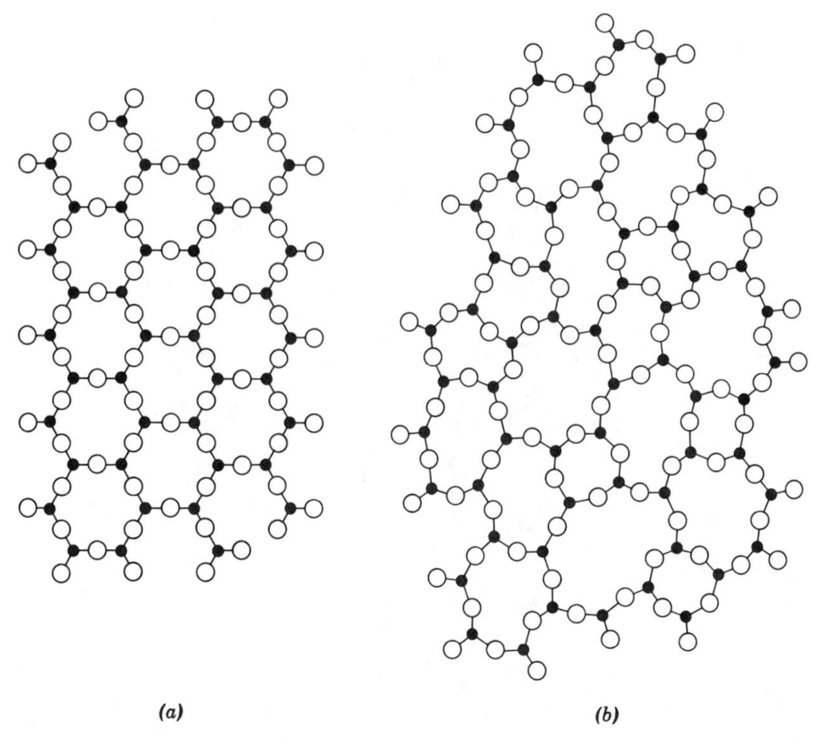

Fig. 3.4. Schematic representation of (*a*) ordered crystalline form and (*b*) random-network glassy form of the same composition.

3. Oxygen polyhedra share corners, not edges or faces.
4. At least three corners of each polyhedron should be shared.

In practice, the glass-forming oxygen polyhedra are triangles and tetrahedra, and cations forming such coordination polyhedra have been termed *network formers.* Alkali silicates form glasses easily, and the alkali ions are supposed to occupy random positions distributed through the structure, located to provide local charge neutrality, as pictured in Fig. 3.5. Since their major function is viewed as providing additional oxygen ions which modify the network structure, they are called *network modifiers.* Cations of higher valence and lower coordination number than the alkalis and alkaline earths may contribute in part to the network structure and are referred to as *intermediates.* In a general way the role of cations depends on valence and coordination number and the related value of single-bond strength, as illustrated in Table 3.1.

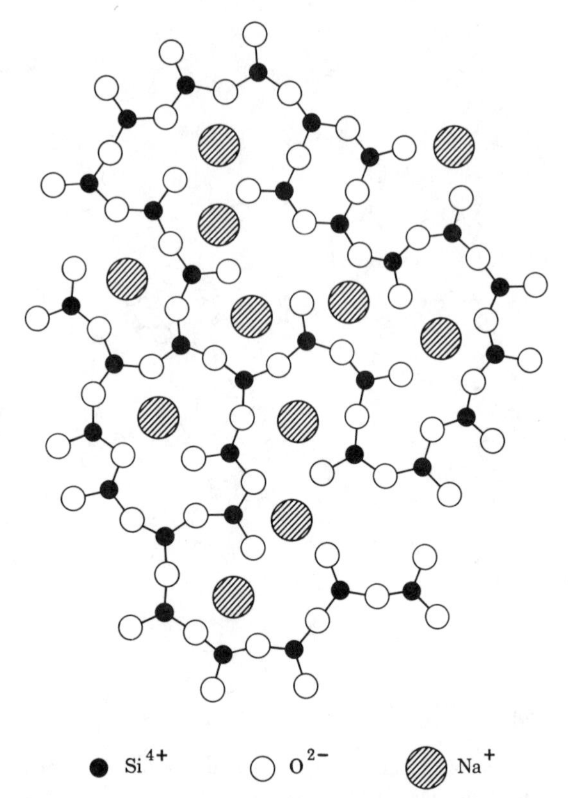

Fig. 3.5. Schematic representation of the structure of a sodium silicate glass.

The random-network model was originally proposed to account for glass formation as resulting from the similarity of structure and internal energy between crystalline and glassy oxides. Although this remains one factor to be considered, we now believe that kinetic considerations preventing crystallization during cooling are more important. The model remains, however, as the best general picture of many silicate glasses and may readily be generalized as a *random-array* model in which the structural elements are randomly arranged and in which no unit of the structure is repeated at regular intervals in three dimensions. In this form, the model may be used to describe a variety of liquid and glass structures, both oxide and nonoxide, in which three-dimensional networks are not possible.

Other Structural Models. Several other models have been suggested to represent the structures of glasses. One of these, termed the pentagonal dodecahedron model, views silicate glasses as composed of pentagonal

Table 3.1. Coordination Number and Bond Strength of Oxides

	M in MO_x	Valence	Dissociation Energy per MO_x (kcal/g-atom)	Coordination Number	Single-Bond Strength (kcal/g-atom)
Glass formers	B	3	356	3	119
	Si	4	424	4	106
	Ge	4	431	4	108
	Al	3	402–317	4	101–79
	B	3	356	4	89
	P	5	442	4	111–88
	V	5	449	4	112–90
	As	5	349	4	87–70
	Sb	5	339	4	85–68
	Zr	4	485	6	81
Intermediates	Ti	4	435	6	73
	Zn	2	144	2	72
	Pb	2	145	2	73
	Al	3	317–402	6	53–67
	Th	4	516	8	64
	Be	2	250	4	63
	Zr	4	485	8	61
	Cd	2	119	2	60
Modifiers	Sc	3	362	6	60
	La	3	406	7	58
	Y	3	399	8	50
	Sn	4	278	6	46
	Ga	3	267	6	45
	In	3	259	6	43
	Th	4	516	12	43
	Pb	4	232	6	39
	Mg	2	222	6	37
	Li	1	144	4	36
	Pb	2	145	4	36
	Zn	2	144	4	36
	Ba	2	260	8	33
	Ca	2	257	8	32
	Sr	2	256	8	32
	Cd	2	119	4	30
	Na	1	120	6	20
	Cd	2	119	6	20
	K	1	115	9	13
	Rb	1	115	10	12
	Hg	2	68	6	11
	Cs	1	114	12	10

rings of SiO_4 tetrahedra. From a given tetrahedron, the rings extend in six directions to include the six edges and form twelve-sided dodecahedral cavities. Because of their fivefold symmetry, these dodecahedral cages cannot be extended in three dimensions without an accompanying strain which ultimately prevents maintenance of the silicon-oxygen bonds. Although pentagonal rings of SiO_4 tetrahedra may indeed exist in the structure of glasses such as fused silica, there is little reason to believe that the structure is composed entirely of such elements.

According to another model, glasses are composed of micelles or paracrystals characterized by a degree of order intermediate between that of a perfect crystal and that of a random array. These paracrystalline grains may themselves be arranged in arrays with differing degrees of order. The degree of order in the grains should be large enough to discern their mutual misorientation in an electron microscope and small enough to avoid sharp Bragg reflections in X-ray diffraction patterns. Although such models seem plausible, the evidence for the existence of such structures, at least in oxide glasses, is marginal.

3.3 The Structure of Oxide Glasses

In discussing the structures of oxide glasses, it should be emphasized that these structures are not known to anything like the confidence with which the crystal structures discussed in Chapter 2 have been determined. Recent advances in experimental techniques and means of analyzing data have opened a new era of glass-structure studies, and the next decade should be marked by significant advances in our knowledge of such structure. Even the best experimental techniques are inadequate, however, for establishing any particular model as the structure of a given glass. Rather, the results of structural investigations of glasses should be regarded as providing information with which any proposed structure must be consistent.

Silica. Early controversies between proponents of the crystallite and random-network models of glass structure were generally decided in favor of the random-network model, based largely on the arguments advanced by B. E. Warren.* From the width of the main broad diffraction peak in the glass diffraction pattern, the crystallite size in the case of SiO_2 was estimated at about 7 to 8 Å. Since the size of a unit cell of cristobalite is also about 8 Å, any crystallites would be only a single unit cell in extent; and such structures seem at variance with the notion of a crystalline array. This remains a powerful argument even if the estimate of crystallite

*B. E. Warren, *J. Appl. Phys.*, **8**, 645 (1937).

size were only accurate to within a factor of 2. Further, in contrast to silica gel, there is no marked small-angle scattering from a sample of fused silica (see Fig. 3.3). This indicates that the structure of the glass is continuous and is not composed of discrete particles like the gel. Hence, if crystallites of reasonable size are present, there must be a continuous spatial network connecting them which has a density similar to that of the crystallites.

A more recent X-ray diffraction study of fused silica was carried out with advanced experimental techniques and means of analyzing data.* In this study, the distribution of silicon-oxygen-silicon bond angles (Fig. 3.6*a*) was determined. As shown in Fig. 3.6*b*, these angles are distributed

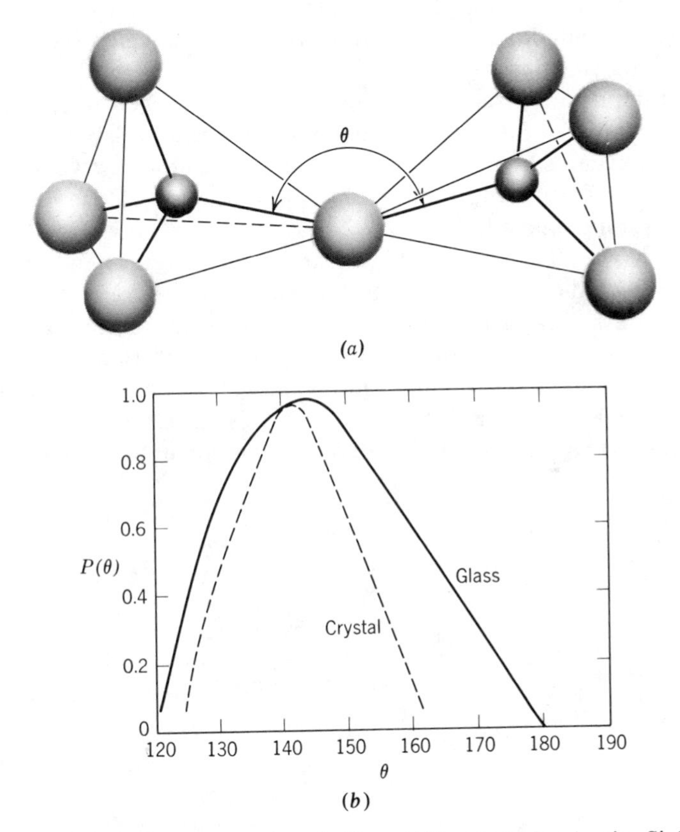

Fig. 3.6. (*a*) Schematic representation of adjacent SiO_4 tetrahedra showing Si–O–Si bond angle. Closed circles = Si; open circles = O. (*b*) Distribution of Si–O–Si bond angles in fused silica and crystalline crystobalite. From R. L. Mozzi, Sc.D. thesis, MIT, 1967.

*R. L. Mozzi and B. E. Warren, *J. Appl. Cryst.*, **2**, 164 (1969).

over a broad range, from about 120 to about 180°, centered about 145°. This range is much broader in the glass than in the corresponding distribution for crystalline cristobalite. In contrast, the silicon-oxygen and oxygen-oxygen distances are nearly as uniform in the glass as in the corresponding crystal.

The essential randomness of the SiO_2 glass structure results, then, from a variation in the silicon-silicon distances (the silicon-oxygen-silicon bond angles). Beyond this direct joining of the tetrahedra over a range of Si–O–Si angles, the structure of fused silica seems to be completely random. X-ray diffraction work by G. G. Wicks* provides strong evidence for a random distribution of rotation angles of one tetrahedron with respect to another. That is, there appears to be no pronounced preference in fused silica for edge-to-face sharing of tetrahedra, which is often found in crystalline silicates.

The structure of fused silica seems, then, to be well described by a random network of SiO_4 tetrahedra, with significant variability occurring in silicon-oxygen-silicon bond angles. Such a random network is, however, not necessarily uniform, and local variations in density and structure are to be expected.

B_2O_3. X-ray diffraction and nuclear magnetic resonance studies of glassy B_2O_3 indicate clearly that the structure is composed of BO_3 triangles. Less clearly defined is the way in which these triangles are linked together in the structure. A random network of triangles provides a poor representation of the diffraction data.† A better description is provided by a model in which the triangles are linked in a boroxyl configuration (Fig. 3.7). A still better representation of the data is obtained

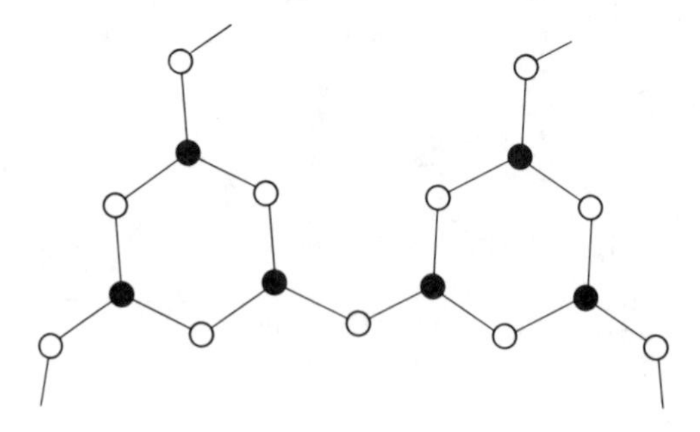

Fig. 3.7. Schematic representation of boroxyl configurations. Filled circles = B; open circles = O.

*Sc.D. thesis, Massachusetts Institute of Technology, 1974.
†R. L. Mozzi and B. E. Warren, *J. Appl. Cryst.*, **3**, 251 (1970).

from a model based on a distorted version of the crystal structure in which the triangles are linked in ribbons. The distortions are such as to destroy the essential symmetry of the crystal, and the notion of discrete crystallites embedded in a matrix is not appropriate.

Silicate Glasses. The addition of alkali or alkaline earth oxides to SiO_2 increases the ratio of oxygen to silicon to a value greater than 2 and breaks up the three-dimensional network with the formation of singly bonded oxygens which do not participate in the network (Fig. 3.8). The structural units found in crystalline silicates are shown for different oxygen–silicon ratios in Table 3.2. For reasons of local charge neutrality, the modifying cations are located in the vicinity of the singly bonded oxygens. With divalent cations, two singly bonded oxygens are required

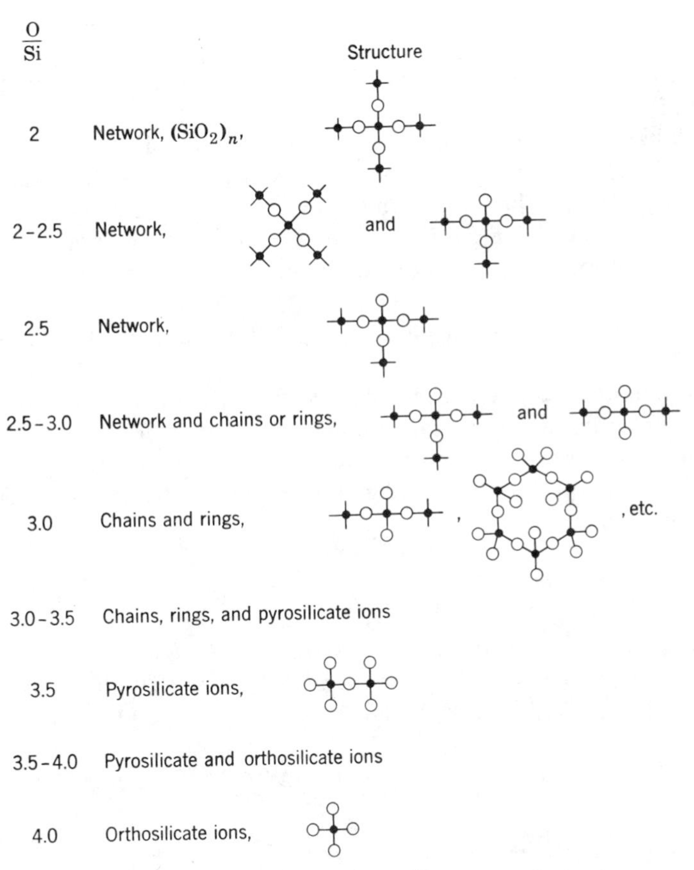

Fig. 3.8. Effect of oxygen-silicon ratio on silicate network structures.

Table 3.2. **Structural Units Observed in Crystalline Silicates**

Oxygen-Silicon Ratio	Silicon-Oxygen Groups	Structural Units	Examples
2	SiO_2	Three-dimensional network	Quartz
2.5	Si_4O_{10}	Sheets	Talc
2.75	Si_4O_{11}	Chains	Amphiboles
3	SiO_3	Chains	Pyroxenes
		Rings	Beryl
3.5	Si_2O_7	Tetrahedra sharing one oxygen ion	Pyrosilicates
4	SiO_4	Isolated orthosilicate tetrahedra	Orthosilicates

for each cation; for univalent alkali ions, only one such oxygen is required.

An X-ray diffraction study of a number of K_2O–SiO_2 glasses by G. G. Wicks* indicates systematic changes in the structure as the alkali oxides are added to SiO_2. The data seem to indicate a random-network structure in which the alkali ions are distributed in pairs at random through the structure but located adjacent to singly bonded oxygens. In the case of a Tl_2O–SiO_2 glass containing 29.4 mole% Tl_2O, Blair and Milberg† suggest clustering of the modifying cations, with an average cluster diameter of about 20 Å.

It is sometimes convenient to describe the network character of silicate glasses in terms of the average number R of oxygen ions per network-forming ion, usually the oxygen-silicon ratio. For example, $R = 2$ for SiO_2; for a glass containing 12 g-atom% Na_2O, 10 g-atom% CaO, and 78 g-atom% SiO_2

$$R = \frac{12 + 10 + 156}{78} = 2.28$$

Following Stevels,‡ for glasses containing only one type of network-forming cation surrounded by Z oxygens ($Z = 3$ or 4), with X nonbridging (i.e., singly bonded) and Y bridging oxygens per polyhedron, one may write

$$X + Y = Z \qquad \text{and} \qquad X + 0.5Y = R \tag{3.3}$$

**Op. cit.*

†*J. Am. Ceram. Soc.*, **57**, 257 (1974).

‡J. M. Stevels, *Handb. Phys.*, **20**, 350 (1957).

For silicate glasses, when the oxygen polyhedra are SiO_4 tetrahedra, $Z = 4$ and Eq. 3.3 becomes

$$X = 2R - 4 \qquad \text{and} \qquad Y = 8 - 2R$$

In the case of silicate glasses containing more alkali and alkaline earth oxides than Al_2O_3, the Al^{3+} is believed to occupy the centers of AlO_4 tetrahedra. Hence the addition of Al_2O_3 in such cases introduces only 1.5 oxygens per network-forming cation, and nonbridging oxygens of the structure are used up and converted to bridging oxygens. This is shown in Table 3.3, in which the values of X, Y, and R are given for a number of glass compositions.

Table 3.3. Values of the Network Parameters *X*, *Y*, and *R* for Representative Glasses

Composition	R	X	Y
SiO_2	2	0	4
$Na_2O \cdot 2SiO_2$	2.5	1	3
$Na_2O \cdot 1/2Al_2O_3 \cdot 2SiO_2$	2.25	0.5	3.5
$Na_2O \cdot Al_2O_3 \cdot 2SiO_2$	2	0	4
$Na_2O \cdot SiO_2$	3	2	2
P_2O_5	2.5	1	3

The parameter Y gives the average number of bridges between the oxygen tetrahedra and their neighbors. For silicate glasses with Y values less than 2, no three-dimensional network is possible, since the tetrahedra have fewer than two oxygen ions in common with other tetrahedra. Chains of tetrahedra of various lengths are then expected as the characteristic structural feature.

In crystalline silicates, the SiO_4 tetrahedra are found in a variety of configurations, depending on the oxygen-to-silicon ratio, as shown in Table 3.2. Such configurations may also occur in glasses of the corresponding compositions, and mixtures of these configurations may occur in glasses of intermediate compositions; occurrence in the crystalline phases indicates that these structural units represent low-energy configurations. However, since glasses are derived from supercooled liquids, in which the greater entropy of more random arrays may be controlling, the analogy between crystalline and glassy structural units should be pursued with caution.

For a variety of glazes and enamels it is typically found that the oxygen–to–network-former ratio is in the range of 2.25 to 2.75, as shown

Table 3.4. Compositions of Some Glazes and Enamels

Composition (mole fraction, with $RO + R_2O = 1.00$)

Description	Firing Temperature (°C)	$\begin{bmatrix} Na \\ K \end{bmatrix}_2 O$	$\begin{bmatrix} Mg \\ Ca \end{bmatrix} O$	PbO	Al_2O_3	B_2O_3	SiO_2	Other	Ratio Oxygen to Network Former[a]
Leadless raw porcelain glaze, glossy	1250	0.3	0.7		0.4		4.0		2.46
Leadless raw porcelain glaze, mat	1250	0.3	0.7		0.6		3.0		2.75
High-temperature glaze, glossy	1465	0.3	0.7		1.1		14.7		2.25
Bristol glaze, glossy	1200	0.35	0.35		0.55		3.30	0.3ZnO	2.65
Aventurine glaze (crystals precipitate)	1125	1.0			0.15	1.25	7.0	$0.75Fe_2O_3$	2.56
Lead-containing fritted glaze, glossy	1080	0.33	0.33	0.33	0.13	0.53	1.73		2.61
Lead-containing fritted glaze, glossy	930	0.17	0.22	0.65	0.12	0.13	1.84		2.25
Lead-containing fritted glaze, glossy	1210	0.05	0.50	0.45	0.27	0.32	2.70		2.30

Raw lead glaze, glossy	1100	0.1	0.3	0.6	0.2		1.6		2.30
Raw lead glaze, mat	1100	0.1	0.35	0.55	0.35		1.5		2.50
Raw lead glaze, opaque	1100	0.1	0.2	0.7	0.2		2.0	$0.33SnO_2$	2.68
Jewelry enamel for copper	950	0.5		0.5	0.1	0.1	1.5		2.40
Sheet-iron ground-coat enamel	850	1.0			0.3	1.0	3.6	$0.3CaF_2$ $0.06CoO + NiO$ $0.08MnO_2$	2.28
Cast-iron cover-coat enamel, opaque	820	0.73		0.27			1.4	$0.18SnO_2$	2.72
Sheet-iron cover-coat enamel, opaque	800	0.76				1.05	3.3	$0.24Na_2SiF_6$ $1.05TiO_2$	2.46

[a] Si + B + 1/3Al + 1/2Pb (this is fairly arbitrary).

in Table 3.4. Usually soda lime silica glasses have an oxygen–to–network-former ratio of about 2.4. The compositions of these and other commercial glasses are shown in Table 3.5.

Borate Glasses. It has been established* that the addition of alkali or alkaline earth oxides to B_2O_3 results in the formation of BO_4 tetrahedra. The variation of the fraction of four-coordinated borons with the concentration of alkali oxide is shown in Fig. 3.9. The smooth curve shown in the

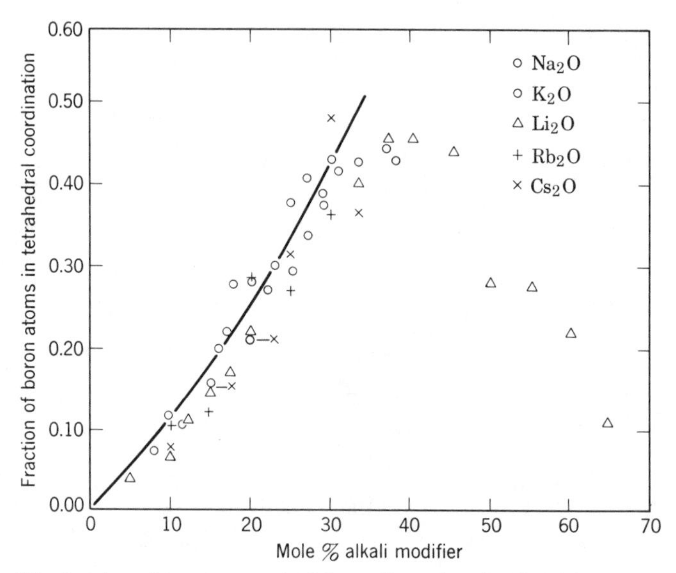

Fig. 3.9. The fraction of boron atoms in BO_4 configurations in alkali borate glasses plotted against the molar percent of alkali oxide. From P. J. Bray, in *Interaction of Radiation with Solids*, Plenum Press, New York, 1967.

figure represents the assumption that each of the oxygens added with the alkali ions converts two triangles to tetrahedra. Up to alkali oxide concentrations of about 30 mole%, nearly all the modifier oxides have the effect of converting BO_3 triangles to BO_4 tetrahedra. Beyond this composition range, the experimentally determined fractions of four-coordinated borons depart significantly from the indicated curve, and this suggests that singly bonded oxygens are produced in appreciable numbers. These singly bonded oxygens are presumably associated with BO_3 triangles rather than with BO_4 tetrahedra, since the requirements for local charge compensation by the modifying cations is simpler in the case of the triangular arrangements.

*P. J. Bray in A. Bishay, Ed., *Interaction of Radiation with Solids*, Plenum Press, Plenum Publishing Corporation, New York, 1967.

Table 3.5. Approximate Composition (wt%) Typical of Some Commercial Glasses

Glass	SiO_2	Al_2O_3	Fe_2O_3	CaO	MgO	BaO	Na_2O	K_2O	SO_3	F_2	ZnO	PbO	B_2O_3	Se	CdO	CuO
Container flint	72.7	2.0	0.06	10.4		0.5	13.6	0.4	0.3	0.2						
Container amber	72.5	2.0	0.1	10.2		0.6	14.4	0.2	S-0.02	0.2						
Container flint	71.2	2.1	0.05	6.3	3.9	0.5	15.1	0.4	0.3	0.1						
Container flint	70.4	1.4	0.06	10.8	2.7	0.7	13.1	0.6	0.2	0.1						
Window green	71.7	0.2	0.1	9.6	4.4		13.1		0.4							
Window	72.0	1.3		8.2	3.5		14.3	0.3	0.3							
Plate	71.6	1.0		9.8	4.3		13.3		0.2							
Opal jar	71.2	7.3		4.8			12.2	2.0		4.2						
Opal illumination	59.0	8.9		4.6	2.0		7.5			5.0	12.0	3.0				
Ruby selenium	67.2	1.8	0.03	1.9	0.4		14.6	1.2	S-0.1	0.4	11.2		0.7	0.3	0.4	
Ruby	72.0	2.0	0.04	9.0			16.6	0.2		Tr.*						0.05
Borosilicate	76.2	3.7		0.8			5.4	0.4					13.5			
Borosilicate	74.3	5.6		0.9		2.2	6.6	0.4					10.0			
Borosilicate	81.0	2.5					4.5						12.0			
Fiber glass	54.5	14.5	0.4	15.9	4.4		0.5			0.3			10.0			
Lead tableware	66.0	0.9		0.7		0.5	6.0	9.5				15.5	0.6			
Lead technical	56.3	1.3					4.7	7.2				29.5	0.6			
Lamp bulb	72.9	2.2		4.7	3.6		16.3	0.2	0.2				0.2			
Heat absorbing	70.7	4.3	0.8	9.4	3.7	0.9	9.8	0.7		Tr.*			0.5			

Source. F. V. Tooley, *Handbook of Glass Manufacture*, Ogden Publishing Co., New York, N.Y.
*Trace.

Germanate and Phosphate Glasses. Glassy GeO_2 is composed of GeO_4 tetrahedra, with a mean germanium-oxygen-germanium bond angle of about 138°. The structural model of a random network of oxygen tetrahedra seems reasonable for this material. In contrast to fused silica, however, the distribution of intertetrahedral angles (Ge–O–Ge in this case) for vitreous germania is quite sharp. The essential randomness of glassy GeO_2 apparently results from a random distribution of the rotation angles of one tetrahedron with respect to another, and this represents a second mode of generating random tetrahedral networks* (in addition to the mode based on the broad distribution of intertetrahedral angles).

Measurements of physical properties such as density suggest that the addition of alkali oxide to GeO_2 may result in the formation of GeO_6 octahedra up to about 15 to 30 g-atom% alkali oxide. For larger additions of alkali oxide, a rapid return to tetrahedral configurations seems indicated, presumably accompanied by the formation of singly bonded oxygens in large numbers. These structural changes remain, however, to be confirmed by diffraction studies.

Information on the structure of phosphate glasses has been determined largely from chromatographic studies. Like silicate and most germanate glasses, phosphate glasses are composed of oxygen tetrahedra; but unlike the silicate and germanate analogs, a PO_4 tetrahedron can be bonded to at most three other similar tetrahedra. The most familiar structural units in phosphate glasses are rings or chains of PO_4 tetrahedra. The results of the chromatographic studies have elucidated the change in average length of chains as the P_2O_5 concentration of phosphate glasses is varied.† With other additions, such as alumina, it is possible to simulate the characteristics of network-based silicate or germanate glasses.

3.4 Submicrostructural Features of Glasses

For several decades after the pioneering work of Warren, glasses were regarded as homogeneous materials, and the conceptual picture of the random network was widely accepted as the best structural model for glasses. Despite this wide acceptance, however, it was known that several glass systems, such as the alkaline earth silicates, exhibited miscibility gaps in their phase diagrams, and it was also known that heterogeneities provided the structural basis for the commercial Vycor process. In this process, a glass containing about 75 wt% SiO_2, 20 wt% B_2O_3, and 5 wt%

*G. G. Wicks, *op. cit.*

†A. E. R. Westman in J. D. Mackenzie, Ed., *Modern Aspects of the Vitreous State*, Vol. 1, Pergamon Press, New York, 1961.

Na_2O is melted, formed into desired shapes, and then heat-treated in the range of 500 to 600°C. Such heat treatment results in the glass separating into two distinct phases, one almost pure SiO_2 and the other rich in Na_2O and B_2O_3. On exposure to a suitable solvent at modest temperatures, the latter phase may be leached out, leaving a SiO_2-rich framework containing a network of pores on a scale of 40 to 150 Å. Subsequent compaction at elevated temperatures (in the range of 900 to 1000°C) results in a transparent glass containing 96 wt% SiO_2.

The introduction of electron microscopy as a tool for investigating materials revolutionized the field of glass structure when submicroscopic features on a scale of a few hundred Å were observed in many glasses, using both replication and direct-transmission electron microscopy. It has now been well established that submicrostructures on a scale of 30 to a few hundred Å are characteristic of many glass systems. Such submicrostructural features have been observed in silicate, borate, chalcogenide, and fused-salt glasses. These submicrostructures have been shown to result from a process of phase separation, in which a liquid which is homogeneous at high temperatures separates into two or more liquid phases on cooling.

To understand this phenomenon, consider the miscibility gap in the MgO–SiO_2 phase diagram shown in Fig. 3.10*a*. As shown in the corresponding free energy versus composition plot in Fig. 3.10*b* at a high temperature such as 2300°C, a homogeneous solution represents the minimum free-energy configuration for all compositions and is the thermodynamically stable phase. At such temperatures, the free energy versus composition curve exhibits positive curvature everywhere. As the temperature is lowered from 2300°C, the free energy increases by an amount proportional to the entropy, since

$$\frac{\partial G}{\partial T} = -S \tag{3.4}$$

For simple solutions, the solution entropy should be greatest in some central region of composition and smallest for pure components and compounds. For this reason it may be expected that with decreasing temperature the free-energy curve flattens. At some lower temperature, such as 2000°C, the free energy versus composition curve develops a region of negative curvature, and the minimum free-energy configuration becomes a mixture of two phases rather than a single phase. These phases are given by the common tangent to the free-energy curve shown in Fig. 3.10*b*.

For composition C_0 at temperature T_2, the lowest free-energy configuration consists of a mixture of two phases of compositions C^α and C^β, in

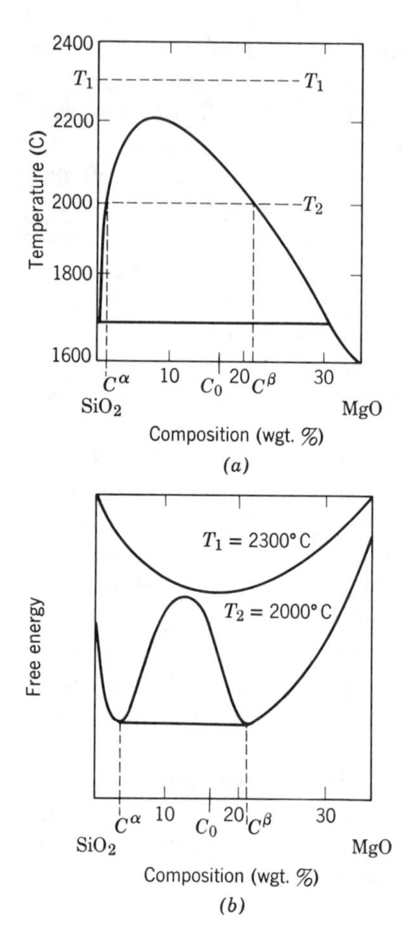

Fig. 3.10. (*a*) Phase diagram for the SiO_2-rich part of the MgO–SiO_2 system showing miscibility gap. From Y. I. Ol'shanskii, *Dokl. Akad. Nauk. SSSR*, **76**, 95 (1951). (*b*) Free energy versus composition relations (schematic) for the temperatures T_1 and T_2 in (*a*).

proportions $X_{\alpha,\beta}$ given by the familiar lever rule (see Chapter 7):

$$\frac{X_\alpha}{X_\beta} = \frac{(C^\beta - C_0)}{(C_0 - C^\alpha)} \tag{3.5}$$

In cases such as that shown for the system sodium tetraborate–silica in Fig. 3.11, the miscibility gaps are metastable. That is, at temperatures such as T_1, the minimum free-energy assemblage of composition C_0 consists of crystal of composition C^C and liquid of composition C^L. If for kinetic reasons the crystal does not form, however, the free energy of a homogeneous liquid of composition C_0 at temperature T_1 can be lowered by its separating into two liquids of composition C^α and C^β. A mixture of these two liquid phases in amounts given by the lever rule represents the liquid configuration of lowest free energy at T_1.

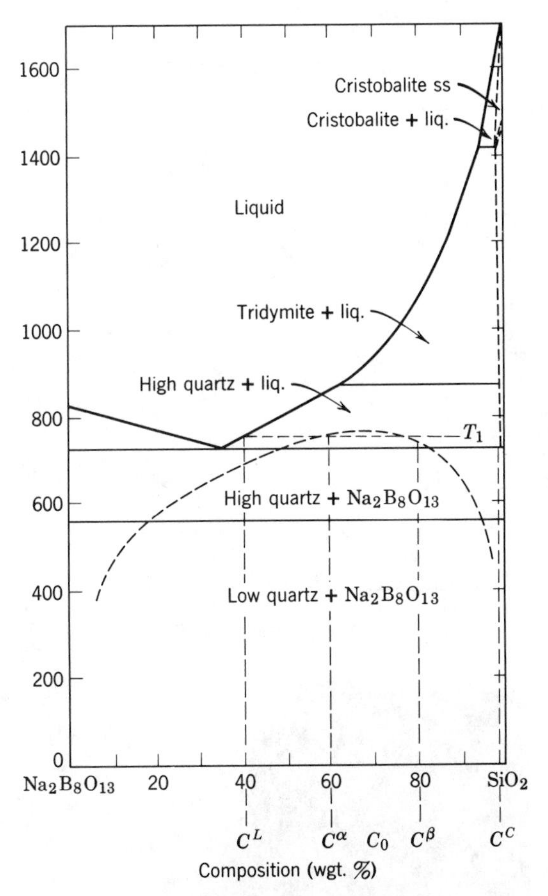

Fig. 3.11. Metastable liquid-liquid immiscibility in the system $Na_2B_8O_{13}$–SiO_2. From T. J. Rockett and W. R. Foster, *J. Am. Ceram. Soc.*, **49**, 31 (1966).

With this background, let us consider the submicrostructural features of glasses in the system BaO–SiO_2. The miscibility gap determined for the system is metastable and is shown in Fig. 3.12*a*. The direct-transmission electron micrographs shown in Fig. 3.12*b* to *d* illustrate features observed for BaO concentrations of 4, 10, and 24 g-atom%. For the 4 g-atom% BaO composition, which lies on the silica-rich side of the miscibility gap, the submicrostructure consists of discrete spherical particles of a BaO-rich phase embedded in a continuous matrix of a SiO_2-rich phase. Similarly, for compositions near the baria-rich side of the miscibility gap, such as the 24 g-atom% BaO composition, the submicrostructure consists of spherical SiO_2-rich particles embedded in a continuous BaO-rich matrix.

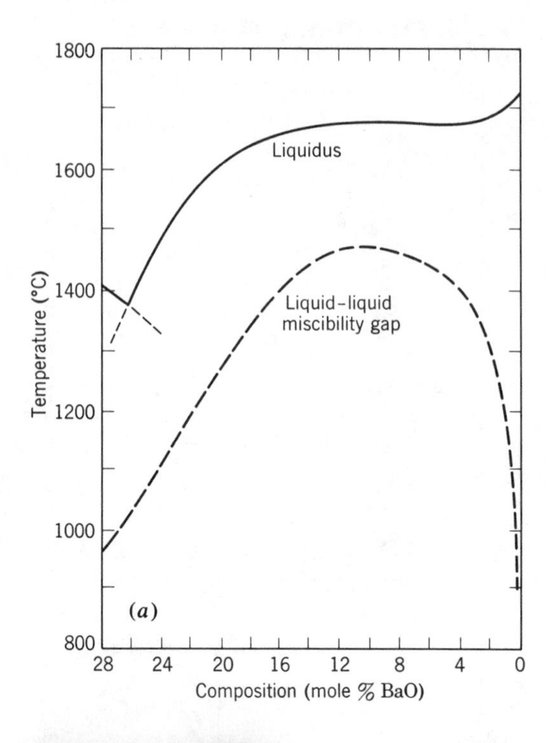

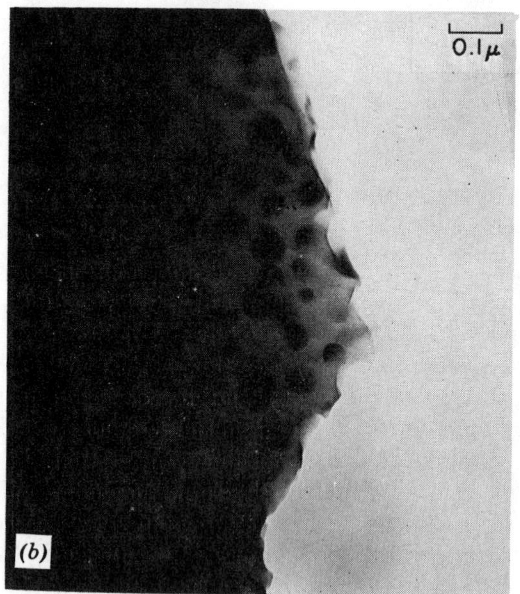

Fig. 3.12. (*a*) Liquidus curve and miscibility gap in the system $BaO–SiO_2$. (*b*) Direct-transmission electron micrograph of $0.04\ BaO–0.96\ SiO_2$ glass.

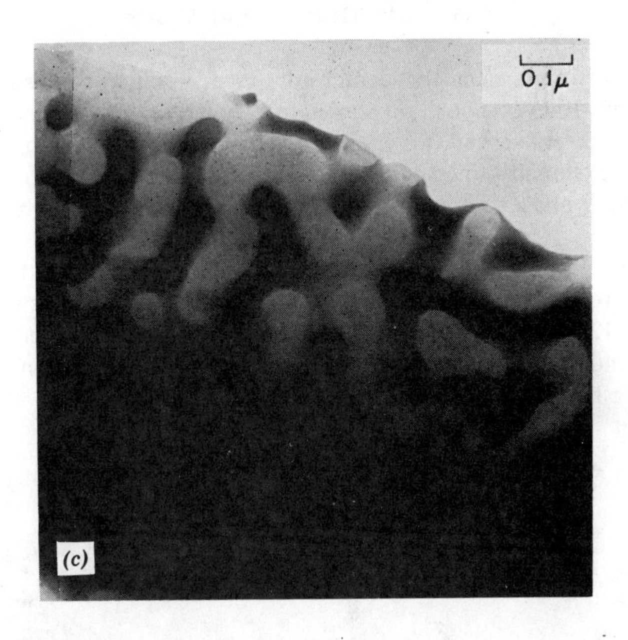

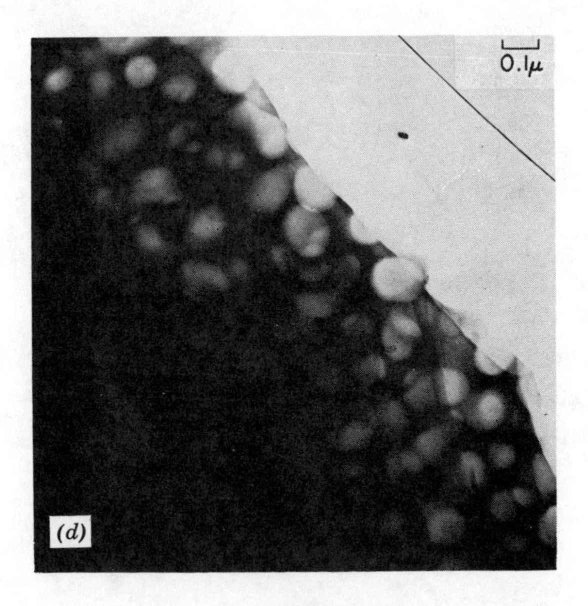

Fig. 3.12 (*contd.*) (*c*) Direct-transmission electron micrograph of 0.10 BaO–0.90 SiO_2 glass. (*d*) Direct-transmission electron micrograph of 0.24 BaO–0.76 SiO_2 glass. From T. P. Seward, D. R. Uhlmann, and D. Turnbull, *J. Am. Ceram. Soc.*, **51**, 278 (1968).

For compositions near the center of the miscibility gap, such as the 10 g-atom% BaO composition shown in Fig. 3.12*c*, the submicrostructure is frequently observed to consist of two phases, each of which is three-dimensionally interconnected. In all the electron micrographs shown, both phases are amorphous, as determined from electron diffraction.

In many respects, the most interesting characteristics on this scale of structure relate to interconnected submicrostructures such as that shown in Fig. 3.12*c*. Similar submicrostructures have been reported for many glasses and seem generally to be characteristic of conditions in which there is a large volume fraction of both phases present. On subsequent

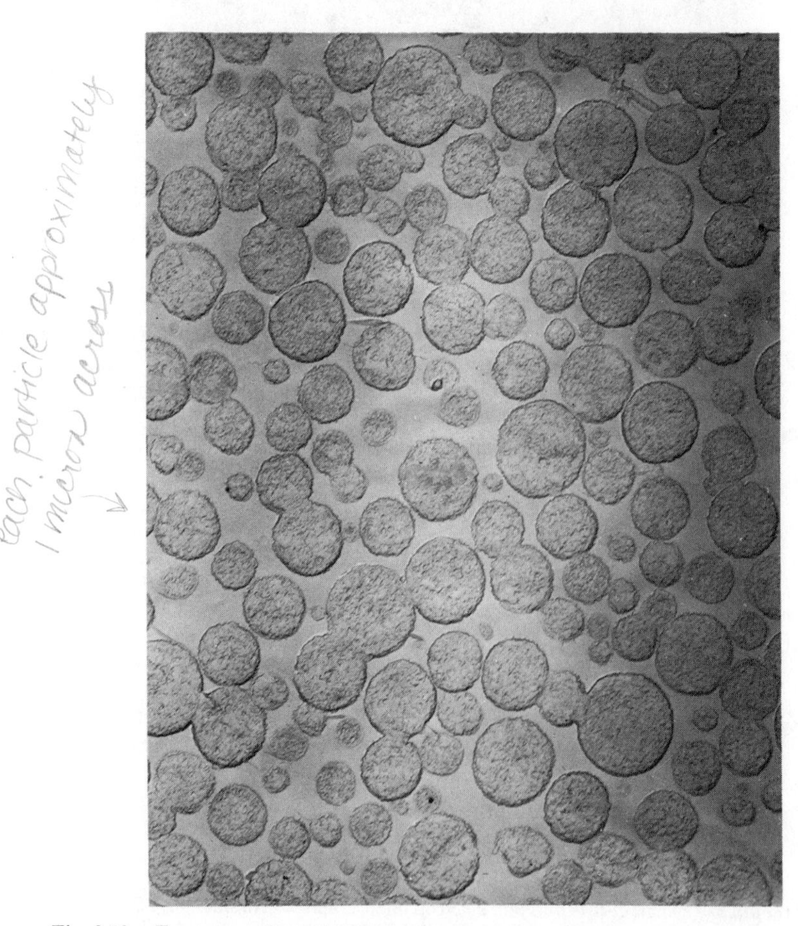

Fig. 3.13. Fracture surface of 11 PbO·89 B_2O_3 sample, showing B_2O_3-rich discrete particles in a PbO-rich matrix. Simultaneous platinum/carbon-shadowed replica. Bar indicates 1 micron. From R. R. Shaw and D. R. Uhlmann, *J. Non-Cryst. Solids*, **1**, 474 (1969).

heating, such interconnected submicrostructures in some cases coarsen while maintaining a high degree of connectivity and in other cases coarsen, neck off, and spheroidize. An example of a system in which discrete-particle structures are observed near the center of a miscibility gap is shown in Fig. 3.13 for the system $PbO–B_2O_3$.

3.5 Miscibility Gaps in Oxide Systems

The addition of modifier oxides to the two most important glass-forming oxides, SiO_2 and B_2O_3, often leads to liquid-liquid immiscibility. Examples of the types of miscibility gaps which result have already been shown for the systems $MgO–SiO_2$ (Fig. 3.10*a*) and $BaO–SiO_2$ (Fig. 3.12*a*). An appreciation of the widespread tendency toward immiscibility in silicate and borate systems may be obtained from Figs. 3.14 to 3.16 and Table 3.6. As shown in Fig. 3.14, miscibility gaps are found when MgO, FeO, ZnO, CaO, SrO, or BaO are added to SiO_2, and only in the case of BaO additions is the miscibility gap metastable. Among the alkali silicates, metastable miscibility gaps are found in the $Li_2O–SiO_2$ and $Na_2O–SiO_2$ systems (Fig. 3.15). The existence of a metastable miscibility gap in $K_2O–SiO_2$ system has been suggested, but the low temperatures of its possible occurrence, in the region of the glass transition and below for all compositions, effectively preclude the observation of phase separation. Stable miscibility gaps are found in all alkaline earth borate systems, and metastable gaps are found in all the alkali borates (Table 3.6).

A large miscibility gap, stable over a wide range of composition, is found in the system $TiO_2–SiO_2$ (Fig. 3.16). The extension of this range of

Table 3.6. Characteristics of Metastable Miscibility Gaps in Alkali Borate Systems

System	Consolute Temp. (°C)	Consolute Composition (mole% alkali oxide)	Approximate Extent of Immiscibility (mole% alkali oxide)
$Li_2O–B_2O_3$	660	10	2–18
$Na_2O–B_2O_3$	590	16	7–24
$K_2O–B_2O_3$	590	10	2–22
$Rb_2O–B_2O_3$	590	10	2–16
$Cs_2O–B_2O_3$	570	10	2–20

Source. R. R. Shaw and D. R. Uhlmann, *J. Am. Ceram. Soc.*, **51**, 377 (1968).

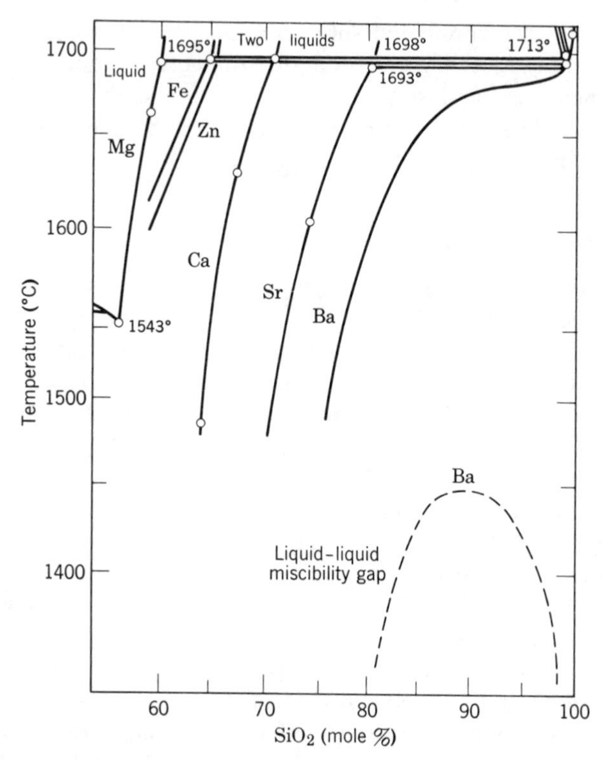

Fig. 3.14. Miscibility gaps in various divalent oxide-silica systems. For the $BaO-SiO_2$ system, in which the gap is metastable, both the liquidus curve and the miscibility gap are shown.

immiscibility into ternary systems is important for the effective use of TiO_2 as a nucleating agent in many glazes, enamels, and glass-ceramic systems. The avoidance of this gap is important for the formation of the very low expansion TiO_2-containing fused silica. A large region of metastable immiscibility is also found in the $Al_2O_3-SiO_2$ system, extending from less than 10 to greater than 50 g-atom% Al_2O_3.

The addition of TiO_2 and alkali oxides tends to enhance immiscibility in complex systems; the addition of Al_2O_3 tends to suppress it. A striking example of the latter behavior is shown in Fig. 3.17 for the system $BaO-SiO_2-Al_2O_3$.

The ranges of immiscibility in two commercially important systems, $Na_2O-B_2O_3-SiO_2$ and $Na_2O-CaO-SiO_2$, have been extensively investigated. In the $Na_2O-B_2O_3-SiO_2$ system, three regions of immiscibility, designated I, II, and III in Fig. 3.18, have been suggested. Glasses of the Pyrex type and the Vycor type occur in different parts of immiscibility

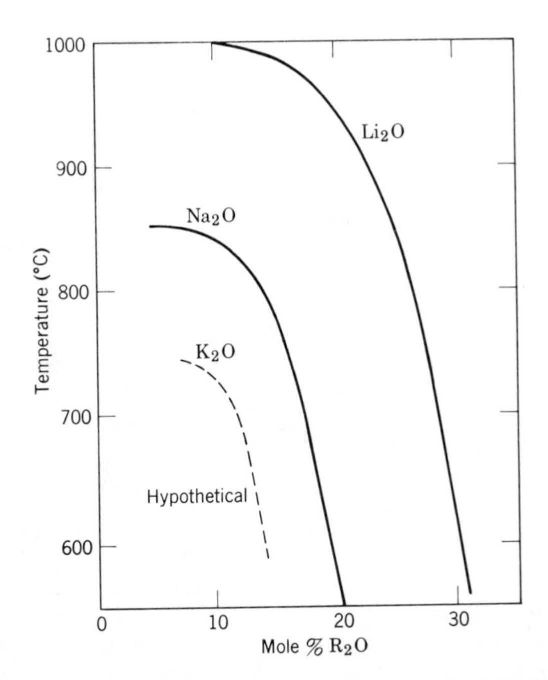

Fig. 3.15. Metastable miscibility gaps in the Li_2O–SiO_2 and Na_2O–SiO_2 systems and hypothecated gap in the K_2O–SiO_2 system. From Y. Moriya, D. H. Warrington, and R. W. Douglas, *Phys. Chem. Glasses*, **1**, 19 (1967).

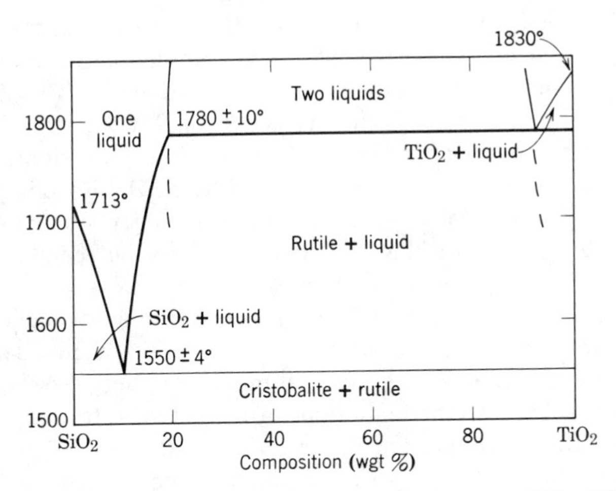

Fig. 3.16. Miscibility gap and its metastable extension in the system TiO_2–SiO_2. From R. C. DeVries, R. Roy, and E. F. Osborn, *Trans. Brit. Ceram. Soc.*, **53**, 531 (1954).

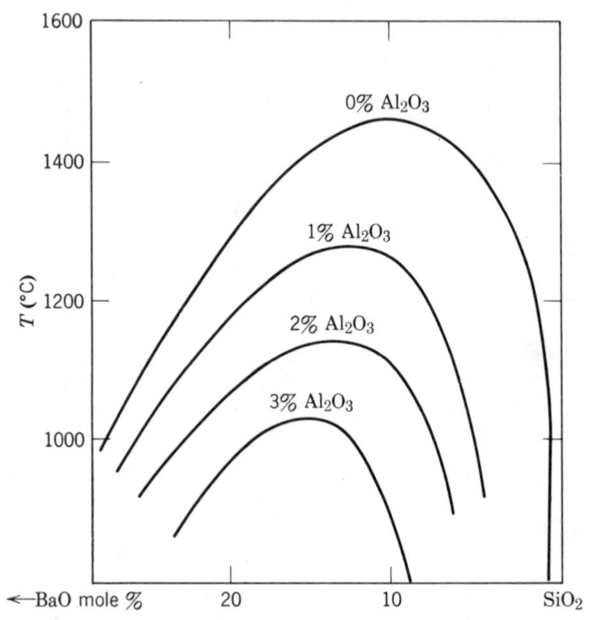

Fig. 3.17. Effects of Al_2O_3 additions on metastable liquid-liquid immiscibility in the system $BaO-SiO_2$. From T. P. Seward and D. R. Uhlmann.

region II in this system. Pyrex glasses exhibit phase separation on a fine scale, typically less than 50 Å; the ability to control the scale and connectivity of immiscible regions forms the very basis of the Vycor process. In the case of $Na_2O-CaO-SiO_2$ glasses, the concentration ratios of the three major constituents in standard commercial compositions lie somewhat within the immiscibility boundary shown in Fig. 3.19. The Al_2O_3 concentration, typically in the range of 2 wt%, very likely has a marked effect in decreasing the extent of immiscibility, similar perhaps to that shown in Fig. 3.17 for Al_2O_3 additions to $BaO-SiO_2$ compositions. The resulting commercial products are then homogeneous glasses.

As discussed relative to Figs. 3.10 and 3.11, the occurrence of immiscibility depends on the relative free energies of the phases which may form in a system. B. E. Warren and A. G. Pincus* originally suggested that liquid-liquid phase separation arises from competition between the cations to surround themselves with a minimum-energy oxygen anion configuration, subject to the limitations of the network-forming tendency of the silica. Modifier and intermediate cations with a limited capability to substitute for silica in the network and with a strong oxygen bond strength

J. Am. Ceram. Soc., **23**, 301 (1940).

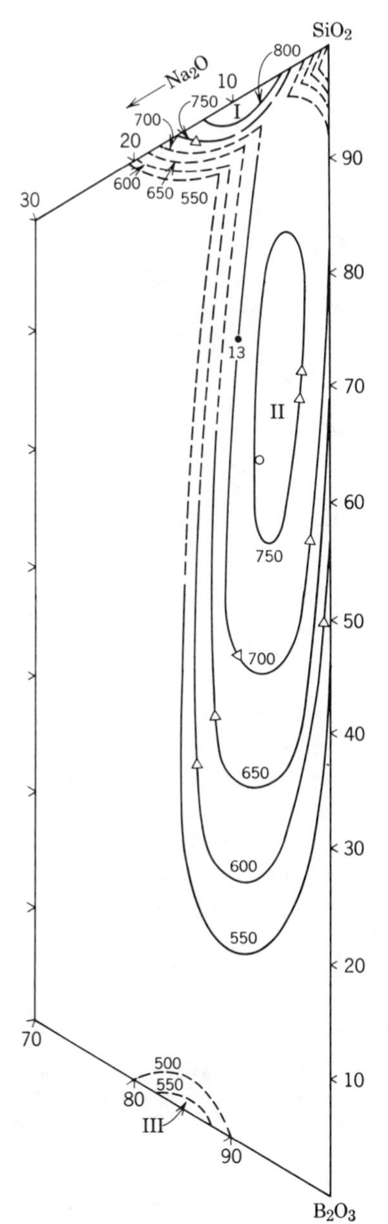

Fig. 3.18. Three regions of liquid-liquid immiscibility in the Na_2O–B_2O_3–SiO_2 system (wt%). From W. Haller, D. H. Blackburn, F. E. Wagstaff, and R. J. Charles, *J. Am. Ceram. Soc.*, **53**, 34 (1970).

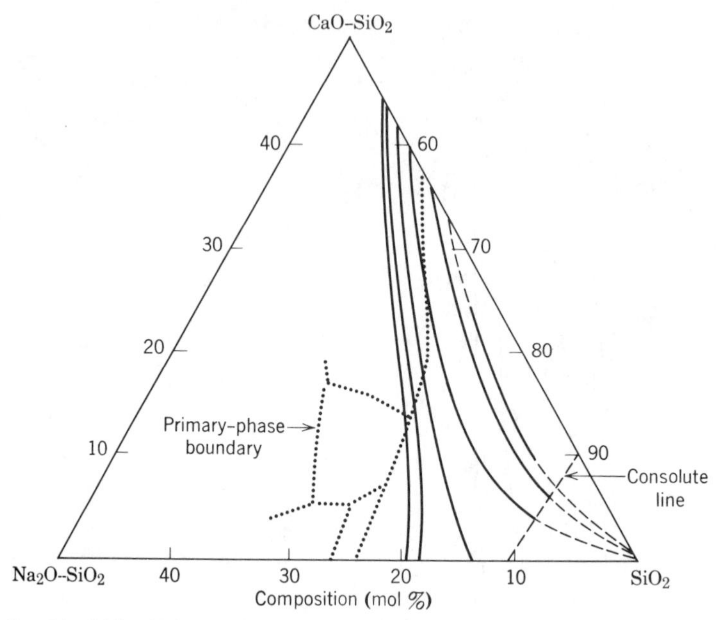

Fig. 3.19. Liquid-liquid immiscibility region in the soda lime–silica system. From D. G. Burnett and R. W. Douglas, *Phys. Chem. Glasses*, **11**, 125 (1970).

cannot be accommodated in large concentrations in a minimum-energy configuration, which would require excessive breakup of the network. The lowest-energy configuration would then come about when the system separates into two liquid phases, one favoring the network (high in silica) and the other favoring the lowest-energy modifier configuration.

This view is best considered as a first approximation which does not adequately describe the range of immiscibility behavior observed in various systems. It does, however, focus attention on some of the salient features of the problem, which will undoubtedly be significant in any detailed treatment of immiscibility. The location of miscibility gaps can, of course, be inferred from thermodynamic activity data when these are available, but the relation between the activities and various structural features of the solutions remains to be elucidated for the oxide systems of interest.

3.6 General Discussion

Some 15 years ago, when the first edition of this text was prepared, it was only possible to warn that the detailed structure of glass was complex and that there was "some evidence" for heterogeneities in structure; equilibrium immiscibility in the alkaline earth oxide–silica systems was, of course, well known. Primarily as a result of direct observation with more powerful electron-microscope techniques, the demonstration that metastable immiscibility and phase separation in glass systems are quite common has clarified and also significantly complicated the problem of characterizing the structure of glasses. In addition to the question of the influence of cooling rate on the glass transition temperature and the few percent range of variation in the specific volume of a single-phase glass, there must be added the question of possible phase separation.

With regard to the possibility of phase separation, it is not sufficient to consider only the phase-equilibrium diagram, since metastable phase separation and suppression of phase separation by rapid cooling often occur. In many cases, glass-in-glass phase separation results in opacity which can easily be associated, incorrectly, with crystallinity. In other cases, the scale of the submicrostructure (tens of Å) is too small to be detected by optical measurements with the much larger (7000-Å) wavelength of light and is only found with rather careful electron-microscope or small-angle X-ray scattering studies.

Clearly, the properties of a glass and any description of its structure must take into account whether it is a single phase or multiphase system. As a result, discussions of structure and correlations of properties with structure, particularly in the older literature, which do not include detailed thermal histories and phase-composition information must be evaluated with caution.

Suggested Reading

1. R. H. Doremus, *Glass Science*, John Wiley & Sons, Inc., New York, 1973.
2. R. W. Douglas and B. Ellis, Eds., *Amorphous Materials*, John Wiley & Sons, Inc., New York, 1972.
3. M. B. Volf, *Technical Glasses*, Pitman, London, 1961.
4. H. Rawson, *Inorganic Glass-Forming Systems*, Academic Press, Inc., New York, 1967.
5. G. M. Bartenev, *The Structure and Mechanical Properties of Inorganic Glasses*, Walters-Noordhoff, Groningen, 1970.
6. G. O. Jones, *Glass*, 2d ed., Chapman & Hall, Ltd., London, 1971.
7. L. D. Pye, H. J. Stevens, and W. C. LaCourse, Eds., *Introduction to Glass Science*, Plenum Press, New York, 1972.
8. J. E. Stanworth, *Physical Properties of Glass*, Oxford University Press, Oxford, 1953.

Problems

3.1. Lead orthosilicate forms a glass of density 7.36 g/cc. What is the oxygen density of this glass? How does this compare with fused silica (density 2.2 g/cc)? Where would you predict the lead ions are going?

3.2. The structure of many multicomponent oxide glasses has been found to be heterogeneous on a scale of 50 to 500 Å.
 (*a*) How might such heterogeneities be detected and observed?
 (*b*) Discuss the relation between such heterogeneities and the random-network model of glass structure.
 (*c*) How might such heterogeneous structures be explained in terms of liquid-liquid immiscibility? Include a hypothetical temperature-composition diagram, as well as free energy-composition diagrams for several temperatures.

3.3. Explain how you can experimentally distinguish among crystalline SiO_2, SiO_2 glass, silica gel, and liquid silica. Explain in terms of the structures of these different forms of the same composition.

3.4. A certain glass found in recent geological formations is high in SiO_2 (70+%) and Al_2O_3 (11.5+%). The alkalis (Na_2O and K_2O) are minor constituents (5%), as are the alkaline earths ($MgO + CaO = 2\%$). There is almost a 10% weight loss on heating above about 900°C, and the volatile constituent associated with the weight loss is water. This glass has the unusual and irreversable property of softening at 850°C, becoming hard at 950 to 1100°C, and then softening at about 1150°C. On cooling the glass has only one fictive temperature range at 1150°C similar to but higher than most commercial glasses. Explain the irreversible softening at about 850°C. For your reference, a soda-lime glass has 72% SiO_2, 1% Al_2O_3, 9% CaO, 4% MgO, 13% Na_2O, 1% K_2O.

3.5. (*a*) What is the fictive temperature of a glass?
 (*b*) Why do glasses have coefficients of expansion more like liquids above the fictive temperature and more like crystalline solids below the fictive temperature.
 (*c*) What would you predict would happen to the fictive temperature of a silica if a function of Na_2O content, NaF content?

3.6. (*a*) If you were asked to obtain a glass (liquid) at 800°C with the highest mole percent silicon dioxide and you were restricted to only one other oxide in addition to silica, what additional material would you choose? Give an explanation.
 (*b*) Why does quartz melt at a temperature below the melting point of crystobalite?

3.7. (*a*) Sketch a possible bond arrangement in B_2O_3 which follows Pauling's rules and has $CN_B = 3$.
 (*b*) Enumerate Zachariasen's rules for glass formation. Does the structure you have sketched fulfill these requirements?
 (*c*) Mention two distinctly different ways that structure could adapt to addition of alkali oxide, say, Na_2O.

3.8. A typical soda-lime-silica window glass has a composition of _____% Na_2O, _____% CaO, _____% SiO_2.

3.9. Classify the following elements as modifiers, intermediates, or network formers (glass formers) in connection with their use in forming oxide type glasses: Si, Na, B, Ca, Al, P, K, Ba.

4

Structural Imperfections

In the previous two chapters we have considered structures of ideal crystals and also structures of some noncrystalline glasses. There are many properties that are markedly dependent on small deviations from ideal structures, and we can best approach these structure-sensitive properties on the basis of departures from the ideal structures.

If we consider a perfect crystal as one with a completely ordered structure having its atoms at rest (except for zero-point oscillation at the absolute zero temperature) and with the electrons distributed in the lowest-energy states, there are several types of deviations or imperfections which may occur. The first is increased amplitude of vibration of the atoms about their equilibrium rest positions as the temperature is increased. (These elastic vibrations are nearly harmonic, since the forces between atoms nearly obey Hooke's law; wavelike solutions for the motions of the atoms can be quantized and amounts of energy $h\nu$, called *phonons*, associated with unit quantum excitation of the elastic vibration. The relationship between phonons and the vibrational frequency is the same as that between light photons and the vibrational frequency of light waves.) Imperfections also occur in the electronic energy levels; *electrons* may be excited into higher energy levels, leaving vacant positions in the normally filled electronic energy-level bands, called *electron holes.* If the excited electron remains closely associated with the electron hole, the electron–electron-hole pair is called an *exciton,* which also may be looked on as an excited state of an atom or ion. Finally there are a number of *atomic defects,* including *substitution* of a wrong atom or a foreign atom for a normal one, *interstitial* atoms, *vacant atom sites,* and line imperfections called *dislocations.* Finally we might also consider the crystal surfaces or boundaries between crystals as imperfections, but these are discussed separately in the next chapter.

At the outset we must admit that there are a large number of combinations, permutations, and interactions among solutes, atomic defects, electronic defects, dislocations, and surfaces. In the present chapter we describe some of these structural imperfections individually and briefly indicate how and when they are likely to occur. Later, as it becomes important for the discussion of particular ceramic processes or properties, we consider their characteristics more fully. There is no ceramic material for which our knowledge is complete. More extensive discussions of structural imperfections are available in the references cited.

4.1 Notation used for Atomic Defects

Several types of structural imperfections are believed to occur in ceramic materials. One departure from ideality involves the motion of an atom from a normal site to an interstitial position, as illustrated in Fig. 4.1*a*. This type of disorder, which results in equal concentrations of

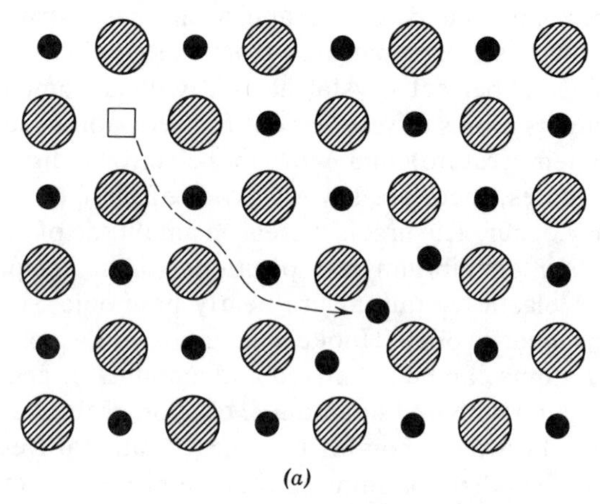

Fig. 4.1. (*a*) Frenkel disorder. Ion leaving normal site forms interstitial ion and leaves a vacancy.

vacant lattice sites and interstitial atoms, is called *Frenkel disorder.** Another kind of disorder, which involves the simultaneous production of both cation and anion vacancies, as illustrated in Fig. 4.1*b*, is referred to as *Schottky disorder.*† Ceramic systems are rarely if ever without impurities, and solute atoms may either substitute for host atoms on normal lattice sites, as in substitutional solid solution, or incorporate on normally unoccupied interstitial sites in the host lattice, as in interstitial solid solution. These two arrangements are illustrated in Fig. 4.1*c* and *d*. In addition to atom locations one must also describe their valence state or, to be more precise, the electronic energy levels in the crystal which also

*J. Frenkel, *Z. Phys.*, **35**, 652 (1926).
†C. Wagner and W. Schottky, *Z. Phys. Chem.*, **B11**, 163 (1931).

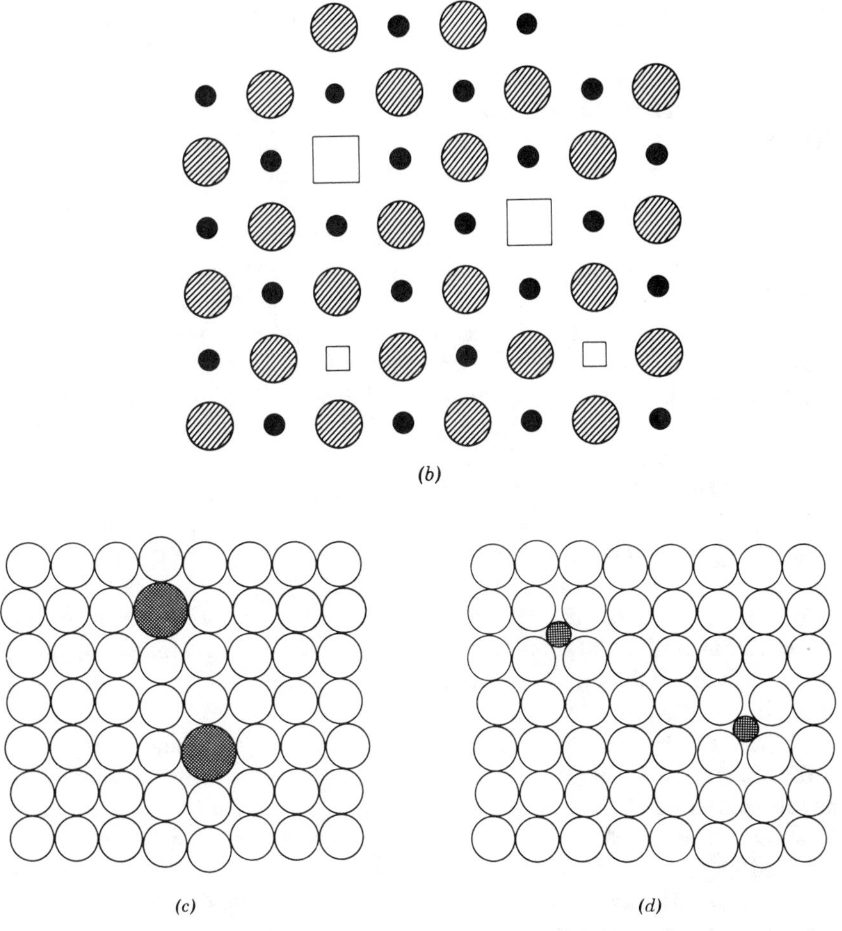

Fig. 4.1. (*contd.*) (*b*) Schottky disorder. Equal numbers of anion and cation vacancies occur. (*c*) Substitutional and (*d*) interstitial solid solutions.

tend to depart from complete order at any temperature above absolute zero. Such deviations are also affected by the presence of atomic imperfections and solute atoms.

In keeping track of the various defects which may exist simultaneously in a single ceramic material, an essential requirement is an adequate system of notation for describing point defects in ceramics. The Kroger-Vink notation (Refs. 1 and 2) is most widely used. In this notation, when we add or subtract elements from the crystal, we do so by adding or subtracting electrically neutral atoms and thus avoid making judgments and decisions about bond types; for application to ionic systems this requires that we separately add or subtract electrons. Let us consider the various conceivable defects and notations for the imaginary binary compound MX.

1. Vacant Lattice Sites (Vacancies). When empty lattice sites occur, they are indicated by the symbols V_M and V_X for the M and X sites, respectively; in the atomic notation, the subscript M indicates a missing M atom. In an ionic lattice such as sodium chloride this would mean the removal of a Na^+ ion together with an electron; similarly, V_{Cl} would indicate the removal of a Cl^- ion with the addition of an electron.

2. Interstitial Atoms. In addition to the normally occupied lattice sites of a crystal structure, there are also interstitial sites. When atoms occupy these interstitial positions, they are denoted by M_i and X_i.

3. Misplaced Atoms. It is also possible in some compounds for M atoms to be on X sites (M_X); the subscript always indicates the position of each particular atom in the lattice.

4. Associated Centers. In addition to single defects, it is also possible for one or more lattice defects to associate with one another, that is, to cluster together. These are indicated by bracketing the components of such a cluster; for example, ($V_M V_X$) or ($X_i X_M$).

5. Solutes. These, if present, are coded as to lattice position in the same way as native defects; L_M and S_X indicate a solute atom L on an M site and S on an X site. L_i indicate that the solute L is on an interstitial site.

6. Free Electrons, Electron Holes. In strongly ionic materials, electrons are normally localized at a particular atom site in a way which can be described in terms of the ion valence. However, as discussed in Chapter 2, this is not always the case, and some fraction of the electrons, denoted e', may not be localized at a particular site; similiarly there may be missing electrons, denoted electron holes $h^{\cdot}$, which are not localized at a particular atom site.

7. Charged Defects. In insulators and semiconductors we usually think of the respective species as ions; for example, sodium chloride is made up of Na^+ and Cl^- ions. If we conceive of removing a positively charged Na^+ sodium ion from the NaCl structure, we remove the sodium atom without one of its electrons; as a result, the vacancy has associated with it an extra electron with a negative charge which we write as e', where the superscript refers to a unit negative charge. If this excess electron is localized at the vacant site, as would normally be the case in NaCl, we write V'_{Na}. Similarly, if we conceive of removing a negatively charged Cl^- chlorine ion, we remove the chlorine atom plus an associated electron, leaving a positive electron hole, which we write $h^{\cdot}$, where the superscript refers to a unit positive charge. If this excess positive charge is localized at the vacant site, as would normally be the case for NaCl, we write $V_{Cl}^{\cdot}$. In some materials less strongly ionic than NaCl these excess or missing electrons, e' or $h^{\cdot}$, may not remain localized at the vacant site, the separation being represented by the reactions

$$V'_{Na} = V_{Na} + e' \tag{4.1}$$

$$V_{Cl}^{\cdot} = V_{Cl} + h^{\cdot} \tag{4.2}$$

By separating the notation for atoms from the notation for electric charge, we avoid the prospect of unintentionally making a priori assumptions about the nature of the defects.

With each of the other defect symbols—V_M, V_X, M_i, M_X, $(V_M V_X)$—an effective charge relative to the host lattice is also possible. Thus $Zn_i^{\cdot\cdot}$ would indicate a Zn^{2+} ion at an interstitial site which is normally unoccupied and without an effective change. Substituting a divalent Ca^{2+} ion for monovalent Na^{+} on a sodium site gives a local electronic structure augmented by one extra positive charge and is represented as $Ca_{Na}^{\cdot}$. Note that the superscripts + and − are used to indicate real charged ions, whereas the superscripts $^{\cdot}$ and $'$ indicate effective positive and negative charges with respect to the host lattice. Other possibilities arise from nonstoichiometry. In FeO, for example, it is possible to have Fe^{3+} ions in addition to the normal Fe^{2+} ions. In this case, the Fe^{3+} ions are indicated as $Fe_{Fe}^{\cdot}$.

4.2 Formulation of Reaction Equations

As each type of defect and its concentration in a material can be described in terms of associated energies of formation and other thermodynamic properties, it is possible to treat all imperfections as chemical entities and treat them in a manner referred to as *defect chemistry*. Defect interactions may be conceptualized in terms of mass-action equilibria, thus enabling the representation of such interactions by means of defect equations. The following rules must be observed:

1. Site Relation. The number of M sites in a compound M_aX_b must always be in correct proportion to the number of X sites (1 : 1 in MgO, 1 : 2 in UO_2, etc.). In maintaining this proportion, however, the total number of each type of site may change.

2. Site Creation. Some defect changes such as introducing or eliminating a vacant M site V_M correspond to an increase or decrease in the number of lattice sites. It is important that this be done in a way that does not change the site relation described in rule 1. Defects indicating site creation are V_M, V_X, M_M, M_X, X_M, X_X, and so on. Nonsite-creating entities are e', $h^{\cdot}$, M_i, L_i, and so on.

3. Mass Balance. As in any chemical equation, a mass balance must be maintained. Here it is helpful to remember that the subscript in the defect symbol indicates the site under consideration and is of no significance for the mass balance.

4. Electrical Neutrality. The crystal must remain electrically neutral. Only neutral atoms or molecules are exchanged with other phases outside the crystal under consideration; within the crystal neutral particles can yield two or more oppositely charged defects. The condition of electrical neutrality requires that both sides of a defect-reaction equation have the same total effective charge, not necessarily zero.

5. Surface Sites. No special indication of surface sites is used. When an atom M is displaced from the bulk of the crystal to its surface, the number of M sites increases.

At this point we anticipate the next section, in which we discuss the solution of $CaCl_2$ in KCl to see how these rules apply.

In KCl there are equal numbers of cation and anion sites; to introduce the two chlorine atoms in $CaCl_2$ on anion sites, we must use potassium sites as well as two chlorine sites; since we have only one Ca, we may tentatively assume that the second K site required for the proper site relation may be vacant. Considering only atomic substitutions, a possible solution process is

$$CaCl_2(s) \xrightarrow{KCl} Ca_K + V_K + 2Cl_{Cl} \tag{4.3}$$

For a strongly ionic material such as $CaCl_2$, we may further assume that the substitutions are fully ionized, which gives us an alternate and more realistic solution process:

$$CaCl_2(s) \xrightarrow{KCl} Ca_K^{\cdot} + V_K' + 2Cl_{Cl} \tag{4.4}$$

which also conserves electrical neutrality, mass balance, and site relation. So also a third possibility, the formation of charged Ca^{2+} interstitials, chlorine ions on chlorine sites, and potassium ion vacancies, for which we write

$$CaCl_2(s) \xrightarrow{KCl} Ca_i^{\cdot\cdot} + 2V_K' + 2Cl_{Cl} \tag{4.5}$$

which also conserves electrical neutrality, mass balance, and site relation. Deciding among these and other possibilities is a matter for defect chemistry and is the main subject of the following sections.

4.3 Solid Solutions

Of the many types of departure from an ideal crystal, the one easiest to visualize is the inclusion of foreign atoms in the host crystal, as illustrated in Fig. 4.1*c* and *d*. If a material crystallizes in the presence of foreign atoms, they may be almost completely rejected by the crystal if they appreciably increase the energy of the crystalline form. On the other hand, if building them into the host structure in an ordered way leads to a large lowering of the system's energy, a new crystalline form develops. In intermediate cases foreign atoms fit into the structure in a random way as the crystal is built up. When this occurs, there is usually a change in the cell size with composition in accordance with *Vegard's law* that the lattice-cell dimensions vary linearly with the concentration of solute added.

Solid solutions are stable when the mixed crystal has a lower free energy than the alternative—building up two crystals of different composition or building up a new structure in which the foreign atoms are put on ordered sites. As previously discussed, the free energy is given by the relation

$$G = E + PV - TS \tag{4.6}$$

where E is largely determined by the structural energy and the entropy is a measure of the randomness (probability) of the structure. If an atom added at random greatly increases the structure energy, the solid solution is unstable and two crystal structures result. On the other hand, if the addition of a foreign atom greatly lowers the structure energy, the system tends to form an ordered new phase. If the energy is not much changed, the entropy is increased by random additions so that the solid solution has the lowest energy and is the stable configuration. Different rules for the likelihood of solid solutions being stable in particular systems are applications of these general principles to specific cases.

Two examples of stable solid solution are shown in Figs. 4.2 and 4.3. For the MgO–NiO system, both end members have the sodium chloride crystal structure, and a complete series of solid solutions occurs. The schematic free-energy–composition diagram for 1500°C is shown in Fig. 4.2*b*. The lowest-free-energy phase for all compositions at 1500°C is the solid solution rather than any ordered structure or the liquid. For the $MgO–Al_2O_3$ system shown in Fig. 4.3, the end members have different crystal structures, and a third possibility is the formation of an intermediate compound. Experimentally it is found that for a 50:50 mixture of MgO and Al_2O_3, the spinel phase has the lowest free energy and at 1750°C in the intervals *a–b* and *c–d* two solid solutions of fixed composition but

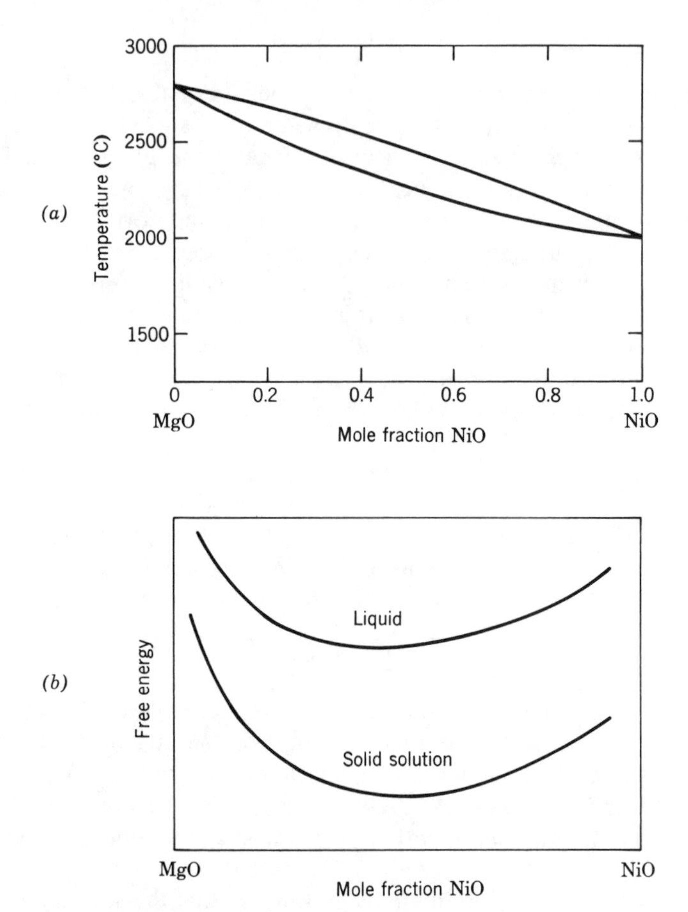

Fig. 4.2. (*a*) Phase diagram of MgO–NiO. (*b*) Schematic representation of free-energy–composition diagram of MgO–NiO system for $T < 2000°C$.

variable amounts are in equilibrium with each other (the compositions in equilibrium have the same values of chemical potential, $\mu = \left(\frac{\partial G}{\partial X}\right)_{T,P}$, that is, an equal slope, a common tangent, for the free-energy–composition curves). In each of the regions MgO–a, b–c, and d–Al_2O_3 a different variable–composition solid solution is the lowest-free-energy equilibrium structure.

Some rules and generalizations useful for predicting solid-solution behavior are discussed in subsequent sections.

Substitutional Solid Solution. Substitution of one ion for another is common in the formation of ceramic crystals. The solid-solution phases

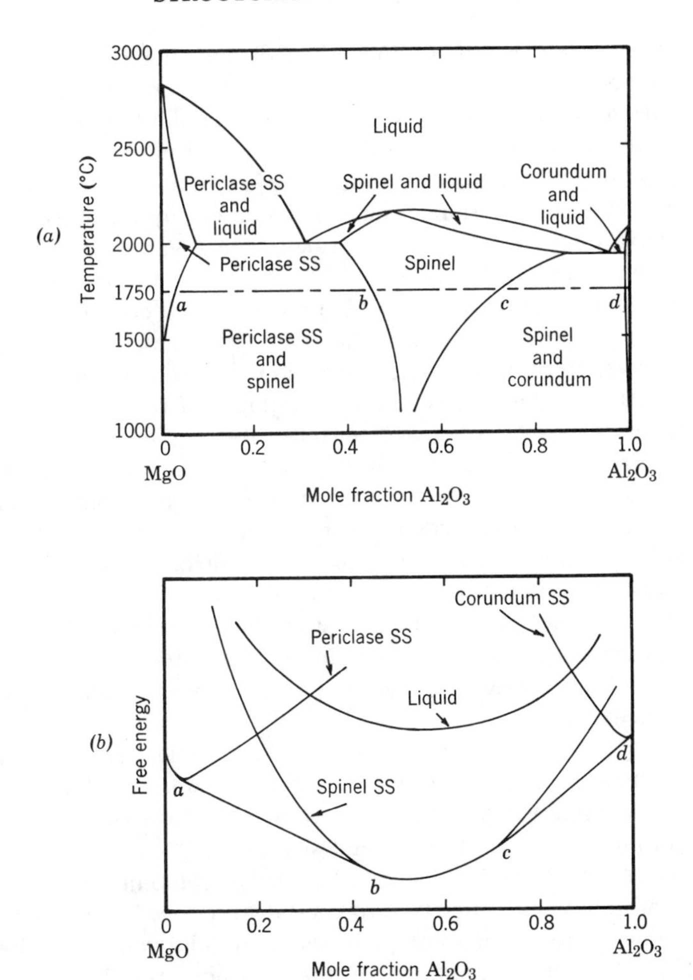

Fig. 4.3. (*a*) Phase diagram of the $MgO–Al_2O_3$ system. (*b*) Schematic representations of free-energy–composition diagram of $MgO–Al_2O_3$ for $T = 1750°C$.

shown in Figs. 4.2 and 4.3 represent this kind of substitution. For example, crystals of magnesium oxide frequently contain appreciable amounts of NiO or FeO with a random distribution of Ni^{2+} or Fe^{2+} ions replacing Mg^{2+} in the crystals so that the final composition of the crystal can be read $Mg_{1-x}Ni_xO$, as shown in the phase diagram in Fig. 4.2. A similar series of solid solutions exists in the system $Al_2O_3–Cr_2O_3$ (rubies are 0.5 to 2% Cr_2O_3 in Al_2O_3), $ThO_2–UO_2$, albite-anorthite, and many of the spinels. In some systems there is a complete series of solid solutions formed between end members (Fig. 4.2). However, in most systems only a limited number of foreign atoms can be added to the substitutional solid solution (Fig. 4.3); an excess above the solubility limit at a given temperature results in the formation of a second phase.

There are several factors determining the extent of substitution that can take place in solid solutions, and a number of rules expressing these factors have been devised. These rules reflect variations in the free energy, which is made up of several terms. Since the free energy is a function of temperature, a series of free-energy vs. composition curves for each temperature can be drawn similar to the ones shown in Figs. 4.2 and 4.3. Lowering of the free energy resulting from an increase in entropy requires that there be at least a small solubility of foreign atoms in any structure. Factors which allow extensive substitution are as follows.

1. SIZE FACTOR. If the sizes of two ions differ by less than about 15% conditions are favorable for the formation of substitutional solid solutions. If the relative ion sizes differ by more than 15%, substitution is generally limited and is usually less than a fraction of 1%. This factor is by far the most important for ionic compounds.

2. VALENCY FACTOR. If the added ion has a valence different from that of the host ion, substitution is limited. It can occur, as indicated below, but other structural changes are also required to maintain overall electrical neutrality.

3. CHEMICAL AFFINITY. The greater the chemical reactivity of the two crystalline materials, the more restricted is solid solubility, since a new phase is usually more stable. For oxides this restriction is usually implicit in factors of ion valency and size.

4. STRUCTURE TYPE. For complete solid solubility the two end members must have the same type of crystal structure. For example, TiO_2 could obviously not form a complete series of solid solutions with SiO_2 (see Chapter 2 for descriptions of these structures). This does not, however, restrict limited solid solution.

On the basis of these factors, an estimate of the extent of substitutional solid solution to be expected can usually be obtained. For oxides, the major factors are the relative ion sizes and valencies. Although different ion sizes definitely preclude extensive solid-solution formation, valence differences can frequently be made up in other ways. For example, extensive solid solution among Mg^{2+}, Al^{3+}, and Fe^{2+} ions is common in clay minerals having the montmorillonite structure. The charge deficiency due to replacing trivalent Al^{3+} with divalent Mg^{2+} or Fe^{2+} is made up by exchangeable ions adsorbed on the surface of the small clay particles. Similarly, substitution of the Si^{4+} ions by Al^{3+} ions in tetrahedral coordination in kaolinite leads to a charge difference which is made up by exchangeable ions adsorbed on the particle surfaces. These substitutional-solid solution-effects are largely responsible for the observed base-exchange properties and the ability to form stable suspensions with the clay minerals. In much the same way, many of the aluminosilicate structures are derivative structures in which an Al^{3+} replaces an Si^{4+} in the crystal and an alkali or alkaline earth ion fits into an expanded interstitial position. Solid solutions frequently occur in which a number of different ion substitutions have taken place.

There is yet another way in which the equal-valency requirement can be circumvented, and this is by leaving an occasional atom site vacant. The magnesium aluminate spinel structure, $MgAl_2O_4$, has been described in Chapter 2. Extensive substitutional solid solution occurs between this material and Al_2O_3. This corresponds to the substitution of Al^{3+} ions for some of the Mg^{2+} ions. In order to maintain electrical neutrality, each two Al^{3+} ions added must replace three Mg^{2+} ions, leaving one vacant lattice site. The end member of this complete series is γ-Al_2O_3, which has a face-centered cubic packing of oxygen ions as in spinel in a structure corresponding to $Al_{8/3}O_4$ with one-ninth of the total cation sites vacant.

This kind of solid solution in which different ion and vacant atom sites are added in the right proportions to give electrical neutrality is not uncommon. For example, additions of CaO to ZrO_2 form a solid solution with the cubic fluorite structure in which Ca^{2+} are substituted for Zr^{4+}. Each time this is done, an oxygen ion site is left vacant to maintain the cation-anion site relationship of 1:2.

$$CaO(s) \xrightarrow{ZrO_2} Ca''_{Zr} + O_O + V_{\ddot{O}} \qquad (4.7)$$

Similarly, additions of La_2O_3 to CeO_2 or ZrO_2 and CdO with Bi_2O_3 give rise to substantial numbers of vacant sites in the anion array. In contrast, additions of $MgCl_2$ to LiCl, Al_2O_3 to $MgAl_2O_4$, and Fe_2O_3 to FeO lead to vacant sites in the cation array. Although in the systems just mentioned these effects are large, similar effects occur in samples for which solid solution is extremely limited; for example, $CaCl_2$ dissolved in KCl. In this system the solid solubility is less than 1%.

In refractory oxides with close-packed crystal structures, the temperature dependence of the solubility is often large (see Fig. 4.3). Although the solubility of Al_2O_3 in MgO is several percent at 2000°C, it decreases to only 0.01% at a temperature of 1300°C. At high temperatures the TS product dominates the free energy for solution; but as the temperature is lowered, the large enthalpy term $(E + PV)$ for the formation of vacant lattice sites dominates (Eq. 4.6).

The most direct evidence for the formation of vacancies is obtained by determining the lattice constant for a structure in order to compare theoretical with measured crystal density. Several examples follow.

A crystal with the composition $Zr_{0.85}Ca_{0.15}O_{1.85}$ which crystallizes in the fluorite structure has an X-ray pattern indicating that in a unit cell having an edge of 5.131 Å there are four cation sites and eight anion sites. If the cation sites are all filled, and oxygen ion vacancies occur to give the proper site relationship, there are $4 \times 0.15 \times 40.08/6.03 \times 10^{23}$ g Ca, $4 \times 0.85 \times 91.22/6.03 \times 10^{23}$ g Zr, and $8 \times 1.85/2 \times 16.0/6.03 \times 10^{23}$ g O in 135.1 A^3 for a total of 5.480 g/cm^3. [The weight of each ion per unit cell is given by (number of sites) (fraction occupied) (atomic weight)/(Avogadro's number).] This is in excellent agreement with the directly measured value of 5.477 g/cm^3.

Figure 4.4*a* shows the change in density calculated from the X-ray lattice parameter as compared with the directly measured density; the upper curve is calculated for another possible model in which the extra cation occupies an interstitial position. The anion vacancy structure (Eq. 4.7) is found to be in accordance with these calculations. However, for samples quenched from 1800°C, the data in Fig. 4.4*b* show that the higher temperature equilibrium corresponds to a different sort of defect structure. (This is perhaps our first clear warning of the possible divergence between observations at room temperature and the equilibrium situation at high temperatures.) When $CaCl_2$ is added to KCl, the density change is found to be in accordance with cation-vacancy formation, as indicated in Eq. 4.4 and Fig. 4.5*a*. When Al_2O_3 is added to MgO, the density change is also found to be in accordance with cation-vacancy formation. These data are shown in Fig. 4.5*b*.

A charge balance can also be achieved by changes in the electronic structure, as discussed in Sections 4.8 and 4.9.

Interstitial Solid Solutions. If added atoms are small, they can go on interstitial sites in the crystal to form solid solutions. This type of solution is particularly common with metallic bonding, in which added H, C, B, and N fit easily into interstitial sites.

The ability to form interstitial solid solutions depends on the same factors, except for structure type, that apply for substitutional solid solutions—size, valency, and chemical affinity. The size effect depends on the original host crystal structure. In face-centered cubic structures, such as MgO, the only available interstitial sites are tetrahedral sites surrounded by four oxygen ions. In contrast, in TiO_2 there are normally vacant octahedral interstices; in the fluorite structure there are larger interstices with eightfold coordination; in some of the network silicate structures such as the zeolites the interstitial positions are very large. Therefore, we expect the order of ease for forming interstitial solid

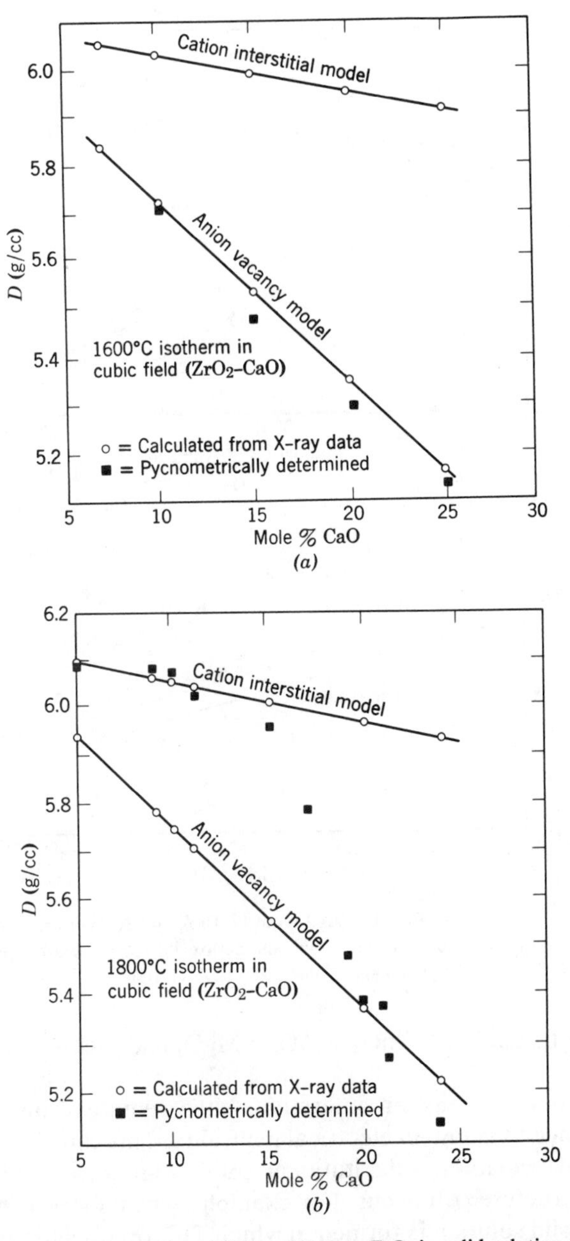

Fig. 4.4. Change in density on the addition of CaO to ZrO_2 in solid solution quenched from (*a*) 1600°C and (*b*) 1800°C. At 1600°C each Ca^{2+} addition is accompanied by the formation of a vacant lattice site. At 1800°C there is apparently a change in defect type with composition. From A. Diness and R. Roy, *Solid State Communication*, **3**, 123 (1965).

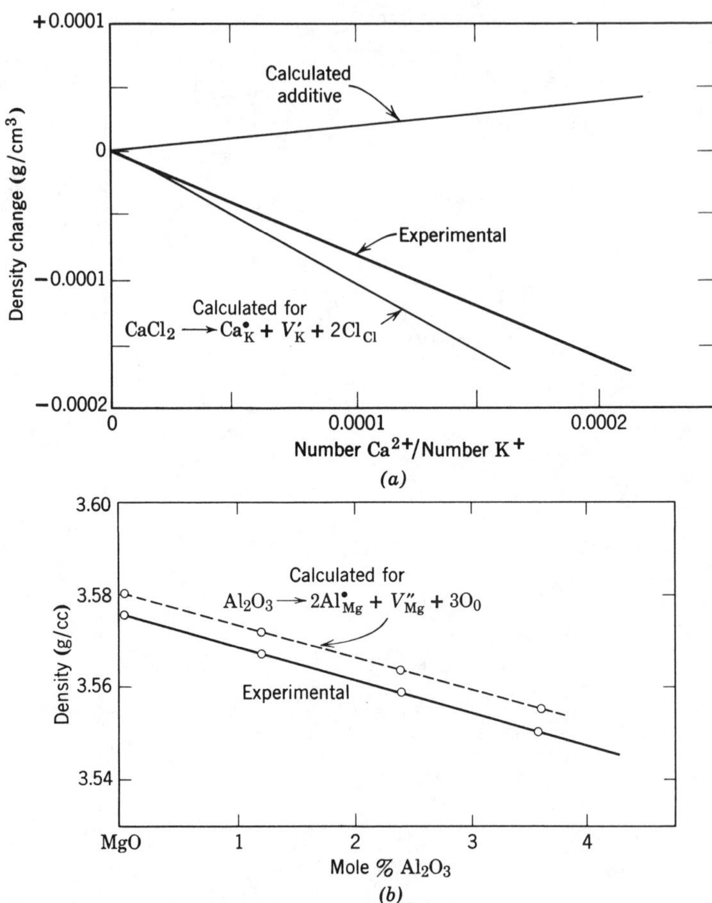

Fig. 4.5. When (*a*) $CaCl_2$ is added to KCl from H. Pick and H. Weber, *Z. Physik*, **128**, 409 (1950) and (*b*) Al_2O_3 is added to MgO, vacant cation lattice sites are created. From V. Stubican and R. Roy, *J. Phys. Chem. Solids*, **26**, 1293 (1965).

solutions to be zeolite > ThO_2 > TiO_2 > MgO, and this is found to be the case.

The addition of ions on interstitial sites requires some associated charge balance to maintain electrical neutrality. This can be accomplished by vacancy formation, substitutional solid solution, or changes in the electronic structure. All occur. For example, when YF_3 or ThF_4 is added to CaF_2, a solid solution is formed in which Th^{4+} or Y^{3+} substitute for Ca^{2+} and at the same time F^- ions are placed on the interstitial sites so that electrical neutrality is maintained (see Fig. 4.6*a*).

Likewise, the addition of ZrO_2 to Y_2O_3 creates oxygen interstitials to

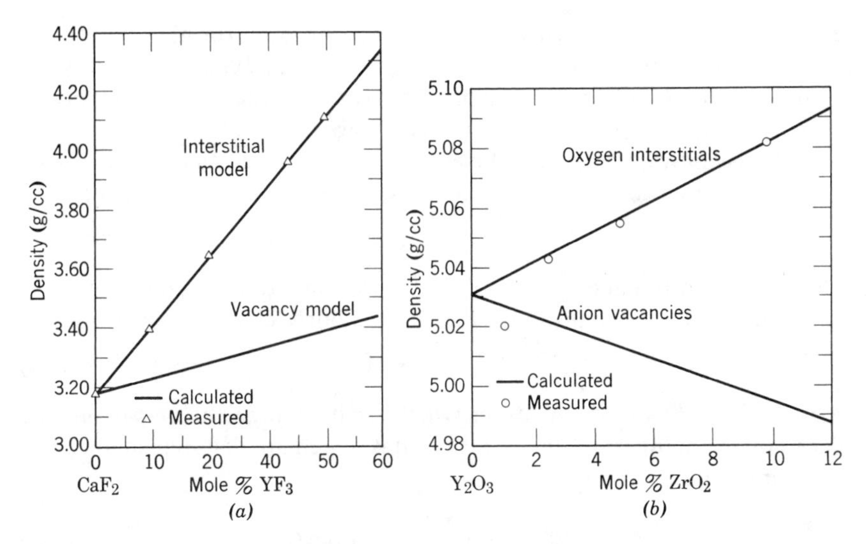

Fig. 4.6. Additions of (*a*) YF_3 to CaF_2 and (*b*) ZrO_2 to Y_2O_3 give rise to the formation of interstitial anions.

maintain the proper site relation and electrical neutrality. The measured density data are given in Fig. 4.6*b* and favor Eq. 4.8 (oxygen interstitials) rather than Eq. 4.9 (yttrium vacancies).

$$\text{Density increase:} \quad 2ZrO_2(s) \xrightarrow{Y_2O_3} 2Zr_Y^{\cdot} + 3O_O + O_i'' \tag{4.8}$$

$$\text{Density decrease:} \quad 3ZrO_2(s) \xrightarrow{Y_2O_3} 3Zr_Y^{\cdot} + 6O_O + V_Y''' \tag{4.9}$$

In many silicate structures the additional charge due to an interstitial Be^{2+}, Li^+, or Na^+ is balanced by the replacement of some of the Si^{4+} by Al^{3+} in solid solution.

4.4 Frenkel Disorder

Several different types of disorder can occur in crystals without the addition of any foreign atoms. The particular type of disorder in which equal numbers of vacant lattice sites and interstitial atoms occur is called Frenkel disorder* and is illustrated in Fig. 4.1*a*. As the extent of this kind of disorder is increased, the structural energy is increased, but at the same time the entropy (randomness of the structure) increases. At higher

*Frenkel, *op. cit.*

temperatures the higher entropy form, the disordered form, is favored to give the minimum free energy required for thermodynamic stability.

The free energy of the crystal can be written as the free energy of the perfect crystal ΔG_0 plus the free energy change $n\,\Delta g$ necessary to create n interstitials and vacancies less the entropy increase ΔS_c that accrues from the different possible ways in which the defects can be arranged:

$$\Delta G = \Delta G_0 + n\,\Delta g - T\Delta S_c \tag{4.10}$$

The configurational entropy ΔS_c is proportional to the number of ways in which the defects can be arranged, W, and is given by

$$\Delta S_c = k \ln W \tag{4.11}$$

For a perfect crystal, the N atoms which are indistinguishable can only be placed in one manner on the N lattice sites which are distinguishable. The configurational entropy is

$$\Delta S_c = k \ln \frac{N!}{N!} = k \ln 1 = 0 \tag{4.12}$$

However, if there are N normal sites and an equal number of interstitial sites, the interstitial atoms n_i can be arranged in $N!/[(N-n_i)!n_i!]$ ways, and the vacant sites n_v can be arranged in $N!/[(N-n_v)!n_v!]$ ways. The configurational entropy for these randomly (noninteracting) arranged defects is

$$\Delta S_c = k \ln \left[\frac{N!}{(N-n_i)!n_i!}\right]\left[\frac{N!}{(N-n_v)!n_v!}\right] \tag{4.13}$$

Since Stirling's approximation for large numbers yields $\ln N! = N \ln N - N$ and since $n_i = n_v = n$, the entropy is

$$\Delta S_c = 2k[N \ln N - (N-n)\ln(N-n) - n \ln n] \tag{4.14}$$

and the total free-energy change is

$$\Delta G = \Delta G_0 + n\,\Delta g - 2kT\left[N \ln\left(\frac{N}{N-n}\right) + n \ln\left(\frac{N-n}{n}\right)\right] \tag{4.15}$$

At equilibrium the free energy is a minimum with respect to the number of defects; thus $\left(\frac{\partial \Delta G}{\partial n}\right)_{T,P} = 0$. By differentiating Eq. 4.15 and setting it equal to zero and then taking $N - n \approx N$, we have $\Delta g = 2kT \ln (N/n)$, or

$$\frac{n}{N} = \exp\left(-\frac{\Delta g}{2kT}\right) = \exp\left(\frac{\Delta s}{2k}\right)\exp\left(-\frac{\Delta h}{2kT}\right) \tag{4.16}$$

In the application of Eq. 4.16 it is sometimes assumed that the entropy change in addition to the configurational entropy is negligible, such that

$$\frac{n}{N} \approx \exp\left(-\frac{\Delta h}{2kT}\right) \tag{4.17}$$

The added entropy Δs results mainly from lattice strains and accompanying changes in vibrational frequencies resulting from the introduction of the defect. There is less than full agreement on theoretical estimates, and experimental observations indicate that although values of exp $(\Delta s/2k)$ between 10 and 100 are most common, values as small as 10^{-4} and as large as 10^4 have been reported. That is, estimates of the absolute values of defect concentrations are subject to much uncertainty. In contrast, the relative change in defect concentration with temperature, and therefore Δh, is much more amenable to measurement.

Silver bromide has been well studied and forms Frenkel defects on the cation sublattice at moderate temperatures:

$$\mathrm{Ag_{Ag}} \leftrightarrows \mathrm{Ag}_i^{\cdot} + V'_{\mathrm{Ag}} \tag{4.18}$$

By defining $[V'_{\mathrm{Ag}}]$ as the fraction of charged vacant silver sites $[V'_{\mathrm{Ag}}] = n_v/N$ and the fraction of silver interstitials as $[\mathrm{Ag}_i^{\cdot}] = n_i/N$,

$$[V'_{\mathrm{Ag}}][\mathrm{Ag}_i^{\cdot}] = \exp\left(-\frac{\Delta g}{kT}\right) \tag{4.19}$$

or since $[V'_{\mathrm{Ag}}] = [\mathrm{Ag}_i^{\cdot}]$ for stoichiometric AgBr,

$$[\mathrm{Ag}_i^{\cdot}] = \exp\left(-\frac{\Delta g}{2kT}\right) \tag{4.20}$$

Another general way in which we can look at the formation of a small fraction of defects is by application of the law of mass action. Creation of a vacancy and an interstitial ion in an ionic crystal can be written as a chemical equation:

(normal ion) + (interstitial site) = (interstitial ion) + (vacant site)

$$\mathrm{Ag_{Ag}} + V_i = \mathrm{Ag}_i^{\cdot} + V'_{\mathrm{Ag}} \tag{4.21}$$

For this equation the mass-action constant is

$$K_F = \frac{[\mathrm{Ag}_i^{\cdot}][V'_{\mathrm{Ag}}]}{[V_i][\mathrm{Ag_{Ag}}]} \tag{4.22}$$

for small concentrations of defects, $[V_i] \approx [\mathrm{Ag_{Ag}}] \approx 1$; thus

$$[\mathrm{Ag}_i^{\cdot}][V'_{\mathrm{Ag}}] = K_F \qquad \text{or} \qquad [\mathrm{Ag}_i^{\cdot}] = \sqrt{K} \tag{4.23}$$

The concentration of Frenkel defects is determined by the energy of forming a vacancy and interstitial ion and by the temperature as given in Eq. 4.16. For energies of formation in the range of 1 to 6 eV and temperatures between 100 and 1800°C, concentration of defects may range between a few percent to 1 part in 10^{41}, as shown in Table 4.1. The equilibrium concentrations at room temperature are always small. They can become appreciable at higher temperatures if the energy of formation for the defects is not too large.

Table 4.1. Defect Concentrations at Different Temperatures

$$\frac{n}{N} = \exp\left[-\frac{\Delta g}{2kT}\right] = \exp\left[\frac{\Delta s}{2k}\right]\exp\left[-\frac{\Delta h}{2kT}\right] \approx \exp\left(-\frac{\Delta h}{2kT}\right)$$

Defect Concentration	1 eV[a]	2 eV	4 eV	6 eV	8 eV
n/N at 100°C	2×10^{-7}	3×10^{-14}	1×10^{-27}	3×10^{-41}	1×10^{-54}
n/N at 500°C	6×10^{-4}	3×10^{-7}	1×10^{-13}	3×10^{-20}	8×10^{-27}
n/N at 800°C	4×10^{-3}	2×10^{-5}	4×10^{-10}	8×10^{-15}	2×10^{-19}
n/N at 1000°C	1×10^{-2}	1×10^{-4}	1×10^{-8}	1×10^{-12}	1×10^{-16}
n/N at 1200°C	2×10^{-2}	4×10^{-4}	1×10^{-7}	5×10^{-11}	2×10^{-19}
n/N at 1500°C	4×10^{-2}	1×10^{-4}	2×10^{-6}	3×10^{-9}	4×10^{-12}
n/N at 1800°C	6×10^{-2}	4×10^{-3}	1×10^{-5}	5×10^{-8}	2×10^{-10}
n/N at 2000°C	8×10^{-2}	6×10^{-3}	4×10^{-5}	2×10^{-7}	1×10^{-9}

[a] 1 eV = 23.05 kcal/mole.

Although the configurational entropy change for forming Frenkel defects can be calculated from statistical mechanics, the energy change in putting an atom into an interstitial position depends to a great extent on the structure and ion characteristics, as discussed in the last section. Calculation of this energy is difficult because there is a large correction term required for the ion polarizabilities, making calculations for alkali halides difficult and for oxides nearly impossible. An example of such a calculation is discussed in the next section.

For alkali halide crystals with a sodium chloride structure the energy required to form an interstitial ion plus a vacancy is of the order of 7 to 8 eV, so that they do not occur in measurable numbers. For crystals with the fluorite structure there is a large interstitial position in the structure, the amount of energy necessary for forming interstitials is lower, $\Delta h =$ 2.8 eV for CaF_2 and $\exp\left(\frac{\Delta s}{2k}\right)$ is about 10^4, thus Frenkel defects are common. They are only prominent in crystals containing ions that have a high polarizability and consequently are better able to be accommodated into interstitial sites. This is true for AgBr, for example, in which substantial numbers of interstitial Ag^+ ions occur along with associated vacancies. The Frenkel formation energy $\Delta h = 1.1$ eV; the preexponential term $\exp(\Delta s/2k)$ is in the range of 30 to 1500.

An oxide system in which Frenkel defects are formed is Y_2O_3, for which we can write

$$O_O + V_i = O_i^{\cdot\cdot} + V_O'' \qquad K_F = [O_i^{\cdot\cdot}][V_O''] \tag{4.24}$$

When this relationship is combined with the solid-solution behavior shown for solid solutions of zirconia in yttria (Eqn. 4.8), that is, the formation of oxygen interstitials in a concentration determined by the solute concentration, we see that the concentration of vacant oxygen sites must be simultaneously diminished. That is, the Frenkel equilibrium, the product of interstitial and vacancy concentrations, remains in force.

4.5 Schottky Disorder

Another particular kind of disorder which occurs in ionic crystals is the presence of both cation and anion vacancies in thermal equilibrium (Fig. 4.1*b*). Just as for Frenkel disorder, energy must be expended to form vacant sites, but the increased entropy makes a finite vacancy concentration favorable for a minimum free energy as the temperature is raised.*

If we consider Schottky disorder in a crystal such as NaCl, we can derive the concentration of vacant sites in exactly the same way as was done for Frenkel disorder in Eqs. 4.10 to 4.16. If Δg is the energy required to form a *pair* of vacant sites by moving two ions to the surface,

$$\frac{n_v}{N} = \exp\left(-\frac{\Delta g}{2kT}\right) = \exp\left(\frac{\Delta s}{2k}\right)\exp\left(-\frac{\Delta h}{2kT}\right) \approx \exp\left(-\frac{\Delta h}{2kT}\right) \tag{4.25}$$

The defect concentration increases exponentially with temperature, and the vacancy concentration is given as shown in Table 4.1.

In principle, the enthalpy to create point defects in ionic crystals can be calculated from a Born-Haber cycle, which involves (1) the creation of charged imperfections in the crystal by removal of ions to the vapor, (2) the transformation of the gaseous ions into atoms, and (3) the formation of the compound from the gaseous atoms. The results for NaCl are in good agreement with experiment:†

$$\Delta h_{\text{calc}} = 2.12 \text{ eV} \qquad \Delta h_{\text{obs}} = 2.02 - 2.19 \text{ eV}$$

*Wagner and Schottky, *op. cit.*
†F. G. Fumi and M. P. Tosi, *Discuss. Faraday Soc.*, **23**, 92 (1957).

Schottky disorder commonly occurs in the alkali halides at elevated temperatures.

For oxides the calculations which involve Coulomb interactions, Born repulsions, and polarization effects are subject to substantial uncertainties. However, the energy for formation of vacancies in oxides is two to three times as large as for the alkali halides, which means that equilibrium Schottky disorder does not become important in oxide crystals until very high temperatures are reached. Thus, the intrinsic number of defects caused by thermal effects is almost always smaller than those from solutes, as discussed in the sections on solid solutions and nonstoichiometry. Table 4.2 contains some experimental data and estimates for Schottky and Frenkel defect formation energies.

Table 4.2. Some Defect Energies of Formation

Compound	Reaction	Energy of Formation, Δh (eV)	Preexponential Term = $\exp(\Delta s/2k)$
AgBr	$Ag_{Ag} \rightarrow Ag_i^{\cdot} + V'_{Ag}$	1.1	30–1500
BeO	$null \rightleftarrows V''_{Be} + V_0^{\cdot\cdot}$	~6	?
MgO	$null \rightleftarrows V''_{Mg} + V_0^{\cdot\cdot}$	~6	?
NaCl	$null \rightleftarrows V'_{Na} + V_{Cl}^{\cdot}$	2.2–2.4	5–50
LiF	$null \rightleftarrows V'_{Li} + V_F^{\cdot}$	2.4–2.7	100–500
CaO	$null \rightleftarrows V''_{Ca} + V_0^{\cdot\cdot}$	~6	?
CaF_2	$F_F \rightleftarrows V_F^{\cdot} + F'_i$	2.3–2.8	10^4
	$Ca_{Ca} \rightleftarrows V''_{Ca} + Ca_i^{\cdot\cdot}$	~7	?
	$null \rightleftarrows V''_{Ca} + 2V_F^{\cdot}$	~5.5	?
UO_2	$O_O \rightleftarrows V_O^{\cdot\cdot} + O''_i$	3.0	?
	$U_U \rightleftarrows V''''_U + U_i^{\cdot\cdot\cdot\cdot}$	~9.5	?
	$null \rightleftarrows V''''_U + 2V_O^{\cdot\cdot}$	~6.4	?

A final but significant principle must be remembered when considering concentrations of interstitial ions and lattice vacancies at a particular temperature. Since the equilibrium defect concentrations were derived for equilibrium conditions, sufficient time must be allowed for equilibrium to be reached. Since this usually involves diffusional processes over many atomic dimensions, equilibrium at low temperatures may in practice never be reached. Thus the high-temperature defect concentrations may be quenched in when the crystal is cooled, as illustrated in Fig. 4.4.

An important distinction between Schottky defects and Frenkel defects is that Schottky defects require a region of lattice perturbation such as a grain boundary, dislocation, or free surface for their formation to occur. For example, in MgO, magnesium ions must leave their lattice positions and migrate to a surface or grain boundary; thus,

$$Mg_{Mg} + O_O \rightleftarrows V''_{Mg} + V_O^{\cdot\cdot} + Mg_{surf} + O_{surf} \tag{4.26}$$

Since magnesium ions and oxygen ions on the surface form a layer over other ions previously located at the surface, this equation is equivalent to the usual form of the Schottky equation:

$$\text{null} \rightleftarrows V''_{Mg} + V_O^{\cdot\cdot} \tag{4.27}$$

and affects only the kinetics, not the equilibrium state.

As is true for Frenkel defects, it is the product of the vacancy concentrations that is fixed by the Schottky equilibrium. From Eq. 4.27,

$$K_s = [V''_{Mg}][V_O^{\cdot\cdot}] \tag{4.28}$$

When Al_2O_3 is added to MgO as a solute, cation vacancies are created, as shown in Fig. 4.5*b*. This effect combined with the Schottky equilibrium (Eq. 4.28) requires that the concentration of anion vacancies be simultaneously diminished.

4.6 Order-Disorder Transformations

In an ideal crystal there is a regular arrangement of atom sites with a periodic arrangement of atoms on all these positions. In real crystals, however, we have seen that foreign atoms, vacant sites, and the presence of interstitial atoms disturb this complete order. Another type of departure from order is the exchange of atoms between different kinds of positions in the structure, leading to a certain fraction of the atoms being on "wrong" sites. This disorder is similar to the other kinds of structural imperfections we have discussed, in that it raises the structure energy but also increases the randomness or entropy so that disorder becomes increasingly important at high temperatures. This leads to an order-disorder transition between the low-temperature form which is mostly ordered and the high-temperature form which is disordered. This kind of transition is commonly observed in metal alloys. It also occurs for ionic systems, but these are more likely to be either completely ordered or completely disordered, and transitions are only infrequently observed. There are some similarities and also differences between order-disorder transitions and high-low polymorphic transitions (Section 2.10).

The degree of order can be described on a long-range basis as the fraction of atoms on "wrong" sites, or on a short-range basis as the fraction of "wrong" atoms in a first or second coordination ring. For our purposes a description of long-range order is sufficient. Let us consider two kinds of atoms, A and B, in a lattice having two kinds of sites, α and β, with the total number of atoms equal to the number of sites N. If R_α is the fraction of α sites occupied by the "right" A atoms and R_β is the fraction of β sites occupied by the B atoms in a perfectly ordered crystal, all the atoms are on the "right" sites and $R_\alpha = R_\beta = 1$. If there are an equal number of A and B atoms and α and β sites, then for a completely random arrangement only half the A atoms are on α sites, $R_a = 1/2$, and only half the B atoms are on β sites, $R_\beta = 1/2$. We can define an order parameter S, which is a measure of how completely the α sites are filled with A atoms, in such a way that for complete order, S equals one, and for complete disorder, S equals zero, as

$$S = \frac{R_\alpha - \frac{1}{2}}{1 - \frac{1}{2}} = \frac{\frac{1}{2} - W_\alpha}{1 - \frac{1}{2}} \tag{4.29}$$

where W_α is the fraction of α sites containing the wrong B atoms.

If only a small degree of disorder occurs, we can derive the dependence of order on temperature and on the energy E_D required for the exchange of a pair of atoms in exactly the same way as was done for Frenkel disorder in Eqs. 4.11 to 4.16, with the result

$$\frac{W_\alpha}{R_\alpha} = \frac{W_\beta}{R_\beta} = \exp\left(-\frac{E_D}{2kT}\right) \tag{4.30}$$

For increasing amounts of disorder, however, more of the neighbors of a "wrong" atom will also be "wrong," so that there is an increasing *ease* of disordering (a lower value for E) as the amount of disorder increases. In the simplest and nearly satisfactory theory of disorder,* it is assumed that the energy required for disorder of a pair of ions is directly proportional to the amount of order, that is,

$$E_D = E_0 S \tag{4.31}$$

This is an oversimplification because the value of E_D depends on the short-range order even when the long-range order is constant. More satisfactory relationships may be derived by considering the effect of short-range ordering on the energy of disorder.† In either case, as the disorder increases with temperature, owing to the cooperative nature of the phenomenon, the rate of disorder also increases until complete

*W. L. Bragg and E. J. Williams, *Proc. R. Soc. (London)*, **145A**, 699 (1934).
†H. A. Bethe, *Proc. R. Soc. (London)*, **15A**, 552 (1935).

disorder is reached at some transition temperature (Fig. 4.7). Generally the number of *A* and *B* atoms are not equal, so that relationships derived must include this variable as well. An excellent review of the entire subject is given by F. C. Nix and W. Shockley.*

Disorder transformations are common in metals in which the nearest neighbors in an *AB* alloy can be ordered or disordered without a large change in energy. In ionic materials exchanging a cation with one of its coordination polyhedra of anions is so unfavorable energetically that it never occurs; all order-disorder phenomena are related to cation positions in the cation substructure or anion positions in the anion substructure. In this case the energy change is one of the second coordination; the first coordination remains unchanged. If the atoms are about the same size and charge, the energy from the second coordination ring of like-charged ions is almost entirely coulombic. If all the cation sites in the structure are equivalent, the energy change of disorder is small, and the disordered form is the only one that occurs; this is true, for example, in solid solutions of NiO–MgO and Al_2O_3–Cr_2O_3. (But at sufficiently low temperatures at which the *TS* product of Eq. 4.6 is sufficiently small, phase separation is to be expected in almost all systems, as discussed in Chapter 8.) In addition, there are many materials which are almost completely disordered, even though the valencies are different as long as only one kind of ion site is involved. For example, both $Li_2Fe_2O_4$ and Li_2TiO_3 have the sodium chloride structure with random distribution of the cations on the cation sites. These two compounds also form a continuous series of solid solutions not only with each other but also with MgO. In a similar way, in the compound $(NH_4)_3MoO_3F_3$ it is impossible to distinguish between the positions of the O^{2-} and F^- ions; that is, in this compound there is disorder on the anion sites. No ordered form of these compounds is known.

The most important examples of order-disorder transformation in ceramic systems occur in materials having two different kinds of cation sites, for example, the spinel structure in which some cations are on octahedral sites and some are on tetrahedral sites (see Fig. 2.25); various degrees of order in the cation positions occur, depending on the heat treatment. It has been found in almost all ferrites having the spinel structure that the cations are disordered at elevated temperatures and the stable equilibrium low-temperature form is ordered. The change of order with temperature follows a relation such as that illustrated in Fig. 4.7.

Another kind of disorder may result when there are unoccupied sites available in the ordered structure. This is the case for Ag_2HgI_4. In the

Rev. Mod. Phys.*, **10, 1 (1938).

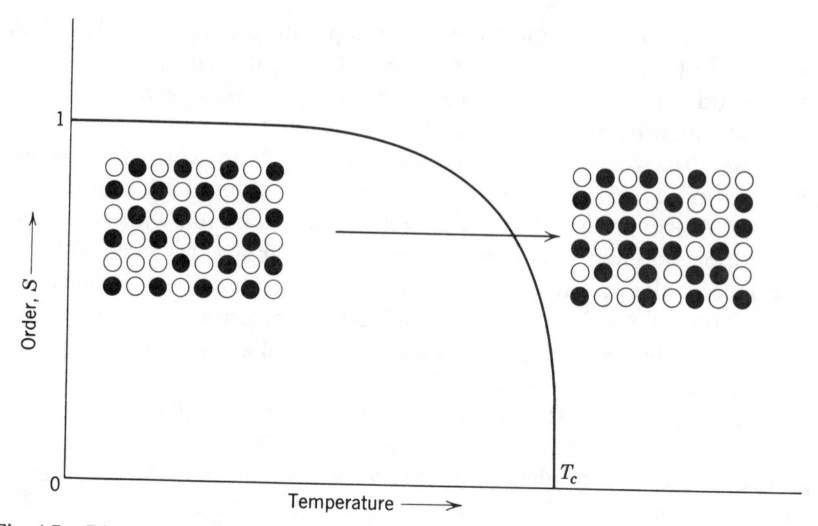

Fig. 4.7. Disorder as a function of temperature. Complete disorder is reached at a critical temperature T_c.

ordered low-temperature form three-quarters of the available sites are filled in an ordered way. A typical order-disorder transformation occurs at a temperature of about 500°C. Above this temperature there is complete disorder, with one Hg and two Ag ions randomly arranged on the four cation sites available.

4.7 Association of Defects

When Schottky or Frenkel defects are present in an ionic crystal, there is a Coulomb force of attraction between the individual defects of opposite effective charge. The electrostatic interaction between defects of opposite charge can be described by the Debye-Huckel theory of electrolytes (Refs. 2 and 7). However, within the precision of available theory to take into account repulsive force, rearrangements of nearby atoms and polarization effects (and considering the paucity of experimental data) it is preferable to focus on the major contribution of the electrostatic interaction at small distances and consider the association as resulting in the formation of a complex defect, for example, a vacancy pair consisting of an anion vacancy and a cation vacancy on nearest-neighbor sites in a material containing Schottky defects. We can write for the formation of such a vacancy pair

$$V'_{\text{Na}} + V^{\bullet}_{\text{Cl}} = (V'_{\text{Na}} V^{\bullet}_{\text{Cl}}) \tag{4.32}$$

The fractional molar concentration of vacancy pairs is given by

$$\frac{[(V'_{Na}V^{\cdot}_{Cl})]}{[V'_{Na}][V^{\cdot}_{Cl}]} = Z \exp\left(\frac{-\Delta g_{vp}}{kT}\right) = Z \exp\left(\frac{\Delta s_{vp}}{k}\right) \exp\left(\frac{-\Delta h_{vp}}{kT}\right) \quad (4.33)$$

Where Z is the distinct number of orientations of the pair which contribute to the configurational entropy ($Z = 6$ for $V^{\cdot}_{Cl}V'_{Na}$ pairs). Since the product of sodium ion and chlorine ion vacancies is fixed by the Schottky equilibrium,

$$[V'_{Na}][V^{\cdot}_{Cl}] = \exp\left(\frac{\Delta s_s}{k}\right) \exp\left(\frac{-\Delta h_s}{kT}\right) \quad (4.34)$$

and
$$[(V'_{Na}V^{\cdot}_{Cl})] = Z \exp\left(\frac{\Delta s_s}{k}\right) \exp\left(\frac{\Delta s_{vp}}{k}\right) \exp\left(-\frac{\Delta h_s + \Delta h_{vp}}{kT}\right) \quad (4.35)$$

the concentration of vacancy pairs is a thermodynamic characteristic of the crystal (a function of temperature) and independent of solute concentrations.

The coulombic energy of attraction between oppositely charged defects is

$$-\Delta h_{vp} \approx \frac{q_i q_j}{\kappa R} \quad (4.36)$$

where $q_i q_j$ are the effective charges (electronic charge $\times$ valence), κ is the static dielectric constant, and R is the separation between defects. This relationship is clearly very approximate, but it gives about the right values and leads to useful insights. For sodium chloride the cation-anion separation is 2.82 Å, the dielectric constant is 5.62 such that the energy required to separate a vacancy pair is

$$-\Delta h_{vp} \approx \frac{(4.8 \times 10^{-10}\ \text{esu})^2 \times 6.24 \times 10^{11}\ \text{eV/esu}^2\text{/cm}}{5.62 \times 2.82 \times 10^{-8}\ \text{cm}} \approx 0.9\ \text{eV} \quad (4.37)$$

where 4.8×10^{-10} esu is the electronic charge. A more precise calculation* gives a somewhat lower value than this, 0.6 eV. For the combination of two vacant sites to form a vacancy pair it is a reasonable assumption, supported by some experimental data, that the preexponential term, $\exp(\Delta s_{vp}/k)$ is near unity.

For oxide materials, in which vacancies have larger effective charges, the energy gained by the formation of vacancy pairs is larger, as illustrated in Table 4.3. Hence vacancy pairs in ceramic oxides should be more important than for the better-studied alkali halides. In Fig. 4.8 we have calculated on a speculative basis the expected concentrations of

*Fumi and Tosi, *op. cit.*

Table 4.3. Approximate Coulombic Defect Association Energies Calculated from Eq. 4.36
(This simple calculation overestimates the correct value by an uncertain amount, perhaps 50 to 100%)

	κ	R (Å)	$-\Delta h^* \approx q_i q_j/\kappa R$ (eV)
NaCl	5.62		
$V'_{Na} - V^{\cdot}_{Cl}$		2.82	0.9
$Ca^{\cdot}_{Na} - V'_{Na}$		3.99	0.6
CaF_2	8.43		
$F'_i - V^{\cdot}_F$		2.74	0.6
$Y^{\cdot}_{Ca} - V''_{Ca}$		3.86	0.9
$Y^{\cdot}_{Ca} - V''_{Ca} - Y^{\cdot}_{Ca}$		3.86	0.4
MgO	9.8		2.8
$V''_{Mg} - V^{\cdot\cdot}_0$		2.11	2.8
$Fe^{\cdot}_{Mg} - V''_{Mg}$		2.98	1.0
$Fe^{\cdot}_{Mg} - V''_{Mg} - Fe^{\cdot}_{Mg}$		2.98	0.5
NiO	12.0		
$V''_{Ni} - V^{\cdot\cdot}_0$		2.09	2.3
$V''_{Ni} - Ni^{\cdot}_{Ni}$		2.95	0.8
$Ni^{\cdot}_{Ni} - V''_{Ni} - Ni^{\cdot}_{Ni}$		2.95	0.4
$Li'_{Ni} - Ni^{\cdot}_{Ni}$		2.95	0.4
UO_2	~15		
$O''_i - V^{\cdot\cdot}_0$		2.09	0.5

*$esu^2/cm \times 6.242 \times 10^{11} = eV$.

Schottky defects and vacancy pairs in sodium chloride and in magnesium oxide.

The electrostatic attraction of oppositely charged defects also leads to association between solutes and lattice defects. For the incorporation of calcium into sodium chloride we have the reaction

$$CaCl_2(s) \xrightarrow{NaCl} Ca^{\cdot}_{Na} + V'_{Na} + 2Cl_{Cl} \tag{4.38}$$

The free energy of the system is decreased by the association reaction

$$Ca^{\cdot}_{Na} + V'_{Na} = (Ca^{\cdot}_{Na} V'_{Na}) \tag{4.39}$$

for which we can write a mass-action constant

$$\frac{[(Ca^{\cdot}_{Na} V'_{Na})]}{[V'_{Na}][Ca^{\cdot}_{Na}]} = Z \exp\left(\frac{\Delta g_a}{kT}\right) = Z \exp\left(\frac{\Delta s_a}{k}\right) \exp\left(\frac{-\Delta h_a}{kT}\right) \approx Z \exp\left(\frac{-\Delta h_a}{kT}\right) \tag{4.40}$$

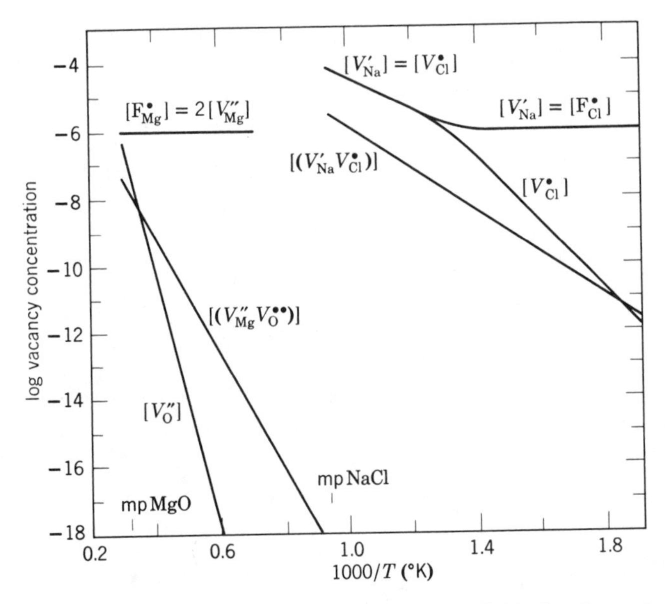

Fig. 4.8. Calculated (for NaCl) and estimated (for MgO) individual and associated defect concentrations for samples containing 1 ppm aliovalent solute.

where Z is the distinct number of solute–vacancy-pair orientations ($Z = 12$ in the NaCl lattice for neighboring cation sites), and it is reasonable to assume that the preexponential term involving vibrational entropy is near unity. An estimation of the energy of association based on the coulombic attraction (Eq. 4.36) gives results for a number of systems, as illustrated in Table 4.3. In contrast to the intrinsic nature of vacancy pairs, the concentration of solute-vacancy associates depends strongly on the solute concentration.

As the temperature of a solute-containing crystal is further lowered, a temperature is reached which corresponds to the solubility limit where precipitation of the solute occurs. At temperatures below this level, the solute concentration remaining in solid solution in the crystal is determined by the free energy of the precipitation reaction. For sodium chloride containing $CaCl_2$, we can write

$$Ca^{\cdot}_{Na} + V'_{Na} + 2Cl_{Cl} \rightleftharpoons CaCl_2(ppt) \tag{4.41}$$

$$\frac{CaCl_2(ppt)}{[V'_{Na}][Ca^{\cdot}_{Na}][Cl_{Cl}]^2} = \exp\left(-\frac{\Delta g_{ppt}}{kT}\right) \propto \exp\left(-\frac{\Delta h_{ppt}}{kT}\right) \tag{4.42}$$

$$\text{or} \quad [V'_{Na}] \propto \exp\left(+\frac{\Delta h_{ppt}}{2kT}\right) \tag{4.43}$$

Similarly, from Fig. 4.3 we see that the solubility of aluminum oxide in MgO decreases from almost 10% at 2000°C to less than 0.1% at 1500°C, corresponding to a heat of solution of about 3 eV. For the precipitation reaction we can write

$$\mathrm{Mg_{Mg}} + 2\mathrm{Al}^{\cdot}_{\mathrm{Mg}} + V''_{\mathrm{Mg}} + 4\mathrm{O_O} \rightleftharpoons \mathrm{MgAl_2O_4(ppt)} \tag{4.44}$$

$$[V''_{\mathrm{Mg}}][\mathrm{Al}^{\cdot}_{\mathrm{Mg}}]^2 \propto \exp\left(+\frac{\Delta h_{\mathrm{ppt}}}{kT}\right) \tag{4.45}$$

$$\text{since } [\mathrm{Al}^{\cdot}_{\mathrm{Mg}}] \approx 2[V''_{\mathrm{Mg}}] \tag{4.46}$$

$$[V''_{\mathrm{Mg}}] \propto \left(\frac{1}{4}\right)^{1/3} \exp\left(+\frac{\Delta h_{\mathrm{ppt}}}{3kT}\right) \tag{4.47}$$

such that the defect concentration in the crystal is approximately determined by the heat of precipitation as defined by these reactions. Since the total solubility also includes defect associates, Δh_{ppt} given in Eqs. 4.43 and 4.47 is not equal to the negative of the heat of solution.

4.8 Electronic Structure

In our ideal crystal, in addition to all atoms being on the right sites with all sites filled, the electrons should be in the lowest-energy configuration. Because of the Pauli exclusion principle, the electron energy levels are limited to a number of energy bands up to some maximum cutoff energy at 0°K which is known as the Fermi energy $E_f(0)$. At higher temperatures thermal excitation gives an equilibrium distribution in some higher energy states so that there is a distribution about the Fermi level $E_f(T)$ which is the energy for which the probability of finding an electron is equal to one-half. Only a small fraction of the total electron energy states are affected by this thermal energy, depending on the electron energy band scheme.

The different temperature effects observed for metals, semiconductors, and insulators are related to the electronic energy band levels (Fig. 4.9). In metals, these bands overlap so that there is no barrier to excite electrons to higher energy states. In semiconductors and insulators a completely filled energy band is separated from a completely empty conduction band of higher electron energy states by a band gap of forbidden energy levels. In intrinsic semiconductors the energy difference between the filled and empty bands is not large compared with the thermal energy, so that a few electrons are thermally excited into the conduction band, leaving empty electron positions (electron holes) in the normally filled band. In perfect insulators the gap between bands is so large that thermal excitation is insufficient to change the electron energy states, and at all temperatures

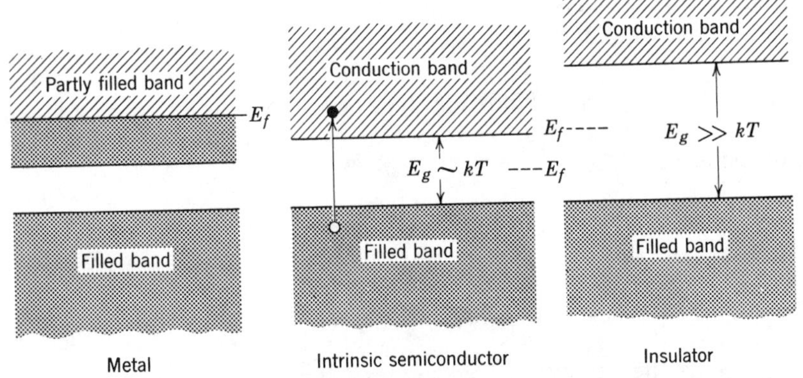

Fig. 4.9. Electron energy band levels for metals with partly filled conduction band, intrinsic semiconductors with a narrow band gap, and insulators with a high value for E_g.

the conduction band is completely devoid of electrons and the next lower band of energy is completely full, with no vacant states.

In an intrinsic semiconductor, each electron whose energy is increased so that it goes into the conduction band leaves behind an electron hole, so that the number of holes equals the number of electrons, $p = n$. The nomenclature usually employed is to indicate the positive electron-hole concentration by p, that is, $p = [h^{\cdot}]$, and the negative excess electron concentration by n, that is, $n = [e']$. In this case the Fermi level E_f is halfway between the upper limit of the filled band and the lower level of the conduction band.

The concentrations of the intrinsic electronic defects can be calculated in a manner analogous to that described for Frenkel and Schottky defect concentrations. In this calculation the thermal randomization of electrons is related to the probability of a valence electron in the full band having enough energy to jump across the energy gap E_g into the conduction band. Because of the Pauli exclusion principle, Fermi statistics are required to calculate the distribution. The concentration of free electrons is

$$n = [e'] = \frac{n_e}{N_c} = \frac{1}{1 + \exp[(E_c - E_f)/kT]} \tag{4.48}$$

where n_e is the number of electrons per cubic centimeter, N_c is the density of available states in the conducton band,

$$N_c = 2\left[\frac{2\pi m_e^* kT}{h^2}\right]^{3/2} \approx 10^{19}/\text{cm}^3 \text{ at } T = 300°\text{K} \tag{4.49}$$

E_c is the energy level at the bottom of the conduction band, and E_f is the

Fermi energy, as illustrated in Fig. 4.9. The Fermi energy represents the chemical potential of the electron and at 0°K is at the center of the band gap.

A similar relationship holds for electron holes in the valence band, and when the concentration of electrons and electron holes is small, these expressions reduce to

$$n = [e'] = \frac{n_e}{N_c} = \exp\left[-\frac{(E_c - E_f)}{kT}\right] \tag{4.50}$$

$$p = [h^{\cdot}] = \frac{n_p}{N_v} = \exp\left[-\frac{(E_f - E_v)}{kT}\right] \tag{4.51}$$

where N_v is the density of electron-hole states in the valence band,

$$N_v = 2\left[\frac{2\pi m_h^* kT}{h^2}\right]^{3/2} \approx 10^{19}/\text{cm}^3 \text{ at } T = 300°\text{K} \tag{4.52}$$

The product of the electron concentration per cubic centimeter times the hole concentration per cubic centimeter is given by

$$\begin{aligned} n_e n_p &= 4\left(\frac{2\pi kT}{h^2}\right)^3 (m_e^* m_h^*)^{3/2} \exp\left(-\frac{E_g}{kT}\right) \\ &\approx 10^{38} \exp\left(-\frac{E_g}{kT}\right) \text{cm}^{-6} \text{ at } 300°\text{K} \end{aligned} \tag{4.53}$$

where $E_g \equiv E_c - E_f$, h is Planck's constant, and m_e^*, m_h^* are the effective masses of free electrons and electron holes in the crystal lattice, usually somewhat larger than the mass of a free electron (in oxides m^* is approximately 2 to $10m$ and in alkalide halides m^* approximately equals $1/2m$). In a pure crystal the concentration of electrons equals the concentration of electron holes. When solutes or nonstoichiometry affects the electron energy levels, the ratio of electrons to holes changes but, as is the case for Frenkel and Schottky equilibrium, their product remains constant.

The magnitude of the energy band gap covers a wide range, varying from as small a value as 0.35 eV for PbS to a value of about 8 eV for stable oxides such as MgO and Al_2O_3. In Table 4.4 some characteristic values of the band gap and the resulting concentrations of electrons and holes in pure materials are illustrated.

Lattice defects, atom vacancies, interstitial atoms, and solute atoms are sites of perturbations to the energy states respresented in the band scheme in Fig. 4.9 and result in localized energy states in the band gap.

If an added electron or hole is loosely associated with an impurity site, we can approximately calculate the energy to add or remove an electron by assuming that the electron is bound to the defect in a way similar to the

Table 4.4. Band Gap[a] and Approximate Concentrations of Electrons and Holes in Pure, Stoichiometric Solids

		$n \approx 10^{19} \exp\left[-\frac{E_g}{2kT}\right]$ electrons/cm^3			
Crystal	E_g (eV)	Room Temp	1000°K	Melting Point	Temp (°K)
KCl	7	10^{-40}	20	150	1049
NaCl	7.3	10^{-43}	4	70	1074
CaF_2	10	10^{-66}	10^{-6}	10^3	1633
UO_2	5.2	10^{-25}	10^6	10^{15}	3150
NiO	4.2	10^{-16}	10^8	10^{13}	1980
Al_2O_3	7.4	10^{-44}	2.0	10^{11}	2302
MgO	8	10^{-49}	0.01	10^{12}	3173
SiO_2	8	10^{-49}	0.01	10^8	1943
AgBr	2.8	10^{-5}	10^{12}	10^9	705
CdS*	2.8	10^{-5}	10^{12}	10^{15}	1773
CdO*	2.1	20	10^{13}	10^{16}	1750
ZnO*	3.2	10^{-8}	10^{11}	10^{14}	1750
Ga_2O_3	4.6	10^{-20}	10^7	10^{13}	2000
LiF	12		10^{-11}	10^{-8}	1143
Fe_2O_3*	3.1	10^{-7}	10^{11}	10^{14}	1733
Si	1.1	10^{10}	10^{16}	10^{17}	1693

[a] Most of the data are based on the optical band gap, which may be larger than the electronic band gap.
*Sublimes or decomposes.

hydrogen atom, except that it has an effective electron mass m_e^* and is immersed in a medium with a dielectric constant κ. The energy is assumed to be proportional to the first excited level in the hydrogen atom:

$$E = 13.6\left(\frac{m_e^*}{m}\right)\left(\frac{z}{\kappa}\right)^2 \text{ eV} \tag{4.54}$$

where z is the ionization state of the defect. For the alkali halides m_e^* is about $1/2m$ and the dielectric constant is about 5, so that the energy required to ionize a sodium or chlorine atom vacancy in NaCl or to excite an electron from a neutral calcium solute atom is estimated as about 0.3 eV. If we assume that the excess electron is located at the nearest-neighbor distance, we can calculate that the energy of ionization as calculated for ion associates (Eq. 4.36) is

$$E = \frac{q_1 q_2}{\kappa R}$$

which gives the energy required to ionize a sodium or chlorine ion vacancy in NaCl as about 0.9 eV. Finally, in the case in which the electron is bound within a narrow orbit, interactions between the valence electron and the impurity center are decisive and the ionization energy of the center is determined by specific quantities such as the ionization energy or electron affinity, polarization terms, the local electrostatic potential, and so forth. In general it is a priori unknown which of these cases applies.

In showing the electron energy levels at defects within the band gap, we always follow the convention of indicating the nature of the level by labeling it as if occupied. Neutral levels near the conduction band may be ionized to free an electron and are called electron donor levels. Neutral levels near the valence band may be ionized by accepting electrons and are called acceptors. In a sample of potassium chloride (Fig. 4.10), vacant chlorine sites may be ionized with the expenditure of about 1.8 eV; calcium atoms substituted on potassium sites may be ionized with the expenditure of about 1 eV. When a neutral site such as a potassium atom vacancy is ionized, approximately 1 eV is required. Potassium chloride is a wide-band-gap material with $E_g \approx 7$ eV. The difference in energy between the lowest donor level and the highest acceptor level is 4.2 eV; this is the energy gained from the ionization of a neutral chlorine atom vacancy and the transfer of its electron to the potassium vacancy, which then has an effective negative charge. Thus we can write for KCl that the Schottky equilibrium for atomic unionized defects is given by

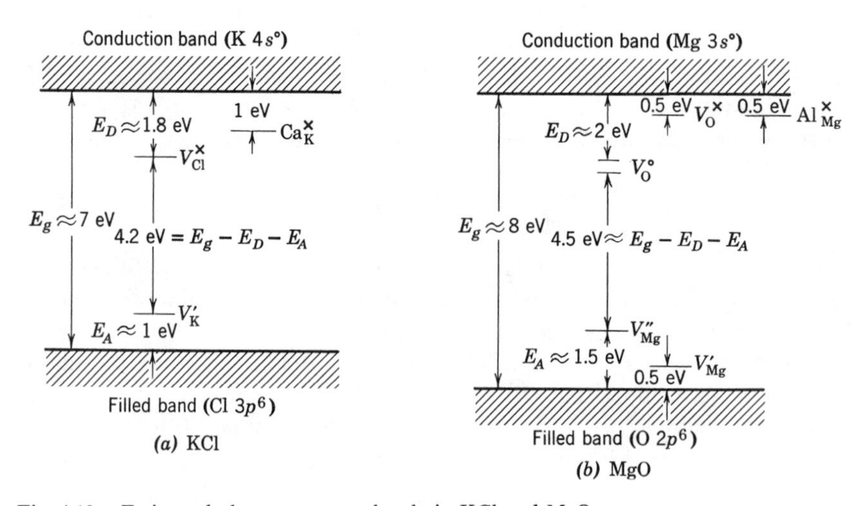

Fig. 4.10. Estimated electron energy levels in KCl and MgO.

$$[V_K][V_{Cl}] \approx \exp\left[-\frac{\Delta G_{s,a}}{kT}\right] \approx \exp\left[-\frac{\Delta G_{s,i}}{kT} + \frac{(E_g - E_A - E_D)}{kT}\right] \quad (4.55)$$

$$\approx \exp\left[-\frac{6.4\text{ eV}}{kT}\right]$$

and

$$[V_K'][V_{Cl}^{\cdot}] \approx \exp\left[\frac{\Delta G_{s,i}}{kT}\right]$$

$$\approx \exp\left[-\frac{2.2\text{ eV}}{kT}\right]$$

That is, for the pure material the ratio of unionized to ionized vacancies is given by

$$\frac{[V_K]}{[V_K']}\frac{[V_{Cl}]}{[V_{Cl}^{\cdot}]} \approx \exp\left[-\frac{4.2\text{ eV}}{kT}\right] \quad (4.56)$$

As a consequence, in wide-band-gap materials the concentration of neutral defects is many orders of magnitude smaller than the concentration of ionized defects, a fact which we have assumed in Sections 4.2 to 4.6. For materials which have a narrower band gap, particularly the transition elements with unfilled *d* orbitals and the higher atomic weight elements, the defect energy levels approach the center of the band gap, are near the Fermi level, unionized or partially ionized defects occur, and the electron energy levels are both more complicated and frequently more controversial.

4.9 Nonstoichiometric Solids

In elementary chemistry and in many analytical chemical techniques we rely on the idea that chemical compounds are formed with constant fixed proportions of constituents. From a consideration of structure vacancies and interstitial ions we have already seen that this is only a special case and that compounds without simple ratios of anions to cations, that is, nonstoichiometric compounds, are not uncommon. An example for which the stoichiometric ratio does not even exist is wüstite, having an approximate composition of $Fe_{0.95}O$. This material has the sodium chloride structure; samples of different compositions were studied by E. R. Jette and F. Foote,* with the results shown in Table 4.5. For samples of different composition, the unit-cell size and the crystal density were determined. The departure from stoichiometry might be accounted for either by oxygen ions in interstitial positions (to give $FeO_{1.05}$, for example) or by vacant cation sites. Since the density increases

*E. R. Jette and F. Foote, *J. Chem. Phys.*, **1**, 29 (1933).

Table 4.5. Composition and Structure of Wüstite[a]

Composition	Atom% Fe	Edge of Unit Cell (Å)	Density (g/cm^3)
$Fe_{0.91}O$	47.68	4.290	5.613
$Fe_{0.92}O$	47.85	4.293	5.624
$Fe_{0.93}O$	48.23	4.301	5.658
$Fe_{0.945}O$	48.65	4.310	5.728

Source. E. R. Jette and F. Foote, *J. Chem. Phys.*, **1**, 29 (1933).

as the oxygen-to-iron ratio decreases, the changing structure must be due to cation vacancies. As more iron vacancies are created, the density decreases, as does the size of the unit cell.

To compensate for the smaller number of cations and consequent loss of positive charge, two Fe^{2+} ions must be transformed into Fe^{3+} ions for each vacancy formed. From a chemical point of view, we may consider this simply as a solid solution of Fe_2O_3 in FeO in which, in order to maintain electrical neutrality, three Fe^{2+} ions are replaced by two Fe^{3+} and a vacant lattice site, that is, $Fe_2^{3+}V_{Fe}O_3$ replaces Fe_3O_3, in which V_{Fe} represents a vacant cation site. To a first approximation the Fe^{2+} ions may be considered as distributed at random. Similar structures are observed for FeS and FeSe, in which ranges of stoichiometry occur corresponding to vacancies in the cation lattice. Other examples are $Co_{1-x}O$, $Cu_{2-x}O$, $Ni_{1-x}O$, γ-Al_2O_3, and γ-Fe_2O_3. Similarly, there are compounds with vacancies in the anion lattice such as ZrO_{2-x} and TiO_{2-x}. Also oxides occur in which there are interstitial cations such as $Zn_{1+x}O$, $Cr_{2+x}O_3$, and $Cd_{1+x}O$. Compounds with interstitial anions are less common, but UO_{2+x} is one.

All these structures can be considered, from a chemical point of view, solid solutions of higher and lower oxidation states, that is, Fe_2O_3 in FeO, U_3O_8 in UO_2, and Zr in ZrO_2. However, the electrons associated with the valency differences are frequently not fixed at one specific ion site but readily migrate from one position to another. The idea that this electron is independent of any fixed ion position can be indicated by representing it separately in the reaction of formation of the nonstoichiometric compound. For the reaction of TiO_2 to form TiO_{2-x} plus $\frac{x}{2}O_2(g)$

$$2Ti_{Ti} + O_O = 2Ti'_{Ti} + V_O^{\cdot\cdot} + \frac{1}{2}O_2(g) \tag{4.57}$$

is equivalent to

$$O_O = V_O^{\cdot\cdot} + \frac{1}{2}O_2(g) + 2e' \tag{4.58}$$

where e' is an added electron in the structure. Similarly, the absence of an electron normally present in the stoichiometric structure corresponds to an electron hole or a missing electron $h^{\cdot}$

$$2Fe_{Fe} + \frac{1}{2}O_2(g) = 2Fe_{Fe}^{\cdot} + O_O + V_{Fe}'' \tag{4.59}$$

$$\frac{1}{2}O_2(g) = O_O + V_{Fe}'' + 2h^{\cdot} \tag{4.60}$$

Oxides in general show a variation of composition with oxygen pressure, owing to the existence of a range of stoichiometry. Stable oxides having a cation with a preference for a single valence state (a high ionization potential) such as Al_2O_3 and MgO have very limited ranges of nonstoichiometry, and in these materials observed nonstoichiometric effects are very often related to impurity content. Oxides of cations having a low ionization potential can show extensive regions of nonstoichiometry. For reactions such as those illustrated in Eqs. 4.57 to 4.60 we can write mass-action expressions and equilibrium constants and relate the atmospheric pressure to the amount of nonstoichiometry observed. For example, cobaltous oxide is found to form cation vacancies:

$$\frac{1}{2}O_2(g) = O_O + V_{Co}'' + 2h^{\cdot} \tag{4.61}$$

For this equation the equilibrium constant is given by

$$K = \frac{[O_O][V_{Co}''][h^{\cdot}]^2}{P_{O_2}^{1/2}} \tag{4.62}$$

Since the concentration of oxygen ions in the crystal is not significantly changed ($[O_O] \approx 1$) and the concentration of electron holes equals twice the concentration of vacancies, $2[V_{Co}''] = [h^{\cdot}]$,

$$[V_{Co}''] \sim P_{O_2}^{1/6}. \tag{4.63}$$

Similarly, when ZnO is heated in zinc vapor, we obtain a nonstoichiometric composition containing excess zinc, $Zn_{1+x}O$, for which we can write

$$Zn(g) = Zn_i^{\cdot} + e' \tag{4.64}$$

$$K = \frac{[Zn_i^{\cdot}][e']}{P_{Zn}} \tag{4.65}$$

$$[Zn_i^{\cdot}] \sim P_{Zn}^{1/2} \tag{4.66}$$

Or similarly for the oxygen pressure dependence ($Zn(g) + 1/2O_2 \rightleftarrows ZnO$)

$$[Zn_i^{\cdot}] \sim P_{O_2}^{-1/4} \tag{4.67}$$

An essential consideration in each case is the nature of the defect (substitutional, interstitial, vacancy) and the degree of ionization. For example, the zinc interstitials in ZnO might be doubly ionized:

$$Zn(g) = Zn_i^{\cdot\cdot} + 2e' \tag{4.68}$$

which would give a different concentration–partial-pressure relationship:

$$[Zn_i^{\cdot\cdot}] \propto [e'] \propto P_{Zn}(g)^{1/3} \propto P_{O_2}^{-1/6} \tag{4.69}$$

The correct model choice requires experimental data. Since the electrical conductivity is proportional to the concentration of free electrons and therefore to the concentration of charged zinc interstitials, the electrical-conductivity data in Fig. 4.11 support our choice of singly charged zinc interstitials (Eqs. 4.64 to 4.67) as the actual defect mechanism.

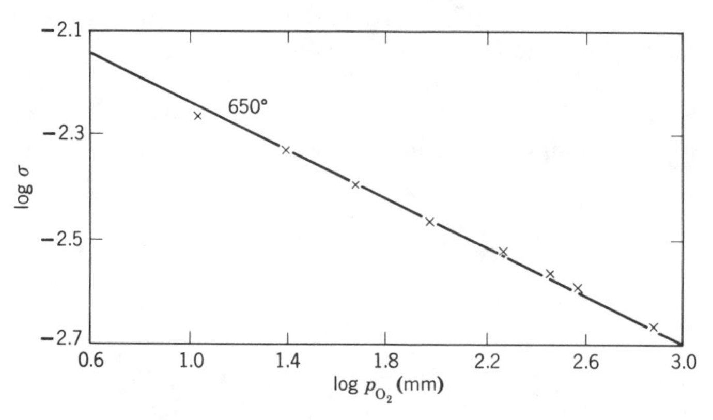

Fig. 4.11. Conductivity of ZnO as a function of oxygen pressure at 650°C. From H. H. Baumbach and C. Wagner, *Z. Phys. Chem.*, **B22**, 199 (1933).

So far we have only considered the major species present over a limited range of stoichiometry. For a more complete description of the defect structure it is necessary to write down all the equilibria expressions involving interactions among vacancies, interstitials, electron energy levels, and chemical composition, including the influence of solutes and impurities, and solve this set of equations together with relations expressing electrical-charge balance, site balance, and mass balance. In an approximate method proposed by Brouwer* we can write the mass-action

Philips Res. Rep.*, **9, 366 (1954).

equations in a logarithmic form such that there is a linear relationship between terms and make the assumption that on each side of the neutrality equation one of the concentrations is so dominant as to make the others negligible. At a given temperature we can then prepare a diagram of the log concentration of each species as a function of the log oxygen pressure; the log concentration of each species appears as a straight line with a slope corresponding to the oxygen pressure dependence within a given neutrality condition.

Let us consider an oxide material in which oxygen Frenkel defects occur, the oxygen content varies over a range of stoichiometry, and the electron and electron-hole concentration is appreciable. We can write

$$\mathrm{O_O} = \mathrm{O}_i'' + V_\mathrm{O}^{\cdot\cdot} \qquad [\mathrm{O}_i''][V_\mathrm{O}^{\cdot\cdot}] = K_F'' \tag{4.70}$$

$$\mathrm{O_O} = \frac{1}{2}\mathrm{O_2}(g) + V_\mathrm{O}^{\cdot\cdot} + 2e' \qquad [V_\mathrm{O}^{\cdot\cdot}][e']^2 P_{\mathrm{O_2}}^{1/2} = K_1 \tag{4.71}$$

$$\text{null} = e' + h^{\cdot} \qquad [e'][h^{\cdot}] = K_i \tag{4.72}$$

$$\frac{1}{2}\mathrm{O_2}(g) = \mathrm{O}_i'' + 2h^{\cdot} \qquad \frac{[\mathrm{O}_i''][h^{\cdot}]^2}{P_{\mathrm{O_2}}^{1/2}} = K_2 \tag{4.73}$$

Actually only three of these four equations are required, since the two representations of the oxygen addition are equivalent; that is, $K_1K_2 = K_i^2K_F''$. The neutrality equation is

$$2[\mathrm{O}_i''] + [e'] = 2[V_\mathrm{O}^{\cdot\cdot}] + [h^{\cdot}]. \tag{4.74}$$

but if the energy gap is such that the concentration of electronic defects at the stoichiometric composition is substantially greater than that of Frenkel defects, we can replace this representation with the simpler requirement that $n = p$. When the concentration of electrons is fixed, the oxygen vacancy concentration is proportional to $P_{\mathrm{O_2}}^{-1/2}$ according to Eq. 4.71. Similarly with the electron-hole concentration fixed, the oxygen interstitial concentration is proportional to $P_{\mathrm{O_2}}^{+1/2}$; at the stoichiometric composition the oxygen interstitial and oxygen vacancy concentrations are equal.

At a sufficiently high oxygen pressure the concentration of oxygen interstitials increases to a point at which the neutrality condition can be approximated by $[\mathrm{O}_i''] = 1/2p$. At a sufficiently low oxygen pressure the concentration of oxygen vacancies with a positive effective charge increases to a point at which the neutrality condition can be approximated by $[V_\mathrm{O}^{\cdot\cdot}] = 1/2n$.

An alternate possibility occurs when the concentration of Frenkel defects is substantially greater than the concentration of intrinsic elec-

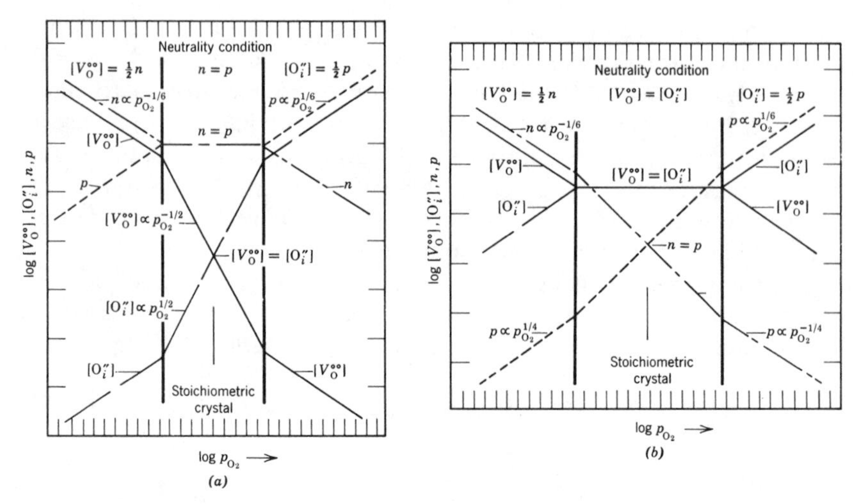

Fig. 4.12. Schematic representation of concentration of oxygen point defects and electronic defects as a function of oxygen pressure in an oxide which, depending on the partial pressure of oxygen, may have an excess or deficit of oxygen. In (*a*) $K_i > K_f''$; in (*b*) $K_f'' > K_i$ (reference 9).

tronic defects; that is, $\Delta g_F'' < E_g$. The relative defect concentrations as a function of oxygen pressure are illustrated in Fig. 4.12.

Relative to the actual situation existing in real ceramic materials, Fig. 4.12 has been simplified by ignoring the presence of associates and the influence of impurities, which are often decisive. For an oxide MO in which Schottky equilibrium is predominant for the pure material, we have shown in Fig. 4.13 a schematic representation of the substantial changes which result from the introduction of impurities. It will be well worthwhile for the reader to apply our earlier discussion to the careful interpretation of Figs. 4.12 and 4.13. Although these Brouwer diagrams clearly indicate the strong influence of nonstoichiometry and the expected oxygen pressure dependence of the defect structure, we should warn again that they are largely schematic; precise values for all the necessary equilibrium constants are not available for any oxide system.

4.10 Dislocations

All the imperfections we have considered thus far are point defects. Another kind of imperfection present in real crystals is the line defect called a dislocation. These are unique in that they are never present as equilibrium imperfections for which the concentration can be calculated by thermodynamics. They may be formed in various ways but are perhaps best visualized by considering the plastic deformation of a crystal, illustrated in Fig. 4.14. Deformation occurs by relative shearing of two parts of a crystal with respect to each other along a plane, the slip plane, parallel to a plane in the lattice. If it were necessary to carry out this shearing process by one simultaneous jump of all the atoms on the slip plane, an excessively large amount of energy would be required and plastic deformation would need much higher stresses (about 10^6 psi) than are actually observed. Instead, it is believed that deformation occurs by a wavelike motion (Fig. 4.14), with the lattice distortion limited to a narrow

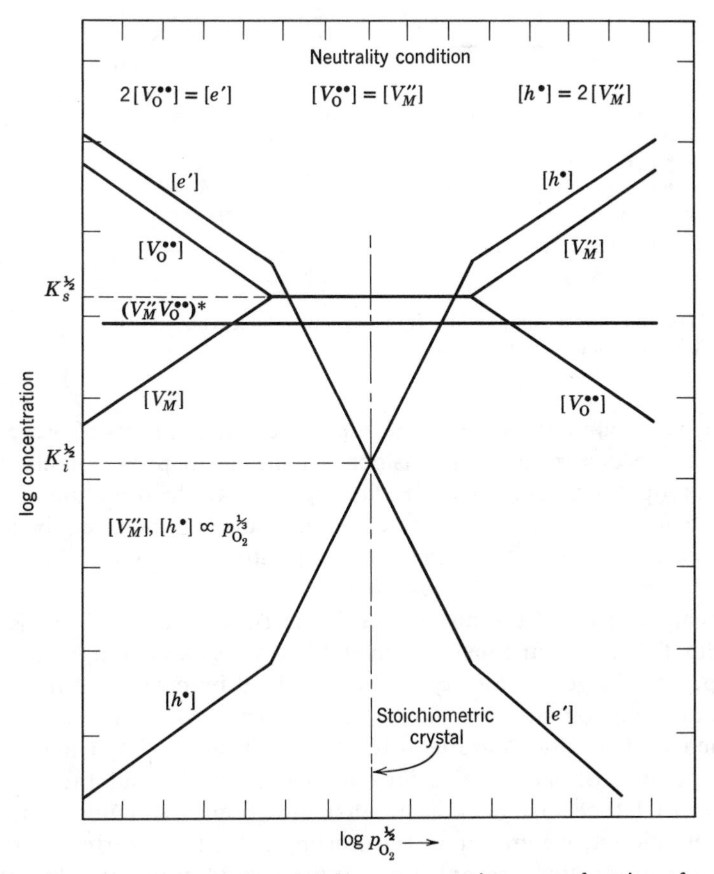

Fig. 4.13. Schematic representation of defect concentrations as a function of oxygen pressure for (*a*) a pure oxide which forms predominantly Schottky defects at the stoichiometric composition.

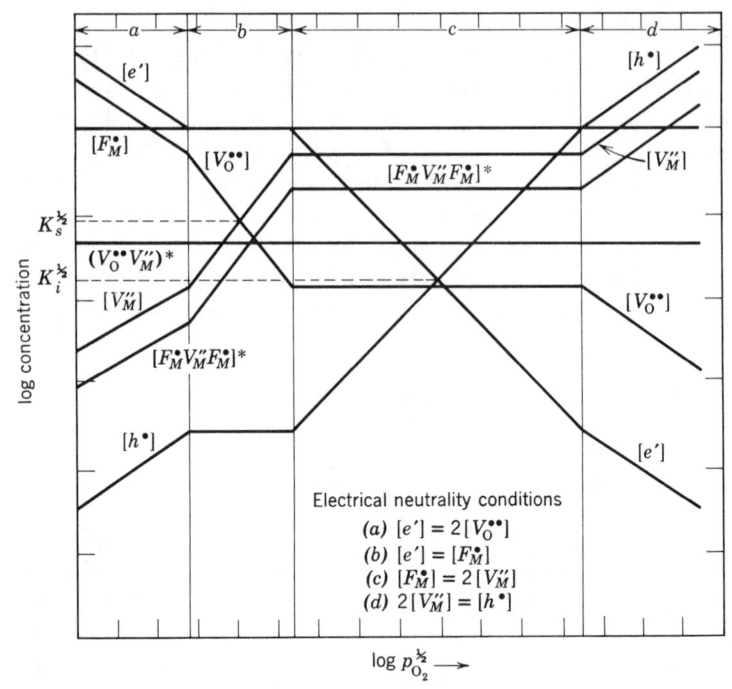

Fig. 4.13 (*contd.*) (*b*) An oxide which forms Schottky defects but contains cation impurities $[F_M^{\cdot}] > K_s^{1/2}$ [modified from (reference 8)].

region. The boundary between the slip and unslipped parts of a crystal is called the dislocation line. The dislocation line can be perpendicular to the direction of slip, an *edge* dislocation, or parallel to the direction of slip, a *screw* dislocation. The structure of an edge dislocation is equivalent to the insertion of an extra plane of atoms into the crystal. This can be illustrated by means of a soap-bubble raft (Fig. 4.15).

A characteristic of the dislocation is the Burgers vector **b**, which is a unit slip distance for the dislocation and is always parallel to the direction of slip. The Burgers vector can be determined by carrying out a circuit count of atoms on lattice positions around the dislocation, as illustrated for the two dislocations in Fig. 4.16. If we start at a point *A* and count a given number of lattice distances in one direction and then another number of lattice distances in another direction, continuing to make a complete circuit, we end up at the starting point for a perfect lattice. If there is a dislocation present, we end up at a different site. The vector between the starting point and the end point of this kind of circuit is the Burgers vector. For an edge dislocation the Burgers vector is always

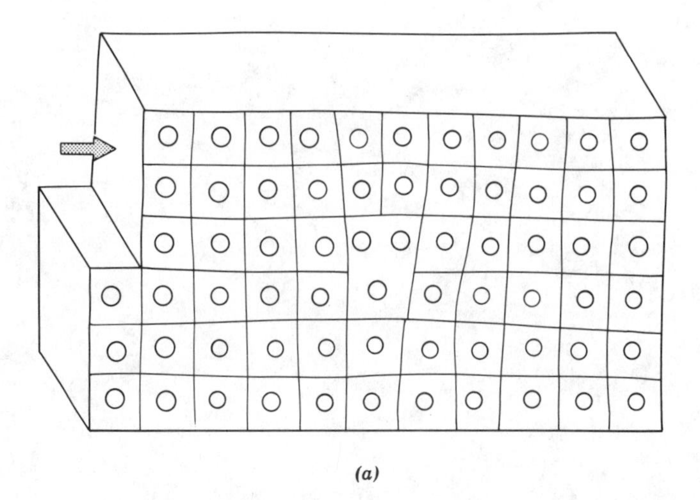

(a)

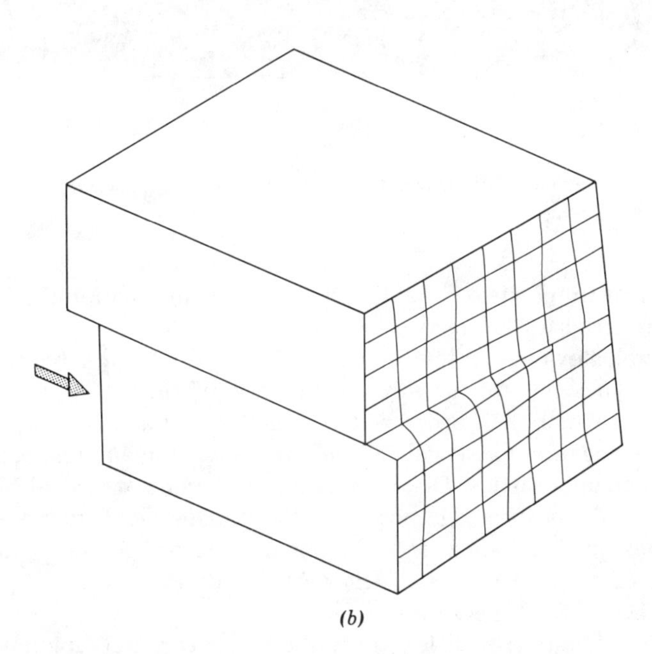

(b)

Fig. 4.14. (a) Pure edge and (b) pure screw dislocations occurring during plastic deformation.

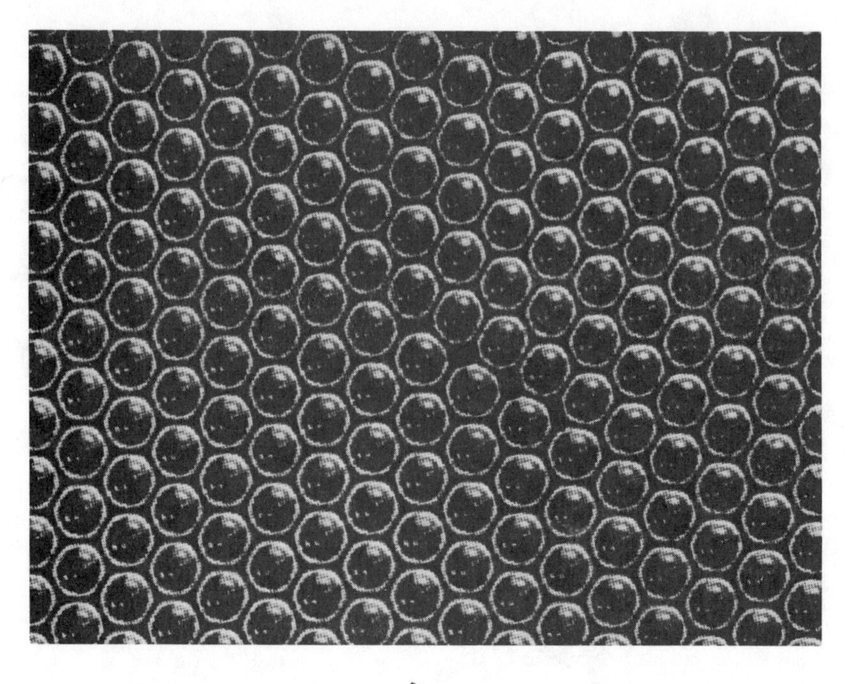

Fig. 4.15. Dislocation in a raft of soap bubbles. From W. L. Bragg and J. F. Nye, *Proc. R. Soc.* (*London*), **A190,** 474 (1947).

perpendicular to the dislocation line. For a screw dislocation the Burgers vector is parallel to the dislocation line.

In general, however, a line defect or dislocation is not restricted to these two types but can be any combination of them (Fig. 4.16). Any dislocation in which the Burgers vector is neither parallel nor perpendicular to the dislocation line is called a *mixed* dislocation and has both edge and screw characteristics. Dislocations can terminate at crystal surfaces but never inside the crystal lattice. Thus they must either form nodes with other dislocations or form a closed loop within the crystal. Such loops and nodes are often observed (Fig. 4.17). At a node the vector sum of the Burgers vectors must be zero.

The original source of dislocations in crystals is not completely clear. No dislocations are present at equilibrium, since their energy is much too great in comparison with the increase in entropy they produce. They must be introduced in a nonequilibrium way during solidification, cooling, or handling. Possible sources include thermal stresses, mechanical stresses,

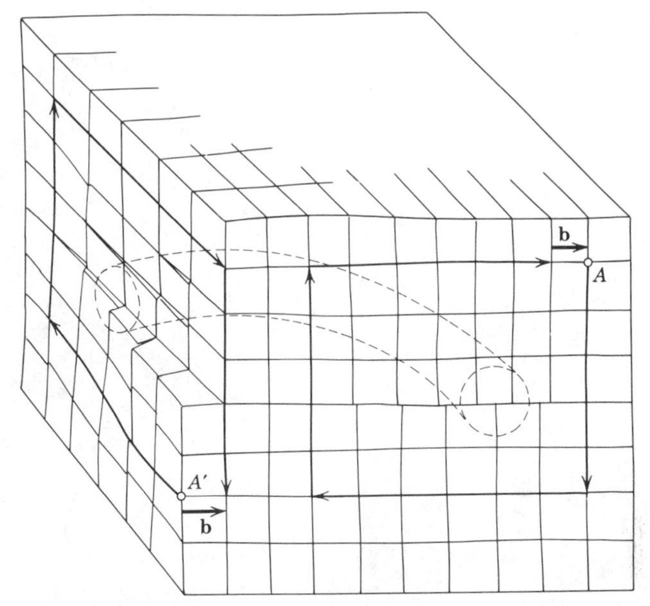

Fig. 4.16. Combination edge and screw dislocation. Burgers vector **b** shown for pure screw and for pure edge. Dislocation line connecting these is shown.

precipitation of vacancies during cooling, and growth over second-phase particles.

Crystal dislocations, which were first postulated independently to account for plastic deformation by Orowan, by Taylor, and by Polanyi in 1934, were not directly observed in real crystals until 1953. In that year precipitates formed along the dislocations (this technique is called *decoration*) were observed in silicon under infrared lighting. Etch pits formed by chemical etching where the dislocation lines touch the crystal surface were also first used to study dislocations in 1953. In the late 1950s various X-ray topographic techniques (Lang and Berg-Barrett) were developed. Transmission-electron-microscope techniques which were developed in the late 1950s provide perhaps the best means for observation. In the transmission electron microscope if the wave vector **g** of the electron beam and the Burgers vector **b** are such that $\mathbf{g} \cdot \mathbf{b} = 0$, one observes the dislocation lines disappear and thereby determines the Burgers vector. Transmission-electron-microscope and etch-pit techniques have been used to characterize dislocations and to measure their velocities resulting from applied stresses.

The concentration of dislocations is measured by the number of dislocation lines which intersect a unit area. Carefully prepared crystals

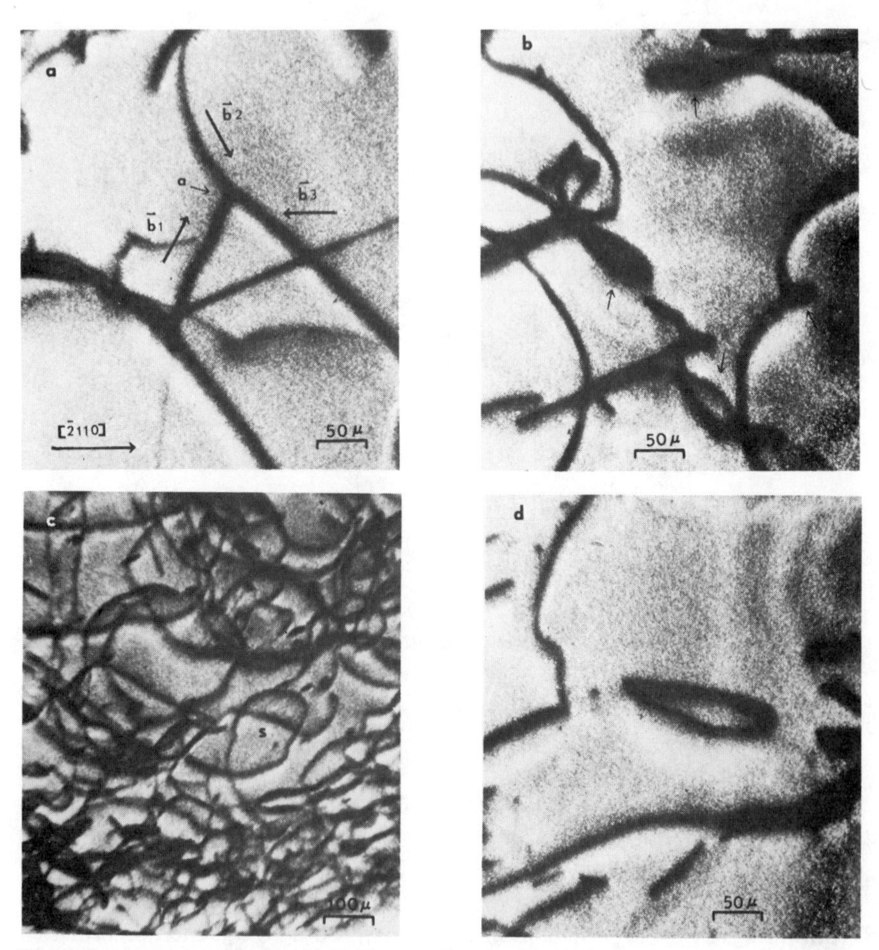

Fig. 4.17. X-ray topographs of sapphire samples representing (*a*) a node formed by three basal dislocations. The directions of Burgers vectors are denoted by arrows. (*b*) Several single helical turns indicated by the arrows. (*c*) A single spiral turn shown around the letter *s*. This type of turn is usually larger than single helical turns shown in (*b*). (*d*) A dislocation loop and cusp dislocation formed by closing a single helical turn. $2\bar{1}\bar{1}0$ reflection; CuKα radiation; traces of (2110) planes are vertical; thickness of sample: (*a*), (*b*), and (*d*) 185 μm; (*c*) 125 μm. From J. L. Caslavsky and C. P. Gazzara, *Philos. Mag.*, **26**, 961 (1972).

may contain 10^2 dislocation lines per square centimeter, and some bulk crystals and crystal whiskers have been prepared nearly free of all dislocations; after plastic deformation the concentration of dislocations increases tremendously, to 10^{10} to 10^{11} per square centimeter for some heavily deformed metals.

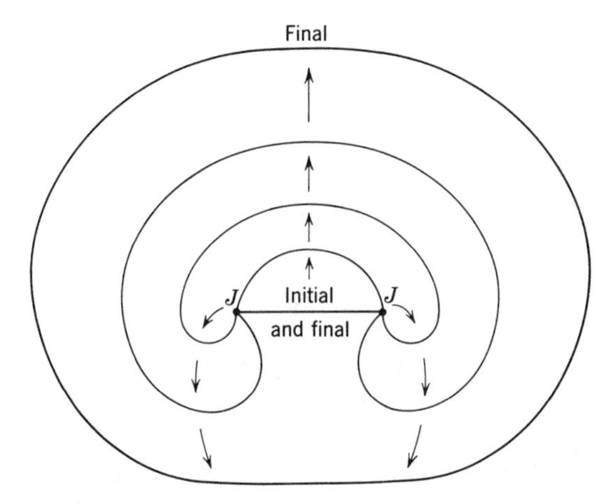

Fig. 4.18. Frank-Read mechanism for multiplying dislocations. Successive stages are shown for the generation of a dislocation loop by the pinned segment *J–J* of a dislocation line. This process can be repeated indefinitely.

Multiplication occurs when dislocations are made to move during deformation. For example, the dislocation in Fig. 4.18, pinned at two points by impurities, boundaries, or other dislocations, may be caused to move out to form a loop by an applied stress and eventually breakaway, forming a new dislocation and the original pinned segment. A segment such as this, pinned at its ends, is called a *Frank-Read source.*

Another multiplication mechanism, multiple cross glide, assumes that Frank-Read sources are generated from cross slip. This is represented schematically for a face-centered cubic crystal in Fig. 4.19. This process assumes that a screw dislocation lying along *AB* can cross glide onto position *CD* on a parallel glide plane. The composite jogs *AC* and *BD* are relatively immovable; however, the segments lying in the two slip planes are free to expand and can operate as a Frank-Read source. Multiple cross glide is a more effective mechanism than a simple Frank-Read source, since it results in more rapid multiplication of dislocations.

Just as we associate an excess energy per unit area with surfaces, an excess energy per unit length can be used to describe dislocations. Analogous to the behavior of soap bubbles in which the total surface area and thus its surface energy are reduced as much as possible, a dislocation containing a bulge straightens out and minimizes its length if free to move; a dislocation loop tends to decrease its radius and ultimately disappear. A dislocation may be considered to have a *line tension* equal to its energy per unit length.

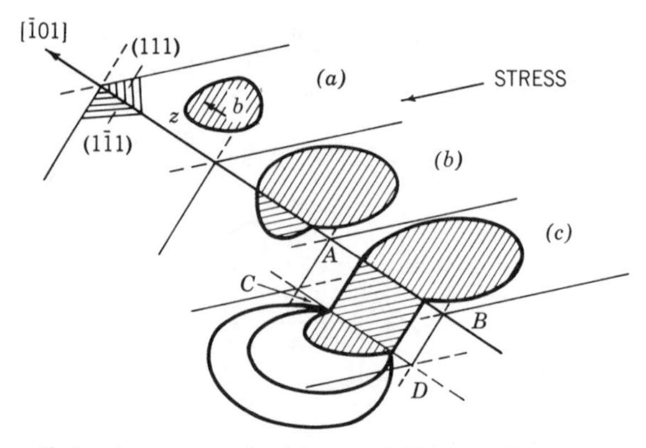

Fig. 4.19. Cross slip in a face-centered cubic crystal. The [101] direction is common to (111) and (11̄1) close-packed planes. A screw dislocation at z is free to glide in either of these planes. Cross slip produces a nonplanar slip surface. In (c) cross slip has caused a dislocation generation source at C–D. (Compare with Fig. 4.18.)

At the center of a dislocation the crystal is highly strained with atoms displaced from their normal sites. This is true to a lesser degree even some distance away from the dislocation center. At distances of more than a few interatomic distances from the dislocation center, elasticity theory can be used to obtain some useful properties of dislocations. We can consider a screw dislocation such as that illustrated in Fig. 4.20 as a distortion of a cylinder of radius r. The shear strain γ is approximately equal to the tangent of γ which is equal to $\mathbf{b}/2\pi r$, as illustrated in Fig. 4.20. If Hooke's law for elastic shear is obeyed, τ equals $G\gamma$, where G is the modulus of elasticity in shear; the shear stress is given by $\tau = G\mathbf{b}/2\pi r$. That is, the magnitude of the shear stress is proportional to $1/r$, where r is the distance from the dislocation center. Inside some limiting value r_0,

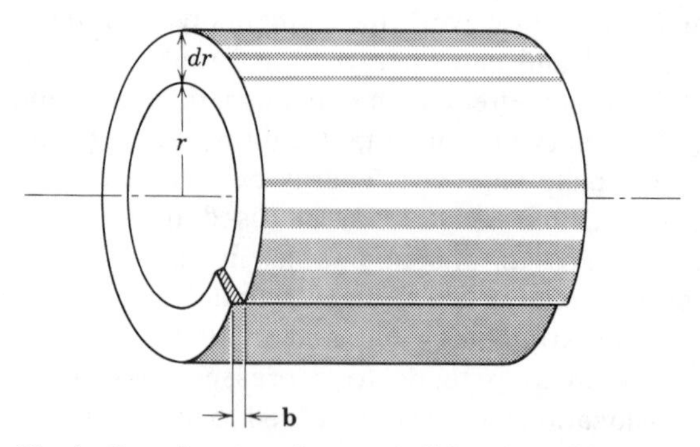

Fig. 4.20. Elastic distortion around a screw dislocation with Burgers vector $\mathbf{b}$.

Hooke's law does not hold, and the shear stress cannot be calculated. The strain energy associated with the strained region is equal to $1/2G\gamma^2$. That is, the strain energy per unit volume is given by $E = 1/2G(\mathbf{b}/2\pi r)^2$. If the distorted cylindrical shell has a thickness dr and a length l, its volume is $2\pi r\,dr\,l$, and

$$\frac{dE'}{l} = \frac{1}{2}G\left(\frac{\mathbf{b}}{2\pi r}\right)^2 \times 2\pi r\,dr = \frac{G\mathbf{b}^2}{4\pi}\frac{dr}{r} \tag{4.75}$$

and

$$E = \int_{r_0}^{r_1} \frac{dE'}{l} = \frac{G\mathbf{b}^2}{4\pi} \ln \frac{r_1}{r_0} \tag{4.76}$$

where E is the strain energy per unit length. Calculations of the strain energy for edge dislocations or mixed dislocations (edge and screw components) yield essentially the same functional dependence. Thus, an approximate relationship for the strain energy per unit length can be written:

$$E = \alpha G\mathbf{b}^2 \tag{4.77}$$

where $\alpha \approx 0.5\text{–}1.0$.

One important result is that the strain energy of a dislocation is proportional to the square of the Burgers vector **b**. This is important because it provides a criterion for what dislocations can be formed in a given crystal. Those with the smallest Burgers vector have the lowest strain energy and consequently are the most likely to form. Similar relationships hold for edge dislocations but are somewhat more complicated, since an edge dislocation is unsymmetrical. In an edge dislocation, as is clear from considering the added layer of atoms, there is a compressive stress above and a tensile stress below the dislocation line.

Many common ceramic systems contain a close-packed array of oxygen atoms. Slip in these oxide systems is usually observed in one of these close-packed directions. This is consistent with the energy required to cause strain (Eq. 4.76) because the Burgers vector in a close-packed direction is smaller; $\mathbf{b}^2$ is smaller and therefore the strain energy, also.

Dislocations in ionic materials are more complex than in elemental or metallic systems. Compare the edge dislocations for a metal and for sodium chloride in Fig. 4.21*a*. Note that in order to maintain the regularity of ions above and below the glide plane, two extra half planes of atoms are required for sodium chloride. Dislocations may also have an effective charge just as do point defects (vacancies, interstitials, impurities). This is illustrated in Fig. 4.21*b*. A jog in the dislocation results in incomplete bonding for the negative ion in the case illustrated and results in an effective charge of $-e/2$.

One place in which dislocation theory has been particularly successful

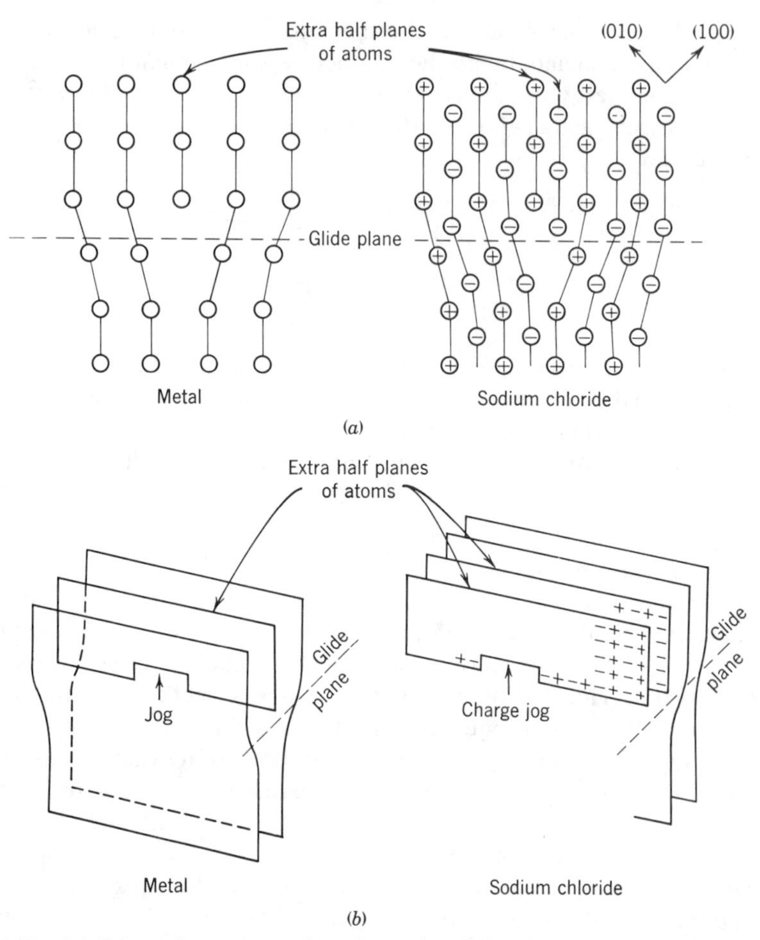

Fig. 4.21. (*a*) Schematic representation of an edge dislocation in sodium chloride; (*b*) demonstration of how dislocation jogs in ionic crystals can have effective charges.

is in describing the structure of low-angle grain boundaries. Just above an edge dislocation, as illustrated in Fig. 4.22, where an extra plane of atoms is inserted, there is a compressive stress, and below the dislocation there is a tensile stress. Consequently, dislocations of the same sign (positive for those with the extra plane inserted above the slip plane) in the slip plane tend to repel one another. Similarly, dislocations of the same sign in different slip planes tend to line up above each other to form low-angle grain boundaries (Fig. 4.22). After annealing, dislocations line up to form networks of low-angle grain boundaries. A mosaic structure results (Fig. 4.23).

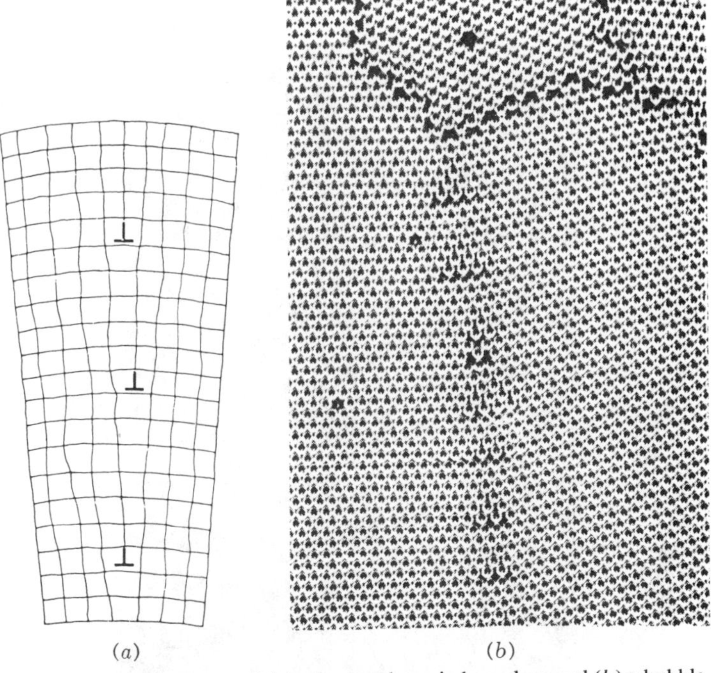

Fig. 4.22. Dislocation structure of (a) a low-angle grain boundary and (b) a bubble model. From C. S. Smith, *Metal. Prog.*, **58**, 478 (1950).

Fig. 4.23. Three-dimensional dislocation network in KCl decorated with silver particles (495×). Courtesy S. Amelinckx.

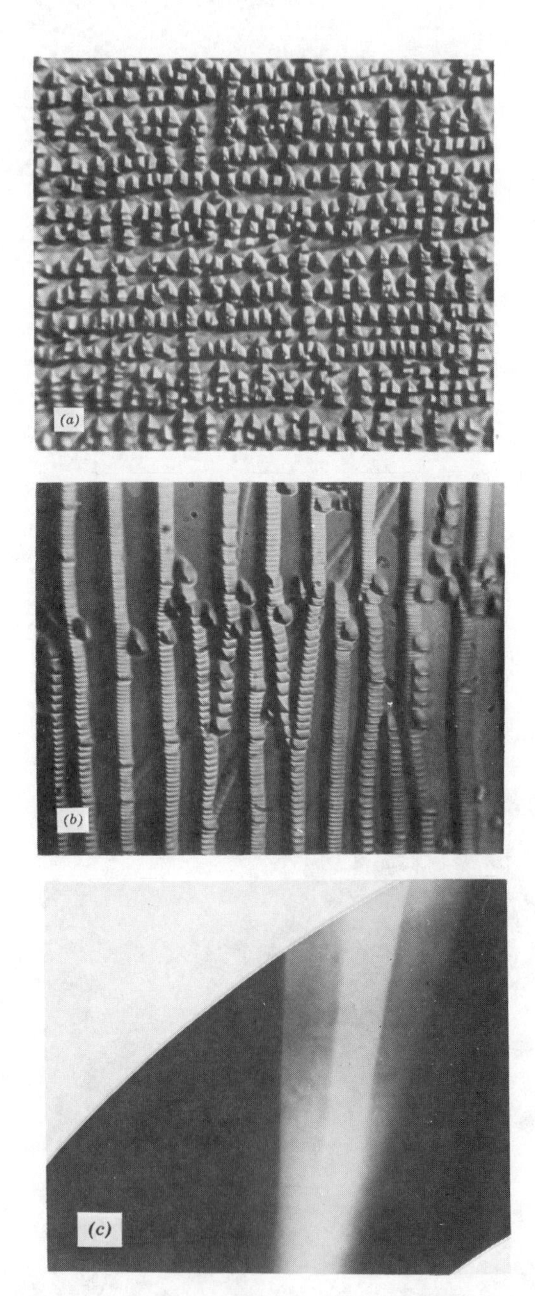

Fig. 4.24. Polygonization of Al_2O_3. (*a*) Etch pits at dislocations in bent rod. Courtesy P. Gibbs. (*b*) Dislocations lined up in polygon boundaries after annealing. Courtesy P. Gibbs. (*c*) Polygons in bent crystal viewed in polarized light. Courtesy M. Kronberg.

When a crystal is plastically deformed and then annealed, some of the dislocations introduced by the deformation process tend to line up in low-angle grain boundaries in a process called *polygonization* which has been observed for Al_2O_3, H_2O, and many metals. In Fig. 4.24 the result of bending a single crystal of aluminum oxide (sapphire) at high temperatures, which forms a greater number of positive than negative dislocations, is illustrated. Annealing leads to the lining up of the excess positive dislocations above each other in the form of low-angle grain boundaries which can be seen either with the etch-pit technique or by the different optical properties illustrated by observation of the bent crystal in polarized light.

Dislocations are particularly important in connection with plastic deformation (Chapter 14) and also in connection with crystal growth (Chapter 8) and are considered in somewhat more detail with these phenomena.

Suggested Reading

1. F. A. Kröger and V. J. Vink, "Relations between the Concentrations of Imperfections in Crystalline Solids," *Solid State Physics*, Vol. 3, F. Seitz and D. Turnbull, Eds., Academic Press, Inc., New York, 1956, pp. 307–435.
2. F. A. Kröger, *The Chemistry of Imperfect Crystals*, North-Holland Publishing Company, Amsterdam, 1964.
3. N. F. Mott and R. W. Gurney, *Electronic Processes in Ionic Crystals*, 2d ed., Clarendon Press, Oxford, 1950.
4. D. Hull, *Introduction to Dislocations*, Pergamon Press, New York, 1965.
5. F. R. N. Nabarro, *Theory of Crystal Dislocations*, Clarendon Press, Oxford, 1967.
6. H. G. Van Bueren, *Imperfections in Crystals*, North-Holland Publishing Company, Amsterdam, Interscience Publishers, Inc., New York, 1960.
7. L. W. Barr and A. B. Lidiard, "Defects in Ionic Crystals," in *Physical Chemistry*, Vol. 10, W. Jost, Ed., Academic Press, New York, 1970.
8. R. J. Brook "Defect Structure of Ceramic Materials," Chapter 3 in *Electrical Conductivity in Ceramics and Glass*, Part A, N. M. Tallen, Ed., Marcel Dekker, Inc., New York, 1974.
9. P. Kofstad, *Nonstoichiometry, Electrical Conductivity, and Diffusion in Binary Metal Oxides*, John Wiley & Sons, Inc., New York, 1972.

Problems

4.1. Assuming no lattice relaxation around vacancies, what would you predict as the P_{O_2} and T dependence of the density of (*a*) $Fe_{1-\delta}O$, (*b*) UO_{2+x}, and (*c*) $Zn_{1+x}O$.

4.2. Estimate the concentration of associates at 1000°C in ZrO_2 doped with 12 m/o CaO ($\kappa \sim 30$).

4.3. Al_2O_3 will form a limited solid solution in MgO. At the eutectic temperature (1995°C), approximately 18 wt% of Al_2O_3 is soluble in MgO. The unit-cell dimensions of MgO decrease. Predict the change in density on the basis of (*a*) interstitial Al^{3+} ions and (*b*) substitutional Al^{3+} ions.

4.4. Make a table listing the structural imperfections that occur in crystalline solids. Do not consider secondary imperfections that result from the interaction of two or more basic imperfections such as F centers. In your table make a one-sentence definition so that the instructor can evaluate your understanding of the imperfection. In a third column designate whether or not the imperfection is thermodynamically stable.

4.5. (*a*) If two parallel edge dislocations of the same sign lie on the same slip plane, that is, their half planes of atoms are parallel and terminate on the same plane perpendicular to the half planes, would there be a force of attraction or repulsion between them?

(*b*) Would this be the result of compressive, tensile, or shear forces?

(*c*) If the distance between them increased 10 times, how much would the force of interaction decrease or increase?

(*d*) If one of the edge dislocations were replaced by a screw dislocation, describe the interaction between the two line defects.

4.6. Estimate the number of free vacancies, interstitials, and associates in 1 cm^3 at 500°C of (*a*) pure AgBr and (*b*) AgBr + 10^{-4} m/o CdBr.

4.7. Construct a diagram similar to Fig. 4.8 for the AgBr data in Problem 4.6.

4.8. Estimate the electron binding energy for the reaction $Ti_{Ti} + e' = Ti'_{Ti}$ in rutile, TiO_2 ($\kappa \sim 100$).

4.9. To the schematic data in Fig. 4.13*b* add the curve for the nonneutral associate $(V''_M F^{\cdot}_M)$. Assuming that cation interstitials form, will $(V''_M F^{\cdot}_M)$ associates be larger in number than $(V''_M M^{\cdot}_i)$?

4.10. The common edge dislocation in rock salt structures is shown in Fig. 4.21. Sketch the edge dislocation in CaF_2.

4.11. Dislocations are observed in most simple oxides, and in fact crystals are difficult to prepare without large concentrations, $>10^4/cm^2$. In more complex oxides, such as garnets (e.g., yttrium aluminum garnet, $Y_3Al_5O_{12}$, gadolinium gallium garnet, $Gd_3Ga_5O_{12}$) single crystals are easily grown dislocationfree. Why?

5

Surfaces, Interfaces, and Grain Boundaries

The surfaces and interfaces between different grains and phases are important in determining many properties and processes. From one point of view these may be regarded as two-dimensional imperfections or departures from the ideal crystal lattice structure. Or the surfaces of a crystal may equally well be included in defining the ideal crystal structure for any particular environment. In either case it is desirable to understand the structure, composition, and properties of the boundaries of a solid or liquid and the interfaces between phases, for they have a strong influence on many mechanical properties, chemical phenomena, and electrical properties.

5.1 Surface Tension and Surface Energy

It is observed experimentally that a force is required to extend a liquid surface. The surface or interfacial tension γ is thus defined as the reversible work w_r required to increase the surface of the liquid by a unit area:

$$dw_r = \gamma \, dA \tag{5.1}$$

Specification of the conditions under which this surface work is done allows the surface tension to be related to other thermodynamic properties of the system. Consider the two-phase multicomponent system shown in Fig. 5.1. According to the first and second laws of thermodynamics, the variation in the internal energy E or the Gibbs free energy G of this system, in changing from one equilibrium state to another, is expressed by

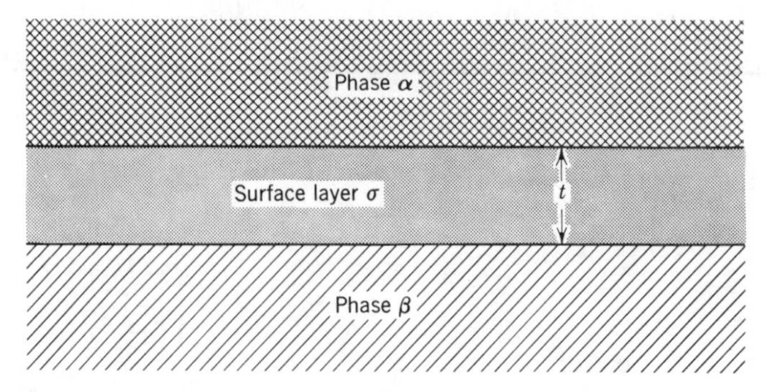

Fig. 5.1. Surface layer σ of thickness t between phases.

$$dE = T\,dS - P\,dV + \gamma\,dA + \Sigma\,\mu_i\,dn_i \tag{5.2}$$

$$dG = -S\,dT + V\,dP + \gamma\,dA + \Sigma\,\mu_i\,dn_i \tag{5.3}$$

These relations apply to systems with a plane interface separating the bulk phases; the effect of surface curvature is discussed in a later section. We can define the surface tension relative to the internal energy and the free energy as

$$\gamma = \left(\frac{\partial E}{\partial A}\right)_{S,\,V,\,n_i} = \left(\frac{\partial G}{\partial A}\right)_{P,\,T,\,n_i} \tag{5.4}$$

where the subscripts refer to the independent variables which have to remain constant during the increase of the surface area by a unit amount. Equation 5.4 may also be applied to systems which include a solid phase. Because a liquid cannot support shear stresses, its surface tension may be obtained by either reversible stretching of the existing surface or by reversible creation of a new surface. However, solids are able to sustain shearing stresses and thus oppose the attempt by surface tension to contract the surface area. The surface tension γ of a solid is therefore defined by the reversible work done in creating new surface by adding additional atoms to the surface; the work required to deform the solid surface is a measure of the surface stress, which may be either compressive or tensile and which is generally unequal to γ. For a liquid, surface tension and surface stress are numerically equal (dynes/cm = ergs/cm^2).

Probably the most familiar manifestation of surface tension is observed in the tendency of liquids to form the low energy state of minimum surface area. Soap bubbles, for example, are always spherical. The source of surface free energy may be seen by comparing the surroundings of atoms on the surface and in the interior, as illustrated in Fig. 5.2. On the surface each atom is only partly surrounded by other atoms. On bringing

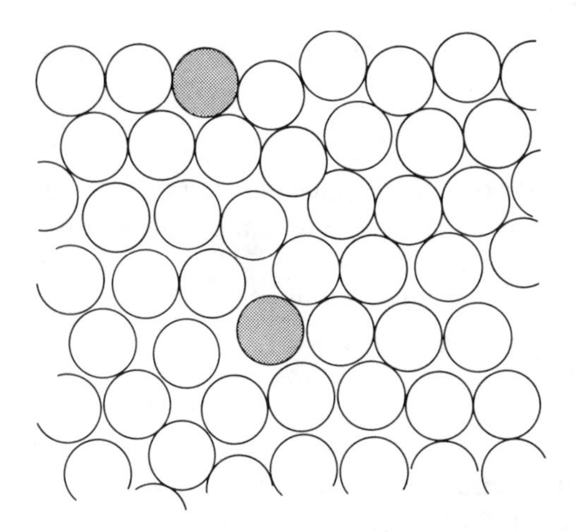

Fig. 5.2. Cross-section of a liquid illustrating the difference in surroundings of a surface atom and an interior atom.

an atom from the interior to the surface, bonds must be broken or distorted and consequently there is an increase in energy. Surface energy is defined as this increase in free energy per unit area of new surface formed (Eq. 5.4). For a finite change in surface area at constant P, T, and n_i,

$$\Delta G = \int_{A_1}^{A_2} \gamma \, dA = \gamma(A_2 - A_1) \tag{5.5}$$

In performing this integration for crystalline solid surfaces, it must be understood that the surface is created without changing the crystallographic orientation of the new surface from that of the existing surface. For crystals, γ is a function of orientation. Although the surface tension and surface free energy of solids are not in general equal, equilibrium conditions for a surface subjected to surface forces can be determined as long as there is no change in the internal strain energy, that is, when the solid is incompressible and no plastic deformation takes place. This is true, for example, when surface configurations are changed by solution, vaporization, surface diffusion, volume diffusion, and grain growth. These processes result in configurations fixed by the surface free energy. For crystalline solids, the surface free energy arises in the same way, as illustrated in Fig. 5.2. Bonds are broken and distorted when a new surface is formed, a process leading to an increase in energy. In general, different crystallographic faces have different surface energies; those surfaces that are planes of densest atomic packing are also planes of lowest surface energy and consequently are usually the most stable.

Thermodynamic Properties of the Interface. If we combine the first and second laws of thermodynamics for the surface layer in Fig. 5.1

instead of the total system, as in Eq. 5.2, we can describe surface quantities (also called excess quantities) as

$$dE = T\,dS + \gamma\,dA - P\,dV + \Sigma\mu_i\,dn_i \tag{5.6}$$

where dV is the volume of the interface layer (thickness $\times\, dA$), dS the excess entropy due to the interface, and dn_i the excess moles of i within the boundary.

Integrating, without change of composition, we have

$$E = TS + \gamma A - PV + \Sigma\mu_i n_i \tag{5.7}$$

and

$$\gamma = \frac{1}{A}(E - TS + PV - \Sigma\mu_i n_i) \tag{5.8}$$

Thus the surface tension of a flat interface is the excess Gibbs free energy per unit area. By differentiating Eq. 5.7 and comparing with Eq. 5.6, we have

$$A\,d\gamma = -S\,dT + V\,dP - \Sigma n_i\,d\mu_i \tag{5.9}$$

and for a unit area

$$d\gamma = -s\,dT + v\,dP - \Sigma\Gamma_i\,d\mu_i \tag{5.10}$$

where Γ_i is the excess moles of i per unit area of the interface layer and s and v are the excess entropy and volume per unit area. For changes at constant temperature and pressure

$$d\gamma = -\Sigma\Gamma_i\,d\mu_i \tag{5.11a}$$

For two components, this becomes

$$d\gamma = -\Gamma_1\,d\mu_1 - \Gamma_2\,d\mu_2 \tag{5.11b}$$

The terms $d\mu_1$ and $d\mu_2$ are not independent but are related through the Gibbs-Duhem equation

$$x_1\,d\mu_1 + x_2\,d\mu_2 = 0 \tag{5.12}$$

where x_1 and x_2 are the mole fractions of the two components in the phase being considered. Hence

$$-d\gamma = \left[\Gamma_2 - \frac{x_2}{x_1}\Gamma_1\right] d\mu_2 \tag{5.13}$$

This is the Gibbs adsorption isotherm and usually is written

$$-d\gamma = \Gamma_{2(1)}\,d\mu_2 \approx \Gamma_{2(1)}RT\,d\ln c_2 \tag{5.14}$$

where $\Gamma_{2(1)}$ is defined as

$$\Gamma_{2(1)} = \left(\Gamma_2 - \frac{x_2}{x_1}\Gamma_1\right) \tag{5.15}$$

and c_2 is the concentration of component 2 in the phase being considered and assumes that the activity coefficient is nearly constant at low concentration. The excess of component 2 in the interface is therefore related to the variation in the surface tension:

$$\Gamma_{2(1)} = -\frac{d\gamma}{d\mu_2} = -\frac{d\gamma}{RT\,d\ln a_2} \approx -\frac{d\gamma}{RT\,d\ln c_2} \tag{5.16}$$

Effects of Impurities. There is a strong tendency for the distribution of material to be such that the minimum surface energy results. If a small amount of a low-surface-tension component is added, it tends to concentrate in the surface layer so that the surface energy is sharply decreased with but small additions. If a high-surface-tension component is added to one of lower surface energy, it tends to be less concentrated in the surface layer than in the bulk and has only a slight influence on the surface tension. Consequently the surface energy does not change linearly with composition between end members. This is illustrated in Fig. 5.3. Measured in units of moles per square centimeter, $\Gamma_{2(1)}$ can be determined at low concentrations by plotting γ versus $\ln c_2$ and measuring the slope. For many materials that have high surface activity, this slope remains constant over a considerable composition range, the maximum of this range corresponds approximately to the adsorption of a monolayer at the surface. Particularly for high-surface-energy materials, such as metals, the effects of surface-active materials are very great. For example, oxygen and sulfur can decrease the surface tension of liquid iron from a value of about 1835 dynes/cm to 1200 dynes/cm with additions as small as 0.05%. The same is true of the effect of oxygen on solid metal, carbide,

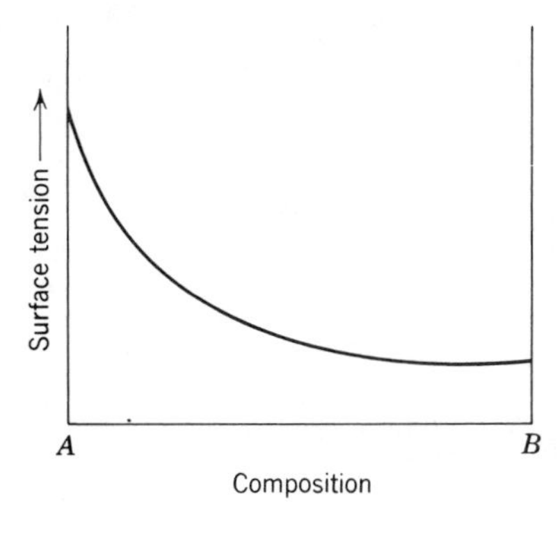

Fig. 5.3. Surface tension in a two-component system.

and nitride surfaces. The large effects of these surface-active components account for many discrepancies in the literature.

In the case of soda-lime-silicate melts, additions of alumina have been shown to increase the surface tension. The surface tensions of molten glasses, including soda-lime silicate glasses, have also been shown to depend on the atmosphere as well as on the bulk glass composition. The magnitude and even the direction of this effect depend on the temperature of measurement, since the temperature coefficient of the surface tension changes with temperature.

Values of Surface Energy. The range of values that has been observed for surface energies is large; it is about 72 ergs/cm^2 for water at room temperature and several thousand for materials such as diamond and silicon carbide. Measured values for a number of materials are given in Table 5.1. The excess free energy of a high-surface-area material is sufficient to provide the driving force for several processes of interest in ceramics. It is the driving force for the sintering of a powdered compact into a dense product.

Gradient Term of the Energy. In Fig. 5.1 we have represented the surface layer as corresponding to a step change in the composition. However, in general we must anticipate that the interface is to a certain extent diffuse, such as is schematically illustrated in Fig. 5.4. Using a bond-counting technique suggested by R. Becker,* we can derive an easy

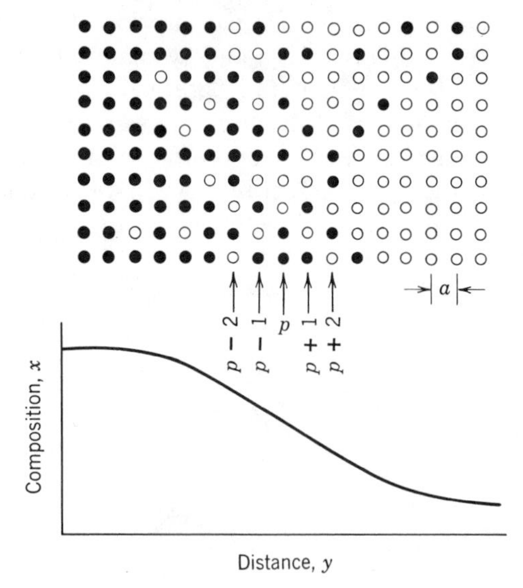

Fig. 5.4. Composition variation across a diffuse interface.

Ann. Phys., **32**, 128 (1938).

Table 5.1. Measured Surface Energies of Various Materials in Vacuo or Inert Atmospheres

Material	Temperature (°C)	Surface Energy (ergs/cm^2)
Water (liquid)	25	72
Lead (liquid)	350	442
Copper (liquid)	1120	1270
Copper (solid)	1080	1430
Silver (liquid)	1000	920
Silver (solid)	750	1140
Platinum (liquid)	1770	1865
Sodium chloride (liquid)	801	114
NaCl crystal (100)	25	300
Sodium sulfate (liquid)	884	196
Sodium phosphate, $NaPO_3$ (liquid)	620	209
Sodium silicate (liquid)	1000	250
B_2O_3 (liquid)	900	80
FeO (liquid)	1420	585
Al_2O_3 (liquid)	2080	700
Al_2O_3 (solid)	1850	905
0.20 Na_2O–0.80 SiO_2	1350	380
0.13 Na_2O–0.13 CaO–0.74 SiO_2 (liquid)	1350	350
MgO (solid)	25	1000
TiC (solid)	1100	1190
CaF_2 crystal (111)	25	450
$CaCO_3$ crystal (1010)	25	230
LiF crystal (100)	25	340

conceptual model for the excess energy at the diffuse interface E_s. This is similar to a regular solution model but applied to the interface. For a system containing components A and B, if the bond energy for A–A bonds is ϵ_{AA}, for B–B bonds is ϵ_{BB}, and for A–B bonds is ϵ_{AB}, then the pure phases are stable if $2\epsilon_{AB} > \epsilon_{AA} + \epsilon_{BB}$. If we define atomic planes parallel to the interface having a composition (atom fraction of A) x_p, x_{p+1}, and so on, we can first calculate the energy of one atomic plane bounded by neighboring planes of the same composition, that is,

$$E_{x_p} = z\left(\frac{m}{2}\right)[x_p x_p \epsilon_{AA} + (1 - x_p)(1 - x_p)\epsilon_{BB} + 2x_p(1 - x_p)\epsilon_{AB}] \tag{5.17a}$$

where z is the number of bonds per atom between planes and m is the atom density per unit interface area. A similar expression holds for the x_{p+1} plane. The summation of bond energy between the planes of differing composition x_p and x_{p+1} is

$$E_{x_p - x_{p+1}} = z\left(\frac{m}{2}\right)\{x_p x_{p+1}\epsilon_{AA} + (1 - x_p)(1 - x_{p+1})\epsilon_{BB} + [x_p(1 - x_{p+1}) + x_{p+1}(1 - x_p)]\epsilon_{AB}\} \tag{5.17b}$$

and the extra energy resulting from the compositional gradient relative to the average homogeneous composition is given by

$$\begin{aligned} E_s &= 2E_{x_p - x_{p+1}} - (E_{x_p} + E_{x_{p+1}}) \\ &= z\left(\frac{m}{2}\right)(2\epsilon_{AB} - \epsilon_{AA} - \epsilon_{BB})(x_p - x_{p+1})^2 \end{aligned} \tag{5.18}$$

If $(x_p - x_{p+1})/a_0$ represents the compositional gradient $\frac{\partial c}{\partial y}$ and $\nu = (2\epsilon_{AB} - \epsilon_{AA} - \epsilon_{BB})$ is the interaction energy, the excess surface energy is given as

$$E_s = z\left(\frac{m}{2}\right)a_0^2\nu\left(\frac{\partial c}{\partial y}\right)^2 \tag{5.19}$$

where a_0 is the interplanar spacing.

A rigorous and more general derivation by Cahn and Hilliard* also gives the free energy of a small volume of nonuniform solution as the sum of two contributions, one the free energy that this volume would have in a homogenous solution g_0 and the other the gradient energy which is a function of the local composition

$$G = N_v \int [g_0 + \kappa(\nabla c)^2 + \cdots]\, dV \tag{5.20}$$

where N_v is the number of molecules per unit volume and:

$$\nabla c = \frac{\partial c}{\partial x} + \frac{\partial c}{\partial y} + \frac{\partial c}{\partial z}$$

and κ is related to derivatives of the free energy with respect to composition. That is, the energy contribution of a diffuse interface increases with the square of the magnitude of the concentration gradient $(\nabla c)^2$.

J. Chem. Phys., **28**, 258 (1958).

5.2 Curved Surfaces

Pressure Difference across a Curved Surface. Many of the important effects of surfaces and interfaces arise from the fact that surface energy causes a pressure difference across a curved surface. This may be seen by considering a capillary inserted in a liquid bath from which a bubble is blown, as illustrated in Fig. 5.5. If the density difference (and consequently the gravitational effect) is negligible, the only resistance to expansion of the bubble is the increased surface area being formed and the increased total surface energy. At equilibrium the work of expansion $\Delta P\, dv$ must equal the increased surface energy $\gamma\, dA$, and

$$\Delta P\, dv = \gamma\, dA \tag{5.21}$$

$$dv = 4\pi r^2\, dr \qquad dA = 8\pi r\, dr \tag{5.22}$$

$$\Delta P = \gamma \frac{dA}{dv} = \gamma \frac{8\pi r\, dr}{4\pi r^2\, dr} = \gamma \left(\frac{2}{r}\right) \tag{5.23}$$

For the more general shape, when the surface is not spherical, similar reasoning leads to the result

$$\Delta P = \gamma \left(\frac{1}{r_1} + \frac{1}{r_2}\right) \tag{5.24}$$

where r_1 and r_2 are the principal radii of curvature.

It is this pressure difference that causes liquid to rise in a capillary. For example, as shown in Fig. 5.6, the pressure difference due to the surface energy is balanced by the hydrostatic pressure of the liquid column. If the

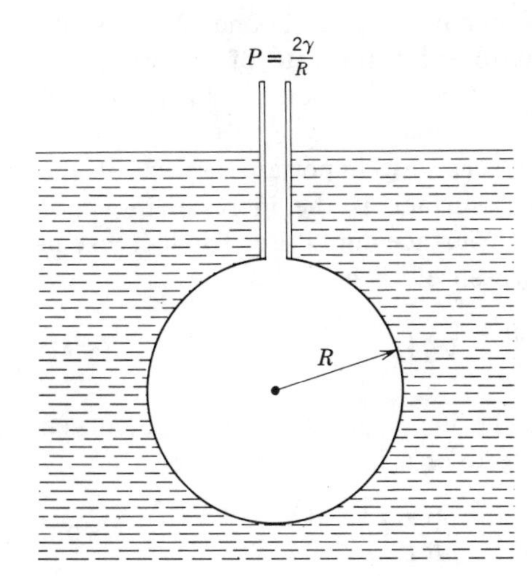

Fig. 5.5. Determination of the pressure at equilibrium to maintain a spherical surface of radius R.

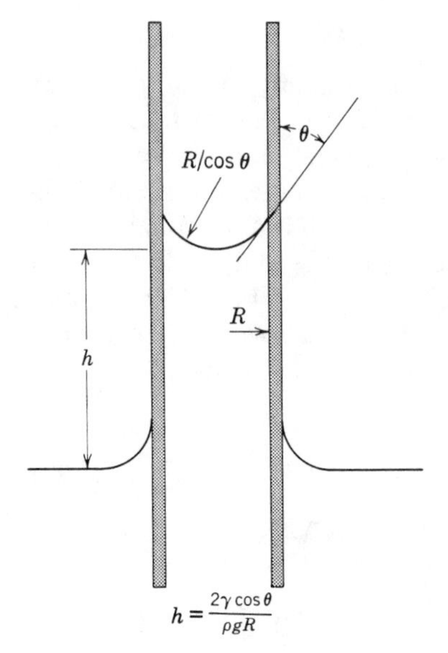

Fig. 5.6. Liquid rise in a capillary. Radius of curvature of liquid surface is $R/\cos\theta$.

capillary radius is R,

$$\Delta P = \gamma\left(\frac{2}{r}\right) = \gamma\left(\frac{2\cos\theta}{R}\right) = \rho g h \tag{5.25}$$

$$\gamma = \frac{R\rho g h}{2\cos\theta} \tag{5.26}$$

and the surface energy can be determined from the capillary rise if the contact angle θ is known. As will be seen in Chapters 11 and 12, this same pressure difference leads to expressions for the rate of sintering and vitrification during the heat treatment of ceramic bodies and is important in connection with grain growth phenomena.

Vapor Pressure of a Curved Surface. An important aspect of the pressure difference across a curved surface is the increase in vapor pressure or solubility at a point of high surface curvature. The increase in vapor pressure due to an applied pressure ΔP is

$$V\,\Delta P = RT\ln\frac{p}{p_0} = V\gamma\left(\frac{1}{r_1}+\frac{1}{r_2}\right) \tag{5.27}$$

where V is the molar volume, p is the vapor pressure over the curved surface, and p_0 is the vapor pressure over a flat surface. Then

$$\ln\frac{p}{p_0} = \frac{V\gamma}{RT}\left(\frac{1}{r_1}+\frac{1}{r_2}\right) = \frac{M\gamma}{\rho RT}\left(\frac{1}{r_1}+\frac{1}{r_2}\right) \tag{5.28}$$

where R is the gas constant, T temperature, M the molecular weight, and ρ the density.

This same relationship can also be derived by considering the transfer of one mole of material from a flat surface through the vapor phase to a spherical surface. The work done must be equal to the surface energy and the change in the surface area. That is,

$$RT \ln \frac{p}{p_0} = \gamma \, dA = \gamma 8\pi r \, dr \tag{5.29}$$

Since the change in volume is $dv = 4\pi r^2 \, dr$, the radius change for a one-mole transfer is $dr = V/4\pi r^2$, and

$$\ln \frac{p}{p_0} = \frac{V\gamma}{RT}\left(\frac{2}{r}\right) \tag{5.30}$$

which is the same result as obtained previously. The pressure which is developed across a curved surface and the resultant increase in vapor pressure or solubility may be substantial for small-particle materials, as indicated in Table 5.2.

The strong effect of particle size in these relations is one of the bases for the use of clay minerals in ceramic technology. Their fine particle size aids in fabrication processes, since it is a source of plasticity. In addition, this fine particle size produces surface-energy forces which cause densifi-

Table 5.2. Effect of Surface Curvature on Pressure Difference and Relative Vapor Pressure across a Curved Surface

Material	Surface Diameter (microns)	Pressure Difference (psi)	Relative Vapor Pressure (p/p_0)
Silica glass	0.1	1750	1.02
(1700°C)	1.0	175	1.002
γ = 300 ergs/cm^2	10.0	17.5	1.0002
Liquid cobalt	0.1	9750	1.02
(1450°C)	1.0	975	1.002
γ = 1700 ergs/cm^2	10.0	97.5	1.0002
Liquid water	0.1	418	1.02
(25°C)	1.0	41.8	1.002
γ = 72 ergs/cm^2	10.0	4.18	1.0002
Solid Al_2O_3	0.1	5250	1.02
(1850°C)	1.0	525	1.002
γ = 905 ergs/cm^2	10.0	52.5	1.0002

cation during the firing process. Nonclay materials which are not naturally fine-grained must be ground or otherwise treated to give them the particle sizes in the micron range that are necessary for satisfactory firing.

5.3 Grain Boundaries

One of the simplest kinds of interface is the boundary between two crystals of the same material. If two crystals of exactly the same orientation are brought together, they fit perfectly. This can be done by splitting mica sheets in a vacuum and then fitting them back together. The resulting crystal cannot be distinguished from one that has not been so treated. However, if the crystals are slightly tilted and brought back together, there is a disregistry at the interface equivalent to the insertion of a row of dislocations, as illustrated in Chapter 4. This row of dislocations can be observed experimentally by etching techniques.

If the angle of disregistry is small (small-angle boundaries), the boundary consists of regions of perfect fit and regions of misfit which result in the formation of dislocations (Fig. 5.7). Examination of Fig. 5.7 shows that the dislocation count should be simply related to the misorientation angle θ. This is true for low-angle tilt boundaries (edge dislocations) or low-angle twist boundaries (screw dislocations). Geometrically for the tilt boundary in Fig. 5.7, this is

$$D = \frac{\mathbf{b}}{\sin\theta} \approx \frac{\mathbf{b}}{\theta} \qquad \text{for small } \theta \tag{5.31}$$

where D is the dislocation spacing, $\mathbf{b}$ the Burgers vector, and θ the angle of disregistry. The strain energy of a dislocation is the sum of the elastic energy and the core energy ($E = E_{el} + E_{core}$). In Chapter 4 we calculated the elastic term, and we add a term for the core energy B

$$E_{edge} = \frac{G\mathbf{b}^2}{4\pi(1-\nu)} \ln\frac{R}{\mathbf{b}} + B \tag{5.32}$$

where E_{edge} is the energy per unit length. Since R is the distance the elastic field extends away from the core, it is equal to the dislocation spacing D. Thus the elastic term of a pure tilt boundary is the energy per unit area E/D

$$\frac{E}{D} = E' = \frac{G\mathbf{b}^2}{D4\pi(1-\nu)} \ln\frac{D}{\mathbf{b}} + \frac{B}{D} = \frac{G\mathbf{b}\theta}{4\pi(1-\nu)} \ln\frac{1}{\theta} + \frac{B\theta}{\mathbf{b}} = E_0[A - \ln\theta]\theta \tag{5.33}$$

where E_0 and A are constants given by

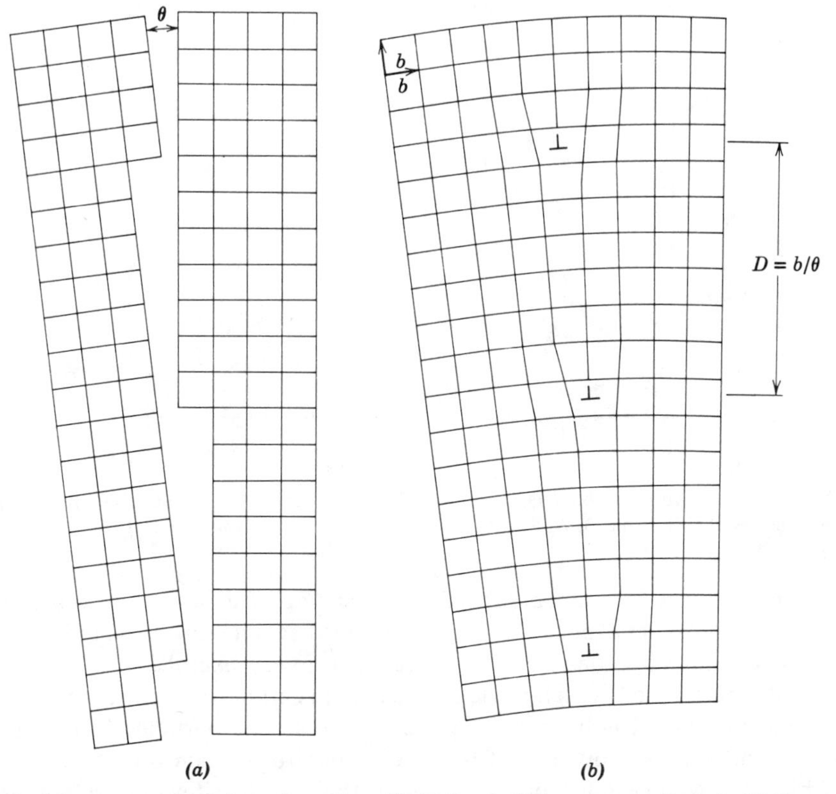

Fig. 5.7. Low-angle tilt boundary. From W. T. Read, *Dislocations in Crystals*, McGraw-Hill Book Company, New York, 1953.

edge ⊥

$$E_0 = \frac{G\mathbf{b}}{4\pi(1-\nu)} \qquad A = \frac{4\pi(1-\nu)B}{G\mathbf{b}^2} \tag{5.34}$$

An equation identical in form to Eq. 5.33 is obtained for twist boundaries (screw dislocations), where E_0 and A become

screw ⊥

$$E_0 = \frac{G\mathbf{b}}{2\pi} \qquad A = \frac{2\pi B}{G\mathbf{b}^2} \tag{5.35}$$

Figure 5.8 is a plot of the relative grain-boundary energy of NiO for various tilt angles. The solid curve represents Eq. 5.33. Up to about 22° the data are reproducible, and the energy increases rapidly with tilt angle; above 22° the energy remains nearly constant. Equation 5.33 was derived for low-angle boundaries, and the model is only applicable where there is an appreciable spacing between dislocations; although higher-angle

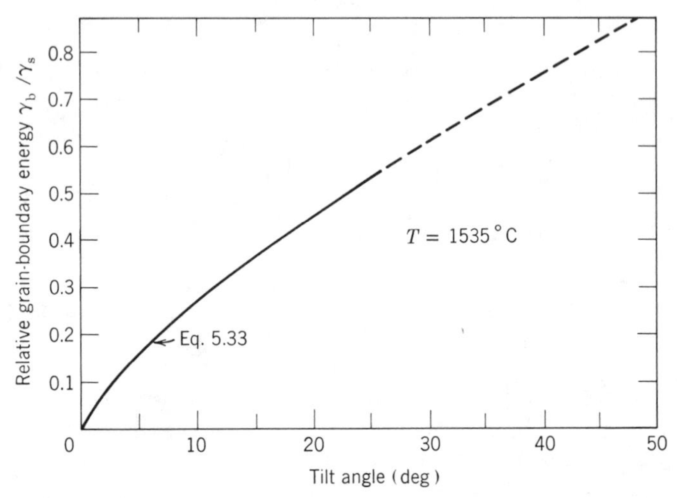

Fig. 5.8. Relative grain-boundary energy versus tilt angle for symmetrical [001] tilt boundaries in NiO. From D. W. Readey and R. E. Jech, *J. Am. Ceram. Soc.*, **51**, 201 (1968).

boundaries can be conceptualized as consisting of dislocations, a set of dislocations spaced only one or two atom distances apart would have such special properties that the model is not very useful.

Data are sparse for high-angle boundaries in ceramic systems, but when magnesium metal is burned in air, the oxide smoke particles form with twist boundaries showing a strong preference for certain orientations in which the two crystals have a fraction of their lattice sites in common, as shown in Fig. 5.9. Grain boundaries between crystals with lattices related in this way are called *coincidence boundaries.* For an ionic crystal such as MgO a possible representation of the tilt coincidence boundary at 36.8°, a {310} twin, is shown in Fig. 5.10. The boundary consists of repeated structural units having dimensions of a few or several atomic distances. Boundaries which deviate from an exact coincidence relationship can be described in terms of the repeated structural coincidence unit combined with dislocations in the coincidence lattice of the grain boundary, as illustrated for a simple cubic lattice in Fig. 5.11.

5.4 Grain-Boundary Potential and Associated Space Charge

J. Frenkel* and K. Lehovec† first showed that in thermodynamic equilibrium the surface and grain boundaries of an ionic crystal may carry

Kinetic Theory of Liquids, Oxford University Press, Fair Lawn, N.J., 1946, p. 37.
†*J. Chem. Phys.*, **21**, 1123 (1953).

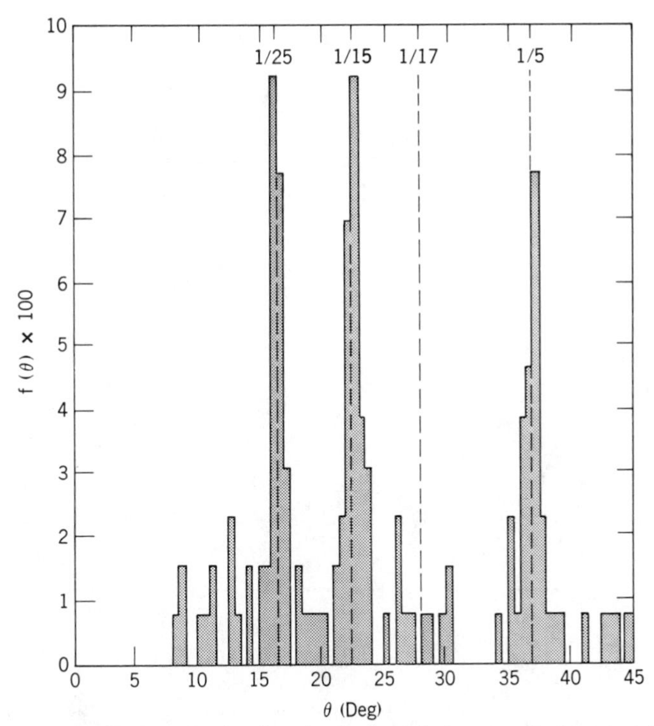

Fig. 5.9. Histogram of $f(\theta)$, the fraction of twist boundaries at angle θ, against θ, for MgO smoke particles collected above hot and rapidly burning magnesium rods. Twist boundaries corresponding to $\theta = 0$ are not included. The histogram is drawn for 1/2° intervals. From P. Chandhari and I. W. Matthews, *J. Appl. Phys.*, **42**, 3063 (1971).

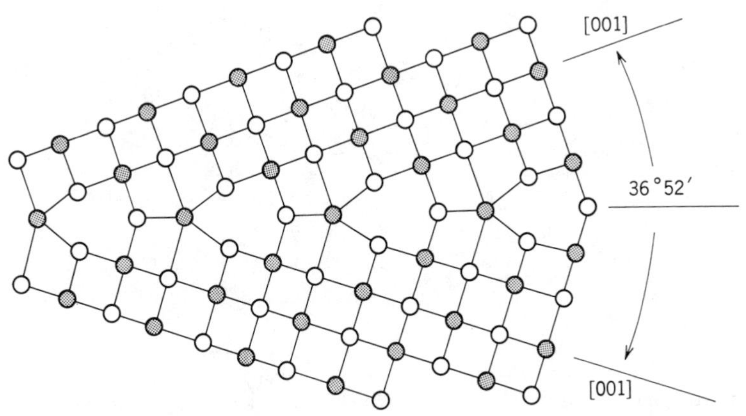

Fig. 5.10. Representation of a possible 36.8° tilt boundary (310) twin in NaCl or MgO.

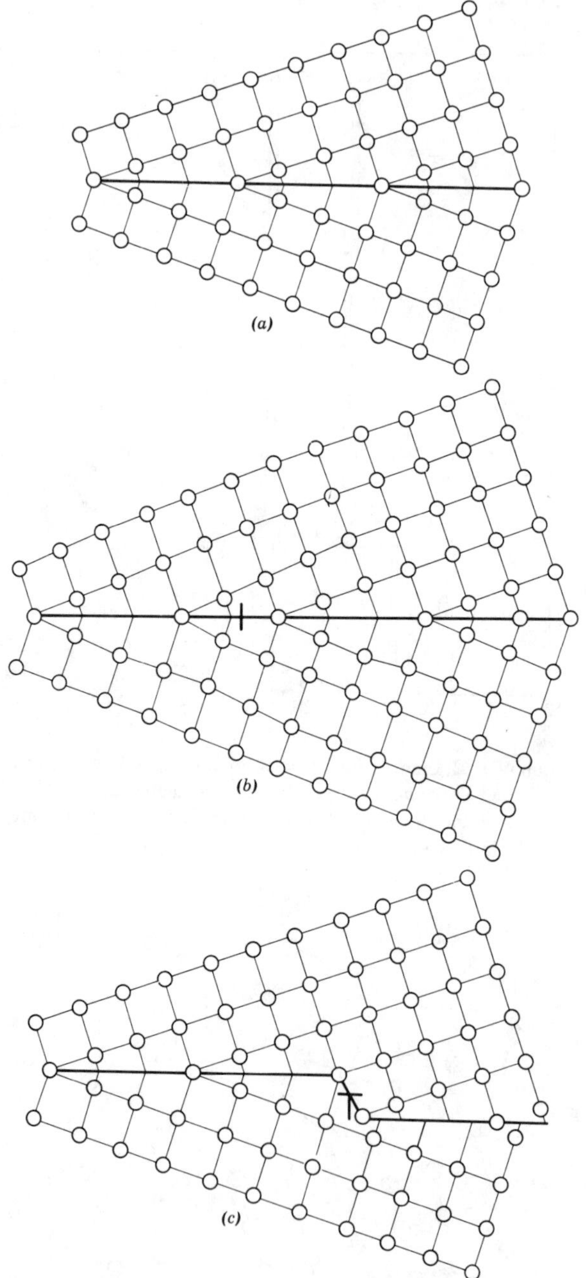

Fig. 5.11. Representation of (*a*) perfect 36.8° tilt boundary (310) twin in a simple cubic crystal and grain-boundary dislocations at (*b*) slightly rotated interface and (*c*) boundary ledge.

an electric charge resulting from the presence of excess ions of one sign; this charge is just compensated by a space-charge cloud of the opposite sign adjacent to the boundary. For a pure material this charge arises if the energies to form anion and cation vacancies or interstitials at the boundary are different; the magnitude and sign of the boundary charge changes if there are aliovalent solutes present which alter the concentrations of lattice defects in the crystal. Many extensions of the theory from various different approaches have been reported; from the point of applying the theory to ceramic oxides, similar conclusions result.

In the simplest formulation, certainly incorrect in detail, it is assumed that surfaces and boundaries act as an infinite uniform source and sink for vacancies. For an ideal pure material containing Schottky defects, there is usually a difference in the free energy of formation at the grain boundary for the anion and cation vacancies. Unfortunately it is not possible to measure these quantities separately or even to estimate them reliably for oxides, but for NaCl the energy required to form a cation vacancy is perhaps two-thirds of the energy required to form an anion vacancy. We can visualize the result as a tendency on heating for an excess of cation vacancies with an effective negative charge to be generated at the grain boundaries and other vacancy sources (surfaces, dislocations); the resulting space charge slows the emission of further cation vacancies and speeds the emission of anion vacancies. At equilibrium (independent of the imagined process) there is electrical neutrality in the bulk of the crystal but a positive charge on the boundary balanced by an equal and opposite negative space-charge cloud penetrating some distance into the crystal.

For the interaction of a lattice ion with the boundary to form a vacancy in a crystal such as NaCl we can write

$$\mathrm{Na_{Na}} = \mathrm{Na}^{\cdot}_{\mathrm{Boundary}} + V'_{\mathrm{Na}} \tag{5.36}$$

$$\mathrm{Cl_{Cl}} = \mathrm{Cl}'_{\mathrm{Boundary}} + V^{\cdot}_{\mathrm{Cl}} \tag{5.37}$$

That is, simultaneously with the formation of an excess of vacant lattice sites of one sign there is a change in the ratio of occupied boundary sites. The assumption that the boundary is a perfect source and sink for vacancies in the following calculation is equivalent to supposing that there are an unlimited number of surface sites of equal energy for the excess sodium and chlorine ions. Various experimental observations, for example, the difficulty of etching pure twist boundaries, clearly indicate that this assumption is not generally true.

At any point in the crystal the number of cation and anion vacancies per lattice site is determined both by the intrinsic energy of formation

$(g_{V_M'}, g_{V_X^{\cdot}})$ and by the effective charge z and electrostatic potential ϕ, that is,

$$[V_M'] = \exp\left[-\frac{(g_{V_M'} - ze\phi)}{kT}\right] \qquad (5.38)$$

$$[V_X^{\cdot}] = \exp\left[-\frac{(g_{V_X^{\cdot}} + ze\phi)}{kT}\right] \qquad (5.39)$$

Far from a surface electrical neutrality requires that $[V_M']_\infty = [V_X^{\cdot}]_\infty$, and the concentration is given by the total energy of formation

$$[V_M']_\infty = [V_X^{\cdot}]_\infty = \exp\left[-\frac{1}{2}\frac{(g_{V_M'} + g_{V_X^{\cdot}})}{kT}\right] \qquad (5.40)$$

$$[V_M']_\infty = \exp\left[-\frac{(g_{V_M'} - ze\phi_\infty)}{kT}\right] \qquad (5.41)$$

$$[V_X^{\cdot}]_\infty = \exp\left[-\frac{(g_{V_X^{\cdot}} + ze\phi_\infty)}{kT}\right] \qquad (5.42)$$

As a result the electrostatic potential at the interior of the crystal is

$$ze\phi = \frac{1}{2}(g_{V_M'} - g_{V_X^{\cdot}}) \qquad (5.43)$$

with a space charge extending a depth that depends on the dielectric constant, typically 20 to 100 Å from the boundary. An estimate for NaCl, taking $g_{V_M'} = 0.65$ eV and $g_{V_X^{\cdot}} = 1.21$ eV gives $\phi_\infty = -0.28$ V; parallel (probably incorrect) assumptions for MgO would give $\phi_\infty \sim -0.7$ V. Thus, the electrostatic potentials being discussed are not trivial. Physically, this corresponds (for NaCl, where $g_{V_M'} < g_{V_X^{\cdot}}$) to an excess of cations located at the boundary giving it a positive charge, together with a space-charge region containing an excess of cation vacancies and a deficit of anion vacancies, such as illustrated in Fig. 5.12*a*. As a result, equilibrium boundaries, even in the purest materials, require vacancy or interstitial ion equilibration within the crystal.

If a concentration C_s of aliotropic solute is present, such as $CaCl_2$ in NaCl or Al_2O_3 in MgO, additional cation vacancies are formed; for additions which are large relative to the thermally generated vacancies, Eq. 5.41 still holds, and

$$\ln C_s \simeq \ln [V_M']_\infty = -\frac{g_{V_M'}}{kT} + \frac{ze\phi_\infty}{kT} \qquad (5.44)$$

As a result, according to Eq. 5.44, the sign and magnitude of the electrostatic potential at a boundary are determined by both the solute content and temperature, as shown in Fig. 5.13*a*.

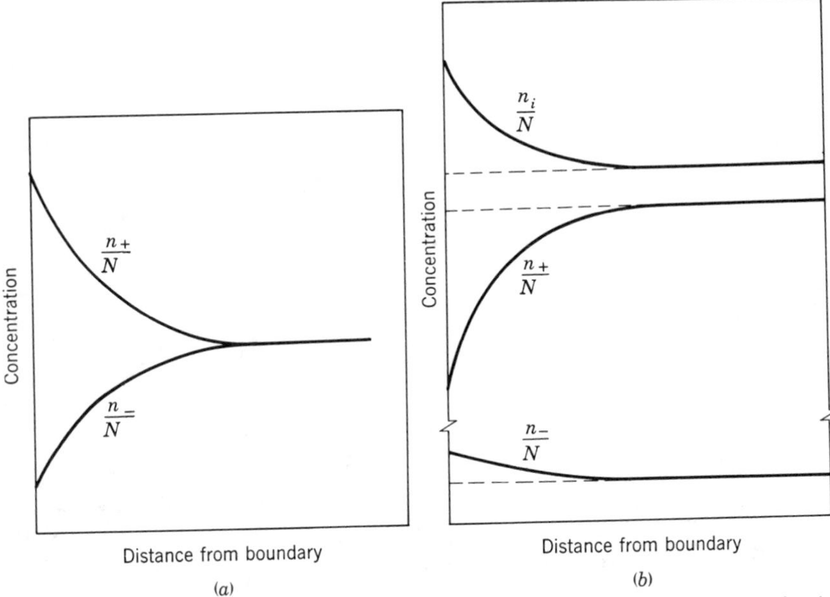

Fig. 5.12. Boundary space charge and associated charged defect concentrations for (*a*) pure NaCl and (*b*) NaCl containing aliovalent solute with a positive effective charge. From K. L. Kliewer and J. S. Koehler, *Phys. Rev.*, **140**, **4A**, 1226 (1965).

This result can also be found, for $CaCl_2$ in NaCl as a typical example, by combining the vacancy concentration generated by aliovalent solution

$$CaCl_2 \xrightarrow[\text{NaCl}]{} Ca^{\cdot}_{Na} + V'_{Na} + 2Cl_{Cl} \tag{5.45}$$

with the Schottky equilibrium

$$\text{null} \rightleftarrows V'_{Na} + V^{\cdot}_{Cl} \tag{5.46}$$

and Eqs. 5.36 and 5.37. Increasing $[V'_{Na}]$ by calcium additions decreases $[V^{\cdot}_{Cl}]$ according to Eq. 5.46, decreases $[Na^{\cdot}_{boundary}]$ according to Eq. 5.36, and increases $[Cl'_{boundary}]$ according to Eq. 5.37, leading to a negative boundary potential (positive ϕ_∞).

Since solute concentrations are much larger in oxide systems than thermally generated vacancy concentrations, even at high temperature, the solute-controlled situation is to be expected for all practical systems. The impurity and vacancy concentration profiles to be expected are illustrated for NaCl in Fig. 5.12*b*. In addition to the change in sign, the boundary charge is temperature-dependent when controlled by impurities and exists independently of the relative values of $g_{V'_M}$ and $g_{V^{\cdot}_X}$.

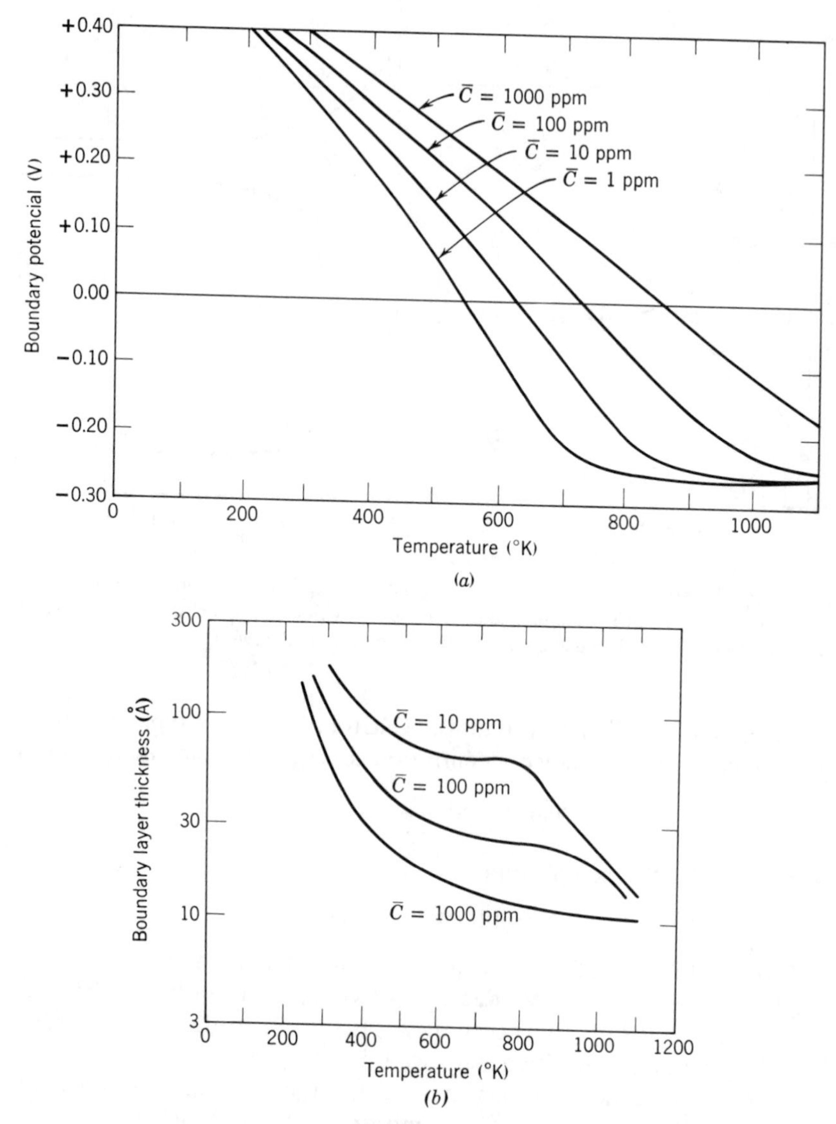

Fig. 5.13. (*a*) Temperature dependence of boundary charge for NaCl containing various concentrations of aliovalent solute; (*b*) temperature dependence of space-charge width for NaCl containing various concentrations of aliovalent solute. From K. L. Kliewer and J. S. Koehler, *Phys. Rev.*, **140**, **4A**, 1226 (1965).

For sodium chloride, assuming a reasonable value of 0.4 eV for the binding energy between a cation vacancy and divalent solute, the temperature dependence of the potential and of the space-charge thickness is shown in Fig. 5.13. Values for the space-charge thickness in the temperature range of boundary-migration processes and in the concentration range appropriate to good-quality ceramics are 20 to 100 Å, increasing at lower temperatures.

Since only charged vacancies and not vacancy-vacancy pairs or vacancies associated with solute atoms to form neutral pairs are affected by the field, the binding energies for vacancy-solute pair formation are important variables. In addition, the free energy of formation of vacancies at surfaces, boundaries, and dislocations of different types and orientations need not be identical, and the charge on an interface or dislocation depends on the difference in properties at a particular interface or dislocation. For oxide materials there are not reliable data or calculations for the separate energy of formation of anion and cation imperfections, there are not reliable data or calculations for solute-vacancy or vacancy-vacancy pair formation, and for most systems studied at elevated temperature electrical measurements leave in doubt the extent of defect ionization; this, of course, affects behavior in any boundary-associated electrical field. As a result, quantitative application of the theory seems far off.

The existence and sign of the boundary potential and associated space charge have been demonstrated for NaCl, Al_2O_3, and MgO. In oxides, because of the low concentrations of thermally induced lattice defects (Tables 4.1 and 4.2) the potential at the boundary and associated space charge are determined by aliovalent solute concentrations. (Positive boundary for Al_2O_3 containing MgO as a solute; negative boundary for MgO containing Al_2O_3 or SiO_2 as a solute.)

5.5 Boundary Stresses

In most ceramic systems a powder composition is heated at a high temperature to develop increased strength and density and then cooled to room temperature where it is used. When two materials of different expansion coefficients are used, stresses are set up between the two phases when the materials are cooled, sometimes causing cracking and separation at the grain boundaries. This has also been observed in single-phase materials such as graphite, aluminum oxide, TiO_2, Al_2TiO_5, and quartz which have different coefficients of thermal expansion in different crystallographic directions. These boundary stresses are used in crushing quartzite rock; the rock is heated and develops sufficient boundary stresses because of the differential thermal expansion that cracks open between grains and allow easy grinding.

Boundary Stresses in Laminates. The source of these stresses can be illustrated by considering the effects in a laminate composed of two materials forming alternate sheets with linear thermal expansion coefficients α_1, α_2, elastic moduli E_1, E_2, and Poisson's ratios μ_1, μ_2. When the temperature is changed from T_0, at which the laminate is assembled stressfree, to a new uniform temperature T, so that $T - T_0 = \Delta T$, one material attempts to expand $\alpha_1 \Delta T$, whereas the second material expands $\alpha_2 \Delta T$. These expansions are not compatible, so that the system must adopt an intermediate overall expansion, depending on the relative elastic moduli and fractions of each component, such that the net compressive force in one component is equal to the net tensile force in the other (Fig. 5.14). If σ is the stress, V the volume fraction (equal to the cross-sectional area fraction), and ϵ the actual strain,

$$\sigma_1 V_1 + \sigma_2 V_2 = 0 \tag{5.47}$$

$$\left(\frac{E_1}{1-\mu_1}\right)(\epsilon - \epsilon_1)V_1 + \left(\frac{E_2}{1-\mu_2}\right)(\epsilon - \epsilon_2)V_2 = 0 \tag{5.48}$$

If $E_1 = E_2$, $\mu_1 = \mu_2$, and $\alpha_1 - \alpha_2 = \Delta\alpha$, then

$$\Delta\alpha \, \Delta T = \epsilon_1 - \epsilon_2 \tag{5.49}$$

and

$$\sigma_1 = \left(\frac{E}{1-\mu}\right) V_2 \, \Delta\alpha \, \Delta T \tag{5.50}$$

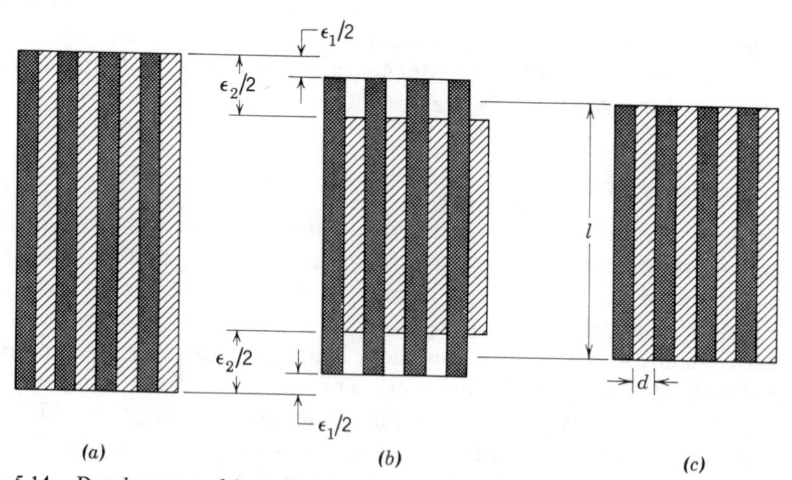

Fig. 5.14. Development of boundary stresses in a laminate (*a*) at high temperature, (*b*) cooled stressfree, and (*c*) cooled with layers bonded together.

These stresses are calculated by setting the total force (equal to the summation of stress in each phase times the cross-sectional area of each phase) equal to zero, since the positive and negative forces in the respective components are in balance. The force might be transferred through the grain boundary; the force transferred to a single layer through the grain boundary is given by $\sigma_1 A_1 = -\sigma_2 A_2$. The total force $\sigma_1 A_1 + \sigma_2 A_2$ produces an average boundary shear stress (τ_{av}), given by

$$\tau_{av} = \frac{(\sigma_1 A_1)_{av}}{\text{Local boundary area}} \tag{5.51}$$

The boundary area is proportional to v/d for laminates, where d is the thickness or edge length and v the volume of the piece. For a laminate we can then write

$$\tau \sim \frac{\left(\dfrac{V_1 E_1}{1-\mu_1}\right)\left(\dfrac{V_2 E_2}{1-\mu_2}\right)}{\left(\dfrac{E_1 V_1}{1-\mu_1}\right) + \left(\dfrac{E_2 V_2}{1-\mu_2}\right)} \Delta\alpha \, \Delta T \frac{d}{l} \tag{5.52}$$

where l is the length of the laminates (Fig. 5.14).

Boundary Stresses in Three-Dimensional Structures. In a three-dimensional equiaxed grain structure the fraction of the total force transferred by boundary shear stresses is smaller than in the laminate, since normal boundary stresses also begin to be important. For the simple case of a spherical particle in an infinite matrix, the sphere is subject to uniform hydrostatic stress, $\bar{\sigma}$:

$$\bar{\sigma} = \frac{(\alpha_m - \alpha_p)\,\Delta T}{(1+\mu_m)/2E_m + (1-2\mu_p)/E_p} \tag{5.53a}$$

where m refers to the matrix and p to the particle. The stresses in the matrix are

$$\sigma_{rr} = \frac{\bar{\sigma} R^3}{r^3} \tag{5.53b}$$

$$\sigma_{\phi\phi} = \sigma_{\theta\theta} = -\frac{\bar{\sigma} R^3}{2r^3} \tag{5.53c}$$

where R is the particle radius and r the distance from the center. For single phase materials with anisotropic thermal expansions the stresses near grain boundaries are more complicated and tend to drop rapidly with distance from the boundary; the maximum stresses are similar in magnitude to those given by Eqs. 5.50 and 5.53 where $\Delta\alpha$ is the difference between the extreme values of the thermal expansion.

These boundary stresses are important in determining many properties

Fig. 5.15. Photomicrograph of large-grain Al_2O_3 showing reflections from separated grain boundaries (30×, transmitted light). Courtesy R. L. Coble.

of polycrystalline ceramics. It is commonly found that in samples having components with different thermal expansion coefficients, such as porcelains, or with a single phase with an anisotropic expansion, such as aluminum oxide, the stresses are high enough in large-grained samples to cause cracking and separation between individual grains. This is illustrated in the photograph of a large-grain sample of aluminum oxide in Fig. 5.15; the reflections from separated grain boundaries show up clearly. Although the stresses are independent of grain size as indicated in Eq. 5.53, spontaneous cracking occurs predominately in large-grain sized samples because the reduction in the internal strain energy is proportional to the cube of the particle size whereas the increased surface energy caused by the fracture is proportional to the square of the particle size. These separations mean that large-grain compositions are weak and in general have poor physical properties because of the substantial grain-boundary stresses.

5.6 Solute Segregation and Phase Separation at and Near Grain Boundaries

Direct observations of ceramic grain boundaries with transmission electron microscopy show that submicroscopic precipitation at the

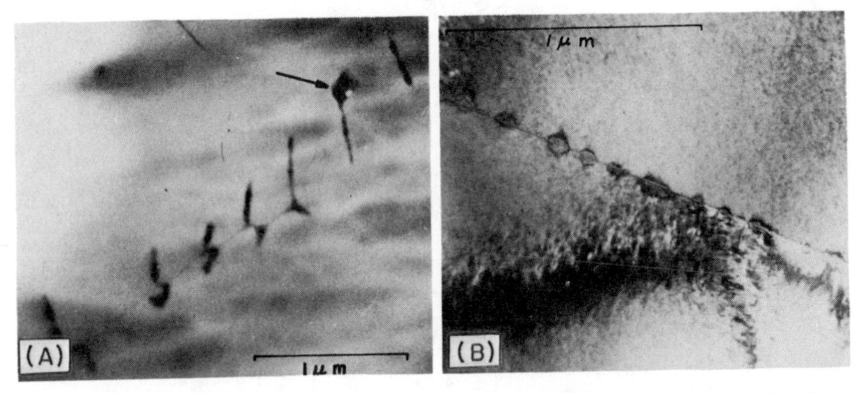

Fig. 5.16. Precipitate particles (*a*) at dislocations in a subboundary in MgO and (*b*) along grain boundary in $ZrO_2 + Y_2O_3$. From N. J. Tighe and J. R. Kreglo, *Bull. Am. Ceram. Soc.*, **49**, 188 (1970).

boundaries is common. Such precipitates at a low-angle dislocation network boundary and at a high-angle boundary are shown in Fig. 5.16. In addition to precipitates the segregation of solutes in the vicinity of a grain boundary is often observed, particularly in samples that are slowly cooled to room temperature, as shown in Fig. 5.17. Recent developments of analytical techniques (such as Auger spectroscopy) for analysing thin layers adjacent to fractured grain boundaries has shown that segregation, as well as precipitation, is very common indeed.

In our discussion of the electrostatic potential and associated space charge at grain boundaries we have already seen that there is an enhancement or diminution of lattice defect concentrations and of aliovalent-solute concentrations near the interface. Figure 5.13*a* shows that the electrostatic charge on the boundary increases as the temperature is lowered, and this same result has been found for the surface concentration of Ca^{+2} in NaCl (Table 5.3).

Table 5.3. Difference in Bulk and Surface Concentration of Ca^{2+} in NaCl

Temperature	Site Fraction Ca^{2+} in Bulk	Site fraction Ca^{2+} in Surface
500°C	1.9×10^{-3}	6.4×10^{-2}
400°C	1.7×10^{-3}	7.1×10^{-2}
300°C	1.8×10^{-3}	1.2×10^{-1}
250°C	1.6×10^{-3}	1.3×10^{-1}

Source. A. R. Allnatt, *J. Phys. Chem.*, **68**, 1763 (1964).

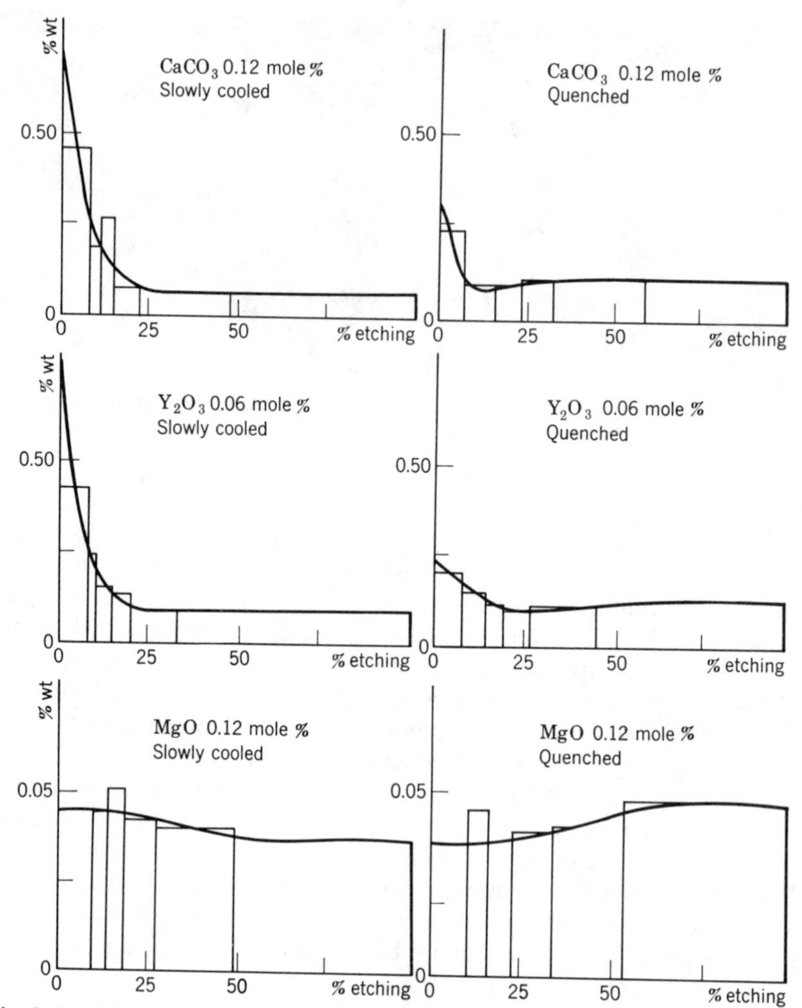

Fig. 5.17. Chemical analysis of calcium, yttrium, and magnesium in a solution obtained by successive chemical etchings on ferrites ground down to a fineness comparable with the average grain size. The ion content is plotted as a function of the etching percentage of the ferrite powder. From M. Paulus, *Materials Science Research*, Vol. 3, Plenum Press, N.Y. p. 31.

In addition to the electrostatic potential there are stress fields associated with boundaries that affect the solute distribution. The heat of solution of many solutes in oxide systems is high, partly because of strain energy and partly because of the energy to form the accompanying vacancies or interstitials necessary for electrical neutrality, as discussed in Chapter 4. If some unknown fraction of the grain boundary consists of already distorted sites for which the added strain energy of inserting a solute atom and accompanying vacancy or interstitial is small, then the overall free energy of the sample is minimized by preferentially filling these low-energy sites. If the fractional filling of these special sites is defined as the grain-boundary concentration C_b and the energy difference of a solute atom at these special sites and in the lattice is $E - e$, then for small solute concentrations C

$$C_b = \frac{AC \exp((E-e)/kT)}{(1 + AC \exp(E - e/kT))} \tag{5.54}$$

where A is a constant which allows for a decrease in vibrational entropy at the boundary.* This treatment is both quite general and quite imprecise but, as shown in Fig. 5.18 it indicates that the tendency toward grain-

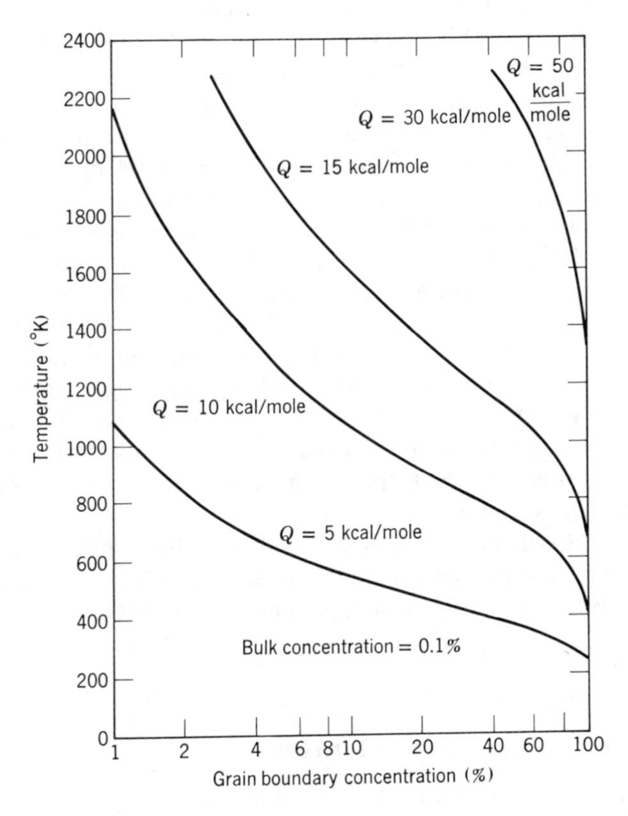

Fig. 5.18. Fraction of low-energy boundary sites occupied by solute shown for different solute concentrations and heats of solution according to Eqn. 5.54.

*Reference 5.

boundary segregation increases as the solute concentration is raised and the temperature is lowered. For heats of solution usual in oxides, 15 to 90 kcal/mole, saturation of the special low energy sites at boundaries is to be expected.

As discussed in the previous section, there are many ceramics in which substantial stresses are induced at the grain boundaries on cooling. For the same reasons that the lowered strain energy induces solute segregation at the boundary core, there is a tendency for solute segregation in the associated stress field during cooling. This effect is believed to be responsible for the influence of silica precipitates in manganese ferrite in enhancing the boundary segregation of calcia and it is probably the reason that segregation effects in alumina are strongly dependent on the relative orientation of the grains. This stress-field effect is probably important in many common oxide ceramics such as alumina, titania, beryllia, zircon, mullite, and quartz which have a high degree of anisotropy in elastic constants and thermal expansion coefficients.

The high heats of solution which lead to grain-boundary segregation also lead to very low crystalline solubilities at moderate temperatures. For example, the solubility of Al_2O_3 in MgO is only 100 ppm at 1300°C, the solubility of MgO in Al_2O_3 is 100 ppm at 1550°C, and the solubility of CaO in MgO is 100 ppm at 1000°C. As a result, supersaturation develops on cooling, and fine precipitates, often submicroscopic, as shown in Fig. 5.16, form at the grain boundaries (a preferred site for nucleation, as discussed in Chapter 8) during cooling or extended heat treatments at moderate temperatures. A related phenomenon occurs when nonstoichiometric oxides undergo a temperature change or change in oxygen pressure which leads to a change in chemical composition, as discussed in Chapter 4. Frequently oxygen diffusion is more rapid at grain boundaries than in the crystalline lattice, so that the composition of the grain boundaries equilibrates more rapidly than that of the bulk.

The boundary-layer thickness that is affected by transient changes may range up to tens of microns in width, depending on time, temperature, and atomic mobility. Discussion of this important question is deferred until Chapter 9.

5.7 Structure of Surfaces and Interfaces

As discussed in Section 5.1, the surface of a phase or the interface between phases is a region of high energy relative to the bulk. In order to maintain the lowest total energy for the system, the configuration of the surface adapts itself to minimize its excess energy. Solutes that lower the surface energy tend to concentrate in the surface, and dipoles orient themselves in such a way as to give a minimum surface energy.

Surface Composition; Adsorption. The composition and structure of surfaces depend a great deal on the conditions of formation and subsequent treatment. For example, it is found that freshly fractured oxide surfaces have high chemical reactivity compared with the same surfaces after they are allowed to stand in the air or are heated at high temperatures. The surface energy of mica freshly cleaved *in vacuo* has been found to be much higher than the same surface cleaved in air. Similarly, an iron bar broken under a surface of liquid mercury is silvered, whereas one broken in air and immediately plunged into mercury is not. Also, the surface of freshly fractured silica is a strong oxidizing agent, but this property disappears with time. These effects are manifestations of the tendency of a surface to adjust its structure to a low energy state by atom migration or adsorption of additional components.

Ions can fit on the surface with relatively low energy only if they are highly polarizable ions, such that the electron shells can be distorted to minimize the energy increase produced by the surface configuration. Consequently, highly polarizable ions tend to form the major fraction of the surface layer (Fig. 5.19), making a crystallographic surface with equal numbers of cations and anions.

When an oxide, such as silica, is fractured at low temperature, the fracture does not follow any particular crystallographic direction, and large numbers of Si–O bonds are broken, leaving Si^{4+} and O^{2-} ions in the surface with unsatisfied valencies. This is a high-energy and very reactive surface, which adsorbs oxygen from the air in order to form a lower-energy surface; this reaction takes place rapidly. Similarly, adsorption of an oxygen layer usually occurs on metal and carbide surfaces. Once formed, these can only be removed with difficulty. When glasses and oxides are heated at high temperatures, there is sufficient ion mobility so

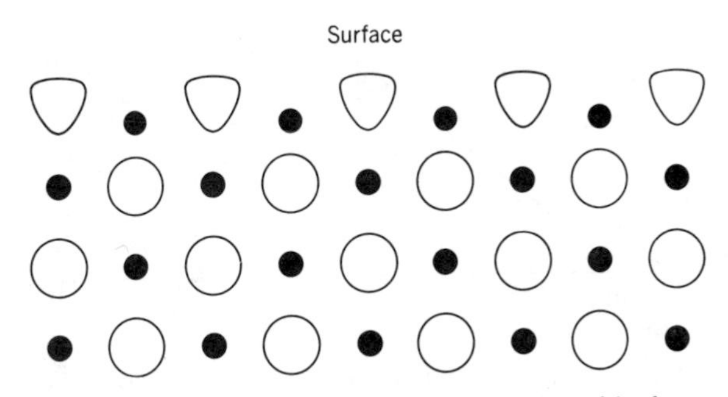

Fig. 5.19. Surface structure of oxide or sulfide dominated by large anions.

that this kind of low-energy surface configuration always results, and this surface configuration is what is normally measured. Even systems without a strong oxygen affinity adsorb oxygen on the surface, as shown for silver in Fig. 5.20. In an atmosphere containing oxygen the surface energy of solid silver is substantially lowered, corresponding to absorption of a monolayer of oxygen. Similar but more pronounced effects occur for higher-surface energy-liquid iron, as illustrated in Fig. 5.21. In liquid iron the addition of oxygen or sulfur forms a monolayer which results in a strongly reduced surface energy. For the smaller ions, such as carbon and nitrogen, the effect is small or absent altogether.

Fused silica has a surface energy of about 300 ergs/cm^2, and the surface is already dominated by oxygen ions in a way similar to that illustrated in Fig. 5.19. Consequently the effect of surface additions on the surface energy of silica is not so strong as that observed for metals. A larger effect occurs with the addition of silica to liquid oxides having a higher surface energy, such as iron oxide and calcium oxide. Surface-energy relationships in some of these two-component systems are illustrated in Fig. 5.22.

Interfaces in Two-Phase Systems. In the same way that free surfaces and the boundaries between two grains of the same material have an associated energy, interfaces between two phases—solid-solid, liquid-liquid, solid-liquid, solid-vapor, liquid-vapor—are characterized by an interface energy corresponding to the energy necessary to form a unit area of new interface in the system. This interfacial energy is always less than the sum of the separate surface energies of the two phases, since

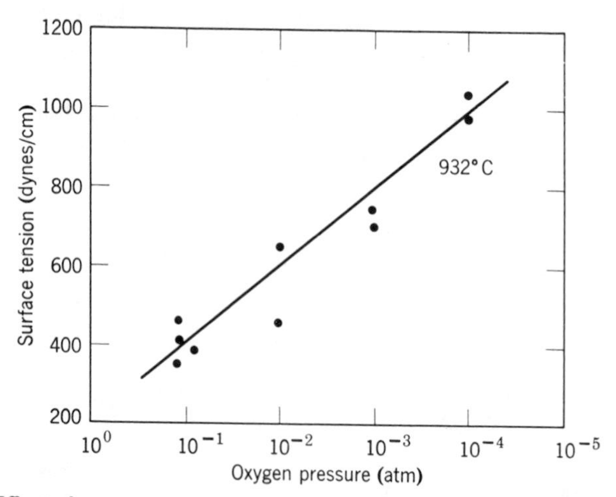

Fig. 5.20. Effect of oxygen pressure on surface energy of solid silver. From F. H. Buttner, E. R. Funk, and H. Udin, *J. Phys. Chem.*, **56**, 657 (1952).

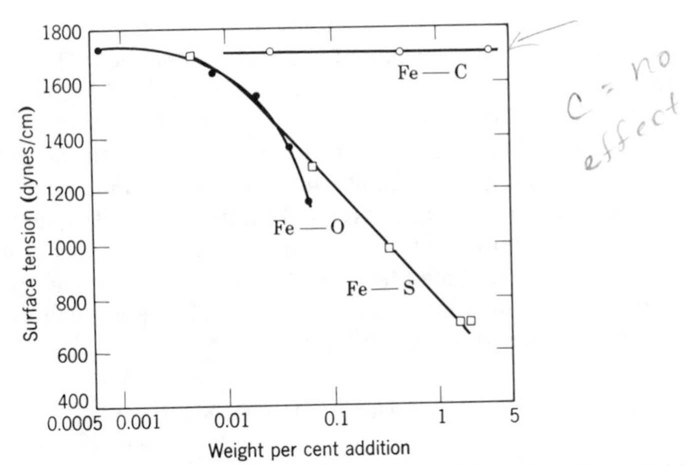

Fig. 5.21. Effect of various additions on the surface tension of liquid iron. From F. H. Halden and W. D. Kingery, *J. Phys. Chem.*, **59**, 557 (1955).

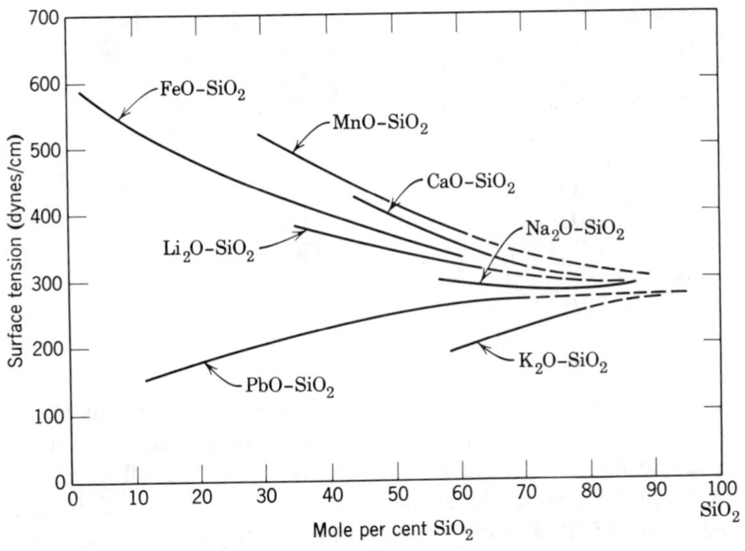

Fig. 5.22. Surface tension in some silicate systems.

there is always some energy of attraction between the phases. It may be any value less than this sum, depending on the mutual attraction of the two phases. Indeed, when two miscible phases are mixed, the surface is sometimes observed to extend itself, that is, increase its area as the first stage of complete mixing (exhibit so-called negative surface energy).

In general, the interfacial energy between chemically similar phases is low compared with the sum of the surface energies. That is, liquid oxides on aluminum oxide or other solid oxides generally have a low interface energy, as do liquid metals on solid metals. Similarly, the interface energy is low when there are strong forces of chemical attraction. In fact, these forces lead to some difficulty in defining interface energy, since some degree of chemical interaction and mutual solubility usually takes place. That is, the initial interfacial energy between unsaturated phases is usually different from the equilibrium interfacial energy between the saturated compositions. Some interfacial energy values for mutually saturated systems are given in Table 5.4.

Table 5.4. Interfacial Energies

System	Temperature (°C)	Interface Energy ($ergs/cm^2$)
$Al_2O_3(s)$–silicate glaze(l)	1000	<700
$Al_2O_3(s)$–Pb(l)	400	1440
$Al_2O_3(s)$–Ag(l)	1000	1770
$Al_2O_3(s)$–Fe(l)	1570	2300
SiO_2(glass)–sodium silicate(l)	1000	<25
SiO_2(glass)–Cu(l)	1120	1370
Ag(s)–$Na_2SiO_3(l)$	900	1040
Cu(s)–$Na_2SiO_3(l)$	900	1500
Cu(s)–$Cu_2S(l)$	1131	90
TiC(s)–Cu(l)	1200	1225
MgO(s)–Ag(l)	1300	850
MgO(s)–Fe(l)	1725	1600

Adsorption phenomena occur at interfaces in the same way as at surfaces. Wetting agents for metals on oxides that are effective in reducing the interfacial energy are widely used. The effect of these is shown in Fig. 5.23, which illustrates the interfacial energy of liquid nickel with additions of titanium in contact with Al_2O_3. The titanium is strongly attracted to the oxide interface and concentrates there because of its high chemical reactivity with oxygen. Frequently interface phenomena are complicated by the presence of an oxide layer on the surface of nonoxides, such as silicon carbide, silicides, nitrides, and metals. The solid oxide layer makes them behave as oxides unless the film is removed by heating in a reducing atmosphere or by using reactive fluxes. The oxide film which forms on silicides makes them behave as oxides. However, in

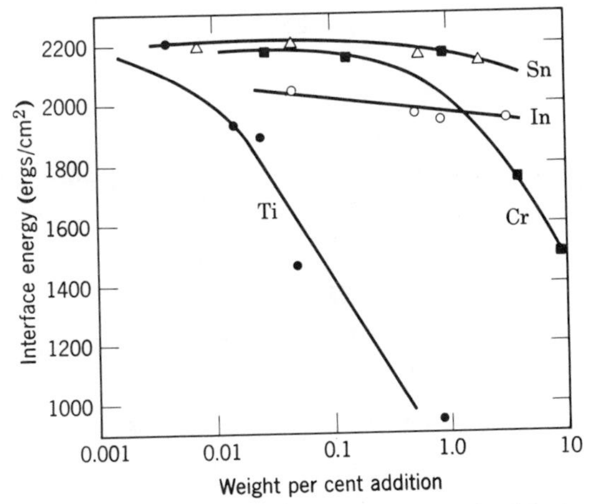

Fig. 5.23. Effect of titanium and chromium additions on Ni–Al_2O_3 interface energy at 1475°C.

vacuo the surface oxide layer vaporizes at about 1200°C, and the metallic nature of the material becomes apparent.

5.8 Wetting and Phase Distribution

The relationship between surface and interfacial energies determines to a large extent the wetting behavior of a liquid on a solid surface and the phase morphology of mixtures of two or more phases.

Wetting of Solid Surface by a Liquid. If we consider the stable configuration of a liquid placed on a solid surface, the equilibrium shape conforms to the minimum total interfacial energy for all the phase boundaries present. If the solid-liquid interfacial energy (γ_{SL}) is high, the liquid tends to form a ball having a small interfacial area, as shown in Fig. 5.24*a*. In contrast, if the solid-vapor interfacial energy (γ_{SV}) is high, the liquid tends to spread out indefinitely to eliminate this interface, as shown in Fig. 5.24*c*. An intermediate drop shape is shown in Fig. 5.24*b*.

The angle between the solid surface and the tangent to the liquid surface at the contact point, the contact angle, may vary between 0 and 180°. This angle specifies the conditions for minimum energy according to the relation

$$\gamma_{LV} \cos \theta = \gamma_{SV} - \gamma_{SL} \tag{5.55}$$

where γ_{SV}, γ_{SL}, and γ_{LV} are the interfacial energies between the phases actually present in the system at the time of measurement, usually not the pure surfaces. We may define $\theta = 90°$ as the boundary between "nonwet-

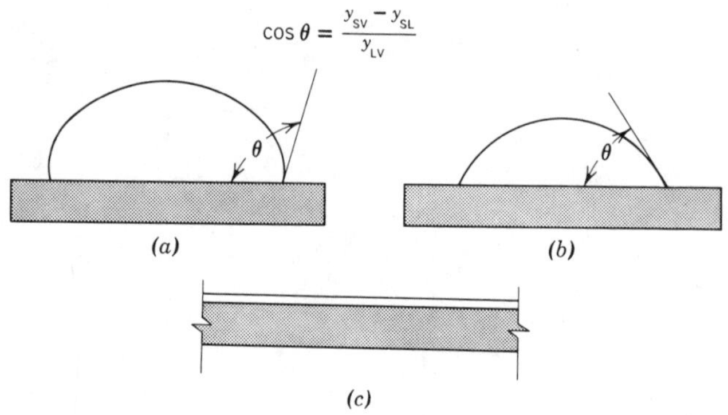

Fig. 5.24. Illustration of (*a*) nonwetting ($\theta > 90°$), (*b*) wetting ($\theta < 90°$), and (*c*) spreading ($\theta = 0$) of liquid on a solid.

ting" ($\theta > 90°$, liquid depression in a capillary, Fig. 5.24*a*) and wetting ($\theta < 90°$, liquid elevation in a capillary, Fig. 5.24*b*). Spreading is the condition in which the liquid completely covers the solid surface ($\theta = 0°$, Fig. 5.24*c*). Relationships between the surface energies determine the wetting behavior and spreading tendency according to a spreading coefficient S:

$$S_{LS} = \gamma_{SV} - (\gamma_{LV} + \gamma_{SL}) \tag{5.56}$$

For spreading to occur it is necessary that S_{LS} be positive. A corresponding necessary, but not sufficient, condition for spreading is that the liquid-vapor interface energy be less than the solid-vapor interface energy ($\gamma_{LV} < \gamma_{SV}$). This condition is sometimes useful for screening purposes.

Table 5.5 shows measured contact angles for several metals and for a

Table 5.5. Measured Contact Angles of Liquids on Single-Crystal MgO

Liquid	Test Temp (°C)	Measured Contact Angles on MgO Crystal Plane (100)	(110)	(111)
Cu	1300	106	159	149
Ag	1300	136	141	147
Co	1600	114	153	144
Fe	1600	59	110	90
Basic slag[a]	1400	9	17	32

[a] 40% SiO_2, 20% Al_2O_3, 40% CaO.

basic slag on MgO single-crystal surfaces. As noted earlier, the surface energy of solids is dependent on the crystallographic orientation, which is demonstrated by the variation in wetting angle with crystallographic planes for MgO.

A complete analysis of wetting behavior is more complex than indicated by Eqs. 5.55 and 5.56, since the composition of the phases normally changes during the process. It is common to speak of *initial* spreading coefficients for the pure phases and *final* spreading coefficients for the mutually saturated phases. A whole series of intermediate coefficients can be of importance under special conditions. In general, all the interface energies can be affected by changing composition, and both *delayed* wetting and initial spreading followed by *dewetting* are known to occur. Equation 5.55 always applies as long as γ_{LV}, γ_{LS}, and γ_{SV} refer to the conditions at the time of measurement. In addition, it is often observed that there is a considerable difference between the angle at which a liquid advances over a solid surface and the angle with which the liquid recedes from a previously wetted surface. Sometimes this difference may be due to surface cleanliness; other times the solid surface may be irreversibly changed after wetting. Usually the angle for the receding liquid is smaller than that for the advancing liquid, and once a surface is wet it tends to remain wetted. This fact is used in the addition of binders to glaze compositions such that the initial liquid is formed with a glaze uniformly distributed over the entire surface. If cracks are allowed to develop initially, the glaze may not completely rewet the surface on fusion.

In general, oxide glaze compositions wet oxide ceramics, although they may not completely spread out over the surface, particularly if the contact angle is advancing. In many cases there is no clearcut definition of wetting, since part of the ceramic tends to dissolve in the liquid glaze layer, illustrated in the microstructure of a glaze-porcelain interface shown in Fig. 5.25. Even though the glaze does wet the surface, viscous resistance to flow may prevent it from spreading out readily if the firing temperature is low.

Oxide liquids have much lower surface energies than solid metals, and consequently oxide layers tend to wet metals on which they are deposited; contact angles vary between 0 and 50°. This means that porcelain enamels flow out readily on iron or copper surfaces, for example. Differences between the adhesion of liquid oxides or porcelain enamels to different samples of metals are not related to the wetting behavior but are determined by other factors. In contrast, liquid metals have much higher surface energies than most ceramic oxides, and the interfacial energy is high so that liquid metals do not wet and spread unless special precautions are taken. Two general approaches have been used for the development

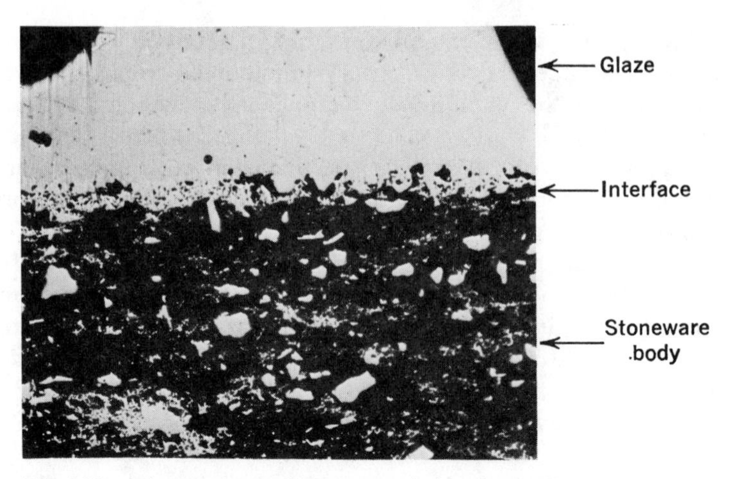

Fig. 5.25. Microstructure of glaze-ceramic interface (150 ×).

of metal brazes for use with oxides. In one method, active metals such as titanium or zirconium are added to the metal; these effectively reduce the interfacial energy by their strong chemical attraction to the oxide and enhance wetting behavior. In the other method, the use of molybdenum-manganese combinations, a reaction occurs, forming a fluid liquid oxide at the interface which wets both the solid metal layer and the underlying oxide ceramic. This gives satisfactory adhesion and allows the formation of sound metalized coatings which are subsequently wet by metallic brazes.

Grain-Boundary Configuration. In the same way that a solid-liquid system reaches an equilibrium configuration determined by the surface energies, the interface between two solid grains reaches equilibrium after a sufficient time at elevated temperatures for atomic mobility or vapor-phase material transfer to occur. The equilibrium between the grain-boundary energy and the surface energy is as shown in Fig. 5.26*a*. At equilibrium

$$\gamma_{SS} = 2\gamma_{SV} \cos \frac{\psi}{2} \tag{5.57}$$

Grooves of this kind are normally formed on heating polycrystalline samples at elevated temperatures, and thermal etching has been observed in many systems. The ratio of grain boundary to surface energies can be determined by measuring the angle of thermal etching. Similarly, if a solid and liquid are in equilibrium in the absence of a vapor phase, the equilibrium condition is as shown in Fig. 5.26*b*,

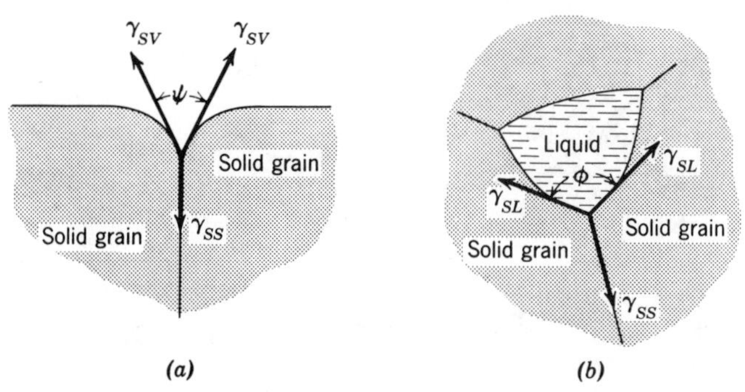

Fig. 5.26. (*a*) Angle of thermal etching and (*b*) dihedral angle for solid-solid-liquid equilibrium.

$$\gamma_{SS} = 2\gamma_{SL} \cos \frac{\phi}{2} \tag{5.58}$$

where ϕ is the dihedral angle. For two-phase systems the dihedral angle ϕ depends on the relationship between the interfacial and grain-boundary energies according to the relation

$$\cos \frac{\phi}{2} = \frac{1}{2} \frac{\gamma_{SS}}{\gamma_{SL}} \tag{5.59}$$

If the interfacial energy γ_{SL} is greater than the grain-boundary energy, ϕ is greater than 120° and the second phase forms isolated pockets of material at grain intersections. If the ratio γ_{SS}/γ_{SL} is between 1 and $\sqrt{3}$, ϕ is between 60 and 120°, and the second phase partially penetrates along the grain intersections at corners of three grains. If the ratio between γ_{SS} and γ_{SL} is greater than $\sqrt{3}$, ϕ is less than 60°, and the second phase is stable along any length of grain edge forming triangular prisms at the intersections of three grains. When γ_{SS}/γ_{SL} is equal or greater than 2, ϕ equals zero, and at equilibrium the faces of all the grains are completely separated by the second phase. These structures are illustrated in Fig. 5.27.

These relationships are important in processes that take place during firing of powder compacts. If a liquid phase is present, it can be effective in speeding the densification process only if ϕ equals zero so that the solid grains are separated by a liquid film. This occurs, for example, with the addition of a small amount of kaolin or talc to magnesium oxide. These relationships are also important in determining properties of resultant compositions. The effect of phase distribution on properties is almost self-evident. The electrical and thermal conductivity, deformation

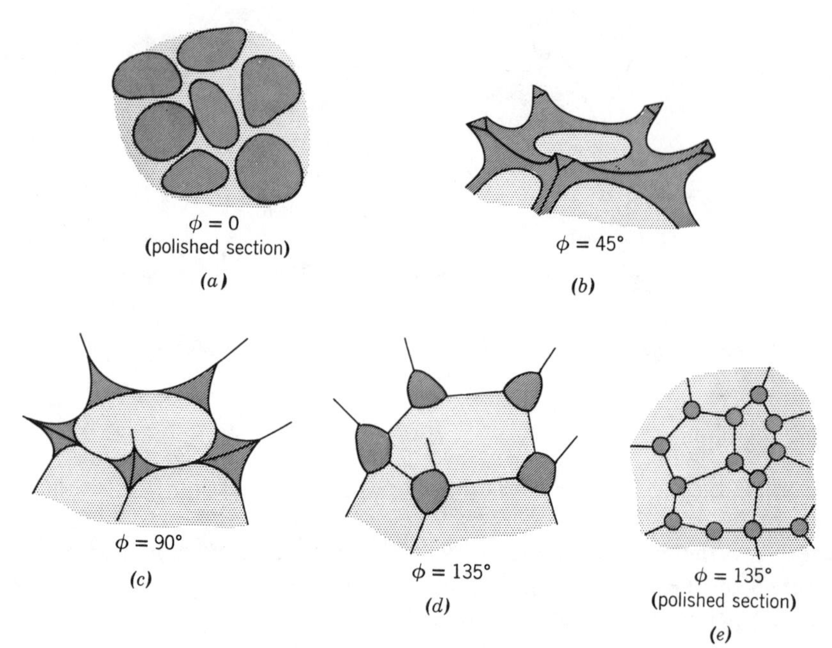

Fig. 5.27. Second-phase distribution for different values of the dihedral angle.

under stress, and chemical reactivity of a complex mixture all depend not only on the properties of the individual phases but also on the relative distribution of the phases.

Suggested Reading

1. N. K. Adam, *The Physics and Chemistry of Surfaces*, Oxford University Press, New York, 1941; also in paperback, Dover Publications, Inc., New York, 1968.
2. W. D. Kingery, "Role of Surface Energies and Wetting in Metal-Ceramic Sealing," *Bull. Am. Ceram. Soc.*, **35**, 108 (1956).
3. C. S. Smith, "Some Elementary Principles of Polycrystalline Microstructures," *Metall. Rev.*, **9**, 1 (1964).
4. W. D. Kingery, "Plausible Concepts Necessary and Sufficient for the Interpretation of Ceramic Grain Boundary Phenomena," *J. Am. Ceram. Soc.*, **57**, 1–8, 74–83 (1974).
5. D. McLean, *Grain Boundaries in Metals*, Clarendon Press, Oxford, 1957.

6. H. Gleiter and B. Chalmers, *Progress in Materials Science*, Vol. 16, Pergamon Press, New York, 1972.
7. H. Hu, Ed., *The Nature and Behavior of Grain Boundaries*, Plenum Press, New York, 1972.
8. P. Chaudhari and J. W. Matthews, Eds., *Grain Boundaries and Interfaces* (*Surface Science*, Vol. 31), North-Holland Publishing Company, Amsterdam, 1972.
9. *Metal Surfaces: Structure, Energetics and Kinetics*, ASM, Metals Park, Ohio, 1963.

Problems

5.1. The surface tension of Al_2O_3 is estimated to be 900 erg/cm^2, for liquid iron it is 1720 erg/cm^2 in a vacuum. Under the same conditions, the interfacial tension (liquid iron–alumina) is about 2300 erg/cm^2. What is the contact angle? Will liquid iron wet alumina? What can be done to lower the contact angle?

5.2. A liquid silicate with surface tension of 500 erg/cm^2 makes contact with a polycrystalline oxide with an angle $\theta = 45°$ on the surface of the oxide. If mixed with the oxide, it forms liquid globules at three grain intersections. The average dihedral angle ϕ is 90°. If we assume the interfacial tension of the oxide-oxide interface, without the silicate liquid is 1000 dyn, compute the surface tension of the oxide.

5.3. The following data are available for the surface tension of iron as a function of composition:

Surface tension (erg/cm^2)	Addition (ppm)	Surface tension (erg/cm^2)	Addition (ppm)
1670	100 sulfur	1710	10 carbon
1210	1000 sulfur	1710	100 carbon
795	10,000 sulfur	1710	1000 carbon
		1710	10,000 carbon

What information can be derived from these data concerning surface excess quantities? Describe the composition of the surface with regard to these two systems (Fe–S and Fe–C).

5.4. Dislocation etch pits measured on the average of 6.87 microns apart on a low-angle grain boundary. The angle between grains amount to 30 sec of arc by X-ray diffraction techniques. What is the length of the Burgers vector? Note: 1 sec = 0.00028°.

5.5. It is determined that, at a distance of 20 Å from an edge dislocation, the shear stress has a certain value S. A line joining the point in question forms an angle of 30° with the slip plane of the material.

(*a*) Describe the shear stress in terms of S at a distance of 100 Å.

(*b*) What will be the maximum tensile stress at a distance of 100 Å?

5.6. Following polygonization, polishing, and etching, etch pits are observed to be spaced at 10-micron distances along a line in a crystal of lithium fluoride. The low-angle grain boundary is observed to move normal to the plane of the boundary under an applied shear stress. How can this happen? If the Burgers vector is 2.83 Å, what is the tilt angle across the boundary?

5.7. A metal is melted on an Al_2O_3 plate at high temperatures.

(*a*) If the surface energy of Al_2O_3 is estimated to be 1000 erg/cm^2, that of the molten metal is similar and the interfacial energy is estimated to be about 300 ergs/cm^2, what will the contact angle be?

(*b*) If the liquid had only half the surface energy of Al_2O_3 but the interfacial energy were twice the surface tension of Al_2O_3, estimate the contact angle.

(*c*) Under conditions described in (*a*), a cermet is formed by mixing 30% metal powder with Al_2O_3 and heated above the melting point of the metal. Describe and show with a drawing the type of microstructure expected between the metal and Al_2O_3.

5.8. At 0°C the solid-liquid interfacial energy in the ice–water system is 28 ergs/cm^2, the grain-boundary energy is 70 ergs/cm^2, the liquid surface energy is 76 ergs/cm^2, and liquid water completely wets an ice surface.

(*a*) Estimate the angle of grain-boundary etching to be expected for ice in air and for ice under water.

(*b*) Alcohol added to the liquid phase is observed to lower the solid-liquid interface energy. How much of a decrease would be required to decrease the angle of grain-boundary etching to zero degrees?

(*c*) How does ice compare with (*a*) SiO_2 (solid)–silicate liquid and (*b*) MgO (solid)–silicate liquid in regard to grain-boundary etching? Explain how and why these two latter systems are similar to or differ from one another.

5.9. A dilute solution of B in A crystallizes in a tetragonal form. On long standing under conditions where vapor-phase transport can occur, the crystals form parallelipipeds with faces perpendicular to the c and a axes, respectively. The ratio of the length of the crystal in the c axis direction to that in the a axis direction is 1.76:1 when the mole fraction of B is 0.05 and 1.78:1 when the mole fraction is 0.07. Calculate the ratio of the surface tension of the a and c faces, and tell which face has a relatively higher density of B atoms.

5.10. One solid sphere of radius R and density ρ supports another by a bridge of liquid. The liquid completely wets the solid, and the effect of gravity on the profile of the liquid surface may be ignored. Find an expression for the vapor pressure of the pendular liquid ring in terms of the physical and geometrical parameters given. You may consider the liquid surface to be a circular arc. Let the vapor pressure of the liquid across a flat surface be P_0 at temperature T.

5.11. Irradiation in a reactor produces interstitial He in a certain metal. If annealed, the He forms bubbles which cause the metal to swell and reduce its density to 0.9 of its original value. Solution of the metal in acid yields 3.95 cm^3 of He gas at STP for every cm^3 of metal dissolved. Microscope examination shows that the bubbles are of uniform size and about 1 micron in radius. What value of surface energy do you calculate for this metal? Is it reasonable? How would you determine if this is an equilibrium value?

6

Atom Mobility

In order for microstructure changes or chemical reactions to take place in condensed phases it is essential that atoms be able to move about in the crystalline or noncrystalline solid. There are a number of possible mechanisms by which an atom can move from one position to another in a crystalline structure. One of these is by the direct exchange of positions between two atoms, or more probably by a ring mechanism in which a closed circle of atoms rotates. (Direct exchange of only two atoms is not energetically probable because of the high strain energy necessary to squeeze the two atoms past one another. This is particularly true in ionic solids, in which we would never expect the exchange of cations and anions.) The ring mechanism illustrated in Fig. 6.1*b* is possible, but it has not been demonstrated to occur in any actual system. Another process which is energetically more favorable is the motion of atoms from a normal position into an adjacent vacant site. As seen in Chapter 4, there are vacant sites in every crystalline solid at temperatures above the absolute zero. The rate at which atom diffusion can occur by this process depends on the ease of moving an atom from a normal site to a vacant site and on the concentration of the vacant sites. Mobility by means of this vacancy mechanism is probably the most common process giving rise to atom motion. It is equivalent to the mobility of the vacancies in the opposite direction, and occasionally we talk of vacancy diffusion. A third process that can occur is the motion of atoms on interstitial sites. If atoms can move from a regular site to an interstitial position, as in the formation of Frenkel defects, the easy movement of these interstitial atoms through the lattice is a mechanism of atom movement. High mobility is also a characteristic of second-component atoms which are in interstitial solid solution. A variant of this process is the *interstitialcy* mechanism in which an interstitial ion moves from its interstitial site onto a lattice site, bumping another atom off the lattice site into a new interstitial position. This kind of process can take place, even though the direct movement from one interstitial site to another is energetically unfavorable. These

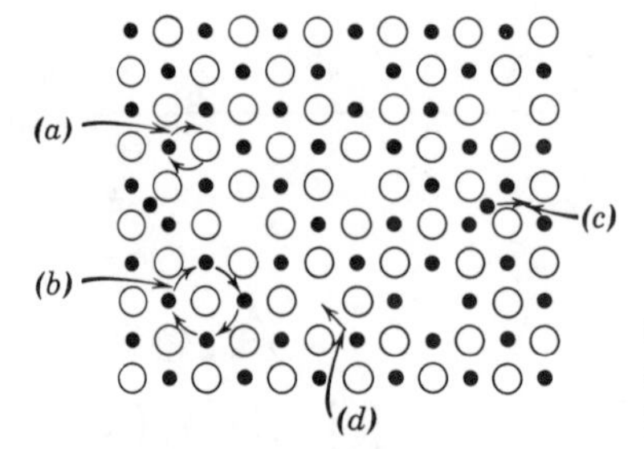

Fig. 6.1. Atomic diffusion mechanisms. (*a*) Exchange; (*b*) ring rotation; (*c*) interstitial; (*d*) vacancy.

mechanisms are illustrated in Fig. 6.1. Which occurs in any particular system depends on the relative energies of the different processes.

On a microscopic scale the effect of atomic mobility and diffusion is illustrated in Fig. 6.2. If two miscible components are brought together, there is a gradual intermingling until an equilibrium structure is reached in which there is a uniform distribution of A and B. The rate of reaching this

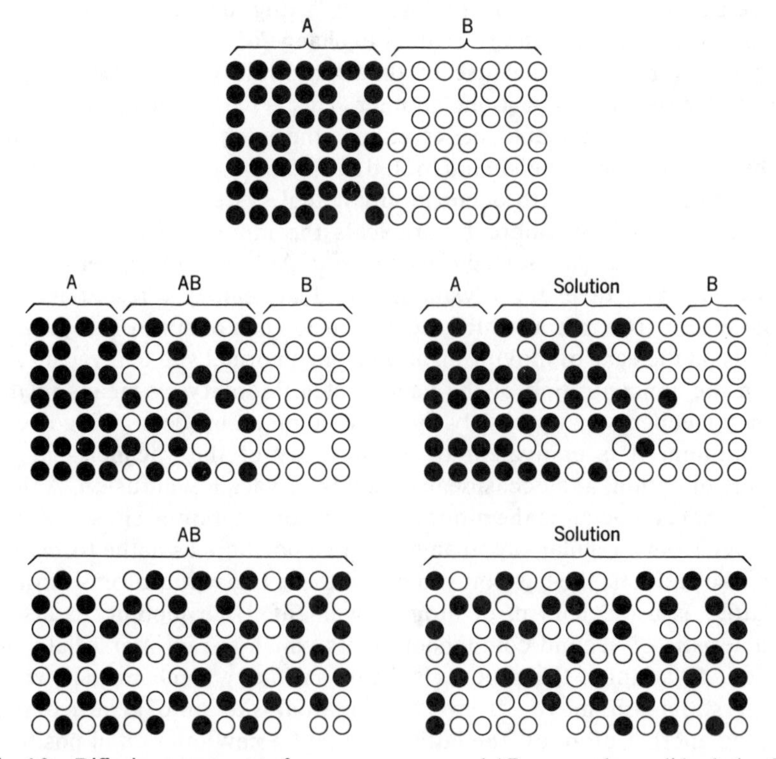

Fig. 6.2. Diffusion processes to form a new compound AB or a random solid solution from pure starting materials A and B.

final state depends on the rate of diffusion of the individual atoms. Similarly, if a new compound is formed between A and B, continuation of the reaction requires that material diffuse through the intermediate layer. The speed of this diffusion process limits the rate of the reaction. In addition to the rate at which uniformity of composition is attained and solid-state reactions proceed, many other processes such as refractory corrosion, sintering, oxidation, and gas permeability are influenced by diffusion properties.

6.1 Diffusion and Fick's Laws

Fick's Laws. If we consider a single-phase composition in which diffusion occurs in one direction under conditions of constant temperature and constant pressure, the transfer of material occurs in such a way that concentration gradients (chemical potential gradients) are reduced. This kind of a system might be represented by the contact between two miscible solids such as MgO and NiO. For such a system Fick's first law states that the quantity of diffusing material which passes per unit time through a unit area normal to the direction of diffusion is proportional to its concentration gradient. This is given by

$$J = -D\frac{\partial c}{\partial x} \tag{6.1}$$

where c is the concentration per unit volume, x is the direction of diffusion, J is the flux (quantity per unit time per unit area). The factor D which is the *diffusion coefficient* comes in as a proportionality factor; it usually is given in dimensions of square centimeters per second. This relationship is similar in form to Ohm's law, in which electrical current is proportional to the gradient of electrical potential, and to Fourier's law, in which the rate of heat flow is proportional to the temperature gradient.

We can determine the change in concentration at any point with time during a diffusion process by determining the difference between the flux into and the flux out of a given volume element. If we consider two parallel planes separated by a distance dx, as illustrated in Fig. 6.3, the flux through the first plane is

$$J = -D\frac{\partial c}{\partial x} \tag{6.2}$$

and the flux through the second plane is

$$J + \frac{\partial J}{\partial x}\,dx = -D\frac{\partial c}{\partial x} - \frac{\partial}{\partial x}\left(D\frac{\partial c}{\partial x}\right)dx \tag{6.3}$$

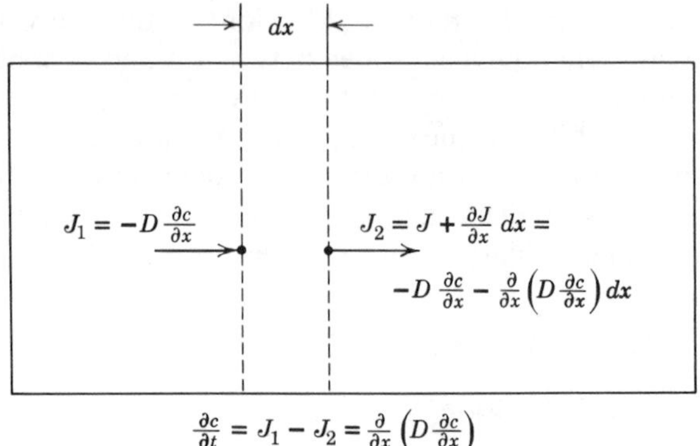

Fig. 6.3. Derivation of Fick's second law.

and by subtraction

$$\frac{\partial J}{\partial x} = -\frac{\partial}{\partial x}\left(D\frac{\partial c}{\partial x}\right) \tag{6.4}$$

The change in flux with distance is equal to $-\partial c/\partial t$ so that we arrive at Fick's second law:

$$\frac{\partial c}{\partial t} = \frac{\partial}{\partial x}\left(D\frac{\partial c}{\partial x}\right) \tag{6.5}$$

If D is constant and independent of the concentration, this can be written

$$\frac{\partial c}{\partial t} = D\frac{\partial^2 c}{\partial x^2} \tag{6.6}$$

The Nernst-Einstein Equation. Although Eqs. 6.1 to 6.6 are written in terms of concentration, it was first suggested by Einstein and has since been confirmed by others that the virtual force which acts on a diffusing atom or ion is the negative gradient of the chemical potential or partial molal free energy. If the absolute mobility of an atom, that is, the velocity v_i obtained under the action of a unit force, is B_i, the virtual force of the chemical potential gradient gives rise to a drift velocity and resultant flux:

$$-B_i = \frac{\text{velocity}}{\text{force}} = \frac{v_i}{\dfrac{1}{N}\dfrac{d\mu_i}{dx}} \tag{6.7}$$

$$J_i = -\frac{1}{N}\frac{d\mu_i}{dx}B_i c_i \tag{6.8}$$

where μ_i is the partial molal free energy or chemical potential of i and N is Avogadro's number. If we assume unit activity coefficient for species i, the change in chemical potential is given by

$$d\mu_i = RT\, d \ln c_i \tag{6.9}$$

Substituting this expression in Eq. 6.7 and comparing with Eq. 6.1, we find that the diffusion coefficient is directly proportional to the atomic mobility:

$$J_i = -\frac{RT\, dc_i}{N\, dx} B_i \tag{6.10}$$

$$D_i = kTB_i \tag{6.11}$$

where k is Boltzmann's constant. In the more general case it is necessary either to define all these equations in terms of activity gradients or to include an activity coefficient term in Eq. 6.9. This expression, called the Nernst-Einstein relation, is particularly useful in considering the mobility of charged particles and the relationship between diffusion coefficients and electrical conductivity. Table 6.1 gives dimensional units of mobility for chemical forces and for electrical forces.

Table 6.1. Dimensional Units for Mobility
(Mobility ≡ velocity/unit force)

$B = \frac{\text{cm/sec}}{\text{dynes}} = \frac{\text{cm/sec}}{\text{dyne}\cdot\text{cm/cm}} = \frac{\text{cm/sec}}{\text{erg/cm}} = \frac{\text{cm/sec}}{10^{-7}\ \text{J/cm}}$	
$B_i = \frac{V_i}{(1/N)(\partial\mu_i/\partial x)} = \frac{\text{cm}^2}{\text{erg}\cdot\text{sec}}$ = absolute mobility	V_i = cm/sec μ_i = ergs/mole = 10^{-7} J/mole N = Avogadro's number = atoms/mole x = cm
$B'_i = \frac{V_i}{\partial\mu_i/\partial x} = \frac{\text{mole}\cdot\text{cm}^2}{\text{J}\cdot\text{sec}}$ = chemical mobility	V_i = cm/sec μ_i = J/mole x = cm
$B''_i = \frac{V_i}{\partial\phi/\partial x} = \frac{\text{cm}^2}{\text{V}\cdot\text{sec}}$ = electrical mobility	V_i = cm/sec ϕ = V x = cm
$B_i = NB'_i$	
$B''_i = z_i F B'_i$	z_i = valence = equiv/mole F = Faraday const = 96,500 J/V · equiv
$\frac{\partial\mu_i}{\partial x} = z_i F \frac{\partial\phi}{\partial x}$	

1 J = C · V = 10^7 ergs = 0.2389 cal = 6.243×10^{18} eV.

Random-Walk Diffusional Processes. Before discussing the mechanisms and mathematics of diffusion, it is helpful to study a simple situation in which no detailed mechanism is assumed. We will discuss a one-dimensional random-walk process to arrive at an approximate value for the diffusion coefficient, which will be related to a jump frequency and a jump distance. Consider a crystal which has a composition gradient along the z axis (Fig. 6.4). We allow atoms to move left or right one jump distance λ along the z axis of the crystal. Let us look specifically at two adjacent lattice planes, designated 1 and 2, a distance λ apart. There are n_1 diffusing solute atoms per unit area in plane 1 and n_2 in plane 2. The jump frequency Γ is the average number of jumps per second that an atom makes out of a plane. Thus in the period of time δt the number of atoms jumping out of plane 1 is $n_1\Gamma\ \delta t$. Half of these atoms will jump to the right into plane 2, and half to the left. Similarly, the number of atoms jumping from plane 2 to plane 1 in the interval δt is $1/2 n_2\Gamma\ \delta t$. A flux from planes 1 and 2 results;

$$J = \frac{1}{2}(n_1 - n_2)\Gamma = \frac{\text{number of atoms}}{\text{(area)(time)}} \tag{6.12}$$

The quantity $(n_1 - n_2)$ can be related to the concentration or a number per unit volume by noting that $n_1/\lambda = c_1$, and $n_2/\lambda = c_2$, and that $(c_1 - c_2)/\lambda = -\partial c/\partial z$. Thus the flux is

$$J = -\frac{1}{2}\lambda^2\Gamma\frac{\partial c}{\partial z} \tag{6.13}$$

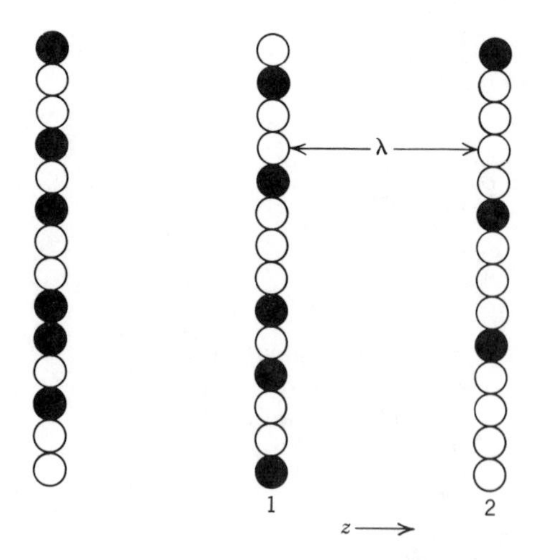

Fig. 6.4. One-dimensional diffusion.

This equation is identical to Fick's first law if the diffusion coefficient is given by

$$D = \frac{1}{2}\lambda^2\Gamma \tag{6.14}$$

If jumps occur in three directions, this value is reduced by one-third; and a rigorous development of the random-walk process in three dimensions gives

$$D = \frac{1}{6}\Gamma\lambda^2 \tag{6.15}$$

It must be remembered that this result is strictly for a random-walk process and no bias or driving force has been assumed to give preferential direction to the overall process. In addition, for a particular diffusion mechanism (vacancy, interstitial) and crystal structure, a geometric factor must be included; this factor γ, which is of the order of unity, includes the number of nearest-neighbor jump sites and the probability that the atom will jump back into its original position. Second, we must consider the availability of vacant sites to which the atoms can jump. If we focus our attention on an interstitial atom, essentially all the neighboring sites are vacant; similarly, if we focus on the motion of a vacant lattice site, its motion corresponds to an exchange with a neighboring occupied site, essentially all of which are occupied and thus available. As a result

$$D_i = \gamma\lambda^2\Gamma \qquad D_v = \gamma\lambda^2\Gamma \tag{6.16}$$

For the diffusion of a lattice atom which moves by jumping into an adjacent vacant site, we must include a term for the probability that an adjacent site will be vacant. This is equal to the fraction of vacant sites n_v, determined as discussed in Chapter 4, and

$$D_l = \gamma\lambda^2 n_v \Gamma \tag{6.17}$$

Boundary Conditions. The measurement and application of diffusion coefficients require solution of the partial differential equations (6.1 and 6.6) for various boundary conditions. If the flux is to be determined, Fick's first law (Eq. 6.1) can be used if the conditions are such that steady-state diffusion with a fixed concentration gradient is maintained. This would be the case for diffusion of a gas through a glass or ceramic diaphragm. The solution of Fick's second law (Eq. 6.6) leads to a determination of the concentration as a function of position and time, that is, $c(x,t)$. In general, these solutions for a constant diffusion coefficient fall in two forms: (1) when the diffusion distance is short relative to the dimension of the initial inhomogeneity, the concentration profile as a function of time and distance can most simply be expressed in terms of error functions, and (2) when complete homogenization is approached, $c(x, t)$ can be represented by the first terms of an infinite trigonometric series. More commonly these two cases are described as short-time and long-time solutions.

Consider first the case of steady-state diffusion with a fixed concentration gradient for the determination of the flux from Eq. 6.1. If a gas is maintained at a pressure p_1 on one side of a thin slab acting as a diaphragm and at some lower uniform pressure p_0 on the other side, a steady state is reached in which the permeation through the diaphragm occurs at a constant rate. The concentration at either surface is determined by its solubility, which is often proportional to the square root of the pressure for many diatomic gases, indicating that the gaseous species, for example, oxygen, dissolves as two independent ions. Thus, the concentration is proportional to the square root of the pressure, and the flux can be expressed in terms of the pressure, since the concentration is proportional to the square root of pressure ($c = b\sqrt{p}$):

$$J = -D\frac{\partial c}{\partial x} = -Db\frac{\sqrt{p_1} - \sqrt{p_0}}{\Delta X} \tag{6.18}$$

where ΔX is the thickness of the thin slab and b is a constant.

One boundary condition which is often approached in practice is that of diffusion into a semi-infinite solid or liquid, that is, one whose dimension in the direction of the diffusion is large. We can consider that the composition is initially uniform, that the surface is brought instantly to some specific surface concentration C_s at time zero, and that the surface concentration is maintained constant during the whole process. This would correspond, for example, to the diffusion of silver from a surface stain into the interior of the glass sample. If, at $x > 0$, C equals C_0 at $t = 0$ and C equals C_s at $x = 0$, the distribution of material at some later time is given by the relation

$$\frac{C(x,t) - C_0}{C_s - C_0} = 1 - \frac{2}{\sqrt{\pi}}\int_0^{x/2\sqrt{Dt}} e^{-\lambda^2}\, d\lambda \tag{6.19}$$

The expression on the right is one minus the error function, $1 - \text{erf } z$, where $z = x/2\sqrt{Dt}$:

$$C(x,t) - C_0 = C_s - C_0\left[1 - \text{erf}\left(\frac{x}{2\sqrt{Dt}}\right)\right] \tag{6.20}$$

The above integral (the error function) often occurs in diffusion and heat-flow problems. It varies from 0 to 1 for $x/2\sqrt{Dt}$ values of 0 to about

3. The compositional variation at a specific time and distance $c(x,t)$ is shown in Fig. 6.5. It can also be shown that $\text{erf}(\infty) = 1$ and that $\text{erf}(-z) = -\text{erf}(z)$.

The method may readily be applied to other boundary conditions for a semi-infinite solid or liquid. For example, if the surface concentration of an initially solute-free specimen is maintained at some composition C_s' for all $t > 0$, solute is added to the specimen with a time and distance dependence:

$$C(x,t) = C_s'\left[1 - \text{erf}\left(\frac{x}{2\sqrt{Dt}}\right)\right] \tag{6.21}$$

or similarly if the ambient was held at $C_s = 0$ and the specimen was initially at C_0, the solution becomes $\text{erf}(-z) = -\text{erf}(z)$:

$$C(x,t) = C_0 \, \text{erf}\left(\frac{x}{2\sqrt{Dt}}\right) \tag{6.22}$$

Let us now consider the long-time solution to Fick's second law, that is, the case when homogenization tends to completion. This could occur for diffusion out of a slab of thickness L, with solute being lost from both faces. If the initial composition is C_0 and the surface composition is maintained at C_s for $t > 0$, the solution can be reliably approximated to give the mean concentration within the specimen, C_m:

$$C_m - C_s = \frac{8}{\pi^2}(C_0 - C_s)\exp\left(-\frac{\pi^2}{L^2}Dt\right) \tag{6.23}$$

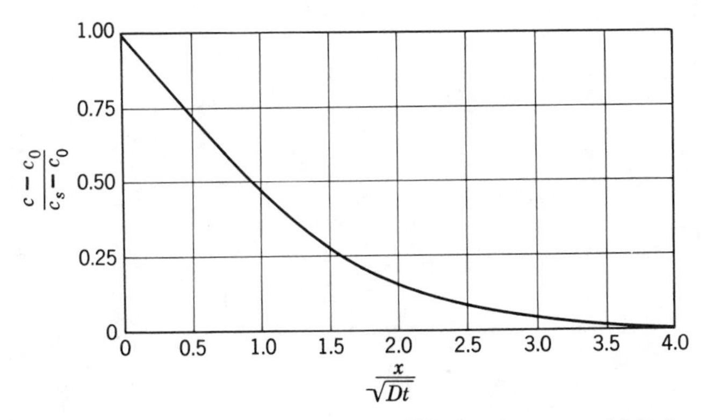

Fig. 6.5. Penetration curve for unidimensional diffusion into a semi-infinite medium of uniform initial concentration c_0 and constant surface concentration c_s; c is the concentration at x and t.

and is valid for $\frac{C_m - C_s}{C_0 - C_s} < 0.8$, that is, long times. The change in the mean concentration with time is shown in Fig. 6.6 for a slab and for other geometries. The dimensionless parameter $\sqrt{Dt}/l$ is used where l is the sphere radius, cylinder radius, or half the slab thickness. For order of magnitude calculations, we note that exchange or homogenization is nearly complete (more than 98%) when $\sqrt{Dt} = 1.5l$ for a slab, $\sqrt{Dt} = 1.0l$ for a cylinder, and $\sqrt{Dt} = 0.75l$ for a sphere. These approximations allow a rapid estimate of the extent to which a diffusion-controlled process will occur under given conditions.

The experimental technique by which most diffusion coefficients are measured involves the application of a thin film of radioactive material on the host material. If the quantity α of a radioactive tracer is diffused into a semi-infinite rod for a time t, the thin-film solution of Fick's law is

$$C = \frac{\alpha}{2\sqrt{\pi Dt}} \exp\left(-\frac{x^2}{4Dt}\right) \tag{6.24}$$

(Initial conditions $t = 0$, $c = 0$ for $|x| > 0$.) Measurement of the relative concentration of radioactive atoms as a function of distance from the surface yields the diffusion coefficient directly (Fig. 6.7). Again the important parameter is $x \approx \sqrt{Dt}$, which indicates the approximate diffusion distance during the time t.

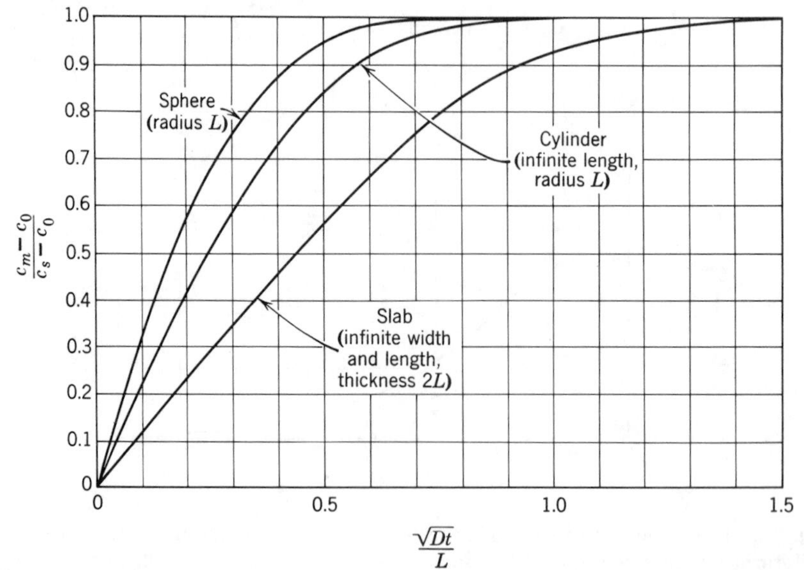

Fig. 6.6. Fractional saturation of sheet, cylinder, and sphere of uniform initial concentration c_0 and constant surface concentration c_s, with c_m the mean concentration at time t.

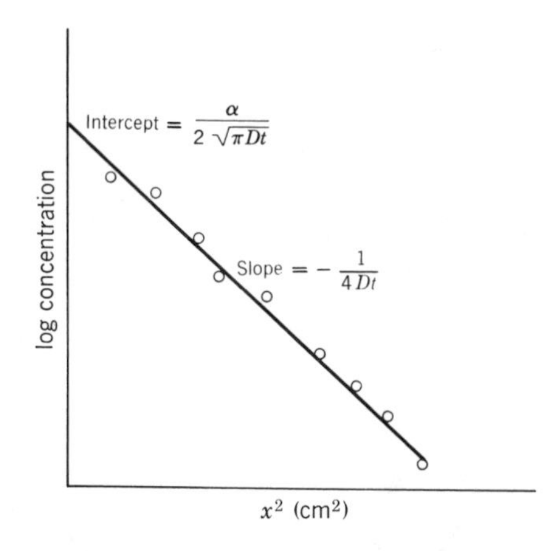

Fig. 6.7. Penetration curve of a radioactive tracter diffusing into a semi-infinite medium for time t.

Often the diffusion coefficient is a function of the distance into the solid, that is, a function of concentration, and Eq. 6.5 rather than Eq. 6.6 must be solved to determine the time-distance concentration relationships. Solutions for many special cases are given in references 1, 3, and 6.

6.2 Diffusion as a Thermally Activated Process

If we consider the change in energy of an atom as it moves from one lattice site to another by a diffusion jump, there is an intermediate position of high energy (Fig. 6.8). Only a certain fraction of the atoms present in the lattice have sufficient energy to overcome this barrier to moving from one site to another. The magnitude of energy which must be supplied in order to overcome this barrier is called the *activation energy* for the process. Diffusion is one of many processes characterized by an energy barrier between the initial and the final states. As the temperature is increased, the fraction of atoms present which have sufficient energy to surmount this barrier increases exponentially, so that the temperature dependence of diffusion can be represented as $D = D_0 \exp(-\Delta G^{\dagger}/RT)$.

Diffusion can be considered a special case of more general reaction-rate theories.*

Two general considerations are the basis for rate studies. The first is that each individual step in a rate process must be relatively simple, such as an individual diffusion jump. Although overall processes are frequently complex and require a series of individual separate unit steps, the individual steps are simple, and the movement of an atom from an

*S. Glasstone, K. J. Laidler, and H. Eyring, *The Theory of Rate Processes*, McGraw-Hill Book Company, Inc., New York, 1941.

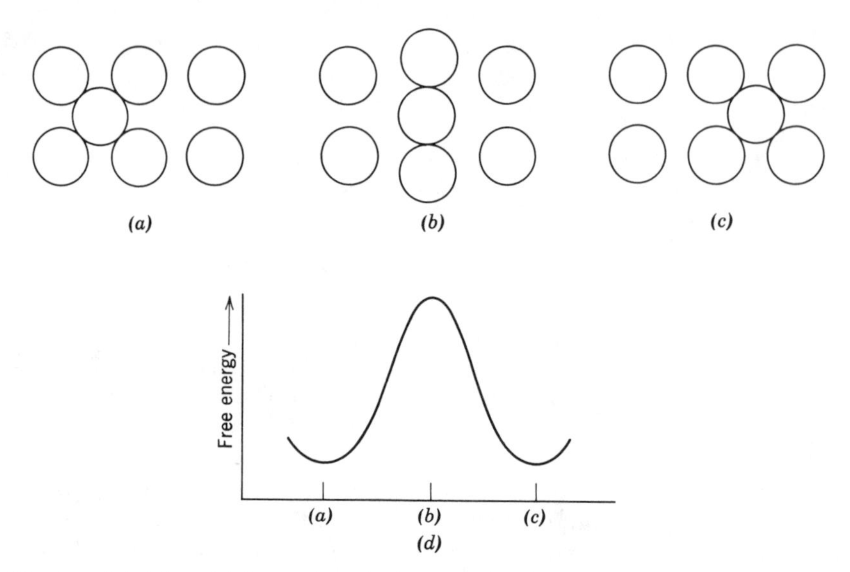

Fig. 6.8. (a), (b), and (c) are schematic drawings showing the sequence of configurations involved when an atom jumps from one normal site to a neighboring one. (d) shows how the free energy of the entire lattice would vary as the diffusing atom is reversibly moved from configuration (a) to (b) to (c).

occupied site to an unoccupied site is typical. Second, the reaction path of each step, such as the individual atom jump in diffusion, a molecular decomposition, or formation of a new chemical bond, involves an *activated complex* or *transition state* of maximum energy along the reaction path. Of all possible parallel paths of reaction, the one with the lowest energy barrier is most rapid and the major contributor to the overall process. This activated-complex theory has provided a general form of the equation for rate processes and is a model which allows semi-empirical calculations for simple processes. The concept of an activated complex corresponding to the energy maximum, actually a saddle point, intermediate between the initial and final atom positions (Fig. 6.8) has been universally accepted as the basis for reaction-rate studies.

There are two main principles which form the basis for reaction-rate theory of activated processes. First, the activated complex can be treated as any other atomic species and is in equilibrium with a reactant, even though its lifetime is short. That is, an equilibrium constant $K^{\dagger}$ can be used for the formation of the activated complex. If the free energy of formation is $\Delta G^{\dagger}$, then $\Delta G^{\dagger}$ equals $-RT \ln K^{\dagger}$. Second, the rate of transition of the activated complex into a product is proportional to a frequency factor ν, which for solids has a value of about 10^{13}/sec. The reaction rate **k** is thus the product of the frequency term and the concentration of activated complexes.

For an individual reaction step to form an activated complex, such as $A + B = AB^{\dagger}$, assuming unit activity coefficients, $K^{\dagger}$ equals $C_{AB^{\dagger}}/C_A C_B$, and the reaction rate or number of activated complexes which decompose per unit time is given by

$$\text{Reaction rate} = \nu C_{AB^{\dagger}} = (\nu K^{\dagger}) C_A C_B \tag{6.25}$$

where the coefficient of concentration terms is the specific reaction constant. Then

$$K^{\dagger} = \exp\left(-\frac{\Delta G^{\dagger}}{RT}\right) = \exp\left(-\frac{\Delta H^{\dagger}}{RT}\right)\exp\frac{\Delta S^{\dagger}}{R} \tag{6.26}$$

and the specific reaction-rate constant is

$$\mathbf{k} = \nu \exp\left(-\frac{\Delta H^{\dagger}}{RT}\right)\exp\frac{\Delta S^{\dagger}}{R} \tag{6.27}$$

To apply this general theory to the diffusion process, the fundamental step during diffusion is the passage of a solute atom from one normal or interstitial position to an adjacent vacant site or interstitial position. The atom midway between two positions, as illustrated in Fig. 6.8, is in the activated state. If the concentration of the diffusing species is c, the rate of passage of the atoms from one site to another is given by $\nu K c^{\dagger}$, and a net atom flux results from a concentration gradient. A more general approach is to consider the driving force as a perturbing force on the activation barrier. The driving force could be a chemical-potential gradient (concentration gradient), electrical field, or the like, and may be treated as follows. Consider the energy-distance plot in Fig. 6.9. The gradient in the free energy (chemical potential) per atom is

$$\chi = -\frac{1}{N}\frac{dG}{dx} = -\frac{1}{N}\frac{d\mu}{dx} \tag{6.28}$$

thus for one jump

$$\Delta\mu = -N\lambda\chi \tag{6.29}$$

The rate in the forward direction is related to the activation energy for that direction ($\Delta G^{\dagger}_{\text{forward}} = \Delta G^{\dagger} - 1/2\chi\lambda$) and is

$$\mathbf{k}_{\text{forward}} = \nu \exp\left(-\frac{\Delta G^{\dagger} - \frac{1}{2}\chi\lambda}{kT}\right) \tag{6.30}$$

The back reaction has a different probability, owing to a different barrier

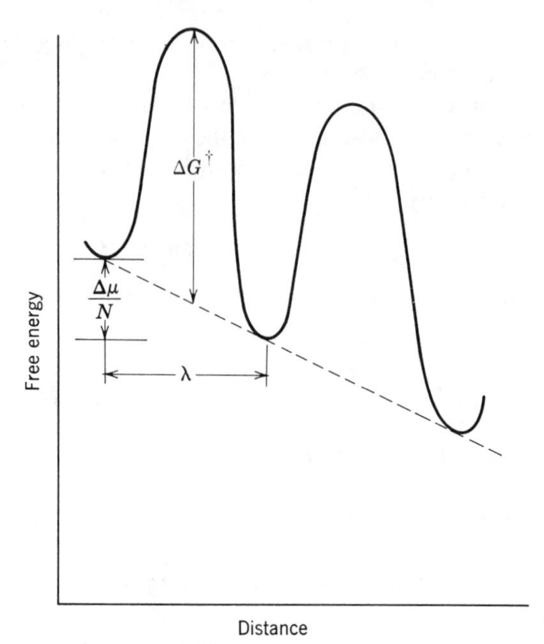

Fig. 6.9. Diffusion in a potential gradient $\Delta\mu$; $\Delta G^{\dagger}$ is the activation energy and λ the jump distance.

height:

$$\mathbf{k}_{\text{backward}} = \nu \exp\left(-\frac{\Delta G^{\dagger} + \frac{1}{2}\chi\lambda}{kT}\right) \tag{6.31}$$

The net rate is

$$\begin{aligned}\mathbf{k}_{\text{net}} = \mathbf{k}_f - \mathbf{k}_b &= \nu \exp\left(-\frac{\Delta G^{\dagger}}{kT}\right)\left[\exp\left(\frac{\frac{1}{2}\chi\lambda}{kT}\right) - \exp\left(-\frac{\frac{1}{2}\chi\lambda}{kT}\right)\right] \\ &= 2\nu \exp\left(-\frac{\Delta G^{\dagger}}{kT}\right)\sinh\left[\frac{1}{2}\frac{\chi\lambda}{kT}\right]\end{aligned} \tag{6.32}$$

In the case of diffusion the chemical-potential gradient is usually small compared with the thermal energy kT $\left(\frac{\Delta\mu}{kT} \ll 1\right)$ and similarly for the electrical potential ($ze\lambda/kT \ll 1$); thus Eq. 6.32 can be approximated as

$$\mathbf{k}_{\text{net}} = 2\nu e^{-\Delta G^{\dagger}/kT}\left[\frac{\chi\lambda}{2kT}\right] \tag{6.33}$$

replacing the gradient term $\chi = -\frac{1}{N}\frac{d\mu}{dx}$

$$\mathbf{k}_{\text{net}} = 2\nu e^{-\Delta G^{\dagger}/kT}\left[-\frac{\lambda\, d\mu/dx}{2NkT}\right] = -\frac{\lambda\nu}{NkT}e^{-\Delta G^{\dagger}/kT}\frac{d\mu}{dx} \tag{6.34}$$

The mass flux is related to the reaction-rate constant by the following substitutions:

$$\lambda\,\mathbf{k}_{\text{net}} = \text{velocity}$$

and

$$J_{\text{flux}}\left(\frac{\text{mole}}{\text{sec-cm}^2}\right) = \lambda\,\mathbf{k}_{\text{net}}c\left(\frac{\text{mole}}{\text{cm}^3}\right) = \lambda\,\mathbf{k}_{\text{net}}c$$

$$= -\frac{\lambda^2\nu c}{NkT}e^{-\Delta G^{\dagger}/kT}\frac{d\mu}{dx} \tag{6.35}$$

Comparison with Eq. 6.8

$$B \equiv \frac{\lambda^2\nu}{kT}e^{-\Delta G^{+}/kT}$$

for an ideal solution $\mu = \mu_0 + NkT\ln c$ and $d\mu/dx = \frac{NkT}{c}\frac{dc}{dx}$, we have

$$J_{\text{flux}} = -\lambda^2\nu e^{-\Delta G^{\dagger}/kT}\frac{dc}{dx} = -D\frac{dc}{dx} \tag{6.36}$$

where

$$D = \lambda^2\nu e^{-\Delta G^{\dagger}/kT} = \lambda^2\nu e^{+\Delta S^{\dagger}/k}e^{-\Delta H^{\dagger}/kT} \tag{6.37}$$

by comparison with Fick's first law. This does not include the geometrical factor or the probability that the adjacent site is vacant; however, comparison with Eq. 6.16

$$D = \gamma\lambda^2\nu e^{-\Delta G^{\dagger}/kT} = \gamma\lambda^2\Gamma \tag{6.38}$$

shows that $\nu\exp[-\Delta G^{\dagger}/kT]$ is the jump frequency.

This analysis of diffusion in terms of general reaction-rate theory and activated processes gives a rational basis for the temperature-dependent and temperature-independent terms (the temperature-independent term includes the entropy of the activated complex). In Eq. 6.38, we have ignored the activity coefficient and changes of the diffusion coefficient with composition. In general, the activity coefficient should be retained, and frequently it is observed that there is a substantial change in diffusion coefficients with composition in a multi-component system. One of the major results of Eq. 6.38 is to show that the diffusion coefficient has an exponential temperature dependence. Diffusion coefficients can almost always be represented within the precision of the experimental measurements by the expression

$$D = D_0 \exp\left(-\frac{Q}{RT}\right) \tag{6.39}$$

where the preexponential value D_0 can be separated into more fundamental terms from comparison with Eq. 6.38. The term Q in Eq. 6.39 is sometimes called the experimental activation energy.

6.3 Nomenclature and Concepts of Atomistic Processes

There are several terminologies in the literature for specifying diffusion coefficients (Table 6.2). The term *self-diffusion* refers to diffusion in the absence of a chemical concentration gradient; the *tracer diffusion* coefficient refers to the same constant which measures only the random motion of a radioactive ion, with no net flow of vacancies or atoms. Strictly speaking, there is always a concentration gradient when A is plated on B or on an AB solid solution, but with radioactive tracers the amount of solute added can be so small that the composition change can be ignored.

The diffusional processes which occur during high-temperature reactions often result from composition gradients. The diffusion coefficient defined by Eq. 6.40,

$$\tilde{D} = -\frac{J}{dc/dx} \tag{6.40}$$

is designated as the *chemical* or *interdiffusion coefficient*. The chemical diffusion coefficient can be obtained by an analysis that allows for $\tilde{D}$ to vary with distance or composition. In the interdiffusion of MgO and NiO, for example, the cations diffuse in a fixed oxygen matrix; thus the effective, chemical, or interdiffusion coefficient in an oversimplified manner represents counterdiffusion of magnesium and nickel ions, which are related to the individual tracer diffusion coefficients by the Darken equation*

$$\tilde{D} = [X_2 D_1^T + D_2^T X_1]\left(1 + \frac{d \ln \gamma_1}{d \ln X_1}\right) \tag{6.41}$$

where X_1 and X_2 are the mole fractions of the diffusing species (e.g., Ni and Mg) and γ_1 is the activity coefficient of component 1. For ideal or dilute solutions the term $\frac{d \ln \gamma_1}{d \ln X_1} \to 0$, and the interdiffusion coefficient is just the weighted average of the tracer diffusion coefficients.†

*L. Darken, *Trans. A.I.M.E.*, **174,** 184 (1948).

†Experimental data often can be represented by a Darken type equation or by a Nernst-Planck equation which couples the flux of the two species through the internal electrical field which results if one ion is more mobile than the other. See Chap. 9 for examples.

Table 6.2. Common Symbols and Terms for Diffusion Coefficients

The *tracer* or *self* diffusion coefficient represents only the *random-walk* diffusion process, i.e., no chemical potential gradients:

$$D^T, D^*, D_{\text{self}}$$

The lattice diffusion coefficient refers to any diffusion process within the *bulk* or *lattice* of the crystal:

$$D_l$$

The *surface* diffusion coefficient measures diffusion along a surface:

$$D_s$$

Boundary diffusion occurs along an *interface* or *boundary*, such as a grain boundary; the term may also include diffusion along *dislocations* (dislocation-pipe diffusion):

$$D_b$$

The *chemical, effective,* or *interdiffusion* coefficient refers to diffusion in a chemical-potential gradient:

$$\tilde{D}$$

Intrinsic diffusion refers to a process when only native point defects (thermally created) are the vehicle for transport.

Extrinsic diffusion refers to diffusion via defects not created from thermal energy, e.g., impurities.

The *apparent* diffusion coefficient includes the contributions from several diffusion paths into one net diffusion coefficient:

$$D_a$$

The *defect* diffusion coefficient refers to the diffusivity of a particular point defect and usually implies more than just the random motion but also the effects of the biasing concentration gradient. Usually the *vacancy* diffusion coefficient is the defect diffusivity of interest:

$$D_v$$

Defect diffusion coefficients are often specified; for example, the interstitial diffusion coefficient refers to species diffusing via an interstitial mechanism; the vacancy diffusion coefficient refers to diffusion of vacant sites. When diffusion occurs by a vacancy mechanism, the tracer diffusion coefficient D_l^T is equal to the diffusion coefficient of the vacancy D_v times the fraction of the vacant lattice sites V_l:

$$[V_l]D_v = D_l^T \tag{6.42}$$

For many nonstoichiometric ceramics, for example, UO_{2+x}, FeO_{1+x}, MnO_{1+x}, the chemical diffusion coefficient in Eq. 6.41 may be determined by measuring oxidation or reduction of the material from one anion-cation ratio to another. For example, if a crystal is oxidized, oxygen ions diffuse into the bulk, and cations simultaneously diffuse to the surface to react with oxygen. Often one of the species (anion or cation) has a much higher diffusivity, for example, O in UO_2, metal ions in FeO, CoO, MnO; in this special case, when diffusion occurs by a vacancy mechanism, the chemical diffusion coefficient is approximately

$$\tilde{D} \approx (1 + Z)D_v \tag{6.43}$$

where Z is the magnitude of the effective charge of the fastest ion.

Other frequently used terms distinguish diffusion within the crystalline lattice from diffusion along line or planar defects. The lattice or bulk diffusion coefficient is used to designate the former and may refer to tracer or chemical diffusion. Other diffusion coefficients are called dislocation diffusion coefficient, grain-boundary diffusion coefficient, and surface diffusion coefficient and refer to the diffusion of atoms or ions within the specified region, which are often found to be *high diffusivity paths* and are discussed in Section 6.6.

In Section 4.7 we discussed the association of defects such as vacancy pairs and also the association between solutes and lattice defects. These associations have a significant influence on the atomistic processes occurring and on the resulting diffusion coefficients. For example, if a solute ion in substitutional solid solution is about the same size as the host lattice ion and randomly distributed relative to defects, it has a diffusion coefficient similar to that of the host ion. However, if the solute is associated with a vacancy, it always has an adjacent site to jump into (see Eq. 6.17) such that the solute diffusion coefficient is similar to that of the vacancy diffusion coefficient rather than the lattice diffusion coefficient, that is, increased by many orders of magnitude. In Section 4.8 association in wustite is considered as one example; for wustite it is experimentally observed that diffusional processes can lead to phase separation at temperatures as low as 300°C.

6.4 Temperature and Impurity Dependence of Diffusion

The diffusion of ions into or out of ceramics is known to be strongly affected by temperature, by the ambient atmosphere and impurities, and by high diffusivity paths discussed in Section 6.6. The motion of ions in condensed matter was shown to be a thermally activated process (Eq. 6.38). We now wish to consider the individual terms for diffusion coefficients, usually written as $D = D_0 e^{-Q/kT}$. For the purpose of discussion we choose potassium chloride, for which careful measurements have been made. The analogy of this system to many important ceramic materials is appropriate, since many oxides also have a close-packed anion lattice.

Diffusion of potassium ions in potassium chloride occurs by interchange of the potassium ion with cation vacancies. From Chapter 4 we know that the concentration of vacancies in a pure crystal is given by the Schottky formation energy

$$[V_K'] = e^{-\Delta G_s/2kT} \tag{6.44}$$

Combining the concentration of defects with the motion term (Eq. 6.38),

$$\begin{aligned} D_K &= [V_K']\gamma\lambda^2 \nu e^{-\Delta G^\dagger/kT} = \gamma\lambda^2 \nu e^{-(\Delta G_s/2kT) - \Delta G^\dagger/kT} \\ &= \gamma\lambda^2 \nu e^{\left(\frac{\Delta S_s}{2} + \Delta S^\dagger\right)/k} e^{[-\Delta H^\dagger - (\Delta H_s/2)]/kT} \end{aligned} \tag{6.45}$$

Thus we see that the random-walk diffusion process can be expressed as $D = D_0 e^{-Q/kT}$. Both D_0 and Q must be of reasonable magnitude before applying diffusion models to specific materials. The preexponential term for a pure stoichiometric compound can be estimated as

$$D_0(\text{vacancy}) = \gamma\lambda^2 \nu \exp\left(\frac{\Delta S^\dagger + \Delta S_s/2}{k}\right) = 10^{-2} - 10^{+1}$$

$$D_0(\text{interstitial}) = \gamma\lambda^2 \nu \exp\left(\frac{\Delta S^\dagger}{k}\right) = 10^{-3} - 10^{+1}$$

The numerical values were obtained by assuming $\lambda \approx 2\ \text{Å} = 2 \times 10^{-8}$ cm, $\gamma \approx 0.1$, $\nu = 10^{13}$/sec, and $\Delta S^\dagger/k$ and $\Delta S_s/k$ as small positive numbers. The activation entropy and enthalpy terms in Eq. 6.45 for KCl are given in Table 6.3.

In most crystals, diffusion is more complex because of impurity content and past thermal history. As shown in Fig. 6.10 for KCl, the high-temperature region represents the intrinsic properties of pure materials. The slope of the $\ln D$ versus $1/T$ plot in this region gives

$$\left(\frac{\Delta H^\dagger}{k} + \frac{\Delta H_s}{2k}\right).$$

In the case of KCl, this represents the enthalpy of potassium ion migration and of potassium vacancy formation. The intercept at $1/T = 0$ gives D_0^{in} for the intrinsic crystal. Table 6.4 gives the formation enthalpy of Schottky defects and the enthalpy of motion for several halides.

In the lower temperature region, impurities within the crystal fix the vacancy concentration. This is the extrinsic region, where the diffusion

Table 6.3. Enthalpy and Entropy Values for Diffusion in KCl

Schottky defect formation:	
Enthalpy ΔH_s (eV)	2.6
Entropy $\Delta S_s/k$	9.6
Potassium ion migration:	
Enthalpy ΔH_1^+(eV)	0.7
Entropy $\Delta S_1^+/k$	2.7
Chlorine ion migration:	
Enthalpy ΔH_2^+(eV)	1.0
Entropy $\Delta S_2^+/k$	4.1

Source. S. Chandra and J. Rolfe, *Can. J. Phys.*, **48**, 412 (1970).

Table 6.4. Values of the Schottky Formation Enthalpy ΔH_s and the Cation Jump Enthalpy $\Delta H^\dagger$ of Several Halides

Substance	ΔH_s (eV)	$\Delta H^\dagger$ (eV)
LiF	2.34	0.70
LiCl	2.12	0.40
LiBr	1.8	0.39
LiI	1.34, 1.06	0.38, 0.43
NaCl	2.30	0.68
NaBr	1.68	0.80
KCl	2.6	0.71
KBr	2.37	0.67
KI	1.60	0.72
CsCl	1.86	0.60
CsBr	2.0	0.58
CsI	1.9	0.58
TlCl	1.3	0.5
$PbCl_2$	1.56	
$PbBr_2$	1.4	

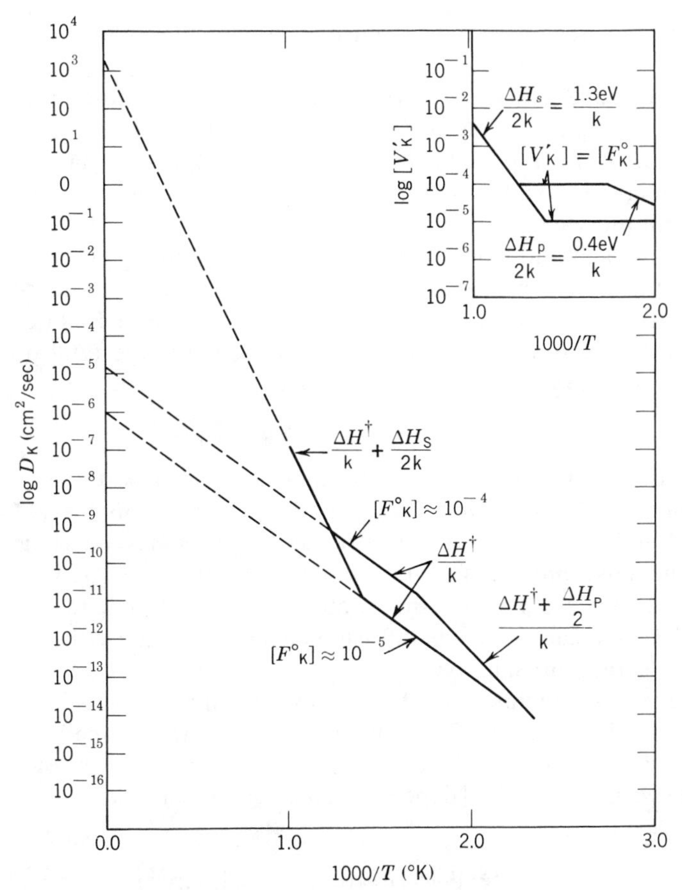

Fig. 6.10. Diffusion-temperature plot for KCl with 10^{-4} and 10^{-5} atom fraction divalent cation impurities. Insert plot shows the variation of the $[V'_K]$ with temperature.

coefficient is given by

$$D_K = \gamma\nu\lambda^2[F^\bullet_K]e^{-(\Delta G^\dagger/kT)} \tag{6.46}$$

where $[F^\bullet_K]$ is the concentration of divalent cation impurities such as calcium and

$$[V'_K] = [Ca^\bullet_K]$$

From Fig. 6.10, we note that $D_0^{ex} \ll D_0^{in}$, consistent with these relationships.

The bend in the curve in Fig. 6.10 occurs in the region where the intrinsic defect concentration is comparable with the extrinsic defect concentration resulting from impurities. When the Schottky formation enthalpy is in the range of 150 kcal/mole (6 eV), as is typical for BeO, MgO, CaO, and Al_2O_3, the crystal must have an aliovalent impurity concentration smaller than 10^{-5} before intrinsic diffusion can be observed at 2000°C. Thus it is unlikely that intrinsic diffusion has ever been observed in these oxides; impurity levels of parts per million are sufficient to control the vacancy concentrations.

When association of vacancies and solutes or precipitation of solutes occurs, there may be a variety of values for the observed activation energy of diffusion. Consider as an example the case of KCl doped with $CaCl_2$. If association between $Ca_K^{\cdot}$ and V_K' occurs, the total potassium vacancy concentration $[V_K']_{Tot}$ is increased

$$[V_K']_{Tot} = [V_K'] + (Ca_K^{\cdot} V_K')^*$$

to include the vacancies which become associated with the impurity according to Eq. 4.39. Interdiffusion measurements in the NiO–Al_2O_3 system* indicate that significant association occurs as Al ions migrate into NiO. The most rapidly diffusing complex appears to be $[Al_{Ni}^{\cdot} V_{Ni}'']'$, which forms with an association enthalpy about 6 to 9 kcal/mole (compare with Table 4.3) and leads to a high diffusion coefficient because the vacancy is coupled to the diffusing ion.

The eventual precipitation of solute as the temperature is lowered also affects the diffusivity. In Fig. 6.10 a second change in slope is shown for the diffusivity of K in KCl containing 10^{-4} mole fraction Ca solute. From Eq. 4.43 we note that $[V_K']$ depends on the heat of solution or precipitation enthalpy,

$$[V_K'] = [CaCl_2]^{1/2}_{\text{solubility limit}} \exp\left(\frac{-\Delta g_P}{2kT}\right) \tag{6.47a}$$

so that the diffusion coefficient becomes

$$D^{ex}_{(ppt)} = \gamma\nu\lambda^2[CaCl_2]^{1/2}_{s.l.} \exp\left(\frac{\Delta s^{\dagger} + \Delta s_P/2}{k}\right) \exp\left(-\frac{\Delta H^{\dagger} + \dfrac{\Delta H_P}{2}}{kT}\right) \tag{6.47b}$$

Similar reasoning applies to the case of Al_2O_3 as an impurity in MgO. The vacancy concentration when Al_2O_3 begins to precipitate as spinel depends on the heat of solution ΔH_P of Al_2O_3 in MgO. Thus, the activation energy for diffusion of magnesium ions would be $Q = \Delta H^{\dagger} + \frac{\Delta H_P}{2}$. Other more complex situations may arise in real materials.

*W. J. Minford and V. S. Stubican, *J. Am. Ceram. Soc.*, **57**, 363, 1974.

6.5 Diffusion in Crystalline Oxides

The diffusion characteristics of crystalline oxides can be classified, first, on the basis of whether they are stoichiometric, that is, free from appreciable electron or electron-hole concentrations; and second, on the basis of whether diffusion characteristics are intrinsic or related to impurity concentration. We shall discuss diffusion characteristics of oxides according to the following scheme:

1. Stoichiometric: (*a*) intrinsic diffusion coefficients, (*b*) impurity-controlled diffusion coefficients.
2. Nonstoichiometric: (*a*) intrinsic diffusion coefficients, (*b*) impurity-controlled diffusion coefficients.

Although this arrangement is a reasonable and inclusive one in the sense that each section is well defined and all the available data should fit in somewhere, many of the experimental data now available are not easy to characterize. Frequently it is not clear whether diffusion characteristics are intrinsic or related to impurities. Other times it is not clear whether compositions are stoichiometric or nonstoichiometric over a limited composition range. Experimental data for diffusion in several oxides are collected in Fig. 6.11.

Stoichiometric Oxides. Measured diffusion coefficients for oxides have not generally shown the break or knee in a curve, corresponding to the change from impurity-controlled diffusion to intrinsic diffusion, that is commonly found for alkali halides (Fig. 6.10), probably because the temperature range of measurements has not been large.

There are a number of data for stoichiometric oxides which clearly correspond to composition-controlled diffusion coefficients. One group of these oxides includes those having the fluorite structure, such as UO_2, ThO_2, and ZrO_2. Additions of divalent or trivalent cation oxides, such as La_2O_3 and CaO, go into solid solution and are known from X-ray and electrical conductivity studies to form structures in which the concentration of oxygen ion vacancies is fixed by composition and is independent of temperature (see Chapter 4). In $Zr_{0.85}Ca_{0.15}O_{1.85}$, for example, the oxygen ion vacancy concentration is high and independent of temperature. Consequently the oxygen ion diffusion coefficient has a temperature dependence fixed solely by the activation energy required for oxygen ion mobility (29 kcal/mole). Similarly, low-temperature oxygen ion diffusion in both stoichiometric and nonstoichiometric UO_2 has been found to occur by an interstitialcy mechanism in which an interstitial ion moves into a regular lattice site, bumping the lattice ion into the neighboring interstitial. Here the activation energy is 28 kcal/mole. In the zirconia-calcia system, the oxygen ion diffusion coefficient increases as the

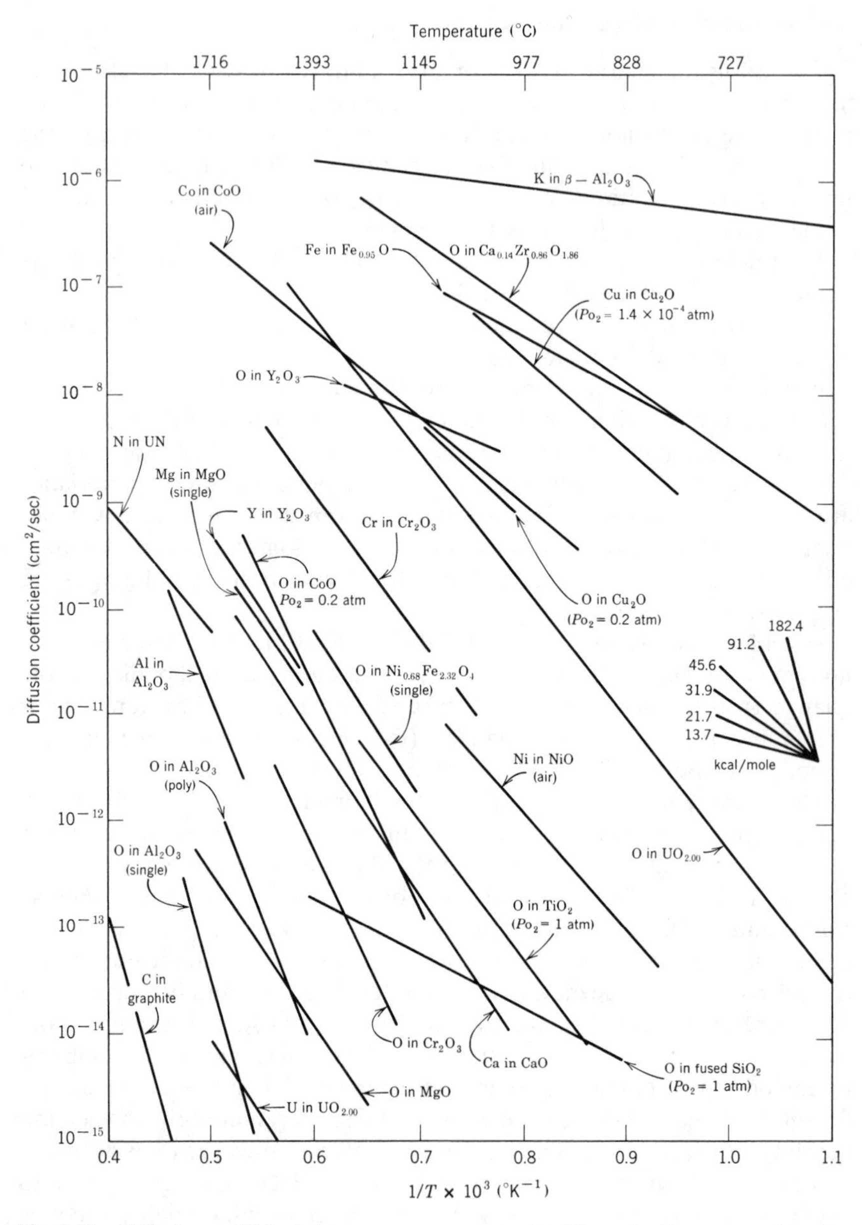

Fig. 6.11. **Diffusion coefficients in some common ceramics. The activation energy Q can be estimated from the slope and the insert, for example, for O in $Ca_{0.14}Zr_{0.86}O_{1.86}$ the $Q \approx 29$ kcal/mole.**

concentration of oxygen ion vacancies increases (oxygen-to-metal ratio decreases). In contrast, for the interstitialcy mechanism in UO_2 the diffusion coefficient for oxygen ion increases as the concentration of interstitial oxygen ions increases, at least for small concentrations of interstitial oxygen ions (oxygen-to-metal ratio increases).

Cation diffusion coefficients for stoichiometric compositions are more difficult to categorize unequivocally. The high activation energies for diffusion in systems such as magnesium oxide, calcium oxide, and spinels suggest that the measured values at the high temperatures studied may be intrinsic and independent of minor impurities. However, lack of specific data for "high"- and "low"-temperature behavior and of other confirming measurements makes this conclusion tentative.

Nonstoichiometric Oxides. The most common behavior for many oxide materials is as intrinsic nonstoichiometric semiconductors at equilibrium with an oxidizing or reducing atmosphere. Typical examples of this type of behavior are the formation of interstitial zinc ions in zinc oxide and the formation of cation vacancies in cobaltous oxide. As discussed in Chapter 4, when zinc oxide is heated in a reducing atmosphere, zinc vapor is maintained at an equilibrium with interstitial zinc ions and excess electrons according to the relationship

$$\mathrm{Zn}(g) = \mathrm{Zn}_i^{\cdot} + e' \tag{6.48a}$$

The concentration of interstitial zinc ions is related to the pressure of the zinc vapor:

$$C_{\mathrm{Zn}_i^{\cdot}} = [\mathrm{Zn}_i^{\cdot}] \approx P_{\mathrm{Zn}}^{1/2} \tag{6.48b}$$

The diffusion of zinc ions occurs by the interstitial mechanism; thus the diffusion coefficient increases with P_{Zn} (Fig. 6.12). A similar type of

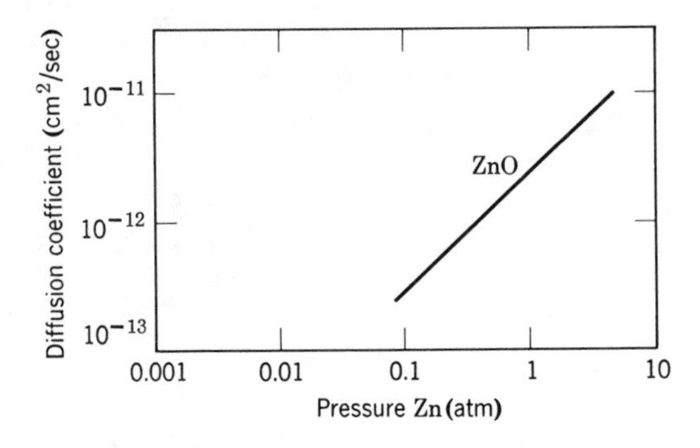

Fig. 6.12. Effect of atmosphere on diffusion in ZnO.

dependence occurs for oxygen interstitial diffusion in nonstoichiometric UO_2.

Diffusion through vacant sites in nonstoichiometric oxides is usually classed as metal-deficient or oxygen-deficient oxides:

(*a*) METAL-DEFICIENT OXIDES (e.g., FeO, NiO, CoO, MnO). The concentration of vacant cation sites in many nonstoichiometric compounds is large, especially in transition metal compounds because of the variable valence cations. For example, $Fe_{1-x}O$ contains from 5 to 15% vacant iron sites. The simple defect reactions are

$$2M_M + \frac{1}{2}O_2(g) = O_0 + V_M'' + 2M_M^{\cdot} \tag{6.49a}$$

where $M_M^{\cdot}$ represents an electron hole localized at a cation (e.g., $M_M^{\cdot} = Co^{+3}, Fe^{+3}, Mn^{+3}$). Equation 6.49a is a solubility reaction for oxygen dissolving into the metal oxide MO and is governed at equilibrium by the free energy of solution ΔG_0:

$$\frac{4[V_M'']^3}{P_{O_2}^{1/2}} = K_0 = e^{-\Delta G_0/kT} \tag{6.49b}$$

The diffusion of cations in the temperature range where the defect concentration is controlled by this reaction is given by

$$\begin{aligned} D_M &= \gamma\nu\lambda^2[V_M''] \exp\left[-\frac{\Delta G^{\dagger}}{kT}\right] \\ &= \gamma\nu\lambda^2\left(\frac{1}{4}\right)^{1/3} P_{O_2}^{1/6} \exp\left[-\frac{\Delta G_0}{3kT}\right] \exp\left[-\frac{\Delta G^{\dagger}}{kT}\right] \end{aligned} \tag{6.50}$$

The effects of pressure and temperature are shown schematically in Fig. 6.13. The actual defect equilibria are more complex; the defects may be effectively neutral, singly charged, or doubly charged. Figure 6.14 shows the possible defect concentrations in CoO at 1100°C with 400 ppm trivalent cation impurity, $[F_M^{\cdot}]$. The measured cobalt tracer diffusion coefficient is observed to conform to the total vacancy concentration $[V_{Tot'}]$. Data for other oxides are given in Fig. 6.15.

(*b*) OXYGEN-DEFICIENT OXIDES. A similar set of relationships exists for defects in structures with anion vacancies:

$$O_0 = 1/2O_2(g) + V_O^{\cdot\cdot} + 2e' \tag{6.51}$$

$$[V_O^{\cdot\cdot}] \approx (1/4)^{1/3} P_{O_2}^{-1/6} \exp\left[-\frac{\Delta G_0}{3kT}\right] \tag{6.52}$$

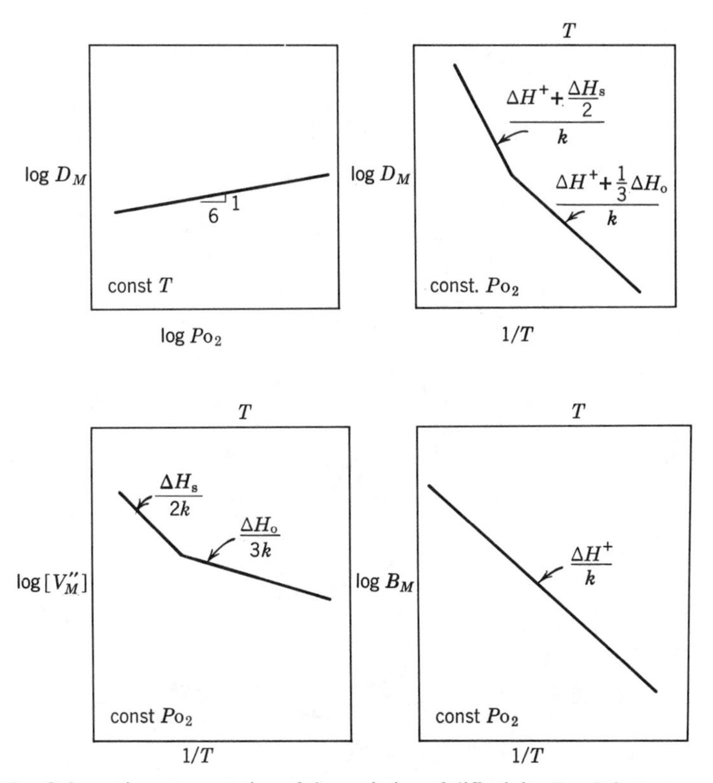

Fig. 6.13. Schematic representation of the variation of diffusivity D_M, defect concentration $[V_M'']$, and mobility B_M with temperature and oxygen pressure.

Thus the oxygen diffusion coefficient is

$$D_o = \gamma\nu\lambda^2[V_O^{\cdot\cdot}]\exp\left[-\frac{\Delta G^\dagger}{kT}\right]$$
$$= \gamma\nu\lambda^2(1/4)^{1/3}P_{O_2}^{-1/6}\exp\left[-\frac{\Delta G_0}{3kT}\right]\exp\left[-\frac{\Delta G^\dagger}{kT}\right] \qquad (6.53)$$

The pressure and temperature effects are shown schematically in Fig. 6.16, and actual data are given in Fig. 6.15 for CdO and Nb_2O_5. Figure 6.16 also shows three possible temperature regimes: (1) low temperature, where the oxygen vacancy concentration is controlled by impurities; (2) intermediate temperature, where the oxygen vacancy concentration changes, owing to variation of the oxygen solubility with temperature (nonstoichiometry); and (3) high temperature, where thermal vacancies become dominant.

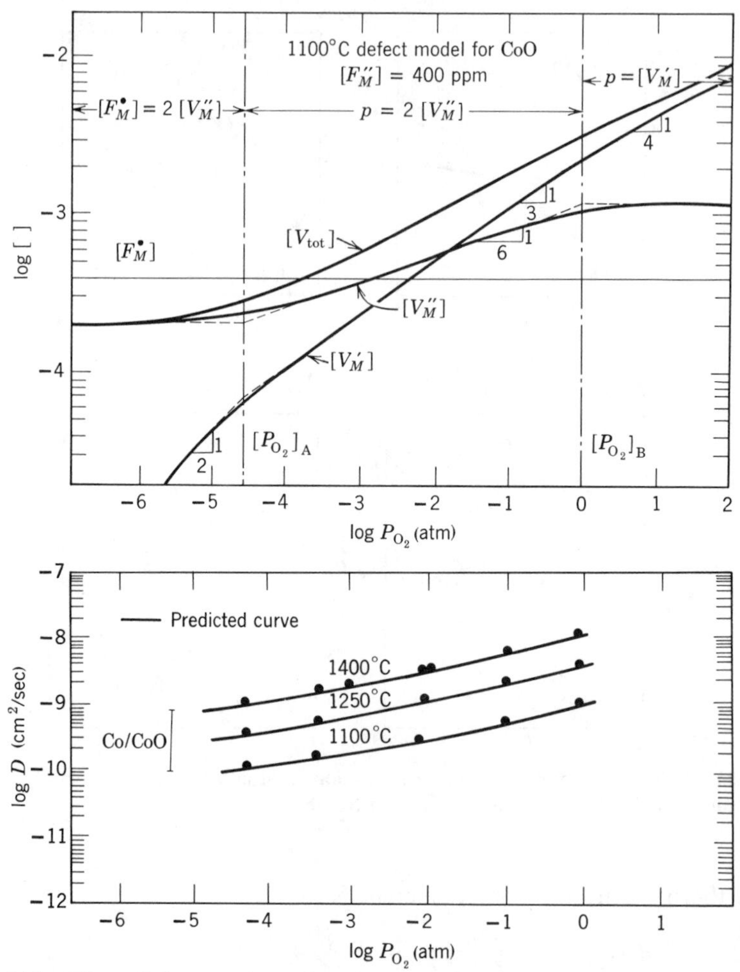

Fig. 6.14. Effect of the oxygen partial pressure on the defect concentration and cobalt tracer diffusivity in CoO. From W. B. Crow, Aerospace Res. Labs.

As a final example of defect equilibria and diffusion in nonstoichiometric oxides, let us consider uranium dioxide, which may exist at high temperatures as oxygen-deficient UO_{2-x} or, with an oxygen excess, UO_{2+x}. The appropriate defect reactions for U and O have been reviewed by H. Matzke*.

J. de Physique, **34**, C9-317 (1973).

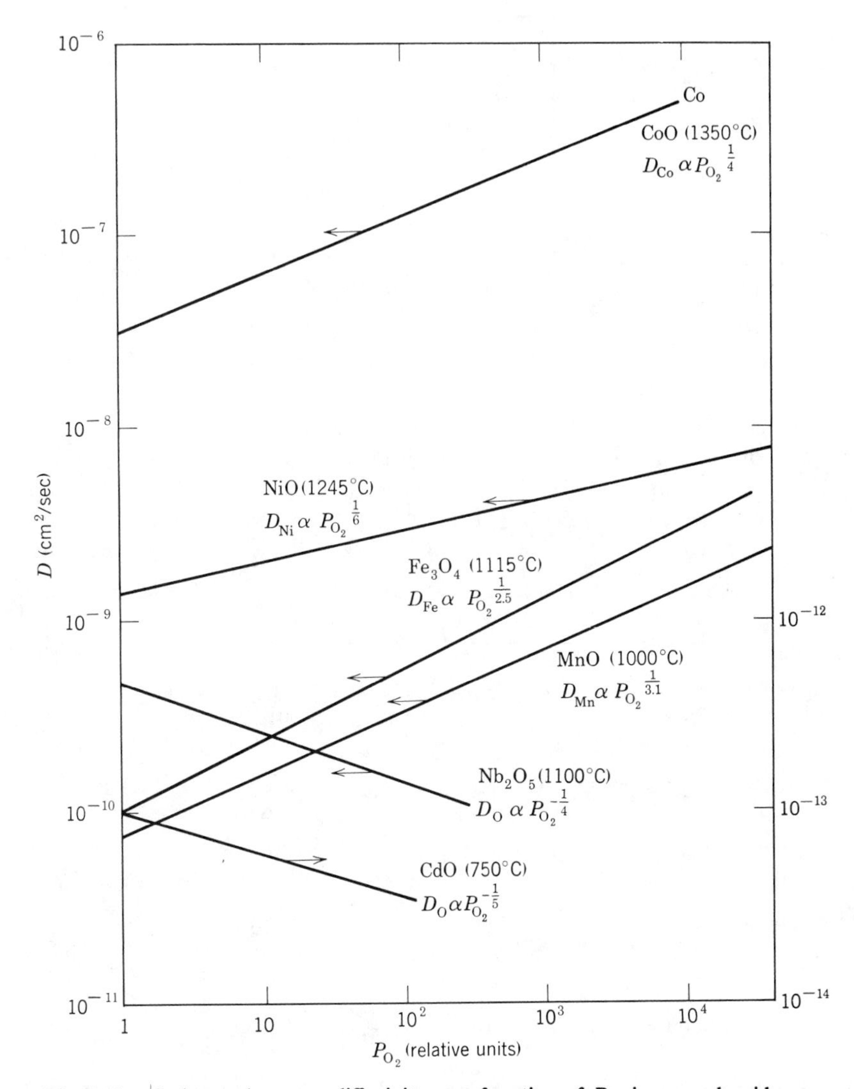

Fig. 6.15. Cation and oxygen diffusivity as a function of P_{O_2} in several oxides.

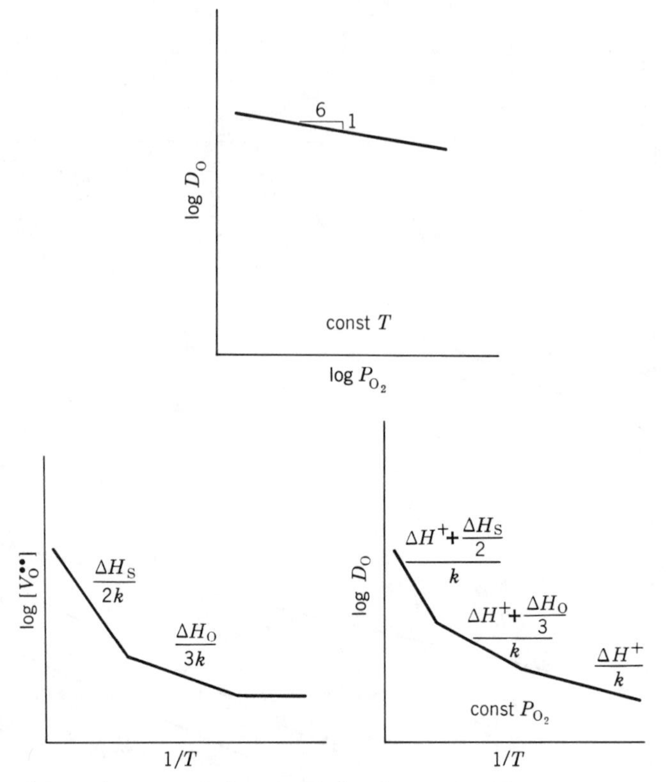

Fig. 6.16. Schematic representation of diffusion in oxygen-deficient oxides.

$$\text{null} = 2V_{\text{O}}^{\cdot\cdot} + V_{\text{U}}^{''''} \qquad K_S \approx \exp\frac{-6.4}{kT} \tag{6.54}$$

$$\text{O}_{\text{O}} = \text{O}_i'' + V_{\text{O}}^{\cdot\cdot} \qquad K_{F_{\text{O}}} \approx \exp\frac{-3.0}{kT} \tag{6.55}$$

$$\text{U}_{\text{U}} = \text{U}_i^{\cdot\cdot\cdot\cdot} + V_{\text{U}}^{''''} \qquad K_{F_{\text{U}}} \approx \exp\frac{-9.5}{kT} \tag{6.56}$$

$$O_2(g) = 2\text{O}_{\text{O}} + V_{\text{U}}^{''''} + 4h^{\cdot} \qquad K_{\text{O}_2} \approx P_{\text{O}_2}{}^{1/4} \exp\frac{-\Delta H_{\text{O}_2}}{4kT} \tag{6.57}$$

The predicted defect concentrations are shown schematically in Fig. 6.17 for interstitials and vacancies of oxygen and uranium at 1600°C as a function of nonstoichiometry O/M. At large deviations from stoichiometry the oxygen defects are insensitive to O/M; that is, the crystal is extrinsic. Near the stoichiometric value O/M = 2 each defect

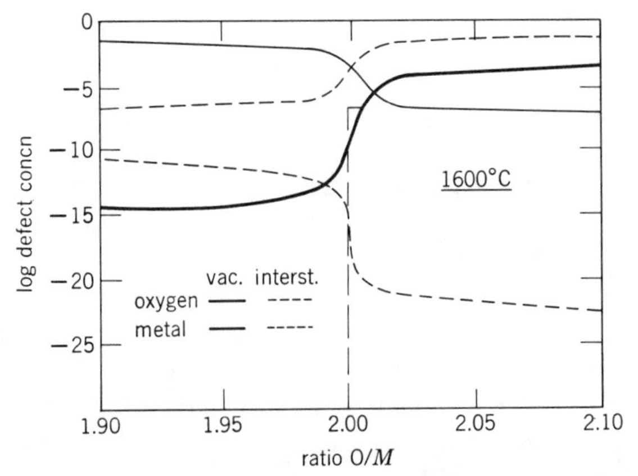

Fig. 6.17. Predicted defect concentrations in UO_2 at 1600°C for different deviations from stoichiometry. The set of formation energies $\Delta G_{FO} = 3.0$ eV, $\Delta G_S = 6.4$ eV and $\Delta G_{FU} = 9.5$ eV was used for the calculations. From H. Matzke.

concentration changes rapidly because the thermal defect generation equations (6.54 to 6.56) become the determining factors. The predominant diffusion process depends on the ease of moving a vacancy relative to an interstitial and on the defect concentrations. In the UO_{2+x} region, $\Delta H^{\dagger}_{O,i}$ is about the same as $\Delta H^{\dagger}_{O,V}$, but interstitials are present in much higher concentrations and the diffusion coefficient is given by

$$D_{O,i} = D_0 \exp \frac{-\Delta H^{\dagger}_{O,i}}{kT} \tag{6.58}$$

In the oxygen-deficient range UO_{2-x}, oxygen vacancies are more prevalent, and

$$D_{O,v} = D_0 \exp \frac{-\Delta H^{\dagger}_{O,v}}{kT} \tag{6.59}$$

Because the uranium defect concentrations are coupled through the Schottky equation to the oxygen defect concentrations, the diffusion of uranium in UO_{2+x} is given by

$$D_{U,v} = D_0 \exp \frac{-(\Delta G_S - 2\Delta G_{F_O} + \Delta H^{\dagger}_{U,v})}{kT} \tag{6.60}$$

and for UO_{2-x}

$$D_{U,v} = D_0 \exp \frac{-\Delta G_S + \Delta H^{\dagger}_{U,v}}{kT} \tag{6.61}$$

Near the stoichiometric composition, where oxygen defects are generated from thermal equilibrium,

$$D_{\mathrm{O},i} = D_0 \exp \frac{-\frac{1}{2}(\Delta G_{F_\mathrm{O}} - \Delta H^{\dagger}_{\mathrm{O},i})}{kT} \tag{6.62}$$

$$D_{\mathrm{U},v} = D_0 \exp \frac{-(\Delta G_S - \Delta G_{F_\mathrm{O}} + \Delta H^{\dagger}_{\mathrm{U},v})}{kT} \tag{6.63}$$

These expressions for diffusion in UO_{2+x} are in general accordance with experimental observations, as shown in Fig. 6.18.

Chemical Effects. The chemical nature of diffusion is illustrated in Fig. 6.19, which shows the measured diffusion coefficients of several divalent and trivalent cations in MgO. In each case the cation diffusion is much more rapid than oxygen diffusion. Second, the diffusion rates of similar cations, for example, Ca, Mg, Ni, Co, are different. If we assume that the

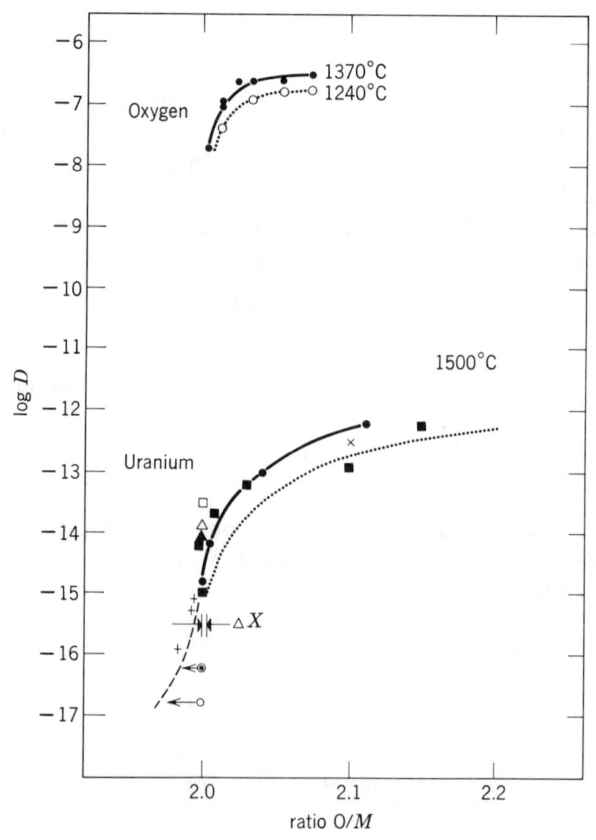

Fig. 6.18. Experimental data for O and U diffusion in UO_{2+x}. From H. Matzke.

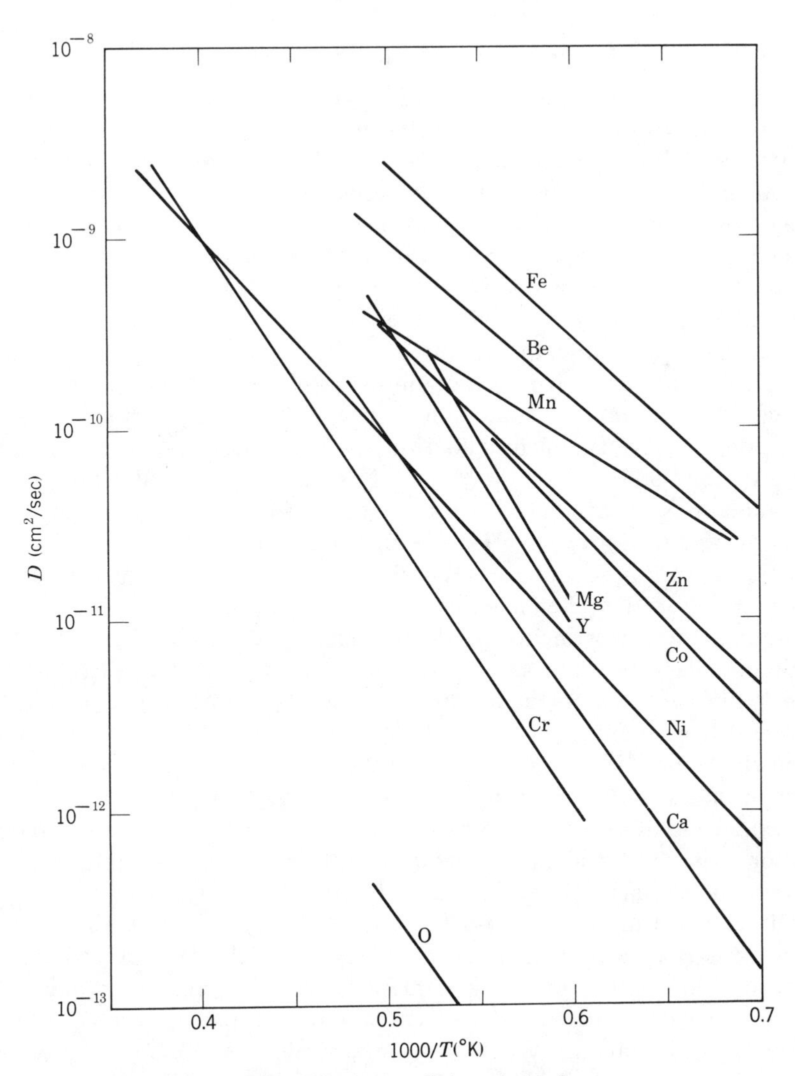

Fig. 6.19. Measured tracer diffusion coefficients in MgO.

individual diffusion experiments were carried out on similar crystals and under similar conditions, we must ascribe the differences to the different chemical nature of the diffusant in the MgO crystal. This can be related to the chemical potential and the activity (e.g., see Eq. 6.41) and to the mobility (jump probability). It is also clear from the atomistic view of point defects that differences in the ionic radii and in the valence contribute to differences in measured diffusion coefficients.

6.6 Dislocation, Boundary, and Surface Diffusion

At the surface of a phase an atom moving from one site to another is not constrained to squeezing between surrounding atoms on all sides, as illustrated in Figs. 6.8 and 6.1. As a result, atomic mobility is greater on the surface and takes place with a lower activation energy. Similarly, but to a lesser extent, the less dense packing of atoms at a grain boundary and at a dislocation core generally leads to greater atomic mobility than in the crystal lattice and a lower activation energy than that for lattice diffusion. The influence of impurities and solutes is often important, however, and available data are subject to various conceptual interpretations. Although the activation energy for surface diffusion is often about half that for lattice diffusion and the activation energy for boundary and dislocation diffusion is intermediate in value, this is sometimes not even approximately true, and caution is required in making generalizations and extrapolations. The most important process for the broadest range of ceramic applications (sintering, deformation, precipitation, oxidation, reactions to form new phases) is grain-boundary diffusion, and we shall concentrate on this phenomenon.

Mathematical treatments of grain-boundary diffusion represent the grain boundary as a uniform isotropic slab of material of width δ within which diffusion occurs in accordance with Fick's laws but at a rate different from the bulk, and the product of the grain-boundary diffusion coefficient and the boundary width ($D_b\delta$) is determined by measuring concentration profiles as a function of time, by measuring the average concentration at different depths from the surface as a function of time, or, for most experiments with oxygen diffusion, by measuring the total amount of exchange as a function of time. The model evaluated and resulting penetration profiles are illustrated in Fig. 6.20. A simple but approximate solution was given by Fisher,* which shows that the log concentration of the diffusing element averaged along the x axis decreased linearly with the distance from the surface when boundary diffusion is dominant. When volume diffusion is dominant, the log concentration decreases with the square of the distance from the surface (Eq. 6.24). The Fisher equation is

$$c(x,y,t) = C_0 \exp\left[\frac{-\sqrt{2}y}{(\pi D_l t)^{1/4}(\delta D_b/D_l)^{1/2}}\right] \operatorname{erf} c\left(\frac{x - \frac{1}{2}\delta}{\sqrt{2Dt}}\right) \tag{6.64}$$

J. Appl. Phys., 22, 74 (1951).

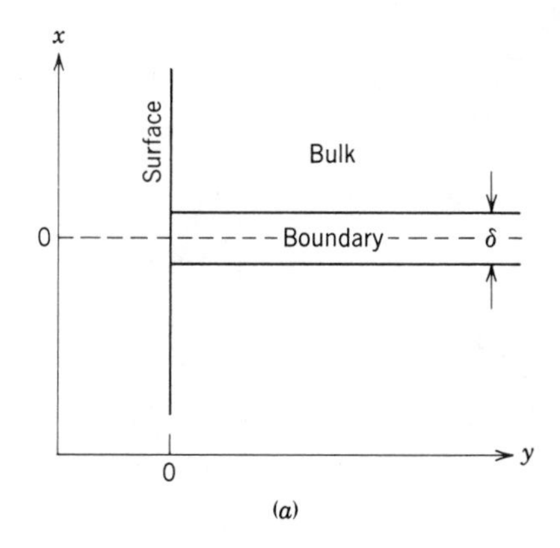

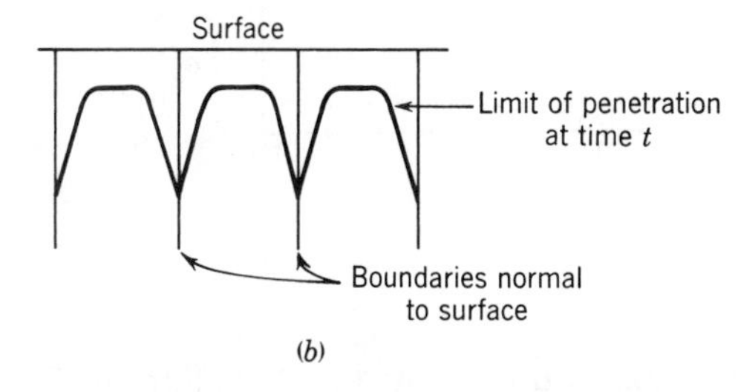

Fig. 6.20. (*a*) Model for grain-boundary diffusion and (*b*) penetration profile resulting from combined grain-boundary and crystal-lattice diffusion.

and the average concentration $\bar{c}$ along the x axis is

$$\ln \bar{c}(y,t) = \frac{-\sqrt{2}}{(\pi D_l t)^{1/4}(\delta D_b / D_l)^{1/2}} \cdot y + \text{const} \tag{6.65}$$

More rigorous but complex solutions were obtained by Whipple* for constant surface concentration and by Suzuoka† for the finite source case. Their solutions also give a log penetration-depth relation close to linearity. With these equations the product of the boundary diffusion coefficient and the boundary thickness $D_b\delta$ can be obtained from diffusion experiments. If we define a nondimensional parameter β as

$$\beta = \frac{D_b}{D_l} \cdot \frac{\frac{1}{2}\delta}{\sqrt{D_l t}} \tag{6.66}$$

boundary diffusion becomes significant when β is larger than unity. In Fig. 6.21 $\ln \bar{c}$ is plotted against y for uranium ion self-diffusion in polycrystalline UO_2. The linear relationship in the deep-penetration

Phil. Mag., **45**, 1225 (1954).

†*Trans. Jap. Inst. Metal.*, **2**, 25 (1961).

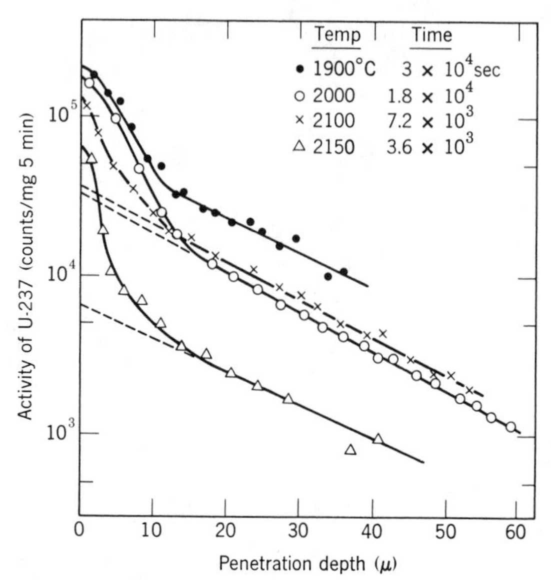

Fig. 6.21. Penetration curve for uranium diffusion into UO_2.

region shows that boundary diffusion is dominant; the deviation in the region near the surface occurs because of the contribution from volume diffusion.

In some cases of polycrystalline ionic crystals, the linear $\ln \bar{c} - y$ relation is not observed; rather $\ln \bar{c}$ varies as y^2, as in volume diffusion, but the apparent diffusion coefficient is larger than in single crystals. This results* when the diffusion penetration in the bulk, Dt, is much larger than the grain size, such that every diffusing atom diffuses sufficiently far to enter, migrate in, and leave a large number of boundary regions. As a result, the boundary enhancement is averaged, and just the apparent diffusion coefficient D_a is increased:

$$D_a = D_l(1-f) + f \cdot D_b \tag{6.67}$$

where f is the average time fraction which a diffusing atom spends in the boundary region.

For dislocation-enhanced diffusion, a solution corresponding to Fisher's was given by Smoluchowski† and one corresponding to Suzuoka's by Mimkes and Wuttig.‡ Hart's apparent diffusion relation also

*Hart, *Acta Met.*, **5**, 597 (1957).
†*Phys. Rev.*, **87**, 482 (1952).
‡*J. Appl. Phys.*, **41**, 3205 (1970).

holds for dislocation-enhanced diffusion when the dislocation spacing is smaller than the diffusion penetration length.

Laurent and Benard* found extensive grain-boundary diffusion for anions but not for cations (except for Cs in CsCl) in polycrystalline alkali halide samples prepared by dry pressing and sintering for 100 hr at a temperature 20 to 30°C below the melting point. The diffusion followed a Fick's law relationship corresponding to volume diffusion ($\ln c \sim x^2$), and the grain-boundary diffusional enhancement was proportional to the inverse grain size (proportional to grain-boundary area), as shown in Fig. 6.22. Further experiments in the same laboratory showed that this effect

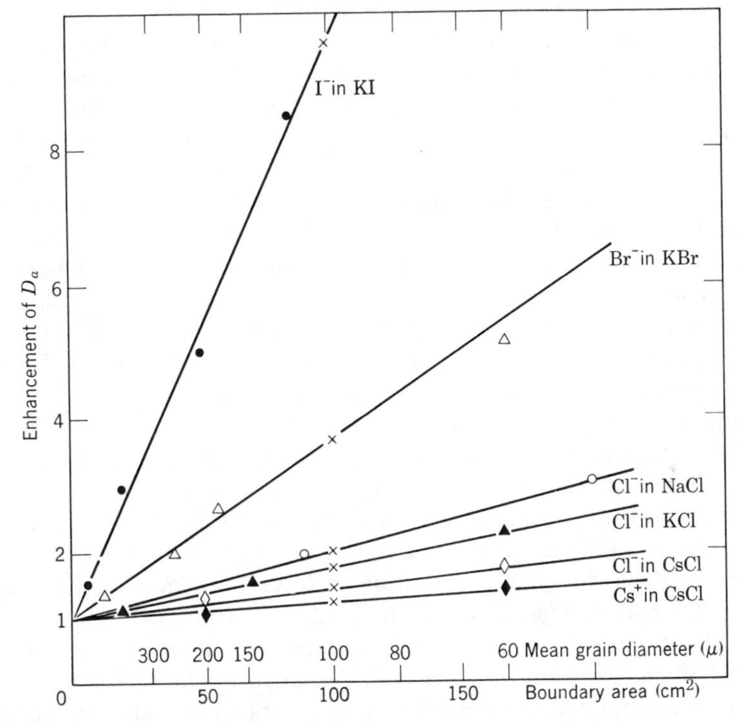

Fig. 6.22. Enhancement of apparent diffusion coefficient in "wet" polycrystalline alkali halides as a function of grain size and grain-boundary area. From J. F. Laurent and J. Benard, *J. Phys. Chem. Solids*, 7, 218 (1958).

was eliminated by carefully preparing moisture-free samples. The kinetics with the "wet" samples cannot be interpreted by any of the above relationships without assuming a very wide boundary thickness (>1 μm).

***J. Phys. Chem. Solids*, 7, 218 (1958).

In contrast to "dry" samples, the "wet" samples exhibited more rapid boundary etching than "dry" samples and a greater residual porosity (about 5% versus 1–1/2%). Wuensch and Tiernan* have similarly found no enhanced grain-boundary diffusion for Tl in "dry" KCl bicrystals but a large enhancement in a "wet" KCl bicrystal. A clearer characterization of the role of water, its distribution and its effect on defect structure, segregation, precipitation (CaF_2 heated in a moist atmosphere and then cooled shows boundary precipitates, presumably of CaO), and internal strains, seems required to establish an explanation for these results. In dry alkali halide samples, boundary diffusion is observed at low temperatures with a lower activation energy than found for volume diffusion.

Among oxide systems, data on cation diffusion in MgO bicrystals and polycrystalline samples over the temperature range of 1000 to 1300°C indicate a wide region of enhanced penetration (Fig. 6.23). In each case of enhanced diffusion there is evidence of solute (usually Ca, Si) precipitation or segregation at the boundaries; in cases with no detectible grain-boundary impurity there was no enhanced diffusion. Without a knowledge of sample impurities, solute solubilities, precipitate morphologies, and detailed thermal histories, none of which are available for the measurements, detailed analysis of diffusion rates and boundary width is meaningless except for the qualitative observation that enhanced diffusion is observed in regions having a width measured in microns. Data for the oxide having a wide range of stoichiometry for which the most measurements are available, ZnO, are characterized by appreciable scatter. In addition to problems associated with impurities it is difficult to assure a uniform chemical composition corresponding to the equilibrium value. Transient compositional changes are expected, which is in accordance with observations that grain-boundary diffusion coefficients can change by an order of magnitude during a long-time anneal.

Most oxygen diffusion measurements have been made by exchange with an O^{18} enriched gas phase, with results similar to those shown in Fig. 6.24. The amount of exchange versus time data is in accordance with Fick's law. In explaining this behavior the most satisfactory conjecture is that the grain-boundary penetration is rapid enough so that the boundaries are close to saturation and the measured exchange is primarily into the grains from the grain boundaries. On this assumption the single-crystal and polycrystal data should be in accordance if the polycrystalline grain size is used in the Fick's law calculation. Examples of this are shown for both Al_2O_3 and MgO in Fig. 6.25. Enhanced oxygen diffusion along grain boundaries or dislocations has also been reported for Fe_2O_3, CoO, and

*Ph.D. thesis, Massachusetts Institute of Technology, 1970.

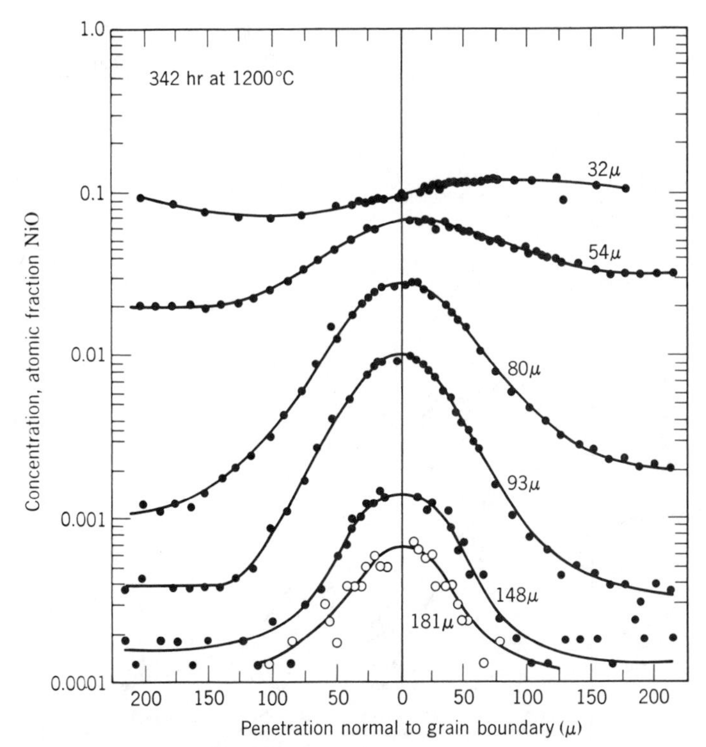

Fig. 6.23. Plot of Ni^{2+} concentration as a function of penetration normal to the grain boundary at various distances from specimen surface. Sandwich type of couple, prepared with a natural bicrystal of MgO. Microbeam probe data. NiK_α fluorescent radiation. From B. J. Wuensch and T. Vasilos, *J. Am. Ceram. Soc.*, **49**, 433 (1966).

$SrTiO_3$ and looked for but not found in BeO, UO_2, Cu_2O, $(ZrCa)O_2$, and yttrium aluminum garnet. Enhanced cation diffusion in oxides has been reported for UO_2, $SrTiO_3$, $(ZrCa)O_2$, and looked for but not found in Al_2O_3, Fe_2O_3, NiO, and BeO. In many cases the reports are discordant and even contradictory, as might be expected from the difficulty of the measurements and a strong expected influence of impurities and nonstoichiometry.

For "wet" alkali halides and for nonequilibrium measurements on other materials with observed phase separation and wide regions of segregation at the grain boundary, interpretation requires a detailed description of the precipitate or segregate distribution, which has yet to be provided. For pure dry samples of alkali halides and for measurements where concentration profiles have been determined in the regime of constant sample composition, the activation energy for boundary diffusion is smaller than that for

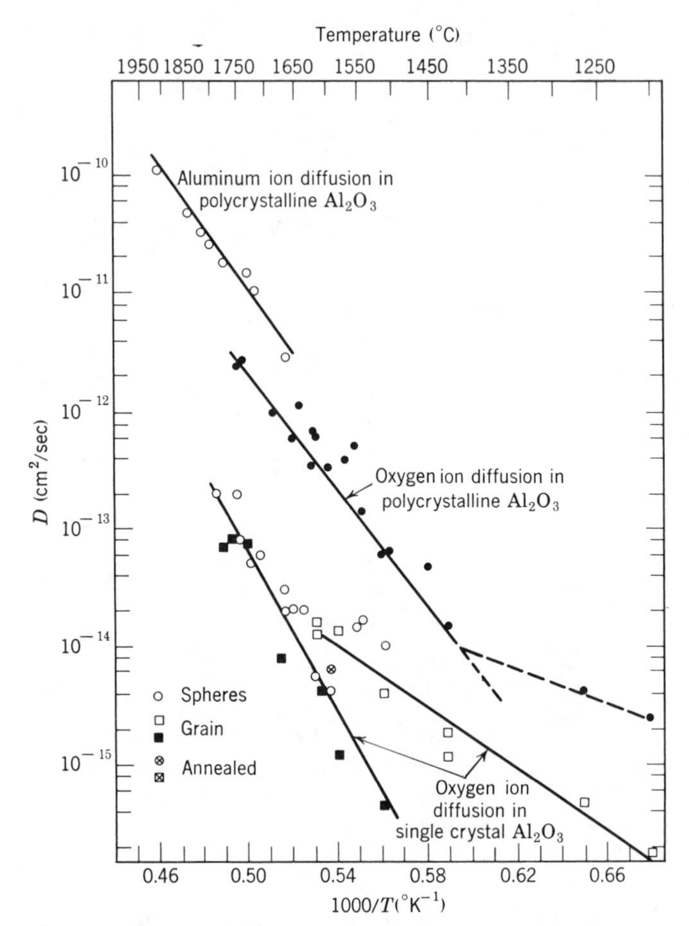

Fig. 6.24. Comparison of aluminum ion self-diffusion coefficients with oxygen ion self-diffusion coefficients in aluminum oxide. From Y. Oishi and W. D. Kingery, *J. Chem. Phys.*, **33**, 905 (1960).

bulk diffusion. This behavior is found with metal systems and is expected from considerations of grain-boundary structure. These data can be adequately interpreted in terms of a boundary diffusion width of a few atom distances and a boundary diffusion coefficient 10^3 to 10^6 greater than the bulk diffusion coefficient.

For at least some materials which have a substantial preference for grain-boundary diffusion of only one ion—O in Al_2O_3 and MgO, U in UO_2, Ca and Zr in $(ZrCa)O_2$—the ion with high grain-boundary mobility has the same sign as the anticipated grain-boundary charge for the composition

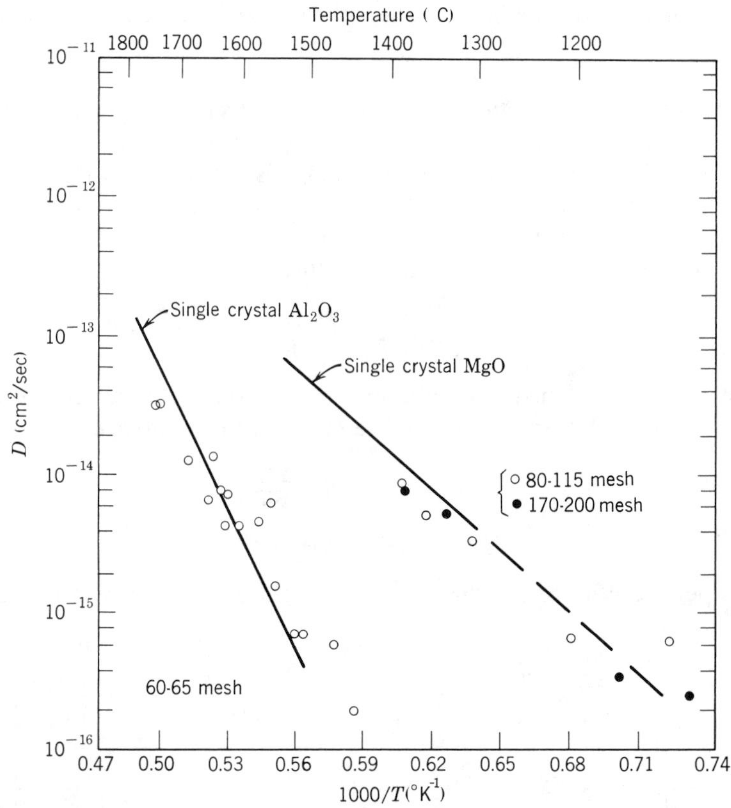

Fig. 6.25. Single-crystal and polycrystalline oxygen diffusion data calculated by taking a as the grain radius. From Kingery, *J. Am. Ceram. Soc.*, (1974).

studied; that is, it is the ion likely to be present in excess at the grain-boundary core. This suggests that a mechanism of excess ion migration at the core may be responsible for enhanced grain-boundary mobility. If this is true, the concentration of aliotropic impurities which affects the charge and hence the added ion concentration on the boundary should have a significant effect.

6.7 Diffusion in Glasses

The physical principles in the preceding sections may also be used in discussing diffusion in noncrystalline solids. The simplest case to be considered in this regard is the diffusion of gases in simple silicate glasses. Much of the data in this area is presented in the form of permeability rather than diffusion coefficient, where the two are related by the relation

$$K = DS \tag{6.68}$$

Here K is the permeability (the volume of gas at s.t.p. passing per second through a unit area of glass, 1 cm in thickness, with a 1-atm pressure difference of the gas across the glass), and S is the solubility (the volume of gas at s.t.p. dissolved in a unit volume of glass per atmosphere of external gas pressure).

The solubility increases with temperature, as

$$[S] = [S_0] \exp\left(-\frac{\Delta H_s}{RT}\right) \tag{6.69}$$

where ΔH_s is the heat of solution. Hence the permeability is also expected to exhibit an exponential temperature dependence:

$$K = K_0 \exp\left(-\frac{\Delta H_\kappa}{RT}\right) \tag{6.70}$$

where

$$\Delta H_\kappa = \Delta H_s + \Delta H^\dagger$$

The permeability of He through various glasses is shown in Fig. 6.26.

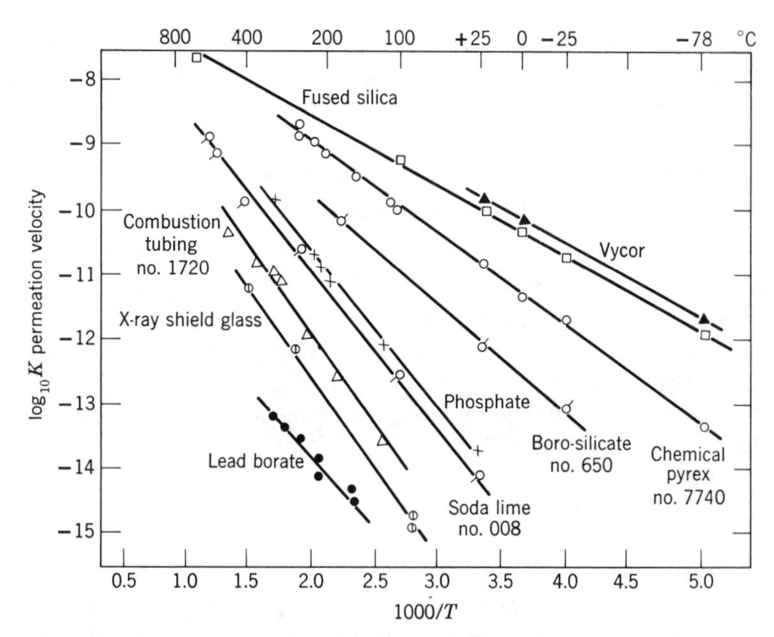

Fig. 6.26. The permeation velocity of helium through various glasses, as measured by Norton. From *J. Am. Ceram. Soc.*, **36**, 90 (1953).

As noted in the figure, marked differences are found in the permeability of He in various glasses at room temperature (comparison of 7740 Pyrex with the X-ray shield glass, for example, indicates a difference of some five or six orders of magnitude). These differences can be important in several applications. For example, although a particular glass is quite adequate for ordinary high-vacuum use, if one requires a vacuum in the range of 10^{-10} torr, or one of 10^{-7} torr for extended periods without pumping, one of the more impermeable glasses should be used.

The differences among the permeabilities of various glasses can be rationalized to a first order of approximation by considering the modifying cations as blocking the holes or openings in the glass networks. On this basis, the permeability would be expected to increase with increasing concentration of the network-forming constituents. Such an increase is shown by the data in Fig. 6.27.

The diffusion coefficient of various gases in fused silica has separately been determined in a number of investigations. The results for He, Ne, H_2, and N_2 all give D_0 values in the range of 10^{-4} to 10^{-3} cm^2/sec. The activation energies increase with increasing size of the gas molecules, as

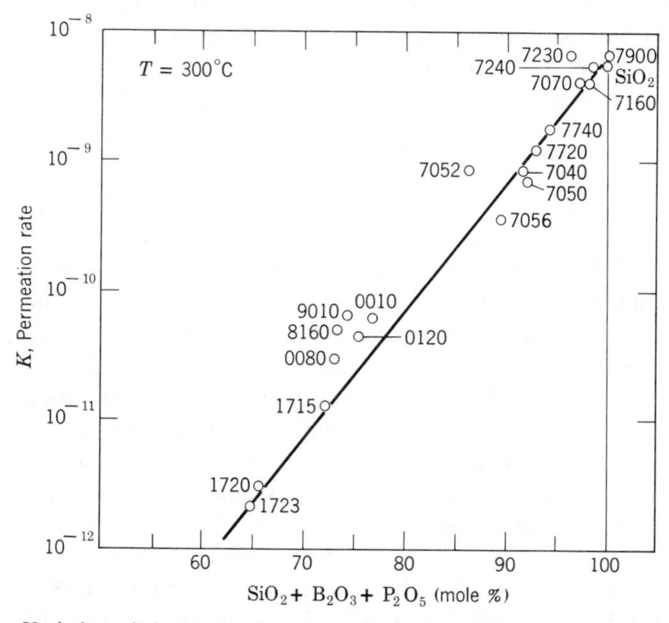

Fig. 6.27. Variation of the permeation rate with the concentration of network formers (SiO_2, B_2O_3, P_2O_5) in the glass. From V. O. Altemose, in *Seventh Symposium on the Art of Glassblowing*, American Scientific Glassblowers Society, Wilmington, 1962.

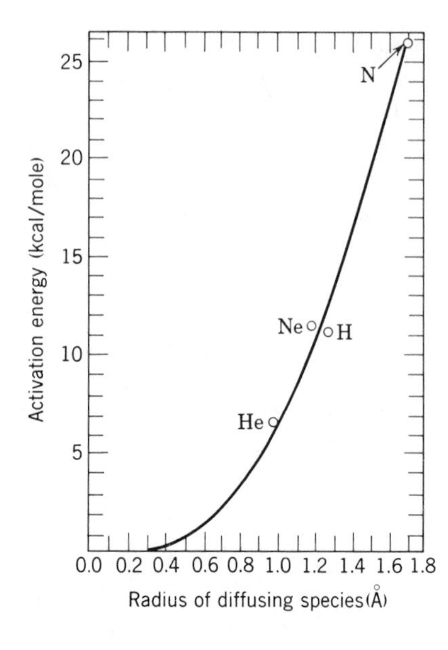

Fig. 6.28. Activation energy for diffusion as a function of the radius of the diffusing species in silica glass. From R. M. Hakim and D. R. Uhlmann, *Phys. Chem. Glasses*, **12**, 132 (1971).

shown in Fig. 6.28. These activation energies should reflect largely the strain energy required to dilate cavities in the glass structure sufficiently to permit transport. Useful information about structural features of fused silica can be obtained from the data in Fig. 6.28, using an expression for the elastic energy required to dilate an ellipsoidal cavity from radius r_d to r

$$E_{st} = 8\pi G(r - r_d)^2 E\left(\frac{c}{a}\right) \tag{6.71}$$

where G is the shear modulus of the surrounding medium and $E(c/a)$ is a factor which depends on the ratio of the minor c to the major a axis of the ellipsoid. Applying this relation to the data in Fig. 6.28, one obtains $r_d = 0.31$ Å and $E(c/a) = 0.4$. The solid curve shown in Fig. 6.27 was drawn using these values of the parameters together with Eq. 6.71. Hence it seems that the holes between interstices in the SiO_2 network are of the order of 0.3 Å in size and depart significantly from sphericity.

In the case of oxygen, transport may take place either as molecular diffusion or as network diffusion. Evidence for the two has been provided by separate studies of oxygen diffusion in fused silica. In one, molecular transport was indicated by a low activation energy: $D = 2.8 \times 10^{-4} \exp(-2.7 \times 10^4/RT)$; in another, network diffusion seems operative: $D = 1.2 \times 10^{-2} \exp(-7.05 \times 10^4/RT)$. At a temperature of 1000°C, then, molecular diffusion apparently takes places some seven orders of magnitude faster than network diffusion.

For diffusion of rare gases (He, Ne, etc.) in glass, there seems to be little change in behavior in the vicinity of the glass transition (for some of the glasses shown in Fig. 6.26, the data bridge the region of the glass transition). In contrast to this, pronounced changes are generally observed in the diffusion of modifying cations. An example of such behavior is shown in Fig. 6.29 for Na^{+1} diffusion in soda-lime-silica glasses of various compositions. The breaks seen in the data correspond to the region of the glass transition and somewhat above in temperature. The activation energies for transport in the liquids are in the range of 23 to 27 kcal/g-atom; those for the corresponding glasses range from 15 to 20 kcal/g-atom. Similar changes in the activation energy for cation diffusion on passing through the glass transition have been observed for other systems as well, but the origin of such changes remains to be elucidated satisfactorily.

As the amounts of modifying cations in silicate glasses are increased, the activation energy for their diffusive transport decreases, and the diffusion coefficient increases. These changes presumably reflect the breaking up of the network and a decrease in the average interionic separation. Divalent cations diffuse much more slowly at a given tempera-

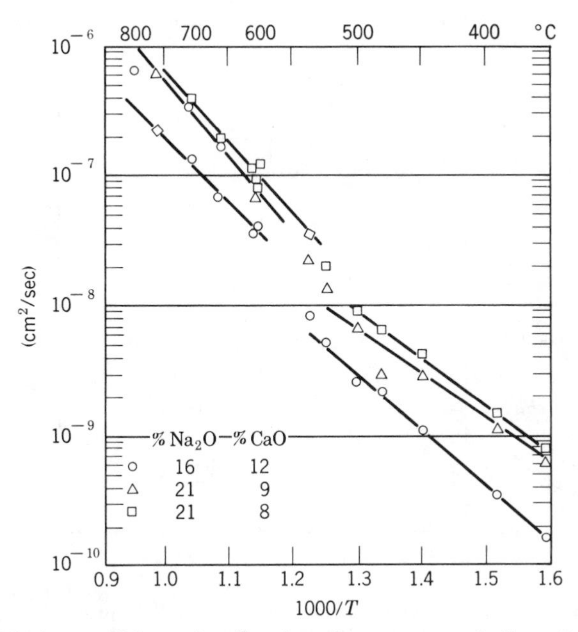

Fig. 6.29. Diffusion coefficients of sodium in various ternary soda-lime-silicate glasses, as measured by Johnson (Thesis, Ohio State University, 1950).

ture than monovalent cations, and their activation energies are generally much larger. In all cases, there seems to be no direct relation between transport of the modifying cations and the flow properties of the materials. For example, the activation energies for viscous flow in soda-lime-silica compositions similar to those shown in Fig. 6.29 are in the range of 100 kcal/g-atom; the activation energy for Na ion diffusion is much smaller, in the range of 25 kcal/g-atom. Such differences are not unexpected in light of the significantly different atomic processes involved in viscous flow and modifying cation diffusion. The viscous-flow data may correlate much better with the diffusion of the network-forming cations (e.g., Si in this case; a comparison of the diffusion of Na, Ca, and Si in glass is given in Fig. 6.30).

Diffusion coefficients of modifying cations in glasses cooled rapidly through the glass transition region are generally higher than those in well-annealed glasses of the same composition. This difference can be as large as an order of magnitude or more and very likely reflects differences

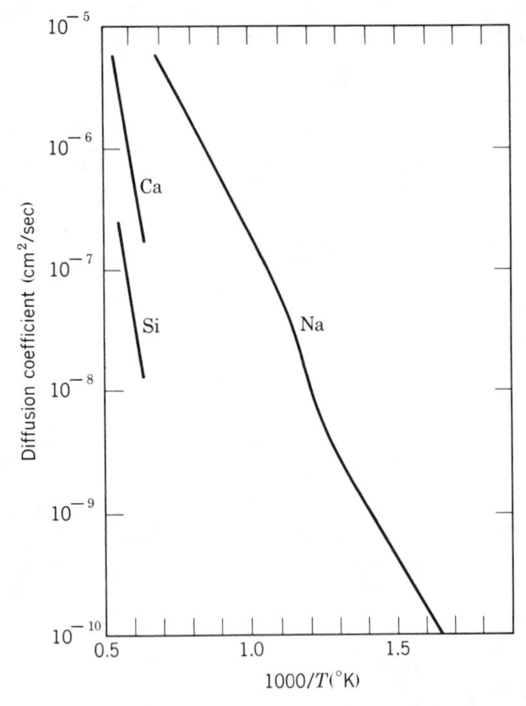

Fig. 6.30. Diffusion of Na^+ in a sodium silicate glass, from J. R. Johnson, R. H. Bristow, and H. H. Blau, *J. Am. Ceram. Soc.*, **34**, 165 (1951), and Ca^{2+} and Si^{4+} in a $40CaO–20Al_2O_3–40SiO_2$ slag, from H. Towers and J. Chipman, *Trans. A.I.M.E.*, **209**, 709 (1957).

in the specific volumes of the glasses, with the larger-volume, more open structure being characterized by the larger diffusivity.

Many of the univalent cations can be substituted for one another in various glass structures. This is exemplified by Ag ions, which can be substituted completely for Na ions even at temperatures below the glass transition region. Li and K ions can be at least partially substituted for Na ions under similar conditions, but in these cases if the exchange is carried too far, stresses can build up in the glass which are large enough to result in its fracture. The substitution of one ion for another is widely used in the technique of ion-exchange strengthening of glasses and is discussed in Chapter 15.

Suggested Reading

1. P. G. Shewmon, *Diffusion in Solids*, McGraw-Hill Book Company, New York, 1963.
2. P. Kofstad, *Nonstoichiometry, Diffusion, and Electrical Conductivity in Binary Metal Oxides*, John Wiley & Sons, Inc., New York, 1972.
3. J. Crank, *Mathematics of Diffusion*, Clarendon Press, Oxford, 1956.
4. W. D. Kingery, Ed., *Kinetics of High-Temperature Processes*, Technology Press, Cambridge, Mass., and John Wiley & Sons, Inc., New York, 1959.
5. R. E. Howard and A. B. Lidiard, "Matter Transport in Solids," *Rep. Prog. Phys.*, **27**, 161–240 (1964).
6. W. Jost, *Diffusion in Solids, Liquids, Gases*, Academic Press, Inc., New York, 1952.
7. R. H. Doremus, "Diffusion in Non-Crystalline Silicates," in *Modern Aspects of the Vitreous State II*, J. D. Mackenzie, Ed., Butterworth, Washington, 1962.

Problems

6.1. (*a*) Describe and discuss fully the experiments you would conduct and the nature of the results you would hope to obtain if, for a given ceramic oxide, you wished to ascertain:

(1) Whether diffusion rates in a given temperature range occurred by an intrinsic or extrinsic mechanism.

(2) Diffusion in a given polycrystalline ceramic was predominantly along grain boundaries or through the lattice.

(3) Whether diffusion occurred via a vacancy mechanism or a ring type of interchange.

(*b*) What concentration of trivalent impurity is required to make the cation diffusivity of Mg^{++} in MgO extrinsic to its melting point. Explain fully all estimates of property values made in your calculation.

6.2. The application of pressure (not necessarily hydrostatic) has been observed to affect several processes which are presumed to be diffusion-controlled. Give several ways in which pressure can affect self-diffusion coefficients and the expected direction of change in D with increasing pressure for (a) vacancy diffusion and (b) interstitial diffusion.

6.3. If diffusion anneal times are doubled, at a given temperature, average penetration depths for the diffusing species will increase by a factor of ______.

6.4. Discuss the influence of zinc chloride addition (10^{-4} mole%) to the diffusivity of all ions (Zn, Na, and Cl) in a single-crystal NaCl from room temperature to the melting temperature.

6.5. From the sintering data on ZnS, diffusion coefficients were measured. At 563°C, a diffusion coefficient of 3×10^{-4} cm^2/sec was measured; at 450°C 1.0×10^{-4} cm^2/sec. (a) Determine the activation energy and D_0. (b) From your knowledge of the structure, predict the nature of the activation energy from the point of view of movement and creation of defects. (c) On the basis of similarity with ZnO predict the change in D with partial pressure of sulfur.

6.6. Figure 6.30 shows diffusion coefficients for ions in an annealed sodium–calcium silicate glass. (a) Why does Na^+ diffuse faster than Ca^{++} and Si^{+4}? (b) What is the nonlinear part of the Na^+ diffusion curve due to? (c) How would quenching the glass change the plot? (d) What is the activation energy (experimental) for Na^+ diffusion in the liquid state of the glass.

6.7. (a) What is the predicted oxygen partial pressure dependence of iron ion diffusion in iron deficient Fe_3O_4? (b) What is the predicted oxygen partial pressure dependence of oxygen diffusion in iron excess Fe_3O_4?

6.8. A student decides to study Ca diffusion in NaCl. It is known that Ca diffuses via a vacancy mechanism on the Na sublattice and that over the range of experimentation $[Ca^{\cdot}_{Na}] = [V'_{Na}]$. Show that D_{Ca} is a function of $[Ca^{\cdot}_{Na}]$; thus $\partial c/\partial t \neq D\partial^2 c/\partial x^2$.

part III

DEVELOPMENT OF MICROSTRUCTURES IN CERAMICS

The properties of ceramics are determined by the properties of each phase present and by the way in which these phases, including porosity and in many cases the interfaces, are arranged. In Part II we have discussed the structure of crystalline materials, the structures of glasses, imperfections in these structures, the characteristics of interfaces, and how the mobility of atoms is related to these structural characteristics. This is the foundation on which we can build an understanding of the properties of each phase present in more complex ceramics. In Part III we want to develop an understanding of the factors which determine the phase distribution and how they operate in ceramic systems.

The development of microstructure proceeds on two fronts. First there are chemical changes and a tendency to form an equilibrium concentration of phases such as to minimize the free energy of the system. Phase-equilibrium diagrams are an economical method for describing the final state towards which the phase composition tends. In our discussion of phase diagrams we have limited ourselves to a maximum of three components and have developed the underlying thermodynamics only to the minimum level necessary. In many actual systems more than three components are important, but the extension of our treatment to this more complicated case uses the same principles which have been described and discussed. The primary difficulty with including a greater number of components is not so much conceptual as in the easy representation of a large body of data in concise diagrammatic form. For ceramic students we have found that the most useful introductory discussion to multicomponent systems is that given by A. Muan and E. F. Osborn, *Phase Equilibria in and among Oxides in Steelmaking.**

In addition to changes in the chemical constitution and amounts of phases present physical factors are also important in determining the direction in which changes proceed during the development of microstructure. A lower free energy of the system is achieved with decreased surface and interface area, which occurs during the processes of sintering, vitrification, and grain growth. In addition there are strain-energy terms and surface-energy terms associated with the formation of a new phase which affect both its morphology and its tendency to appear. These aspects of the driving forces toward minimizing the system's free energy during microstructural development are discussed in Chapter 8 in relationship to phase transformations and in Chapter 10 in relationship to grain growth and sintering. The physical changes occurring, such as the decrease in porosity, the distribution in porosity, and the morphology of the phases present, are equally as important as the chemical processes related to phase equilibria discussed in Chapter 7 and chemical equilibria discussed in Chapter 9.

Only a small percentage of real ceramic systems are treated under conditions such that equilibrium is achieved. Particularly with regard to the small driving forces associated with surface and interface energy and for systems in which the mobility of atoms is small, including many silicate systems and almost all systems at moderate and low temperatures, the way in which equilibrium is approached and the rate at which it is approached are equally as important as the equilibrium being approached. In the condensed phase systems with which we are mostly concerned, material transfer processes may take place by lattice, boundary, or surface diffusion, by viscous flow, or by vapor transport processes. The rate and kinetics by which these processes are important in affecting the development of microstructure are discussed in Chapter 8 with regard to phase transformations, in Chapter 9 with regard to solid-state reactions, and in Chapter 10 with regard to grain growth and sintering. A thorough understanding of the way in which systems modify their microstructure in the approach toward equilibrium is absolutely essential for understanding the microstructure and therefore the properties of ceramic products.

In Chapter 11 some characteristic measurements necessary to describe microstructure together with typical examples of ceramic microstructures in a variety of real systems are discussed and described. In addition to the specific systems described in Chapter 11 we have been implicitly or

*Addison-Wesley Publishing Company, Inc., Reading, Mass., 1965.

explicitly concerned with microstructure development throughout this book. Consequently, microstructure characteristics are described throughout. Indeed, the development of microstructure, its influence on the properties of ceramics, and its control by composition and processing changes are a central theme.

7

Ceramic Phase-Equilibrium Diagrams

At equilibrium a system is in its lowest free energy state for the composition, temperature, pressure, and other imposed conditions. When a given set of system parameters is fixed, there is only one mixture of phases that can be present, and the composition of each of these phases is determined. Phase-equilibrium diagrams provide a clear and concise method of graphically representing this equilibrium situation and are an invaluable tool for characterizing ceramic systems. They record the composition of each phase present, the number of phases present, and the amounts of each phase present at equilibrium.

The time that it takes to reach this equilibrium state from any arbitrary starting point is highly variable and depends on factors other than the final equilibrium state. Particularly for systems rich in silica the high viscosity of the liquid phase leads to slow reaction rates and very long times before equilibrium is established; equilibrium is rarely achieved. For these systems and for others, metastable equilibrium, in which the system tends to a lower but not the lowest free energy state, becomes particularly important.

It is obvious that the phases present and their composition are an essential element in analysing, controlling, improving, and developing ceramic materials. Phase diagrams are used for determining phase and composition change occurring when the partial pressure of oxygen or other gases is changed, for evaluating the effects of heat treatments on crystallization and precipitation processes, for planning new compositions, and for many other purposes. We have already seen the importance of thermodynamic equilibrium in our discussions of single-phase systems: crystalline solid solutions (Chapter 2), crystalline imperfections (Chapter 4), structure of glasses (Chapter 3), and surfaces and interfaces (Chapter 5). In this chapter we concentrate our attention on equilibria involving two or more phases.

7.1 Gibbs's Phase Rule

When a system is in equilibrium, it is necessary that the temperature and pressure be uniform throughout and that the chemical potential or vapor pressure of each constituent be the same in every phase. Otherwise there would be a tendency for heat or material to be transferred from one part of the system to some other part. In 1874 J. Willard Gibbs* showed that these equilibrium conditions can occur only if the relationship

$$P + V = C + 2 \tag{7.1}$$

is satisfied. This is known as the *phase rule*, with P being the number of phases present at equilibrium, V the variance or number of degrees of freedom, and C the number of components. This relationship is the basis for preparing and using phase-equilibrium diagrams.

A phase is defined as any part of the system which is physically homogeneous and bounded by a surface so that it is mechanically separable from other parts of the system. It need not be continuous; that is, two ice cubes in a drink are one phase. The number of degrees of freedom or the variance is the number of intensive variables (pressure, temperature, composition) that can be altered independently and arbitrarily without bringing about the disappearance of a phase or the appearance of a new phase. The number of components is the smallest number of independently variable chemical constituents necessary and sufficient to express the composition of each phase present. The meaning of these terms will become clearer as they are applied to specific systems in the following sections.

Deduction of the phase rule follows directly from the requirement that the chemical potential μ_i of each constituent i be the same in every phase present at equilibrium. The chemical potential is equal to the partial molar free energy $\bar{G}_i$,

$$\bar{G}_i = \left(\frac{\partial G}{\partial n_i}\right)_{T, P, n_1, n_2, \ldots}$$

which is the change in free energy of a system at constant temperature and pressure resulting from the addition of one mole of constituent i to such a large quantity of the system that there is no appreciable change in the concentration. In a system with C components we have an independent equation for each component representing the equality of chemical potentials. For a system containing P phases, we have

$$\mu_1{}^a = \mu_1{}^b = \mu_1{}^c = \cdots = \mu_1{}^P \tag{7.2}$$

Collected Works, Vol. 1, Longmans, Green & Co., Ltd., London, 1928.

$$\mu_2^{\ a} = \mu_2^{\ b} = \mu_2^{\ c} = \cdots = \mu_2^{\ P} \tag{7.3}$$

etc.

which constitute $C(P-1)$ independent equations which serve to fix $C(P-1)$ variables. Since the composition of each phase is defined by $C-1$ concentration terms, completely defining the composition of P phases requires $P(C-1)$ concentration terms, which together with the imposed conditions of temperature and pressure give

$$\text{Total number of variables} = P(C-1)+2 \tag{7.4}$$

$$\text{Variables fixed by equality of chemical potentials} = C(P-1) \tag{7.5}$$

$$\text{Variables remaining to be fixed} = P(C-1)+2-C(P-1) \tag{7.6}$$

$$V = C - P + 2 \tag{7.7}$$

which is Gibbs's phase rule (Eq. 7.1).

The main limitation on the phase rule is that it applies only to equilibrium states, requiring homogeneous equilibrium within each phase and heterogeneous equilibrium between phases. Although a system in equilibrium always obeys the phase rule (and nonconformance proves that equilibrium does not exist), the reverse is not always true. That is, conformation with the phase rule is not a demonstration of equilibrium.

7.2 One-Component Phase Diagrams

In a single-component system the phases that can occur are vapor, liquid, and various polymorphic forms of the solid. (The energy of different polymorphic forms as related to temperature and crystallographic structure has been discussed in Section 2.10, and might well be reviewed by the reader, since it is closely related to the present section.) The independent variables that cause appearance or disappearance of phases are temperature and pressure. For example, when we heat water, it boils; if we cool it, it freezes. If we put it in an evacuated chamber, the water vapor pressure quickly reaches some equilibrium value. These changes can be diagrammatically represented by showing the phases present at different temperatures and pressures (Fig. 7.1).

Since this is a one-component system, even the air phase is eliminated, and different phase distributions correspond to Fig. 7.2*a* to *c*. In actual practice measurements in which the vapor phase is unimportant are usually made at constant atmospheric pressure in a way similar to Fig. 7.2*d*. Although this is not an ideal closed system, it closely approximates one as long as the vapor pressure is low compared with atmospheric pressure (so that we can ignore the insignificant vapor phase which would

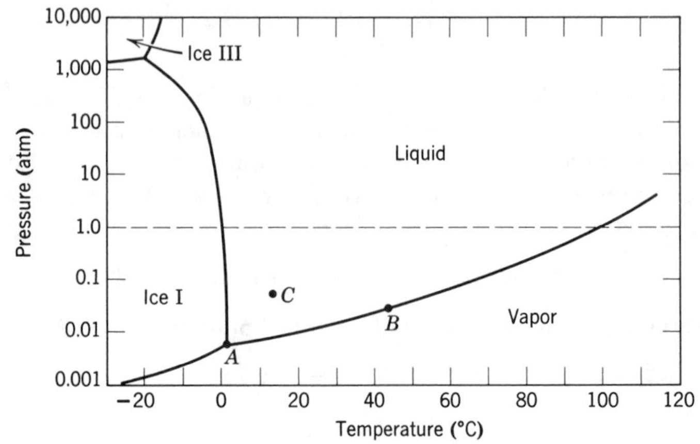

Fig. 7.1. Pressure-temperature diagram for H_2O.

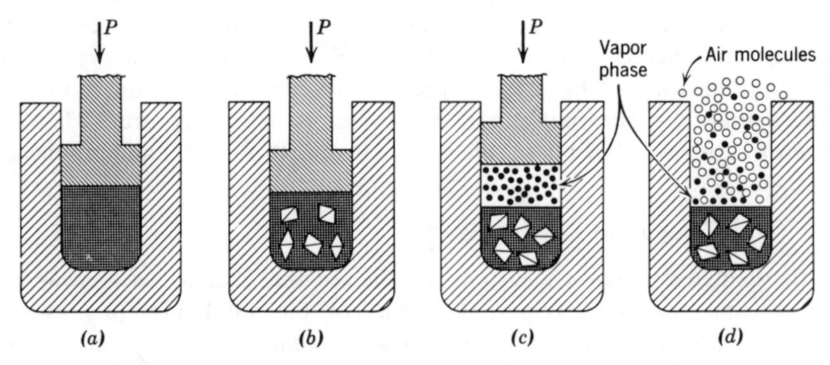

Fig. 7.2. Experimental conditions for a single-component system with (*a*) one phase, (*b*) two phases, (*c*) three phases, and (*d*) common conditions, with the condensed phase exposed to a gas atmosphere.

not exist at all in a closed system) or is equal to or greater than atmospheric pressure (so that the vapor phase has the partial pressure predicted by the phase diagram). For many condensed systems of interest, the first criterion is satisfied.

In a one-component system the largest number of phases that can occur at equilibrium is given when the variance is zero: $P + V = C + 2$, $P + 0 = 1 + 2$, $P = 3$. When three phases are present at equilibrium (ice, water, vapor), as at point A in Fig. 7.1, any change in pressure or temperature causes the disappearance of a phase. The lines on the diagram represent conditions for two phases to exist together at equilibrium; for example, when liquid and vapor are present, as at point B, $P + V = C + 2$,

$2 + V = 1 + 2$, $V = 1$, and the variance is one. This means that either pressure or temperature, but not both, can be changed arbitrarily without the disappearance of a phase. If we change T_1 to T_2, P_1 must also change to P_2 if both phases are to remain present. If only one phase is present, as at C,

$$P + V = C + 2, 1 + V = 1 + 2, V = 2,$$

and both pressure and temperature can be arbitrarily changed without the appearance of a new phase.

At 1 atm pressure, as shown in Fig. 7.1, equilibrium between the solid and liquid occurs at 0°C, the freezing point. Equilibrium coexistence of liquid and vapor occurs at 100°C, the boiling point. The slope of these phase-boundary curves is at any point determined by the Clausius-Clapeyron equation

$$\frac{dp}{dT} = \frac{\Delta H}{T\,\Delta V} \tag{7.8}$$

where ΔH is the molar heat of fusion, vaporization, or transformation, ΔV is the molar volume change, and T is the temperature. Since ΔH is always positive and ΔV is usually positive on going from a low-temperature to a high-temperature form, the slopes of these curves are usually positive. Since ΔV is usually small for condensed-phase transformations, lines between solid phases are often almost vertical.

There are a number of applications of one-component phase diagrams in ceramics. Perhaps the most spectacular of these is the development of the commercial production of synthetic diamonds from graphite. High temperatures and high pressures are necessary, as shown in Fig. 7.3. In addition, the presence of a liquid metal catalyst or mineralizer such as nickel is required for the reaction to proceed at a useful rate. Another system which has been extensively studied at high pressure and temperature is SiO_2. At pressures above 30 to 40 kilobars a new phase, *coesite*, appears which has been found to occur in nature as a result of meteorite impacts. At even higher pressures, above 100 kilobars, another new phase, *stishovite*, has been found.

Of greater interest for ceramic applications are the low-pressure phases of silica, still subject to some dispute as to the role of minor impurities, but illustrated schematically in Fig. 7.4. There are five condensed phases which occur at equilibrium—α-quartz, β-quartz, β_2-tridymite, β-cristobalite, and liquid silica. At 1 atm pressure the transition temperatures are as shown. As discussed in Section 2.10, the α-quartz–β-quartz transition at 573° is rapid and reversible. The other transformations shown are sluggish, so that long periods of time are required to reach equilib-

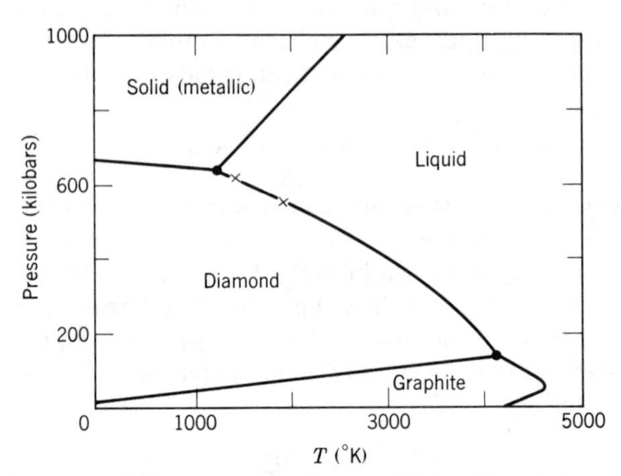

Fig. 7.3. High-pressure, high-temperature phase equilibrium diagram for carbon. From C. G. Suits, *Am. Sci.*, **52**, 395 (1964).

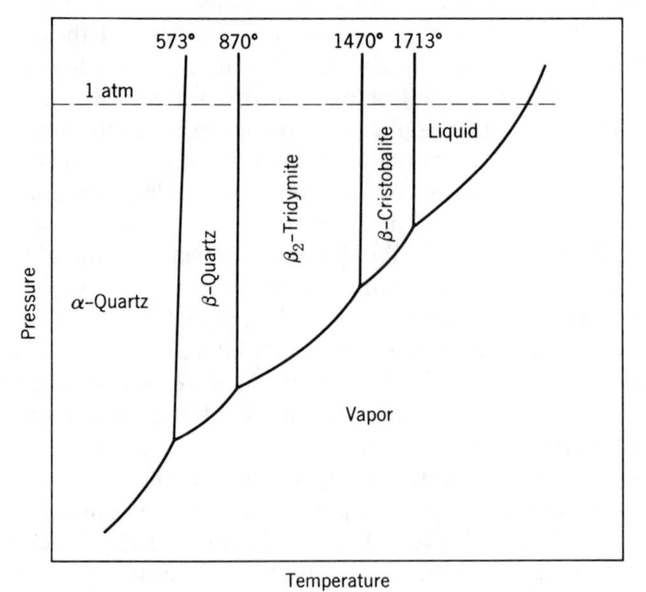

Fig. 7.4. Equilibrium diagram for SiO_2.

rium. The vapor pressure shown in the diagram is a measure of the chemical potential of silica in the different phases, and this same kind of diagram can be extended to include the metastable forms of silica which may occur (Fig. 7.5). The phase with the lowest vapor pressure (the heavy lines in the diagram) is the most stable at any temperature, the equilibrium phase. However, once formed, the transition between cristobalite and quartz is so sluggish that β-cristobalite commonly transforms on cooling into α-cristobalite. Similarly, β_2-tridymite commonly transforms into α- and β-tridymite rather than into the equilibrium quartz forms. These are the forms present in the refractory silica brick, for example. Similarly, when cooled, the liquid forms silica glass, which can remain indefinitely in this state at room temperature.

At any constant temperature there is always a tendency to transform into another phase of lower free energy (lower vapor pressure), and the reverse transition is thermodynamically impossible. It is not necessary, however, to transform into the lowest energy form shown. For example, at 1100° silica glass could transform into β-cristobalite, β-quartz, or β_2-tridymite. Which of these transformations actually takes place is

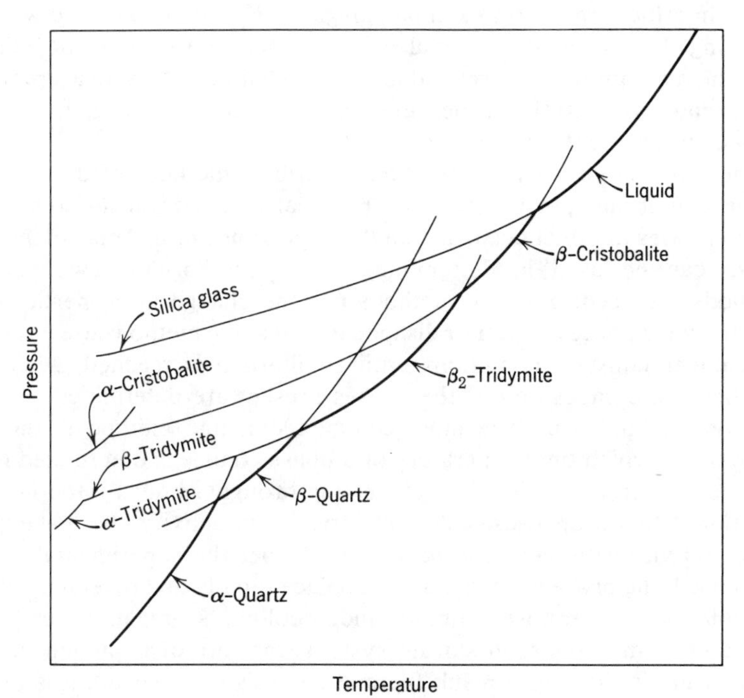

Fig. 7.5. Diagram including metastable phases occurring in the system SiO_2.

determined by the kinetics of these changes. In practice, when silica glass is heated for a long time at this temperature, it crystallizes, or devitrifies, to form cristobalite, which is not the lowest energy form but is structurally the most similar to silica glass. On cooling, β-cristobalite transforms into α-cristobalite.

The silica system illustrates that the phase-equilibrium diagram graphically represents the conditions for minimum free energy in a system; extension to include metastable forms also allows certain deductions about possible nonequilibrium behavior. Almost always, however, a number of alternative nonequilibrium courses are possible, but there is only one equilibrium possibility.

7.3 Techniques for Determining Phase-Equilibrium Diagrams

The phase-equilibrium diagrams discussed in the last section and in the rest of this chapter are the product of experimental studies of the phases present under various conditions of temperature and pressure. In using phase-equilibrium diagrams it is important to remember this experimental basis. In critical cases, for example, diagrams should not be used without referring directly to the original experimenter's description of exactly how the diagram was determined and with what detail the measurements were made. As additional measurements are carried out, diagrams are subject to constant revision.

There is a large body of literature describing methods of determining phase equilibrium. In general, any physical or chemical difference between phases or effect occurring on the appearance or disappearance of a phase can be used in determining phase equilibrium. Two general methods are used: dynamic methods use the change in properties of a system when phases appear or disappear, and static methods use a sample held under constant conditions until equilibrium is reached, when the number and composition of the phases present are determined.

Dynamic Methods. The most common dynamic method is thermal analysis, in which the temperature of a phase change is determined from changes in the rate of cooling or heating brought about by the heat of reaction. Other properties such as electrical conductivity, thermal expansion, and viscosity have also been used. Under the experimental conditions used, the phase change must take place rapidly and reversibly at the equilibrium temperature without undercooling, segregation, or other nonequilibrium effects. In silicate systems the rate of approach toward equilibrium is slow; as a result thermal-analysis methods are less useful for silicates than they are for metals, for example.

Dynamic methods are suitable for determining the temperature of phase changes but give no information about the exact reactions taking place. In addition to the measurements of temperature changes then, phase identification before and after any phase change is required. This analysis is usually carried out by chemical determination of composition, determination of optical characteristics, X-ray determination of crystal structure, and microscopic examination of phase amounts and phase distribution.

Static Methods. In contrast to dynamic measurements, static measurements often consist of three steps. Equilibrium conditions are held at elevated temperatures or pressures, the sample is quenched to room temperature sufficiently rapidly to prevent phase changes during cooling, and then the specimen is examined to determine the phases present. By carrying out these steps at a number of different temperatures, pressures, and compositions, the entire phase diagram can be determined. Sometimes high-temperature X-ray and high-temperature microscopic examinations can determine the phases present at high temperatures, making quenching unnecessary.

For silicate systems the major problem encountered in determining phase-equilibrium diagrams is the slow approach toward equilibrium and the difficulty in ensuring that equilibrium has actually been reached. For most systems this means that static measurements are necessary. A common technique is to mix together carefully constituents in the correct ratio to give the final composition desired. These are held at a constant temperature in platinum foil; after rapid cooling, the mixture is reground in a mortar and pestle and then heated for a second time and quenched. The phases present are examined, the sample mixture remixed, reheated, and quenched again. The resulting material is then reexamined to ensure that the phase composition has not changed.

This process requires much time and effort; since several thousand individual experiments, such as those just described, may be necessary for one ternary diagram, we can understand why only a few systems have been completely and exhaustively studied.

Reliability of Individual Diagrams. In general, the original experimenter investigating a particular phase diagram is usually concerned with some limited region of composition, temperature, and pressure. His effort is concentrated in that area, and the other parts of the phase diagram are determined with much less precision and detail. As reported in summarizing descriptions (such as those given in this chapter), the diagram is not evaluated as to which parts are most reliable. As a result, although the general configuration of diagrams given can be relied on, the exact temperatures and compositions of individual lines or points on the diagram should only be accepted with caution. They represent the results of difficult experimental techniques and analysis.

These cautions are particularly applicable to regions of limited crystalline solution at high temperatures, since for many systems exsolution occurs rapidly on cooling and for many systems this was not a feature of the experimenters' interest. Similarly, phase separation at moderate and low temperatures often results in submicroscopic phases which are not recognized without the use of electron microscopy and electron diffraction, which have not as yet been widely applied to crystalline solid solutions.

7.4 Two-Component Systems

In two-component systems one additional variable, the composition, is introduced so that if only one phase is present, the variance is three: $P + V = C + 2$, $1 + V = 2 + 2$, $V = 3$. In order to represent the pressure, temperature, and composition region of the stability of a single phase, a three-dimensional diagram must be used. However, the effect of pressure is small for many condensed-phase systems, and we are most often concerned with the systems at or near atmospheric pressure. Consequently, diagrams at constant pressure can be drawn with temperature and composition as variables. A diagram of this kind is shown in Fig. 7.6.

If one phase is present, both temperature and composition can be arbitrarily varied, as illustrated for point A. In the areas in which two phases are present at equilibrium, the composition of each phase is

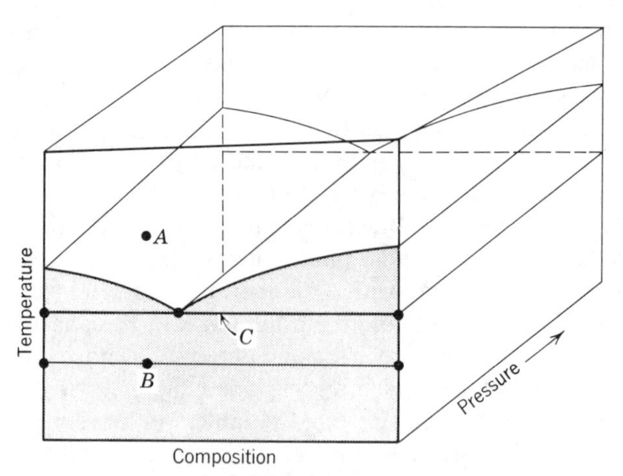

Fig. 7.6. Simple binary diagram.

indicated by lines on the diagram. (In binary diagrams two-phase regions will often be shaded, single-phase regions not.) The intersection of a constant-temperature "tie line" with the phase boundaries gives the compositions of the phases in equilibrium at temperature T. With two phases present, $P + V = C + 2$, $2 + V = 2 + 2$, $V = 2$. At an arbitrarily fixed pressure, any arbitrary change in either temperature or composition of one of the phases present requires a corresponding change in the other variable. The maximum number of phases that can be present where pressure is arbitrarily fixed ($V = 1$) is

$$P + V = C + 2, P + 1 = 2 + 2, P = 3.$$

When three phases are present, the composition of each phase and the temperature are fixed, as indicated by the solid horizontal line at C.

Systems in Which a Gas Phase Is Not Important. Systems containing only stable oxides in which the valence of the cations is fixed comprise a large fraction of the systems of interest for ceramics and can adequately be represented at a constant total pressure of 1 atm. At equilibrium the chemical potential of each constituent must be equal in each phase present. As a result the variation of chemical potential with composition is the underlying thermodynamic consideration which determines phase stability. If we consider a simple mechanical mixture of two pure components, the free energy of the mixture G^M is

$$G^M = X_A G_A + X_B G_B \tag{7.9}$$

For the simplest case, an ideal solution in which the heat of mixing and changes in vibrational entropy terms are zero, random mixing gives rise to a configurational entropy of mixing ΔS_m which has been derived in Eq. 4.14; the free energy of the solution is

$$G^{id,S} = G^M - T\,\Delta S_m \tag{7.10}$$

and under all conditions the free energy of the solution is less than that of a mechanical mixture; the free energy curves for the solid and liquid solutions and the resulting phase-equilibrium diagram are similar to those already illustrated in Fig. 4.2. Since very dilute solutions approach ideal behavior, Eq. 7.10 requires that there is always at least some minute solubility on the addition of any solute to any pure substance.

Most concentrated solutions are not ideal, but many can be well represented as *regular* solutions in which the excess entropy of the solution is negligible, but the excess enthalpy or heat of mixing ΔH^{xs} is significant. In this case the free energy of the regular solution is

$$G^{r,S} = G^M + \Delta H^{xs} - T\,\Delta S_m \tag{7.11}$$

The resulting forms of typical free-energy–composition curves for an ideal solution and for regular solutions with positive or negative excess enthalpies are shown in Fig. 7.7. In Fig. 7.7*c* the minimum free energy for the system at compositions intermediate between α and β consists of a mixture of α and β in which these two solution compositions have the same chemical potential for each component and a lower free energy than intermediate single-phase compositions; that is, phase separation occurs. When differences of crystal structure occur (as discussed in Chapter 2), a complete series of solid solutions between two components is not possible, and the free energy of the solution increases sharply after an initial decrease required by the configurational entropy of mixing. This

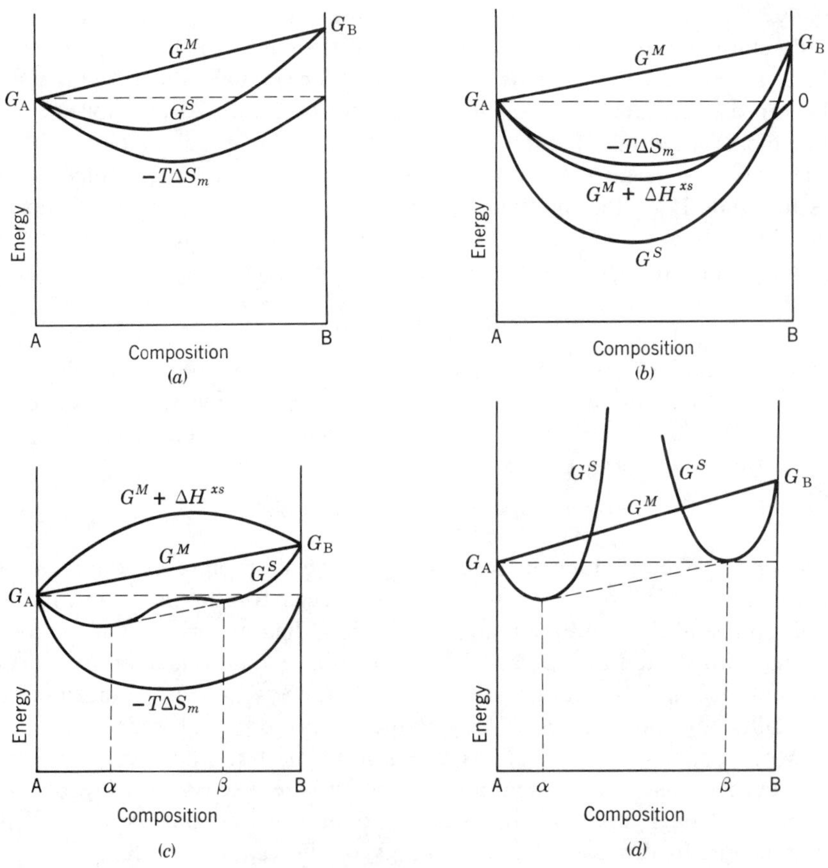

Fig. 7.7. Free-energy–composition diagrams for (*a*) ideal solution, (*b*) and (*c*) regular solutions, and (*d*) incomplete solid solution.

situation is illustrated in Fig. 7.7*d*, in which the minimum system free energy again consists of a mixture of the two solutions α and β.

When, for any temperature and composition, free-energy curves such as shown in Fig. 7.7 are known for each phase which may exist, these phases actually occur at equilibrium which give the lowest system free energy consistent with equal chemical potentials for the components in each phase. This has been illustrated for an ideal solution in Fig. 4.2, compound formation in Fig. 4.3, and phase separation in Fig. 3.10 and is illustrated for a series of temperatures in a eutectic system in Fig. 7.8.

Systems in Which a Gas Phase Is Important. In adjusting the oxygen pressure in an experimental system, it is often convenient to use the equilibria

$$CO + \frac{1}{2}O_2 = CO_2 \qquad (7.12)$$

$$H_2 + \frac{1}{2}O_2 = H_2O. \qquad (7.13)$$

In this case, with no condensed phase present, $P + V = C + 2$, $1 + V = 2 + 2$. $V = 3$, and it is necessary to fix the temperature, system total pressure, and the gas composition, that is, CO_2/CO or H_2/H_2O ratio, in order to fix the oxygen partial pressure. If a condensed phase, that is, graphite, is in equilibrium with an oxygen-containing vapor phase, $P + V = C + 2, 2 + V = 2 + 2, V = 2$, and fixing any two independent variables completely defines the system.

The most extensive experimental data available for a two-component system in which the gas phase is important is the Fe–O system, in which a number of condensed phases may be in equilibrium with the vapor phase. A useful diagram is shown in Fig. 7.9, in which the heavy lines are boundary curves separating the stability regions of the condensed phases and the dash-dot curves are oxygen isobars. In a single condensed-phase region (such as wüstite) $P + V = C + 2, 2 + V = 2 + 2, V = 2$, and both the temperature and oxygen pressure have to be fixed in order to define the composition of the condensed phase. In a region of two condensed phases (such as wüstite plus magnetite) $P + V = C + 2, 3 + V = 2 + 2, V = 1$, and fixing either the temperature or oxygen pressure fully defines the system. For this reason, the oxygen partial-pressure isobars are horizontal, that is, isothermal, in these regions, whereas they run diagonally across single condensed-phase regions.

An alternative method of representing the phases present at particular oxygen pressures is shown in Fig. 7.9*b*. In this representation we do not show the O/Fe ratio, that is, the composition of the condensed phases, but only the pressure–temperature ranges for each stable phase.

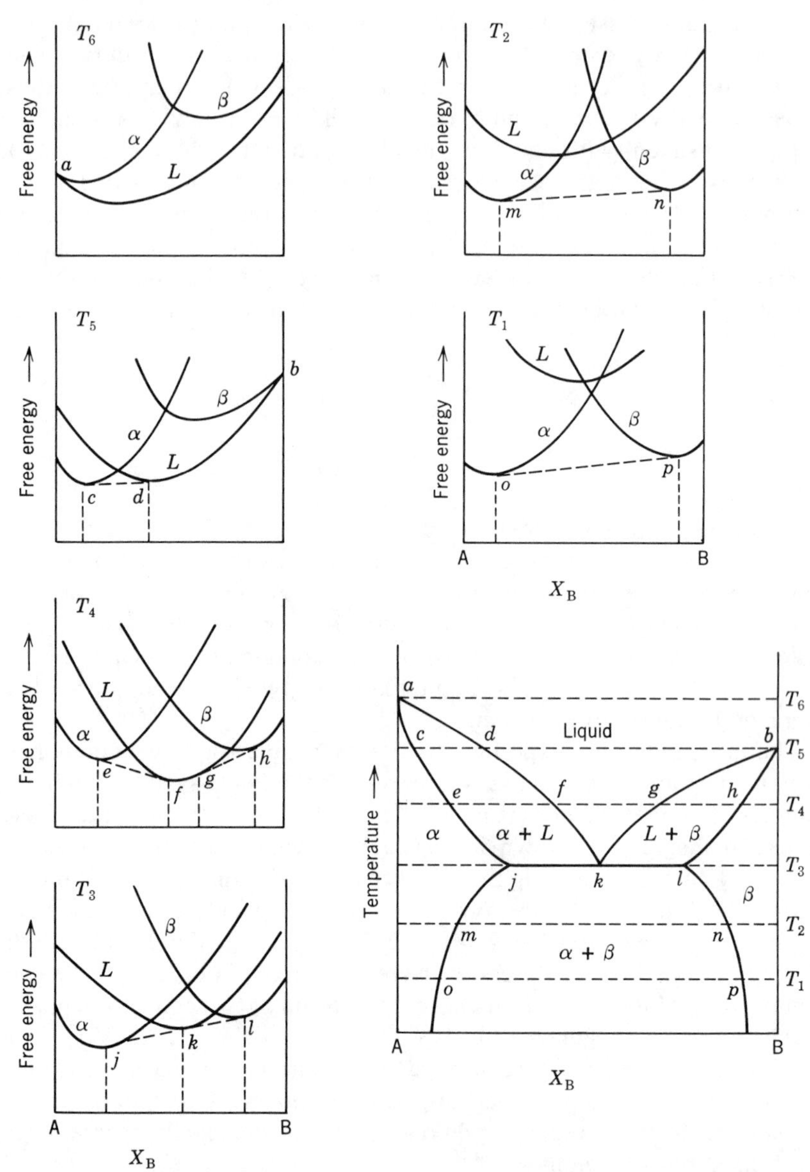

Fig. 7.8. Free-energy–composition curves and the temperature-composition equilibrium diagram for a eutectic system. From P. Gordon, *Principles of Phase Diagrams in Materials Systems*, McGraw-Hill Book Company, New York, 1968.

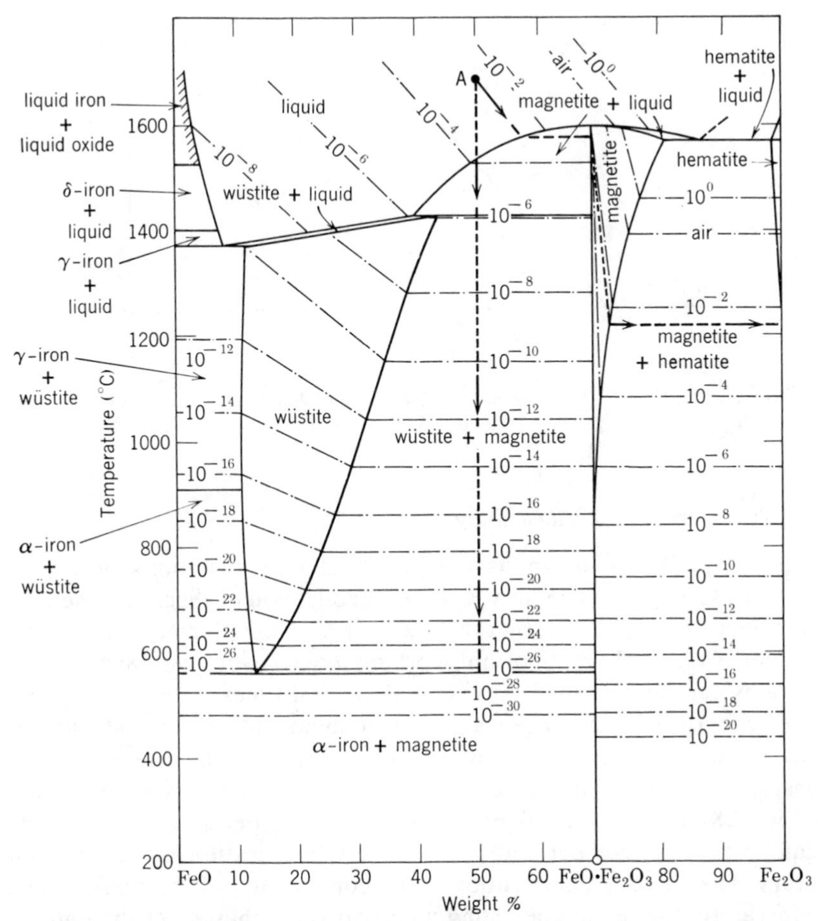

Fig. 7.9. (*a*) Phase relations in the FeO–Fe_2O_3 system. Dash-dot lines are oxygen isobars. Alternate solidification paths for composition A are discussed in text. From A. Muan and E. F. Osborn, *Phase Equilibria among Oxides in Steelmaking*, Addison-Wesley Publishing Company, Inc., Reading, Mass., 1965.

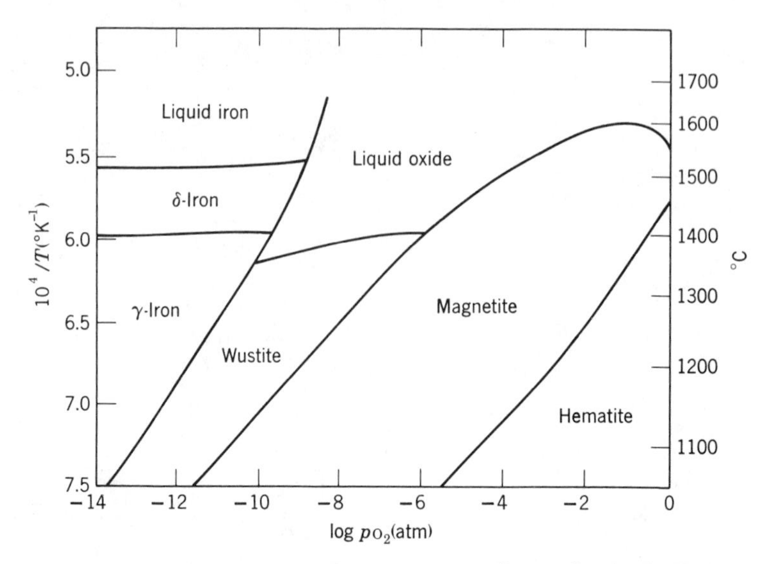

Fig. 7.9 (*continued*). (*b*) Temperature-oxygen pressure diagram for the $Fe–Fe_2O_3$ system. From J. B. Wagner, *Bull. Am. Cer. Soc.*, **53**, 224 (1974).

7.5 Two-Component Phase Diagrams

Phase-equilibrium diagrams are graphical representations of experimental observations. The most extensive collection of diagrams useful in ceramics is that published by the American Ceramic Society in two large volumes, which are an important working tool of every ceramist.* Phase diagrams can be classified into several general types.

Eutectic Diagrams. When a second component is added to a pure material, the freezing point is often lowered. A complete binary system consists of lowered liquidus curves for both end members, as illustrated in Fig. 7.8. The *eutectic temperature* is the temperature at which the liquidus curves intersect and is the lowest temperature at which liquid occurs. The eutectic composition is the composition of the liquid at this temperature, the liquid coexisting with two solid phases. At the eutectic temperature three phases are present, so the variance is one. Since pressure is fixed, the temperature cannot change unless one phase disappears.

In the binary system $BeO–Al_2O_3$ (Fig. 7.10) the regions of solid solution that are necessarily present have not been determined and are presumed

*E. M. Levin, C. R. Robbins, and H. F. McMurdie, *Phase Diagrams for Ceramists*, American Ceramic Society, Columbus, 1964; *Supplement*, 1969.

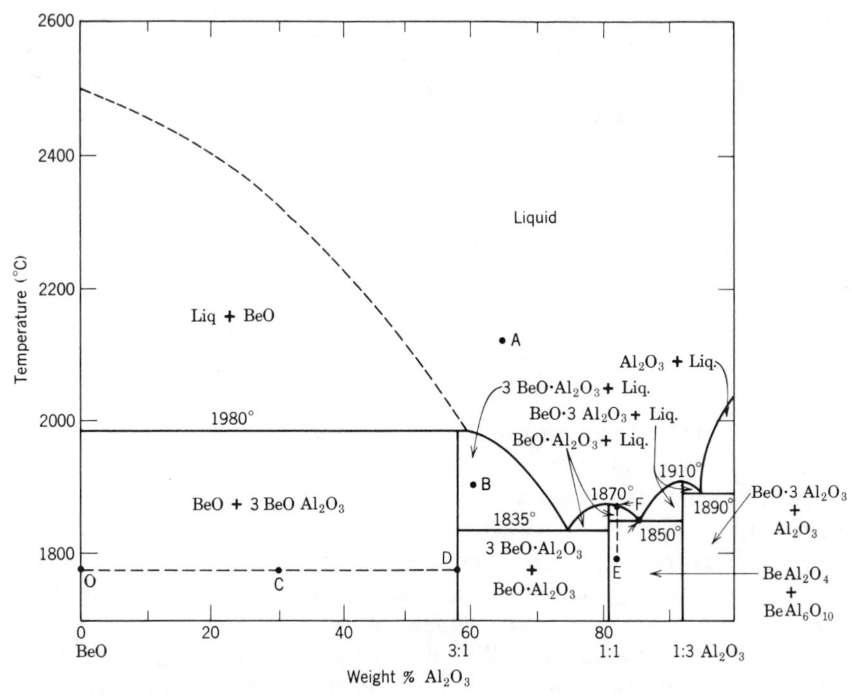

Fig. 7.10. The binary system BeO–Al_2O_3.

to be of limited extent, although this is uncertain, and are not shown in the diagram. The system can be divided into three simpler two-component systems (BeO–$BeAl_2O_4$, $BeAl_2O_4$–$BeAl_6O_{10}$, and $BeAl_6O_{10}$–Al_2O_3) in each of which the freezing point of the pure material is lowered by addition of the second component. The BeO–$BeAl_2O_4$ subsystem contains a compound, $Be_3Al_2O_6$, which melts incongruently, as discussed in the next section. In the single-phase regions there is only one phase present, its composition is obviously that of the entire system, and it comprises 100% of the system (point *A* in Fig. 7.10). In two-phase regions the phases present are indicated in the diagram (point *B* in Fig. 7.10); the composition of each phase is represented by the intersection of a constant temperature tie line and the phase-boundary lines. The amounts of each phase can also be determined from the fact that the sum of the composition times the amount of each phase present must equal the composition of the entire system. For example, at point *C* in Fig. 7.10 the entire system is composed of 29% Al_2O_3 and consists of two phases, BeO (containing no Al_2O_3) and 3BeO·Al_2O_3 (which contains 58% Al_2O_3). There must be 50% of each phase present for a mass balance to give the correct overall composition. This can be represented graphically in the diagram by the *lever principle*, in which the distance from one phase boundary to the overall system composition, divided by the distance from that boundary to the second phase boundary, is the fraction of the second phase present. That is, in Fig. 7.10,

$$\frac{OC}{OD}(100) = \text{Per cent } 3BeO \cdot Al_2O_3$$

A little consideration indicates that the ratio of phases is given as

$$\frac{DC}{OC} = \frac{BeO}{3BeO \cdot Al_2O_3}$$

This same method can be used for determining the amounts of phases present at any point in the diagram.

Consider the changes that occur in the phases present on heating a composition such as *E*, which is a mixture of $BeAl_2O_4$ and $BeAl_6O_{10}$. These phases remain the only ones present until a temperature of 1850°C is reached; at this eutectic temperature there is a reaction, $BeAl_2O_4 + BeAl_6O_{10}$ = Liquid (85% Al_2O_3), which continues at constant temperature to form the eutectic liquid until all the $BeAl_6O_{10}$ is consumed. On further heating more of the $BeAl_2O_4$ dissolves in the liquid, so that the liquid composition changes along *GF* until at about 1875°C all the $BeAl_2O_4$ has disappeared and the system is entirely liquid. On cooling this liquid, exactly the reverse occurs during equilibrium solidification.

As an exercise students should calculate the fraction of each phase present for different temperatures and different system compositions.

One of the main features of eutectic systems is the lowering of the temperature at which liquid is formed. In the BeO–Al_2O_3 system, for example, the pure end members melt at temperatures of 2500°C and 2040°C, respectively. In contrast, in the two-component system a liquid is formed at temperatures as low as 1835°C. This may be an advantage or disadvantage for different applications. For maximum temperature use as a refractory we want no liquid to be formed. Addition of even a small amount of BeO to Al_2O_3 results in the formation of a substantial amount of a fluid liquid at 1890°C and makes it useless as a refractory above this temperature. However, if high-temperature applications are not of major importance, it may be desirable to form the liquid as an aid to firing at lower temperatures, since liquid increases the ease of densification. This is true, for example, in the system TiO_2–UO_2, in which addition of 1% TiO_2 forms a eutectic liquid, which is a great aid in obtaining high densities at low temperatures. The structure of this system, shown in Fig.

7.11, consists of large grains of UO_2 surrounded by the eutectic composition.

The effectiveness of eutectic systems in lowering the melting point is made use of in the Na_2O–SiO_2 system, in which glass compositions can be melted at low temperatures (Fig. 7.12). The liquidus is lowered from 1710°C in pure SiO_2 to about 790° for the eutectic composition at approximately 75% SiO_2–25% Na_2O.

Formation of low-melting eutectics also leads to some severe limitations on the use of refractories. In the system CaO–Al_2O_3 the liquidus is strongly lowered by a series of eutectics. In general, strongly basic oxides such as CaO form low-melting eutectics with amphoteric or basic oxides, and these classes of materials cannot be used adjacent to each other, even though they are individually highly refractive.

Incongruent Melting. Sometimes a solid compound does not melt to form a liquid of its own composition but instead dissociates to form a new solid phase and a liquid. This is true of enstatite ($MgSiO_3$) at 1557°C (Fig. 7.13); this compound forms solid $Mg_2S{:}O_4$ plus a liquid containing about 61% SiO_2. At this *incongruent melting point* or *peritectic temperature* there

Fig. 7.11. Structure of 99% UO_2–1% TiO_2 ceramic (228X, HNO_3 etch). UO_2 is the primary phase, bonded by eutectic composition. Courtesy G. Ploetz.

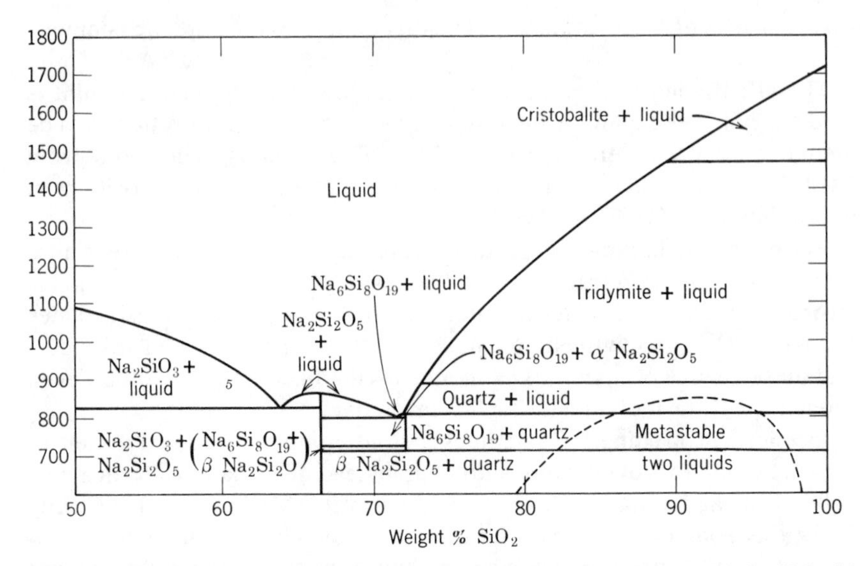

Fig. 7.12. The binary system Na_2SiO_3–SiO_2. The dashed line shows metastable liquid-liquid phase separation.

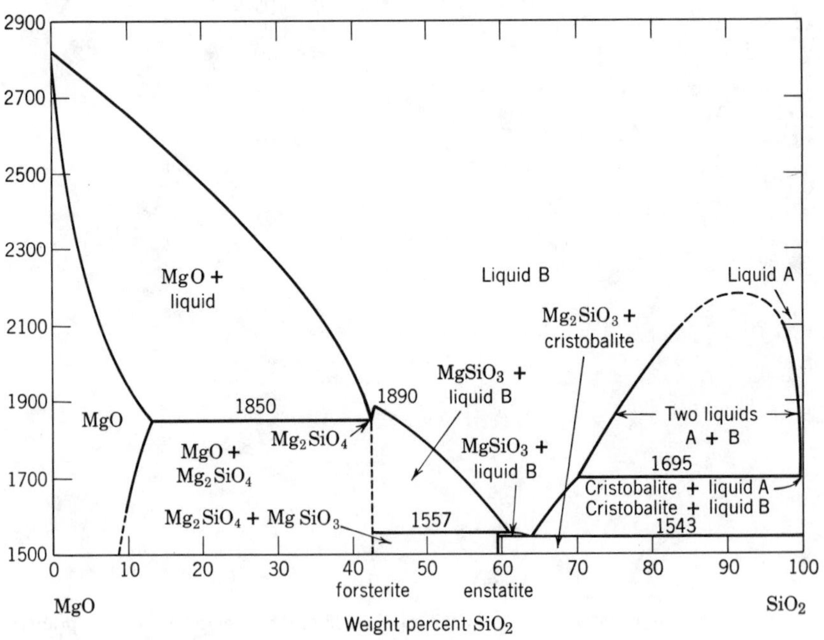

Fig. 7.13. The binary system MgO–SiO_2.

are three phases present (two solids and a liquid), so that the temperature remains fixed until the reaction is completed. Potash feldspar (Fig. 7.14) also melts in this way.

Phase Separation. When a liquid or crystalline solution is cooled, it separates into two separate phases at the *consolute temperature* as long as the excess enthalpy is positive (see Fig. 7.7). This phenomenon is particularly important relative to the development of substructure in glasses, as discussed in Chapter 3 (Figs. 3.11, 3.12, 3.14 to 3.19). Although it has been less fully investigated for crystalline oxide solid solutions, it is probably equally important for these systems when they are exposed to moderate temperatures for long periods of time. The system CoO–NiO is shown in Fig. 7.15.

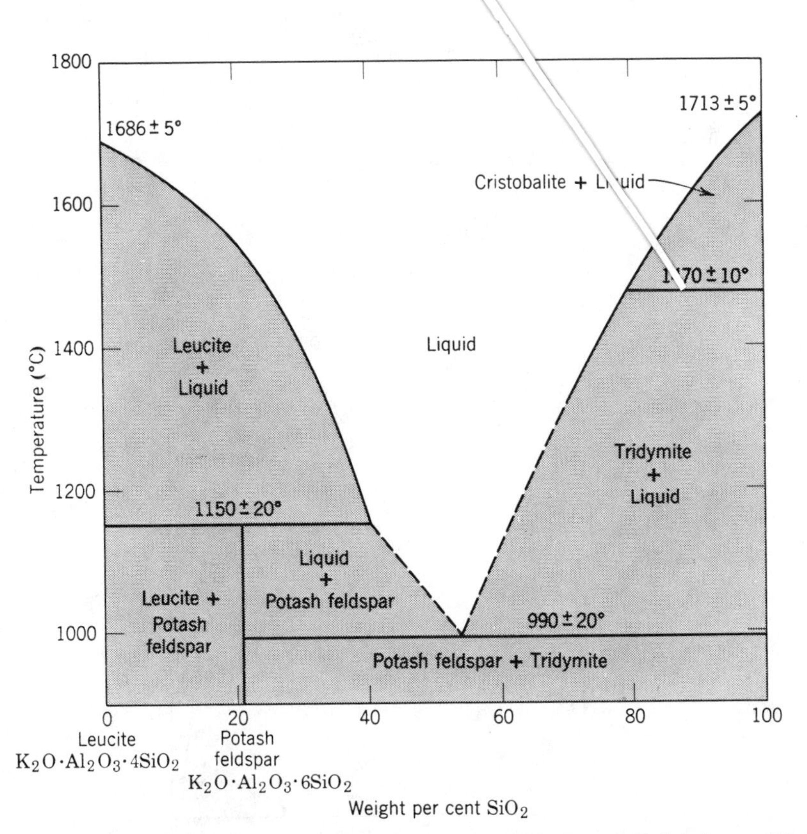

Fig. 7.14. The binary system $K_2O \cdot Al_2O_3 \cdot 4SiO_2$ (leucite)–SiO_2. From J. F. Schairer and N. L. Bowen, *Bull. Soc. Geol. Finl.*, **20**, 74 (1947). Two-phase regions are shown shaded in this diagram.

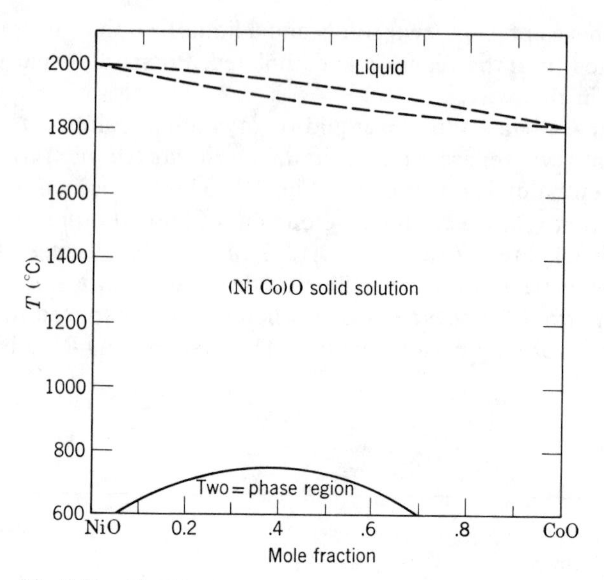

Fig. 7.15. The binary system NiO–CoO.

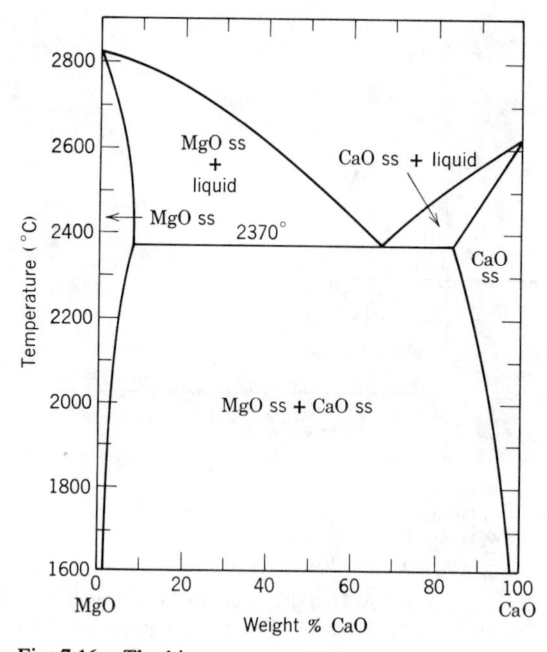

Fig. 7.16. The binary system MgO–CaO.

Solid Solutions. As discussed in Chapter 4 and in Section 7.4, a complete series of solid solutions occurs for some systems such as illustrated in Fig. 4.2 and Fig. 7.15, and some minute or significant limited solid solution occurs for all systems, as shown in Figs. 4.3, 7.13, and 7.15.

It has only been in the last decade or so that careful experimentation has revealed the wide extent of solid solubility, reaching several percent at high temperatures in many systems, as shown in Figs. 4.3, 7.13, and 7.15 and for the MgO–CaO system in Fig. 7.16 and the MgO–Cr_2O_3 system in Fig. 7.17. For steel-plant refractories directly bonded magnesia–chromite brick is formed when these materials are heated together at temperatures above 1600°C as a result of the partial solubility of the constituents; exsolution occurs on cooling. Almost all open-hearth roofs are formed of either direct-bonded, rebonded fine-grain, or fusion-cast magnesia-chromite refractories. In the basic oxygen-furnace process for steel making MgO–CaO refractories bonded with pitch are widely used, and the solid solubility at high temperatures forms a high-temperature bond. In magnesia refractories the lower solid solubility of SiO_2 as compared

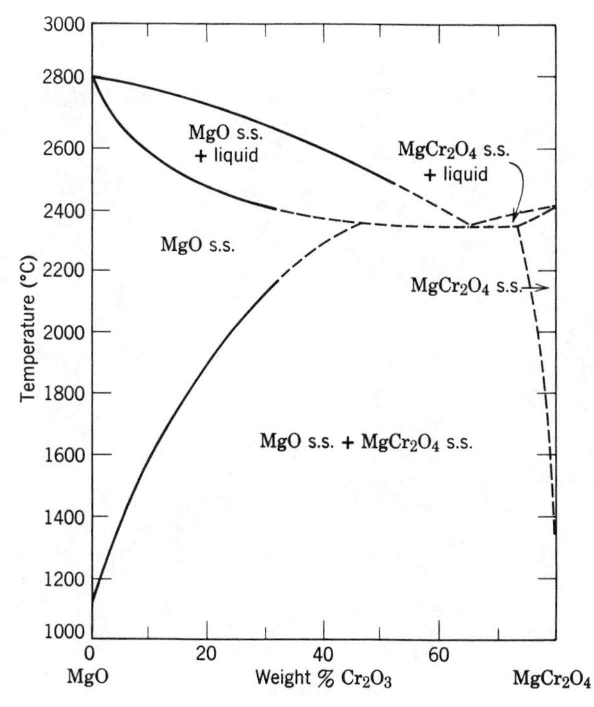

Fig. 7.17. The binary system MgO–$MgCr_2O_3$.

with CaO in MgO requires that excess CaO be added to prevent the formation of low-melting intergranular silicates.

In the MgO–Al_2O_3 system (Fig. 4.3) there is extensive solubility of MgO and of Al_2O_3 in spinel. As spinel in this composition range is cooled, the solubility decreases, and corundum precipitates as a separate solid phase (Fig. 7.18).

This same sort of limited solid solution is observed in the CaO–ZrO_2 system (Fig. 7.19); in this system there are three different fields of solid solution, the tetragonal form, the cubic form, and the monoclinic form. Pure ZrO_2 exhibits a monoclinic tetragonal phase transition at 1000°C, which involves a large volume change and makes the use of pure zirconia impossible as a ceramic material. Addition of lime to form the cubic solid solution, which has no phase transition, is one basis for *stabilized zirconia*, a valuable refractory.

Complex Diagrams. All the basic parts of binary phase-equilibrium diagrams have been illustrated; readers should be able to identify the number of phases, composition of phases, and amounts of phases present at any composition and temperature from any of these diagrams with ease and confidence. If they cannot, they should consult one of the more extensive treatments listed in the references.

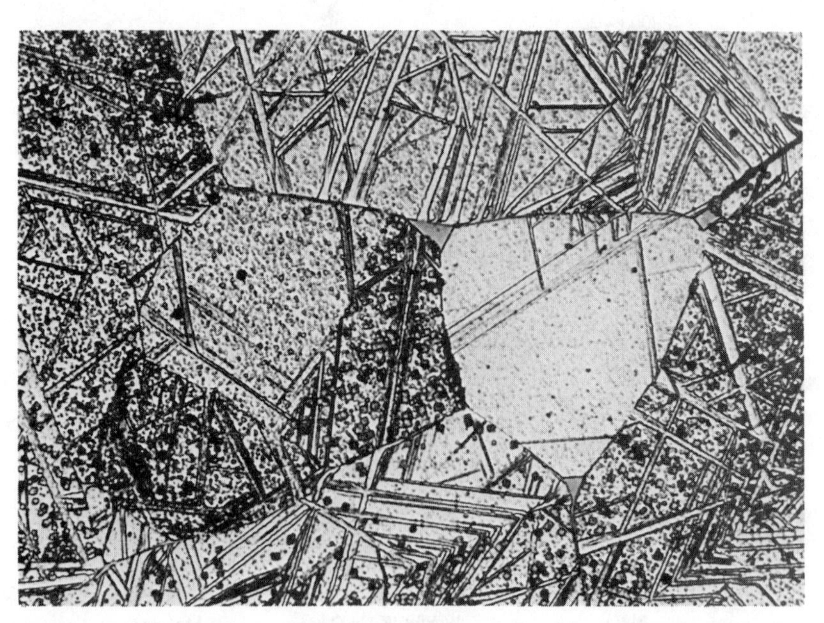

Fig. 7.18. Precipitation of Al_2O_3 from spinel solid solution on cooling (400× H_2SO_4 etch). Courtesy R. L. Coble.

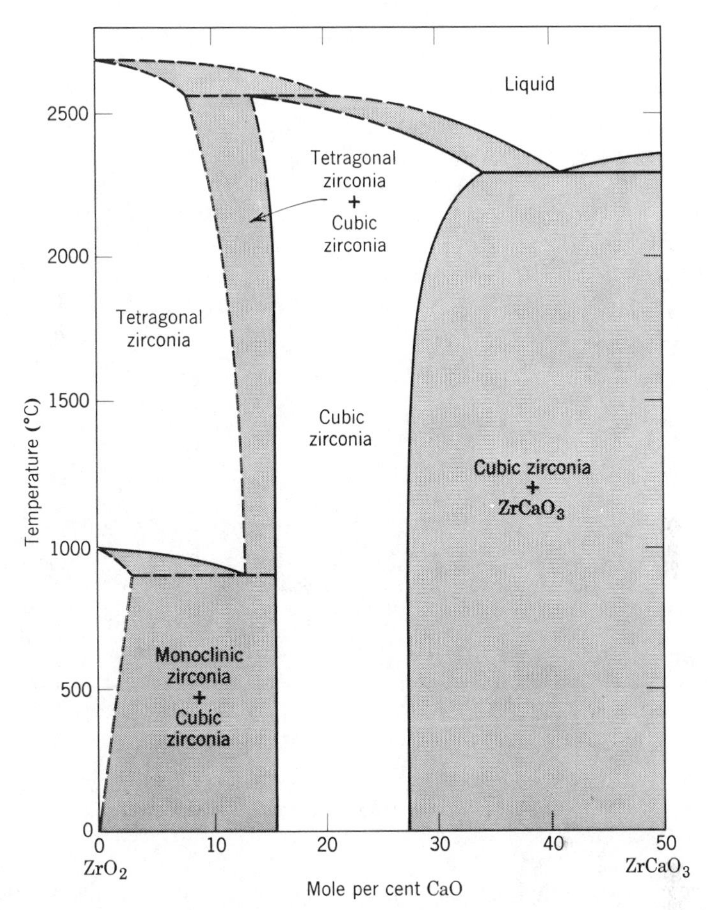

Fig. 7.19. The binary system $CaO–ZrO_2$. From P. Duwez, F. Odell, and F. H. Brown, Jr., *J. Am. Ceram. Soc.*, **35**, 109 (1952). Two-phase regions are shown shaded in this figure.

Combinations of simple elements in one system sometimes appear frightening in their complexity but actually offer no new problems in interpretation. In the system $Ba_2TiO_4–TiO_2$ (Fig. 7.20), for example, we find two eutectics, three incongruently melting compounds, polymorphic forms of $BaTiO_3$, and an area of limited solid solution. All of these have already been discussed.

Generally phase diagrams are constructed at a total pressure of 1 atm with temperature and composition as independent variables. Since the interesting equilibrium conditions for many ceramics involve low oxygen partial pressures, phase diagrams at a fixed temperature but with oxygen

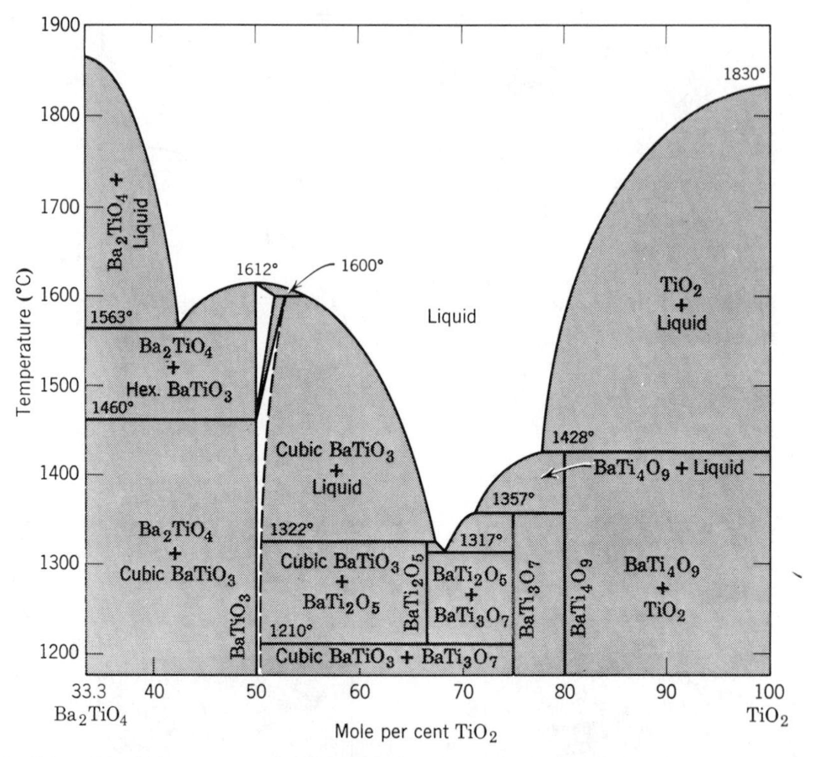

Fig. 7.20. The binary system Ba_2TiO_4–TiO_2. From D. E. Rase and R. Roy, *J. Am. Ceram. Soc.* **38**, 111 (1955). Two-phase regions are shown shaded in this figure.

pressure and composition as variables become a useful alternative for describing phase equilibria, for example, Fig. 7.9*b*. Figure 7.21(*a*-1) shows such a diagram for Co–Ni–O at 1600°K. The lens-shaped two-phase region between (Co,Ni)O and the NiCo alloy is similar to that between the liquid oxides and (Co,Ni)O in a temperature-composition plot (Fig. 7.15). Figure 7.21(*a*-2) shows the oxygen isobar tie lines between the metal alloy and the oxide solid solution; for example, the dotted line represents the equilibrium at $P_{O_2} = 1.5 \times 10^{-7}$ atm between $Ni_{0.62}Co_{0.38}O$ and $Ni_{0.9}Co_{0.1}$. (A tie line connects phases in equilibrium and designates the composition of each phase. For example, a constant temperature tie line in Fig. 7.17 at 2600°C specifies the composition of the solid solution, 10 w/o Cr_2O_3, in equilibrium with the liquid, which contains 40 w/o Cr_2O_3.) A plot of the nickel activity as a function of P_{O_2} is shown in Fig. 7.21(*a*-3). In systems which form intermediate compounds, such as spinels, the diagrams become more complex. The Fe–Cr–O ternary system at 1573°K is shown in Fig. 7.21*b*. At an oxygen pressure of $P_{O_2} = 10^{-10}$ atm, the stable phases may be FeO, FeO + $(Fe,Cr)_3O_4$, $(Fe,Cr)_3O_4$ + $(Fe,Cr)_2O_3$, or $(Fe,Cr)_2O_3$, depending on the concentration of chromium. The oxygen isobars shown in Fig. 7.21(*b*-2) are tie lines between the compositions in equilibrium at 1573°K.

7.6 Three-Component Phase Diagrams

Three-component systems are fundamentally no different from two-component systems, except that there are four independent variables—pressure, temperature, and the concentrations of two components (which fix the third). If pressure is arbitrarily fixed, the presence of four phases gives rise to an invariant system. A complete graphical representation of ternary systems is difficult, but if the pressure is held constant, compositions can be represented on an equilateral triangle and the temperature on a vertical ordinate to give a phase diagram such as Fig. 7.22. For two-dimensional representation the temperatures can be projected on an equilateral triangle, with the liquidus temperatures represented by

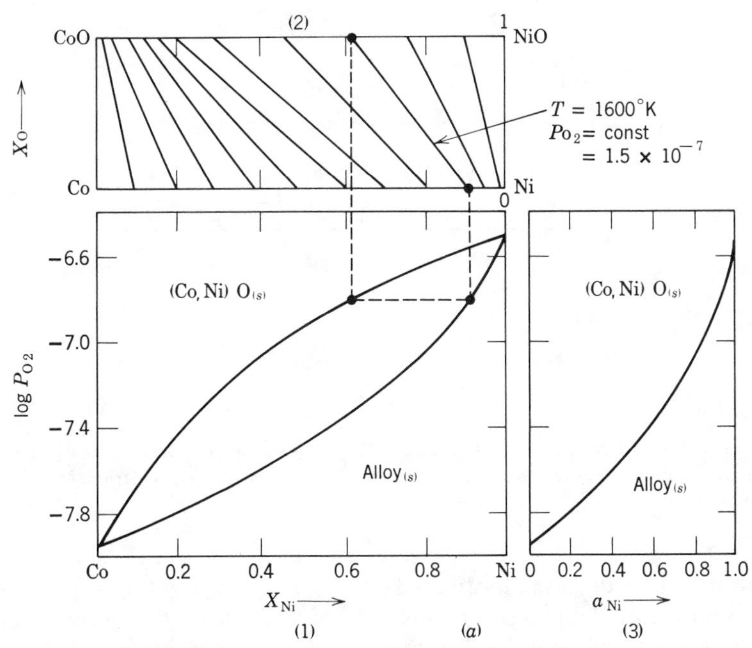

Fig. 7.21. (*a*) Co–Ni–O system. (1) Composition of condensed phases as a function of P_{O_2}; (2) oxygen isobars for equilibrium between the oxide solid solution and the alloy solution; (3) nickel activity as a function of P_{O_2}.

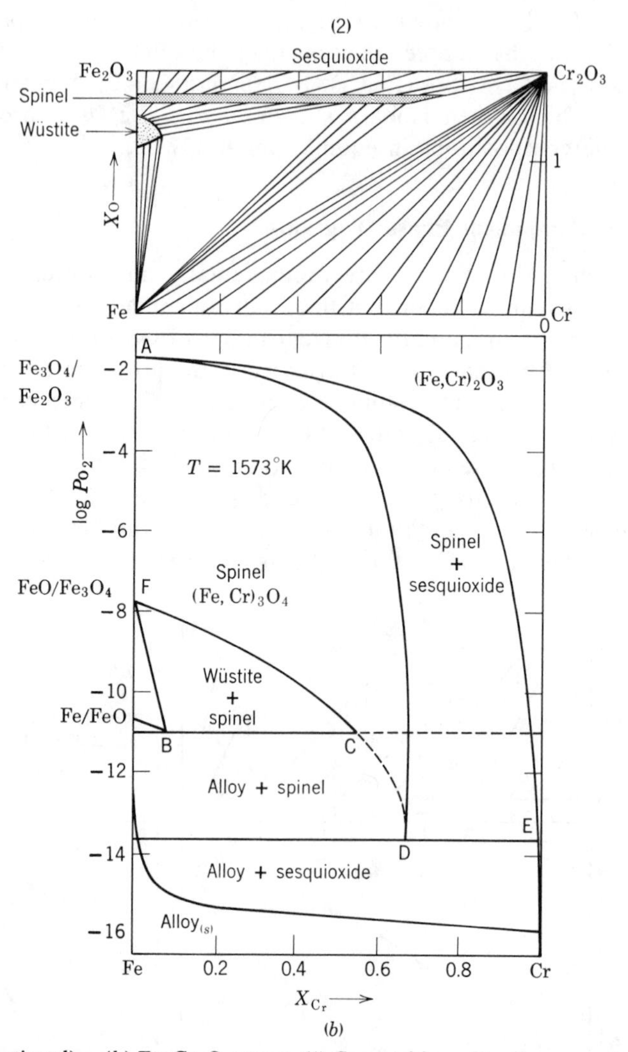

Fig. 7.21 (*continued*). (*b*) Fe–Cr–O system. (1) Composition—P_{O_2} diagram and (2) oxygen isobars for equilibrium between two phases. From A. Pelton and H. Schmalzried, *Met. Trans.*, **4**, 1395 (1973).

isotherms. The diagram is divided into areas representing equilibrium between the liquid and a solid phase. Boundary curves represent equilibrium between two solids and the liquid, and intersections of three boundary curves represent points of four phases in equilibrium (invariant points in the constant-pressure system). Another method of two-

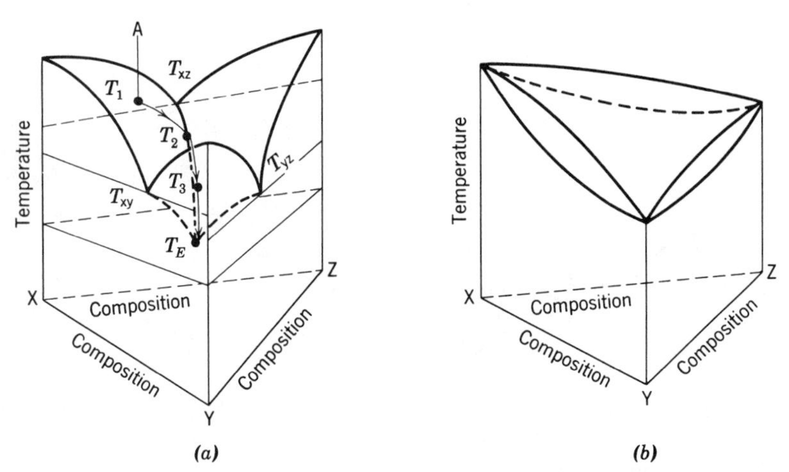

Fig. 7.22. Space diagram of (*a*) ternary eutectic and (*b*) complete series of solid solutions.

dimensional representation is to take a constant-temperature cut through the diagram, indicating the phases at equilibrium at some fixed temperature.

Interpretation of ternary diagrams is not fundamentally different from that of binary diagrams. The phases in equilibrium at any temperature and composition are shown; the composition of each phase is given by the phase-boundary surfaces or intersections; the relative amounts of each phase are determined by the principle that the sum of the individual phase compositions must equal the total composition of the entire system. In Fig. 7.22 and Fig. 7.23, for example, the composition A falls in the primary field of X. If we cool the liquid A, X begins to crystallize from the melt when the temperature reaches T_1. The composition of the liquid changes along AB because of the loss of X. Along this line the lever principle applies, so that at any point the percentage of X present is given by 100(BA/XB). When the temperature reaches T_2 and the crystallization path reaches the boundary representing equilibrium between the liquid and two solid phases X and Z, Z begins to crystallize also, and the liquid changes in composition along the path CD. At L, the phases in equilibrium are a liquid of composition L and the solids X and Z, whereas the overall composition of the entire system is A. As shown in Fig. 7.23*b*, the only mixture of L, X, and Z that gives a total corresponding to A is $x\text{A}/x\text{X}$ (100) = Per cent X, $z\text{A}/z\text{Z}$ (100) = Per cent Z, $l\text{A}/l\text{L}$ (100) = Per cent L. That is, the smaller triangle XZL is a ternary system in which the composition of A can be represented in terms of its three constituents.

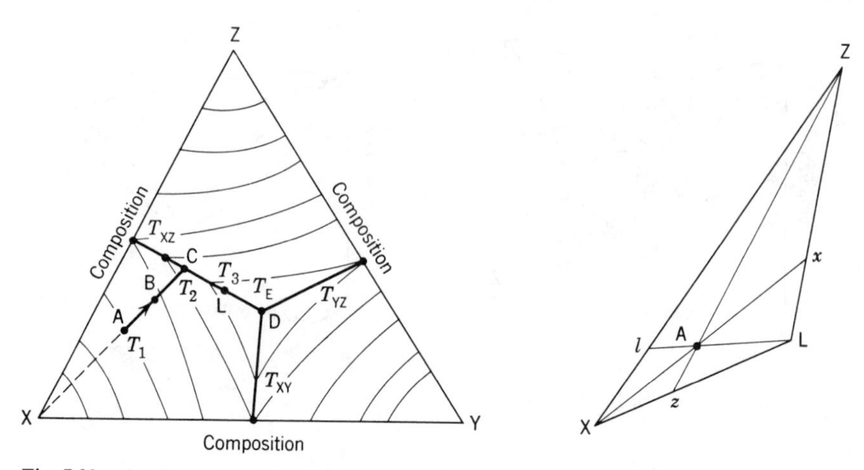

Fig. 7.23. (*a*) Crystallization path illustrated in Fig. 7.22*a* and (*b*) application of center of gravity principle to a ternary system.

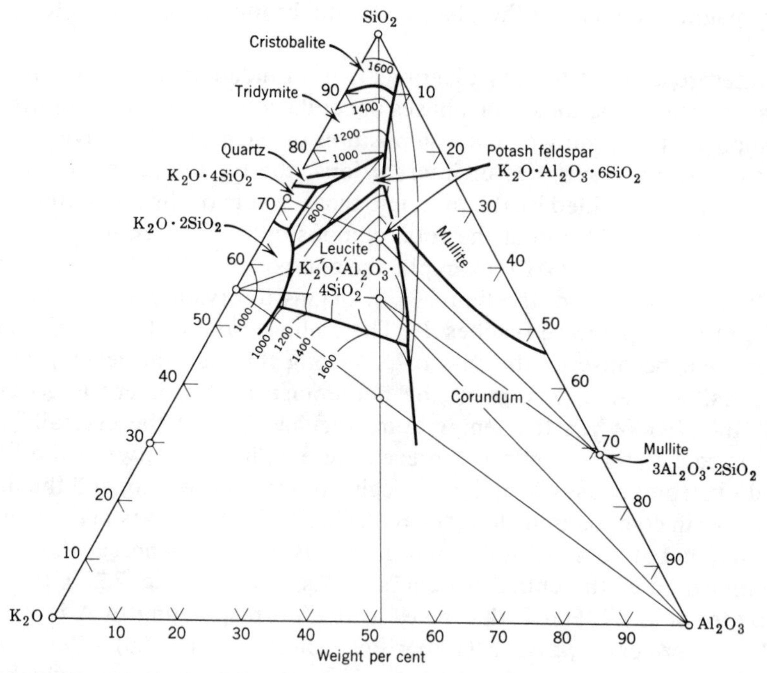

Fig. 7.24. The ternary system K_2O–Al_2O_3–SiO_2. From J. F. Schairer and N. L. Bowen, *Am. J. Sci.*, **245**, 199 (1947).

Many ternary systems are of interest in ceramic science and technology. Two of these, the K_2O–Al_2O_3–SiO_2 system and the Na_2O–CaO–SiO_2 system, are illustrated in Figs. 7.24 and 7.25. Another important system, the MgO–Al_2O_3–SiO_2 system, is discussed in Section 7.8. The K_2O–Al_2O_3–SiO_2 system is important as the basis for many porcelain compositions. The eutectic in the subsystem potash–feldspar–silica–mullite determines the firing behavior in many compositions. As discussed in Chapter 10, porcelain compositions are adjusted mainly on the basis of (*a*) ease in forming and (*b*) firing behavior. Although real systems are usually somewhat more complex, this ternary diagram provides a good description of the compositions used. The Na_2O–CaO–SiO_2 system forms the basis for much glass technology. Most compositions fall along the border between the primary phase of devitrite, $Na_2O{\cdot}3CaO{\cdot}6SiO_2$, and silica; the liquidus temperature is 900 to 1050°C.

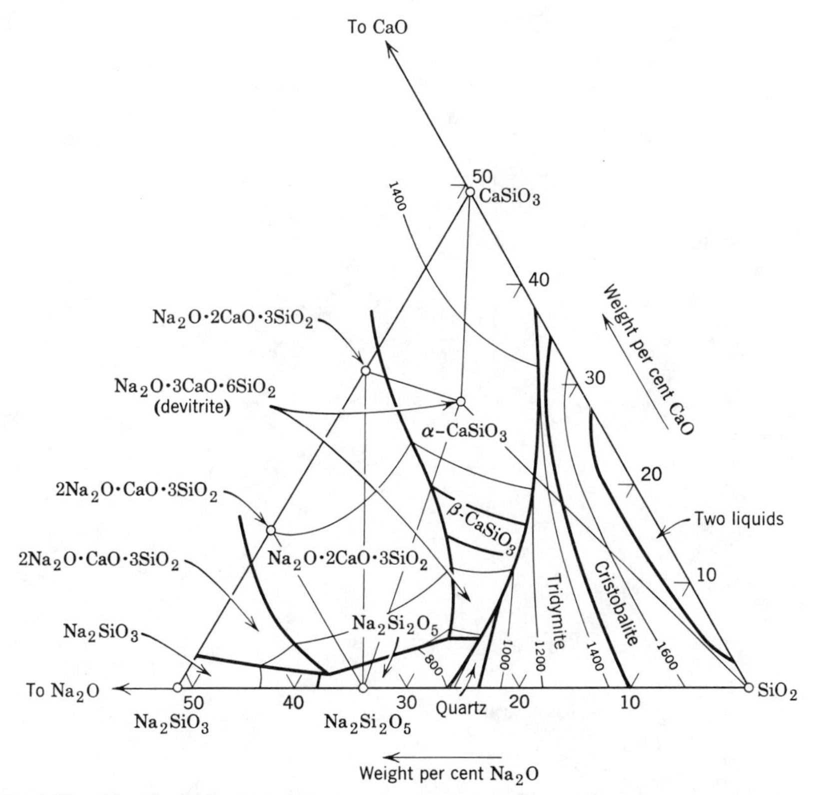

Fig. 7.25. The Na_2O–O–CaO–SiO_2 system. From G. W. Morey and N. L. Bowen, *J. Soc. Glass Technol.*, **9**, 232 (1925).

This is a compositional area of low melting temperature, but the glasses formed contain sufficient calcium oxide for reasonable resistance to chemical attack. When glasses are heated for extended times above the transition range, devitrite or cristobalite is the crystalline phase formed as the devitrification product.

Very often constant-temperature diagrams are useful. These are illustrated for subsolidus temperatures in Figs. 7.24 and 7.25 by lines between the forms that exist at equilibrium. These lines form composition triangles in which three phases are present at equilibrium, sometimes called compatibility triangles. Constant-temperature diagrams at higher temperatures are useful, as illustrated in Fig. 7.26, in which the 1200° isothermal

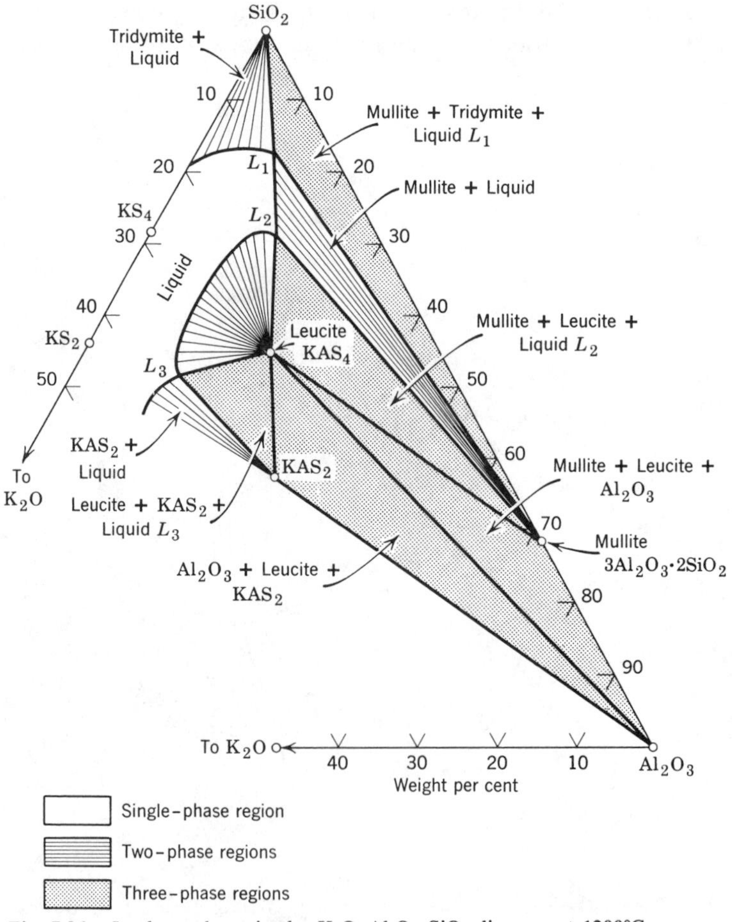

Fig. 7.26. Isothermal cut in the K_2O–Al_2O_3–SiO_2 diagram at 1200°C.

plane is shown for the K_2O–Al_2O_3–SiO_2 diagram. The liquids formed in this system are viscous; in order to obtain vitrification, a substantial amount of liquid must be present at the firing temperature. From isothermal diagrams the composition of liquid and amount of liquid for different compositions can be easily determined at the temperature selected. Frequently it is sufficient to determine an isothermal plane rather than an entire diagram, and obviously it is much easier.

Although our discussion of three-component diagrams has been brief and we do not discuss phase-equilibrium behavior for four or more component systems at all, students would be well advised to become familiar with these as an extra project.

7.7 Phase Composition versus Temperature

One of the useful applications for phase equilibrium diagrams in ceramic systems is the determination of the phases present at different temperatures. This information is most readily used in the form of plots of the amount of phases present versus temperature.

Consider, for example, the system MgO–SiO_2 (Fig. 7.13). For a composition of 50 wt% MgO–50 wt% SiO_2, the solid phases present at equilibrium are forsterite and enstatite. As they are heated, no new phases are formed until 1557°C. At this temperature the enstatite disappears and a composition of about 40% liquid containing 61% SiO_2 is formed. On further heating the amount of liquid present increases until the liquidus is reached at some temperature near 1800°C. In contrast, for a 60% MgO–40% SiO_2 composition the solid phases present are forsterite, Mg_2SiO_4, and periclase, MgO. No new phase is found on heating until 1850°C, when the composition becomes nearly all liquid, since this temperature is near the eutectic composition. The changes in phase occurring for these two compositions are illustrated in Fig. 7.27.

Several things are apparent from this graphical representation. One is the large difference in liquid content versus temperature for a relatively small change in composition. For compositions containing greater than 42% silica, the forsterite composition, liquids are formed at relatively low temperatures. For compositions with silica contents less than 42% no liquid is formed until 1850°C. This fact is used in the treatment of chromite refractories. The most common impurity present is serpentine, $3MgO{\cdot}2SiO_2{\cdot}2H_2O$, having a composition of about 50 wt% SiO_2. If sufficient MgO is added to put this in the MgO–forsterite field, it no longer has a deleterious effect. Without this addition a liquid is formed at low temperatures.

Another application of this diagram is in the selection of compositions

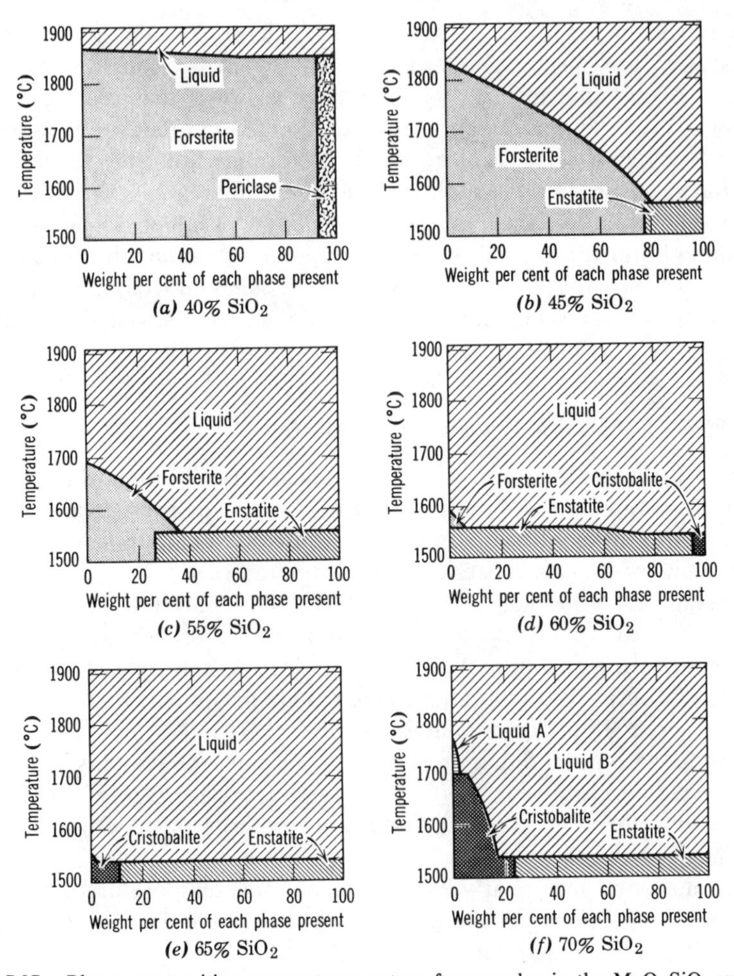

Fig. 7.27. Phase composition versus temperature for samples in the $MgO–SiO_2$ system.

that have desirable firing characteristics. It is necessary to form a sufficient amount of liquid for vitrification, but not so much that ware slumps or warps during firing. The limits of liquid required vary with the properties of the liquid but are in the range of 20 to 50 wt%. To have a sufficient range of firing temperature, it is desirable that the liquid content not change much with temperature. Forsterite compositions cannot be fired until very high temperatures if the composition is exactly 42% SiO_2, since no liquid is formed below 1850°C. Compositions in the forsterite-enstatite field which are mainly forsterite form a liquid at 1557°C, and since the liquidus curve is steep, the amount of liquid present changes but slightly with temperature, as shown in Fig. 7.27. Consequently, these compositions have a good firing range and are easy to vitrify. In contrast, compositions that are mostly enstatite (55, 60, 65% SiO_2) form large amounts of liquid at low temperature, and the amount of liquid present changes rapidly with temperature. These materials have a limited firing range and pose difficult control problems for economic production.

For systems in which the gas phase is important the way in which condensed phases appear and their compositional changes on cooling depend on the conditions imposed. Referring back to the Fe–O system illustrated in Fig. 7.9, if the total condensed-phase composition remains constant, as occurs in a closed nonreactive container with only a negligible amount of gas phase present, the composition A solidifies along the dotted line with a corresponding decrease in the system oxygen pressure. In contrast, if the system is cooled at constant oxygen pressure, the solidification path is along the dashed line. In one case the resulting product at room temperature is a mixture of iron and magnetite; in the second case the resulting product is hematite. Obviously in such systems the control of oxygen pressure during cooling is essential for the control of the products formed.

For detailed discussions of crystallization paths in ternary systems the references should be consulted. The following summary* can serve as a review.

1. When a liquid is cooled, the first phase to appear is the primary phase for that part of the system in which the composition of the melt is represented.
2. The crystallization curve follows to the nearest boundary the extension of the straight line connecting the composition of the original liquid with that of the primary phase of that field. The composition of the liquid within the primary fields is represented by points on the crystallization curve. This curve is the intersection of a plane (perpendicular to the base triangle and passing through the compositions of original melt and the primary phase) with the liquidus surface.
3. At the boundary line a new phase appears which is the primary phase of the adjacent field. The two phases separate together along this boundary as the temperature is lowered.
4. The ratio of the two solids crystallizing is given by the intersection of the tangent to the boundary curve with a line connecting the composi-

*After E. M. Levin, H. F. McMurdie, and F. P. Hall, *Phase Diagrams for Ceramists*, American Ceramic Society, Cleveland, Ohio, 1956.

tions of the two solid phases. Two things can occur. If this tangent line runs between the compositions of the two solid phases, the amount of each of these phases present increases. If the tangent line intersects an extension of the line between solid compositions, the first phase decreases in amount (is resorbed; Reaction A + Liquid = B) as crystallization proceeds. In some systems the crystallization curve leaves the boundary curve if the first phase is completely resorbed, leaving only the second phase. Systems in which this occurs may be inferred from a study of the mean composition of the solid separating between successive points on the crystallization path.

5. The crystallization curve always ends at the invariant point which represents equilibrium of liquid with the three solid phases of the three components within whose composition triangle the original liquid composition was found.

6. The mean composition of the solid which is crystallizing at any point on a boundary line is shown by the intersection at that point of the tangent with a line joining the composition of the two solid phases which are crystallizing.

7. The mean composition of the total solid that has crystallized up to any point on the crystallization curve is found by extending the line connecting the given point with the original liquid composition to the line connecting the compositions of the phases that have been separating.

8. The mean composition of the solid that has separated between two points on a boundary is found at the intersection of a line passing through these two points with a line connecting the compositions of the two solid phases separating along this boundary.

7.8 The System Al_2O_3–SiO_2

As an example of the usefulness of phase diagrams for considering high-temperature phenomena in ceramic systems, the Al_2O_3–SiO_2 system illustrates many of the features and problems encountered. In this system (Fig. 7.28), there is one compound present, mullite, which is shown as melting incongruently. (The melting behavior of mullite has been controversial; we show the metastable extensions of the phase boundaries in Fig. 7.28. For our purposes this is most important as indicative of the fact that experimental techniques are difficult and time consuming; the diagrams included here and in standard references are summaries of experimental data. They usually include many interpolations and extrapolations and have been compiled with greater or lesser care, depending on the needs of the original investigator.) The eutectic between mullite and

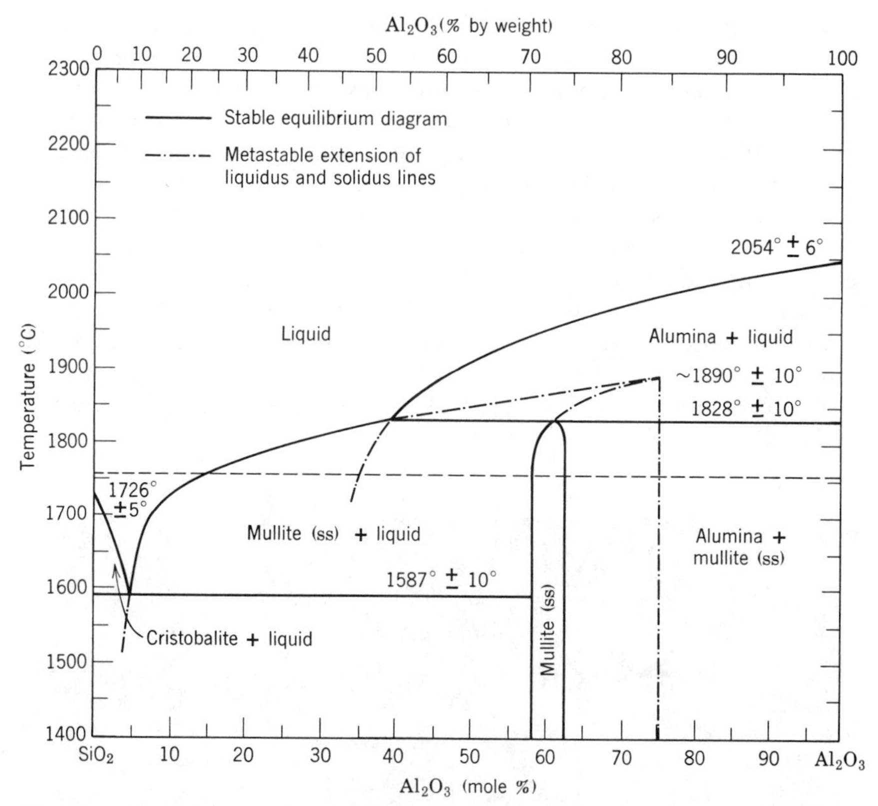

Fig. 7.28. The binary system Al_2O_3–SiO_2. From Aksay and Pask, *Science*, **183**, 69 (1974).

cristobalite occurs at 1587°C to form a liquid containing about 95 mole% SiO_2. The solidus temperature between mullite and alumina is at 1828°C.

Factors affecting the fabrication and use of several refractory products can be related to this diagram. They include refractory silica brick (0.2 to 1.0 wt% Al_2O_3), clay products (35 to 50 wt% Al_2O_3), high-alumina brick (60 to 90 wt% Al_2O_3), pure fused mullite (72 wt% Al_2O_3), and pure fused or sintered alumina (>90 wt% Al_2O_3).

At one end of the composition range are silica bricks widely used for furnace roofs and similar structures requiring high strength at high temperatures. A major application was as roof brick for open-hearth furnaces in which temperatures of 1625 to 1650°C are commonly used. At this temperature a part of the brick is actually in the liquid state. In the development of silica brick it has been found that small amounts of aluminum oxide are particularly deleterious to brick properties because the eutectic composition is close to the silica end of the diagram. Consequently, even small additions of aluminum oxide mean that substantial amounts of liquid phase are present at temperatures above 1600°C. For this reason *supersilica* brick, which has a lower alumina content through special raw-material selection or treatment, is used in structures that will be heated to high temperatures.

Fire-clay bricks have a composition ranging from 35 to 55% aluminum oxide. For compositions without impurities the equilibrium phases present at temperatures below 1587°C are mullite and silica (Fig. 7.29). The relative amounts of these phases present change with composition, and there are corresponding changes in the properties of the brick. At temperatures above 1600°C the amount of liquid phase present is sensitive to the alumina-silica ratio, and for these high-temperature applications the higher-alumina brick is preferred.

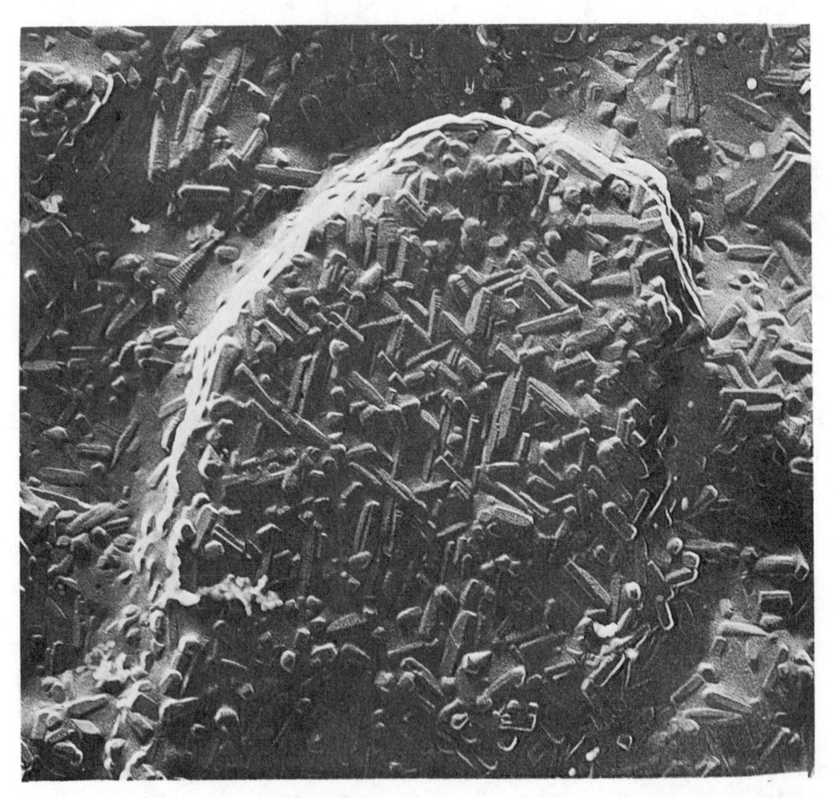

Fig. 7.29. Mullite crystals in silica matrix formed by heating kaolinite (37,000×). Courtesy J. J. Comer.

Refractory properties of brick can be substantially improved if sufficient alumina is added to increase the fraction of mullite present until at greater than 72 wt% alumina the brick is entirely mullite or a mixture of mullite plus alumina. Under these conditions no liquid is present until temperatures above 1828°C are reached. For some applications fused mullite brick is used; it has superior ability to resist corrosion and deformation at high temperatures. The highest refractoriness is obtained with pure alumina. Sintered Al_2O_3 is used for laboratory ware, and fusion-cast Al_2O_3 is used as a glass tank refractory.

7.9 The System MgO-Al_2O_3-SiO_2

A ternary system important in understanding the behavior of a number of ceramic compositions is the MgO–Al_2O_3–SiO_2 system, illustrated in Fig. 7.30. This system is composed of several binary compounds which

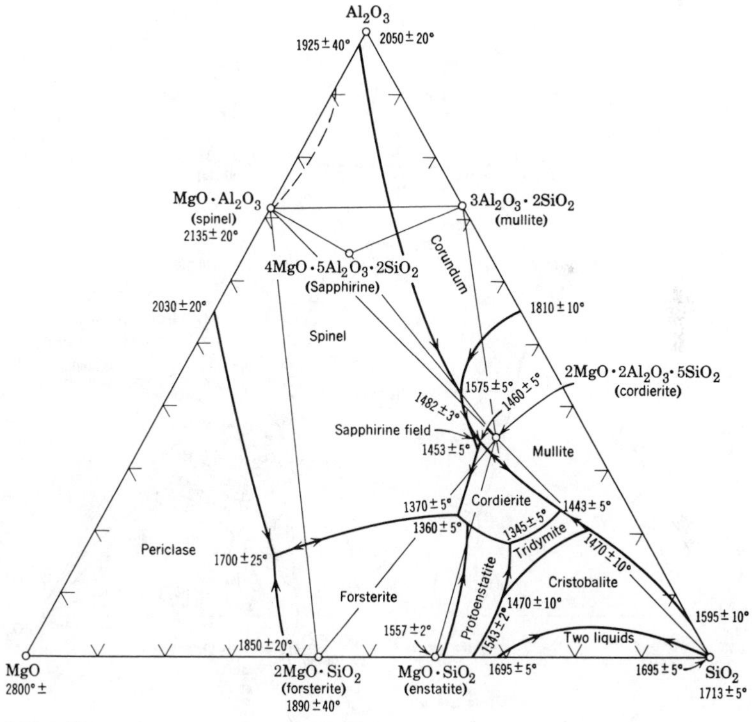

Fig. 7.30. The ternary system MgO–Al_2O_3–SiO_2. From M. L. Keith and J. F. Schairer, *J. Geol.*, **60**, 182 (1952). Regions of solid solution are not shown; see Figs. 4.3 and 7.13.

have already been described, together with two ternary compounds, cordierite, $2MgO{\cdot}2Al_2O_3{\cdot}5SiO_2$, and sapphirine, $4MgO{\cdot}5Al_2O_3{\cdot}2SiO_2$, both of which melt incongruently. The lowest liquidus temperature is at the tridymite–protoenstatite–cordierite eutectic at 1345°C, but the cordierite–enstatite–forsterite eutectic at 1360°C is almost as low-melting.

Ceramic compositions that in large part appear on this diagram include magnesite refractories, forsterite ceramics, steatite ceramics, special low-loss steatites, and cordierite ceramics. The general composition areas of these products on the ternary diagram are illustrated in Fig. 7.31. In all but magnesite refractories, the use of clay and talc as raw materials is the basis for the compositional developments. These materials are valuable in large part because of their ease in forming; they are fine-grained and platey and are consequently plastic, nonabrasive, and easy to form. In addition, the fine-grained nature of these materials is essential for the

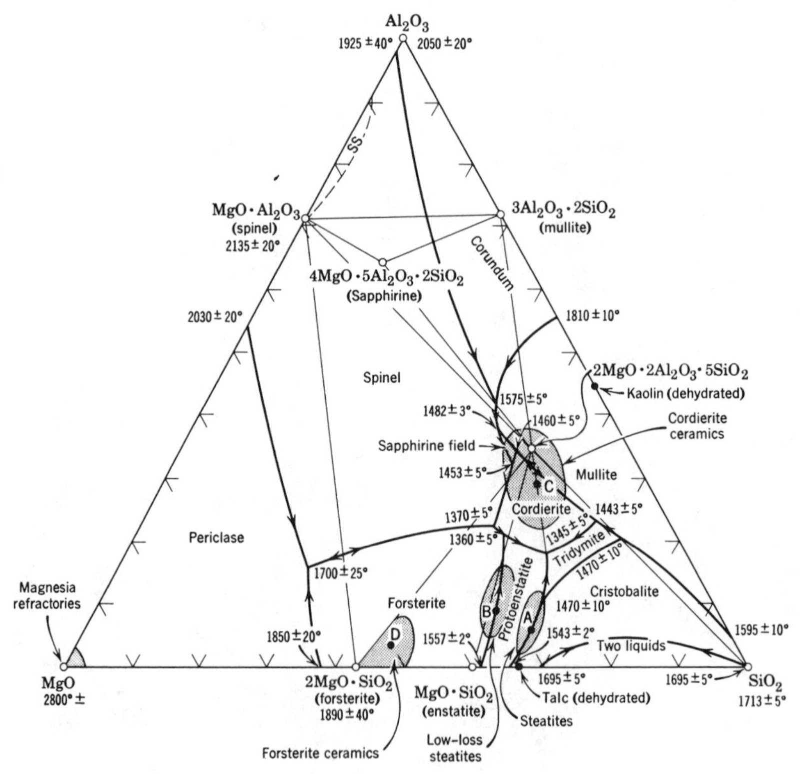

Fig. 7.31. Common compositions in the ternary system $MgO–Al_2O_3–SiO_2$. See text for other additives.

firing process, which is described in more detail in Chapter 12. On heating, clay decomposes at 980°C to form fine-grained mullite in a silica matrix. Talc decomposes and gives rise to a similar mixture of fine-grained protoenstatite crystals, $MgSiO_3$, in a silica matrix at about 1000°C. Further heating of clay gives rise to increased growth of mullite crystals, crystallization of the silica matrix as cristobalite, and formation of a eutectic liquid at 1595°C. Further heating of pure talc leads to crystal growth of the enstatite, and liquid is formed at a temperature of 1547°C. At this temperature almost all the composition melts, since talc (66.6% SiO_2, 33.4% MgO) is not far from the eutectic composition in the MgO–SiO_2 system (Fig. 7.13).

The main feature which characterizes the melting behavior of cordierite, steatite porcelain, and low-loss steatite compositions is the limited firing range which results when pure materials are carried to partial fusion. In general, for firing to form a vitreous densified ceramic about 20 to 35% of a viscous silicate liquid is required. For pure talc, however, as indicated in Fig. 7.32, no liquid is formed until 1547°C, when the entire composition liquifies. This can be substantially improved by using talc-clay mixtures. For example, consider the composition A in Fig. 7.31 which is 90% talc–10% clay, similar to many commercial steatite compositions. At this composition about 30% liquid is formed abruptly at the liquidus temperature, 1345°C; the amount of liquid increases quite rapidly with temperature (Fig. 7.32), making close control of firing temperature necessary, since the firing range is short for obtaining a dense vitreous

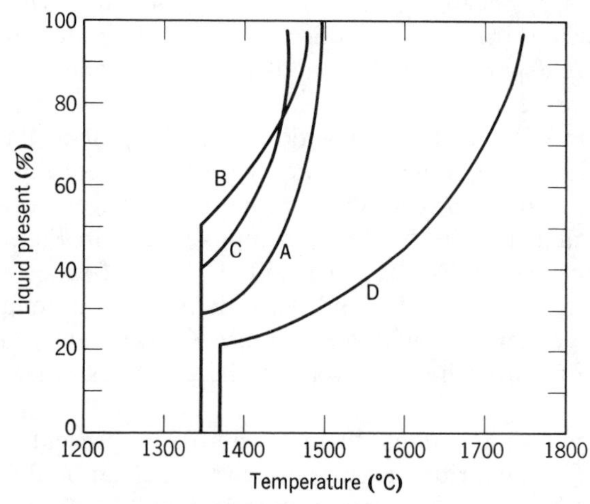

Fig. 7.32. Amount of liquid present at different temperatures for compositions illustrated in Fig. 7.31.

body (this composition would be fired at 1350 to 1370°C). In actual fact, however, the raw materials used contain Na_2O, K_2O, CaO, BaO, Fe_2O_3, and TiO_2 as minor impurities which both lower and widen the fusion range. Additions of more than 10% clay again so shorten the firing range that they are not feasible, and only limited compositions are practicable. The addition of feldspar greatly increases the firing range and the ease of firing and has been used in the past for compositions intended as low-temperature insulators. However, the electrical properties are not good.

For low-loss steatites, additional magnesia is added to combine with the free silica to bring the composition nearer the composition triangle for forsterite–cordierite–enstatite. This changes the melting behavior so that a composition such as B in Fig. 7.31 forms about 50% liquid over a temperature range of a few degrees, and control in firing is very difficult (Fig. 7.32). In order to fire these compositions in practice to form vitreous bodies, added flux is essential. Barium oxide, added as the carbonate, is the most widely used.

Cordierite ceramics are particularly useful, since they have a very low coefficient of thermal expansion and consequently good resistance to thermal shock. As far as firing behavior is concerned, compositions show a short firing range corresponding to a flat liquidus surface which leads to the development of large amounts of liquid over a short temperature interval. If a mixture consisting of talc and clay, with alumina added to bring it closer to the cordierite composition, is heated, an initial liquidus is formed at 1345°C, as for composition C in Fig. 7.31. The amount of liquid rapidly increases; because of this it is difficult to form vitreous bodies. Frequently when these compositions are not intended for electrical applications, feldspar (3 to 10%) is added as a fluxing medium to increase the firing range.

Magnesia and forsterite compositions are different in that a eutectic liquid is formed of a composition widely different from the major phase with a steep liquidus curve so that a broad firing range is easy to obtain. This is illustrated for the forsterite composition D in Fig. 7.31 and the corresponding curve in Fig. 7.32. The initial liquid is formed at the 1360°C eutectic, and the amount of liquid depends mainly on composition and does not change markedly with temperature. Consequently, in contrast to the steatite and cordierite bodies, forsterite ceramics present few problems in firing.

In all these compositions there is normally present at the firing temperature an equilibrium mixture of crystalline and liquid phases. This is illustrated for a forsterite composition in Fig. 7.33. Forsterite crystals are present in a matrix of liquid silicate corresponding to the liquidus

Fig. 7.33. Crystal-liquid structure of a forsterite composition (150×).

composition at the firing temperature. For other systems the crystalline phase at the firing temperature is protoenstatite, periclase, or cordierite, and the crystal size and morphology are usually different as well. The liquid phase frequently does not crystallize on cooling but forms a glass (or a partly glass mixture) so that the compatibility triangle cannot be used for fixing the phases present at room temperature, but they must be deduced instead from the firing conditions and subsequent heat treatment.

7.10 Nonequilibrium Phases

The kinetics of phase transitions and solid-state reactions is considered in the next two chapters; however, from our discussion of glass structure in Chapter 3 and atom mobility in Chapter 6 it is already apparent that the lowest energy state of phase equilibria is not achieved in many practical systems. For any change to take place in a system it is necessary that the free energy be lowered. As a result the sort of free-energy curves illustrated in Figs. 3.10, 4.2, 4.3, 7.7, and 7.8 for each of the possible phases that might be present remain an important guide to metastable equilibrium. In Fig. 7.8, for example, if at temperature T_2 the solid solution α were absent for any reason, the common tangent between the liquid and solid solution β would determine the composition of those phases in which the constituents have the same chemical potential. One of the common types of nonequilibrium behavior in silicate systems is the slowness of crystallization such that the liquid is supercooled. When this happens, metastable phase separation of the liquid is quite common, discussed in Chapter 3.

Glasses. One of the most common departures from equilibrium behavior in ceramic systems is the ease with which many silicates are cooled from the liquid state to form noncrystalline products. This requires that the driving force for the liquid-crystal transformation be low and that the activation energy for the process be high. Both of these conditions are fulfilled for many silicate systems.

The rate of nucleation for a crystalline phase forming from the liquid is proportional to the product of the energy difference between the crystal and liquid and the mobility of the constituents that form a crystal, as discussed in Chapter 8. In silicate systems, both of these factors change so as to favor the formation of glasses as the silica content increases. Although data for the diffusion coefficient are not generally available, the limiting mobility is that of the large network-forming anions and is inversely proportional to the viscosity. Thus, the product of $\Delta H_f/T_{mp}$ and $1/\eta$ can be used as one index for the tendency to form glasses on cooling, as shown in Table 7.1.

Table 7.1. Factors Affecting Glass-Forming Ability

Composition	T_{mp}(°C)	$\Delta H_f/T_{mp}$ (cal/mole/°K)	$(1/\eta)_{mp}$ (poise^{-1})	$(\Delta H_f/T_{mp}) \times (1/\eta)_{mp}$	Comments
B_2O_3	450	7.3	2×10^{-5}	1.5×10^{-4}	Good glass former
SiO_2	1713	1.1	1×10^{-6}	1.1×10^{-6}	Good glass former
$Na_2Si_2O_5$	874	7.4	5×10^{-4}	3.7×10^{-3}	Good glass former
Na_2SiO_3	1088	9.2	5×10^{-3}	4.5×10^{-2}	Poor glass former
$CaSiO_3$	1544	7.4	10^{-1}	0.74	Very difficult to form as glass
NaCl	800.5	6.9	50	345	Not a glass former

Metastable Crystalline Phases. Frequently in ceramic systems crystalline phases are present that are not the equilibrium phases for the conditions of temperature, pressure, and composition of the system. These remain present in a metastable state because the high activation energies required for their conversion into more stable phases cause a low rate of transition. The energy relationships among three phases of the same composition might be represented as given in Fig. 7.34. Once any one of these phases is formed, its rate of transformation into another more stable phase is slow. In particular, the rate of transition to the lowest energy state is specially slow for this system.

The kinetics of transformation in systems such as those illustrated in Fig. 7.34 are discussed in Chapter 9 in terms of the driving force and energy barrier. Structural aspects of transformations of this kind have been discussed in Chapter 2. In general, there are two common ways in which metastable crystals are formed. First, if a stable crystal is brought into a new temperature or pressure range in which it does not transform into the more stable form, metastable crystals are formed. Second, a precipitate or transformation may form a new metastable phase. For example, if phase 1 in Fig. 7.34 is cooled into the region of stability of phase 3, it may transform into the intermediate phase 2, which remains present as a metastable crystal.

The most commonly observed metastable crystalline phases not undergoing transformation are the various forms of silica (Fig. 7.5). When a porcelain body containing quartz as an ingredient is fired at a temperature of 1200 to 1400°C, tridymite is the stable form but it never is observed; the quartz always remains as such. In refractory silica brick, quartz used as a raw material must have about 2% calcium oxide added to it in order to be transformed into the tridymite and cristobalite forms which are desirable. The lime provides a solution-precipitation mechanism which essentially eliminates the activation energy barrier, shown in Fig. 7.34, and allows

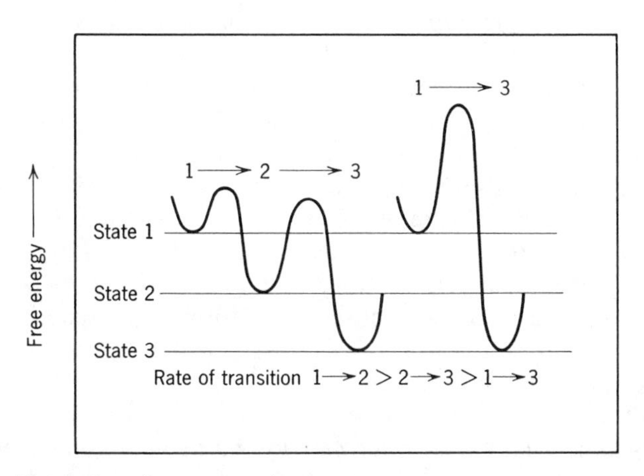

Fig. 7.34. Illustration of energy barriers between three different states of a system.

the stable phase to be formed. This is, in general, the effect of mineralizers such as fluorides, water, and alkalies in silicate systems. They provide a fluid phase through which reactions can proceed without the activation energy barrier present for the solid-state process.

Frequently, when high-temperature crystalline forms develop during firing of a ceramic body, they do not revert to the more stable forms on cooling. This is particularly true for tridymite and cristobalite, which never revert to the more stable quartz form. Similarly, in steatite bodies the main crystalline phase at the firing temperature is the protoenstatite form of $MgSiO_3$. In fine-grained samples this phase remains as a metastable phase dispersed in a glassy matrix after cooling. In large-grain samples or on grinding at low temperature, protoenstatite reverts to the equilibrium form, clinoenstatite.

A common type of nonequilibrium behavior is the formation of a metastable phase which has a lower energy than the mother phase but is not the lowest-energy equilibrium phase. This corresponds to the situation illustrated in Fig. 7.34 in which the transition from the highest-energy phase to an intermediate energy state occurs with a much lower activation energy than the transition to the most stable state. It is exemplified by the devitrification of silica glass, which occurs in the temperature range of 1200 to 1400°C, to form cristobalite as the crystalline product instead of the more stable form, tridymite. The reasons for this are usually found in the structural relationships between the starting material and the final product. In general, high-temperature forms have a more open structure than low-temperature crystalline forms and consequently are more nearly like the structure of a glassy starting material. These factors tend to favor crystallization of the high-temperature form from a supercooled liquid or glass, even in the temperature range of stability of a lower-temperature modification.

This phenomenon has been observed in a number of systems. For example, J. B. Ferguson and H. E. Merwin* observed that when calcium-silicate glasses are cooled to temperatures below 1125°C, at which wollastonite ($CaSiO_3$) is the stable crystalline form, the high-temperature modification, pseudowollastonite, is found to crystallize first and then slowly transform into the more stable wollastonite. Similarly, on cooling compositions corresponding to $Na_2O \cdot Al_2O_3 \cdot 2SiO_2$, the high-temperature crystalline form (carnegieite) is observed to form as the reaction product, even in the range in which nephelite is the stable phase; transformation of carnegieite into nephelite occurs slowly.

In order for any new phase to form, it must be lower in free energy than the starting material but need not be the lowest of all possible new phases.

*_Am. J. Science, Series 4_, **48**, 165 (1919).

This requirement means that when a phase does not form as indicated on the phase equilibrium diagram, the liquidus curves of other phases on the diagram must be extended to determine the conditions under which some other phase becomes more stable than the starting solution and a possible precipitate. This is illustrated for the potassium disilicate–silica system in Fig. 7.35. Here, the compound $K_2O{\cdot}4SiO_2$ crystallizes only with great difficulty so that the eutectic corresponding to this precipitation is frequently not observed. Instead, the liquidus curves for silica and for potassium disilicate intersect at a temperature about 200° below the true eutectic temperature. This nonequilibrium eutectic is the temperature at which both potassium disilicate and silica have a lower free energy than the liquid composition corresponding to the false eutectic. Actually, for this system the situation is complicated somewhat more by the fact that cristobalite commonly crystallizes from the melt in place of the equilibrium quartz phase. This gives additional possible behaviors, as indicated by the dotted line in Fig. 7.35.

Extension of equilibrium curves on phase diagrams, such as has been

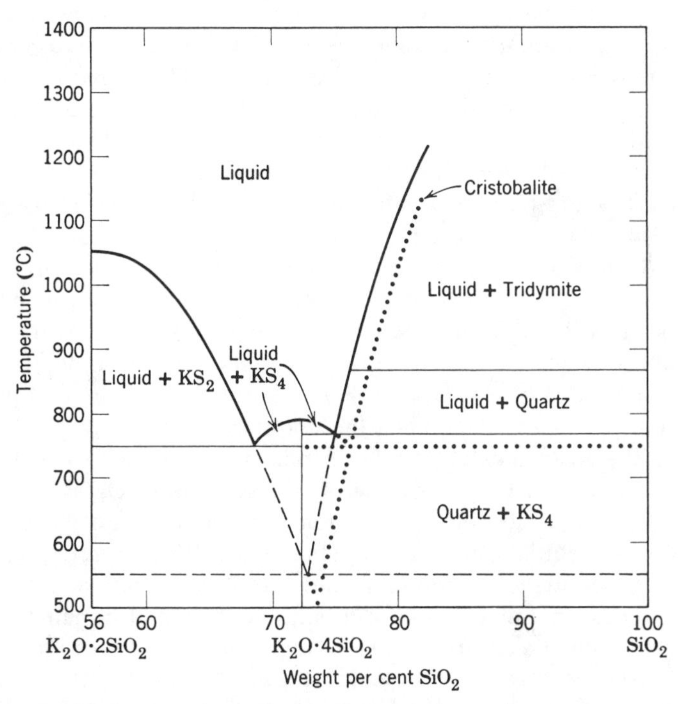

Fig. 7.35. Equilibrium and nonequilibrium liquidus curves in the potassium disilicate–silica system.

shown in Fig. 7.35 and also in Fig. 7.5, provides a general method of using equilibrium data to determine possible nonequilibrium behavior. It provides a highly useful guide to experimental observations. The actual behavior in any system may follow any one of several possible courses, so that an analysis of the kinetics of these processes (or more commonly experimental observations) is also required.

Incomplete Reactions. Probably the most common source of nonequilibrium phases in ceramic systems are reactions that are not completed in the time available during firing or heat treatment. Reaction rates in condensed phases are discussed in Chapter 9. The main kinds of incomplete reactions observed are incomplete solution, incomplete solid-state reactions, and incomplete resorption or solid-liquid reactions. All of these arise from the presence of reaction products which act as barrier layers and prevent further reaction. Perhaps the most striking example of incomplete reactions is the entire metallurgical industry, since almost all metals are thermodynamically unstable in the atmosphere but oxidize and corrode only slowly.

A particular example of incomplete solution is the existence of quartz grains which are undissolved in a porcelain body, even after firing at temperatures of 1200 to 1400°C. For the highly siliceous liquid in contact with the quartz grain, the diffusion coefficient is low, and there is no fluid flow to remove the boundary layer mechanically. The situation is similar to diffusion into an infinite medium, illustrated in Fig. 6.5. To a first approximation, the diffusion coefficient for SiO_2 at the highly siliceous boundary may be of the order of 10^{-8} to 10^{-9} cm^2/sec at 1400°C. With these data it is left as an exercise to estimate the thickness of the diffusion layer after 1 hr of firing at this temperature.

The way in which incomplete solid reactions can lead to residual starting material being present as nonequilibrium phases will be clear from the discussion in Chapter 9. However, new products that are not the final equilibrium composition can also be formed. For example, in heating equimolar mixtures of $CaCO_3$ and SiO_2 to form $CaSiO_3$, the first product formed and the one that remains the major phase through most of the reaction is the orthosilicate, Ca_2SiO_4. Similarly, when $BaCO_3$ and TiO_2 are reacted to form $BaTiO_3$, substantial amounts of Ba_2TiO_4, $BaTi_3O_7$, and $BaTi_4O_9$ are formed during the reaction process, as might be expected from the phase-equilibrium diagram (Fig. 7.20). When a series of intermediate compounds is formed in a solid reaction, the rate at which each grows depends on the effective diffusion coefficient through it. Those layers for which the diffusion rate is high form most rapidly. For the $CaO–SiO_2$ system this is the orthosilicate. For the $BaO–TiO_2$ system the most rapidly forming compound is again the orthotitanate, Ba_2TiO_4.

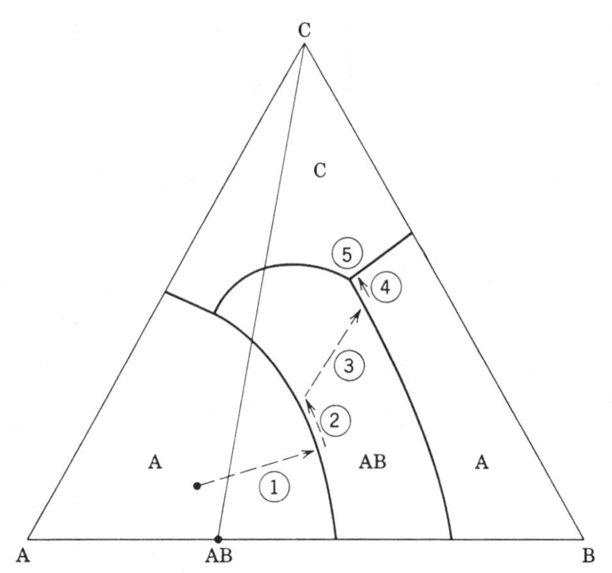

Fig. 7.36. Nonequilibrium crystallization path with (1) Liquid → A, (2) A + liquid → AB, (3) Liquid → AB, (4) Liquid → AB + B, (5) Liquid → AB + B + C.

A final example of nonequilibrium conditions important in interpreting phase-equilibrium diagrams is the incomplete resorption that may occur whenever a reaction, A + Liquid = AB, takes place during crystallization. This is the case, for example, when a primary phase reacts with a liquid to form a new compound during cooling. A layer tends to build up on the surface of the original particle, forming a barrier to further reaction. As the temperature is lowered, the final products are not those anticipated from the equilibrium diagram. A nonequilibrium crystallization path for incomplete resorption is schematically illustrated in Fig. 7.36.

Suggested Reading

1. E. M. Levin, C. R. Robbins, and H. F. McMurdie, *Phase Diagrams for Ceramists*, American Ceramic Society, Columbus, Ohio, 1964.
2. E. M. Levin, C. R. Robbins, H. F. McMurdie, *Phase Diagrams for Ceramists, 1969 Supplement*, American Ceramic Society, Columbus, Ohio, 1969.
3. A. M. Alper, Ed., *Phase Diagrams: Materials Science and Technology*, Vol. I, "Theory, Principles, and Techniques of Phase Diagrams," Academic Press, Inc., New York, 1970; Vol. II, "The Use of Phase Diagrams in Metal, Refractory, Ceramic, and Cement Technology," Academic Press, Inc., New

York, 1970; Vol. III, "The Use of Phase Diagrams in Electronic Materials and Glass Technology," Academic Press, Inc., New York, 1970.
4. A. Muan and E. F. Osborn, *Phase Equilibria among Oxides in Steelmaking*, Addison-Wesley, Publishing Company, Inc., Reading, Mass., 1965.
5. A. Reisman, *Phase Equilibria*, Academic Press, Inc., New York, 1970.
6. P. Gordon, *Principles of Phase Diagrams in Materials Systems*, McGraw Hill Book Company, New York, 1968.
7. A. M. Alper, Ed., *High Temperature Oxides*, Part I, "Magnesia, Lime and Chrome Refractories," Academic Press, Inc., New York, 1970; Part II, "Oxides of Rare Earth, Titanium, Zirconium, Hafnium, Niobium, and Tantalum," Academic Press, Inc., New York, 1970; Part III, "Magnesia, Alumina, and Beryllia Ceramics: Fabrication, Characterization and Properties," Academic Press, Inc., New York; Part IV, "Refractory Glasses, Glass–Ceramics, Ceramics," Academic Press, New York, Inc., 1971.
8. J. E. Ricci, *The Phase Rule and Heterogeneous Equilibrium*, Dover Books, New York, 1966.

Problems

7.1. A power failure allowed a furnace used by a graduate student working in the K_2O–CaO–SiO_2 system to cool down over night. For the fun of it, the student analyzed the composition he was studying by X-ray diffraction. To his horror, he found β–$CaSiO_3$, $2K_2O{\cdot}CaO{\cdot}3SiO_2$, $2K_2O{\cdot}CaO{\cdot}6SiO_2$, $K_2O{\cdot}3CaO{\cdot}6SiO_2$, and $K_2O{\cdot}2CaO{\cdot}6SiO_2$ in his sample.
   (*a*) How could he get more than three phases?
   (*b*) Can you tell him in which composition triangle his original composition was?
   (*c*) Can you predict the minimum temperature above which his furnace was operating before power failure?
   (*d*) He thought at first he also had some questionable X-ray diffraction evidence for $K_2O{\cdot}CaO{\cdot}SiO_2$, but after thinking it over he decided $K_2O{\cdot}CaO{\cdot}SiO_2$ should not crystallize out of his sample. Why did he reach this conclusion?

7.2. According to Alper, McNally, Ribbe, and Doman,* the maximum solubility of Al_2O_3 in MgO is 18 wt% at 1995°C and of MgO in $MgAl_2O_4$ is 39% MgO, 51% Al_2O_3. Assuming the NiO–Al_2O_3 binary is similar to the MgO–Al_2O_3 binary, construct a ternary. Make isothermal plots of this ternary at 2200°C, 1900°C, and 1700°C.

7.3. You have been assigned to study the electrical properties of calcium metasilicate by the director of the laboratory in which you work. If you were to make the material synthetically, give a batch composition of materials commonly obtainable in high purity. From a production standpoint, 10% liquid would increase the rate of sintering and reaction. Adjust your composition accordingly. What would be the expected firing temperature? Should the boss ask you to explore the possibility of lowering the firing temperature and maintain a white body, suggest the direction to procede. What polymorphic transformations would you be conscious of in working with the above systems?

J. Am. Ceram. Soc.* **45(6), 263–268 (1962).

7.4. Discuss the importance of liquid-phase formation in the production and utilization of refractory bodies. Considering the phase diagram for the $MgO-SiO_2$ system, comment on the relative desirability in use of compositions containing $50MgO-50SiO_2$ by weight and $60MgO-40SiO_2$ by weight. What other characteristics of refractory bodies are important in their use?

7.5. A binary silicate of specified composition is melted from powders of the separate oxides and cooled in different ways, and the following observations are made:

Condition	Observations
(*a*) Cooled rapidly	Single phase, no evidence of crystallization
(*b*) Melted for 1 hr, held 80°C below liquidus for 2 hr	Crystallized from surface with primary phases SiO_2 plus glass
(*c*) Melted for 3 hr, held 80°C below liquidus for 2 hr	Crystallized from surface with primary phases compound $AO \cdot SiO_2$ plus glass
(*d*) Melted for 2 hr, cooled rapidly to 200°C below liquidus, held for 1 hr, and then cooled rapidly	No evidence of crystallization but resulting glass is cloudy

Are all these observations self consistent? How do you explain them?

7.6. Triaxial porcelains (flint–feldspar–clay) in which the equilibrium phases at the firing temperature are mullite and a silicate liquid have a long firing range; steatite porcelains (mixtures of talc plus kaolin) in which the equilibrium phases at the firing temperature are enstiatite and a silicate liquid have a short firing range. Give plausible explanations for this difference in terms of phases present, properties of phases, and changes in phase composition and properties with temperature.

7.7. For the composition $40MgO-55SiO_2-5Al_2O_3$, trace the equilibrium crystallization path in Fig. 7.30. Also, determine the crystallization path if incomplete resorption of forsterite occurs along the forsterite-protoenstatite boundary. How do the compositions and temperatures of the eutectics compare for the equilibrium and nonequilibrium crystallization paths? What are the compositions and amounts of each constituent in the final product for the two cases?

7.8. If a homogeneous glass having the composition $13Na_2O-13CaO-74SiO_2$ were heated to 1050°C, 1000°C, 900°C, and 800°C, what would be the possible crystalline products that might form? Explain.

7.9. The clay mineral kaolinite, $Al_2Si_2O_5(OH)_4$, when heated above 600°C decomposes to $Al_2Si_2O_7$ and water vapor. If this composition is heated to 1600°C and left at that temperature until equilibrium is established, what phase(s) will be present. If-more than one is present, what will be their weight percentages. Make the same calculations for 1585°C.

8

Phase Transformations, Glass Formation, and Glass-Ceramics

Phase-equilibrium diagrams graphically represent the ranges of temperature, pressure, and composition in which different phases are stable. When pressure, temperature, or composition is changed, new equilibrium states are fixed, as indicated in the phase-equilibrium diagrams, but a long time may be required to reach the new lower-energy conditions. This is particularly true in solid and liquid systems in which atomic mobility is limited; indeed, in many important systems equilibrium is never attained. In general, the rate at which equilibrium is reached is just as important as knowledge of the equilibrium state.

As discussed by J. W. Gibbs a century ago,* there are two general types of processes by which one phase can transform into another: (*a*) changes which are initially small in degree but large in spatial extent and in the early stages of transformation resemble the growth of compositional waves, as illustrated schematically in Fig. 8.1*a*; and (*b*) changes initially large in degree but small in spatial extent, as illustrated schematically in Fig. 8.1*b*. The first type of phase transformation is called *spinodal decomposition*; the latter is termed *nucleation and growth.*

The kinetics of either type of process may be rapid or slow, depending on factors such as the thermodynamic driving force, the atomic mobility, and heterogeneities in the sample.

During transformation by a nucleation and growth process, either the nucleation or growth step may limit the rate of the overall process to such an extent that equilibrium is not easily attained. For example, to induce precipitation from supersaturated cloud formations, nucleation is the stumbling block and *seeding* of clouds leads to precipitation in the form of rain or snow. In contrast, on cooling a gold-ruby glass, nuclei are formed

**Scientific Papers*, Vol. 1, Dover Publications, Inc., New York, 1961.

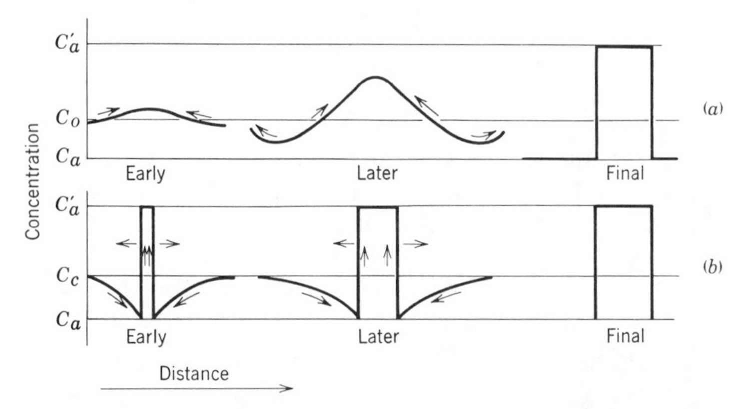

Fig. 8.1. Schematic evaluation of concentration profiles for (*a*) spinodal decomposition and (*b*) nucleation and growth. From J. W. Cahn, *Trans. Met. Soc. AIME*, **242**, 166 (1968).

during cooling but do not grow until the glass is reheated, forming the beautiful ruby color.

In the present chapter, both types of phase transformation are discussed and applied to phase separation in glass-forming materials, unidirectional solidification, glass formation, the development of desired microstructures in glass-ceramic materials, photosensitive glasses, opacified enamels, and photochromic glasses.

8.1 Formal Theory of Transformation Kinetics

In considering phase transformations taking place by a nucleation and growth process, it is often useful to describe the volume fraction of a specimen which is transformed in a given time. Consider a specimen brought rapidly to a temperature at which a new phase is stable and maintained at this temperature for a time τ. The volume of the transformed region present is designated V^{β} and that of the remaining original phase V^{α}. In a small time interval $d\tau$, the number of particles of the new phase which form is

$$N_{\tau} = I_v V^{\alpha}\, d\tau \tag{8.1}$$

where I_v is the nucleation rate, that is, the number of new particles which form per unit volume per unit time. If the growth rate per unit area of the interface, u, is assumed isotropic (independent of orientation), the transformed regions are spherical in shape. If u is then taken to be independent of time, the volume at time t of the transformed material which originated at τ is

$$V_\tau^\beta = \frac{4\pi}{3} u^3(t-\tau)^3. \tag{8.2}$$

During the initial stages of the transformation, when the nuclei are widely spaced, there should be no significant interference between neighboring nuclei, and $V^\alpha \approx V$, the volume of the sample. Hence the transformed volume at time t resulting from regions nucleated between τ and $\tau + dt$ is

$$dV^\beta = N_\tau V_\tau^\beta \approx \frac{4\pi}{3} V I_v u^3(t-\tau)^3\, dt \tag{8.3}$$

and the fractional volume transformed at time t is

$$\frac{V^\beta}{V} = \frac{4\pi}{3} \int_0^t I_v u^3(t-\tau)^3\, d\tau. \tag{8.4}$$

When the nucleation rate is independent of time

$$\frac{V^\beta}{V} = \frac{\pi}{3} I_v u^3 t^4 \tag{8.5}$$

A more exact treatment, first carried out by M. Avrami,* includes the effects of impinging transformed regions and excluded nucleation in already transformed material. The corresponding relations to Eq. (8.4) and (8.5) are

$$\frac{V^\beta}{V} = 1 - \exp\left[-\frac{4\pi}{3} u^3 \int_0^t I_v(t-\tau)^3\, d\tau\right] \tag{8.6}$$

and, with I_v constant,

$$\frac{V^\beta}{V} = 1 - \exp\left[-\frac{\pi}{3} I_v u^3 t^4\right] \tag{8.7a}$$

These reduce to Eqs. (8.4) and (8.5) for small fractions transformed, where interference between growth centers is not expected. Other variations with time of the nucleation frequency and growth rate have also been analyzed† and lead to expressions of the form

$$\frac{V^\beta}{V} = 1 - \exp(-at^n) \tag{8.7b}$$

where the exponent n is often referred to as the Avrami n. These expressions describe fraction transformed versus time curves of sigmoid shape.

J. Chem. Phys., **7**, 1103 (1939).

†See J. W. Christian, *Phase Transformations in Metals*, Pergamon Press, New York, 1965.

It is apparent that the volume fraction of new phase formed in a given time depends on the individual kinetic constants describing the nucleation and growth processes I_v and u. These, in turn, can be related to a variety of thermodynamic and kinetic factors such as the heat of transformation, the departure from equilibrium, and the atomic mobility.

8.2 Spinodal Decomposition

Spinodal decomposition refers to a continuous type of phase transformation in which the change begins as compositional waves that are small in amplitude and large in spatial extent (Fig. 8.1). In a phase diagram containing a miscibility gap, the free energy versus composition relations for several temperatures are shown in Fig. 8.2. The free energy versus composition relations for temperatures below the consolute temperature are characterized by regions of negative curvature ($\partial^2 G/\partial C^2 < 0$). The

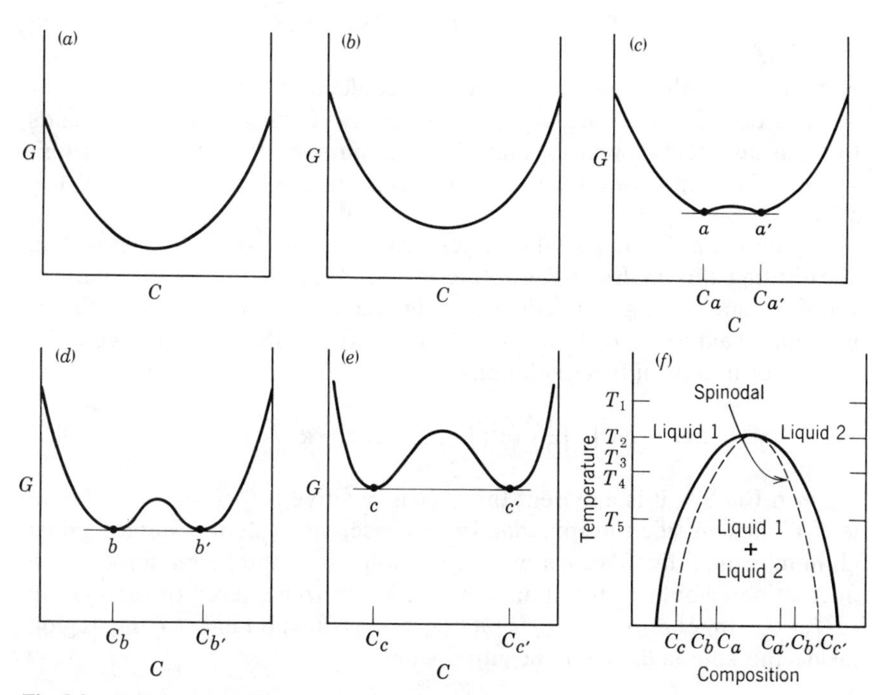

Fig. 8.2. Two-liquid immiscibility. (*a*) to (*e*) show a sequence of Gibbs free energy curves corresponding to the phase diagram shown in (*f*). (*a*) $T = T_1$; (*b*) $T = T_2$; (*c*) $T = T_3$; (*d*) $T = T_4$; (*e*) $T = T_5$. From T. P. Seward, in *Phase Diagrams*, Vol. 1, Academic Press, Inc., New York, 1970.

inflection points ($\partial^2 G/\partial C^2 = 0$) are termed the spinodes, and their locus as a function of temperature defines the spinodal curve shown in Fig. 8.2.

The spinodal represents the limit of chemical stability. For compositions outside the spinodal, the chemical potential of the given component increases with the density of the component, and a homogeneous solution is stable or metastable, depending on whether the given composition lies inside or outside the miscibility gap. Within the gap but outside the spinodal, a homogeneous solution is stable against infinitesimal fluctuations in composition but can separate into an equilibrium two-phase system by a nucleation and growth process. In contrast, for compositions within the spinodal, a homogeneous solution is unstable against infinitesimal fluctuations in density or composition, and there is no thermodynamic barrier to the growth of a new phase.

Thermodynamics of Spinodal Decomposition. For an infinite, incompressible, isotropic binary solution, the free energy of a nonuniform solution may be expressed to first order by a relation

$$G = N_v \int_v [g(C) + \kappa(\nabla C)^2]\, dV \tag{8.8}$$

Here $g(C)$ is the free energy per molecule of a uniform solution of composition C, κ is a constant which is positive for a solution which tends to separate into two phases, and N_v is the number of molecules per unit volume. The gradient term $\kappa(\nabla C)^2$ has been discussed in Chapter 5 (Eq. 5.20).

Expanding $g(C)$ in a Taylor series about C_0, the average composition, substituting this in Eq. 8.8, subtracting the free energy of the uniform solution, and noting that odd terms in the expansion must vanish for isotropic solutions, one has the free energy difference between the nonuniform and uniform solutions:

$$\Delta G = N_v \int_v \left[\frac{1}{2}\left(\frac{\partial^2 g}{\partial C^2}\right)_{C_0} (C - C_0)^2 + \kappa(\nabla c)^2\right] dV \tag{8.9}$$

From Eq. 8.9, it is apparent that ΔG is positive if $(\partial^2 g/\partial C^2)_{C_0} > 0$, that is, if C_0 lies outside the spinodal. In this case, the system is stable against all infinitesimal fluctuations in composition, since the formation of such fluctuations would result in an increase in the free energy of the system ($\Delta G > 0$). In contrast, if $(\partial^2 g/\partial C^2)_{C_0} < 0$, corresponding to the region inside the spinodal, ΔG is negative when

$$\frac{1}{2}\left|\left(\frac{\partial^2 g}{\partial C^2}\right)_{C_0}\right| (C - C_0)^2 > \kappa(\nabla C)^2 \tag{8.10}$$

The formation of fluctuations can be accompanied by a decrease in the free energy of the system within the spinodal provided the scale of the fluctuations is large enough so that the gradients are sufficiently small so that Eq. 8.10 is satisfied. Hence the system is always unstable against some fluctuations within the spinodal region.

In proceeding further, consider a composition fluctuation of the form

$$C - C_0 = A \cos \beta x \tag{8.11}$$

Substituting this in Eq. 8.9, one obtains for the change in free energy on forming fluctuations

$$\frac{\Delta G}{V} = \frac{A^2}{4}\left[\left(\frac{\partial^2 g}{\partial C^2}\right)_{C_0} + 2\kappa\beta^2\right] \tag{8.12}$$

The solution is then unstable ($\Delta G < 0$) for all fluctuations of wave number β smaller than a critical wave number β_c:

$$\beta_c = \left[-\frac{1}{2\kappa}\left(\frac{\partial^2 g}{\partial C^2}\right)_{C_0}\right]^{1/2} \tag{8.13a}$$

or for all fluctuations of wavelength $\lambda = 2\pi/\beta$ longer than a critical wavelength λ_c

$$\lambda_c = \left[\frac{-8\pi^2\kappa}{\left(\dfrac{\partial^2 g}{\partial C^2}\right)}\right]^{1/2} \tag{8.13b}$$

From these expressions, as from Eq. 8.10, it is seen that the incipient surface energy, reflected in the gradient energy term, prevents the solution from decomposing on too small a scale. The more negative the factor $(\partial^2 g/\partial C^2)_{C_0}$, the more gradient energy can be accommodated and still have a negative ΔG, and hence the smaller the scale of the continuous decomposition which can take place. For decomposition on a scale finer than λ_c, ΔG is positive, and the solution is stable against the fluctuations. In this case, the system might still separate into two phases, but not by the spinodal decomposition mechanism by which the free energy of the system continuously decreases (separation would have to take place by a nucleation and growth process).

In summary, for compositions which lie at a given temperature within the miscibility gap but outside the spinodal, phase separation can take place only by a nucleation and growth process, since the formation of infinitesimal compositional fluctuations of all wavelengths are accompanied by an increase in the free energy of the system. Such a system is described as being metastable. Within the spinodal, the system is unstable against compositional fluctuations on some sufficiently large scale, since such fluctuations can decrease the free energy of the system. For fluctuations on a smaller scale, however, the system is effectively metastable because of the incipient surface energy involved in forming regions differing in composition.

Spinodal Decomposition Kinetics. The expected kinetics of the spinodal decomposition process are obtained by deriving and solving a generalized diffusion equation for the free-energy expression of Eq. 8.8. Considering the early stages of isothermal decomposition of an initially uniform solution, one obtains for the sinusoidal composition fluctuations of Eq. 8.11*

$$A(\beta,t) = A(\beta,0)\exp[R(\beta)t] \tag{8.14}$$

where the amplification factor $R(\beta)$ is given by

$$R(\beta) = -\frac{\tilde{M}\beta^2}{N_v}\left[\left(\frac{\partial^2 g}{\partial C^2}\right)_{C_0} + 2\kappa\beta^2\right] \tag{8.15}$$

Here $A(\beta,t)$ is the amplitude of a fluctuation of wave number β at time t, $A(\beta,0)$ is its initial amplitude, and $\tilde{M}$ is the mobility.

It is seen from Eqs. 8.14, 8.15, and 8.12 that when a solution is stable with respect to fluctuations of a given wavelength, $R(\beta)$ is negative, and the fluctuations decay with time. In contrast, when the solution is unstable with respect to such fluctuations, $R(\beta)$ is positive, and the fluctuations grow rapidly with time. The amplification factor is expected to vary with wave number, as shown in Fig. 8.3. The maximum value of $R(\beta)$, designated R_m, corresponds to the wavelength λ_m which grows most

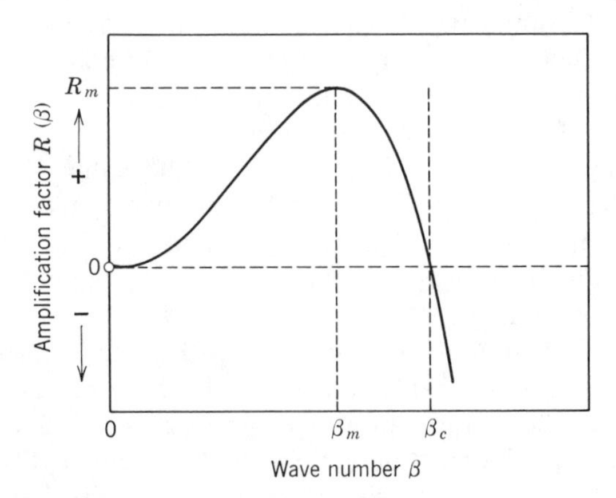

Fig. 8.3. Amplification factor versus wave number (schematic).

*See J. W. Cahn, *Acta Met.*, **9**, 795 (1961); *Trans. Met. Soc. AIME*, **242**, 166 (1968).

rapidly with time. From Eq. 8.15 these are

$$\lambda_m = \sqrt{2}\lambda_c \tag{8.16}$$

and

$$R_m = -\frac{1}{2}\tilde{M}\left(\frac{\partial^2 f}{\partial C^2}\right)_{C_0} \qquad \beta_m{}^2 = 2\tilde{M}\kappa\beta_m{}^4 \tag{8.17}$$

When a specimen is quenched from a temperature above a miscibility gap to a temperature within the gap, the initial distribution of fluctuations, which can be determined from studies of small-angle X-ray scattering and light scattering, should change with time, as indicated by Eq. 8.14. After a sufficient period of time, the decomposition is dominated by fluctuations of wavelength λ_m. The microstructure predicted for this case for the central region of a miscibility gap is shown in Fig. 8.4. It indicates a phase morphology in which two phases are both three-dimensionally interconnected. Such calculated microstructures are reminiscent of those seen in

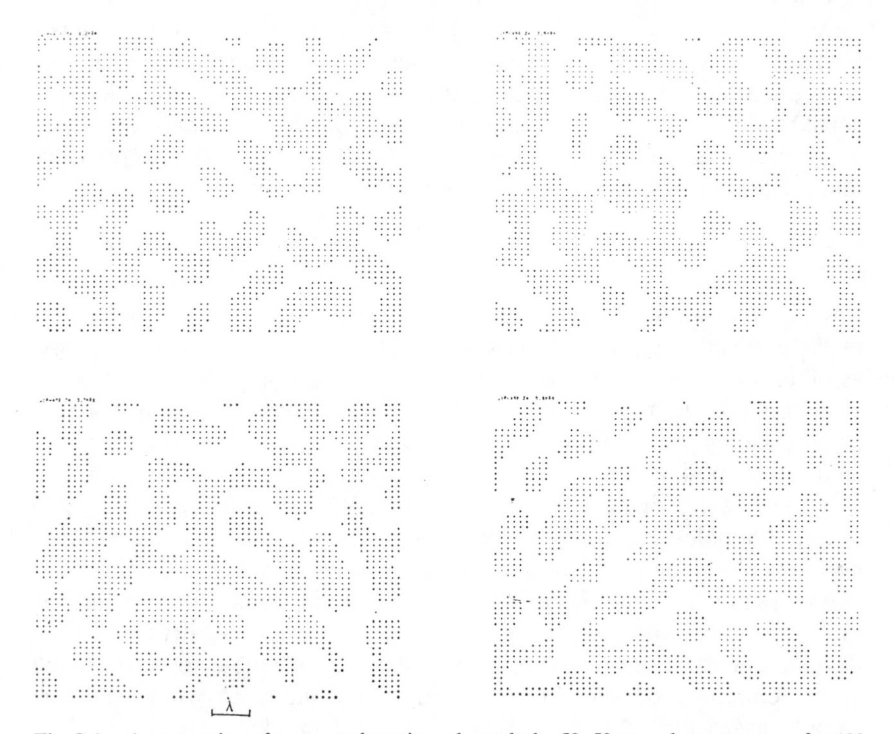

Fig. 8.4. A succession of computed sections through the 50:50 two-phase structure for 100 random sine waves. Note that all particles are interconnected. The spacing between sections is $1.25/\beta_m$. From J. W. Cahn, *J. Chem. Phys.*, **42**, 93 (1965).

many experimental studies of phase separation in which both phases occupy large volume fractions (see Fig. 3.12).

These morphological predictions are based on the simple linearized theory of spinodal decomposition presented above, which should be appropriate for the early stages of phase separation near the central regions of miscibility gaps. The theory neglects, however, higher-order terms in the diffusion equation which led to Eqs. (8.14) and (8.15). Inclusion of the most important of these terms seems to result in a sharpening of interfaces and a breaking off of connectivity. Beyond this, although extensive interconnectivity is predicted by the linear theory for the case in which both phases are present in high-volume fractions, when the volume fraction is smaller than about 15 to 20%, such connectivity is not expected.

8.3 Nucleation

When a new phase is formed by a nucleation and growth process, it must start as a very small region and then increase in size. Initially it has a high surface-to-volume ratio, which tends to make it unstable because of its high surface energy. This can perhaps be illustrated by considering a small bubble of water vapor being formed at the boiling point at which the vapor pressure is just 1 atm. As discussed in Chapter 5, there is a pressure difference across the surface of the bubble given by

$$\Delta P = \frac{2\gamma}{r} \tag{8.18}$$

Formation of a stable bubble requires a vapor pressure equal to the sum of the atmospheric pressure plus the pressure due to the surface energy, as given in Eq. 8.18. To exist stably then in water ($\gamma = 72$ dynes/cm), a 1-micron bubble must have a vapor pressure equal to about 4 atm, corresponding to a temperature of about 145°C, 45° higher than the equilibrium boiling point.

In view of the high degree of supersaturation indicated as necessary to nucleate bubbles at the boiling point, we may ask how it is that water is normally observed to boil at a temperature very close to 100°C. There are two ways in which bubbles probably originate in liquids. One is by means of vortices resulting from agitation; a very high negative pressure develops at the center of the vortex, allowing a bubble to start. Much more commonly bubbles start growing at an interface where there is already a gas film. As the bubble grows and breaks loose, it leaves a starting nucleus of gas for the next bubble. Streams of bubbles issuing from the same point are commonly observed in boiling water or carbonated beverages. In order to prevent *bumping* during boiling in glass vessels, it is a common laboratory practice to add a porous porcelain chip which provides sites at which nucleation is favored.

Nucleation from a homogeneous phase is called homogeneous nucleation; when surfaces, grain boundaries, second-phase particles, and other discontinuities in the structure serve as favorable sites, catalyzing nucleation, the process is called heterogeneous nucleation. Heterogeneous nucleation is by far the more commonly observed.

Homogeneous Nucleation. The formation of nuclei requires the formation of an interface between the two phases. Because of this, the formation of very small particles usually requires an increase in the free energy of the system. Once the particle has reached a sufficiently large size, the interface energy is small compared with the volume energy decrease so that the overall change in free energy on forming the new phase becomes negative. The local increase in free energy which gives rise to the formation of small regions of a new phase must come from fluctuations in a homogeneous system. The kinetics of formation of nuclei involves both the free energy of formation of the small region, including its surface energy, and the rate of atom transport at the interface boundary.

Let us consider a system in which a phase α is the stable phase above a transition temperature T_0 and a phase β is the stable phase below this temperature. The transition can be solidification, precipitation, or another phase transition. We wish to consider the formation of a small region of β within the homogeneous phase α. The overall change in free energy can be considered to consist of two major parts. One arises from the formation of the interface and another from the change in volume free energy connected with the phase change $\alpha \rightarrow \beta$. (In reactions in a solid phase another term may be necessary to include elastic strains arising from a volume change.) If we assume for simplicity that the new region of β is spherical and has a radius r, the free-energy change is given by

$$\Delta G_r = 4\pi r^2 \gamma + \frac{4}{3}\pi r^3 \Delta G_v \tag{8.19}$$

where γ is the interface energy and ΔG_v is the free-energy change per unit volume for the phase transition, neglecting interface energy.

For very small particles the first term in Eq. 8.19 predominates; as the size of an embryo (a particle smaller than a stable nucleus) increases, the free energy required to form it also increases. On further growth, however, the second term tends to predominate, so that once embryos reach some critical size (Fig. 8.5), the second term predominates and further growth leads to an increasingly lower free energy and a stabler

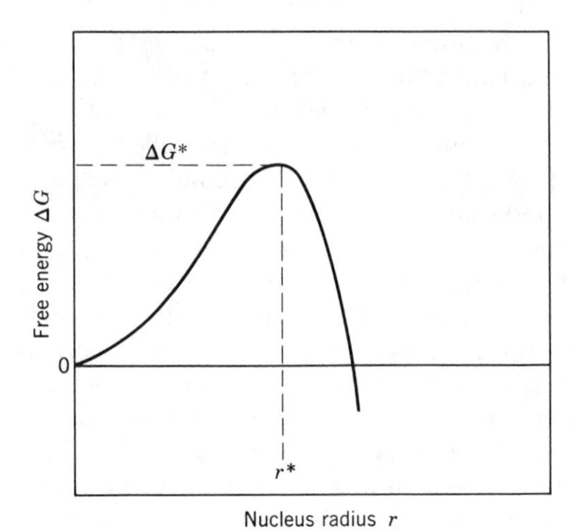

Fig. 8.5. Free energy of nucleus as a function of size. Some critical size must be exceeded before the nucleus becomes stable.

system. There will be, however, an equilibrium concentration of embryos smaller in size than the critical nucleus. By minimizing the free energy of the system with respect to the number of embryos (there is a gain of entropy from the mixing of embryos with unassociated molecules), the equilibrium concentration of embryos of size r may be written

$$\frac{n_r}{n_0} = \exp\left(-\frac{\Delta G_r}{kT}\right) \tag{8.20}$$

and the concentration of critical size nuclei

$$\frac{n^*}{n_0} = \exp\left(-\frac{\Delta G^*}{kT}\right) \tag{8.21}$$

where n_r, n^*, and n_0 are, respectively, the number of embryos of size r, critical nuclei, and single molecules per unit volume and ΔG_r and ΔG^* are the free energy of formation of an embryo of size r and of critical size.

The size of the cluster which has the maximum free energy and which on further growth leads to a continuous decrease in free energy is determined from the relation $\partial(\Delta G_r)/\partial r = 0$ and is given by

$$r^* = \frac{-2\gamma}{\Delta G_v} \tag{8.22}$$

$$\Delta G^* = \frac{4}{3}\pi r^{*2}\gamma = \frac{16\pi\gamma^3}{3(\Delta G_v)^2} \tag{8.23}$$

Regions of the new phase β having a radius less than r^* are called subcritical embryos; those of radius r^* and larger are called critical and supercritical nuclei. Nuclei generally grow larger, but those of the critical size may grow indefinitely large or shrink back and disappear. Either process decreases the free energy of the critical nucleus.

The number of molecules in a nucleus of critical size is generally in the range of 100 molecules. This size of nucleus is much too large to result from a single fluctuation, and the nucleation process is assumed to consist of the addition of molecules one by one to an embryo. If it is assumed that the concentration of critical-size nuclei is that characteristic of equilibrium and that all nuclei which become just supercritical grow to a large size, the equilibrium rate of nucleation per unit volume $(I_v)_{eq}$ may be written

$$(I_v)_{eq} = \nu n_s n^* \tag{8.24}$$

where ν is the collision frequency of single molecules with nuclei and n_s is the number of molecules on the periphery of the critical-size nucleus. That is, the nucleation rate per unit volume equals the number of critical-size clusters per unit volume times the number of molecules in contact with the nucleus of critical size times the frequency with which single molecules are attached to the nucleus of critical size.

When the phase α is a perfect vapor, the collision frequency ν can be expressed

$$\nu = \frac{\alpha_c p}{n_A (2\pi MkT)^{1/2}} \tag{8.25}$$

where α_c is the condensation coefficient for molecules impinging on the clusters (the fraction of incident molecules which condense on the cluster), p is the vapor pressure, n_A is the number of atoms per unit area of the cluster, and M is the molecular weight. Combining this with Eqs. 8.24 and 8.21, the equilibrium rate of nucleation is

$$(I_v)_{eq} = \frac{\alpha_c p}{(2\pi MkT)^{1/2}} A^* n_0 \exp\left(\frac{-\Delta G^*}{kT}\right) \tag{8.26}$$

where A^* is the area of the critical nucleus.

For nucleation in condensed phases, the collision frequency may often be expressed

$$\nu = \nu_0 \exp\left(\frac{-\Delta G_m}{kT}\right) \tag{8.27}$$

where ν_0 is the molecular jump frequency and ΔG_m is the activation energy for transport across the nucleus-matrix interface. The equilibrium rate of nucleation is then given by

$$(I_v)_{eq} = \nu_0 n_s n_0 \exp\left(\frac{-\Delta G^*}{kT}\right) \exp\left(\frac{-\Delta G_m}{kT}\right) \tag{8.28}$$

The equilibrium treatment outlined above neglects the back flux of supercritical embryos becoming subcritical as well as the decrease in embryo population caused by growth of nuclei to finite size. These may be included in a so-called steady-state analysis in which a steady-state current of single molecules through the range of subcritical clusters is treated. The resulting expressions for the steady-state nucleation frequency are

$$I_v \approx \frac{\alpha_c p}{n_A (2\pi MkT)^{1/2}} n_0 \exp\left(\frac{-\Delta G^*}{kT}\right) \tag{8.29}$$

and

$$I_v \approx n_0 \nu_0 \exp\left(\frac{-\Delta G_m}{kT}\right) \exp\left(\frac{-\Delta G^*}{kT}\right) \tag{8.30}$$

If we consider the effect of increasing supersaturation at a constant temperature, the value of the free-energy change of the phase transition is determined by the relation

$$\Delta G_v = -\frac{kT}{V_\beta} \ln \frac{p}{p_0} \tag{8.31}$$

where p is the actual vapor pressure, p_0 is the equilibrium vapor pressure, and V_β is the molecular volume of the β phase. Substituting this value into Eq. 8.23 and substituting this in turn into Eq. 8.29, the nucleation rate is

$$I_v = I_0 \exp\left[-\frac{16\pi\gamma^3}{3kT(kT/V_\beta \ln p/p_0)^2}\right] \tag{8.32}$$

Since the logarithm of the supersaturation ratio (p/p_0) enters into the exponential term of Eq. 8.32 as a factor determining ΔG^*, the rate of nucleation is sensitive to the degree of supersaturation. In fact, the rate of nucleation is very small except as the supersaturation approaches some critical value; then I changes by many orders of magnitude for small changes about the critical value of the supersaturation ratio.

For nucleation in condensed phases, the free-energy change per unit volume, ΔG_v, on transforming α phase to β phase material can be estimated as follows: From the definition of the Gibbs free energy

$$\Delta G_v = \Delta H_v - T\,\Delta S_v \tag{8.33a}$$

where ΔH_v and ΔS_v are respectively the differences between phases of the enthalpy and entropy per unit volume. At the equilibrium transformation temperature T_0, $\Delta G_v = 0$. Hence

$$\Delta H_v = T_0 \, \Delta S_v \tag{8.33b}$$

At temperatures not too far from T_0, where ΔH_v and ΔS_v are close to their values at T_0, Eq. 8.33*a* becomes

$$\Delta G_v \approx \frac{\Delta H_v (T_0 - T)}{T_0} \tag{8.33c}$$

where ΔH_v is now identified as the heat of transformation per unit volume. Substituting Eq. 8.33*c* into Eqs. 8.23 and 8.30, we obtain for the temperature dependence of nucleation

$$I = I_0 \exp \frac{-16\pi\gamma^3 T_0^2}{3kT \, \Delta H_v^2 (T_0 - T)^2} \exp\left(-\frac{\Delta G_m}{kT}\right) \tag{8.34}$$

This variation of nucleation rate with temperature is so sharp for crystallization from fluid liquids that experimentally the temperature at which homogeneous nucleation takes place, T^*, may be considered a characteristic property of the substance. For highly viscous liquids, the steady-state rate of homogeneous nucleation is sufficiently small so that the formation of crystals in the interiors of specimens can often not be identified as homogeneous or heterogeneous.

The results of experimental studies on a variety of fluid liquids crystallizing from the melt are summarized in Table 8.1. As shown there, homogeneous nucleation is observed for most liquids at relative undercoolings, $\Delta T^*/T_0$, between 0.15 and 0.25. From these values, the magnitudes of the respective crystal-liquid interfacial energies γ_{SL} can be evaluated from Eqs. 8.32 and 8.34. The values of γ_{SL} obtained in this way have been related to the respective heats of fusion, as

$$\gamma_{SL} = \beta \, \Delta H_f \tag{8.35}$$

where γ_{SL} and ΔH_f are respectively the crystal-liquid surface energy per atom and the heat of fusion per atom, and $\beta \approx 1/2$ for metals and $\approx 1/3$ for nonmetals.

Heterogeneous Nucleation. The nucleation of most phase transformations takes place heterogeneously on container walls, impurity particles, or structural imperfections. The general action of such nucleating substrates is to reduce the barrier to nucleation represented by the surface energy. When a nucleus forms on a substrate, in addition to the creation of the nucleus-matrix interface, some high-energy substrate-matrix surface is replaced by a lower-energy substrate-nucleus surface, thereby resulting in a smaller overall surface-energy contribution.

When a nucleus of phase β forms on a flat substrate characterized by

Table 8.1. Experimental Nucleation Temperatures

	T_0 (K)	T^* (K)	$\Delta T^*/T_0$
Mercury	234.3	176.3	0.247
Tin	505.7	400.7	0.208
Lead	600.7	520.7	0.133
Aluminum	931.7	801.7	0.140
Germanium	1231.7	1004.7	0.184
Silver	1233.7	1006.7	0.184
Gold	1336	1106	0.172
Copper	1356	1120	0.174
Iron	1803	1508	0.164
Platinum	2043	1673	0.181
Boron trifluoride	144.5	126.7	0.123
Sulfur dioxide	197.6	164.6	0.167
CCl_4	250.2	200.2 ± 2	0.202
H_2O	273.2	232.7 ± 1	0.148
C_5H_5	278.4	208.2 ± 2	0.252
Naphthalene	353.1	258.7 ± 1	0.267
LiF	1121	889	0.21
NaF	1265	984	0.22
NaCl	1074	905	0.16
KCl	1045	874	0.16
KBr	1013	845	0.17
KI	958	799	0.15
RbCl	988	832	0.16
CsCl	918	766	0.17

T_0 (K) is the melting point, T^* (K) the lowest temperature to which the liquid could be supercooled. $\Delta T^*/T_0$ is the maximum supercooling in reduced temperature units. Note that $\Delta T^*/T_0$ is approximately constant.
Source. K. A. Jackson in *Nucleation Phenomena*, American Chemical Society, Washington, 1965.

contact angle θ (Fig. 8.6), the free energy of forming a critical-size cluster having the shape of a spherical cap is given by

$$\Delta G_s^* = \Delta G^* f(\theta) \tag{8.36}$$

where

$$f(\theta) = \frac{(2 + \cos\theta)(1 - \cos\theta)^2}{4} \tag{8.37}$$

and ΔG^* is the free energy for homogeneous nucleation, given by Eq. 8.23. Equation 8.36 shows that the thermodynamic barrier for nucleation

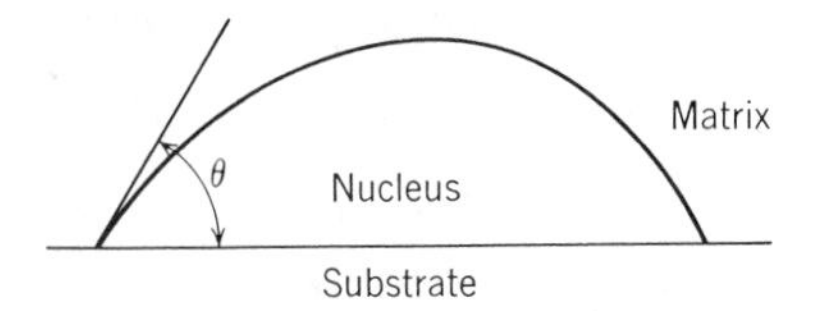

Fig. 8.6. Spherical cap model of heterogeneous nucleation.

on a substrate should decrease with decreasing θ and approach zero as θ approaches zero. Potent nucleation catalysis is favored by similar configurations of atoms in the interface planes in nucleus and substrate.

The steady-state heterogeneous nucleation rate per unit area of substrate in condensed phases may be written

$$I_s = K_s \exp\left(\frac{-\Delta G_s^*}{kT}\right) \tag{8.38}$$

where

$$K_s \approx N_s^0 \nu_0 \exp\left(\frac{-\Delta G_m}{kT}\right) \tag{8.39}$$

This expression is similar to its counterpart for homogeneous nucleation, but with ΔG_s^* in place of ΔG^* in the exponential and with the number of molecules per unit area in contact with the substrate N_s^0 replacing the number of molecules per unit volume in the matrix.

Studies of the heterogeneous nucleation of crystals from the melt have often produced results which must be regarded as erratic. On the one hand, rather good correlations are generally found between the lattice match across the substrate-nucleus interface and the potency of resulting nucleation catalysis in cases of close lattice match (disregistry of less than perhaps 5%). In contrast, rather poor correlations are often found in cases in which the lattice match is rather poor (disregistries in the range of 15%).

More generally, it should be noted that little can be said à priori about the number or potency of nucleating heterogeneities associated with a given sample of a given material. Experience with glass-forming liquids has indicated that crystal nucleation almost invariably takes place at external surfaces and sometimes but not always at interior bubble surfaces. Nucleation at the external surfaces has been associated with superficial condensed-phase impurities, and the nucleation sometimes observed at interior bubble surfaces may well have a similar origin. In oxide glasses to which a nucleating agent has not been added, internal nucleation is seldom observed, and most of the reported cases of such nucleation have involved heat treatment at relatively low temperatures. The resulting crystallization is often observed in the form of rosettes and elongated striae and may often be associated with nonuniform concentrations of impurities in the melt.

When many glass-forming materials are cooled into the glassy state and then reheated to a temperature T between the glass transition and the melting point, copious crystallization is frequently observed. In contrast, when a sample of the same material is cooled directly to T from a temperature above the melting point, the sample may remain free of visible crystallization for an extended period. The origin of this difference in behavior can be associated with nuclei which form during cooling and reheating in the first type of heat treatment. In no case has it been established whether such nucleation is homogeneous or heterogeneous.

The ability to produce large densities of internal nuclei in glass-forming liquids by the addition of selected nucleating agents to the melt has important applications in forming glass-ceramic materials, as well in producing glazes and enamels having desired degrees of crystallinity. This is discussed in Sections 8.6 and 8.8.

8.4 Crystal Growth

After a stable nucleus has been formed, it grows at a rate fixed by conditions of temperature and the degree of supersaturation. The rate of growth is determined by the rate at which material reaches the surface and the rate at which it can be built into the crystal structure. It is convenient to consider crystal growth from dilute solutions or from the vapor phase separately from growth from the melt.

In both cases, the nature of the interface between a crystal and the phase from which it is growing is expected to have a decisive influence on the kinetics and morphology of crystallization. Each of the models used to describe the crystallization process is based on a different assumption concerning the interface and the nature of the sites on the interface where atoms are added and removed.

The nature of the interface has been related* to a bulk thermodynamic property, the entropy of fusion. For crystallization processes involving small entropy changes ($\Delta S < 2R$), even the most closely packed interface planes should be rough on an atomic scale, and the growth rate anisotropy, the differences in growth rate for different orientations, should be small. In contrast, for crystallization involving large entropy changes ($\Delta S > 4R$), the most closely packed faces should be smooth, the less closely packed faces should be rough, and the growth rate anisotropy

*K. A. Jackson in *Progress in Solid State Chemistry*, Vol. 3, Pergamon Press, New York, 1967.

should be large. This difference in structural features between the small entropy change and large entropy change cases is illustrated by the two-dimensional interfaces shown in Fig. 8.7.

Large entropy changes are expected for growth from the vapor or from dilute solution, as well as for the growth of most organic and inorganic compounds, including most silicates and borates, from the melt. Small entropy changes are expected for the growth of metals, SiO_2, and GeO_2 from the melt.

The predictions based on Jackson's model have been confirmed by many experimental observations. Crystallization processes involving large entropy changes generally are characterized by faceted interface morphologies (Fig. 8.8*a*); crystallization involving small entropy changes is characterized under similar conditions by the nonfaceted interface morphologies which are typical of nearly isotropic growth (Fig. 8.8*b*). For materials with rough interfaces on an atomic scale (nonfaceted interfaces), the fraction of growth sites on the interface is expected to be on the order of unity, and although it depends in general on orientation, it should not vary strongly with undercooling. For materials with smooth interfaces on an atomic scale (faceted interfaces), growth is expected to be affected significantly by defects. For sufficiently perfect crystals, nucleation barriers to the formation of new layers may well be noted.

Crystal Growth Processes. A number of models have been proposed to describe the crystal growth process. Each is based on a different view of the interface and the sites available for growth.

1. NORMAL (ROUGH SURFACE) GROWTH. When atoms can be added to or removed from any site on the interface, the growth rate may be

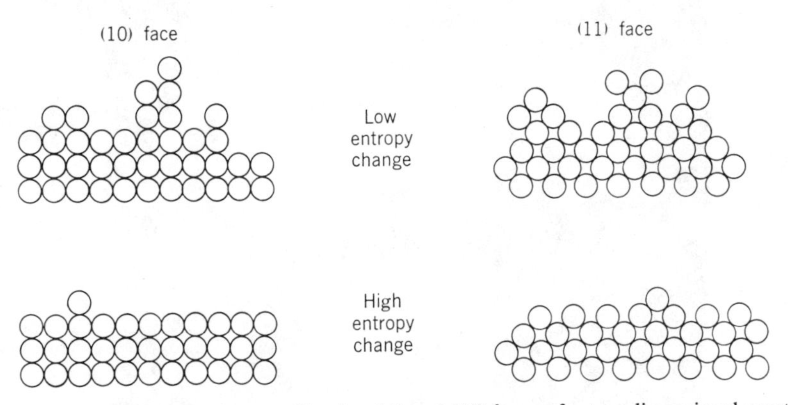

Fig. 8.7. Calculated interface profiles for (10) and (11) faces of a two-dimensional crystal, showing the effects of entropy change on interface structure. From K. A. Jackson, private communication.

expressed

$$u = \nu a_0\left[1 - \exp\left(-\frac{\Delta G}{kT}\right)\right] \tag{8.40}$$

where u is the growth rate per unit area of the interface, ν is the frequency factor for transport at the crystal–liquid interface, a_0 is the distance advanced by the interface in a unit kinetic process, approximately a molecular diameter, and ΔG is the free-energy change accompanying crystallization. As indicated by Eq. 8.33, ΔG may be taken as proportional to the undercooling. The frequency factor ν may be expressed by Eq. 8.27 or by the alternative form

$$\nu = \frac{kT}{3\pi a_0^{\ 3}\eta} \tag{8.41}$$

where η is the viscosity.

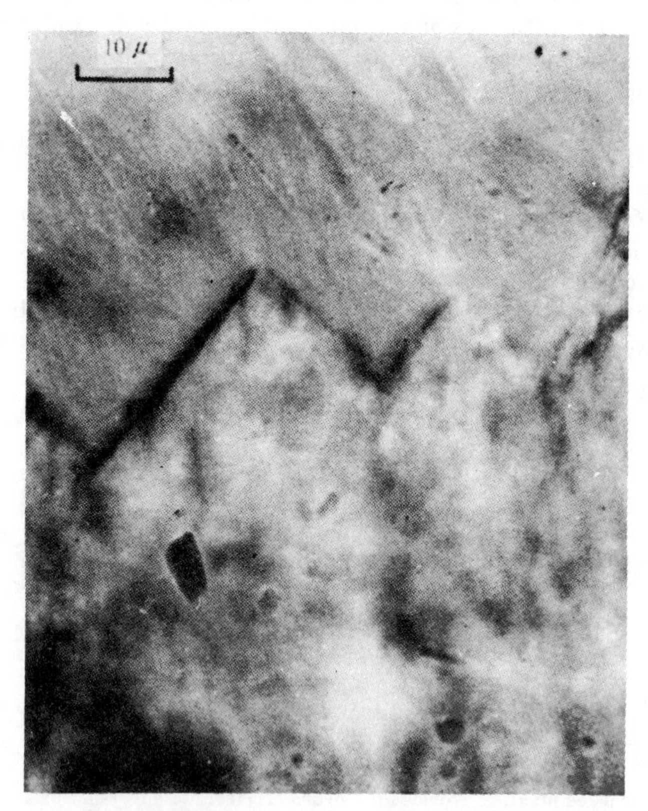

Fig. 8.8. (*a*) Interface morphology of $Na_2O{\cdot}SiO_2$. From G. S. Meiling and D. R. Uhlmann, *Phys. Chem. Glasses*, **8**, 62 (1967).

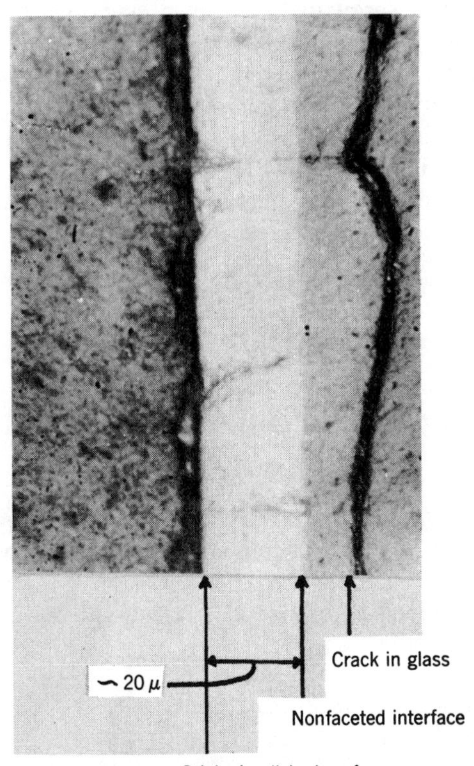

(b)

Fig. 8.8 (*continued*) (*b*) Interface morphology of GeO_2. From P. J. Vergano and D. R. Uhlmann, *Phys. Chem. Glasses*, **11**, 30 (1970).

For small departures from equilibrium ($\Delta G \ll kT$), this model predicts a linear relation between growth rate and driving force or undercooling. For large departures from equilibrium ($\Delta G \gg kT$), the model predicts a limiting growth rate

$$u \rightarrow \nu a_0 \tag{8.42}$$

This limiting growth rate for fluid materials is calculated to be in the range of 10^5 cm/sec.

For the normal growth model to correspond to reality, the interface must be rough on an atomic scale and be characterized by a large fraction of step sites where atoms can preferentially be added and removed.

2. Screw Dislocation Growth. Crystal growth may take place at step sites provided by screw dislocations intersecting the interface. Such dislocations provide a self-perpetuating source of steps as molecules are added to the crystal. The emergence point of the dislocation acquires a higher curvature until a steady state is reached in which the form of the spiral remains constant and the whole spiral rotates uniformly around the dislocation. This is illustrated in Fig. 8.9.

Assuming that growth takes place only at the dislocation ledges, the fraction f of preferred growth sites on the interface in growth from the melt is approximately*

$$f \approx \frac{\Delta T}{2\pi T_E} \tag{8.43}$$

and the growth rate is

$$u = f\nu a_0 \left[1 - \exp\left(-\frac{\Delta G}{kT}\right)\right] \tag{8.44}$$

The increase of f with undercooling reflects the winding of the dislocation into a tighter spiral, with an accompanying decrease in the separation

*W. B. Hillig and D. Turnbull, *J. Chem. Phys.*, **24**, 914 (1956).

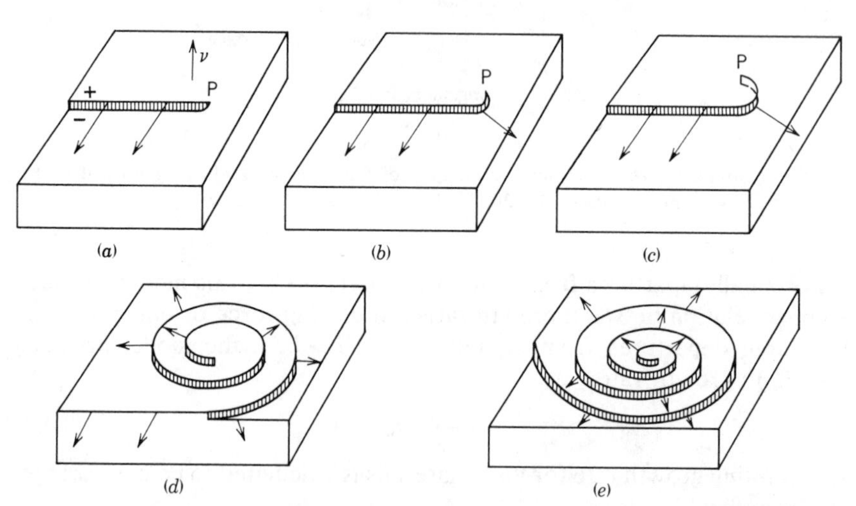

Fig. 8.9. (*a*) Step attached to the emergence point P of a dislocation with a Burgers vector which is not parallel to the surface. The step height $h = \bar{b} \cdot \bar{\nu}$, where $\bar{\nu}$ is the unit normal on the surface. (*b*) to (*e*) Under the influence of supersaturation the step in (*a*) winds up into a spiral centered on P.

between the dislocation ledges. For small departures from equilibrium, this model predicts a growth rate which varies as the square of the undercooling or driving force. For the model to provide a useful representation of growth, the interface must be quite smooth on an atomic scale and must be imperfect, with growth taking place only at steps provided by screw dislocations.

3. SURFACE NUCLEATION GROWTH. In this process, growth takes place at step sites provided by two-dimensional nuclei formed on the interface, and the growth rate may be written

$$u = A\nu \exp\left(-\frac{B}{T\,\Delta T}\right) \tag{8.45}$$

The exponential constant B depends on the model used in the analysis but in all cases is proportional to the square of the edge surface energy of the two-dimensional nucleus.

The growth rate predicted by Eq. 8.45 should vary exponentially with driving force and for a small driving force should be unobservably low. For the model to correspond to reality, the interface must be smooth on an atomic scale and must also be perfect (free of intersecting screw dislocations).

Crystal Growth from Vapor or Dilute Solution. When atoms are adsorbed from the vapor on a crystal-vapor interface, they diffuse a considerable distance across the surface before they reevaporate. If a diffusing atom reaches a step, it is relatively tightly held, and the step moves across the surface; as growth proceeds, the step finally reaches the edge of the crystal. When this happens, the step is eliminated, and in a perfect crystal it is necessary to nucleate an additional layer for growth to proceed. The growth kinetics expected for this case are given by Eq. 8.45.

Experimentally it is found that the growth of crystals from the vapor phase often occurs at measurable rates, even at low values of supersaturation. This has been associated with growth taking place at steps provided by screw dislocations intersecting the interface.

A variety of spiral growth forms, suggestive of screw dislocation growth, has been observed. An example is shown in Fig. 8.10. Growth spirals frequently have the same symmetry as the face on which they are growing. Double growth spirals, similar in appearance to Frank-Read sources, are often observed. The step heights in many studies in which spiral growth forms are observed are estimated to be of the order of microns or larger. The origin of such large step heights remains to be satisfactorily explained.

The rate of growth of crystals from dilute solutions or the vapor phase is determined by the rate of forming steps on the surface and by the rate

Fig. 8.10. Growth spiral on cadmium iodide crystals growing from water solution. Interference contrast micrograph at 1025 ×. Courtesy Dr. K. A. Jackson, Bell Telephone Laboratories.

of migration of the steps across the surface. As the supersaturation increases, an increasing number of sources of steps become available. For very low supersaturation dislocations are probably the only step sources; at high degrees of supersaturation steps are also formed by surface nucleation. Growth can be radically affected by the presence of impurities; the effects of impurities increase with decreasing supersaturation and with decreasing temperature.

Where only one dislocation serves to form a single spiral ramp, crystal whiskers, which are perfect crystals except for the central dislocation core, are grown. These are interesting, since they have exceptional mechanical properties. Many give elastic deformations corresponding to a strength of several million psi. They have been grown from a variety of materials, including oxides, sulfides, alkali halides, and many metals.

Growth from the Melt. The experimental relationship between the growth rate and the undercooling at the interface is often difficult to obtain because of the difficulty in measuring the interface temperature. In many situations, the growth rate is not limited by interface kinetics but by the rate at which the latent heat of fusion can be removed from the freezing front (see discussion in Section 8.6).

The large viscosities of many glass-forming materials result in growth rates, and hence rates of latent heat evolution, which are small even at large undercoolings. The interface temperature can then be well taken as the bath or furnace temperature, and the growth rate can be measured over a wide range of undercooling.

The form of the growth-rate–temperature relation for a typical glass-forming system is shown in Fig. 8.11. The growth rate is zero at the melting point, increases with increasing undercooling, rises to a maximum, and then decreases, as the fluidity decreases, with further increases in undercooling. Information about the nature of the interface process during growth can be obtained from the reduced growth rate U_R:

$$U_R \equiv \frac{u\eta}{[1 - \exp(-\Delta G/kT)]} \tag{8.46}$$

Provided only that the transport process at the interface varies with temperature in the same way as the viscosity ($\nu \propto 1/\eta$), the U_R versus ΔT relation should represent the temperature dependence of the interface site factor. For normal (rough surface) growth, this relation would be a horizontal line (see Eqs. 8.40 and 8.41); for screw dislocation growth, a line of positive slope (see Eqs. 8.43, 8.44, and 8.41); and for surface nucleation growth, a concave upward curve (see Eqs. 8.45 and 8.41). In the last case a plot of $\log(u\eta)$ versus $1/T\,\Delta T$ should be a straight line of negative slope (see Eqs. 8.45 and 8.41).

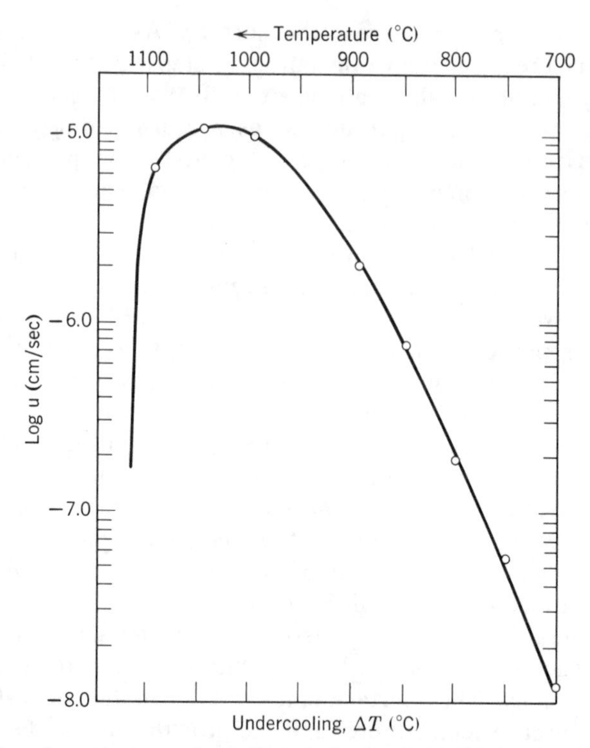

Fig. 8.11. Growth rate versus undercooling relation for GeO_2. From P. J. Vergano and D. R. Uhlmann, *Phys. Chem. Glasses*, **11**, 30 (1970).

Materials with small entropies of fusion such as GeO_2 grow and melt with nonfaceted interface morphologies (see Fig. 8.8*b*); their growth rate anisotropy is small. As shown in Fig. 8.12*a*, the forms of their crystallization and melting kinetics are those predicted by the normal growth model (Eq. 8.40), and their rates of crystallization and melting, corrected for the variation of viscosity with temperature, are equal at equal small departures from equilibrium (Fig. 8.12*b*). In contrast, materials with large entropies of fusion such as $Na_2O \cdot 2SiO_2$ have faceted interface morphologies in growth (see Fig. 8.8*a*) and nonfaceted morphologies in melting; their growth rate anisotropy is large. The forms of their crystallization kinetics are not well described by Eq. 8.44 or 8.45 (see Fig. 8.13*a*), and pronounced asymmetry is observed in their rates of crystallization and melting in the vicinity of the melting point, with melting taking place more rapidly than growth at equal small departures from equilibrium (Fig. 8.13*b*).

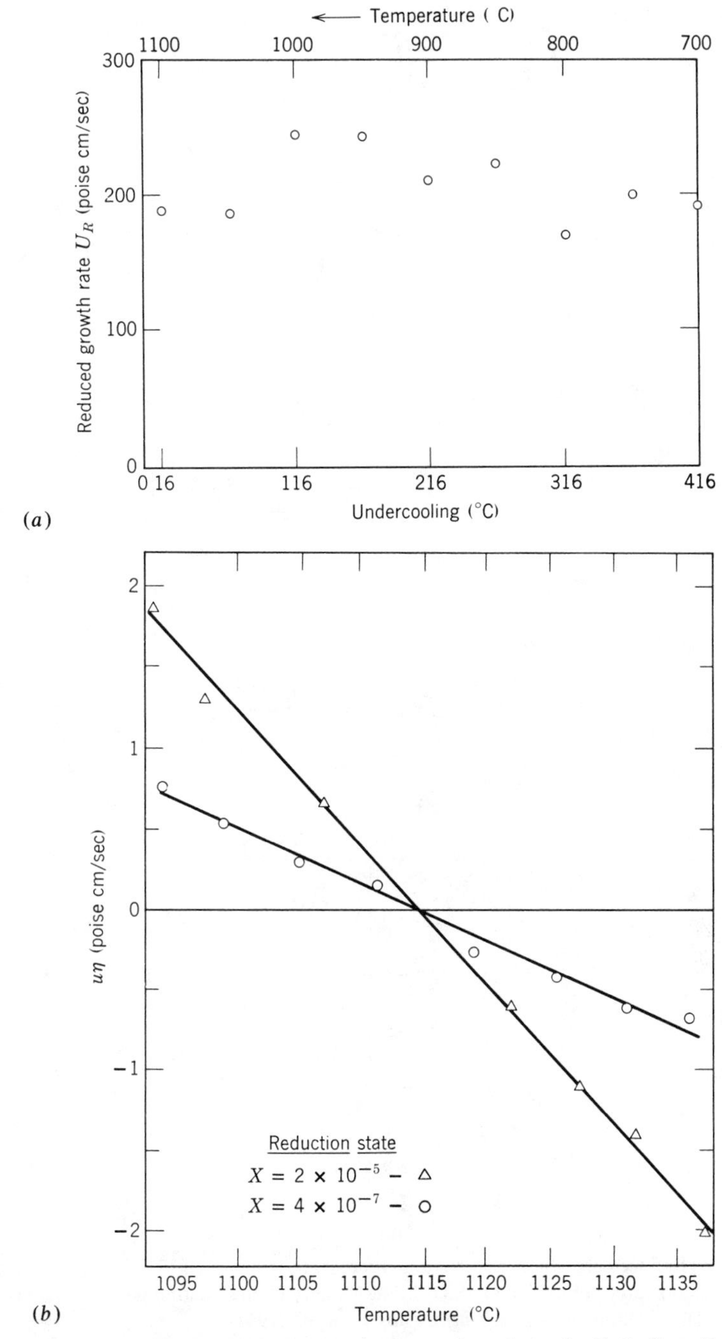

Fig. 8.12. (*a*) Reduced growth rate versus undercooling relation for GeO_2. From P. J. Vergano and D. R. Uhlmann, *Phys. Chem. Glasses*, **11**, 30 (1970). (*b*) Growth and melting rate times viscosity versus temperature relation for GeO_2 glasses having different states of reduction, represented by x in GeO_{2-x}. From P. J. Vergano and D. R. Uhlmann, *Phys. Chem. Glasses*, **11**, 39 (1970).

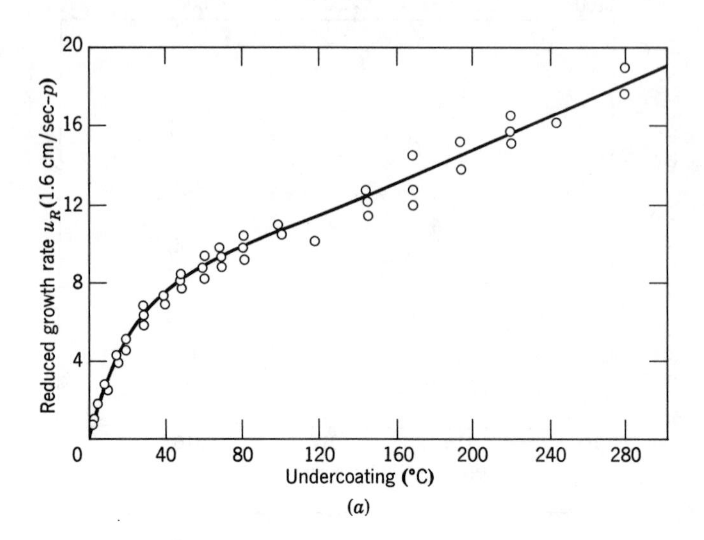

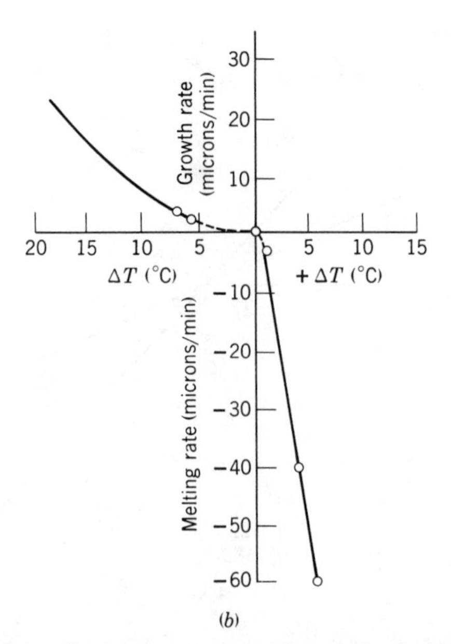

Fig. 8.13. (*a*) Reduced growth rate versus undercooling relation for $Na_2O{\cdot}2SiO_2$. (*b*) Melting rate and crystallization rate of $Na_2O{\cdot}2SiO_2$ in the vicinity of the melting point. From G. S. Meiling and D. R. Uhlmann, *Phys. Chem. Glasses*, **8**, 62 (1967).

8.5 Glass Formation

Criteria for the formation of oxide glasses were developed by W. H. Zachariasen,* who considered the structural conditions necessary for forming an oxide liquid with an energy similar to that of the corresponding crystal. These criteria, which have been discussed in Chapter 3, are rather specific to the case of oxide glasses, but the basic concept of forming a liquid with an energy similar to that of the corresponding crystal has quite general use.

However, it is now well established that glass formers are found in every category of material and that any glass former crystallizes if held for a sufficiently long time in the temperature range below the melting point. For this reason, it seems more fruitful to consider how fast must a given liquid be cooled in order that detectable crystallization be avoided, rather than whether a given liquid is a glass former. In turn, the estimation of a necessary cooling rate reduces to two questions: (1) how small a volume fraction of crystals embedded in a glassy matrix can be detected and identified and (2) how can the volume fraction of crystals be related to the kinetic constants describing the nucleation and growth processes.

For crystals which are distributed randomly through the bulk of the liquid, a volume fraction of 10^{-6} can be taken as a just detectable concentration. Concern with crystals distributed throughout the liquid provides an estimate of the necessary rather than a sufficient cooling rate for glass formation.

In treating the problem we make use of Eq. 8.5:

$$\frac{V^{\beta}}{V} \approx \frac{\pi}{3} I_v u^3 t^4$$

In using this relation we neglect heterogeneous nucleation events and are thus concerned with minimum cooling rates capable of leading to glass formation.

The cooling rate necessary to avoid a given volume fraction crystallized can be estimated from Eq. 8.5 by the construction of so-called T-T-T (time-temperature-transformation) curves, an example which is shown in Fig. 8.14. In constructing such curves, a particular fraction crystallized is selected, the time required for the volume fraction to form at a given temperature is calculated with nucleation rates calculated from Eq. 8.30 and growth rates measured experimentally or calculated from Eqs. 8.40, 8.44, or 8.45, and the calculation is repeated for a series of temperatures (and possibly other fractions crystallized).

J. Am. Chem. Soc., **54**, 3841 (1932).

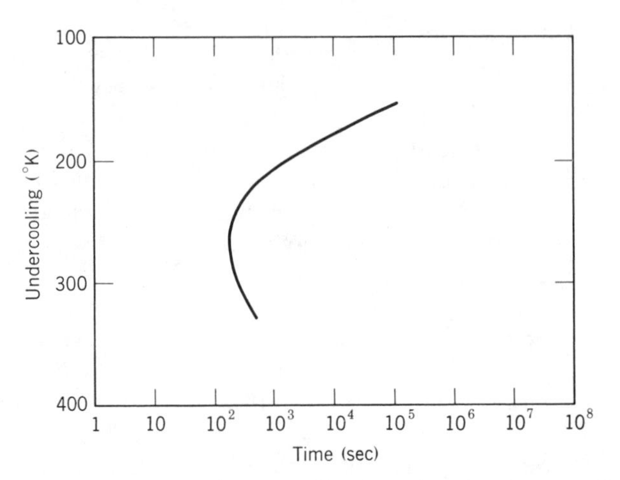

Fig. 8.14. Calculated T-T-T curve for $Na_2O{\cdot}2SiO_2$ for volume fraction crystallized of 10^{-6}. Curve constructed from calculated nucleation rates and the growth rate and viscosity data of G. S. Meiling and D. R. Uhlmann, *Phys. Chem. Glasses*, **8**, 62 (1967).

The nose in a T-T-T curve, corresponding to the least time for the given volume fraction crystallized, results from a competition between the driving force for crystallization, which increases with decreasing temperature, and the atomic mobility, which decreases with decreasing temperature. The cooling rate required to avoid a given fraction crystallized may be roughly estimated from the relation

$$\left(\frac{dT}{dt}\right)_c \approx \frac{\Delta T_n}{\tau_n} \tag{8.47}$$

where $\Delta T_n = T_0 - T_N$, T_N is the temperature at the nose of the T-T-T curve, and τ_n is the time at the nose of the T-T-T curve. More accurate estimates of the critical cooling rate for glass formation may be obtained by constructing continuous cooling curves from the T-T-T curves, following the approach outlined by Grange and Kiefer.*

From the form of Eq. 8.5 it is apparent that the critical cooling rate for glass formation is insensitive to the assumed volume fraction crystallized, since the time at any temperature on the T-T-T curve varies only as $(V^\beta/V)^{1/4}$.

In calculating T-T-T curves for a given material, one can in principle use measured values of the kinetic quantities. In practice, however, information on the temperature dependence of the nucleation frequency

Trans. ASM* **29, 85 (1941).

is almost never available, and for only a few cases of interest are adequate data available on the variation of growth rate with temperature. In nearly all cases, therefore, it is necessary to estimate the nucleation frequency.

In estimating nucleation frequencies, ΔG^* is generally taken as 50 to 60 kT at a relative undercooling $\Delta T/T_0$ of 0.2, in accordance with experimental results on a wide variety of materials. In estimating the growth rate when data are not available, a normal growth model (rough surface) (Eq. 8.40) can be assumed for materials having small entropies of fusion, and a screw dislocation growth model (Eq. 8.44) for materials characterized by large entropies of fusion.

The analysis can readily be extended to include time-dependent growth rates, such as those characteristic of diffusion-controlled growth, as well as heterogeneous nucleation in bulk liquid or at external surfaces. Such nucleation is included by replacing $I_v t$ in the various relations with N_v, the number of effective heterogeneous nuclei per unit volume. This last quantity has in general a significant temperature dependence, as nucleating particles of different potency become active in different ranges of temperature. In cases in which the nuclei are primarily associated with the external surfaces or with the center line of a glass body, a criterion of minimum observable crystal size may be preferable in some applications to the present criterion of minimum detectable fraction crystallinity.

The effects of nucleating heterogeneities can be explored with the model of a spherical-cap nucleus (Fig. 8.6). The number of nucleating heterogeneities per unit volume characterized by a given contact angle θ can be estimated as follows: Experiments on a variety of materials, discussed in Section 8.3, indicate that division of a sample into droplets having sizes in the range of 10 microns in diameter is sufficient to ensure that most droplets, perhaps 99%, do not contain a nucleating heterogeneity. These results indicate a density of nucleating particles in the range of 10^9 per cubic centimeter. Using this value together with an assumed heterogeneity size (e.g. 500 Å), one can obtain the total area of nucleating surface per unit volume.

The nucleation rate associated with heterogeneities is then obtained with Eqs. 8.36 to 8.39. The effect of the contact angle of nucleating heterogeneities on glass formation may be evaluated by calculating T-T-T curves for different θ's. Typical results, for $Na_2O \cdot 2SiO_2$, are shown in Fig. 8.15, in which ΔG^* has been taken as 50 kT at $\Delta T/T_0 = 0.2$. As seen there, heterogeneities characterized by modest contact angles ($\theta \leq 80°$) can have a pronounced effect on glass-forming ability; heterogeneities characterized by large contact angles ($\theta \geq 120°$) have a negligible effect.

Similar calculations have been carried out for a variety of materials,

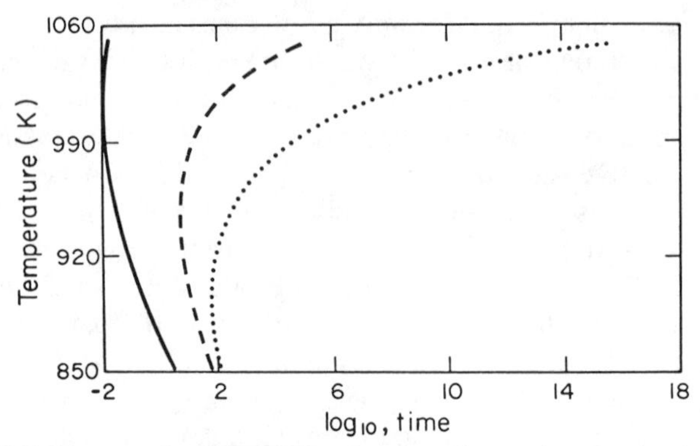

Fig. 8.15. *T-T-T* curves for $Na_2O{\cdot}2SiO_2$ showing effects of nucleating heterogeneities. Volume fraction crystallized $= 10^{-6}$. ———, contact angle = 40°; -----, contact angle = 80°; ········, homogeneous nucleation or contact angle = 120° and 160°. From P. Onorato and D. R. Uhlmann.

including oxides, metals, organics and water, which indicate that nucleating heterogeneities with $\theta > 90$ to 100° quite generally have a negligible effect on glass-forming ability. Table 8.2 compares the critical cooling rates estimated, assuming only homogeneous nucleation, and those estimated with 10^9 heterogeneities per cubic centimeter, 500 Å in size, all characterized by a contact angle of 80°. The results shown in the table for water and particularly the metal are subject to considerable uncertainty because the viscosity data had to be extrapolated over a wide range to carry out the calculations. The results indicate that it is highly unlikely, with or without heterogeneities, that a pure metal can be formed as a glass by cooling from the liquid state.

The effects on glass formation of changes in the barrier to nucleation are also shown in Table 8.2, in which critical cooling rates are compared

Table 8.2. Estimated Cooling Rates for Glass Formation

Material	dT/dt (°K/sec) Homogeneous nucleus $\Delta G^* = 50kT$ $T_r = 0.2$	dT/dt (°K/sec) Heterogeneous nucleus $\theta = 80°$ $\Delta G^* = 50kT$ $T_r = 0.2$	dT/dt (°K/sec) Homogeneous nucleus $\Delta G^* = 60kT$ $T_r = 0.2$
$Na_2O{\cdot}2SiO_2$	4.8	46	0.6
GeO_2	1.2	4.3	0.2
SiO_2	7×10^{-4}	6×10^{-3}	9×10^{-5}
Salol	14	220	1.7
Metal	1×10^{10}	2×10^{10}	2×10^{9}
H_2O	1×10^{7}	3×10^{7}	2×10^{6}

for $\Delta G^* = 50\,kT$ and $\Delta G^* = 60\,kT$ at $\Delta T/T_0 = 0.2$. As seen there, these effects can also be substantial. When the calculated rates for various oxides are compared with experience in the laboratory, the difficulty of forming glasses is generally overestimated by assuming $\Delta G^* = 50\,kT$ at $\Delta T/T_0 = 0.2$. That is, the calculated cooling rates, even neglecting nucleating heterogeneities, are consistently too high. Reasonable agreement between calculated rates and laboratory experience can be obtained by taking somewhat larger values for ΔG^* (in the range of 60 to 65 kT at $\Delta T/T_0 = 0.2$).

When this analysis is applied to a variety of liquids, it is found that the material characteristics most conducive to glass formation are a high viscosity at the melting point and a viscosity which increases strongly with falling temperature below the melting point. For materials with similar viscosity-temperature relations, glass formation is then favored by low melting points or liquidus temperatures. The observed ease of glass formation in regions of composition near many eutectics can be related to the composition redistribution required for crystallization to proceed as well as to the lower liquidus temperatures.

By emphasizing the importance of the crystallization rate and viscosity in glass formation, attention is in turn focused on those characteristics of materials which determine their flow behavior and the nature of their crystal–liquid interfaces. In this view, various correlations which have been suggested between glass transition temperatures and liquidus temperatures (e.g., $T_g \sim 2/3\,T_M$) for particular systems must be regarded as of limited generality.

Materials such as Al_2O_3, H_2O, and $Na_2O \cdot SiO_2$, which are quite fluid over a range of temperature below their melting points, can only be obtained as glasses by achieving very rapid cooling. This is effected by techniques such as splat cooling or condensation from the vapor onto a cold substrate. With splat cooling, cooling rates in the range of 10^6 °K/sec can be obtained, and even higher effective cooling rates, for material in thin-film form, can be obtained by condensation from the vapor.

8.6 Composition as a Variable, Heat Flow, and Precipitation from Glasses

In considering nucleation and growth with composition as a variable, consider the schematic free energy versus composition diagram in Fig. 8.16. As shown there, the equilibrium compositions are given by the contact points of the free-energy curve with the common tangent, designated as C^α and C^β. For a composition C_0, the change in free energy on crystallization to the equilibrium distribution of phases and hence the driving force for crystallization is given by ΔG_0.

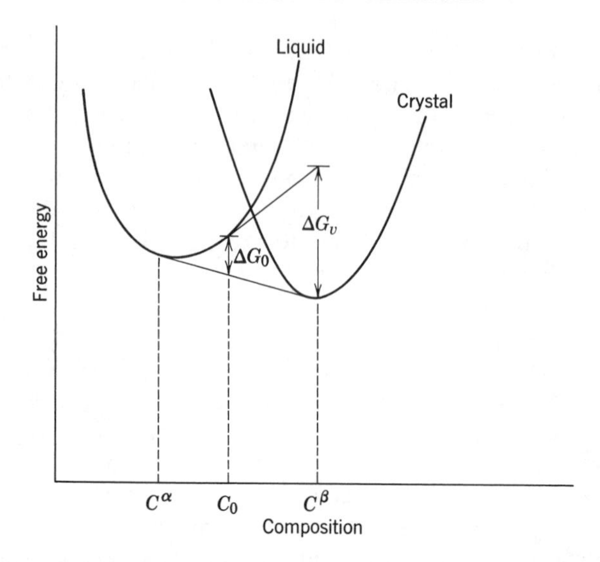

Fig. 8.16. Free energy versus composition relation for solutions (schematic), showing driving force for crystallization (ΔG_0) and driving force for nucleation (ΔG_v).

In considering nucleation, however, one must note that as a cluster of composition C^β forms, the composition of the remaining liquid need change only slightly from the initial composition C_0 to conserve mass. For this reason, the free-energy change driving the nucleation process may be given:

$$\Delta G_v = G(C^\beta) - G(C_0) - (C^\beta - C_0)\left(\frac{dG}{dC}\right)_{C_0} \tag{8.48}$$

Hence, the driving force for nucleation is given by the difference between the tangent line drawn to C_0 and the free-energy curve of the phase being formed at the composition of interest. If the nucleus is assumed to form with composition C^β, the formalism of Section 8.3 may be used directly, with ΔG_v given now by Eq. 8.48 and n_0 replaced by the number of clustering molecules per unit volume.

For most materials other than the familiar oxide glasses, the presence of impurities is expected to decrease the rate of crystallization, since part of the driving force must be used in the diffusional process required for crystallization or since the preferred growth sites on the interface can be poisoned by the impurities. High-entropy-of-fusion materials should be more sensitive to impurity effects than low-entropy-of-fusion materials.

In contrast, impurities, including atmospheric impurities and departures from stoichiometry, can appreciably increase the growth rate in many oxide materials and certainly in many of the important glass-forming materials. Examples of this have been observed with water, oxygen, and Na_2O in SiO_2, water and Na_2O with GeO_2, and water with B_2O_3, and the use of water and other mineralizing agents to increase the rate of crystallization in oxide systems has been known for many years. It is expected that these effects should be most important for compositions having a network character and for pure materials.

It is well known that the redistribution of solutes takes place as a crystal grows into an impure melt. This results from the equilibrium concentration of solute in the crystal at the interface C_s being different from that in the adjacent liquid C_L. This is usually expressed in terms of the distribution coefficient k_0:

$$k_0 = \frac{C_s}{C_L} \tag{8.49}$$

Most solutes in most systems are less soluble in the crystal than in liquid, and for these k_0 is less than unity (see the solidus and liquidus curves in Fig. 8.17).

For cases in which convection in the liquid can be neglected, the impurity distributions in the crystal and liquid can be estimated for different stages of growth. This is shown schematically in Fig. 8.18 for the case of constant growth rate in which k_0 is less than unity and Eq. 8.49 pertains, that is, in which the interface is at the solidus temperature for the local compositions C_L and C_s. For an initial impurity concentration of C_0 in the liquid, the crystals first formed should have a concentration of k_0C_0. Under conditions of steady-state growth, the concentration in the liquid at the interface should be C_0/k_0 and that in the crystal C_0.

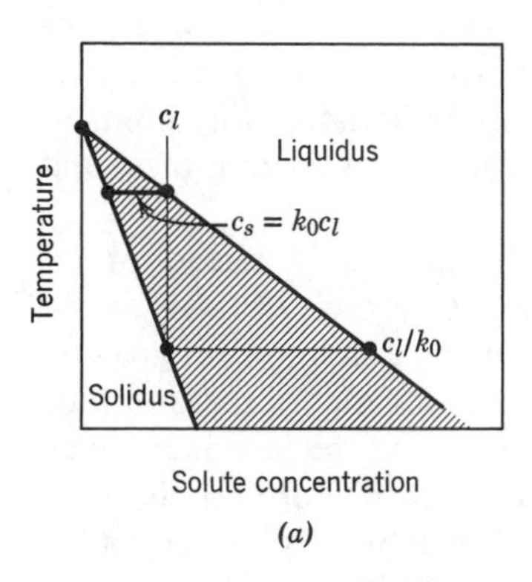

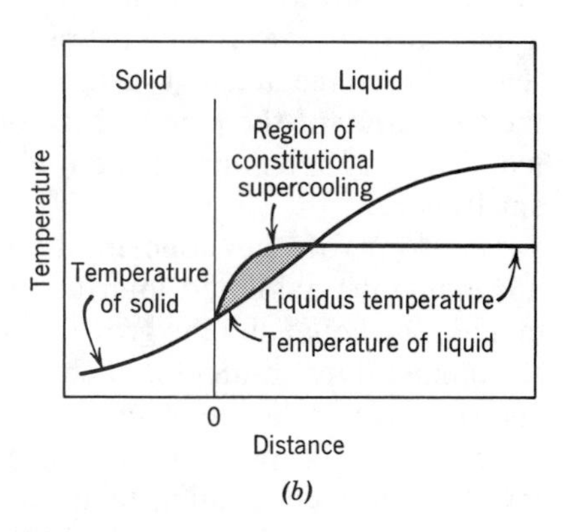

Fig. 8.17. (*a*) Solidus and liquidus curves and (*b*) tendency for constitutional supercooling.

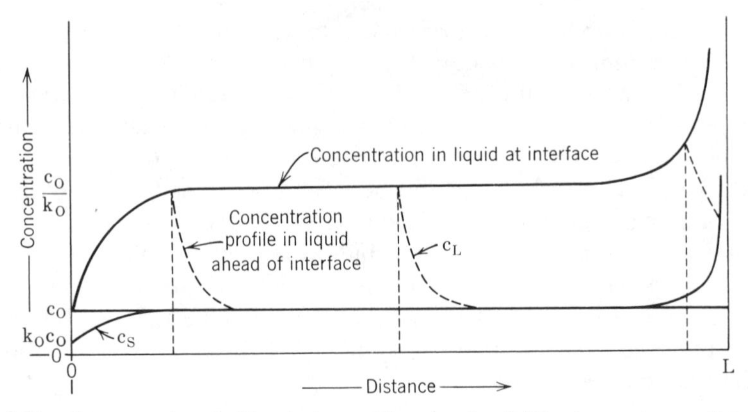

Fig. 8.18. Concentration profiles during unidirectional solidification under conditions in which Eq. 8.49 is valid. C_{LP} = concentration profile in liquid ahead of interface. From W. A. Tiller, K. A. Jackson, J. W. Rutter, and B. Chalmers, *Acta Met.*, **1**, 428 (1953).

In normal solidification the nucleation at the wall of the container is rapid, and a large number of crystals of different orientations form a rim around the mold interface. Gradually these crystals tend to be oriented in their fastest-growing directions to form large columnar crystals growing perpendicular to the face of the mold. This leads to a large grain size in fusion-cast materials, which has a considerable effect on resulting properties. In practical mold design, a main problem is the change in volume occurring on solidification. In general there is a 10 to 25% volume decrease during the liquid-solid transition. This causes the formation of a *pipe* in the casting which must be controlled by suitable mold design. In fusion-cast ceramic ware this is even more of a problem than for metals because the low thermal conductivity of the liquid allows the formation of a solid surface layer so that the pipe forms with a large interior void rather than being connected to a surface opening. In order to form sound blocks, the formation of the pipe and its location must be satisfactorily controlled. This has been one of the difficulties in the fusion casting of sound small pieces.

The form of the concentration profile in the liquid ahead of an advancing crystal-liquid interface is determined by the diffusion coefficient in the liquid and the growth rate. Under many conditions of growth, the temperature gradient is such that the material in front of the interface shows a greater degree of supercooling than occurs at the interface; once an irregularity grows into this region, it tends to grow more rapidly than the flat interface, leading to instability. This instability gives rise to a cellular structure determined by the greater rate of diffusion away from a rounded interface and the tendency for constitutional supercooling to segregate the solute between growing perturbances. This is what happens, for example, in the growth of sea ice. Concentrated brine solution forms along boundaries between growing crystals and becomes mechanically trapped as growth proceeds. During melting, this trapped material is the first to melt and is one cause of *rotten ice* during the melt season. The brine distribution has a strong effect on the properties of sea ice.

When a fluid liquid is supercooled before nucleation of the solid phase, the rate of isothermal crystal growth is rapid, but the heat of solidification tends to warm the interface above the surrounding liquid temperature. For solidification to continue, this heat must diffuse away through the lower-temperature liquid phase. This occurs most rapidly for a small-radius-of-curvature interface; consequently there is dendritic growth with the small-radius-of-curvature end of a dendrite spike advancing at a rapid rate. Side dendrites begin to form when thermal conditions along the side of the initial dendrite are such that growth can take place.

Growth Controlled by Diffusion of Solutes. Treatments of diffusion-controlled growth generally indicate dimensions of crystals which increase proportionally with the square root of time (growth rates which decrease as the square root of time). This time variation is associated with the fact that matter must be transported over progressively larger distances as growth proceeds. For cases in which the scale of the growing particle is large relative to the characteristic diffusion distance, however, the interface should advance at a rate which is independent of time. For the specific case of a rod of constant diameter to which atoms are added only near the ends, the length and volume are expected to increase linearly with time.

For cases in which the crystal size is comparable with the scale of the diffusion field, the crystal dimensions should increase as the square root of time for diffusion-controlled growth and linearly with time for interface-controlled growth. A transition from interface-controlled growth to diffusion-controlled growth is anticipated as progressively large concentrations of impurities are added to materials such as SiO_2 and GeO_2.

For solute species which increase the growth rate, their rejection and buildup at the interface may result in a greatly enhanced growth rate which increases with time. In such cases, specimens may be completely crystallized before anything like steady-state growth conditions are maintained. Such autocatalytic effects should be more likely to initiate in regions between two advancing interfaces, where the buildup of solutes is most significant. In other cases, the transport of crystallizing material through a solute region of high mobility at the interface may result in a crystallization process which resembles growth from a concentrated solution. For small concentrations of solutes which do not result in diffusion-controlled growth or autocatalytic growth effects, the effects on kinetics should be describable in terms of their combined effects on the viscosity and the liquidus temperature.

The crystallization behavior of glasses in the Na_2O–CaO–SiO_2 system illustrates further the principles outlined above. The growth rates of devitrite ($Na_2O \cdot 3CaO \cdot 6SiO_2$) crystals in a glass of percentage composition $17Na_2O$, $12CaO$, $2Al_2O_3$, $69SiO_2$ were found to be fairly constant until the specimens were almost completely devitrified (Fig. 8.19*a*). As shown in Fig. 8.19*b*, the rates of growth and dissolution appear to vary smoothly and continuously through the liquidus temperature of 1007°C.

The lack of dependence of the growth rate on time might a priori have been unexpected, considering the large changes in CaO concentration involved in the crystallization of devitrite. The explanation for this behavior can, however, be seen from the morphologies of the growing crystals, which are fibrillar in character, and related to the crystal length being large in comparison with the relevant diffusion length.

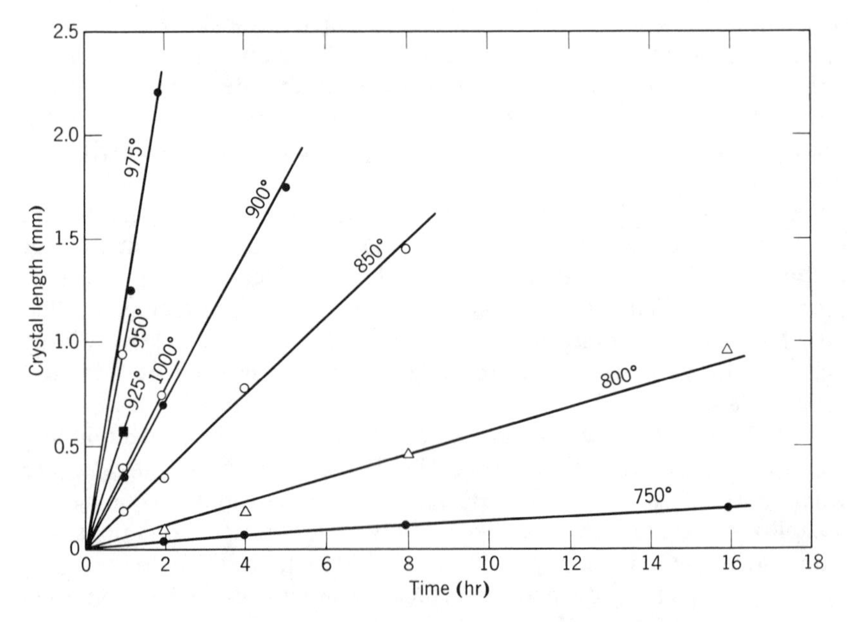

Fig. 8.19. (*a*) Growth of devitrite crystals in a glass of composition $17Na_2O$–$12CaO$–$2Al_2O_3$–$69SiO_2$ showing the effects of time and temperature.

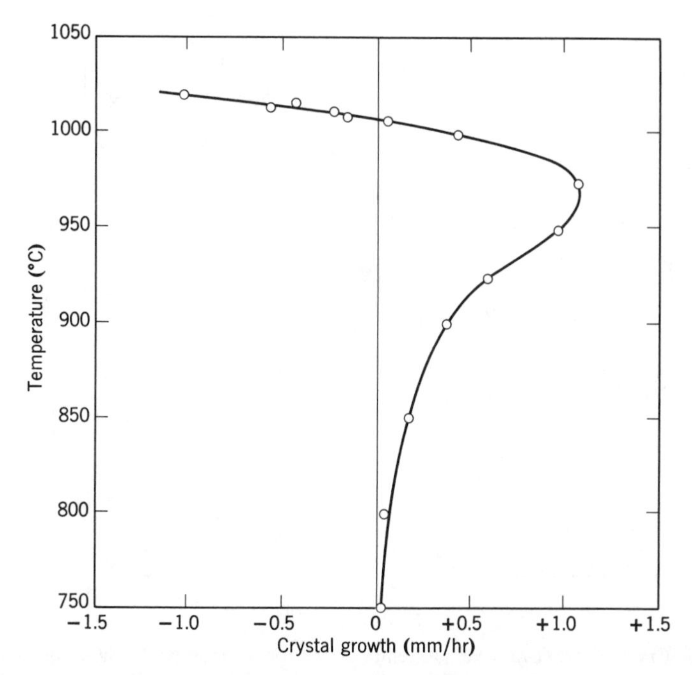

Fig. 8.19 (*continued*) (*b*) Rate of growth of devitrite crystals in the same glass as shown in Fig. 8.19*a*. From H. R. Swift, *J. Am. Ceram. Soc.*, **30**, 165 (1947).

Crystal growth rates have been determined by Scherer* over wide ranges of temperature for several compositions in the Na_2O–SiO_2 and K_2O–SiO_2 systems. The results for compositions containing 1.5, 10, 15, 20, 25, and 30 mole% Na_2O are shown in Fig. 8.20. For comparison, the phase diagram for this system is shown in Fig. 8.21. Compositions containing 1.5, 10, 15, and 20% Na_2O crystallized with a dendritic morphology. For the first three compositions, cristobalite was the only crystallization product; the last material crystallized to cristobalite and $Na_2O \cdot 3SiO_2$, the latter being a metastable crystalline phase. In all cases, the growth rates were independent of time.

Liquid-liquid phase separation was observed in the glasses containing 10 and 15% Na_2O but was found to have no discernible effect on the crystallization kinetics. This was related to the small scale of phase separation compared with the scale of the diffusion fields of the dendrites. The material containing 30% Na_2O crystallized with spherulitic crystals of

*Sc.D. thesis, Massachusetts Institute of Technology, 1974.

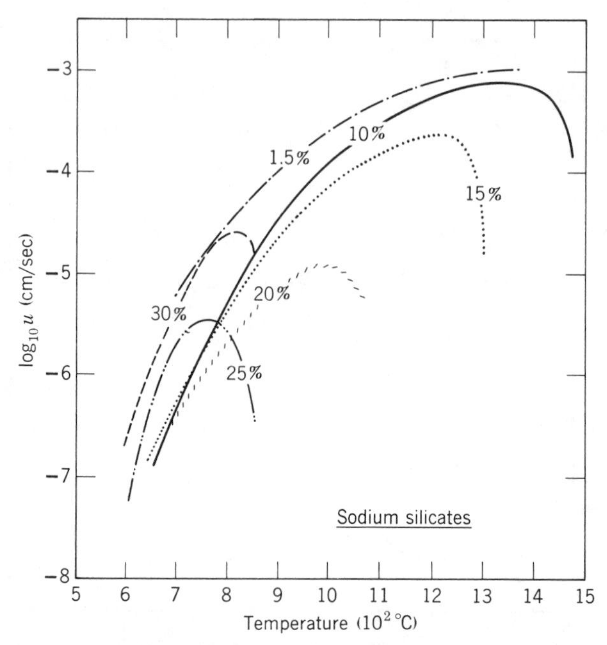

Fig. 8.20. Crystallization rate versus temperature for various Na_2O–SiO_2 compositions. Compositions indicated in mole percent Na_2O. From G. W. Scherer, Sc.D. thesis, MIT, 1974.

$Na_2O \cdot 2SiO_2$; the 0.25 Na_2O–0.75 SiO_2 composition crystallized with a faceted interface morphology. The crystal growth rate of the latter composition was apparently limited by interface attachment kinetics rather than by diffusion in the melt, and the reduced growth rate versus undercooling relation (Fig. 8.22) is a straight line of positive slope passing through the origin. The form of this relation is suggestive of growth by a screw dislocation mechanism (see discussion in Section 8.4). Examination of the phase diagram for the Na_2O–SiO_2 system (Fig. 8.21) indicates that the observed crystallization of the 0.25 Na_2O–0.75 SiO_2 composition takes place to a metastable crystalline phase of the same composition, rather than to the equilibrium-phase assemblage of $SiO_2 + Na_2O \cdot 2SiO_2$.

Crystallization of compositions in the K_2O–SiO_2 system containing 10 and 15 mole% K_2O resulted in the formation of dendritic crystals of cristobalite. At all temperatures studied, the growth rates of these crystals were independent of time. The measured growth rate as a function of temperature is shown in Fig. 8.23 for the 0.10 K_2O–0.90 SiO_2 composition. Other work on the same composition by Christensen, Cooper, and Rawal*

J. Am. Ceram. Soc., 56, 557 (1973).

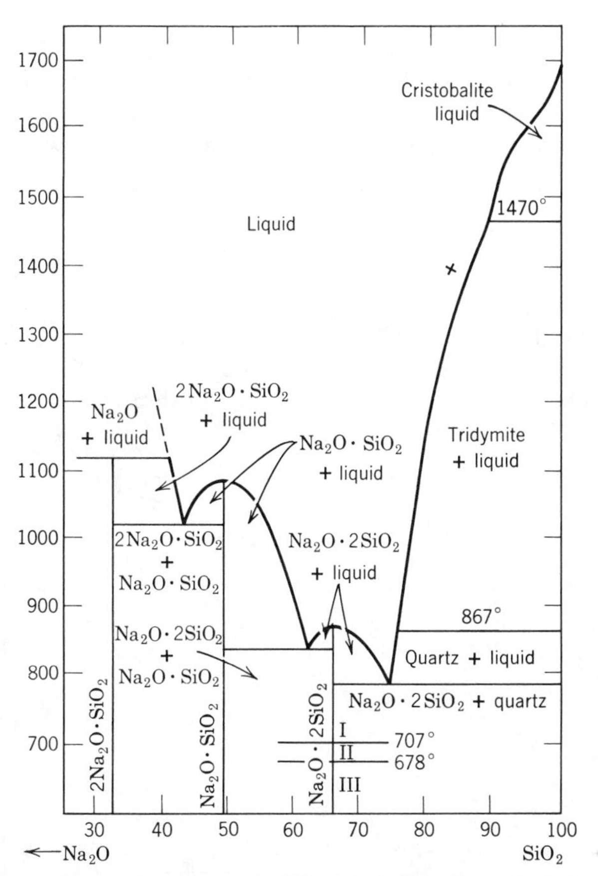

Fig. 8.21. Phase diagram for the system SiO_2–$2Na_2O{\cdot}SiO_2$. From F. C. Kracek, *J. Phys. Chem.*, **34**, 1588 (1930), and *J. Am. Chem. Soc.*, **61**, 2869 (1939).

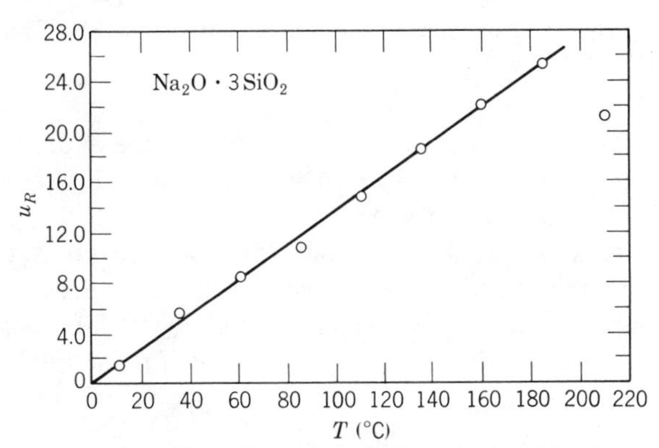

Fig. 8.22. Reduced growth rate versus undercooling relation for the crystallization of $Na_2O{\cdot}3SiO_2$ from a melt of the same composition. From G. Scherer, Sc.D. thesis, MIT, 1974.

demonstrated the existence of a K-rich boundary layer adjacent to the dendrites, which is suggestive that the growth is diffusion controlled. If the composition gradient at the interface is approximated by its average value over the thickness of the boundary layer, the growth rate under conditions of diffusion control can be approximated:*

$$u = \frac{D}{\delta}\frac{(C_0 - C_s)}{1 - C_s} \tag{8.50}$$

where D is the chemical interdiffusion coefficient, δ is the effective boundary-layer thickness adjacent to the interface, C_0 is the concentration in bulk liquid, and C_s is the equilibrium (liquidus) concentration for the temperature of growth. The dotted curve drawn in Fig. 8.23 was

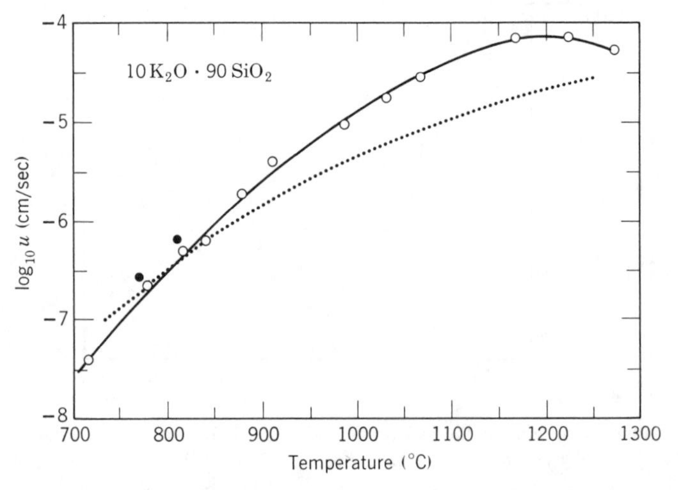

Fig. 8.23. Variation of growth rate with temperature for $0.10K_2O$–$0.90SiO_2$ composition. Solid curve = experimental data; dotted curve = predictions of Eq. 8.50. From G. Scherer, Sc.D. thesis, MIT, 1974.

constructed with a constant boundary-layer thickness of 3 microns (a boundary layer of this thickness was measured by Christensen and his coworkers for crystallization at 810°C).

The predictions of the simple model of Eq. 8.50 are seen to be in reasonable agreement with the experimental data. Improved agreement can be obtained by directly solving the problem of diffusion-controlled growth of dendrites. This can be done exactly for isolated dendrites,†

*Ibid.

†G. Horvay and J. W. Cahn, *Acta Met.*, **9**, 695 (1961).

with results which agree well with experimental data under conditions in which the growing dendrites are in fact isolated. An approximate analysis by G. Scherer* provides improved agreement with experimental results for the more common case in which the dendrites are growing in the presence of neighbors. In this case, the effects of overlapping diffusion fields from neighboring dendrites must be included in the analysis.

In most studies of crystallization from silicate liquids, it is found that the growth rate becomes negligible at a temperature about 300 to 500°C below the liquidus. For most commercial glasses, this temperature is above the annealing range, and when the glasses are held at temperatures above the transition range for long periods of time, crystallization (devitrification) takes place.

Crystalline Glazes. Glassy silicates are often used as glazes for ceramic bodies and as porcelain enamels on iron and aluminum. The compositions of some of these have been given in Table 3.4. They are usually prepared from presmelted frit plus some clay and other constituents as null additions that assist in forming a stable suspension and give good properties for smooth application. Lead glasses are particularly low-melting; additions of calcium fluoride also contribute to low-melting systems. In art ceramics it is sometimes desirable to prepare glazes having a controlled crystalline content.

Mat glazes are those in which crystals are formed during heat treatment. The crystalline products are usually anorthite, $CaO \cdot Al_2O_3 \cdot 2SiO_2$, or wollastonite, $CaO \cdot SiO_2$. These glazes characteristically have a higher oxygen-silicon ratio than the clear glazes which do not crystallize on cooling. Many can be formed as clear glazes without the precipitation of a crystal phase by overfiring and rapid cooling. For a fixed alkali and alkaline earth content, the ratio of Al_2O_3 to SiO_2 determines whether a porcelain glaze is clear (Fig. 8.24).

The liquid phase which forms at the firing temperature during vitrification of clay and triaxial porcelain compositions (clay-flint-feldspar) has a sufficiently high oxygen-silicon ratio to form a glassy phase on cooling. In contrast, steatite compositions, particularly those with barium carbonate as an added flux, form a more fluid liquid phase at the firing temperature and frequently precipitate at least some crystalline phases on cooling. There is usually a residual glassy matrix, however. Similarly, the liquid phase present during the firing of basic refractories also normally crystallizes on cooling or during the firing process.

For mat glazes the crystals should be uniform in size and relatively small. For other effects large crystals are desirable. One of the systems

**Op. cit.*

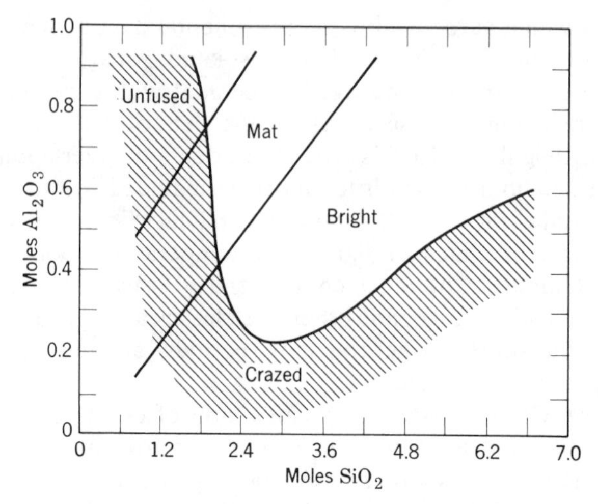

Fig. 8.24. Porcelain glaze field for constant RO ($0.3K_2O$–$0.7CaO$) and variable SiO_2 and Al_2O_3 fired at cone 11, 1285°C. From R. T. Stull and W. L. Howat, *Trans. Am. Ceram. Soc.*, **16**, 454 (1914).

found particularly effective for the development of matte-glaze compositions are glazes having a sufficient lime content for calcium silicate crystals to form.

For crystals to grow to a large size in a thin glaze layer, the glaze composition must be such that the crystals grow in a planar habit. Willemite crystals, Zn_2SiO_4, form hexagonal plates and can be grown to large sizes in thin glaze layers. In a glaze composition described by F. H. Norton* the nucleation and growth regions are sufficiently separated for large crystals to be grown. The rate of nucleation is high below the critical nucleation temperature (Fig. 8.25). In order to grow large crystals, the composition must be heated to a temperature of about 1200°C, which is above the liquidus temperature, for a sufficient time to dissolve nearly all the nuclei formed during the initial heating. Then, on cooling into the region of rapid growth, large crystals grow on nuclei which remained undissolved at special heterogeneous nucleation sites, and crystals having sizes in the centimeter range can easily be obtained.

These crystals are particularly attractive, since cobalt and other coloring additives have a tendency to concentrate in the crystalline phase rather than in the glass, so that brightly colored crystals can be grown against a clear background.

J. Am. Ceram. Soc.*, **20, 217 (1937).

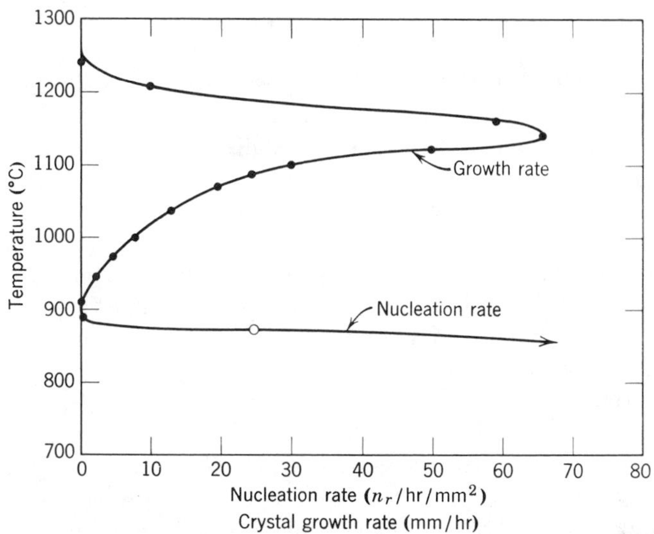

Fig. 8.25. Nucleation rate and growth rate of crystals in a glaze composition suitable for growing large decorative crystals. From F. H. Norton, *J. Am. Ceram. Soc.*, **20**, 217 (1937).

Opacified Enamels. When porcelain enamels are applied over steels, a ground coat which contains cobalt and is dark blue is often used. Sometimes no ground coat is applied. In either event, a satisfactory overcoat must have high reflectance, high opacity, and good whiteness. These properties are obtained by the precipitation of a second phase which gives rise to light scattering in a glassy matrix. In order to be effective, the particle size of the precipitate must be very small, and the difference in index of refraction between the precipitate and the glassy matrix should be large. Some of the most satisfactory of these enamels are prepared using TiO_2 for opacification. Various investigators have found that for satisfactory opacification particle sizes in the range of from 0.05 to 0.5 micron (0.1 to 1 times the wavelength range of visible light) are most suitable. The opacification results from a scattering of light by crystals of TiO_2 dispersed in a glass matrix, and the effectiveness of TiO_2 as an opacifier results from its high index of refraction and the controlled range of particle sizes. These particles are generally prepared by a process of internal nucleation and crystal growth.

Titania is quite soluble in the borosilicate liquids used in enamel application at the smelting temperature (in the range of 1100 to 1200°C). After melting, the glass is usually quenched into water; the resulting frit is combined with small amounts of suspending media and with H_2O to form the enamel slip, which is sprayed onto the ground-coated metal substrate.

The metal part is then fired at temperatures in the range of 700 to 900°C. During firing, the frit particles flow together to form a coherent glass, and because the solubility of TiO_2 in these glasses decreases significantly with decreasing temperature, there is a high supersaturation at the firing temperature, and much of the TiO_2 crystallizes as a fine-grained precipitate.

Some typical enamel compositions have been given in Table 3.4. Because of the relative positions of the nucleation and growth curves as functions of temperature, it is desirable to fire the enamel at a low temperature for which the nucleation rate is rapid and the growth rate is low in order to obtain the fine-grained dispersion of titania particles most effective as an opacifier.

For forming opal glasses it is more common to add phosphates or fluorides as the opacifying agent. Like TiO_2, sodium fluoride, which is formed in fluoride opals, has an appreciable solubility at the melting and working temperatures, but this solubility decreases rapidly as the temperature is lowered. After forming as a glass, reheating into the nucleation and growth range allows the formation of NaF nuclei at a temperature at which the growth rate is low so that a fine dispersion can be obtained. This gives the maximum light scattering, resulting in the formation of desirable optical characteristics. The high degree of supersaturation at the nucleation and growth temperature frequently produces dendritic *snowflake* crystals.

8.7 Colloidal Colors, Photosensitive Glasses, and Photochromic Glasses

The development of gold-ruby or copper-ruby glasses depends on the controlled nucleation and growth of metal particles from a glass matrix. When such colloidal colors in glasses are prepared, compounds of the coloring metal are added to the batch, and the metal initially dissolves as an ion. Also added to the batch are reducing agents such as the oxides of antimony, tin, selenium, or lead. When the glass containing these reducing agents is cooled, the gold or copper ions are reduced to neutral atoms, and a large number of metal particle nuclei may be formed. On subsequent reheating to an appropriate temperature range, growth of these nuclei takes place to a colloidal size.

In order for good colors to be formed, it is essential that the particles be present in small (colloidal) sizes and significant concentrations. This requires that a large number of nuclei be formed, both to prevent excessive particle growth and to provide the necessary concentration of scattering centers. A typical gold-ruby glass contains 0.01 to 0.1 wt% gold.

Photosensitive Glasses. Photosensitive glasses can be made by substituting an optical sensitizer (cerium ions) for the reducing ions present in normal ruby glasses. These glasses are colorless and transparent when they are first made and cooled to room temperature. Exposure to ultraviolet light or X-rays results in the absorption of photons by the sensitizing cerium ions converting the Ce^{3+} to Ce^{4+}:

$$Ce^{3+} + h\nu \rightarrow Ce^{4+} + e^{-} \tag{8.51}$$

While the glass remains at ambient temperature, the electrons are believed to be trapped close to the parent Ce ions. When the glass is subsequently heated, these electrons can migrate to nearby gold ions and convert them to gold atoms, which then aggregate to form small metallic particles. The density of these particles can be controlled, by control of the incident radiation, from very small values to values in excess of 10^{12} per cubic centimeter. A similar process takes place with glasses containing copper and silver, but in these cases the metal ions can act as their own sensitizers.

The phenomenon of photosensitivity has been used commercially in a number of applications. One of these involves the use of the small metal particles as heterogeneous nuclei for further crystal growth of other compositions. In particular, lithium silicate glasses are suitable for the precipitation of a crystalline phase, Li_2SiO_3, which is much more soluble in hydrofluoric acid than the surrounding glass. This has made it possible to prepare a glass which can be *chemically machined.* The photosensitive glass is exposed to an ultraviolet or X-ray light pattern which forms metal nuclei and is then heated to a temperature at which the lithium silicate crystals grow to form a pattern in the glassy matrix. Exposure to hydrofluoric acid dissolves out these crystalline parts of the glass and provides a method of forming desired structures, such as fine grids of holes, which cannot be obtained in other ways. With this technique, arrays of holes with densities as large as 5×10^4 per square centimeter can be produced.

Photochromic Glasses. Another development which depends on the control of crystallization in glass-forming systems is the achievement of stable photochromic materials. Such materials darken when exposed to sunlight or other radiation of appropriate wavelength ranges and regain their original color when the light is removed.

A large number of photochromic materials are known, including many organic compounds as well as some inorganic compounds of Zn, Cd, Hg, Cu, and Ag. All these materials involve atoms or molecules which can exist in two states having different molecular or electronic configurations and different absorption coefficients for light in the visible range. In their normal state the molecules have one color (or perhaps are colorless); on exposure to light or other radiation of the appropriate wavelength range, they switch to a second state in which they exhibit a second color; and on removal of the light they revert to the original state.

Although photochromic materials have a variety of applications, particularly in the opthalmic area, most suffer from fatigue when subject to repeated light-and-dark cycling. This fatigue usually results from an irreversible chemical reaction between the active species produced in the photochemical process and water or oxygen or other chemicals present in the material.

Problems associated with such fatigue can be avoided with photochromic glasses. Their photochromic properties result from small, dispersed crystals of silver halides which form during the initial cooling of the glass or on subsequent heat treatment at temperatures between the strain and softening points of the host glass. These silver halide particles may contain sizable concentrations of impurities such as alkali ions which are present in the glass, and the photochromic behavior is affected significantly by both the composition and thermal history of the glass.

The base glasses are usually alkali borosilicates; the concentration of silver is between 0.2 and 0.7 wt% for the commercially important transparent glasses and between 0.8 and 1.5% for opaque glasses. The halogen may be Cl, Br, or I or combinations thereof and is usually added in concentrations of a few tenths of a weight percent. Also added to the batch are small concentrations, of the order of 0.01 wt%, of a sensitizer, most notably CuO, which significantly enhances the sensitivity and increases the photochromic darkening. The Cu ions serve as hole traps and prevent the recombination of holes and electrons produced on exposure to the light. In typical cases, the silver halide particles may have an average diameter in the range of 100 Å and are formed with an average spacing between particles of about 500 to 800 Å, corresponding to a bulk concentration of particles between 10^{15} and 10^{16} per cubic centimeter.

The wavelengths which cause darkening depend primarily on the halide particles present. Glasses containing AgCl particles are sensitive to the violet and ultraviolet; that is, they darken when exposed to such radiation. Additions of Br or I generally shift the sensitivity toward longer wavelengths. In addition to the darkening process which occurs on exposure to sensitizing radiation, bleaching processes, both optical and thermal, also occur. The steady-state degree of coloration achieved for a given temperature and light intensity results from a competition between the darkening and bleaching processes.

The kinetics of both darkening and clearing of a given glass are generally slower than those needed for the establishment of steady-state darkening. The rate of darkening is relatively insensitive to temperature; the clearing rate increases significantly with increasing temperature. Consequently, the steady-state darkening decreases with increasing temperature. This is shown in Fig. 8.26, where it is also seen that glasses with more rapid clearing rates at a given temperature are apparently more sensitive to changes in temperature than those with slower clearing rates.

The photochromic process in these glasses seems closely similar to the photolytic dissociation of silver halide and the formation of neutral silver atoms in the familiar photographic-film process. In photochromic glasses this process is reversible, whereas in photographic emulsions it is irreversible and leads to the formation of stable silver particles. This difference in behavior results primarily from differences in the impermeability and chemical inertness of the host material (unlike the gelatine of

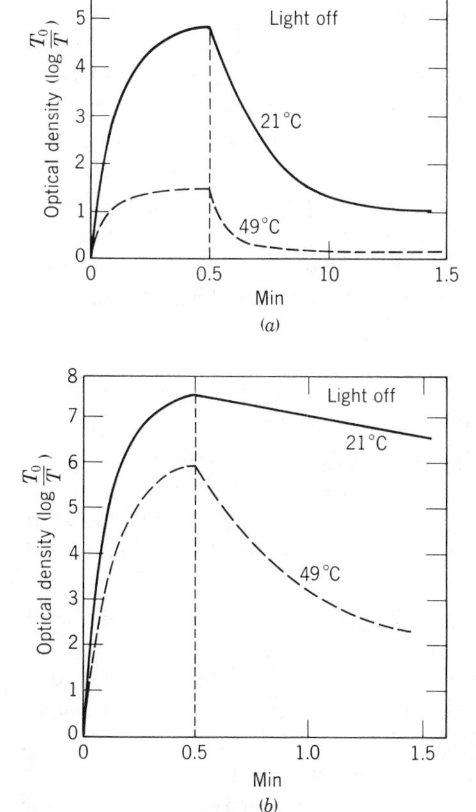

Fig. 8.26. Effect of temperature on rates of darkening and clearing in (*a*) fast-clearing glass and (*b*) more slowly clearing glass. The darkening is more sensitive to temperature in the fast-clearing glass. From W. D. Armistead and S. D. Stookey, *Science*, **144**, 150 (1964).

the emulsion, the glass prevents loss of halogen from the silver halide particles).

8.8 Glass-Ceramic Materials

Glass-ceramic materials are produced by the controlled crystallization of appropriate glasses. They consist of a large proportion, typically 95 to 98 vol%, of very small crystals, generally smaller than 1 micron, with a small amount of residual glass phase making up a porefree composite.

When these materials are fabricated, bodies of desired shapes are formed with conventional glass-forming techniques. As discussed in Section 8.3, the conventional crystallization of glasses is almost invariably observed to initiate at the external surfaces, followed by the crystals growing into the amorphous phase and producing a nonuniform body of large grain size. For a variety of reasons, it is desirable that the crystals be small (less than 1 micron) and uniform in size. To obtain such small crystals occupying a large volume fraction of the material, a uniform density of nuclei of the order of 10^{12} to 10^{15} per cubic centimeter is required. Such copious nucleation is produced by adding selected nucleating agents to the batch during the melting operation and carrying out a controlled heat treatment.

The most commonly used nucleating agents are TiO_2 and ZrO_2, but P_2O_5, the Pt group and noble metals, and fluorides are also used. TiO_2 is often used in concentrations of 4 to 12 wt%; ZrO_2 is used in concentrations near its solubility limit (4 to 5 wt% in most silicate melts). In some cases, ZrO_2 and TiO_2 are used in combination to obtain desired properties in the final crystallized bodies.

The Pt group and noble metal nucleating agents seem to function by directly forming a crystalline nucleating phase in a precipitation process. The major crystalline phase or phases subsequently grow on particles of the nucleating phase. Such a process could also be operative in the case of oxide nucleating agents, but in many cases these melt additions seem to be effective in promoting a phase-separation process. The separation can provide a fine dispersion of second-phase material, which can then form a crystalline nucleant phase.

The role of phase separation in the nucleation process has been associated with a number of factors, including (1) the formation of an amorphous phase of relatively high mobility in a temperature range in which the driving force for crystallization is large. From such a phase, crystal nucleation can occur rapidly. (2) The introduction of second-phase boundaries between the phase-separated regions on which the nucleation of the first crystalline phase may take place. (3) The provision of a driving force for crystallization in cases in which no such driving force exists for the homogeneous solution. Of these possibilities, the first seems to be the most important and the most generally applicable.

Direct experimental evidence on the precise role of oxide nucleating agents is rather meager. In one study of a TiO_2-nucleated Li_2O–Al_2O_3–SiO_2 glass-ceramic, the nucleation stage consisted of phase separation on a scale of about 50 Å followed by the formation of a crystalline TiO_2-rich nucleating phase. This phase was estimated to contain about 35 wt% Ti and about 20 wt% Al; the starting material contained less than 5 wt% TiO_2. In other similar systems, however, there is no evidence of structural heterogeneities, detectable by either electron-microscope or light-scattering observations, prior to the appearance of the crystals of the major phases. This could reflect different modes of nucleation being effective in different systems or merely indicate the difficulties of detecting small-scale heterogeneities. More generally, it should be noted that the important oxide nucleating agents are of great use in a number of systems containing sizable concentrations of SiO_2 and often significant amounts of Al_2O_3 as well. In contrast, for many other systems—for example, many phosphate-based systems—they are not at all effective. The differences seem most likely associated with differences in immiscibility behavior but might also reflect differences in lattice match of possible crystalline phases as well.

The steps used in processing a glass-ceramic body are illustrated schematically by the temperature-time cycle shown in Fig. 8.27. The material is melted and formed at elevated temperatures and then often

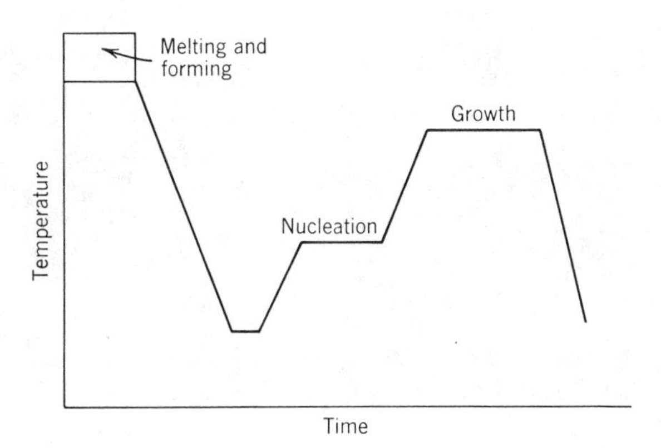

Fig. 8.27. Schematic temperature-time cycle for the controlled crystallization of a glass-ceramic body.

cooled to ambient, at which additional processing steps may be carried out. The material at this stage may be largely homogeneous, or it may contain some phase-separated domains or some very small crystals of the nucleant phase.

The sample is then heated at a rate limited by the avoidance of thermal shock to a holding temperature at which nucleation of the major phases is effected. The sample is typically held at this nucleation temperature, at which the melt viscosity is often in the range of 10^{11} to 10^{12} P, for 1 to 2 hr. The scale of formation of the initial nuclei is often in the range of 30 to 70 Å (see Fig. 8.28 as an example).

After nucleation is completed, the material is heated further to effect the growth of the major crystalline phases. The maximum temperature for growth is generally chosen to maximize the kinetics of crystal growth, subject to the constraints of obtaining the desired combination of phases and avoiding deformation of the sample or unwanted transformations within the crystalline phases or redissolution of some of the phases. This temperature and the time for which the material is held at the temperature, which can be very brief, depend on the system and composition as well as on the phases and properties desired in the final body.

In most cases the crystallization is carried out to a fraction crystallized exceeding 90% and often exceeding 98%. The final grain size is typically in the range of 0.1 to 1 micron. This is considerably smaller than the grain size of conventional ceramic bodies, illustrated by the micrographs in Chapter 11.

The volume fractions of the various phases, both crystalline and glassy, in a glass-ceramic body are determined by the composition of the initial

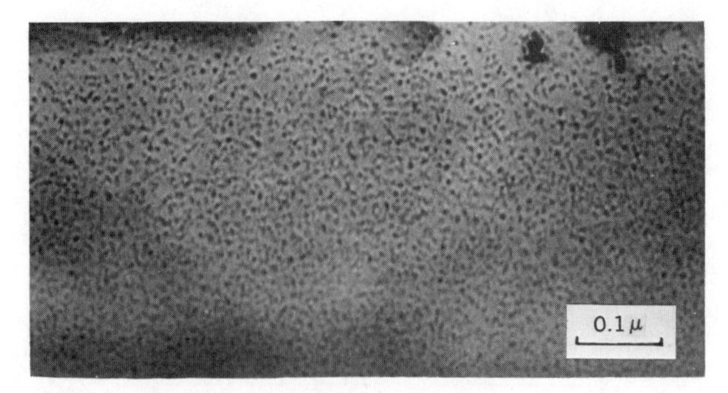

Fig. 8.28. Crystalline nuclei in Li_2O–Al_2O_3–SiO_2 composition with 4% TiO_2 addition. From G. H. Beall in L. L. Hench and S. W. Freiman, Eds., *Advances in Nucleation and Crystallization in Glasses*, American Ceramic Society, 1972, pp. 251–261.

glass, the stoichiometry of the crystalline phases, and the temperatures and times of the crystallization treatment. Metastable phases are often encountered, however, and both the final assemblage of phases and the sequence of microstructural development depend in general on details of the crystallization heat treatment and its relation to the kinetics of the nucleation and crystal growth processes.

As examples of the variation in crystal size which can be achieved by variations in the heat-treatment cycle, consider the composition studied by Doherty and his associates:* SiO_2, 70 wt%; Al_2O_3, 18%; MgO, 3%; Li_2O, 3%; and TiO_2, 5%. The TiO_2-rich nuclei, estimated to contain about 35 wt% Ti, begin to form at temperatures about 725°C; their rate of formation reaches a maximum between 800 and 825°C and becomes quite small again at temperatures in the range of 850°C. The major crystalline phase, *β-eucryptite,* forms on the TiO_2-rich nuclei and subsequently transforms to *β-spodumene* at temperatures above 1000°C. The growth rate of the eucryptite crystals becomes significant at temperatures about 825°C and increases with increasing temperature over a range above this. This knowledge of the kinetics of nucleation and growth can be used to control the microstructure by controlling the heat treatment. Examples of such control are given in Figs. 8.29 and 8.30. Figure 8.29 shows a sample which was heated rapidly to 875°C and held for 25 min. Because it was heated rapidly through the region of rapid nucleation, relatively few nuclei formed, and large eucryptite crystals, in the range of several microns, result. The same composition held at 775°C for 2 hr and then heated to 975°C for 2 min is characterized by an appreciably smaller crystal size, in the range of 0.1 micron (Fig. 8.30).

Important Glass-Ceramic Systems. Among the systems in which technologically important glass-ceramic materials have been produced, the following seem most noteworthy:

(1) Li_2O–Al_2O_3–SiO_2. This system is by far the most important commercial system. It is used for glass-ceramic materials having very low thermal expansion coefficients and hence very high resistance to thermal shock. Among the trade names for materials in this system and Corning's Corning Ware, Owens-Illinois' Cer-Vit, and PPG's Hercuvit. The very low expansion coefficients in this system, which in some cases are appreciably lower than that of fused silica, are associated with the presence in the crystallized materials of crystalline β-spodumene ($Li_2O \cdot Al_2O_3 \cdot 4SiO$), which has a low expansion coefficient, or β-eucryptite ($Li_2O \cdot Al_2O \cdot 2SiO_2$), which has an expansion coefficient that is larger in

*P. E. Doherty in R. M. Fulrath and J. A. Pask, Eds., *Ceramic Microstructures,* John Wiley & Sons, Inc., New York, 1968.

Fig. 8.29. Submicrostructure in Li_2O–Al_2O_3–SiO_2 glass-ceramic heated rapidly to 875°C and held for 25 min. From P. E. Doherty in R. M. Fulrath and J. A. Pask, Eds., *Ceramic Microstructures*, John Wiley & Sons, Inc., New York, 1968, pp. 161–185.

magnitude and negative. Commercial compositions are found in several ranges in this system. Among them are the following (in weight percent): Li_2O (2.6), Al_2O_3 (18), SiO_2 (70), and TiO_2 (4.5). Different phase assemblages, characterized by different sets of properties, can result from using TiO_2 and ZrO_2 in varying proportions as the nucleating agent.

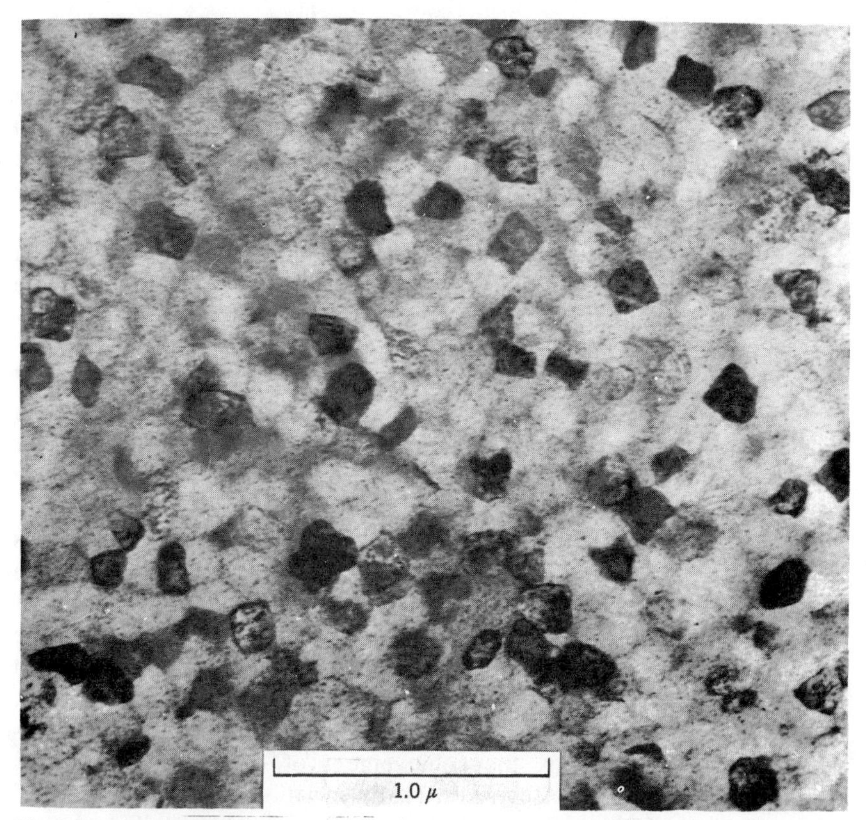

Fig. 8.30. Submicrostructure in Li_2O–Al_2O_3–SiO_2 glass-ceramic, having identical composition with that shown in Fig. 8.29, held at 775°C for 2 hr before heating to 975°C for 2 min. From P. E. Doherty in R. M. Fulrath and J. A. Pask, Eds., *Ceramic Microstructures*, John Wiley & Sons, Inc., New York, 1968, pp. 161–185.

(2) MgO–Al_2O_3–SiO_2. This system forms glass-ceramic materials having high electrical resistivity and high mechanical strength. The high strength has been associated with the presence in the crystallized materials of crystalline β-cordierite ($2MgO \cdot 2Al_2O_3 \cdot 5SiO_2$). The compositions of some useful glass-ceramic materials in this system cover a range about MgO (13), Al_2O_3 (30), SiO_2 (47), and TiO_2 (10).

(3) Na_2O–BaO–Al_2O_3–SiO_2. Commercial glass-ceramic materials in this system have thermal expansion coefficients in the range of 80×10^{-7} per degree centigrade and cover a range of composition about Na_2O (13), BaO (9), Al_2O_3 (29), SiO_2 (41), and TiO_2 (7). The important crystalline phases are nepheline ($Na_2O \cdot Al_2O_3 \cdot 2SiO_2$) and celsian ($BaO \cdot Al_2O_3 \cdot 2SiO_2$). In practice, the commercial products based on this system, most notably

Corning's Centura Ware, use a compressive glaze to achieve desired mechanical strengths. The use of such glazes is discussed in Chapter 16.

(4) Li_2O–MgO–Al_2O_3–SiO_2. Glass-ceramic materials in this system are noted for their variable, in some cases low or negative, thermal expansion coefficients, transparency (in some cases), and the ease with which they can be chemically strengthened. The important crystalline phase is a stuffed β-quartz solid solution.

(5) K_2O–MgO–Al_2O_3–SiO_2. Glass-ceramic materials in this system are melted with small concentrations of fluoride and are noted for the quality of machinability. This characteristic is associated with the presence of mica phases having large aspect ratios in the crystallized bodies.

Properties of Glass-Ceramic Materials. The principal advantages of glass-ceramic materials over conventional ceramics are associated with the economy and precision of the forming operations, with the absence of porosity in the materials, and with the occurrence of well-dispersed, very small crystals having desirable properties in the crystallized bodies. The absence of porosity is related to the relatively small volume changes involved in crystallizing these systems and to the fact that changes in volume can be accommodated by flow. The small crystal size is a result of copious nucleation achieved by adding the nucleant to the melt. The particular material properties can to a significant extent be programmed by the selection of a suitable composition and crystallization treatment. This last factor constitutes perhaps the essential feature of the glass-ceramic concept, namely, the achievement of desired properties or combinations of properties by a systematic variation of the chemistry and microstructure of materials containing both glassy and crystalline phases. This variation in turn is effected by means of a controlled crystallization treatment; its systematic feature is made possible by the continuous and wide-ranging variation of phase assemblages which can be achieved in glass-and-ceramic bodies (in contrast to the more restricted range possible with crystalline ceramics).

In some cases, a glass-ceramic material is not designed for its properties in the as-crystallized state but rather for the ease with which postcrystallization treatments can be effected. An example is the development of glass-ceramic materials whose surfaces can easily be strengthened by cladding or by an ion-exchange process. The latter development can be particularly complex, since the ion-exchange treatment of a glass-ceramic body includes not only the direct strengthening of the glass phase or phases by a stuffing process, as discussed in Chapter 16, but also more important effects which involve the crystalline phases as well: phase transformations to phases of different volumes and expansion coefficients, and solid solution in the already existing crystalline phases, changing their volumes and expansion coefficients.

8.9 Phase Separation in Glasses

As discussed in Chapter 3, phase separation is a widespread phenomenon in glass-forming systems. Of particular interest has been the origin of the interconnected structures observed in many systems in which both phases occupy large volume fractions. Before Cahn's calculations based on the linearized theory of spinodal decomposition, which indicated an interconnected structure (see Fig. 8.4), phase-separation phenomena were customarily explained on the basis of nucleation and growth processes. For example, in accounting for the two-phase leachable submicrostructures in the alkali borosilicates and other glasses, the continuous second phase was assumed to arise from a high concentration of second-phase droplets which grew rapidly and linked with one another to form a continuous network.

This view has received support from calculations in which a random array of point nuclei begin simultaneously to grow at a constant rate until they (the minor phase) occupy a given volume fraction of the material. At 50% occupied volume fraction, for example, it was found that each sphere intersects on average 5.5 other spheres and only 0.4% of the spheres are unattached. A calculated cross section through such an assemblage of spheres is shown in Fig. 8.31. A high degree of connectivity can be seen in the figure, and allowing for the smoothing effect of surface tension in the regions of the necks between particles, this morphology closely resembles those seen in many phase-separated glasses.

Overlapping diffusion fields in the region of contact between two growing particles should, however, reduce the diffusive flux to the region and curtail further growth. When surface-tension effects are neglected, such interparticle interference effects can completely prevent coalescence (two spheres growing by a diffusion-controlled process should not coalesce but merely flatten in the region of their mutual approach). The critical parameter in such interference is the ratio of particle separation to

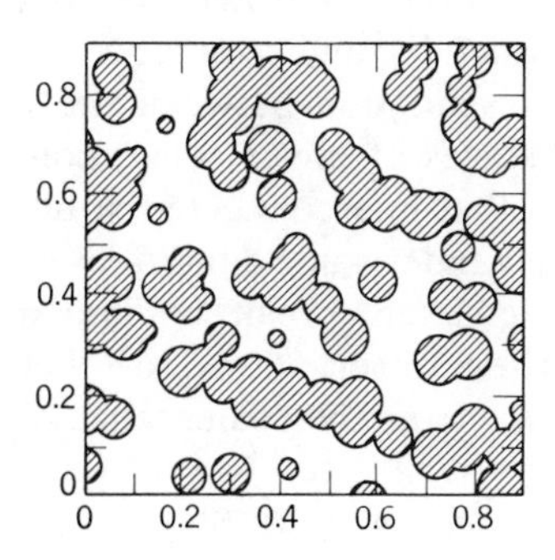

Fig. 8.31. Cross section through three-dimensional array of equal spheres (radius $r = 0.05$). From W. Haller, *J. Chem. Phys.*, **42**, 686 (1965).

radius, with large particles being affected by interference at relatively large separation.

A number of mechanisms have been considered which might permit particle coalescence despite the interference effects.* These include (1) enhanced diffusion to the flattened region of approach, driven by capillarity, either through or external to the particles, (2) nucleation of a neck joining the two particles, and (3) diffusion in the region between the particles driven by a surface energy which decreases with decreasing interparticle separation. Each of these mechanisms can lead to coalescence in reasonable times, but each should be effective only when the particles have approached within some small separation (of the order of 10 Å). Such close approach may result simply from diffusion-controlled growth of the particles, but in many cases the growth process can be assisted in this regard by Brownian motion of the second-phase particles. This motion is only likely to be significant when copious nucleation leads to many small second-phase particles (30 to 100 Å in diameter) occupying a large volume fraction in fluid systems.

Interconnected submicrostructures in the later stages of phase transformation, such as those shown in Chapter 3, can therefore result either from spinodal decomposition or from the coalescence of discrete particles. Direct experimental evidence that interconnected structures can result from the formation, growth, and coalescence of discrete second-phase particles has been provided by an electron-microscope study of phase separation in the $BaO-SiO_2$ system. Several stages in this development are shown in Fig. 8.32. A similar coalescence process is implied by data on $Al_2O_3-SiO_2$ glasses, in which discrete second-phase particles were seen in the parts of a sample which were most rapidly quenched, whereas an interconnected submicrostructure was noted in the less rapidly quenched areas.

The significant feature of these results is the observation of coalescence as the origin of interconnectivity. The discrete particles seen prior to coalescence could result either from a nucleation and growth process or from a spinodal decomposition process in which the higher-order terms in the diffusion equation are important. Differentiation between these possibilities requires data on the very early stages of phase separation. Other data on $BaO-SiO_2$ glasses have shown that interconnected structures coarsen and remain highly interconnected under some conditions of annealing but under other conditions they coarsen, break up into a discrete particle structure, and spheroidize. These observations, together with those of interconnected structures resulting from the coalescence of

*R. W. Hopper and D. R. Uhlmann, *Discuss. Faraday Soc.*, **50**, 166 (1971).

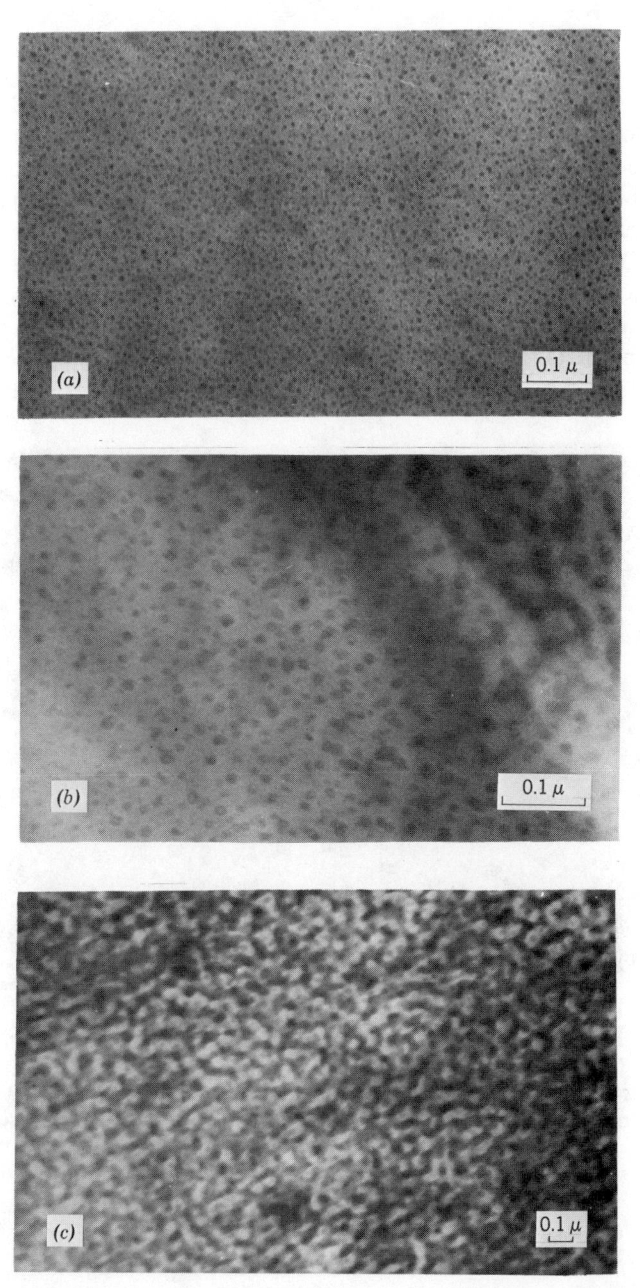

Fig. 8.32. Stages in the development of interconnected submicrostructures 0.12BaO–0.88SiO_2 thin films beam-heated in the electron microscope. (*a*) Early stage of separation showing isolated, discrete particles; (*b*) intermediate stage of separation showing coalescence; (*c*) final stage of separation showing high degree of interconnectivity. From T. P. Seward, D. R. Uhlmann, and D. Turnbull, *J. Am. Ceram. Soc.*, **51**, 634 (1968).

discrete particles, emphasize the importance of obtaining data on the chronology of microstructural development and not inferring mechanisms of phase separation from morphological observations of the later stages.

Another striking example of thermal-history effects on phase separation is provided by a study of a borosilicate glass containing 67.4 SiO_2, 25.7 B_2O_3, and 6.9 wt% Na_2O.* A sample of this glass which had been cooled to ambient and heat-treated for 3 hr at 750°C exhibited pronounced interconnectivity (Fig. 8.33*a*); the same composition cooled directly to 750°C and held for 3 hr exhibited a much smaller degree of connectivity (Fig. 8.33*b*). The difference in behavior probably reflects a difference in

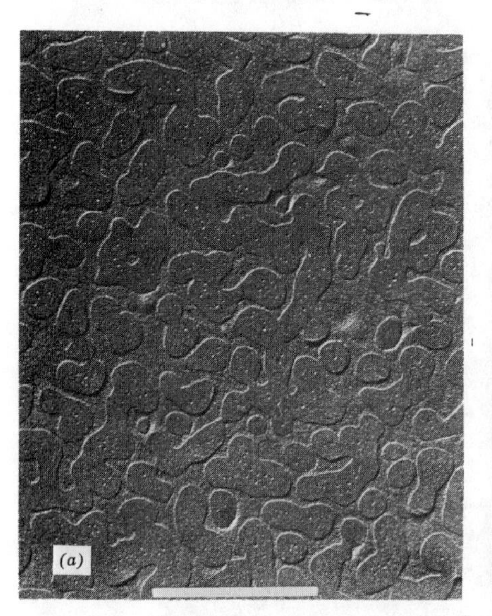

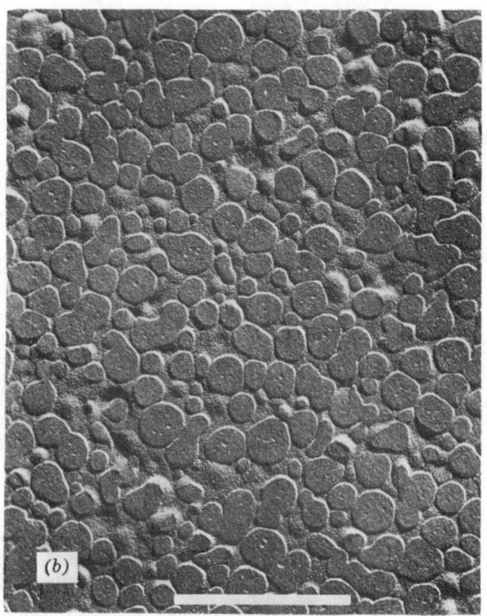

Fig. 8.33. (*a*) Submicrostructure in glass of composition 67.4SiO_2, 25.7B_2O_3, and 6.9 wt% Na_2O which has been cooled to ambient and heat-treated for 3 hr at 750°C. Note the pronounced interconnectivity. (*b*) Same composition cooled directly from the melt to 750°C and heat-treated for 3 hr. Note the smaller degree of interconnectivity. From T. H. Elmer, M. E. Nordberg, G. B. Carrier, and E. J. Korda, *J. Am. Ceram. Soc.*, **53**, 171 (1970).

the density of second-phase particles formed at 750°C and at various temperatures during cooling to ambient, as well as the increased likelihood of coalescence for the case of small particles occupying a large volume fraction. From the published electron micrographs, however, the structural differences could also reflect different mechanisms of initial separation at different temperatures.

*T. H. Elmer, M. E. Nordberg, G. B. Carrier, and E. J. Korda, *J. Am. Ceram. Soc.*, **53**, 171 (1970).

Suggested Reading

1. D. Turnbull in *Solid State Physics*, Vol. 3, Academic Press, New York, 1956.
2. K. A. Jackson in *Progress in Solid State Chemistry*, Vol. 3, Pergamon Press, New York, 1967.
3. D. R. Uhlmann in *Advances in Nucleation and Crystallization in Glasses*, American Ceramic Society, Columbus, 1972.
4. R. J. Araujo in *Photochromism*, Wiley, New York, 1971.
5. P. W. McMillan, *Glass Ceramics*, Academic Press, New York, 1964.
6. J. W. Cahn, *Trans. Met. Soc. AIME*, **242**, 166 (1968).
7. J. E. Hilliard in *Phase Transformations*, American Society for Metals, Metals Park, 1970.
8. J. W. Christian, *The Theory of Transformations in Metals and Alloys*, Pergamon Press, New York, 1965.
9. R. D. Shannon and A. L. Friedberg, *Univ. Ill. Eng. Exp. Sta. Bull.*, **456** (1960).

Problems

8.1. Preferential nucleation of crystals at external surfaces is observed for the crystallization of most glass-forming liquids. It is also observed that melting occurs heterogeneously at free surfaces at negligible departures from equilibrium. Combine these two pieces of information with your knowledge of thermodynamics to comment on the source of crystal nucleation (e.g., is it associated with the free surfaces of the liquids themselves?).

8.2. Discuss the effects of the phase transformations exhibited by SiO_2 which affect the processing and the resulting properties or use limitations for:
(*a*) SiO_2 brick.
(*b*) A conventional porcelain.

8.3. Compare and contrast the two processes of phase transformation, spinodal decomposition, and homogeneous nucleation and growth. Discuss thermodynamic and kinetic aspects and the influence of undercooling and time on the resulting structures. How could they be distinguished experimentally?

8.4. Derive a relationship for the rate of homogeneous nucleation of crystals from a glassy matrix near the liquidus temperature. How would this differ from heterogeneous nucleation? What additional factors would have to be included to calculate the nucleation rate from glasses at room temperature? Would this latter calculation be appropriate to samples of glass uncovered in an archeological investigation? Explain.

8.5. The viscosity of liquid glapium oxide varies with temperature as:

T(°C)	(*poise*)
1400	10^2
1200	10^3
1000	10^5

The melting point of this material is about 1300°C, and its entropy of fusion is about 2 cal/mole°C. Consider that over the indicated range of undercooling the principal source of crystal nuclei is condensed impurities, present in concentrations of $10^6/cm^3$, which may be assumed effective at undercoolings greater than 100°C. For how long could a 1-cm^3 sample be held for processing at 1000°C without sensible bulk crystallization?

8.6. Discuss the variation of viscosity, density, melting temperature, and tendency for devitrification of sodium silicate glasses as a function of sodium content. Do the same for viscosity and density variations of sodium borates. (Be sure to explain the reason for different behavior of silicates and borates.)

8.7. Read D. Turnbull, *J. Chem. Phys.*, **18**, 768 (1950), and comment on his method for determination of the crystal-liquid surface energy of mercury.

8.8. If the measurable nucleation rate $I_v = 10^{-1}$/droplet-sec is observed for 20-μm diameter droplets, calculate the crystal-liquid surface energy for germanium if it can be undercooled by 227°C:

$$T_{mp} = 1231°\text{K}, \qquad \Delta H_f = 8.3\ \text{kcal/mole}, \qquad \rho = 5.35\ \text{gm/cm}^3$$

8.9. The AO–BO system has a miscibility gap at 1000°K that ranges from 4 to 98 m/o BO. What is the free-energy change for complete precipitation of a mole of 6% material? What is the free-energy change per mole of precipitate formed? What is the free-energy change per mole of precipitate during the first stage of precipitation? What pressure on the precipitate (but not on the matrix) is required to prevent the initial precipitate from growing if the molar volume is 10 cc?

9

Reactions with and between Solids

In heterogeneous reactions there is a reaction interface between the reacting phases, such as nucleus and matrix or crystal and melt. In order for the reaction to proceed, three steps must take place in series—material transport to the interface, reaction at the phase boundary, and sometimes transport of reaction products away from the interface. In addition, reactions at the phase boundary liberate or absorb heat, changing the boundary temperature and limiting the rate of the process. Any of these steps may determine the overall rate at which a heterogeneous reaction takes place, since the overall reaction rate is determined by the slowest of these series steps.

In this chapter we consider these rate-determining steps as applied to changes taking place in ceramic systems. Decomposition of hydrates and carbonates, solid-state reactions, oxidation, corrosion, and many other phenomena must be considered on the basis of limitations imposed by the rates of phase-boundary reactions, material transport, and heat flow.

9.1 Kinetics of Heterogeneous Reactions

Reaction Order. Classical chemical-reaction kinetics has been mainly concerned with homogeneous reactions and cannot be directly applied to many phenomena of particular interest in ceramics, but it provides the basis for understanding rate phenomena. Reaction rates are frequently classified as to molecularity—the number of molecules or atoms formally taking part in the reaction. Overall reactions are also commonly classified as to reaction order—the sum of the powers to which concentrations c_1, c_2, and so on, must be raised to give empirical agreement with a rate equation of the form

$$\frac{-dc}{dt} = Kc_1^{\alpha}c_2^{\beta}c_3^{\gamma} \cdots \tag{9.1}$$

In a first-order reaction, for example,

$$\frac{-dc}{dt} = Kc \tag{9.2}$$

On integration this gives

$$\ln \frac{c}{c_0} = K(t - t_0) \tag{9.3}$$

where K is a constant and c_0 is the initial concentration at time t_0. For the simplest overall reactions which involve but one elementary step, the order and molecularity are the same. For more complex reactions which consist of several consecutive elementary steps involving different species and for heterogeneous reactions in general, the molecularity and order are quite different, and characterization by reaction order is a purely formal empirical method. In fact, zero and fractional reaction orders are sometimes found. Although the reaction-order concept is useful as a means of representing data for heterogeneous reactions, these cannot usually be interpreted simply in terms of molecular interactions.

Activation Energy and Reaction Rate. The effect of temperature on the rate of processes taking place is frequently great. The historical basis for its understanding is the Arrhenius equation, in which it was found that for many processes the specific reaction-rate constant could be related to temperature by the relation $\log K \sim 1/T$ or, alternatively, $K = A \exp(-Q/RT)$, where Q is the experimental activation energy. The basis of this relationship in a general theory of rate processes has been discussed in Chapter 6 in connection with diffusion as an activated process. In general, an activation energy is required for each of the steps involved in an overall rate process (Fig. 9.1).

Two general considerations are the basis for interpreting most kinetic data. The first of these is that each individual step in a rate process must be relatively simple and that the reaction path of each step, such as an individual atom jump in diffusion, a molecular decomposition, or a new chemical bond being formed, involves an *activated complex* or *transition state* of maximum energy along the reaction path. Of all possible parallel paths of reaction, the one with the lowest energy barrier is the most rapid and the major contributor to the overall process. This activated-complex theory has provided a general form of equation for rate processes and a model that allows semiempirical calculations for simple processes. The second general consideration has been that the overall rate of a complex process involving a series of consecutive steps is fixed by the rate of the slowest individual step.

If we plot energy along a distance coordinate corresponding to the

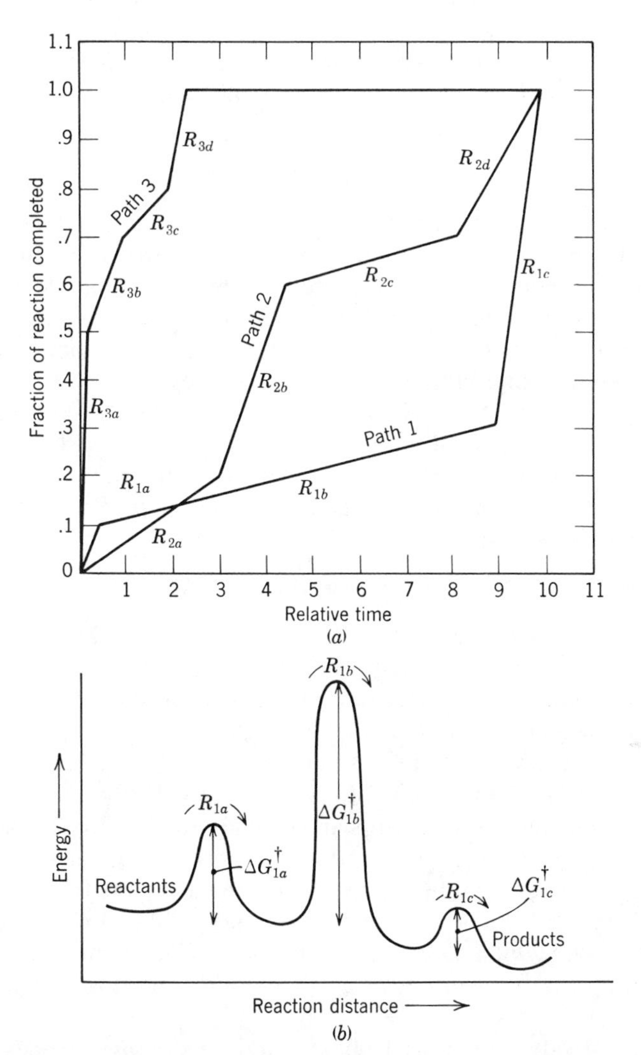

Fig. 9.1. Schematic representation of (*a*) multipath process in which each path contains several steps; the process is dominated by the fastest path (path 3). (*b*) Multistep path in which each step has an activation energy; the overall rate along this path is determined by the slowest step.

reaction path of lowest energy between the reactant and product, there is an energy maximum, actually a saddle point, corresponding to the activated complex or transition state, such as discussed for diffusion in Chapter 6. This concept of an activated complex has been generally accepted as the basis for reaction-rate studies and, as discussed in Chapter 6, leads to a specific reaction-rate constant given by

$$\mathbf{k} = \frac{kT}{h} \exp\left(-\frac{\Delta H^{\dagger}}{RT}\right) \exp \frac{\Delta S^{\dagger}}{R} \tag{9.4}$$

where k is the Boltzmann constant, h the Planck constant, and $\Delta H^{\dagger}$ and $\Delta S^{\dagger}$ the enthalpy and entropy of activation, respectively. Individual reaction steps in an overall reaction process are usually simple and may be designated as monomolecular or bimolecular. Semiempirical treatments of the unit steps on the basis of activated-complex theory allow a rational theoretical approach to reaction processes.

Complex Processes. Overall processes are frequently complex and require a series of individual separate unit steps. In such a sequence the rate of any individual step depends on the specific reaction-rate constant and the concentration of the reactants for this step. For a series of consecutive steps,

$$A_1 = A_2 = \cdots A_i = A_{i+1} = \cdots A_n \tag{9.5}$$

We can define a *virtual maximum rate* for each step as the rate that would be found if equilibria were established for all previous and following steps. Under these conditions the reaction with the lowest virtual maximum rate controls the overall rate if it is much lower than the rates of other steps. Under these conditions equilibrium will have been virtually established for all previous steps but will not necessarily be established for the following steps. As shown schematically in Fig. 9.1, the reaction rate for path 1 is determined by step $1b$ of the process; it has the slowest rate and the largest activation energy barrier and accounts for 85% of the reaction time; steps $1a$ and $1c$ occur more rapidly. Reaction step R_{1a} will be slowed to a rate giving a virtual equilibrium concentration of products; reaction R_{1c} will be slowed because R_{1b} is producing few reactants for step R_{1c}.

We have already noted that most condensed-phase processes of interest in ceramics involve heterogeneous systems; changes take place at a phase boundary. The overall process involves (1) transport of reactants to the phase boundary, (2) reaction at the phase boundary, (3) transport of products away from the phase boundary. This series of reaction steps has relatively simple kinetics, provided the virtual maximum rate of one step is much slower than that of any of the other steps. If we assume this to be the case, we have two general classes of heterogeneous reactions: (1) those controlled by transport rate and (2) those controlled by phase-boundary reaction rate. In general, both the transport process and the phase-boundary process involve a number of individual steps, one of which has the lowest virtual maximum rate. In going from reactants to products, there may be several possible reaction paths for transport processes and for phase-boundary reactions. There are three different possible reaction paths shown in Fig. 9-1*a*.

9.2 Reactant Transport through a Planar Boundary Layer

Slip Casting. As an example of the usefulness of determining the rate-limiting step for deriving kinetic equations, we begin with the ceramic processing technique of slip casting, in which a slurry containing clay particles dispersed in water is poured into a plaster of paris (gypsum) mold which contains fine capillaries (see Fig. 11.36) that absorb water from the slip. This causes a compact layer of clay particles to form at the mold-slip interface (Fig. 9.2). The rate of the process is determined by the

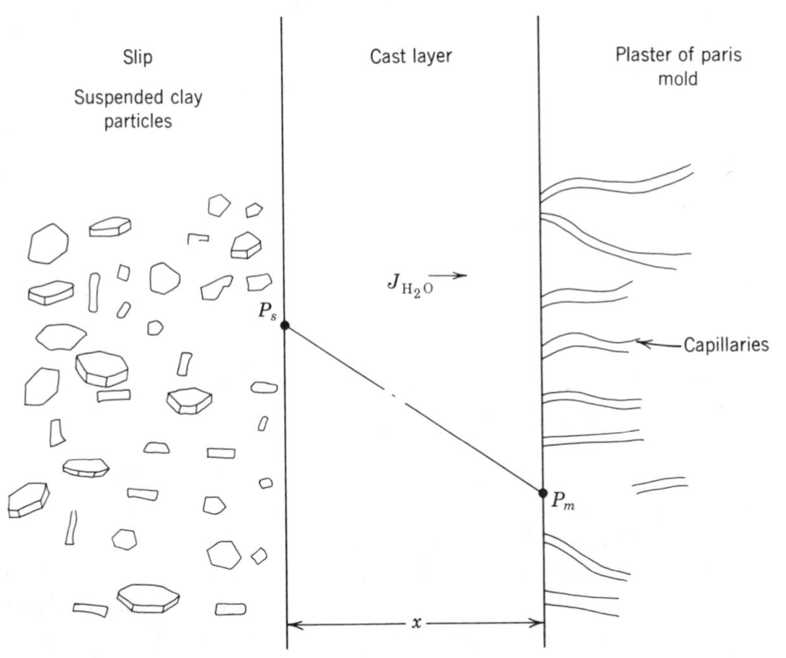

Fig. 9.2. Schematic representation of the formation of a slip-cast layer formed by the extraction of water by capillary action of a plaster of paris mold.

transport of water out of the slip and into the capillaries; the rate-limiting step is the flow of water through the compact clay layer. As the layer thickness increases, the overall rate of material transport decreases because of the increased permeation distance (similar to gas permeation through glasses, discussed in Chapter 6).

We begin by writing the flux equation for water,

$$J = -K\frac{dP}{dx} \tag{9.6}$$

where we assume a planar deposit (unidirectional flow) and that the water flux J_{H_2O} is proportional to the pressure gradient resulting from the capillaries of the plaster mold. The permeation coefficient K depends on the clay particle size, particle packing, and the viscosity of water and is temperature-sensitive. The water pressure in the slip, P_s, is 1 atm; in the mold, P_m, it is determined by capillarity, $\Delta P = P_s - P_m \approx 2\gamma/r$ (Chapter 5). The surface tension is a function of the deflocculating agents used. Until the capillaries become filled with water, ΔP is approximately constant, and the flux can be related to the change of the layer thickness dx/dt,

$$J = \left(\frac{1}{\kappa\rho}\right)\frac{dx}{dt} = -K\frac{dP}{dx} = -K\frac{\Delta P}{x} \cong -K\frac{2\gamma}{rx} \tag{9.7}$$

where ρ is the density of the cast layer and κ is a factor for converting the volume of water removed to the volume of clay particles deposited. Integration of Eq. 9.7 gives

$$x = \left(2K\rho\kappa\frac{2\gamma}{r}\right)^{1/2} t^{1/2}$$

or in the general parabolic form

$$x = (K't)^{1/2} \tag{9.8}$$

That is, the wall thickness of a planar casting should increase with the square root of time (Fig. 9.3).

This parabolic rate law is commonly observed for kinetic processes in which the limiting step is mass transport through a reaction layer.

Interdiffusion between Solids. In Section 6.3, we discussed the chemical diffusion coefficient and its formulation in terms of the tracer diffusion coefficients for the case of interdiffusion. If we measure the rate at which two ceramics interdiffuse, this too can be considered the formation of a reaction product which is a solid solution rather than a distinct or separate phase. Let us consider the interdiffusion between crystals of NiO and CoO at a high temperature. The solid solution that forms is nearly ideal;

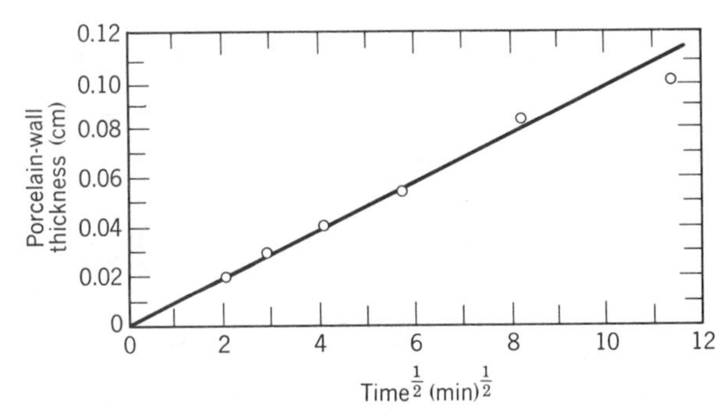

Fig. 9.3. Parabolic dependence of the slip-casting rate of a porcelain slip in a plaster of paris mold.

thus the chemical potential is related directly to the composition (concentration) by $\mu_i = \mu_i^\circ + RT \ln \gamma c_i$ where the activity coefficient γ is equal to one. Thus Eq. 6.41 becomes

$$\tilde{D} = [D_1{}^T X_2 + D_2{}^T X_1]\left(1 + \frac{d \ln \gamma_1}{d \ln X_1}\right) = D_{\text{Co}}{}^T X_{\text{Co}} + D_{\text{Ni}}{}^T (1 - X_{\text{Co}}) \quad (9.9)$$

This is the familiar Darken equation and assumes local equilibrium everywhere in the interdiffusion zone and is not strictly valid in ceramics. As will be seen later, ambipolar coupling will decrease the value of $\tilde{D}$ through electrochemical fields which arise if one charged specie moves faster than another.

The measured interdiffusion data for the CoO–NiO system are shown in Fig. 9.4*a*. The curves on the plot were calculated from the tracer diffusivities (Fig. 9.4*b*) and Eq. 9.9, assuming ideal solution behavior. In the case of interdiffusion of NiO–MgO, Eq. 9.9 is not directly applicable because the tracer diffusivities are a function of the nickel concentration. The experimental interdiffusion coefficient (Fig. 9.5*a*) has an exponential dependence on the concentration of nickel. Trivalent nickel ions and cation vacancies become associated (see Sections 4.7 and 6.4) and increase the transport rate of nickel into MgO. The measurements were made in air, so that there is sufficient trivalent nickel to dominate the cation vacancy formation process. That is, most of the vacancies arise from the presence of $Ni_{Ni}^{\cdot}$. As discussed in Sections 4.7 and 6.4, a significant fraction of the trivalent nickel ions and cation vacancies are coupled by association. Data for some other systems are shown in Fig. 9.5*b*.

Next let us consider a reaction in which a compound is formed as the reaction layer, for example, the formation of nickel aluminate spinel ($NiAl_2O_4$) from NiO and Al_2O_3. There are many possible reaction paths; five are shown schematically in Fig. 9.6. The rate of spinel formation

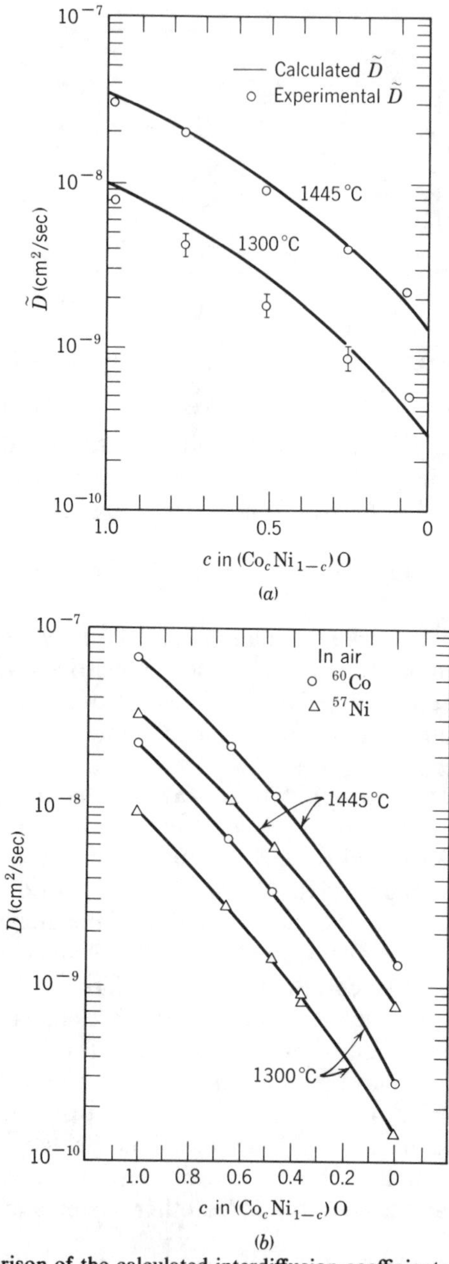

Fig. 9.4. (*a*) Comparison of the calculated interdiffusion coefficients $\tilde{D}$ and experimental values at 1445 and 1300°C in air. (*b*) Tracer diffusion coefficients of ^{60}Co and ^{57}Ni in $(Co_cNi_{1-c})O$ crystals at 1445 and 1300°C in air. Plotted as log D versus c. From W. K. Chen and N. L. Petersen, *J. Phys. Chem. Soc.*, **34**, 1093 (1973).

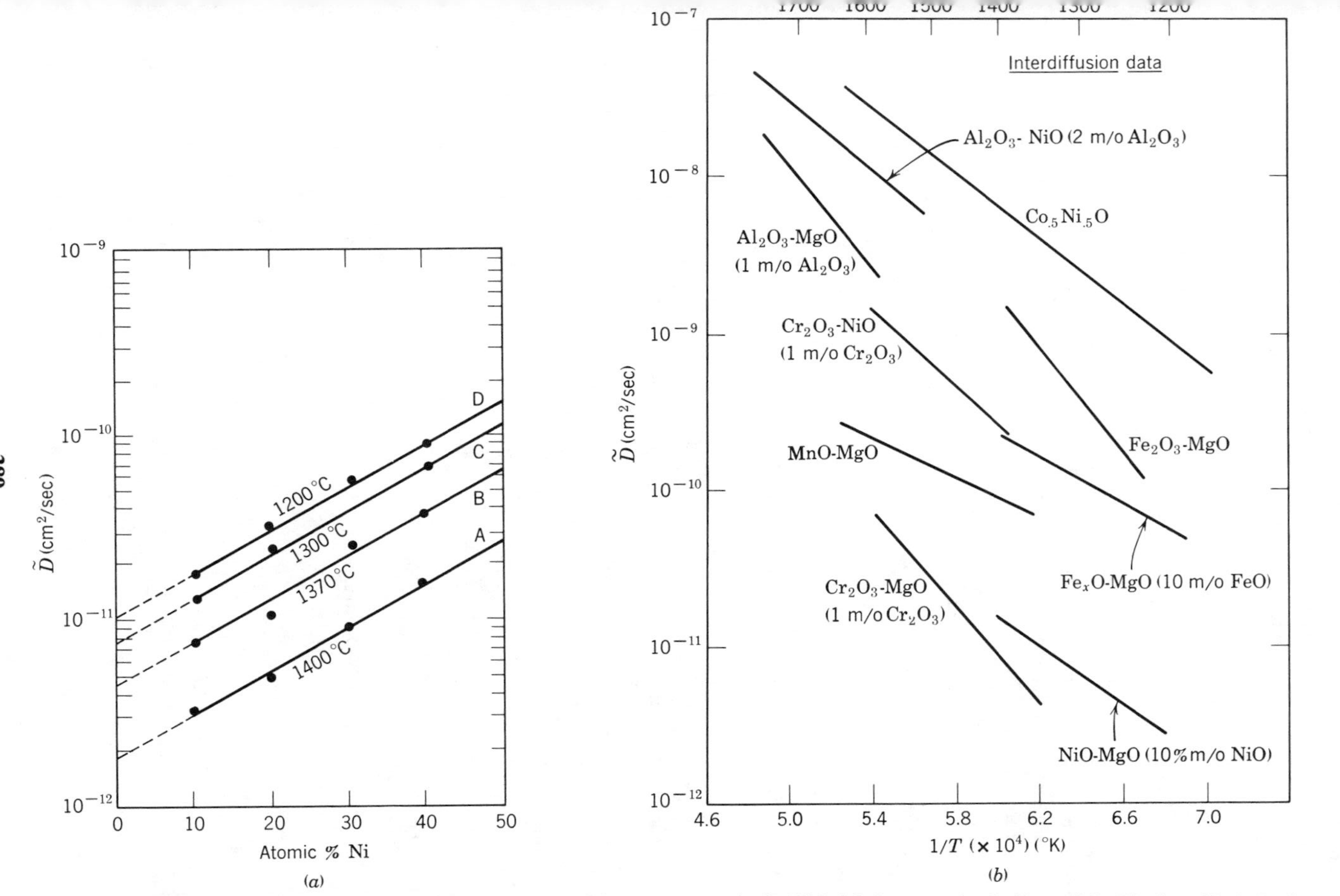

Fig. 9.5. (*a*) Diffusivity versus nickel concentration at several temperatures in the NiO–MgO system in air. From S. L. Blank and J. A. Pask, *J. Am. Ceram. Soc.*, **52**, 669 (1969). (*b*) Interdiffusion coefficients in several oxides for specific compositions.

Reaction occurs at AB_2O_4–B_2O_3 interface: oxygen gas phase transport with A^{2+} ion and electron transport through AB_2O_4:

$$A^{2+} + 2e^- + \frac{1}{2}O_2 + B_2O_3 = AB_2O_4$$

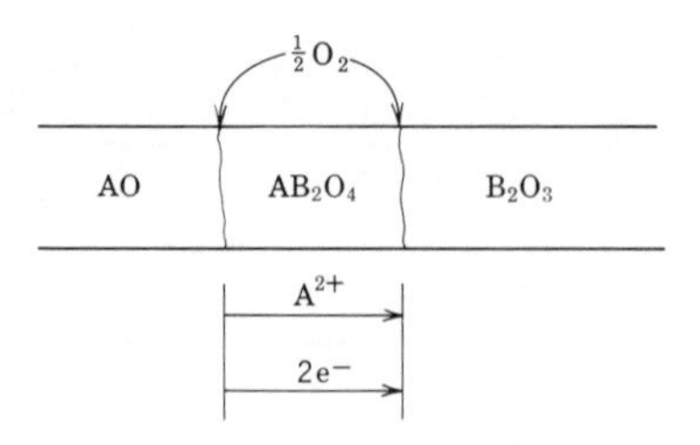

Reaction occurs at AO–AB_2O_4 interface: oxygen gas phase transport with B^{3+} ion and electron transport through AB_2O_4:

$$AO + 2B^{3+} + 6e^- + \frac{3}{2}O_2 = AB_2O_4$$

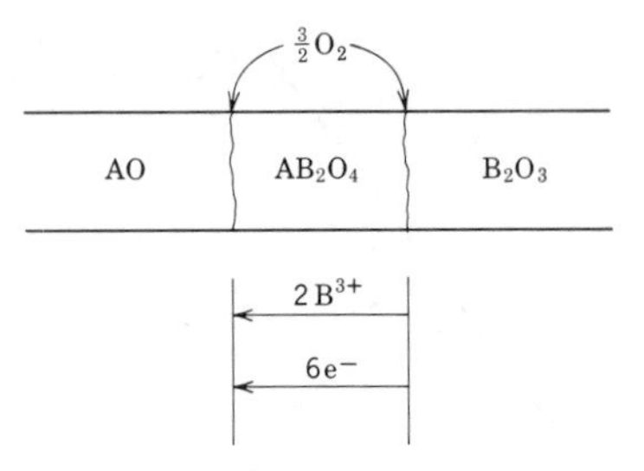

Oxygen and cation transport through AB_2O_4:

(1) Both cations diffuse $\left(J_{B^{3+}} = \frac{2}{3}J_{A^{2+}}\right)$.

Reactions occur at

AO–AB_2O_4 interface

$$2B^{3+} + 4AO = AB_2O_4 + 3A^{2+}$$

and at

AB_2O_4–B_2O_3 interface

$$3A^{2+} + 4B_2O_3 = 3AB_2O_4 + 2B^{3+}$$

(2) A^{2+} and O^{2-} diffuse.
Reaction at

AB_2O_4–B_2O_3 interface

$$A^{2+} + O^{2-} + B_2O_3 = AB_2O_4$$

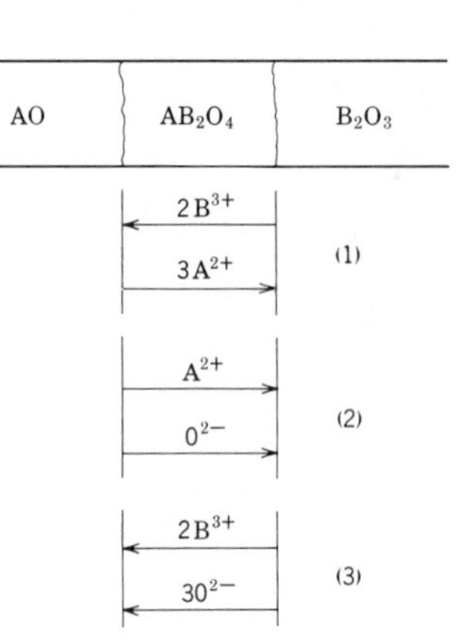

(3) B^{3+} and O^{2-} diffuse.
Reaction at

AO–AB_2O_4 interface

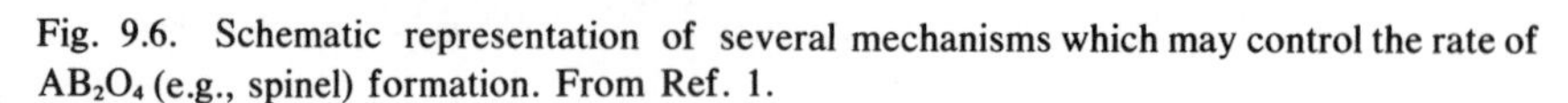

$$AO + 2B^{3+} + 3O^{2-} = AB_2O_4$$

Fig. 9.6. Schematic representation of several mechanisms which may control the rate of AB_2O_4 (e.g., spinel) formation. From Ref. 1.

might be controlled by the diffusion of A^{2+} ions; B^{3+} ions or O^{2-} ions, by the transport of electrons (holes), by the transport of O_2 gas, or by the interface reactions at AO–AB_2O_4 or AB_2O_4–B_2O_3.

When the rate of reaction-product formation is controlled by diffusion through the planar product layer, the parabolic rate law is observed (Eq. 9.8), in which the diffusion coefficient is that for the rate-limiting process. Figure 9.7 shows the parabolic time dependence for $NiAl_2O_4$ formation at two different temperatures, and Fig. 9.8 is a photomicrograph of the planar spinel reaction product on Al_2O_3. (More complex situations arise when several phases are formed as reaction products. These are discussed in reference 1 and by C. Wagner.*)

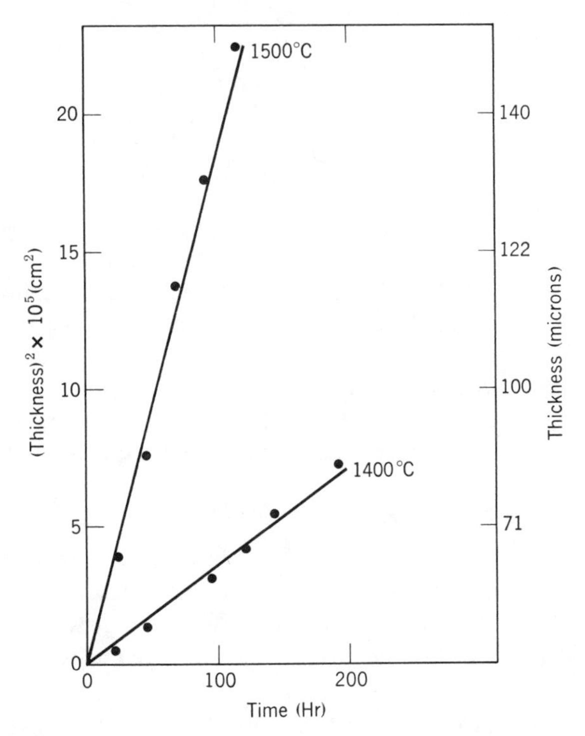

Fig. 9.7. Thickness of $NiAl_2O_4$ formed in NiO–Al_2O_3 couples as a function of time for couples heated in argon at 1400 and 1500°C. From F. S. Pettit et al., *J. Am. Ceram. Soc.*, **49**, 199 (1966).

The Electrochemical Potential in Ionic Solids. When considering point defects (Chapter 4) and atom mobility (Chapter 6), we noted that a

Acta Met.*, **17, 99 (1969).

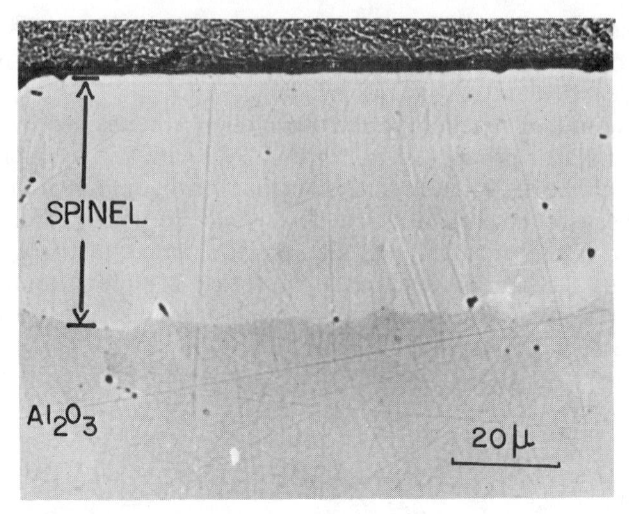

Fig. 9.8. Cross-sectional view of a typical $NiAl_2O_4$ layer formed in an NiO–Al_2O_3 couple after 73 hr at 1400°C. From F. S. Pettit, et al., *J. Am. Ceram. Soc.*, **49**, 199 (1966).

distinguishing feature of ionic crystals is the effective charge that an atomic specie may have within the crystal lattice. When there is mass transport in a ceramic, the transport of one charged specie is usually coupled to the transport of an ion or defect of the opposite charge. We must therefore consider the electrochemical potential as the motivating force for mass transport rather than just the chemical potential or concentration gradient. The electrochemical potential of the ith specie η_i is the sum of the chemical potential μ_i and the electrical potential ϕ acting on it:

$$\eta_i = \mu_i + Z_i F \phi \tag{9.10}$$

where Z_i is the effective charge and F is the Faraday constant. We have already noted in Table 6.1 the interrelationship between the mobilities expressed in terms of electrical and chemical driving forces. The flux due to an electrochemical potential gradient is thus given by

$$j_i = c_i v_i = -c_i B_i \frac{\partial \eta_i}{\partial x} = -c_i B_i \left[\frac{\partial \mu_i}{\partial x} + Z_i F \frac{\partial \phi}{\partial x} \right] \tag{9.11}$$

Examination of the two gradient terms in this equation shows the importance of the ionic nature of ceramics. For example, a concentration gradient (chemical-potential gradient) in one direction may be offset by an electrical-field gradient that motivates the ion in the opposite direction. Another kind of effect results from the local electrical field between oppositely charged species. For example, the cations in most close-packed oxides diffuse more rapidly than oxygen, as for the NiO–MgO and NiO–CoO interdiffusion already discussed. If this begins to happen in the case in which there is a net mass flow (not for the case of diffusion coefficient measurements using radioactive tracers), for example, Al^{3+} ions in alumina, a net electrical field results and thereby couples the motion of Al^{+3} ions and O^{-2} ions. Several solid reactions based on Eq. (9.11) are now considered.

Oxidation of a Metal. The most extensive studies of the parabolic rate law in which the process is controlled by diffusive transport through the reaction product are investigations into the formation of oxide layers on metals. The analysis techniques were developed by Carl Wagner which begin with Eq. (9.11). They are described here in some detail because the results extend to many ceramic problems. Consider the formation of a coherent oxide layer on a metal where the ambient oxygen pressure is $P_{O_2}{}^{g}$ and the effective oxygen pressure at the oxide-metal interface $P_{O_2}{}^{i}$ is determined by the temperature and the standard free energy of formation of the oxidation reaction (see Fig. 9.9):

$$2\text{Me} + \text{O}_2 = 2\text{MeO} \qquad \Delta G^\circ_{\text{formation}} \tag{9.12}$$

$$P_{O_2}{}^{i} = e^{+\Delta G^\circ/RT}$$

The oxygen concentration gradient (chemical potential) across the oxide layer (Fig. 9.10) provides the driving force for oxygen diffusion towards the metal-oxide interface. A gradient of the chemical potential of the metal ion in the opposite direction produces metal-ion diffusion toward the oxygen atmosphere. If one atomic flux is larger than the other, there is also a net flux of electrons or electron holes. The net transport, which determines the rate of oxide growth, is the sum of flux of anions and cations and electrons or holes. First we must consider each of these fluxes and then we shall look for circumstances when one specie is rate-determining and the complex relationships reduce to more simple forms.

The flux of the atomic and electronic species given by Eq. 9.11 can be changed to the flux of charged particles by multiplying by the valence:

$$
\begin{aligned}
J_0 &= -|Z_0| c_0 B_0 \frac{\partial \eta_0}{\partial x} = |Z_0| j_0 \\
J_{\text{Me}} &= -|Z_{\text{Me}}| c_{\text{Me}} B_{\text{Me}} \frac{\partial \eta_{\text{Me}}}{\partial x} = |Z_{\text{Me}}| j_{\text{Me}} \\
J_{e'} &= -\text{n} B_{e'} \frac{\partial \eta_{e'}}{\partial x} = |-1| j_{e'} \\
J_{h^\cdot} &= -p B_{h^\cdot} \frac{\partial \eta_{h^\cdot}}{\partial x} = |+1| j_{h^\cdot}
\end{aligned}
\tag{9.13}
$$

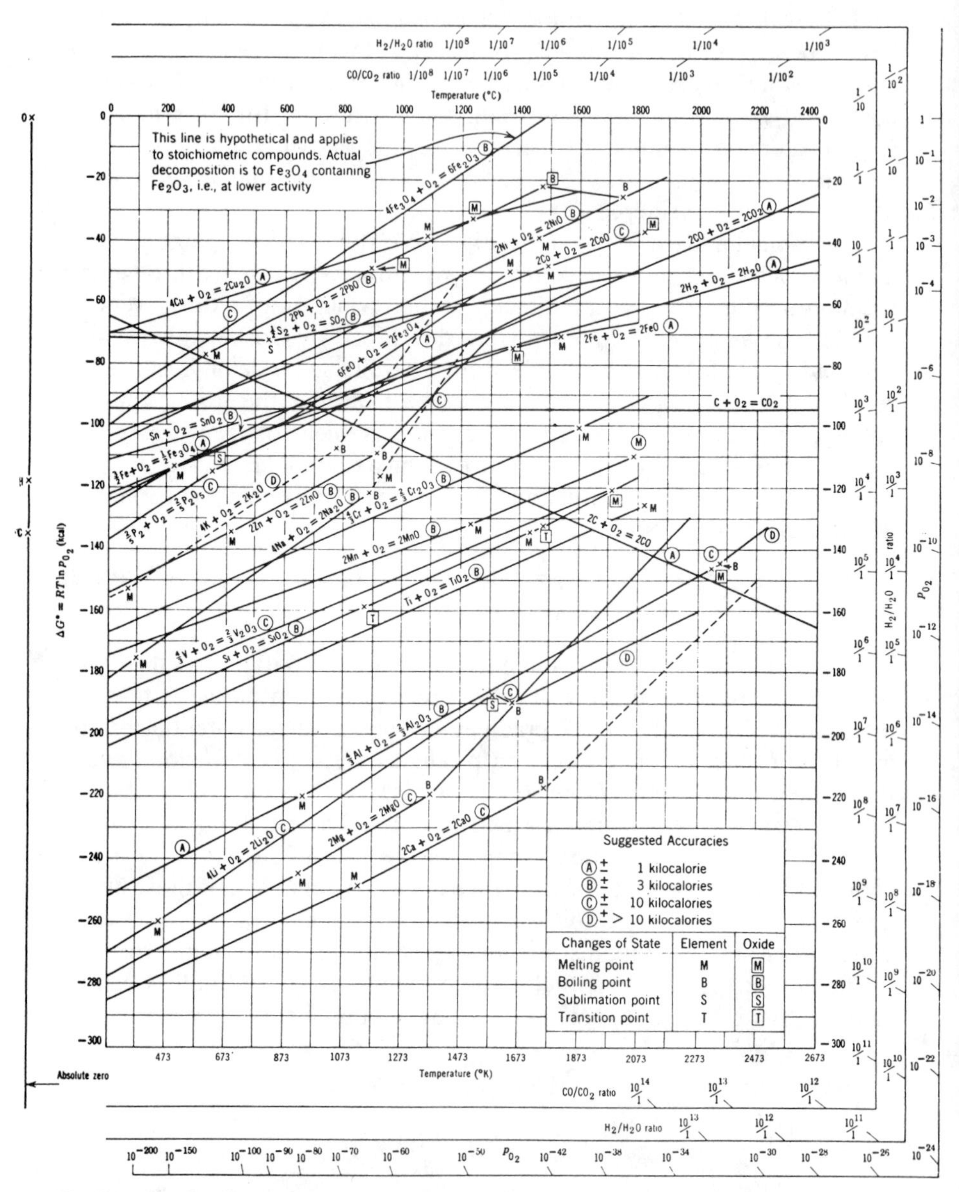

Fig. 9.9. Standard free energy of formation of oxides as a function of temperature. From F. D. Richardson and J. H. E. Jeffes, *J. Iron Steel Inst.*, **160**, 261 (1948); modified by L. S. Darken and R. W. Gurry, *Physical Chemistry of Metals*, McGraw-Hill Book Company, New York, 1953.

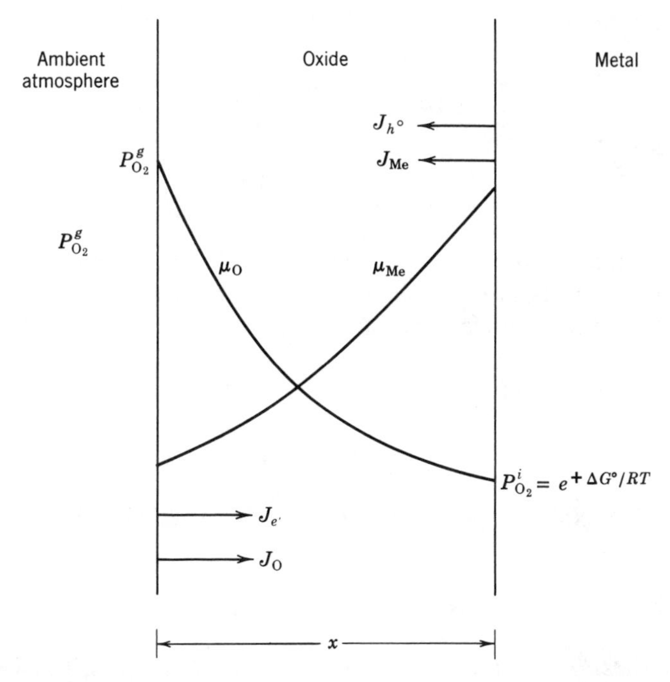

Fig. 9.10. Chemical-potential gradients across an oxide layer on a metal.

For a given oxide layer, either electrons or holes are predominant, so that only one of the last two equations is necessary. The constraint on the net flux is electrical neutrality. If we assume electrons to be the important electronic defect, this constraint requires that

$$J_O + J_{e'} = J_{Me} \tag{9.14}$$

The net flux and therefore the rate of oxidation is the sum $J_{ox} = |J_O| + |J_{Me}|$. The general result can be expressed in terms of the conductivity σ and the transference number t_i (see also Chapter 17), which represents the fraction of the total charge flux carried by a particular specie;

$$\begin{aligned} J_{ox} &= \frac{\sigma t_{e'}}{|Z_{Me}|F^2}(t_O + t_{Me})\left|\frac{\partial \mu_{Me}}{\partial x}\right| \\ &= \frac{\sigma t_{e'}}{|Z_O|F^2}(t_O + t_{Me})\left|\frac{\partial \mu_O}{\partial x}\right| \end{aligned} \tag{9.15}$$

Although the composition varies through the layer, average values can be assumed for t_i and σ to simplify the result, which yields a form of the parabolic rate law;

$$\frac{dx}{dt} = \frac{1}{x}\left[\frac{\bar{\sigma}\bar{t}_{e'}}{|Z_{Me}|F^2}(\bar{t}_0 + \bar{t}_{Me})|\Delta\mu_{Me}|\right] = \frac{K}{x} \tag{9.16}$$

Recalling that,

$$t_i\sigma = \sigma_i = eZ_ic_i\mu_i = \frac{c_iZ_i^2e^2F^2}{RT}D_i \tag{9.17}$$

we can see that the oxidation rate is governed by the atomic mobilities or diffusivities. Let us now consider specific rate limiting cases:

1. The electrical current is carried primarily by electronic defects, $t_{e'} \approx 1$ or $t_{h^{\cdot}} \approx 1$:

a. If $D_0 \gg D_{Me}$, then

$$K = \frac{\bar{\sigma}\bar{t}_0}{|Z_{Me}|F^2}|\Delta\mu_{Me}| \tag{9.18}$$

which for an oxide for predominant $V_0^{\cdot\cdot}$ defects reduces to the approximation

$$K \approx \frac{c_0}{2|Z_{Me}|}\int_{\mu_0{}^i}^{\mu_0{}^g} D_0\, d\mu_0 \tag{9.19}$$

since $\mu_0 = 1/2\mu_{O_2} = 1/2(\mu_{O_2}^{\circ} + RT \ln P_{O_2})$. If we assume that $[V_0^{\cdot\cdot}] \propto P_{O_2}^{-1/6}$, as discussed in Chapter 6, and that the diffusion coefficient varies similarly;

$$D_0 = [V_0^{\cdot\cdot}]D_{V_0^{\cdot\cdot}} = \frac{(K_{V_0^{\cdot\cdot}})^{1/3}}{4^{1/3}}D_{V_0^{\cdot\cdot}}P_{O_2}^{-1/6} \tag{9.20}$$

and the rate constant becomes

$$K \approx \frac{3c_0}{4^{1/3}|Z_{Me}|}(K_{V_0^{\cdot\cdot}})^{1/3}D_{V_0^{\cdot\cdot}}\{(P_{O_2}{}^g)^{-1/6} - (P_{O_2}{}^i)^{-1/6}\} \tag{9.21}$$

b. If $D_{Me} \gg D_0$ and we assume singly charged metal vacancies V'_{Me}, the rate constant is

$$K \approx 2(K_{V'_{Me}})^{1/2}D_{V'_{Me}}\{(P_{O_2}{}^g)^{1/4} - (P_{O_2}{}^i)^{1/4}\}c_{Me} \tag{9.22}$$

note that $D_{V'_{Me}}[V'_{Me}] = D_{Me}$.

2. If the electral current is carried primarily by the ions, $(t_0 + t_{Me}) \approx 1$, the rate constant from Eq. 9.16 becomes

$$K \approx \frac{kT}{8|Z_{Me}|e^2}\int_{P_{O_2}{}^i}^{P_{O_2}{}^g} \sigma_{e1}\, d\ln P_{O_2} \tag{9.23}$$

where σ_{e1} is the conduction due to electrons and holes which have mobilities μ_e and μ_h, respectively (see Table 6.1 $\mu_i = B''_i$).

$$\sigma_{e1} = \mathrm{en}\mu_e + \mathrm{ep}\mu_h \tag{9.24}$$

If we assume that the defect concentration does not have a large variation over the oxide layer

$$K \approx \frac{\sigma_{e1}kT}{8|Z_{\mathrm{Me}}|e^2}[\ln P_{O_2}{}^g - \ln P_{O_2}{}^i] \tag{9.25}$$

An example of the applicability of this relationship is the diffusive transport of oxygen through calcia-stabilized zirconia. The oxygen diffusion coefficient plotted in Fig. 6.11 is very large and accounts for $t_0 \approx 1$. Thus, the slower-moving specie, the electron hole, becomes rate-limiting for oxygen permeation (Eq. 9.25), as shown in Fig. 9.11.

3. If the metal undergoing oxidation has an impurity with a different oxidation state, for example, Li in Ni, the defect concentration in the oxide may be determined by the impurity concentration. As an example, consider the analogous case to Eq. 9.22 for which $D_{\mathrm{Me}} \gg D_0$ but where $[V'_{\mathrm{Me}}] = [F^{\cdot}_{\mathrm{Me}}]$. The thickness of this extrinsic layer is again determined by the parabolic rate law, Eq. 9.16, but with the reaction constant,

$$K_{\mathrm{ex}} = 2D_{V'_{\mathrm{Me}}}[V'_{\mathrm{Me}}][\ln P_{O_2}{}^g - \ln P_{O_2}{}^i]c_{\mathrm{Me}} \tag{9.26}$$

If the impurity concentration and oxygen pressure are such that the defect concentrations are in an intermediate range, an intrinsic layer may

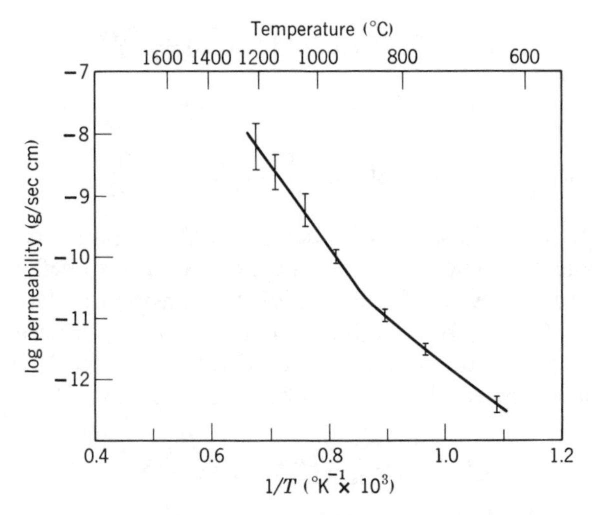

Fig. 9.11. Oxygen permeation through calcia-stabilized zirconia as a function of temperature. The oxygen transport is controlled by the concentration and mobility of electron holes, Eq. 9.25. From K. Kitazawa, Ph.D. thesis, MIT, 1972.

form on the oxygen-rich side (external) and an extrinsic oxide layer on the metal-rich side (at the oxide-metal interface).

Short-Circuit Diffusion Paths. In each of the examples of metal oxidation, lattice diffusion D_l was assumed to be the rate-determining transport process. In Section 6.6 the importance of other more rapid diffusion paths was discussed. The effects of short-circuit paths can be incorporated into the parabolic rate equations. For example, an apparent diffusivity D_a from Eq. 6.67 can be used in Eq. 9.16 to include the contributions from lattice D_l and boundary diffusion D_b;

$$D_a = D_l(1-f) + fD_b$$
$$\frac{dx}{dt} = \frac{K'D_a}{x} \tag{9.27}$$

where the diffusion coefficient has been extracted from the rate constant to give another constant K'. Low-temperature oxidation and oxide layers with fine grain sizes are expected to form by boundary diffusion.

Chemical Diffusion in Nonstoichiometric Oxides. The chemical diffusion coefficient for the counter diffusion of cations and anions can also be determined from the Wagner analysis. If we assume that electrical conduction is mainly electronic ($t_{el} \approx 1$) that is, movement of electrical charge is not the rate-limiting step for mass transport (ions), the chemical diffusion coefficient $\tilde{D}$ can be determined. In terms of diffusion coefficients rather than transference numbers Eq. 9.15 becomes

$$J_{ox} = \frac{c_0}{|Z_{Me}|}(|Z_{Me}|D_0 + |Z_0|D_{Me})\frac{1}{kT}\frac{d\mu_0}{dx} \tag{9.28}$$

In terms of Fick's first law this can be rewritten

$$J_{ox} = \left[(|Z_{Me}|D_0 + |Z_0|D_{Me})\frac{c_0}{|Z_{Me}|}\left(\frac{1}{kT}\frac{d\mu_0}{d\tilde{c}}\right)\right]\frac{d\tilde{c}}{dx} = \tilde{D}\frac{d\tilde{c}}{dx} \tag{9.29}$$

where $\tilde{c}$ represents the excess (or deficit) of the metal or oxygen in the nonstoichiometric compound. The chemical diffusion coefficient is the bracketed term. Consider, for example, the transition metal monoxides ($Fe_{1-\delta}O$, $Ni_{1-\delta}O$, $Co_{1-\delta}O \cdots$) for which $\tilde{c} \propto [V_{Me}^{\alpha'}]$, where α is the effective charge on the vacancy and where $D_{Me} \gg D_0$. The chemical diffusion coefficient can be written from Eq. 9.29 in the form

$$\tilde{D} = \frac{1}{2}\frac{c_{Me}}{[V_{Me}^{\alpha'}]}D_{Me}\frac{d \ln P_{O_2}}{d \ln [V_{Me}^{\alpha'}]} \tag{9.30}$$

for which the substitution $d\mu_0 = 1/2kTd \ln P_{O_2}$ has been made. From the defect equilibrium reaction, the mass action law gives

$$[h^{\cdot}]^{\alpha}\,[V_{Me}^{\alpha'}] = K_{V_{Me}^{\alpha'}}P_{O_2}^{1/2} \tag{9.31}$$

The derivative in Eq. 9.30 can now be determined;

$$\frac{d \ln P_{O_2}}{d \ln [V_{Me}^{\alpha'}]} = 2(\alpha + 1) \tag{9.32}$$

Substituting this into Eq. 9.30 and recalling that $c_{Me}D_{Me} = c_V D_V$, the chemical diffusion coefficient is given by

$$\tilde{D} = (\alpha + 1) D_{V_{Me}^{\alpha'}} \tag{9.33}$$

Thus for singly charged vacancies, $\tilde{D} = 2D_{V_{Me}'}$, and for doubly charged vacancies, $\tilde{D} = D_{V_{Me}''}$.

If the oxygen pressure is changed from one value to another, a new O/Me value is established in a nonstoichiometric oxide, and the oxidation-reduction rate is determined by a diffusion coefficient of the type in Eq. 9.33. This value is larger than the diffusivity of the cation or the anion. Figure 9.12 shows the chemical diffusion coefficient determined in $Fe_{1-\delta}O$ by step changes in the oxygen pressure which cause diffusion-controlled changes in the composition. The value of the chemical diffu-

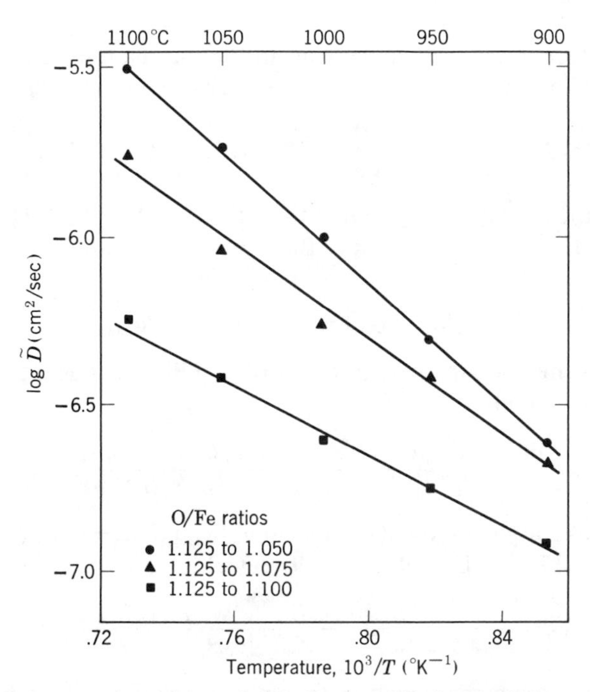

Fig. 9.12. Temperature dependence of the chemical-diffusion coefficient of wustite for several composition increments. From R. L. Levin and J. B. Wagner, *Trans. AIME*, **233**, 159 (1965).

sion coefficient correlates with the tracer value (Eq. 9.30) when the defect equilibrium relationships are known.

Ambipolar Diffusion. The formality used to derive Eqs. 9.11 and 9.13 also allows us to determine the effective diffusion constants when cations and anions are flowing in the same direction. Referred to as ambipolar diffusion, a description of the atomistic process must again consider the coupling between the oppositely charged species when the transport of electrons and holes is slower than ion transport. If the flux of cations becomes excessive, a local internal electric field builds up to "drag" along the anions. This behavior is important in processes involving reactions which cause product formation, in processes which are in response to an applied electric field, and in processes which result in a shape change due to mechanical or surface tension forces such as sintering and creep.

As an example, consider a pure oxide for which $t_{el} \approx 0$. Equation 9.11 can be written for anion and cation transport as in Eq. 9.13. Since the transport of each ion is in the same direction, electrical neutrality is maintained when

$$J_T = \frac{J_O}{|Z_{Me}|} = \frac{J_{Me}}{|Z_O|} \tag{9.34}$$

where J_T refers to the total molecular flux. Equating the anion and cation charge flux allows for the solution of the internal electric field, $\partial\phi/\partial x$,

$$|Z_O|c_O B_O \left[\frac{\partial \mu_O}{\partial x} + Z_O F \frac{\partial \phi}{\partial x}\right] = |Z_{Me}| c_{Me} B_{Me} \left[\frac{\partial \mu_{Me}}{\partial x} + Z_{Me} F \frac{\partial \phi}{\partial x}\right] \tag{9.35}$$

in terms of the chemical potential of the oxide, $\mu(Me_{Z_O}O_{Z_{Me}})$. The chemical potential of the oxide is the sum of the chemical potentials of cations and anions,

$$d\mu(Me_{Z_O}O_{Z_{Me}}) = Z_{Me}\, d\mu_O + Z_O\, d\mu_{Me} \tag{9.36}$$

which yields the coupling field in terms of mobilities, concentrations, and the chemical potential,

$$\frac{\partial \phi}{\partial x} = \frac{[|Z_O|c_O B_O - |Z_{Me}|c_{Me}B_{Me}]}{F[Z_{Me}|Z_{Me}|c_{Me}B_{Me} - Z_O|Z_O|c_O B_O]} \frac{\partial \mu_{Me}}{\partial x} \tag{9.37}$$

where we have assumed local equilibrium, $|Z_{Me}|\, dc_{Me} = |Z_O|\, dc_O$. Substitution of Eq. 9.37 into Eqs. 9.34 and 9.13 yields

$$J_T = \frac{-c_{Me}B_{Me}c_O B_O}{Z_{Me}|Z_{Me}|c_{Me}B_{Me} - Z_O|Z_O|c_O B_O} \frac{\partial \mu(Me_{Z_O}O_{Z_{Me}})}{\partial x} \tag{9.38}$$

This term is the correction due to ambipolar effects to the diffusion transport resulting from a chemical potential gradient. Consider as an example of the applicability of Eq. 9.38 to sintering of pure MgO for which the values of $Z_{Mg} = |Z_0| = 2$ and $c_{Mg} = c_0 = c$:

$$J_T = -\frac{cB_{Mg}B_0}{[B_{Mg} + B_0]} \frac{\partial \mu_{MgO}}{\partial x} \tag{9.39}$$

Since $\mu_{Mg} = \mu_{Mg}^{\circ} + RT \ln a \approx \mu_{Mg}^{\circ} + RT \ln c$, Eq. (9.39) can be expressed as

$$J_T = -\frac{c_{MgO}B_{Mg}B_0RT}{[B_{Mg} + B_0]} \frac{d \ln c_{MgO}}{dx} = -\frac{B_{Mg}B_0RT}{[B_{Mg} + B_0]} \frac{dc_{MgO}}{dx} \tag{9.40}$$

where $\frac{dc_{MgO}}{dx}$ is the concentration gradient due to curvature (Chapter 10). Recalling that the tracer diffusion coefficient and mobility are related by Eq. (6.11),

$$J_T \approx -\frac{D_{Mg}{}^T D_0{}^T}{[D_{Mg}{}^T + D_0{}^T]} \frac{dc_{MgO}}{dx} \tag{9.41}$$

Thus the total molecular transport may be governed by the slowest-moving specie if there is a large difference in diffusivities (e.g., $D_{Mg} \gg D_0$; $J_T \propto D_0$) or by an intermediate value when they are not too disimilar (e.g., $D_{Mg} = 3D_0$; $J_T \propto D_{Mg}/4$).

Since some ions transport more rapidly in boundaries or along dislocations, a relationship for ambipolar diffusion can be derived when paths other than the lattice are assumed. A simple case has been derived for steady-state grain boundary and lattice transport.* The effective area of transport in the lattice A^l and boundary A^b must be incorporated in the equation for total mass flow. For the case of a pure material MO the effective diffusion coefficient is similar in form to Eq. 9.41 and given by

$$D_{\text{effective}} \approx \frac{(A^l D_{Me}{}^l + A^b D_{Me}{}^b)(A^l D_0{}^l + A^b D_0{}^b)}{(A^l D_{Me}{}^l + A^b D_{Me}{}^b) + (A^l D_0{}^l + A^b D_0{}^b)} \tag{9.42}$$

where D^l refers to lattice diffusion and D^b refers to boundary diffusion. In many oxides, it has been observed that $A^b D_0{}^b \gg A^l D_0{}^l$ and that $A^l D_{Me}{}^l > A^b D_{Me}{}^b$; thus Eq. 9.42 reduces to

$$D_{\text{effective}} \approx \frac{A^l D_{Me}{}^l A^b D_0{}^b}{A^l D_{Me}{}^l + A^b D_0{}^b} \tag{9.43}$$

Diffusive transport in real materials is more complex, owing to impurities and imperfections, but relationships like these can be derived to include more complex situations.†

*R. S. Gordon, *J. Am. Ceram. Soc.*, **56**, 147 (1973).
†D. W. Readey, *J. Am. Ceram. Soc.*, **49**, 366 (1966).

9.3 Reactant Transport through a Fluid Phase

As discussed in Section 9.1, heterogeneous reactions at high temperatures require, first, material transfer to the reaction interface, second, reaction at the phase boundary, and in some cases diffusion of products away from the reaction site. Any of these steps can have the lowest virtual reaction rate and be rate-controlling for the overall process. Generally, once a reaction is initiated, material-transfer phenomena determine the overall rate in the high-temperature systems of importance in ceramics. As discussed in the previous section, the diffusion of ions and electrons through a stable oxide film on the surface of a metal determines the reaction rate. If, however, the film forms with cracks and fissures, the rate may be determined by gaseous diffusion through these channels. In this section we wish to consider several important examples of the way ceramic materials interact with gases and liquids and to determine the rate-limiting kinetic equations.

Gas-Solid Reactions: Vaporization. The simplest kind of solid-gas reactions are those related to vaporization or thermal decomposition of the solid. Section 9.4 contains a discussion of the decomposition of a solid to a gas and another solid; in this section we are primarily concerned with reactions in which the solid forms only gaseous products. The rate of decomposition is dependent on the thermodynamic driving forces, on the surface-reaction kinetics, on the condition of the reaction surface, and on the ambient atmosphere; for example, at high temperatures oxides volatilize much more rapidly in a vacuum than in air.

The loss of silica from glass and refractories in reducing atmospheres is an important factor which limits the usefulness of these ceramic products. Consider the following reaction which can cause the volatilization of SiO_2:

$$2SiO_2(s) = 2SiO(g) + O_2(g) \tag{9.44}$$

At 1320°C, the equilibrium constant is

$$K_{eq} = \frac{P_{SiO}^2 P_{O_2}}{a_{SiO_2}^2} = 10^{-25} \tag{9.45}$$

Assuming unit activity for the silica, it is apparent that the ambient oxygen partial pressure controls the pressure of $SiO(g)$ and therefore the rate of vaporization. Under reducing conditions (inert atmosphere, H_2 or CO atmosphere) of $P_{O_2} = 10^{-18}$ atm, the SiO pressure is 3×10^{-4} atm (0.23 torr).

The rate of evaporation near equilibrium is given by the Knudsen equation:*

$$\frac{dn_i}{dt} = \frac{AP_i\alpha_i}{\sqrt{2\pi M_i RT}} \tag{9.46}$$

where $\frac{dn_i}{dt}$ is the loss of component i in moles per unit time, A is the sample area, α_i is the evaporation coefficient ($\alpha_i \leq 1$), M_i is the molecular weight of i, and P_i is the pressure of i above the sample. If there is a high gas flow rate over the sample or if the evaporation is into a vacuum, the sample is not able to maintain its equilibrium vapor pressure P_i, and the evaporation rate is controlled by the interface reaction rate. For the gas to be in equilibrium with the solid, the gas flow rate S (moles/sec) and the total pressure P^T (atm) must satisfy the inequality

$$\frac{A\alpha_i P^T}{S(M_i T)^{1/2}} \gg 2.3 \times 10^{-9} \tag{9.47}$$

where A is in square centimeters and T in degrees Kelvin.

When the oxygen partial pressure in the gas phase is controlled by gas mixtures $P_{O_2}{}^{ext}$, the equation (9.46) becomes

$$J_{O_2} = \frac{(P_{O_2} - P_{O_2}{}^{ext})\alpha_{O_2}}{(2\pi M_{O_2} RT)^{1/2}} \tag{9.48}$$

where P_{O_2} is calculated from the standard free energy of the decomposition reaction (e.g. Eq. 9.45).

For the vaporization of SiO_2 by reaction (9.44), Eq. 9.46 predicts a loss rate of about 5×10^{-5} moles SiO_2/cm^2 sec at 1320°C. Figure 9.13 shows actual SiO_2 loss rates from various silica-containing refractories annealed in hydrogen. The overall decomposition reaction in this case is

$$H_2(g) + SiO_2(s) = SiO(g) + H_2O \tag{9.49}$$

The effect of a few mole percent water vapor in the gas stream is evident from Fig. 9.14. As predicted from Eq. 9.49, the $SiO(g)$ pressure is decreased by an increase in the $H_2O(g)$ pressure.

Chemical Vapor Transport. Next let us consider the reaction of an active transport gas with a ceramic. The net effect is to increase the vapor-phase transport. Some high-temperature ceramics and many thin-film electronic devices are prepared by chemical vapor deposition. By controlling the chemical potential (concentration) of reaction gases, the rate of deposition can be controlled. Generally the rate of deposition and the temperature of deposition determine the reaction kinetics and rates at

*M. Knudsen, *Ann. Phys.*, **47**, 697 (1915).

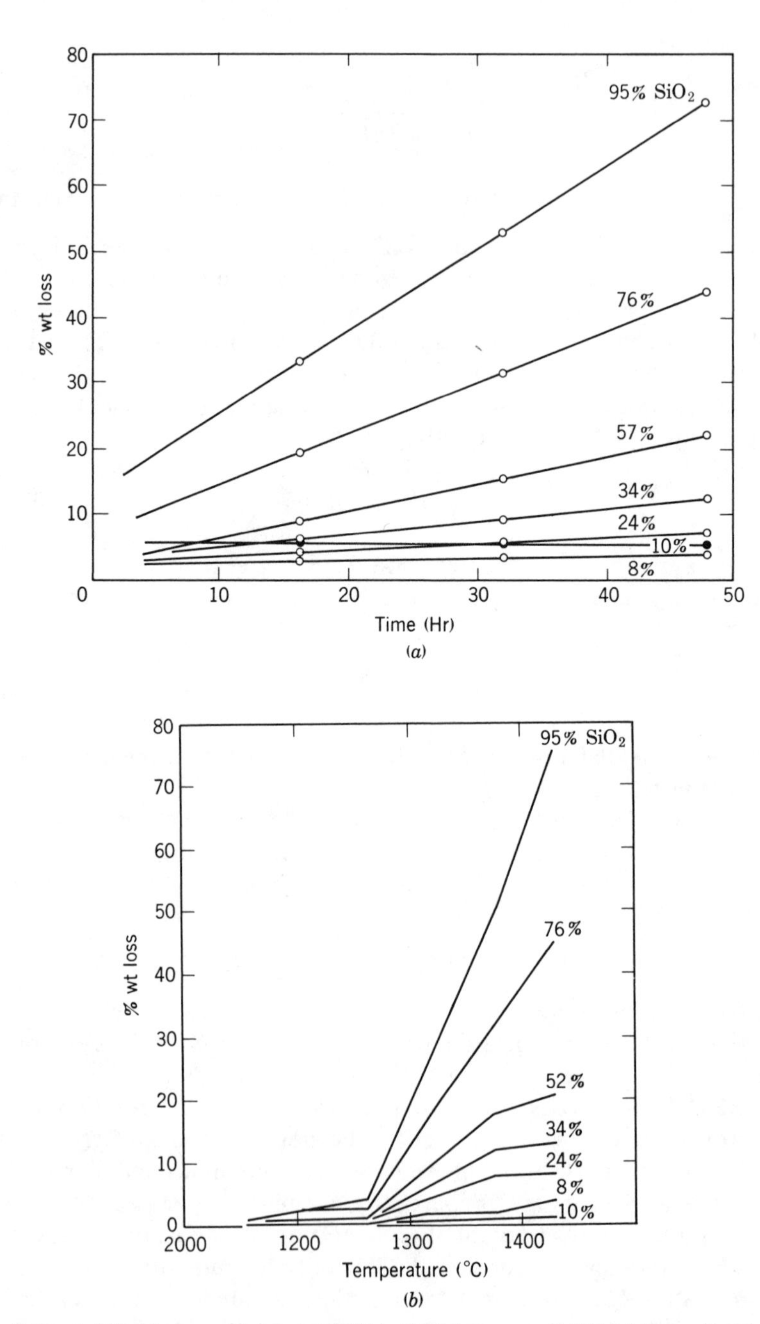

Fig. 9.13. (*a*) Weight loss of brick at 1425°C in 100% hydrogen. (*b*) Weight loss of brick after 50 hr in 100% hydrogen. From M. S. Crowley, *Bull. Am. Ceram. Soc.*, **46**, 679 (1967).

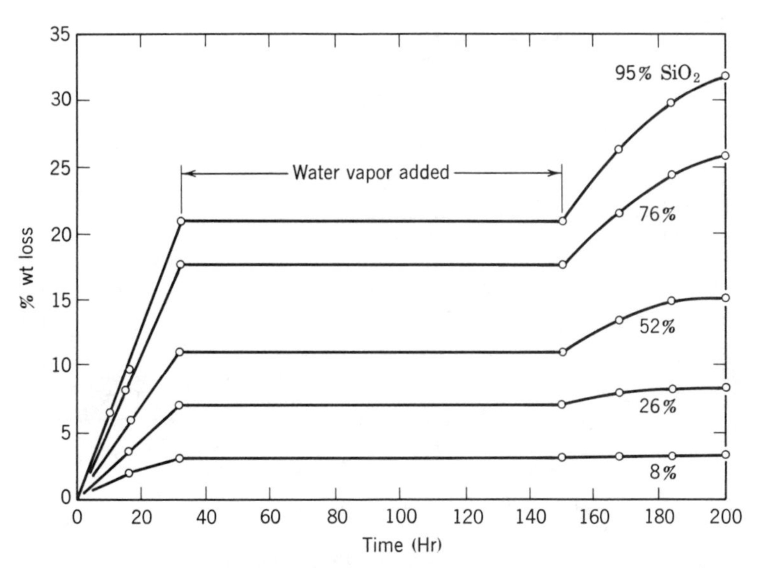

Fig. 9.14. Weight loss of brick at 1370°C in 75% H_2–25% N_2 atmosphere. After 32 hr water vapor was added for 150 hr. From M. S. Crowley, *Bull. Am. Ceram. Soc.*, **46**, 679 (1967).

which the decomposition products can "crystallize" on the reaction surface. If the supersaturation is large, homogeneous gas-phase nucleation occurs; that is, a heterogeneous surface is not needed. As the supersaturation is reduced, the gases react in the vicinity of a surface, and a polycrystalline deposit is formed. The perfection of the deposit, porosity, preferred grain orientation, and so on, depend on the particular material and the rate of deposition; usually slower deposition and higher temperatures result in a more perfect reaction product. Finally, when a single-crystal substrate is used as the heterogeneous reaction surface, epitaxial deposition occurs. In the latter case, a single crystal with an orientation determined by the substrate is formed.

To understand the kinetics of chemical vapor deposition fully requires a knowledge of all of the thermodynamic equilibria involved and the respective kinetic processes for the generation of reactants, mixing of reactant gases, diffusion through the boundary layers, molecular combinations at the interface, exsolution of gaseous products, surface diffusion of the solid products, and so on. We have chosen, as an example, a simple system for which the rate-determining step is diffusion in the gas phase. Consider the closed system shown in Fig. 9.15 in which two chambers are held at thermal equilibrium. Assume that the chemical reactions in each chamber reach thermodynamic equilibrium such that the diffusion flux of

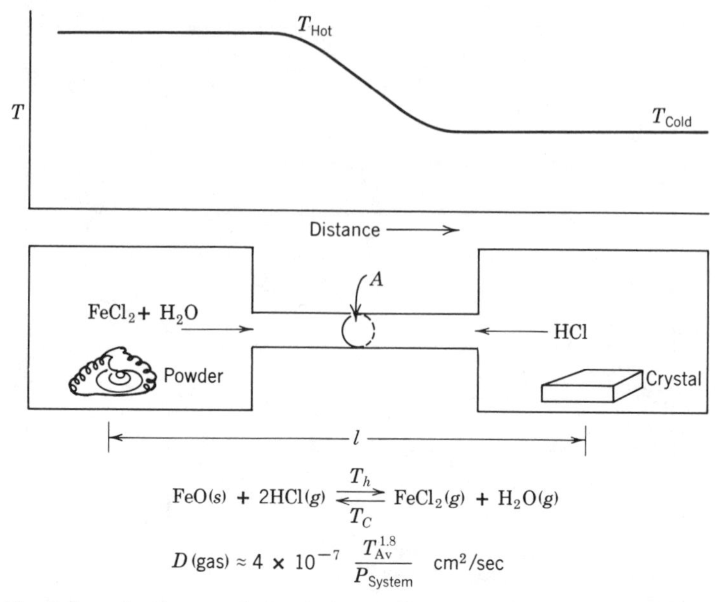

$$FeO(s) + 2HCl(g) \underset{T_C}{\overset{T_h}{\rightleftarrows}} FeCl_2(g) + H_2O(g)$$

$$D\,(\text{gas}) \approx 4 \times 10^{-7}\,\frac{T_{Av}^{1.8}}{P_{System}}\ \text{cm}^2/\text{sec}$$

Fig. 9.15. Schematic diagram of chemical vapor transport of iron oxide in a temperature gradient.

matter is from the hot chamber to the cooler chamber because of the concentration gradient (the direction of transport is determined by the sign of the enthalpy of the reaction).

The kinetics of mass transport as determined by the diffusion of the rate-limiting specie—for example, diffusion of $FeCl_2(g)$—is given by Fick's law:

$$\frac{dn}{dt} = -AD\frac{\partial c}{\partial x} = -AD\frac{\Delta c}{l} = -AD\frac{(c_h - c_c)}{l} \tag{9.50}$$

where n is the number of moles transported, A the cross-sectional area of the connecting tube (cm^2), D the diffusion coefficient of the rate-limiting specie, and c the concentrations in the respective isothermal chambers. For an ideal gas

$$c_h = \frac{n_h}{V} = \frac{P_h}{RT_h} \tag{9.51}$$

and the composition difference is

$$(c_h - c_c) \approx \frac{(P_h - P_c)}{RT_{av}} \tag{9.52}$$

Thus, the transport rate is determined by

$$\frac{dn}{dt} = -\frac{AD}{lRT_{av}}(P_h - P_c) \tag{9.53}$$

The equilibrium pressures can be determined by the standard free energy of formation at each temperature; for example, at the higher temperature

$$\Delta G_h^\circ = -RT_h \ln \frac{P_{FeCl_2}P_{H_2O}}{P_{HCl}^2 a_{FeO}} \tag{9.54}$$

In a closed system such as a quartz ampoule an initial HCl pressure of B atm results in the adjustment of the formation reaction by the formation of equivalent numbers of moles $FeCl_2$ and H_2O. The expression (9.54) reduces to

$$\Delta G_h^\circ = -RT_h \ln \frac{P_{FeCl_2}^2}{(B - 2P_{FeCl_2})^2} \tag{9.55}$$

which can be solved for each temperature and therefore leads to a prediction of the transport rate from Eq. 9.53.

In general, the rate-limiting gas-phase transport step is a function of the total pressure of the system. At very low pressures ($P_{total} < 10^{-4}$ atm) gas-phase molecular collisions are infrequent and thus transport becomes line-of-sight. At intermediate pressures ($10^{-4} < P_{total} < 10^{-1}$ atm), the diffusion-limited case discussed above becomes important. At higher pressures ($P_{total} > 10^{-1}$ atm) convective mass transport is more rapid. If convection or forced flow becomes rapid, gas-phase diffusion through the boundary layer may become the rate-determining process.

Liquid–Solid Reactions: Refractory Corrosion. An important example of the kinetics of liquid–solid reactions is the rate of dissolution of solids in liquids, particularly important in connection with refractory corrosion by molten slags and glasses, with the rate of conversion of solid batch components to glass in the glass-making process, and with the firing of a ceramic body in which a liquid phase develops. No nucleation step is required for the dissolution of a solid. One process that can determine the rate of the overall reaction is the phase-boundary reaction rate which is fixed by the movement of ions across the interface in a way equivalent to crystal growth (Section 8.4). However, reaction at the phase boundary leads to an increased concentration at the interface. Material must diffuse away from the interface in order for the reaction to continue. The rate of material transfer, the dissolution rate, is controlled by mass transport in the liquid which may fall into three regimes: (1) molecular diffusion, (2) natural convection, and (3) forced convection.

For a stationary specimen in an unstirred liquid or in a liquid with no fluid flow produced by hydrodynamic instabilities, the rate of dissolution is governed by molecular diffusion. The kinetics are similar to those discussed in Chapter 6 on diffusion. The effective diffusion length over which mass is transported is proportional to $\sqrt{Dt}$, and therefore the change in thickness of the specimen, which is proportional to the mass dissolved, varies with $t^{1/2}$. Even in a system which may undergo convection due to hydrodynamic instabilities from density gradients which arise from thermal gradients or from concentration gradients (due to dissolution), the initial dissolution kinetics should be governed by molecular diffusion.

The diffusion coefficient for dissolution kinetics must be considered in the same light as in Section 9.2; the electrical and chemical effects of the various possible species must be accounted for. For example, the dissolution of Al_2O_3 in a silicate slag may be controlled by any of the cations or anions in the Al_2O_3 or slag or more probably a combination (e.g., Eq. 9.41). An example of dissolution controlled by molecular diffusion is shown in Fig. 9.16 for the dissolution of sapphire in a $CaO–Al_2O_3–SiO_2$ melt containing 21 wt% Al_2O_3.

Natural or free convection occurs, owing to hydrodynamic instabilities in the liquid which give rise to fluid flow over the solid. This enhances the dissolution kinetics. It has often been observed in metals processing that the amount of dissolution is dependent on whether or not the ceramic is totally immersed in the liquid. Generally, a partially submerged sample undergoes more extensive dissolution near the liquid-gas interface, called

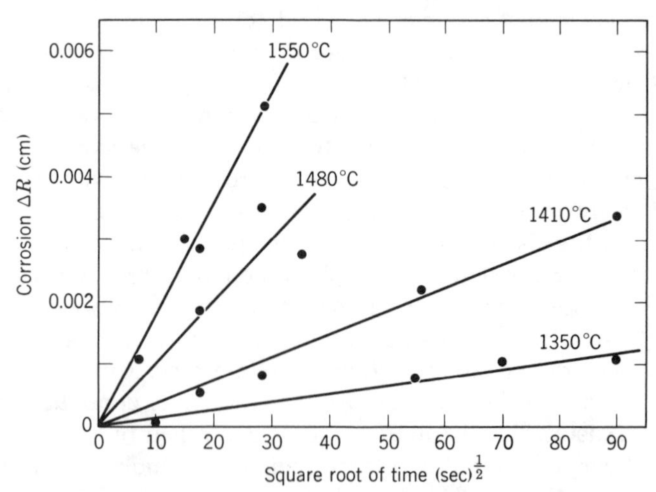

Fig. 9.16. Dissolution of sapphire cylinder in $CaO–Al_2O_3–SiO_2$ with 21 wt % Al_2O_3 versus square root of time. From Ref. 6.

the metal line. Below this interface the corrosion kinetics can be analyzed from free convection principles. It is clear that after a relatively short induction period during which molecular diffusion kinetics prevail the rate of dissolution becomes nearly independent of time. The general expression for mass transport during convection is

$$j = \frac{dn/dt}{A} = \frac{D(c_i - c_\infty)}{\delta(1 - c_i\bar{V})} \tag{9.56}$$

where j is the number of moles per second per square centimeter removed, c_∞ is the concentration in the bulk liquid, c_i is the concentration at the interface (saturation concentration), δ is the boundary layer thickness, D is the effective diffusion coefficient through the boundary layer, and $\bar{V}$ is the partial molar volume. The boundary layer is shown in Fig. 9.17 and defined by

$$\delta = \frac{c_i - c_\infty}{(dc/dy)} \tag{9.57}$$

where (dc/dy): is the concentration gradient at the interface. The boundary-layer thickness is determined by the hydrodynamic conditions of fluid flow. Viscous liquids form thicker boundary layers and cause slower material transfer. Higher liquid velocities form thinner boundary layers and permit more rapid material transfer. For refractory dissolution in glasses and silicate slags, the high viscosity and slow fluid velocity combine to give relatively thick boundary layers. The thickness of the boundary layer may be a centimeter. In comparison, for rapidly stirred aqueous solutions the boundary-layer thickness is a fraction of a millimeter. Also the diffusion rate is much slower in viscous silicate liquids than

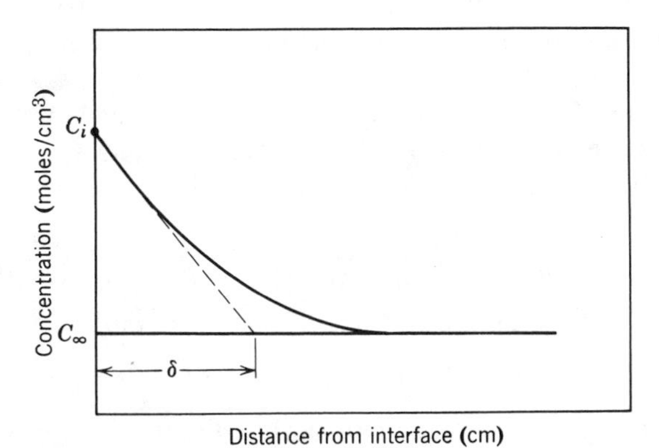

Fig. 9.17. Concentration gradient through diffusion layer at a solution interface.

in aqueous solutions, so that there is more of a tendency for the reaction process to be controlled by material-transfer phenomena rather than interface reactions.

Values for the boundary-layer thickness have been derived for special cases in fluid flow. The boundary-layer thickness for mass transport from a vertical slab with natural convection caused by density-difference driving forces is

$$\delta = 1.835 \times \left[\frac{D\nu\rho_\infty}{gx^3(\rho_i - \rho_\infty)}\right]^{1/4} \tag{9.58}$$

where x is the distance from the leading edge of the plate, ν is the kinematic viscosity η/ρ, g is the gravitational constant, ρ_∞ is the density of the bulk liquid, and ρ_i is the density of the saturated liquid (the liquid at the interface). Thus the average dissolution rate for a plate of height h is given by

$$J = \frac{dn/dt}{A} = 0.726D\left(\frac{g(\rho_i - \rho_\infty)}{\nu D h \rho_\infty}\right)^{1/4}(c_i - c_\infty) \tag{9.59}$$

The boundary-layer thickness for mass transport from a rotating disc is

$$\delta = 1.611\left(\frac{D}{\nu}\right)^{1/3}\left(\frac{\nu}{\omega}\right)^{1/2} \tag{9.60}$$

where ω is the angular velocity (rad/sec). The mass transfer for a rotating disc is proportional to the square root of the angular velocity:

$$j = \frac{dn/dt}{A} = 0.62\,D^{2/3}\nu^{-1/6}\omega^{1/2}\frac{(c_i - c_\infty)}{(1 - c_i\bar{V})} \tag{9.61}$$

Figure 9.18 shows the dissolution kinetics of sapphire into CaO–Al_2O_3–SiO_2 for the free convection kinetics and in Fig. 9.19 for forced flow. In each case the kinetics are time-independent, as predicted by Eqs. 9.59 and 9.61.

Comparison of the data for sapphire dissolution at 1550°C for kinetics limited by molecular diffusion, free convection, and forced convection (126 rad/sec) show the dimensional change ΔR (cm) to be related to time as

$$\begin{aligned} \Delta R\ \text{(molecular diffusion)} &= (1.77 \times 10^{-4}\ \text{cm/sec}^{1/2})t^{1/2} \\ \Delta R\ \text{(free convection)} &= (3.15 \times 10^{-6}\ \text{cm/sec})t \\ \Delta R\ \text{(forced convection)} &= (9.2 \times 10^{-5}\ \text{cm/sec})t \end{aligned} \tag{9.62}$$

The important parameters for convective dissolution are fluid velocity, kinematic viscosity, the diffusivity, and the composition gradient.

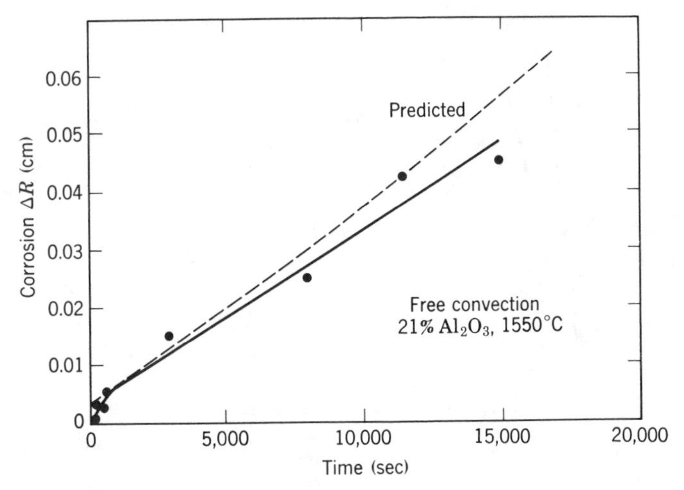

Fig. 9.18. Dissolution at relatively long times of sapphire cylinders in $CaO–Al_2O_3–SiO_2$ with 21 wt % Al_2O_3 versus time. From Ref. 6.

It is clear from the data in Figs. 9.16 and 9.19 that the dissolution rate is extremely temperature-sensitive. Since we have assumed transport-limited kinetics, the temperature dependence is largely determined by the exponential temperature dependence of diffusion (Eq. 6.39). The dependence of the corrosion rate of several ceramics on temperature is shown in Fig. 9.20.

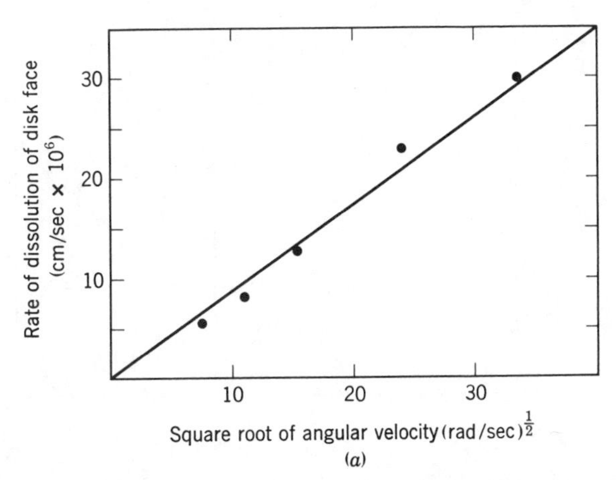

Fig. 9.19. (*a*) Dependence of rate of dissolution of face of sapphire disk on square root of angular velocity.

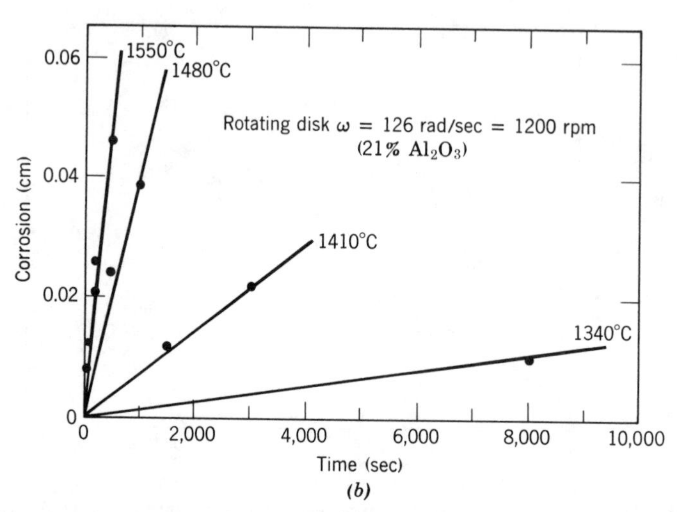

(*b*)

Fig. 9.19 (*continued*). (*b*) Rate of dissolution of face of sapphire disk rotating at 126 rad/sec on $CaO–Al_2O_3–SiO_2$ with 21 wt% Al_2O_3. From Ref. 6.

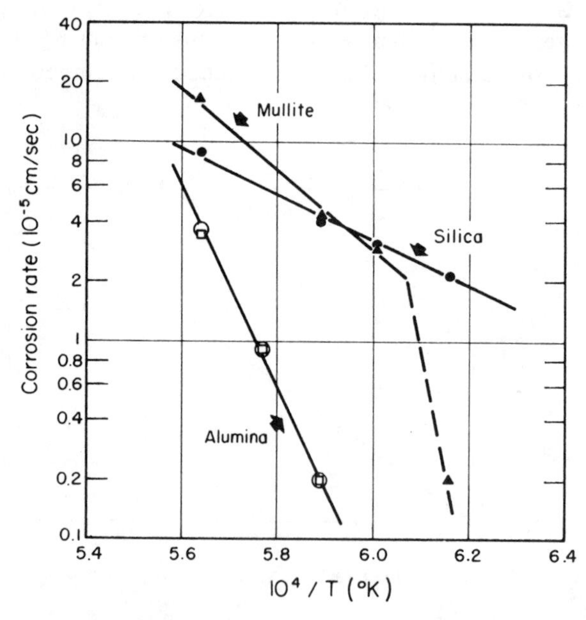

Fig. 9.20. Temperature dependence of forced convection corrosion in the $40CaO–20Al_2O_3–40SiO_2$ slag of alumina, mullite, and fused silica. From Ref. 6.

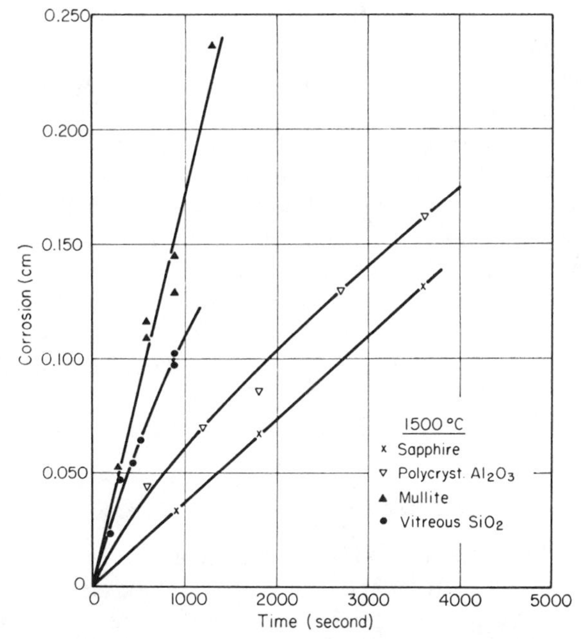

Fig. 9.21. Corrosion rate under forced convection conditions in the $40CaO–20Al_2O_3–40SiO_2$ slag of indicated specimens of sapphire, polycrystalline alumina, mullite, and vitreous silica. From Ref. 6.

Refractory corrosion is often much more complex. Besides complexities in the hydrodynamics of a molten bath, refractories seldom have ideal surfaces and are usually not of uniform composition. Multiphase bodies and brick with extensive porosity provide centers for accelerated corrosion, spalling, and penetration by the liquid. In dense single-phase ceramics, corrosion may be greatest at grain boundaries. This can be seen from the data in Fig. 9.21, in which the corrosion of polycrystalline Al_2O_3 is about 40% greater than sapphire after 2500 sec.

9.4 Reactant Transport in Particulate Systems

Of particular interest to ceramists is the large number of transformations which occur with granular or powdered raw materials; for example, the dehydration of minerals, decarbonization of carbonates, and polymorphic transformations. In general, the minerals and reaction products involved are used in large volumes; thus even though the nature of these reactions is complex, study of a few examples is important and elucidates the important kinetic parameters and illustrates the concept of the rate-limiting step.

Calcination and Dehydration Reactions. Calcination reactions are common for the production of many oxides from carbonates, hydroxides, sulfates, nitrates, acetates, oxalates, alkoxides, and so on. In general the reactions produce an oxide and a volatile reaction product (e.g., CO_2, SO_2, H_2O, . . .). The most extensively studied reactions are the decomposition of $Mg(OH)_2$, $MgCO_3$, and $CaCO_3$. Depending on the particular conditions of temperature, time, ambient pressure, particle size, and so on, the process may be controlled (1) by the reaction rate at the reaction surface, (2) by gas diffusion or permeation through the oxide product layer, or (3) by heat transfer. The kinetics of each of these rate-limiting steps is considered.

Let us first consider the thermodynamics of decomposition, for example, the calcination of $CaCO_3$:

$$CaCO_3(s) \rightarrow CaO(s) + CO_2(g) \qquad \Delta H^{298}_{\text{react}} = 44.3 \text{ kcal/mole} \quad (9.63)$$

The standard heat of reaction is 44.3 kcal/mole, that is, strongly endothermic, which is typical for most decomposible salts of interest. This means that heat must be supplied to the decomposing salt.

The standard free energy for the decomposition of $CaCO_3$, $MgCO_3$, and $Mg(OH)_2$ is plotted in Fig. 9.22. The equilibrium partial pressure of the gas for each of the reactions is also plotted in Fig. 9.22. Note, for example, that when $\Delta G°$ becomes zero, P_{CO_2} above $MgCO_3$ and $CaCO_3$ and P_{H_2O} above $Mg(OH)_2$ have become 1 atm. The temperatures at which this occurs are 1156°K ($CaCO_3$), 672°K ($MgCO_3$), and 550°K ($Mg(OH)_2$). The P_{CO_2} normally in the atmosphere and the range of P_{H_2O} (humidity) in air are also shown in Fig. 9.22. From these values we can determine the temperature at which the salt becomes unstable when fired in air. For example, $CaCO_3$ becomes unstable over 810°K, $MgCO_3$ above 480°K. Depending on the relative humidity, $Mg(OH)_2$ becomes unstable above 445 to 465°K. Because acetates, sulfates, oxalates, and nitrates have essentially zero partial pressure of product gases in the ambient atmosphere, it is clear that they are unstable at room temperature. That they exist as salts to a decomposition temperature of about 450°K indicates that their decomposition is governed by atomistic kinetic factors and not by thermodynamics.

The kinetics, as noted above, may be limited by the reaction at the surface, the flow of heat from the furnace to the reaction surface, or the diffusion (permeation) of the product gas from the reaction surface to the ambient furnace atmosphere. This is shown schematically in Fig. 9.23, which also includes the appropriate heat and mass flow equations. The

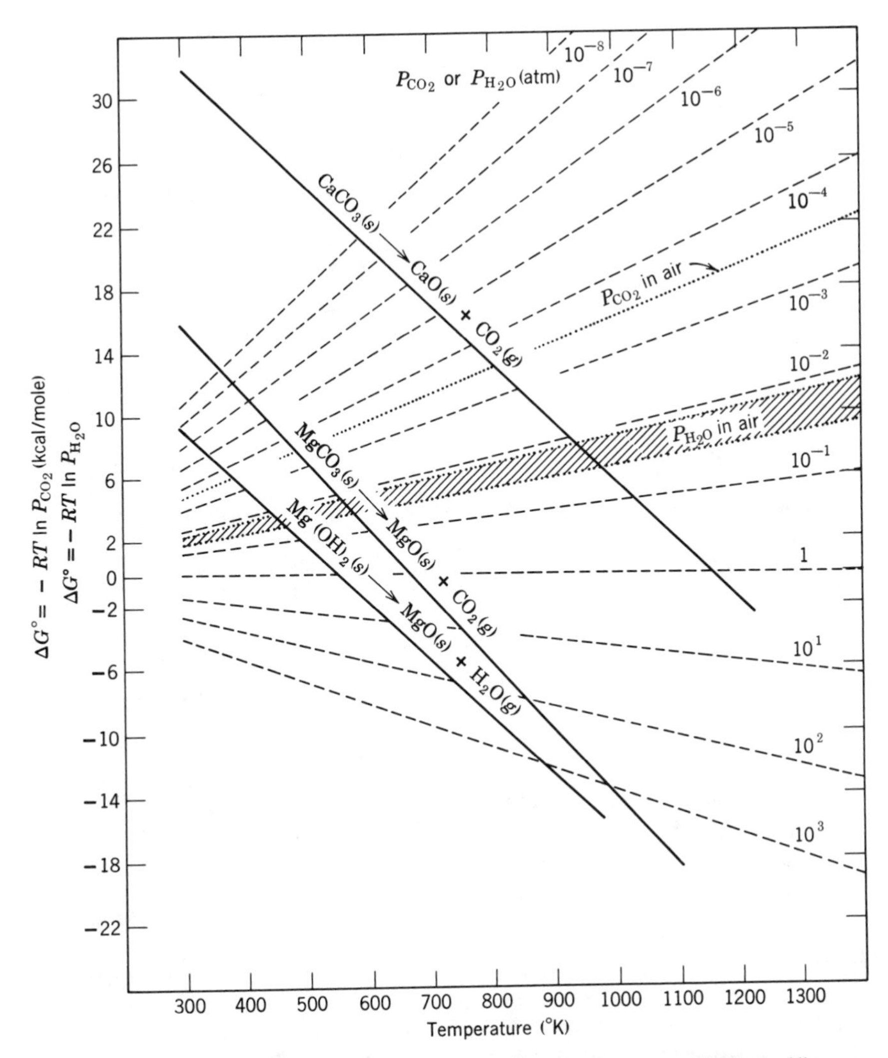

Fig. 9.22. Standard free energy of reaction as a function of temperature. The dashed lines are the equilibrium gas pressure above the oxide and carbonate (hydroxide).

rate-limiting step depends on the particular substance which is undergoing decomposition and the relative temperature. For example, at low temperatures the existence of unstable salts which decompose at higher temperatures suggests that the initial decomposition must be controlled by atomistic processes because there is no reaction-product interference in

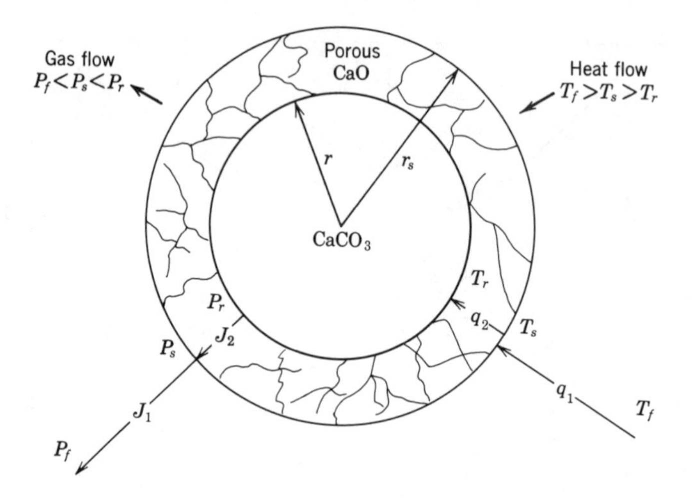

CO₂ flow to furnace

$$J_{\text{interface}} = k_r 4\pi r^2 (P_e - P_r)$$

$$J_2 = 4\pi \frac{D'_{CO_2}(P_r - P_s) r r_s}{r_s - r}$$

$$J_1 = 4\pi r_s^2 \frac{D''_{CO_2}}{\delta}(P_s - P_f)$$

$$P_e = e^{-\Delta G^\circ / RT_f}$$

δ = boundary-layer thickness

Heat flow to reaction interface

$$q_{\text{interface}} = \frac{4\pi r^2 \rho}{M} \Delta H^\circ_{T_r} \frac{dr}{dt}$$

$$q_1 = h_s 4\pi r_s^2 (T_f - T_s)$$

$$q_2 = \frac{4\pi k (T_s - T_r) r r_s}{r_s - r}$$

ρ = density of $CaCO_3$

M = molecular weight

h_s = heat-transfer coefficient

k = thermal conductivity of CaO

Fig. 9.23. Schematic representation of the decomposition of a spherical particle (e.g., $CaCO_3$) of a salt which yields a porous oxide product (e.g., CaO) and a gas (CO_2). The reaction is endothermic, requiring heat transfer. The driving forces for heat and mass transport for steady-state decomposition are expressed as temperatures and pressures in the furnace (T_f, P_f), at the particle surface (T_s, P_s), and at the reaction interface (T_r, P_r).

the transport of heat to the reaction interface or gaseous product away from the interface.

The reaction shown schematically in Fig. 9.23 is heterogeneous; that is, the reaction occurs at a sharply defined reaction interface. Figure 9.24 shows this interfacial area for $MgCO_3$ for which the reaction proceeds from nucleation sites on the surface of the $MgCO_3$ platelets. The

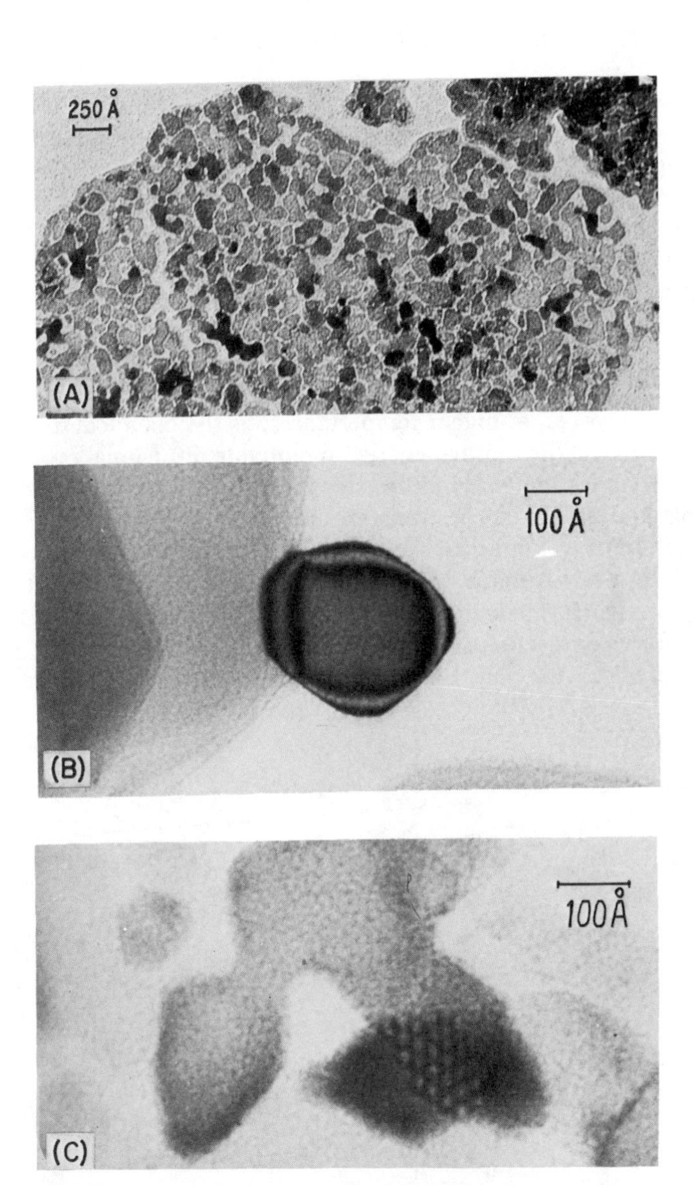

Fig. 9.24. Transmission electron micrographs of MgO prepared by thermal decomposition of basic magnesium carbonate. (*a*) Pseudomorphed MgO (550°C calcine); (*b*) crystallite approaching cube form (900°C calcine); (*c*) two-dimensional moire pattern from overlapped crystallites (550°C calcine). From A. F. Moodie, C. E. Warble, and L. S. Williams, *J. Am. Ceram. Soc.*, **49**, 676 (1966).

decomposition kinetics for cylindrical geometry is

$$(1-\alpha)^{1/2} = 1 - kt/r_0 \tag{9.64}$$

where α is the fraction decomposed, k is the thermally activated kinetic constant, t is the time (assumed constant temperature), and r_0 is the initial particle radius. The first-order kinetics (Eq. 9.2) for this reaction at several temperatures is shown in Fig. 9.25 for decomposition of $Mg(OH)_2$.

The importance of the surface on the decomposition rate is indicated by the time to decompose (700°C) a cleaved calcite crystal ($CaCO_3$), 60 hr, compared with an equivalent mass of the same material in powder form, 4 hr.

At low temperatures the crystallite size strongly affects the decomposition rate; however, at higher temperatures, as the chemical driving force increases and as the thermal energy to motivate diffusional processes and reaction kinetics increases, other steps may become rate-controlling, for example the rate of heat transfer. Figure 9.26 shows the center-line temperature of a cylindrical sample of pressed $CaCO_3$ powder which was thrust into a hot furnace. The sample temperature increases to a maximum, at which nucleation of CaO finally occurs. The decrease in temperature represents the endothermic heat absorbed by the reaction.

The effect of varying the ambient CO_2 pressure is illustrated in Fig.

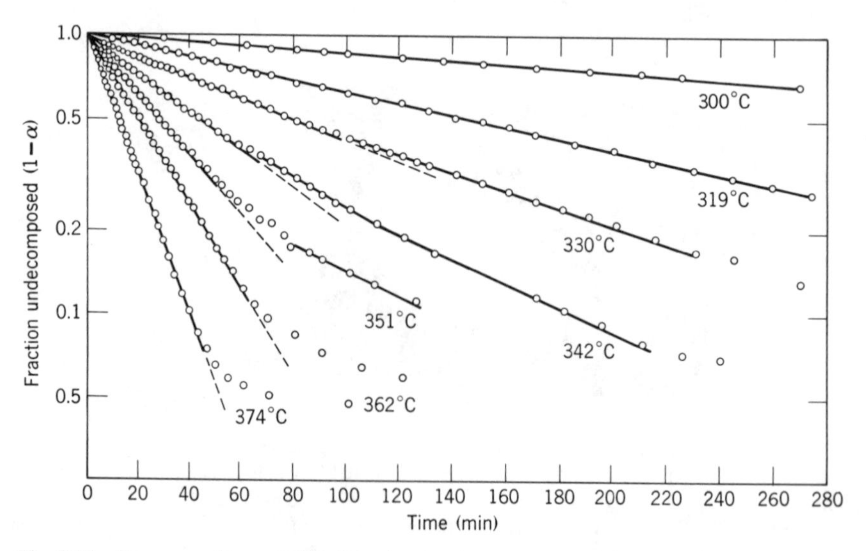

Fig. 9.25. Decomposition of $Mg(OH)_2$ showing first-order kinetics. From R. S. Gordon and W. D. Kingery, *J. Am. Ceram. Soc.*, **50**, 8 (1967).

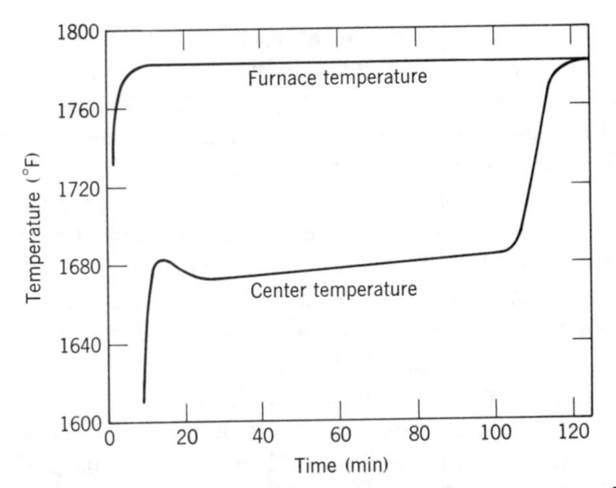

Fig. 9.26. Comparison of the furnace temperature to center-line temperature of a cylindrical sample of $CaCO_3$ thrust into a preheated furnace. From C. N. Satterfield and F. Feales, *A.I.CH.E.J.*, **5**, 1 (1959).

9.27. As the P_{CO_2} is increased, the driving potential for the reaction decreases, and thus the reaction rate is decreased.

Some of the clay minerals, kaolin in particular, do not decompose in the manner shown in Fig. 9.23; that is, they do not have a heterogeneous

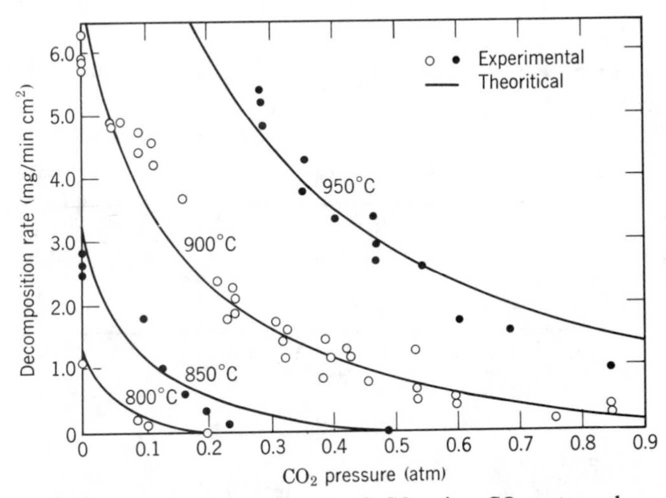

Fig. 9.27. Rate of decomposition of $CaCO_3$ in CO_2 atmosphere; $R_{\text{theor}} = (1 - P_{CO_2}/P^{\cdot}_{CO_2})/(BP_{CO_2} + 1/R_0)$, where $P^{\cdot}_{CO_2}$ = equilibrium CO_2 pressure, B = constant, and R_0 = decomposition rate in a pure neutral atmosphere. From E. P. Hyatt, I. B. Cutler, and M. E. Wadsworth, *J. Am. Ceram. Soc.*, **41**, 70 (1958).

reaction interface or a reaction product which breaks up into small crystallites. Above 500°C the water of crystallization is evolved, and a pseudomorphic structure remains until 980°C. The pseudomorph is a matrix of the original crystal structure containing large concentrations of vacant anion sites. Above 980°C the structure collapses irreversibly into crystalline mullite and silica, which releases heat (see Fig. 9.28).

The reaction kinetics is controlled by the diffusion of hydroxyl ions in the bulk rather than the heterogenous surface decomposition illustrated in Fig. 9.23. The kinetics is thus homogeneous and controlled by diffusion in the solid, which gives a parabolic rate law. The dehydration kinetics of kaolinite is given (1) in Fig. 9.29 for size fractions. A similar situation is observed for the decomposition of $Al(OH)_3$.

Powder Reactions. In most processes of interest in ceramic technology, solid-state reactions are carried out by intimately mixing fine powders. This changes the geometry from that considered in Fig. 9.6, and the actual reaction is more like that illustrated in Fig. 9.30.

If the reaction is carried out isothermally, the rate of formation of the reaction zone depends on the rate of diffusion. For the initial parts of the reaction the rate of growth of the interface layer is given to a good

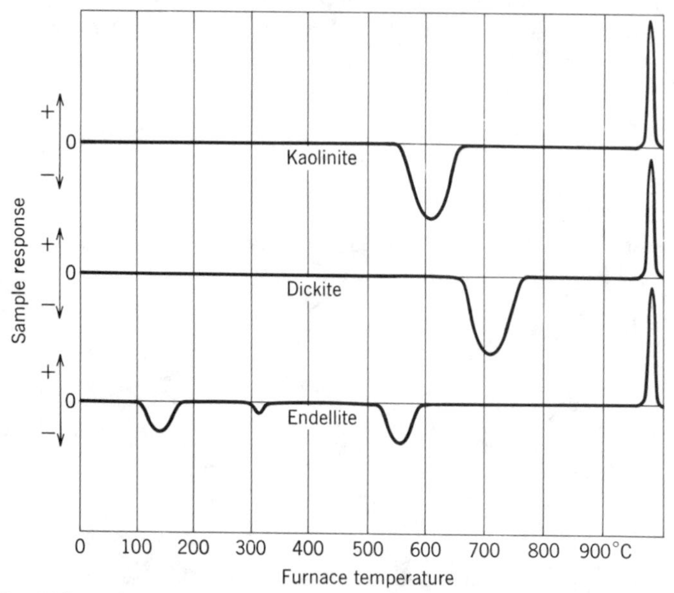

Fig. 9.28. Differential thermal analysis curves of kaolin clays. The sample temperature leads (+) or lags (−) the furnace temperature at levels at which heat is evolved or absorbed by chemical changes.

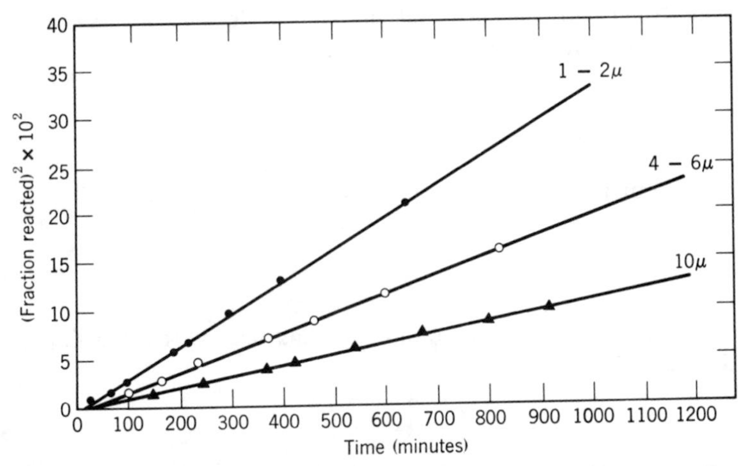

Fig. 9.29. Parabolic plots for three-size fractions of kaolinite at 400°C in vacuum. From J. B. Holt, I. B. Cutler, and M. E. Wadsworth, *J. Am. Ceram. Soc.*, **45**, 133 (1962).

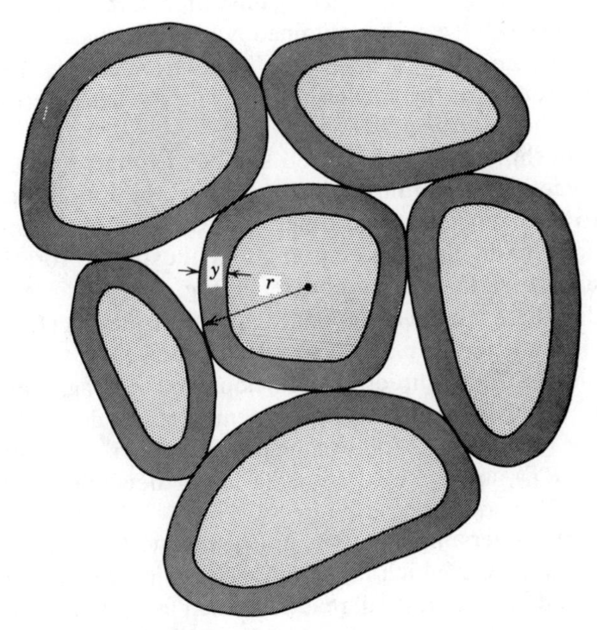

Fig. 9.30. Schematic representation of reaction-product layers forming on surface of particles in powder mixture.

approximation by the parabolic relationship in Eq. 9.8. If V is the volume of material still unreacted at time t, then

$$V = \frac{4}{3}\pi(r - y)^3 \tag{9.65}$$

The volume of unreacted material is also given by

$$V = \frac{4}{3}\pi r^3(1 - \alpha) \tag{9.66}$$

where α is the fraction of the volume that has already reacted. Combining Eqs. 9.65 and 9.66,

$$y = r(1 - \sqrt[3]{1 - \alpha}) \tag{9.67}$$

Combining this with Eq. 9.8 gives for the rate of reaction

$$(1 - \sqrt[3]{1 - \alpha})^2 = \left(\frac{KD}{r^2}\right)t \tag{9.68}$$

Note that this is for spherical geometry where Eq. 9.64 is for cylindrical geometry. By plotting $(1 - \sqrt[3]{1 - \alpha})^2$ against time, a reaction-rate constant equivalent to KD/r^2 can be obtained which is characteristic of the reaction conditions. The constant K is determined by the chemical-potential difference for the species diffusing across the reaction layer and by details of the geometry.

The relationship given in Eq. 9.68 has been found to hold for many solid-state reactions, including silicate systems, the formation of ferrites, reactions to form titanates, and other processes of interest in ceramics. The dependence on different variables is illustrated for the reaction between silica and barium carbonate in Fig. 9.31. In Fig. 9.31*a* it is observed that there is a linear dependence of the function $(1 - \sqrt[3]{1 - \alpha})^2$ on time. The dependence on particle size illustrated in Fig. 9.31*b* shows that the rate of the reaction is directly proportional to $1/r^2$ in agreement with Eq. 9.68: in 9.31*c* it is shown that the temperature dependence of the reaction-rate constant follows an Arrhenius equation, $K' = K'_0 \exp(-Q/RT)$, as expected from its major dependence on diffusion coefficient.

There are two oversimplifications in Eq. 9.68 which limit its applicability and the range over which it adequately predicts reaction rates. First, Eq. 9.68 is valid only for a small reaction thickness, Δy; and second, there was no consideration of a change in molar volume between the reactants and the product layer. The time dependence of the fraction reacted

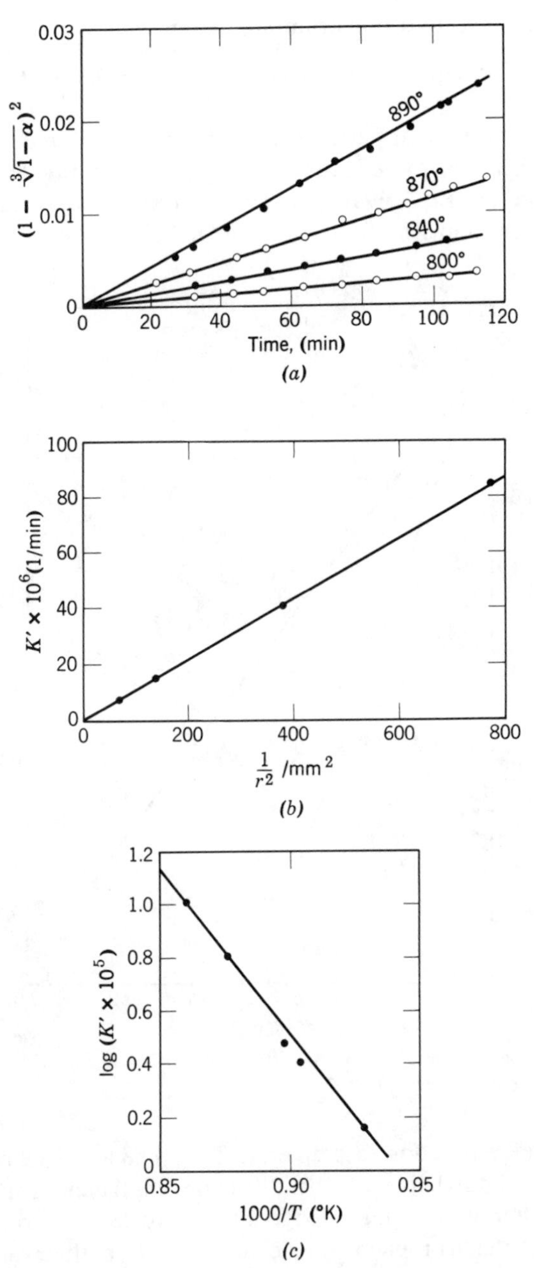

Fig. 9.31. Solid reaction between silica and barium carbonate showing (*a*) time dependence, (*b*) particle-size dependence, and (*c*) temperature dependence of reaction rate. From W. Jander, *Z. Anorg. Allg. Chem.*, **163**, 1 (1927).

corrected for these two constraints is given as*

$$[1+(Z-1)\alpha]^{2/3}+(Z-1)(1-\alpha)^{2/3}=Z+(1-Z)\left(\frac{KD}{r^2}\right)t \qquad (9.69)$$

where Z is the volume of particle formed per unit volume of the spherical particle which is consumed, that is, the ratio of equivalent volumes. A demonstration that Eq. 9.69 is valid even to 100% reaction is shown in Fig. 9.32 for the reaction $ZnO + Al_2O_3 = ZnAl_2O_4$.

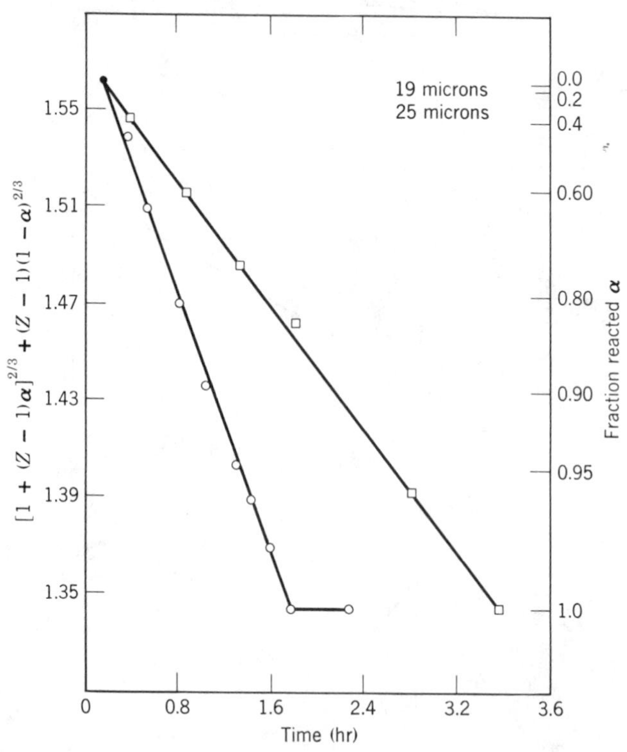

Fig. 9.32. Reaction between ZnO and Al_2O_3 to form $ZnAl_2O_4$ at 1400°C in air (two spherical particle sizes). See reference 1, p. 102.

Calculating the reaction rate given in Eqs. 9.68 and 9.69 on an absolute basis requires knowledge of the diffusion coefficient for all the ionic species together with a knowledge of the system's geometry and the chemical potential for each specie as related to their position in the reaction-product layer. The diffusing species which control the reaction

*R. E. Carter, *J. Chem. Phys.*, **34**, 2010 (1961); **35**, 1137 (1961).

rate are the most rapidly moving ions, or ions plus electrons, capable of arriving at a phase-boundary interface. All the constraints discussed in Section 9.2 must be considered.

Another difficulty in detailed quantitative calculations is the strong dependence of reaction rates on the structure of the reaction product. In many cases the reaction product is formed in such a way that it is not coherent with the reactants. Because of volume changes it may be formed with many defects and fissures. Consequently, there is extensive opportunity for surface and boundary diffusion, and the diffusion coefficient indicated in Eqs. 9.68 and 9.69 is not necessarily identical with diffusion through a single-crystal or dense polycrystalline body; these values set a lower limit for the actual diffusion coefficient and the possible reaction rate. When new phases are formed, as by carbonate decomposition at low temperatures, there is a strong tendency for the initial lattice parameter to be some nonequilibrium value corresponding to a coherent interface and structure with the reactant, as discussed in connection with nucleation and growth in Sections 8.3 and 8.4. Diffusion coefficients for this nonequilibrium lattice are normally larger than for the final equilibrium product. For example, an increase in solid-state reaction rate is frequently observed at a polymorphic transition temperature (the Hedvall effect). This effect is related to lattice strains and fissures formed by volume changes at the transition point; these lattice strains and fissures occur extensively in quartz, for example, in which the volume change is large. Also, at the transition temperature equilibrium between two polymorphic forms tends to occur with a coherent interface giving rise to lattice strains which increase the diffusion coefficients and the opportunity for material transfer. At present there are no data available for putting these effects on a quantitative basis.

Coarsening of Particles. After a solid has precipitated, the particles may undergo a coarsening effect because the variation in particle size represents a variation in the chemical activity from particle to particle. Generally termed Ostwald ripening, the principles apply to precipitates dispersed in solids or liquids. For the system of dispersed particles of varying size in a medium in which they have some solubility, the smaller ones dissolve and the larger ones grow. The driving force is the reduction of the interfacial free energy. The Thompson-Freundlich equation (Chapter 5) relates the increased solubility of the precipitate c_a to the curvature a relative to that for a planar interface $c_{\text{p.i.}}$:

$$RT \ln \frac{c_a}{c_{\text{p.i.}}} = \frac{2\gamma}{a} \frac{M}{\rho} \tag{9.70}$$

where γ is the interfacial energy (ergs/cm^2), M the molecular weight, and

ρ the density of the precipitate particle. This relation also assumes that the activity is given by the concentration. If $\frac{2\gamma M}{RTa\rho} < 1$, the increased solubility is given by

$$c_a = c_{\text{p.i.}}\left(1 + \frac{2\gamma M}{RTa\rho} + \cdots\right) = c_{\text{p.i.}} + \frac{2\gamma M c_{\text{p.i.}}}{RTa\rho} \tag{9.71}$$

For simplicity, consider a system of two particle sizes a_1 and a_2 where $a_1 > a_2$. The a_2 particles are more soluble in the matrix and thus tend to dissolve because of the concentration driving force:

$$c_{a_1} - c_{a_2} = \frac{2M\gamma c_{\text{p.i.}}}{RT\rho}\left(\frac{1}{a_1} - \frac{1}{a_2}\right) \tag{9.72}$$

From Fick's law we can determine the rate of growth of these particles if we assume the rate is controlled by diffusion in the matrix, that is, solvent (see Fig. 9.33a). The rate of mass gain by a_1 is

$$\frac{dQ}{dt} = -D\left(\frac{A}{x}\right)(c_{a_1} - c_{a_2}) \tag{9.73}$$

where A/x is a representative area-to-length ratio for diffusion between two dissimilar particles. Substitution of Eq. 9.72 into 9.73 yields

$$\frac{dQ}{dt} = -D\left(\frac{A}{x}\right)\frac{2M\gamma c_{\text{p.i.}}}{RT\rho}\left(\frac{1}{a_1} - \frac{1}{a_2}\right) \tag{9.74a}$$

As we have assumed spherical particles and must conserve mass,

$$-\frac{dQ}{dt} = \rho 4\pi a_2{}^2\frac{da_2}{dt} = -\rho 4\pi a_1{}^2\frac{da_1}{dt} \tag{9.74b}$$

the growth of a_1 is

$$\rho 4\pi a_1\frac{da_1}{dt} = -D\left(\frac{A}{x}\right)\frac{2M\gamma c_{\text{p.i.}}}{RT\rho}\left(\frac{1}{a_1} - \frac{1}{a_2}\right) \tag{9.75}$$

Equation 9.75 can be integrated under various approximations, however, the same solution results by considering the following approximate solution. If we assume that the small particles contribute solute to the matrix faster than the solute is precipitated onto the large particles, the growth of large particles can be treated as a diffusion-limited-growth problem. The rate-limiting step is assumed to be diffusion of matter to the large particle from the matrix. Assume a diffusion field of $r(r \gg a_1)$

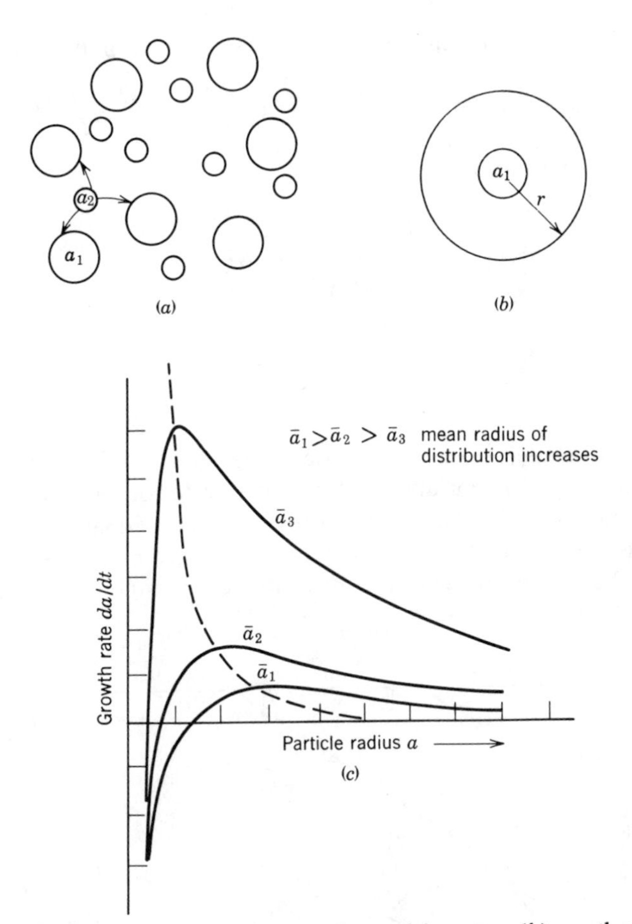

Fig. 9.33. (*a*) Coarsening of particles in a two-size particle system; (*b*) growth of particle a_1 in a diffusion field of radius r; (*c*) variation in the particle growth rate with particle radius.

around the growing particle (Fig. 9.33*b*); thus Fick's first law of spherical symmetry is

$$J = 4\pi D \Delta c \left(\frac{a_1 r}{r - a_1}\right) \tag{9.76}$$

where

$$\Delta c = \frac{2M\gamma c_{\text{p.i.}}}{RT\rho a_1}$$

Recalling that for dispersed particles $r \gg a_1$, the flux is given by

$$J = \frac{4\pi D 2M\gamma c_{\text{p.i.}}}{RT\rho a_i}\left(\frac{a_1 r}{r - a_1}\right) \approx \frac{4\pi D c_{\text{p.i.}} 2M\gamma}{RT\rho} = \text{const} \tag{9.77}$$

The flux is a constant, independent of the growing particle radius,

$$J = \text{constant} = \rho \frac{dV}{dt} = \rho 4\pi a_1^2 \frac{da_1}{dt} = \frac{4\pi D c_{\text{p.i.}} 2M\gamma}{\rho RT} \tag{9.78}$$

which after integration becomes

$$a_f^3 - a_i^3 = \frac{6Dc_{\text{p.i.}}M\gamma}{\rho^2 RT} \tag{9.79}$$

or

$$\left(\frac{a(t)}{a_i}\right)^3 = 1 + \frac{t}{\tau}$$

where

$$\tau = \frac{6Dc_{\text{p.i.}}M\gamma}{\rho^2 RT a_i^3} \tag{9.80}$$

More rigorous analyses give essentially the same result for a distribution of precipitates.* The variation in the growth rate for varying particle size and for increases in the mean radius is illustrated in Fig. 9.33*c*. The diffusion-limited growth of precipitates and of grains during liquid-phase sintering have been observed to have this cubic time dependence (Figs. 9.34 and 9.35).

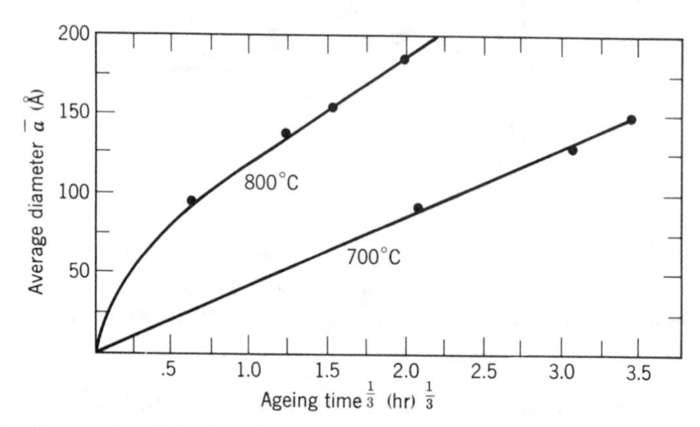

Fig. 9.34. Coarsening of $Mg_{1.2}Fe_{1.8}O_{3.9}$ precipitates in MgO. From G. P. Wirtz and M. E. Fine, *J. Am. Ceram. Soc.*, **51**, 402 (1968).

The coarsening relationships discussed above assumed spherical particles. The following discussion demonstrates that faceted particles and even those with different surface energies can be included in the growth expressions by properly defining Δc, the concentration difference.

*C. Wagner, *Z. Electrochem.*, **65**, 581–591 (1961); G. W. Greenwood, *Acta Met.*, **4**, 243–248 (1956).

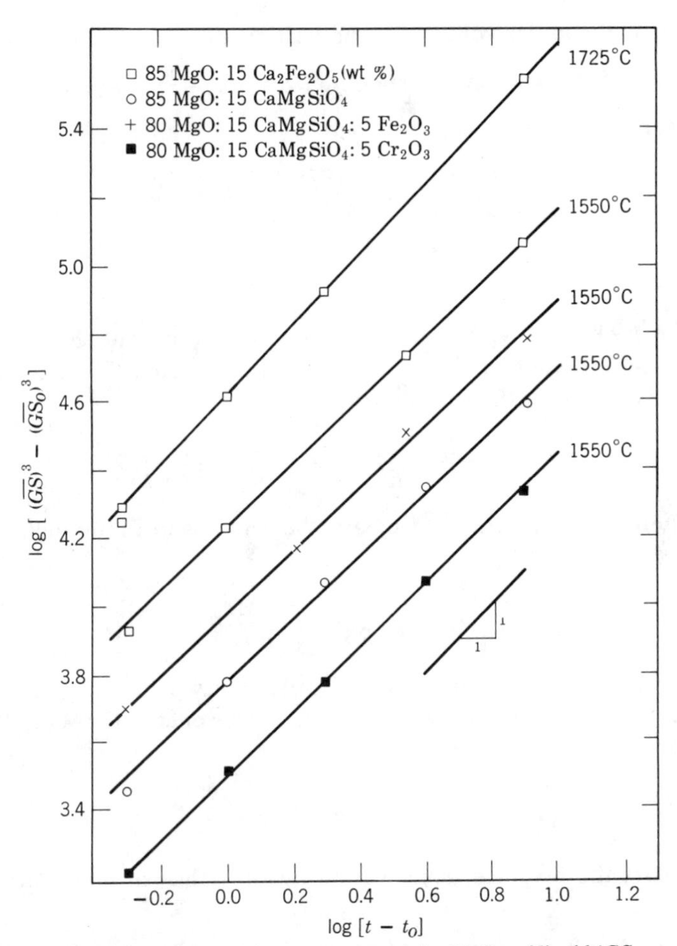

Fig. 9.35. Isothermal grain growth in systems containing MgO and liquid (GS = grain size in microns, t = time in hours) $(\overline{GS})^3 \alpha t$. From J. White in *Materials Sci. Research*, Vol. 6, Plenum Publishing Corporation, New York, 1973, p. 81.

Consider a size class of particles of constant shape but with a surface-to-volume ratio $S_V = S/V$. The surface contribution to the free energy is

$$G - G_{\text{p.i.}} = \gamma S = \gamma S_V V \tag{9.81}$$

The chemical potential difference between the faceted particle (μ) and the planar interface ($\mu_{\text{p.i.}}$) is

$$\mu - \mu_{\text{p.i.}} = \frac{d}{dn}(G - G_{\text{p.i.}}) = \gamma \bar{V}\left(\frac{dS}{dV}\right)_{S_V} \tag{9.82}$$

where $\bar{V}$ is the molar volume for the particle phase. Let x be some linear parameter of the particle size such that $S = ax^2 = \frac{\alpha V}{x}$, where a and α are characteristic shape constants. Then

$$\left(\frac{dS}{dV}\right)_{S_V} = S_V \frac{dS/S}{dV/V} = \frac{2}{3} S_V = \frac{2}{3}\frac{\alpha}{x} \tag{9.83}$$

and

$$\mu - \mu_{\text{p.i.}} = \gamma \bar{V} \frac{2}{3}\frac{\alpha}{x} \tag{9.84}$$

Equations 9.81 to 9.84 hold for any system of constant-shape particles, irrespective of whether they are spherical or faceted. If we assume the activity is given by the concentration,

$$\Delta c = \frac{(\mu - \mu_{\text{p.i.}})c_{\text{p.i.}}}{RT} = \frac{\gamma \bar{V} c_{\text{p.i.}}}{RT}\left(\frac{2}{3}\frac{\alpha}{x}\right) \tag{9.85}$$

For spheres $\alpha = 3$ and $x = r$, we have the Thompson-Freundlich equation (9.71):

$$\Delta c = \frac{2\gamma \bar{V} c_{\text{p.i.}}}{rRT} = \frac{2\gamma M c_{\text{p.i.}}}{RTr\rho}$$

where $\bar{V}$ is the molar volume, M the molecular weight and ρ the density. For faceted interfaces with varying surface free energy the Wulff theorem is applicable:

$$\mu - \mu_{\text{ref}} = \frac{2}{3} \bar{V} \sum \left(\frac{\alpha_i \gamma_i}{x_i}\right) \tag{9.86}$$

where x_i is the distance from the ith facet to the particle center.

9.5 Precipitation in Crystalline Ceramics

The nucleation and growth of a new phase has been discussed in Chapter 8 and applied there to processes occurring in a liquid or glass matrix. Polymorphic phase transformations in crystalline solids are discussed in Chapter 2. Precipitation processes from a crystalline matrix in which the precipitate has a composition different from the original crystal are important in affecting the properties of many ceramic systems, and as techniques such as transmission electron microscopy capable of observing and identifying fine precipitate particles are more fully applied, the widespread occurrence and importance of precipitation is becoming more fully recognized. Initiation of the process may occur by a spinodal process or by discrete particle nucleation (Chapter 8) when a driving force for phase separation occurs (Chapter 7); growth rates are limited by atom mobility (Chapter 6).

For nucleation in solids, strain energy resulting from differences in volume between precipitate and matrix must be included in evaluating the free-energy change on forming a nucleus. In these cases, Eq. 8.19 is replaced by

$$\Delta G_r = 4\pi r^2 \gamma + \frac{4}{3}\pi r^3(\Delta G_v + \Delta G_\epsilon) \tag{9.87}$$

where the strain energy per unit volume is given by $\Delta G_\epsilon = b\epsilon^2$, ϵ is the strain, and b is a constant which depends on the shape of the nucleus and can be calculated from elasticity theory. The presence of ΔG_ϵ in the expression for ΔG_r results in a free energy on forming the critical nucleus, ΔG^*, which corresponds to a definite crystallographic relation of the α and β structures and the boundary between them when both are crystalline phases. Inclusion of ΔG_ϵ can affect greatly the morphology of stable nuclei and increase the tendency for nucleation at heterogeneous sites. The strain energy typically causes the formation of parallel platelets when a decomposition (precipitation) reaction occurs. The configuration of precipitates as parallel platelets allows growth to take place with the minimum increase in strain energy. In general, the formation of thick or spherical particles produces large values of strain; since the strain energy is proportional to ϵ^2, precipitates with a platelike habit are preferred when the volume change on precipitation is appreciable, as is often the case. Strain energy also effects spinodal decomposition by increasing the energy of the inhomogeneous solution and depressing the temperature at which phase separation occurs and by causing separation to occur as lamellae in preferred crystallographic directions.

The energy for nucleation of a new phase depends on the interface structure and orientation, as discussed in Chapter 5. We can define two general kinds of precipitate. In a *coherent* precipitate, as in Fig. 9.36, planes of atoms are continuous across the interface so that only the second coordination of individual atoms is changed, similar to a twin boundary. In contrast, a *noncoherent* precipitate is one in which the planes of atoms, or some of them, are discontinuous across the interface, giving rise to dislocations or a random structure in the boundary layer, as described in Chapter 5. The interface energy of a coherent boundary is an order of magnitude less than that of an incoherent boundary, so that formation of new phases with definite structural relationships to the mother matrix is strongly preferred. In addition, the oxygen ions are commonly the more slowly moving in oxide structures, so that a transformation in which these ions must migrate to new positions is bound to be

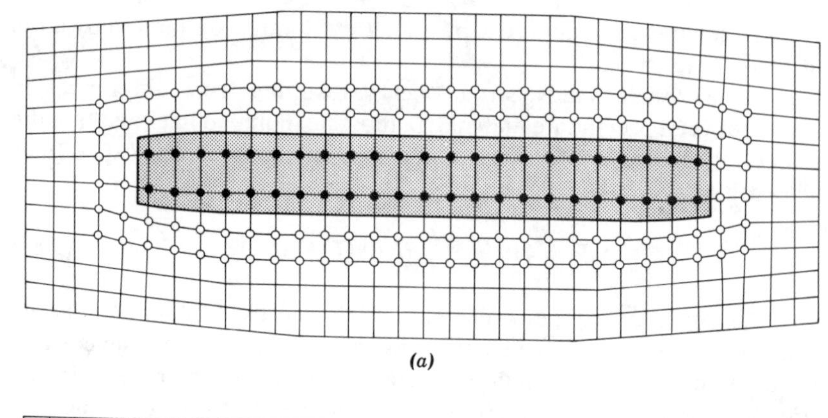

(a)

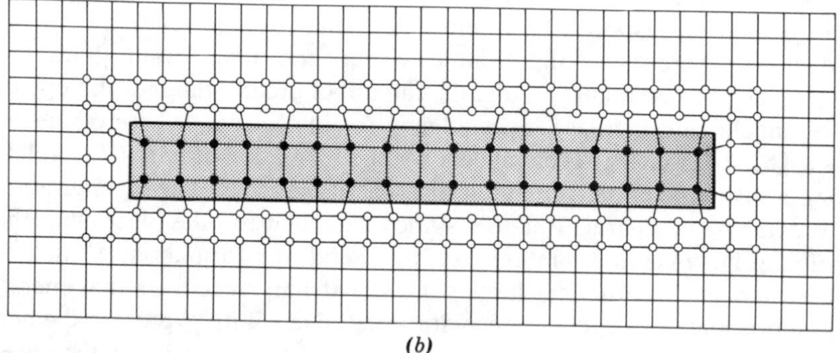

(b)

Fig. 9.36. (a) Coherent precipitate with continuous planes of atoms across the interface; (b) noncoherent precipitate with discontinuous planes of atoms across the interface.

relatively slow. Therefore, coherency of the oxygen ion lattice is favorable both for the driving force of nucleation and for the rate of nucleation and crystal growth.

Precipitation Kinetics. The kinetics of precipitation in a crystalline solid depend on both the rate of initiation or nucleation of the process and the rate of crystal growth, as discussed in Chapter 8. When the precipitation process consists of a combination of nucleation and growth, the sigmoidal curve characteristic of the Johnson-Mehl or Arrami relations (Chapter 8) results in an apparent incubation time period, as illustrated in Fig. 9.37. For precipitation processes far from an equilibrium phase boundary, which is the most usual case, both the nucleation rate and the growth rate increase with temperature, as illustrated in Fig. 9.38 such that the incubation time is decreased and the transformation time for formation of the new phase is decreased at higher temperatures, such as occurs for the process illustrated in Fig. 9.37. In many cases, however, the

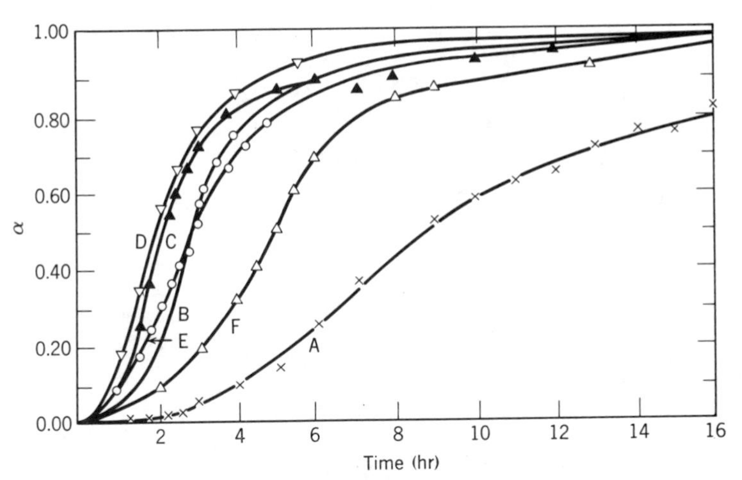

Fig. 9.37. Effect of temperature on the decomposition of solid solutions of 80 mole % ZrO_2–20 mole % MgO. Solid solutions prepared at 1520°C for 1 hr and then decomposed at A, 1000°C; B, 1075°C; C, 1150°C; D, 1250°C; E, 1350°C; F, 1375°C. From V. S. Stubican and D. J. Viechnicki, *J. Appl. Phys.*, **37**, 2751 (1966).

nucleation process occurs rapidly during cooling, such that a large number of nuclei are available for growth. This is particularly the case in which heterogeneous nucleation sites are available in a not very perfect matrix crystal or in which the interface energy term is low for a coherent precipitate. When this occurs the overall precipitation process, as measured by the fraction of material transformed, corresponds to growth of existing nuclei and no incubation period is observed, as illustrated for precipitation of $MgAl_2O_4$ spinel from MgO in Fig. 9.39. For precipitation of magnesium ferrite, $MgFe_2O_4$, from MgO, application of superparamagnetic measurements capable of identifying newly formed crystals having an average diameter of about 15 Å has shown no indication of an incubation period; that is, the critical nucleus size is very very small in accordance with a low energy for the coherent interface.*

Precipitate Orientation. The influence of strain energy and coherent interfaces leads to a high degree of precipitate orientation for many oxide precipitation processes. These relationships are particularly strong for the many oxide structures based on close-packed arrangements of oxygen ions described in Chapter 2. In the case of magnesium aluminate spinel containing excess aluminum oxide in solid solution, the influence of strain energy and coherency relationships leads to precipitation of a metastable

*G. P. Wirtz and M. E. Fine, *J. Am. Ceram. Soc.*, **51**, 402 (1968).

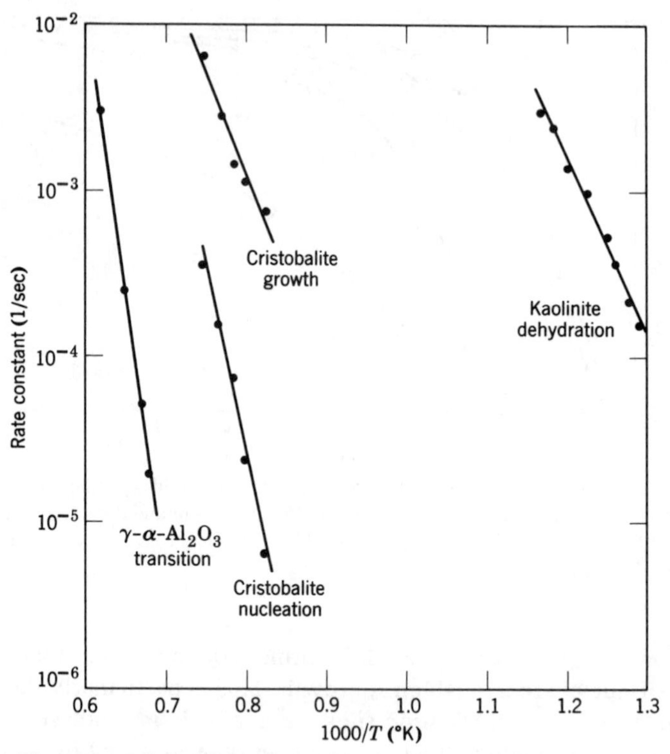

Fig. 9.38. Temperature dependence of rate of (*a*) nucleation, (*b*) growth of cristobalite from silica gel, (*c*) formation of α-Al_2O_3, and (*d*) dehydration of kaolinite.

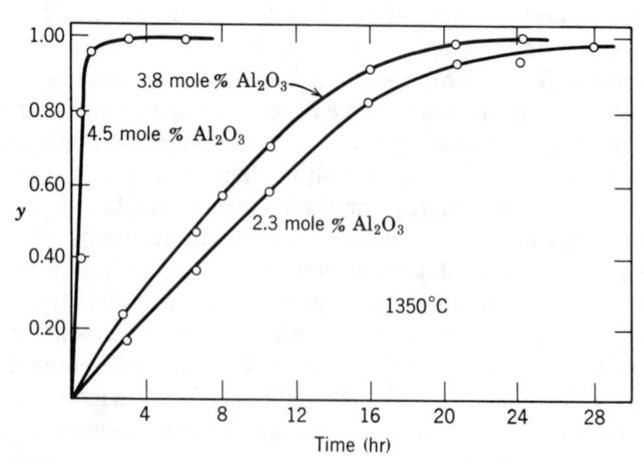

Fig. 9.39. The fractional precipitation y of spinel as a function of time at 1350°C. Specimens prepared at 1950°C. From V. S. Stubican and D. J. Viechnicki, *J. Appl. Phys.*, **37**, 2751 (1966).

intermediate with a structure similar to spinel as the first precipitation product* which is more easily nucleated than the stable equilibrium product, α-alumina. In fact, as shown in Fig. 9.40, two different types of metastable precipitates initially form, plus a smaller amount of α-alumina. After long annealing at 850°C the α-alumina particles grow at the expense of the metastable intermediate precipitates.

Synthetic star sapphires are produced by precipitating an alumina-rich titaniferous precipitate from single crystals of sapphire containing 0.1 to 0.3% TiO_2. When viewed in the direction of the c-axis stellate opalescence causes the reflected light to form a well-defined six-ray star. Aging times for precipitation range from approximately 72 hr at 1100°C to 2 hr at 1500°C. The lath-shaped precipitates formed are illustrated in Fig. 9.41. As for precipitation from spinel, the precipitate particle formed is not the equilibrium phase (Al_2TiO_5) but a metastable product.

Strong orientation effects are also observed in systems which are believed to exhibit spinodal decomposition, shown in Fig. 9.42, for the SnO_2–TiO_2 system in which a lamellar microstructure is formed after a 5-min anneal at 1000°C. The electron diffraction pattern at the lower corner of Fig. 9.42 shows streaking of the diffraction spots perpendicular to the 001 direction, which is to be expected for the periodic structure formed by spinodal decomposition. Other crystalline systems such as Al_2O_3–Cr_2O_3 and $CoFe_2O_4$–Co_3O_4 are also believed to phase separate in this manner. A similar structure, Fig. 9.43, is found for precipitation of the spinel phase from an FeO–MnO solid solution at low temperature. The large metal deficit in this highly nonstoichiometric system (discussed in Chapter 4) is believed to result in defect association on cooling; defect agglomerates may serve as nucleation sites for the precipitation reaction forming the spinel phase. Because of the high defect concentration and the resulting high diffusivity of the cations, precipitation processes occur in this and related systems at quite low temperatures, in this case about 300°C. On cooling a sample, it is not possible to prevent the formation of defect clusters, even with the most rapid quench.

When growth is rapid or occurs at low temperatures with a composition change, the rate of flow of heat or material limits the growth rate and fixes the morphology. Under these conditions the rate at which heat is dissipated or material added to a growing precipitate is proportional to the inverse radius of curvature of the growing tip of the crystal. As a result, dendritic forms result, with the radius of curvature of the growing tip remaining small and side arms developing to form a treelike structure.

* H. Jagodzinski, *Z. Krist.*, **109**, 388 (1957), and H. Saalfeld, *Ber. Deut. Keram Ges.*, **39**, 52 (1962).

Fig. 9.40. (*a*) Metastable precipitates I and II first form along with α-Al_2O_3 (A) on annealing spinel at 850°C; (*b*) after long times the α-Al_2O_3 particles coarsen at the expense of the intermediate precipitates. Courtesy G. K. Bansal and A. Heuer.

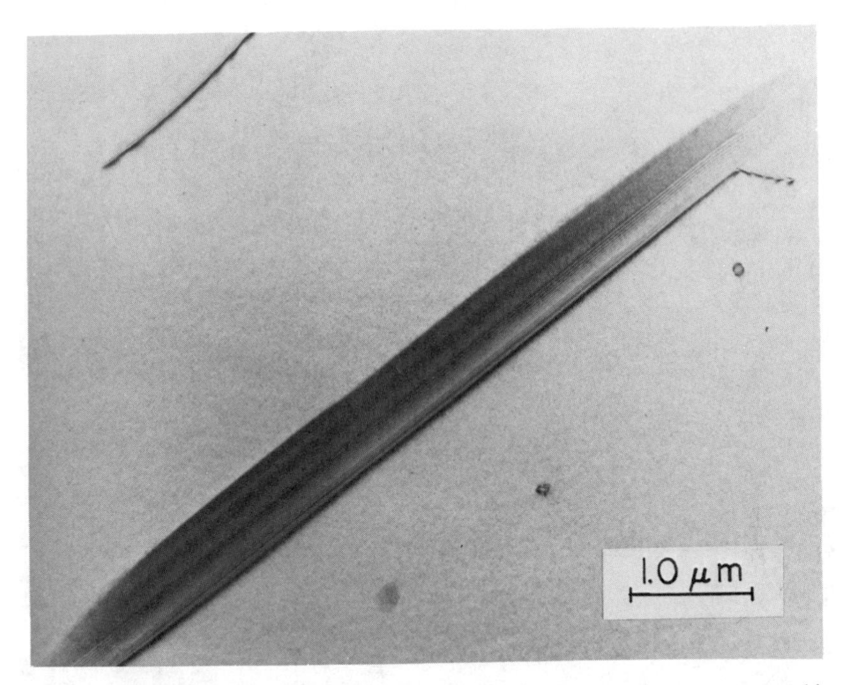

Fig. 9.41. Lath-shaped crystallographically oriented precipitate particles in star sapphire. Courtesy B. J. Pletka and A. Heuer.

Depending on the conditions of formation, different structures arise, as illustrated for the precipitation of magnesioferrite from magnesia in a basic refractory brick (Fig. 9.44). Sometimes a crystallographic orientation of the precipitate occurs in which the platelets of $MgFe_2O_4$ form along (100) planes in the parent magnesia phase. On precipitation during long periods of time at a lower temperature at which diffusion is probably rate-determining, dendritic precipitates form which still have crystallographic orientations with the matrix but in which the rate of growth is limited, so that starlike crystals result (Fig. 9.45*b*); finally, after long periods at the higher temperature levels, there is a tendency for a spheroidal precipitate to develop in which the total surface energy is a minimum and the strain energy may be relieved by plastic flow.

Heterogeneous Precipitation. It is frequently observed (Fig. 9.45*a*) that precipitation of a new phase occurs primarily along grain boundaries; when more extensive precipitation occurs (Fig. 9.45*b*) grain boundaries may show precipitates surrounded by an area of material which is nearly precipitation free. This can result from heterogeneous nucleation at the grain boundary, although in the case of precipitation of Fe_3O_4 from wüstite the microstructure observed at low magnifications results primar-

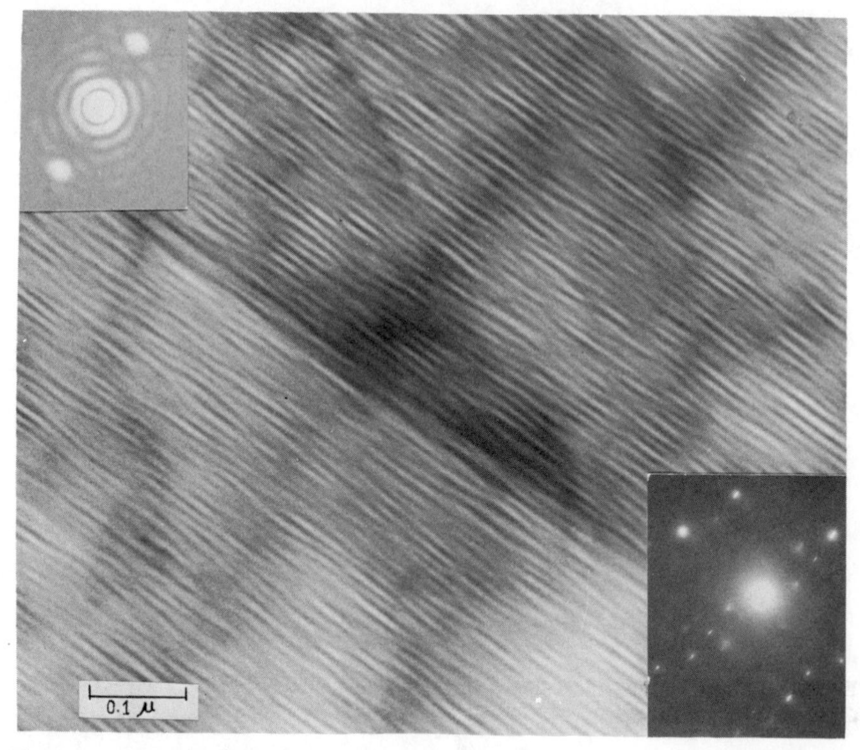

Fig. 9.42. Equimolar TiO_2–SnO_2 crystal homogenized at 1600°C and annealed 5 min at 1000°C. Electron diffraction pattern in lower right; optical diffraction pattern in upper left. Courtesy M. Park and A. Heuer.

ily from differences in the growth rate adjacent to grain boundaries rather than from a nucleation process. In this system the grain boundaries act as high diffusivity paths, discussed in Chapter 6, which allow nuclei at the grain boundary to grow initially at a faster rate than nuclei in the bulk, which tends to denude the area adjacent to the grain boundary of solute; at later stages in the precipitation process (Fig. 9.45*b*) there is an area adjacent to the grain boundaries which tends to be precipitate free. In this system, as for many of those previously described, the precipitate particles are coherent with the matrix crystal, and all have the same orientation in each grain of wüstite.

For samples in which solubility is small, direct observation of grain boundaries and dislocations indicates that second-phase precipitation at these sites is very common indeed. Particularly for many systems containing silicates as minor impurities, coherency is not to be expected;

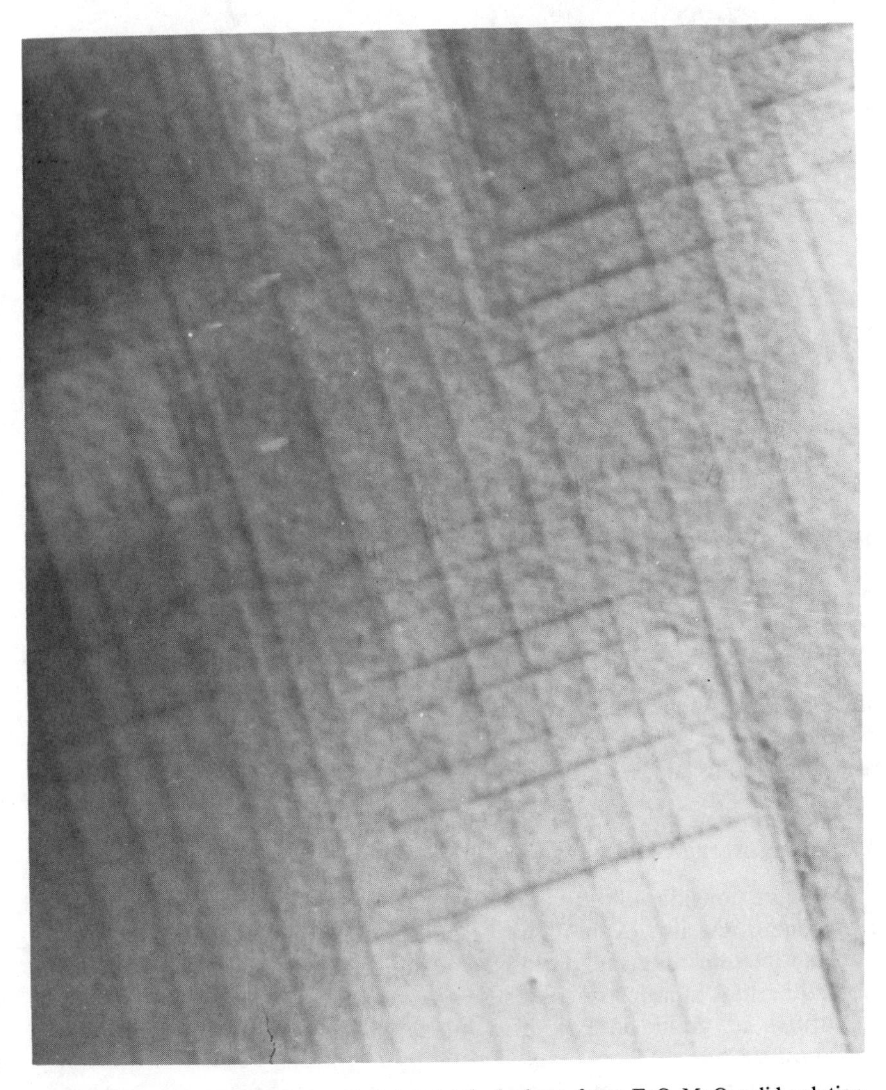

Fig. 9.43. Low-temperature precipitate of spinel phase from FeO–MnO solid solution. Courtesy C. A. Goodwin, Ph.D. Thesis, MIT, 1973.

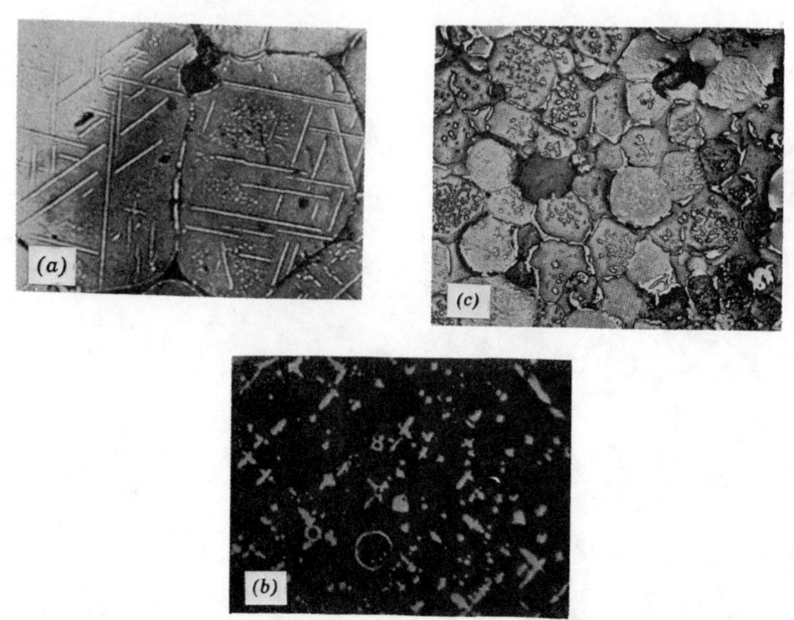

Fig. 9.44. Precipitation of $MgFe_2O_4$ from MgO in basic refractory brick as (*a*) platelets parallel to (100) planes in MgO (500 ×); (*b*) dendritic precipitate (975 ×); (*c*) spheroidal morphology (232 ×). Courtesy F. Trojer and K. Konopicky, *Radex Rdsch.*, **7**, **8**, 149 (1948), and B. Tavasci, *Radex Rdsch.*, **7**, 245 (1949).

the smaller driving force and larger energy barrier to nucleation enhances the importance of heterogeneous nucleation and growth at dislocations and grain boundaries, as illustrated in Fig. 9.46.

9.6 Nonisothermal Processes

We have considered diffusional processes as they occur under isothermal conditions; however, many ceramic processing procedures include substantial nonisothermal periods. One important example of the effect of nonisothermal kinetic processes is the segregation and precipitation of impurities at grain boundaries while a specimen cools from a high temperature.*

If we assume that over the temperature range of interest, the diffusion coefficient is given by

$$D = D_0 e^{-Q/RT}$$

*See W. D. Kingery, *J. Am. Ceram. Soc.*, **57**, 1 (1974).

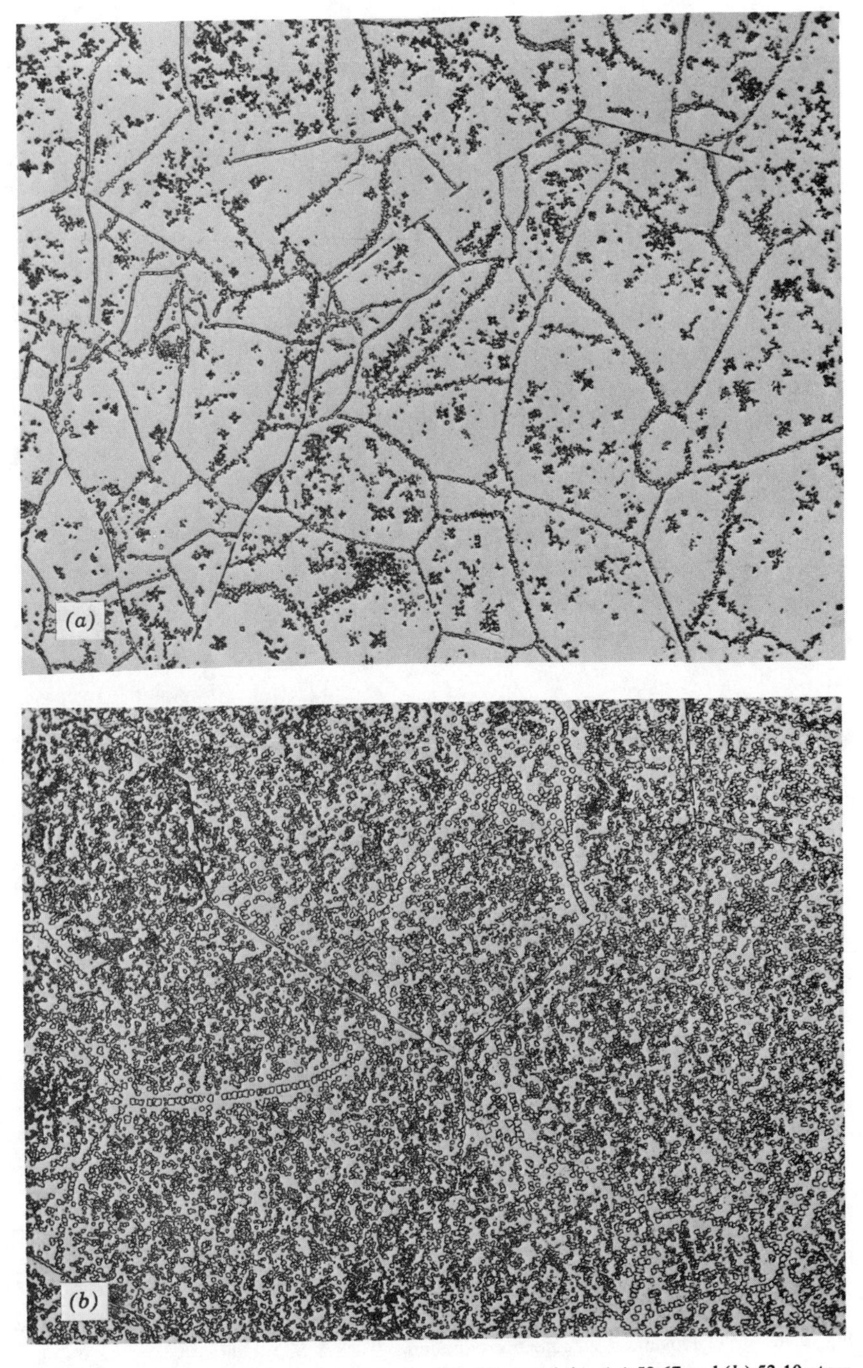

Fig. 9.45. Precipitate of Fe_3O_4 from wustite (Fe_xO) containing (*a*) 52.67 and (*b*) 53.10 atom % oxygen (95 ×). Courtesy L. Himmel.

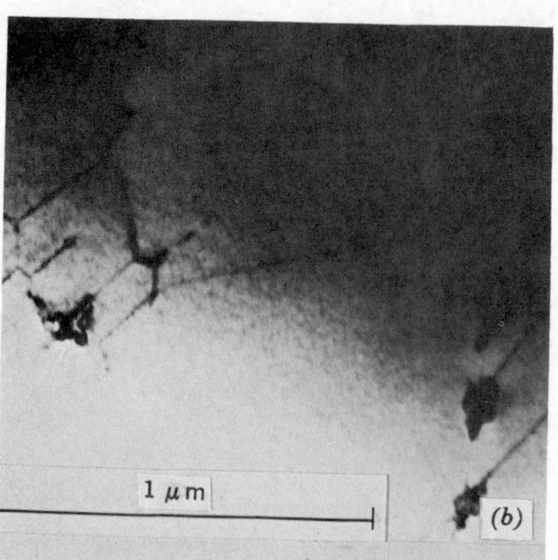

Fig. 9.46. (*a*) Manganese zinc ferrite containing 0.042% SiO_2 heated 4 hr at 1200°C in $N_2 + 1\%\ O_2$ showing SiO_2 inclusions at grain-boundary fracture. Courtesy M. Paulus. (*b*) Precipitate particles at dislocations in MgO. Courtesy N. J. Tighe.

and we assume the temperature to vary from T_1 to T_2 at a linear rate of α, the time-dependent diffusion coefficient is

$$D = D_0 \exp\left[\frac{Q/R}{T_1 - \alpha T}\right] \tag{9.88}$$

An approximate diffusion length l may be estimated from the integral;

$$l^2 = D_0 \int_0^t \exp\left(\frac{-Q/R}{T_1 - \alpha t}\right) dt \tag{9.89}$$

$$l^2 \approx \frac{D_0 R}{\alpha Q}(T_1^2 e^{-Q/RT_1} - T_2^2 e^{-Q/RT_2})$$
$$l^2 \approx \frac{R}{\alpha Q}[D_1 T_1^2 - D_2 T_2^2] \tag{9.90}$$

Let us consider, as an example of the use of Eq. 9.89, Al_2O_3 impurities in MgO. The diffusion of supersaturated aluminum ions from within a grain to the grain boundary is essentially that for the defect diffusion (vacancy) because of the impurity-vacancy pair which tends to form (see Section 6.4). From the data of impurity diffusion into MgO a value of 2 to 3 eV (50 to 75 kcal/mole) seems a reasonable activation energy for vacancy diffusion. Assume a sample of MgO annealed at high temperature contains 100 ppm Al_2O_3. If the sample is cooled at 0.1°C/sec, the solubility limit at 1300°C produces the onset of grain-boundary precipitation. For an assumed diffusivity of 10^{-8} cm^2/sec at 1300°C and $Q = 2$ eV, Eq. 9.89 yields a value of 30 microns for the effective diffusion distance. A similar calculation for 100 ppm MgO in Al_2O_3 ($T_s = 1530$°C, $Q \simeq 3$ eV, and $D \simeq 5 \times 10^{-8}$ cm^2/sec) yields a segregation thickness of 60 microns.

There are many other examples of ceramic processes which occur during nonisothermal annealing. As porcelain or refractories are processed in production kilns, much of the densification and reaction between granular components takes place during the heating cycle. We consider finally two examples of nonisothermal kinetic processes which are described in detail in Section 9.4 and Section 10.3 for isothermal conditions.

First, let us consider the nonisothermal decomposition reaction (Eq. 9.63) in which $CaCO_3$ decomposes to CaO and CO_2. The reaction rate is determined by decomposition at the surface and obeys linear kinetics. The reaction rate R is equal to the change in weight per unit area of the $CaCO_3$ with time, $d(\omega/a)/dt$. Thus Eq. 9.4 can be rewritten

$$\frac{d(\omega/a)}{dt} = R = \frac{kT}{h}\exp\left(\frac{\Delta S^{\dagger}}{R}\right)\exp\left(-\frac{\Delta H^{\dagger}}{RT}\right) = A \exp\left(-\frac{\Delta H^{\dagger}}{RT}\right) \tag{9.91}$$

If the temperature of the $CaCO_3$ is changed at a constant rate, $T = \alpha t$, the weight change as a function of temperature is obtained from

$$\frac{d(\omega/a)}{dT} = \frac{A}{\alpha}\exp\left(-\frac{\Delta H^{\dagger}}{RT}\right) \tag{9.92}$$

The integration of Eq. 9.92, assuming that A is not a strong function of temperature, yields the approximate solution

$$\frac{\Delta\omega}{\omega_0} \approx \frac{ART^2}{\alpha\,\Delta H^{\dagger}}\exp\left(-\frac{\Delta H^{\dagger}}{RT}\right) \tag{9.93}$$

The form of the equation is similar to Eq. 9.89. A plot of the nonisothermal decomposition in vacuum of a single crystal of $CaCO_3$ is given in Fig. 9.47. For this reaction and for several other endothermic decomposition reactions the activation energy for decomposition is identical with the heat of reaction (Eq. 9.63).

As a final example of nonisothermal kinetic processes consider the sintering of glass spheres (discussed in Chapter 10). The shrinkage rate $\frac{d(\Delta L/L_0)}{dt}$, which is a function of the surface tension γ, the viscosity $\eta = B\,e^{Q/RT}$, and the particle radius a, can be determined from nonisother-

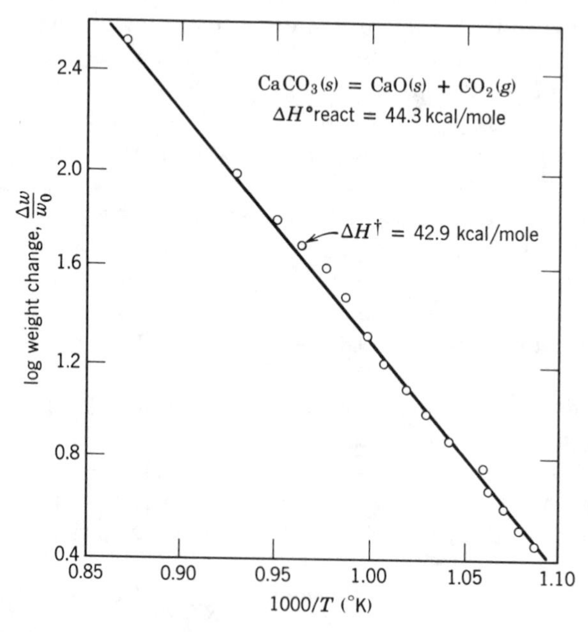

Fig. 9.47. Nonisothermal decomposition of $CaCO_3$ in vacuum.

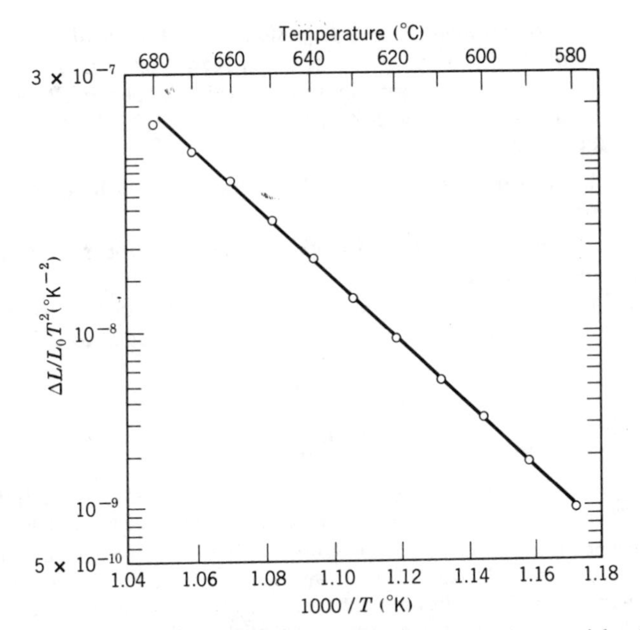

Fig. 9.48. Kinetic data for nonisothermal sintering of 0.25-μm glass particles (soda-lime-silica). From I. B. Cutler, *J. Am. Ceram. Soc.*, **52**, 14 (1969).

mal sintering from*

$$\frac{\Delta L}{L_0} \approx \left(\frac{\gamma R T^2}{2a\alpha BQ}\right) \exp\left(-Q/RT\right) \tag{9.94}$$

Kinetic data illustrating Eq. 9.94 are given in Fig. 9.48 for the sintering of 25-micron soda-lime-silica particles in an atmosphere of oxygen and water vapor.

Suggested Reading

1. H. Schmalzried, *Solid State Reactions*, Academic Press, New York, 1974.
2. G. C. Kuczynski, N. A. Hooton, and C. F. Gibbon, Eds., *Sintering and Related Phenomena*, Gordon and Breach, New York, 1967.
3. G. C. Kuczynski, Ed., "Sintering and Related Phenomena," *Materials Science Research*, Vol. 6, Plenum Press, New York, 1973.
4. T. J. Gray and V. D. Frechette, Eds., "Kinetics of Reaction in Ionic Systems," *Materials Science Research*, Vol. 4, Plenum Press, New York, 1969.

*I. B. Cutler, *J. Am. Ceram. Soc.*, **52**, 14 (1969).

5. P. Kofstad, *Nonstoichiometry, Diffusion, and Electrical Conductivity in Binary Metal Oxides*, John Wiley & Sons, New York., 1972.
6. For a discussion of dissolution kinetics see A. R. Cooper, Jr., B. N. Samaddar, Y. Oishi, and W. D. Kingery, *J. Am. Ceram. Soc.*, **47**, 37 (1964); **47**, 249 (1964); and **48**, 88 (1965).
7. G. M. Schwab, Ed., *Reactivity of Solids*, Elsevier Publishing Company, New York, 1965.
8. W. D. Kingery, Ed., *Kinetics of High Temperature Processes*, John Wiley & Sons, New York, 1959.
9. M. E. Fine, *Introduction to Phase Transformations in Condensed Systems*, McGraw-Hill, New York, 1964; *Bull. Am. Ceram. Soc.*, **51**, 510 (1972).

Problems

9.1. Rates of solution can be controlled by (*a*) diffusion in the liquid, (*b*) diffusion through a reaction layer, or (*c*) phase-boundary reaction. How would you distinguish these?

9.2. While measuring the rate of decomposition of alumina monohydrate, a student finds the weight loss to increase linearly with time up to about 50% reacted during an isothermal experiment. Beyond 50%, the rate of the weight loss is less than linear. The linear isothermal rate increases exponentially with temperature. An increase of temperature from 451 to 493°C increases the rate tenfold. Compute the activation energy. Is this a diffusion-controlled reaction, a first-order reaction, or an interface-controlled reaction?

9.3. Consider formation of $NiCr_2O_4$ from spherical particles of NiO and Cr_2O_3 when the rate is controlled by diffusion through the product layer.
 (*a*) Carefully sketch an assumed geometry, and then derive a relation for the rate of formation early in the process.
 (*b*) What governs the particles on which the product layer forms?
 (*c*) At 1300°C, $D_{Cr} > D_{Ni} > D_O$ in $NiCr_2O_4$. Which controls the rate of formation of $NiCr_2O_4$? Why?

9.4. Polymorphic transformations in solids result in polycrystalline materials of small size (fine-grained) or large size (coarse-grained), depending on the rates of nucleation and nuclei growth. How do these rates vary to produce fine-grained and coarse-grained products? Draw a time versus size for an individual grain illustrating the growth of a fine grain compared to a coarse grain. Start the time on the time axis at the moment of transformation.

9.5. According to Alper *et al.* [*J. Am. Ceram. Soc.*, **45**(6) 263–66 (1962)], Al_2O_3 is soluble in MgO to the extent of 3% by weight at 1700°C, 7% at 1800°C, 12% at 1900°C, and 0% at 1500°C. They observed crystallization of spinel crystals from the solid solution region on slow cooling. Fast quenching retained the solid solution as a single phase at room temperature. The exsolved spinel appeared uniformly without regard to grain boundaries within the periclase grains but on specific planes. (*a*) Is the nucleation of spinel homogeneous or heterogeneous within the periclase grains? (*b*) Account for the appearance of spinel crystals along specific planes of periclase crystals. Predict the shape of the rate of crystallization versus temperature for nucleated periclase solid solution containing 5% Al_2O_3 over the temperature range 0°C to 1850°C.

9.6. In the previous problem, we described a solid solution of Al_2O_3 in MgO. Assuming a manufacturer of basic refractories uses MgO contaminated with 5 to 7% Al_2O_3, what microstructure differences will exist in slow-cooled refractory compared to fast-cooled material? Would you predict sintering by self-diffusion (bulk), grain growth, and cation diffusion in this material would be different than in pure MgO? Why?

9.7. Suppose that the formation of mullite from alumina and silica powder is a diffusion-controlled process. How would you prove it? If the activation energy is 50 kcal/mole and the reaction proceeds to 10% of completion at 1400°C in 1 hr, how far will it go in 1 hr at 1500°C? in 4 hr at 1500°C?

9.8. An amorphous SiO_2 film on SiC builds up, limiting further oxidation. The fraction of complete oxidation was determined by weight gain measurements and found to obey a parabolic oxidation law. For a particular-particle-sized SiC and pure O_2 the following data were obtained. Determine the apparent activation energy in kcal/mole. How can it be shown that this is a diffusion-controlled reaction?

Temp (°C)	Fraction Reacted	Time (hr)
903	2.55×10^{-2}	100
1135	1.47×10^{-2}	10
	4.26×10^{-2}	100
1275	1.965×10^{-2}	10
	6.22×10^{-2}	100
1327	1.50×10^{-2}	5
	4.74×10^{-2}	50

9.9. The slow step in the precipitation of $BaSO_4$ from aqueous solution is the interface addition of the individual Ba^{++} and $SO_4^{=}$. Diffusion to the surface is assumed sufficiently fast that we may neglect any concentration differences in the solution. Assume that the rate of addition is first-order in both Ba^{++} and $SO_4^{=}$.

(*a*) Derive an expression for the approach to equilibrium in terms of the rate constants for the forward and back reaction and the surface area.

(*b*) What is the effect of an excess of Ba^{++}?

(*c*) Why can you assume the surface area to be constant?

(*d*) How would you modify your approach to include a correction for diffusion?

9.10. One-micron spheres of Al_2O_3 are surrounded by excess MgO powder in order to observe the formation of spinel. Twenty percent of the Al_2O_3 was reacted to form spinel during the first hour of a constant-temperature experiment. How long before all the Al_2O_3 will be reacted? Compute the time for completion on the basis of (*a*) no spherical geometry correction and (*b*) the Jander equation for correction of spherical geometry.

9.11. In fired chrome ore refractories, an R_2O_3 phase precipitates as platelets in the spinel phase matrix. Write the chemical equation for this reaction, and explain why it occurs. The precipitate is oriented so that the basal plane in the R_2O_3 phase is parallel to the (111) plane in the spinel. Explain why this should occur in terms of crystal structure.

10

Grain Growth, Sintering, and Vitrification

We have previously discussed phase changes, polymorphic transformations, and other processes independent of, or subsequent to, the fabrication of ceramic bodies. Phenomena that are of great importance are the processes taking place during heat treatment before use; these are the subject of this chapter.

During the usual processing of ceramics, crystalline or noncrystalline powders are compacted and then fired at a temperature sufficient to develop useful properties. During the firing process changes may occur initially because of decomposition or phase transformations in some of the phases present. On further heating of the fine-grained, porous compact, three major changes commonly occur. There is an increase in grain size; there is a change in pore shape; there is change in pore size and number, usually to give a decreased porosity. In many ceramics there may be solid-state reactions forming new phases, polymorphic transformations, decompositions of crystalline compounds to form new phases or gases, and a variety of other changes which are frequently of great importance in particular cases but are not essential to the main stream of events.

We shall be mainly concerned with developing an understanding of the major processes taking place. There are so many things which can happen, and so many variables that are occasionally important, that no mere cataloging of phenomena can provide a sound basis for further study. In general, we shall be concerned first with recrystallization and grain-growth phenomena, second with the densification of single-phase systems, and finally with more complex multiphase processes. There are many important practical applications for each of these cases.

10.1 Recrystallization and Grain Growth

The terms recrystallization and grain growth have had a very broad and indefinite usage in much of the ceramic literature; they have sometimes been used to include phase changes, sintering, precipitation, exsolution, and other phenomena which produce changes in the microstructure. We are mainly concerned with three quite distinct processes. *Primary recrystallization* is the process by which nucleation and growth of a new generation of strainfree grains occurs in a matrix which has been plastically deformed. *Grain growth* is the process by which the average grain size of strainfree or nearly strainfree material increases continuously during heat treatment without change in the grain-size distribution. *Secondary recrystallization*, sometimes called abnormal or discontinuous grain growth, is the process by which a few large grains are nucleated and grow at the expense of a fine-grained, but essentially strain-free, matrix. Although all these processes occur in ceramic materials, grain growth and secondary recrystallization are the ones of major interest.

Primary Recrystallization. This process has as its driving force the increased energy of a matrix which has been plastically deformed. The energy stored in the deformed matrix is of the order of 0.5 to 1 cal/g. Although this is small compared with the heat of fusion, for example (which is a 1000 or more times this value), it provides a sufficient energy change to effect grain-boundary movement and changes in grain size.

If the isothermal change in grain size of strainfree crystals in a deformed matrix is measured after an initial induction period, there is a constant rate of grain growth for the new strainfree grains. If the grain size is d,

$$d = U(t - t_0) \tag{10.1}$$

where U is the growth rate (cm/sec), t is the time, and t_0 is the induction period. This is illustrated in Fig. 10.1 for recrystallization of a sodium chloride crystal which had been deformed at 400°C and then annealed at 470°C. The induction period corresponds to the time required for a nucleation process, so that the overall rate of recrystallization is determined by the product of a nucleation rate and a growth rate.

The nucleation process is similar to those discussed in Chapter 8. For a nucleus to be stable, its size must be larger than some critical diameter at which the lowered free energy of the new grain is equal to the increased surface free energy. The induction period corresponds to the time required for unstable embryos present to grow to the size of a stable nucleus. If an unlimited number of sites is available, the rate of nucleation increases to some constant rate after an initial induction period. In

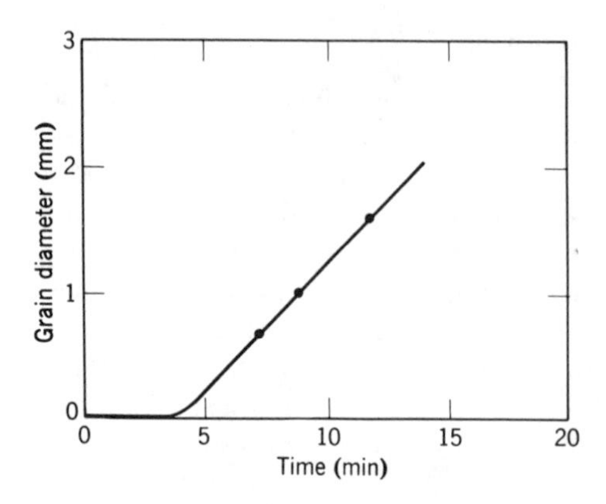

Fig. 10.1. Recrystallization of NaCl deformed at 400°C (stress = 4000 g/mm²) and recrystallized at 470°C. From H. G. Muller, *Z. Phys.*, **96**, 279 (1935).

practice the number of favorable sites available is limited, and the rate of nucleation passes through a maximum as they are used up. H. G. Müller* observed that nuclei in sodium chloride tended to form first at grain corners, for example. As the temperature is increased, the rate of nucleation increases exponentially:

$$\frac{N}{dt} = N_0 \exp\left(-\frac{\Delta G_N}{RT}\right) \tag{10.2}$$

where N is the number of nuclei and ΔG_N is the experimental free energy for nucleation. Consequently, the induction period, $t_0 \sim 1/(dN/dt)$, decreases rapidly as the temperature level is raised.

As indicated in Eq. 10.1, the growth rate remains constant until the grains begin to impinge on one another. The constant in growth rate results from the constant driving force (equal to the difference in energy between the strained matrix and strainfree crystals). The final grain size is determined by the number of nuclei formed, that is, the number of grains present when they finally impinge on one another. The atomistic process necessary for grain growth is the jumping of an atom from one side of a boundary to the other and is similar to a diffusional jump in the boundary. Consequently the temperature dependence is similar to that of diffusion:

$$U = U_0 \exp\left(-\frac{E_U}{RT}\right) \tag{10.3}$$

where the activation energy E_U is normally intermediate between that for boundary and lattice diffusion. The growth-rate-temperature curve for

***Z. Phys.*, **96**, 279 (1935).

recrystallization of sodium chloride has a knee similar to that observed for diffusion and conductivity data, as discussed in Chapter 6.

Since both the nucleation rate and the growth rate are strongly temperature-dependent, the overall rate of recrystallization changes rapidly with temperature. For a fixed holding time, experiments at different temperatures tend to a show either little or nearly complete recrystallization. Consequently, it is common to plot data as the amount of cold work or the final grain size as a function of the *recrystallization temperature.* Since the final grain size is limited by impingement of the grains on one another, it is determined by the relative rates of nucleation and growth. As the temperature is raised, the final grain size is larger, since the growth rate increases more rapidly than the rate of nucleation. However, at higher temperatures recrystallization is completed more rapidly, so that the larger grain size observed in constant-time experiments (Fig. 10.2) may be partly due to the greater time available for grain growth following recrystallization. The growth rate increases with increasing amounts of plastic deformation (increased driving force), whereas the final grain size decreases with increasing deformation.

In general, it is observed that (1) some minimum deformation is required for recrystallization, (2) with a small degree of deformation a higher temperature is required for recrystallization to occur, (3) an increased annealing time lowers the temperature of recrystallization, and

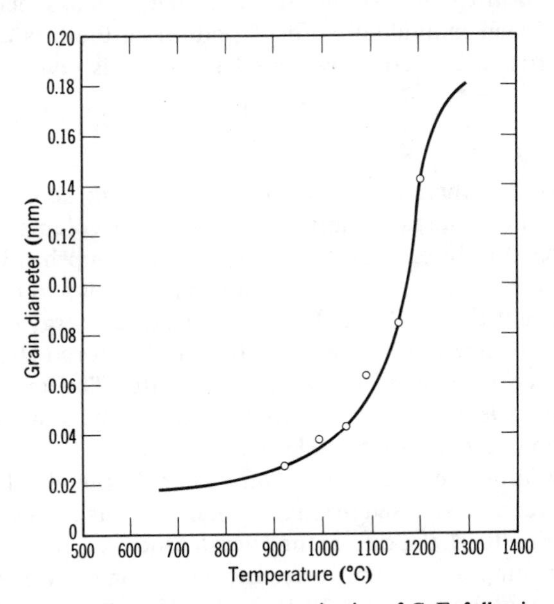

Fig. 10.2. Effect of annealing temperature on grain size of CaF_2 following compression at 80,000 psi and 10 hr at temperature. From M. J. Buerger, *Am. Mineral.*, **32**, 296 (1947).

(4) the final grain size depends on the degree of deformation, the initial grain size, and the temperature of recrystallization. In addition, continued heat after recrystallization is completed leads to the continuation of grain growth.

Primary recrystallization is particularly common in metals which are extensively deformed in normal processing techniques. Ceramic materials are seldom plastically deformed during processing, so that primary recrystallization is not commonly observed. For relatively soft materials, such as sodium chloride or calcium fluoride, deformation and primary recrystallization do occur. It has also been observed directly in magnesium oxide; also, the polygonization process described in Chapter 4 (see Fig. 4.24) for aluminum oxide has many points of similarity.

Grain Growth. Whether or not primary recrystallization occurs, an aggregate of fine-grained crystals increases in average grain size when heated at elevated temperatures. As the average grain size increases, it is obvious that some grains must shrink and disappear. An equivalent way of looking at grain growth is as the rate of disappearance of grains. Then the driving force for the process is the difference in energy between the fine-grained material and the larger-grain-size product resulting from the decrease in grain-boundary area and the total boundary energy. This energy change corresponds to about 0.1 to 0.5 cal/g for the change from a 1-micron to a 1-cm grain size.

As discussed in Chapter 5, an interface energy is associated with the boundary between individual grains. In addition, there is a free-energy difference across a curved grain boundary which is given by

$$\Delta G = \gamma \bar{V}\left(\frac{1}{r_1} + \frac{1}{r_2}\right) \tag{10.4}$$

where ΔG is the change in free energy on going across the curved interface, γ is the boundary energy, $\bar{V}$ is the molar volume, and r_1 and r_2 are the principal radii of curvature. (This relationship has been derived and discussed in Chapter 5. That part of Chapter 5 should be reviewed if its meaning is not clear.) This difference in the free energy of material on the two sides of a grain boundary is the driving force that makes the boundary move toward its center of curvature. The rate at which a boundary moves is proportional to its curvature and to the rate at which atoms can jump across the boundary.

Grain growth provides an opportunity to apply the absolute-reaction-rate theory already discussed in Chapter 6. If we consider the structure of a boundary (Fig. 10.3), the rate of the overall process is fixed by the rate at which atoms jump across the interface. The change in energy with an atom's position is shown in Fig. 10.3b, and the frequency of atomic jumps

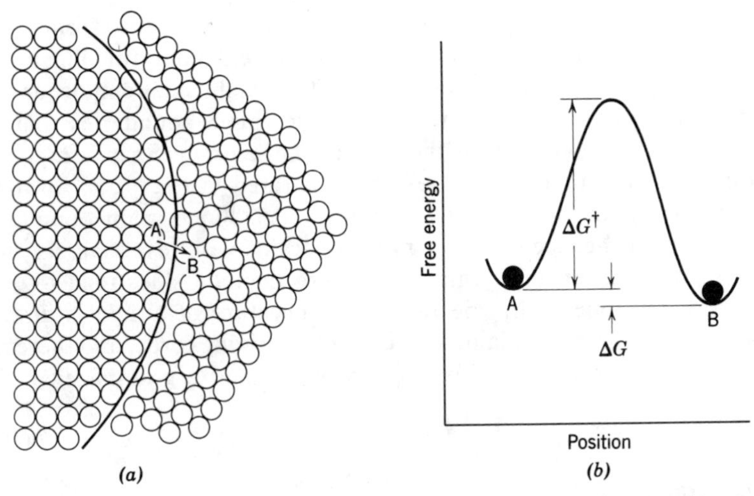

Fig. 10.3. (*a*) Structure of boundary and (*b*) energy change for atom jump.

in the forward direction is given by

$$f_{AB} = \frac{RT}{Nh} \exp\left(-\frac{\Delta G^{\dagger}}{RT}\right) \tag{10.5}$$

and the frequency of reverse jumps is given by

$$f_{BA} = \frac{RT}{Nh} \exp\left(-\frac{\Delta G^{\dagger} + \Delta G}{RT}\right) \tag{10.6}$$

so that the net growth process, $U = \lambda f$, where λ is the distance of each jump is given by

$$U = \lambda f = \lambda (f_{AB} - f_{BA}) = \frac{RT}{Nh} (\lambda) \exp\left(-\frac{\Delta G^{\dagger}}{RT}\right)\left(1 - \exp\frac{\Delta G}{RT}\right) \tag{10.7}$$

and since $1 - \exp\frac{\Delta G}{RT} \cong \frac{\Delta G}{RT}$, where $\Delta G = \gamma \bar{V}\left(\frac{1}{r_1} + \frac{1}{r_2}\right)$ and $\Delta G^{\dagger} = \Delta H^{\dagger} - T\,\Delta S^{\dagger}$,

$$U = \left(\frac{RT}{Nh}\right)(\lambda)\left[\frac{\gamma \bar{V}}{RT}\left(\frac{1}{r_1} + \frac{1}{r_2}\right)\right] \exp\frac{\Delta S^{\dagger}}{R} \exp\left(-\frac{\Delta H^{\dagger}}{RT}\right) \tag{10.8}$$

which is equivalent in form to Eq. 10.3 given previously. That is, the rate of growth increases exponentially with temperature. The unit step involved is the jump of an atom across the boundary, so that the activation energy should correspond approximately to the activation energy for boundary diffusion.

If all the grain boundaries are equal in energy, they meet to form angles of 120°. If we consider a two-dimensional example for illustrative purposes, angles of 120° between grains with straight sides can occur only for six-sided grains. Grains with fewer sides have boundaries that are concave when observed from the center of the grain. Shapes of grains having different numbers of sides are illustrated in Fig. 10.4; a sample with uniform grain size is shown in Fig. 10.5. Since grain boundaries migrate toward their center of curvature, grains with less than six sides tend to grow smaller, and grains with more than six sides tend to grow larger. For any one grain, the radius of curvature of a side is directly proportional to the grain diameter, so that the driving force, and therefore the rate of grain growth, is inversely proportional to grain size:

$$\dot{d} = \frac{\mathrm{d}(d)}{\mathrm{d}t} = \frac{k}{d} \tag{10.9}$$

and integrating,

$$d - d_0 = (2k)^{1/2} t^{1/2} \tag{10.10}$$

where d_0 is the grain diameter at time zero. Experimentally it is found that when $\log d$ is plotted versus $\log t$, a straight line is obtained (Fig. 10.6). Frequently the slope of curves plotted in this way is smaller than one-half,

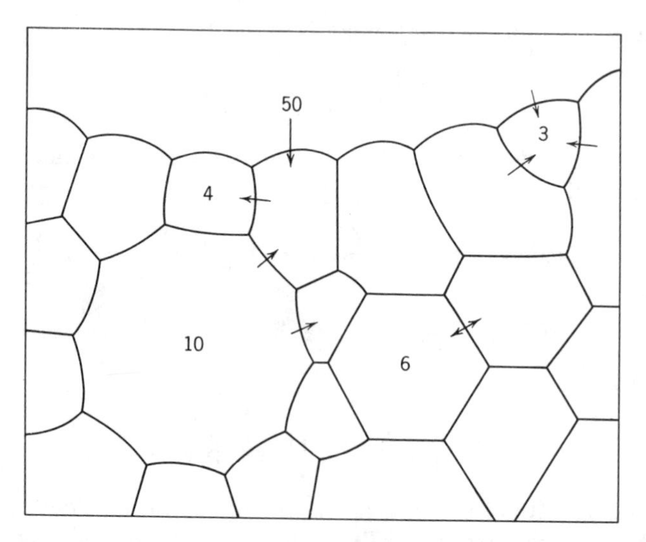

Fig. 10.4. Schematic drawing of polycrystalline specimen. The sign of curvature of the boundaries changes as the number of sides increases from less than six to more than six, and the radius of curvature is less, the more the number of sides differs from six. Arrows indicate the directions in which boundaries migrate. From J. E. Burke.

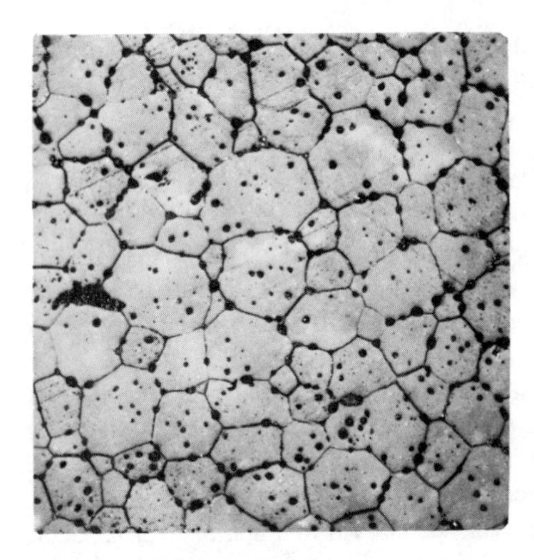

Fig. 10.5. Polycrystalline CaF_2 illustrating normal grain growth. Average angle at grain junctures is 120°.

usually falling between 0.1 and 0.5. This may occur for several reasons, one being that d_0 is not a large amount smaller than d; another common reason is that inclusions or solute segregation or sample size inhibits grain growth.

A somewhat different approach is to define a grain-boundary mobility B_i such that the boundary velocity v is proportional to the applied driving force F_i resulting from boundary curvature:

$$v = B_i F_i \tag{10.11a}$$

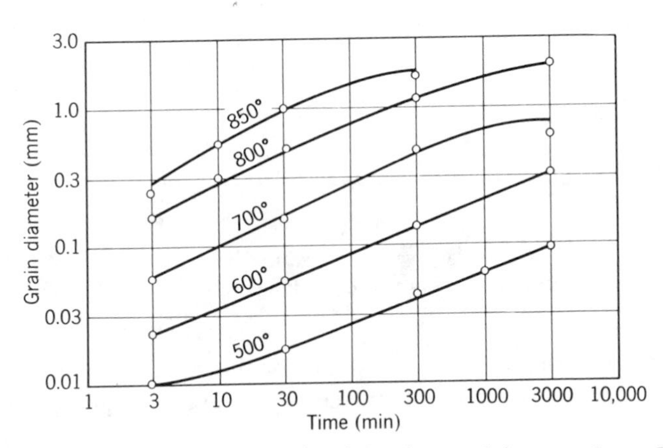

Fig. 10.6. Log grain diameter versus log time for grain growth in pure α-brass. From J. E. Burke.

For the atomic-jump mechanism illustrated in Fig. 10.3, the boundary mobility is given by the atomic mobility divided by the number of atoms involved, n_a:

$$B_i = \frac{B_a}{n_a} = \left(\frac{D_b}{kT}\right)\left(\frac{\Omega}{Sw}\right) \tag{10.11b}$$

where D_b is the grain-boundary diffusion coefficient, Ω is the atomic volume, S is the boundary area, and w is the boundary width. Since the average boundary velocity is equal to v and the driving force is inversely proportional to grain size, a grain-growth law of the form of Eqs. 10.9 and 10.10 results. However, as discussed in Chapter 5, the actual structure of a ceramic grain boundary is not quite so simple as pictured in deriving Eqs. 10.8 and 10.11*b*. Even for a completely pure material there is a space-charge atmosphere of lattice defects associated with the boundary and usually solute segregation as well, as shown in Figs. 5.11, 5.12, 5.17, and 5.18. The effect of this lattice defect and impurity atmosphere is to sharply reduce the grain-boundary velocity at low driving forces, as shown in Fig. 10.7 and analysed by J. Cahn* and K. Lücke and H. D. Stuwe.† The influence of this atmosphere becomes stronger as the grain

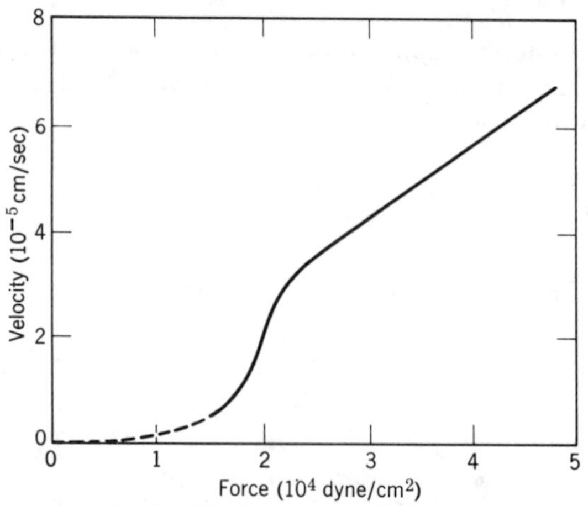

Fig. 10.7. Variation of boundary velocity v with driving force F at 750°C for a 20° tilt boundary in NaCl. From R. C. Sun and C. L. Bauer, *Acta Met.*, **18**, 639 (1970).

size increases, the solute segregate concentration increases, and the average boundary curvature decreases. Additions of MgO to Al_2O_3, $CaCl_2$

***Acta Met.*, **10**, 789 (1962).
†*Acta Met.*, **19**, 1087 (1971).

to KCl and of ThO_2 to Y_2O_3 in amounts below the solubility limit have proved effective as grain-growth inhibitors.

When grains grow to such a size that they are nearly equal to the specimen size, grain growth is stopped. In a rod sample, for example, when the grain size is equal to the rod diameter, the grain boundaries tend to form flat surfaces normal to the axis so that the driving force for boundary migration is eliminated and little subsequent grain growth occurs. Similarly, inclusions increase the energy necessary for the movement of a grain boundary and inhibit grain growth. If we consider a boundary such as the one illustrated in Fig. 10.8, the boundary energy is decreased when it reaches an inclusion proportional to the cross-sectional area of the inclusion. The boundary energy must be increased again to pull it away from the inclusion. Consequently, when a number of inclusions are present on a grain boundary, its normal curvature becomes insufficient for continued grain growth after some limiting size is reached. It has been found that this size is given by

$$d_l \approx \frac{d_i}{f_{d_i}} \tag{10.12}$$

where d_l is the limiting grain size, d_i is the particle size of the inclusion, and f_{d_i} is the volume fraction of inclusions. Although this relationship is only approximate, it indicates that the effectiveness of inclusions increases as their particle size is lowered and the volume fraction increases.

For the process illustrated in Fig. 10.8, the boundary approaches, is attached to, and subsequently breaks away from a second-phase particle. Another possibility is that the grain boundary drags along the particle

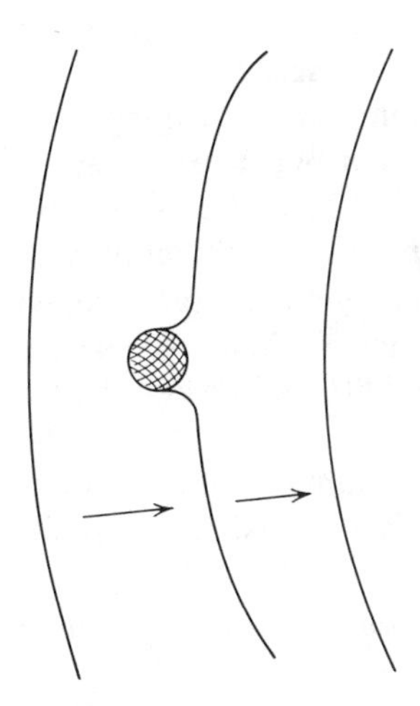

Fig. 10.8. Changing configuration of a boundary while passing an inclusion.

which remains attached to the boundary as it moves. This requires material transport across the particle, which may occur by interface or surface or volume diffusion, by viscous flow, or by solution (precipitation in a liquid or glass inclusion), or by evaporation (condensation in a gas inclusion). We can define an inclusion particle mobility B_p relating the driving force and particle velocity $v_p = B_p F_p$ in the same way as has been done for the boundary (Eq. 10.11*b*) and for atomic diffusion in Chapter 6. When the inclusion is dragged by the boundary, their velocities are identical; in the case in which $B_p \ll B_b$ we can neglect the intrinsic boundary mobility, and the resulting grain-boundary velocity is controlled by the driving force on the boundary together with the mobility and number of inclusions per grain boundary, p:

$$v_b = \frac{B_p F_b}{p} \tag{10.13}$$

The inclusion particle moves along with the boundary, gradually becoming concentrated at boundary intersections and agglomerating into larger particles as grain growth proceeds. This is illustrated for the special case of pore agglomeration in Figs. 10.9 and 10.10.

Thus, second-phase inclusions can either (1) move along with boundaries, offering little impedance; (2) move along with boundaries, with the inclusion mobility controlling the boundary velocity; or (3) be so immobile that the boundary pulls away from the inclusion, depending on the relative values of the boundary driving force (inversely proportional to grain size), the boundary mobility (Fig. 10.7), and the inclusion particle mobility, which, depending on the assumed mechanism and particle shape, may be proportional to r_p^{-2}, r_p^{-3}, or r_p^{-4}.* As grain growth proceeds, the driving force diminishes, and any inclusions dragged along by the boundary increase in size so that their mobility decreases. As a result, the exact way in which second-phase inclusions inhibit grain growth not only depends on the properties of the particular system but also can easily change during the grain-growth process. Sorting out these effects requires a careful evaluation of the microstructure evolution in combination with the kinetics of grain growth and a detailed knowledge of system properties. Inhibition of grain growth by solid second-phase inclusions has been observed for MgO additions to Al_2O_3, for CaO additions to ThO_2, and in other systems.

A second phase that is always present during ceramic sintering and in almost all ceramic products prepared by sintering is residual porosity

*P. G. Shewmon, *Trans. A.I.M.E.*, **230**, 1134 (1964); M. F. Ashby and R. M. A. Centamore, *Acta Met.*, **16**, 1081 (1968).

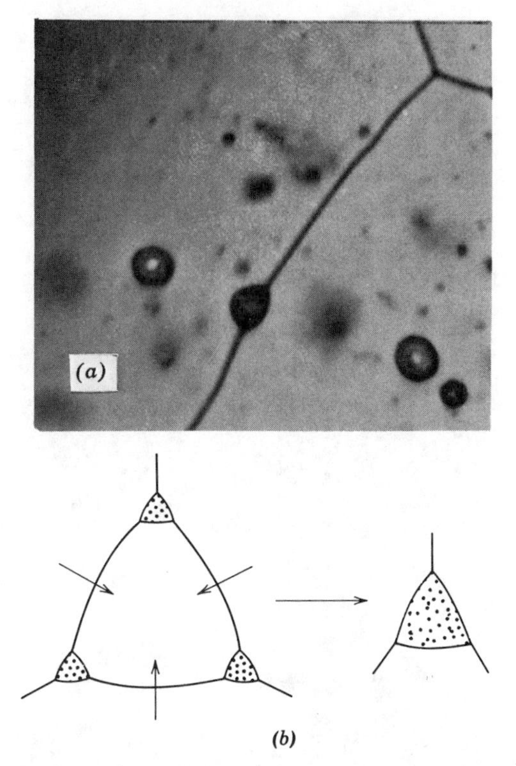

Fig. 10.9. (*a*) Pore shape distorted from spherical by moving boundary and (*b*) pore agglomeration during grain growth.

remaining from the interparticle space present in the initial powder compact. This porosity is apparent both on the grain boundaries (intergranular) and within the grains (intragranular) in the sintered CaF_2 sample shown in Fig. 10.5. It is present almost entirely at the grain corners (intergranular) in the sintered UO_2 samples shown in Fig. 10.10. As with particulate inclusions, pores on the grain boundaries may be left behind by the moving boundary or migrate with the boundary, gradually agglomerating at grain corners, as illustrated in Figs. 10.9 and 10.10. In the early stages of sintering, when the boundary curvature and the driving force for boundary migration are high, pores are often left behind, and a cluster of small pores in the center of a grain is a commonly observed result (see Fig. 10.5). In the later stages of sintering, when the grain size is larger and the driving force for boundary migration is lower, it is more usual for pores to be dragged along by the boundary, slowing grain growth.

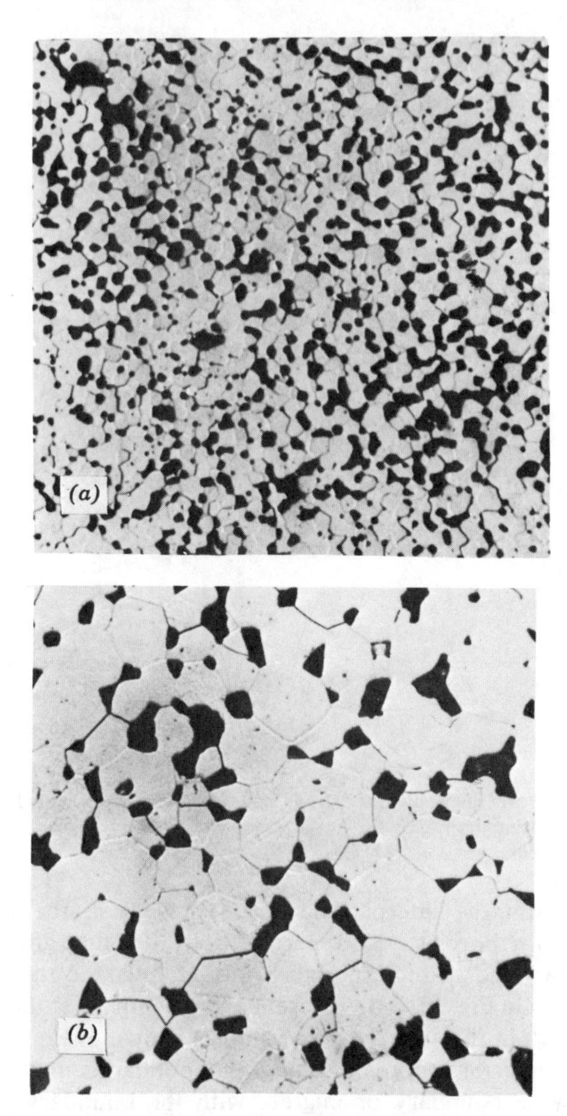

Fig. 10.10. Grain growth and pore growth in sample of UO_2 after (a) 2 min, 91.5% dense, and (b) 5 hr, 91.9% dense, at 1600°C (400×). From Francois and Kingery.

Another factor that may restrain grain growth is the presence of a liquid phase. If a small amount of a boundary liquid is formed, it tends to slow grain growth, since the driving force is reduced and the diffusion path is increased. There are now two solid-liquid interfaces, and the driving force is the difference between them, that is, $(1/r_1 + 1/r_2)_A - (1/r_1 + 1/r_2)_B$, which is smaller than either alone; in addition, if the liquid wets the boundary, the interface energy must be lower than the pure-grain-boundary energy. Also, the process of solution, diffusion through a liquid film, and precipitation is usually slower than the jump across a boundary. However, this case is more complex in that grain growth may be enhanced by the presence of a reactive liquid phase during the densification process, as discussed in Section 10.4. In addition, a very small amount of liquid may enhance secondary recrystallization, as discussed later, whereas larger amounts of liquid phase may give rise to the grain-growth process described in Chapter 9. In practice, it is found that addition of a moderate amount of silicate liquid phase to aluminum oxide prevents the extensive grain growth which frequently occurs with purer materials.

Secondary Recrystallization. The process of secondary recrystallization, sometimes called discontinuous or exaggerated grain growth, occurs when some small fraction of the grains grow to a large size, consuming the uniform-grain-size matrix. Once a single grain grows to such a size that it has many more sides than the neighboring grains (such as the grain with fifty sides illustrated in Fig. 10.4), the curvature of each side increases, and it grows more rapidly than the smaller grains with fewer sides. The increased curvature on the edge of a large grain is particularly evident in Fig. 10.11, which shows a large alumina crystal growing at the expense of a uniform-particle-size matrix.

Secondary crystallization is particularly likely to occur when continuous grain growth is inhibited by the presence of impurities or pores. Under these conditions the only boundaries able to move are those with a curvature much larger than the average; that is, the exaggerated grains with highly curved boundaries are able to grow, whereas the matrix material remains uniform in grain size. The rate of growth of the large grains is initially dependent on the number of sides. However, after growth has reached the point at which the exaggerated grain diameter is much larger than the matrix diameter, $d_g \gg d_m$, the curvature is determined by the matrix grain size and is proportional to $1/d_m$. That is, there is an induction period corresponding to the increased growth rate and the formation of a grain large enough to grow at the expense of the constant-grain-size matrix. Therefore, the growth rate is constant as long as the grain size of the matrix remains unchanged. Consequently, the kinetics of secondary recrystallization is similar to that of primary recrystallization, even though the nature of the nucleation and driving force is different.

Secondary recrystallization is common for oxide, titanate, and ferrite ceramics in which grain growth is frequently inhibited by minor amounts of second phases or by porosity during the sintering process. A typical resultant structure is illustrated for barium titanate in Fig. 10.12, and the

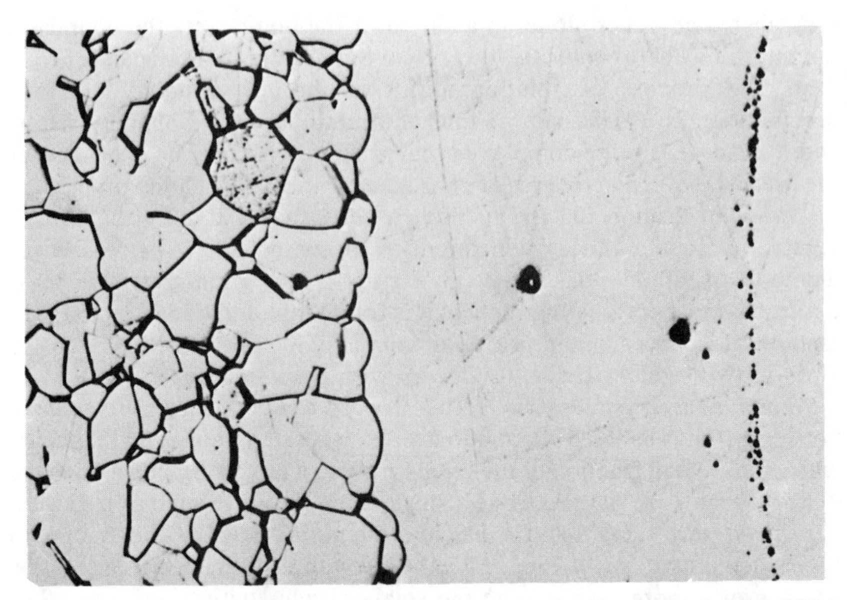

Fig. 10.11. Growth of a large Al_2O_3 crystal into a matrix of uniformly sized grains (495×). Compare with Fig. 10.4. Courtesy R. L. Coble.

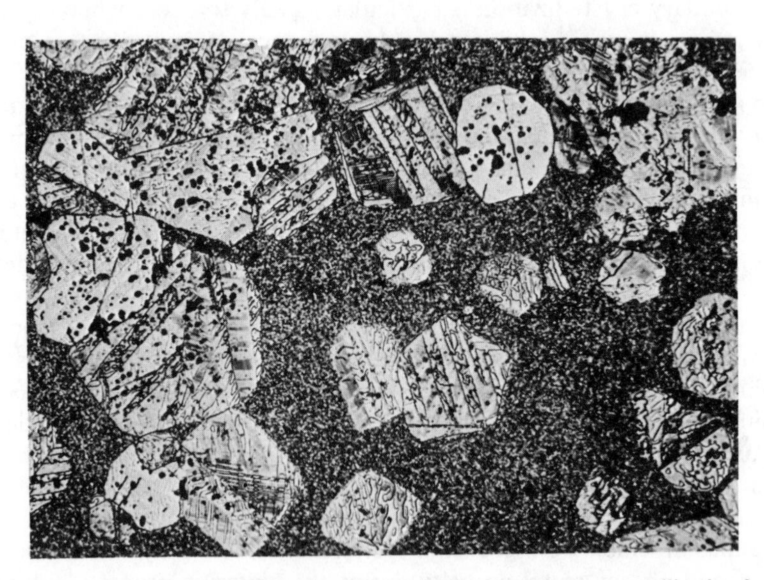

Fig. 10.12. Large grains of barium titanate growing by secondary recrystallization from a fine-grained matrix (250×). Courtesy R. C. DeVries.

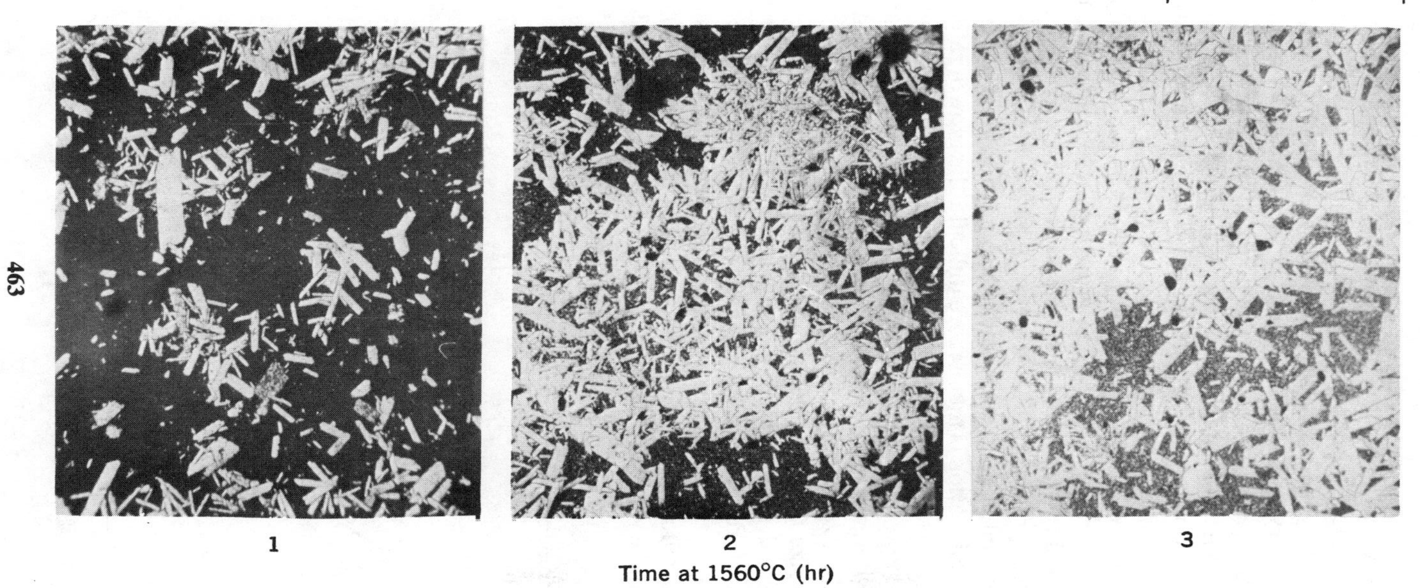

Fig. 10.13. Large grains of Al_2O_3 growing by secondary recrystallization from a fine-grained matrix. Courtesy I. B. Cutler, in reference 5.

progressive growth of aluminum oxide crystals during secondary recrystallization is illustrated in Fig. 10.13.

When polycrystalline bodies are made from fine powder, the extent of secondary recrystallization depends on the particle size of the starting material. Coarse starting material gives a much smaller relative grain growth, as illustrated in Fig. 10.14 for beryllia. This is caused by both the rate of nucleation and the rate of growth. There are almost always present in the fine-grained matrix a few particles of substantially larger particle size than the average; these can act as embryos for secondary recrystallization, since already $d_g > d_m$, and growth proceeds to a rate proportional to $1/d_m$. In contrast, as the starting particle size increases, the chances of grains being present which are much larger in particle size than the average are much decreased, and consequently the nucleation of secondary recrystallization is much more difficult; the growth rate, proportional to $1/d_m$, is also smaller. In the data shown in Fig. 10.14, material having a starting particle size of 2 microns grows to a final particle size of about 50 microns, whereas material with an initial particle size of 10 microns shows a final grain size of only about 25 microns. This result of a much larger final grain size for a smaller initial particle size would be very puzzling if the process of secondary recrystallization was not known to occur.

Secondary recrystallization has been observed to occur with the boundaries of the large grains apparently perfectly straight (Fig. 10.15). Here the previous discussion of the surface tension and curvature of the phase boundary does not apply directly. That is, the boundary energy is

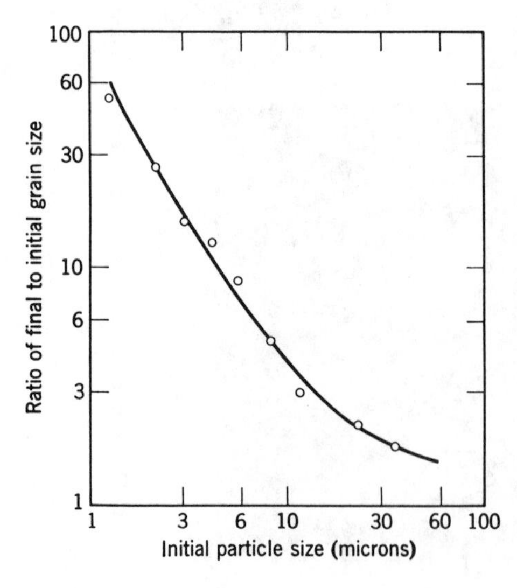

Fig. 10.14. Relative grain growth during secondary recrystallization of BeO heated $2\frac{1}{2}$ hr at 2000°C. From P. Duwez, F. Odell, and J. L. Taylor, *J. Am. Ceram. Soc.*, **32**, 1 (1949).

not independent of crystal directions, and the growth planes are those of low surface energy. These structures all seem to occur in systems having a small concentration of impurity which gives rise to a small amount of a boundary phase. The driving force for secondary recrystallization is the lower surface energy of the large grain compared with the high-surface-energy faces or small radius of curvature of adjacent grains. Transfer of material under these conditions can only occur when there is an intermediate boundary phase separating the surfaces of the small and large grains. The amount of second phase present tends to increase at the boundaries of the large crystals compared with that at other boundaries in the system, and a large grain continues to grow once it is initiated. If the amount of boundary phase is increased, however, normal grain growth and this kind of secondary recrystallization are both inhibited, as discussed previously.

Secondary recrystallization affects both the sintering of ceramics and resultant properties. Excessive grain growth is frequently harmful to mechanical properties (see Sections 5.5 and 15.5). For some electrical and magnetic properties either a large or a small grain size may contribute to improved properties. Occasionally grain growth has been discussed in the literature as if it were an integral part of the densification process. That this is not true can best be seen from Fig. 10.16. A sample of aluminum oxide with an initial fine pore distribution was heated to a high temperature so that secondary recrystallization occurred. The recrystallization has left almost the same amount of porosity as was present in the initial

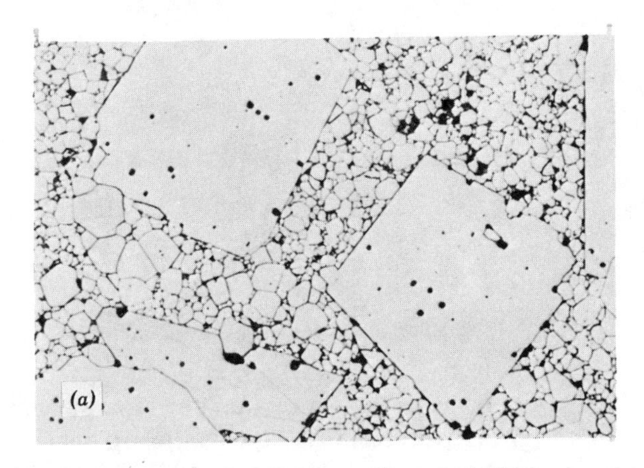

Fig. 10.15. (*a*) Idiomorphic grains in a polycrystalline spinel. The large grain edges appear straight, whereas the shape of the small grains is controlled by surface tension (350×). Courtesy R. L. Coble.

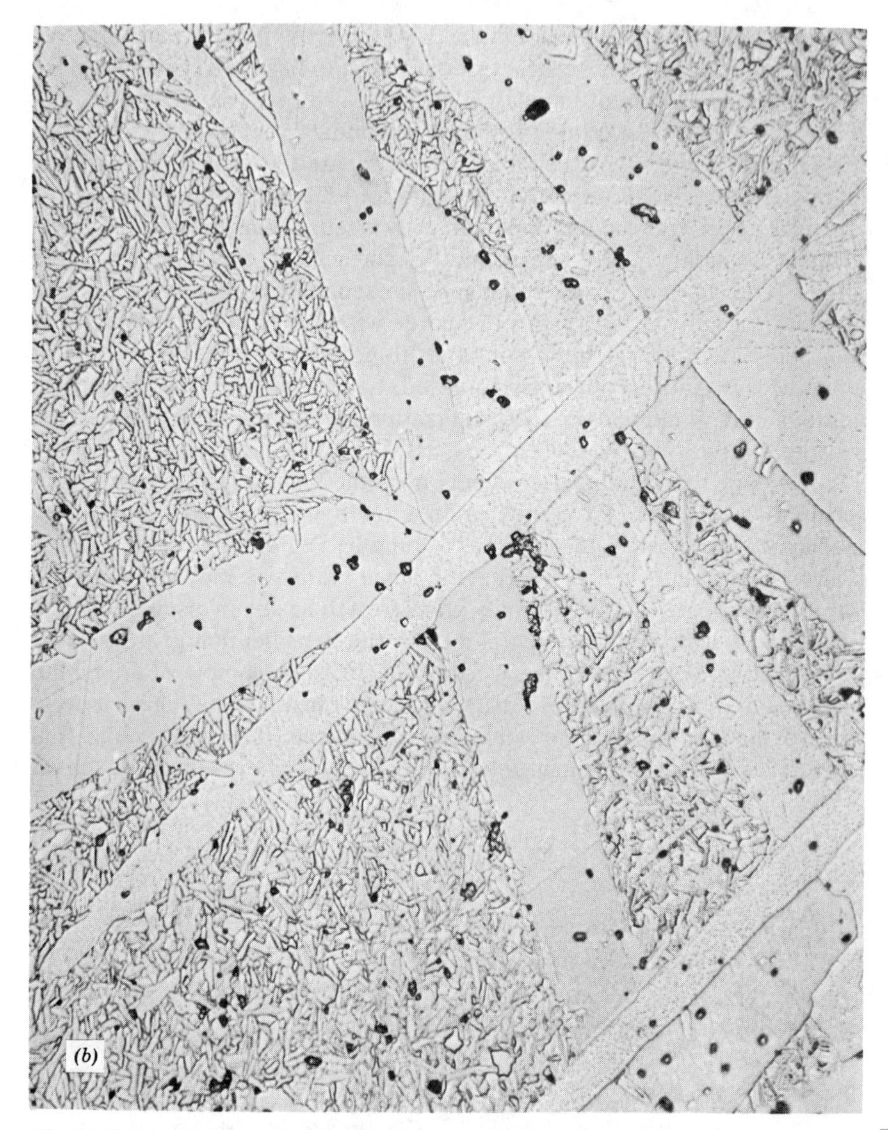

Fig. 10.15 (*Continued*). (*b*) Idiomorphic grains of α-6H SiC in a β-SiC matrix (1000×).

compact. Elimination of porosity is a related but separate subject and is considered in following sections. An application in which secondary recrystallization has been useful is in the development of preferred orientation on firing of the magnetically hard ferrite, $BaFe_{12}O_{19}$.* For this

*A. L. Stuijts, *Trans. Brit. Ceram. Soc.*, **55**, 57 (1956).

Fig. 10.15 (*Continued*). (*c*) Detail of boundary (75,000×). Courtesy S. Prochazka.

magnetic material it is desirable to obtain a high density as well as a high degree of preferred orientation in the sintered product. Particles of the powdered material can be oriented to a considerable extent by subjecting them to a high magnetic field while forming. On sintering there was a 57% alignment after heating at 1250°C. On further heating at 1340°C the

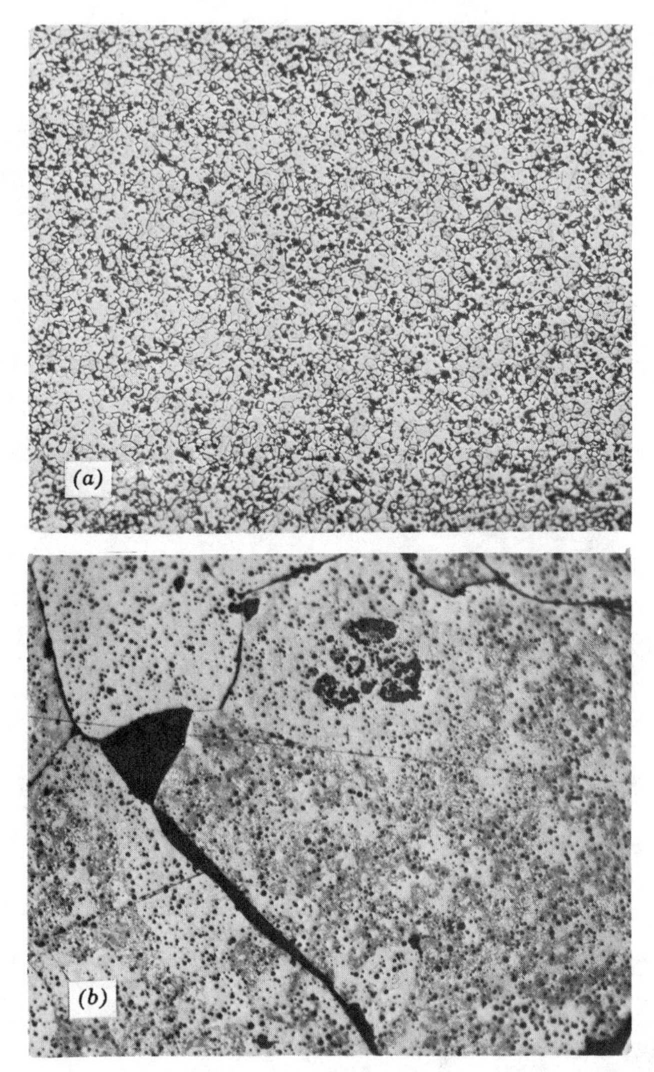

Fig. 10.16. A specimen of alumina (*a*) sintered 1 hr at 1800°C and (*b*) heated 1 hr at 1900°C to give secondary recrystallization. Note that the pore spacing has not changed. Courtesy J. E. Burke.

preferred orientation increased to 93% alignment, corresponding to the structural change brought about by secondary recrystallization. It seems apparent that the few large grains in the starting material are more uniformly aligned than the fine surrounding material. These grains serve as nuclei for the secondary recrystallization process and give rise to a highly oriented final product.

10.2 Solid-State Sintering

Changes that occur during the firing process are related to (1) changes in grain size and shape, (2) changes in pore shape, and (3) changes in pore size. In Section 10.1 we concentrated on changes in grain size; in this and the following section we are mainly concerned with changes in porosity, that is, the changes taking place during the transformation of an originally porous compact to a strong, dense ceramic. As formed, a powder compact, before it has been fired, is composed of individual grains separated by between 25 and 60 vol% porosity, depending on the particular material used and the processing method. For maximizing properties such as strength, translucency, and thermal conductivity, it is desirable to eliminate as much of this porosity as possible. For some other applications it may be desirable to increase this strength without decreasing the gas permeability. These results are obtained during firing by the transfer of material from one part of the structure to the other. The kind of changes that may occur are illustrated in Fig. 10.17. The pores initially present can change shape, becoming channels or isolated spheres, without necessarily changing in size. More commonly, however, both the size and shape of the pores present change during the firing process, the pores becoming more spherical in shape and smaller in size as firing continues.

Driving Force for Densification. The free-energy change that gives rise to densification is the decrease in surface area and lowering of the surface free energy by the elimination of solid-vapor interfaces. This usually takes place with the coincidental formation of new but lower-energy

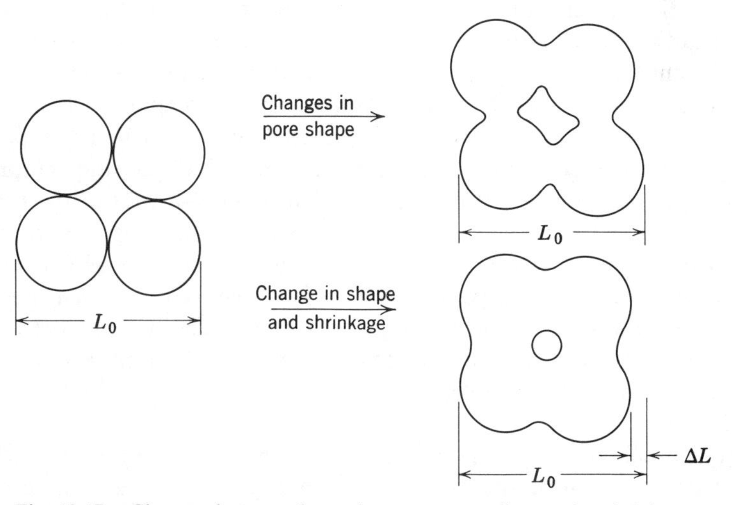

Fig. 10.17. Changes in pore shape do not necessarily require shrinkage.

solid-solid interfaces. The net decrease in free energy occurring on sintering a 1-micron particle size material corresponds to an energy decrease of about 1 cal/g. On a microscopic scale, material transfer is affected by the pressure difference and changes in free energy across a curved surface. These changes are due to the surface energy and have been discussed in Chapter 5 and referred to in Section 10.1. If the particle size, and consequently the radius of curvature, is small, these effects may be of a substantial magnitude. As indicated in Chapter 5, they become large when the radius of curvature is less than a few microns. This is one of the major reasons why much ceramic technology is based on and depends on the use of fine-particle materials.

Most of the insight into the effect of different variables on the sintering process has come from considering simple systems and comparing experimental data with simple models. Since our major aim is to be sure we understand the importance of different variables in traditional or new systems, we use this method here. Since the driving force is the same (surface energy) in all systems, considerable differences in behavior in various types of systems must be related to different mechanisms of material transfer. Several can be imagined—evaporation and condensation, viscous flow, surface diffusion, grain-boundary or lattice diffusion, and plastic deformation are among those that occur to us. Of these, diffusion and viscous flow are important in the largest number of systems; evaporation-condensation is perhaps the easiest to visualize.

Evaporation-Condensation. During the sintering process there is a tendency for material transfer because of the differences in surface curvature and consequently the differences in vapor pressure at various parts of the system. Material transfer brought about in this way is only important in a few systems; however, it is the simplest sintering process to treat quantitatively. We derive the sintering rate in some detail, since it provides a sound basis for understanding more complex processes.

Let us consider the initial stages of the process when the powder compact is just beginning to sinter and concentrate on the interaction between two adjacent particles (Fig. 10.18). At the surface of the particle there is a positive radius of curvature so that the vapor pressure is somewhat larger than would be observed for a flat surface. However, just at the junction between particles there is a neck with a small negative radius of curvature and a vapor pressure an order of magnitude lower than that for the particle itself. The vapor-pressure difference between the neck area and the particle surface tends to transfer material into the neck area.

We can calculate the rate at which the bonding area between particles increases by equating the rate of material transfer to the surface of the

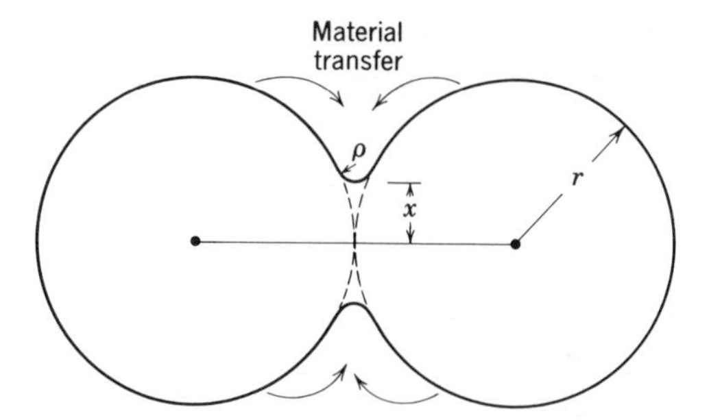

Fig. 10.18. Initial stages of sintering by evaporization-condensation.

lens between the spheres with the increase in its volume. The vapor pressure over the small negative radius of curvature is decreased because of the surface energy in accordance with the Thomson-Freundlich (Kelvin) equation discussed in Chapter 5:

$$\ln \frac{p_1}{p_0} = \frac{\gamma M}{dRT}\left(\frac{1}{\rho} + \frac{1}{x}\right) \tag{10.14}$$

where p_1 is the vapor pressure over the small radius of curvature, M is the molecular weight of the vapor, and d is the density. In this case the neck radius is much larger than the radius of curvature at the surface, ρ, and the pressure difference $p_0 - p_1$ is small. Consequently, to a good approximation, $\ln p_1/p_0$ equals $\Delta p/p_0$, and we can write

$$\Delta p = \frac{\gamma M p_0}{d\rho RT} \tag{10.15}$$

where Δp is the difference between the vapor pressure of the small negative radius of curvature and the saturated vapor in equilibrium with the nearly flat particle surfaces. The rate of condensation is proportional to the difference in equilibrium and atmospheric vapor pressure and is given by the Langmuir equation to a good approximation as

$$m = \alpha\, \Delta p \left(\frac{M}{2\pi RT}\right)^{1/2} \qquad \text{g/cm}^2\text{/sec} \tag{10.16}$$

where α is an accommodation coefficient which is nearly unity. Then the rate of condensation should be equal to the volume increase. That is,

$$\frac{mA}{d} = \frac{dv}{dt} \qquad \text{cm}^3\text{/sec} \tag{10.17}$$

From the geometry of the two spheres in contact, the radius of curvature at the contact points is approximately equal to $x^2/2r$ for x/r less than 0.3;

the area of the surface of the lens between spheres is approximately equal to π^2x^3/r; the volume contained in the lenticular area is approximately $\pi x^4/2r$. That is,

$$\rho = \frac{x^2}{2r}: \qquad A = \frac{\pi^2 x^3}{r}: \qquad v = \frac{\pi x^4}{2r} \tag{10.18}$$

Substituting values for m in Eq. 10.16, A and v in Eq. 10.18 into Eq. 10.17 and integrating, we obtain a relationship for the rate of growth of the bond area between particles:

$$\frac{x}{r} = \left(\frac{3\sqrt{\pi}\gamma M^{3/2} p_0}{\sqrt{2}R^{3/2}T^{3/2}d^2}\right)^{1/3} r^{-2/3} t^{1/3} \tag{10.19}$$

This equation gives the relationship between the diameter of the contact area between particles and the variables influencing its rate of growth.

The important factor from the point of view of strength and other material properties is the bond area in relation to the individual particle size, which gives the fraction of the projected particle area which is bonded together—the main factor in fixing strength, conductivity, and related properties. As seen from Eq. 10.19, the rate at which the area between particles forms varies as the two-thirds power of time. Plotted on a linear scale, this decreasing rate curve has led to characterizations of *end point* conditions corresponding to a certain sintering time. This concept of an end point is useful, since periods of time for sintering are not widely changed; however, the same rate law is observed for the entire process (Fig. 10.19*b*).

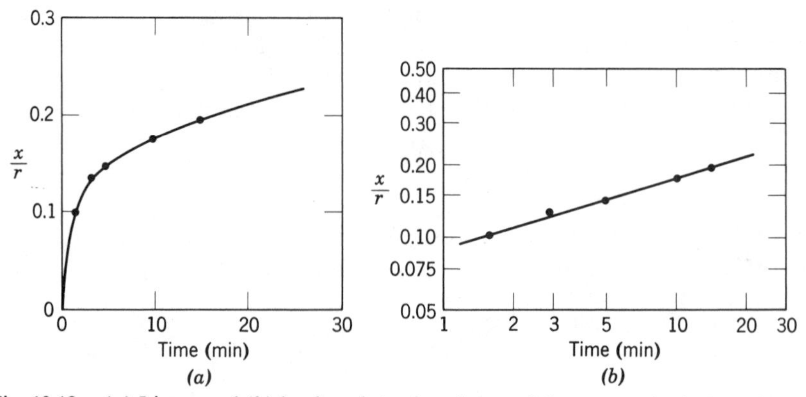

Fig. 10.19. (*a*) Linear and (*b*) log-log plots of neck growth between spherical particles of sodium chloride at 725°C.

If we consider the changes in structure that take place during a process such as this, it is clear that the distance between centers of spherical particles (Fig. 10.18) is not affected by the transfer of material from the particle surface to the interparticle neck. This means that the total shrinkage of a row of particles, or of a compact of particles, is unaffected by vapor-phase-material transfer and that only the shape of pores is changed. This changing shape of pores can have an appreciable effect on properties but does not affect density.

The principal variables in addition to time that affect the rate of pore-shape change through this process are the initial particle radius (rate proportional to $1/r^{2/3}$) and the vapor pressure (rate proportional to $p_0^{1/3}$). Since the vapor pressure increases exponentially with temperature, the process of vapor-phase sintering is strongly temperature-dependent. From a processing point of view, the two main variables over which control can be exercised for any given material are the initial particle size and the temperature (which fixes the vapor pressure). Other variables are generally not easy to control, nor are they strongly dependent on conditions of use.

The negligible shrinkage corresponding to vapor-phase-material transfer is perhaps best illustrated in Fig. 10.20, which shows the shape changes that occur on heating a row of initially spherical sodium chloride particles. After long heating the interface contact area has increased; the

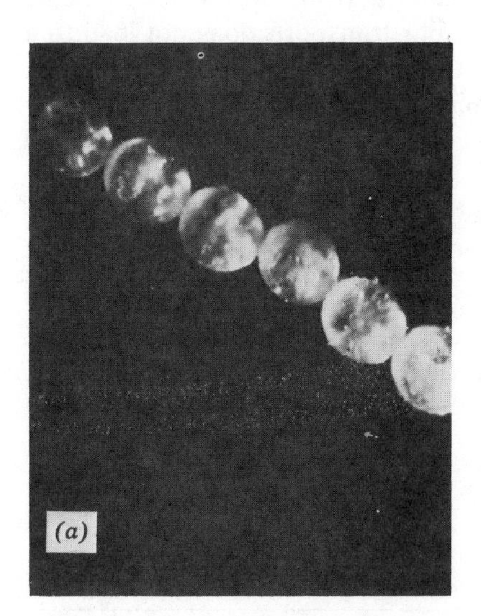

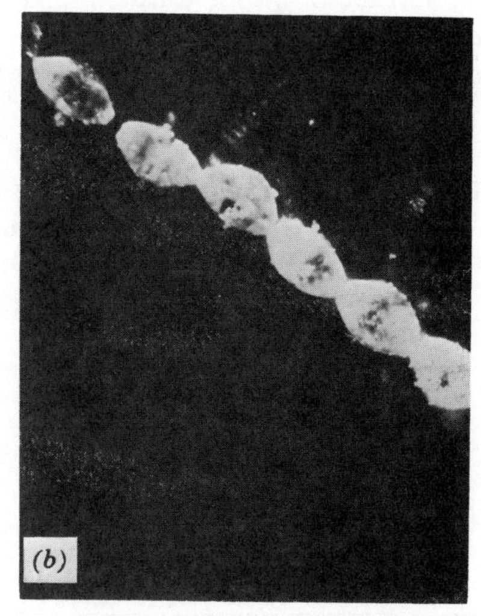

Fig. 10.20. Photomicrographs of sintering sodium chloride at 750°C: (*a*) 1 min; (*b*) 90 min.

particle diameter has been substantially decreased, but the distance between particle centers, that is, the shrinkage, has not been affected.

Vapor-phase-material transfer requires that materials be heated to a temperature sufficiently high for the vapor pressure to be appreciable. For micron-range particle sizes this requires vapor pressures in the order of 10^{-4} to 10^{-5} atm, a pressure higher than those usually encountered during sintering of oxide and similar phases. Vapor-phase transfer plays an important part in the changes occurring during treatment of halides such as sodium chloride and is important for the changes in configuration observed in snow and ice technology.

Solid-State Processes. The difference in free energy or chemical potential between the neck area and the surface of the particle provides a driving force which causes the transfer of material by the fastest means available. If the vapor pressure is low, material transfer may occur more readily by solid-state processes, several of which can be imagined. As shown in Fig. 10.21 and Table 10.1, in addition to vapor transport (process 3), matter can move from the particle surface, from the particle bulk, or from the grain boundary between particles by surface, lattice, or grain-boundary diffusion. Which one or more of these processes actually contributes significantly to the sintering process in a particular system depends on their relative rates, since each is a parallel method of lowering the free energy of the system (parallel reaction paths have been discussed in Chapter 9). There is a most significant difference between these paths for matter transport: the transfer of material from the surface to the neck by surface or lattice diffusion, like vapor transport, does not lead to any decrease in the distance between particle centers. That is, these processes do not result in shrinkage of the compact and a decrease in porosity. Only

Table 10.1. Alternate Paths for Matter Transport During the Initial Stages of Sintering[a]

Mechanism Number	Transport Path	Source of Matter	Sink of Matter
1	Surface diffusion	Surface	Neck
2	Lattice diffusion	Surface	Neck
3	Vapor transport	Surface	Neck
4	Boundary diffusion	Grain boundary	Neck
5	Lattice diffusion	Grain boundary	Neck
6	Lattice diffusion	Dislocations	Neck

[a] See Fig. 10.21.

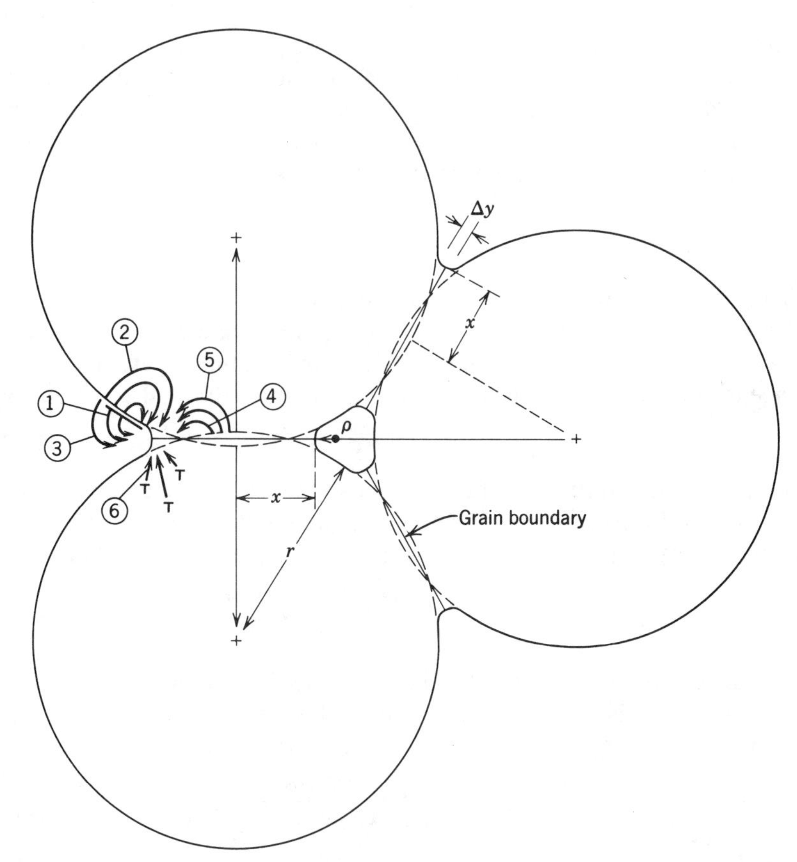

Fig. 10.21. Alternate paths for matter transport during the initial stages of sintering. Courtesy M. A. Ashby. (See Table 10.1.)

transfer of matter from the particle volume or from the grain boundary between particles causes shrinkage and pore elimination.

Let us consider mechanism **5**, matter transport from the grain boundary to the neck by lattice diffusion. Calculation of the kinetics of this process is exactly analogous to determination of the rate of sintering by a vapor-phase process. The rate at which material is discharged at the surface area is equated to the increase in volume of material transferred. The geometry is slightly different:

$$\rho = \frac{x^2}{4r}: \qquad A = \frac{\pi^2 x^3}{2r}: \qquad V = \frac{\pi x^4}{4r} \tag{10.20}$$

The process can be visualized most easily by considering the rate of migration of vacancies. In the same way that there are differences in vapor pressure between the surface of high negative curvature and the nearly flat surfaces, there is a difference in vacancy concentration. If c is the concentration of vacancies and Δc is the excess concentration over the concentration on a plane surface c_0, then, equivalent to Eq. 10.15,

$$\Delta c = \frac{\gamma a^3 c_0}{kT\rho} \tag{10.21}$$

where a^3 is the atomic volume of the diffusing vacancy and k is the Boltzmann constant. The flux of vacancies diffusing away from the neck area per second per centimeter of circumferential length under this concentration gradient can be determined graphically and is given by

$$J = 4D_V \, \Delta c \tag{10.22}$$

Where D_V is the diffusion coefficient for vacancies, D_V equals D^*/a^3c_0 if D^* is the self-diffusion coefficient. Combining Eqs. 10.22 and 10.21 with the continuity equation similar to Eq. 10.17, we obtain the result

$$\frac{x}{r} = \left(\frac{40\gamma a^3 D^*}{kT}\right)^{1/5} r^{-3/5} t^{1/5} \tag{10.23}$$

With diffusion, in addition to the increase in contact area between particles, there is an approach of particles centers. The rate of this approach is given by $d(x^2/2r)/dt$. Substituting from Eq. 10.23, we obtain

$$\frac{\Delta V}{V_0} = \frac{3\,\Delta L}{L_0} = 3\left(\frac{20\gamma a^3 D^*}{\sqrt{2}kT}\right)^{2/5} r^{-6/5} t^{2/5} \tag{10.24}$$

These results indicate that the growth of bond formation between particles increases as a one-fifth power of time (a result which has been experimentally observed for a number of metal and ceramic systems) and that the shrinkage of a compact densified by this process should be proportional to the two-fifths power of time. The decrease in densification rate with time gives rise to an apparent end-point density if experiments are carried out for similar time periods. However, when plotted on a log-log basis, the change in properties is seen to occur as expected from Eq. 10.24. Experimental data for sodium fluoride and aluminum oxide are shown in Fig. 10.22.

The relationships derived in Eqs. 10.23 and 10.24 and similar relationships for the alternate matter transport processes, which we shall not derive, are important mainly for the insight that they provide on the variables which must be controlled in order to obtain reproducible processing and densification. It is seen that the sintering rate steadily decreases with time, so that merely sintering for longer periods to obtain

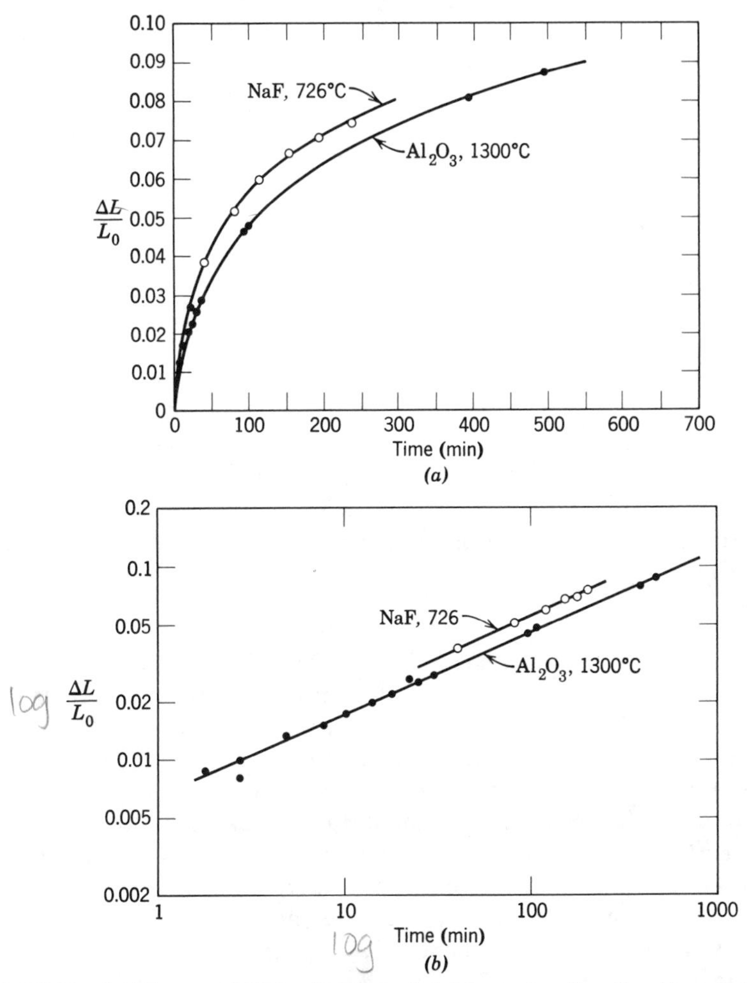

Fig. 10.22. (*a*) Linear and (*b*) log-log plots of shrinkage of sodium fluoride and aluminum oxide compacts. From J. E. Burke and R. L. Coble.

improved properties is impracticable. Therefore, time is not a major or critical variable for process control.

Control of particle size is very important, since the sintering rate is roughly proportional to the inverse of the particle size. The interface diameter achieved after sintering for a period of 100 hr at 1600°C is illustrated in Fig. 10.23 as a function of particle size. For large particles even these long periods do not cause extensive sintering; as the particle size is decreased, the rate of sintering is raised.

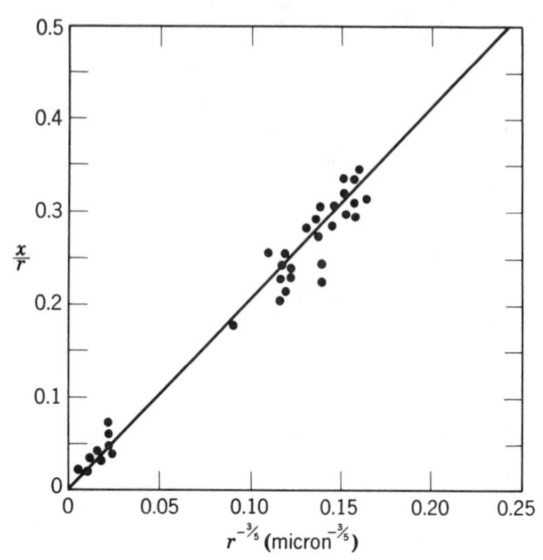

Fig. 10.23. Effect of particle size on the contact area growth in Al_2O_3 heated 100 hr at 1600°C. From R. L. Coble.

The other variable appearing in Eqs. 10.22 and 10.24 that is subject to analysis and some control is the diffusion coefficient; it is affected by composition and by temperature; the relative effectiveness of surfaces, boundaries, and volume as diffusion paths is affected by the microstructure. A number of relationships similar to Eqs. 10.23 and 10.24 have been derived, and it has been shown that surface diffusion is most important during early stages of sintering (these affect the neck diameter between particles but not the shrinkage or porosity); grain-boundary diffusion and volume diffusion subsequently become more important. In ionic ceramics, as discussed in Chapter 9, both the anion and the cation diffusion coefficients must be considered. In Al_2O_3, the best studied material, oxygen diffuses rapidly along the grain boundaries, and the more slowly moving aluminum ion at the boundary or in the bulk controls the overall sintering rate. As discussed in Chapter 5, the grain-boundary structure, composition, and electrostatic charge are influenced strongly by temperature and by impurity solutes; as discussed in Chapter 6, the exact mechanism of grain-boundary diffusion remains controversial. Estimates of the grain-boundary-diffusion width from sintering data range from 50 to 600 Å. These complications require us to be careful not to overanalyze data in terms of specific numerical results, since the time or temperature dependence of sintering may be in accordance with several plausible models. In general the presence of solutes which enhance either boundary or volume diffusion coefficients enhance the rate of solid-state sintering. As discussed in Chapter 6, both boundary and volume diffusion coefficients are strongly temperature-dependent, which means that the sintering rate is strongly dependent on the temperature level.

In order to effectively control sintering processes which take place by solid-state processes, it is essential to maintain close control of the initial particle size and particle-size distribution of the material, the sintering temperature, the composition and frequently the sintering atmosphere.

As an example of the influence of solutes, Fig. 10.24 illustrates the effect of titania additions on the sintering rate of a relatively pure alumina in a region of volume diffusion. (Both volume and boundary diffusion processes are enhanced.) It is believed that Ti enters Al_2O_3 substitutionally as Ti^{+3} and Ti^{+4} (Ti_{Al} and $Ti^{\cdot}_{Al}$). At equilibrium

$$3Ti_{Al} + \frac{3}{4}O_2(g) = 3Ti^{\cdot}_{Al} + V'''_{Al} + \frac{3}{2}O_O \qquad (10.25)$$

from which

$$K_1 = \frac{[Ti^{\cdot}_{Al}]^3[V'''_{Al}]}{[Ti_{Al}]^3[P_{O_2}]^{3/4}} \qquad (10.26)$$

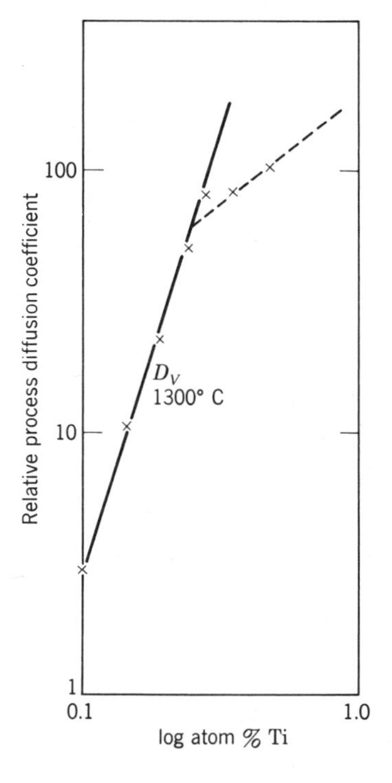

Fig. 10.24. Data for the relative sintering process diffusion coefficient with Ti additions to Al_2O_3. $D \alpha [Ti]^3$. From R. D. Bagley, I. B. Cutler, and D. L. Johnson, *J. Am. Ceram. Soc.*, **53**, 136 (1970); R. J. Brook, *J. Am. Ceram. Soc.*, **55**, 114 (1972).

In the powders used, divalent impurities such as magnesium exceed in concentrations the intrinsic defect levels, so that overall charge neutrality at moderate titania levels is achieved by

$$[Ti_{Al}^{\cdot}] = [Mg_{Al}'] \tag{10.27}$$

and at constant impurity and oxygen pressure levels, combining Eqs. 10.26 and 10.27 gives

$$[V_{Al}'''] = K_2[Ti_{Al}]^3 \tag{10.28}$$

Since the total Ti addition ($Ti_{Al} + Ti_{Al}^{\cdot}$) is much greater than the impurity levels, $[Ti]_{Total} \approx [Ti_{Al}]$ and $[V_{Al}'''] \approx K_2[Ti]_{Total}^3$. The dependence of lattice defect concentrations on titania concentration is shown in Fig. 10.25 for the proposed model. As discussed in Chapter 6, the diffusion coefficient is proportional to the vacancy concentration; as a result the effect of this model is to anticipate an increase in the sintering rate proportional to the third power of titania concentration as experimentally observed (Fig. 10.24). At higher concentrations the dependence on titania concentration should become less steep, which is suggested by the sintering data.

Thus far our discussion of the variables influencing the sintering process has been based on the initial stages of the process, in which models are based on solid particles in contact. As the process continues, an intermediate microstructure forms in which the pores and solid are both continuous, followed by a later stage in which isolated pores are separated from one another. A number of analytical expressions have

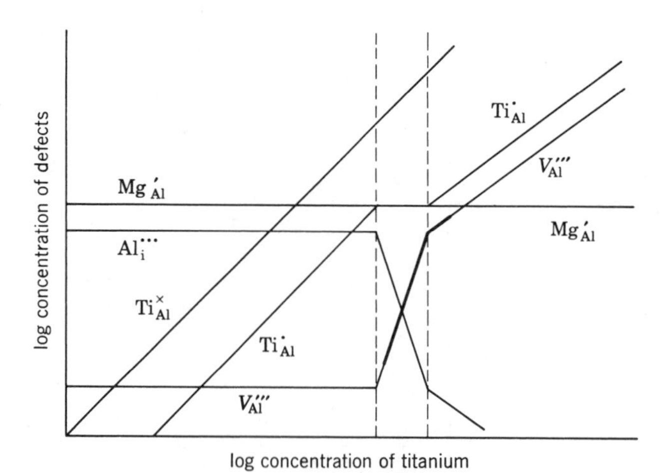

Fig. 10.25. Model for the dependence of defect concentrations on the Ti concentration in Al_2O_3. From R. J. Brook, *J. Am. Ceram. Soc.*, **55**, 114 (1972).

been derived from specific microstructural models for the transport processes listed in Table 10.1. In the later stages of the process only two mechanisms are important: boundary diffusion from sources on the boundary and lattice diffusion from sources on the boundary. For a nearly spherical pore the flux of material to a pore can be approximated as

$$J = 4\pi D_V \, \Delta c \left(\frac{rR}{R - r} \right) \qquad (10.29)$$

where D_V is the volume diffusion coefficient, Δc is the excess vacancy concentration (Eq. 10.21), r is the pore radius, and R is the effective-material-source radius. The importance of microstructure in applying this sort of analysis to specific systems is illustrated in Fig. 10.26. For a sample

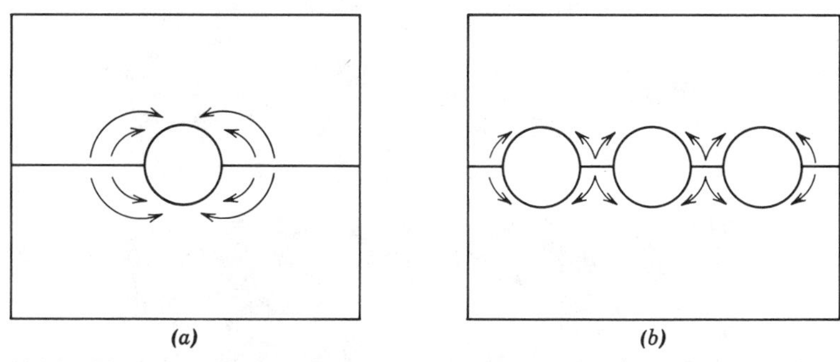

Fig. 10.26. The mean diffusion distance for material transport is smaller when there are more of the same size of pores in a boundary.

with a larger number of pores, all the same size, on a boundary the mean diffusion distance is smaller when there are more pores, and pore elimination is accomplished more quickly for the sample with the higher porosity. Thus, although the terms which influence the rate of sintering—volume or boundary diffusion coefficient (and therefore temperature and solute concentration) surface energy and pore size—are well established, the geometrical relationship of grain boundaries to the pores may have a variety of forms and is critical in determining what actually occurs.

With fine-grained materials such as oxides, it is usual to observe an increase in both grain size and pore size during the early stages of heat treatment, as illustrated for Lucalox alumina in Fig. 10.27. This partially results from the presence of agglomerates of the fine particles which sinter rapidly, leaving interagglomerate pores, and is partly due to the rapid grain growth during which pores are agglomerated by moving with the boundaries, as illustrated in Fig. 10.9. In cases in which agglomeration

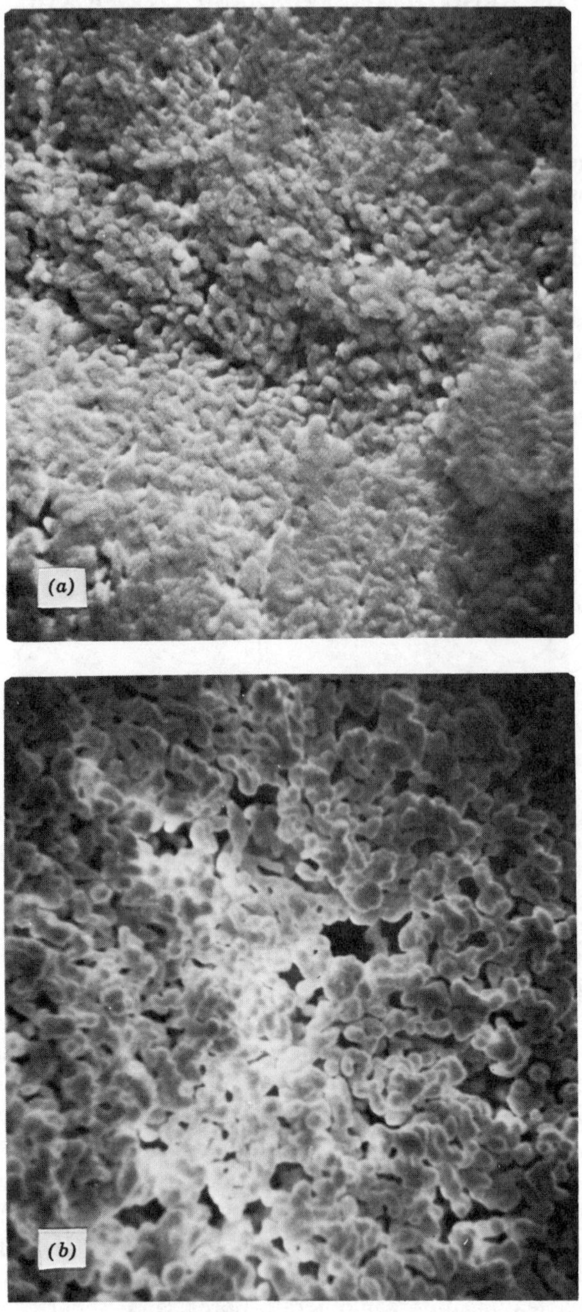

Fig. 10.27. Progressive development of microstructure in Lucalox alumina. Scanning electron micrographs of (*a*) initial particles in the compact (5000×), (*b*) after 1 min at 1700°C (5000×).

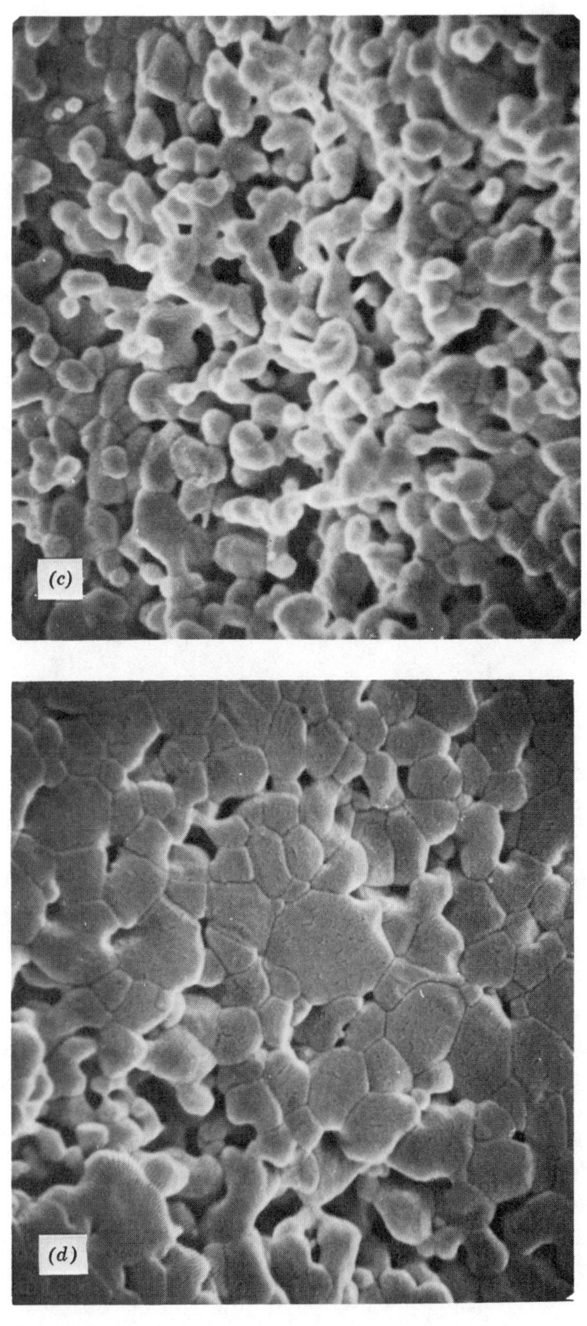

Fig. 10.27 (*Continued*) (*c*) Scanning electron micrographs after $2\frac{1}{2}$ min at 1700°C (5000×), and (*d*) after 6 min at 1700°C (5000×). Note that pores and grains increase in size, that there are variations in packing and in pore size, and that pores remain located between dense grains.

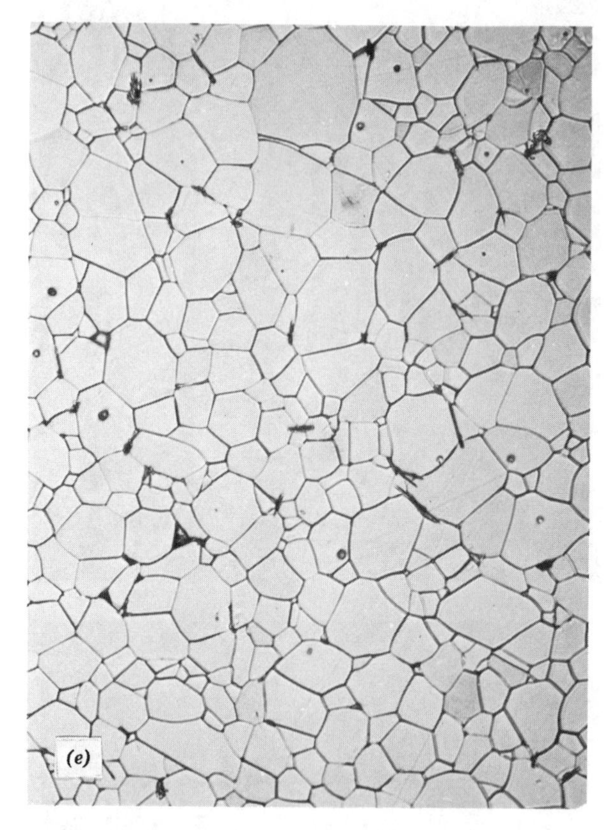

Fig. 10.27 (*Continued*) (*e*) The final microstructure is nearly porefree, with only a few pores located within grains (500×). Courtesy C. Greskovich and K. W. Lay.

of fine precipitated particles into clumps is severe, ball milling to break up the agglomerates leads to a remarkable increase in the sintering rate. Even minor variations in the original particle packing are exaggerated during the pore growth process; in addition, spaces between agglomerates and occasional larger voids resulting from the bridging of particles or agglomerates are present. As a result, during intermediate stages of the sintering process there is a range of pore sizes present, and the slower elimination of the larger pores leads to variations in pore concentration in the later stages of the sintering process, as illustrated in Fig. 10.28*c*.

In addition to local agglomerates and packing differences, pore-concentration variations in the later stages of sintering can result from particle-size variations in the starting material, from green density varia-

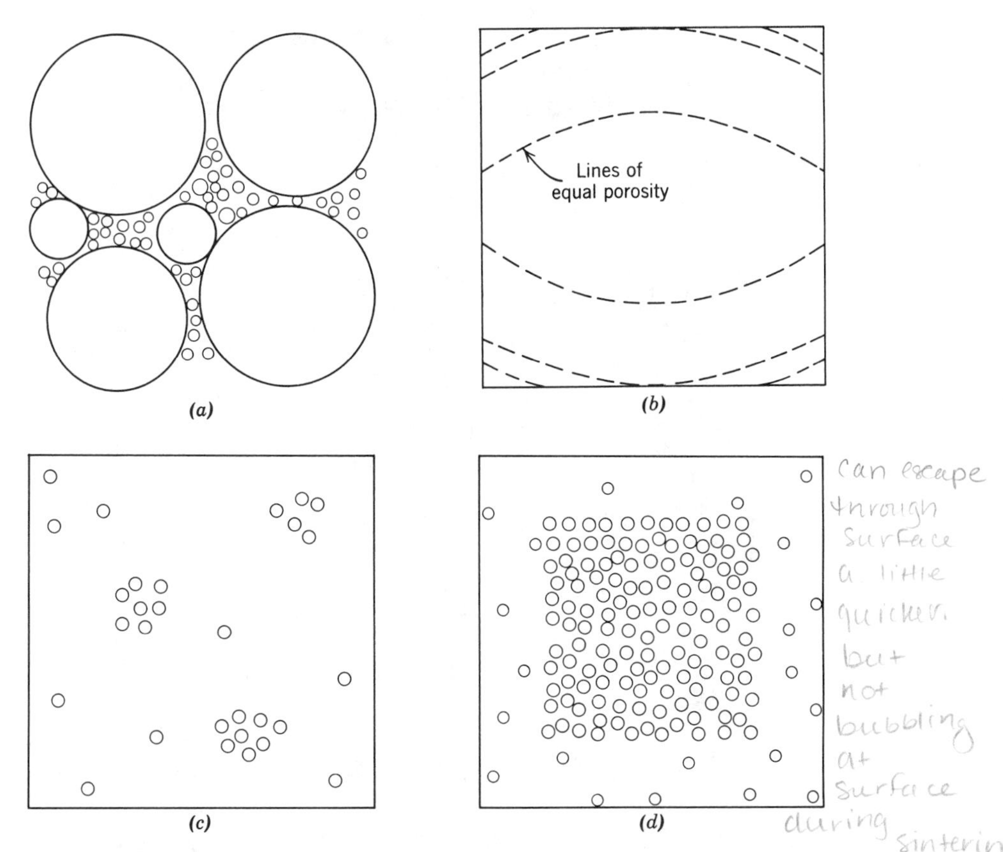

Fig. 10.28. Pore-concentration variations resulting from (*a*) a variation in grain sizes, (*b*) die friction, (*c*) local packing and agglomeration differences, and (*d*) more rapid pore elimination near surfaces.

tions caused by die-wall friction during pressing, and from the more rapid elimination of porosity near surfaces caused by temperature gradients during heating, as shown in Fig. 10.28. The importance of local variations in pore concentration results from the fact that the part of the sample containing pores tends to shrink but is restrained by other porefree parts. That is, the effective diffusion distance is no longer from the pore to an adjacent grain boundary but a pore-pore or pore-surface distance many orders of magnitude larger. An example of residual pore clusters in a sintered oxide is shown in Fig. 10.29.

Not only the kinetics of pore elimination can lead to "stable" and residual porosity, but it is also possible in some cases to have a thermodynamically metastable equilibrium pore configuration. In Fig.

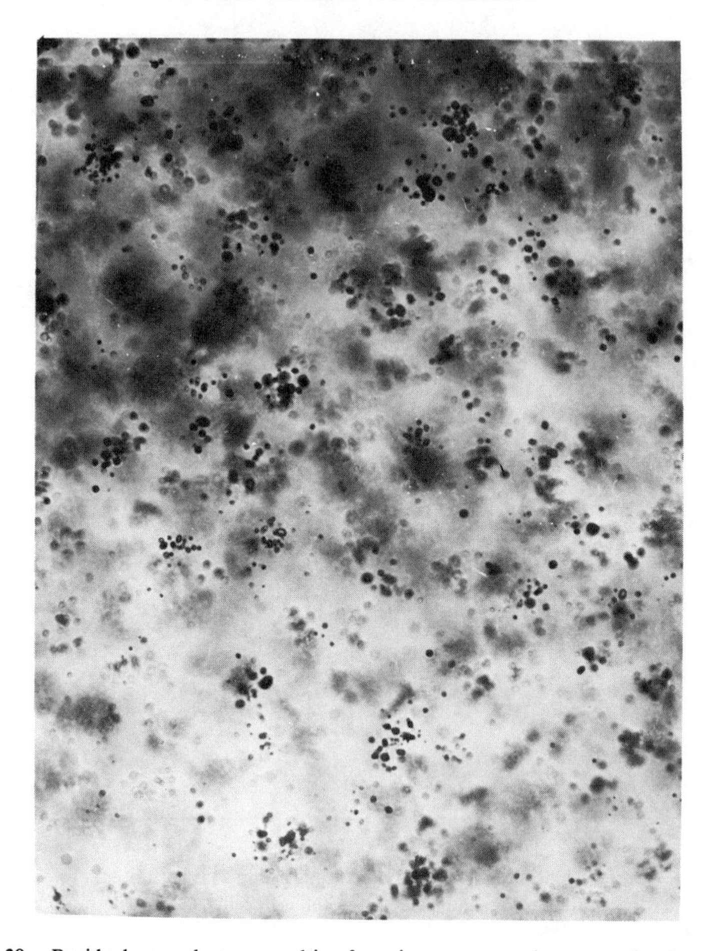

Fig. 10.29. Residual pore clusters resulting from improper powder processing in a sample of 90 mole % Y_2O_3–10 mole % ThO_2. Transmitted light, 137×. Courtesy C. Greskovich and K. N. Woods.

10.26 we have drawn spherical pores located on a grain boundary, the usual model description, but we know from our discussion of interface energies in Chapter 5 that there is a dihedral angle ϕ at the pore-boundary intersection determined by the relative interface energies;

$$\cos\frac{\phi}{2} = \frac{\gamma_{gb}}{2\gamma_s} \tag{10.30}$$

In most cases the dihedral angle for pure oxides is about 150°, and the spherical pore approximation is quite good; but for $Al_2O_3 + 0.1\%$ MgO the value is 130°, for $UO_2 + 30$ ppm C the value is 88°, and for impure boron carbide the value is about 60°. For these materials the consequences of nonspherical pores have to be considered.

As discussed for discontinuous grain growth and illustrated in Figs. 10.4 and 10.11, the boundary curvature between grains or phases depends both on the value of the dihedral angle and on the number of surrounding grains. If we take r as the radius of a circumscribed sphere around a polyhedral pore surrounded by grains, the ratio of the radius of curvature of the pore surfaces ρ to the spherical radius depends both on the dihedral angle and on the number of surrounding grains, as shown in Fig. 10.30*a*. When r/ρ decreases to zero, the interfaces are flat and have no tendency

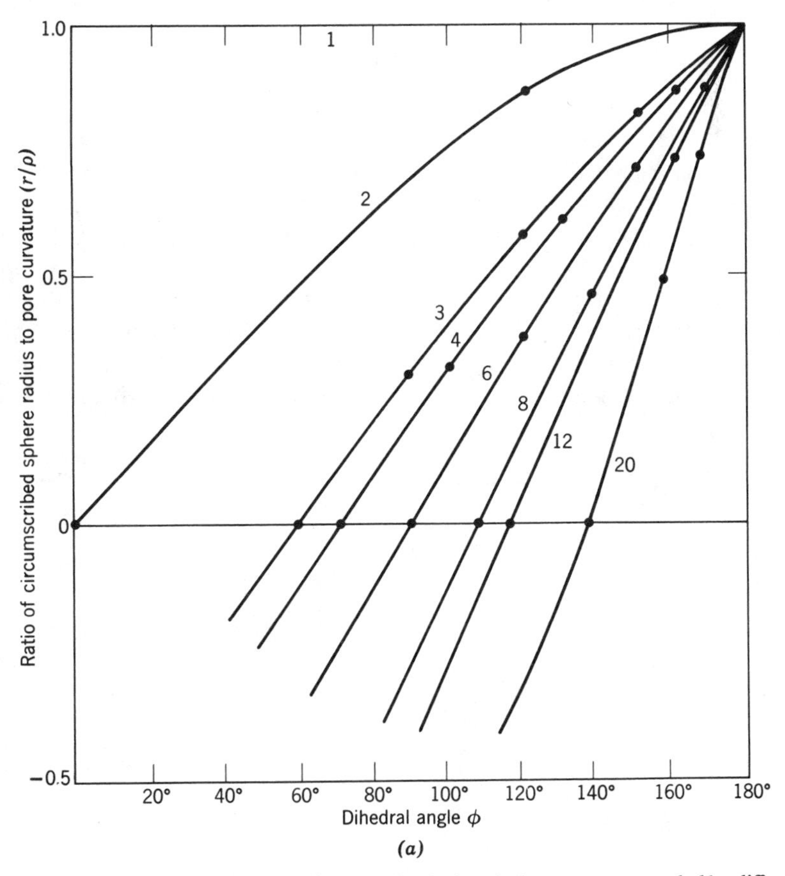

(*a*)

Fig. 10.30. (*a*) Change in the ratio (r/ρ) with dihedral angle for pores surrounded by different numbers of grains as indicated on individual curves.

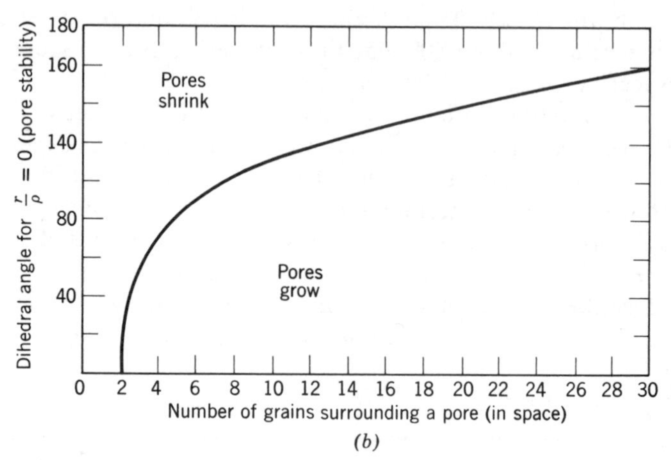

Fig. 10.30 (*Continued*). (*b*) Conditions for pore stability.

for shrinkage; when r/ρ is negative, the pore tends to grow. This is illustrated in Fig. 10.30*b*. For a uniform grain size the space-filling form is a tetrakaidecahedron with 14 surrounding grains. From an approximate relationship between the number of surrounding grains and the pore-diameter to grain-diameter ratio we can derive a relationship for pore stability as a function of dihedral angle and the ratio of pore size to grain size, as shown in Fig. 10.31. From this figure we can see why large pores present in poorly compacted powder such as shown in Fig. 10.32 not only remain stable but grow. It is also seen that an enormous disparity between

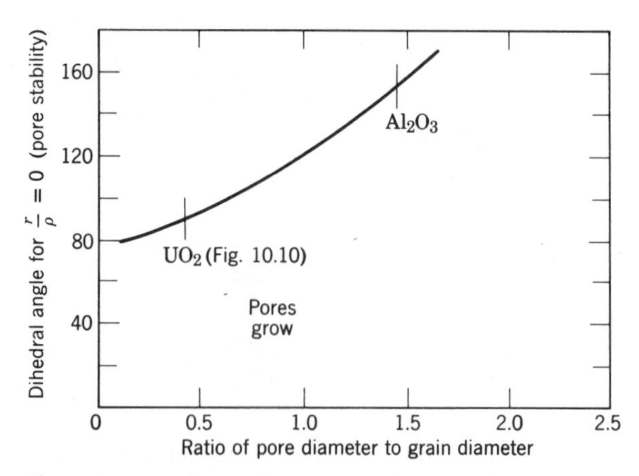

Fig. 10.31. Conditions for pore stability.

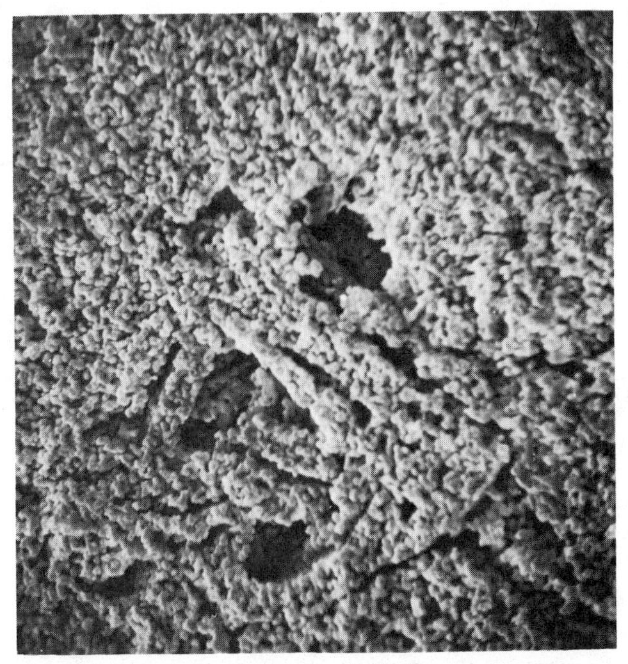

Fig. 10.32. Large voids formed by bridging of agglomerates in fine Al_2O_3 powder viewed with scanning electron microscope at 2000×. Courtesy C. Greskovich.

grain size and pore size is not necessary for pore stability. That is, the site and size of the porosity relative to the grain-boundary network not only affects the necessary distance for diffusion but also the driving force for the process.

The interaction of grain boundaries and porosity is, of course, a two-way street. When many pores are present during the initial stages of sintering, grain growth is inhibited. However, as discussed in Section 10.1, once the porosity has decreased to a value such that secondary grain growth can occur, extensive grain growth may result at high sintering temperatures. When grain growth occurs, many pores become isolated from grain boundaries, and the diffusion distance between pores and a grain boundary becomes large, and the rate of sintering decreases. This is illustrated in Fig. 10.16*b*, in which extensive secondary recrystallization has occurred, with the isolation of pores in the interior of grains and a reduction in the densification rate. Similarly, the sample of aluminum oxide shown in Fig. 10.33 has been sintered at a high temperature at which discontinuous grain growth occurred. Porosity is only removed near the grain boundaries, which act as the vacancy sink. The importance of

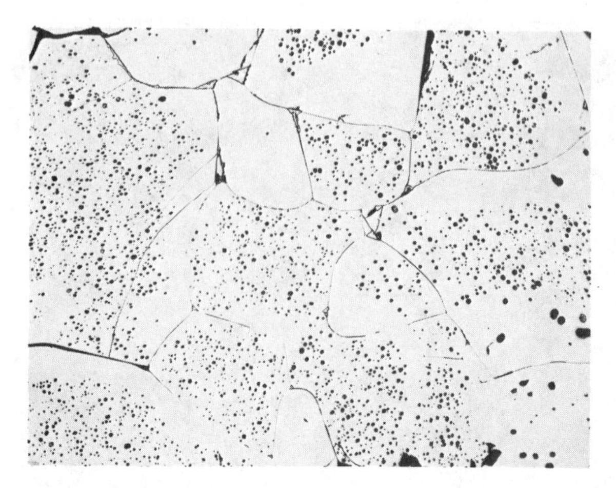

Fig. 10.33. Sintered Al_2O_3 illustrating elimination of porosity adjacent to grain boundaries with residual porosity remaining at grain centers. Courtesy J. E. Burke.

controlling grain growth as an integral part of controlling sintering phenomena cannot be overestimated. Consequently, the grain-growth processes discussed in Section 10.1 must be actively prevented in order to obtain complete densification. Usually densification continues by a diffusion process until about 10% porosity is reached; at this point rapid grain growth occurs by secondary recrystallization, and the rate of densification is sharply reduced. In order to obtain densification much beyond this level, prevention of secondary recrystallization is essential. The most satisfactory way of doing this is with additives which prevent or slow down boundary migration to a point at which it is possible to obtain pore elimination. Additions of MgO to Al_2O_3, ThO_2 to Y_2O_3, and CaO to ThO_2, among others, have been found to slow boundary migration and allow complete pore elimination by solid-state sintering in these systems. The porefree microstructure of a polycrystalline ceramic having optical transparancy suitable for use as a laser material is shown in Fig. 10.34.

10.3 Vitrification

To vitrify is to make glasslike and the vitrification process—densification with the aid of a viscous liquid phase—is the major firing process for the great majority of silicate systems. (In some current glossaries vitrification is defined as being identical to densification on firing, but the more specific usage is preferred.) A viscous liquid silicate is formed at the firing temperature and serves as a bond for the body. For

Fig. 10.34. Polished section of $Y_2O_3 + 10$ mole % ThO_2 sintered to porefree state. 100×. Courtesy C. Greskovich and K. N. Woods.

satisfactory firing the amount and viscosity of the liquid phase must be such that densification occurs in a reasonable time without the ware slumping or warping under the force of gravity. The relative and absolute rates of these two processes (shrinkage and deformation) determine to a large extent the temperature and compositions suitable for satisfactory firing.

Process Kinetics. If we consider two particles initially in contact (Fig. 10.21), there is a negative pressure at the small negative radius of curvature ρ compared with the surface of the particles. This causes a viscous flow of material into the pore region. By an analysis similar to that derived for the diffusion process, the rate of initial neck growth is given as*

$$\frac{x}{r}=\left(\frac{3\gamma}{2\eta\rho}\right)^{1/2}t^{1/2} \tag{10.31}$$

The increase in contact diameter is proportional to $t^{1/2}$; the increase in area between particles is directly proportional to time. Factors of most importance in determining the rate of this process are the surface tension, viscosity, and particle size. The shrinkage which takes place is determined by the approach between particle centers and is

$$\frac{\Delta V}{V_0}=\frac{3\,\Delta L}{L_0}=\frac{9\gamma}{4\eta r}\,t \tag{10.32}$$

That is, the initial rate of shrinkage is directly proportional to the surface tension, inversely proportional to the viscosity, and inversely proportional to the particle size.

The situation after long periods of time can best be represented as small spherical pores in a large body (Fig. 10.35). At the interior of each pore

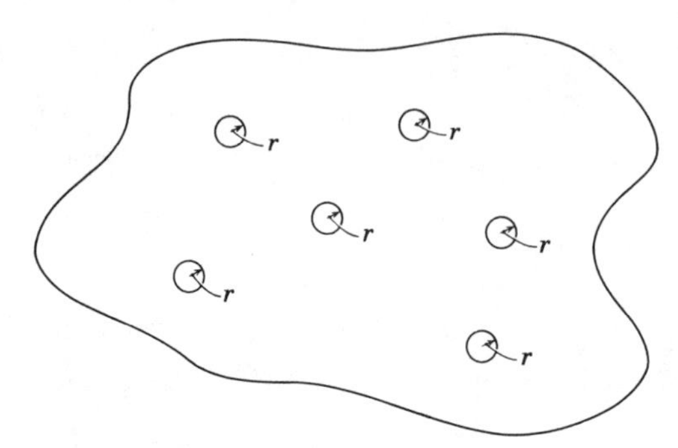

Fig. 10.35. Compact with isolated spherical pores near the end of the sintering process.

there is a negative pressure equal to $2\gamma/r$; this is equivalent to an equal positive pressure on the exterior of the compact tending to consolidate it. J. K. Mackenzie and R. Shuttleworth† have derived a relation for the rate of shrinkage resulting from the presence of isolated equal-size pores in a viscous body. The effect of surface tension is equivalent to a pressure of $-2\gamma/r$ inside all pores or, for an incompressible material, to the applica-

*J. Frenkel, *J. Phys (USSR)*, **9**, 385 (1945).
†*Proc. Phys. Soc. (London)*, **B62**, 833 (1949).

tion of a hydrostatic pressure of $+2\gamma/r$ to the compact. The real problem is to deduce the properties of the porous material from the porosity and viscosity of the dense material. The method of approximation used gives an equation of the form

$$\frac{d\rho'}{dt} = \frac{2}{3}\left(\frac{4\pi}{3}\right)^{1/3} n^{1/3} \frac{\gamma}{\eta} (1-\rho')^{2/3} \rho'^{1/3} \tag{10.33}$$

where ρ' is the relative density (the bulk density divided by the true density or the fraction of true density which has been reached) and n is the number of pores per unit volume of real material. The number of pores depends on the pore size and relative density and is given by

$$n\frac{4\pi}{3}r^3 = \frac{\text{Pore volume}}{\text{Solid volume}} = \frac{1-\rho'}{\rho'} \tag{10.34}$$

$$n^{1/3} = \left(\frac{1-\rho'}{\rho'}\right)^{1/3}\left(\frac{3}{4\pi}\right)^{1/3}\frac{1}{r} \tag{10.35}$$

By combining with Eq. 10.33,

$$\frac{d\rho'}{dt} = \frac{3\gamma}{2r_0\eta}(1-\rho') \tag{10.36}$$

where r_0 is the initial radius of the particles.

The general course of the densification process is best represented by a plot of relative density versus nondimensional time, illustrated in Fig. 10.36 following Eq. 10.33. Spherical pores are formed very quickly to reach a relative density of about 0.6. From this point until the completion of the sintering process about one unit of nondimensional time is

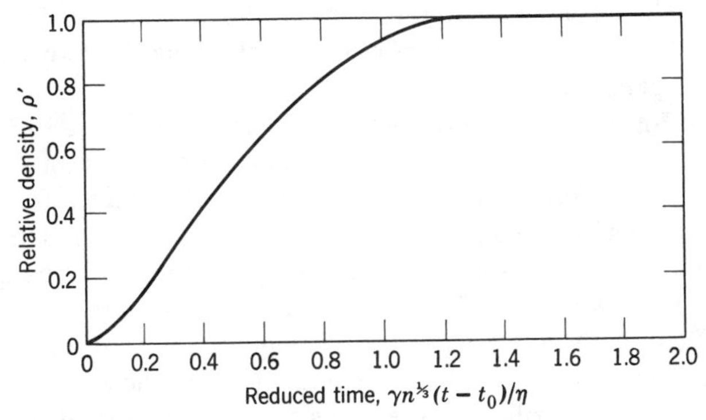

Fig. 10.36. Increase in relative density of compact with reduced time for a viscous material. From J. K. Mackenzie and R. Shuttleworth, *Proc. Phys. Soc.* (*London*), **B62**, 833 (1949).

required. For complete densification

$$t_{\text{sec}} \sim \frac{1.5 r_0 \eta}{\gamma} \tag{10.37}$$

Some experimental data for the densification of a viscous body are shown in Fig. 10.37, in which the strong effect of temperature, that is, the viscosity of the material, is illustrated by the rapid change in sintering rates. The solid lines in Fig. 10.37 are calculated from Eq. 10.33. The

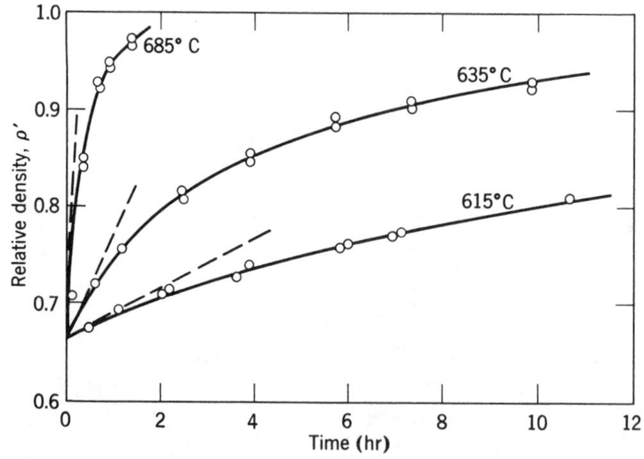

Fig. 10.37. Densification of a soda-lime-silica glass.

initial rates of sintering indicated by the dashed curves are calculated from Eq. 10.32. The good agreement of these relationships with the experimental results gives us confidence in applying them to vitrification processes in general.

Important Variables. The particular importance of Eqs. 10.31 to 10.37 is the dependence of the rate of densification on three major variables—the particle size, viscosity, and surface tension. For silicate materials the surface tension is not changed much by composition, although there are some systems for which surface energy is particularly low, as illustrated in Chapter 5. However, surface tension is not a variable that normally causes difficulty during the design of compositions or the control of processing. The particle size has a strong effect on the sintering rate and must be closely controlled if the densification process is going to be controlled. In changing from a 10-micron to 1-micron particle, the rate of sintering is increased by a factor of 10. Even more important for control purposes is the viscosity and its rapid change with temperature. For a typical soda-lime-silica glass the viscosity changes by a factor of 1000 over an interval of 100°C; the rate of densification changes by an equal factor over the temperature range. This means that the temperature must be closely controlled. Viscosity is also much changed by composition, as discussed in Chapter 3. The rate of densification, then, can be increased by changing the composition to lower the viscosity of the glassy material. The relative values of viscosity and particle size are also important; the viscosity must not be so low that appreciable deformation takes place under the forces of gravity during the time required for densification. This makes it necessary for the particle size to be in such a range that the stresses due to surface tension are substantially larger than the stresses due to gravitational forces. Materials sintered in a fluid state must be supported so that deformation does not occur. The best means of obtaining densification without excessive deformation is to use very fine-grained materials and uniform distribution of materials. This requirement is one of the reasons why successful compositions in silicate systems are composed of substantial parts of talc and clays that are naturally fine-grained and provide a sufficient driving force for the vitrification process.

Silicate Systems. The importance of the vitrification process lies in the fact that most silicate systems form a viscous glass at the firing temperature and that a major part of densification results from viscous flow under the pressure caused by fine pores. Questions that naturally arise are how much liquid is present and what are its properties. Let us consider Fig. 7.26, which shows an isothermal cut at 1200°C in the K_2O–Al_2O_3–SiO_2 system; this is the lower range of firing temperatures used for semivitreous porcelain bodies composed of about 50% kaolin (45% Al_2O_3, 55% SiO_2), 25% potash-feldspar, and 25% silica. This and similar compositions are in the primary field of mullite, and at 1200°C there is an equilibrium between mullite crystals and a liquid having a composition approximately $75SiO_2$, $12.5K_2O$, $12.5Al_2O_3$, not much different in composition from the eutectic liquid in the feldspar-silica system (Fig. 7.14). In actual practice only a small part of the silica present as flint enters into the liquid phase, and the composition of the liquid depends on the fineness of the grinding as well as on the overall chemical composition. However, the amount of silica which dissolves does not have a large effect on the amount and composition of the liquid phase present. The liquid is siliceous and has a high viscosity; the major effect of compositional changes is to alter the relative amounts of mullite and liquid phases present. Since mullite is very fine-grained, the fluid flow properties of the body correspond to those of a liquid having a viscosity greater than the pure liquid phase. For some systems the overall flow process corresponds to plastic flow with a yield point rather than to true viscous flow. This changes the kinetics of the vitrification process by introducing an additional term in Eqs. 10.33 and 10.36 but does not change the relative effects of different variables.

Although phase diagrams are useful, they do not show all the effects of small changes in composition. For example, a kaolinite composition should show equilibrium between mullite and tridymite at 1400°C with no glassy material. However, it is observed experimentally that even after 24 hr about 60 vol% of the original starting material is amorphous and deforms as a liquid. The addition of a small amount of lithium oxide as Li_2CO_3 has been observed to give a larger content of glass than additions of the same composition as the fluoride. Similar small amounts of other mineralizers can also have a profound effect in the firing properties of particular compositions. That fine grinding and intimate mixing reduce the vitrification temperature follows from the analysis in Eqs. 10.31 to 10.37. S. C. Sane and R. L. Cook* found that ball milling for 100 hr reduced the final porosity of a clay-feldspar-flint composition from 17.1 to 0.3% with the same firing conditions. This change is caused in part by increased tendencies toward fusion equilibrium and uniform mixing of constituents and in part by the smaller initial particle and pore size. In contrast to triaxial (flint-feldspar-clay) porcelains, which frequently do not reach fusion equilibrium, many steatite bodies and similar compositions which are prepared with fine-particle, intimately mixed material and form a less siliceous liquid reach phase equilibrium early in the firing process.

The time-temperature relationship and the great dependence of vitrification processes on temperature can perhaps be seen best in the experimental measurements illustrated in Fig. 10.38. As shown there, the time required for a porcelain body to reach an equivalent maturity changes by almost an order of magnitude with a 50° temperature change. There are changes in both the amount and viscosity of the glassy phase during firing, so that it is difficult to elucidate a specific activation energy for the process with which to compare the activation energy for viscous flow. However, the temperature dependence of the vitrification rate of a composition such as this (a mixture of clay, feldspar, and flint) is greater than the temperature dependence of viscosity alone. This is to be expected from the increased liquid content at the higher firing temperatures.

In summary, the factors determining the vitrification rate are the pore size, viscosity of the overall composition (which depends on amount of liquid phase present and its viscosity), and the surface tension. Equivalent

J. Am. Ceram. Soc., **34**, 145 (1951).

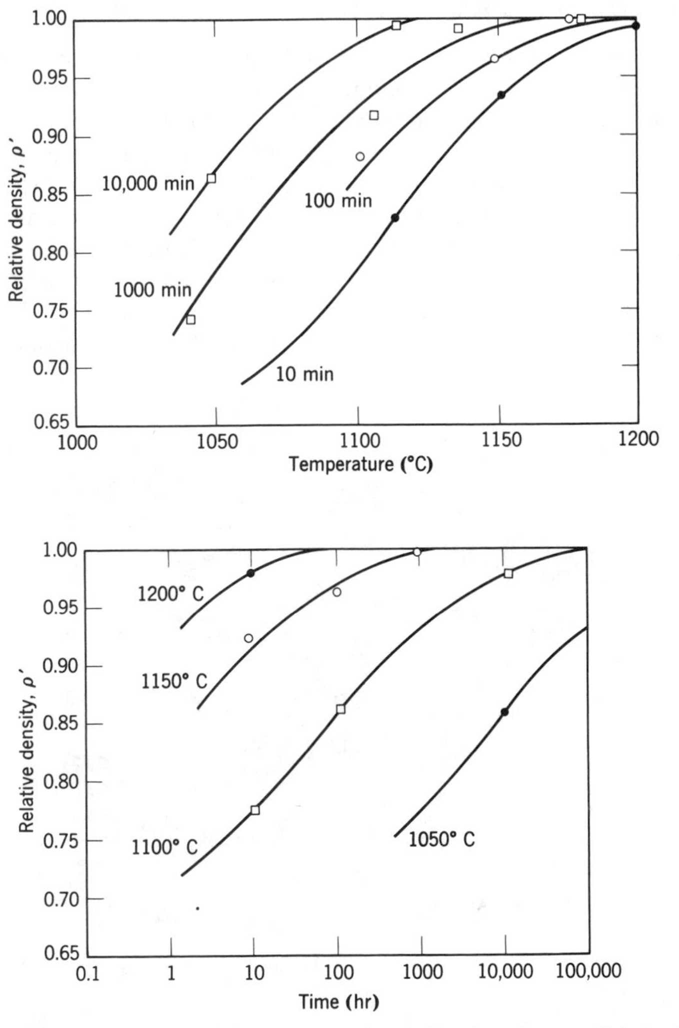

Fig. 10.38. Effect of time and temperature on the vitrification of a porcelain body. Data from F. H. Norton and F. B. Hodgdon, *J. Am. Ceram. Soc.*, **14**, 177 (1931).

densification results from longer periods of time at the same temperature. In controlling the process, the temperature dependence is great because of the increase in liquid content and lowered viscosity at higher temperatures. Changes in processing and changes in composition affect the vitrification process as they affect these parameters.

10.4 Sintering with a Reactive Liquid

Another quite different process which leads to densification is sintering in the presence of a reactive liquid. Here we are referring to systems in which the solid phase shows a certain limited solubility in the liquid at the sintering temperature; the essential part of the sintering process is the solution and reprecipitation of solids to give increased grain size and density. This kind of process occurs in cermet systems such as bonded carbides and also in oxide systems when the liquid phase is fluid and reactive, such as magnesium oxide with a small amount of liquid phase present (Fig. 10.39), UO_2 with the addition of a small amount of TiO_2 (Fig. 7.11), and high-alumina bodies which have an alkaline earth silicate as a bonding material.

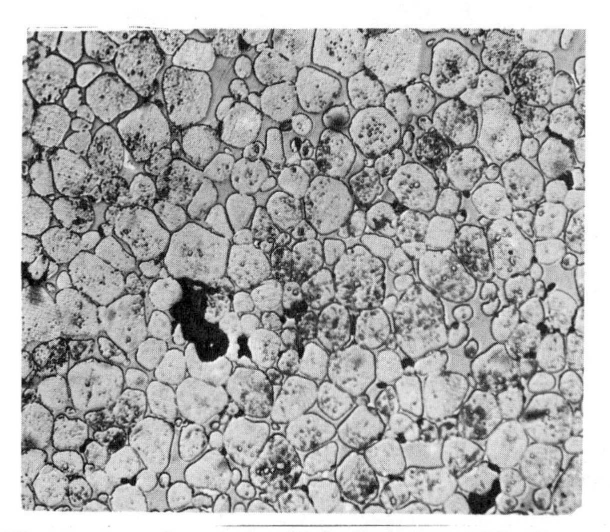

Fig. 10.39. Microstructure of magnesia—2% kaolin body resulting from reactive-liquid sintering (245×).

Studies of a large number of systems indicate that for densification to take place rapidly it is essential to have (1) an appreciable amount of liquid phase, (2) an appreciable solubility of the solid in the liquid, and (3) wetting of the solid by the liquid. The driving force for densification is derived from the capillary pressure of the liquid phase located between the fine solid particles, as illustrated in Fig. 10.40. When the liquid phase wets the solid particles, each interparticle space becomes a capillary in which a substantial capillary pressure is developed. For submicron particle sizes, capillaries with diameters in the range of 0.1 to 1 micron

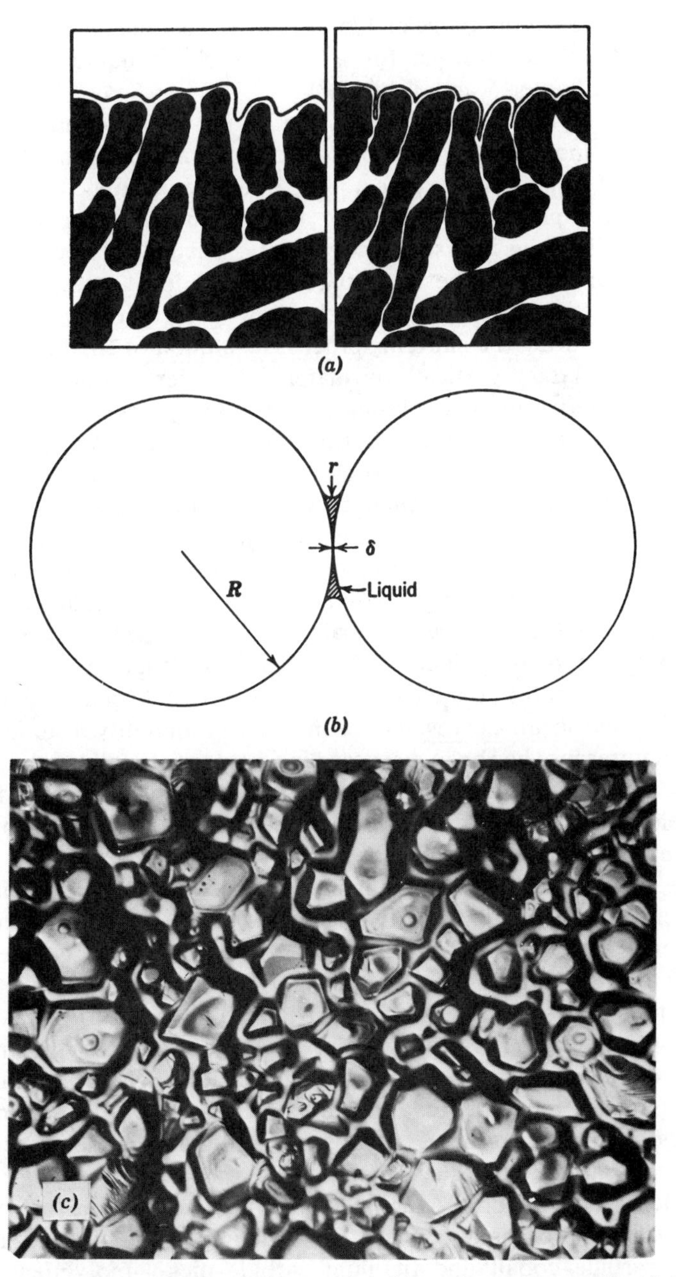

Fig. 10.40. (*a*) Surface of solid-liquid composite with varying amounts of liquid phase. (*b*) Drop of liquid between two solid spheres exerts pressure to pull them together. (*c*) Surface of forsterite ceramic showing liquid capillary depression between crystals.

develop pressures in the range of 175 to 1750 psi for silicate liquids and in the range of 975 to 9750 psi for a metal such as liquid cobalt (see discussion in Chapter 5 and Table 5.2).

The capillary pressure results in densification by different processes which occur coincidentally. First, on formation of a liquid phase there is a rearrangement of particles to give a more effective packing. This process can lead to complete densification if the volume of liquid present is sufficient to fill in the interstices completely. Second, at contact points where there are bridges between particles high local stresses lead to plastic deformation and creep, which allow a further rearrangement. Third, there is during the sintering process a solution of smaller particles and growth of larger particles by material transfer through the liquid phase. The kinetics of this solution-precipitation process have already been discussed in Chapter 9. Because there is a constantly imposed capillary pressure, additional particle rearrangement can occur during grain-growth and grain-shape changes and give further densification. (As discussed for vapor transport and surface diffusion in solid-state sintering, mere solution-precipitation material transfer without the imposed capillary pressure would not give rise to densification). Fourth, in cases in which liquid penetrates between particles the increased pressure at the contact points leads to an increased solubility such that there is material transfer away from the contact areas so that the particle centers approach one another and shrinkage results; the increase in solubility resulting from the contact pressure has been discussed in Chapter 5. Finally, unless there is complete wetting, recrystallization and grain growth sufficient to form a solid skeleton occur, and the densification process is slowed and stopped.

Perhaps even more than for the solid-state process, sintering in the presence of a liquid phase is a complex process in which a number of phenomena occur simultaneously. Each has been shown to occur, but experimental systems in which a single process had been isolated and analysed during sintering have not been convincingly demonstrated. Clearly, the process requires a fine-particle solid phase to develop the necessary capillary pressures which are proportional to the inverse capillary diameter. Clearly, the liquid concentration relative to the solid particle packing must be in a range appropriate for developing the necessary capillary pressure. Clearly, if and when a solid skeleton develops by particle coalescence, the process stops.

A critical and still controversial question is the degree of wetting required for the process to proceed. In some important systems such as tungsten carbide–cobalt and titanium carbide–nickel–molybdenum the dihedral angle is zero. In other systems such as iron-copper and magnesia-silicate liquids this is not the case at equilibrium; but the dihedral angle is low, and the solid is wetted by the liquid phase, as required to develop the necessary capillary pressure. For grain growth of periclase particles in a silicate liquid, the dihedral angle has a large effect on the grain-growth process, as illustrated in Fig. 10.41. Although zero dihedral angle is not essential for liquid-phase sintering to occur, the process becomes more effective as this ideal is approached.

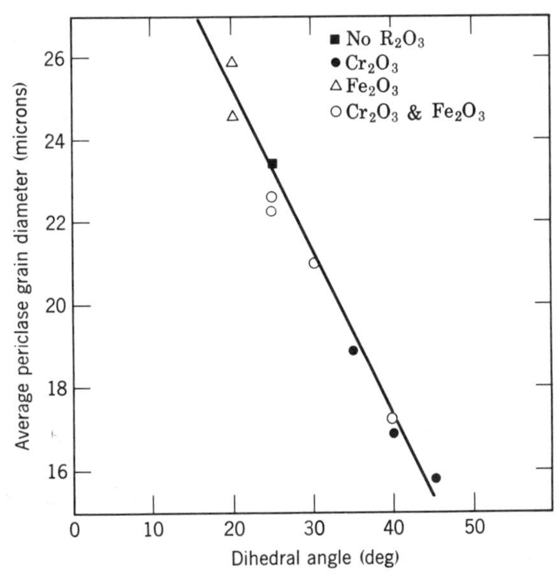

Fig. 10.41. Grain growth of periclase particles in liquid-phase-sintered periclase-silicate compositions as a function of dihedral angle. From B. Jackson, W. F. Ford, and J. White, *Trans. Brit. Ceram. Soc.*, **62**, 577 (1963).

10.5 Pressure Sintering and Hot Pressing

The sintering processes thus far discussed depend on the capillary pressures resulting from surface energy to provide the driving force for densification. Another method is to apply an external pressure, usually at elevated temperature, rather than relying entirely on capillarity.* This is desirable in that it eliminates the need for very fine-particle materials and also removes large pores caused by nonuniform mixing. An additional advantage is that in some cases densification can be obtained at a temperature at which extensive grain growth or secondary recrystalliza-

*R. L. Coble, *J. Appl. Phys.* **41**, 4798 (1970).

tion does not occur. Since the mechanical properties of many ceramic systems are maximized with high density and small grain size, optimum properties can be obtained by hot-pressing techniques. The effect of added pressure on the densification of a beryllium oxide body is illustrated in Fig. 10.42. The main disadvantages of hot pressing for oxide

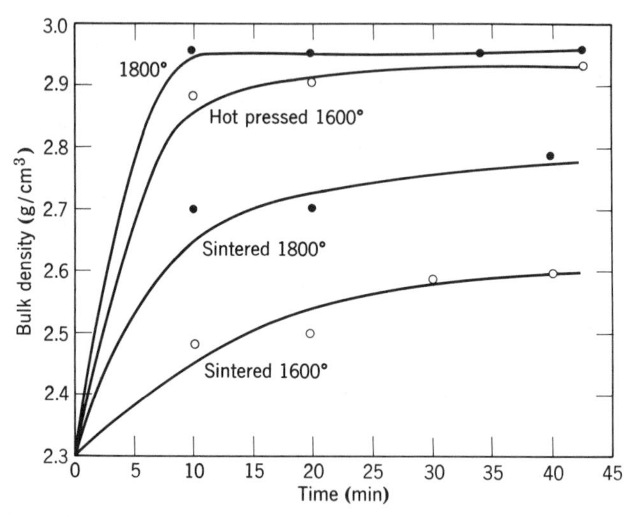

Fig. 10.42. Densification of beryllia by sintering and by hot pressing at 2000 psi.

bodies are the unavailability of inexpensive and long-life dies for high temperatures and the difficulty in making the process into an automatic one to achieve high-speed production. Both factors make the hot-pressing process an expensive one. For oxide materials which have to be pressed at temperature above 1200 or 1300°C (often at 1800 to 2000°C) graphite is the most satisfactory die material available; the maximum stress is limited to a few thousand pounds per square inch, and the life of dies is usually limited to seven or eight pieces. The entire die must be heated and cooled with the formation of each piece. Techniques for using high temperatures in a process in which the die is maintained cool with the material heated have shown some promise in laboratory tests but have not been developed for production.

For lower-temperature materials, such as glasses or glass-bonded compositions which can be pressed in metal dies at temperatures below 800 to 900°C, the hot-pressing process can be developed as an automatic and inexpensive forming method. This is similar to the normal pressing of glass as a glass-forming method in which it is used to obtain the desired shape rather than as a means of eliminating porosity.

Densification during pressure sintering can occur by all the mechanisms which have been discussed for solid-state sintering, vitrification, and liquid-phase sintering. In addition, particularly during the early stages, when high stresses are present at the particle contact points, and for soft materials, such as the alkali halides, plastic deformation is an important densification mode. Since the grain-growth process is insensitive to pressure, pressure-sintering oxides at high pressures and moderate temperatures allows the fabrication of high-density–small-grain samples with optimum mechanical properties and with sufficiently low porosity to be nearly transparent. Covalent materials such as boron carbide, silicon carbide, and silicon nitride can be hot-pressed to nearly complete density. It is often advantageous to add a small fraction of liquid phase (i.e., LiF to MgO, B to silicon carbide, MgO to silicon nitride) to allow pressure-induced liquid phase, or liquid-film, sintering to occur.

10.6 Secondary Phenomena

The primary processes which occur on heating and are important in connection with the firing behavior of all ceramic compositions are grain growth and densification, as discussed in previous sections. In addition to these changes, there are a large number of other possible effects which occur during the firing of some particular compositions. These include chemical reactions, oxidation, phase transitions, effects of gas trapped in closed pores, effects of nonuniform mixing, and the application of pressure during heating. Although they are not processes of the most general importance, they frequently cause the main problems and the major phenomena observed during firing. Although we cannot discuss them in great detail, we should at least be familiar with some of the possibilities.

Oxidation. Many natural clays contain a few percent organic matter which must be oxidized during firing. In addition, varnishes or resins used as binders, as well as starches and other organic plasticizers, must be oxidized during firing, or difficulties result. Under normal conditions organic materials char at temperatures above 150°C and burn out at temperatures ranging from 300 to 400°C. Particularly with low-firing-temperature compositions, it is necessary to heat at a slow enough rate for this process to be completed before shrinkage becomes substantial. If the carbonaceous material is sealed off from the air by vitrification occurring before oxidation is completed, it acts as a reducing agent at higher temperatures. Sometimes this may merely affect the color, giving rise to *black coring* of brick and heavy clay products whose interiors are in a reduced state, black in color. A typical example of a stoneware heated too rapidly for oxidation to be completed is illustrated in Fig. 10.43, which shows the central black core. Very often impurities present, particularly sulfides, may cause difficulties unless oxidized before vitrification. Sulfides in general react with oxygen in the temperature range of 350 to 800°C, forming SO_2 gas which escapes through open pores.

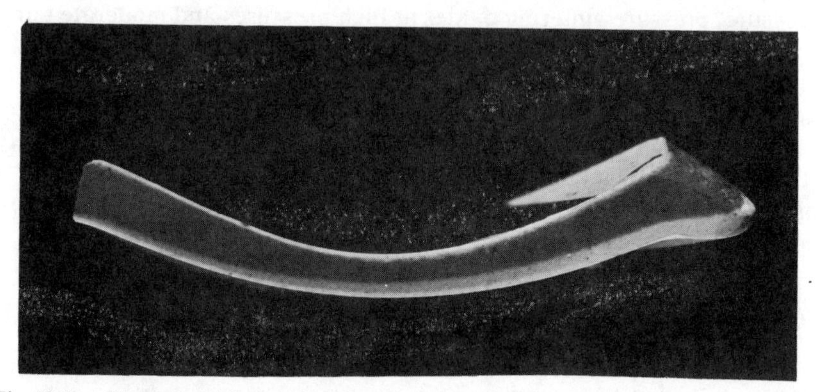

Fig. 10.43. Example of black core produced when time allowed for oxidation reactions was insufficient for completion of the reaction.

In ferrite and titania compositions control of oxidation reactions during firing is particularly important. As illustrated for the Ti–TiO_2 and Fe–O_2 systems (Chapter 7), the phases present depend on the oxygen pressure. In addition, as discussed in Chapter 4, the composition of these phases covers a substantial range of stoichiometry and depends on the oxygen pressure. It is common practice in the manufacture of ferrites to control the oxygen pressure during firing so that the composition of each phase present, and the overall phase composition of the body, is maintained to give the best magnetic properties.

Decomposition Reactions. Many of the constituents used in ceramic bodies are in the form of carbonates or hydrated compounds; these decompose during firing to form the oxide plus a gaseous product (CO_2, H_2O). Many impurities are also incorporated as carbonates, hydrates, and sulfates and decompose during firing (see Section 9.4).

Hydrates decompose over a wide temperature range between 100 and 1000°C, depending on the particular composition. Carbonates decompose over a temperature range from 400 to 1000°C, also depending on the particular composition. For each temperature there is, of course, an equilibrium pressure of the gaseous product; if this pressure is exceeded, further decomposition does not take place, leading to the major problem encountered, the sealing of pores before complete dissociation. As the temperature is raised, the decomposition pressure increases and forms large pores, blistering, and bloating. (This is, of course, the method used to form cellular glass products in which the surface is intentionally sealed off before chemical reaction or decomposition takes place to form a gas phase that expands and produces a foamed product.) This kind of defect is particularly common when high heating rates are used, for then there is a temperature gradient between the surface and interior of the ware, and the surface layer vitrifies, sealing off the interior. This temperature gradient and the time required for oxidation of constituents or impurities are the two most important reasons for limiting the rate of heating during firing.

Sulfates create a particular problem in firing because they do not decompose until a temperature of 1200 to 1300°C is reached. Therefore they remain stable during the firing process used for burning many clay bodies. In particular, $CaSO_4$ is stable but slightly soluble in water, so that a high sulfate content leads to a high concentration of soluble salts in the burned brick. This causes efflorescence—the transport of slightly soluble salts to the surface, forming an undesirable white deposit. Addition of barium carbonate prevents the deposit from forming by reacting with calcium sulfate to precipitate insoluble barium sulfate.

Decomposition also occurs in some materials to form new solid phases. A particular example used in refractory technology is the decomposition of kyanite, $Al_2O_3 \cdot SiO_2$, to form mullite and silica at a temperature of 1300 to 1450°C. This reaction proceeds with an increase in volume, since both mullite and the silica glass or cristobalite formed have lower densities than kyanite. The reaction is useful, since the addition of kyanite to a composition can counteract a substantial part of the firing shrinkage if the other constituents are carefully selected. Similarly, reaction of MgO with Al_2O_3 to form spinel occurs with a decrease in volume. By incorporating magnesia and alumina in a refractory mix, or more commonly in a high-temperature ramming mix or cement, the shrinkage taking place on heating can be decreased.

Phase Transformations. Polymorphic transformations may be desirable or undesirable, depending on the particular composition and the anticipated use. If a large volume change accompanies the polymorphic transformation, difficulties result, owing to the induced stresses. Refractories cannot be made containing pure zirconium oxide, for example, since the tetragonal monoclinic transformation at about 1000° involves such a large volume change that the ware is disrupted. The source of these stresses has been discussed in Chapter 5 in connection with boundary stresses caused by differential thermal expansion or contraction of different grains. The expansion or contraction of a crystal in a matrix leads to the same sort of stresses that may give rise to actual cracking, illustrated for quartz grains in a porcelain body in Fig. 10.44. The stresses in individual grains can be reduced if the grain size is reduced; properties of porcelains are improved if fine-grained flint is used rather than coarse material.

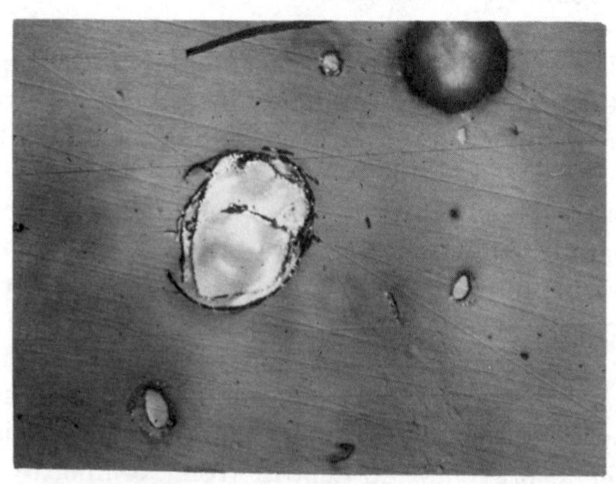

Fig. 10.44. Cracked quartz grain and surrounding matrix in a porcelain body. Differential expansion due mainly to the α-β quartz transition leads to cracking of larger grains but leaves small grains intact (500×).

Sometimes desirable phase transformations only occur sluggishly. This is what happens, for example, with the firing of refractory silica brick. The transition from quartz, the starting material, to tridymite and cristobalite, the desired end constituents, occurs only slowly. In order to increase the rate of transformation, calcium oxide is added as a mineralizer. The calcium oxide forms a liquid in which silica is soluble. Consequently the quartz dissolves and precipitates as tridymite, which is the more stable phase (Chapter 7). Some of the quartz transforms directly to cristobalite during the process as well. In general, mineralizers help in achieving equilibrium conditions by providing a mechanism of material transfer—solution or vaporization—that circumvents energy barriers to direct transformations. In silicate systems the addition of fluorides or hydroxyl ions is particularly helpful in this regard, since they greatly increase the fluidity of the liquid phase present.

Trapped Gases. In addition to the bloating occasioned by decomposition reactions, trapping of gases within closed pores imposes a limitation on the ultimate density that can be reached during firing. Gases such as water vapor, hydrogen, and oxygen (to a lesser extent) are able to escape from closed pores by solution and diffusion. In contrast, gases such as carbon monoxide, carbon dioxide, and particularly nitrogen have a lower solubility and do not normally escape from closed pores. If, for example, spherical pores are closed at a total porosity of 10% and a partial pressure of 0.8 atm nitrogen, the pressure has increased to 8 atm (about 110 psi) when they have shrunk to a total porosity of 1%, and further shrinkage is limited. At the same time that the gas pressure is increasing, however, the negative radius of curvature of the pore becomes small so that the negative pressure produced by surface tension is increased proportional to $1/r$; the gas pressure builds up proportional to $1/r^3$. For sintering in air this factor usually limits densification; where very high densities are required, as for optical materials or dental porcelains requiring high translucency, vacuum or hydrogen atmosphere is preferred.

Nonuniform Mixing. Although not mentioned in most discussions of sintering, the most important reason why densification and shrinkage stop short of complete elimination of pores is that gross defects caused by imperfect mixing and compact consolidation prior to firing are usually present. Examination of typical production ceramics shows that they commonly contain upward of 10% porosity in the millimeter size range (that is, pores much larger than the particle size of the raw materials introduced in the composition). These pores are caused by local variations induced during forming, and there is no tendency for elimination of these pores during firing. Corrective treatment must be taken in the forming method.

Overfiring. Ware is commonly referred to as overfired if for any of a variety of reasons a higher firing temperature leads to poorer properties or a reduced shrinkage. For solid-state sintering, such as ferrites and titanates, a common cause is secondary recrystallization occurring at the higher temperature before the elimination of porosity. Consequently, there is some maximum temperature at which the greatest density or optimum properties are obtained. For vitreous ceramics the most common cause of overfiring is the trapping of gases in pores or the evolution of gases which cause bloating or blistering.

10.7 Firing Shrinkage

As formed, green ware contains between 25 and 50 vol% porosity. The amount depends on the particle size, particle-size distribution, and forming method (Chapter 1). During the firing process this porosity is removed; the volume firing shrinkage is equal to the pore volume eliminated. This firing shrinkage can be substantially decreased by addition of nonshrinking material to the mix; fire-clay brick is commonly manufactured with grog (prefired clay) additions which serve to decrease firing shrinkage. Similarly, this is one of the functions of the flint in the porcelain body; it provides a nonshrinking structure which reduces the shrinkage during firing. Terra-cotta compositions, composed of mixtures of fired grog and clay, can be made in large shapes because a large part of the raw material has been prefired and the firing shrinkage is low.

If firing is carried to complete densification, the fractional porosity originally present is equal to the shrinkage taking place during firing. This commonly amounts to as much as 35% volume shrinkage or 12 to 15% linear shrinkage and causes difficulty in maintaining close tolerances. However, the main difficulties are warping or distortion caused by different amounts of firing shrinkage at different parts of the ware. Nonuniform shrinking can sometimes even cause cracks to open.

Warping. A major cause of warping during firing is density variations in the green ware. There are many reasons for differences in porosity in the green ware. The density after firing is nearly uniform, and there is higher shrinkage for the parts that had a low density than for the parts that had a high density in the green ware. In pressed ware, pressure variations in the die (Chapter 1) cause different amounts of compaction at different parts of a pressed piece; usually the shrinkage at the center is larger than the shrinkage at the ends, and an hourglass shape results from an initially cylindrical sample (Fig. 10.45*a*).

Another source of warping during firing is the presence of temperature gradients. If ware is laid on a flat plate and heated from above, there is a temperature difference between the top and bottom of the ware that may cause greater shrinkage at the top than at the bottom and a corresponding warping. In some cases the gravitational stresses may be sufficient to make the ware lie flat, even though shrinkage is nonuniform. The relationship between temperature distribution, warpage, and deformation under the stresses developed is complicated and difficult to analyze quantitatively. Another source of warpage in firing is preferred orientation of the platey clay particles during the forming process. This causes the drying and firing shrinkage to have directional properties.

Vitreous ware is also warped by flow under forces of gravity. This is especially true for large heavy pieces in which substantial stresses are developed. In the forming of vitreous sanitary ware, the upper surface of a closet bowl (Fig. 10.45*c*) or a lavatory (Fig. 10.45*d*) must be designed with a greater curvature than is desired in the end product so that the settling which occurs on firing produces a final shape that is satisfactory. A final contributor to warpage during firing is the frictional force or *drag*

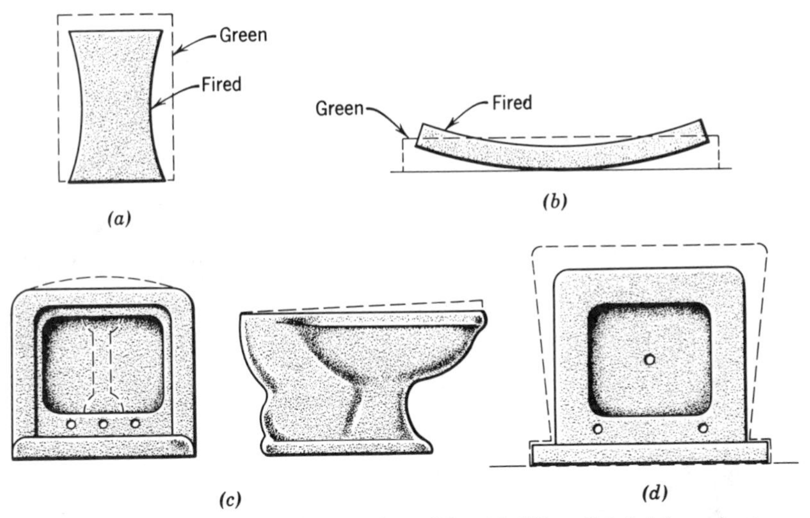

Fig. 10.45. Firing shrinkage of (*a*) pressed crucible with differential shrinkage due to green density variations, (*b*) tile with differential shrinkage due to temperature gradients, (*c*) ware with differential shrinkage due to gravity settling, and (*d*) differential shrinkage due to frictional force of setting.

of the ware against the setter. This means that the bottom surface tends to shrink less than the upper surface (Fig. 10.45*d*). Ware must be designed so that the final shape, including shrinkage, comes out to be rectangular.

Difficulties caused by differential firing shrinkage, resulting distortion, and warping can be eliminated in three ways: first, altering the forming method to minimize the causes of warping; second, designing shapes in a way that compensates for warping; and third, using setting methods in firing that minimize the effects of warping. One obvious improvement in forming methods is to obtain uniformity of the structure during initial forming. This requires elimination of pressure gradients, segregation, and other sources of porosity variation. Pressing samples that have long ratios of length to die diameter cause density variations. Extruded and pressed mixes that have low plasticity are particularly prone to large pressure variations and green density differences. Slip casting and extrusion both cause a degree of segregation and density differences during firing. Some settling may occur during the casting process, causing structural variations. During extrusion pressure differences at various parts of the die or an unsymmetrical setting for the die can cause variations.

Sometimes variations in firing shrinkage and difficulties from warping can be overcome by compensating the shapes. This is true, for example, in Fig. 10.45, in which the closet bowl and lavatory are designed in such a

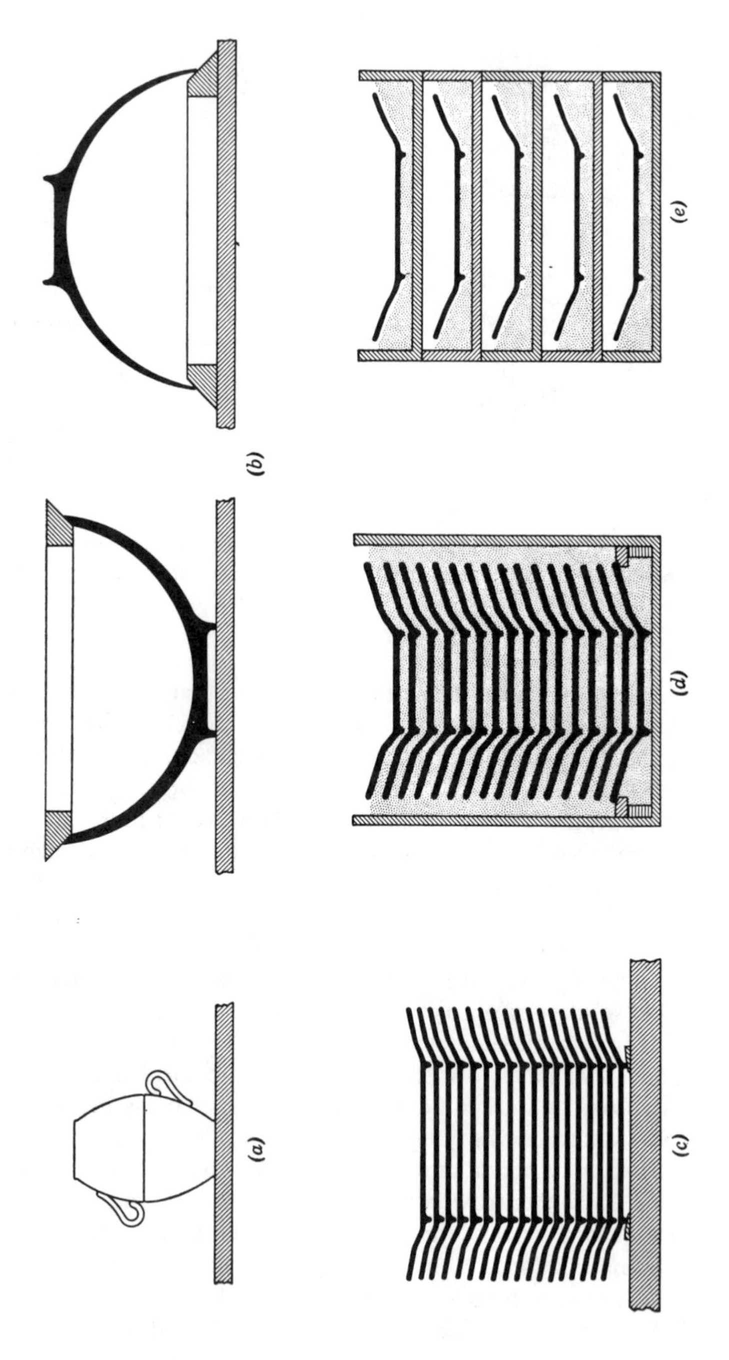
(a)
(b)
(c)
(d)
(e)

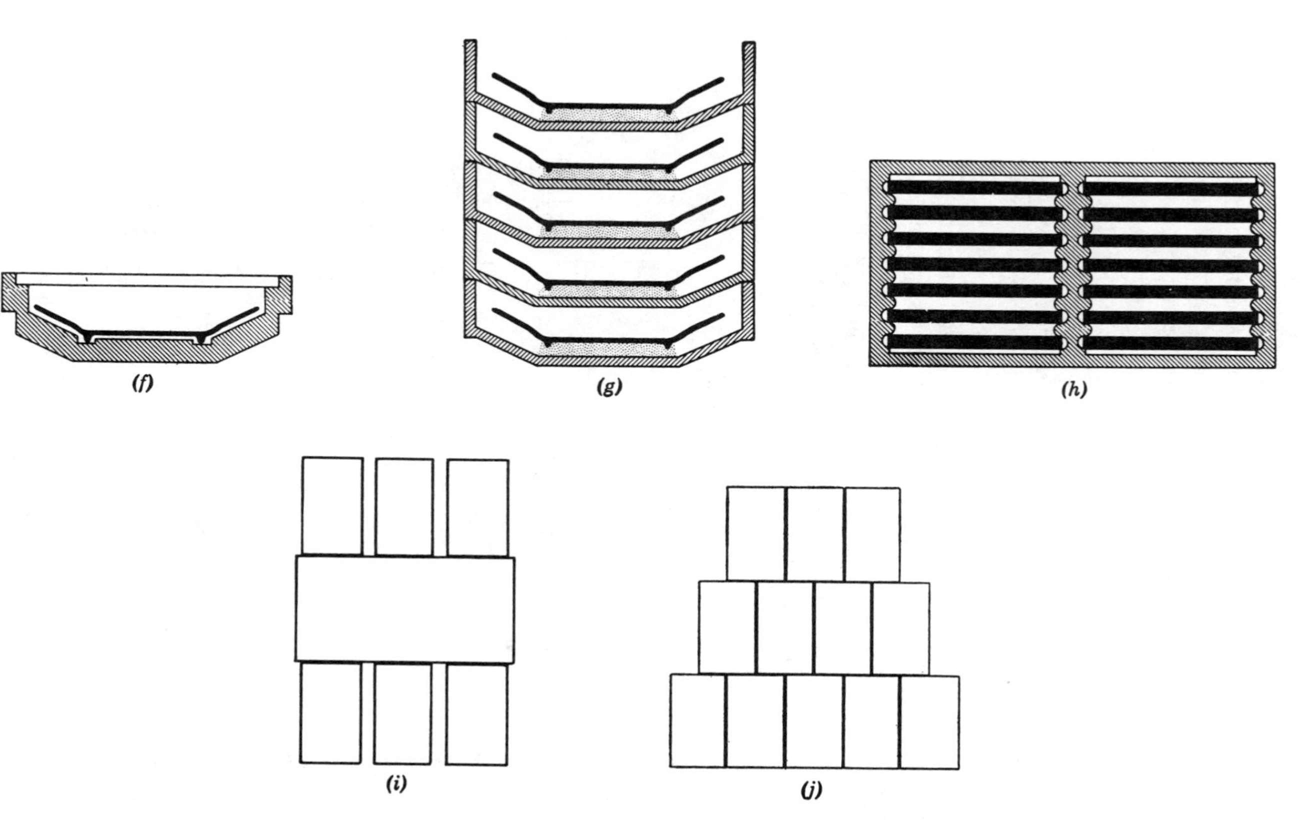

Fig. 10.46. Setting methods for (*a*) cups and bowls, (*b*) large bowls, (*c*) earthenware, (*d*) hotel china plates, (*e*) bone china plates, (*f*) frit porcelain plates, (*g*) hard porcelain plates, (*h*) tile, (*i*) brick checkerwork, (*j*) brick bench setting. From F. H. Norton.

way that the final shape is satisfactory. In the same way, when plates are fired in the horizontal position there is a tendency for the rims to settle; this can be compensated for by adjusting the shape of the initial piece.

Correct setting methods are important in eliminating difficulties caused by firing-shrinkage variations. These have been most extensively developed for porcelain compositions in which complete vitrification is desired and high shrinkages result. Some of the standard setting methods are illustrated in Fig. 10.46. Cups and bowls are commonly boxed as indicated in Fig. 10.46*a*. This keeps the rim circular, since warpage of one restricts warpage of the other; in addition, it prevents the thin rims from being too rapidly heated. For larger pieces, unfired setters are necessary as a means of controlling shrinkage and maintaining circular rims. A variety of methods is used for setting different kinds of plate compositions, depending on the amount of shrinkage expected. For ware fired to complete vitrification individual setting and support are essential. For ware fired to partial densification, plates can be stacked with no ill effects. In general, large tiles and brick do not cause much difficulty.

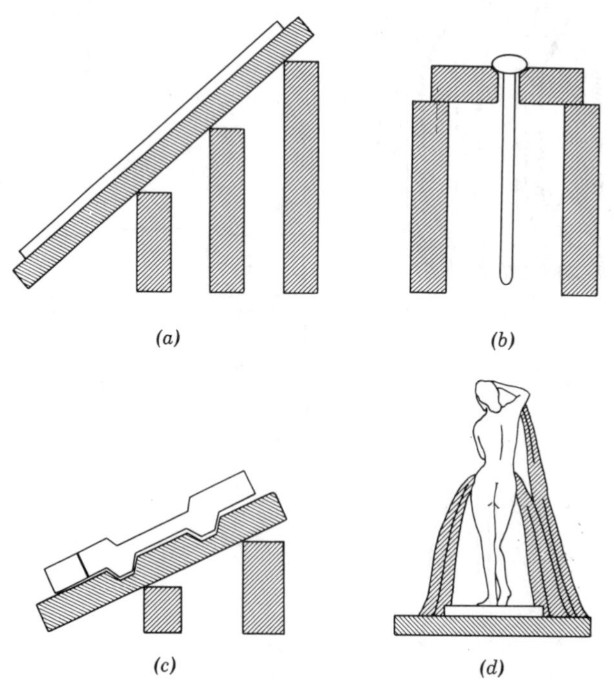

Fig. 10.47. Setting methods for special shapes. (*a*) Large tiles set an angle of repose; (*b*) slender rod supported by collar; (*c*) special shape; (*d*) sculptured piece. From F. H. Norton.

Special shapes may require special setting methods to eliminate adverse effects of firing shrinkage. Large refractory tile can be set at an angle of repose on a flat surface (Fig. 10.47*a*). This allows the tile to shrink without much stress. In the same way rods or tubes may be set in an inclined V groove or supported by a collar from the upper end (Fig. 10.47*b*). Gravitational forces keep the tubes straight up to lengths of several feet. Unique shapes can always be supported on special setters designed for the particular sample. Some experience is necessary to handle unique shapes efficiently. Small pieces of sculptured vitrified ware are particularly difficult. The safest setting provides complete support from unfired struts (Fig. 10.47*d*).

Suggested Reading

1. G. C. Kuczyuski, N. A. Hooton, and C. F. Gibson, Eds., *Sintering and Related Phenomena*, Gordon and Breach, New York, 1967.
2. G. C. Kuczyuski, Ed., "Sintering and Related Phenomena", *Materials Science Research*, Vol. 6, Plenum Press, New York, 1973.
3. R. L. Coble and J. E. Burke, *Progress in Ceramic Science*, Vol. III, J. E. Burke, Ed., Pergamon Press, 1963.
4. W. D. Kingery, Ed., *Ceramic Fabrication Process*, Part IV, Technology Press, Cambridge, Mass., and John Wiley & Sons, New York, 1958.
5. W. D. Kingery, Ed., *Kinetics of High-Temperature Process*, Part IV, Technology Press, Cambridge, Mass., and John Wiley & Sons, New York, 1959.
6. J. E. Burke and D. Turnbull, "Recrystallization and Grain Growth in Metals," *Prog. Met. Phys.*, **3**, 220 (1952).
7. E. Schramm and F. P. Hall, "The Fluxing Effect of Feldspar in Whiteware Bodies," *J. Am. Ceram. Soc.*, **15**, 159 (1936).
8. For additional papers on sintering see: R. L. Coble, *J. Appl. Phys.*, **41**, 4798 (1970); D. L. Johnson and I. B. Cutler, *J. Am., Ceram. Soc.* **46**, 541 (1963).

Problems

10.1. Distinguish between primary recrystallization, grain growth, and secondary recrystallization as to (*a*) source of driving force, (*b*) magnitude of driving force, and (*c*) importance in ceramic systems.

10.2. Explain why the activation energy for grain-boundary migration corresponds approximately with that for boundary diffusion, even though no concentration gradient exists in the former case.

10.3. Can grain growth during sintering cause compaction of ceramics? Explain. Can grain growth affect the sintering rate? Explain.

10.4. Which of the following processes can contribute increased strength to sintered articles without causing compaction? Explain.
(*a*) Evaporation condensation.
(*b*) Volume diffusion.
(*c*) Viscous flow.
(*d*) Surface diffusion.
(*e*) Solution reprecipitation.

10.5. Assuming the surface energy of $NiCr_2O_4$ is 600 erg/cm^2 and estimating diffusion data from Cr_2O_3 and NiO data given in Chapter 6, what would be the initial rate of densification for a compact of 1-micron particles at 1300°C? at 1400°C? at 1200°C?

10.6. If pores of 5-micron diameter are sealed off containing nitrogen at a pressure of 0.8 atm in a glass having a surface tension of 280 dyne/cm and a relative density of 0.85, what will be the pore size at which the gas pressure just balances the negative pressure due to the surface tension? What will be the relative density at this point?

10.7. Explain the mechanism of reactive-liquid sintering, such as occurs for Co–WC compositions. Identify two critical solid-liquid interaction characteristics in a system showing this behavior, and describe how you would quantitatively measure them.

10.8. From data collected during sintering of powder compacts of nominally pure simple phase materials at variable heating rates, the observed rates of density change analyzed on an Arrhenius plot frequently give activation energies higher in value than that for lattice self-diffusion. There are three sets of assumptions on which this behavior can be rationalized. Give two examples of suitable assumptions, and explain the behavior on a mechanistic basis.

10.9. During the normal grain growth of MgO at 1500°C, crystals were observed to grow from 1 micron diameter to 10 micron diameter in 1 hr. Knowing that the grain-boundary diffusion energy is 60 kcal/mole, predict the grain size after 4 hr at 1600°C. What effect would you predict impurities will have on the rate of grain growth of MgO? Why?

10.10. Suppose that in order to reduce sintering shrinkage you were to mix enough fine particles (about 30%), 1 micron in diameter, with coarse particles, 50 microns in diameter, so that all the interstices between the coarse particles were filled with fine particles. What would be the rate of shrinkage of this compact? Make a plot of $\log(\Delta L/L_0)$ versus $\log t$, and place the 1-micron powder and 50-micron powder shrinkage lines in their relative positions; then place the shrinkage curves for the composite material in its proper position with respect to the 1-micron and 50-micron curves. Justify your answer.

10.11. A certain magnetic oxide material is believed to follow the normal grain-growth equation. Magnetic-strength properties deteriorate when grain size increases beyond an average of 1 micron. The original grain size before sintering is 0.1 micron. Sintering for 30 min triples the grain size. Because of warping of large pieces, the superintendent of production wants to increase the sintering time. What is the maximum time you would recommend?

10.12. Alumina with MgO is sintered to nearly theoretical density in hydrogen to the point that the optical transmission in the visible range is almost 100%. Actually the material (Lucalox) is not transparent but translucent because of the hexagonal crystal structure of alpha alumina. It is used to contain sodium vapor (at pressures above atmospheric) for street lamps. An alternative candidate for this application is

CaO which is cubic and could be transparent if sintered to theoretical density. Outline your research program if you were to seek to make CaO transparent through sintering.

10.13. The time required to shrink 5% for a compact of 30-micron glass spheres is 209.5 min at 637°C and 5.8 min at 697°C. Compute the activation energy and viscosity of the glass on the basis that surface energy is 300 ergs/cm^2.

11

Microstructure of Ceramics

One of the main principles on which this introduction to ceramics is based is that the properties of ceramic products are determined not only by the composition and structure of the phases present but also by the arrangement of the phases. The phase distribution or microstructure in the final ware depends on the initial fabrication techniques, raw materials used, phase-equilibrium relations, and kinetics of phase changes, grain growth, and sintering. In this chapter the resulting structures for a number of different systems are discussed.

11.1 Characteristics of Microstructure

The observation and interpretation of microstructure have a long history in geology, metallurgy, and ceramics. Many of the most fruitful analyses of ceramic microstructures were carried out in the period 1910–1930, so that this is by no means a new subject. However, new developments in techniques and the understanding of factors affecting microstructure development have provided a more complete picture of what the structure is actually like and better interpretation of its origin.

The characteristics of microstructure that can be determined are (1) the number and identification of phases present, including porosity, (2) the relative amounts of each phase present, and (3) characteristics of each phase, such as size, shape, and orientation.

Techniques of Studying Microstructure. Many different techniques of studying microstructure have been used; the two most widely used optical methods are observations of thin sections with transmitted light and observations of polished sections with reflected light. Thin-section techniques use light passing through a section that is 0.015 to 0.03 mm thick. The section is prepared by cutting a thin slice of material, polishing one side, cementing this side to a microscope slide, and then grinding and polishing the other side to obtain a section of the required uniform thickness. The method is advantageous in that the optical properties of each phase present can be determined and the phases thus identified. It suffers from two main disadvantages: first, specimens are difficult to prepare, and, second, the individual grains in many fine-grained ceramic materials are smaller in size than the section thickness, which leads to confusion, particularly for those who are not experts. In general, considerable experience is necessary to use thin-section techniques to best advantage.

Polished sections are usually prepared by mounting a cut specimen in Bakelite or Lucite plastic and then grinding and polishing one face smooth, using a series of abrasive papers followed by abrasive powders on cloth wheels. Polished surfaces can be observed directly with reflected light in a metallurgical microscope to distinguish pore structure and, in some ceramics, to distinguish differences in relief or reflectivity between different phases. Usually, however, different phases are best distinguished after chemical etching. When the sample is subjected to a chemical reagent, some phases are more rapidly attacked than others; grain boundaries are usually the most rapidly dissolved. Resulting differences in relief and surface roughness allow a distinction among the phases present. Polished sections are advantageous in that they are relatively easy to prepare and are simpler to interpret, particularly for a beginner. Phases present cannot be identified by their optical properties but can be distinguished and sometimes identified by their etching characteristics. Suitable etchants must be developed for each class of material. As discussed in Chapter 13, the reflectivity of a transparent material depends on its index of refraction; for a silicate with a refractive index of 1.5 only about 4% of normal incident light is reflected. For these materials a xenon or arc-light source is preferred; as an alternative a thin layer of highly reflective gold may be evaporated from a tungsten filament onto the sample surface.

The resolution obtainable in a microscope is limited by the wavelength of light used, and in practice optical microscopy is limited to a magnification of about 1000×. By using an electron beam with a wavelength measured in angstroms, the resolution can be improved to some tens of angstroms (or less with special techniques), and magnifications of 50,000× can be readily achieved. As with light microscopy, thin sections can be viewed with transmitted electrons. In this mode a sample thickness of less than about 10,000 Å (1 micron) is required, preferably less than 1000 Å (0.1 micron). Samples this thin cannot usually be prepared by grinding but require in addition either chemical thinning or thinning by argon ion bombardment. For survey studies it is often possible to shatter a specimen and then use the thin edges of the resulting particles for transmission electron microscopy. One of the powerful advantages of electron microscopy is that selected area electron diffraction patterns can serve to identify and characterize phases just as they are seen in the microstructure.

Scanning electron microscopy uses an electron beam which scans the surface, causing the emission of secondary electrons suitable for viewing. A wide range of magnifications is possible, 20× to 50,000×, and a smooth surface is not required. To prevent electrostatic charging of the surface of ceramic insulators, they must be coated with a thin layer of gold evaporated from a tungsten filament onto the sample surface. Scanning electron microscopy is particularly useful for observing surface features and fracture surfaces but is widely applicable to general microstructure observations as well. A special advantage of this method is that by analysis of the electron emission energy spectrum from a point on the sample, its approximate chemical composition can be determined in situ.

Many other techniques are useful for determining phase distribution, morphology, and characteristics in ceramic materials. Phase microscopy allows greatly increased depth resolution. Use of polarized light with polished or thin sections aids in phase identification. Stereoscopic microscopy, X-ray microscopy, dark-field electron microscopy, and many other special methods are useful and should be considered as possible techniques for each particular problem. The best methods vary from sample to sample, and any general rule would be subject to many exceptions. For most purposes preparation and examination of polished sections by optical or scanning electron microscopy together with X-ray diffraction or microscopic phase identification constitutes a good basic procedure.

Porosity. A phase that is almost always present in ceramics prepared by powder compaction and heat treatment is porosity. Porosity can be characterized by the volume fraction of pores present and their size, shape, and distribution compared with other phases. The amount of porosity can vary from zero to more than 90% of the total volume; this is the basic measurement required, but it is not sufficient. Many properties are also strongly dependent on pore shape and distribution. For example, the direct-current electrical conductivity and also thermal conductivity change with porosity between wide limits (Fig. 11.1). These limits can perhaps be best visualized by considering an idealized case of parallel slabs, as shown in Fig. 11.2. Here the low conductivity (high resistance) of the porosity is in series with the solid for electrical or heat flow normal to the slabs and

$$\frac{1}{k_t} = \frac{f_s}{k_s} + \frac{f_p}{k_p} \tag{11.1}$$

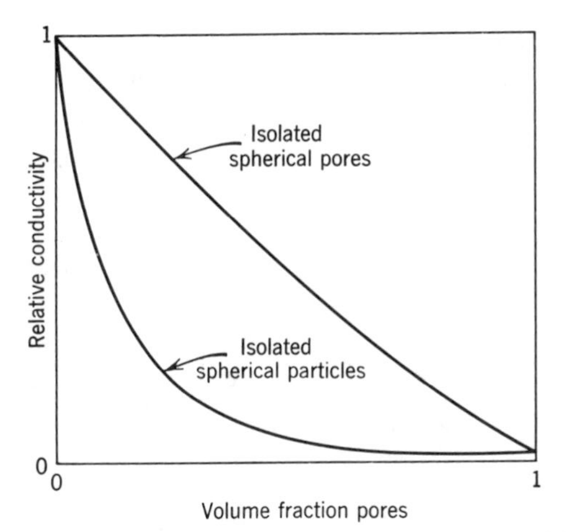

Fig. 11.1. Effect of porosity on direct-current electrical conductivity or on thermal conductivity. Upper curve, isolated spherical pores in a continuous solid matrix; lower curve, isolated solid particles in a continuous pore matrix.

where k_p and k_s are the conductivity of the pore and solid phases and f is the volume fraction for each phase. If k_p is much less than k_s, then k_t nearly equals k_p/f_p. In contrast, for flow parallel to the slabs each conducts heat or electricity with the same thermal or potential gradient and

$$k_t = f_s k_s + f_p k_p \tag{11.2}$$

If k_p is much less than k_s, then k_t nearly equals $f_s k_s$.

In general, samples with low porosity nearly approach a continuous

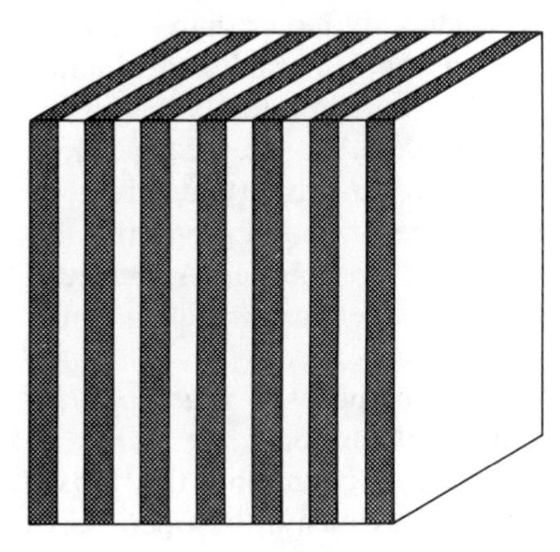

Fig. 11.2. Parallel slabs of pore space and solid material.

solid phase, and samples with high porosity tend to approach a continuous pore phase to give an S-shaped property-variation curve (Fig. 11.1). However, this does not always happen. Although a high-fired porcelain such as that shown in Fig. 11.3 closely approaches having isolated

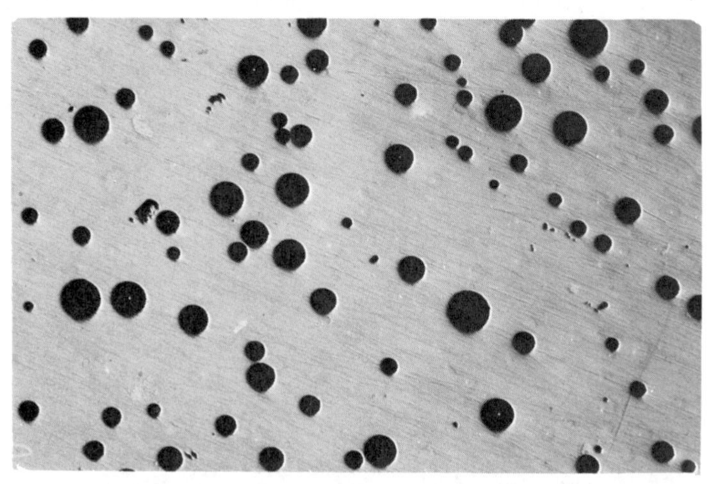

Fig. 11.3. Pore structure in high-fired Japanese hard porcelain (95×, unetched).

spherical pores, the low-porosity large-grain Al_2O_3 sample shown in Fig. 5.15 has flat cracks along many grain boundaries that closely approach a continuous pore phase, even though the fraction porosity is low. Similarly, a foam glass structure is essentially a continuous-solid-phase–isolated-pore-phase structure, even though the porosity is high.

Porosity can also be characterized by its relation to other phases. The nearly spherical pores in Fig. 11.3 are considerably larger than the other constituents. In contrast, the fine pores in the recrystallized alumina shown in Fig. 10.16*b* are similar in shape but are nearly all inside individual grains. The same-size pores in Fig. 10.16*a* are nearly all on grain boundaries. These distinctions can have an important effect on sintering, grain growth, and high-temperature deformation properties. As already mentioned, the shape of pores also affects properties. An interesting example of porosity with crystallographic orientation and crystallographic faces developed is given in Fig. 11.4.

One of the common methods of characterizing porosity is as *apparent porosity*—those pores connected to the surface, or *open pores.* In contrast the *total porosity* includes both the open and the *closed pores*—those not connected to the surface. Obviously, the open pores

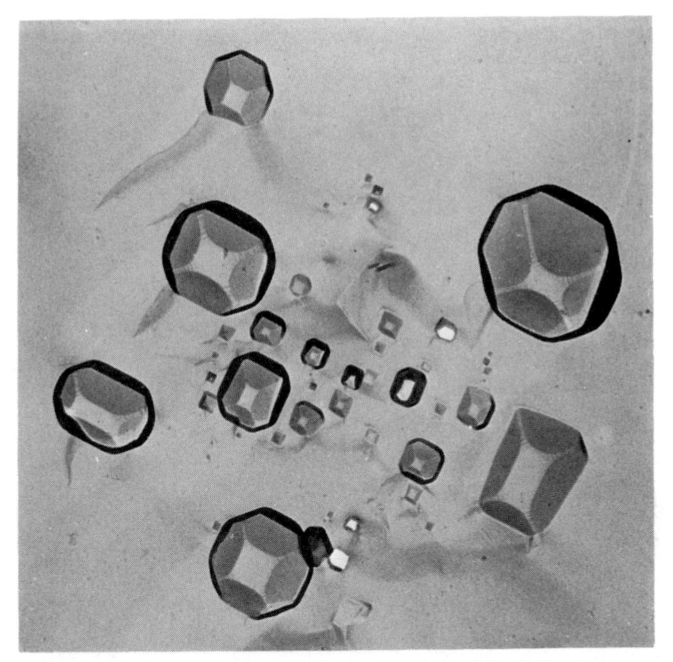

Fig. 11.4. Pores which have formed "negative" crystals in UO_2; (100) planes are parallel to surface (18,000×). Courtesy T. R. Padden.

directly affect properties such as permeability, vacuum tightness, and surface available for catalytic reactions and chemical attack, whereas closed pores have little effect on these properties.

Before firing, almost the entire porosity is present as open pores. During firing, the volume fraction porosity decreases, as discussed in Chapter 10. Although some open pores are eliminated directly, many are transformed into closed pores. As a result, the volume fraction of closed pores increases initially and only decreases toward the end of the firing process. Open pores are generally eliminated when the porosity has decreased to 5%. This is illustrated for the permeability of ceramics to gas flow in Fig. 11.5. By the time 95% of theoretical density is reached, the ware is gastight.

Single-Phase Polycrystalline Ceramics. In addition to porosity it is necessary to determine the amount, size, shape, and distribution of other constituents present in order to characterize a microstructure completely. In the simplest case this is a single phase. The microstructure of a polycrystalline ceramic normally develops as grains that meet at *faces* whose intersections form angles of 120° (discussed in Chapters 5 and 10).

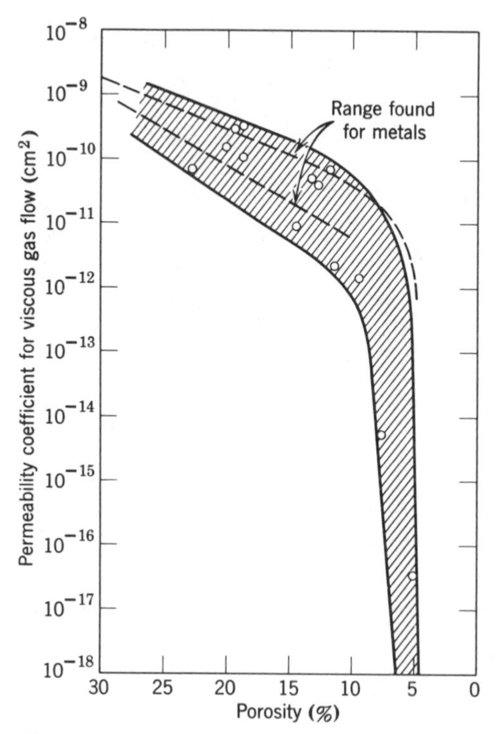

Fig. 11.5. Permeability coefficient for viscous gas flow in beryllia ware of differing porosity. Permeability drops rapidly when the porosity is below 5 to 8%. After J. S. O'Neil, A. W. Hey, and D. T. Livey, U.K. AERE-R3007.

Viewed in a polished section we cannot see the three-dimensional form which is apparent in Fig. 5.15. However, there are only a limited number of ways in which grains satisfying the 120°-angle requirement at the intersection of three grains to form a common boundary can be shaped to fit together and fill space. It is found that grains of many different materials usually have nine to eighteen faces, each face with four to six edges, as discussed in the next section. However, in some materials a duplex structure is developed in which some large grains grow in a fine-grained matrix (Figs. 10.12, 10.13, and 10.15). Also, grains are sometimes columnar, prismatic, cubic, spheroidal, or acicular or have some other special habit of growth; these shapes can give rise to special properties such as those of the oriented ferrites.

In single-phase ceramics the nature and composition of dislocations, subgrain boundaries, and grain boundaries (Figs. 4.23 and 4.24) are frequently important. These often have definite structural characteristics such as specific orientation, concentration, and impurity segregations. However, they are more difficult to investigate than the gross microstructure.

Multiphase Ceramics. In multiphase compositions we must also consider the relationships among the amount, distribution, and orientation of separate phases. Perhaps the most common structure is one or more phases dispersed in a continuous matrix. These can be prismatic crystals in a glass, such as the forsterite-glass ceramic shown in Fig. 7.33, or they can be crystals precipitated on a crystalline matrix, as shown in Figs. 9.40 to 9.43. The crystalline habit may be cubic, prismatic, columnar, dendritic, acicular, or parallel plates or be formed in other specific shapes. They can be orientated with regard to the matrix and in some ceramics are preferentially found along grain boundaries, subgrain boundaries, or other preferred sites. Another common structure is large particles held together with a bond forming lenses between the grains.

The multiplicity of possible phase relations in complex systems makes the microstructure of ceramics a fascinating and valuable study, essential for understanding the effects of processing or environmental variables on properties.

11.2 Quantitative Analysis

Space Filling. Much of our understanding and interpretation of microstructures are based on the interface energy relationships discussed in Chapter 5 and the rate processes discussed in Chapters 8, 9, and 10. However, even the effects of these are limited by geometrical restrictions to the way in which an area or volume can be filled. If we consider a plane surface, for example, the only regular polygons that can fill an area are triangles, rectangles, or hexagons. Various irregular polygons can also fill space. In this case there is a definite relationship between the number of polygons P, edges between polygons E, and corners C which is given by Euler's law:

$$C - E + P = 1 \tag{11.3}$$

This law applies equally well to grain boundaries seen in a polished section and craze marks on a teacup. If the number of sides on an average polygon is $n = 2E/P$, from Eq. 11.3

$$n = 2\frac{C}{P} + 2 - \frac{2}{P} \tag{11.4}$$

The most common corners are those where three edges meet ($C/P = 1/3n$); much less common are corners where four edges meet ($C/P =$

$1/4n$). If we restrict the structure to one with the simplest corners with three grains meeting, the *average polygon* must be a hexagon.

As discussed in Chapter 5, a balance of the grain-boundary energies where three grain corners come together requires that the boundaries meet at an angle of 120°. If this requirement is met, the only equilibrium shape is a hexagon. Polygons with fewer or more than six sides, have curved faces and a pressure difference across these curved faces, and the boundaries tend to migrate. Grains with fewer than six sides shrink, and grains with more than six sides grow (Chapter 10).

In three dimensions there are similar geometric restrictions to filling space with regular polyhedra. For any system the number of corners, edges, faces, and polyhedra must obey a relation:

$$C - E + F - P = 1 \tag{11.5}$$

In addition, if interface energy relationships are to be fulfilled, there must be an angle of 120° between faces where three grains meet along a common boundary and angles of $109\frac{1}{2}°$ where four faces meet at a point. These relationships are nearly met, and space is filled by the packing of truncated octahedra (tetrakaidecahedra) (Fig. 11.6). This shape has six square faces and eight hexahedron faces with twenty-four vertices, each having two angles of 120° and one of 90°. These can be distorted to give a lower total interface energy by having all corners meeting at 109° and all edges at 120°. This sort of ideal grain is not observed exactly; studies of soap foams, plant cells, and polycrystalline metals, using stereoscopic microradiography, have shown that the number of faces in a three-dimensional grain ranges from nine to eighteen. The number of edges on

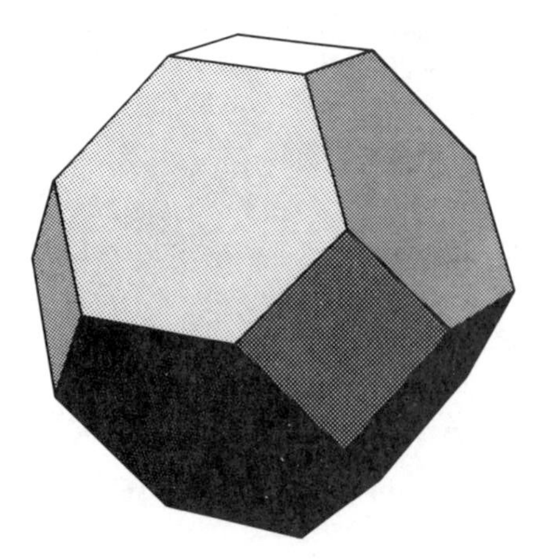

Fig. 11.6. Truncated octahedron (tetrakaidecahedron).

each face is most commonly five and is seldom less than four or more than six (Fig. 11.7).

As illustrated in Fig. 5.27, when a small amount of second phase is added, it can either be dispersed as isolated spherical particles or ideomorphic inclusions (Figs. 11.3 and 11.4), completely penetrate between and separate grains of the major phase (Fig. 7.11), or form between grain junctions with some equilibrium dihedral angle (observe the triangles of boundary phase with ϕ approximately 60° in Fig. 7.18). In a transparent section these angles can be measured with stereoscopic techniques or proper sample orientation. With plane sections, which often are all that are available, the plane of the polished section may lie at any angle to the line of the boundary. Consequently, a number of apparent angles is observed for any true angle present. From geometrical consider-

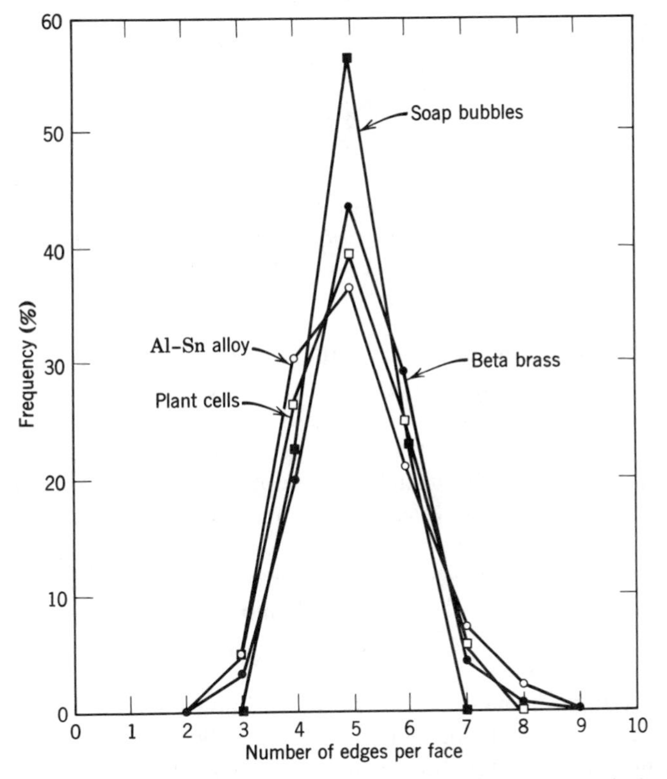

Fig. 11.7. Experimentally observed frequency distribution of the number of edges per face in various systems (reference 7).

ations D. Harker and E. R. Parker* have derived relationships between observed and true angles. A main result is that the most frequently observed angle is always equal to the average true angle. For quantitative measurements a large number of angles must be observed. From a histogram such as that shown in Fig. 11.8 the true angle can be determined; this dihedral angle characterizes one of the special relationships between phases.

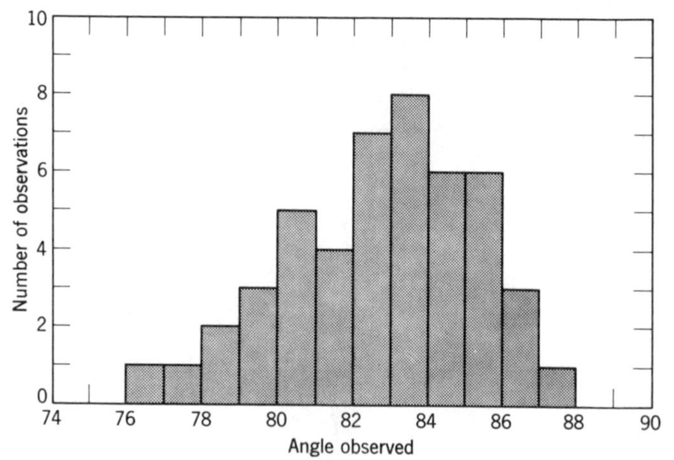

Fig. 11.8. Distribution of observed boundary angles in a polished section of the system Ni–Al_2O_3.

Relative Amounts of Phases Present. The information most frequently desired from microscopic analysis is the relative amounts of different phases present. For a random sample it can be shown that the volume fraction of a phase is equal to the cross-sectional area fraction of the phase in a random plane through the sample; it is also equal to the linear fraction of the phase intersecting a random line drawn through the sample; it is also equal to the fraction of points randomly distributed along a line over a cross-sectional area or throughout a volume that falls within the phase. Using the symbols defined in Table 11.1,

$$V_V = A_A = L_L = P_P \tag{11.6}$$

The experimental measurements on a ceramographic sample, which is a thin section (relative to particle size) or a polished plane taken from a three-dimensional sample, are usually done by a form of area, lineal, or

Trans. Am. Soc. Met., **34**, 156 (1945).

Table 11.1. Basic Symbols Used in Microstructure Analysis and Their Definitions

Symbol	Dimensions[a]	Definition
P		Number of point elements, or test points
P_P		Point fraction. Number of points (in areal features) per test point
P_L	mm^{-1}	Number of point intersections per unit length of test line
P_A	mm^{-2}	Number of points per unit test area
P_V	mm^{-3}	Number of points per unit test volume
L	mm	Length of lineal elements, or test line length
L_L	mm/mm	Lineal fraction. Length of lineal intercepts per unit length of test line
L_A	mm/mm^2	Length of lineal elements per unit test area
L_V	mm/mm^3	Length of lineal elements per unit test volume
A	mm^2	Planar area of intercepted features, or test area
S	mm^2	Surface or interface area (not necessarily planar)
A_A	mm^2/mm^2	Area fraction. Area of intercepted features per unit test area
S_V	mm^2/mm^3	Surface area per unit test volume
V	mm^3	Volume of three-dimensional features, or test volume
V_V	mm^3/mm^3	Volume fraction. Volume of features per unit test volume
N		Number of features (as opposed to points)
N_L	mm^{-1}	Number of interceptions of features per unit length of test line
N_A	mm^{-2}	Number of interceptions of features per unit test area
N_V	mm^{-3}	Number of features per unit test volume
$\bar{L}$	mm	Average lineal intercept, L_L/N_L
$\bar{A}$	mm^2	Average areal intercept, A_A/N_A
$\bar{S}$	mm^2	Average surface area, S_V/N_V
$\bar{V}$	mm^3	Average volume, V_V/N_V

[a] Arbitrarily shown in millimeters.

Source. E. E. Underwood, *Quantitative Stereology*, Addison-Wesley Publishing Company, Inc., Reading, Mass., 1970.

point analysis. *Point counting* is done by randomly distributing a grid of points over a sample in the microscope or on a photomicrograph or randomly moving a microscope cross hair and counting the fraction of the total point count that falls in a given phase P_P, which gives the volume fraction of the phases directly, as indicated in Eq. 11.6. In lineal analysis, a line in the microscope objective or one drawn across a micrograph is used, and the fractional length intercepting each phase L_L is measured. This can be done manually or with an integrating stage micrometer, which is more convenient. Areal analysis to determine the relative area of each phase A_A in the planar section is less commonly done, because it is slower than point counting or lineal analysis. Relative areas can be determined from tracings with a planimeter, by cutting out the separate phases from a micrograph and weighing them, or by counting squares on a grid placed over the micrograph. Modern developments in electronic scanning instrumentation may well bring back the importance of areal analysis.

In addition to determining the volume fraction of the phases present, relatively simple measurements and statistical analyses give a range of additional information about microstructure characteristics important for special purposes. The basic measurements with a known length of test line L, for which the magnification must be known, are the number of intersections of a particular feature such as phase boundaries along the line P_L and the number of interceptions of objects such as particles along the line N_L. Related measurements can be made of the number of points, such as three-grain intersections observed in a known area A, which is P_A, or the number of objects, such as grains, in a known area N_A. From these measurements, one can calculate directly the surface area per unit volume S_V, the length of lineal elements per unit volume L_V, and the number of point elements per unit volume P_V:

$$S_V = \left(\frac{4}{\pi}\right) L_A = 2P_L \tag{11.7}$$

$$L_V = 2P_A \tag{11.8}$$

$$P_V = \left(\frac{1}{2}\right) L_V S_V = 2P_A P_L \tag{11.9}$$

Sizes and Spacing of Structure Constituents. In observing microstructures, it is frequently found that there are two or more levels of structure. There is first a structure associated with the distribution of relatively large pores, grog (prefired clay) particles, or other grains in a matrix phase. Second, there is a structure of phase distribution within the large grains and within the bond phase. Also we are often concerned with the distribution of dislocations and the separation between various structure constituents. It is often necessary to use different techniques for a study of each of these characteristics. For example, we may use a thin or polished section to study the larger-scale structure together with scanning electron microscopy to study a fine-clay matrix. Which level of structure is most important depends on the particular ceramic and the particular property of interest, and it is necessary to be selective about what is critical for a particular problem.

As a particle parameter, the mean intercept length $\bar{L} = L_L/N_L$ is a simple and convenient measurement to characterize particle size. For spherical particles or rods of uniform size, the mean intercept length gives a measure of the particle radius r:

$$\bar{L} = \frac{4}{3} r \qquad \text{sphere} \tag{11.10}$$

$$\bar{L} = 2r \qquad \text{rod} \tag{11.11}$$

and the thickness t of a plate-shaped phase

$$\bar{L} = 2t \qquad \text{plate} \tag{11.12}$$

In addition, for space-filling grains, since $\bar{L} = 1/N_L$ and $P_L = N_L$, the surface area per unit volume is given by

$$S_V = 2P_L = 2N_L = \frac{2}{\bar{L}} \tag{11.13}$$

For separated particles, $P_L = 2N_L$, and

$$S_V = 2P_L = 4N_L = \frac{4(V_V)}{\bar{L}} \tag{11.14}$$

Another important parameter in determining spatial distributions is the mean free distance between particles λ, the mean edge-to-edge distance along a straight line between the particles or phases. It is given by

$$\lambda = \frac{1 - V_V}{N_L} \tag{11.15}$$

This value is related to the mean intercept length:

$$\lambda = \frac{\bar{L}(1 - V_V)}{V_V} \tag{11.16}$$

and if the mean spacing between particle centers is S,

$$S = \frac{1}{N_L} \tag{11.17}$$

and

$$\bar{L} = S - \lambda \tag{11.18}$$

Thus, from relatively simple methods of point counting and lineal analysis, not only the volume fraction of phases present but a good deal of information about their size and spacial distribution can be determined. From a histogram of the mean intercept lengths, a calculation of the particle-size distribution can be made (reference 1). Special precautions are necessary in evaluating nonrandom-oriented and anisometric samples (references 1 and 2).

Porosity. A quantitative measurement of the porosity present is one of the characteristics of ceramics formed by compacting a powder and firing. The pores can occur in widely different sizes, shapes, and distribution, and quantitative characterization is not always simple. In many cases the best method is to use polished sections with lineal or area analysis. This is an effective method for glazes, enamels, porcelains, refractories, and abrasives and probably should be more widely used than it has been in the past. The main difficulties with this technique have arisen from (1) relatively soft specimens that are difficult to polish and (2) samples in which grains tend to pull out during polishing so that misleading high values result. Polishing soft specimens can be done satisfactorily by impregnating the sample with a resin. Pullout of individual particles is mainly caused by differences in constituent hardness or by microfissures or microstresses already present in the sample. Hardness differences cause high relief that encourages pullouts; this can be corrected by using a suitably hard abrasive for polishing (diamond powder, for example) and a hard, flat polishing surface. Microstresses or microfissures are flat cracks along grain boundaries that allow grains to pop out with very small added stresses. Pullouts that occur in spite of careful polishing technique are one of the best indications that microstresses and microfissures are present.

Density. The size, shape, distribution, and amount of the total porosity can be determined from the microstructure. The total porosity can also be measured by determining the bulk density ρ_b of a sample (total weight/total volume, including pores) and comparing this with the true density ρ_t (total weight/volume of solids). Then

$$f_P = \frac{\rho_t - \rho_b}{\rho_t} = 1 - \frac{\rho_b}{\rho_t} \qquad (11.19)$$

Sometimes it is convenient to measure bulk density as the fraction of theoretical density achieved, $\rho_b/\rho_t = 1 - f_P$.

The true density can be determined readily for a single-phase material but not so easily for a polyphase material. For a crystalline solid the density can be calculated from the crystal structure and lattice constant, since the atomic weight for each constituent is known. This calculation has been illustrated in Chapter 4. True density can also be determined by comparing porefree samples with a liquid having a known density. For glasses and single crystals this can be done by weighing the material in air and then suspended in a liquid, determining the volume by Archimedes' method; it can be done more precisely by adjusting the composition or temperature of a liquid column just to balance the density of the solid so that it neither sinks nor rises but remains suspended in the liquid. For complex mixtures and porous solids the sample must be pulverized until there are no residual closed pores, and the density is then determined by the pycnometer method. The sample is put in a known-volume pycnometer bottle and weighed; then liquid is added to give a known volume of liquid plus solid and another weight is taken. To ensure penetration of the solid among all particles, the sample and liquid should be boiled or heated under vacuum. The differences in weights obtained give the liquid volume; this is subtracted from the total pycnometer volume to give the solid sample volume from which the density can be calculated.

The bulk density of porous bodies requires determination of the total volume of solid plus pores. For samples such as bricks this can be done by measuring the sample dimensions and calculating the volume. For smaller samples bulk density can be determined by measuring the weight of mercury (or of any other nonwetting liquid that does not penetrate the pores) displaced by the sample with a mercury volumeter, or the force required to submerge the sample (Archimedes' method). For small samples bulk density can also be determined by coating the sample with an impermeable film such as paraffin. The weight of the film is measured by difference so that the film volume is known. Then the volume of the sample plus film can be determined by Archimedes' method and the sample volume measured by difference. The total porosity can be determined at the same time the open-pore volume is measured by first weighing a sample in air W_a and then heating in boiling water for 2 hr to fill the open pores completely with water. After cooling, the weight of the saturated piece is determined (1) suspended in water W_{sus} and (2) in air W_{sat}. The difference between these last two values gives the sample volume and allows calculation of the bulk density. The difference between saturated and dry weights gives the open-pore volume.

Open Porosity. Various methods have been devised for characterizing the open pores by regarding them as capillaries and determining their equivalent diameter from the rate of fluid flow through them or the extent to which liquid mercury can be forced into them. In the mercury method, for example, the sample is placed in a container and evacuated, and then mercury is admitted and pressure applied. The pressure necessary to force mercury into a capillary depends on the contact angle and surface tension (discussed in Chapter 7) and, according to L. Lecrivain* is given by

$$P = \frac{4\gamma \cos\theta}{d} \simeq \frac{14}{d} \tag{11.20}$$

where P is the pressure (kg/cm^2), d is the pore diameter (micron), γ is the surface tension of mercury, and θ is the contact angle (140° for most oxides). As the pressure is increased, smaller-size pores are permeated. Their amount is measured by the decrease of the apparent volume of mercury plus sample. The result attained is a distribution of open-pore sizes characteristic of a particular material. Average pore sizes for open porosity vary from less than a micron for clay and other fine-grained bodies to the millimeter range for some coarse-textured refractories.

11.3 Triaxial Whiteware Compositions

A wide range of traditional ceramic compositions, the basis for much of the whiteware industry, is mixtures of clay, feldspar, and flint. These compositions include hard porcelain for artware, tableware, vitreous sanitary ware, electrical porcelain, semivitreous tableware, hotel china, dental porcelain, and others (Table 11.2). A typical composition could be considered equal parts of china clay, ball clay, feldspar, and flint. In these

Table 11.2. Composition of Triaxial Whiteware Compositions

Type Body	China Clay	Ball Clay	Feldspar	Flint	Other
Cone 16 hard porcelain	40	10	25	25	–
Cone 14 electrical insulation ware	27	14	26	33	–
Cone 12 vitreous sanitary ware	30	20	34	18	–
Cone 12 electrical insulation	23	25	34	18	–
Cone 10 vitreous tile	26	30	32	12	–
Cone 9 semivitreous whiteware	23	30	25	21	–
Cone 10 bone china	25	–	15	22	38 bone ash
Cone 10 hotel china	31	10	22	35	2 $CaCO_3$
Cone 10 dental porcelain	5	–	95	–	–
Cone 9 electrical insulation	28	10	35	25	2 talc

*_Trans. Brit. Ceram. Soc._, **57**, 687 (1958).

compositions the clays serve a dual purpose of (1) providing fine particle sizes and good plasticity for forming, as discussed in Chapter 1, and (2) forming fine pores and a more or less viscous liquid essential to the firing process, as discussed in Chapter 10. The feldspar acts as a flux, forming a viscous liquid at the firing temperature, and aids in vitrification. The flint is mainly an inexpensive filler material which during firing remains unreactive at low temperature and at high temperature forms a high-viscosity liquid.

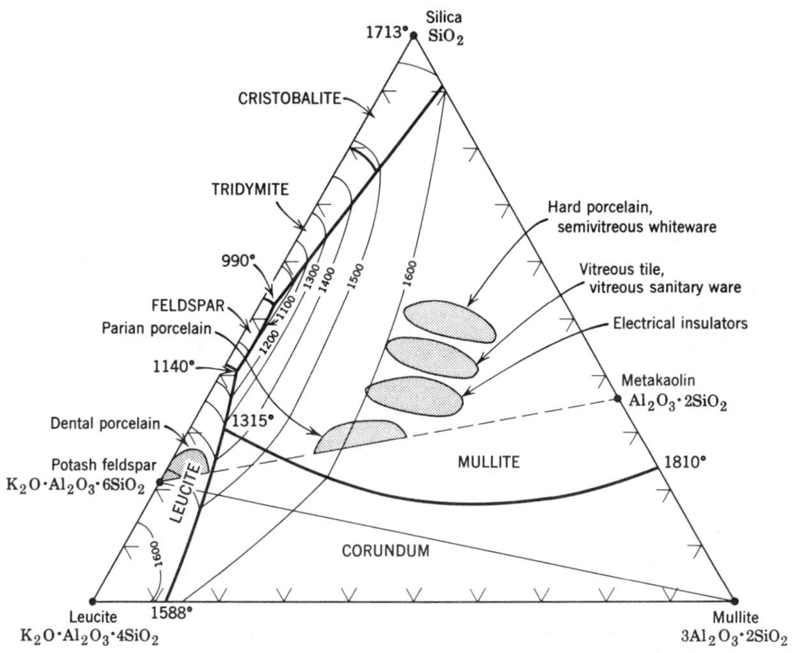

Fig. 11.9. Areas of triaxial whiteware compositions shown on the silica-leucite-mullite phase-equilibrium diagram.

Typical Compositions. Typical composition ranges for different types of bodies are illustrated in the silica-leucite-mullite phase diagram in Fig. 11.9 (these compositions can be readily visualized as flint-clay-feldspar mixtures by drawing in the feldspar-metakaolin join on the diagram). On this diagram the clays have all been lumped together. The main differences among compositions are in the relative amounts and kinds of feldspar and clay used. As an increasing amount of feldspar is added, the amount of liquid formed at the eutectic temperature increases, vitrification proceeds at a lower temperature, there is more liquid present, and greater vitrification and higher translucency are obtained. As feldspar is replaced by clay, higher temperatures are required for vitrification, and the firing process becomes more difficult and expensive. However, the forming processes become easier, and the mechanical and electrical properties of the resulting body are improved. The amount and kinds of clay used are determined for the most part by requirements with regard to forming; as more difficult forming techniques are used, a larger clay content is required.

Considering the different compositions illustrated in Table 11.2 and Fig. 11.9, dental porcelains require higher translucency and are formed in small simple shapes so that high-feldspar–low-clay compositions are clearly indicated. In contrast, hard porcelain artware and tableware must be formed into complicated thin-wall shapes by hand throwing, jiggering, and slip casting. For successful forming it must have a substantial clay content. China clay has relatively large particles (medium plasticity, drying shrinkage, and dry strength) and is white-burning; ball clay has fine particles (high plasticity, drying shrinkage, dry strength) and contains larger amounts of impurities which make it less translucent and usually not white-burning. A balance is struck by using as much china clay as will permit successful forming. For automatic forming machines, such as those used for American hotel china and semivitreous ware, larger amounts of ball clay are used. Compositions such as low-tension electrical porcelain are not critical as regards either forming or firing operations. A balance is desired which leads to the most economical ware. Since the largest individual manufacturing cost is labor, a body is used which is easy to form but at the same time can be fired without difficulty. Firing is not carried to complete vitrification, thus avoiding problems of high firing temperature, high firing shrinkage, and warping.

Changes during Firing. The changes that occur in the structure of a triaxial porcelain during firing depend to a great extent on the particular composition and conditions of firing. As shown in Figs. 11.9 and 7.24, the ternary eutectic temperature in the system feldspar-clay-flint is at 990°C, whereas the temperature at which feldspar grains themselves form a liquid phase is 1140°C. At higher temperatures an increasing amount of liquid is formed which at equilibrium would be associated with mullite as a solid phase. In actual practice equilibrium is not reached during normal firing because diffusion rates are low and the free-energy differences are small between the various phases present. The general equilibrium conditions do not change at temperatures above about 1200°C (see Fig. 7.26) so that long firing times at this temperature give results that are very similar to shorter times at higher temperatures (the relative times required can be determined from the temperature dependence of diffusion and viscosity, as discussed in Chapter 9). Also fine grinding to reduce diffusion paths gives equivalent results in shorter firing times or at lower temperatures.

The initial mix is composed of relatively large quartz and feldspar grains in a fine-grained clay matrix. During firing the feldspar grains melt at about 1140°C, but because of their high viscosity there is no change in shape until above 1200°C. Around 1250°C feldspar grains smaller than about 10 microns have disappeared by reaction with the surrounding clay, and the larger grains have interacted with the clay (alkali diffuses out of the feldspar, and mullite crystals form in a glass). The clay phase initially shrinks and frequently fissures appear. As illustrated in Fig. 7.29, fine mullite needles appear at about 1000°C but cannot be resolved with an optical microscope until temperatures of at least 1250° are reached. With further increases of temperature mullite crystals continue to grow. After firing at temperatures above 1400°C, mullite is present as prismatic crystals up to about 0.01 mm in length. No change is observed in the quartz phase until temperatures of about 1250°C are reached; then rounding of the edges can be noticed in small particles. The solution rim of high-silica glass around each quartz grain increases in amount at higher temperatures. By 1350°C grains smaller than 20 microns are completely dissolved; above 1400°C little quartz remains, and the porcelain consists almost entirely of mullite and glass.

The heterogeneous nature of the product is illustrated in Fig. 11.10, in which quartz grains surrounded by a solution rim of high-silica glass, the outlines of glass-mullite areas corresponding to the original feldspar grains, and the unresolved matrix corresponding to the original clay can be clearly distinguished. Pores are also seen to be present. Although mullite is the crystalline phase in both the original feldspar grains and in the clay matrix, the crystal size and development are quite different (Fig. 11.11). Large mullite needles grow into the feldspar relicts from the surface as the composition changes by alkali diffusion. A quartz grain and the surrounding solution rim of silica-rich glass is shown in Fig. 11.12. Mullite needles extend into the outer edge of the solution rim, and a typical microstress crack is seen; the crack is caused by the greater contraction of the quartz grain compared with that of the surrounding matrix. Usually the quartz forms only glass, but for some compositions fired at high temperatures there is a transformation into cristobalite which starts at the outer surface of the quartz grain (Fig. 11.13). The overall structure of quartz grains, microfissures, solution rims, feldspar relicts of glass and mullite, and fine mullite-glass matrices is shown with great clarity in Fig. 11.14.

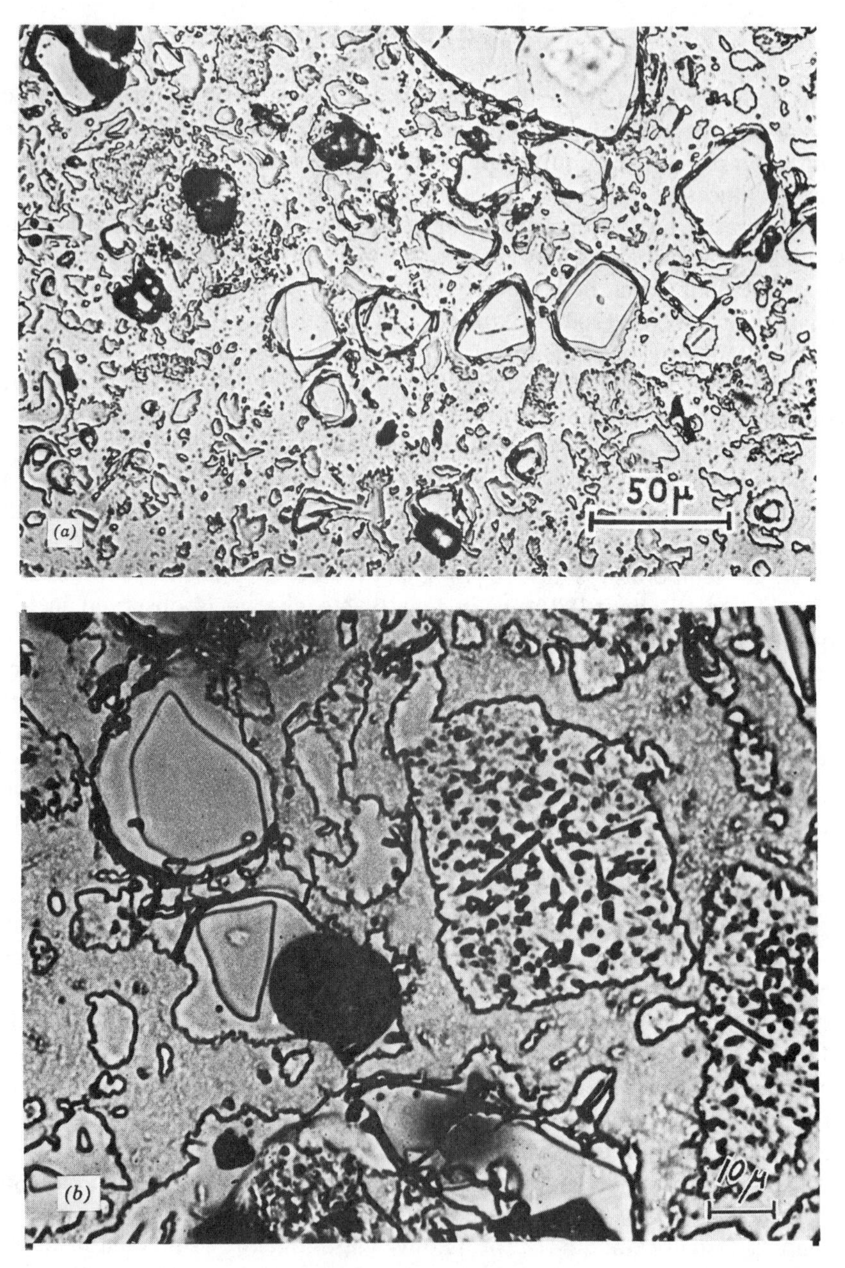

Fig. 11.10. Photomicrographs of electrical insulator porcelain (etched 10 sec, 0°C, 4% HF) showing liquid quartz grains with solution rim, feldspar relicts with indistinct mullite, unresolved clay matrix, and dark pores. Courtesy S. T. Lundin.

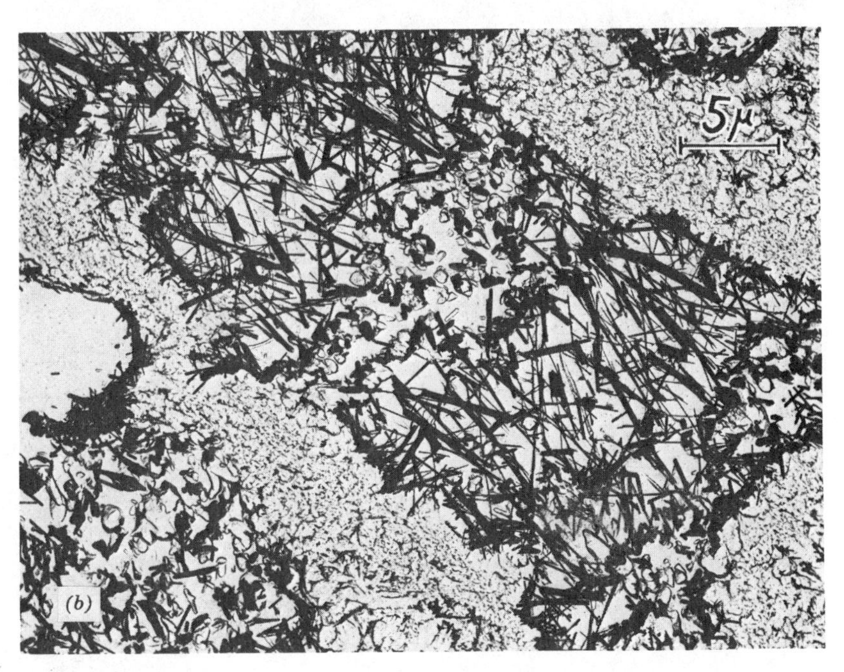

Fig. 11.11. Mullite needles growing into a feldspar relict (etched 10 sec, 0°C, 40% HF). Courtesy S. T. Lundin.

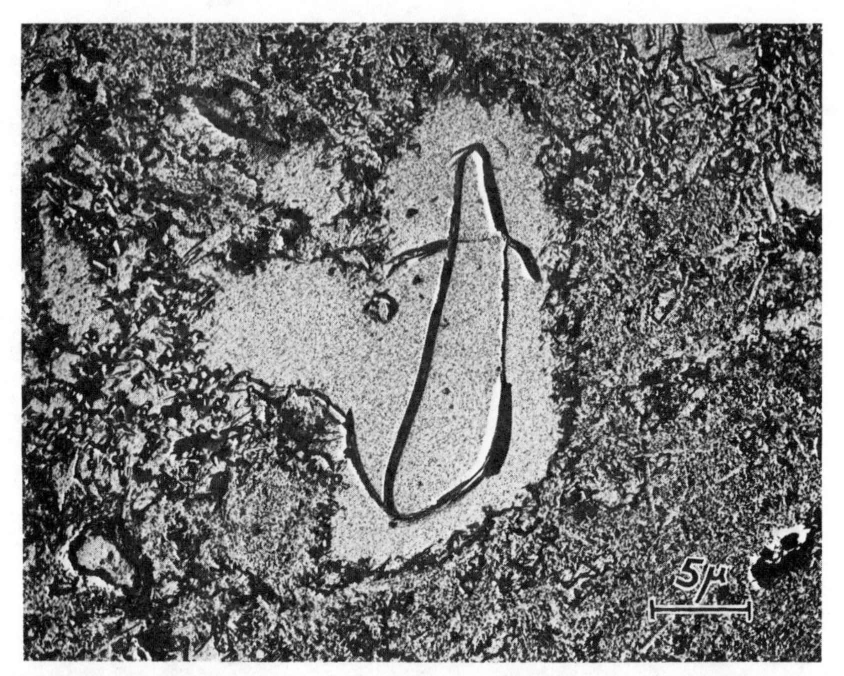

Fig. 11.12. Partly dissolved quartz grain in electrical insulator porcelain (etched 10 sec, 0°C, 40% HF, aluminum replica, 2750×. Courtesy S. T. Lundin.

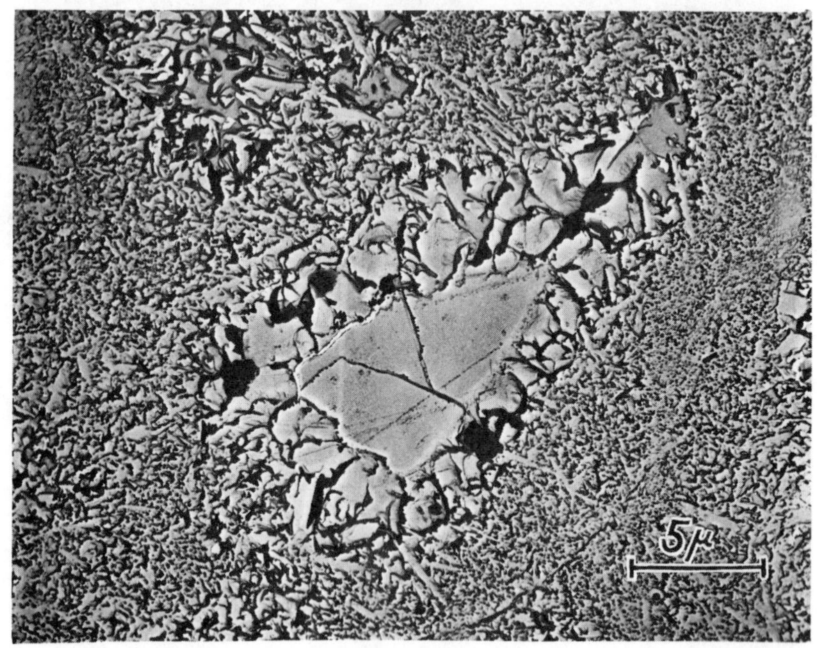

Fig. 11.13. Quartz grain with cristobalite formed at surface (etched 20 min, 100°C, 50% NaOH, silica replica. Courtesy S. T. Lundin.

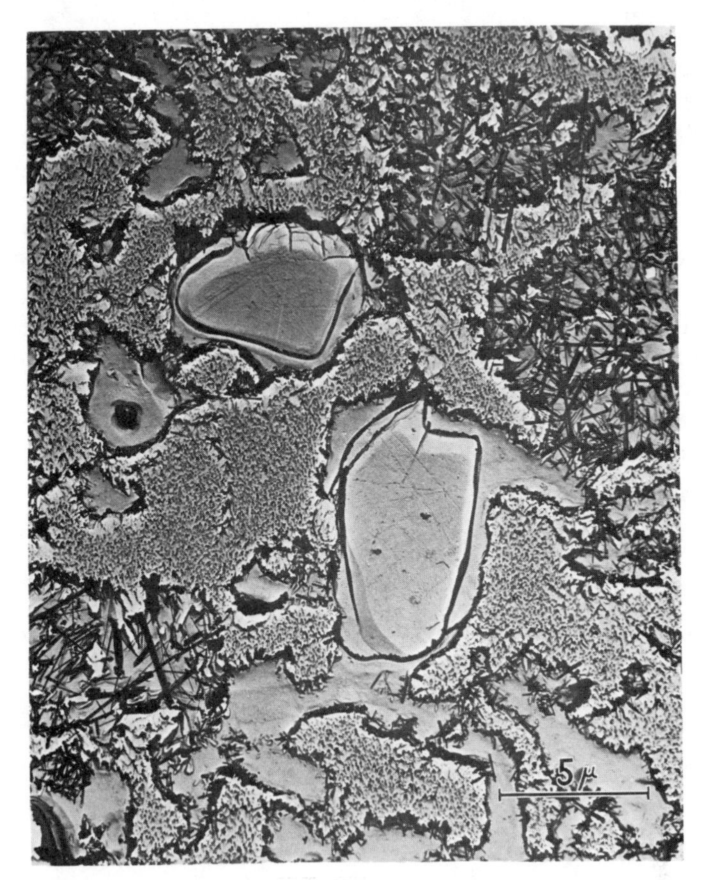

Fig. 11.14. Electron micrograph of electrical insulator porcelain (etched 10 sec, 0°C, 40% HF, silica replica. Courtesy S. T. Lundin.

The changes taking place during firing occur at a rate depending on time, temperature, and particle size. The slowest process is the quartz solution. Under normal firing conditions equilibrium at the firing temperature is only achieved at temperatures above 1400°C, and the structure consists of a mixture of siliceous liquid and mullite. In all cases the liquid at the firing temperature cools to form a glass so that the resulting phases present at room temperature are normally glass, mullite, and quartz in amounts depending on the initial composition and conditions of firing treatment. Compositions with a larger feldspar content form larger amounts of siliceous liquid at lower temperatures and correspondingly vitrify at lower temperatures than the compositions with larger clay contents.

Advantages of Triaxial Porcelains. One of the great advantages of quartz-clay-feldspar bodies is the fact that they are not sensitive to minor changes in composition, fabrication techniques, and firing temperature. This adaptability results from the interaction of the phases present to increase continuously the viscosity of the fluid phase as more of it is formed at higher temperatures. If we consider the phase-equilibrium diagram in Fig. 11.9, the eutectic liquid is increased in amount but also in viscosity (silica content) as the feldspar relict interacts with the clay. At 1300°C, equilibrium is reached between the clay and feldspar to give mullite in a glassy matrix (Fig. 11.15). Further interaction with the quartz grains forms more liquid, but it is of continuously increasing silica content and is consequently more viscous. As a result of these reactions, the body has an unusually long firing range and low sensitivity to compositional variations.

11.4 Refractories

Refractory materials cover a wide range of compositions and structures and are difficult to characterize easily, particularly since the structure is

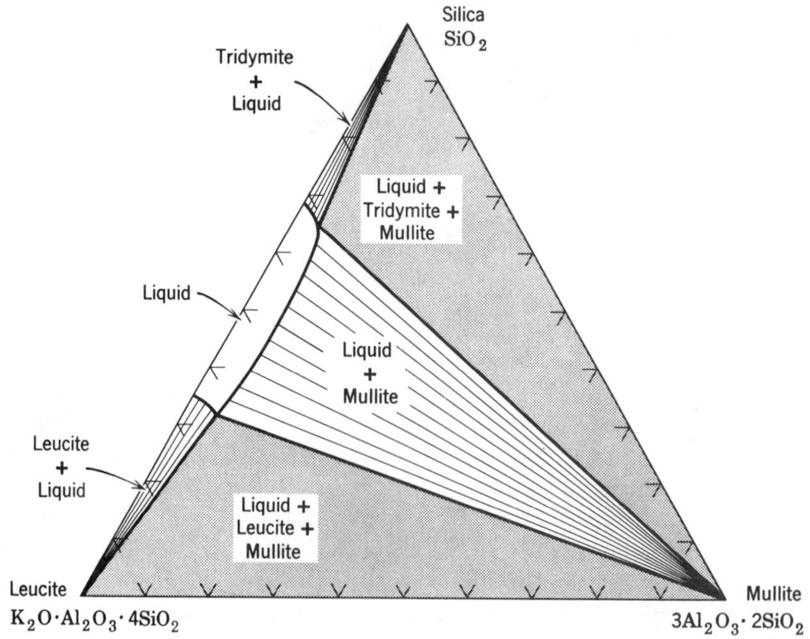

Fig. 11.15. Isothermal section of silica-leucite-mullite phase-equilibrium diagram at 1300°C.

frequently affected by service conditions and changes during the lifetime of the refractory. Generally, refractories are composed of large grog or refractory-grain particles held together with a fine-grained bond. Both the bond material and the refractory grains have a fine structure, and both are often multiphase. Many variations occur among bricks made from different raw materials and by different manufacturers. Some typical compositions are given in Table 11.3.

Fire-Clay Refractories. The largest group of fire-clay refractories is based on mixtures of plastic fire clay, flint clay, and fire-clay grog. Large numbers of brick are also made with increased alumina content by adding sillimanite, kyanite, topaz, diaspore, or bauxite. All these materials tend to form mullite on heating. In addition, quartz is often present as an impurity in the plastic fire clay and is sometimes added to reduce firing and drying shrinkage. The fine structure in the grog (prefired clay) or flint-clay particles and in the plastic fire-clay bond is difficult to resolve with an optical microscope but consists of fine mullite crystals in a siliceous matrix (Fig. 7.29). Alkali, alkaline earth, iron, and similar impurities that are present largely combine with the siliceous material to form a low-melting glass and decrease the refractoriness of the brick.

The brick texture—relation of grog or flint-clay particles to bond clay and porosity—is important to developing good properties. Large amounts of plastic clay, which is less refractory, decrease the refractoriness of the brick. In general, the coarse grog particles form an interlocking mass of bonded particles with numerous cracks and pores also present. These pores tend to prevent the extension of cracks or fissures and contribute resistance to spalling (breaking off of pieces of the brick caused by thermal or mechanical stresses), even though surface checks and cracks may form more readily than in a denser structure.

As discussed in Chapter 1, the volume fraction porosity in a graded series of particle sizes to give dense packing is about 25 to 30%. This is the space available for bond clay during the forming process. During firing the plastic fire-clay bond has a higher shrinkage than the grog or flint clay; if it just fits in between the particles, it contributes little to the overall firing shrinkage; if more than this amount is used, the shrinkage of the body increases. However, if the grog particles are in contact before firing, the shrinkage of the bond clay frequently causes fissures to open up during firing, contributing to stresses and reduced property values. Similarly, if the grog particles have larger contractions during firing than the plastic bond clay, boundary stresses are set up, as discussed in Chapter 5, and cracks tend to form (Fig. 11.16). These result in lower strength and poorer resistance to thermal stresses.

Silica Refractories. Silica brick have a high load-bearing capacity at

Table 11.3. Composition of Some Typical Refractory Brick

Type	Composition								Apparent Porosity (%)	Main Phases in Order of Amount Normally Present
	Al_2O_3	SiO_2	MgO	Cr_2O_3	Fe_2O_3	CaO	TiO_2	Total		
Chrome ore	30.0	5.3	19.0	30.5	13.5	0.7	···	99.0	22	$(MgFe)(AlCr)_2O_4$, R_2O_3 solid solution, $MgFe_2O_4$
Chrome ore–periclase	19.0	6.0	40.0	22.0	11.0	1.2	···	99.2	25	$(MgFe)(AlCr)_2O_4$, MgO, $MgFe_2O_4$, Mg_3SiO_4
Periclase–chrome ore	9.0	5.0	73.0	8.2	2.0	2.2	···	99.4	21	MgO, $(MgFe)(AlCr)_2O_4$, $MgFe_2O_4$, $MgCaSiO_4$
Periclase	1.0	3.0	90.0	0.3	3.0	2.5	···	99.8	22	MgO, $MgFe_2O_4$, $MgCaSiO_4$, $MgSiO_4$
Pitch-bonded low-flux dolomite	0.3	0.4	40.0	0.3	0.3	56.0	···	97.3	20	MgO, CaO, $MgCaSiO_4$, 2.7% residual carbon
Forsterite	1.0	33.3	54.5	0.7	9.1	1.0	···	99.6	23	Mg_2SiO_4, $MgFe_2O_4$, MgO, $MgAl_2O_4$
Silica	0.2	96.3	0.6	···	···	2.2	···	99.3	25	Tridymite, cristobalite
Fire clay	25–45	70–50	0–1	···	0–1	0–1	1–2	···	10–25	Mullite, siliceous phase, quartz
High-alumina fire clay	90–50	10–45	0–1	···	0–1	0–1	1–4	···	18–25	Mullite, siliceous phase

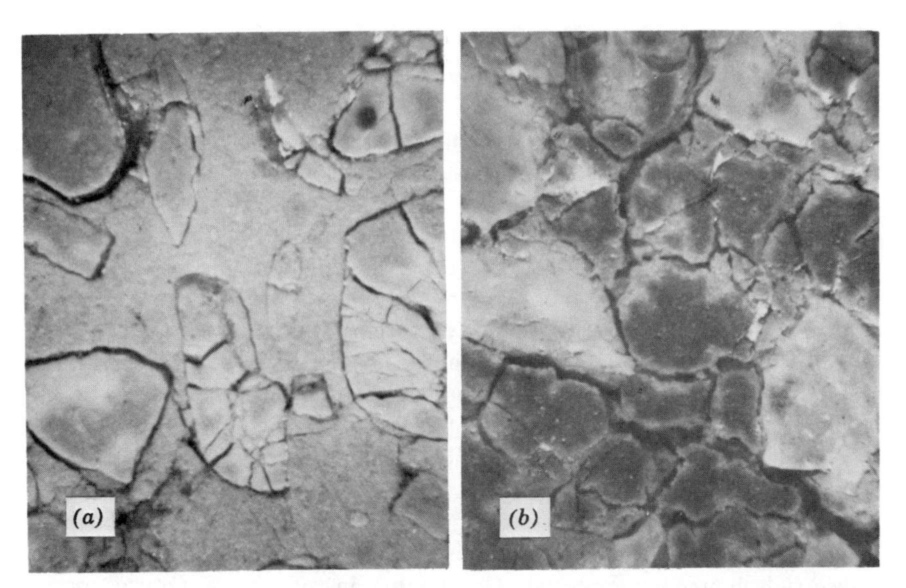

Fig. 11.16. Fissures in test fire-clay brick (50 grain, 50 bond) caused by (*a*) shattering of grains because they shrink more than matrix does and (*b*) cracking of matrix and separation from form because of greater shrinkage of matrix (8×). Courtesy C. Burton Clark.

elevated temperatures and consequently have been widely used for sprung arches as roofs for open-hearth furnaces, glass tanks, kilns, and coke ovens. Bricks are manufactured from ground ganister rock (quartzite) containing about 98% SiO_2 to which is added some 2% CaO as milk of lime. The added lime serves as a mineralizer during firing. Fired brick consist of shattered quartz grains that have been almost completely transformed into cristobalite (starting at the edges of the grain, as illustrated in Fig. 11.13) in a matrix of fine tridymite, cristobalite, and glass. Small amounts of unconverted quartz (about 10%) normally remain with nearly equal amounts of cristobalite and tridymite formed. Excessive quartz, (frequently present in brick at one time) is deleterious because of the large volume change accompanying the α to β transformation at 573°C.

Long firing times normally tend to favor tridymite development during firing. The quartz initially present transforms into cristobalite, starting at the grain edges. This dissolves in the calcium silicate liquid phase and precipitates as tridymite. During firing the amount of quartz present continually decreases, the amount of cristobalite initially increases and then later decreases, and the amount of tridymite continuously increases.

Basic Refractories. In the class of basic refractories are included brick manufactured from chrome ore [$(MgFe)(AlCr)_2O_4$], periclase (MgO), calcined dolomite (CaO, MgO), olivine [$(MgFe)_2SiO_4$], and mixtures of these materials.

Chrome ore contains serpentine and other silicates as impurities which are low-melting and deleterious. If sufficient magnesia is added, it reacts with this material to form forsterite, which is refractory, thus improving the properties of the brick. Brick consist of large chromite grains which usually contain a precipitate of $(Fe, Al, Cr)_2O_3$ resulting from iron oxidation during firing; the bond phase consists of fine chrome ore, magnesioferrite, and forsterite. A typical structure is illustrated in Fig. 11.17. When more than a small amount of magnesite is added, we enter a range of chrome-ore–periclase compositions in which there are usually large grains of chromite in a matrix of fine MgO, $MgFe_2O_4$, and Mg_2SiO_4.

Periclase brick are formed mainly from magnesite ore or seawater magnesia and contain large MgO grains in a bond phase of fine MgO together with some Mg_2SiO_4 and $MgAl_2O_4$. The large magnesia grains are normally made up of smaller crystals separated by thin films of iron- and silica-rich boundary material. On heating in air, the FeO present is oxidized to form a precipitate of magnesioferrite in the periclase crystals (see Fig. 9.44). The morphology of the precipitate depends on its mode of formation. Chrome ore is sometimes added as a bond phase in periclase–chrome-ore brick. The bond is a complex mixture of MgO, $(MgFe)(AlCr)_2O_4$, and Mg_2SiO_4.

A minor class of basic brick are those manufactured from olivine, $(MgFe)_2SiO_4$. The major phase present is forsterite.

Recent years have seen better control of raw-material composition, firing procedures, and methods of production to control microstructure and improve properties as well as a wider range of products available for special purposes. As a result, the compositions and structures of basic brick cover a wide range. A matrix of pure chrome ore usually contains high percentages of metasilicate materials, principally $MgSiO_3$ and $CaMg(SiO_3)_2$, which are low-melting and cause low load-bearing capacity at high temperatures. Additions of magnesia, as in chrome-ore–magnesite brick, remedy this and improve the load-bearing capacity at high temperature. Chemically bonded brick, unfired before service, have this reaction take place during use. Low-silica chrome ore has a high iron oxide content, which results in friability under the cyclic oxidation-reduction atmosphere in open-hearth use and is therefore unsuited for chemically bonded brick. However, the low-silica ore gives an improved product when it is first fired with magnesia to a high enough temperature to form a stable spinel structure and has given a much improved product. Sufficiently high temperatures to form a diffusion bond between phases and

Fig. 11.17. Basic brick. (*a*) Chrome ore–magnesite brick at low magnification. Light grains are chrome ore, gray phase is periclase, and dark gray areas are porosity ($5\frac{1}{4}\times$). Courtesy F. Trojer. (*b*) Different chrome ore–magnesite brick at higher magnification. Angular chrome ore grain and magnesite grain (rounded periclase particles in silicate matrix) in fine-grained bond (150×). Courtesy G. R. Eusner. (*c*) Higher magnification of chromite grain showing lamellar $(CrFe)_2O_3$ precipitate parallel to the (111) plane in spinel (1380×). Courtesy F. Trojer.

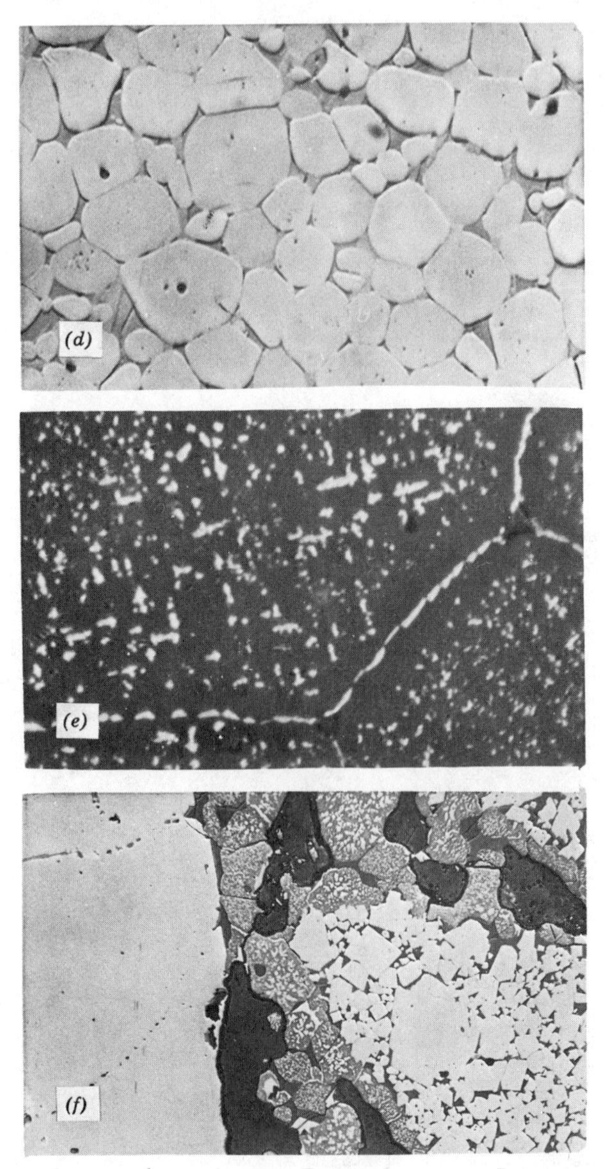

Fig. 11.17 (*Continued*). (*d*) Structure of magnesite grain showing rounded periclase crystals in silicate matrix of forsterite (Mg_2SiO_4) and monticellite ($MgCaSiO_4$) (275×). Courtesy F. Trojer. (*e*) Interior of oxidized periclase crystal showing precipitate of magnesioferrite ($MgFe_2O_4$), the light phase, dicalcium silicate, the dark triangular grains (1180×). Courtesy F. Trojer. (*f*) Complex structure of chromemagnesite brick showing large chromite grains on left, angular spinel grains, rounded periclase with magnesioferrite precipitate, and gray forsterite-monticellite matrix; dark gray is porosity (106×). Courtesy F. Trojer.

formation of large magnesia grains without penetration of the silicate liquid between grains improve performance, and much of the chrome-ore–magnesite brick used for severe service is now either fusion-cast as brick or formed from fused grain or fired at temperatures above the anticipated service temperature to give *direct-bonded* brick, in which the controlled microstructure leads to superior properties.

In pure periclase brick formed from seawater magnesia, the matrix tends to include relatively low-melting calcium-magnesium orthosilicates such as montecellite, $CaMgSiO_4$, and merwinite, $Ca_3Mg(SiO_4)_2$, which give rise to low load-bearing capacity at high temperatures. Not only the total amount of impurities but also the CaO to SiO_2 ratio is significant; when the weight ratio of calcia to silica is greater than 1.86, dicalcium silicate becomes the second phase rather than merwinite and montecellite, and the melting temperature is much increased. At the same time, the wetting behavior of the matrix phase is lessened such that there is enhanced bonding between the periclase grains. Higher firing temperatures to give greater grain growth and an increased CaO to SiO_2 ratio to decrease wetting result in a microstructure that markedly improves properties.

Dolomite ($MgCO_3$–$CaCO_3$) is a raw material of high refractoriness, low cost, and wide availability. Calcined dolomite consists of a mixture of CaO and MgO solid solutions, as shown in Fig. 7.16, plus a minor amount of silicate phase. For use as large blocks in basic open-hearth furnaces and to improve hydration resistance, these brick are formed by warm pressing with 5 to 8% pitch, which dissociates on first use, leaving a residual 2 to 3% carbon. As with other refractory brick, microstructure control is critical. It is found that by tempering the formed brick at 200 to 500°C, the pitch phase distributes itself uniformly in the interstices between grains and leads to greatly improved performance. Until recently, dolomites with up to 10% $Fe_2O_3 + Al_2O_3 + SiO_2$ were used; however, it has been found that low-flux raw materials containing less than 1% $Fe_2O_3 + Al_2O_3 + SiO_2$ with a smaller amount of silicate phase have greatly improved performance. Use of the low-flux material was made possible by developing a two-stage calcination process in which lightly calcined material is pressed to high-density briquettes and subsequently fired to a high temperature sufficient to give a large enough grain size and low enough grain porosity to provide adequate resistance to hydration without the protective silicate or ferrite phase previously thought necessary.

In almost all refractory brick, control of the porosity present is an important aspect of microstructure control. Lower porosity improves strength, load-bearing capacity, and corrosion resistance, but it also leads to catastrophic failure from thermal shock, since the pores present act as crack stoppers in a more porous brick. As a result, control of porosity is a compromise for which the optimum depends on specific service conditions. Users of refractory brick, particularly steelmakers, are very conscious of the refractory cost per ton of product. As a result, first cost of the brick is carefully weighed against useful life and replacement cost.

Special Refractories. Many special refractory compositions are used that are more expensive than either fire clay or basic brick; these are used only where required. For exceptional resistance to corrosion by liquid slag and glasses, porefree fusion-cast refractories of various compositions are used. Fused alumina brick of 99.5% Al_2O_3 plus Na_2O consists of nearly close-packed crystals of β-Al_2O_3 with almost no bond phase present. Other compositions consist of α-alumina crystals, fusion-cast mullite, fusion-cast basic brick, and a special glass tank block containing 45 to $50Al_2O_3$, 30 to $40ZrO_2$, 12 to $13SiO_2$, $2Fe_2O_3$, and $2Na_2O$ for use in glass-melting furnaces. The principal phases present are corundum and baddeleyite, ZrO_2, together with a glassy phase. In all these compositions the porosity is small and the crystallite size large.

Silicon carbide refractories are used where high-temperature load-bearing capacity and high thermal conductivity are desirable, such as in tiles for muffles, kiln furniture, and heat exchanges. The major difficulty is a tendency toward oxidation at high temperatures. For refractory purposes dense packing of the silicon carbide grains is required. Each silicon carbide grain normally develops a thin siliceous coating during firing, and the individual particles are bonded together with a glassy phase or with a mixture of glass and crystals.

A special class of materials are the thermal-insulating refractories, which have a variety of structures, depending on the particular composition and use. For relatively low-temperature brick, porosity is obtained from plaster of paris mixtures that include extremely fine pores which remain after firing. Also for relatively low-temperature use, diatomaceous earth brick are prepared in which the porosity is derived from the extremely high and fine-grained residual porosity in the siliceous skeletons of diatoms; this provides particularly effective thermal insulation. For higher-temperature-insulating firebrick the porosity is normally induced by incorporating combustible materials such as sawdust in the mix; these burn out during firing and form a large fraction of interconnected pores with a firebrick matrix material, similar to other firebrick compositions, as the solid phase. To obtain stability at the higher temperatures, the pore size is increased, and the amount of porosity is subsequently reduced. This means that the effective conductivity is higher for higher-temperature brick, and they should only be used when temperature conditions require it.

11.5 Structural Clay Products

Structural clay products include materials such as building brick, sewer pipe, drainpipe, and various kinds of tile; their manufacture is characterized by inexpensive raw materials, efficient material handling methods, and a low cost for the resultant product, necessary if structural clay products are to compete with other construction materials. The main raw materials are locally available clays having a variety of compositions and structures, depending on the particular locality. The high cost of transportation requires that local clays be used; similarly, the distribution radius for products is usually not large. The clays used are commonly glacial clays, shale or alluvial clays that contain considerable amounts of impurities. The clay minerals present are normally mixed with quartz, feldspar, mica, and other impurities. Secondary impurities include dolomite, rutile, and ferruginous materials.

During regular firing the larger-grain quartz and other accessory minerals are normally not affected. The clay used contains sufficient impurities for a glassy phase to form readily. The resulting structure normally consists of large grains of secondary constituents embedded in a matrix of fine-grained mullite and glass. The fine grain size of the matrix material makes its resolution in normal microscopic observation difficult. A typical structure observed for a stoneware is shown in Fig. 11.18. A glaze body interface for a similar composition has been illustrated in Fig. 5.25.

The details of the resultant structure depend on the particular composition of clays and the firing procedure used, but a structure similar to that in Fig. 11.18 is normally reached. In material that is underfired the clay phase contains many small pores, leading to low strength, poor resistance to frost and freezing, and generally unsatisfactory properties. Material that has been fired to an excessively high temperature is mostly a glass. The lack of porosity gives high strength but leads to failure of the entire brick when mechanical and thermal stresses are applied and prevents obtaining a satisfactory bond when used with mortars. The coloring constituents are mainly iron and TiO_2; these give products varying from yellow to salmon to buff to dark gray or black, depending on the particular impurities present.

11.6 Glazes and Enamels

As discussed in Chapter 8, glazes for ceramic ware and porcelain enamels for iron, aluminum, and jewelry metals are usually silicate glasses that may or may not include dispersed crystals or bubbles. As a

Fig. 11.18. Microstructure of stoneware. Ion-bombardment-etched (100×).

glaze is fired, the structure changes continuously. The materials initially present decompose and fuse, bubbles are formed and rise to the surface, and reaction takes place with the body under the glaze. The interface between the glaze and body is rough (Fig. 5.25) because of differences in solubility in the body constituents. Mullite crystals develop at the interface of porcelain bodies fired at high temperatures.

In a clear, glossy glaze the glass phase has no dispersed particles or porosity, and the surface is ideally perfectly smooth. Gloss is most commonly lost because bubbles rise and burst at the surface, forming small craters. Although the bubble content is initially high, the larger bubbles are rapidly eliminated during firing; the smaller bubbles are only slowly removed unless fluid glazes, which tend to run, are used. In mat glazes the surface texture and low gloss result from the extensive development of fine crystals (Fig. 11.19). For different glazes the composition of these crystals is variable, but anorthite, $CaO \cdot Al_2O_3 \cdot 2SiO_2$, is most common; mullite and wollastonite are also frequently observed. Large crystals in the glaze are sometimes desired for art products, as discussed in Chapter 8.

Perhaps the most common glaze defect is crazing; this occurs when the contraction of the glaze during cooling is greater than that of the underlying body so that tensile stresses develop in the glaze. Fissures appear, as illustrated in Fig. 12.22. In order to eliminate crazing, the composition of the glaze must be adjusted to decrease its thermal expansion coefficient, discussed in Chapter 12.

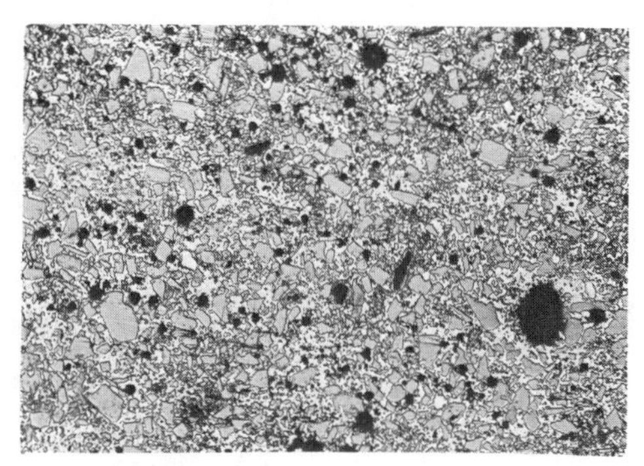

Fig. 11.19. Fine crystals dispersed in a mat glaze (polished section, H_3PO_4 etch, 78×).

Porcelain enamels on metal are similar to glazes in that they are basically a silicate glass coating. However, they are normally fired for shorter periods and at lower temperatures, and a wide range of defects is common. During firing, a ground coat containing nickel or cobalt oxide or a nickel dip (of the metal in a nickel sulfate solution) is frequently used to improve adherence. The main effect is to cause galvanic corrosion of the base metal; certain grains corrode more rapidly than others, providing a rough interface and improved adherence. This interface is similar to the rough interface obtained in glaze-porcelain structures as a result of variable solubility. A typical cross section of an interface boundary is illustrated in Fig. 11.20.

For good results with porcelain enamels it is found that the *bubble structure* is particularly important. Clays containing some organic impurities form fine bubbles adjacent to the metal surface. These provide reservoirs for hydrogen evolved from the metal on cooling and prevent *fish scaling*, the breaking out of pieces of the coating on cooling. Overfiring that eliminates the bubble structure is deleterious. A typical satisfactory bubble structure is illustrated in Fig. 11.20. Many defects in porcelain enamels result from defects in the metal used. Laminations in sheet iron for enameling give off gas during firing and cause blisters, as do surface inclusions such as slag or oxide. Large bubbles formed at the firing temperature may expose the metal underneath and allow rapid oxidation to occur, with the formation of Fe_2O_3 spots called copper heads.

Cover-coat enamels and opaque enamels are prepared with TiO_2, SnO_2, or other additives as opacifiers. These materials form a fine-grained second-phase dispersion, as discussed in Chapter 8. In most glazes and porcelain enamels colors can be obtained by using either colorants in solution to form a colored single-phase glass or pigments which remain

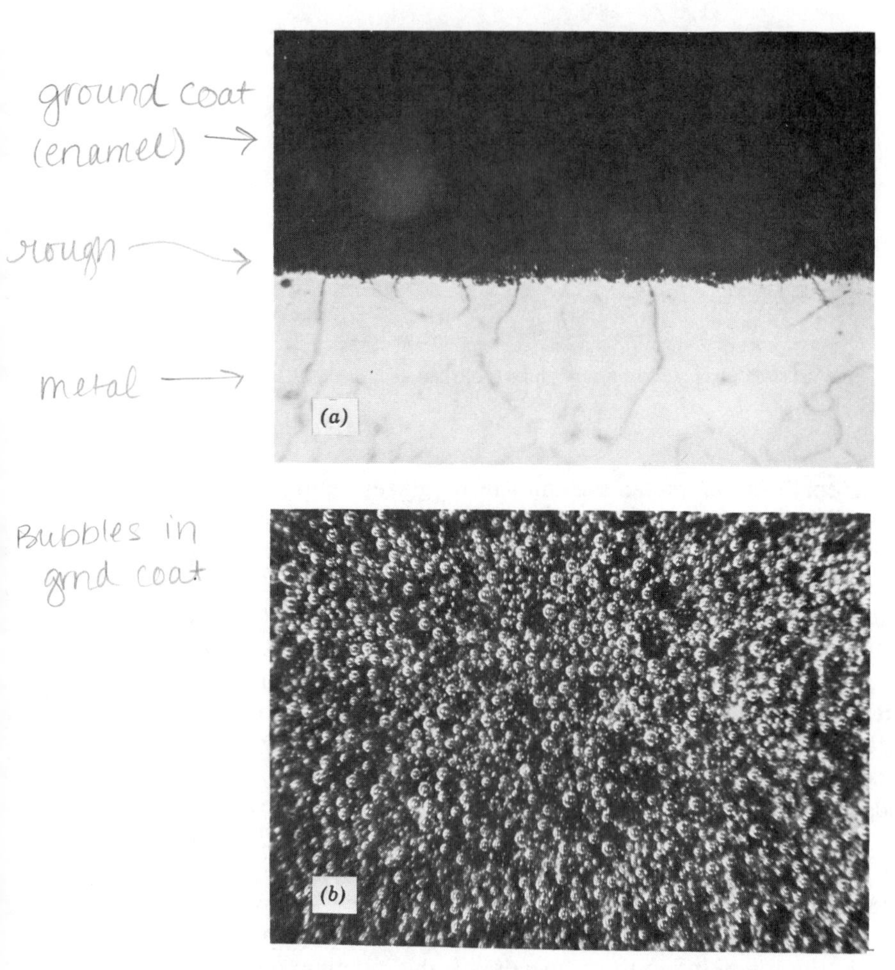

Fig. 11.20. Microstructure of porcelain enamel. (*a*) Cross section of enamel-metal interface showing rough boundary (525×). (*b*) Top view of bubble structure at metal-enamel interface provides space for gas evolution from metal on cooling (38×).

undissolved as a second phase. Optical properties of both glazes and enamels are discussed in Chapter 13.

11.7 Glasses

As discussed in Chapter 3, liquid-liquid immiscibility is widespread in glass-forming systems, and many glasses which appear optically homogeneous may be phase-separated on a scale of 30 to 50 Å up to a few

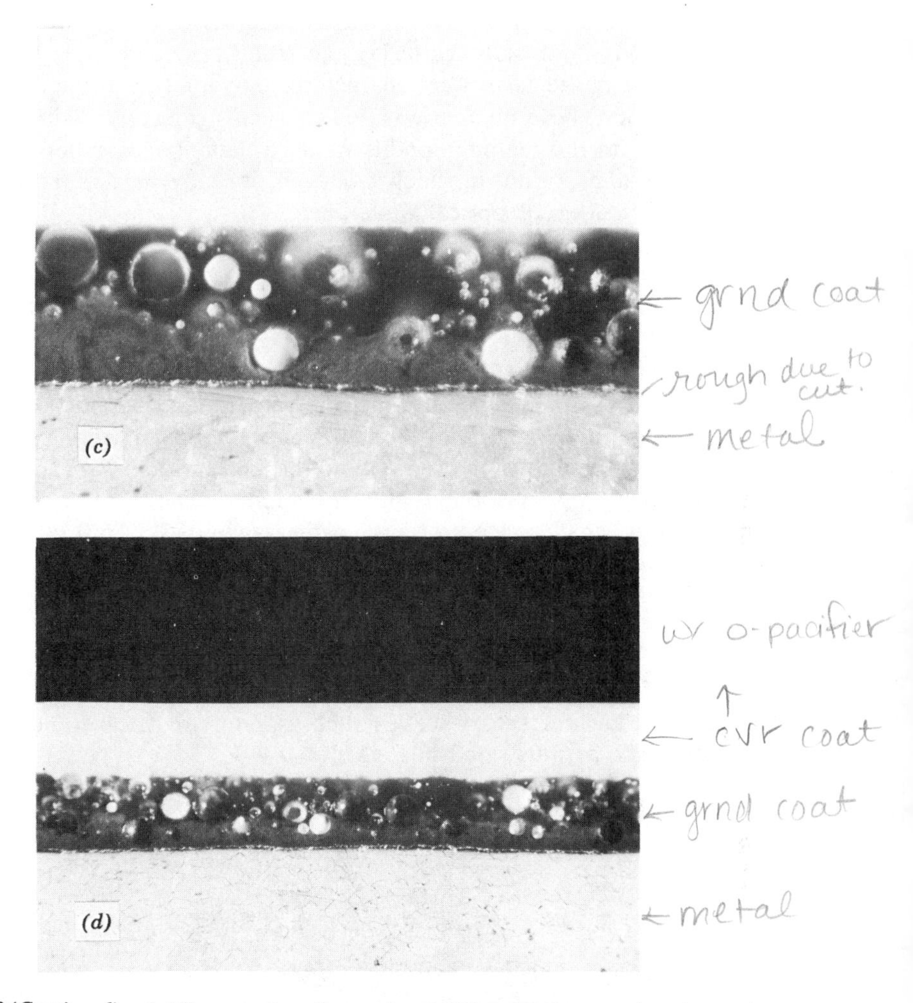

Fig. 11.20 (*Continued*). (*c*) Cross section of ground coat (192×). (*d*) Cross section of ground coat and cover coat (96×). Courtesy G. H. Spencer-Strong.

hundred angstroms. The characteristic forms of structure on this scale have been illustrated in Chapters 3 and 8. The sensitivity of these structural features to thermal history has been discussed in Chapter 8. Interconnected structures may form by spinodal decomposition or by the growth and coalescence of discrete particles, and such structures coarsen with time, retaining a high degree of connectivity in some cases and necking off and spheroidizing in others. Such two-phase submicrostructures may be developed to a sufficiently coarse scale so that they scatter light strongly, and such scattering can result in opalescence or opacity. Even single-phase glasses are characterized by fluctuations in density and composition. These fluctuations, which vary significantly with composition and with the melting conditions used, have become important in a number of applications in which low levels of scattering are of concern, as in the fabrication of optical wave guides.

On the scale of microns to millimeters and larger, a number of types of defects are found in glasses.

1. **Seed.** These are small gaseous bubbles which were not removed during the melting operation. The process by which such gaseous inclusions are removed from the molten glass is termed *fining*, and the time required for fining represents the rate-limiting part of many glass-melting operations. The large bubbles are eliminated from the melt by rising under the influence of gravity to the free surface; the small bubbles are removed by dissolving into the molten glass. To facilitate this dissolution, small concentrations, in the range of 0.1 to 0.3%, of fining agents such as Sb_2O_3 or As_2O_3 are added to the batch.

2. **Stone.** These are small crystalline imperfections found in the glass. These usually result from melting for periods of time which are too short to permit complete dissolution in the molten glass. The refractory linings of the glass tanks are a frequent source of such crystalline fragments, and refractory oxides such as ZrO_2 are often found as stone constituents, as illustrated in Fig. 11.21*a*.

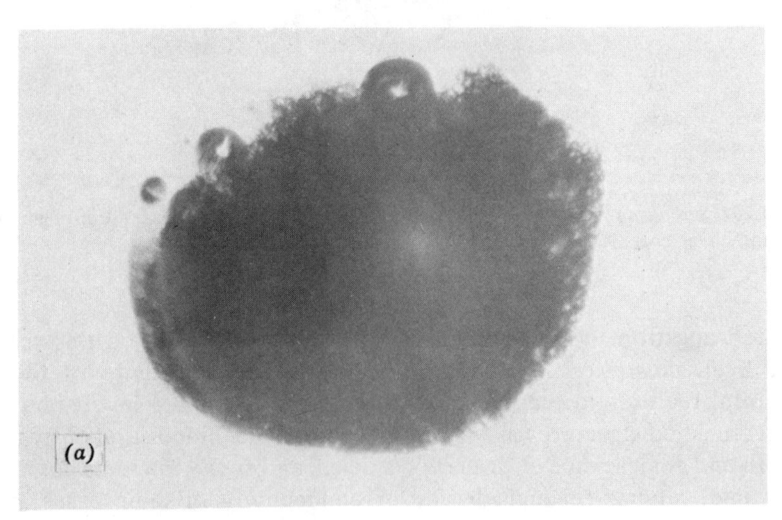

Fig. 11.21. (*a*) Refractory stone.

3. Cord and Striae. Cord are attenuated amorphous inclusions in the glass which have properties, in particular an index of refraction, different from those of the surrounding glass (Fig. 11.21*b*). These result from incomplete homogenization of the melt (inadequate mixing). Striae are cord of low intensity, which are of greatest concern in the preparation of optical glasses. To minimize the occurrence of cord and striae, a stirring step, generally carried out with platinum paddles, is often introduced after fining and before forming these glasses.

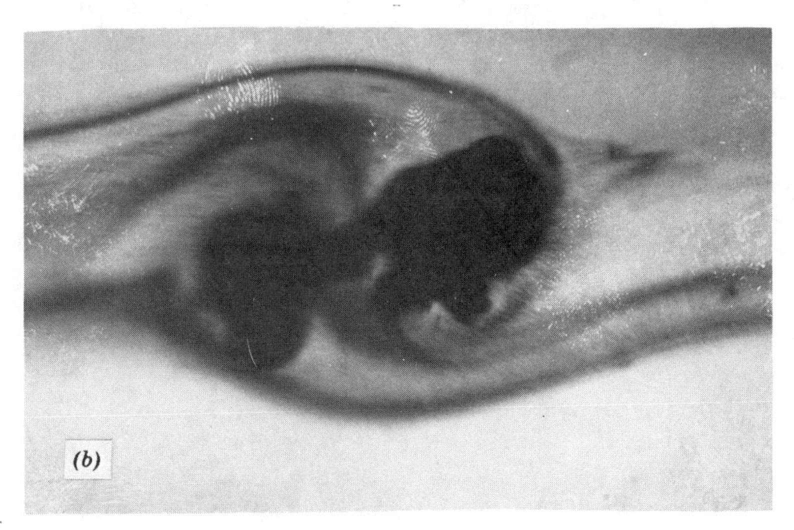

Fig. 11.21 (*Continued*). (*b*) Cord in soda-lime-silica glass.

Among commercial glasses, the occurrence of microstructural features such as seed, stone, and cord is most frequent in operations based on high throughput (output of a tank relative to its capacity) such as the production of glass containers, particularly near the end of the campaign (service life) of the refractories. These features occur with smallest frequency in optical glasses, which are melted with relatively high-purity-batch raw materials, often in platinum-lined tanks or pots, at relatively small throughputs.

11.8 Glass-Ceramics

As discussed in Chapter 8, the formation of glass-ceramic bodies often occurs with phase separation into two amorphous phases as the first step, which is followed by crystallization of the minor nucleant-rich phase,

Fig. 11.22. (*a*) Microstructure of transparent, low-expansion Li_2O–Al_2O_3–SiO_2 glass-ceramic body; (*b*) microstructure of highly crystalline glass-ceramic body. From R. H. Redwine and M. A. Conrad in R. M. Fulrath and J. A. Pask, Eds., *Ceramic Microstructures*, John Wiley & Sons, Inc., New York, 1968, pp. 900–922.

which serves as precursor for the formation of the major crystalline phase or phases. The amount and characteristics of the final-phase assemblage depend both on the composition and on the heat treatment. An example of differences resulting from heat treatment of the same composition has been shown in Fig. 8.32. By controlled heat treatment $Li_2O \cdot Al_2O_3 \cdot SiO_2$ glass-ceramics can be prepared in a highly crystalline form with a particle size as small as 0.05 micron (Fig. 11.22*a*). Since this particle size is less than a tenth of the wavelength of light and the phases have similar indices of refraction, this material is transparent. Modifying the heat treatment of the same material gives particle sizes in the micron range (Fig. 11.22*b*), and this product is opaque. As discussed in Chapter 8, heat treatment at relatively low temperatures at which the ratio of nucleation rate to growth rate is greater leads to a smaller-particle product. The crystalline content of the sample illustrated in Fig. 11.22 is about 85 vol% for the fine-particle product and about 95 vol% for the larger-particle material.

Many glass-ceramic bodies are composed of more than one phase. Figure 11.23 shows a Li_2O–Al_2O_3–SiO_2 glass-ceramic nucleated with TiO_2, in which small rutile crystals, several hundred angstroms in diameter, are observed with the β-spodumene grains. Figure 11.24*a* shows a MgO–Al_2O_3–SiO_2 glass-ceramic nucleated with ZrO_2, in which ZrO_2 crystals are observed in the boundary regions between the β-quartz grains. In this case, the ZrO_2 crystals were rejected at the advancing crystal-melt interface during crystallization of the major phase. Prolonged heat treatment of this same composition results in a solid-state transformation in which spinel forms at the expense of the quartz grains (Fig. 11.24*b*). Such solid-state transformations are often encountered in systems used for glass-ceramic applications in which a metastable phase first crystallizes or in which initial crystallization at a low temperature is followed by a higher-temperature heat treatment.

Compositional variations in the crystalline phase are characteristic of many glass-ceramic microstructures. As discussed in Chapter 2, eucryptite, $LiAlSiO_4$, is a *stuffed derivative* of β-quartz, and these phases are ones commonly occurring in glass-ceramics. Variations in the amount of aluminum and other ion substitution in the silica structure with accompanying changes in the alkali ion content are common.

Special properties of glass-ceramics can be developed when sheet silicates, primarily those of the fluorine mica family, are precipitated as the major phase in the glass-ceramic body. Such glass-ceramics containing more than 65 to 70 vol% mica can be machined to close tolerances if the aspect ratios of the crystals are large enough to cause contact between them. The desirable machinability is associated with basal cleavage of the mica crystals, because of relatively weak bonding between the layers, and

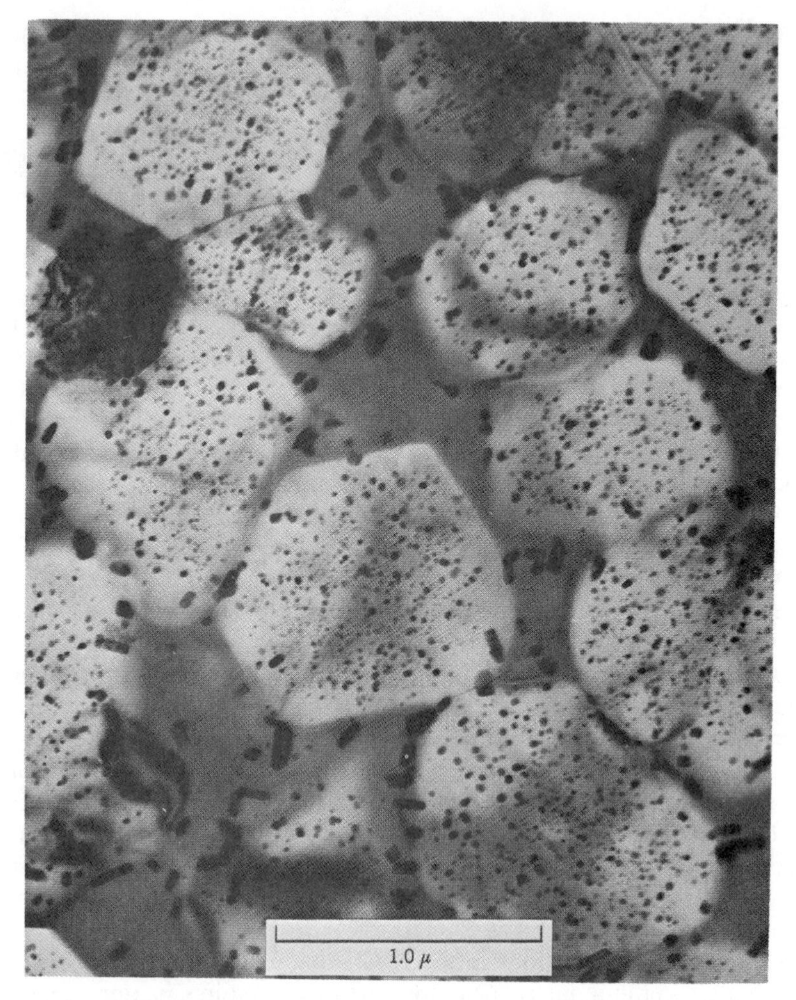

Fig. 11.23. Submicrostructure of Li_2O–Al_2O_3–SiO_2 glass-ceramic nucleated with TiO_2, approximately 80% crystalline. Small rutile crystals shown in larger crystals of β-spodumene. From P. E. Doherty in R. M. Fulrath and J. A. Pask, Eds., *Ceramic Microstructures*, John Wiley & Sons, Inc., New York, 1968, pp. 161–185.

with the difficulty of fracture propagation across the basal planes. Fractures therefore follow the crystal boundaries, causing detachment of individual crystals or small groups of crystals. Figure 11.25 shows mica crystals having low and high aspect ratios in glass-ceramic bodies. The microstructure of high aspect ratio, referred to as an interlocking *house of cards* structure, has the more desirable properties.

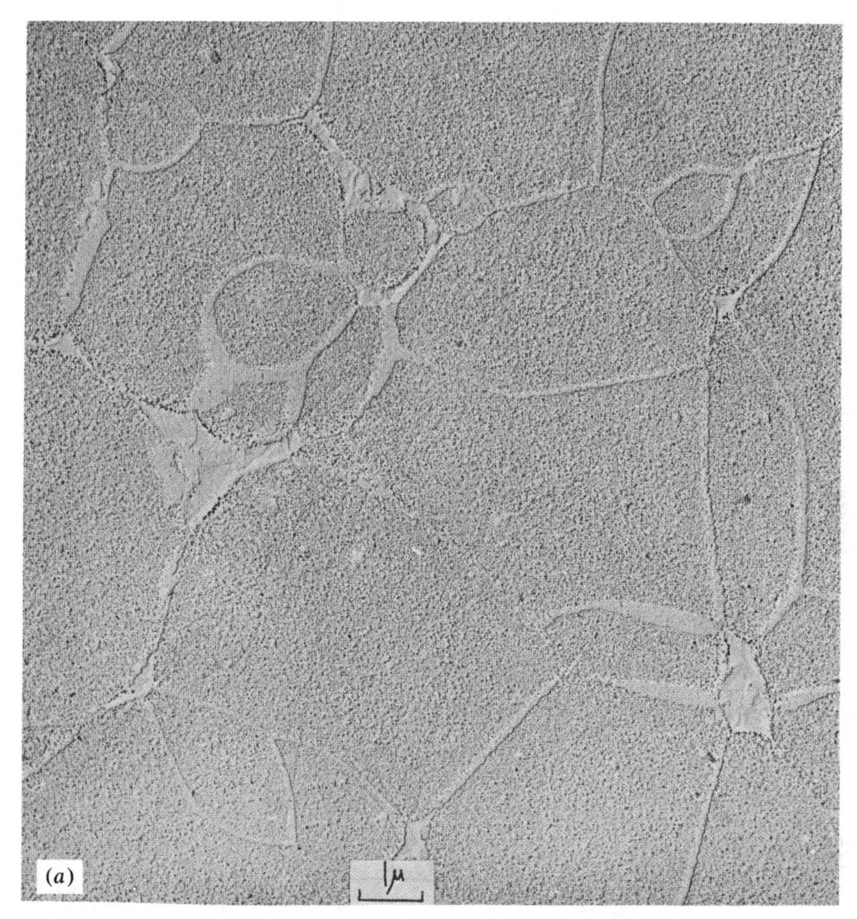

Fig. 11.24. (a) Submicrostructure of $MgO \cdot Al_2O_3 \cdot 3SiO_2$ glass-ceramic nucleated with ZrO_2. Small zirconia crystals at grain boundaries between high-quartz crystals.

Finally, the process of ion-exchange strengthening, discussed in Chapter 8, when applied to glass-ceramic bodies can result in phase transformations in the near-surface regions which have been ion-exchanged. Such transformations can have a pronounced effect on mechanical strength. An example of the microstructures which result from ion-exchange treatments of this type is shown in Fig. 11.26. A glass-ceramic body in the MgO–Al_2O_3–SiO_2 system containing cordierite and cristobalite as the stable crystalline phase assembly has been ion-exchanged in a Li_2SO_4 bath at 1000°C. In the $2LI^+ \rightleftharpoons Mg^{2+}$ exchange, the phase assemblage in the near-surface region transforms successively to a β-quartz solid solution and then to a β-spodumene solid solution as the Li^+ concentration increases.

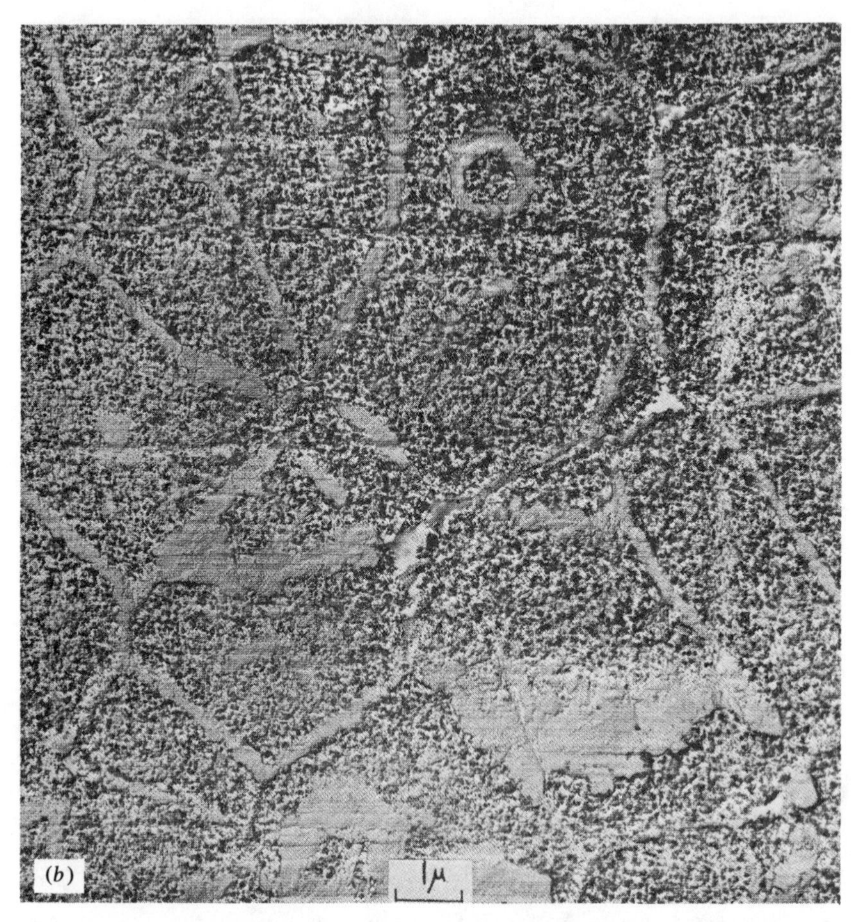

Fig. 11.24 (*Continued*). (*b*) Same composition held at 977°C for 20 hr. Spinel forms at the expense of previously crystallized high-quartz phase. From R. H. Redwine and M. A. Conrad in R. M. Fulrath and J. A. Pask, Eds., *Ceramic Microstructures*, John Wiley & Sons, Inc., New York, 1968, pp. 900–922.

11.9 Electrical and Magnetic Ceramics

The general category of electrical and magnetic ceramics is one that includes a wide range of compositions and structures. Most ceramic materials can be used as electrical insulators or for other electrical purposes. The composition that is most widely used for low-tension insulators are triaxial porcelains, already discussed in Section 11.3. Glasses are also widely used for electrical-insulation purposes. For low-loss and high-frequency applications steatite, forsterite, and alumina ceramics are

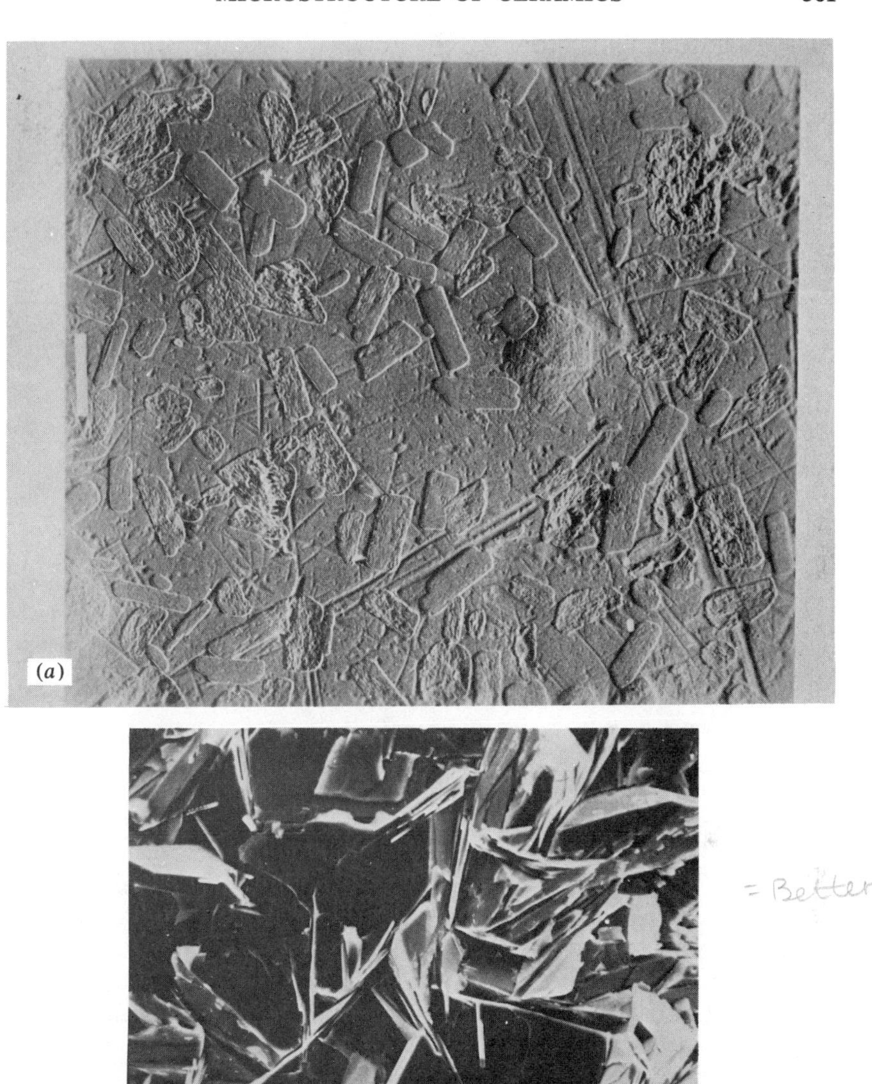

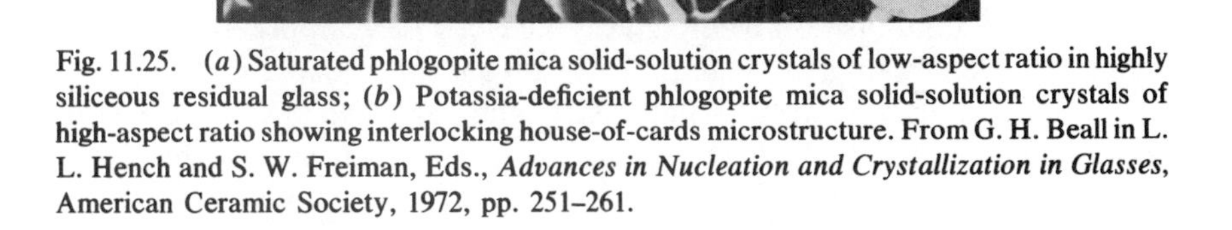

Fig. 11.25. (*a*) Saturated phlogopite mica solid-solution crystals of low-aspect ratio in highly siliceous residual glass; (*b*) Potassia-deficient phlogopite mica solid-solution crystals of high-aspect ratio showing interlocking house-of-cards microstructure. From G. H. Beall in L. L. Hench and S. W. Freiman, Eds., ***Advances in Nucleation and Crystallization in Glasses***, American Ceramic Society, 1972, pp. 251–261.

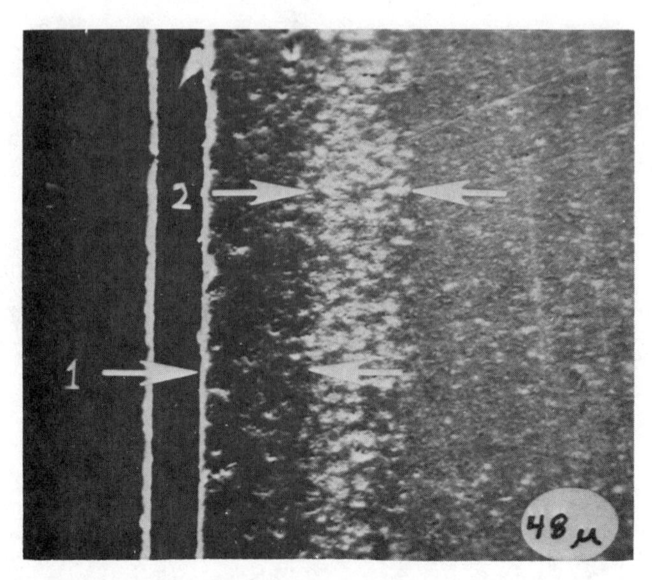

Fig. 11.26. Scanning electron micrograph of $2Li^{+} \rightleftharpoons Mg^{2+}$ ion exchanged cordierite + cristobalite glass-ceramic body showing quartz and spodumene solid solutions developed in the near-surface region. After G. H. Beall in L. L. Hench and S. W. Freiman, Eds., *Advances in Nucleation and Crystallization in Glasses*, American Ceramic Society, 1972, pp. 251–261.

generally used. Reactions occurring during firing and the phase composition of these materials have been discussed in Chapter 7.

Steatite compositions are a general class of dielectrics which contain steatite, or talc, as a major constituent. They are extensively used for high-frequency insulators because of their good strength, relatively high dielectric constant, and low dielectric losses. Two main phases are present in the fired body (Fig. 11.27). The crystalline phase is enstatite, which appears as small discrete prismatic crystallites in a glassy matrix. The high-temperature equilibrium form is protoenstatite, which converts to clinoenstatite on cooling. The conversion is inhibited by the glassy phase, which isolates the individual crystals, and large crystals are converted more rapidly than small ones. The presence of large crystals is harmful to properties, since they tend to crack, owing to differences in expansion coefficient between the crystals and glassy matrix, illustrated in Fig. 10.44.

Electrical properties of steatite ceramics are largely determined by the amount and composition of the glassy phase. Triaxial porcelains contain considerable amounts of alkalies derived from the feldspar used as a flux. This leads to high electrical conductivity and high dielectric loss. Steatite porcelains that have feldspar added as an aid in firing also have high

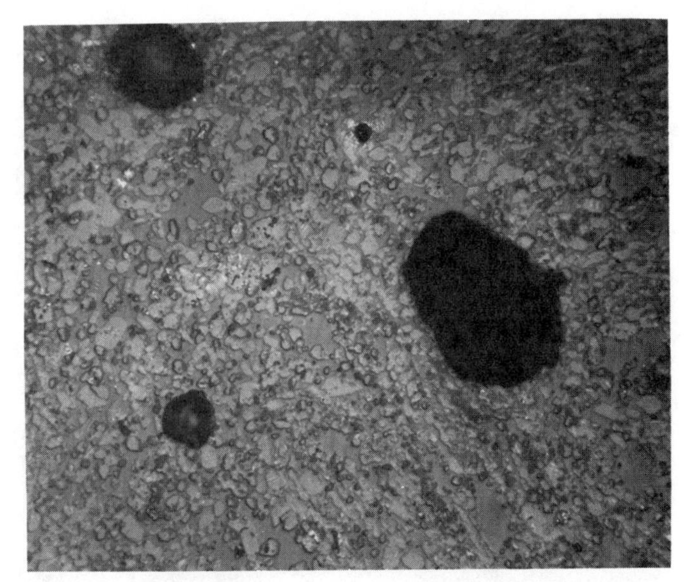

Fig. 11.27. Microstructure of steatite porcelain (500×).

dielectric loss. The low-loss compositions are nearly alkalifree, using alkaline earth oxides as fluxing constituents.

Forsterite ceramics have this material, Mg_2SiO_4, as the major crystalline phase bonded with a glassy matrix. The crystals are prismatic and usually larger in size than the enstatite crystals present in steatite. A typical structure has been illustrated in Fig. 7.33. Differences among various compositions depend for the most part on the amount of glassy phase present. Forsterite ceramics are particularly useful as low-loss dielectrics in designs in which the thermal expansion coefficient must be suitable for metal-ceramic bonding.

Alumina ceramics have Al_2O_3 as the crystalline phase bonded with a glassy matrix. In some bodies the alumina has a prismatic habit, whereas in others particles are nearly spheroidal. The factors controlling crystal habit are not completely understood; a typical microstructure is illustrated in Fig. 11.28. The properties obtained depend in large part on the amount and properties of the glassy phase, which is usually alkalifree, being compounded from mixtures of clay, talc, and alkaline earth fluxes. The firing temperature of alumina ceramics is relatively high. The body must be carefully compounded for satisfactory results. The main imperfection is excessive porosity; the pore size is usually larger than that of the individual grains of raw materials used and results from poor forming or firing techniques. The main advantages of alumina ceramics usually given are

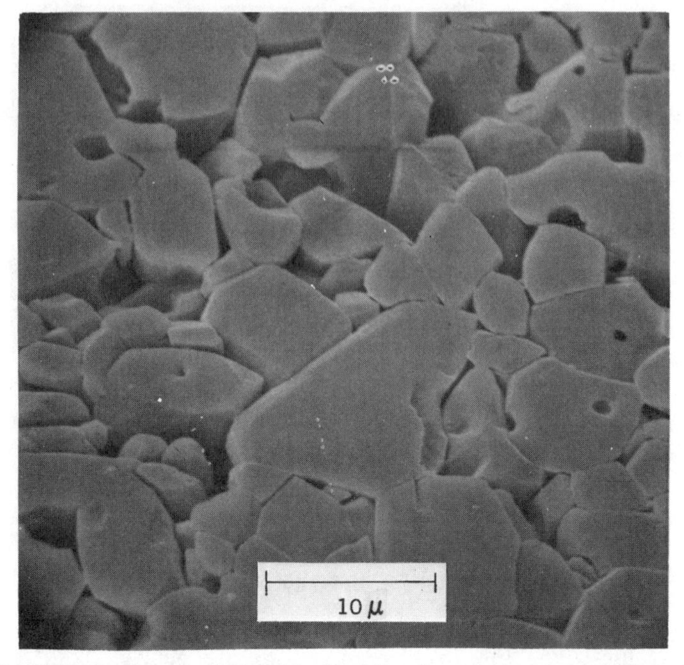

Fig. 11.28. High-alumina porcelain polished and heavily etched to remove silicate bonding phase (2300×).

relatively high dielectric constant and low dielectric losses. Its main usefulness in fact arises mostly from its high strength and resistance to thermal stresses. These properties allow it to be used in automatic forming machinery without excessive breakage or special handling. Alumina is widely used as a substrate for electronic-device applications in which surface resistivity and dielectric losses dictate use of a material containing 99% or more Al_2O_3. Surface smoothness depends in large part on grain size and a small-grain material such as illustrated in Fig. 11.29 is much preferred.

Cordierite ceramics, as discussed in Chapter 7, are useful because of their very low thermal expansion and consequent high resistance to thermal shock. Bodies are manufactured with a variety of fluxes; the cordierite phase develops as prismatic-habit crystals and is associated with a glassy phase, often together with some mullite, corundum, spinel, forsterite, or enstatite.

For ultralow-loss applications, particularly where large energy transfer through a ceramic is required, as in windows for high-powered electronic tubes, it is desirable to eliminate the glassy phase entirely. This can be done

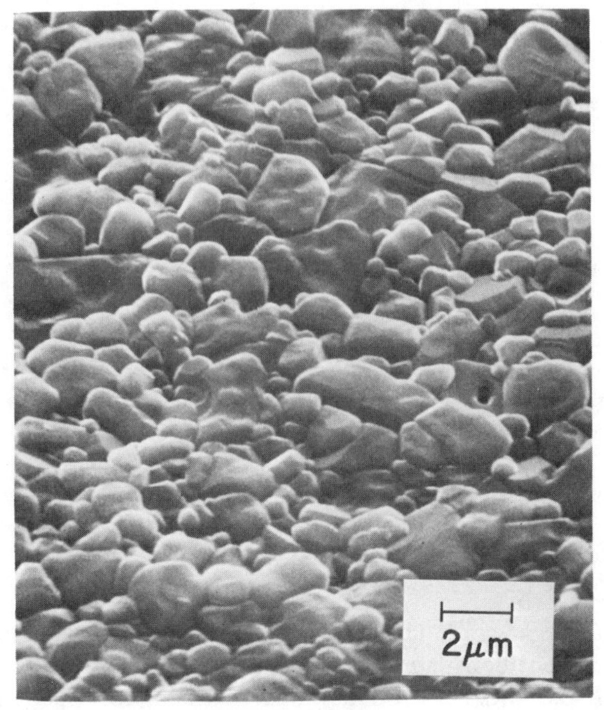

Fig. 11.29. Surface replica of an as-fired fine-grained 99% Al_2O_3 substrate surface. Courtesy R. Mistler.

with alumina by sintering pure fine-grained materials in the solid state at high temperatures. Structures obtained are ideally fine-grained with little porosity (Fig. 10.27*e*). Frequently, secondary crystallization occurs at the high firing temperatures used so that structures such as those illustrated in Fig. 10.13 are obtained. The main effect of increased grain size is to reduce the strength; it has little effect on electrical properties.

For applications in which very high dielectric constants are required, titania (dielectric constant about 100) or barium titanite (dielectric constant about 1500) ceramics are used. In titania ceramics TiO_2 is the major crystalline phase with small amounts of fluxes such as zinc oxide added to form a liquid phase at the firing temperatures. The resulting structure is similar to that of Fig. 7.11. Barium titanite bodies normally consist entirely of crystalline $BaTiO_3$. The individual crystals in a polycrystalline sample contain multiple domains of different ferroelectric orientations (see Chapter 18) which are clearly distinguished on etching (Fig. 11.30). Frequently, some secondary crystallization occurs during firing (Fig. 10.12).

Fig. 11.30. Microstructure of barium titanate ceramic. Different ferroelectric domain orientations are brought out by etching (500×). Courtesy R. C. DeVries.

Magnetic ceramics are ideally composed of a single crystalline phase having a composition determined by the magnetic properties desired ($FeNiFeO_4$, $BaFe_{12}O_{19}$, $FeMnFeO_4$, and so forth; see Chapter 19) and usually with as high density and fine grain size as can be obtained. A typical structure is illustrated in Fig. 11.31. As with other materials formed by solid-phase sintering, secondary recrystallization may occur (illustrated in Fig. 10.15).

As shown for the Fe–O system in Fig. 7.9, the desired single phase of magnetite, Fe_3O_4, occurs only over a limited range of oxygen content corresponding to a limited range of oxygen pressure. This is also true of other magnetic ferrite phases, and in production all manufacturers control the oxygen pressure during firing to ensure obtaining the desired magnetic properties normal to a particular composition of the single-phase ferrite. Where this is not done and two phases occur, the result is frequently similar to that found by R. E. Carter for $NiFe_2O_4$ in which an (MgFe)O phase forms (Fig. 11.32). Differences in the expansion coefficient of the two phases lead to cracking of the ferrite and have an adverse effect on the magnetic properties obtained.

11.10 Abrasives

Abrasive products have as their essential constituent a hard phase that provides many individual particles with sharp cutting edges; a bonding

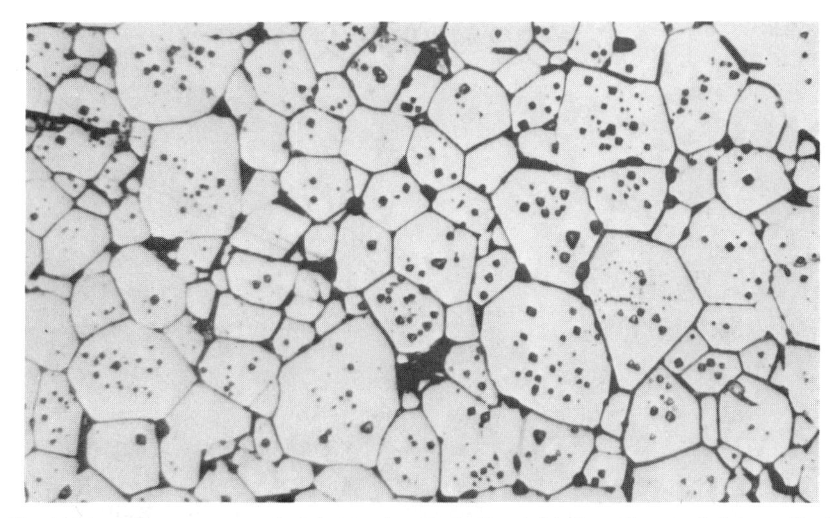

Fig. 11.31. Microstructure of nickel ferrite. Etch pits visible in grains result from sulfuric acid–oxalic acid etchant (638×). Courtesy S. L. Blum.

Fig. 11.32. Microstructure of two-phase ferrite ceramic. Light phase is $MgFe_2O_4$, dark phase (MgFe)O. Cracks due to microstresses are evident (500×). Courtesy R. E. Carter.

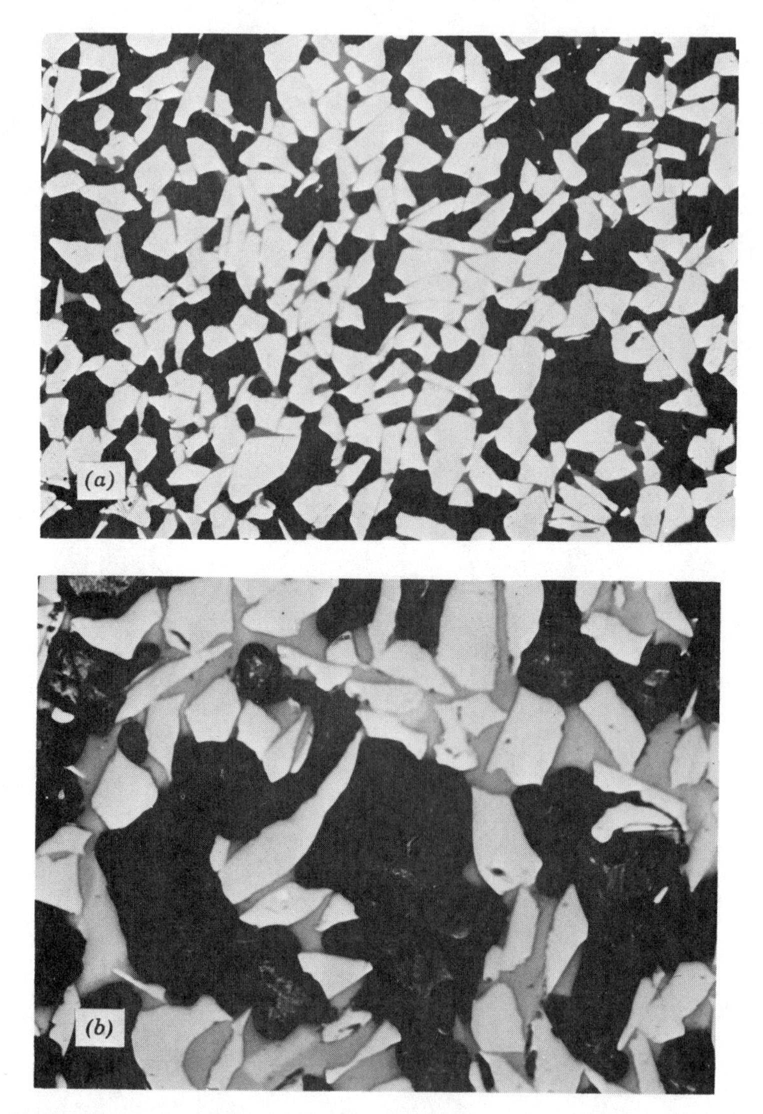

Fig. 11.33. Abrasive products. (*a*) Section of silicon carbide wheel, unetched (50×): (*b*) aluminum oxide wheel, unetched (100×). In both light area is grain, gray area bond, and dark area porosity. Courtesy A. Sidhwa.

phase holds these particles in a more or less tight grip, and a certain amount of porosity provides channels for air or liquid flow through the structure. For the hard abrasive grain either aluminum oxide or silicon carbide is usually used. Aluminum oxide grains are tougher than silicon carbide grains and wear more slowly, but they are not quite so hard. Silicon carbide grains are harder and more satisfactory for grinding hard materials but tend to fracture in use so that the life of the abrasive is shorter. In either abrasive individual grains are bonded to a wheel or paper or cloth with a strength depending on their proposed use. It is desirable that grains break out of the bond material once they become dull. Bond materials include fired ceramic bonds and a variety of organic resins and rubbers. Fired ceramic bonds are relatively hard, provide for a long life, can be used at high speeds satisfactorily, and account for the major part of grinding-wheel production.

No matter what hard grain and bond material are used, the overall structure is similar to those illustrated in Fig. 11.33 in which a section of a silicon carbide wheel and an aluminum oxide wheel are shown. In both the abrasive grains are held together with a glassy bond which determines the relative hardness of the individual wheel. The alumina product illustrated in Fig. 11.33*b* has a greater strength and hardness than the silicon carbide wheel, which has a larger proportion of abrasive grain to bond. In both the structure of the wheel is open with a large void fraction to provide for efficient cooling, either by air currents or liquid coolants during the grinding operation. This also allows grains to fracture and break off as they become worn.

11.11 Cement and Concrete

A wide variety of cementitious materials are used for different purposes. The structures of the products resulting from most of these have not been studied in much detail. The one that is of most economic importance and is most widely used is portland cement; high-alumina cement, *ciment fondu*, is used for refractory purposes. Portland cement is manufactured in rotary kilns, using various raw materials to give an overall composition in which the major resulting constituents are tricalcium silicate, $3CaO \cdot SiO_2$, and dicalcium silicate, $2CaO \cdot SiO_2$. The product is made in a rotary furnace and sintered under such conditions that a fraction of the charge becomes a liquid phase. The transformation occurring in the dicalcium silicate on rapid cooling causes a sufficient volume change to give rise to *dusting*, that is, to break up the particles. Polished sections of portland cement clinker are shown in Fig. 11.34. In addition to the major dicalcium silicate and tricalcium silicate phases, which are the basic materials present, there are frequently smaller amounts of tricalcium aluminate, $3CaO \cdot Al_2O_3$, brownmillerite (approximately $4CaO \cdot Al_2O_3 \cdot Fe_2O_3$), some CaO, some MgO, and glass. The amounts, composition, and morphology of the minor phases present depend a great deal on the raw materials used and conditions for sintering. Aluminous cement has a much higher alumina content (approximately $40Al_2O_3$, $40CaO$, $7SiO_2$, $7Fe_2O_3$, $5FeO$, and five other oxides) and

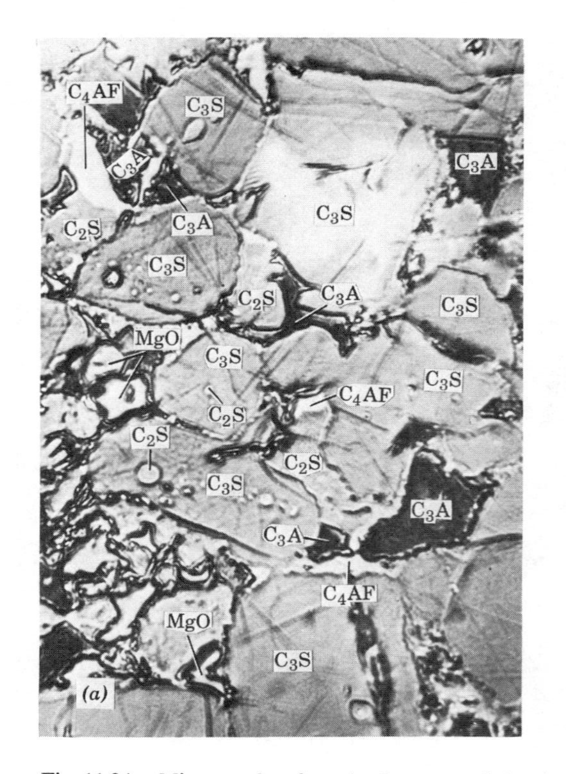

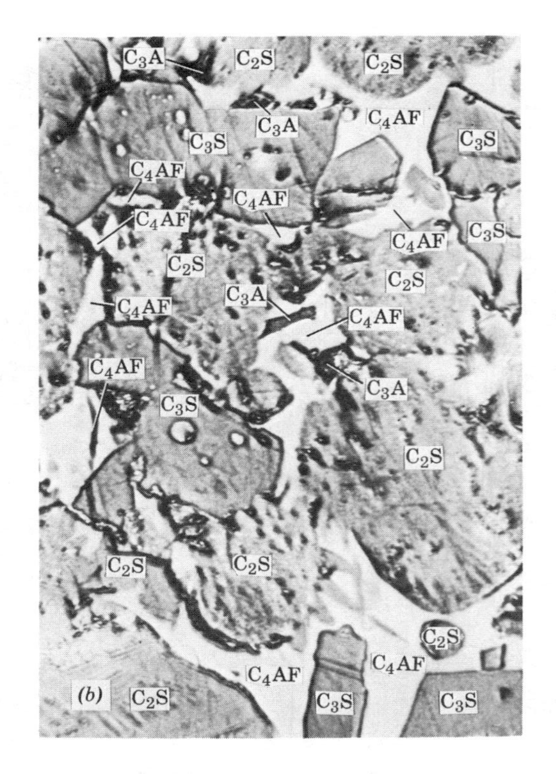

Fig. 11.34. Micrographs of portland cement clinker (835×). (*a*) Type I, high in $3CaO \cdot SiO_2$ (major gray phase is C_3S, dark gray phase C_3A, light gray phase C_2S, white phase mainly C_4AF); (*b*) type II, containing nearly equal parts of $3CaO \cdot SiO_2$ and $2CaO \cdot SiO_2$ (gray C_3S, light gray C_2S, black C_3A, white C_4AF). Courtesy Portland Cement Association.

during the firing process forms a much more fluid liquid phase which gives rise to the name *melted* cement. The main constituents in the clinker are *unstable* $5CaO\cdot3Al_2O_3$, monocalcium aluminate, $CaO\cdot Al_2O_3$, and calcium dialuminate, $CaO\cdot2Al_2O_3$. The unstable $5CaO\cdot3Al_2O_3$ probably has the approximate composition $6CaO\cdot4Al_2O_3\cdot FeO\cdot SiO_2$.

On reaction with water the clinker forms a complex hydrated product which is a cementitious material. The main cementitious product that forms is a noncrystalline calcium silicate gel resulting from the tricalcium silicate and dicalcium silicate present in the clinker material. Along with the calcium silicate hydrate, calcium hydroxide is formed as a by-product and occurs as small hexagonal plates. The calcium hydroxide reacts with carbon dioxide in the air or water available to form calcium carbonate. The resulting structure of the gel as formed after reaction is illustrated in simplified form in Fig. 11.35. In the gel phase itself there are pore spaces between the individual gel particles; in addition, there are large residual capillary pores remaining from the excess water content required to form the cement and place it satisfactorily. This excess capillary porosity should be kept as low as possible if the optimum mechanical properties from the

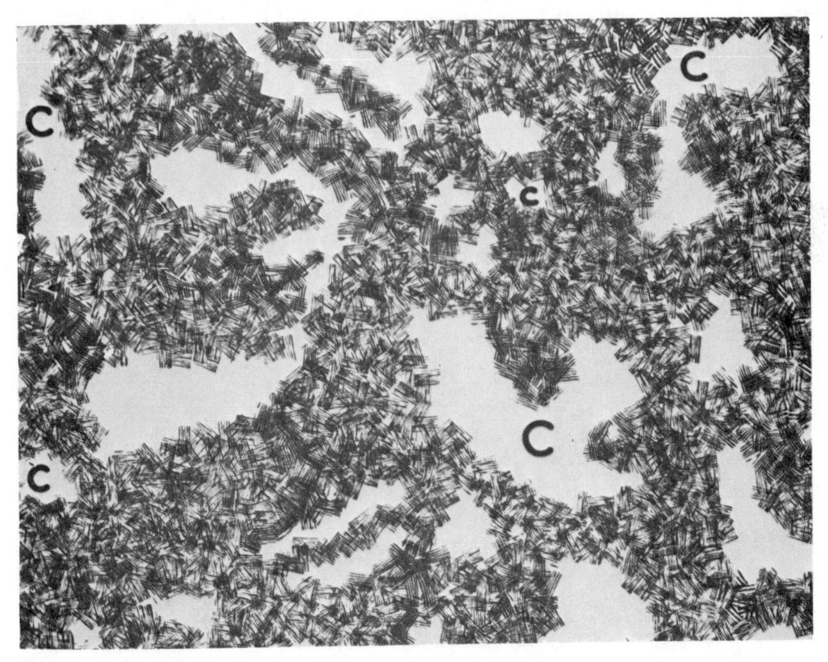

Fig. 11.35. Simplified model of portland cement paste structure showing needle or platelet gel particles and capillary cavities C. From T. C. Powers, *J. Am. Ceram. Soc.*, **41**, 1 (1958).

cement are to be obtained. In addition to the cement gel, there is present in concrete an aggregate of crushed stone that acts as a filler material. The portland cement paste serves to bond together the aggregate particles in much the same way as a bond material present in a refractory brick or an abrasive wheel. The properties of the concrete formed depend on the amount of porosity present, particularly as capillary pores, the strength of the aggregate material, and the properties of the cement paste gel.

In contrast to portland cement, which forms a noncrystalline cement paste having good adhesive properties to bond together the aggregate materials when completely set, plaster of paris, which is widely used as a cementing and mold material in ceramics, has little strength and poor adhesive properties. The structure of set plaster is a highly crystalline reaction product, as shown in Fig. 11.36. Plaster is formed by calcining gypsum and then is used by rehydration:

Calcination: $$CaSO_4{\cdot}2H_2O = CaSO_4{\cdot}\tfrac{1}{2}H_2O + \tfrac{3}{2}H_2O$$

Rehydration: $$CaSO_4{\cdot}\tfrac{1}{2}H_2O + \tfrac{3}{2}H_2O = CaSO_4{\cdot}2H_2O$$

The individual crystals present are in the form of fine needles so that the resulting structure corresponds to a feltlike arrangement in which there are very fine pores, and the interlocking of the crystal needles provides sufficient strength. The amount of residual porosity depends on the amount

Fig. 11.36. Interlocking crystallite network in pottery plaster, $CaSO_4{\cdot}2H_2O$. Courtesy W. Gourdin.

of water present in the original mix. As the water content is increased, the volume fraction porosity after drying is also increased; the absorption capacity of the mold is raised, but at the same time its strength and durability are reduced.

A variety of other cementitious materials are used for various applications of ceramics. These range from tars and sugar solutions to reactions that form oxychlorides or acid phosphates and include the use of fine-grained clay materials as bonding agents. In all cases the resultant structure obtained is similar to that illustrated in Fig. 11.33, in which the bonding material is distributed at the contact points between grains, holding them together. More or less bonding material may be required, depending on the strength desired and the properties of the particular bond used. The general characteristic that all these materials have is a tendency to form noncrystalline products. In many cases this corresponds to the opportunity for extensive development of hydrogen bonds, which provide a mechanism for adherence. However, for many applications great strengths are not particularly required, and as long as a liquid phase distributes itself in positions between solid grains and then is solidified, there is sufficient adherence developed for satisfactory results. The major concern with regard to adhesion is first wetting and then the possibility of deleterious side effects such as gas evolution, volume changes, and adsorption of impurities at the interface; as long as these effects are absent, sufficient adherence is obtained.

11.12 Some Special Compositions

In addition to the classes of materials discussed thus far, in which the great majority of all ceramic production can be included, there are many materials that are difficult to introduce into any of these specific product groups. Although they do not contribute much to the overall volume of the ceramics industry, they frequently supply critical needs or provide properties that cannot be obtained elsewhere; consequently, they are particularly important from the point of view of developing new materials and the understanding of material properties.

Cermets. A group of materials that come under the general classification of refractories are the combinations of metals and ceramic materials called cermets. The compositions of most significance for their practical or potential application are carbides having high-temperature strength (Ni–TiC) and, in particular, great hardness. Also, oxide-base cermets have some valuable properties as high-temperature, high-strength materials that are reasonably stable in air; the most extensively investigated and only commercially available ones are mixtures primarily composed of aluminum oxide and chromium suitably alloyed to give useful high-temperature properties. The carbide-metal compacts consist of either spheroidal or prismatic carbide grains completely enclosed by the metal phase. The bond phase is usually liquid at the firing temperature and completely wets and flows between the carbide particles, forming thin films of metal. These compositions have excellent high-temperature strength and also satisfactory toughness. The carbides used are hard so that they can be used in cutting tools. In the aluminum oxide–chromium system, in contrast, there is a continuous phase of both the oxide and chromium. This gives rise to high-temperature strength and resistance to thermal stresses. During cooling to room temperature there is a tendency for boundary stresses to develop, and the strength and other properties are less favorable at room temperature than at higher temperatures. Microstructures of these compositions are illustrated in Fig. 11.37.

Coatings. The only kinds of coating that have been discussed thus far are glazes and enamels based primarily on forming a glassy liquid which flows over and covers the surface. A variety of compositions has been used as enamel or glass coatings, in much the same way. In addition, however, coating of nonmetals can be applied by reactions from the vapor phase which deposit a coating on the surface, by flame-spraying oxide material through a high-intensity heat source so that it fuses in the flame and solidifies when it strikes the relatively cool surface, or by spraying a suspension on a hot surface so that a fine-particle dispersion is formed on hitting the surface to develop a suitable coating. The microstructures of these different types of coating vary substantially, depending on the particular method of application. Flame-sprayed coatings normally have a porosity of 7 to 10% and frequently show evidences of some layered structure during the buildup, although this depends on the particular techniques used. Layers formed by the evaporation of a solvent have very small crystals in the resultant coating. In contrast, coatings developed from the vapor phase by reaction frequently show large crystals, owing to the nucleation of new crystals in the surface and then subsequent growth in the coating phase. Very often the structure of the coating is parallel to the underlying structure, in that new crystals are nucleated and grow on sites of crystals of the underlying material. Graphite coolings, for example, can be formed by passing hot CH_4, a gas, over a hot surface. In this process the new *pyrolytic* graphite crystals form a deposit with their c axis normal to the underlying surface, in parallel bundles consisting of individual crystallites of nearly the same orientation (Fig. 11.38).

Sintered Oxides. Another kind of material which we have mentioned is the pure sintered single-phase oxide for uses requiring high strength, high-temperature capabilities, good electrical properties, or great hard-

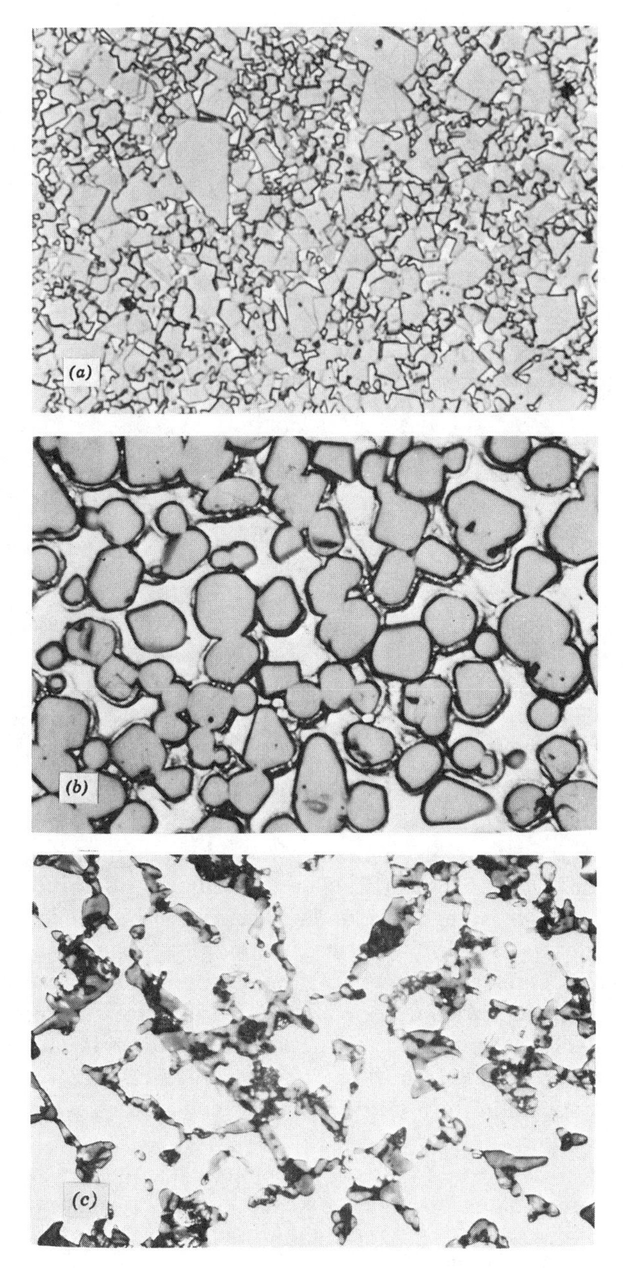

Fig. 11.37. Metal-ceramic compositions. (*a*) 96WC–6Co (1500×). Courtesy M. Humenik. (*b*) 70TiC–30Ni (1580×). Courtesy M. Humenik. (*c*) $30Al_2O_3$–70Cr (545×). Courtesy F. R. Charvat.

Fig. 11.38. Pyrolytic graphite coating deposited on graphite rod at bottom from a methane-hydrogen atmosphere (164×). Courtesy Avco Research and Advanced Development Division.

ness. For example, Al_2O_3 has been used as a tool material, its great hardness and low friction with metal combined with its high strength and its high-temperature capability making it particularly effective. Resulting structures in these ceramic compounds have been discussed in connection with solid-state sintering in Chapter 10, and some structures have been illustrated there. One of the interesting recent applications for this class of material is the use of uranium oxide as a nuclear-reactor-fuel material. It is particularly useful in that a large fraction of the uranium atoms can be fissioned without degradation of the structure. Other oxides which have been used for special or refractory applications include BeO, MgO, ThO_2, ZrO_2, and $MgAl_2O_4$.

Single Crystals. For many special applications, single crystals of oxide and other ceramic materials have been used. Single crystals of Al_2O_3 have been used as windows for heat resistance and good infrared transmission, as high-intensity light-bulb enclosures, and as electronic-device substrates. They have also been used in the form of rods and other special shapes as high-temperature refractory materials. Single crystals of rutile (TiO_2), spinel ($MgAl_2O_4$), strontium titanate ($SrTiO_4$), ruby (Al_2O_3 with some Cr_2O_3 in solid solution), and others have been used as synthetic jewel materials. Lithium niobate ($LiNbO_3$) is used as a laser host and as a substrate. The optical properties of single crystals of magnesium oxide have been of potential interest; the use of alkali and alkaline earth halide crystals for prisms and windows in optical equipment has been widespread for many years. Single crystals of calcium fluoride, lithium fluoride, sodium chloride, and many others are commercially available.

Whiskers. An area of interest from a research point of view has been the structure and properties of whiskers of ceramic materials in which extremely high-strength values have been observed. Under certain conditions of growth, crystals form in which the growth is rapid in one direction, developing filamentary crystals that are presumably free from gross imperfection and have strengths up to several million pounds per square inch. Whiskers of this type of aluminum oxide, several sulfides, several alkali halides, graphite, and others have been grown in the laboratory.

Graphite. A ceramic material that has been widely used but not extensively described from a structural point of view is graphite. Graphite is normally made from mixtures of coke and pitch which are formed and heat-treated to develop a graphite crystal structure. The graphite crystals formed are highly anisometric, forming platelets, as would be expected from the crystal structure. The general microstructure of graphite consists of grains of highly graphitized material in a matrix of very fine-grained material which is more or less graphitized and more or less strongly crystallized, depending on the particular heat treatment. A microstructure of one sample is illustrated in Fig. 11.39. Details of the structure depend a great deal on the structure of the original coke, which again depends on the original petroleum, coal, or tar used for its formation, together with its distribution and heat treatment during graphitization. The properties of the resulting graphite are strongly dependent on the details of the structure; this is indicated by the fact that the properties of graphite products are strongly directional and depend on details of forming techniques. However, exact relationships between structure and properties have not been worked out for any detailed system.

High-Porosity Structures. Another group of materials, mentioned earlier but not discussed in much detail, is highly porous compositions for various insulating purposes. These include fibrous products such as glass wool, powdered insulated grain, and strong insulating firebrick. The common characteristic of all these materials is high porosity. In general,

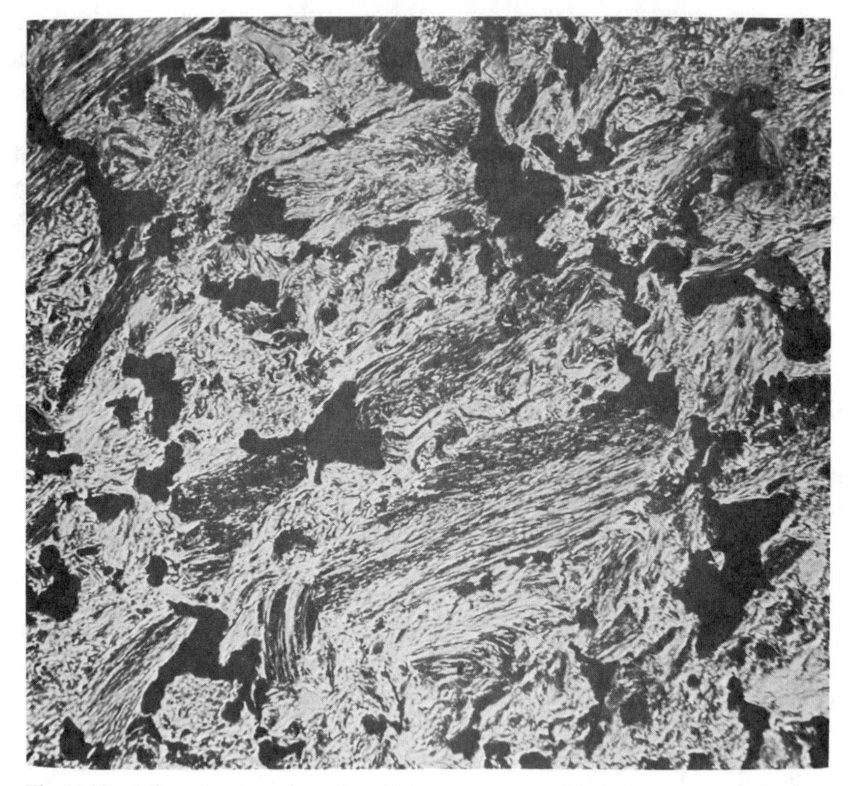

Fig. 11.39. Microstructure of graphite (104×). Courtesy A. Tarpinian.

as the pore size decreases and the amount of porosity increases, materials become more effective as thermal insulators. However, for use at high temperatures, the minimum pore size that can be present without inducing sintering and instability of volume is limited. Consequently, materials used in brick form for high-temperature insulation have a much larger pore size than materials used at temperatures below the sintering range. This is the main reason why powdered carbon has been useful as an extremely high-temperature insulation; the mobility of carbon atoms is very low, as discussed in Chapter 8, so that sintering does not take place, and a very fine particle size material can be used satisfactorily.

Suggested Reading

1. E. E. Underwood, *Quantitative Stereology*, Addison-Wesley Publishing Company, Inc., Reading, Mass., 1970.
2. R. M. Fulrath and J. A. Pask, Eds., *Ceramic Microstructures*, John Wiley & Sons, Inc., New York, 1968.
3. H. Insley and V. D. Fiechette, *Microscopy of Ceramics and Cements*, Academic Press, Inc., New York, 1955.
4. American Society Testing Materials, *Symposium on Light Microscopy, A.S.T.M. Spec. Publ.* 143, 1952.
5. G. R. Rigby, *Thin Section Mineralogy of Ceramic Materials*, 2d ed., British Ceramic Resin Association, Stoke-on-Trent, England, 1953.
6. A. A. Klein, "Constitution and Microstructure of Porcelain," *Natl. Bur. Std., Tech. Pap.* 80, 1916–1917.
7. C. S. Smith, "The Shape of Things," *Sci. Am.*, **190**, 58 (January, 1954).
8. S. T. Lundin, "Electron Microscopy of Whiteware Bodies," *Transactions of the IVth International Ceramic Congress*, Florence, Italy, 1954.

Problems

11.1. A typical porcelain body has the composition 50 clay–25 feldspar–25 quartz. Sketch an expected microstructure of such a body, indicating scale, when (*a*) fired to achieve phase equilibrium (1450°C for 6 hr) and (*b*) fired to 1300°C for 1 hr. Explain how and why these two different firings would affect mechanical, optical, thermal, and electrical properties.

11.2. Suppose that the 10% porosity existing in a sintered alumina ceramic is due to a uniform distribution of pores trapped at the interstices of particles during the sintering process (grain boundary or bulk diffusion). Considering the initial powder compact to be an ideal packing of spheres 1 micron in radius with sixfold coordination for pores as well as spheres, that is, one pore per particle, what is the average size of pore in microns viewed in the microsection? How many pores per square centimeter in a microsection? Consider the theoretical density to be 4.00 g/cc and the atomic weight 102.

11.3. In a triaxial porcelain fired at 1200°C, feldspar grains ($K_2O \cdot Al_2O_3 \cdot 6SiO_2$) melt at the firing temperature to form a blob of viscous liquid surrounded by the product formed by heating clay ($Al_2O_3 \cdot 2SiO_2 \cdot 2H_2O$). In a microstructure of a fired porcelain, needlelike crystals of mullite ($3Al_2O_3 \cdot 2SiO_2$) are observed to extend into the feldspar *pseudomorphs*. Discuss the kinetics of the mullite crystal growth including (*a*) your choice of the most probable rate-limiting step, (*b*) your reasons for that choice, and (*c*) how you would experimentally or analytically confirm or negate that choice.

11.4. List the following in order of importance for control of microstructure during sintering of a typical oxide (like UO_2). Justify the order with suitable numerical approximations for the range of value for each variable over which control can be exercised:

Surface energy
Temperature
Atmosphere
Distribution of particle size
Heating rate
Bulk density (prior to firing)
Inhomogeneity in bulk density
Time

11.5. How would you go about making a large-grain *controlled-orientation* (i.e., all or most crystals oriented parallel to one another) polycrystalline ceramic? Explain the principle on which you base your proposed procedure.

11.6. Draw clear sketches of the microstructure, showing pores, solid phases, and grain boundaries, paying particular care to clearly illustrate the relationship among pores, different phases present, and grain boundaries, for:
(*a*) A triaxial porcelain sintered to maximum density in an atmosphere in which the gas would not diffuse at an appreciable rate in the solid.
(*b*) A single-phase crystalline refractory material such as MgO of large initial particle size (10 microns) sintered at a low temperature (1500°C).
(*c*) A single-phase crystalline material of small initial particle size (0.5 micron) sintered to a high temperature in high vacuum:
(1) In which discontinuous grain growth has occurred.
(2) In which the firing time is much longer than that required for discontinuous grain growth.
(3) In which discontinuous grain growth has been inhibited.

11.7. Describe how you would experimentally determine the fractional porosity, fractional glass content, and fractional crystal content in a steatite porcelain containing three phases (pores, glass, and $MgSiO_3$ crystals).

11.8. From a lineal analysis, estimate the fraction of porosity present in the porcelain in Fig. 11.3. What is the true pore radius?

11.9. How would you determine the surface profile of a fired ceramic? What governs the surface profile at equilibrium?

11.10. On examining a polished section, what characteristics of the microstructure enable one to decide whether it is single-phase or polyphase?

11.12. You are placed in charge of the production control of a ferrite processing line which has been recently set up without quality control. (What you are producing is a soft ferrite for a transformer yoke at intermediate frequencies.) The process involves the use of copper oxide as an addition to the batch which facilitates sintering because of the formation of a liquid phase at elevated temperatures. Draw the microstructure that you would expect to result; then list several microstructural features which you expect to significantly affect magnetic properties, and indicate which properties are most strongly affected by each. For quality control, what microstructural measurements would you have set up. Give the relations between those measurements and the microstructural features of interest. Also indicate what magnetic and other characteristics you would have measured for quality control.

part IV

PROPERTIES OF CERAMICS

The aim of a ceramist in selecting or modifying a particular composition, fabrication method, firing process, or heat treatment is to obtain a product having certain useful properties. This requires a good understanding of material properties per se, which is of course a very broad subject. In this book we concentrate on the ceramic aspects of material properties—how properties can be usefully controlled or improved by the proper selection of composition, forming methods, firing techniques, and application. To do this, and at the same time avoid an excessively long book, we do not give quantitative derivations of different parameters used to describe properties. We do not attempt to present complete tabulations of material properties.

The properties to be considered and the methods of presentation are in large measure arbitrary; no attempt has been made to be exhaustive or completely consistent. In general, we consider the properties of crystals, the properties of glasses, and the properties of mixtures of these phases. The topics presented are based on either their general importance, the state of current development, or their particular significance in illustrating the ceramic parameters involved.

12

Thermal Properties

Physical properties that determine much of the usefulness of ceramic materials are those properties directly related to temperature changes. These properties are important for all ceramics no matter what their use; for applications such as thermal insulators or under conditions in which good thermal stress resistance is required, they are critical.

12.1 Introduction

The properties with which we are mainly concerned are the heat capacity (amount of thermal energy required to change the temperature level), coefficient of thermal expansion (fractional change in volume or linear dimension per degree of temperature change), and thermal conductivity (amount of heat conducted through the body per unit temperature gradient). Heat capacity and thermal conductivity determine the rate of temperature change in a ceramic during heat treatment in fabrication and in use. They are fundamental in fixing the resistance to thermal stresses (Chapter 16) and also determine operating temperatures and temperature gradients. A low thermal conductivity is essential for materials used as thermal insulators. Differential expansion of different constituents of a ceramic body or structure with temperature changes can lead to substantial stresses. Many of the most common difficulties occurring in development of ceramic compositions, development of suitable coatings, glazes, and enamels, and using ceramics in conjunction with other materials result from dimensional changes with temperature.

Heat Capacity. Heat capacity is a measure of the energy required to raise the temperature of a material; from another point of view it is the increase in energy content per degree of temperature rise. It is normally measured as the heat capacity at constant pressure c_p, but theoretical calculations are frequently reported in terms of the heat capacity at constant volume c_v:

$$c_p = \left(\frac{\partial Q}{\partial T}\right)_p = \left(\frac{\partial H}{\partial T}\right)_p \qquad \text{cal/mole °C} \tag{12.1}$$

$$c_v = \left(\frac{\partial Q}{\partial T}\right)_v = \left(\frac{\partial E}{\partial T}\right)_v \qquad \text{cal/mole °C} \tag{12.2}$$

$$c_p - c_v = \frac{\alpha^2 V_0 T}{\beta} \tag{12.3}$$

Q is the heat exchange, E the internal energy, H the enthalpy, $\alpha = dv/(v\,dT)$ the volume thermal expansion coefficient, $\beta = -dv/(v\,dp)$ the compressibility, and V_0 the molar volume. Frequently heat-capacity values are given as the specific heat capacity, calories per gram per degree centigrade. For condensed phases the difference between c_p and c_v is negligibly small for most applications but may become significant at elevated temperatures.

Thermal Expansion. The length and volume changes associated with temperature changes are important for many applications. At any particular temperature, we can define a coefficient of linear expansion:

$$a = \frac{dl}{l\,dT} \tag{12.4}$$

and a coefficient of volume expansion:

$$\alpha = \frac{dv}{v\,dT} \tag{12.5}$$

In general, these values are a function of temperature, but for limited temperature ranges an average value is sufficient. That is,

$$\bar{a} = \frac{\Delta l}{l\,\Delta T} \qquad \bar{\alpha} = \frac{\Delta v}{v\,\Delta T} \tag{12.6}$$

Heat Conduction. Frequently one of the main uses of a ceramic is as a thermal insulator or a thermal conductor. Its usefulness for these applications is largely fixed by the rate of heat transfer through it under a particular temperature gradient. The basic equation for thermal conductivity which serves to define this term is

$$\frac{dQ}{d\theta} = -kA\frac{dT}{dx} \tag{12.7}$$

where dQ is the amount of heat flowing normal to the area A in time $d\theta$. The heat flow is proportional to the temperature gradient, $-dT/dx$, the proportionality factor being a material constant, the thermal conductivity k.

Application of Eq. 12.7 under steady state conditions in which the heat flux, $q = dQ/d\theta$, and temperature at each point are independent of time requires an integration for the particular shape of interest. For heat flux through a flat slab this is

$$q = -kA\frac{T_2 - T_1}{x_2 - x_1} \tag{12.8}$$

For radial heat flow out through a cylinder of length l, inner diameter D_1 and outer diameter D_2,

$$q = -k(2\pi l)\frac{T_2 - T_1}{\ln D_2 - \ln D_1} \tag{12.9}$$

Similar relationships can be derived for many other simple shapes. Complex shapes generally require approximation methods.

If the temperature is not constant, its rate of change with time depends on the ratio of the thermal conductivity to the heat capacity per unit volume ρc_p. This ratio is called the thermal diffusivity $k/\rho c_p$ and has the

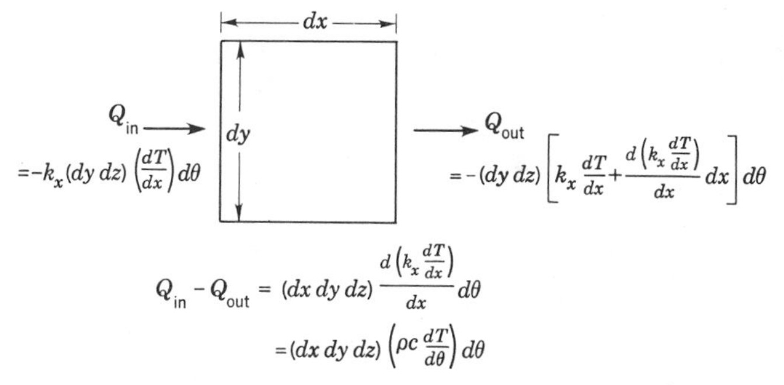

Fig. 12.1. Transient heat flow through a cube element $dx\,dy\,dz$.

same units as material diffusivity, square centimeters per second. In Fig. 12.1, the rate of change of temperature of a volume element $dx\,dy\,dz$ is derived as equal to

$$\frac{dT}{d\theta} = \frac{d\left(\frac{k}{\rho c_p}\frac{dT}{dx}\right)}{dx} \tag{12.10}$$

The similarity between Eqs. 12.10 and 12.7 and equivalent equations for the diffusion of material as discussed in Chapter 8 is apparent.

12.2 Heat Capacity

The energy required for raising the temperature of a material from its minimum energy state at the absolute zero goes into (1) vibrational energy by which atoms vibrate around their lattice positions with an amplitude and frequency that depend on temperature, (2) rotational energy for molecules in gases, liquids, and crystals having rotational degrees of freedom, (3) raising the energy level of electrons in the structure, and (4) changing atomic positions (such as forming Schottky or Frenkel defects, disordering phenomena, magnetic orientation, or altering the structure of glasses at the transformation range). All these changes correspond to an increase in internal energy and are accompanied by an increase in configurational entropy.

The classical kinetic theory of heat requires that each atom have an average kinetic energy of $\frac{1}{2}kT$ and an average potential energy of $\frac{1}{2}kT$ for each degree of freedom. Therefore the total energy for an atom with three degrees of freedom is $3kT$, the energy content per gram atom is $3NkT$, and

$$c_v = \left(\frac{dE}{dT}\right)_v = 3Nk = 5.96 \text{ cal/g-atom °C} \tag{12.11}$$

This is a good representation of the actual value observed at high temperatures. However, at low temperatures this must be multiplied by a function of the dimensionless number $h\nu/kT$, where h is Planck's constant and ν is the vibrational frequency. In the Debye theory of specific heat (reference 1), for a maximum frequency of lattice vibration ν_{max},

$$\frac{c_v}{3Nk} = f\left(\frac{h\nu_{max}/k}{T}\right) = f\left(\frac{\theta_D}{T}\right) \tag{12.12}$$

where $h\nu_{max}/k$ has the dimensions of temperature and is called the Debye temperature or characteristic temperature θ_D. At low temperatures the heat capacity is proportional to $(T/\theta_D)^3$, whereas at higher temperatures $f(\theta_D/T)$ approaches unity so that the heat capacity becomes independent of temperature, as indicated in Eq. 12.11.

The temperature at which the heat capacity becomes constant or only slightly varying with temperature depends on the bond strength, elastic constants, and melting point of the material and varies widely for different materials. Some typical values and experimental heat capacity curves are illustrated in Fig. 12.2. As shown there, the characteristic temperature is of the order of one-fifth to one-half the melting point in degrees absolute. In practice, the characteristic temperature is usually determined from heat capacity data. More detailed and precise discussions of the heat

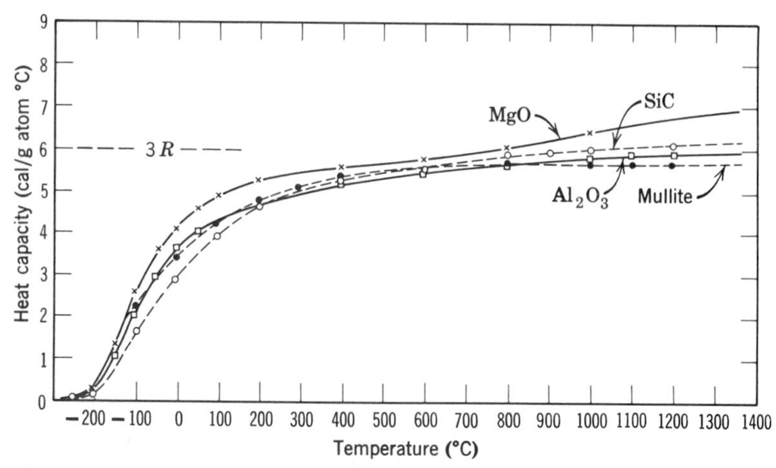

Fig. 12.2. Heat capacity of some ceramic materials at different temperatures.

capacities of materials and their relation to the vibrational spectra are given by de Launay (reference 2).

The main result of heat-capacity theory for ceramic systems is that the heat capacity increases from a low value at low temperature to a value near 5.96 cal/g atom °C at temperatures in the neighborhood of 1000°C for most oxides and carbides. Further increases in temperature do not strongly affect this value, and it is not much dependent on the crystal structure. This is illustrated in Fig. 12.3, in which data for crystalline quartz (SiO_2), CaO, and $CaSiO_3$ are presented. An abrupt change in the heat capacity occurs at the α-β quartz transition. In general, the value approached near 1000°C can be estimated from Eq. 12.11 and the molar composition. Exact values and the temperature dependence must be determined experimentally but are not structure-sensitive with regard to the crystal structure or ceramic microstructure.

As illustrated in Fig. 12.2, the heat capacity continues increasing at a modest rate at temperature above the characteristic termperature θ_D. The constant value indicated in Eq. 12.11 corresponds to the vibrational contribution to the heat capacity, which is the major factor at low temperatures. At higher temperatures the heat capacity at constant pressure also increases more rapidly and deviates to a greater extent from the constant-volume values. Development of Frenkel and Schottky defects, magnetic disorder, electronic energy contributions, and so on, contribute to the increased value of heat capacity at higher temperature. The total value of this contribution depends on the particular structure and energy increase for the higher-energy form formed at higher tempera-

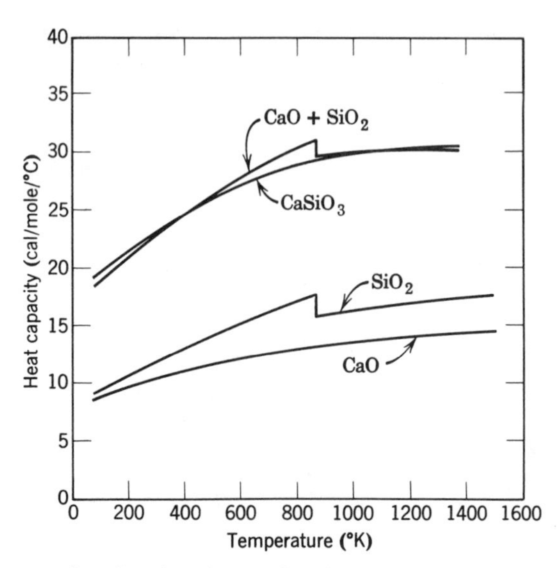

Fig. 12.3. Heat capacity of various forms of $CaO + SiO_2$ in 1 : 1 molar ratio.

tures. In general, the heat capacity at constant pressure can be adequately represented within experimental precision of measurement at higher temperatures as linearly increasing with temperature.

Values of the heat capacity increase particularly rapidly during the course of a cooperative process such as the order-disorder transformation discussed in Chapter 4, in which most of the transformation from the ordered to the disordered structure occurs rapidly over a limited temperature range. The heat capacity corresponding to an order-disorder transformation is shown in Fig. 12.4. Similar changes in heat capacity occur at magnetic and ferroelectric transformations. The cooperative nature of these transformations is discussed in more detail in Chapter 18. They are, in principle, similar to the order-disorder transformations which have already been discussed.

Whereas the heat content changes discontinuously at a polymorphic transformation, the change in heat capacity, although also discontinuous, is usually not large (see Fig. 12.3).

Although it is true that the molar heat capacity for crystalline materials is not structure-sensitive, the value reported as volume heat capacity does depend on the porosity, since the mass of material in a unit volume is decreased in proportion to the pore spaces present. Consequently, the heat energy required to raise the temperature of an insulating firebrick is much lower than that required to raise the temperature of a dense

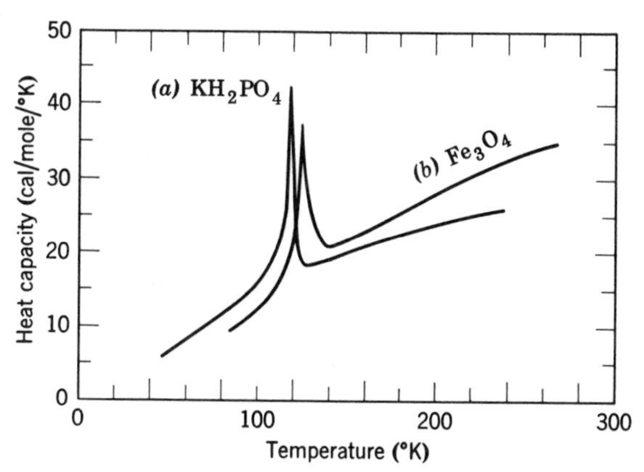

Fig. 12.4. Heat capacity at order–disorder transformations of (*a*) hydrogen bonds in KH_2PO_4 and (*b*) Fe^{3+} ions in Fe_3O_4.

firebrick. This is one of the valuable and useful properties of insulating materials for the manufacture of furnaces which must be periodically heated and cooled. Similarly, for laboratory furnaces that must be rapidly heated or cooled, radiation shielding of molybdenum sheet or low-density fiber or powder insulation has a low solid content and consequently a low heat capacity per unit volume; this allows rapid heating and cooling to be achieved.

The heat capacities of most oxide glasses approach 0.7 to 0.95 of the $3R$ value at the low-temperature end of the glass transition. On passing through the glass transition to the liquid state, the heat capacity generally increases by a factor of 1.3 to 3 (Fig. 12.5). The increased heat capacity reflects the increase in configurational entropy which becomes possible in the liquid state, in which the time for molecular rearrangement is short with respect to the experimental time scale.

12.3 Density and Thermal Expansion of Crystals

Density and Crystal Structure. The volume of crystalline materials and their volume changes with temperature are closely related to the crystal structures discussed in Chapter 2. The density is directly determined by the crystal structure, that is, the efficiency of atomic packing. Oxides such as those discussed in Section 2.6 can be considered as having a basic structure of oxygen ions with interstices filled or partly filled by cations. Structures with close-packed oxygen ions have a high density of atomic packing and high values of atoms per cubic centimeter. The density, as

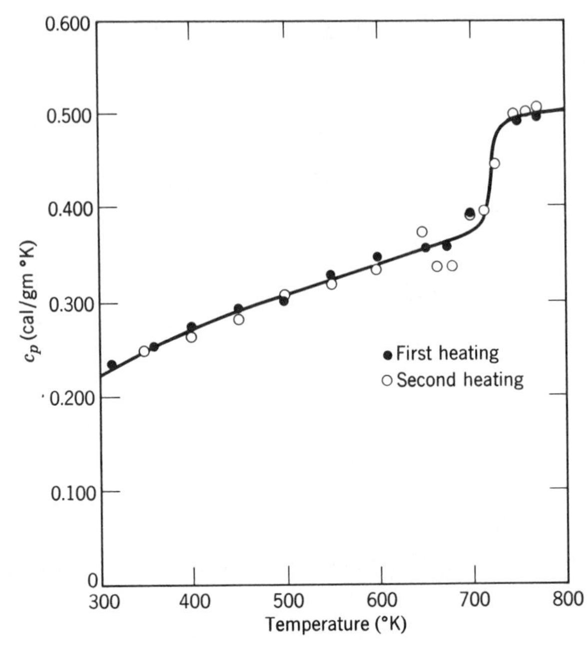

Fig. 12.5. Variation of heat capacity with temperature for $0.15Na_2O–0.85B_2O_3$ glass. From D. R. Uhlmann, A. G. Kolbeck, and D. L. de Witte, *J. Non-Cryst. Solids*, **5**, 426 (1971).

usually measured (grams per cubic centimeter), depends on this value and on the atomic weight of the constituents. Structures with simple cubic packing of the oxygen atoms, such as UO_2, have relatively low density of atoms per cubic centimeter; with high-atomic-weight cations, such as uranium, the measured specific gravity is high.

In contrast to structures based on close packing of oxygen ions, many silicates, such as those discussed in Section 2.7, have low densities of atomic packing. Aluminum and silicon have similar atomic weights, but the density of Al_2O_3 (hexagonal close packing of oxygen ions) is 3.96 g/cm^3 compared with a value of 2.65 for quartz, the common form of silica. This low density for silicates results from the network structures required by the low coordination number and high valence of the silicon ion. Higher-temperature forms of silica have an even lower density (cristobalite 2.32, tridymite 2.26), corresponding to more open structures, as discussed in Section 2.10. As illustrated in Fig. 12.6, this is generally the case; that is, high-temperature polymorphic forms have a higher specific volume (lower density) than the low-temperature forms. This corresponds to a discontinuous increase in volume at the transformation temperature.

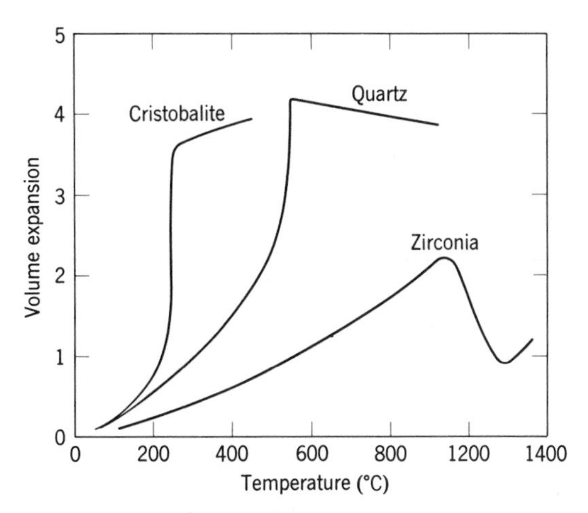

Fig. 12.6. Volume changes at polymorphic transformations.

The direct relationship between density and crystal structure is clear. This, of course, includes crystal imperfections such as those discussed in Chapter 4. These become important in some systems at high temperatures.

Thermal Expansion. The specific volume of any given crystal increases with temperature, and the crystal tends to become more symmetrical. The general increase in volume with temperature is mainly determined by the increased amplitude of atomic vibrations about a mean position. The repulsion term between atoms changes more rapidly with atomic separation than does the attraction term. Consequently, the minimum-energy trough is nonsymmetrical (Fig. 12.7); as the lattice energy increases, the increased amplitude of vibration between equivalent energy positions leads to a higher value for the atomic separation corresponding to a lattice expansion. Thermodynamically, the structure energy increases but the entropy decreases.

The change in volume due to lattice vibrations is closely related to the increase in energy content. Consequently, changes in the thermal expansion coefficient, $\alpha = dv/v\,dT$, with temperature are parallel to changes in heat capacity (Fig. 12.8). The thermal expansion coefficient increases rapidly at low temperature and reaches a nearly constant value above the Debye characteristic temperature θ_D. Normally there is a continued increase observed above this temperature, resulting from the formation of Frenkel or Schottky defects. The concentration of these, as discussed in Chapter 4, can be directly translated into expansion behavior. Some typical expansion coefficient curves are illustrated in Fig. 12.9.

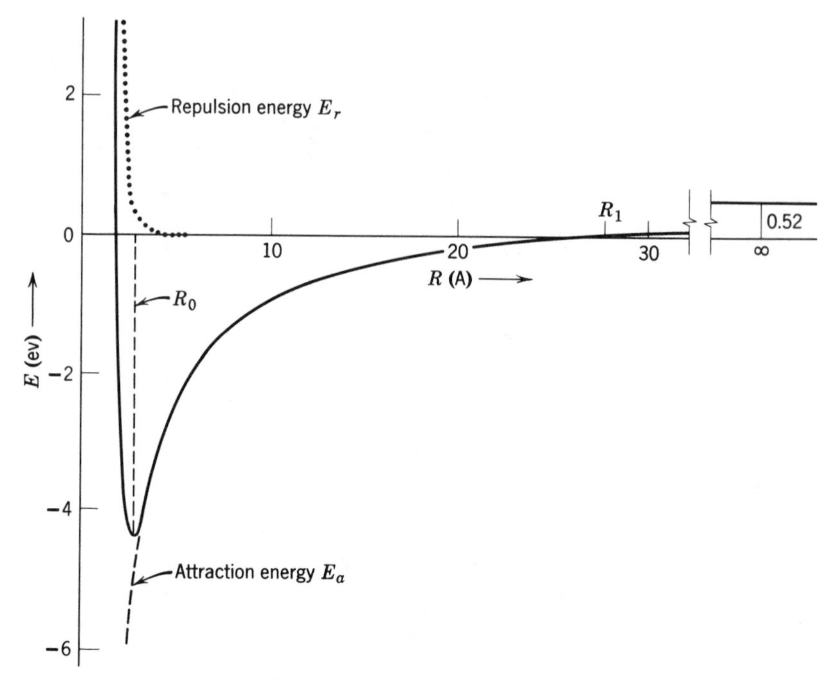

Fig. 12.7. Lattice energy as a function of atomic separation.

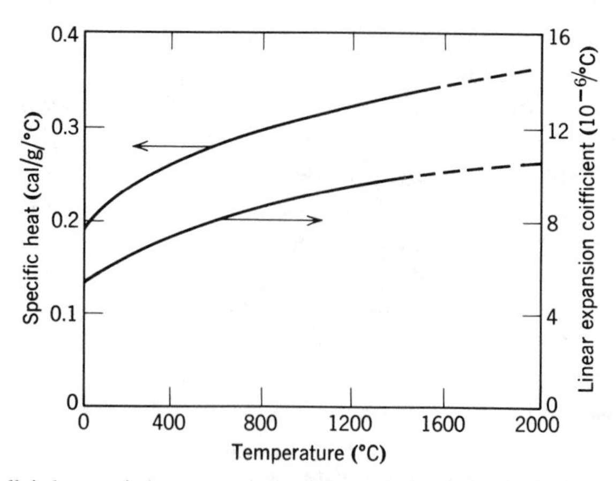

Fig. 12.8. Parallel changes in heat capacity and thermal expansion coefficient of Al_2O_3 over a wide temperature range.

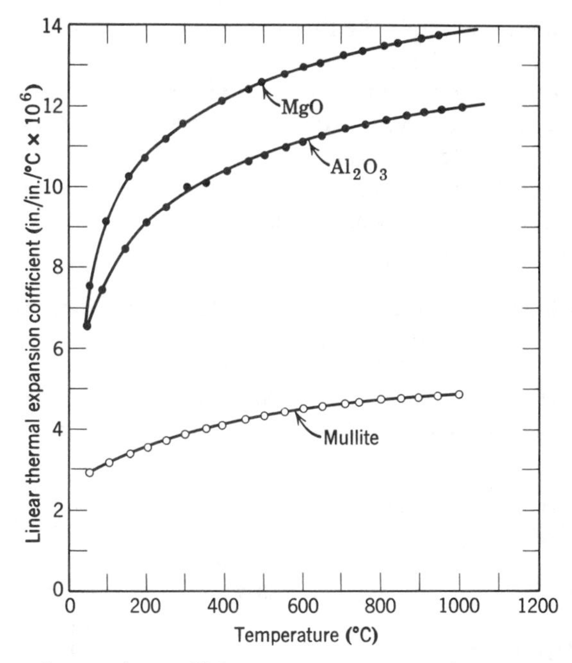

Fig. 12.9. Thermal expansion coefficient versus temperature for some ceramic oxides.

An important practical consequence of the temperature dependence of the expansion coefficient is that for many oxides it is erroneous to use data from room-temperature expansion coefficients, the ones most frequently tabulated, for application over a wide temperature range or to a different temperature range. This error is one commonly encountered in the ceramic literature.

For cubic crystals the expansion coefficients along different crystalline axes are equal, and the changes in dimensions with temperature are symmetrical. Consequently, the linear expansion coefficient $\bar{a}$ is the same measured in any direction. For isotropic materials the average volume expansion coefficient $\bar{\alpha}$ over a limited temperature range is related to the linear expansion coefficient by

$$1 + \bar{\alpha}\,\Delta T = 1 + 3\bar{a}\,\Delta T + 3\bar{a}^2\,\Delta T^2 + \bar{a}^3\,\Delta T^3$$

$$\bar{\alpha} = 3\bar{a} + 3\bar{a}^2\,\Delta T + \bar{a}^3\,\Delta T^2 \qquad (12.13)$$

For most cases, since $\bar{a}$ is small, to a good approximation for a limited temperature range,

$$\bar{\alpha} = 3\bar{a} \qquad (12.14)$$

For nonisometric crystals the thermal expansion varies along different crystallographic axes. The variation is such that it almost always results in a more symmetrical crystal at higher temperatures, for the same reasons as those discussed in Section 2.10. That is, in tetragonal crystals the c/a ratio decreases as the temperature is increased. At the same time the ratio of the expansion coefficients a_c/a_a tends to decrease as the temperature is raised. Data for some anisometric materials are given in Table 12.1.

Table 12.1. Thermal Expansion Coefficients for Some Anisometric Crystals ($\bar{a} \times 10^6/°C$)

Crystal	Normal to c-Axis	Parallel to c-Axis
Al_2O_3	8.3	9.0
Al_2TiO_5	−2.6	+11.5
$3Al_2O_3 \cdot 2SiO_2$	4.5	5.7
TiO_2	6.8	8.3
$ZrSiO_4$	3.7	6.2
$CaCO_3$	−6	25
SiO_2 (quartz)	14	9
$NaAlSi_3O_8$ (albite)	4	13
C (graphite)	1	27

Perhaps the most striking examples of anisometric expansion are related to layer crystalline structures such as graphite, in which the bonding is strongly directional and expansion is much lower in the plane of the layer than normal to it. For strongly anisometric crystals the expansion coefficient in one direction may be negative, and the resulting volume expansion may be very low. Materials such as this are useful for thermal-shock applications. Extreme examples are aluminum titanate, cordierite, and various lithium aluminum silicates. In the highly interesting case of β-eucryptite, the overall volume expansion coefficient is negative. In these materials the small or negative volume expansion is related to highly anisotropic structures. Consequently in polycrystalline bodies the grain boundaries are under such high stresses that the materials are inherently weak, as discussed in Sections 5.5 and 12.4.

The absolute value of the expansion coefficient is closely related to the crystal structure and bond strength. Materials with high bond strength, such as tungsten, diamond, and silicon carbide, have low coefficients of thermal expansion. However, these are materials with high characteristic temperatures so that comparisons of expansion coefficients with room temperature values are not completely satisfactory. It is preferable to compare materials at their characteristic temperature when discussing structural effects.

Typical values for oxide structures with dense packing of oxygen ions are in the range of 6 to 8×10^{-6} in./in. °C at room temperature (linear coefficient) and increase to 10 to 15×10^{-6} at temperatures near the characteristic temperature. A number of silicates have much lower values than this, related to the ability of these open structures to absorb vibrational energy by transverse modes of vibration and by adjustment of bond angles. Some typical values for expansion coefficients are collected in Table 12.2.

Table 12.2 Mean Thermal Expansion Coefficients for a Number of Materials

Material	Linear Expansion Coefficient, 0–1000°C (in./in. °C × 10^6)	Material	Linear Expansion Coefficient, 0–1000°C (in./in. °C × 10^6)
Al_2O_3	8.8	ZrO_2 (stabilized)	10.0
BeO	9.0	Fused silica glass	0.5
MgO	13.5	Soda-lime-silica glass	9.0
Mullite	5.3	TiC	7.4
Spinel	7.6	Porcelain	6.0
ThO_2	9.2	Fire-clay refractory	5.5
UO_2	10.0	Y_2O_3	9.3
Zircon	4.2	TiC cermet	9.0
SiC	4.7	B_4C	4.5

12.4 Density and Thermal Expansion of Glasses

As discussed in Chapter 3 and particularly in Section 3.3, the volume of a glass is largely determined by the nature of the vitreous network. The density is a minimum value for the pure network former and increases as modifier ions, which increase the number of atoms present without changing the network much, are added. The effect of introducing the additional ions into the structure thus outweighs the influence of the modifiers in loosening the network. Because of the structural considerations discussed in Chapter 3, glasses usually but not always have a lower density than corresponding crystalline compositions. The variations of density with composition in binary oxide glass systems are reviewed by

Shaw and Uhlmann.* A typical example of this variation for a silicate glass system is shown in Fig. 12.10. Data of similar form are obtained for other simple silicate systems.

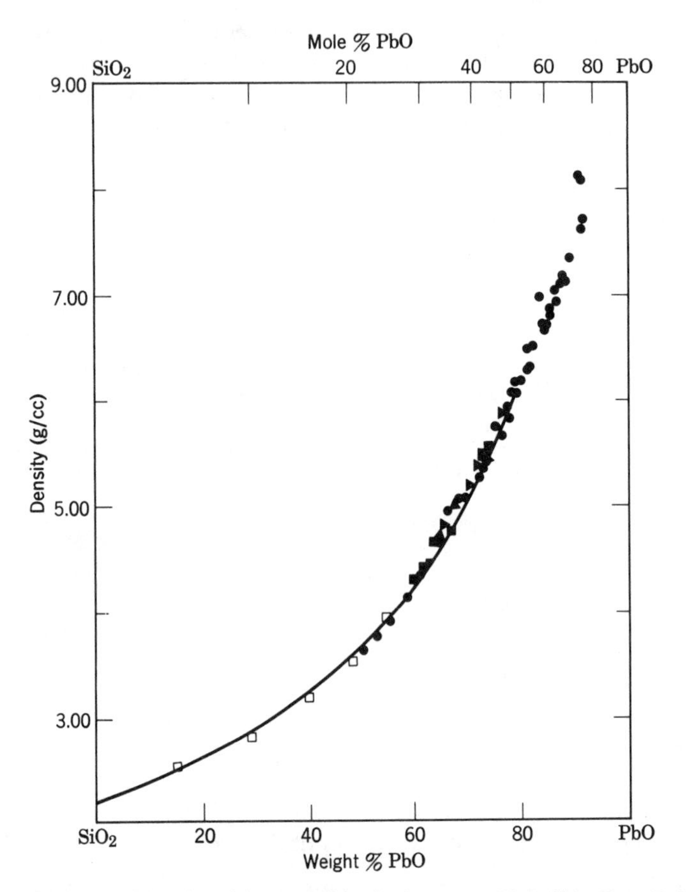

Fig. 12.10. Variation of density with composition in the system $PbO-SiO_2$. From R. R. Shaw and D. R. Uhlmann, *J. Non-Cryst. Solids*, **1**, 474 (1969).

Effect of Heat Treatment. In addition to composition, the room-temperature density of glasses depends on prior heat treatment (Fig. 3.1(*b*)); a rapidly cooled glass has a higher specific volume than the same composition cooled more slowly. The transformation from a supercooled liquid, in which structural rearrangement occurs with temperature

J. Non-Cryst. Solids, **1**, 474 (1969).

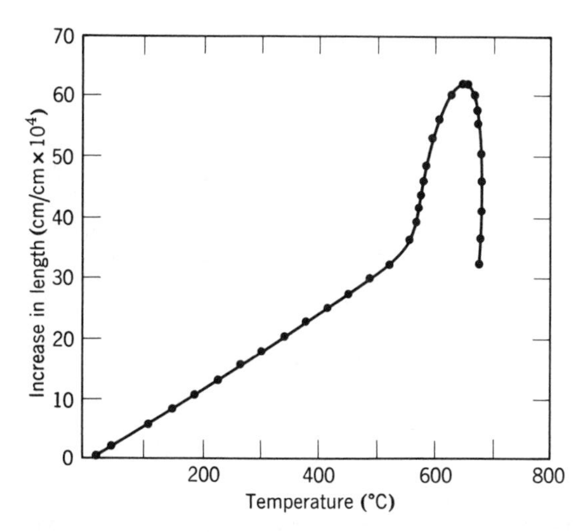

Fig. 12.11. Typical data for thermal expansion of a glass.

changes, to the glassy solid, in which the structure is fixed and independent of temperature, is clearly illustrated in the normal measurement of thermal expansion coefficient which gives a result such as that shown in Fig. 12.11. A sudden increase in the expansion coefficient occurs at a temperature of 500 to 600°C for commercial silicate glasses; this is sometimes referred to as the transformation temperature but is better referred to as a transformation *range*, since its value depends on the rate of heating and prior treatment of the glass. (The decrease in length at about 700° corresponds to viscous flow of the sample under the stresses imposed by the measuring device.)

The dependence of expansion behavior on prior heat treatment is illustrated in Fig. 12.12. Curve *a* corresponds to the length observed after the sample is held at temperature long enough to come to equilibrium. The time required is longer at the lower temperatures. The upper curve *b* corresponds to a sample of the same composition which has been rapidly cooled. With the rate of heating used, the same expansion coefficient (slope of the curve) is found up to about 400°; above this temperature the glass contracts until at about 560° it reaches the equilibrium structure and again expands. Curve *c* shows a sample which has been annealed for a long time at 500°. On reheating, the expansion coefficient remains constant even above this temperature, corresponding to a time lag required for structural equilibrium to be established. In curve *d* an intermediate result is found for a sample with a slower rate of cooling. These changes correspond to variations in the expansion coefficient (Fig. 12.13). All samples approach a constant value, corresponding at low

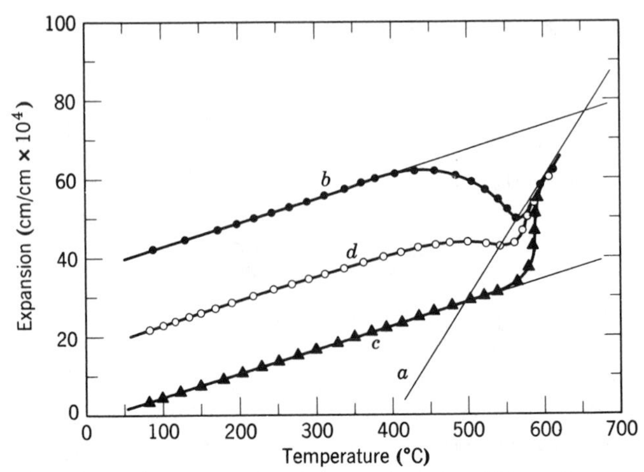

Fig. 12.12. Linear dimension changes for the same glass after different heat treatments. (*a*) Ideally slow heating to equilibrate at each temperature above 400°; (*b*) rapidly cooled; (*c*) long anneal at 500°; (*d*) slowly cooled. Samples *b*, *c*, and *d* were measured at heating rate of 10°C/min.

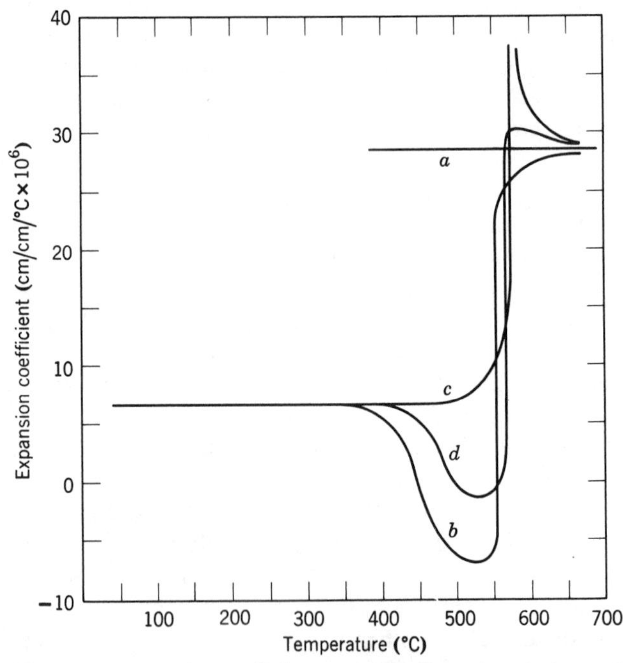

Fig. 12.13. Variation in expansion coefficient corresponding to length changes illustrated in Fig. 12.12.

temperature to the expansion coefficient of the glass and at high temperature to the supercooled liquid. Behavior in the transition range is variable, depending on the heat treatment.

In many glasses containing B_2O_3, striking variations have been reported in some of the property versus composition relations; the classic example of this is the pronounced minimum in the thermal expansion coefficient of glasses in the Na_2O–B_2O_3 system, originally reported at about 15 g-atom% Na_2O. This has been referred to as the boric oxide anomaly and was originally attributed to the conversion of BO_3 triangles to BO_4 tetrahedra for alkali concentrations of less than about 15% and the formation of singly bonded oxygens for larger alkali concentrations. Subsequent NMR results (see Fig. 3.9) have indicated, however, that the fraction of boron atoms in tetrahedral coordination continues to increase with increasing alkali oxide content up to about 30 to 35 g-atom% alkali oxide. Although several workers have reported minima in the thermal expansion versus composition relation for Na_2O–B_2O_3 glasses, they often disagree on the location of the minima (Fig. 12.14). Much of the variation seen in Fig. 12.14 is likely associated with variations in experimental procedure, sample purity, composition intervals selected, and the temperature range of the measurements.

This question has been clarified by an investigation of the thermal-expansion behavior of glasses in all five alkali borate systems, carried out over a common temperature interval (−196° to +25°C) at narrow (1 or 2 g-atom%) composition intervals. The results, shown in Fig. 12.15, indicate the absence of any sharp minima in the thermal expansion versus composition relations. Rather, these relations are characterized by broad, spread-out minima, with no single composition uniquely identifiable with a minimum. The pronounced increase in thermal expansion occurs in the range of composition, around 30 g-atom% alkali oxide, at which the NMR results indicate a cessation in the process of each added oxygen converting two boron atoms from BO_3 to BO_4 configurations. Beyond this range, singly bonded oxygens are presumably formed in appreciable concentrations, resulting in a decrease in coherence of the network and an increase in expansion coefficient. The broad minimum shown in Fig. 12.15 reflects a competition between two processes: the formation of BO_4 tetrahedra, tending to decrease the expansion coefficient, and the introduction of modifying cations, tending to increase it. The larger the size of the cation, the larger its effect on the expansion coefficient.

Fictive Temperature. The changes that occur in density and other properties after cooling and during annealing in the region of the glass transition are dependent on time in a way that depends on the thermal history of the specimen as well as on the temperature (Fig. 12.16). One

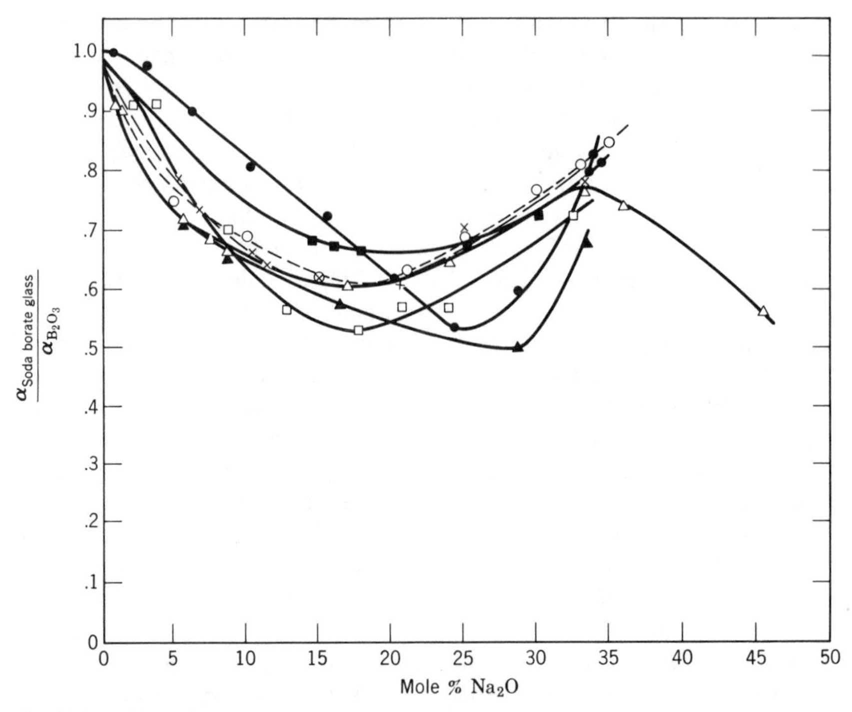

Fig. 12.14. Normalized thermal expansion coefficients of sodium borate glasses reported by different investigators. From R. R. Shaw and D. R. Uhlmann, *J. Non-Cryst. Solids*, **1**, 347 (1969).

widely used approach for discussing these structural changes is that advanced by Tool,* who introduced the notion of a fictive temperature as that temperature at which the glass structure would be in equilibrium if brought infinitely rapidly to that temperature. This approach is perhaps best illustrated by comparing the fictive temperatures plotted in Fig. 12.16*b* with the samples illustrated in Fig. 12.12. On this basis, the total linear expansion can be represented as the sum of two terms, one related to changes in actual temperature with a fixed structure and another caused by changes in structure at a fixed temperature. If a_1 is the low-temperature value in Fig. 12.13 and a_2 is the high-temperature value,

$$\frac{dl}{l} = a_1\,dT + a_2\,d\tau \tag{12.15}$$

J. Am. Ceram. Soc.*, **29, 240 (1946).

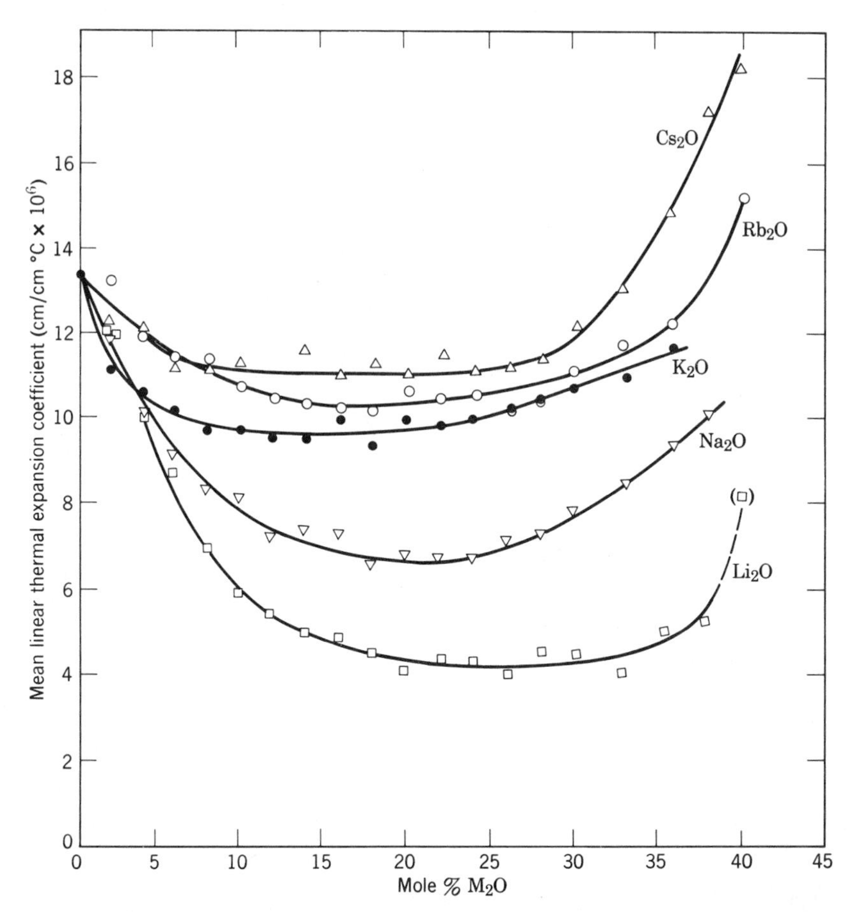

Fig. 12.15. Thermal expansion coefficients of alkali borate glasses as a function of composition. From R. R. Shaw and D. R. Uhlmann, *J. Non-Cryst. Solids*, **1**, 347 (1969).

The rate of change of the fictive temperature is the product of a driving force corresponding to the free-energy difference between the structures of the actual material and that of the equilibrium or fictive temperature and a term representing the barrier to atomic rearrangements. The driving force is proportional to $(T - \tau)$. The dependence of the activation energy for ion movement $(\Delta G^{\dagger})$ on structure is less clear, but Tool empirically found that his results could be represented with this term proportional to $\exp(T/A) \exp(\tau/B)$, so that

$$\frac{d\tau}{d\theta} = K(T - \tau) \exp \frac{T}{A} \exp \frac{\tau}{B} \tag{12.16}$$

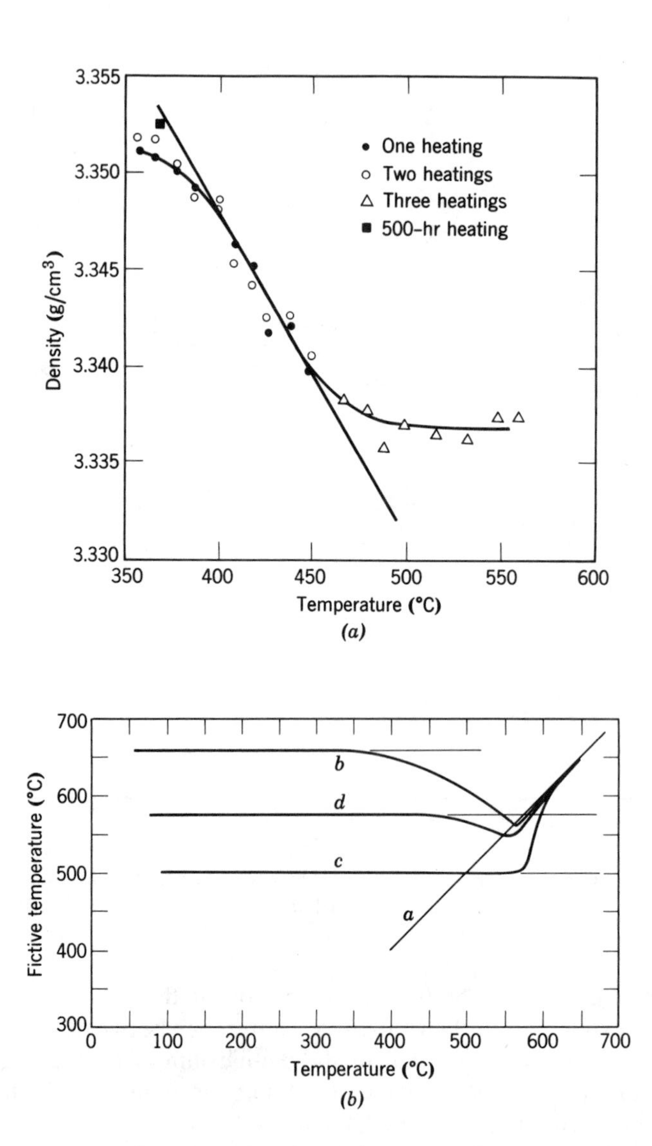

Fig. 12.16. (*a*) Density variation at room temperature of medium flint glass rapidly cooled after long heat treatment at temperatures shown. From A. Q. Tool and E. E. Hill, *J. Soc. Glass Technol.*, **9**, 185 (1925). (*b*) Change in fictive temperature of samples illustrated in Figs. 12.12 and 12.13.

This corresponds to a viscous resistance to relaxation η which is proportional to $\exp(-T/A)\exp(-\tau/B)$. For the equilibrium liquid, $T=\tau$, and for a 1° change in the temperature of the equilibrium liquid, η^{-1} changes by a factor $\exp(1/A+1/B)$. The part $\exp(1/A)$ represents the change due to temperature alone, and the part $\exp(1/B)$ that due to the change in internal state of the liquid. Hence the relative values of $1/A$ and $1/B$ indicate the relative importance of temperature and structure in the same units.

The data of Tool on a borosilicate crown glass were describable by Eq. 12.16 with $1/A = 0.050$ and $1/B = 0.023$ and suggest that temperature is about twice as important as τ. In contrast, data on a different borosilicate glass obtained by Collyer* indicate nearly equal importance of the two parameters, and subsequent work by Ritland† suggests that Eq. 12.16 be replaced by

$$\frac{d\tau}{d\theta} = K_0[|\tau - T| + K_1|\tau - T|^2] \exp\frac{T}{A} \exp\frac{\tau}{B}. \qquad (12.17)$$

The addition of the quadratic term in the brackets complicates the linear dependence on the driving force suggested by Tool. The results can be interpreted in terms of a spectrum rather than a single relaxation time. For more detailed discussion of these points see reference 3.

12.5 Thermal Expansion of Composite Bodies

When a polycrystalline body, a mixture of crystalline phases, or a mixture of crystals and glasses is heated at the firing temperature, a dense coherent structure results from viscous flow, diffusion, or solution and precipitation. If the expansion coefficients in the different crystalline directions are not the same or if the various phases present have different coefficients of thermal expansion, different grains present have different amounts of contraction on cooling and cracks open up between grains if stressfree contraction occurs. In practice, however, each grain is restrained by the surrounding grains so that, instead of grain separation, microstresses are developed which are proportional to the difference between the stressfree contraction and the actual contraction.

Expansion Coefficient of Composite. The resulting expansion coefficient for the composite can be calculated if we assume that no cracks develop, that the contraction of each grain is the same as the overall

J. Am. Ceram. Soc., **30**, 338 (1947).
†*J. Am. Ceram. Soc.*, **37**, 370 (1954).

contraction, and that all microstresses are pure hydrostatic tension and compression (interfacial shear is negligible). Then the stresses on each particle are given by

$$\sigma_i = K(\alpha_r - \alpha_i)\,\Delta T \tag{12.18}$$

where α_r and α_i are the average volume expansion coefficient and the volume expansion coefficient for particle i, ΔT is the temperature change from the stressfree state, and K is the bulk modulus. ($K = -P/(\Delta V/V) = E/3(1-2\mu)$, where P is the isotropic pressure, V the volume, E the elastic modulus, and μ Poisson's ratio.) If the stresses are nowhere large enough to disrupt the structure, the summation of stresses over an area or volume is zero. Consequently, if V_1 and V_2 are the fractional volumes for a mixture of materials,

$$K_1(\alpha_r - \alpha_1)V_1\,\Delta T + K_2(\alpha_r - \alpha_2)V_2\,\Delta T + \cdots = 0 \tag{12.19}$$

and

$$V_1 + V_2 + \cdots = V_r \tag{12.20}$$

$$V_i = \frac{F_i\rho_r V_r}{\rho_i} \tag{12.21}$$

where F_i is the weight fraction of phase i, and ρ_r and ρ_i are the mean and individual phase densities. Substituting in Eq. 12.19 and eliminating ΔT, ρ_r, and V_r gives an expression for the coefficient of volume expansion of the aggregate which was originally obtained by Turner*

$$\alpha_r = \frac{\alpha_1 K_1 F_1/\rho_1 + \alpha_2 K_2 F_2/\rho_2 + \cdots}{K_1 F_1/\rho_1 + K_2 F_2/\rho_2 + \cdots}. \tag{12.22}$$

An alternative model for the expansion behavior of composite materials takes into account shear effects at the boundaries between the grains or phases. It is here assumed that the overall dilation of a composite body is $\alpha_r\,\Delta T$, where α_r is the overall expansion coefficient and ΔT is the temperature difference between the initial unstressed state and the final stressed state of the body. Analyzing the displacement of the individual components and applying continuity relations at the interfaces, the overall expansion coefficient can be expressed by the relation first obtained by Kerner†

$$\alpha_r = \alpha_1 + V_2(\alpha_2 - \alpha_1)\frac{K_1(3K_2 + 4G_1)^2 + (K_2 - K_1)(16G_1^2 + 12G_1K_2)}{(4G_1 + 3K_2)[4V_2G_1(K_2 - K_1) + 3K_1K_2 + 4G_1K_1]} \tag{12.23}$$

where G_i is the shear modulus of phase i.

*J. Res. NBS, **37**, 239 (1946).

†Proc. Phys. Soc. (Lond.), **B69**, 808 (1956).

For comparison, the predictions of Eqs. 12.22 and 12.23 are plotted in Fig. 12.17 for a two-phase composite body whose end-member properties are

$$\alpha_1 = 12 \times 10^{-6}/°C$$
$$K_1 = 1.5 \times 10^{11}\ \text{dyne/cm}^2$$
$$G_1 = 0.8 \times 10^{11}\ \text{dyne/cm}^2$$
$$\rho_1 = 1.86\ \text{g/cm}^3$$
$$\alpha_2 = 4.5 \times 10^{-6}/°C$$
$$K_2 = 3.6 \times 10''\ \text{dyne/cm}^2$$
$$G_2 = 2 \times 10^{11}\ \text{dyne/cm}^2$$
$$\rho_2 = 2.09\ \text{g/cm}^3$$

These values are typical of glasses in the Li_2O–B_2O_3 system for compositions containing less than 20 g-atom% Li_2O, which have been shown to exhibit phase separation (see Table 3.6). Figure 12.17 shows that the predictions of the Turner relation (Eq. 12.22) lie below those of the Kerner expression (Eq. 12.23) by as much as 12% for this system. The two Kerner curves shown in the figure were obtained by reversing the role of matrix and inclusion; with decreasing difference between the end-member properties, the difference between the two Kerner curves decreases. Finally, it is seen that the curve predicted by the Kerner relation is

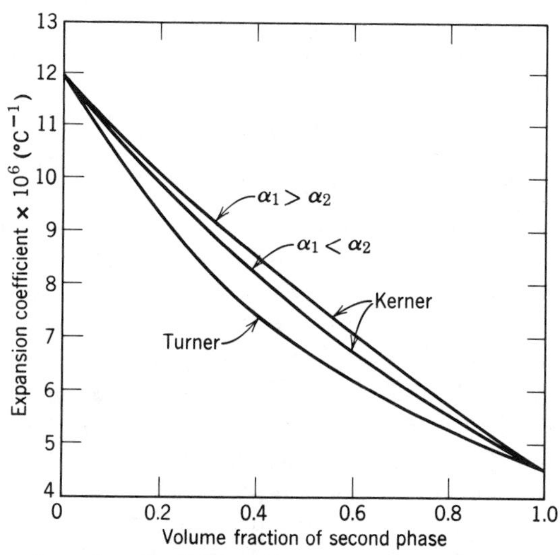

Fig. 12.17. Comparison of predicted thermal expansion coefficients of a two-phase material. From R. R. Shaw, Ph.D. thesis, MIT, 1967.

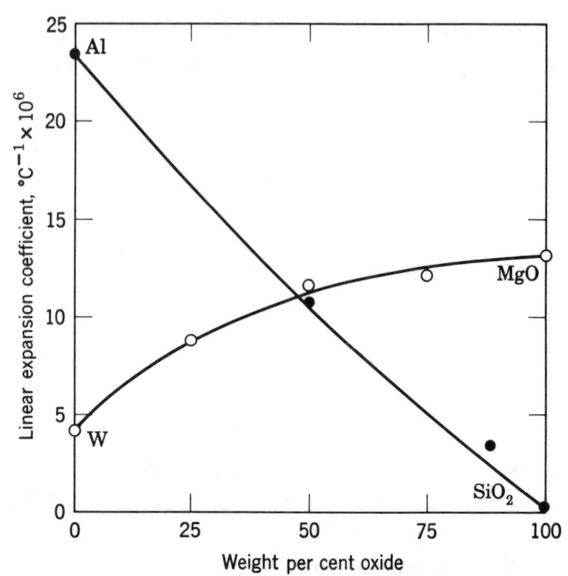

Fig. 12.18. Thermal expansion coefficients of end members and intermediate compositions in the systems MgO–W and Al–SiO_2 glass. The solid lines are calculated from Eq. 12.22. After W. D. Kingery, *J. Am. Ceram. Soc.*, **40**, 351 (1957).

considerably flatter in its dependence on volume fraction than is that predicted by the Turner model.

In a number of cases, the Turner relation has been found to provide a useful representation of experimental data. Examples of this are shown in Fig. 12.18 for two metal-matrix composites. In other cases, however, the experimental results lie closer to the Kerner predictions than to those of Eq. 12.22. In still other cases, the agreement with both models leaves much to be desired, although the experimental data generally fall between the predictions of the two models.

Effect of Polymorphic Transformations. Relations similar to Eqs. 12.22 and 12.23 can be derived for compositions in which one component in a composite body undergoes a polymorphic transformation, for example, porcelain containing quartz or cristobalite as one constituent. These constituents undergo sharp volume changes at polymorphic transformations (Fig. 12.6). When they are combined in a porcelain body, the expansion coefficient of the porcelain is increased at the transformation point. This is illustrated for two porcelain compositions in Fig. 12.19. In one the material present is mostly cristobalite. In another there is some cristobalite and also some quartz present. (These expansion curves should be compared with those given in Fig. 12.6.) An equivalent

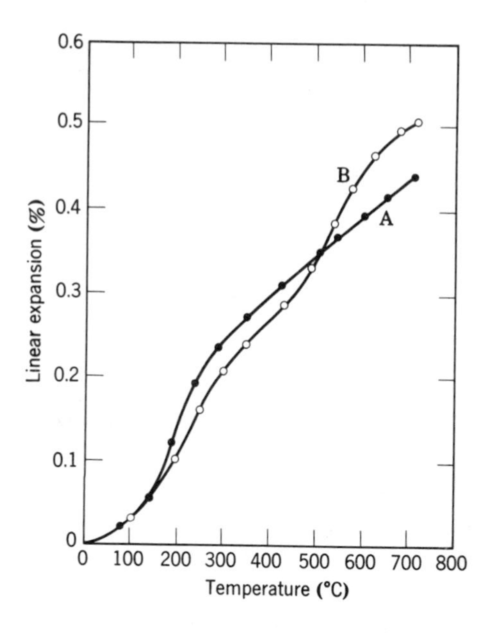

Fig. 12.19. Thermal expansion of two porcelain compositions. Body A contains cristobalite as the silica phase; body B contains both cristobalite and quartz.

expansion coefficient for use in Eqs. 12.22 and 12.23 can be derived by substituting $(\Delta V / V_0 \Delta T)$ in place of α for each phase for the temperature range concerned. This confirms that substantial stresses and changes in the overall expansion coefficient are to be expected at the transformation temperatures.

Microstresses. The stresses developed as a result of large differences between the expansion coefficients of two materials or in crystallographic directions in a composite are sufficient to cause microcracks to occur in the body. These fissures are of great importance for understanding many practical properties of real ceramics. We have already illustrated them for a number of systems. One important result of these microfissures is hysteresis in the thermal expansion coefficient measured for a polycrystalline aggregate or a composite body in which large stresses are developed. This occurs, for example, in some TiO_2 compositions (Fig. 12.20). Polycrystalline titania, when cooled from the firing temperature, develops microcracks, and as these form, the overall observed expansion coefficient is lower than the expansion coefficient of the individual crystals. On heating, these cracks tend to close, and at low temperatures abnormally low expansion coefficients are observed. This kind of expansion hysteresis occurs particularly in polycrystalline compositions when the expansion coefficient is markedly different in different crystalline directions. It also occurs for mixtures of materials having different expansion coefficients.

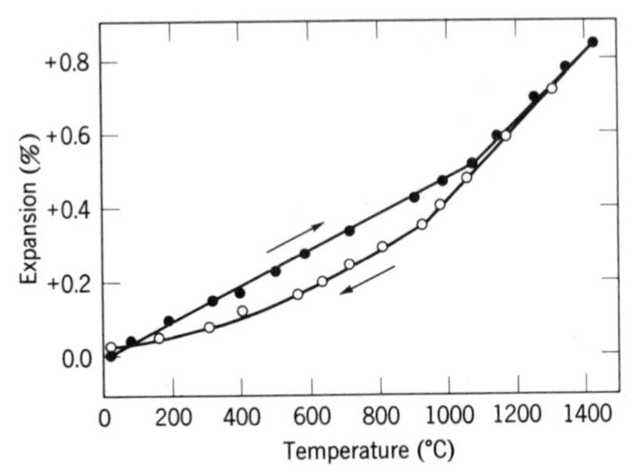

Fig. 12.20. Thermal expansion hysteresis of polycrystalline TiO_2 caused by presence of microfissures.

A particularly striking example of grain-boundary fracture and its resultant effect on measured thermal expansion coefficients is graphite. The expansion coefficient normal to the c axis is about 1×10^{-6}/°C; the expansion coefficient parallel to the c axis is about 27×10^{-6}/°C. Here the observed linear expansion coefficient for polycrystalline samples in the range is $1\text{–}3 \times 10^{-6}$/°C.

Although microstress fractures occur both within grains and along grain boundaries, they are most commonly observed at grain boundaries. As discussed in Chapter 5, the boundary stresses developed are independent of grain size (Eq. 5.53); but grain boundary cracking and thermal expansion hysteresis occur predominately in large-grain samples.

In the same way that microstresses can lead to microcracks and failure, microstresses are developed in bodies that are restrained from expansion by being held in a fixed support or by being attached to a material of different expansion coefficient. If a bar of material is completely restrained from expanding, by application of restraining forces due to the

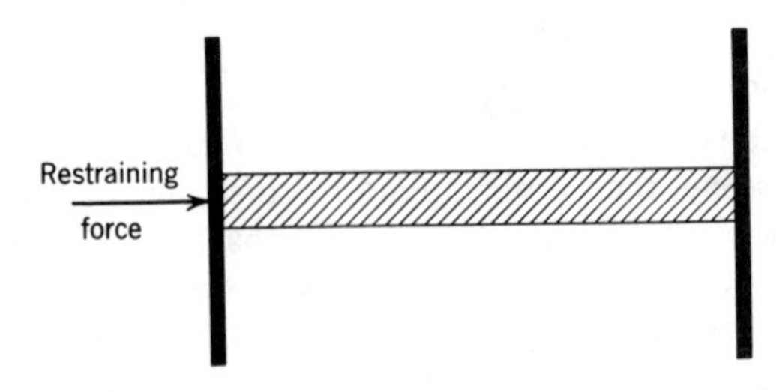

Fig. 12.21. Expansion restraint by fixed supports.

design of the part, as shown in Fig. 12.21, the stress is given by

$$\sigma = \frac{Ea\,\Delta T}{1-\mu} \tag{12.24}$$

where E is Young's modulus, ΔT is the temperature change, and μ is Poisson's ratio.

Glaze Stresses. Stresses are similarly caused by the difference between the expansion coefficient of a glaze or enamel and that of the underlying ceramic or metal. If stressfree at T_0, the stresses depend on the new temperature T', on the elastic properties of the material, and on the coefficients of expansion. For a thin glaze on an infinite slab the stresses are given by Eqs. 12.25 and 12.26 for the simple case in which the elastic properties of glaze and body are the same. This is usually a good approximation for glazes on ceramics.

$$\sigma_{gl} = E(T_0 - T')(a_{gl} - a_b)(1 - 3j + 6j^2) \tag{12.25}$$

$$\sigma_b = E(T_0 - T')(a_b - a_{gl})(j)(1 - 3j + 6j^2) \tag{12.26}$$

where, following the usual convention, a positive stress denotes tension and j is the ratio of glaze to body thickness. In usual practice, T_0 is taken as the setting point of the glaze, generally in its annealing range.

For a thin glaze on a cylindrical body, the corresponding expressions are

$$\sigma_{gl} = \frac{E}{1-\mu}(T_0 - T')(a_{gl} - a_b)\frac{A_b}{A} \tag{12.27}$$

$$\sigma_b = \frac{E}{1-\mu}(T_0 - T')(a_b - a_{gl})\frac{A_{gl}}{A} \tag{12.28}$$

where A, A_b, and A_{gl} are, respectively, the cross-sectional areas of the overall cylinder, body and glaze.

In order to obtain satisfactory fit between a glaze and body or porcelain-enamel and metal, it is desirable that after cooling to room temperature the glaze be in a condition of compression. This is necessary because if tensile stresses develop, the glaze tends to craze (Fig. 12.22). Here the tensile stress has developed to a point at which the tensile strength of the glaze has been exceeded. If the glaze is put under compressive stress, this type of failure does not occur, and it is only with substantial stresses that the reverse, *shivering*, or failure under compressive stresses takes place. Typical glaze and enamel stresses in cooled ware are about 10,000-psi compression (Fig. 12.23).

Even when compressive stress has been developed during cooling, delayed crazing failures may occur in service. Silicate bodies tend to

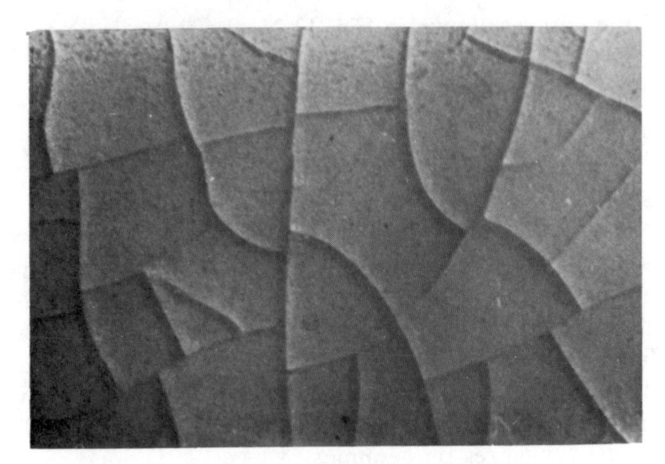

Fig. 12.22. Crazed glaze showing tensile cracks in glass.

increase in volume because of moisture absorption. The expansion of the body decreases the compression of the glaze and transforms the stress into tension, and after sufficient time the ware tends to craze. In order to prevent crazing, it is desirable to have substantial initial compressive stress. In addition, compositions should be adjusted so that moisture expansion is a minimum. This can be done with more vitreous compositions or by using alkalifree compositions. Steatite tile, for example, is much superior to feldspar-clay-flint tile in this regard.

The stresses obtained on cooling from the firing temperature, where the glaze or enamel is sufficiently fluid to relieve stresses, begin to increase when the transformation range is passed. The actual temperature at which stresses begin to increase depends on the rate of cooling, as indicated in Section 12.2. The change in stresses with temperature during cooling depends on details of the cooling curves (Fig. 12.23). However, the overall stress developed at room temperature depends only on the difference between the total expansion of the glaze and body between the stress point and room temperature and is independent of the changes in stress during the cooling process. This is illustrated for two bodies in Fig. 12.23, in which the change in stress with temperature is quite different for the two but the resulting stress is similar.

For many centuries, glazes have been used for decorative effects and to render ceramic bodies impermeable to liquids. In recent years, they have also been used to provide large surface compressions and hence to strengthen ceramic bodies. A noteworthy example of this has been discussed in Chapter 8; moderate-expansion glazes are used to strengthen

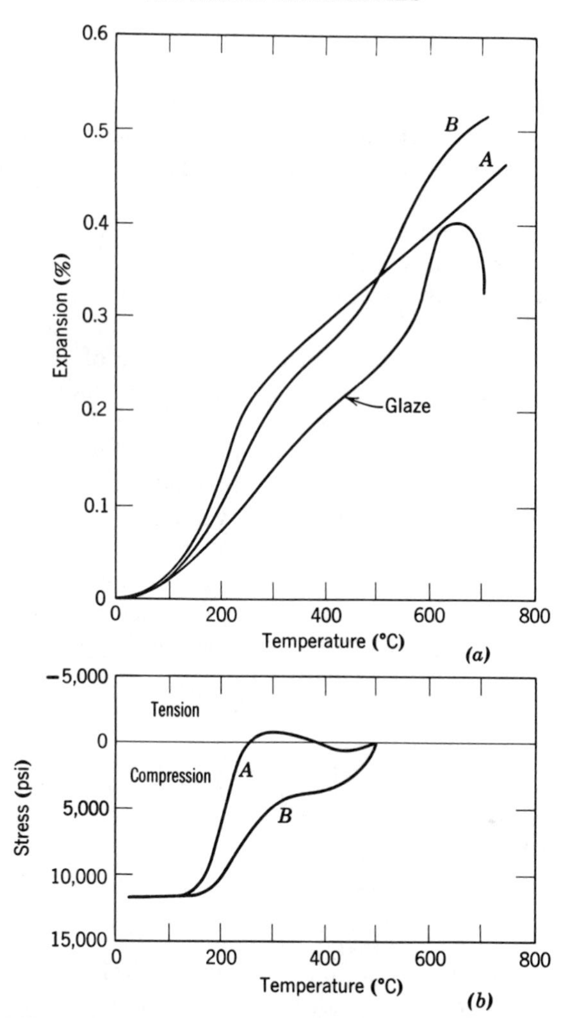

Fig. 12.23. (*a*) Expansion of glaze and porcelain bodies and (*b*) stresses developed on cooling.

high-expansion glass-ceramic bodies in the Na_2O–BaO–Al_2O_3–SiO_2 system. In this case, compressive stresses in the range of 25,000 psi and overall strengths in the range of 38,000 psi can be obtained.* Such strengths permit the tableware based on these materials to be guaranteed for 3 years against breakage.

*See D. A. Duke, J. E. Megles, J. F. MacDowell, and H. F. Bopp, *J. Am. Ceram. Soc.*, **51**, 98 (1968).

12.6 Thermal Conduction Processes

The conduction process for heat-energy transfer under the influence of a temperature gradient depends on the energy concentration present per unit volume, its velocity of movement, and its rate of dissipation with the surroundings. Each of these factors must be understood in order to predict the resulting thermal conductivity. In gases, for example, individual atoms or molecules exchange kinetic energy by collision; the heat energy present is simply equal to the heat capacity per unit volume, the velocity of molecular motion can be calculated from kinetic theory, and the rate of energy dissipation depends on the rate of collision between atoms or molecules. If we consider a temperature gradient in which the concentration of molecules is N and their average velocity is v, the average rate at which molecules pass a unit area in the x direction is equal to $1/3Nv$. If energy equilibrium is obtained by collisions between molecules and the average distance between collisions, the mean free path, is l, molecules moving parallel to the x axis have an energy $E_0 + l\partial E/\partial x$, where E_0 is the mean energy at $x = 0$, l is the mean free path, and $\partial E/\partial x$ is the energy gradient in the x direction. Combining these relations, the net energy flux in the x direction is given by

$$\frac{q}{A} = k\frac{\partial T}{\partial x} = \frac{1}{3}Nvl\frac{\partial E}{\partial x} \tag{12.29}$$

Since

$$N\left(\frac{\partial E}{\partial x}\right) = N\left(\frac{\partial E}{\partial T}\right)\left(\frac{\partial T}{\partial x}\right) = c\left(\frac{\partial T}{\partial x}\right) \tag{12.30}$$

the conductivity must be given by

$$k = \frac{1}{3}cvl \tag{12.31}$$

where c is the heat capacity per unit volume. This relationship is found to represent the behavior of an ideal gas satisfactorily and explain the behavior of different gases at different pressures and temperatures.

Phonon Conductivity. The conduction of heat in dielectric solids may be considered either the propagation of anharmonic elastic waves through a continuum or the interaction between quanta of thermal energy called phonons. The frequency of these lattice waves covers a range of values, and scattering mechanisms or wave interactions may depend on the frequency. The thermal conductivity can be represented in a general form equivalent to Eq. (12.31) by the relation

$$k = \frac{1}{3}\int c(\omega)vl(\omega)\,d\omega \tag{12.32}$$

where $c(\omega)$ is the contribution to the specific heat per frequency interval for lattice waves of that frequency and $l(\omega)$ is the attenuation length for the lattice waves.

The major process giving rise to a finite thermal conductivity and energy dissipation from thermal elastic waves is phonon-phonon interactions corresponding to phonon scattering called *Umklapp processes.* In addition to phonon interaction, which is important over a wide temperature range, various lattice imperfections give rise to anharmonicities and result in phonon scattering, which further decreases the mean free path and affects the conductivity. At sufficiently high temperatures, generally above room temperatures, the imperfection scattering is independent of temperature and vibrational frequency for all types of imperfection. At low temperatures a variety of different scattering mechanisms resulting from the lattice imperfections gives rise to a number of specific results. If more than one process is operative, the effective value of $1/l$ is found by adding $1/l$ for each process; this corresponds approximately to additivity of the corresponding thermal resistances for operative processes. Effects of impurity defects and microstructure on the thermal conductivity of dielectrics have been experimentally investigated with results in general agreement with theoretical predictions. The mechanism and temperature dependence of phonon conductivity in dielectric solids are quite well understood.

Photon Conductivity. In addition to the vibrational energy in solids, a much smaller fraction of the energy content results from higher-frequency electromagnetic radiation energy. Because this fraction of the total energy is so small, it is usually neglected in discussing heat capacity and thermal conductivity, but it becomes important at high temperatures because it is proportional to the fourth power of temperature. The energy per unit volume of blackbody radiation at temperature T is given by

$$E_T = \frac{4\sigma n^3 T^4}{c} \tag{12.33}$$

The volume heat capacity corresponding to the energy necessary to raise the temperature level of this radiation is given by

$$C_R = \left(\frac{\partial E}{\partial T}\right) = \frac{16\sigma n^3 T^3}{c} \tag{12.34}$$

where σ is the Stefan-Boltzmann constant (1.37×10^{-12} cal/cm^2 sec °K^4), c is the velocity of light (3×10^{10} cm/sec), and n is the refractive index. Since the velocity of this radiation is $v = c/n$, substituting in Eq. 12.31, we obtain for the radiant-energy conductivity

$$k_r = \frac{16}{3} \sigma n^2 T^3 l_r \tag{12.35}$$

where l_r is the mean free path of the radiant energy.

This result in Eq. 12.34 is more commonly reached by considering the exchange of energy between two volume elements in a partially absorbing medium. The intensity of radiation passing through an isothermal medium varies with distance according to the Lambert-Beer law, $I_x = I_0 \exp(-ax)$, where a is the absorption coefficient. The rate at which radiation is emitted in all directions by a unit volume of material with refractive index n is given by

$$j = 4a\sigma n^2 T^4 \tag{12.36}$$

Under steady-state conditions the amount of energy emitted by a volume element must be equal to that absorbed. In a temperature gradient, the amount of energy emitted is larger from the higher-temperature regions, leading to a net flux of radiant energy which, for any volume element, is equal to the difference in energy absorbed and energy emitted. A number of authors have shown that analysis of this process leads to the same result as given in Eq. 12.31. For photon conductivity both the energy distribution and the mean free path are strongly dependent on wavelength so that a relationship of the form of Eq. 12.32 should be used for quantitative analysis.

Since the magnitude of the radiant-energy density is small compared with the vibrational energy, the effectiveness of this energy-transfer process depends critically on the mean free path of radiant-energy transmission. For opaque materials ($l \approx 0$) energy transfer by this process is negligible. Similarly, if the mean free path is long compared with the size of the system, the interaction of energy with material is negligible, and radiation-energy transfer is a surface or boundary phenomenon as classically discussed in heat-transfer texts. It is only when the mean free path reaches macroscopic dimensions which are small compared with sample size that the photon-conduction energy-transfer process within the material is significant. This is the practical case with many silicate glasses and also with single crystals at moderate temperature levels. It becomes important for translucent ceramic materials such as sintered oxides at higher temperature levels. As long as the distances involved are large, the photon conductivity is a material characteristic, even though the mean free path is large, as attested to by its importance in astrophysics.

12.7 Phonon Conductivity of Single-Phase Crystalline Ceramics

A variety of processes may limit the mean free path of phonons and participate in fixing the thermal conductivity. The most fundamental of these is the phonon-phonon interaction leading to phonon scattering (the Umklapp process). At low temperatures, the mean free path corresponding to this process becomes large, such that a variety of other effects becomes important. However, for most ceramic materials at temperatures near room temperature and above, phonon-phonon interaction and scattering resulting from lattice imperfections are the processes of major importance and are our sole concern.

Temperature Dependence. The temperature dependence of phonon conduction in dielectric crystals is illustrated for a single crystal of aluminum oxide in Fig. 12.24. At very low temperatures the phonon mean free path becomes of the same magnitude as the sample size, boundary effects predominate, and the conductivity decreases to zero at 0°K. At some low temperature the thermal conductivity reaches a maximum, and phonon-phonon interactions lead to $k \sim \exp(-\theta/\alpha T)$. This exponential temperature dependence changes to $k \sim 1/T$ as the temperature level is raised above the Debye temperature. If the temperature is raised to a sufficiently high level, the mean free path decreases to a value near the lattice spacing, and the conductivity is expected to be independent of temperature.

The temperature dependence of conductivity for several oxides shows the thermal conductivity to be proportional to the inverse temperature above the Debye temperature, as illustrated in Fig. 12.25. To estimate the magnitude of the phonon mean free path in these materials, l was calculated as a function of temperature. Wave-velocity values were determined from the modulus of elasticity by the relation $v = \sqrt{(E/\rho)}$; it was assumed that static measurements on polycrystalline oxides, which indicate a rapid decrease in elastic modulus at temperatures above 700 to 1000°C, result from grain-boundary relaxation and creep; this is in agreement with dynamic elasticity measurements. As a result, the low-temperature linear relationship between E and T was extrapolated to obtain high-temperature wave-velocity values. Values of the mean free path at room temperature vary from a few to more than a hundred angstroms. The temperature variation of l for several materials is illustrated in Fig. 12.26.

Three general kinds of behavior are observed. At temperatures below the Debye temperature, as in Al_2O_3, BeO, and MgO, the inverse mean free path increases more rapidly than linearly with temperature. This corresponds to the exponential increase such as illustrated in Fig. 12.24 and is to

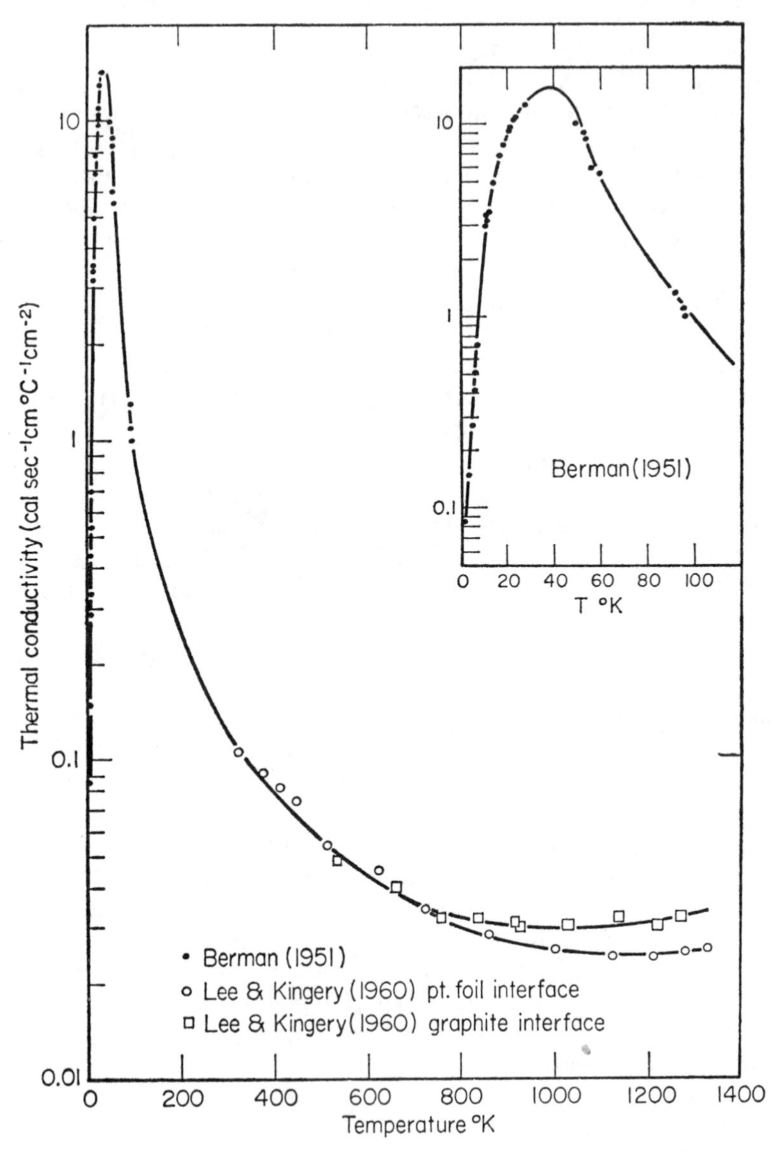

Fig. 12.24. Thermal conductivity of single-crystal aluminum oxide over a wide temperature range.

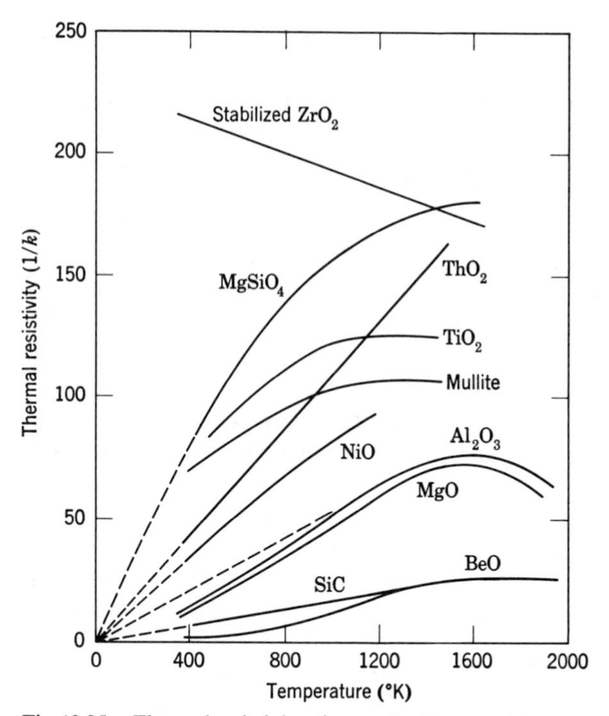

Fig. 12.25. **Thermal resistivity of several oxide materials.**

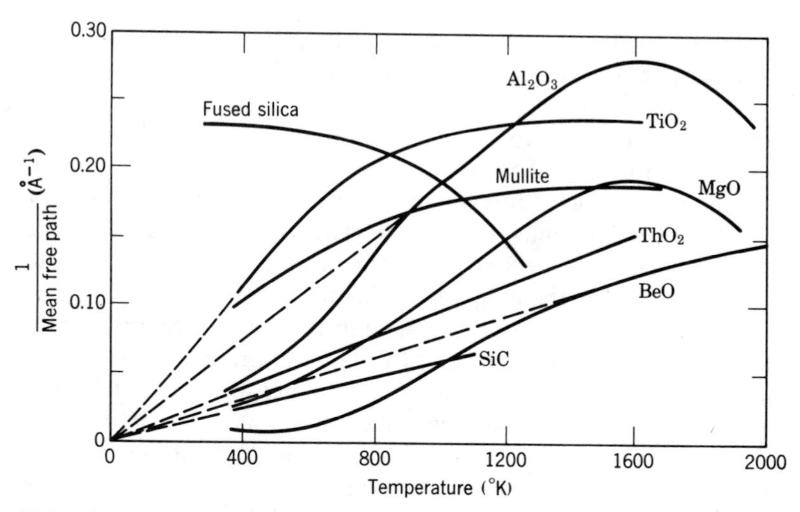

Fig. 12.26. **Inverse phonon mean free path for several crystalline oxides and for vitreous silica (Kingery, 1955).**

be expected at room temperature and above for several oxides. Over a wide temperature range near and above the Debye temperature, a linear increase in inverse mean free path with temperature is observed. At high temperatures, where the mean free path has decreased to a few angstroms, its value becomes fixed independent of temperature, as observed for TiO_2 and mullite. At temperatures above about 500°K for single crystals and about 1600°K for polycrystalline samples, an increase in the apparent mean free path is observed as a result of photon conductivity.

Influence of Structure and Composition of Pure Materials. Although the variation of the mean free path, and consequently the thermal conductivity, with temperature has a sound theoretical basis and experimental measurements are in good agreement with theoretical predictions, the absolute value of the mean free path can be estimated with much less certainty or precision. The extent of phonon scattering and the value of the mean free path should depend on the anharmonicity of the lattice vibrations. In this case, as for the parallel case of thermal expansion coefficients, theoretical calculations of absolute values are difficult. Although the general vibrational energy spectrum is well established, variations from completely harmonic vibrations are only imperfectly known.

Materials with complex structures have a greater tendency toward thermal scattering of lattice waves and consequently a lower thermal conductivity. Magnesium aluminate spinel, for example, has a lower conductivity than either Al_2O_3 or MgO, although each of these has similar structure, expansion coefficient, heat capacity, and elasticity. In the same way, mullite, which has a complex structure, has a much lower thermal conductivity than magnesium aluminate spinel. These relationships also affect the temperature dependence, since, as illustrated in Fig. 12.26, the mean free path tends to approach lattice dimensions at high temperatures in complex structures. The thermal conductivity for anisotropic crystal structures is also found to vary with the direction in the crystal (Table 12.3). This variation tends to decrease as the temperature level is raised. This result is to be expected, since anisotropic crystals always become more symmetrical as the temperature level is raised; the thermal conductivity is greatest in the direction having the lowest thermal expansion coefficient for SiO_2 (quartz), TiO_2 (rutile), and graphite.

Anharmonicities in lattice vibrations increase as the difference in atomic weight of the constituents increases. As a result, the thermal conductivity is a maximum for simple elementary structures and decreases as the atomic weights of the components become more different. Data for oxides and carbides are illustrated in Fig. 12.27. As one result of this analysis, we can be confident that the maximum conductivity to be

Table 12.3. Thermal Conductivity of Quartz (Birch and Clark, 1940) and Rutile (Charvat and Kingery, 1957)

Material	Temperature (°C)	Thermal conductivity (cal/sec °C cm)		
		Normal to c-Axis	Parallel to c-Axis	Ratio
SiO_2	0	0·0016	0·0027	1·69
	100	0·0012	0·0019	1·58
	200	0·0010	0·0015	1·50
	300	0·0084	0·0012	1·43
	400	0·0074	0·0010	1·35
TiO_2	200	0·0240	0·0158	1·52

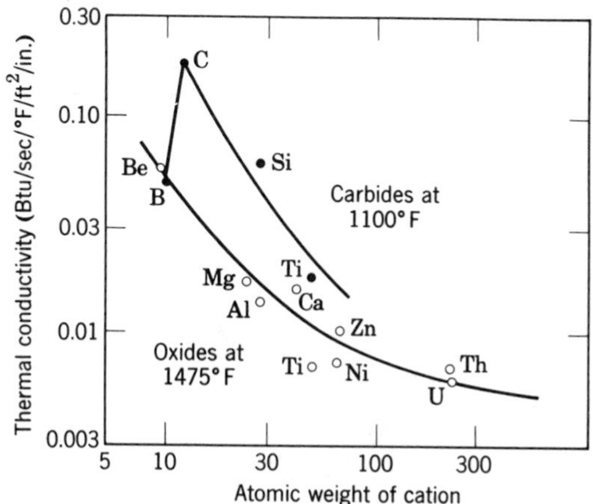

Fig. 12.27. Effect of cation atomic number on thermal conductivity of some oxides and carbides.

expected in ceramic oxides has been found in beryllia. At low temperatures, below room temperature, this effect leads to observable scattering resulting from the presence of different weight isotopes.

Boundary Effects. The scattering of phonons at grain boundaries results in a maximum for thermal conductivity at very low temperatures, as illustrated in Fig. 12.24. For materials at room temperature, the phonon mean free path has decreased to values somewhat lower than 100 Å, as illustrated for several materials in Fig. 12.26. At higher temperatures these values are even lower, and the crystallite size which is necessary for phonon scattering at boundaries to become important relative to other scattering processes is extremely small. It is perhaps of importance for thin-film technology but has not been studied. Comparisons between single crystals and polycrystalline samples of different grain size in the micron range ($\gg l$) are illustrated for aluminum oxide, calcium fluoride, and titania in Fig. 12.28. For each of these materials, the thermal conductivity of polycrystalline and single-crystal samples was identical at temperatures below about 200°C. At higher temperatures the single-crystal values depart from the polycrystalline sample values as a result of photon conductivity.

One boundary effect which should be mentioned is porosity appearing in the form of flat grain-boundary fissures in highly anisotropic materials or in mixed phases having different coefficients of thermal expansion. Under these conditions the heat flow is seriously disrupted with a result that the thermal conductivity may depend critically on the mode of heating, even though the total porosity is small.

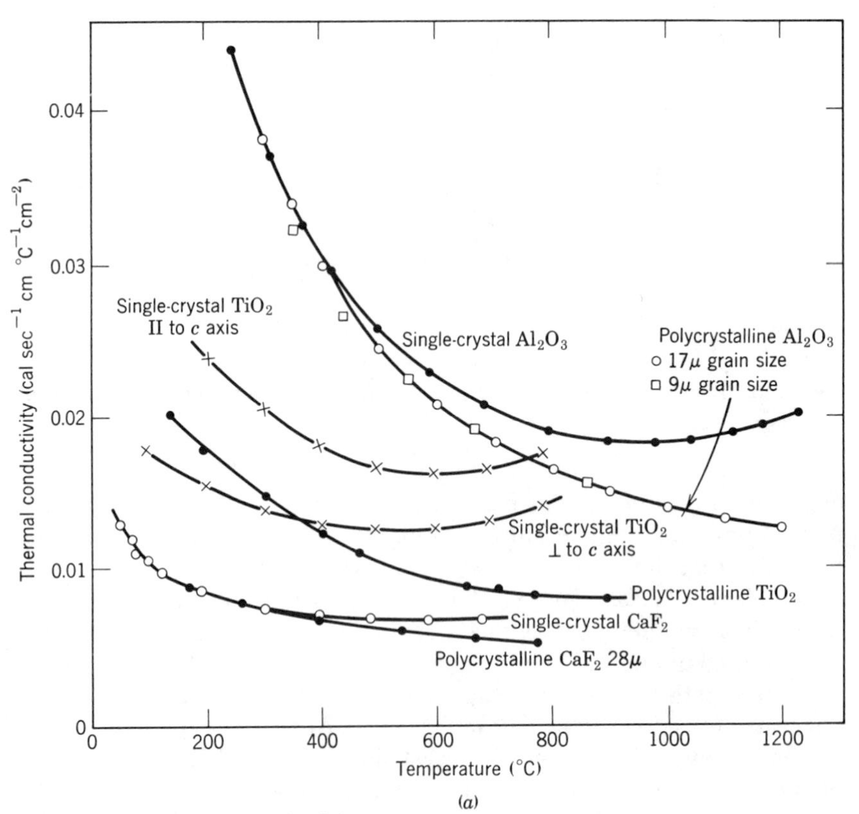

Fig. 12.28. (*a*) Thermal conductivity.

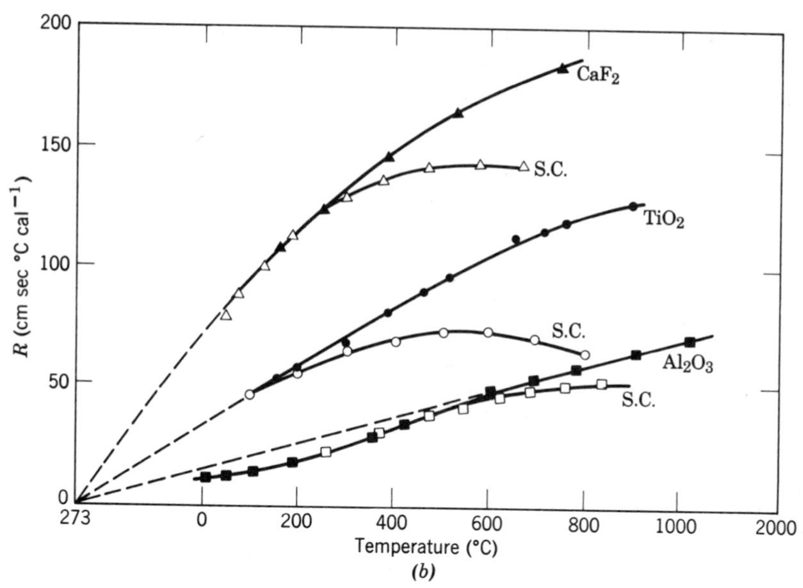

Fig. 12.28 (*Continued*). (*b*) Resistivity of single-crystal and polycrystalline Al_2O_3, TiO_2, and CaF_2. From Charvat and Kingery, 1957.

Impurities and Solid Solutions. In much the same way that complex structures and atoms of different size lead to increased anharmonicity and a lower conductivity, the presence of impurity atoms in solid solution leads to a decrease in the thermal conductivity. The effect of impurities can be treated in terms of the mean free path related to a scattering coefficient for the impurity center. The effects causing phonon scattering arise from differences in mass of an element substituted in the lattice, differences in binding force of the substituted atom compared with the original structure, and the elastic strain field around the substituted atom. Impurity scattering increases as the temperature is raised at very low temperatures, but it becomes independent of temperature at temperatures greater than about half the Debye temperature, as might be expected, since the average wavelength is comparable with or less than the size of the point imperfection for this temperature range. The numerical factors required to apply this theory quantitatively to ceramic systems are not well known, and the results may be uncertain by an order of magnitude.

The inverse mean free paths for different scattering processes are additive:

$$\frac{1}{l_{\text{total}}} = \frac{1}{l_{\text{thermal}}} + \frac{1}{l_{\text{impurity}}} + \cdots. \tag{12.37}$$

Consequently, the effect of solid-solution impurity scattering is greatest in simple lattices and at low temperatures, at which the thermal scattering mean free path is large. As shown in Fig. 12.29, the additional scattering

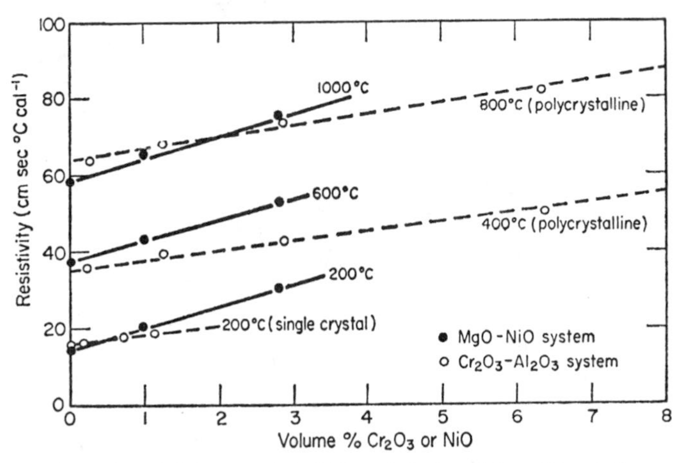

Fig. 12.29. Thermal resistivity of MgO–NiO and Cr_2O_3–Al_2O_3 solid solutions.

for low concentrations of impurities is directly proportional to the volume fraction added. As indicated by the constancy of slope at different temperatures, the effect of impurities on the thermal resistivity is independent of temperature; that is, the mean free path for impurity scattering is independent of temperature. The effectiveness of solid-solution impurities in decreasing the thermal conductivity depends on the mass difference, size difference, and binding-energy difference of the impurity added. For both Ni^{2+} in MgO and Cr^{3+} in Al_2O_3, a 1 vol% addition corresponds to a mean free path of 80 to 100 Å. This is equivalent to a scattering cross section for each point imperfection of the same order of magnitude as the atomic size. This result has also been found for *F* centers in NaCl and calcium additions in KCl. Since the mean free path caused by thermal scattering decreases rapidly with temperature, as illustrated in Fig. 12.26, the importance of the effect of impurities on the overall conductivity is greatly dependent on the temperature level. This is illustrated for the MgO–NiO solid-solution system in Fig. 12.30. At temperature below room temperature, the effectiveness of solid solution in strongly decreasing thermal conductivity is even more impressive.

Compositional variations can have a particularly large effect on nonstoichiometric materials for which solid solutions occur. One system that

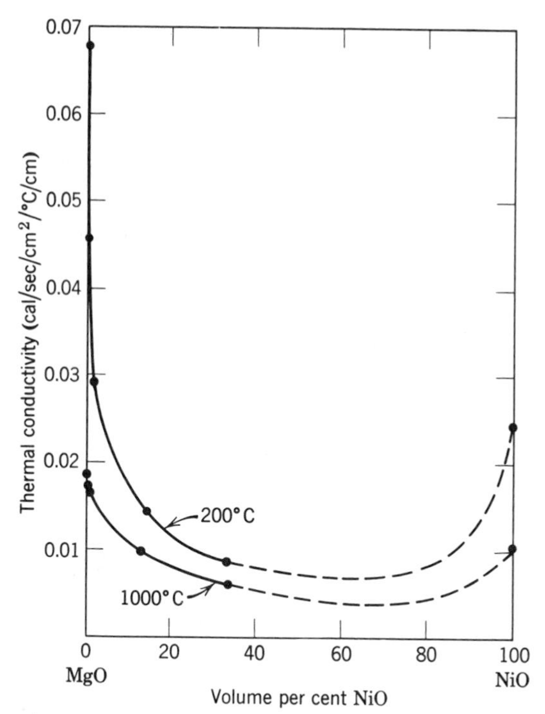

Fig. 12.30. Thermal conductivity in the solid-solution system MgO–NiO.

has been investigated is the UO_2–ThO_2 system. Data for a number of compositions are collected in Fig. 12.31. When UO_2 is oxidized to increase the oxygen content, the conductivity decreases to about a quarter of the value found for the stoichiometric material. (In this case, however, the results are complicated by the possibility that a second phase may precipitate in some samples.) When thorium oxide is substituted for uranium, the conductivity is further decreased. The lowest observed conductivity for an oxygen-deficient thorium-uranium composition has a value of about 0.003 cal/cm °C sec. In this system the specific heat is of the order of 0.66 cal/cm^3 °C, and the wave velocity is about 4×10^5 cm/sec. From these data we can calculate a limiting lower value for conductivity if we assume that the mean free path is limited to about 4 Å by the lattice dimensions. From these assumptions an estimate of 0.0035 is obtained for the minimum conductivity; this is in reasonable agreement with the lowest value actually observed.

One of the most useful high-temperature refractory materials having a low thermal conductivity is stabilized zirconia. In this material a cubic

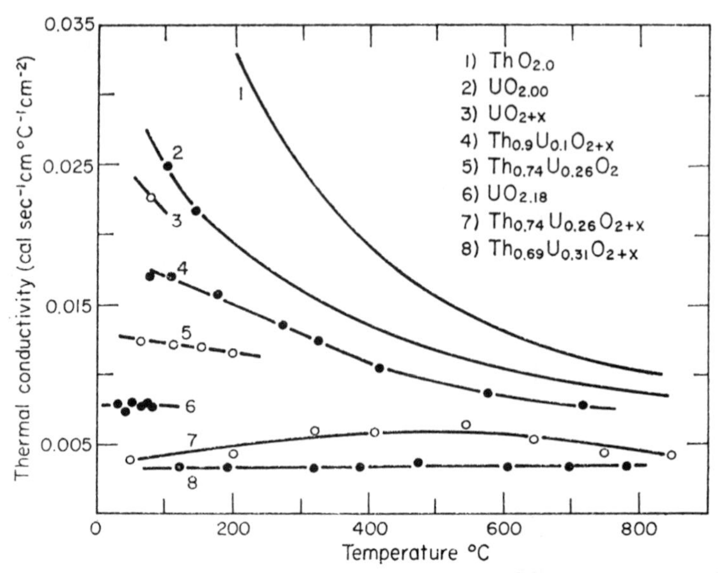

Fig. 12.31. Thermal conductivity data for various compositions in the UO_2–ThO_2–O_2 system.

solid solution is formed in which Ca^{2+} and Hf^{4+} substitute for Zr^{4+}; O^{2-} vacancies are formed to balance the charge deficiency. For compositions approximately $92ZrO_2 \cdot 4HfO_2 \cdot 4CaO$, a calculated phonon mean free path of 3.6 Å is in general agreement with the complex solid solution formed.

Neutron Irradiation. When a crystalline material is irradiated with neutrons, the resultant structure contains displaced atoms and lattice strains corresponding to the presence of impurities but with larger associated strain energies. As a result, neutron irradiation leads to a decrease in the thermal conductivity that is particularly important at low temperatures, as illustrated in Fig. 12.32.

12.8 Phonon Conductivity of Single-Phase Glasses

In the same way that the thermal conductivity of highly disordered crystals such as the $(Th,U)O_{2+x}$ composition shown in Fig. 12.31 has a low value, glasses, with their completely noncrystalline structure, are found to have a phonon mean free path that is limited to the order of interatomic distances by the random structure. This fixing of the mean free path by the structure leads to a much more limited range of thermal-conductivity values for glasses than is found for crystals.

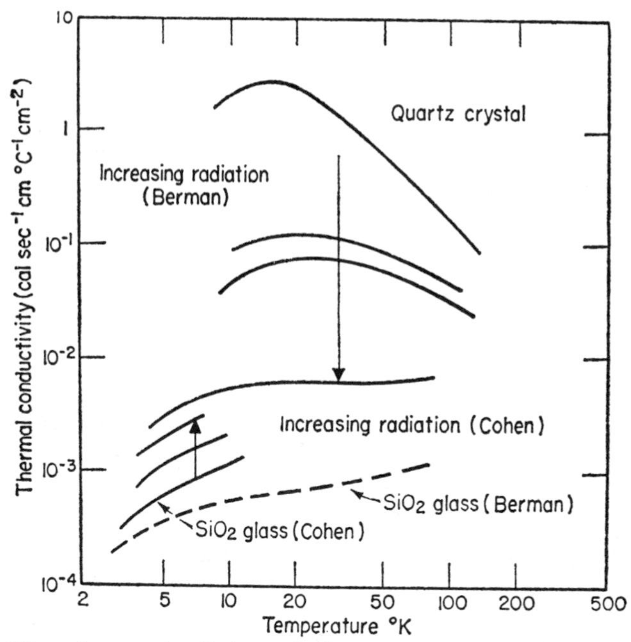

Fig. 12.32. Effect of neutron irradiation on the thermal conductivity of crystalline quartz and vitreous silica.

Temperature Dependence of Glass Conductivity. The glass for which the largest number of reliable thermal-conductivity measurements have been carried out over the widest temperature range is fused silica, SiO_2, for which the conductivity changes with temperature, as illustrated in Fig. 12.33. Since the mean free path is limited to a fixed value independent of temperature by the random-network structure, the thermal conductivity parallels the volume heat capacity. The conductivity (and heat capacity) increases at low temperatures and then reaches a nearly constant value for temperatures above a few hundred degrees centigrade. As for the single crystals discussed previously, high-temperature measurements normally show an increase corresponding to photon conductivity. When the photon contribution is excluded, the conductivity remains sensibly constant at temperatures above about 800°K.

Few other glass compositions have been measured over a wide temperature range, but the general temperature-dependence behavior is similar to that for fused silica, as illustrated in Fig. 12.34.

Effects of Composition. Although the effect of composition on the thermal conductivity of noncrystalline solids is less than that seen in crystals, since the mean free path is limited by the random structure,

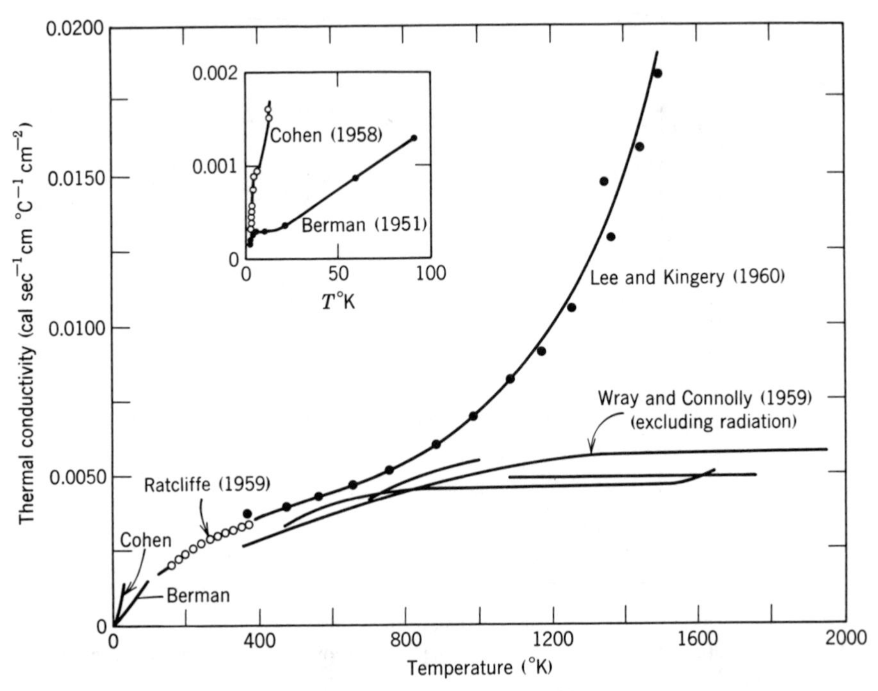

Fig. 12.33. Thermal conductivity of fused silica over a wide temperature range.

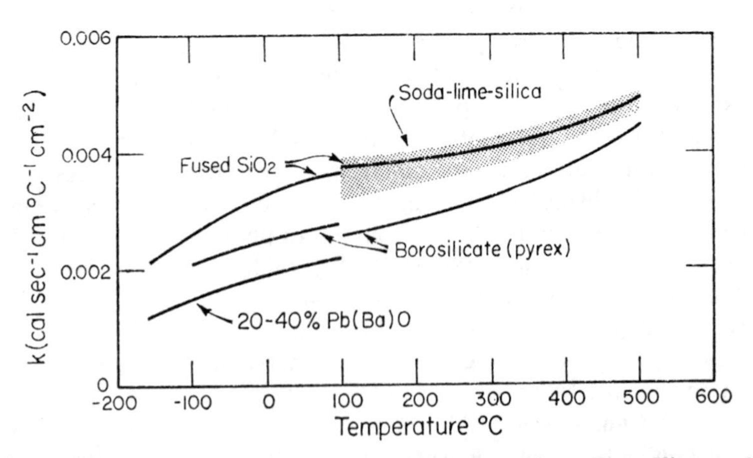

Fig. 12.34. Thermal conductivity of several glass compositions. From Kingery, 1949; Ratcliff, 1960.

significant variations do occur among different compositions. Experimental data for a variety of compositions, some of which are illustrated in Fig. 12.34, range between a value of about 0.0033 cal/sec °C cm^2 at room temperature for fused silica to a value of 0.0013 for a glass containing 80% lead oxide. In general, the thermal conductivity of fused silica is similar to that found for soda-lime-silica glass, and these values are larger than those observed for borosilicate glasses which in turn are higher than values found for high-index optical glasses containing large fractions of heavy metal ions.

The mean free path of phonons in fused silica, soda-lime-silica glass, and Pyrex glass has been estimated from the elastic wave velocity, volume heat capacity, and thermal conductivity values, as illustrated in Table 12.4. The experimental finding that the mean free path of silica is greater than that of a soda-lime-silica glass and is still greater than that of a Pyrex composition is established beyond experimental error and suggests that the more complex structure in the borosilicate and soda-lime glass more severely limits the mean free path.

Table 12.4. Calculated Phonon Mean Free Path in Glasses

Glass	k (cal cm/sec °C cm^2)	ρ (g cm^3)	c (cal cm^{-3} °C^{-1})	$v = \sqrt{(E/\rho)}$ (cm sec^{-1})	$l = k/\frac{1}{3}cv$ Å
Fused silica	0·0037	2·20	0·39	$5{\cdot}5 \times 10^5$	5·2
Corning 0080	0·0037	2·47	0·50	$5{\cdot}2 \times 10^5$	4·3
Pyrex-brand 7740	0·0026	2·23	0·42	$5{\cdot}5 \times 10^5$	3·3

12.9 Photon Conductivity

As discussed in Section 12.6, the conduction process in which photons transfer energy through dielectric solids becomes important as the temperature level is raised and when the photon mean free path is appreciable. As indicated in Eq. 12.35, the photon conductivity can be given as equal to

$$k_r = \frac{16}{3}\frac{\sigma n^2 T^3}{a} = \frac{16}{3}\sigma n^2 T^3 l_r \qquad (12.38)$$

when a, the absorption coefficient, and n, the index of refraction, are assumed independent of temperature and wavelength. However, these depend on frequency, particularly the absorption coefficient.

In general, single crystals of dielectric materials are quite transparent in the visible region of the spectrum, becoming opaque as a result of electron excitation in the ultraviolet and showing absorption bands in the infrared as a result of atomic vibrational phenomena. In addition, certain ions such as the transition elements show strong absorption in the visible spectrum as a result of electronic transitions. Most ceramics are more transparent, that is, have a longer mean free path in the visible and near infrared than at longer wavelengths. Since the peak of the blackbody emission spectrum is between 2 and 3 microns in the temperature range of 700 to 1500°C and shifts to shorter wavelengths as the temperature is raised, the effective mean free path increases at the higher temperatures. As a result, the photon conduction increases more rapidly with increasing temperature than the T^3 relationship indicated in the simplified analysis of Eq. 12.38.

The Photon Mean Free Path. The absorption and scattering of photons in the visible and near infrared regions of the spectrum are the basic material characteristics that fix photon conductivity. For materials with low values of the absorption coefficient, photon conduction becomes important at temperatures of a few hundred degrees centigrade. For materials with high values of the absorption coefficient or substantial scattering, photon conductivity does not become significant until very high temperatures are reached. Typical absorption coefficients for a few characteristic materials over the range of wavelength of interest and at different temperatures are illustrated in Fig. 12.35. As shown there, the absorption coefficient is low up to a wavelength of 2 to 4 microns, at which absorption increases strongly. For a number of different glasses, the mean free path measured at a wavelength of 2 microns at room temperature is nearly proportional to the integrated mean free path.

The change in integrated mean free path with temperature depends on the characteristics of the material and tends to be larger for materials having good transmission in the visible and near infrared, that is, clear glasses and single crystals. Two factors are important: first, the distribution of radiant energy shifts to shorter wavelengths as the temperature is raised; second, the absorption edge usually moves to shorter wavelengths as the temperature is raised. As a result, the change in integrated mean free path with temperature can be considerable, as illustrated in Fig. 12.36.

Aside from glasses, few ceramics are used in transparent form such as single crystals, and the major form of photon attentuation in most ceramics results from light scattering. The main scattering process results from pores acting as scattering centers. Because of the large difference in index of refraction between the pores and the solid and because of the

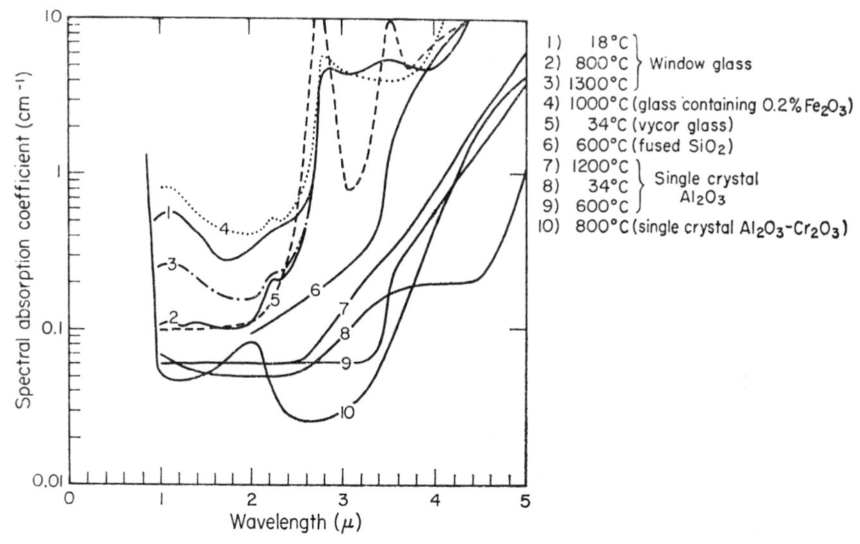

Fig. 12.35. Absorption coefficient for single crystals and glasses at different wavelengths and temperature levels. From Lee and Kingery, 1960; Neuroth, 1952; Grove et al., 1960.

small particle size of the pores normally present, the transmission is markedly reduced by as little as 1/2% porosity. Calculated and experimental observations for alumina based on well-established scattering theory are illustrated in Fig. 12.37. Since almost all ceramics contain a few percent porosity, the effective mean free path is substantially less than for glasses or single crystals. Some experimental measurements of the photon conductivity for aluminum oxide, sintered Vycor glass, and the sintered calcium fluoride, all of which are highly translucent as compared with most ceramics, are illustrated in Fig. 12.38. These values are from one to three orders of magnitude smaller than values for single crystals or glasses. As a result, the photon energy-transfer process only becomes important for sintered materials at quite high temperatures (greater than 1500°C).

Temperature Dependence. The change in photon conductivity with temperature depends on the integrated mean free path which normally increases as some low power of temperature, as indicated in Fig. 12.36, or is nearly independent of temperature for scattering phenomena. The result of a combination of these is that the photon conductivity is proportional to T^{3+x}. Experimental results usually show the temperature exponent to be in the range of 3.5 to 5.

In order for the prior analysis to apply, the mean free path must be

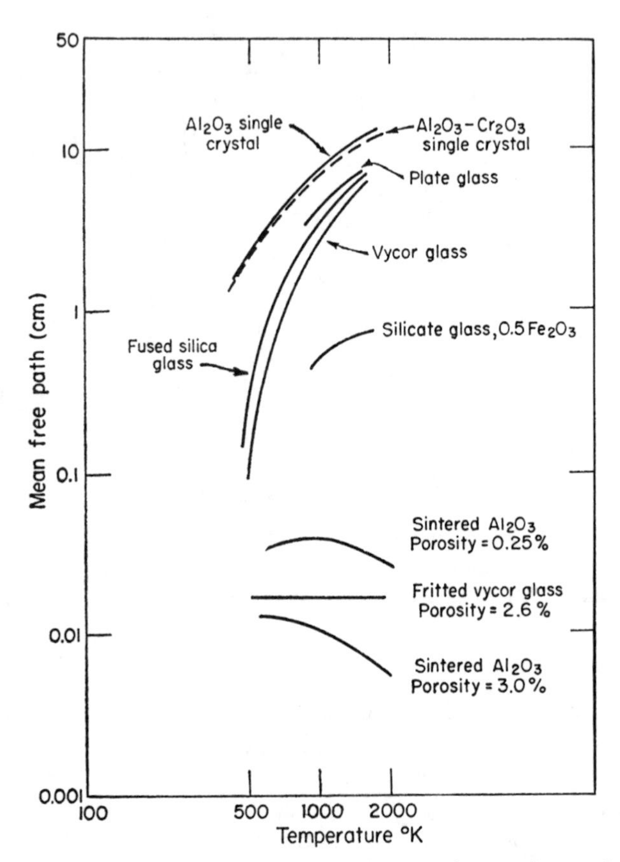

Fig. 12.36. The temperature dependence of the integrated mean free path for various materials. Precision of high values is about ±100%.

small compared with the sample size. For single crystals and for most measurements of glass conductivity, this is not the case; as a result, the measured conductivity is actually less than would be observed for a large specimen. This is illustrated in Fig. 12.38, in which data for single crystals of alumina and for silica glass are compared with calculated values. The effects of boundaries are discussed subsequently.

For samples in which radiation conductivity is appreciable, the temperature dependence of the overall heat transfer changes from a negative to a positive exponent, leading to a minimum in the experimental conductivity.

Effects of Boundaries. Just as the thermal conductivity of crystals is limited at low temperatures when the sample dimensions become the

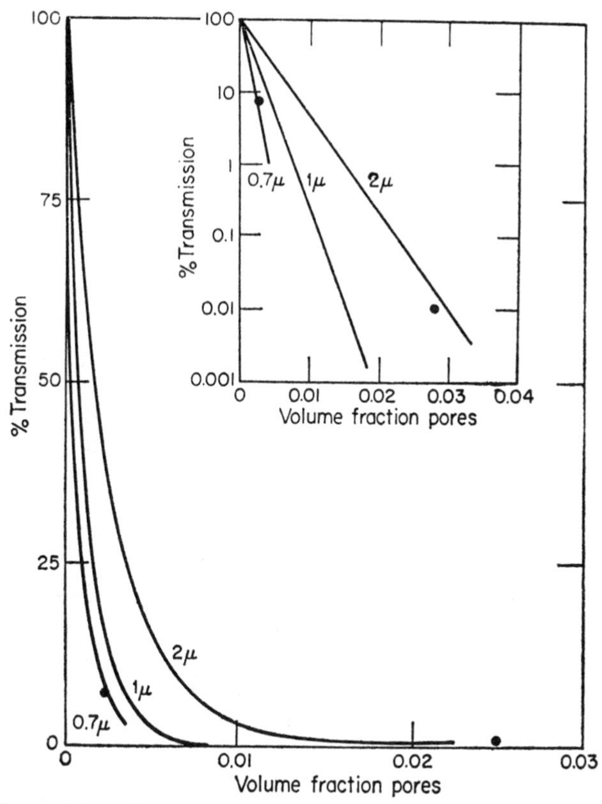

Fig. 12.37. Transmission of polycrystalline alumina containing small amounts of residual porosity (equivalent thickness 0.3 mm).

same order of magnitude as the phonon mean free path, and boundary effects predominate in gas conduction at low pressures when the molecular mean free path is large, boundary effects are found to predominate for photon conductivity when the sample size is similar to the photon mean free path. This boundary effect has led some authors to emphasize the limitations of assigning a conductivity value to the photon process and to treat photon conductivity as something quite different from other heat-transfer processes. We prefer instead to emphasize that boundary effects are important for all conduction mechanisms and that the development of photon conduction theory is exactly parallel to that for other energy-transfer mechanisms.

There are three photon energy-transfer processes that can be separately identified for energy transfer between two boundaries separated by

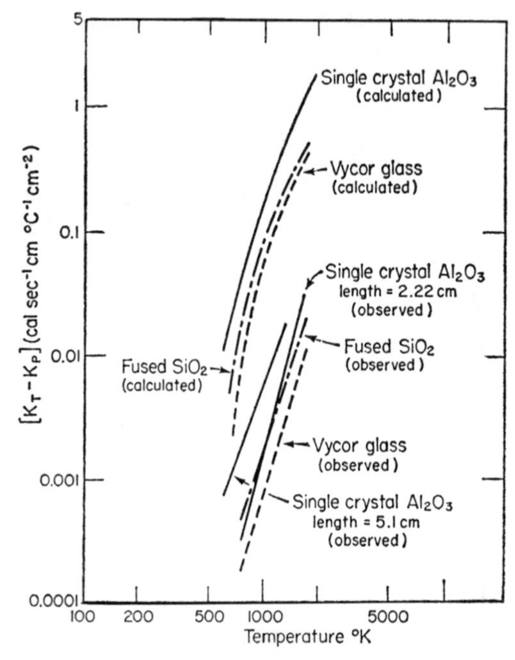

Fig. 12.38. Calculated and observed temperature dependence of photon conduction in single-crystal Al_2O_3, fused silica glass, and Vycor glass. In all cases the integrated mean free path is greater than sample size. From Lee and Kingery, 1960.

any material. First, energy may be transferred directly from one boundary to the other without interaction with the intervening material. In this case, the only effect of the material is to alter the photon velocity, and consequently the rate of heat transfer, but its own temperature is not affected by and has no influence on the rate of heat transfer. Second, energy may be transferred by photon energy exchange between the boundary and the material. This transfer is limited to a region having a thickness of the order of magnitude of the mean free path of the photons, and the rate of energy exchange is determined by the temperature of the boundary and the temperature gradients in the material. Third, energy may be transferred by photon processes within the material, independent of the boundaries. This last is obviously the only process which can be described as a material property; it has been discussed in previous sections.

Boundary effects are important as a practical matter because the values of photon mean free paths commonly found for ceramic materials, 0.1 to 10 cm, are the same order of magnitude as common sample sizes.

Experimentally, this is noted in the effects of sample size as measured *apparent photon conductivity*, that is, including boundary effects.

When radiation does not interact with the medium and photon conduction is the only energy-transfer process, there is a temperature discontinuity at the interface, the temperature gradient in the material present is independent of the rate of heat transfer, and thermal conductivity as a material property has no meaning. However, if the distance between boundaries is d, a value of *effective conductivity* for the heat transfer due to pure boundary processes can be defined in terms of the effective emissivity $e/(2-e)$ as

$$q = \left(\frac{e}{2-e}\right) 4n^2\sigma T^3 \,\Delta T \tag{12.39}$$

$$k_b = q\left(\frac{d}{\Delta T}\right) \tag{12.40}$$

$$k_b = \left(\frac{e}{2-e}\right) 4\sigma n^2 T^3\, d \tag{12.41}$$

The ratio of the observed radiant-energy transfer when the mean free path is large compared with the sample size and the maximum photon conductivity for an infinitely large sample is, to a first approximation, given by

$$\frac{k_b}{k_r} = P = \frac{4\sigma n^2 T^3\{e/(2-e)\}}{16/3n^2T^3 l_r} = \frac{3}{4}\frac{e}{2-e}\frac{d}{l_r} \qquad \text{for } P < 1 \tag{12.42}$$

which only applies when the ratio P is much less than unity. Under these conditions the observed conductivity, that due to boundary conditions, is only a fraction of the expected conductivity based on an infinite sample size. The ratio of these values is given for different optical thicknesses and boundary emissivities for single-crystal alumina, clear fused silica, and Vycor glass in Fig. 12.39. Increasing the ratio d/l_r by increasing the sample thickness increases the apparent conductivity. The importance of the emissivity of the boundary as well as the sample thickness can be seen from the increased apparent conductivity of samples bounded with graphite rather than platinum.

At the boundary regions, where radiation interacts with the material over a finite photon mean free path, the energy emission from an opaque boundary or uniform temperature region such as a furnace enclosure is given by

$$J = en^2\sigma T^4 \tag{12.43}$$

where e is the emissivity of the surface. If photon energy transfer is an

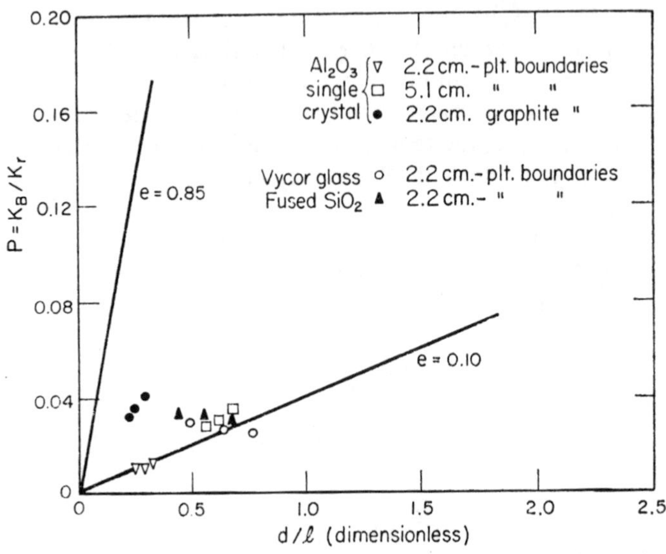

Fig. 12.39. Effect of sample size and boundary material on the fractional transfer resulting from direct exchange between boundaries. From Lee and Kingery, 1960.

important process, the change in properties, both optical and lattice conductivity, at the interface requires a change in temperature gradient in the region of the material near the interface. The gradient change occurs over a distance of the order of magnitude of a mean free path for photons, and the gradient increases or decreases toward the surface, depending on whether photon conductivity is being converted to phonon conductivity or vice versa. Calculated temperature gradients in a glass tank are given in Fig. 12.40.

12.10 Conductivity of Multiphase Ceramics

Most ceramic materials are composed of mixtures of one or more solid phases together with a pore phase. The resulting conductivity of the body depends on the amounts and arrangement of each phase present as well as their individual conductivity.

Since the resulting conductivity of a mixture depends on its arrangement, an understanding of the microstructure is essential in interpreting thermal-conductivity data. Three idealized kinds of phase distribution are illustrated in Fig. 12.41. Of these, the parallel-slab arrangement (Fig. 12.41*a*) is not commonly observed, but a continuous major phase (Fig. 12.41*b*) with a minor amount of a discontinuous second phase is typical of

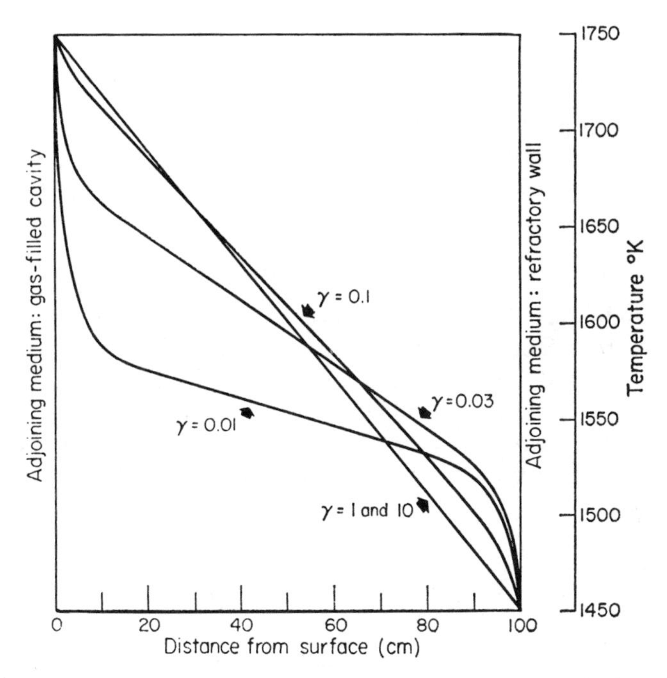

Fig. 12.40. Calculated changes in temperature gradients near boundaries in a glass tank. After Walther et al., *Glastech. Ber.*, **26**, 193 (1953).

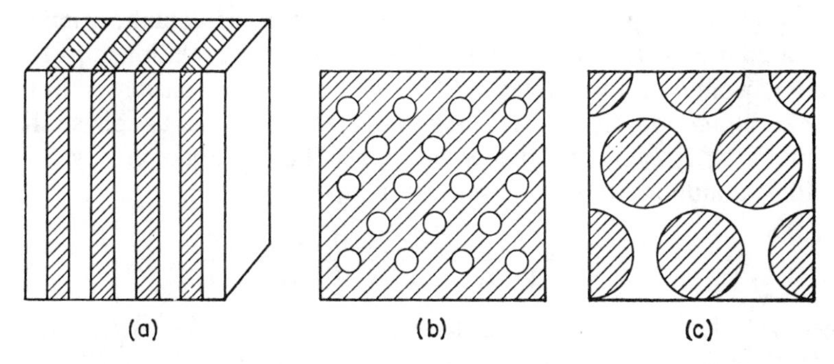

Fig. 12.41. Idealized phase arrangements. (*a*) Parallel slabs; (*b*) continuous major phase; (*c*) continuous minor phase.

many microstructures. It is characteristic of the form of porosity normally present. A discontinuous major phase (Fig. 12.41*c*) with a minor phase appearing as a continuous boundary material, for example, is typical of the glass bond distribution in many ceramics, the metal distribution in metal-bonded carbides, and the porosity in insulating powders.

For the simplest geometry, a series of slabs, it can be easily seen that if the heat flow is parallel to the plane of the slabs, they are equivalent to a parallel electrical circuit, each slab has the same thermal gradient, and most of the heat flow is through the better conductor. The thermal conductivity is given by

$$k_m = v_1 k_1 + v_2 k_2 \tag{12.44}$$

where v_1 and v_2 are the volume fraction of each component, equal to the fraction of cross-sectional area. Under these conditions, the heat conduction is dominated by the better conductor; if $k_1 \gg k_2$, $k_m \simeq v_1 k_1$. In contrast, for heat flow perpendicular to the plane of the slabs, they are equivalent to an electrical series circuit, the heat flow through each slab is equal, but the temperature gradients are different. The total conductivity is given by

$$\frac{1}{k_m} = \frac{v_1}{k_1} + \frac{v_2}{k_2} \tag{12.45}$$

or

$$k_m = \frac{k_1 k_2}{v_1 k_2 + v_2 k_1} \tag{12.46}$$

Here the heat conduction is dominated by the poorer conductor, and if $k_1 > k_2$, $k_m = k_2/v_2$.

More realistic approaches toward the structure of actual ceramics are illustrated in Fig. 12.41*b* and *c*. For these structures, relationships for the resultant conductivity derived by Maxwell have been discussed in terms of thermal conductivity by Eucken. If the continuous phase has a conductivity k_c and the dispersed phase has a conductivity k_d, the resultant conductivity of the mixture is given by

$$k_m = k_c \frac{1 + 2v_d(1 - k_c/k_d)/(2k_c/k_d + 1)}{1 - v_d(1 - k_c/k_d)/(k_c/k_d + 1)}. \tag{12.47}$$

When $k_c > k_d$, the resultant conductivity is $k_m \simeq k_c[(1 - v_d)/(1 + v_d)]$. In contrast, if $k_d > k_c$, then $k_m \simeq k_c[(1 + 2v_d)/(1 - v_d)]$.

For sintered mixtures of MgO with BeO and for MgO with MgO–SiO_2 which have a minor phase dispersed in a continuous major phase, the conductivity changes along an S-shaped curve, as illustrated in Fig. 12.42. The light lines correspond to the end members being the continuous phase. Above about 40% forsterite, the forsterite phase is continuous with dispersed MgO grains within it; in samples containing smaller amounts of

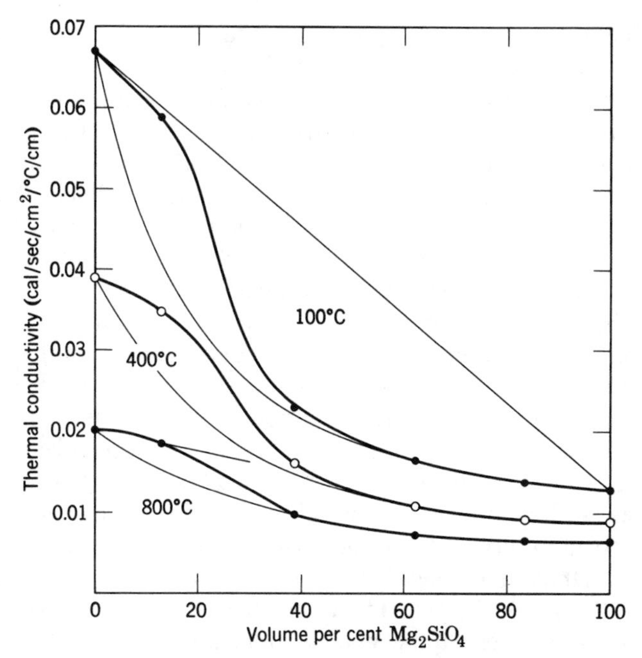

Fig. 12.42. Thermal conductivity in the two-phase system MgO–$MgSiO_4$.

forsterite, it is present as dispersed grains in a continuous MgO matrix. Just where any individual composition falls, in terms of end members being dispersed or continuous, depends on the particular system and the interface energy relationships and sintering conditions. For alumina porcelains, both phases, corundum and glass, are found to be continuous at a composition corresponding to about 9 vol% glass, leading to a thermal conductivity intermediate between calculations for extreme structures. For silicon-silicon carbide mixtures formed by infiltration, the silicon phase is found to be continuous, even though present in amounts up to 30 vol%. In general, the glass phase is continuous in vitreous ceramics, so that the conductivity of porcelains and fire clay is closer to that of the glass contained than the conductivity of the crystalline phase.

In most ceramic systems an important constituent is the porosity which is almost always present. One effect of porosity has been discussed in the previous section, in which the influence of pores as scattering centers for photons was discussed, and it was pointed out that even small fractions of porosity strongly reduce the photon mean free path and severely limit this conduction mechanism.

The effect of porosity on the phonon conductivity of ceramic systems is in approximate agreement with Eq. 12.47. Other models give derivations somewhat differing from Eq. 12.47 but of greater concern that the specific model assumed is the value of effective conductivity to be taken for the pores. At low temperatures the porosity has a low conductivity as compared with any solid phases present, and a nearly linear decrease in conductivity with increasing porosity is found for dispersed pores in a solid. However, in addition to true conduction, radiation across pores contributes to heat transfer at high temperature.

Although for single crystals, single crystals with uniform scattering centers, or opaque materials, the photon conductivity and the effects of porosity on photon conductivity can be treated on the basis of the previous discussions, for other situations the analysis is much more complex. Examples such as insulating firebrick and powder insulations are cases in which regions of solids with small scattering pores surround large pores. Although transfer through the solid sections can be and has been discussed in terms of phonon conductivity and an effective photon conductivity, the latter depending on absorption, re-emission, and scattering within the solid, the transfer across the large pores, those between particles of a powder or the macroscopic pores in firebrick, must be analyzed by other methods.

If the material surrounding the pores is opaque, an *effective* radiation conductivity of a pore can be defined when the temperature gradient is small, since the radiant-energy transfer across a pore is proportional to the temperature difference between pore surfaces:

$$q = n^2 \sigma e_{\text{eff}} A (T_1^4 - T_2^4) \tag{12.48}$$

In this equation $(T_1^4 - T_2^4)$ can be factored, so that

$$q = 4n^2 \sigma e_{\text{eff}} A T_m^3 \,\Delta T \tag{12.49}$$

Considering heat transfer across a flat cavity, with parallel sides and thickness d_p, an effective conductivity is defined by

$$q = -k_{\text{eff}} A \frac{\Delta T}{d_p} \tag{12.50}$$

An effective conductivity is thus defined which gives the proper heat flux when inserted in the normal thermal conductivity relations:

$$k_{\text{eff}} = 4 d_p n^2 \sigma e_{\text{eff}} T_m^3 \tag{12.51}$$

Consequently the effect of radiation on pore conductivity is proportional to the pore size and to the third power of temperature. Pores of larger size

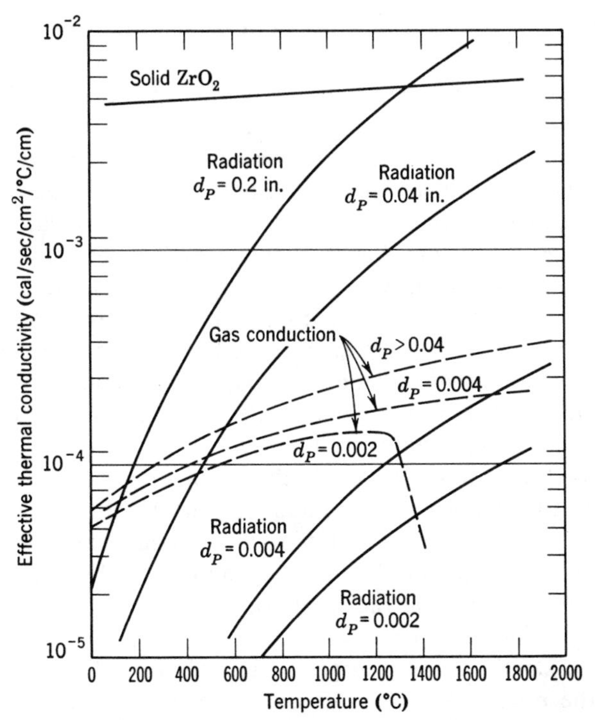

Fig. 12.43. Effective thermal conductivity of pore spaces for conduction and radiation heat transfer.

contribute to increasingly high conductivity at high temperatures; small-size porosity remains a good barrier to heat flow. The calculated conductivity of different-size pores over a wide temperature range is illustrated in Fig. 12.43.

Although the above discussion is valid for opaque materials, most ceramic materials are more or less translucent and transmit a considerable fraction of the incident radiation, especially in thin sections; the latter effect must be considered in real cases. If we think of the material surrounding the pores as radiation shields which diminish the radiant transfer, we can see the physical significance of the previous sections: the smaller the pore size and the larger the porosity, the more pores or shields there are across the radiant flux, and the radiant transfer, or the effective photon conductivity, is decreased. On the other hand, the higher the emissivity, the greater the transfer between surfaces and the larger the effective conductivity.

The effect of translucency is to diminish the efficiency of these radiation shields and thereby increase the effective conductivity. This effect also depends on the thickness of the layers, since thinner layers have greater transmission. In order to treat the effective conductivity of translucent materials, we must add another factor to Eq. 12.51; this factor, a function of the product of the actual thickness and the extinction coefficient, called the optical thickness, is unity for opaque layers, larger for translucent layers, and finally becomes proportional to the reciprocal of the optical thickness for very transparent materials. That is, pores have little effect on the radiant-energy transfer in highly transparent materials. Although decreasing the pore size, while holding other factors constant, increases the number of radiation shields, it also diminishes the efficiency of each shield and therefore does not decrease the conductivity as much as it would if the material were opaque.

In contrast to solids, in which the main effect of the presence of porosity is to decrease the conductivity nearly in proportion to the fraction porosity, powders and fibrous materials have a much lower conductivity, even though the volume fraction solids may be considerable. This results from the fact that the pore phase is continuous, there being no sintering between solid-solid contacts. Under these conditions, the resulting conductivity is largely determined by the effective conductivity of the pore phase (Fig. 12.43). For gas systems having different conductivities from air, the effective conductivity of the overall system increases in proportion to the conductivity of the gas phase present and also depends on the gas pressure. When radiant-energy transfer in powders is significant relative to gas conductivity, such as powders in a vacuum or at high temperatures, the situation is as summarized in Fig. 12.43. A more rigorous treatment of radiant transfer through powders and fibrous insulations necessitates a much more complex heat-transfer calculation which takes into account the transmission through the particles as well as absorption, emission, and scattering in them.

In investigations of polycrystalline TiO_2 it was found that heat treatment had a large effect on conductivity, even though the porosity and grain size were not much changed (Fig. 12.44). Similarly, some compositions in the two-phase systems Al_2O_3–ZrO_2, MgO–$MgAl_2O_4$, and Al_2O_3–mullite have conductivities lower than those found for either end member, without solid solution or other impurity effects. These results occur in systems in which the expansion coefficient is different in different crystallographic directions or between different phases. Microstresses arise which cause flat microcracks to open along boundaries between individual grains. Even though the fraction porosity thus introduced is small, it occurs as a continuous phase in series with the heat-flow path and

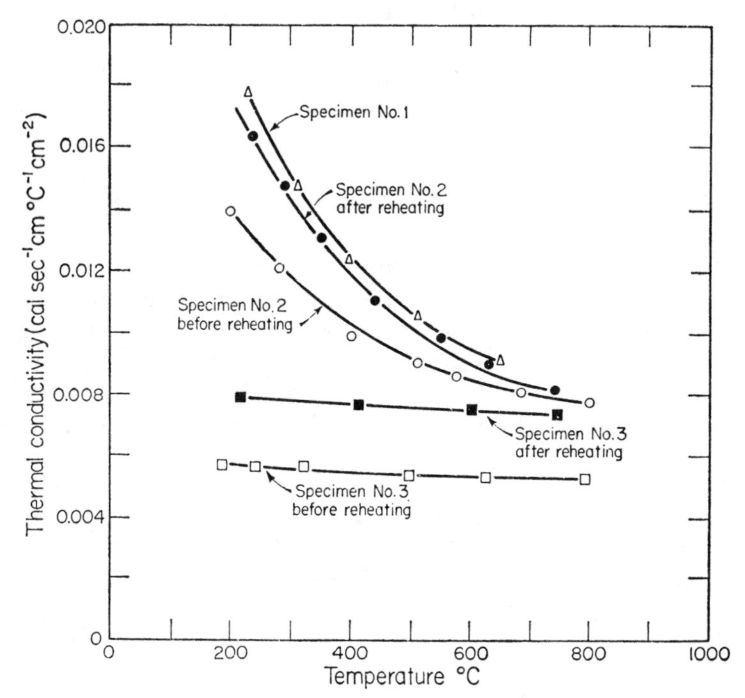

Fig. 12.44. Thermal conductivity of polycrystalline titania samples after various heat treatments. From Charvat and Kingery, 1957.

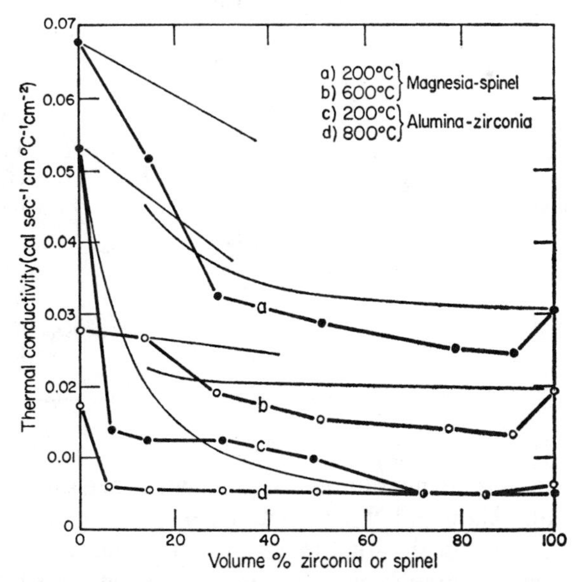

Fig. 12.45. Thermal conductivity for MgO–$MgAl_2O_4$ and Al_2O_3–ZrO_2 two-phase systems in which grain-boundary microcracks occur.

as a result has a large influence on the resulting conductivity. Data for the $MgO-MgAl_2O_4$ system and the $Al_2O_3-ZrO_2$ system are illustrated in Fig. 12.45. In both cases, microstructure studies indicated that cracks formed along grain boundaries.

The effect of heating and reheating results from increased cracking on cooling or by annealing of cracks previously present. The opening and closing of such cracks also give a hysteresis loop in measurements; this phenomenon has also been observed experimentally in aluminum titanate and in anisotropic metals. The much lower conductivity of marble than that of single crystals of calcite is attributed to this phenomenon.

Typical thermal conductivity values for a number of ceramic systems are collected in Table 12.5 and Fig. 12.46. In general, materials with a high

Table 12.5. Thermal Conductivity of Some Ceramic Materials

	Thermal Conductivity (cal/sec/cm²/°C/cm) at	
Material	100°C	1000°C
Al_2O_3	0.072	0.015
BeO	.525	.049
MgO	.090	.017
$MgAl_2O_4$	.036	.014
ThO_2	.025	.007
Mullite	.014	.009
$UO_{2.00}$	.024	.008
Graphite	.43	.15
ZrO_2 (stabilized)	.0047	.0055
Fused silica glass	.0048	.006
Soda-lime-silica glass	.004	–
TiC	.060	.014
Porcelain	.004	.0045
Fire-clay refractory	.0027	.0037
TiC cermet	.08	.02

conductivity at low temperatures have a large negative temperature coefficient; materials with low conductivity have a positive temperature coefficient. As a result, the total range of values observed for different materials decreases as the temperature level is raised. This means that in

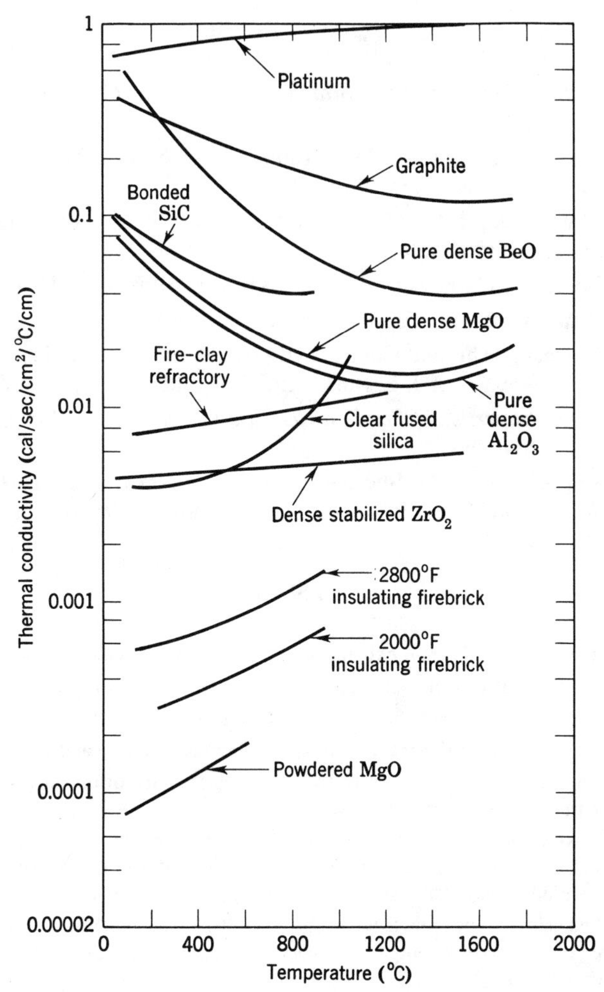

Fig. 12.46. Conductivity (log scale) of a variety of ceramic materials over a wide temperature range.

addition to microstructure determination, the effect of temperature must be considered in comparing different materials. Comparisons made at room temperature, for example, may not be useful in understanding the behavior of materials at 1300 to 1400°C.

Suggested Reading

1. C. Kittel, *Introduction to Solid State Physics*, 3d ed., John Wiley & Sons, Inc., New York, 1968.
2. De Launay in F. Seitz and D. Turnbull, Eds., *Solid State Physics*, Vol. 2, Academic Press, Inc., New York, 1956.
3. J. D. Mackenzie, Ed., *Modern Aspects of the Vitreous State*, Vol. 3, Butterworth, London, 1965.
4. R. H. Doremus, *Glass Science*, John Wiley & Sons, Inc., New York, 1973.
5. W. D. Kingery, *J. Am. Ceram. Soc.*, **42**, 617 (1959).
6. W. D. Kingery, "Thermal Conductivity of Ceramic Dielectrics," *Progress in Ceramic Science*, Vol. 2, J. E. Burke, Ed., Pergamon Press, New York, 1961.
7. *Engineering Properties of Selected Ceramic Materials*, The Am. Ceram. Soc., Columbus, Ohio, 1966.
8. I. E. Campbell and E. M. Sherwood, Eds., *High-Temperature Materials and Technology*, John Wiley & Sons, Inc., New York. (1967).

Problems

12.1. Hysteresis in the thermal expansion of a ceramic is usually due to __________.

12.2. Sketch the heat capacity of a ferrite over a wide range of temperatures.

12.3. Compare the molar specific heat of NaCl and MgO at 700°C.

12.4. A good value for the thermal expansion coefficient for a clay at 0°K is __________.

12.5. Explain why the thermal conductivities of glasses are often orders of magnitude lower than those of crystalline solids.

12.6. For glasses in the 0 to 50% K_2O, 100 to 50% SiO_2 composition range, predict the thermal expansion. Explain the results in terms of (*a*) the melting point of the compositions and (*b*) the structure of the glasses.

12.7. On the basis of your knowledge of the Al_2O_3–Cr_2O_3 system predict the nature of the curve of thermal conductivity versus composition for single crystals and polycrystals.

12.8. Draw a typical thermal conductivity versus temperature curve (10 to 2000°K) for a glass and a crystalline material, and explain the similarities and differences.

12.9. A porcelain of composition 25% quartz (−200 mesh), 25% potash feldspar (−325 M), 15% ball clay (air-floated), and 35% kaolin (water-washed) was fabricated into specimens by slip casting and divided into three groups. Each group was fired at one of three temperatures for an hour (1200°C, 1300°C, 1400°C) but was not labeled, and the student lost his record. He had available a recording dilatometer for measuring thermal expansion. How can he tell at which temperature each group was fired?

12.10. If dense polycrystalline samples of several compositions across the MgO–Cr_2O_3 binary were examined for thermal conductivity as a function of their composition, describe their expected relative thermal conductivity with an appropriate plot. Assume that samples are quenched rapidly from 1800°C and tested at (*a*) room temperature and (*b*) 400°C for thermal conductivity.

12.11. Calculate the effect on the thermal conductivity of ThO_2 (90% dense) at 1500°C of varying the pore size from 0.2 in. in diameter to 0.02 in. in diameter when the pores are isolated.

12.12. (*a*) Estimate the thermal conductivity of fused silica at room temperature.
(*b*) At what temperature would heat transfer by radiation equal phonon thermal conductivity for a glass of average optical absorption coefficient 7 cm within a piece 1 m thick? (Velocity of elastic waves = 5000 m/sec, specific heat = 0.22 cal/g, density = 2.2 g/cc, refractive index = 1.5.)

12.13. Discuss the effects of porosity on thermal diffusivity in pure MgO at (*a*) low temperature, where lattice conduction is predominant, and (*b*) at high temperature, where radiation conductivity is predominant.

13

Optical Properties

Many different optical properties of ceramic products are of concern in different applications. Perhaps most important are those optical glasses and crystals used as windows, lenses, prisms, filters, or in other ways requiring useful optical properties as the primary function of the material. Also, however, much of the value and usefulness of products such as tile, ceramic tableware and artware, porcelain enamels, and sanitary ware depends on properties such as color, translucency, and surface gloss. Consequently, optical properties are important in one way or another for most ceramics.

The variety and complexity of optical properties of potential interest make it essential that we restrict our attention to a few of the major aspects of the problem, rather than attempting any comprehensive treatment. In doing this we first consider properties important for use in optical systems—the index of refraction and dispersion. These properties form the basis for extensive applications of optical glasses. Then we consider various aspects of light reflection, scattering, reflectance, translucency, and gloss. Finally, we consider light absorption and color development and control in ceramic systems and some more recent applications.

13.1 Introduction

Electromagnetic Waves in Ceramics. A dielectric material reacts to electromagnetic radiation differently from free space because it contains electrical charges that can be displaced, as discussed in some detail in Chapter 18. For a sinusoidal electromagnetic wave, there is a change in wave velocity and intensity described by the complex coefficient of refraction:

$$n^* = n - ik \tag{13.1}$$

where n is the index of refraction and k is the index of absorption. The coefficient of refraction is related to the complex dielectric constant (Chapter 18), $n^{*2} = \kappa^*$. Since $\kappa^* = \kappa' - i\kappa''$, where κ' is the relative dielectric constant and κ'' is the relative dielectric loss factor,

$$n^{*2} = n^2 - k^2 - 2ink \tag{13.2}$$

and

$$\kappa' = n^2 - k^2 \qquad \kappa'' = 2nk \tag{13.3}$$

The optical properties of dielectric materials are generally of interest because of their good transmission in the optical part of the spectrum as compared with other classes of materials (Fig. 13.1). At short wavelengths this good transmission is terminated at the ultraviolet absorption edge, which corresponds to radiation energies and frequencies ($E = h\nu = hc/\lambda$) where absorption of energy arises from electronic transitions between levels in the valence band to unfilled states in the conduction band. (Actually, existence of multiple peaks near the ultraviolet absorption edge shows that the details of this process are quite complex.) At long wavelengths the relatively good transmission of dielectrics is terminated by elastic vibration of the ions in resonance with the imposed radiation which gives a frequency of maximum absorption:

$$\nu^2 = 2v\left[\frac{1}{M_c} + \frac{1}{M_a}\right] \tag{13.4}$$

where v is a force constant and M_c and M_a are the ionic masses of cation

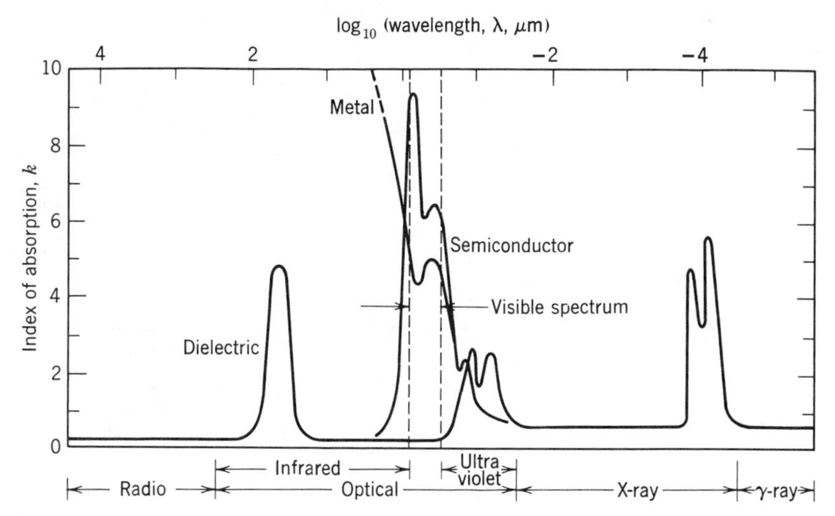

Fig. 13.1. Frequency variation of the index of absorption for metals, semiconductors, and dielectrics.

Table 13.1. Useful Transmission (Exceeding 10%) Regions of Materials for 2-mm Thickness

Wavelength scale (microns): .1, .2, .3, .4, .5, .6, .7, .8, .9, 1, 2, 3, 4, 5, 6, 7, 8, 9, 10, 20, 30, 40, 50, 60, 80, 100 Microns

Ultraviolet (to .4) | Visible (.4–.75) | Near IR (.75–3) | Middle IR (3–6) | Far IR (6–15) | Extreme IR (15 and up)

Material	Lower limit (microns)	Upper limit (microns)
Fused Silica (SiO_2)	.16	4
Fused Quartz (SiO_2)	.18	4.2
Calcium Aluminate Glass ($CaAl_2O_4$)	.4	5.5
Lithium Metaniobate ($LiNbO_3$)	.35	5.5
Calcite ($CaCO_3$)	.2	5.5
Titania (TiO_2)	.43	6.2
Strontium Titanate ($SrTiO_3$)	.39	6.8
Alumina (Al_2O_3)	.2	7
Sapphire (Al_2O_3)	.15	7.5
Lithium Fluoride (LiF)	.12	8.5
Magnesium Fluoride (Polycrystalline) (MgF_2)	.45	9
Yttrium Oxide (Y_2O_3)	.26	9.2
Magnesium Oxide (Single Crystal) (MgO)	.25	9.5
Magnesium Oxide (Polycrystalline) (MgO)	.3	9.5
Magnesium Fluoride (Single Crystal) (MgF_2)	.15	9.6
Calcium Fluoride (Polycrystalline) (CaF_2)	.13	11.8
Calcium Fluoride (Single Crystal) (CaF_2)	.13	12
Barium Fluoride/Calcium Fluoride (BaF_2/CaF_2)	.75	12
Arsenic Trisulfide Glass (As_2S_3)	.6	13
Zinc Sulfide (ZnS)	.6	14.5

Material	Lower limit	Upper limit
Sodium Fluoride (NaF)	.14	15
Barium Fluoride (BaF_2)	.13	15
Silicon (Si)	1.2	15
Lead Fluoride (PbF_2)	.29	15
Cadmium Sulfide (CdS)	.55	16
Zinc Selenide (ZnSe)	.48	22
Germanium (Ge)	1.8	23
Sodium Iodide (NaI)	.25	25
Sodium Chloride (NaCl)	.2	25
Potassium Chloride (KCl)	.21	25
Silver Chloride (AgCl)	.4	30
Thallium Chloride (TlCl)	.42	30
Cadmium Telluride (CdTe)	.9	31
Thallium Chloro Bromide (Tl(Cl, Br))	.4	35
Potassium Bromide (KBr)	.2	38
Silver Bromide (AgBr)	.45	40
Thallium Bromide (TlBr)	.38	40
Potassium Iodide (KI)	.25	47
Thallium Bromo Iodide (Tl(Br, I))	.55	50
Cesium Bromide (CsBr)	.2	55
Cesium Iodide (CsI)	.25	70

and anion. For a wide range of transparency it is desirable to have a high value for the electronic energy gap (Chapters 2 and 4) together with weak interatomic bonding and large ionic mass. These conditions are optimal for the high-atomic-weight monovalent alkali halides, as illustrated in Table 13.1 and Fig. 13.2.

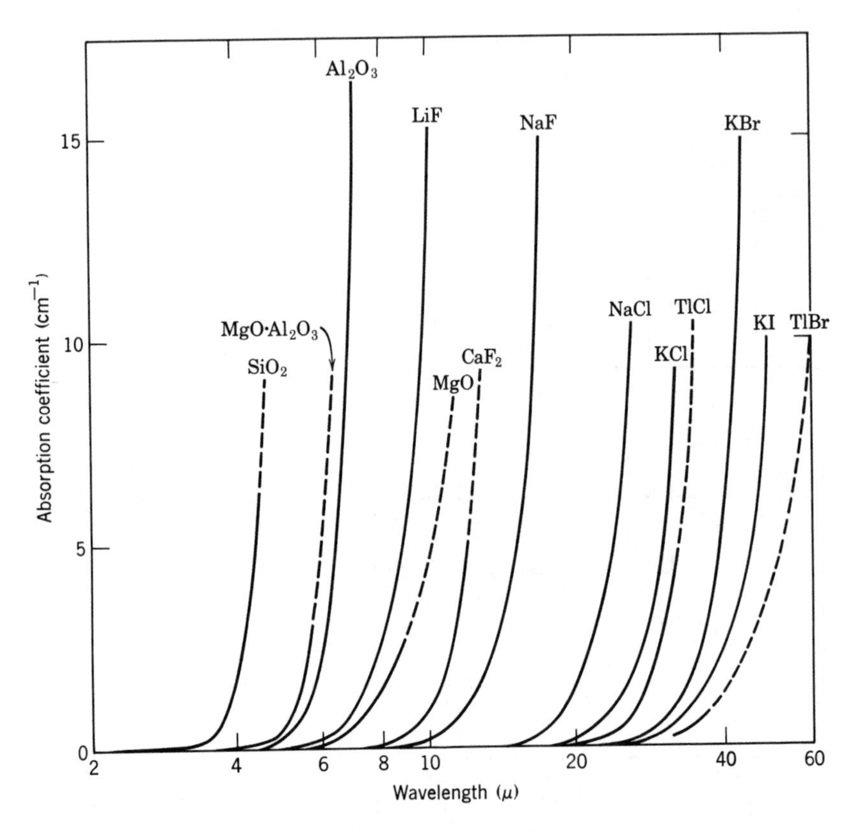

Fig. 13.2. Infrared absorption edges of ionic crystals.

Refractive Index and Dispersion. When light passes from a vacuum into a denser material, its velocity is decreased. The ratio between these velocities determines the index of refraction n:

$$n = \frac{v_{\text{vacuo}}}{v_{\text{material}}} \tag{13.5}$$

The refractive index is a function of the frequency of the light, normally decreasing as the wavelength increases (Fig. 13.3). This change with wavelength is known as dispersion of the refractive index and has been defined in various ways. At any wavelength of interest the dispersion can be most directly given as

$$\text{Dispersion} = \frac{dn}{d\lambda} \tag{13.6}$$

and this value can be determined directly from Fig. 13.3 and is shown in Fig. 13.4. However, most practical measurements are made by using the refractive index at fixed wavelengths rather than by determining the complete dispersion curve. Values are most commonly reported as reciprocal relative dispersion, or ν value:

$$\nu = \frac{n_D - 1}{n_F - n_C} \tag{13.7}$$

where n_D, n_F, and n_C refer to the refractive indices measured with sodium D light, the hydrogen F line, and hydrogen C line (5893 Å, 4861 Å, and 6563 Å—these are indicated in Fig. 13.3*a*). Typical values of the refractive index vary from 1.0003 for air to the range of 1.3 to 2.7 for solid oxides. Silicate glasses have refractive indices of about 1.5 to 1.9.

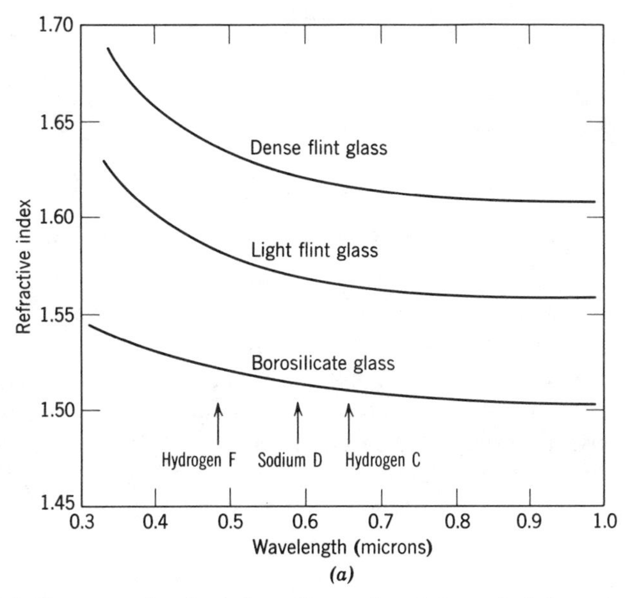

Fig. 13.3. (*a*) Change in refractive index with wavelength for typical glasses in the visible spectrum.

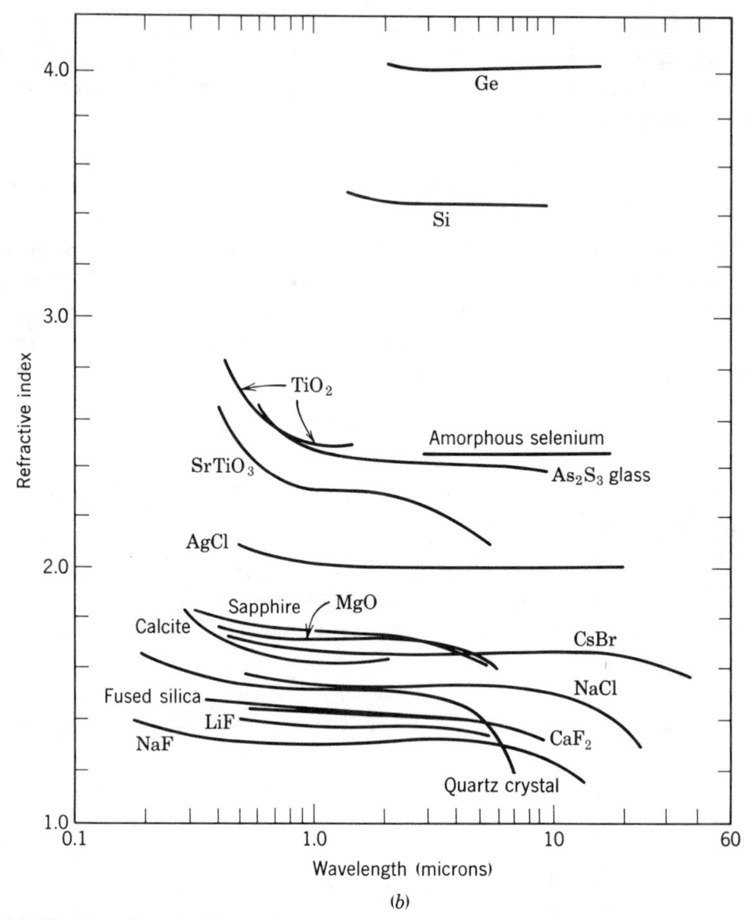

(*b*)

Fig. 13.3 (*Continued*). (*b*) Change in refractive index with wavelength for several crystals and glasses.

Reflection and Refraction. The relative index of refraction between phases (ratio of refractive indices) determines the reflectance and refractive properties of a phase boundary. The change in velocity causes light to bend on passing through the interface. If the angle of incidence from a normal to the surface is i and the angle of refraction is r (Fig. 13.5), these angles are related, when one medium is air or vacuo, by

$$n = \frac{\sin i}{\sin r} \tag{13.8}$$

Some of the light is reflected at an angle equal to the incident angle,

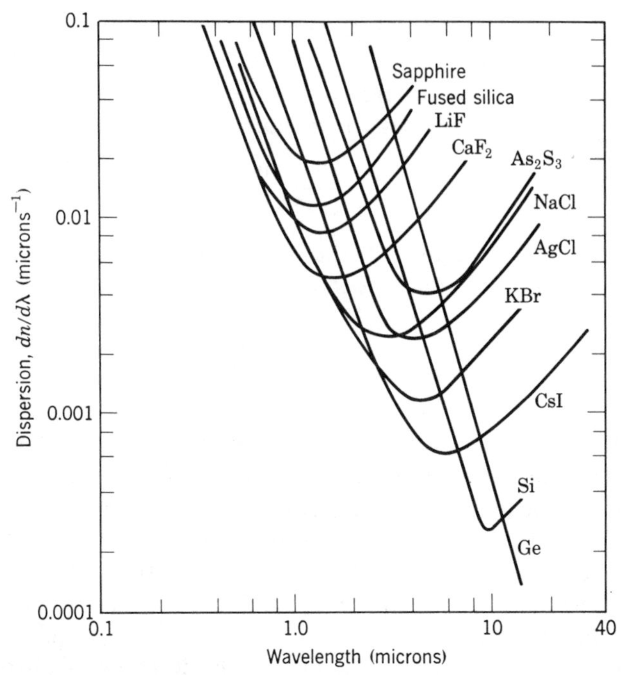

Fig. 13.4. Dispersion as a function of wavelength for several ceramics shown in Fig. 13.3*b*.

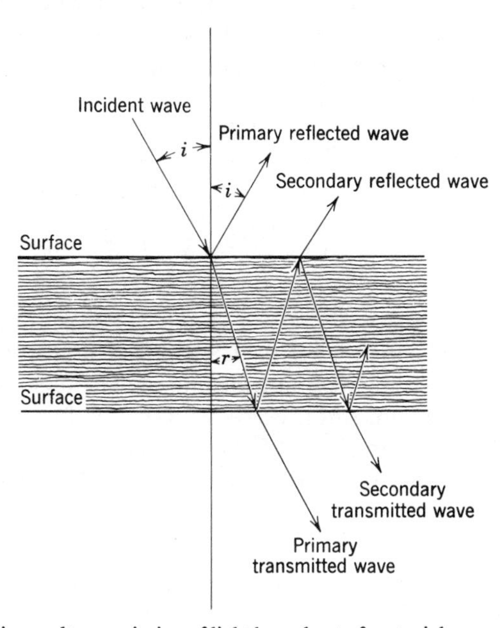

Fig. 13.5. Reflection and transmission of light by a sheet of material.

specular reflection. The fraction of light reflected in this way is given for normal incidence by Fresnel's formula:

$$R = \frac{(n-1)^2 + k^2}{(n+1)^2 + k^2} \tag{13.9}$$

Within the optical region of the spectrum (Fig. 13.1) the index of absorption is much less than the index of refraction; thus Eq. 13.9 reduces to the useful form

$$R = \left(\frac{n-1}{n+1}\right)^2 \tag{13.10}$$

This surface reflection may be desirable, for example, when observing microscopic polished sections or for cut glass, or it may be undesirable, as is frequently true in optical systems. It increases for normal incident light from a value of about 4% at $n = 1.5$ to about 10% at $n = 1.9$. This means that high-index samples are easy to observe microscopically in polished sections and that they require special precautions to avoid excess reflection losses in optical systems.

Absorption. As shown in Eq. 13.1, both the index of absorption and the refractive index are necessary to describe the optical properties of a ceramic. The absorption index is a function of wavelength (Fig. 13.1) and is most directly related to the absorption coefficient $\beta = 4\pi k/\lambda$. For a single-phase material, the fraction of light transmitted is given by the absorption coefficient and sample thickness;

$$\frac{dI}{I_0} = -\beta\, dx \tag{13.11}$$

or

$$T = \frac{I}{I_0} = \exp(-\beta x) \tag{13.12}$$

$$\ln \frac{I}{I_0} = -\beta x \tag{13.13}$$

where I_0 is the initial density, I is the transmitted intensity, x is the optical path length, and T is the fraction transmitted. The absorption coefficient is usually measured in units of inverse centimeters. For a plate of material such as that illustrated in Fig. 13.5, the overall transmission depends on both reflectance losses and absorption. This overall transmission is given for normal incidence by

$$T' = \frac{I_{\text{in}}}{I_{\text{out}}} = (1-R)^2 \exp(-\beta x) \tag{13.14}$$

where R is the reflectivity given by Eq. 13.9.

The absorption coefficient varies greatly with wavelength; selected absorption in the visible range of the spectrum between wavelengths of 0.3 to 0.7 micron is of course the source of color. In addition, it is obvious that incoming light must all be accounted for by reflection, transmission, or absorption. That is, $A' + R' + T' = 1$ if A' is the total fraction of the incident light absorbed in the sample, R' is the summation of the light reflected from the surface, and T' is the summation of light transmitted. These quantities change with frequency or wavelength, as shown in Fig. 13.6, which illustrates a glass with a green color in the visible range and an absorption cutoff at approximately 3 microns in the infrared and about 0.3 micron in the ultraviolet.

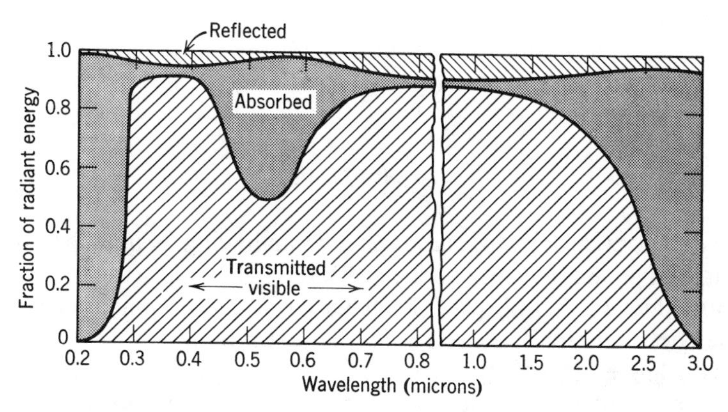

Fig. 13.6. Change in reflection, absorption, and transmission of light with wavelength for a green glass.

The light absorption illustrated in Fig. 13.6 is frequently due to a particular ion in solution, or chromophore; in addition to Lambert's law (Eq. 13.12), the absorption coefficient is also proportional to the concentration of the absorbing ion. That is, β equals ϵc where ϵ is the extinction coefficient or absorption observed per unit concentration. Then

$$\frac{I}{I_0} = \exp(-\epsilon c x) \tag{13.15}$$

which is known as Beer's law or the Beer-Lambert equation.

Scattering. For systems that are optically heterogeneous, such as a transparent medium containing small particles, part of the beam is scattered, and the intensity of the beam is reduced. For a single particle of projected area πr^2, the fraction of beam intensity lost is given by

$$\frac{\Delta I}{I} = -K\frac{\pi r^2}{A} \tag{13.16}$$

where r is the particle radius, A is the beam area, and K is a scattering factor that varies between 0 and 4. If we neglect multiple scattering,

$$\frac{I}{I_0} = \exp(-Sx) = \exp(-KN\pi r^2 x) \tag{13.17}$$

where S is the scattering coefficient, sometimes called the turbidity coefficient, and N is the number of particles per unit volume. If the volume fraction of second-phase particles present is V_p,

$$V_p = \frac{4}{3}\pi r^3 \cdot N \tag{13.18}$$

$$S = \frac{3}{4}KV_p r^{-1} \tag{13.19}$$

$$\frac{I}{I_0} = \exp\left(-3KV_p\frac{x}{4r}\right) \tag{13.20}$$

The energy loss during attenuation of a beam having an initial intensity I_0 is measured as the intensity of the scattered light i at the distance $\mathbf{r}$ from the individual scattering particle, where $\mathbf{r}$ is large compared with the wavelength of the radiation. The energy distribution is a function of the scattering angle between the scattered wave and the incident beam, since the scattered intensity varies as a function of $\mathbf{r}^2$. The scattering can be measured in terms of the Rayleigh ratio:

$$\mathbf{R} = \frac{i\mathbf{r}_\theta^2}{I} \tag{13.21}$$

where θ is the angle to the incident beam. The scattered intensity of the scattered light is a maximum parallel to the beam and a minimum normal to the incident beam. The scattering coefficient can be determined by the integration of the scattered intensity at all angles to the incident beam. In terms of the scattering normal to the beam, it is given by

$$S = \frac{16\pi}{3}R_{90^\circ} = \frac{8\pi}{3}R_0. \tag{13.22}$$

The value of the scattering factor K is mainly dependent on (1) the ratio of the particle diameter to the wavelength of the incident light, usually measured in terms of $\alpha = 2\pi r/\lambda$, and (2) the relative refractive index of the particle and the medium ($m = n_{\text{particle}}/n_{\text{medium}}$). In addition, the amount of energy scattered depends on the solid angle subtended by the beam of the incident, measured radiation as indicated in Eq. 13.22, and also the shape of the particle and its orientation in the beam. The greatest effect in ceramic systems is related to the relative refractive index in which there is increasing scattering with an increasing difference in index between the particle and the medium. In addition, the scattering is strongly dependent on particle size, so that the maximum scattering occurs at a particle size of the same magnitude as the radiation wavelength. For particle sizes much smaller than the wavelength of incident radiation, the scattering constant K increases with particle size and is inversely proportional to the fourth power of wavelength. The scattering coefficient reaches a maximum when the particle size is about equal to the wavelength of incident radiation and decreases at larger particle-size values. The scattering constant K reaches a constant value for particle sizes substantially larger than the wavelength of the incident radiation, so that for a fixed concentration of second phase the measured scattering coefficient is inversely proportional to particle size, as indicated in Eqs. 13.19 and 13.20. Effects of particle size and relative refractive index are illustrated in Fig. 13.7.

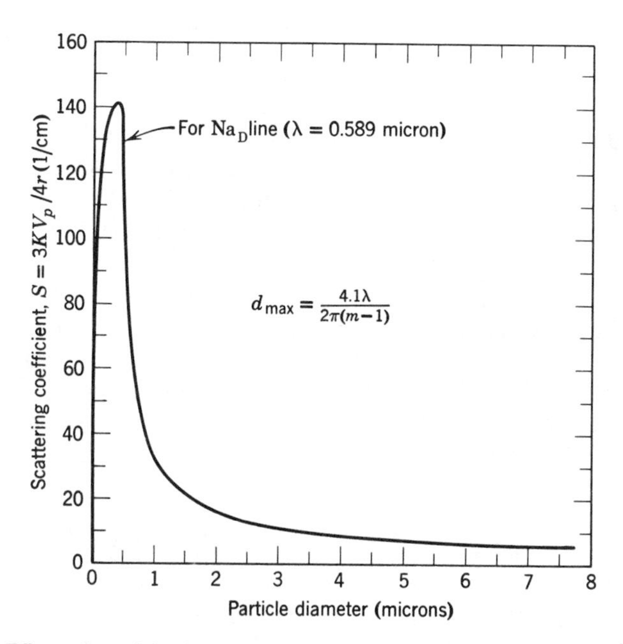

Fig. 13.7. Effect of particle size on the scattering coefficient ($S = 3KV_p/r4$) of a fixed volume of particles (1.0 vol%) for a relative refractive index of 1.8 (TiO_2 in glass).

13.2 Refractive Index and Dispersion

Electric Dipole Moments. A dielectric material reacts with and affects electromagnetic radiation because it contains charged carriers that can be displaced. The light waves are retarded because of the interaction of the electromagnetic radiation and the electronic systems of the atoms. The relationship between the applied field and the medium can be considered as resulting from the presence of elementary electric dipoles having an average dipole moment $\bar{\mu}$. If the dipole is represented by two charges of the opposite polarity, + and $-Q$, separated by a distance d, then $\bar{\mu}$ equals Qd. Over the range of optical frequencies the source of this dielectric polarization is the shift of the electron cloud around the atomic nucleus (Fig. 13.8).

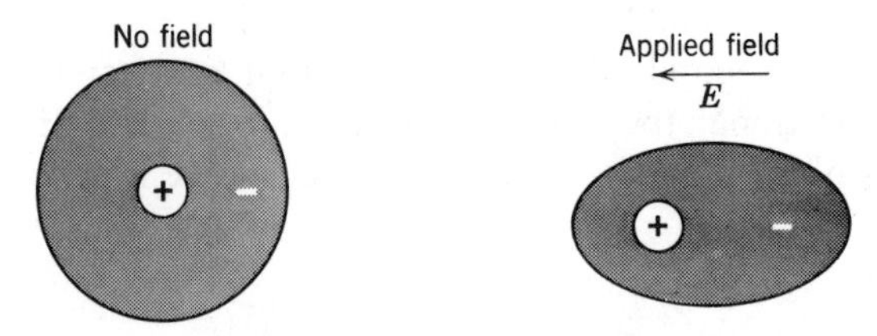

Fig. 13.8. Shift of electron cloud giving rise to dipole moment and electronic polarization.

The average dipole moment $\bar{\mu}$ is proportional to the local electric field strength that acts on the particle; the proportionality constant α is called the polarizability and measures the average dipole moment per unit field strength, $\bar{\mu} = \alpha E$. (The polarizability is measured in units of cubic centimeters in the esu system or $\sec^2\ C^2/kg = \epsilon\, m^3$ in the mks system of units.) The summation of all the elementary dipoles gives the total dipole moment per unit volume or the polarization P; that is, if there are N particles per unit volume, $P = N\bar{\mu} = N\alpha E$. The Lorentz-Lorenz equation connects the polarizability α of the atoms of an elementary monatomic gas with the refractive index for waves of infinite wavelength,

$$\alpha = \frac{3\epsilon_0}{N_0}\frac{n^2-1}{n^2+2}\frac{M}{\rho} = \frac{3\epsilon_0}{N_0}R_\infty \tag{13.23}$$

where ϵ_0 is the dielectric constant of a vacuum ($\epsilon_0 = 1/4\pi$ in the esu system; 8.854×10^{-12} F/m in the mks system), N_0 is Avogadro's number, M is the molecular weight, ρ is the density, and R_∞ is the molar refractivity and is directly proportional to the atomic polarizability.

Polarizability. In the simplest case of a monatomic gas, the molar refractivity can be directly and simply rationalized by considering that the electron density is uniform within the atomic radius r_0 (this is incorrect of course, as discussed in Chapter 2, but is a suitable simple approximation for the present purpose). If we consider an atom of atomic number Z, the nucleus has a positive charge $+Ze$ and is surrounded by an electron atmosphere of the charge $-Ze$. An external field E exerts a force on the electron atmosphere, $F = -ZeE$, and its center is displaced by a distance d. When displaced, it is in equilibrium with the coulomb attraction exercised by the positive nucleus $+Ze$. If the displacement occurs with spherical symmetry (Fig. 13.9) and the displacement of the center is d, the

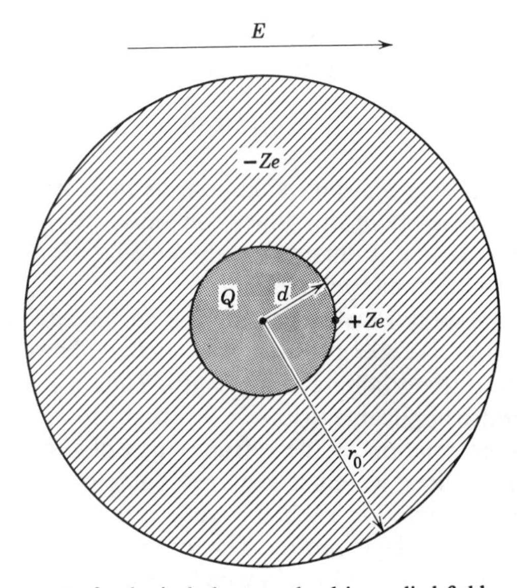

Fig. 13.9. Displacement of spherical electron cloud in applied field.

part of the charge Q_d that contributes to the asymmetry is given by

$$Q_d = -Ze\frac{d^3}{r_0^3} \tag{13.24}$$

This charge acts as if it were concentrated at a distance d from the nucleus, so that the coulomb force between the nucleus $+Ze$ and the negative charge Q_d is

$$F_C = -(Ze)^2\frac{d^3/r_0^3}{4\pi\epsilon_0 d^2} \tag{13.25}$$

At equilibrium this coulomb force is balanced by the force of the external field, $F_C = F$,

$$-ZeE = \frac{-(Ze)^2 d^3/r_0^3}{4\pi\epsilon_0 d^2} \tag{13.26}$$

From this,

$$d = 4\pi\epsilon_0 r_0^3 E/Ze \tag{13.27}$$

and

$$\bar{\mu} = Zed = 4\pi\epsilon_0 r_0^3 E \tag{13.28}$$

so that the electronic polarizability of the atom is given by

$$\alpha = 4\pi\epsilon_0 r_0^3 \tag{13.29}$$

In the unrationalized esu system of units, $4\pi\epsilon_0$ equals 1 so that the electronic polarization is equal to r_0^3. That is, for helium,

$$\alpha_{\mathrm{He}} = r_0^3 = 0.74 \times 10^{-24}\ \mathrm{cm}^3 \tag{13.30}$$

This calculation is only an approximation, as indicated previously, and actually gives a value about five times too small. Its main interest from our point of view is that it illustrates the close relationship between the atomic volume and atomic polarizability.

The molar refractivity R_∞ of gases can be determined experimentally by measuring the refractive index for various wavelengths, as in Fig. 13.3, and then extrapolating to an infinite wavelength. This procedure, with some modification, is also applicable to ions in solution. However, the value thus obtained is the summation of the refractivities of the separate ions; to obtain individual values, we must assume a value of some one particular ion. A reasonable assumption is that the molar refractivity of the H^+ is zero, since this consists of a single proton. The molar refractivities of a number of ions are collected in Table 13.2, in which they are grouped according to the total number of orbital electrons. As indicated before, ion size has a considerable effect, but increasing negative charge for any given number of orbital electrons also has a substantial effect. This is understandable, since the outer electrons are less firmly bound in a negative ion than in a positve one; thus, in addition to size effects, the outer electrons are expected to contribute the major share to the polarizability.

In general, then, we expect that the ionic polarizability increases with the size of the ion and with the degree of negative charge on isoelectric ions. Since the index of refraction increases along with the polarizability, as indicated in Eq. 13.23, we obtain high refractive indices with large ions ($n_{\mathrm{PbS}} = 3.912$) and low indices with small ions ($n_{\mathrm{SiCl_4}} = 1.412$). However, the immediate surroundings and arrangement of ions also affect the refractive

Table 13.2. Molar Refractivities of a Number of Atoms and Ions

Number of Electrons	Charge					
	−2	−1	0	+1	+2	+3
10	O	F	Ne	Na	Mg	–
	7.0	2.4	1.0	0.5	0.3	
18	–	Cl	A	K	Ca	Sc
	–	9.0	4.2	2.2	1.3	0.9
36	–	Br	Kr	Rb	Sr	–
	–	12.6	6.3	3.6	2.2	
54	–	I	Xe	Cs	Ba	La
		19.0	10.3	6.1	4.2	3.3

index; only in glasses and in cubic crystals is the index independent of crystallographic direction. In other crystal systems the index of refraction is high in directions that are close-packed in the structure. This follows directly from Eq. 13.23. Similarly, the more open structures of high-temperature polymorphic forms have lower refractive indices than the low-temperature forms, and glasses have lower indices than crystals of the same composition. For SiO_2, for example, $n_{glass} = 1.46$, $n_{tridymite} = 1.47$, $n_{cristobalite} = 1.49$, $n_{quartz} = 1.55$.

In the same way that densely packed directions in crystals have the highest refractive index, the application of a tensile stress to an isotropic material such as a glass increases the index normal to the direction of the stress and decreases the index along the stressed direction. Uniaxial compression has the reverse effect. Since the incremental changes in light velocity (refractive index) are directly proportional to the applied stress, the birefringence (difference in indices in different directions) can be used as a method of measuring stress.

The refractive index of a typical soda-lime-silicate glass is about 1.51. In the great majority of practical glasses, constituents having the greatest effect in raising the index of the fraction are lead (atomic number 82) and barium (atomic number 56) additions. Glasses containing 90 wt% PbO have an index of refraction of about 2.1. The total range of values obtainable in practical optical glasses is from about 1.4 to 2.0 (Fig. 13.10). Some compositions and refractive indices are collected in Table 13.3.

Dispersion. The dispersion of the refractive index results from the fact that the visible spectrum is adjacent to the natural frequency of electronic

Table 13.3. Refractive Indices of Some Glasses and Crystals

	Average Refractive Index	Birefringence
Glass composition:		
From orthoclase ($KAlSi_3O_8$)	1.51	
From albite ($NaAlSi_3O_8$)	1.49	
From nepheline syenite	1.50	
Silica glass, SiO_2	1.458	
Vycor glass (96% SiO_2)	1.458	
Soda-lime-silica glass	1.51–1.52	
Borosilicate (Pyrex) glass	1.47	
Dense flint optical glasses	1.6–1.7	
Arsenic trisulfide glass, As_2S_3	2.66	
Crystals:		
Silicon chloride, $SiCl_4$	1.412	
Lithium fluoride, LiF	1.392	
Sodium fluoride, NaF	1.326	
Calcium fluoride, CaF_2	1.434	
Corundum, Al_2O_3	1.76	0.008
Periclase, MgO	1.74	
Quartz, SiO_2	1.55	0.009
Spinel, $MgAl_2O_4$	1.72	
Zircon, $ZiSiO_4$	1.95	0.055
Orthoclase, $KAlSi_3O_8$	1.525	0.007
Albite, $NaAlSi_3O_8$	1.529	0.008
Anorthite, $CaAl_2Si_2O_8$	1.585	0.008
Sillimanite, $Al_2O_3 \cdot SiO_2$	1.65	0.021
Mullite, $3Al_2O_3 \cdot 2SiO_2$	1.64	0.010
Rutile, TiO_2	2.71	0.287
Silicon carbide, SiC	2.68	0.043
Litharge, PbO	2.61	
Galena, PbS	3.912	
Calcite, $CaCO_3$	1.65	0.17
Silicon, Si	3.49	
Cadmium telluride, CdTe	2.74	
Cadmium sulfide, CdS	2.50	
Strontium titanate, $SrTiO_3$	2.49	
Lithium niobate, $LiNbO_3$	2.31	
Yttrium oxide, Y_2O_3	1.92	
Zinc selenide, ZnSe	2.62	
Barium titanate, $BaTiO_3$	2.40	

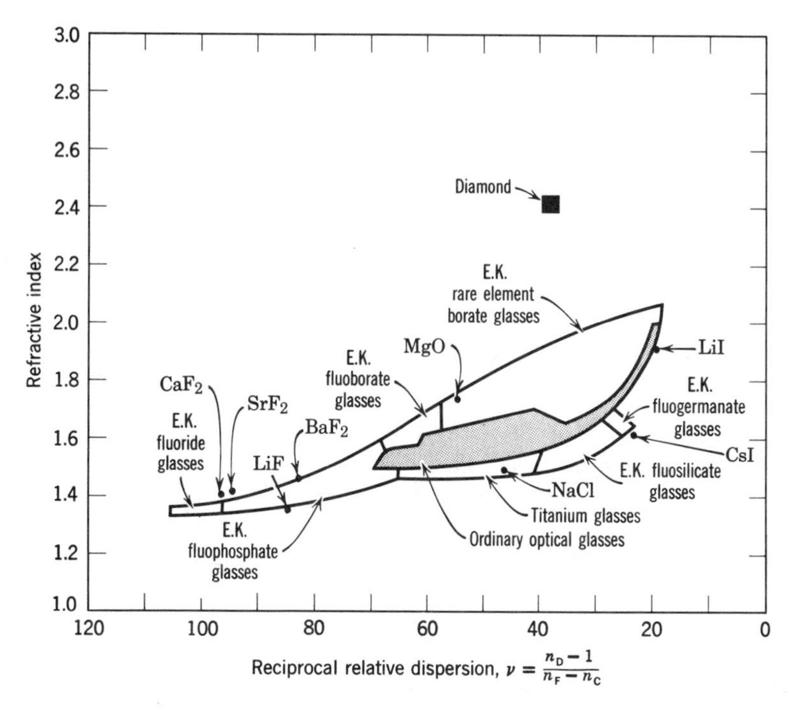

Fig. 13.10. Range of optical properties obtained with crystals and ordinary optical glasses and with Eastman Kodak Company fluoride and rare earth glasses.

oscillators in the ultraviolet. There is strong interaction with electromagnetic radiation of the natural frequency causing resonance, or reinforcement of the natural oscillations, and a high absorption at this resonant frequency (compare with Chapter 18). At frequencies far from the resonant frequency for a single electron,

$$n^2 = 1 + \frac{Ne^2/\pi m}{\nu_0^2 - \nu^2} \tag{13.31}$$

where N is the number of atoms per unit volume, e is the electronic charge, m is the electronic mass, ν_0 is the natural frequency, and ν is the frequency of the incident radiation. In general, there is a sum of terms of the form of Eq. 13.31, for solids the interaction between adjacent ions must also be considered, but the same general shape of the dispersion characteristic results (Fig. 13.11). The index increases with decreasing wavelength in the visible range, and this is referred to as normal dispersion. One of the

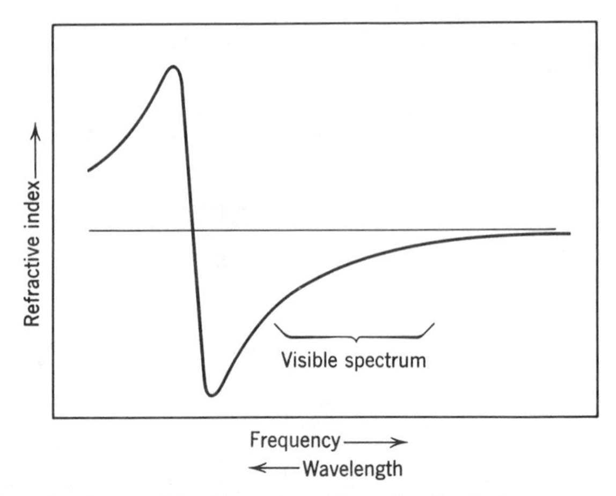

Fig. 13.11. Normal and anomalous dispersion of the refractive index.

common empirical relations for this behavior is the Cauchy formula:

$$n = A + \frac{B}{\lambda^2} + \frac{C}{\lambda^4} \tag{13.32}$$

It can easily be shown that Eq. 13.31 leads to this result if ν is much less than ν_0. In the region of the natural frequency where resonance occurs, the index decreases with decreasing wavelength, and this is referred to as anomalous dispersion.

13.3 Boundary Reflectance and Surface Gloss

As indicated in Eq. 13.10, Fresnel's formula shows that the specular reflectance from a perfectly smooth surface is determined by the index of refraction, $R = (n - 1)^2/(n + 1)^2$. This means that in optical systems the reflectance losses increase as the index of refraction is raised. In contrast, high reflectivity is desirable along with strong refraction as the basis for cut-glass "crystal," which has a high lead content, a high refractive index, and consequently about twice as high reflectivity as ordinary soda-lime-silica glass. Similarly, the high index of refraction of gem stones gives them desirable strong refraction as well as high reflectivity. Glass fibers are used as light pipes for illumination and communications and depend on total internal reflection of the light beam. This is accomplished with a variable-refractive-index glass or by coatings.

For optical-engineering applications it is desirable to combine strong refraction with low reflectivity. This can be done by applying to optics a coating with a thickness of one-fourth of a wavelength of light, usually in the middle of the visible spectrum, and an intermediate index of refraction such that the primary reflected wave illustrated in Fig. 13.5 is just canceled by the secondary reflected wave of equal magnitude and opposite phase. Coated objectives of this kind are used in most microscopes and many other optical systems. This same system is used to make "invisible" windows.

Most surfaces in ceramic systems are not perfectly smooth, and consequently there is considerable diffuse reflection from the surface. If the amount of energy reflected at different angles from a single incident beam is measured for an opaque material, results such as those shown in Fig. 13.12 are obtained. In opaque materials the diffuse reflection arises from the surface roughness, as illustrated in Fig. 13.13, (for nonopaque materials subsurface reflections discussed in the next section are the most important).

The *gloss* of a surface is difficult to define exactly, but it is intimately related to the relative amounts of specular and diffuse reflection. It has been found to be most closely related to the sharpness and perfection of the

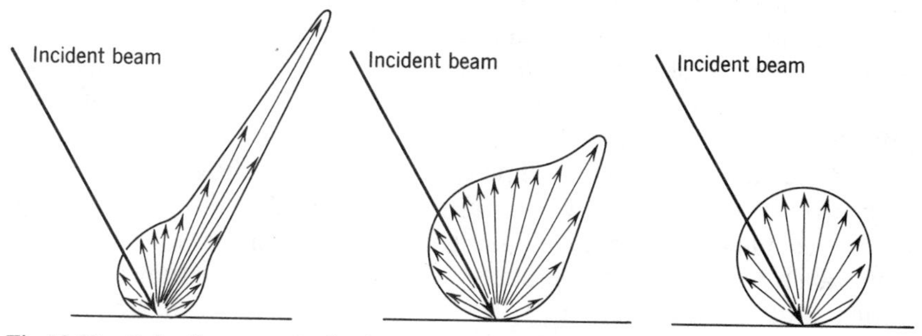

Fig. 13.12. Polar diagrams of reflection from surfaces of increasing roughness.

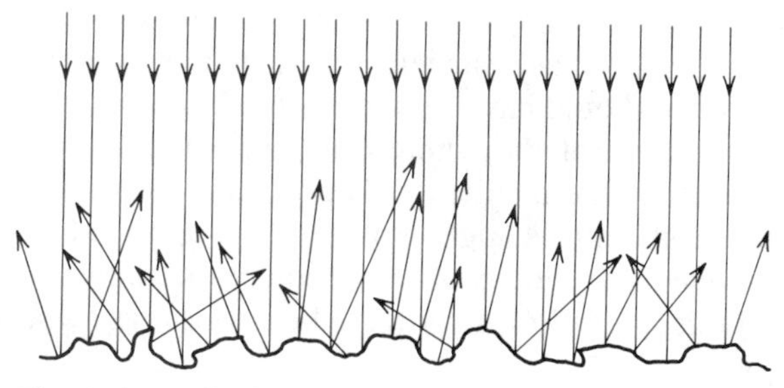

Fig. 13.13. Reflections from a rough surface.

reflected image, that is, to the narrowness of the specular reflection band and its intensity.* These factors are primarily determined by the index of refraction and by the surface smoothness. In order to obtain a high surface gloss, it is desirable to use a lead-base glaze or enamel composition fired to a temperature at which it flows out to form a completely smooth surface. The surface gloss can be decreased by using a lower-index glass phase or by creating surface roughness. Surface roughness can be produced by grinding or sand blasting, by chemical etching an initially smooth surface, and by the deposition of particulate material from a suspension, solution, or vapor phase. Difficulties in obtaining high gloss with glazes and enamels usually result from surface roughness caused by crystal formation, a wavy surface, or craters created by bursting bubbles.

13.4 Opacity and Translucency

The appearance and application of glazes, enamels, opal glasses, and porcelains depend in good part on reflectance and transmission properties. These are strongly influenced by the light-scattering characteristics of the multiphase systems normally used. The overall effect of small-particle scattering is illustrated for a glaze or enamel and a glass sheet or porcelain in Fig. 13.14. The important optical characteristics are the fraction of the light specularly reflected, which determines the gloss, the fraction of the light directly transmitted, the fraction of the incident light diffusely reflected, and the fraction of the incident light diffusely transmitted. Polar-reflection diagrams for some typical glazes are shown in Fig. 13.15. High opacity or covering power requires that the light be diffusely reflected before reaching the underlayer having variable optical characteristics. For high translucency the light should be scattered, so that the transmission is diffuse, but a large fraction of the incident light should be transmitted rather than diffusely reflected.

Opacification. As illustrated in Fig. 13.7, the main factors determining the overall scattering coefficient and consequently affecting the opacity of a two-phase system are the particle size, relative refractive index, and volume of the second-phase particles present. For maximum scattering power the particles should have an index of refraction far different from that of the matrix material, they should have a particle size nearly the same as the wavelength of the incident radiation, and the volume fraction of particles present should be high. For opacifying porcelain-enamel and silicate-glass systems, the index of refraction of the glasses used is limited to a range of about 1.49 to 1.65. In order to be effective as a scattering agent,

*See, for example, A. Dinsdale and F. Malkin, *Trans. Brit. Ceram. Soc.*, **54**, 94 (1955).

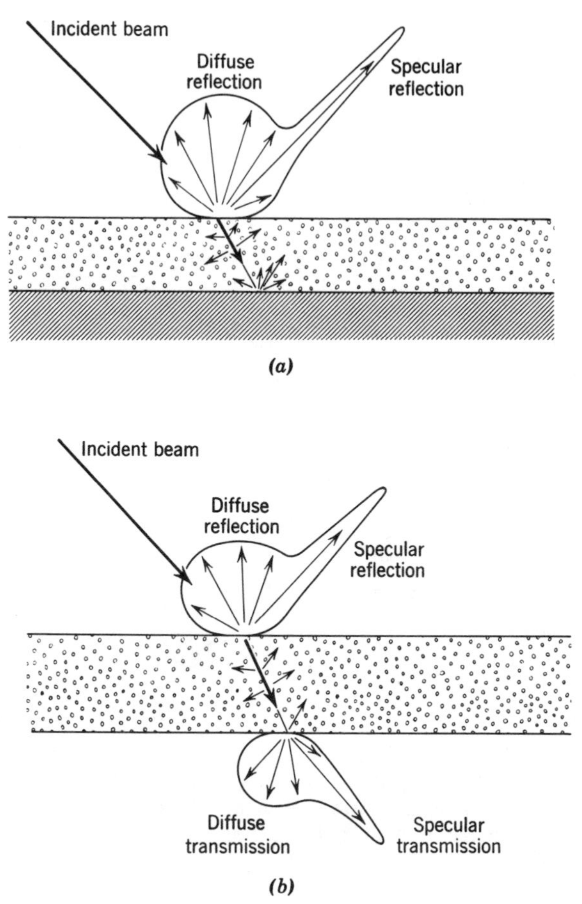

Fig. 13.14. Specular and diffuse reflection and transmission from (*a*) surface coating of glaze or enamel and (*b*) plate of translucent glass or porcelain.

the opacifier must have an index of refraction substantially different from this value, and this limits the number of materials available. In addition, opacifiers must be able to be formed as small particles in a silicate liquid matrix; this further limits the materials that can be used.

Opacifiers can be materials that are completely inert with the glass phase (exactly analogous to pigments in paints), they can be inert products formed during melting, or they can be crystallized from the melt during cooling or reheating, as discussed in Chapter 8. This last is the most effective method of obtaining the fine particle sizes desired and is consequently the most commonly used when a high degree of opacification

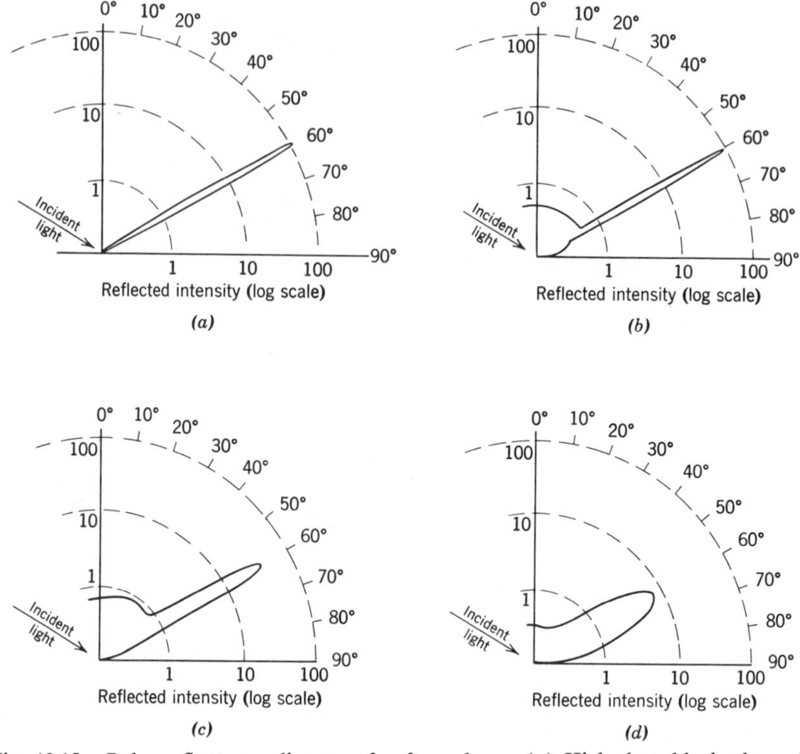

Fig. 13.15. Polar reflectance diagrams for four glazes. (*a*) High-gloss black glaze; (*b*) high-gloss china glaze; (*c*) medium-low-gloss sanitary ware glaze; (*d*) low-gloss semimat glaze. From A. Dinsdale and F. Malkin, *Trans. Brit. Ceram. Soc.*, **54**, 94 (1955).

is required. Several materials commonly used as opacifiers or present as crystalline phases in glasses and glazes are collected in Table 13.4. As seen there, the most effective opacifier is titanium dioxide. Since this is also a material that can be nucleated and crystallized to give very fine particle sizes, it is widely used as an opacifier for porcelain enamels in which a high degree of opacification is required.

The fractions of incident light that are reflected, absorbed, and transmitted depend on the thickness of the specimen as well as the scattering and absorption characteristics. The reflectivity R_∞, which is equal to the reflectance (fraction of incident light diffusely and specularly reflected) of an infinitely thick specimen, is unity for a material in which there is no light absorption. For materials with a high absorption coefficient the reflectivity is low; consequently, a good opacifier must have a low absorption coefficient, that is, good transmission characteristics on a microscale.

Table 13.4. Classification of Opacifiers Suitable for Silicate Glass Media ($n_{glass} = 1.5$)

Opacifiers	n_D	$n_{crystal}/n_{glass}$
Inert additions		
SnO_2	2.0	1.33
$ZrSiO_4$	2.0	1.33
ZrO_2	2.2	1.47
ZnS	2.4	1.6
TiO_2 (anatase)	2.52	1.68
TiO_2 (rutile)	2.76	1.84
Inert products of reactions on melting		
Air Pores	1.0	0.67
As_2O_5	2.2	1.47
$PbAs_2O_6$	2.2	1.47
$Ca_4Sb_4O_{13}F_2$	2.2	1.47
Nucleated and crystallized from the glass		
NaF	1.3	0.87
CaF_2	1.4	0.93
$CaTiSiO_5$	1.9	1.27
ZrO_2	2.2	1.47
$CaTiO_3$	2.35	1.57
TiO_2 (anatase)	2.52	1.68
TiO_2 (rutile)	2.76	1.84

For practical applications of opacifiers we are concerned with the light reflectance of a specimen in contact with a backing of some reflectance R' such as the metal surface under a porcelain enamel or a porcelain body under a glaze surface. If R_0 is defined as the light reflectance of a specimen in contact with a backing of zero reflectance (a material that either completely absorbs or completely transmits the incident radiation), then $R_{R'}$, the light reflectance of a specimen in contact with a backing of reflectance R', is given by P. Kubelka and F. Munk* as

$$R_{R'} = \frac{(1/R_\infty)(R' - R_\infty) - R_\infty(R' - 1/R_\infty)\exp[Sx(1/R_\infty - R_\infty)]}{(R' - R_\infty) - (R' - 1/R_\infty)\exp[Sx(1/R_\infty - R_\infty)]} \tag{13.33}$$

This equation is difficult to solve analytically but indicates that the reflectance increases as the backing reflectance R' increases, as the

Z. Tech. Phys., **12**, 593 (1931).

scattering coefficient increases, as the thickness of coating increases, and as the reflectivity R_∞ increases.

The reflectivity R_∞ is determined by the ratio of the absorption coefficient β and the scattering coefficient S. It is given by

$$R_\infty = 1 + \frac{\beta}{S} - \left(\frac{\beta^2}{S^2} + \frac{2\beta}{S}\right)^{1/2} \tag{13.34}$$

That is, the reflectance of a thick coating is determined equally as much by the absorption coefficient as by the scattering characteristics.

The covering power of an opaque coating is related to the fraction of light reflected from adjacent areas in the underlying material, one area having a high backing reflectance R' and the other a low backing reflectance. This is measured in terms of a contrast ratio or opacifying power $C_{R'} = R_0/R_{R'}$. It is convenient to take $R' = 0.80$, so that $C_{0.80} = R_0/R_{0.80}$, that is, the ratio of reflectance with a completely absorbing backing and the reflectance with a fairly good reflectance backing. This ratio can be related to the reflectance of a specimen having $R' = 0$ from Eq. 13.33. This equation is graphically represented for any value of the reflectivity by the family of curves in Fig. 13.16. For a specific contrast ratio R_∞, scattering coefficient, coating thickness, and value of R_0, the opacity or contrast ratio can be determined. Good opacification is obtained with high values of reflectivity, thick coating, and high values of the scattering coefficient or some combination of these.

The particular combination of opacifiers used depends on the system being opacified and on the method of fabrication. For sheet-steel porcelain enamels a thin coating applied by spraying or brushing is desired. The reflectance of the base is low. For good covering power with a low value of the backing reflectance in thin layers, the highest possible value of the scattering coefficient must be obtained. For this purpose titania-opacified enamels, in which the relative refractive index is high and particle sizes equal to the wavelength of light can be obtained by nucleation and precipitation, are preferred. In contrast, cast-iron enamels are applied on the heavy metal sections by a dry process in which the coating is put on the hot metal as a powder. This technique and the roughness of the casting require that a heavy coat of enamel be used. Coating thicknesses are on the order of 0.070 in. compared with a value of 0.007 in. for sheet steel, an increase of a factor of 10. Consequently, a lower opacifying power can be satisfactorily used on cast iron than for sheet steel applications, and a more economical opacifier than titania is indicated. (Also, the long cooling cycle which varies with the casting thickness makes control of a nucleation process very difficult.) Antimony base opacifiers have been found satisfactory.

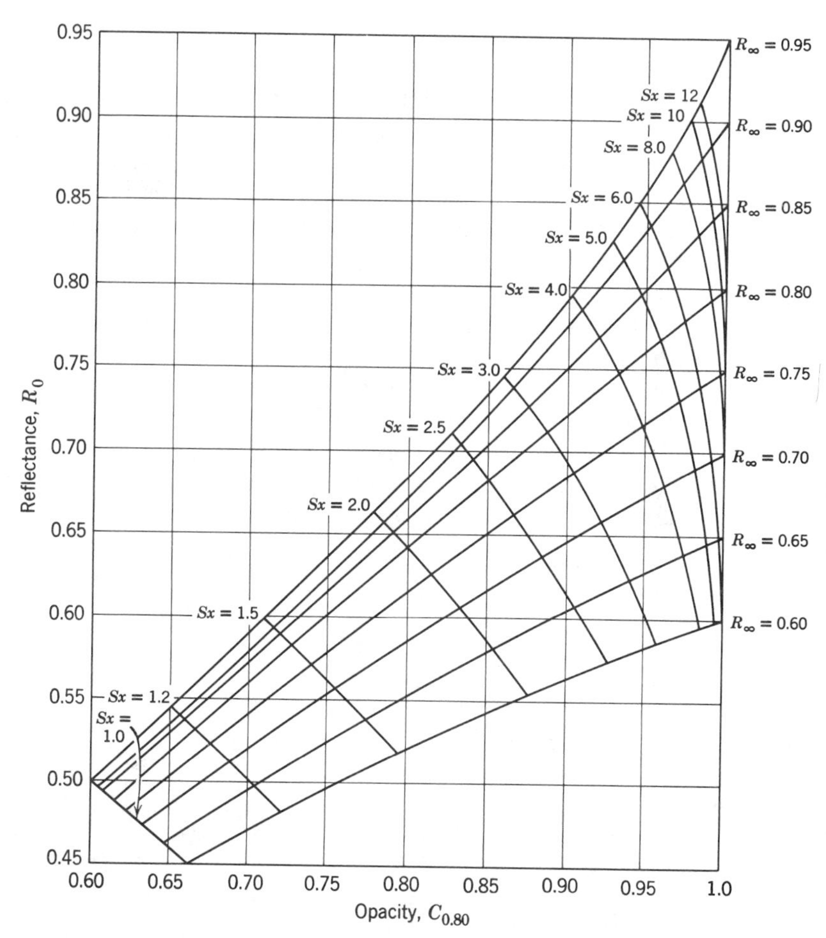

Fig. 13.16. Reflectance-opacity chart. From D. B. Judd, W. N. Harrison, and B. J. Sweo, *J. Am. Ceram. Soc.*, **21**, 16 (1938).

For a porcelain body, a fairly heavy coating of glaze is required to cover surface defects adequately; coating thicknesses of 0.020 in. are typical. In addition, the reflectance of the white bodies ordinarily used is quite high. Consequently, requirements for opacification are not so rigid as they are for porcelain enamels on either sheet steel or cast iron. Since firing the glaze requires a longer period of time, it is more difficult to control precipitation processes in the glaze than it is in enamels; consequently, inert insoluble opacifiers that can be added to the glaze batch, such as zircon and tin oxide, have been found to be the most generally suitable.

In designing particular opacifying compositions, the reflectivity of an infinitely thick coating is determined by the absorption characteristics in relation to the light scattering. If the absorption is zero, the reflectivity must become unity when a sufficiently thick coating is applied. The rate at which opacity increases with the concentration of opacifying phase depends on the scattering coefficient, that is, the relative index of refraction and the particle size of the material. Consequently, even though the reflectivity of an infinitely thick coating may be equivalent for two opacifiers, the volume fraction of added material required to obtain a specified level of reflectance at a particular coating thickness may be quite different. Consequently, both factors must be known in order to make a reasonable selection of a particular system for a particular application. Typical variations of total reflectance for enameled coatings of different thicknesses are illustrated in Fig. 13.17. Equivalent opacification can be obtained by choosing different thicknesses or different opacifier concentrations. Some typical examples of glass-opacifier systems are illustrated in Table 13.5.

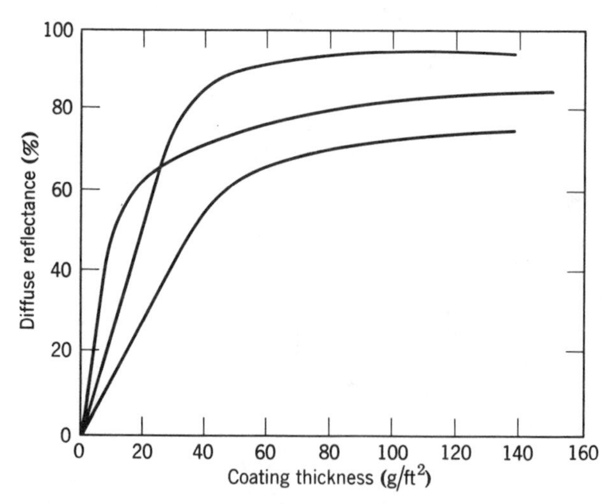

Fig. 13.17. Relation between diffuse reflectance and thickness for three different coatings.

Translucency. In addition to the diffuse reflection produced by internal scattering, the fraction of incident light which emerges as diffuse transmission is important for applications such as opal glasses and translucent porcelains. For opal glasses it is desirable to have appreciable scattering with a minimum absorption so that there is a maximum of diffuse transmission. In this glass a dispersed phase with an index not much

Table 13.5. Typical Examples of Glass-Opacifier Systems[a]

Glass Composition	Fluoride Opal Glass	Zircon Glaze, Cone 11	Tin Oxide Glaze, Cone 11	Zircon Glaze, Cone 06	Antimony Cast-Iron Enamel	Titanium Sheet Steel Enamel
SiO_2	71.7	73.1	61.1	43.0	48.3	64.1
Al_2O_3	5.3	14.6	12.1	3.1		4.7
Na_2O	4.7	0.8	1.0	4.2	16.4	9.1
K_2O	3.5	3.2	5.2	1.6		0.8
B_2O_3	6.4			8.5	5.1	16.2
CaO		2.0	12.1	2.5		1.3
ZnO	6.4		8.4			1.3
PbO	2.0			37.2	23.0	
MgO		6.3				2.5
TiO_2					7.2	
	100.0	100.0	99.9	100.1	100.0	100.0
Composition						
Opacifiers (weight)	7.7% NaF 8.7% CaF_2	13.6$ZrSiO_4$	4.1SnO_2	14.0$ZrSiO_4$	5.4NaF 16.5$Ca_4F_2Sb_4O_{13}$	11.9NaF 17.0TiO_2
Composition by volume	88.2% glass 5.9% NaF 5.9% CaF_2	92.8% glass 7.2% $ZrSiO_4$	98.4% glass 1.6% SnO_2	90.8% glass 9.2% $ZrSiO_4$	87.4% glass 4.7% NaF 7.9% $Ca_4F_2Sb_4O_{13}$	82.1% glass 8.2% NaF 9.7% TiO_2
Refractive indices						
Glass	1.5	1.5	1.5	1.6	1.6	1.5
Opacifiers	NaF = 1.3 CaF_2 = 1.4	$ZrSiO_4$ = 2.0	SnO_2 = 2.0	$ZrSiO_4$ = 2.0	NaF = 1.3 $Ca_4F_2Sb_4O_{13}$ = 2.2	NaF = 1.3 TiO_2 = 2.5

[a] After W. W. Coffeen, *Ceram. Ind.*, **70**, 77 (May 1938).

different from that of the matrix is the most satisfactory; sodium and calcium fluoride are very commonly used, as indicated in Table 13.5.

Frequently the translucency of single-phase oxide ceramics is cited as an indication of their overall quality. This index of quality works reasonably well because the pore size is similar in different bodies and the translucency depends almost exclusively on the pore concentration. (Strength and other properties are closely related to the porosity.) In aluminum oxide, for example, the index of refraction of the solid is relatively high ($n_D = 1.8$), whereas the index of refraction of the pore phase is near unity, giving a relative index of 1.8, which is very high. The pore size for these bodies usually corresponds to the original particle size of the starting material (frequently 0.5 to 2 microns) and is nearly the wavelength of the incident radiation so that the scattering is a maximum. Consequently, as indicated in Fig. 13.18, the transmission is found to be reduced to 0.01% by the addition of about 3% porosity. Even when the porosity is reduced to a value of 0.3%, the transmission is still only about 10% that of a completely dense sample. That is, for high-density single-phase ceramics containing fine porosity the translucency is a sensitive measure of the residual porosity and consequently a good indication of the quality of the ware.

The aesthetic value of many porcelain compositions is judged by the translucency. This together with good mechanical properties is the basis for compositions such as bone china and hard porcelain having high translucency. For porcelain bodies the phases present are normally a glass having an index of refraction of approximately 1.5, mullite ($n_D = 1.64$), and quartz ($n_D = 1.55$). As discussed in Chapter 11, in the normal microstructure of dense vitrified porcelain the mullite phase appears as fine needle crystals in a glassy matrix with larger quartz crystals which are undissolved or partially dissolved (see Fig. 11.11). Consequently, although the particle size of the mullite is in the micron range, the particle size of the quartz is much larger. Both because of the difference in particle size and because of the greater difference in the index of refraction, the mullite phase is the main contributor to scattering and decreased translucency in porcelain bodies. The primary method by which the translucency can be increased is to increase the glass content, decreasing the amount of mullite present. This can be accomplished, for example, by increasing the ratio of feldspar to clay, as discussed in Chapters 7 and 11. Frit porcelains and compositions, such as dental porcelains having a high feldspar content, are given high translucency by increasing the amount of glass at the expense of mullite development. For other purposes, however, this is deleterious, since it lowers the strength which the presence of mullite adds to the body.

In the same way that porosity greatly decreases the translucency of single-phase alumina, the presence of voids ($n_D = 1$) is deleterious in

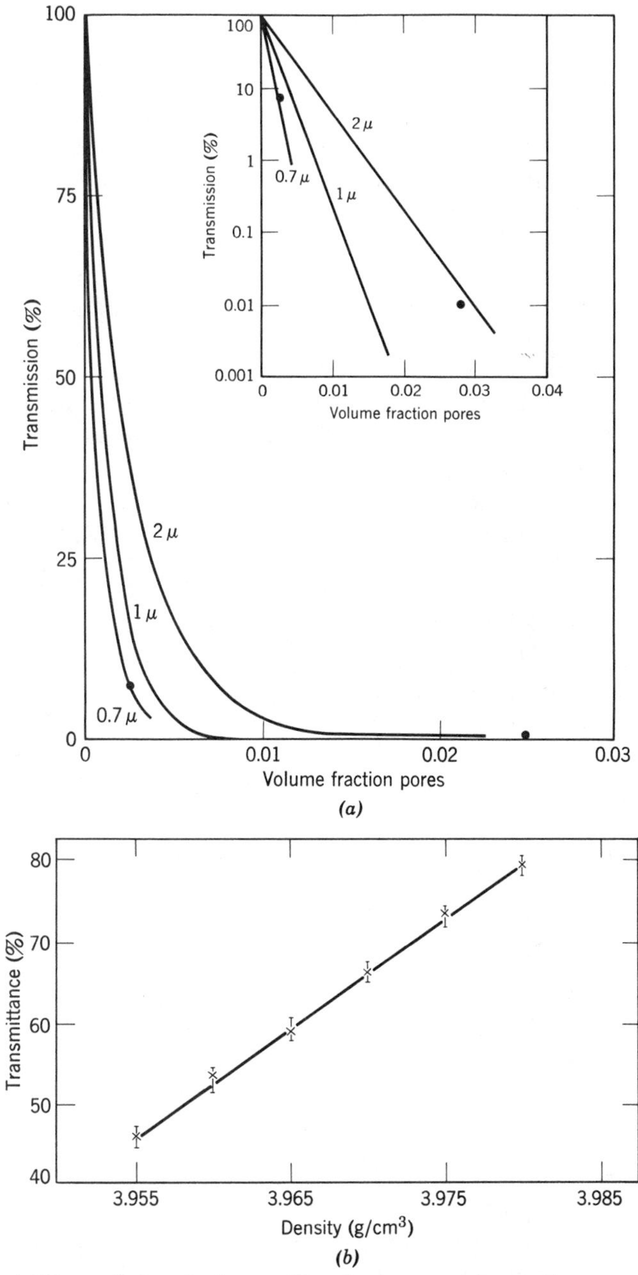

Fig. 13.18. (*a*) Transmission of polycrystalline alumina containing small amounts of residual porosity (equivalent thickness 0.5 mm). From D. W. Lee and W. D. Kingery, *J. Am. Ceram. Soc.*, **43**, 594 (1960). (*b*) Effect of density on transmittance at 4.5 microns of Al_2O_3, grain size 27 ± 3 microns, thickness 0.5 mm, surface finish <5 μ in. From H. Grimm et al., *Bull. Am. Ceram. Soc.*, **50**, 962 (1971).

porcelain bodies. Translucent bodies are only obtained when the ware is fired to a sufficient temperature for the fine pores resulting from the interstices between clay particles to be completely eliminated. This occurs in ware containing high feldspar or frit content which develops a large glassy phase or in ware fired to a sufficiently high temperature so that the densification process can go to completion. Typically, in high-fired translucent ware, the residual porosity is limited to pores of large size resulting from inhomogeneities in the mixing process. Even these pores contribute substantially to decrease the translucency; as much as 5% porosity is found equivalent to 50% mullite.

Another way of achieving a highly translucent body is to adjust the index of refraction of the various phases present so that a better match is obtained than is normally found. For a typical body in which quartz and mullite are both present this is not possible because the quartz and mullite have different indices of refraction. The mullite which is present in a fine particle size is the phase which the glass should be closest to in order to obtain maximum translucency (minimum scattering). In practical compositions this approach is best realized in English bone china having liquid with an index of refraction of about 1.56, which brings it into a region nearly equivalent to the crystal phases present. This, together with its low porosity, gives English bone china its exceptionally good translucency. The effectiveness of changing the index of refraction of the liquid phase is illustrated for a series of tests compositions in Fig. 13.19.

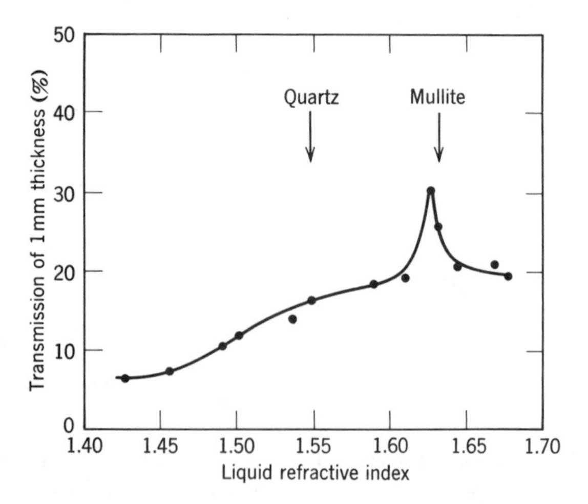

Fig. 13.19. Effect of changing liquid index on translucency of 20% quartz, 20% mullite, 60% liquid mixture. From G. Goodman, *J. Am. Ceram. Soc.*, **33**, 66 (1950).

13.5 Absorption and Color

The absorption coefficient has been defined by the relation $\ln I/I_0 = -\beta x$ in Eq. 13.13. The value of the absorption coefficient determines the transmission of light through the glass sheet and also fixes the reflectivity for an infinite thickness of a two-phase system, as discussed in Section 13.4. The absorption coefficient for many materials varies with wavelength, and this variation gives rise to color. The importance of color in providing usefulness and sales appeal for many ceramic products cannot be overemphasized.

Absorption Bands. The physical description of absorption, and reflection and transmission, characteristics is best done in terms of the changes in absorption coefficient or transmission with wavelength. In order for appreciable absorption to occur in the visible range, there must be transitions in the electronic structure of the atoms present. Referring back to the descriptions of ionic structures in Chapter 2, in ions with closed noble gas electron shells, all the electronic energy levels are filled; as a result these materials are transparent and colorless. It is only when additional unfilled energy levels are available, as in the transition and rare earth elements or when impurity additions create new energy levels, as discussed in Chapter 4, that light absorption occurs in the visible part of the spectrum.

For gaseous atoms the electronic transitions correspond to fixed changes in energy levels, and sharp spectral lines are observed when these transitions take place. In liquids and solids the ions are closer together and interact to form bands of allowed states rather than discrete separate energy levels. Consequently the transitions observed also correspond to a range of energies so that more or less broad absorption bands are observed rather than discrete spectral lines.

Impurity color centers are commonly observed whenever acceptor or donor levels, which allow electronic transitions to take place, are present. Perhaps the best studied of these are the F centers in alkali halides, which result from heating in an alkali vapor; similar bands are found in MgO heated in magnesium vapor. The resulting crystals are characteristically blue in color, corresponding to the presence of an associated electron-anion vacancy impurity center. Color centers are also produced by X-ray or neutron bombardment. In oxide materials impurity centers are commonly associated with oxygen-deficient materials. Clear transparent TiO_2, ZrO_2, and similar materials rapidly darken in color as they become nonstoichiometric. These changes in color with impurity additions or departures from stoichiometry are common causes of difficulty in obtaining good white colors and maintaining satisfactory color control. They are seldom sufficiently controllable to be used as reliable sources of color in ceramic systems.

Color. The most commonly used coloring constituents in ceramic systems are the transition elements characterized by an incomplete *d* shell, particularly V, Cr, Mn, Fe, Co, Ni, Cu, and to a lesser extent the rare earth elements characterized by an incomplete *f* shell. In addition to the individual ion and its oxidation state, absorption phenomena are markedly affected by the ionic environment.

When most oxide crystals and glasses are exposed to ionizing radiation (X rays, γ rays, ultraviolet light), optical absorption bands in the visible and ultraviolet part of the spectrum are produced. In oxide crystals the absorption centers from holes trapped at cation vacancy sites and cation vacancy–solute associate sites have been useful in studying the characteristics of these defect centers. Three bands in silicate glasses are of technological significance, since they give rise to absorption in the visible spectrum. The phenomenon of darkening under a radiation field is known as *solarization* and is undesirable in many applications such as television face plates and laser rods.

J. S. Stroud* was able to show that the bands produced in the visible, with maxima about 4400 and 6200 Å, are associated with trapped holes; trapped electrons cause an absorption band centered in the near ultraviolet which extends into the visible. The intensity of these bands increases with increasing alkali content in alkali silicate glasses, which suggests that the formation of the color centers is associated with singly bonded oxygens in the structure. This suggestion is reinforced by the observed high resistance of phosphate glasses to solarization effects.

In most technological applications, the formation of color centers on irradiating silicate glasses is prevented by adding cerium to the glasses. Stroud showed that Ce^{3+} inhibits the formation of the two-hole bands in the visible by trapping holes $[Ce^{3+} + h^{\cdot} \rightarrow (Ce^{3+}h^{\cdot})]$; Ce^{4+} inhibits the formation of the electron band in the near ultraviolet by capturing electrons $[Ce^{4+} + e' \leftarrow (Ce^{4+}\, e')]$.

Ligand-Field Chemistry. As indicated above, ions with incomplete electronic configurations, notably the transition metals such as iron, cobalt, nickel, chromium, and molybdenum and the rare earths such as neodymium, erbium, and holmium, absorb light in characteristic ranges of wavelength. In a crystal or glass there is a polarization of each ion present which has a substantial effect on the exact energy distribution of the outer electrons. These are the electron energy levels that contribute to color formation in the transition elements, and the colors of these materials are

J. Chem. Phys., **37**, 836 (1962).

particularly subject to change in coordination numbers and the nature of adjacent ions. These changes give rise to the description of colors as resulting from specific chromophores—complex ions which produce particular absorption effects. In contrast, the rare earth colorants depending on transitions in the inner *f* shell are much less subject to environmental changes.

A quantitative description of these effects of environment can be provided by crystal-field and ligand-field chemistry. In the original crystal-field approach, the bonding was treated as electrostatic (ionic), and the splittings of the energy levels were inferred from the symmetry of the electric field acting on the metal ion. An alternative approach, based on consideration of molecular orbitals (see below), includes factors such as covalent bonding, nonspherical ions, and distortions due to surrounding charges. The electrostatic and molecular-orbital treatments of these phenomena form the basis of ligand-field chemistry. The term ligand refers to any ion or molecule which is immediately adjacent to a metal ion and can be regarded as bonded to it. The most common ligands are monatomic or polyatomic negative ions such as $O^{=}$ and Cl^{-} or neutral polar molecules such as NH_3, H_2O, and CO.

Essential to understanding the behavior of transition metal ions are the shapes of the *d* orbitals (the distributions in space of the charge density for the *d* electrons). As shown in Fig. 13.20, there are five *d* orbitals, corresponding to the five allowed values of the magnetic quantum number (five orientations of the angular momentum vector). In three of these states, there are four lobes at 45° to the principal axes (for a magnetic field in the *Z* direction, these would lie respectively in the *XY*, *YZ* and *XZ* planes); in one, the four lobes lie along the *X* and *Y* axes; and in one, there is a charge distribution about the origin together with two lobes along the *Z* axis.

In a fieldfree environment (a free ion) the five *d* orbitals are degenerate; that is, they have the same energy. In the presence of a crystal field, however, as in a crystal, all the *d* orbitals no longer have the same energy but are split into groups. This can be illustrated by the simple case of a Ti^{3+} ion, which has only one *d* electron, surrounded by six negative ions arranged at the corners of an octahedron. Such an ion is described as being in an octahedral field. In the presence of such a field, the electronic energy in the two orbitals which lie along the axes, d_{Z^2} and $d_{X^2-Y^2}$ in Fig. 13.20, are raised relative to the free ion because of repulsion from the surrounding anions. In contrast, the electronic energy in the other three orbitals, which do not lie along the axes, is raised by a smaller amount. This splitting of the *d* levels thus reflects the tendency of the electron to avoid regions in which the field of the ligands is largest.

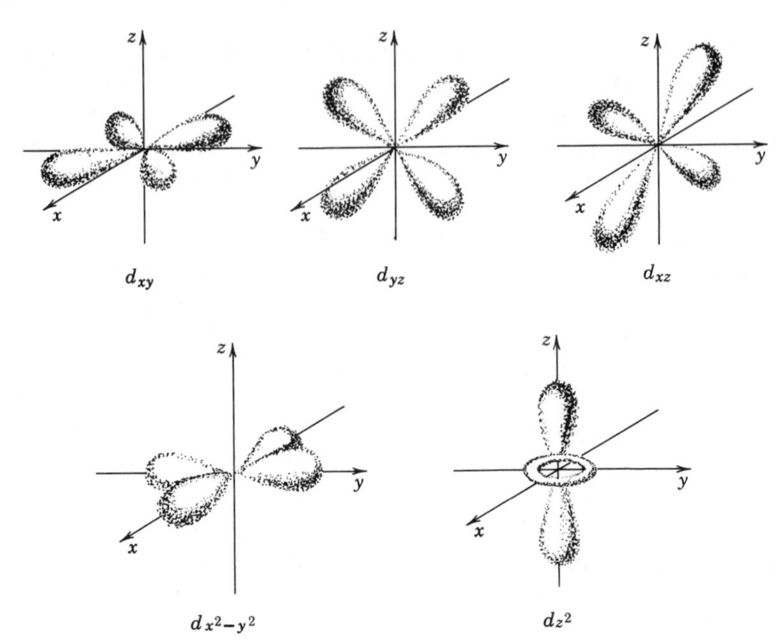

Fig. 13.20. The five d orbitals.

If the Ti^{3+} ion is surrounded by four negative ions arranged at the corners of a tetrahedron (a tetrahedral field), the electronic energy in the d_{Z^2} and $d_{X^2-Y^2}$ orbitals is less than that in the d_{XY}, d_{YZ} and d_{XZ} orbitals, and the magnitude of the splitting in this case is smaller than in the octahedral field. These splittings are illustrated schematically in Fig. 13.21.

The total energy difference between the higher and lower energy levels is denoted by $10Dq$. For d orbitals, this is generally in the range of 1 to 2 eV. Hence absorption of light, associated with electronic transitions between the lower and upper d levels, generally occurs in the visible or near infrared regions of the spectrum. Such absorption is used, in fact, to determine values for the energy splitting $10Dq$.

For transition metal ions with more than a single d electron, two factors must be considered in assessing the distribution of the d electrons among the available orbitals. These are the tendency toward a maximum number of parallel spins and the preference for orbitals of lower energy in the ligand field. In many cases, these influences are in conflict with each other (e.g., for ions having 4 to 7 d electrons placed in an octahedral field). In such cases, the state having the maximum number of parallel spins is preferred when the $10Dq$ splitting is small relative to the interaction causing parallel spins; the state having the maximum number of electrons

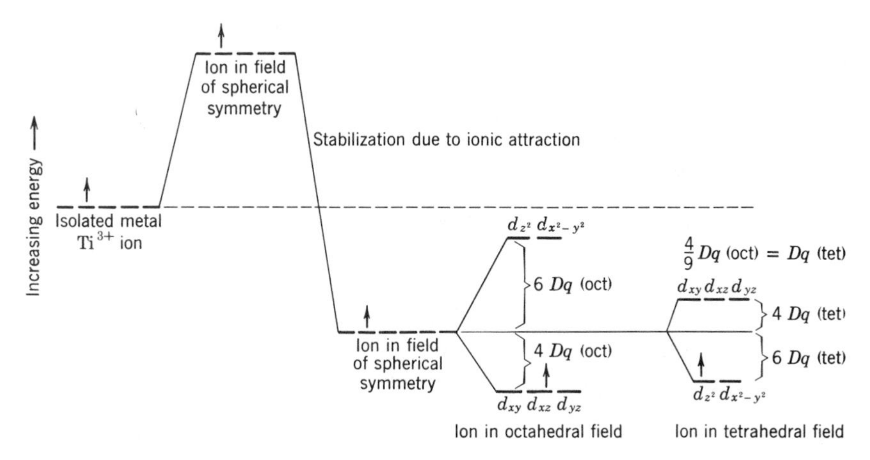

Fig. 13.21. Crystal field energy relationships for Ti^{3+} ion in an octahedral and tetrahedral field.

in the low-energy orbitals is preferred when $10Dq$ is large. The nature of the ligand can have a marked effect on the splitting; for a given transition metal ion, $10Dq$ increases in the order $I^- < Br^- < Cl^- < F^- < H_2O \sim O^= < NH_3 < NO_2^- < CN^-$. Magnitudes of the ligand-field splitting for various transition metal ions in tetrahedral and octahedral fields in hydrates and oxides are given in Table 13.6.

Colorants. As indicated in the preceding section, the environment or ligand field around a transition metal ion can have a sizable effect on its absorption characteristics and hence on the color produced. Often, and this is particularly true in organic dyestuffs, the particular color obtained can be identified with a certain ion combination or chromophore. This is true for the yellow color of cadmium sulfide in glasses; neither Cd^{2+} nor S^{2-} alone causes visible absorption, but the combination gives a yellow color corresponding to the Cd–S chromophore. Other and technologically more important examples of such chromphores are provided by the familiar amber glasses. It has been established by Douglas* and by Harding and Ryder† that the absorption center responsible for the amber color comprises a Fe^{3+} ion in tetrahedral coordination with one S^{2-} ion substituted for one of the surrounding oxygen ions.

As might be expected, electronic transitions are particularly probable when two oxidation states are present in the same compound. Materials

Phys. Chem. Glasses*, **10, 125 (1969).
†*J. Can. Ceram. Soc.*, **39**, 59 (1970).

Table 13.6. Crystal-Field Data for Transition Metal Ions

		Hydrates		Oxide
Number of *d* electrons	Ion	Octahedron 10 *Dq*/cm	Tetrahedron 10 *Dq*/cm	Octahedron 10 *Dq*/cm
1	Ti^{3+}	20,300	9000	6000
2	V^{3+}	18,000	8400	
3	V^{2+}	11,800	5200	
	Cr^{3+}	17,600	7800	16,800
4	Cr^{2+}	14,000	6200	
	Mn^{3+}	21,000	9300	
5	Mn^{2+}	7500	3300	7300–9800
	Fe^{3+}	14,000	6200	12,200
6	Fe^{2+}	10,000	4400	9520
	Co^{3+}	18,100	7800	
7	Co^{2+}	10,000	4400	
8	Ni^{2+}	8600	3800	8600
9	Cu^{2+}	13,000	5800	
10	Zn^{2+}	0	0	

1 eV = 8066 cm = 23.06 kcal/mole.

of this kind are usually semiconductors and are always deeply colored (Fe_3O_4, black; Fe_2O_3, red; $TlCl_3 \cdot 3TlCl$, red; Au_2Cl_4, dark red; Ti_3O_5, dark blue).

To be used in ceramics, stability at elevated temperatures is required, and this limits the palette of available colors. As the temperature level is increased, the number of stable colors diminishes so that underglaze colors for high-temperature porcelains (1400 to 1500°C) are limited and marking colors for 1800°C are not available; overglaze colors and colors for low-temperature glazes and enamels are numerous. Colors in glazes and enamels are normally formed either by solution of ions in the glass phase or by dispersion of solid colored particles in the same way as colored particles are dispersed in paints.

The colors of ions in silicate glasses depend primarily on the oxidation state and coordination number. Coordination number corresponds to being in a network former or network modifier position. For example, in normal silicate glasses Cu^{2+} replaces Na^{+} in a modifier position and is surrounded by six or more oxygen ions; Fe^{3+} and Co^{2+} usually replace Si^{4+}, forming CoO_4 and FeO_4 groups in the network. However, as the basicity of a host glass changes, there is a change in the function of these ions, which fall in the general classification of intermediates in terms of structural participation. At the same time there is frequently a change in oxidation state so that a wide range of colors may be produced by the same ion in different host glasses. Many individual colors have been discussed by W. A. Weyl.* The absorption characteristics of some are shown in Fig. 13.22.

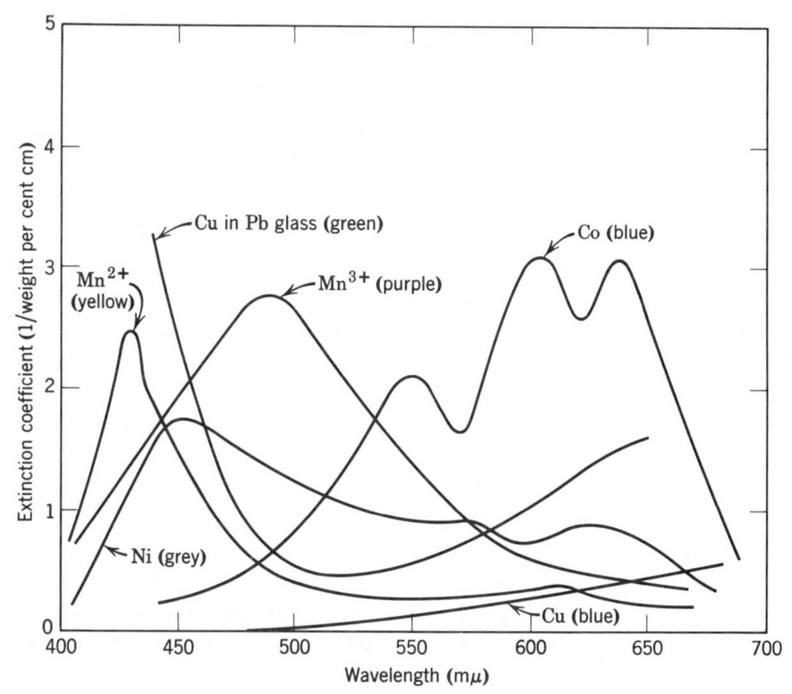

Fig. 13.22. Extinction coefficients for several ions in glasses.

Another source of structural variation in glasses is the replacement of oxygen ions with other anions such as S^{2-}, F^-, Cl^-, or I^-. This in itself does not give rise to absorption in a visible range but can have a substantial modifying effect on other ions present and lead to the formation of specific chromophores such as the Cd–S and Fe–S combinations already discussed. Another example is the green cobalt glass formed when K^+ is added to a cobalt containing glass melt. Similarly a polysulfide S–S–S

**Colored Glasses*, Society of Glass Technology, Sheffield, England (1951).

chromophore is produced when carbon or other reducing agents are added to form an amber glass.

Ceramic Stains. In addition to solution colors, associated with the presence of absorbing species, and scattering colors, associated with the presence of fine particles precipitated from the matrix, there are many applications in which ceramic stains are most suitable. The use of such stains involves the dispersion in a matrix material of pigmentary particles of a second phase which are colored and insoluble in the matrix. Stains are widely used in imparting color to glazes on wall tile and other whiteware bodies.

The stains must be colored compounds which are stable at high temperatures and inert in silicate systems. Many different compositions have been used for stains. With the notable exception of the CdS–CdSe red pigments, most ceramic stains are oxide materials. In some cases, the crystals are colored even in the absence of doping. Examples of such stains are $CoAl_2O_4$ (blue), $3CaO \cdot Cr_2O_3 \cdot 3SiO_2$ (green), and $Pb_2Sb_2O_7$ (yellow). In other cases, the host crystals of the stain are colorless in the absence of doping but are doped to achieve the desired color. Examples of this type include the family of zircon stains (see below) and Al_2O_3 doped with Cr.

One of the most successful crystal structures for the development of a wide variety of stable colors is the spinel structure described in Chapter 2. In this structure, which is stable and chemically inert, ions in different valence states and in both tetrahedral and octahedral coordination are present. In addition there are a number of different ways of arranging the cations on the tetrahedral and octahedral sites (inverse and regular spinels). These characteristics led to the development of many different intense and stable colors suitable for use as stains and are illustrated by the cobalt aluminates and chromates. Such stains are widely used in porcelain enamels, which are fired at temperatures in the range of 750 to 850°C.

For bodies which are fired at more elevated temperatures, such as 1000 to 1250°C, host crystals of ZrO_2 and $ZrSiO_4$ (zircon) have found wide use. These offer increased resistance to attack by the glass phase and added versatility in their accommodation of rare earth ions in solid solution. Examples of dopants used in such stains include vanadium (blue), praseodymium (yellow), and iron (pink).

Spinels and other colored crystals can be prepared by grinding together fine powders and calcining in the temperature range of 1000 to 1400°C. Additional B_2O_3, $Na_2B_4O_7$, NaCl, or NaF is frequently added as a flux to increase the reaction rate and allow use of lower reaction temperatures. The intensity of a color can be greatly affected by the flux used. The product is ground, after leaching with HCl if a B_2O_3 flux is used, to a particle size of 1 to 5 microns. Particle-size control is of great importance. As illustrated in Fig. 13.7, maximum reflectance is obtained when the particle is near the wavelength of light (0.4 to 0.7 micron), and this size is preferred for opacifiers. Smaller particle sizes rapidly lose their effectiveness and also tend to go into solution more rapidly in the glaze or enamel. The maximum color brilliance is obtained with somewhat larger particle sizes; about 5 microns is frequently optimum.

In the development and application of colored glasses, glazes, and enamels, all the optical properties thus far discussed come into play. That is, the final result depends on specular and diffuse reflectance, direct and diffuse transmission, and selective absorption characteristics. These must be adjusted to give desired results by manipulation of the properties of each phase present and their distribution. In clear colored glasses the only consideration is refractive index and absorption characteristics. Colored opaque glazes and enamels can be developed with an absorbing glass phase and a transparent dispersed-phase opacifier, or a clear glass can be opacified with colored particles, or some mixture of these can be used.

In general, the widest range of color and color control can be exercised when dispersed particles are used as pigments in a clear glaze, and this is usually done for low-temperature overglaze colors and porcelain enamels. For higher temperatures, such as those required for glasses and porcelain glazes, the number of stable nonsoluble crystals suitable for use in stains is limited, and solution colors are generally used.

Color Specification. There is an added difficulty in color development in that the human eye is extremely sensitive to color matches. The eye's sensitivity varies greatly with wavelength, however, as illustrated for the typical observer in Fig. 13.23. The curves indicate the response of the eye under both dark-adapted and light-adapted conditions. Curve *B*, corresponding to light-adapted conditions, is of primary importance for lighting purposes, since the eye is light-adapted under most conditions of artificial illumination. Because of this variation of eye sensitivity with wavelength, visual effects depend not only on the spectral distribution of absorption coefficients but also on the sensitivity of the receptor, usually the human eye. This sensitivity varies from individual to individual, but Fig. 13.23 is a good average.

The adaptability of the eye to color distribution also means that a pure color may not be the most satisfactory one. For example, we are accustomed to a reddish tinge in incident light; this is particularly true when artificial light is used for viewing. Consequently, a white porcelain with transmission completely independent of wavelength is described as bluish by a typical observer; a sample having increased transmission in

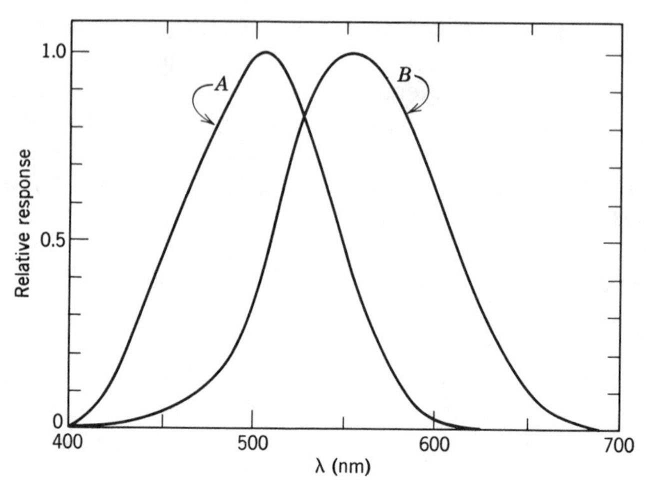

Fig. 13.23. Color's sensitivity of typical observer under dark-adapted (curve A) and light-adapted (curve B) conditions.

the red part of the spectrum, which gives it an off-white color from a physical point of view, is, on comparison, described by an observer as a whiter color.

In addition to the eye's sensitivity and bias, color depends to a great extent on the light source in which it is viewed. Shadows on snow are blue because the incident light from the sky is that color. Similarly, we commonly find that colors appear quite different under daylight, incandescent light, and fluorescent light. The widespread nature of this phenomenon is perhaps less well recognized. As an example, light entering a room diffused from a large tree is mostly green, whereas sunlight reflected to surroundings in a room with a red-carpeted area is mostly red. Colored objects viewed under these two situations may appear quite different. This causes difficulty in color matching, since colors appearing the same under standard conditions or under a particular light source may not match at all when viewed in some other surroundings. The spectral distribution of some common light sources is shown in Fig. 13.24.

In view of these variables it is obvious that the specification of color in meaningful terms such as red or green or blue and the shades of differences among them is not easy. A variety of systems has been proposed. In general there are three fundamental variables of color as mental phenomenon. The most important for most purposes is *hue,* which may be described as the main quality factor in color. It is the essential

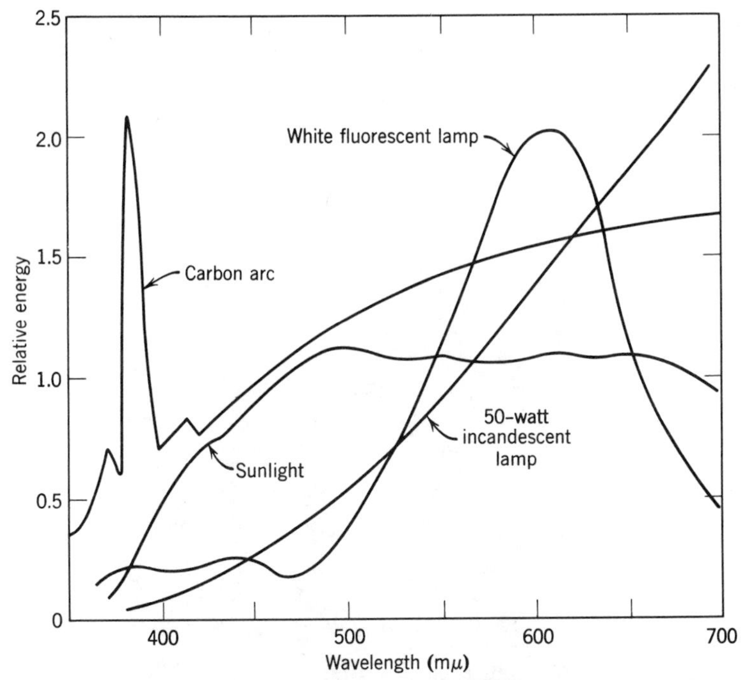

Fig. 13.24. Spectral energy distribution in common light sources.

element that leads us to describe a color as yellow or blue; it is the main factor that changes in the spectrum as the wavelength of light changes, and for this reason it is the quality that permits us to describe color in terms of the wavelength of monochromatic light which it matches. Under good conditions the eye can distinguish about 200 different hues. In addition to this factor, the *saturation* can be defined as the percentage of hue present in a color. This is a characteristic indicated by terms such as pale or weak, deep or strong in connection with some particular hue. Saturation and hue together define the quality aspect of the color. Finally, *brightness* can be defined as the quantitative aspect of color which describes it in terms of its apparent amount. Of these three variables, brightness is the most difficult one for an observer to define accurately without reference to some auxiliary standard. This is well known in connection with night vision, for example, in which marked adaption of brightness sensitivity is observed.

The three variables, hue, saturation, and brightness, can be used in various ways to define and measure color. For example, all colors can be considered as distributed through the interior of a solid in which brightness varies vertically, hue varies with position about the center in a horizontal plane, and saturation varies with the distance outward normal to the center vertical. This color cylinder is illustrated in Fig. 13.25 and provides the basis for the arrangement of color chips in the Muncel color system.

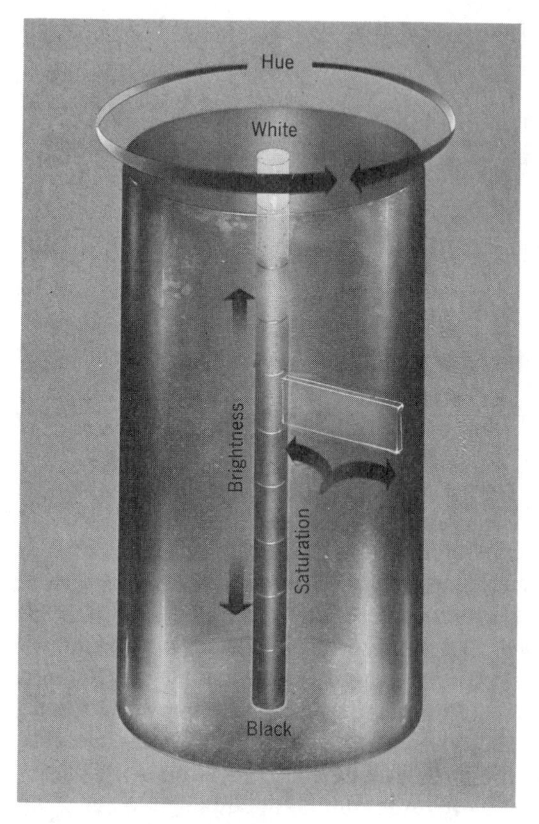

Fig. 13.25. Diagram of color cylinder. Brightness varies along axis, hue along circumference, and saturation along radius.

Rather than comparing a color with standard samples in which surface characteristics differ and colors may change with time, it is desirable to be able to measure color in terms of spectrometer curves (Fig. 13.22). In this procedure the absorption for each wavelength must be multiplied by a factor corresponding to the sensitivity of a typical observer (Fig. 13.23) to determine that wavelength contribution toward the color impression.

Resulting curves can be given as color specifications expressed as the dominant wavelength (hue), fraction of white light that must be mixed with monochromatic light of the dominant wavelength (saturation), and intensity of the color (brilliance). Standard methods for carrying out these comparisons have been agreed upon and specified by the International Commision of Illumination (the ICI system, called the CIE system, Commission International d'Eclairage, in England and Europe).

13.6 Applications

In addition to their usefulness in a number of large-volume applications such as optical and opthalmic glasses, glazes, and enamels, the optical properties of ceramics have importance in several applications involving smaller tonnages in which the development of new technologies have depended on improved understanding of the chemistry and physics of materials.

Phosphors. The decay of an electron from an excited energy level to a lower-lying level may be accompanied by the transfer of heat to the lattice or by radiative emission. The emission of light in such a process is termed fluorescence or phosphorescence, depending on the time between excitation and emission. Excitation is most often provided by electrons or photons, but in some applications, chemical reactions or electric fields can be used as well.

Phosphors are widely used in fluorescent lamps, screens in cathode-ray tubes, and scintillation counters. The light emission of ordinary phosphors depends critically on the presence of impurities, which can have an important effect even at small concentrations and may act as activators, coactivators, or poisons. In some cases, the emission depends on electronic transitions of the impurity; in others, it depends on the recombination of electrons and holes at the impurity. The activator impurities may serve as electron or hole traps; their energy levels can be shifted by interaction with nearby coactivator centers, which are often used to enhance the solubility of the activator impurities in the crystal. The poison impurities are centers which provide alternative and undesired routes for electron-hole recombination in the material. Because of the sensitivity of phosphors to small amounts of impurities, it is not surprising that their development has depended on the ability to obtain high-purity materials and materials with controlled additions of impurities.

The operation of a fluorescent lamp depends on an electrical discharge through a mixture of mercury vapor and a rare gas, usually argon, which converts a large part of the electrical-energy input to radiation in a single line of the mercury spectrum (at 2537 Å). This mercury radiation in the ultraviolet excites a broad-band emission in the visible from a phosphor coating on the wall of the discharge tube. The phosphor is prepared as a powder, dispersed in a liquid, and applied as a film or coating. In early fluorescent lamps (1940s), Mn-activated Zn_2SiO_4 was used as the phosphor. These were replaced with mixed phosphates and eventually with halogen phosphates: $Ca_5(PO_4)_3$ (Cl, F): $Sb^{3+}Mn^{2+}$. The antimony and manganese dopants are activators which provide two overlapping emission bands in the visible. Table 13.7 lists some common lamp phosphors, the color of their emitted light, and some applications.

In the operation of cathode-ray tubes, excitation of the phosphor is provided by an electron beam. In the color-television application, different phosphors are used for their emission in frequency ranges corresponding to each of the primary colors. The decay time of the phosphor is vital in these applications, with the relevant time scale imposed by the time for the electron beam to sweep the face of the tube. Table 13.8 lists several important phosphors used in cathode-ray tubes, their characteristics, and their applications.

Lasers. Many ceramic materials have found use as hosts for solid-state lasers and as window materials for gas lasers. The solid-state laser is a luminescent solid in which the light emitted in the fluorescence of one excited center stimulates other centers to emit in phase with the light from the first center. The difference between this stimulated emission and the light-emitting processes of conventional light sources is that the lasing ions in lasers have other excited-state energy levels such as that shown in Fig. 13.26*a* at which the excited ions can reside before being stimulated to return to the ground state. The characteristics of this intermediate state (fluorescence level) are important in laser operation, and the light emitted when the atoms return to the ground level from this state has the same wavelength as the light which stimulated them into leaving the state. The light emitted when the atoms return from the fluorescence level to the ground state can in turn stimulate further emission from other atoms in the lasing level.

For laser action to occur, it is essential for the lasing ions to be excited to the higher energy levels, with the population in the fluorescence level exceeding that in the ground state. This condition represents a nonequilibrium inversion of the usual populations of ions in the two states. The excitation is generally accomplished by means of an external flash lamp which emits light that is absorbed by the lasing ions. For laser action to occur, a minimum of three energy levels is required. If only two levels are present, optical excitation can at most produce equal populations in the ground state and the excited state. A laser such as that shown in Fig. 13.26*a*, in which the terminal state is the ground state, is termed a

Table 13.7. Lamp Phosphors

Matrix	Activators	Color of Emission	Comments
Calcium tungstate, $CaWO_4$	. . .	Deep Blue	Mainly in blue lamps
	Pb	Pale blue	
Barium disilicate, $BaSi_2O_5$	Pb	UV peak at 350 nm	For long UV emission
Zinc orthosilicate, Zr_2SiO_4	Mn	Green	Mainly in green lamps
Calcium metasilicate, $CaSiO_3$	Pb, Mn	Yellow to orange	In "deluxe" color lamps
Cadmium borate, $Cd_2B_2O_5$	Mn	Orange-red	Mainly in red lamps
Barium pyrophosphate	Ti	Blue-white	
Strontium pyrophosphate	Sn	Blue	
Calcium halophosphate, $Ca_5(PO_4)_3(Cl, F)$	Sb, Mn	Blue to orange and white	Main group of lamp phosphors; also strontium halophosphates
Strontium orthophosphate (containing Zn or Mg), $(Sr, Zn)_3(PO_4)_2$	Sn	Orange	In "deluxe" color lamps, and also high-efficiency high-pressure lamps for color correction
Magnesium arsenate, $Mg_6As_2O_{11}$	Mn	Red	In "deluxe" color lamps; emission attributed to Mn^{4+}
Magnesium fluorogermanate $3MgO \cdot MgF_2 \cdot GeO_2$	Mn	Red	Comment as for strontium orthophosphate; emission attributed to Mn^{4+}
Yttrium vanadate, YVO_4	Eu	Red	Color correction in high-pressure lamps
Magnesium Gallate $MgGa_2O_4$ (aluminium-substituted)	Mn	Green	Photoprinting

Source. Reference 11.

Table 13.8. Phospors for Cathode-Ray Tubes

Phosphor Material	Color of Emission	Decay Time (to 10%)	Uses
Zn_2SiO_4: Mn	YG; $\lambda_m = 530$ nm	$2 \cdot 45 \cdot 10^{-2}$ sec	Radar, oscillography
$CaSiO_3$: Pb, Mn	O; $\lambda_m = 610$ nm	$4 \cdot 6 \cdot 10^{-2}$ sec	Radar, because of long persistence
(Zn, Be)SiO_4:Mn	W; $\lambda_m = 543$ and 610 nm	$1 \cdot 3 \cdot 10^{-2}$ sec	Projection TV
(Ca, Mg)SiO_4: Ti	$\lambda_m = 427$ nm	$5 \cdot 5 \cdot 10^{-5}$ sec	
Zn_2SiO_4: Mn, As	G; $\lambda_m = 525$ nm	$1 \cdot 5 \cdot 10^{-1}$ sec	Integrating phosphor for low-repetition-rate displays
$Ca_2MgSi_2O_7$: Ce	BP; $\lambda_m = 335$ nm	$1 \cdot 2 \cdot 10^{-7}$ sec	Flying-spot scanner tubes, photography
$Zn_3(PO_4)_2$: Mn	R; $\lambda_m = 640$ nm	$2 \cdot 7 \cdot 10^{-2}$ sec	Old standard red for color TV
ZnO	UV; $\lambda_m = 390$ nm G; $\lambda_m = 500$ nm	$\lesssim 5 \cdot 10^{-8}$ sec $2 \cdot 8 \cdot 10^{-6}$ sec	Flying-spot scanners, photography
ZnO	G; $\lambda_m = 510$ nm	$1 \cdot 5 \cdot 10^{-6}$ sec	Flying-spot scanners
YVO_4: Eu	OR; $\lambda_m = 618$ nm	$9 \cdot 10^{-3}$ sec	Color TV

R—red, O—orange, Y—yellow, G—green, B—blue, P—purple, W—white.
Source. Reference 11.

three-level system. In the alternative four-level system (Fig. 13.26*b*), the lasing transition takes place between the fluorescence level and some lower level which is above the ground state.

Ruby lasers consist of single-crystal sapphire (Al_2O_3) rods doped with a small concentration of Cr, typically in the range of 0.05%. The end faces are highly polished, planar, and parallel. Mirrors are placed adjacent to the end faces to cause some of the spontaneously emitted light to reflect back and forth through the rod. One of the mirrors is almost completely reflecting; the other is only partially reflecting. The rod is excited along its length by a flash lamp. Most of the energy of the flash is dissipated as heat, but a small fraction is absorbed by the rod and serves to excite the

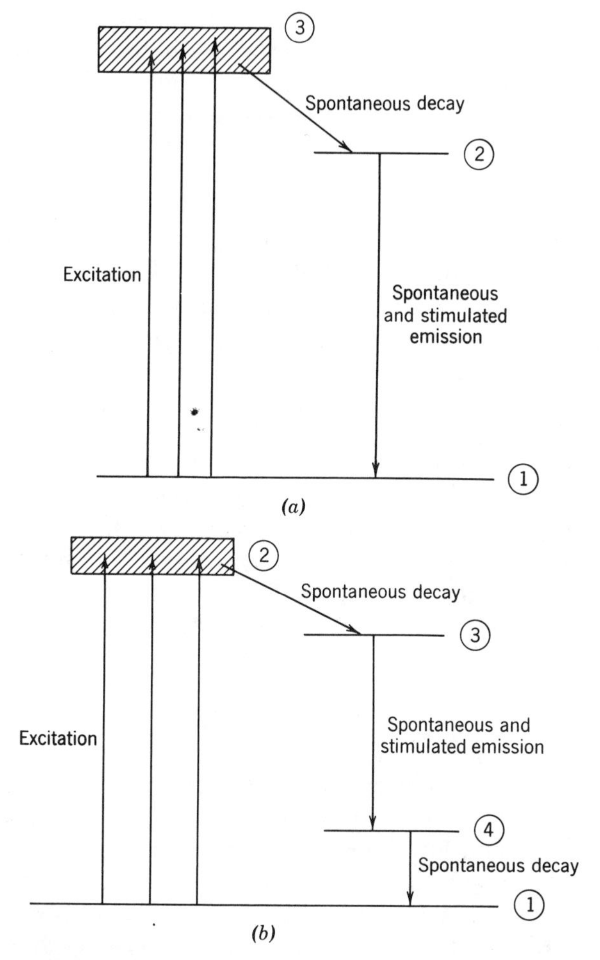

Fig. 13.26. Simplified energy-level diagrams for (*a*) a three-level and (*b*) a four-level lasing system.

Cr ions to elevated energy levels. The exciting energy is absorbed in a broad band, and the emission takes place in a narrow line of the trivalent Cr ion at 6943 Å, corresponding to the transition between level 2 and level 1 in Fig. 13.26*a*. The output radiation generated when stimulated emission occurs passes out through the partially reflecting end of the laser rod.

Another important crystalline laser is Nd-doped YAG ($Y_3Al_5O_{12}$), which is a four-level system. The wavelength at which the ruby laser emits radiation is about 0.69 microns; that of the Nd–YAG laser is about

1.06 microns. In both of these cases, the appropriately doped host materials are prepared in single-crystal form to avoid scattering or absorption of energy. Another four-level crystal laser, Nd-doped Y_2O_3, has been prepared by sintering to theoretical density to yield a polycrystalline laser.

Glasses are also widely used as host materials for lasers. Relative to crystal lasers, they offer improved flexibility in size and shape and may readily be obtained as large, homogeneous, isotropic bodies with high optical quality. The indices of refraction n of glass hosts can be varied between about 1.5 and 2.0, and both the temperature coefficient of n and the strain-optic coefficient can be adjusted by changes in the composition. Glasses have, however, lower thermal conductivities than the Al_2O_3 or YAG, which imposes limitations on their use in continuous- and high-repetition-rate applications.

The principal differences between the behavior of glass and crystal lasers are associated with the greater variation in the environments of lasing ions in glasses. This leads to a broadening of the fluorescent levels in the glasses. As an example of this broadening, the width of the Nd^{3+} emission in YAG is about 10 Å; that in oxide glasses is typically about 300 Å. The broadened fluorescent lines in glasses make it more difficult to obtain continuous (CW) laser operation, relative to the same lasing ions in crystal hosts.

The broadened fluorescent lines in glasses are used to advantage in so-called Q-switched operation. In this case, the completely reflecting mirror of the previous example is replaced by a reflector of low reflectivity while pumping. After an inversion in the population of the fluorescent level and lower levels has been obtained, the reflector is rapidly increased to a high value of reflectivity, and the energy stored by the excited ions is rapidly converted to a pulse of light whose peak power can be orders of magnitude greater than power levels reached in continuous or *long-pulsed* operation. The duration of such Q-switched pulses is typically in the range of 10 to 50×10^{-9} sec.

A number of ions have been made to lase in glass. These include Nd^{3+}, Yb^{3+}, Er^{3+}, Ho^{3+}, and Tm^{3+}. The most important is neodymium, a four-level system, because it can be operated at room temperature with high efficiency. Interest has developed as well, however, in the erbium glass laser, a three-level system, because its emission at about 1.54 microns provides greatly improved eye safety relative to neodymium lasers (the transmission of light through the eye to the retina is smaller by many orders of magnitude at 1.5 microns than at the emission wavelengths of other common solid-state lasers). To obtain useful efficiencies with erbium lasers, the glasses are also doped with ytterbium.

The latter ion serves to sensitize the fluorescence of the Er by absorbing light from the flash lamp and subsequently transferring energy to the fluorescent ion.

Although very high-power densities can be achieved with glass lasers operated in a Q-switched mode, they are inappropriate when high-power levels are desired under continuous conditions. For the latter applications, gas lasers are used. Many of these lasers, which use a glow discharge to produce the population inversion, emit in the visible part of the spectrum (e.g., He–Ne at 6328 Å and Ar at 4880 and 5145 Å), but the two most important for very high-power applications emit energy in the infrared region (at about 5 microns for the CO laser and about 10.6 microns for the CO_2 laser).

Window materials used with the lasers which emit in the infrared must be highly transmitting in this region. The transmission of various ceramic materials in the infrared has been given in Table 13.1. From the data presented there, it is seen that for gas lasers which operate about 1 to 2 microns, a number of oxide materials such as Al_2O_3 would be suitable as windows. In the region around 5 microns, the alkaline earth halides such as CaF_2 may be used; in the region around 10 microns, the alkali halides or various II–VI compounds, such as ZnSe or CdTe, seem indicated. In addition to very high optical quality, materials for use as such windows must have reasonable mechanical strength.

Fiber Optics. As indicated by Fig. 13.5 and Eq. 13.8, when a light ray emerges from a glass into air, it is bent away from the normal to the surface. Consequently, for some value of the angle r, the angle i reaches 90°, which corresponds to the emergent light ray traveling parallel to the surface. For any larger value of r, the light rays are totally internally reflected back within the glass. From Eq. 13.8, the critical angle for total internal reflection may be expressed:

$$\sin r_{\text{crit}} = \frac{1}{n}. \tag{13.35}$$

For a typical glass with $n = 1.50$, the critical angle is about 42°.

Because of total internal reflection, a glass rod can transmit light around corners. An image incident on one end of the rod, however, is seen at the other end as an area of approximately uniform intensity, representing the average of the intensities incident on the initial end. If the single rod is replaced by a bundle of smaller fibers, each fiber transmits only the light incident on it, and an image can be transmitted with a resolution equal to the individual fiber diameters.

Despite this potential for transmitting images, fiber optic devices were limited for many years by problems of light loss. Among the sources of loss are points of contact between fibers in the bundle, where light is transmitted from one fiber to the other rather than reflected; scratches in the surface of the fibers, which change locally the angle of incidence of the light rays on the surface; grease on the surface, which changes the critical angle for total internal reflection; and dust particles on the surface, which cause scattering losses, since the light propagates about a half wavelength into the surrounding medium. These problems can be overcome by cladding the fibers with a glass of lower refractive index. In this way, the reflections take place predominantly at the interface between the two fibers, which is protected by the cladding, rather than at the external surface of the fiber-cladding body. In manufacturing such clad fibers, care must be directed to the relative thermal-expansion and viscous-flow behaviors of the core and cladding, as well as to the relative softening points and optical properties of the respective glasses. The overall diameters of such fibers are typically in the range of about 50 microns. The thickness of cladding required to prevent losses is in the range of twice the wavelength of the light. In some cases, the cladding glass in a fiber bundle can be fused at elevated temperatures to provide vacuum-tight fiber optic assemblies. In all cases, the quality of the core-cladding interface is an important parameter affecting the performance of the device.

In addition to cladding techniques in which a step change in refractive index between core and cladding is produced, a continuous variation in refractive index across the fiber radius can also be used in fabricating fiber optic devices. In the latter application, a gradient in refractive index between a high-index axis and a lower-index surface is produced by an ion-exchange process. The exchange is typified by the substitution of K^+ ions for Tl^+ ions in a Na_2O–Tl_2O–PbO–SiO_2 glass. The resulting refractive-index distribution is approximately radially parabolic, and such distributions have the effect of focusing the light rays. That is, rather than being totally reflected at the surface of the fiber, the light rays are bent in a continuous sinusoidal path about the fiber axis.

Optical Waveguides. One of the early rationales for developing laser devices was the potential for using light as the carrier wave in communications systems. Laser-generated light in the visible or near-infrared regions would, because of its higher frequency than the microwave radiation now widely used, provide greater bandwidth and hence higher rates of information transmittal. It has been estimated, for example, that a single optical system could simultaneously carry 500 million telephone conversations.

The principal elements of a laser communications system are shown schematically in Fig. 13.27. It consists of a laser source, modulator, transmission line, and detector. The laser sources, detectors, and tech-

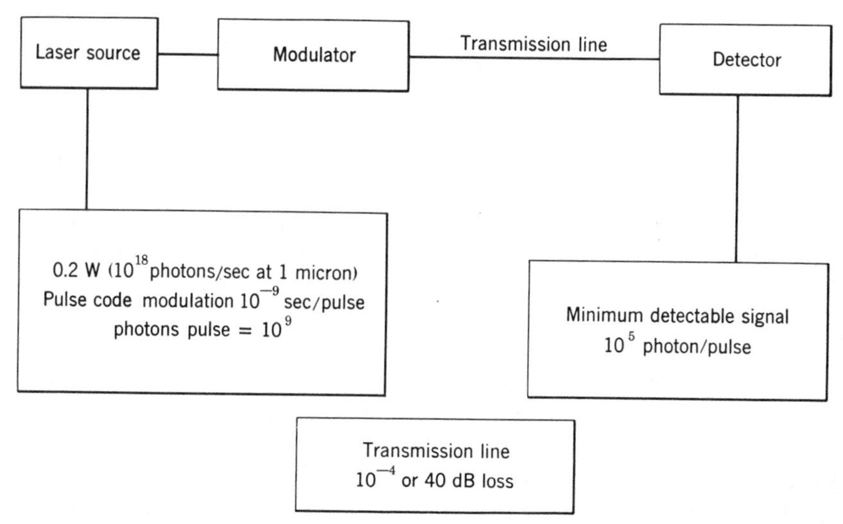

Fig. 13.27. Schematic diagram of a laser communications system. From E. Snitzer, *Bull. Am. Ceram. Soc.*, **52**, 516 (1973).

niques for modulation have all been the subject of considerable technical attention. Since lasers cannot be used for direct transmission through the atmosphere, because of high attenuation in rain, fog, or snow, much activity has also been directed to transmission lines based on fiber optic technology.

In discussing fiber optic wave guides for communications, reference is often made to single-mode and multimode fibers. Here the term *mode* describes the possible discrete distributions of electric field intensity across the aperture of the fiber which can propagate. Each mode is characterized by specific phase and group velocities. Conventional fiber optic configurations, with a step change in index of refraction between core and cladding, have core diameters (10 to 100 microns) which are large enough to support many modes. With such multimode fibers, the rate of transmitting information would be limited by the spreading of a pulse because of differences in the velocity of propagation in the different modes.

For single-mode operation, core diameters in the range of a few microns are required. Since most fiber handling requires outer (cladding) diameters in the range of at least 100 microns, single-mode fibers would have cladding-core-diameter ratios in the range of 20 or 30 to 1; such fibers represent a most critical test of fabrication techniques and the ability to control form.

With an optical communication system, it is anticipated that repeater stations will be required along the length of the system, at spacings determined by the attenuation of the transmission line. The permissible loss in the transmission line has been estimated by E. Snitzer* as follows: A laser with an output power of 0.2 W, emitting at 1 micron, would generate 10^{18} photons/sec. Assuming each pulse lasts for 10^{-9} sec, an effective limiting time for electronic detectors, the information per bit would be launched with an energy pulse of 10^9 photons. For reliable detection, the pulse must contain of the order of 10^5 photons. These values suggest that the transmission line cannot have an attenuation larger than 10^4, that is, a loss of 40 dB.

This value of tolerable loss places severe constraints on the transmission line. For perspective, a typical high-quality optical glass would be characterized by an absorption coefficient in the range of 0.25 %/cm, or a loss of 1000 dB/km. If only 40-dB loss can be tolerated, repeater stations would be required at 40-m intervals. Use of fiber optic wave guides for a practical transmission line will therefore require fibers with exceptionally low loss, as in the range of 20 dB/km or less. Materials with losses in this low range can be produced by working with raw materials of high purity, using a vapor-phase transport process to obtain increased purity, and by directing great care to various phases of the fabrication process. The principal sources of loss in the wave guide are scattering by composition and density fluctuations in the glass and by irregularities at the core-cladding interface and absorption by transition metal ions, particularly Fe, Cu, and Co, and hydroxyl ions. In general, the concentrations of the transition metal ions in the glasses must be less than 1 ppm, and in some cases, for example, the Fe^{2+} concentration for transmission in the range of 1 micron, the impurity content must be in the range of 10s of ppb.

The importance of impurities on the attenuation of optical wave guides is illustrated by the data in Fig. 13.28. All the peaks shown in the figure are associated with the presence of OH^- ions in the glass (about 10^{-4} atom% for the case shown). The total loss for this wave guide was as low as 4 dB/km over some ranges of wavelength. The ultimate lower limit of attenuation would be that associated with scattering by density and compositional fluctuations in the glass, which is estimated to be in the range of 1 to 2 dB/km.

Electrooptic and Acoustooptic Materials. Systems based on laser technology will require much additional hardware in addition to the lasers and wave guides. For example, devices will be required to modulate, switch, deflect, translate in frequency, and otherwise modify the optical signal in a predictable and controllable manner. Needs in this area have prompted the

Bull. Am. Ceram. Soc., **52**, 516 (1973).

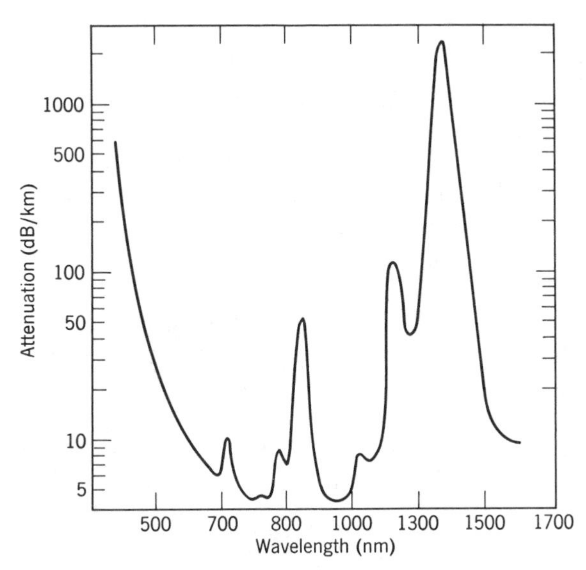

Fig. 13.28. Total attenuation of low-loss optical wave guide as a function of wavelength. All peaks can be associated with absorption due to OH^- ions. From R. D. Maurer, *Proc. IEEE,* **61**, 452 (1973).

development of materials which are capable of optical transmission with low loss, which have optical properties that can be modified by electric or magnetic fields or by externally applied stresses, and whose properties will interact in a specified manner with the optical signal. Important among these are the so-called electrooptic and acoustooptic crystals.

The electrooptic effect occurs whenever an applied electric field produces a change in the optical dielectric properties. The applied field may be a static, microwave, or an optical electromagnetic field. In some crystals, the electrooptic actions are largely electronic in origin; in others they are principally associated with vibrational modes. In some cases, the electrooptic effects may vary linearly with the applied electric field; in others it may vary quadratically with the field.

In terms of a single electron-oscillator description of the index of refraction, the effect of a low-frequency field E is to shift the characteristic frequency from ν_0 to ν:

$$\nu^2 - \nu_0^{\,2} = \frac{2ve(\kappa_0 + 2)E}{3mv_0^{\,2}} \tag{13.36}$$

where v is the anharmonic force constant, e is the electronic charge, m is the mass of an electron, and κ_0 is the low-frequency dielectric constant.

The index of refraction n varies as $[\nu^2 - \nu_0^2]^{-1}$, and hence Eq. 13.36 directly expresses the linear variation of the index with electric field.

The effect of an applied electric field may also be represented in terms of the polarizability. In the derivation of Eq. 13.28, it was assumed that the dipole strength μ was a linear function of the applied field, $\mu = \alpha E$, and for N dipoles the polarization is $P = N\mu = N\alpha E$. For crystals which lack a center of symmetry, the polarization is not a linear function of the electrical field. Rather

$$P = \alpha' E + \beta' E^2 + \gamma' E^3 + \cdots \tag{13.37}$$

where $\alpha' = N\alpha$ and β' and γ' are material constants associated with the nonlinear terms.

The principal electrooptic effect is describable in terms of the half-wave field-distance product $[E \cdot l]_{\lambda/2}$, where E is the electric field strength and l is the optical path length. This product represents the voltage required to produce half-wave retardation in a geometry $l/d = 1$, where d is the crystal thickness in the direction of applied field.

The optical phase retardation Γ, in radians, is given by

$$\Gamma = \frac{2\pi l}{\lambda_0}[n_1(E) - n_2(E)] \tag{13.38}$$

where λ_0 is the wavelength of the light in vacuum and $n_1(E)$ and $n_2(E)$ are the field-dependent indices of refraction. The form taken by $n_1(E) - n_2(E)$ depends on the crystal symmetry and on the direction of application of the electric field as well as the propagation and polarization directions of the optical beam.

Among the important electrooptic materials are $LiNbO_3$, $LiTaO_3$, $Ca_2Nb_2O_7$, $Sr_xBa_{1-x}Nb_2O_6$, KH_2PO_4, $K(Ta_xNb_{1-x})O_3$, and $BaNaNb_5O_{15}$. In many of these crystals, the basic structural unit is a Nb or Ta ion coordinated by an octahedron of oxygen ions. Because of the change in the refractive index with electric field, electrooptic crystals can be used in a variety of electronic applications such as optical oscillators, frequency doublers, voltage-controlled switches in laser cavities, and modulators for optical communications systems.

In addition to applied electric fields, changes in the index of refraction of crystals can also be induced by strain (the so-called acoustooptic effect). The strain acts to alter the internal potential of the lattice, and this changes the shape and size of the orbitals of the weakly bound electrons and hence causes changes in the polarizability and refractive index. The effect of strain on the refractive indices of a crystal depends on the direction of the strain axes and the direction of the optical polarization with respect to the crystal axes.

When a plane elastic wave is excited in a crystal, a periodic strain pattern is created with a spacing equal to the acoustic wavelength. The strain pattern produces an acoustooptic variation in the index of refraction equivalent to a volumetric diffraction grating. Acoustooptic devices are based on the fact that light incident on an acoustooptic grading at an appropriate angle is partially diffracted. The use of crystals in such devices depends in general on the piezoelectric coupling, the ultrasonic attenuation, and the acoustooptic coefficients. Among the important acoustooptic crystals are $LiNbO_3$, $LiTaO_3$, PbM_0O_4, and PbM_0O_5. All these crystals have refractive indices about 2.2 and are highly transparent in the visible.

The coming decades will see expanded attention directed to the development of ceramic materials for use in integrated optics technology. This technology will require significant improvements in the available ceramic materials for use elements such as lasers, wave guides, couplers, modulators, diffractors, and detectors (references 14 to 17).

Suggested Reading

1. F. S. Sears, *Principles of Physics, III, Optics*, Addison-Wesley Publishing Company, Inc., Reading, Mass., 1946.
2. H. C. van de Hulst, *Light Scattering by Small Particles*, John Wiley & Sons, Inc., New York, 1957.
3. R. M. Evans, *An Introduction to Color*, John Wiley & Sons, Inc., New York, 1948.
4. W. A. Weyl, *Coloured Glasses*, Society of Glass Technology, Sheffield, England, 1951.
5. W. W. Coffeen, "How Enamels, Glasses and Glazes are Opacified," *Ceram. Ind.*, **70**, 120 (April, 1958); 77 (May, 1958).
6. G. Goodman, "Relation of Microstructure to Translucency of Porcelain Bodies," *J. Am. Ceram. Soc.*, **33**, 66 (1950).
7. C. W. Parmelee, *Ceramic Glazes*, Industrial Publications, Chicago, 1948.
8. D. B. Judd, *Color in Business, Science, and Industry*, John Wiley & Sons, Inc., New York, 1952.
9. L. E. Orgel, *An Introduction to Transition-Metal Chemistry*, 2d ed., Methuen & Co., Ltd., London, 1966.
10. B. A. Lengyel, *Introduction to Laser Physics*, John Wiley & Sons, Inc., New York, 1966.
11. B. Cockayne and D. W. Jones, Eds., *Modern Oxide Materials*, Academic Press, Inc., New York, 1972.
12. A. Smakula, "Synthetic Crystals and Polarizing Materials," *Opt. Acta*, **9**, 205 (1962).

13. E. Snitzer, "Lasers and Glass Technology," *Bull. Am. Ceram. Soc.*, **52**, 516 (1973).
14. R. D. Maurer, "Glass Fibers for Optical Communication," *Proc. IEEE*, **61**, 452 (1973).
15. E. G. Spencer, P. V. Lenzo, and A. A. Ballman, "Dielectric Materials for Electrooptic, Elastooptic and Ultrasonic Device Applications," *Proc. IEEE*, **55**, 2074 (1967).
16. S. E. Miller, "Integrated Optics: An Introduction," *Bell Syst. Tech. J.*, **48**, 2059 (1969).
17. D. Marcuse, Ed., *Integrated Optics*, IEEE Press, New York, 1973.

Problems

13.1. Windows for CO_2 lasers (10.6 microns) and CO lasers (5 microns) require low absorption yet high strength and ease of fabrication. Contrast the properties and requirements of oxides and halides.

13.2. For the following compounds, decide whether (*a*) $n^2 \gg K$, (*b*) $n^2 = K$, or (*c*) $n^2 \ll K$:

$MgAl_2O_4$	InP	Ge	FeO
SiC	GaAs	ZrO_2	NiO
CsCl	CdS	UO_2	
SiO_2 (fused)	Na_2O–CaO–SiO_2 (glass)	Al_2O_3	

13.3 Determine for MgO the polarizability and polarizability per mole at 10 kHz and the molar refraction for light at $\lambda = 0.590$ micron. Compare these values and comment. What would the difference be if the complex permittivity were used?

13.4 Infrared-transmitting optics are becoming of great importance. Arsenic trisulfide glasses are suitable for this purpose. In terms of their absorption characteristics, explain why silicate glasses are not used. What common impurities might you expect to be harmful in As_2S_3? Why?

13.5 What differences in index of refraction and dispersion would you expect between LiF and PbS? List reasons for your answer.

13.6 In the production of porcelain it is desirable to have a high degree of translucency, which is frequently not achieved. How would you define translucency as a measurable characteristic? Discuss the factors which contribute to translucency in porcelains, and explain techniques used in (*a*) composition selection, (*b*) fabrication methods, and (*c*) firing procedures which would enhance the translucency.

13.7. Which material transmits infrared radiation of the longest wavelength: MgO, SrO, or BaO?

13.8. Titania is extensively used to opacify porcelain enamels. What are the light-scattering particles? What particle characteristics give these enamels a high opacifying quality? Explain the relative and absolute importance of refractive index, particle size, crystal structure, color, transparency, etc. Explain how the light-scattering particles are formed in the enamel.

13.9. Zinc sulfide is an important phosphor with a band gap of 3.64 eV for the cubic (zinc blende) structure. Under proper excitation zinc blende doped with Cu^{++} (0.01 atom%) emits radiation of 6700 Å. When zinc vacancies are produced in the zinc blende lattice by the incorporation of Cl^{-} ions, the radiation emitted is centered around 4400 Å. (*a*) Compute the longest wavelength capable of producing fluorescence on the assumption that excitation is independent of impurity levels. (*b*) Locate the impurity levels in the band gap in relation to the valence band (illustrate with a drawing).

14

Plastic Deformation, Viscous Flow, and Creep

In this chapter we are concerned with modes of deformation under an applied stress that lead to permanent changes in shape. Atomic mechanisms by which these changes occur have been known in broad outline for a long time. The processes are varied and complex, however, and details on an atomic scale or even a microscale are still not completely understood. As the subject of intensive research at the present time, this area is definitely one of rapidly expanding knowledge.

The plastic deformation of crystals and the viscous flow of liquids and glasses are important for fabrication processes and also for many applications of ceramics. As discussed in the next chapter, there is increasing evidence that fracture in many materials is initiated by prior plastic deformation. Since fracture is one of the main limitations to a wider usefulness of ceramic materials, new development and understanding in this area are particularly important. Viscous flow, plastic deformation, and creep are all major criteria for structural applications of ceramics at high temperatures. This is true of traditional refractories and furnaces and also of new applications such as nose cones for space vehicles, nuclear fuel elements, and ceramics in high-temperature gas turbines.

14.1 Introduction

Since many different engineering and scientific disciplines have been concerned with mechanical properties of ceramic materials, a variety of test methods, terminologies, and points of view has developed. In this chapter we are concerned with deformation that results in permanent changes in shape. Depending on the time rate of the deformation process, these are conveniently described either as the deformation *resulting* from a given applied stress or as the *rate of deformation* caused by a stress.* In this introductory section we describe the macroscopic observations; in later sections the mechanisms of deformation are discussed and the effects of important variables.

Plastic Deformation. Plastic deformation is most commonly measured by increasing the load at a constant rate and observing the deformation. Some measurements have been made at a constant rate of strain, observing the corresponding stress. At low temperatures and at low rates of load application, measurements are usually not highly sensitive to these variables, and equivalent results are obtained. (But this is not always true.) The result obtained from this kind of a test is a stress-strain curve (Fig. 14.1). Characteristics necessary to describe the curve and compare it with

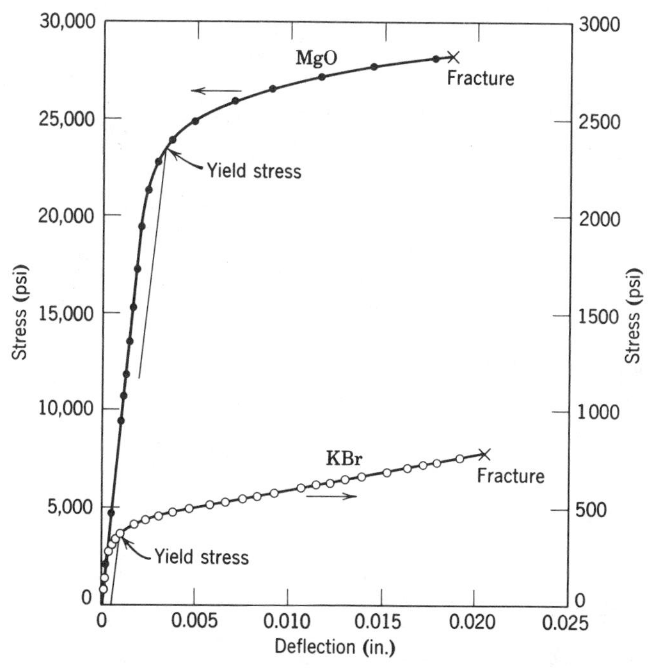

Fig. 14.1. Stress-strain curves for crystals of KBr and MgO tested in bending. From A. E. Gorum, E. R. Parker, and J. A. Pask, *J. Am. Ceram. Soc.*, **41**, 161 (1958).

other material are the elastic modulus E, the yield strength Y, and the fracture stress S. The elastic modulus is the ratio of stress to strain during the initial extension that is completely recovered when the stress is

*Units of stress include meganewton/m^2 = 1450 psi = 101 Kg/cm^2 = 1.01 kg/mm^2 = 10^6 bar.

removed, $E = \sigma/\epsilon$. Values of 5 to 90×10^6 psi are common for ceramic materials, as discussed in Chapter 15. The yield strength Y is the stress causing some small permanent deformation. It is usually determined by drawing a line parallel to the elastic part of the curve at some fixed strain, frequently 0.05%, as illustrated in Fig. 14.1. The fracture stress S is the stress at fracture, as shown in Fig. 14.1.

Several other ways of describing plastic deformation have been used from time to time. For the tensile tests usually used for metals, there is a substantial decrease in cross-sectional area of the specimen as the test proceeds, so that the actual stress on the sample is greater than the nominal stress calculated from the original dimensions. Frequently *tensile strengths* are reported as the maximum stress calculated on the basis of original dimensions; these are useful to engineers but are not related to fracture criteria. For most ceramic materials of low ductility transverse bending tests are used, and the problem of decrease in area does not arise. Sometimes the *proportional limit* is referred to. This is defined as the highest stress for which the stress-strain curve is strictly linear. Since it is really nowhere strictly linear, this value depends mainly on the sensitivity of testing equipment and is better avoided.

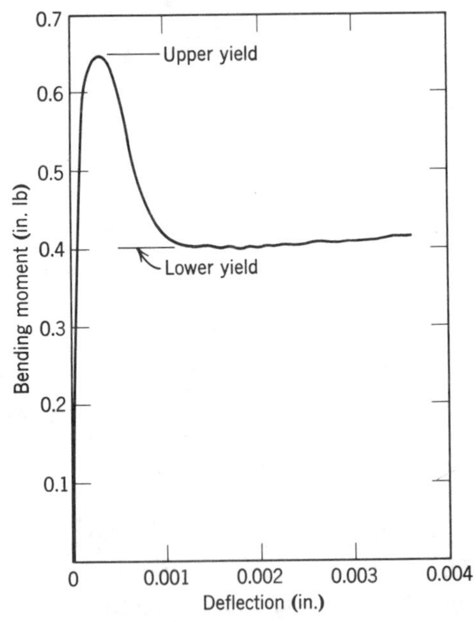

Fig. 14.2. Stress-strain curve for crystal of LiF illustrating yield point. From J. J. Gilman and W. G. Johnston, reference 6.

The term *yield point* has been used with various more or less distinct meanings. It is best reserved for deformation in which the stress drops off immediately after deformation is initiated, giving a definite maximum (Fig. 14.2), called the *upper* yield point; the following lower value required for continued deformation is the *lower* yield point. This kind of deformation is commonly observed for low-carbon steels; it is found in some samples of lithium fluoride and also for aluminum oxide and magnesium oxide at high temperatures.

Creep Deformation. When measurements are made at constant stress over an extended time period at elevated temperatures, a technical *creep* curve is observed, corresponding to continuing deformation with time (Fig. 14.3). After initial elastic extension there is a period during which the deformation rate decreases (primary creep or transient creep); this is followed by a short or long period of a minimum or constant rate of creep (steady-state creep or secondary creep); finally there is frequently a period when deformation rate increases because of impending fracture (accelerating creep or tertiary creep).

The shape of creep curves (Fig. 14.3) varies, depending on particular conditions of test and the material tested. Frequently the initial part of the curve can be represented by an expression

$$\dot{\epsilon} = (\text{constant})\theta^{-n}. \tag{14.1}$$

At low temperatures data can often be represented by $n = 1$, $\epsilon = (\text{constant}) \log \theta$.

Both the temperature and the stress affect the shape of the constant-temperature creep curve (Fig. 14.4). When the temperature is raised,

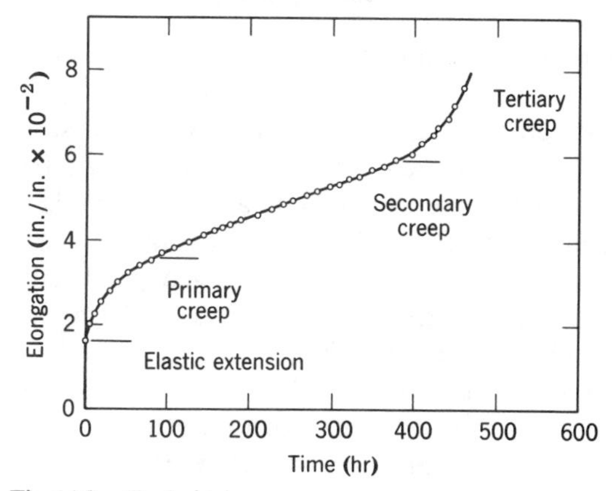

Fig. 14.3. Technical creep curve.

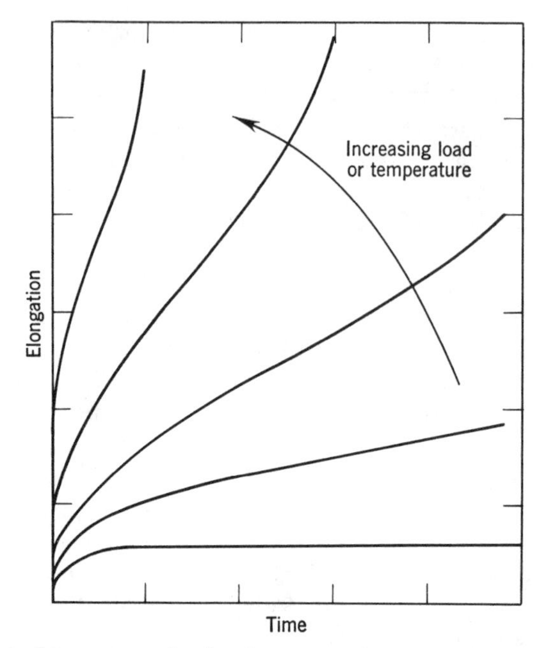

Fig. 14.4. Effect of temperature level and stress on the creep curve.

deformation is more rapid, and the duration of the constant-creep period is shortened. The same sort of change in the shape of the curve is observed with increasing stress. Frequently the deformation rate is proportional to some power of the applied stress:

$$\dot{\epsilon} = (\text{constant})\, \sigma^n \tag{14.2}$$

where n varies between 2 and 20, with a value of 4 being most commonly observed.

The strong temperature dependence of the constant-stress deformation rate has led to the development of various refractory *load tests.* In these measurements a fixed load is applied, and the sample temperature is increased at a constant rate. The resulting deformation is a complex sum of thermal expansion and creep of a sample which usually contains considerable temperature gradients; the overall result is similar to that shown in Fig. 14.5. Either a fixed deformation (often 10%) or rate of deformation (often 1% per 10° temperature rise) is taken as an end-point temperature, indicating the *load-bearing capacity.* Although these tests do not discriminate between deformation mechanisms, they do indicate in a qualitative way the difference in load-bearing capacity between samples of the same material and between different materials.

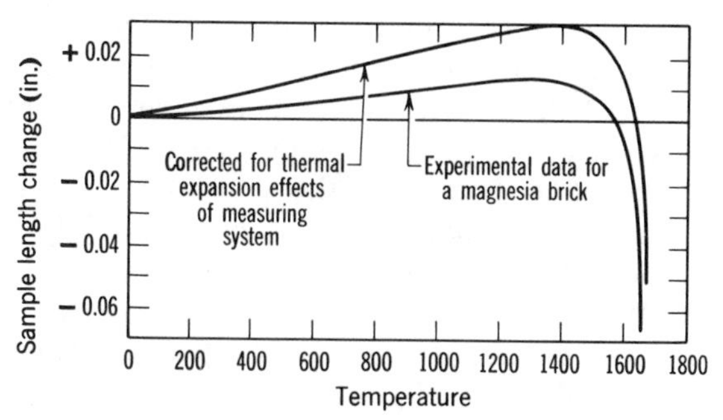

Fig. 14.5. Deformation of sample with fixed load and temperature increasing at a constant rate.

The deformation encountered in service by a ceramic is generally complex and far removed from laboratory test situations. Nevertheless, selection of materials for structural use is based on the mechanical-property data generated in the laboratory tests. The amount of deformation which is equivalent to a strain ϵ is not only a function of the stress σ, time t, and temperature T, as indicated above, but also of the structure S:

$$\epsilon = f(\sigma,t,S,T). \tag{14.3}$$

Since the deformation is strongly dependent on the constitution of the material, a structure term is required which involves both the macrostructure (i.e., grain size, porosity, phase distribution) and microstructure (i.e., crystal structure, point defects, dislocation tangles, vacancy clusters, etc.). Most of this chapter deals with the effect of structure on the deformation behavior of ceramics.

Viscous Deformation. For simple liquids the deformation rate is directly proportional to the shearing stress. At low velocities the liquid moves in parallel lines (Fig. 14.6); the viscosity is defined as the ratio of shear stress and velocity gradient:

$$\eta = \frac{\tau}{dv/dx} \tag{14.4}$$

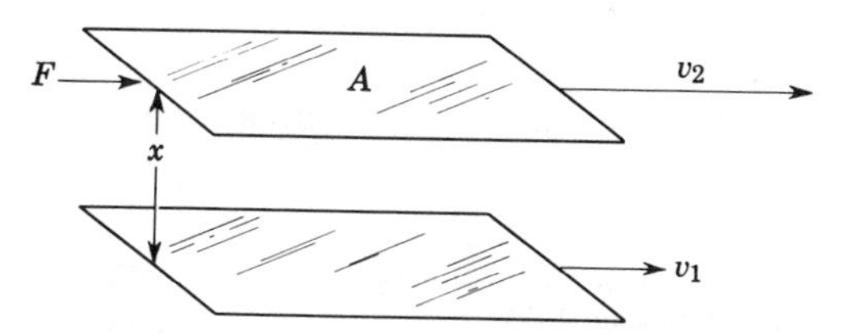

Fig. 14.6. Force per unit area τ and velocity gradient are related by viscosity.

where τ is the force exerted on a unit area of a plane parallel to the direction of flow and dv/dx is the normal velocity gradient. The viscosity coefficient η has units of shear stress per velocity gradient (g/cm-sec, or poise, ρ); the reciprocal of viscosity is called the fluidity ($\phi = 1/\eta$). Sometimes it is convenient to discuss fluid-flow properties in terms of the ratio of viscosity to density; this is called the kinematic viscosity ($\nu = \eta/\rho$ cm^2/sec).

The viscosity varies over wide limits. For water at room temperature and for liquid metals the value is of the order of 0.01 P (1 cP). For soda-lime-silica glasses at the liquidus temperature the value is about 1000 P; in the annealing range the viscosity of glasses is about 10^{14} P.

At the lower temperatures the deformation curve for a glass sample with a fixed load is similar to the creep curve shown in Fig. 14.3. It is distinguished from plastic flow in that the rate of deformation during the constant-rate period shown in Fig. 14.3 is directly proportional to the applied stress; this is generally not true for plastic deformation and creep of complex materials.

14.2 Plastic Deformation of Rock Salt Structure Crystals

The first studies of crystal plasticity were those of E. Reusch,* who discovered and investigated the plastic deformation of sodium chloride. The plastic deformation of sodium chloride and other alkali halide crystals was extensively studied during the 1930s.† These studies and the investigation of oxide crystals, particularly MgO, have been renewed and intensified under the stimulus of developments in the theory of dislocations.

It is clear, as has already been discussed in Chapter 4, that the plastic deformation of crystals takes place by the movement of dislocations through the crystal structure. (Dislocations have been discussed in Section 4.10 and are illustrated in Figs. 4.14 to 4.24.) The importance of diffusion in these processes is discussed in Section 14.5. Crystallographically the deformation process consists of crystal elements gliding over one another (called slip) or being homogeneously sheared (called twinning). These processes are illustrated in Fig. 14.7. The slip mechanism is simpler and is of very widespread importance; we restrict our discussion almost completely to this process. Macroscopically it is observed that slip tends to take place discontinuously in bands such as those shown in Fig. 14.7. In addition, the slip direction and usually the plane of slip have a definite crystallographic orientation. The stress required for plastic deformation of

Pogg. Ann., **132**, 441 (1867).

†See, for example, E. Schmidt and W. Boas, *Plasticity of Crystals*, Chapman & Hall, Ltd., London, 1968 (originally published 1935).

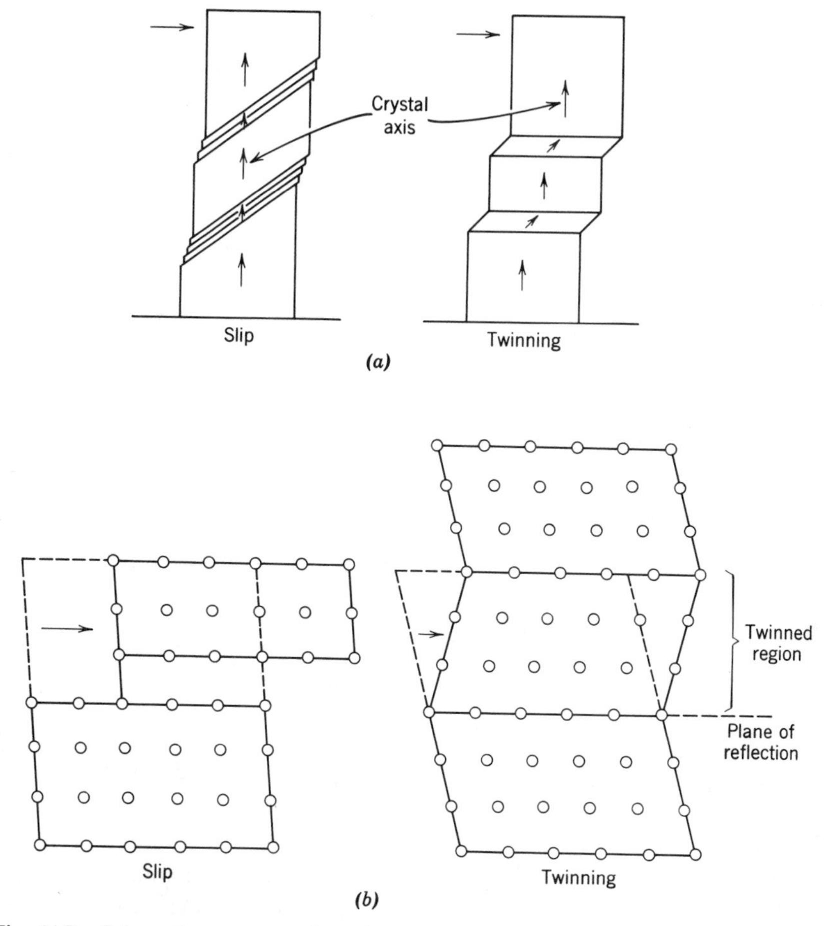

Fig. 14.7. Schematic representation of (*a*) macroscopic and (*b*) microscopic slip and twinning.

a single crystal is the resolved shear stress in the direction of slip on the area of the slip plane (Fig. 14.8). If the normal to the slip plane is at an angle ϕ to the applied stress, the stress in this plane is $(F/A)\cos\phi$; if the direction of slip is at an angle ψ to the direction of loading, the critical shear stress is

$$\tau_{\text{crit}} = \frac{F}{A}\cos\phi\cos\psi. \qquad (14.5)$$

The particular planes and directions in which slip takes place are dictated

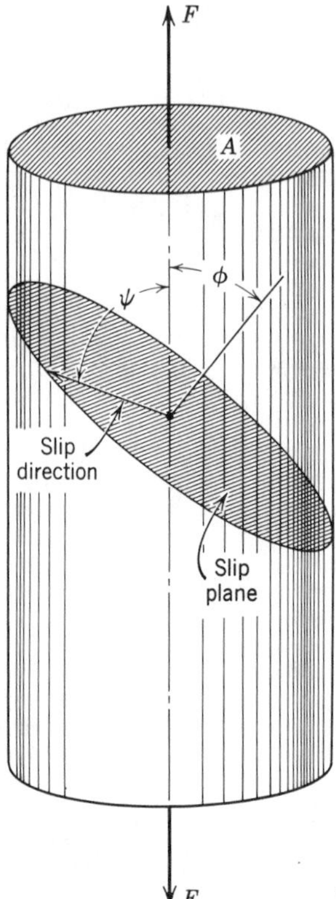

Fig. 14.8. Determination of critical shear stress.

by crystallographic considerations. In general, there are wide differences between the critical shear stress for different slip systems; often only one slip system is observed. In the sodium chloride structure ionic crystals, slip occurs most easily on {110} planes and in the [1$\bar{1}$0] direction at low temperatures. Slip lines in MgO are illustrated in Fig. 14.9.

Restrictions on slip systems and slip directions result from both geometric and electrostatic considerations. In sodium chloride structure crystals the direction of gliding, ⟨110⟩, is the shortest translation vector of the crystal structure and requires the smallest amount of displacement across the glide plane to restore the structure (Fig. 14.10). Also, translation in the [110] direction does not require any nearest-neighbor ions of the same polarity to become juxtaposed during the glide process, and no large electrostatic repulsive forces develop. The preference of strongly ionic crystals like NaCl and MgO to glide in the ⟨110⟩ direction is related to the greater electrostatic energy at half a unit translation distance for {100} glide, where like-charged ions would be brought into nearest-neighbor

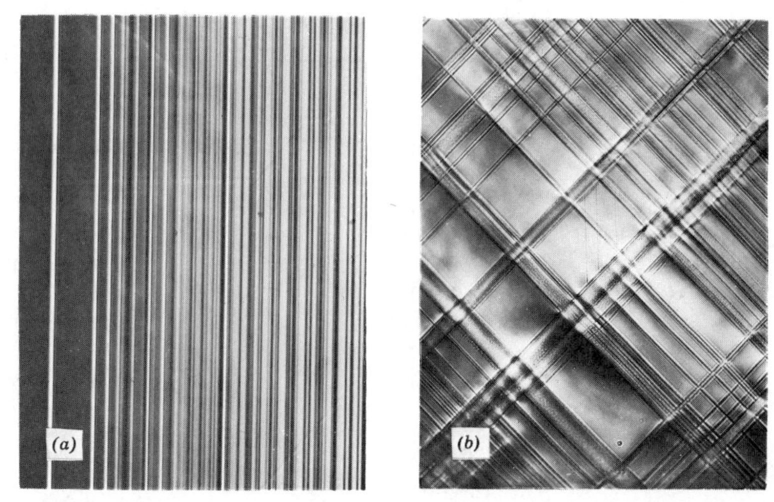

Fig. 14.9. (*a*) Deformation markings on surface of bent MgO crystal showing group of (110) deformation offsets in region of increasing deformation, going from left to right (175×). Courtesy M. L. Kronberg, J. E. May, and J. H. Westbrook. (*b*) Etched cross section of (110) deformation bands in bent MgO crystal (130×). Courtesy T. L. Johnston, R. J. Stokes, and C. H. Li.

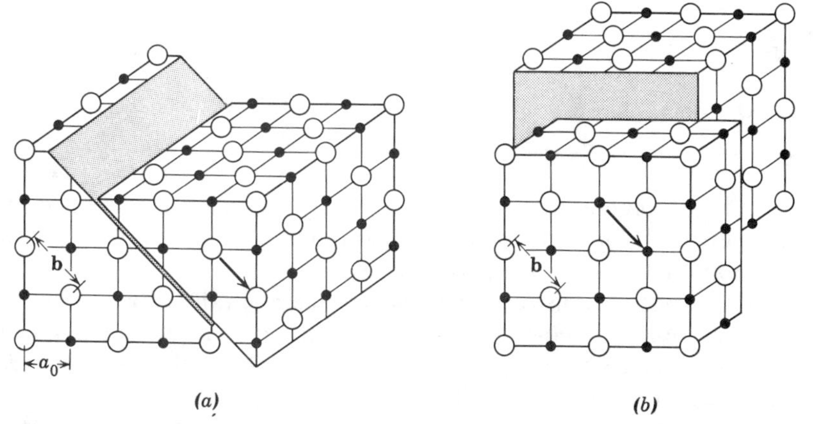

Fig. 14.10. Translation gliding in the ⟨110⟩ direction and on (*a*) the {110} plane and (*b*) the {100} plane for crystals with the rock salt structure. {110}⟨110⟩ glide is preferred.

positions. At high temperatures {100} ⟨110⟩ glide is observed for these materials. The slip systems for several ceramics are given in Table 14.1. Figure 14.11 illustrates that slip due to compression in the ⟨111⟩ direction is more difficult than in the ⟨110⟩ or ⟨100⟩ direction for MgO single crystals

Table 14.1. Slip Systems in Some Ceramic Crystals

Crystal	Slip System	Number of Independent Systems	Comments
C (diamond), Si, Ge	$\{111\}\langle 1\bar{1}0\rangle$	5	At $T > 0.5T_m$
NaCl, LiF, MgO, NaF	$\{110\}\langle 1\bar{1}0\rangle$	2	At low temperatures
NaCl, LiF, MgO, NaF	$\{110\}\langle 1\bar{1}0\rangle$ $\{001\}\langle 1\bar{1}0\rangle$ $\{111\}\langle 1\bar{1}0\rangle$	5	At high temperatures
TiC, UC	$\{111\}\langle 1\bar{1}0\rangle$	5	At high temperatures
PbS, PbTe	$\{001\}\langle 1\bar{1}0\rangle$ $\{110\}\langle 001\rangle$	3	
CaF_2, UO_2	$\{001\}\langle 1\bar{1}0\rangle$	3	
CaF_2, UO_2	$\{001\}\langle 1\bar{1}0\rangle$ $\{110\}$ $\{111\}$	5	At high temperatures
C (graphite), Al_2O_3, BeO	$\{0001\}\langle 11\bar{2}0\rangle$	2	
TiO_2	$\{101\}\langle 10\bar{1}\rangle$ $\{110\}\langle 001\rangle$	4	
$MgAl_2O_4$	$\{111\}\langle 1\bar{1}0\rangle$ $\{110\}$	5	

even at high temperatures. In weakly ionic crystals such as PbTe and PbS slip occurs on {100} planes at low temperatures because the polarizability of the ions reduces the repulsive forces.

In ionic crystals a dislocation must also maintain the cation-anion site ratio. As a result, an edge dislocation for slip on a $(1\bar{1}0)$ plane in a $[1\bar{1}0]$ direction in magnesium oxide requires the removal of a plane of molecules (two planes of atoms) with a Burgers vector larger than the elementary Mg–O separation. Atom pairs must be removed for the crystal to come back to the correct structure, as illustrated in Fig. 14.12.

For macroscopic deformation to take place (experimentally observed yielding), it is necessary to cause dislocations to start moving. If there are no dislocations present, some must be created; if existing dislocations are pinned by impurities, some must be freed. Once these initial dislocations are moving, they accelerate and cause increasing multiplication and

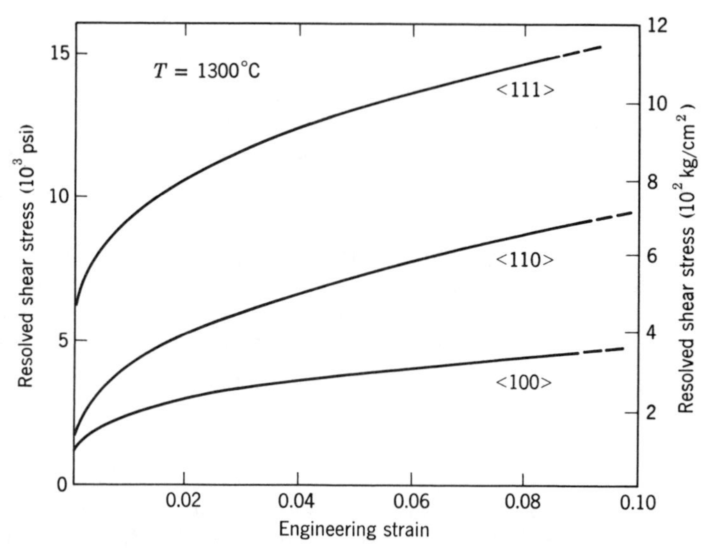

Fig. 14.11. In single-crystal MgO slip occurs on a more difficult system for the ⟨111⟩ axis than it does for the ⟨110⟩ and ⟨100⟩ axes. From S. M. Copley and J. A. Pask in *Materials Science Research*, Vol. 13, W. W. Kriegel and H. Palmorr III, Eds., Plenum Press, New York, 1966, pp. 189–224.

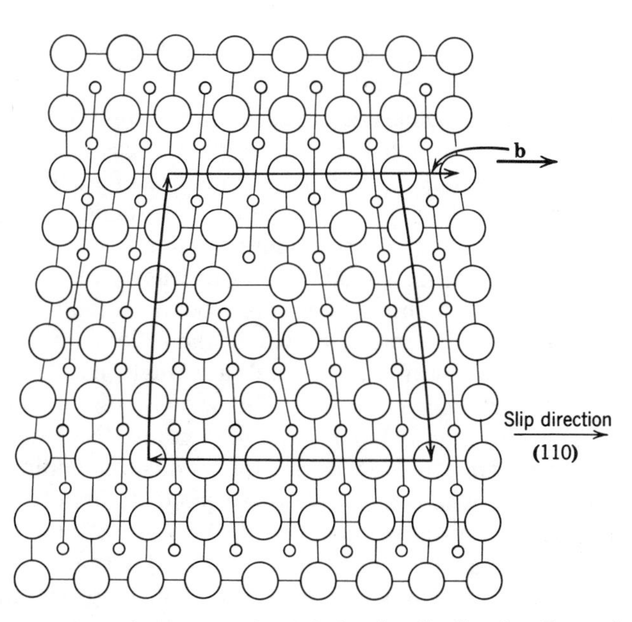

Fig. 14.12. Edge dislocation structure in MgO showing slip direction, Burgers' circuit, and Burgers' vector **b**.

macroscopic yielding. Plastic-deformation characteristics are related both to the energy required to form dislocations or initiate their movement and to the force required to keep them moving at any specified velocity. Either can be a restriction on plastic deformation. It is found that for filamentary dislocationfree whiskers a very large stress is necessary to initiate plastic deformation; however, once slip is initiated, it continues at a much lower stress level.

The microscopic theory of plastic deformation was founded by Orowan, who first interpreted plastic flow as a dynamic process by which the rate of glide is given by the mobile dislocation density N_m and their average velocity $\bar{v}$. Thus, the plastic-strain rate is given by the product of these two terms and the magnitude of the Burgers vector.

$$\dot{\epsilon} = \frac{d\epsilon}{dt} = N_m \mathbf{b} \bar{v}. \tag{14.6}$$

Gilman and Johnston* first clearly tested this relationship; they were able to evaluate the velocity and the number of mobile dislocations separately in LiF crystals. As shown in Fig. 14.13, etch-pit techniques can be used to distinguish the motion of an individual dislocation loop. The large flat-bottom etch pits correspond to the initial position of the dislocation, whereas the pointed-bottom pits show their positions after an applied stress has caused them to move.

Stress-strain curves for the same lithium fluoride crystals with different surface treatment are shown in Fig. 14.14. In as-cleaved crystals mechanical stresses present during cleavage lead to formation of dislocations and easy slip. After chemical polishing a high yield stress and definite yield point are observed, corresponding to the elimination of mobile dislocations from previous handling and the requirement for forming new ones. After polishing and sprinkling with carborundum powder, impacts cause formation of many new dislocation loops and again easy plastic deformation. That is, generation and multiplication of dislocations require more force than their subsequent movement; however, the ratio of these stresses is seldom more than a factor of 2, and crystals as actually used almost always have surface imperfections present. The growth of an initial single-dislocation loop into a glide band containing many dislocations is illustrated in Fig. 14.15. The stress necessary for dislocation multiplication and formation of a glide band (Fig. 14.14) corresponds to about 800 g/mm^2 in this lithium fluoride crystal compared with the value of about 600 g/mm^2 necessary to enlarge a single half-loop (Fig. 14.13) without multiplication taking place. The stability of small dislocation half-loops and the stresses

*See reference 6.

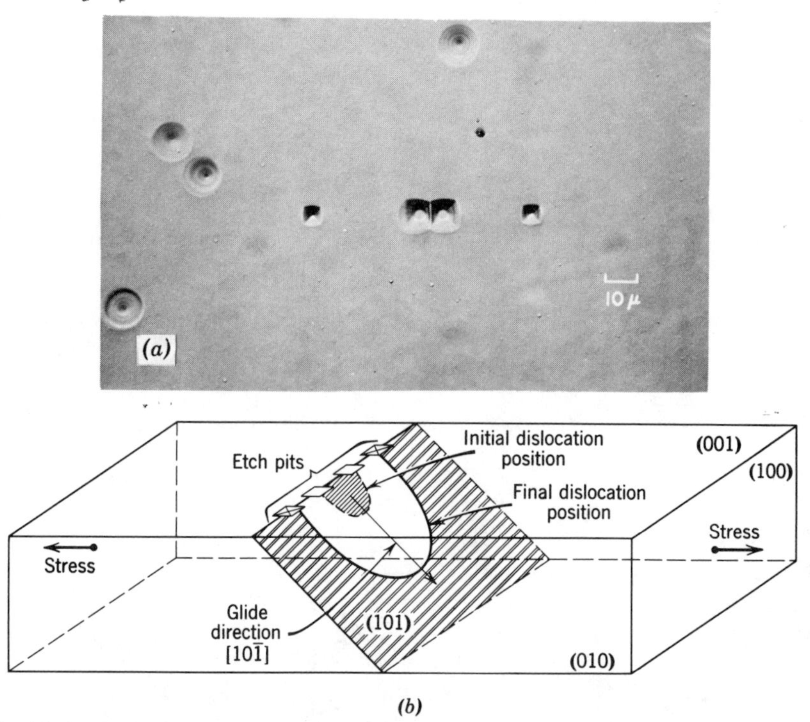

Fig. 14.13. Etch pits showing motion of individual dislocation loop in a lithium fluoride crystal. From J. J. Gilman and W. G. Johnston, reference 6.

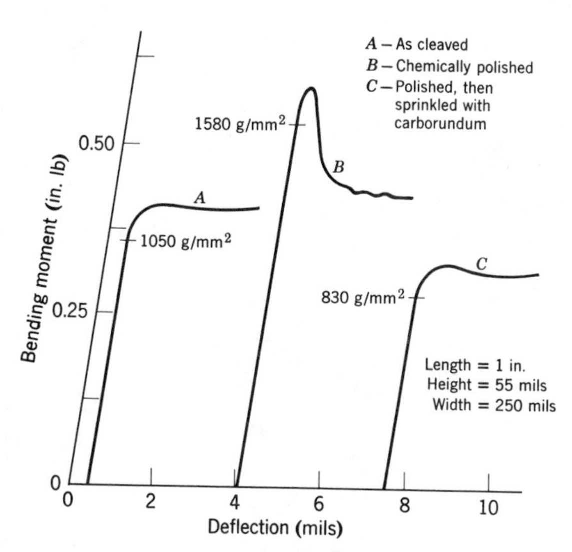

Fig. 14.14. Stress-strain curves showing effects of surface treatment for LiF crystals. From J. J. Gilman and W. G. Johnston, reference 6.

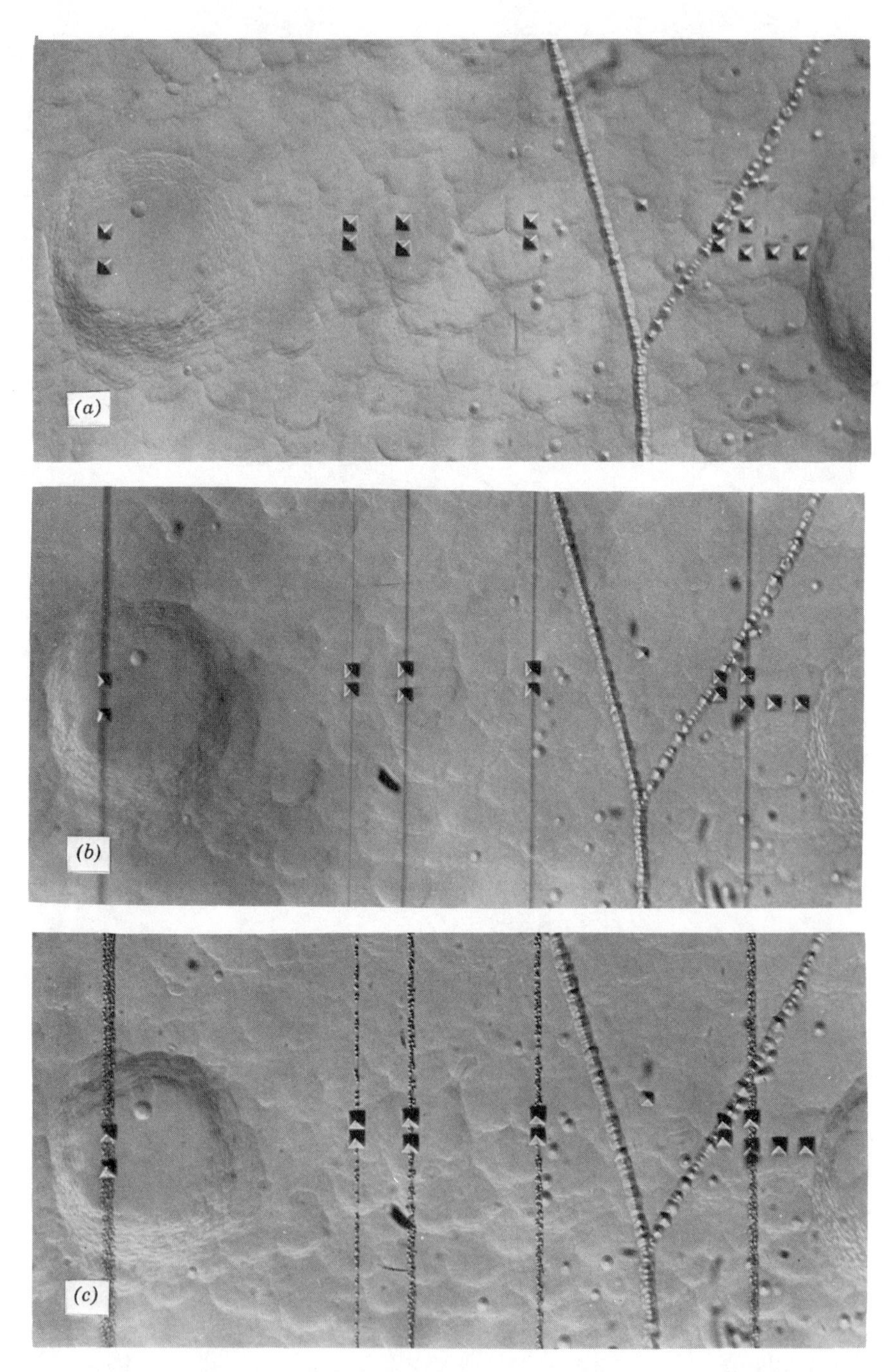

Fig. 14.15. Growth of glide bands from dislocation half-loops. Strain rate 2×10^{-5} sec. (*a*) Single-dislocation half-loops at the surface of a LiF single crystal; (*b*) same crystal after bending shows glide bands passing through five of the half-loops; (*c*) a light etch reveals many dislocations in each glide band. From J. J. Gilman and W. G. Johnston, reference 6.

necessary for their motion correspond to solute hardening or a viscous drag on dislocation movement in sodium chloride structure crystals.

Dislocation velocities varying between 10 and 10^{12} atom distances per second have been measured in lithium fluoride by W. G. Johnston and J. J. Gilman* by applying stress pulses to crystals. They found, first, that a certain shear stress must be applied before dislocations move through the crystal at an observable rate. Second, the velocity increases very rapidly with small increases of stress. Third, the edge components move more rapidly than the screw components. Finally, the velocity of sound appears to be a limiting velocity. Results for a typical crystal are shown in Fig. 14.16. As-grown samples tested at low temperatures require a greater stress for dislocation movement; samples softened by heat treatment require a

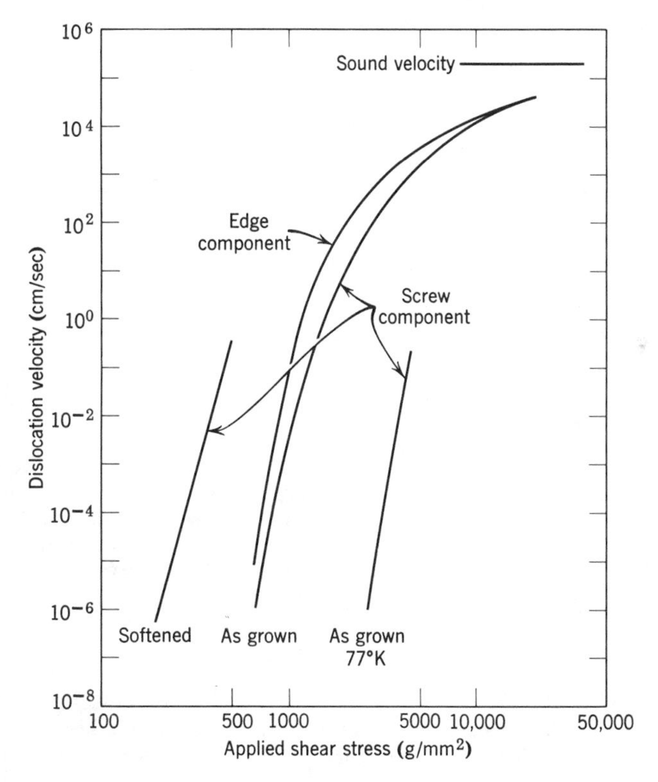

Fig. 14.16. Dislocation velocities in a typical LiF crystal as grown, heat-treated, and at 77°K. From W. G. Johnston and J. J. Gilman, *J. Appl. Phys.*, **30**, 192 (1959).

***J. Appl. Phys.*, **30**, 129 (1959).

smaller stress for dislocation movement. In all samples a rapid increase in dislocation velocity with increased stress is observed.

The rate at which a crystal deforms plastically depends on how many dislocations are moving as well as their velocity (Eq. 14.6). Once deformation is started, all but a few of the dislocations are formed through regenerative multiplication. The general way in which this happens when ends of a dislocation segment are pinned has been discussed in Chapter 4, and two forms of this process, the Frank-Read source and *multiple cross glide* are described there. They are similar, but the latter is probably more common for most alkali halide types of crystals. As a result of these processes the average dislocation density increases with the amount of strain (Fig. 14.17).

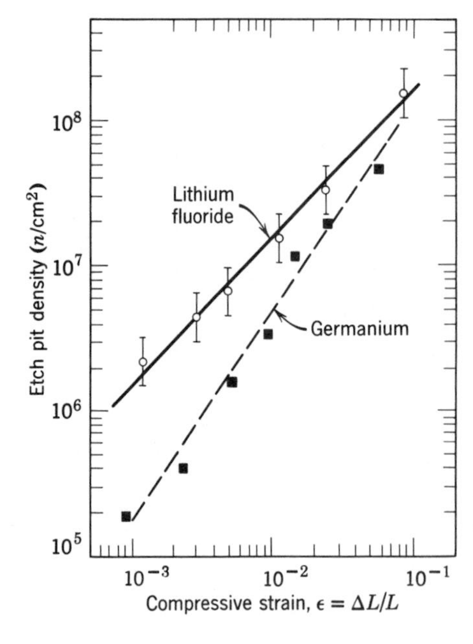

Fig. 14.17. Average etch-pit density versus plastic strain in LiF, from W. G. Johnston and J. J. Gilman, *J. Appl. Phys.*, **30**, 129 (1959), and in germanium, from J. R. Patel and B. H. Alexander, *Acta Met.*, **4**, 385 (1956).

In general, the stress-strain curve can have various shapes at the beginning of the yielding process (Fig. 14.18). In curve *A*, a large number of dislocations are initially present, and at a stress slightly higher than the stress at which dislocation motion starts, the product of dislocation concentration and velocity is sufficient to allow the crystal to deform at the applied strain rate. In curve *C* the initial number of dislocations is much lower, and a stress sufficient to nucleate dislocations or pull them away

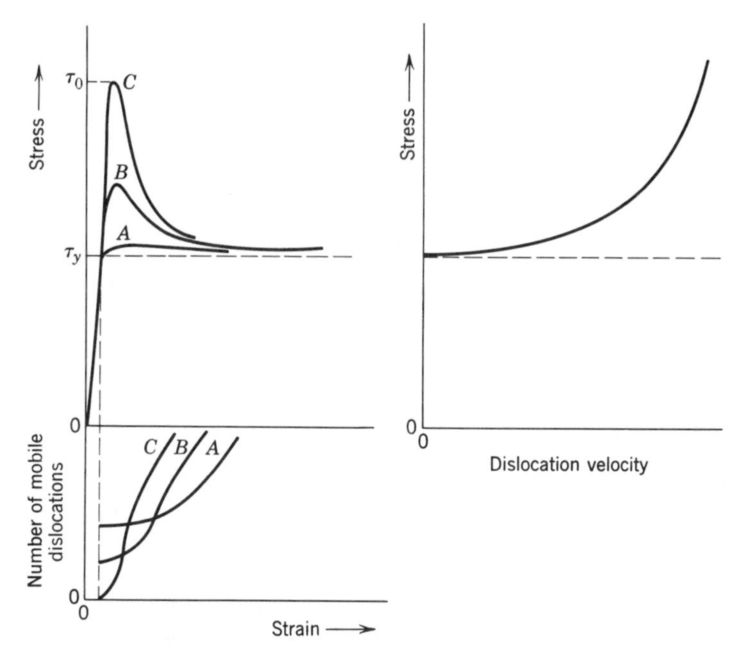

Fig. 14.18. Different shapes of the stress-strain curve at the beginning of the yielding process and interpretation in terms of dislocation density and velocity. From J. J. Gilman, reference 8.

from pinning impurities is necessary to initiate deformation. Once sufficient dislocations are moving, the stress drops. In curve *B* an intermediate condition is shown.

It is apparent from the nature of the process that many variables can affect the plastic-deformation properties of sodium chloride structure crystals. In addition to temperature, prior history, and stress level, the strain rate is important, as implied in the dislocation velocities shown in Fig. 14.16. Most metals, which are less sensitive to strain rate than ionic crystals, must have an even steeper slope than LiF. Some partially ionic crystals such as Al_2O_3, which apparently are more sensitive to strain rate, must have a flatter slope than LiF (Table 14.2).

When the strain fields of dislocations begin to interact, it becomes more difficult for the dislocations to glide. Thus, the yield stress is sensitive to the total strain, and this sensitivity is called strain hardening. G. I. Taylor* first showed that the flow stress for an incremental change in the plastic strain is proportional to the square root of the dislocation density *N* or the square

Proc. R. Soc.* (*London*), **A145, 362 (1934).

Table 14.2. Stress-Sensitivity Exponent of the Dislocation Velocity ($v \sim \tau^m$)

Material	Crystal Structure	m (at R.T.)[a]
LiF	Rock salt	13.5–21
NaCl	Rock salt	7.8–29.5
NaCl (ultra high purity)	Rock salt	3.9
KCl	Rock salt	20
KBr	Rock salt	65
MgO	Rock salt	2.5–6
CaF_2	Flourite	7.0
UO_2	Fluorite	4.5–7.3
Ge	Diamond	1.35–1.9[b]
Si	Diamond	1.4–1.5[b]
GaSb	Diamond	2.0[b]
InSb	Diamond	1.87[b]

[a] The values of m may be more sensitive to impurities and prior thermal history than to crystal structure. In addition the homologous temperature (here $298°/T_{mp}$) is important.
[b] Above R.T.

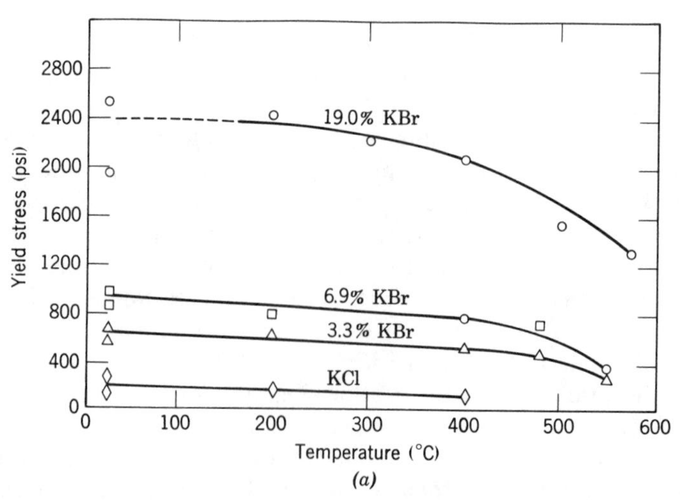

Fig. 14.19. (*a*) Yield stress of KCl–KBr solid solutions as a function of temperature, compressed along ⟨100⟩ direction. From N. S. Stoloff et al., *J. Appl. Phys.*, **34**, 3315 (1963).

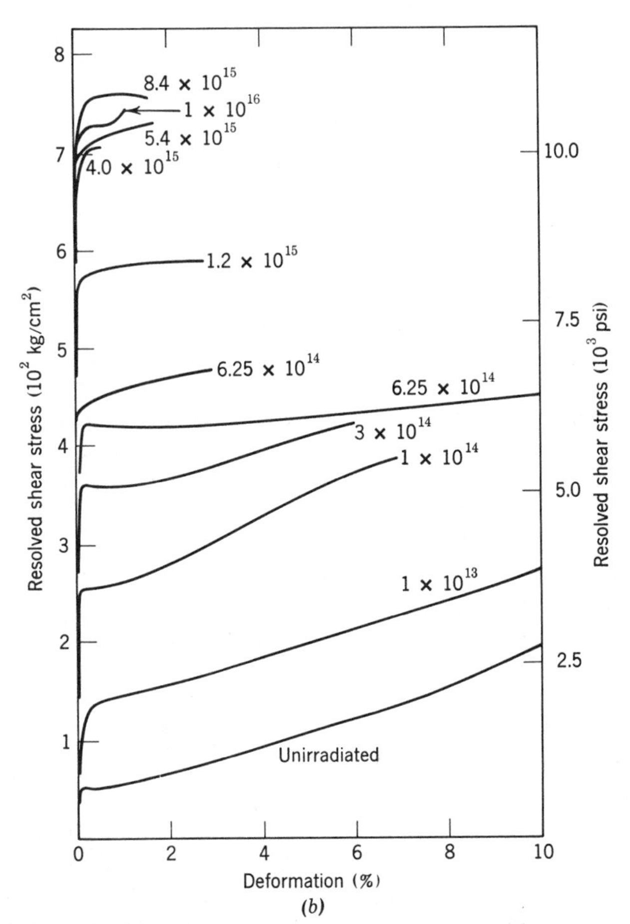

Fig. 14.19 (*Continued*). (*b*) At low temperatures, irradiation hardens ceramic materials as shown here for LiF at room temperature. The dosages are in electrons/cm^2. From A. D. Whapham and M. J. Makin, *Phil. Mag.*, **5**, 237–250 (1960).

root of the plastic strain

$$\sigma_{\text{yield}} \propto N^{1/2} \propto \epsilon_p^{1/2}. \qquad (14.7)$$

The strain hardening for MgO compressed in three orientations was shown in Fig. 14.11 (slope of σ versus ϵ) to also be a function of crystallographic orientation.

The motion of dislocations may also be inhibited by impurities in the lattice. Substitutional impurities with the same valence but with different ionic radii increase the yield strength because of the associated strain field around the impurity (Fig. 14.19*a*). In fact the dislocations tend to become pinned by the impurity. Lattice defects (vacancies and interstials) caused by irradiation similarly increase the yield strength (Fig. 14.19*b*). Substitutional impurities with a different valence require additional lattice defects for compensation; for example, $SrCl_2$ in KCl requires that additional potassium vacancies be formed. Dipoles are formed from the associated defects ($V_K' Sr_K^{\cdot}$) which more effectively inhibit glide. As a consequence the strengthening effects are generally larger. Figure 14.19*a* shows the increase in the yield for KCl–KBr solid solutions, and Fig. 14.20*a* shows the dramatic effects due to small additions of $SrCl_2$ to KCl. The replacement of potassium ions by 0.084% strontium ions increases the yield stress by more than a factor of 10.

The yield stress depends on the concentration of the solute. Often this has been shown to depend on the square root of the concentration:

$$\sigma_{\text{yield}} \propto C^{1/2}. \tag{14.8}$$

Data in Fig. 14.20*b* show this dependence for several alkali halide crystals. However more complex behavior has been observed (Fig. 14.21 and 14.22) and must be related to impurity segregation, impurity oxidation state, etc.

In many cases the solute may form precipitates which also act as inhibitors to dislocation motion. The precipitation-hardening mechanism is generally more effective than solid-solution strengthening. The AgCl–NaCl system has limited solid-solid solubility below 175°C, which leads to precipitation in single crystals and results in large increases in the compressive yield strength over the pure crystals and over the solid-solution-strengthened crystals (Fig. 14.21).

Increasing the temperature at which deformation tests are made decreases the effects of all these strengthening techniques (strain, solution, precipitation). The lattice planes are more "flexible" because of absorbed thermal energy, which causes lattice vibrations; thus the impedence of these obstacles is less. In addition, slip which may be impeded on one slip system may be accommodated on another system. Higher temperatures also increase the amount of soluble impurities and can lead to coarsening of precipitates (Chapter 9), which decreases the number of obstacles to dislocation motion. A. Joffe and others* indicated that sodium chloride was brittle in air but remained ductile when immersed in water so that a new surface was continually being formed. This was believed to result from the removal of surface microcracks by the solubility of the surface in water. Recently it has been found by A. E. Gorum, E. R. Parker, and J. A. Pask† and A. Aerts and W. D. Dekeyser‡ that freshly cleaved samples of NaCl,

*A. Joffe, M. W. Kupitschewa, and N. A. Levitsky, *Z. Phys.*, **22**, 286 (1924).
†*J. Am. Ceram. Soc.*, **41**, 161 (1958).
‡*Acta Met.*, **4**, 557 (1956).

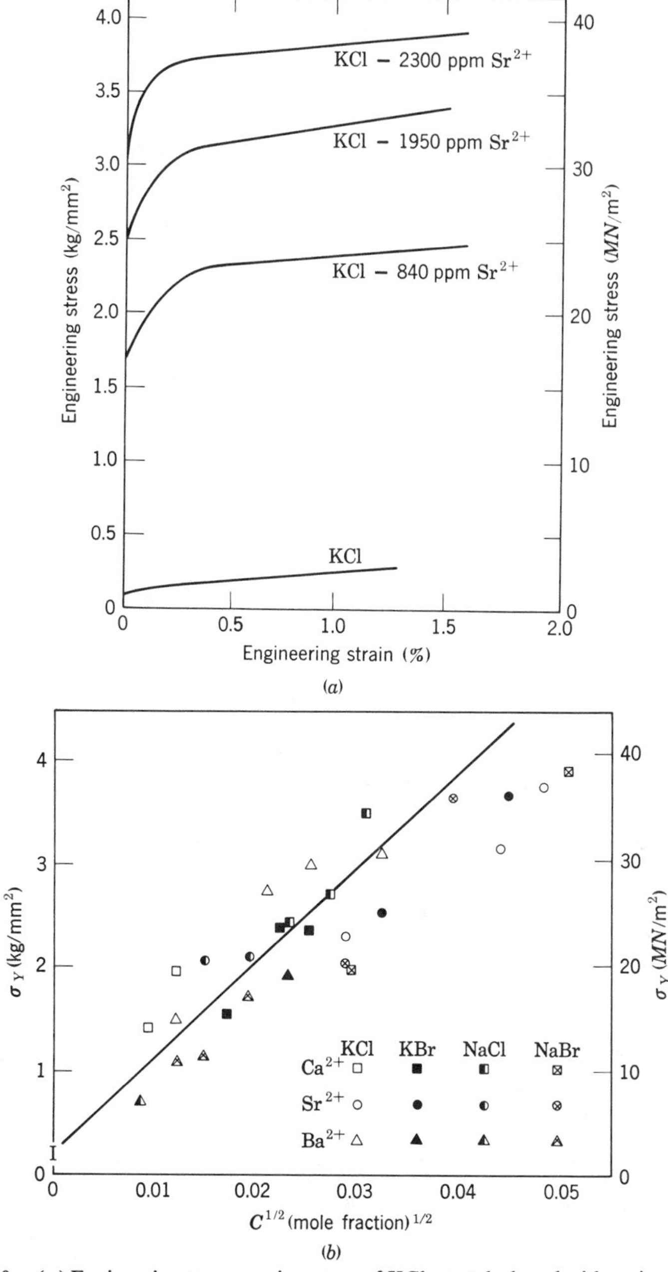

Fig. 14.20. (*a*) Engineering stress-strain curves of KCl crystals doped with various amounts of Sr^{2+}. Crystals tested in compression in a ⟨100⟩ direction after air-cooling following a half hour anneal at 725°C. (*b*) Values of yield strength versus the square root of concentration of M^{2+} for various dopants in alkali halide crystals. From G. Y. Chin et al., *J. Am. Ceram. Soc.*, **56**, 369 (1973).

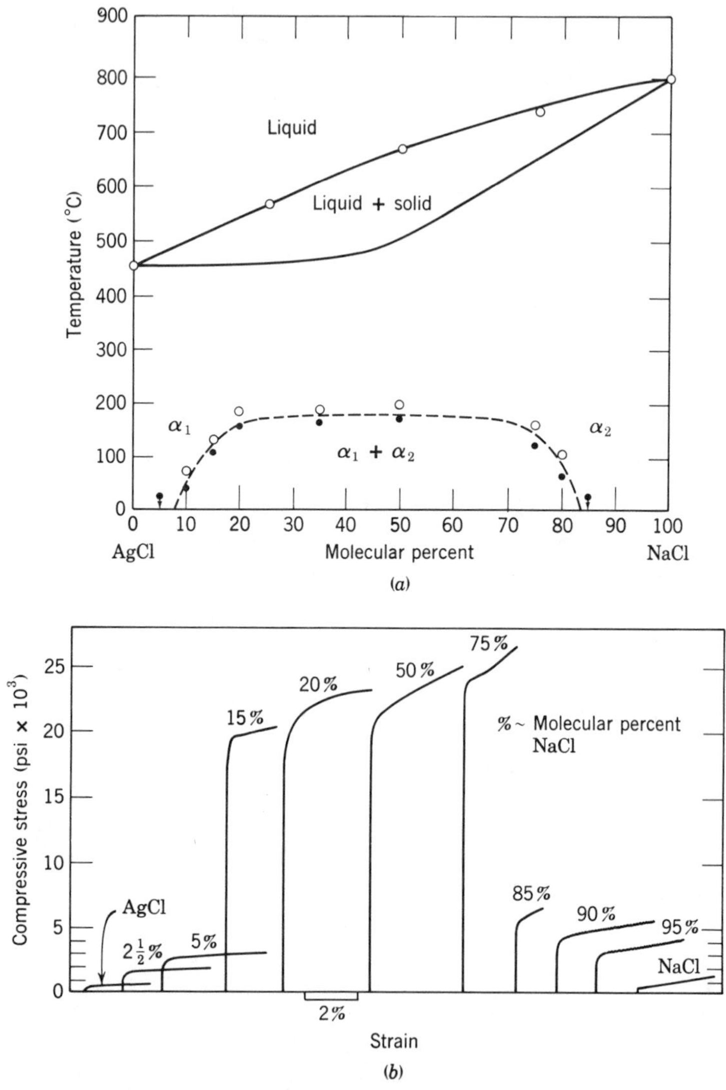

Fig. 14.21. (*a*) Solid-solution and (*b*) precipitation hardening in the silver chloride–sodium chloride alloy system. From R. J. Stokes and C. H. Li, *Acta Met.*, **10**, 535 (1962).

KCl, and other ionic crystals remain ductile for long periods of time if stored under conditions such that surface damage cannot occur. The several mechanisms by which surface contamination can prevent ductility have not been completely elucidated. Various mechanisms by which

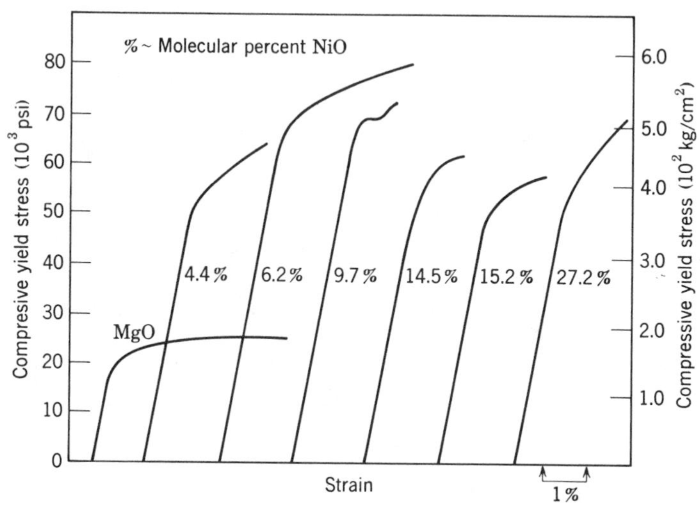

Fig. 14.22. Alloying single-crystal MgO results in hardening, but the effect is limited with NiO. From T. S. Liu, R. J. Stokes and C. H. Li, *J. Am. Ceram. Soc.*, **47**, 276–279 (1964).

microcracks can be initiated are discussed in Chapter 15. A review of surface-sensitive mechanical properties of ionic crystals is given by A. R. C. Westwood.*

14.3 Plastic Deformation of Fluorite Structure Crystals

Most studies on dislocation motion in fluorite structure crystals have been done with CaF_2, although some studies have been made with other fluorides and with UO_2. The deformation behavior of fluorite structure crystals is different from the rock salt structure crystals in that low-temperature slip occurs on systems of the type $\{100\}\langle 110\rangle$. At high temperatures, five independent systems are active, due to the secondary $\{111\}\langle 110\rangle$ and $\{110\}\langle 110\rangle$ systems. (The importance of five independent slip systems is discussed in Section 14.5.) The mechanisms for multiplication of dislocations appear to be the same, but the dislocation velocity in fluorite appears to be less stress-sensitive than in rock salt (compare Figs. 14.16 and 14.23). An interesting observation is that screw-dislocation velocities in CaF_2 are higher than edge dislocations at the same temperature and stress (Fig. 14.23). This allows for easy cross glide (Fig. 4.19) at elevated temperatures and a transition from distinct slip bands to wavy

***Mater. Sci.*, **1**, 114 (1963).

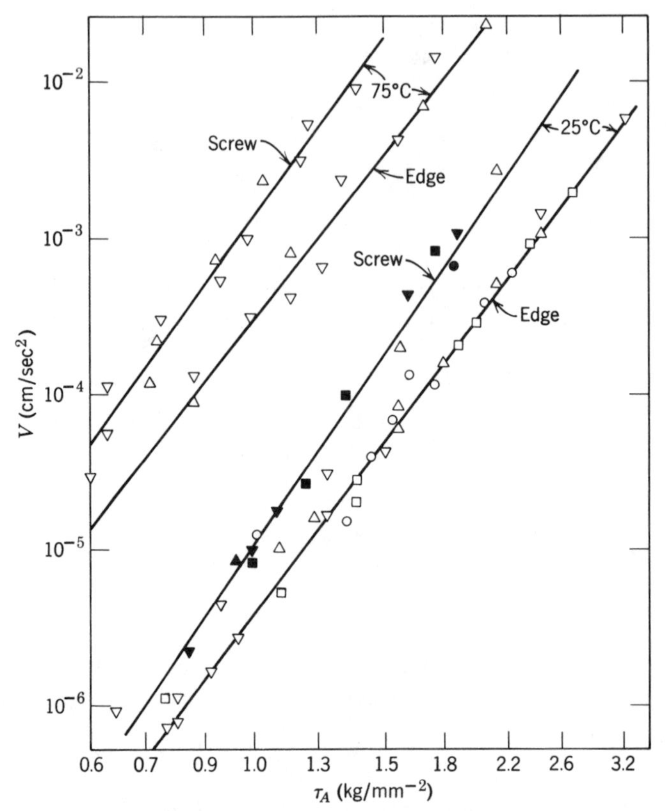

Fig. 14.23. The stress dependence of the dislocation velocity. From A. G. Evans and P. L. Pratt, *Phil. Mag.*, **20**, 1213 (1969).

slip. The variation of the yield stress with temperature and strain rate is shown in Fig. 14.24.

Impurities in the lattice cause hardening, as discussed for rock salt structures. Dopants such as NaF, YF_3, or NdF_3 cause a dipole to be formed between the impurity and the lattice defect (vacancy or interstial), which gives a very effective obstacle to dislocation motion. The effects of 0.002% Nd in CaF_2 can be seen in the yield characteristics at 160°C (Fig. 14.25), more than doubling the value from 1500 to 3500 kg/mm².

14.4 Plastic Deformation of Al_2O_3 Crystals

The plastic-deformation characteristics of aluminum oxide are of particular interest, since alumina is a widely used ceramic material and the deformation of this noncubic, strongly anisotropic crystal probably

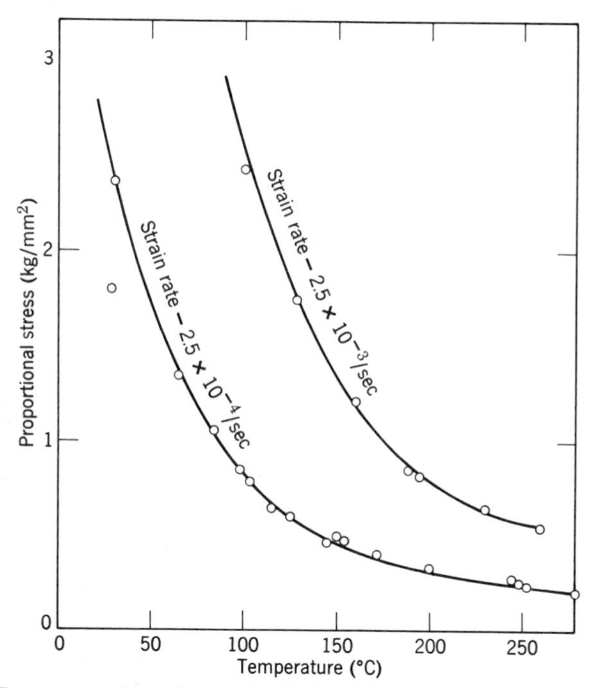

Fig. 14.24. Temperature dependence of critical resolved shear stress of CaF_2. From P. L. Pratt, C. Roy and A. G. Evans, *Mat. Sci. Res.*, **3**, 225 (1966).

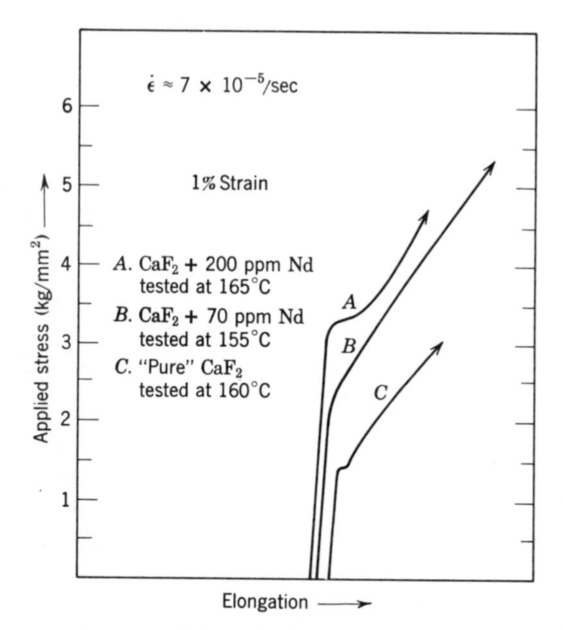

Fig. 14.25. Stress-strain curves of CaF_2 single crystals of various dopant levels tested in compression along the $[11\bar{2}]$ direction. From R. N. Katz and R. L. Coble.

represents an extreme in behavior. The deformation characteristics are directly related to the crystal structure. Single crystals deform plastically above 900°C by basal slip on $(0001)\langle 11\bar{2}0\rangle$ systems, giving rise to anisotropic deformation. At higher temperatures slip may occur on the prismatic planes $\{1\bar{2}10\}$ in the $\langle 10\bar{1}0\rangle$ or $\langle 10\bar{1}1\rangle$ directions and on the pyramidal systems $\langle 01\bar{1}1\rangle\{1\bar{1}02\}$ and $\langle 01\bar{1}1\rangle\{10\bar{1}1\}$; slip on these nonbasal systems can also occur at lower temperatures at very high stresses. Even at 1700°C the stress to initiate nonbasal slip is 10 times that for basal slip.

At temperatures above 900°C, the characteristics of plastic deformation are shown in Fig. 14.26 and may be summarized as (1) a strong temperature dependence, (2) a large strain-rate dependence, and (3) a definite yield point in constant-strain rate-tests, or an equivalent induction period in long-time low-stress creep tests (Fig. 14.27). Both the upper and lower yield stresses in Fig. 14.26*a* are temperature-sensitive and show an approximately exponential decrease with increasing temperature (Fig. 14.26*b*). The sharp yield point and yield-point drop observed both in tension and compression

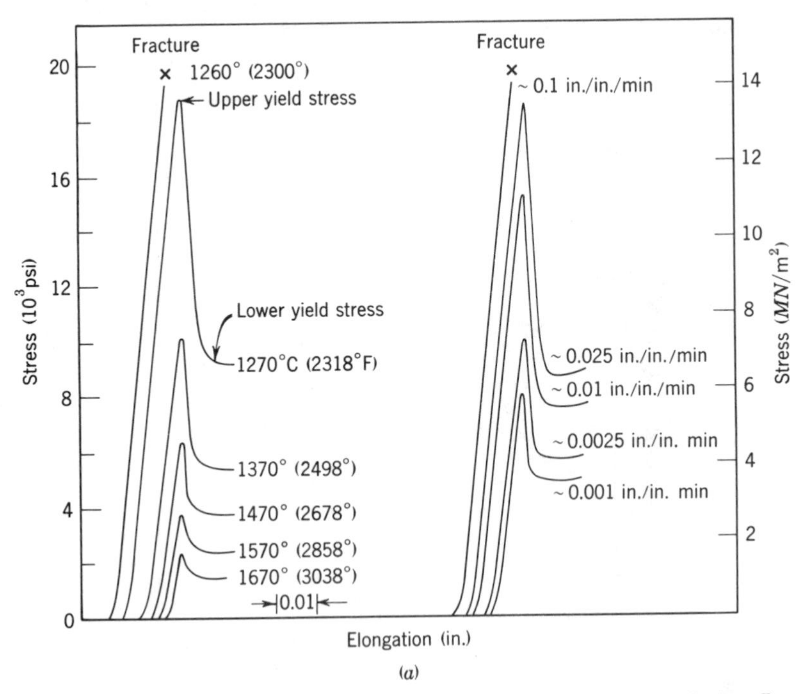

Fig. 14.26. (*a*) Deformation behavior of single-crystal alumina. On the left the effect of temperature and on the right the effect of strain rate. A sharp yield stress and a large yield drop are apparent.

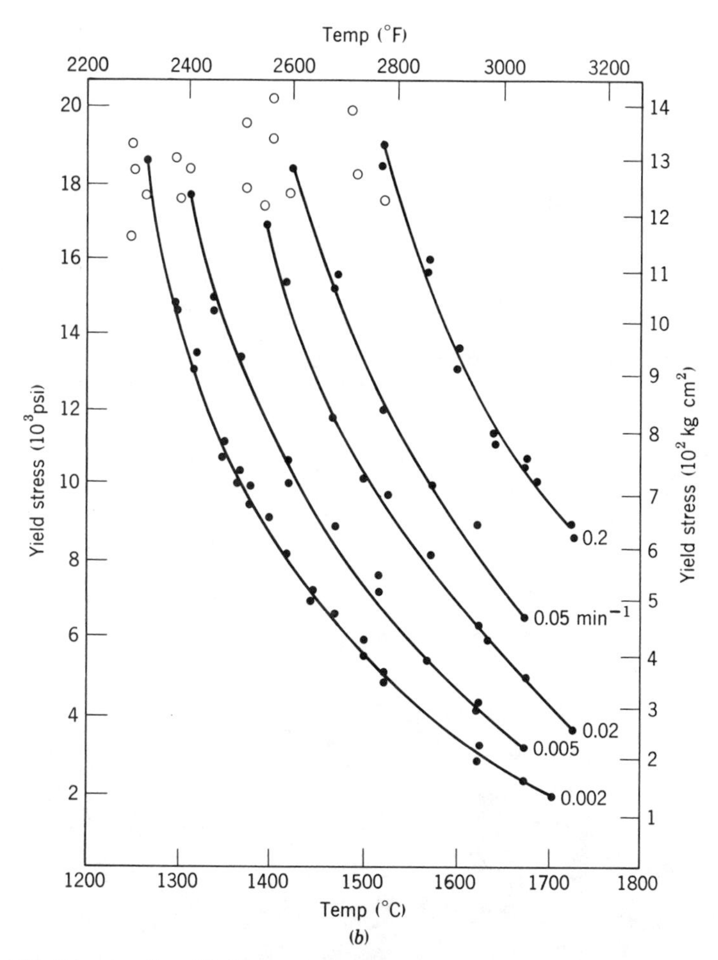

Fig. 14.26 (*Continued*). (*b*) Illustration of the strong dependence of yield stress on temperature and strain rate for single-crystal Al_2O_3. From M. L. Kronberg, *J. Am. Ceram. Soc.*, 45, 274–279 (1962).

are explained by the need for dislocation multiplication rather than unpinning of dislocations. A tenfold increase in the strain rate doubles the yield stress.

As discussed in Chapter 2, Al_2O_3 has a rhombohedral hexagonal structure with close-packed oxygen ions; two-thirds of the octahedral interstices are filled with Al^{3+} ions. Two layers of oxygen ions with the intermediate aluminum ions are shown in Fig. 14.28. On that figure the minimum translation on the basal plane that is necessary to give reregistry

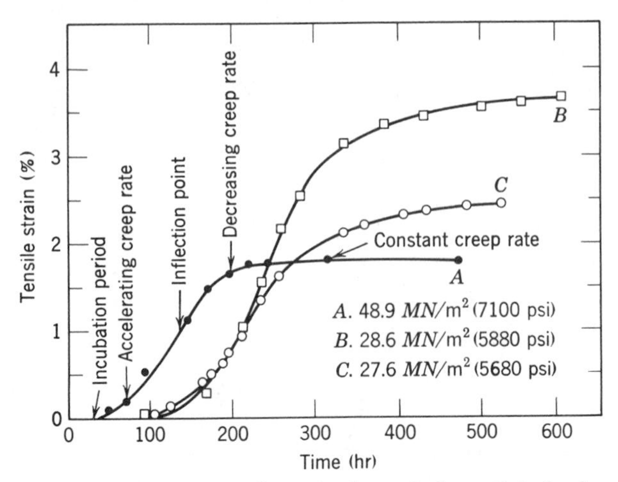

Fig. 14.27. Sigmoidal creep curves determined on single-crystal alumina. From J. B. Wachtman and L. M. Maxwell, *J. Am. Ceram. Soc.*, **40**, 377 (1957).

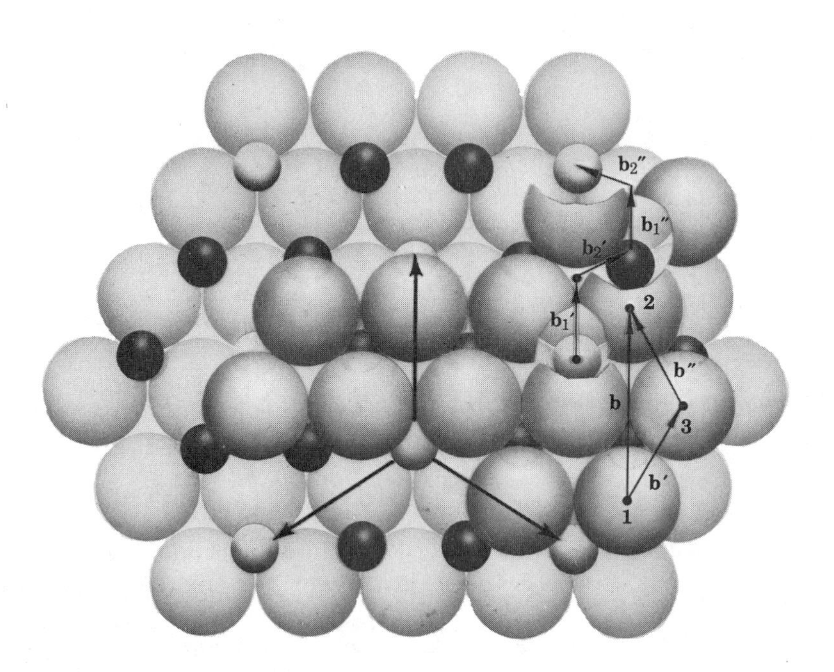

Fig. 14.28. Structure of Al_2O_3 showing two layers of large oxygen ions with hexagonal array of Al^{3+} and vacant octahedral interstices. Slip directions and Burgers' vector **b** for basal plane slip are indicated.

of the structure is also indicated. These are the slip directions in this structure and correspond to the minimum Burgers vector of a dislocation required for basal plane slip. This Burgers vector is substantially larger than the one that would be required for a close-packed hexagonal metal (the oxygen-oxygen distance in Fig. 14.28).

The large Burgers vector and the fact that the ion motion which corresponds to moving the dislocation through the structure is a jump from site **1** to site **2**, directly over an oxygen ion in the next layer below, certainly would require a large amount of energy for dislocation motion. Hence it seems more likely that the dislocation will slip as two *partial* dislocations with the Burgers vectors $\mathbf{b}'$ and $\mathbf{b}''$. The energy dependence of a dislocation on $\mathbf{b}^2$, as discussed in Chapter 4, means that separation into partials corresponds to a lesser total energy for the dislocation, but the region between the two partial dislocations is not crystallographically perfect. Instead, it corresponds to a stacking fault or error in the order in which atomic layers are added onto the basal plane. In this particular example of a partial dislocation, the Burgers circuit around one of the partial dislocations causes the oxygen ions to be in satisfactory registry but not the aluminum ions and cation vacancies. The ion motion to move the dislocations from sites **1** to **3** and then from **3** to **2** can take place along saddle points in the structure and should be a much lower-energy process than the direct **1** to **3** movement.

A further reduction in the total dislocation line energy and in the energy necessary to move the dislocation through the lattice can be achieved by also moving the aluminum ions along through troughs in the structure by the paths $\mathbf{b}_1'$–$\mathbf{b}_2'$–$\mathbf{b}_1''$–$\mathbf{b}_2''$. This corresponds to splitting the total dislocation into four partials separated by areas of stacking faults. The plastic-deformation process, as illustrated in Fig. 14.29, then consists of moving these associated partial dislocations through the structure.

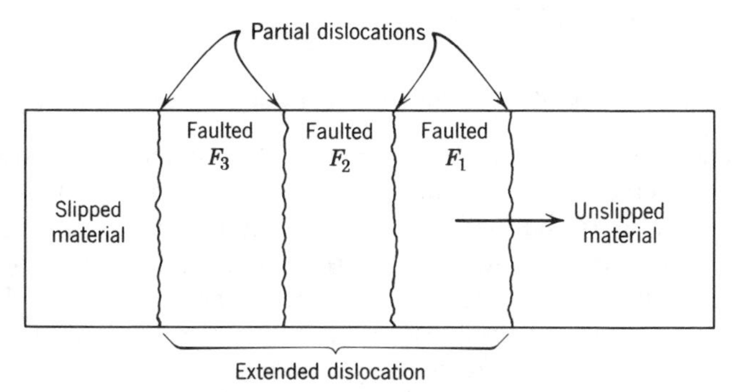

Fig. 14.29. Schematic diagram of extended dislocation composed of quarter partials separated by strips of faulted material.

Although this complex dislocation structure corresponds to a crystallographically low-energy arrangement with relatively little energy required for movement through the structure, the Al^{3+} motion and the O^{2-} motion in the dislocation core must be synchronized. When ion mobility is high at high temperatures and when strain rates are low, these motions can be synchronized and the stress required for plastic deformation is low. At higher strain rates and at lower temperatures (lower ionic mobility) synchronization is a more difficult process to achieve so that the slip deformation is highly sensitive to both temperature and strain rate. The high yield point observed may be caused in part from *self-pinning* of dislocations resulting from the energy required to form the extended dislocation capable of movement through the structure; the observed high yield point may also be related to impurities that act to pin the dislocations; the yield point may also be related to the energy needed for multiplication and acceleration of dislocations. Some requisite combination of velocity and number of dislocations must be satisfied in order for the sample deformation to keep up with the applied strain rate.

As is the case with most of the mechanical properties of ceramic materials, the plastic properties of Al_2O_3 are sensitive to surface treatment. Figure 14.30*a* show the effects of flame polishing and annealing on the plastic yield of a single crystal. The yield stress (at the same strain rate) is greater in the case of the Verneuil-grown (flame fusion) crystals than for the more strainfree Czochralski crystals.

Additions of different-sized ions or different valency impurities can cause solid-solution strengthening. Figure 14.30*b* indicates that Fe, Ni, Cr, Ti, and Mg increase the compressive yield strength. Because of the low solubility of all cations except Cr in Al_2O_3, the data shown in Fig. 14.30 probably reflect solid solution and precipitation hardening. For example, hardening of Al_2O_3 by Ti in solid solution is much less effective than aging the crystals which, causes precipitation of needle-shaped precipitates (star sapphire).

14.5 Creep of Single-Crystal and Polycrystalline Ceramics

The general shape of creep curves and their change with stress and temperature have been described in Figs. 14.3, 14.4, and 14.27. The time-dependent deformation becomes increasingly important as the temperature level is raised; in the use of refractories as structural materials at high temperatures, creep and creep-rupture properties are the main mechanical criteria for usefulness. As it becomes more important to use ceramics in high-temperature applications, especially in energy-conversion systems, the understanding of creep becomes more important.

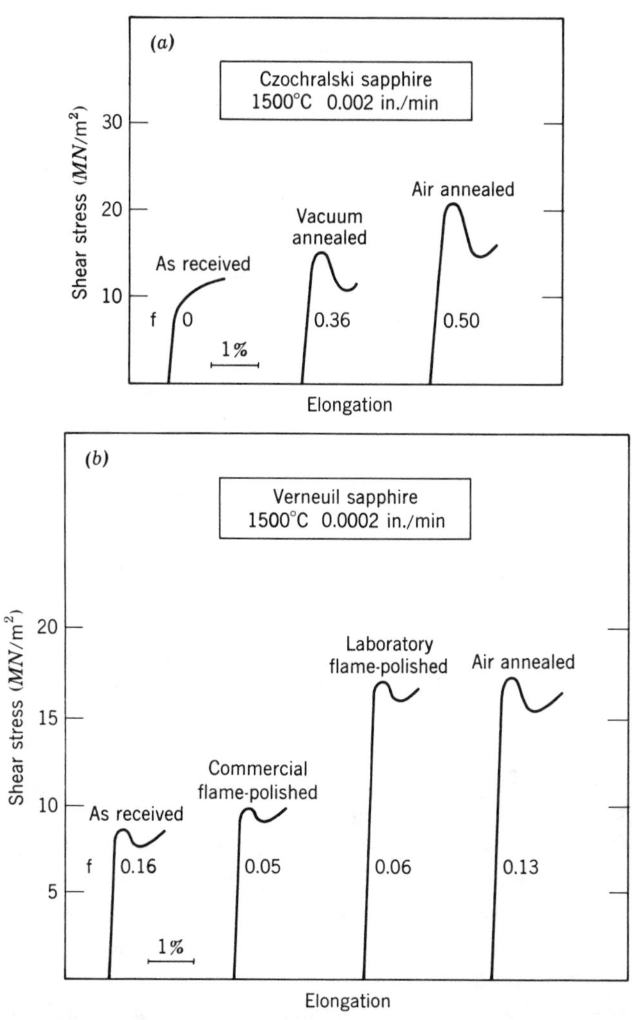

Fig. 14.30. (*a*) Effect of thermal treatments on initial yielding of (*A*) Czochralski and (*B*) Verneuil sapphire crystals. *f* is the yield drop factor. From R. F. Firestone and A. H. Heuer, *J. Am. Ceram. Soc.*, **56**, 136 (1973).

Deformation characteristics are also important for processing ceramics (hot-pressing, hot-working, sintering, etc.).

In imperfect crystals, plastic deformation does not proceed simply by glide of dislocations. Instead, a wide variety of obstacles to dislocation motion exists which must be surmounted for glide to occur. These obstacles have been mentioned in the above discussion and can be

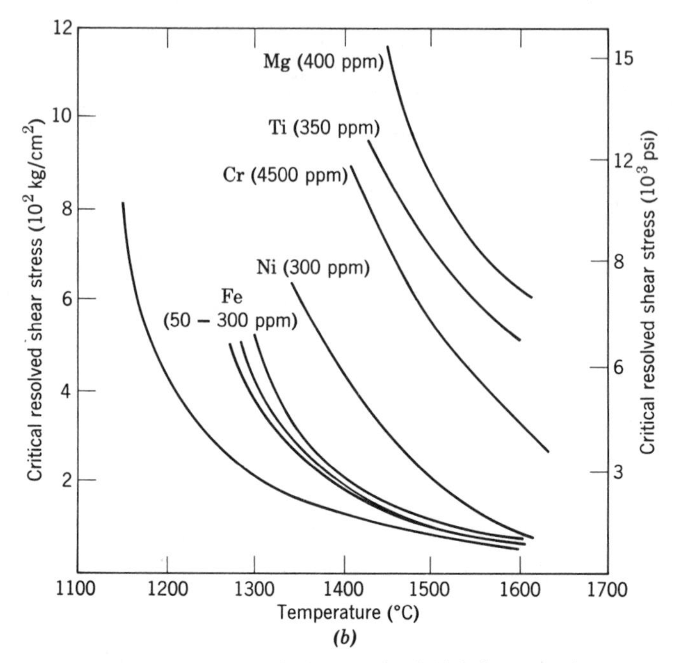

Fig. 14.30 (*Continued*). (*b*) Alloying single crystals of Al_2O_3 increases its strength (tested in air). From K. C. Radford and P. L. Pratt, *Proc. Brit. Ceram. Soc.*, **15**, 185–202 (1970).

classified in two categories: (1) Obstacles that possess long-range stress fields (>10 atomic diameters) such as large precipitates and dislocations. These are known as *athermal* because their magnitude is such that thermal fluctuations cannot directly assist the stress in overcoming them. (2) Obstacles that possess short-range stress fields, called *thermal* obstacles, since thermal energy can assist the stress. Examples are solid-solution ions and point defects.

Generally the plastic strain rate can be fitted to an equation similar to Eq. 14.2 but with a thermal activation term:

$$\dot{\epsilon} = A\sigma^n \exp\left(-\frac{\Delta H}{RT}\right) \tag{14.9}$$

where A is a constant and ΔH represents the activation enthalpy to overcome an obstacle. The effects of temperature on the plastic deformation of KCl–KBr, CaF_2, Al_2O_3, and Al_2O_3 with dopants are shown in Figs. 14.19, 14.24, 14.26*b*, and 14.30*b*.

The importance of rate-controlling mechanisms for kinetic processes was discussed in Chapter 9. The two rate-controlling deformation processes which describe the deformation of a wide range of ceramic materials are dislocation climb and diffusional creep. Grain-boundary sliding may also be important under some conditions.

Dislocation Climb. Dislocation climb, the movement of dislocations out of their slip planes, requires a diffusional kind of jump of an atom above the dislocation line into the dislocation; this is equivalent to the motion of the dislocation into an adjacent plane, just as volume diffusion is equivalent to a vacancy migration. Thus the climb process depends on the diffusion of lattice vacancies and the rate of deformation is controlled by diffusion (Fig. 14.31). The steady-state strain rate for small stresses

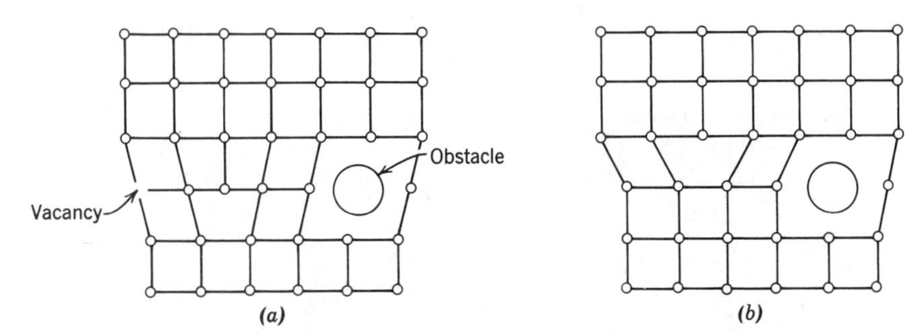

Fig. 14.31. By absorbing vacancies, a dislocation can climb out of its slip plane to where its glide is not hindered by an obstacle.

derived by Weertman* is given by

$$\dot{\epsilon} \approx \frac{\pi^2 D \sigma^{4.5}}{b^{0.5} G^{3.5} N^{0.5} kT} \tag{14.10}$$

where D is the diffusion coefficient of the rate-limiting species, G is the shear modulus, b the Burgers vector, and N the density of dislocation sources. Clear visual evidence that this process does take place is shown in Fig. 14.32 for a sample of lithium fluoride that has been plastically bent, etched to indicate dislocation sites, annealed, and then etched again. Many of the dislocations are found to be displaced from their original sites, moving away from the slip planes. One result of this process, polygonization, has already been discussed in Chapter 4. The movement of dislocations out of slip planes into an alignment along small-angle grain boundaries is shown in Fig. 4.24. The strong temperature dependence of

J. Appl. Phys., **26**, 1213 (1955), and **28**, 362 (1957).

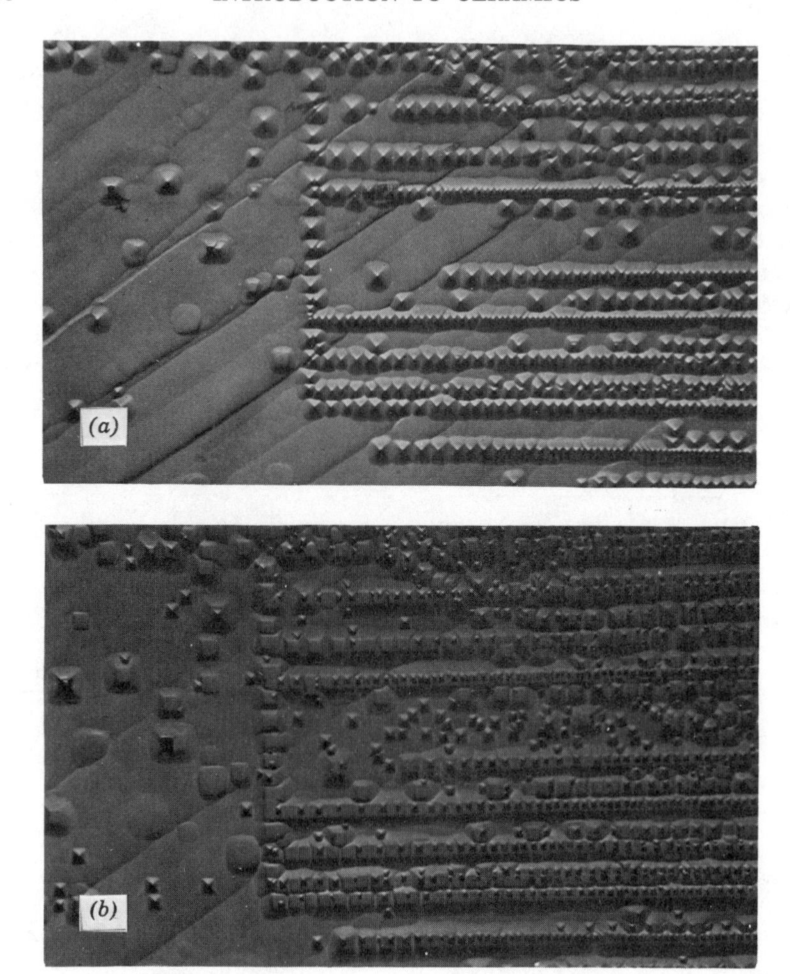

Fig. 14.32. Effect of annealing on glide bands in plastically bent LiF crystal (500×). (*a*) Etched before anneal; (*b*) annealed for 16 hr at 400°C and re-etched. Large flat-bottomed pits are original sites; small pointed pits are sites after anneal. From J. J. Gilman and W. G. Johnston, *J. Appl. Phys.*, 7, 1018–1022 (1956).

this process is illustrated in Fig. 14.33. The polygonization rate has an apparent activation energy of 140 kcal/mole, similar to that found for oxygen ion volume diffusion.

In the case of polycrystalline ceramics, grain boundaries act as barriers to the glide of dislocations. Some grains are poorly oriented with respect to the stress axis and block the shear of others with the result that the

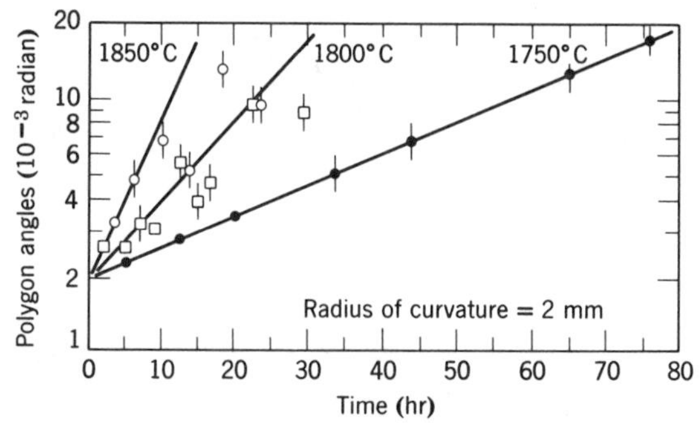

Fig. 14.33. Rate of polygonization of plastically deformed Al_2O_3 at three different temperatures. From J. E. May, *Kinetics of High-Temperature Processes*, W. D. Kingery, Ed., Technology Press, Cambridge, Mass., and John Wiley & Sons, Inc., New York, 1959, p. 35.

aggregate is not ductile. Von Mises* and Taylor† determined that five independent slip systems are necessary for ductility of a polycrystalline material. From Table 14.1, it can be seen that secondary (high-temperature) slip systems must become operative for ceramics to meet this criterion.

The grain size is important in determining the yield strength and the fracture strength of ceramics. The Petch equation‡ shows the relationship between the yield strength σ and the grain size d for a material which deforms by dislocation glide:

$$\sigma = \sigma_i + B/d^{1/2} \tag{14.11}$$

where B is a constant and σ_i, the friction stress, is a measure of the lattice resistance to deformation. This strengthening can also result from sub-grains and low-angle grain boundaries.

Figure 14.34 shows this dependence for press-forged (polycrystalline with high degree of texture) KCl and Sr-doped KCl. Note in the latter case that the strengthening effects from grain boundaries and from the solute are additive.

Diffusional Creep. Diffusional, or Nabarro-Herring, creep has already been discussed in Chapter 10 in connection with the sintering of crystalline solids. In this process self-diffusion within the grains of a polycrystalline solid allows the solid to yield to an applied stress. Deformation results from diffusional flow within each crystal grain away from those boundaries where there is a normal compressive force (high chemical potential)

Z. Angew. Math. Mech., **62**, 307 (1938).
†*J. Inst. Met.*, **62**, 307 (1938).
‡N. J. Petch, *Prog. Met. Phys.*, **5**, 1–52 (1954).

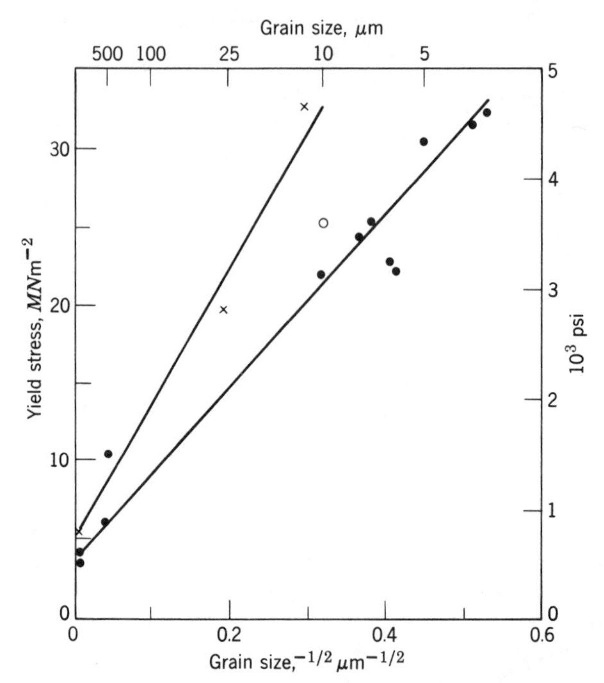

Fig. 14.34. Grain-size dependence of yield strength in hot-forged KCl materials (●—pure, ⟨100⟩ forgings, ○—pure, ⟨111⟩ forgings, and ×—0.1 m/o $SrCl_2$ doped, ⟨100⟩ forgings). From Roy Rice.

toward boundaries having a normal tensile stress (Fig. 14.35). For example, a tensile stress on a boundary increases the vacancy concentration to $c = c_0 \exp(\sigma\Omega/kT)$, where Ω is the vacancy volume, and c_0 is the equilibrium concentration; a compressive stress reduces the concentration to $c = c_0 \exp(-\sigma\Omega/kT)$. The resultant deformation is always accompanied by grain-boundary sliding. Under steady-state conditions, the creep rate calculated by F. R. N. Nabarro* and by C. Herring† is

$$\dot{\epsilon} = \frac{13.3\Omega D\sigma}{kTd^2} \tag{14.12}$$

where d is the grain size.

If the rate-limiting diffusion occurs along the grain boundaries, R. L.

**Report of Conference on Strength of Solids*, University of Bristol, Bristol, England, 1947, pp. 75–90, and *Phys. Soc. London* (1948).

†*J. Appl. Phys.*, **21**, 437 (1950).

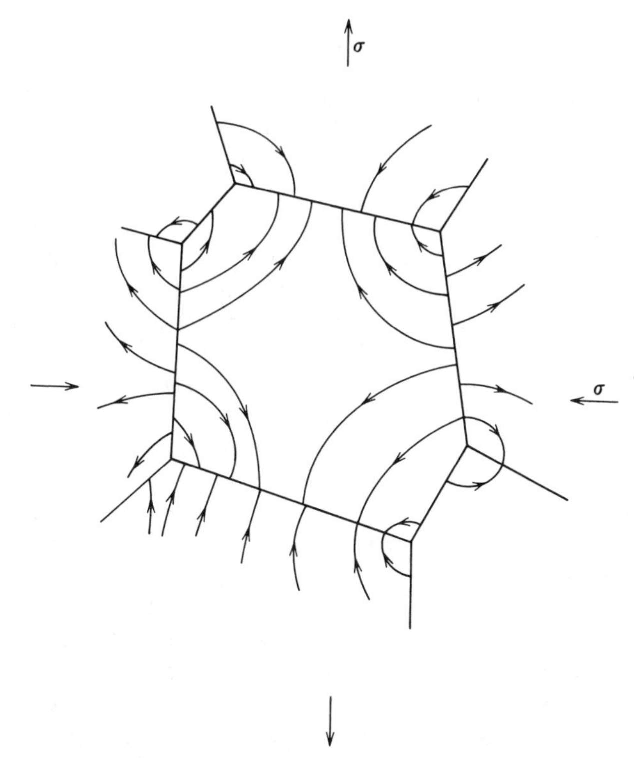

Fig. 14.35. Migration of atoms to grain boundaries parallel to the compressive stress results in elongation of the grain and consequent strain.

Coble* has calculated the relationship

$$\dot{\epsilon} = \frac{47\Omega\delta D_b\sigma}{kTd^3} \tag{14.13}$$

where δ is the grain-boundary width (see Section 6.6) and D_b the diffusion coefficient in the grain boundary.

As discussed in Chapter 6 and in Chapter 9, diffusional processes in ceramics are often complex because of the requirement of charge neutrality and because cations and anions often diffuse at different rates. The effects of ambipolar diffusion and of stoichiometry and impurities must be considered when ascertaining the controlling mechanisms of creep. Generally, the measured stress dependence and grain-size dependence are used to infer the operative mechanism, that is, Eqs. 14.10, 14.12,

J. Appl. Phys., **34**, 1679 (1963).

or 14.13. If grain growth occurs during creep deformation, the time dependence of this process must also be considered.

Grain-Boundary Creep. Grain boundaries have two important effects on the creep rate. First, at high temperatures the grain boundaries can slide relative to one another, which relaxes the shear stress but increases the stress within the grain at any point where there is a restriction to sliding, particularly at triple points, where three grains meet. Second, the grain boundary itself can be a source or sink for dislocations so that the dislocations within about one obstacle-spacing distance from the grain boundary annihilate rather than contributing to strain hardening; where the grain size decreases to be of the order of the obstacle spacing there is a significant increase in the steady-state creep rate.

High-angle grain boundaries (Chapter 5) are areas of poor lattice match and to a first approximation can be thought of as areas of noncrystalline structure between the crystal grains. It has been shown by T. S. Ke* that grain boundaries behave viscously (strain rate proportional to stress) when a shearing stress is applied. The creep rate, however, remains limited by the shape change of the individual grains. If that shape change is limited by slip, then the creep rate is only increased because there is an increase in the stress within the grain. Recent considerations of the effects of grain-boundary sliding† have shown that if the accommodation is by diffusion, the resulting creep rate is the same as given in Eqs. 14.12 and 14.13. In fact, if creep occurs by diffusional processes, grain-boundary sliding is required in order to keep the grains together; on the other hand, if creep results from grain-boundary sliding, diffusional processes are required for accommodation.

Substructure Formation. As a result of dislocation climb during high-temperature plastic deformation (creep), dislocations are able to align themselves into low-energy configurations to form three-dimensional cells of low-angle grain boundaries. Figure 14.36*a* shows the variation of strain rate, dislocation density, and subgrain (cell) size as a function of strain in MgO crystals deformed at 1200°C. At about 40% strain a dynamic equilibrium sets in between dislocations generated and those annihilated at the boundaries (cell walls). Once formed, these boundaries may act as inhibitors to dislocation motion if the material is subsequently strained at lower temperatures (see Fig. 14.34).

The actual subgrain size at the steady-state strain rate (creep rate) has been shown to vary inversely with the steady-state creep stress for many materials. Figure 14.36*b* shows data for NaCl and KCl. The steady-state

Phys. Rev.*, **71, 533 (1947).

†R. Raj and M. F. Ashby, *Met. Trans.*, **2**, 1113 (1971).

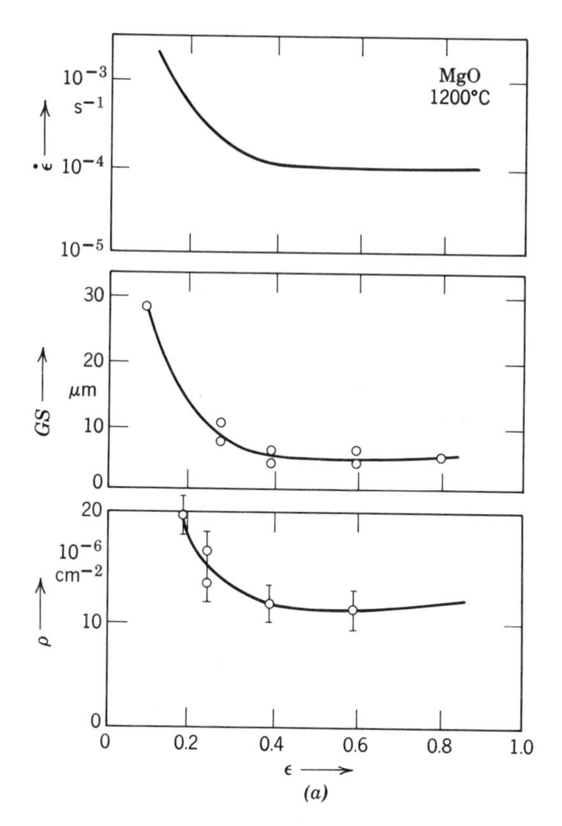

Fig. 14.36. (*a*) Creep substructure formation in MgO crystals deformed at 1200°C. The strain rate $\dot{\epsilon}$, subgrain size *GS*, and dislocation density ρ, are shown as a function of strain. From B. Illschner, *High Temperature Plasticity*, Springer-Verlag, Berlin, 1973.

stress values for various materials are generally proportional to the shear modulus and the Burgers vector (subgrain size $\propto G\mathbf{b}/\sigma$).

Deformation Maps. M. Ashby* has introduced the *deformation-mechanism map* as a method of displaying a material's deformation behavior. Diagrams are constructed using the homologous temperature T/T_m, where T_m is the absolute melting-point temperature, and the normalized stress σ/G, where G is the shear modulus, as the axes; various steady-state deformation mechanisms which are dominant in a particular stress-temperature region are depicted as fields. The maps are constructed from data for a given material using the strain-rate equations (e.g., Eqs. 14.9, 14.12, and 14.13). Figure 14.37 shows the tentative map for magnesia.

Acta Met.*, **20, 887 (1972).

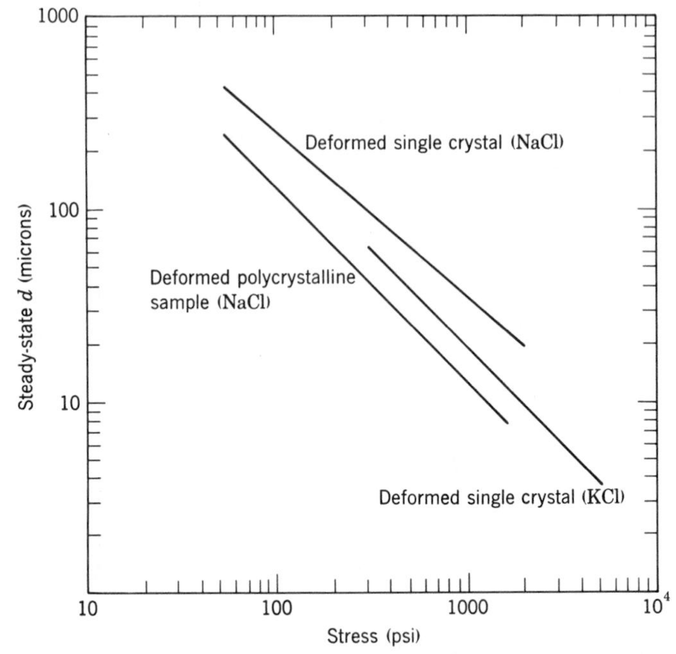

Fig. 14.36 (*Continued*). (*b*) Steady-state subgrain diameter-stress dependence for NaCl and KCl. From M. F. Yan, et al., reference 2.

The field boundaries are calculated by equating strain rates predicted by two different deformation mechanisms and computing the stress at various temperatures. For example, equating the strain rates in Eqs. 14.10 and 14.12 would produce the stress boundaries between Weertman and Nabarro-Herring creep. Since deformation is sensitive to structural properties such as grain size, obstacle spacing, dislocation density, and so on, these maps are dependent both on good experimental data and also on microstructure, impurities, and thermal history. As these data become more available, the use of deformation maps as guides in predicting engineering applicability of a material may become more common.

Creep of Polycrystalline Ceramics. In addition to temperature and stress, the most important variables that affect the creep behavior of ceramics are the microstructure (grain size and porosity), composition and stoichiometry, lattice perfection, and environment.

In general, a considerable enhancement of deformation occurs with increasing porosity. For example, MgO with 12% porosity deforms six times faster than with 2% porosity. Figure 14.38 shows similar behavior for Al_2O_3. A proposed relationship between the creep rate and porosity is

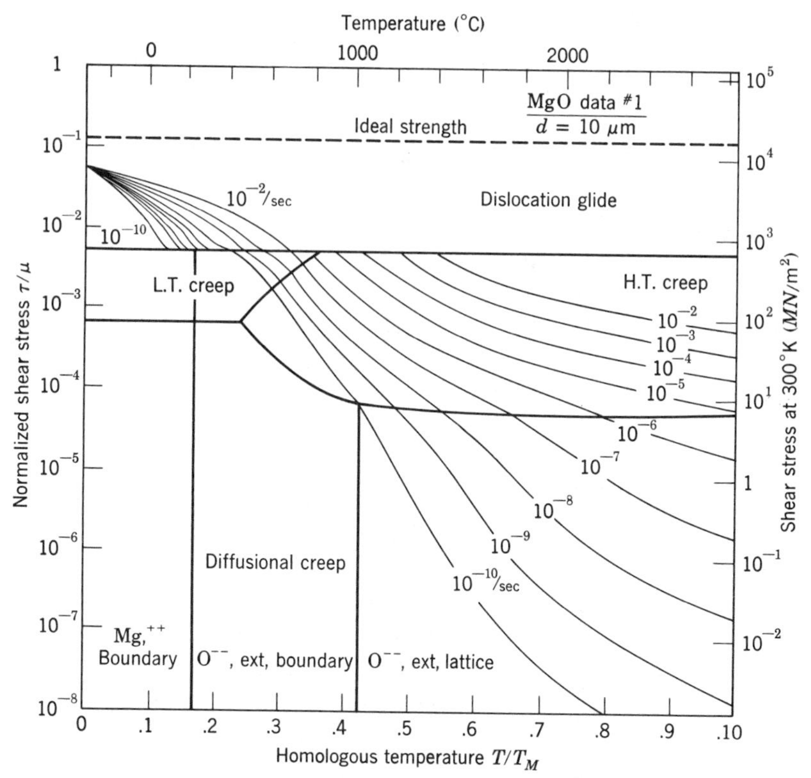

Fig. 14.37. Deformation-mechanism map for MgO. From M. Ashby.

based on the assumption that the porosity P reduces the cross-sectional area available to resist creep:

$$\dot{\epsilon} \propto (1 - P^{2/3})^{-1} \tag{14.14}$$

Deformation studies of MgO bicrystals have revealed that dislocations intersecting grain boundaries have difficulty penetrating into adjacent grains. Thus in fine-grained material mechanisms other than conservative dislocation motion become rate-controlling. Samples prepared by hot-pressing or sintering may have pores or second phases at boundaries which cause grain-boundary sliding which leads to crack initiation and eventual failure before appreciable plasticity occurs. Creep of MgO containing Fe^{3+} is completely diffusional creep at low stresses (~ 4000 psi) because the diffusivity is enhanced by the solute additions and glide is impeded by them. This results in a linear strain-rate–stress dependence and a $(d)^{-2}$ strain rate dependence predicted by the Nabarro-Herring

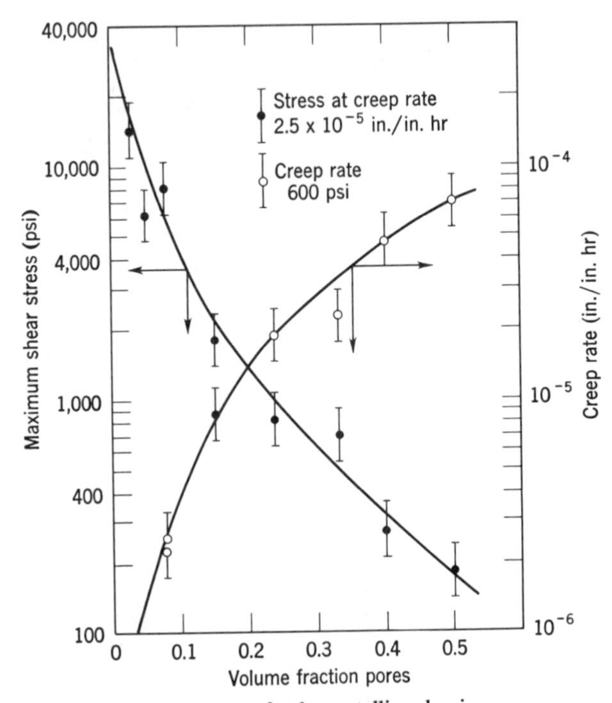

Fig. 14.38. Effect of porosity on creep of polycrystalline alumina.

theory (Fig. 14.39). The rate-controlling diffusion species appears to be magnesium ions, which is contrary to expectation, since oxygen diffuses more slowly than magnesium in the MgO lattice (Fig. 6.11). This suggests that oxygen diffuses along the grain boundaries faster than magnesium diffuses through the bulk. The effect of P_{O_2} is consistent with decreasing the cation vacancy concentration with decreasing oxygen pressure, thus decreasing magnesium diffusion and the creep rate. At higher stresses, the stress exponent ($\dot{\epsilon} \propto \sigma^n$) increases, to conform more with a dislocation climb mechanism.

For Al_2O_3, the von Mises criterion is only satisfied if nonbasal slip systems are activated; thus below 2000°C and at stresses less than 20,000 psi ($\sigma/G < 10^{-3}$) other mechanisms than conservative dislocation motion must contribute to and control the creep behavior. Figure 14.40*a* shows the tensile stresses required to activate slip on the basal, prismatic, and pyramidal systems. Generally in polycrystalline alumina these stresses are not reached prior to formation of grain-boundary fissures and sample failure. Using Eqs. 14.12 and 14.13, the rate-limiting diffusion coefficient was calculated from the strain-rate data for Al_2O_3, as shown in

Fig. 14.40*b*. For 5 to 70 microns grain-size material, tests in the range of 1400 to 2000°C indicate aluminum ion diffusion through the lattice to be rate-controlling (Nabarro-Herring creep). (As for MgO, this requires faster migration of oxygen along the grain boundaries.)

At lower temperatures (<1400°C) and finer grain sizes (1 to 10 microns), the data shown in Fig. 14.40*c* indicate that aluminum ion diffusion along the grain boundary is rate-limiting (Coble creep). The fact that much of the measured creep-rate data is sensitive to the grain size also indicates diffusional creep. However, large-grained material (> 60 microns) appears to deform with significant contributions from dislocation mechanisms.

Most of the work on polycrystalline UO_2 has shown two stress regimes for steady-state creep (Fig. 14.41). At low stresses, the strain rate is proportional to the stress; at high stresses the stress exponent is between 4 and 5.

The effect of nonstoichiometry is important in both Nabarro-Herring and dislocation climb mechanisms as it relates to the diffusivity. Theoretically the diffusivity of the slower-diffusing uranium ion, by uranium vacancies, should be proportional to the square of the nonstoichiometry ($D_u \propto x^2$ in UO_{2+x}). This behavior is observed in single crystals, but a linear dependence is seen in polycrystalline material. As stoichiometry increases, it appears that the slip system for dislocation creep changes. One of the most important effects on the creep of UO_2 is that of fission. Fission is relatively unimportant above 1200°C, but from 200 to 1200°C the creep rate under irradiation is always greater than for normal conditions.

14.6 Creep of Refractories

There are many factors contributing to the deformation of classical ceramics which make their behavior extremely difficult to analyze. The presence of many phases, particularly glass phases and incompletely reacted phases, render theoretical studies difficult. However, from a practical point of view the general effect of certain variables can be determined sufficiently so that modifications to the stress-response behavior can be made.

The glass phase present in most refractories plays a very important role in determining the deformation behavior (see Section 14.7). Its effect depends on the degree to which it wets the crystalline phases present. If the glass does not wet the grains, a large degree of self-bonding occurs, whereas the further the glass penetrates into the grain boundaries, the smaller the amount of self-bonding. When the glass completely penetrates into the grain boundaries, no self-bonding exists, and complete wetting of

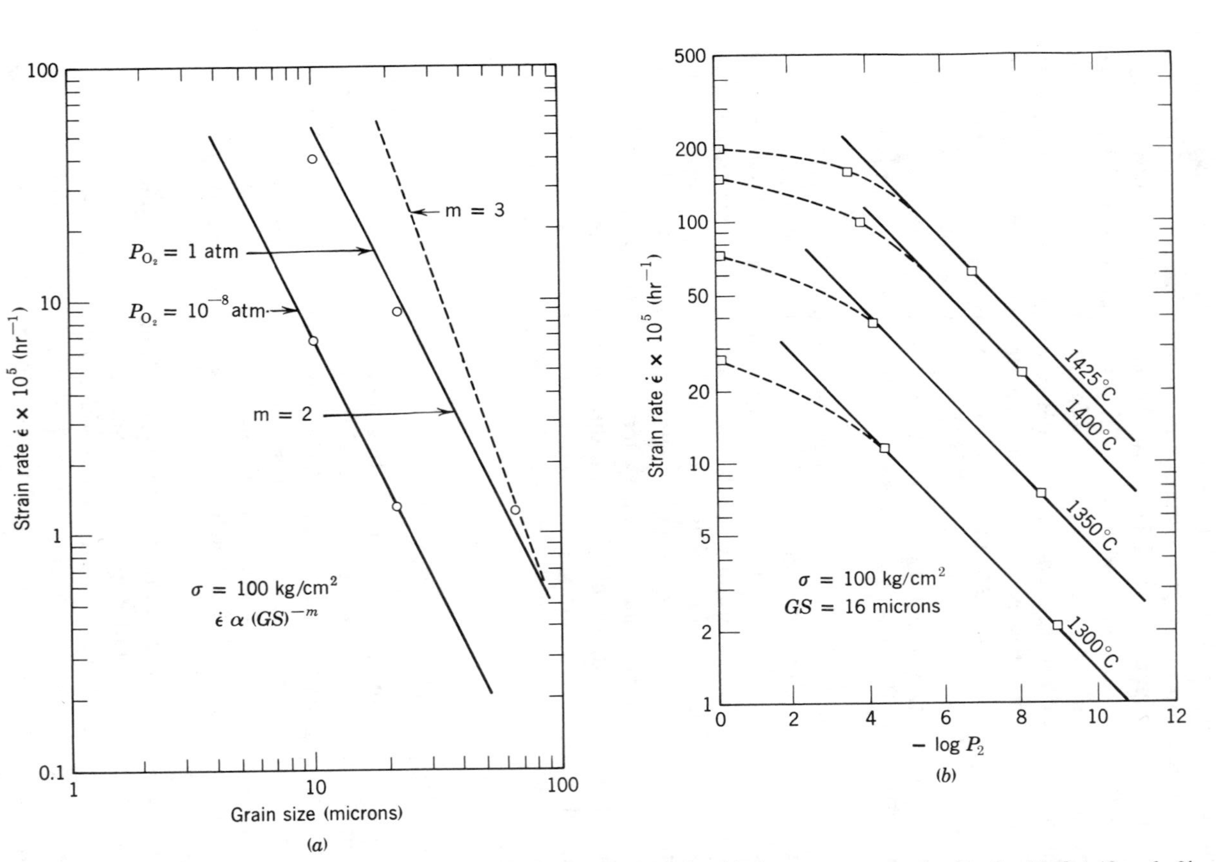

Fig. 14.39. (*a*) Creep rate versus grain size for MgO—1.0 mole % Fe_2O_3. (*b*) Creep rate versus log P_{O_2} for MgO + 10 mole % Fe_2O_3. Courtesy R. S. Gordon.

the crystalline phase by the glass occurs, resulting in the weakest structure. For high-strength refractories, complete elimination of the glass phase is necessary, but because this is not usually feasible, the second approach is to minimize the wetting characteristics. This is possible by firing the ceramic at temperatures at which very little wetting occurs or by modifying the glass composition so that it does not wet the crystalline phases. This is not easily accomplished, since these same grain-boundary phases make possible the sintering of ceramics to higher densities at lower temperatures. Another method of strengthening refrac-

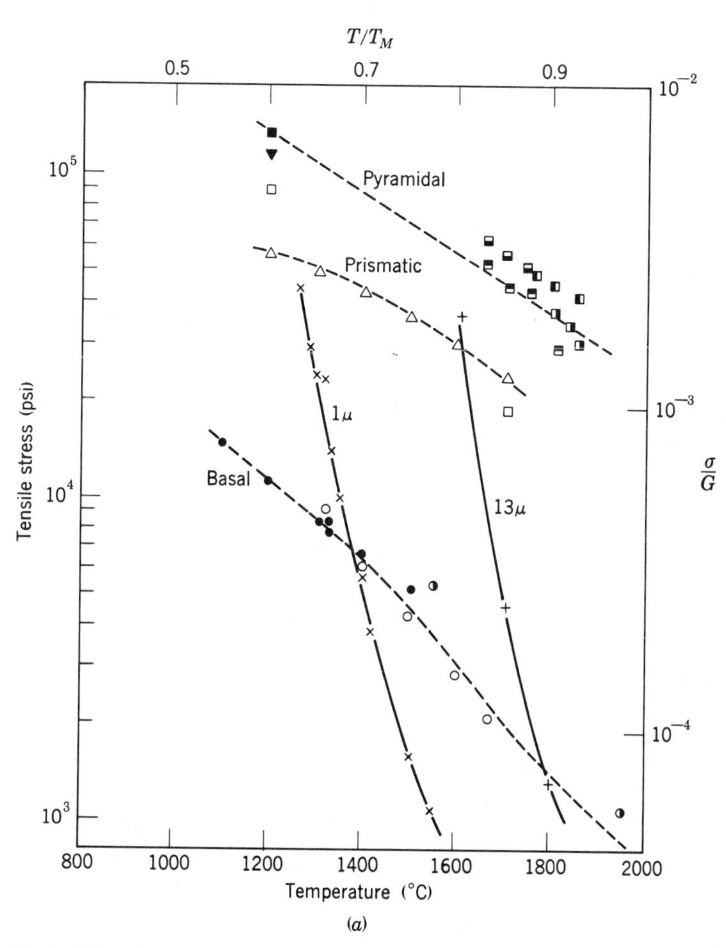

(a)

Fig. 14.40. (*a*) Tensile stress to activate various slip systems in single-crystal Al_2O_3 and flow stress for fine-grained alumina indicating diffusional creep rather than plastic flow.

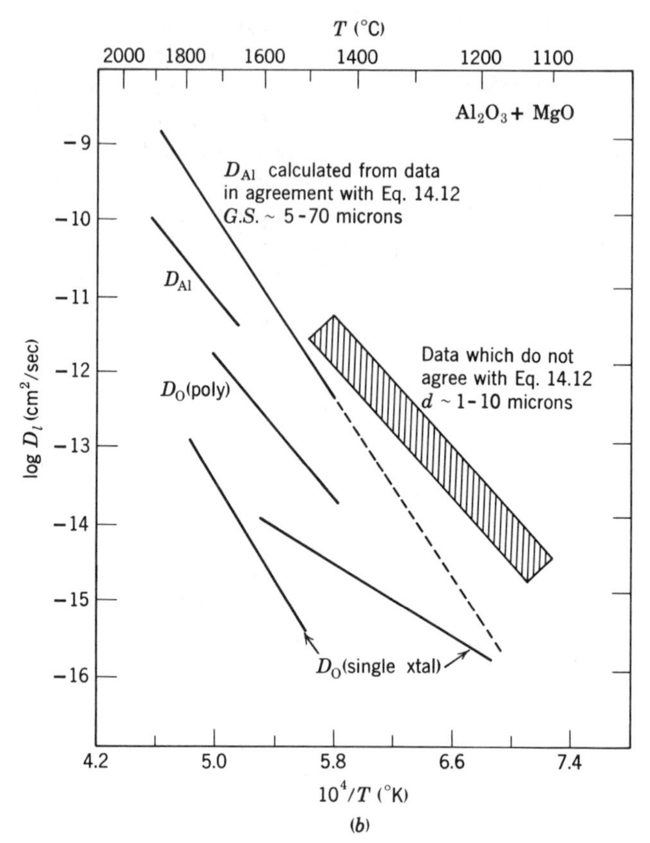

Fig. 14.40 (*Continued*). (*b*) Calculated diffusion coefficient for MgO saturated Al_2O_3 indicating Al ion lattice diffusion controlled creep at $T > 1400°C$. The aluminum and oxygen tracer diffusion coefficients are also shown.

tories is to alter the viscosity of the glass phase by temperature control or by compositional changes. Magnesia refractories, often called magnesite because they are derived from $MgCO_3$, were found to be more resistant to deformation with Cr_2O_3 additions because the wetting of the grains by the silicate phase was reduced, thereby increasing the crystalline bonding. Fe_2O_3 additions increase wetting and thus reduce strength.

The degree of reaction among the various phases is also important in the deformation behavior. Different firing conditions result in different phase development. In many instances service temperatures exceed fabrication temperatures, resulting in changes that can markedly affect the deformation behavior. For example, alumino-silicates held at high temperature (~ 1200°C) develop elongated mullite ($3Al_2O_3{\cdot}2SiO_2$) crystals

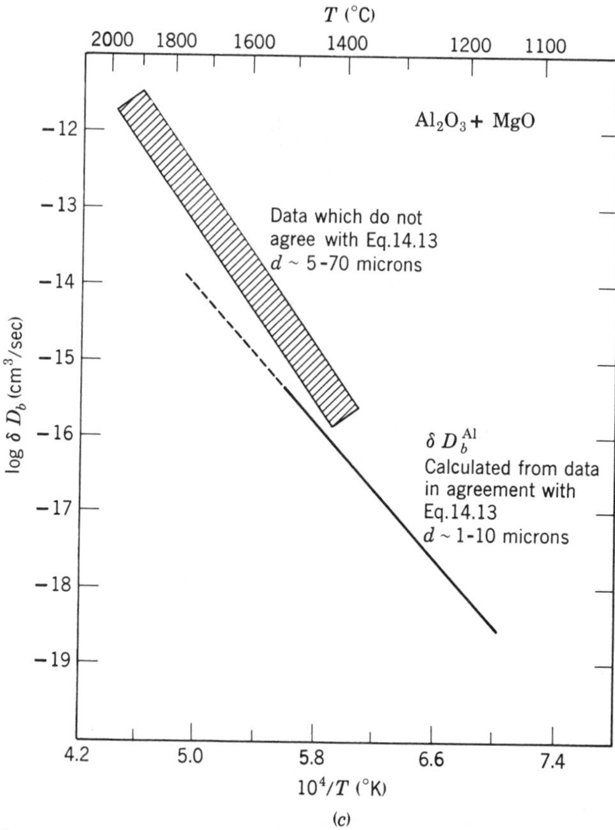

Fig. 14.40 (*Continued*). (*c*) Calculated grain-boundary diffusion (Al) controlled creep at $T < 1500°C$ in MgO saturated Al_2O_3. From R. M. Cannon and R. L. Coble, reference 2.

which form an interlocking network of high strength. The presence of small amounts of sodium oxide (~ 0.5%) increases the rate of mullite formation, resulting in higher creep strengths. Because of incomplete reactions in ceramics, composition is not a completely reliable indicator of strength. High-alumina refractories (~ 60% Al_2O_3) generally increase in strength with increasing alumina content, but reactions during test can change this behavior, as evidenced in the creep rates of alumina-silica refractories which decrease with increasing Al_2O_3 content at 1300°C; at higher temperatures, the formation of mullite at the expense of the SiO_2 and Al_2O_3 changes the resistance to deformation. On the other hand, magnesite bricks exhibit higher strengths with increasing firing temperature because the amount of glassy bonding is reduced.

The glass phase substantially controls the deformation behavior when

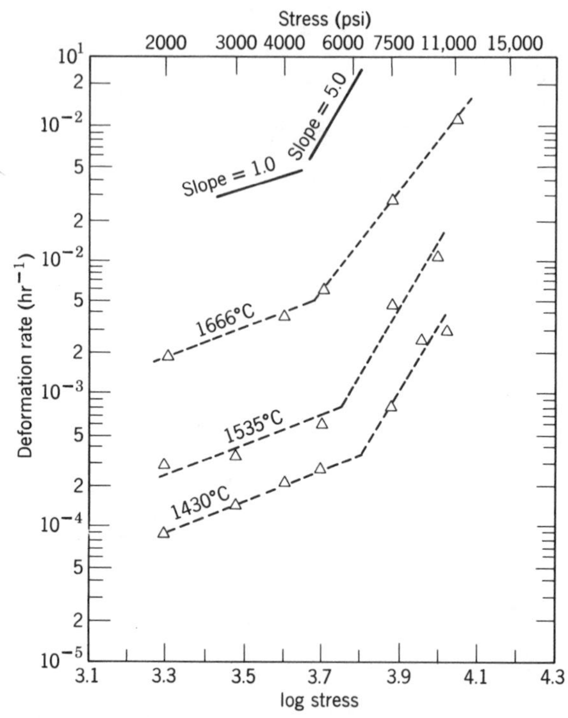

Fig. 14.41. Steady-state creep behavior of UO_2 can be divided into two regimes with different stress exponents. The transition between the regimes decreases with increasing temperatures. From L. E. Poteat and C. S. Yust in *Ceramic Microstructures*, R. M. Fulrath and J. A. Pask, Eds., John Wiley & Sons, Inc., New York., 1968.

present in large amounts. The crystalline materials present reduce the deformation rate, but the temperature and stress dependencies are similar to those of a viscous medium: the deformation rate depends linearly on stress and has an activation energy similar to that for the viscosity of glasses. Higher purity sometimes results in better performance. Fireclay brick has a lower creep resistance than mullite and alumina. With increasing purity, mechanisms other than the shear of glassy phases can contribute to creep. Grain-boundary sliding has been suggested for high-alumina refractories and a dislocation plastic flow mechanism for high-periclase (~ 95% MgO) refractories.

As the crystal structure becomes more covalent, diffusion and dislocation mobility decrease. Thus in carbides and nitrides, the pure materials are very creep-resistant; however, second phases at grain boundaries which are introduced to increase sinterability also increase the creep rate or decrease the yield strength. Figure 14.42 shows the effect of temperature on the yield strength of several carbides. Note also the importance of

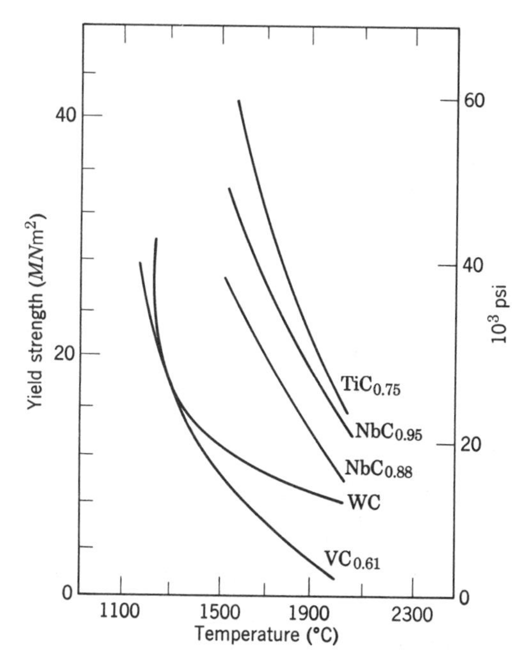

Fig. 14.42. The temperature dependence of the yield stress for a number of transition metal carbides. High temperatures are required for plastic deformation, and the strength is greatly affected by stoichiometry. From G. E. Hollox, *NBS Special Publication* 303, 1969, pp. 201–215.

nonstoichiometry. In Fig. 14.43 creep data for SiAlON (a Si_3N_4–Al_2O_3 alloy) and Si_3N_4 are shown. Since these data are strong functions of the processing parameters (i.e., grain-boundary phases), these measured rates for carbides and nitrides decrease for commercial materials as new processing techniques are learned.

Diffusional creep of SiC at very high temperatures (1900 to 2200°C) has been documented by P. L. Farnsworth and R. L. Coble.* However, careful studies of many of the monoxide ceramics and polyphase refractories have yet to be carried out. There is much variability between samples and difficulty in specifying the microstructure.

Creep rates for a number of crystalline and noncrystalline materials are compared at one temperature and stress in Table 14.3. As seen there,

J. Am. Ceram. Soc., **49**, 264 (1966).

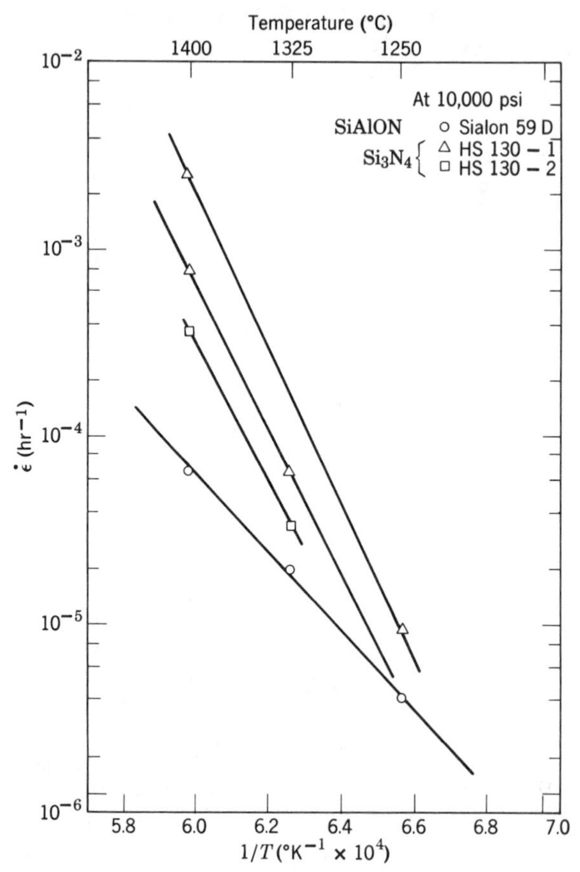

Fig. 14.43. Steady-state creep rate versus reciprocal absolute temperature for SiAlON 59D and $S_{13}N_4$ HS130-1 tested in air and for HS130-2 tested in argon; data normalized to 10,000 psi. From M. S. Seltzer, A. H. Clauer, and B. A. Wilcox, Battelle, Columbus, Ohio, 1974.

these materials are broadly divided into two groups: the noncrystalline glasses are much more readily deformed than the crystalline oxide materials. If we consider variations resulting from porosity, illustrated in Fig. 14.38, and variations resulting from differences in grain size, we conclude that most of the differences reported between different materials are probably unrelated to changes in the composition or crystal structure but rather are caused by changes in microstructure. This conclusion is illustrated in Fig. 14.44, which shows the high-temperature, low-stress creep rates for a number of polycrystalline oxides. Variations among different materials can all be included in a band of values; the differences among individual polycrystalline oxides are best correlated with microstructure variations.

Table 14.3. Torsional Creep of Several Materials.

Material	Creep Rate at 1300°C, 1800 psi (in./in./hr)
Polycrystalline Al_2O_3	0.13×10^{-5}
Polycrystalline BeO	(30×10^{-5})*
Polycrystalline MgO (slip cast)	33×10^{-5}
Polycrystalline MgO (hydrostatic pressed)	3.3×10^{-5}
Polycrystalline $MgAl_2O_4$ (2–5μ)	26.3×10^{-5}
Polycrystalline $MgAl_2O_4$ (1–3 mm)	0.1×10^{-5}
Polycrystalline ThO_2	(100×10^{-5})*
Polycrystalline ZrO_2 (stabilized)	3×10^{-5}
Quartz glass	$20{,}000 \times 10^{-5}$
Soft glass	$1.9 \times 10^{9} \times 10^{-5}$
Insulating firebrick	$100{,}000 \times 10^{-5}$
	Creep Rate at 1300°C, 10 psi (in./in./hr)
Quartz glass	0.001
Soft glass	8
Insulating firebrick	0.005
Chromium magnetite brick	0.0005
Magnesite brick	0.00002

* Extrapolated.

14.7 Viscous Flow in Liquids and Glasses

In sharp contrast to the strong dependence of plastic flow in crystals on crystallography, viscous deformation of liquids and glasses is completely isotropic, depending only on the applied stresses. It is not, however, independent of the atomic structure and composition of liquid or noncrystalline glass.

In a gas viscous drag results from the transfer of gas molecules and their momentum from one laminar flow layer to another. Therefore, the difference in viscosity among different gases is small. The viscosity of a gas increases with temperature, and isothermal changes in density have little effect on the viscosity. Exactly the opposite is found for liquids; different liquids have widely different viscosities, the viscosity decreases as the temperature increases, and the effects of density changes can be considerable.

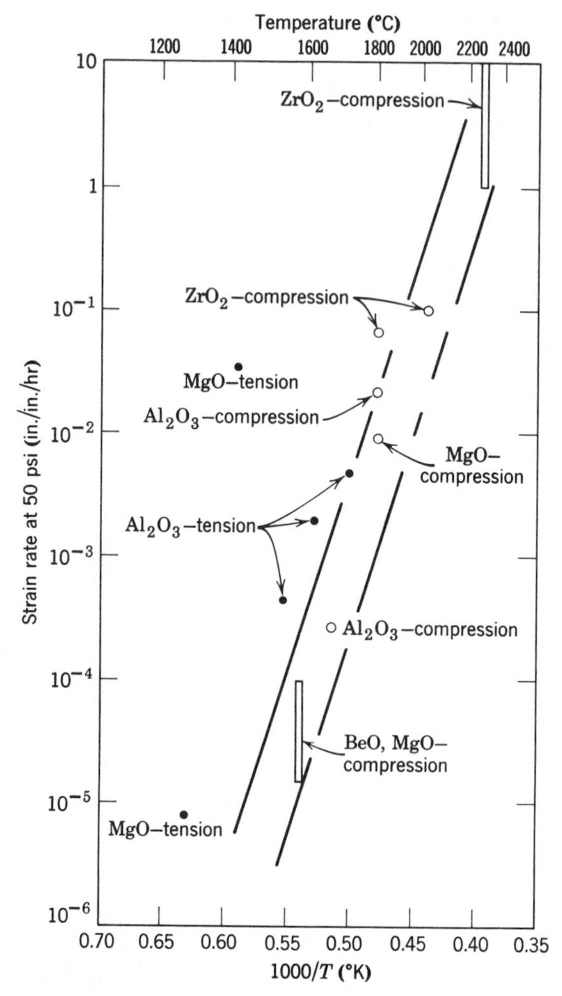

Fig. 14.44. Creep rate of several polycrystalline oxides with an applied stress of 50 psi.

Under most conditions, the flow of oxide liquids may be described as Newtonian; that is, the strain rate is a linear function of the applied stress. A relation of this type is indicated by Eq. 14.4. Phenomenologically, such relations between stress and strain rate result in a stability against necking, which is important in many forming operations. This may be seen by considering a volume element of cross section A and length L of a sample subjected to a normal load F. From conservation of mass

$$\dot{\epsilon} = \frac{\dot{L}}{L} = -\frac{\dot{A}}{A} \tag{14.15}$$

where the dots denote the time derivatives of the respective quantities. Since for a normal stress σ

$$\sigma = 3\eta\dot{\epsilon} \tag{14.16}$$

it follows that
$$\dot{A} = -A\dot{\epsilon} = -\frac{\sigma A}{3\eta} = -\frac{F}{3\eta} \tag{14.17}$$

Hence all cross sections decrease in area at a rate which depends not on their area but only on the load and the viscosity.

At sufficiently high stress levels, the viscosity is expected to decrease with increasing stress. Such stress-dependent viscosities are familiar to the polymer scientist but are noted less frequently for oxide liquids, presumably because of the higher critical stresses for non-Newtonian behavior. For a homogeneous Rb_2O–SiO_2 glass tested in a temperature range in which the low-stress viscosities were between $10^{13.5}$ and $10^{17.1}$ P, the critical stress level for non-Newtonian flow was found* to be about $10^{9.1}$ dyne/cm^2. At higher stresses, the viscosity decreases markedly with increasing stress. The development of two-phase liquid structures by a phase-separation process apparently increases the critical stress for non-Newtonian flow and results in a viscosity which increases over long periods of time at a given temperature.

When high pressure is applied to a liquid, the viscosity at any given temperature is observed to increase. Among oxides, this effect has been most extensively studied for B_2O_3.† Pressures as small as 1000 atm applied to this material increase its viscosity by as much as a factor of 4. The effect of pressure on viscosity is larger at lower temperatures.

Models for Flow. A number of models have been proposed to describe the flow behavior of liquids.

1. ABSOLUTE-RATE THEORY. According to this model, viscous flow can be viewed as a rate process dominated by a transition state of high energy. Transport over the energy barriers is biased by the applied stress, and the standard treatment‡ yields an expression for the viscosity

$$\eta = \frac{\tau \exp(\Delta E/kT)}{2\nu_0 \sinh(\tau V_0/2kT)} \tag{14.18}$$

where τ is the shear stress, ΔE is the height of the energy barrier in the absence of stress, ν_0 is the number of times per second the barrier is

*J. H. Li and D. R. Uhlmann, *J. Non-Cryst. Solids*, **3**, 127 (1970).

†L. L. Sperry and J. D. Mackenzie, *Phys. Chem. Glasses*, **9**, 91 (1968).

‡S. K. Glasstone, K. J. Laidler, and H. Eyring, *The Theory of Rate Processes*, McGraw-Hill Book Company, New York, 1941).

attempted, and V_0 is the flow volume. For small stress ($\tau V_0 \ll 2kT$). The viscosity on this model should be independent of stress; at large stresses, the viscosity should decrease strongly with increasing temperature. For the small-stress case appropriate in most experimental situations, Eq. 14.18 becomes

$$\eta \approx \frac{kT}{\nu_0 V_0} \exp\left(\frac{\Delta E}{kT}\right) = \eta_0 \exp\left(\frac{\Delta E}{kT}\right) \tag{14.19}$$

The close relation of this expression to similar functions for diffusion (see Chapter 6) and other processes treated by absolute-rate theory is apparent.

2. Free-Volume Theory. According to this model, the critical step in flow is opening a void of some critical volume to permit molecular motion. The void is viewed as forming by the redistribution of free volume V_f in the system. The free volume is defined as

$$V_f = V - V_0 \tag{14.20}$$

where V is the molecular volume at a given temperature and V_0 is the effective *hard-core* volume of the molecule. Under most conditions where flow is observed, the average free volume is a small fraction of the hard-core volume.

The most familiar free-volume treatment* yields an expression for the viscosity

$$\eta = B \exp\left(\frac{KV_0}{V_f}\right) \tag{14.21}$$

where B is a constant and K is a constant of the order unity. The temperature dependence of the viscosity is represented here by the temperature dependence of the free volume. By assuming that V_f falls to some small value in the vicinity of the glass transition T_g, the familiar Williams-Landel-Ferry (WLF) relation is obtained. Applied to viscosity this relation is

$$\eta = B \exp\left[\frac{b}{f_g + \Delta\alpha(T - T_g)}\right] \tag{14.22}$$

Here f_g is the fractional free volume at the glass transition, taken as ~ 0.025 for many materials; $b \sim 1$; and $\Delta\alpha$ is the difference in thermal expansion coefficients between liquid and glass ($\sim 5 \times 10^{-4}$/°K for many organic materials but generally smaller for oxides).

*D. Turnbull and M. H. Cohen, *J. Chem. Phys.*, **34**, 120 (1961).

3. Excess-Entropy Theory. According to this model, the decrease in configurational entropy of a liquid with falling temperature results in an increased difficulty of deformation. Considering the size of the smallest region of a system which can change to a new configuration without simultaneous external changes in configuration and relating this to the configurational entropy, an expression for the viscosity can be obtained:*

$$\eta = C \exp\left[\frac{D}{TS_c}\right] \tag{14.23}$$

where C is a constant, S_c is the configurational entropy of the sample, and D should be nearly constant, proportional to the potential-energy barrier to molecular rearrangement. Over a range of temperature near T_g, this expression is effectively indistinguishable from the WLF relation.

The predictions of the free-volume and excess-entropy models can be expressed in the form of the empirical Vogel-Fulcher relation:

$$\eta = E \exp\left[\frac{F}{T - T_0}\right] \tag{14.24}$$

where E and F are constants. Depending on the magnitudes of the respective constants, this expression can be equivalent to the WLF relation.

Temperature Dependence. The temperature dependence of viscosity varies widely for different groups of materials. The large temperature dependence for a typical soda-lime-silica glass is illustrated in Fig. 14.45. This large variation with temperature is one of the bases for glass-forming techniques such as drawing, blowing, and rolling. In the melting range the viscosity is 50 to 500 P; in the working range the viscosity is higher, being 10^4 to 10^8 P; in the annealing range the viscosity is still higher, being $10^{12.5}$ to $10^{13.5}$ P. Since the viscosity is the primary property determining the temperature level at which glass working and the annealing of internal stresses can take place, it is a major factor in the manufacture and working of glasses. These practical operating points are designed on the basis of viscosity and are determined by measuring the viscosity. The two most widely used defined points are the *annealing point* which is the temperature at which internal stresses are substantially reduced in 15 min—equivalent to a viscosity of $10^{13.4}$ poises—and the *Littleton softening point* determined by a fixed procedure equivalent to a viscosity of $10^{7.6}$ P.

According to Eq. 14.19, a plot of log viscosity versus $1/T$ should give a

*G. Adam and J. H. Gibbs, *J. Chem. Phys.*, **43**, 139 (1965).

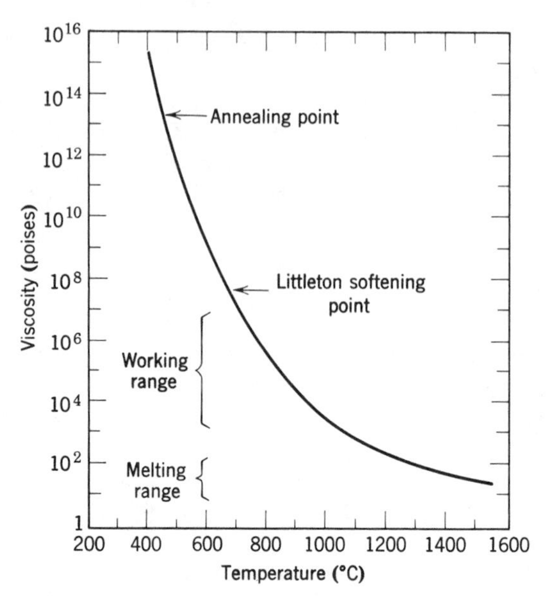

Fig. 14.45. Change in viscosity with temperature of a typical soda-lime-silica glass.

straight line, the slope of which determines the activation energy. The strong temperature dependence of an oxide glass-forming material is shown in Fig. 14.46. The apparent activation energy for flow is higher at low temperatures than at high temperatures. For most oxide glass formers, excluding SiO_2 and GeO_2, which exhibit Arrhenian behavior over the full range, the apparent activation energy at low temperatures is larger by a factor of 2 or 3 than that at high temperatures. In contrast, for most organic glass formers $\Delta E_{\text{low temperature}}$ exceeds $\Delta E_{\text{high temperature}}$ by an order of magnitude or more. This change in apparent activation energy with temperature, together with the magnitudes of the ΔE's obtained from the low-temperature data, indicates that the flow cannot be regarded as a simply activated process, as suggested by absolute-rate theory, but that it involves cooperative motion of more than one atom or molecule.

Significant differences exist in the detailed forms of the viscosity-temperature relations of different liquids. Some materials, like SiO_2 and GeO_2, exhibit an Arrhenian or nearly Arrhenian temperature dependence over the full range of temperature in which data are available. Other materials, like B_2O_3, salol, and possibly the alkali silicates, exhibit Arrhenian behavior at low temperatures, as the glass transition is approached; an intermediate region, in which appreciable curvature is found in the log η versus $1/T$ relation; and a high-temperature region, corres-

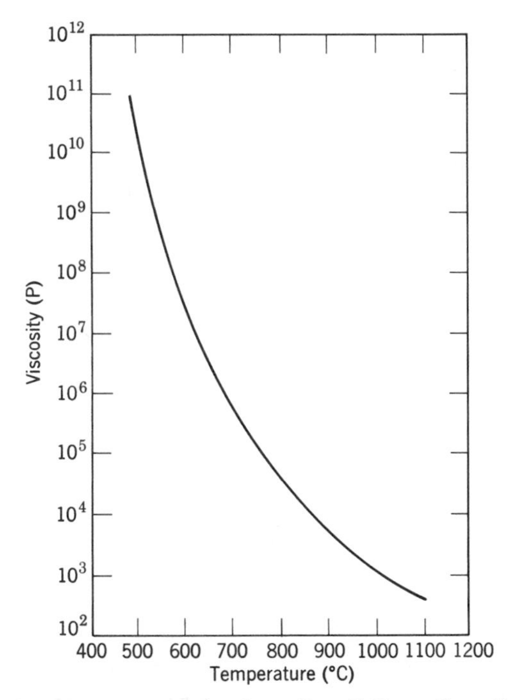

Fig. 14.46. Viscosity-temperature relation for sodium disilicate. From G. S. Meiling and D. R. Uhlmann, *Phys. Chem. Glasses*, **8**, 62 (1967).

ponding to the fluid range, of much smaller and in some cases negligible curvature. For other materials, like anorthite ($CaO \cdot Al_2O_3 \cdot 2SiO_2$) and O-terphenyl, curvature in the log η versus $1/T$ relation is observed over the full range of viscosity, with the curvature being small at high and low temperatures and large at intermediate temperatures. For others, like glycerin, more gradual curvature is observed over the entire range of temperature. These differences in flow behavior are illustrated by the data on a number of organic liquids shown in Fig. 14.47.

Such complexity in form of the temperature dependence of viscosity is beyond description by any of the standard theoretical models, each of which is based on a relatively simple picture of the flow process. It seems, however, that free-volume models can provide a useful representation of flow behavior in the fluid range ($\eta < 10^{3.5} - 10^4$ P), and it is suggested that they be applied in this range and not in the vicinity of the glass transition, the range in which significant variability in form is observed.

Time Dependence. In the region of the glass transition, the viscosity of glass-forming liquids is observed to depend on time. This is illustrated by

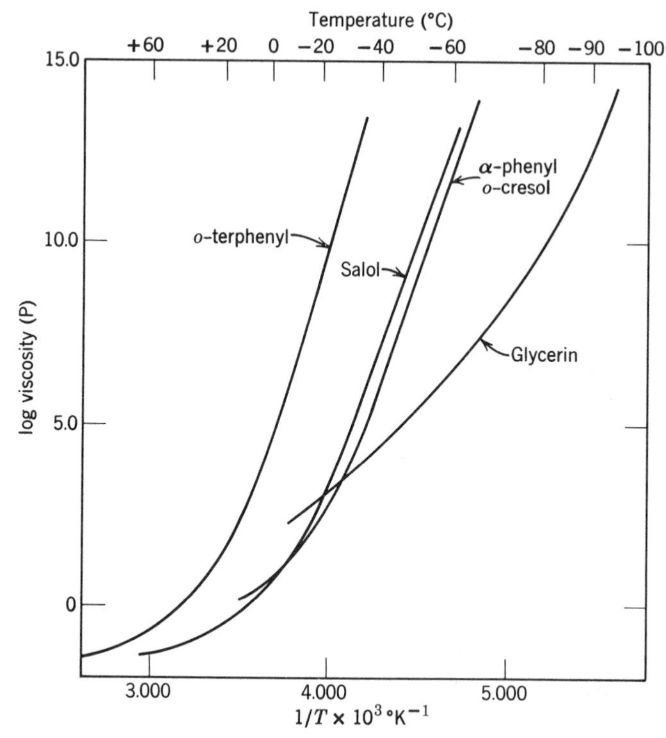

Fig. 14.47. Viscosity-temperature relations for glycerin, α-phenyl, o-cresol, salol, and o-terphenyl. From W. T. Laughlin and D. R. Uhlmann, *J. Phys. Chem.*, **76**, 2317 (1972).

the data shown in Fig. 14.48. These data indicate that the viscosity increases with time for specimens cooled to the annealing temperature from higher temperatures and decreases with time for specimens initially held at temperatures below the annealing temperature. In both cases, the viscosity approaches an equilibrium value which is characteristic of the annealing temperature.

The variations shown in Fig. 14.48 can qualitatively be related to the accompanying changes of the volume with time. For samples cooled to the holding temperature from higher temperatures, the volume decreases with time toward its equilibrium value. This decrease in volume, and hence in free volume, is accompanied by an increase in the viscosity. In detail, the form of the data in the figure is unexpected. The top curve for the specimen which had a higher initial viscosity indicates more rapid relaxation to the equilibrium viscosity than the lower curve for the sample with the smaller initial viscosity. In contrast to this behavior, the

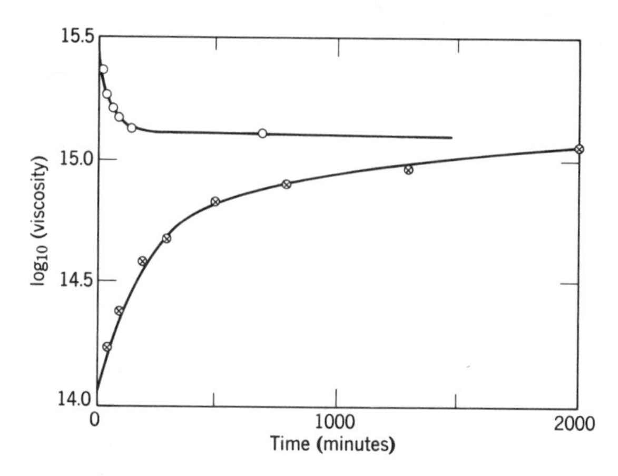

Fig. 14.48. Viscosity-time curves for two soda-lime-silicate glass samples at 486.7°C. Upper curve, sample previously heated at 477.8°C for 64 hr; lower curve, sample in freshly drawn condition. From J. E. Stanworth, *Physical Properties of Glasses*, Oxford University Press, 1953.

characteristic times of the relaxation process are expected to increase with increasing viscosity.

For a given time scale of measurement, viscosity versus temperature curves of the form shown in Fig. 14.49 for different fused silicas are observed. The pronounced decrease in slope of the $\log \eta$ versus $1/T$ relations in the region of the glass transition reflects the time scale for the viscosity to reach its equilibrium value becoming longer than the measurement time. The apparent viscosities determined in this way increase with increasing time of annealing in the glass-transition region, such as shown by the lower curve in Fig. 14.48.

Compositional Dependence. Among inorganic oxide materials, the viscosity is often found to be a strong function of the composition as well as temperature. In the case of silicates, the viscosity is almost invariably found to decrease with increasing concentration of modifying cations. In many cases, this variation is quite pronounced. For example, at a temperature of 1700°C, the viscosity of fused silica can be decreased by about four orders of magnitude by the addition of as little as 2.5 mole% K_2O, and the effect of larger concentrations of modifiers is illustrated in Fig. 14.50. The temperature dependence of the viscosity is correspondingly decreased by the addition of modifying oxides.

In detail, there is no good picture relating viscosity to molecular structure in these systems. It seems clear, however, that an important effect of adding the modifier is the introduction of singly bonded oxygens,

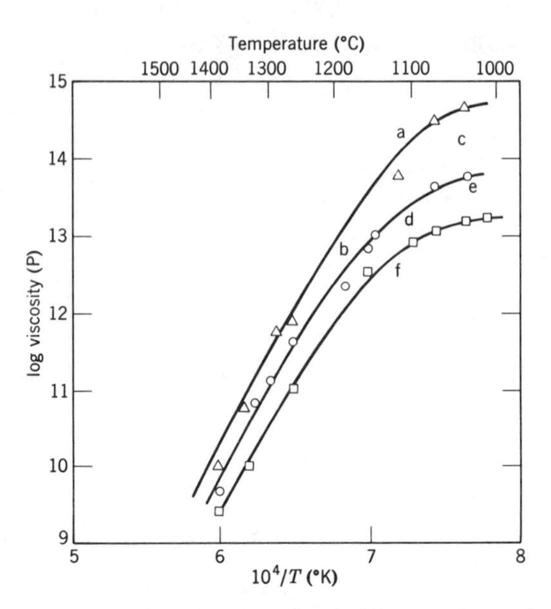

Fig. 14.49. Measured viscosities of various fused silicas. △, I. R. Vitreosil; ○, O. G. Vitreosil; □, Spectrosil. From G. Hetherington, K. H. Jack, and J. C. Kennedy, *Phys. Chem. Glasses*, **5**, 130 (1964).

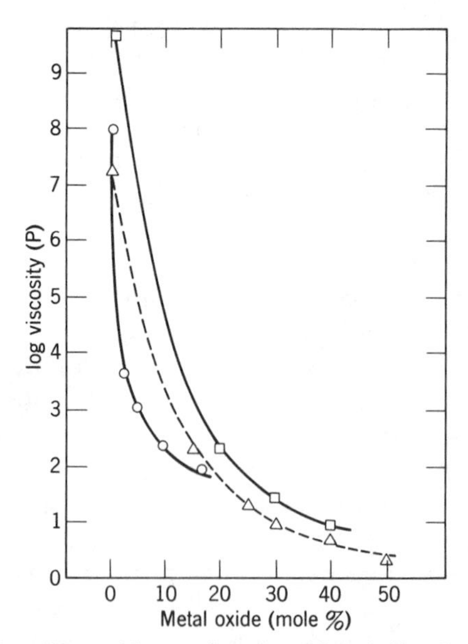

Fig. 14.50. Effect of modifier oxides on viscosity of fused silica. □, Li_2O–SiO_2, 1400°C; ○, K_2O–SiO_2, 1600°C; △, BaO–SiO_2, 1700°C. From J. O'M. Bockris, J. D. Mackenzie, and J. A. Kitchener, *Faraday Soc.*, **51**, 1734 (1955).

which serve as weak links in the Si–O network. For a typical soda-lime-silicate glass at high temperatures, the effectiveness of various divalent cations in decreasing the viscosity seems to correlate with their ionic radii. This is illustrated by the data shown in Fig. 14.51.

In borate glasses, the viscosity at high temperatures is observed to decrease with increasing concentration of alkali oxide. At intermediate temperatures the viscosity decreases with small additions of alkali oxide, then increases to a maximum, and then decreases again with increasing modifier concentration; at low temperatures the viscosity increases with alkali oxide additions. This behavior is not satisfactorily understood at the present time.

In complex oxide glasses, the addition of modifying cations generally decreases the viscosity at any given temperature; the addition of silica or alumina usually increases it. Beyond this, there is a general mixing effect, in which the addition of more than one type of alkali or alkaline earth ion results in a higher viscosity than would be obtained with the same total

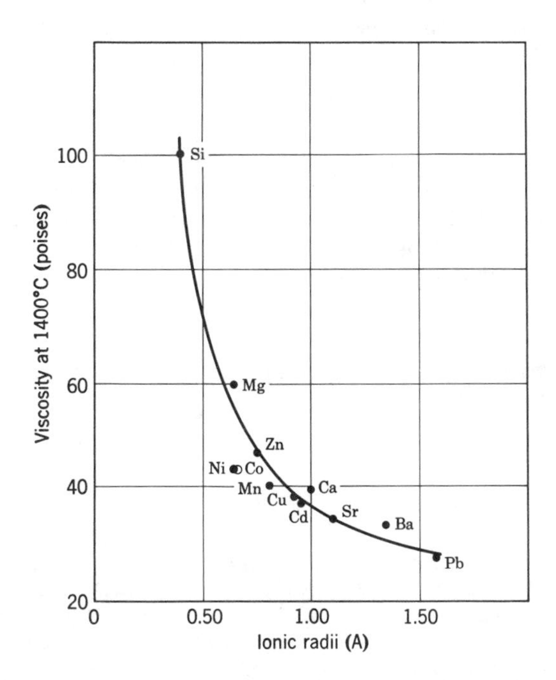

Fig. 14.51. Effect of viscosity of substitution of 8% of silica in a $74SiO_2$–$10CaO$–$16Na_2O$ glass by other divalent oxides on a cation-for-cation basis. From A. F. G. Dingwall and H. Moore, *J. Soc. Glass Technol.*, **37**, 337 (1953).

concentration of a single modifier. In low-melting glazes, the ions commonly used to develop high fluidity are F^- (which substitutes for O^{2-}), Pb^{2+}, and Ba^{2+}.

Suggested Reading

1. G. R. Terwilliger and K. C. Radford, "High Temperature Deformation of Ceramics," *Bull. Am. Ceram. Soc.*, **53**, 173 and 465 (1974).
2. R. E. Tressler and R. C. Bradt, Eds., *Plastic Deformation of Ceramic Materials*, Plenum Press, New York, 1975.
3. R. J. Stokes and C. H. Li, "Dislocations and the Strength of Polycrystalline Ceramics," in *Materials Science Research*, Vol. 1, H. H. Stadelmaier and W. W. Austin, Eds., Plenum Press, New York, 1963.
4. R. J. Stokes, "Basic Mechanisms of Strain-Hardening in Ceramics," in *Strengthening Mechanisms: Metals and Ceramics*, J. J. Burke, N. L. Reed, and V. Weiss, Eds., Syracuse University Press, Syracuse, N.Y., 1966.
5. E. Schmid and W. Boas, *Plasticity of Crystals*, Chapman & Hall, Ltd., London, 1968, from the 1935 German version.
6. J. C. Fisher, W. G. Johnston, R. Thomson, and T. Vreeland, Eds., *Dislocations and Mechanical Properties of Crystals*, John Wiley & Sons, Inc., New York, 1957.
7. A. H. Cottrell, *Dislocations and Plastic Flow in Crystals*, Clarendon Press, Oxford, 1953.
8. J. J. Gilman, "Mechanical Behavior of Ionic Crystals," *Progress in Ceramic Science*, Vol. 1, J. E. Burke, Ed., Pergamon Press, New York, 1960.
9. D. Turnbull and M. H. Cohen, "Crystallization Kinetics and Glass Formation," in *Modern Aspects of the Vitreous State*, J. D. Mackenzie, Ed., Butterworth, Washington, 1960.
10. J. A. Stanworth, *The Physical Properties of Glass*, Clarendon Press, Oxford, 1950.
11. W. Eitel, *The Physical Chemistry of the Silicates*, The University of Chicago Press, Chicago, 1954.

Problems

14.1. Compare the high-temperature creep behavior of (*a*) glass, (*b*) single-crystal Al_2O_3, and (*c*) fine-grain polycrystalline Al_2O_3, explaining any differences in deformation mechanism. Illustrate with curves of deformation versus time and stress versus deformation rate for each material. How would grain size affect the behavior of the polycrystalline alumina?

14.2. The application of pressure, not necessarily hydrostatic, has been observed to affect several processes which are presumed to be diffusion-controlled. Give several ways in which pressure can affect self-diffusion coefficients and the expected direction of change in D with increasing pressure for (*a*) vacancy diffusion and (*b*) interstitial diffusion.

14.3. Sapphire rods are readily available with the c axis forming a 30° angle with the rod axis. (*a*) The critical resolved shear stress is what fraction of the applied stress for plastic deformation? (*b*) Why is plastic deformation of sapphire very temperature dependent?

14.4. The steady-state creep rate of aluminum oxide has been tested at 1750°C. It has been observed that a polycrystalline material creeps at a rate of 3×10^{-6} in./in. sec, whereas the rate for a single crystal is 8×10^{-10} in./in. sec. Why the difference? Would you expect the activation energy to be the same or different for two forms of Al_2O_3? Why?

14.5 The viscosity of SiO_2 glass is 10^{15} P at 1000°C and 10^8 at 1400°C. What is the activation energy for viscous flow of SiO_2 glass? These data were taken at constant pressure. Would you expect a difference in the activation energy if the data were obtained at constant volume? Why?

14.6. (*a*) What are the essential similarities and differences in the plastic deformation of metals, ionic solids, and covalent solids?

(*b*) Assuming hardness characteristics are related to plasticity as well as bond strength, would you predict a hexagonal modification of SiC to be harder or softer than a cubic modification? Why?

14.7. For CaF_2, give the (1) crystal structure, (2) cleavage plane, (3) primary slip plane, (4) Burger's vector, (5) dominant lattice defect type, and (6) influence on plastic flow, at 1200°C by additions of:

(*a*) YF_3

(*b*) CaO

(*c*) NaF

and discuss the mechanism for each case.

14.8. The force of interaction between a pair of parallel dislocation lines on adjacent slip planes is given by $\gamma = G\mathbf{b}/l$, where l is the spacing between lines. If a slip band widens by a multiple-cross-glide mechanism, how would the dislocation densities in slip bands in different crystals change with the yield stress of the crystals?

14.9. It has been reported that polycrystalline UO_{2+x} creeps by a Nabarro-Herring mechanism. What is the rate-controlling specie? How does the creep rate depend on the oxygen partial pressure?

14.10. Drawing of glass rods and fibers is possible, whereas attempts to form metals the same way result in necking down during drawing of rods and balling during formation of fibers.

(*a*) Show that glass rods can be drawn without necking.

(*b*) Explain the nature and cause of surface-tension instability during fiber formation.

(*c*) Explain how it is that glass fibers can be formed in spite of this instability but metal fibers cannot.

15

Elasticity, Anelasticity, and Strength

One of the main reasons why ceramics are not used more widely is the fact that they fail with "glasslike" brittle fracture. They do not normally exhibit appreciable plastic deformation, as discussed in Chapter 14, and their impact resistance is low. The use of ceramics for many applications is limited by these relatively poor mechanical properties. Alumina porcelains, for example, are widely used as dielectric materials. Their choice is often dictated, not by their electrical superiority, but by the fact that they are mechanically stronger than other available materials. Consequently they can be used in automatic machinery without as much chipping or breakage as possible competitors would be subject to. Similarly, even though the high-temperature creep strength and deformation properties of ceramics are the best available, their poor impact resistance limits their use in jet engines, where one impact failure is catastrophic. Many similar examples can be cited.

At the same time, the difference between actual strength levels commonly obtained and the potential theoretical strength, which has been demonstrated in several cases, is as great or greater for fracture characteristics as for any other property. Consequently, this is a particularly attractive area for future developments.

15.1 Introduction

In this chapter we are concerned with the various factors influencing the elasticity and the resistance of ceramic materials to fracture, which is measured quantitatively by some critical value of stress. Analysis and study of these phenomena are complicated by the fact that fracture does not occur by one simple process for all materials and conditions, but rather there are a number of quite different mechanisms that lead to a material breaking as a result of mechanical stresses. To add further complication, a given material may fail by different mechanisms, depending on the stress level, strain rate, previous history, environmental conditions, and temperature level.

Fracture Processes. Most ceramics fail in a brittle manner, that is to say, by a process in which fracture occurs with little or no plastic deformation. Noncrystalline materials such as glass, which are a major component of most ceramics, are always brittle below the softening temperature, and the appearance of the fracture surface is termed *conchoidal.* For the crystalline components brittle fracture generally occurs by *cleavage* over particular crystallographic planes. At high temperatures the crystalline component can fail *intergranularly.* This occurs when grain-boundary shearing takes place and cracks open up between the grains, causing a local stress concentration and ultimate fracture.

In contrast to the brittle fracture of most ceramics, ductile metals and some ceramics fail as a consequence of *necking* or a continual thinning of one section. In extreme cases necking can proceed until separation occurs along a sharp edge or point. No critical fracture stress can be quoted for this process. In general ductile metals fracture within the neck to leave a *cup and cone type of fracture* surface. At the bottom of the cup, where the fracture is normal to the tensile stress, the surface is jagged, and the mode is termed *fibrous fracture*; the side of the cup and cone follows the surface of maximum shear stress, it is quite smooth, and the mode is termed *shear fracture.*

The local decrease in area (necking) during the application of a tensile stress makes the nominal stress, calculated from the total load and initial sample dimensions, less than the actual stress. Consequently, the *engineering tensile strength,* or maximum stress recorded on the basis of initial sample dimensions, is less than the true fracture stress (Fig. 15.1). From the change in cross-sectional area during the test the true fracture stress can be determined.

Fatigue fracture occurs in metals under repeated cyclic stresses by the nucleation and extension of a crack within an intensely cold-worked area at the specimen surface; it is rare in ceramic materials. However, *static fatigue* or *delayed fracture* is common in ceramics; in this case preferential stress corrosion occurs at the tip of a crack under a static applied stress so that fracture occurs at a time after the load is applied. This kind of fracture is particularly sensitive to environmental conditions.

The analysis of failures resulting from mechanical stresses and occurring under different environments depends critically on the particular type

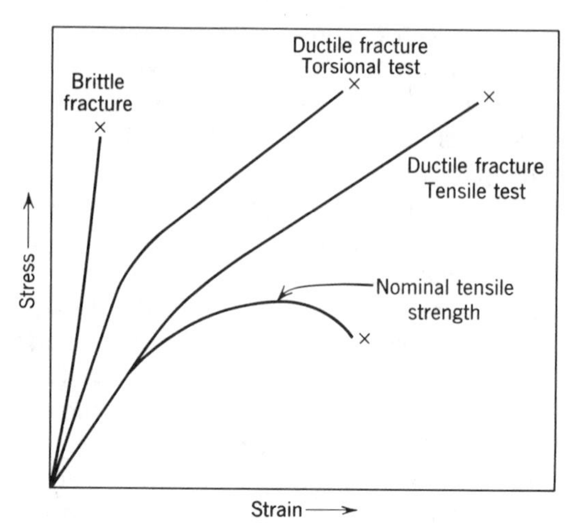

Fig. 15.1. Brittle and ductile fracture.

of fracture observed. In ceramic materials brittle fracture is of greatest importance and is our major concern.

Elastic Deformation. There are many important applications of ceramics which arise from control or manipulation of the elastic deformation under an applied stress. Up to the *proportional limit*, stress is directly proportional to the strain (Hooke's law):

$$\sigma = E\epsilon \tag{15.1}$$

where σ is the normal (tensile) stress, E is Young's modulus, and ϵ is the normal strain. Similarly, the shear stress τ is directly proportional to the shear strain γ:

$$\tau = G\gamma \tag{15.2}$$

where G is the modulus of rigidity or the modulus of elasticity in shear. When a sample is extended in tension, there is an accompanying decrease in thickness; the ratio of the thickness decrease to the length increase is Poisson's ratio:

$$\mu = \frac{\Delta d/d}{\Delta l/l} \tag{15.3}$$

For plastic flow, viscous flow, and creep, the volume remains constant so that $\mu = 0.5$. For elastic deformation Poisson's ratio is found to vary between 0.2 and 0.3, with most materials having values of approximately 0.2 to 0.25. Poisson's ratio relates the modulus of elasticity and modulus of rigidity by the following equation:

$$\mu = \frac{E}{2G} - 1 \tag{15.4}$$

This relationship is only applicable to an isotropic body in which there is only one value for the elastic constant independent of direction. Generally this is not the case for single crystals, it is a good approximation for glasses and for most polycrystalline ceramic materials.

Under conditions of isotropic pressure, the applied pressure P is equivalent to a stress of $-P$ in each of the principal directions. In each principal direction we have a relative strain:

$$\epsilon = -\frac{P}{E} + \mu\frac{P}{E} + \mu\frac{P}{E} = \frac{P}{E}(2\mu - 1) \tag{15.5}$$

The relative volume change is given by

$$\frac{\Delta V}{V} = 3\epsilon = \frac{3P}{E}(2\mu - 1) \tag{15.6}$$

The bulk modulus K, defined as the isotropic pressure divided by the relative volume change, is given by

$$K = \frac{-P}{\Delta V/V} = \frac{E}{3(1-2\mu)} \tag{15.7}$$

The stresses and strains corresponding to these relationships are illustrated in Fig. 15.2.

Anelasticity. In a number of applications, such as glass-forming liquids in the vicinity of the glass transition and polycrystalline materials at high temperature, the elastic moduli cannot be taken as constant but exhibit a significant dependence on time. This behavior is termed anelastic or viscoelastic and characterizes deformation which is recoverable, but not instantaneously, after removal of a stress.

Anelastic behavior is frequently represented by mechanical models consisting of arrays of elastic springs and viscous dash pots, an example of which is shown in Fig. 15.3*a*. The response of such a model to stress is described by

$$\sigma + \tau_R \frac{d\sigma}{dt} = E_1\epsilon + (E + E_1)\tau_R \frac{d\epsilon}{dt} \tag{15.8}$$

where the relaxation time τ_R is given by

$$\tau_R = \frac{\eta}{E} \tag{15.9}$$

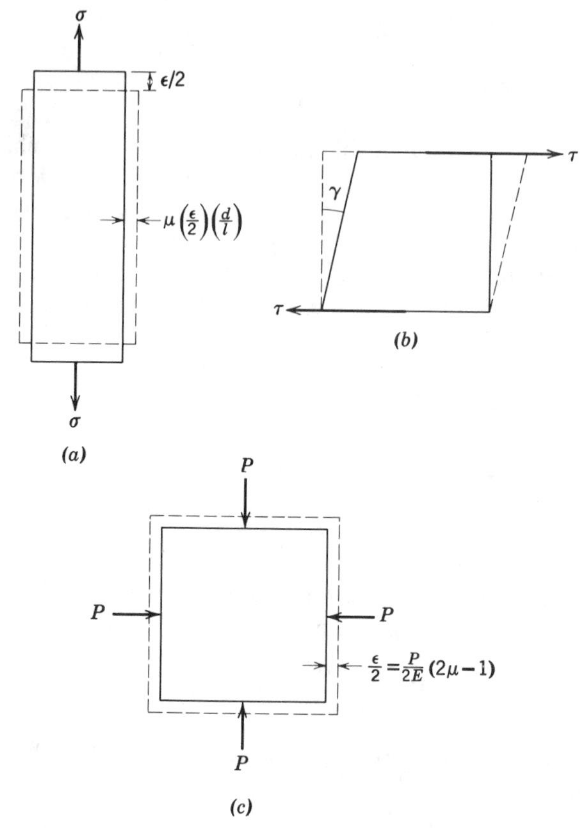

Fig. 15.2. Definitions of elastic constants. (*a*) Young's modulus; (*b*) shear modulus; (*c*) bulk modulus.

For a constant stress σ_0 applied over a time interval t_0 and then released, the resulting strain should have the time dependence shown in Fig. 15.3*b*.

This model is known as the standard linear solid and is generally attributed to Zener.*

For such a solid, the relaxed modulus is the ratio of stress to the total deformation after a long time; the instantaneous deformation on application of stress defines the unrelaxed modulus, $E_U = \sigma_0/\epsilon_0$. If periods of time that are short compared with the relaxation time are used (high frequencies), the measured modulus is the unrelaxed modulus, since there

*C. Zener, *Elasticity and Anelasticity of Metals*, The University of Chicago Press, Chicago, 1948.

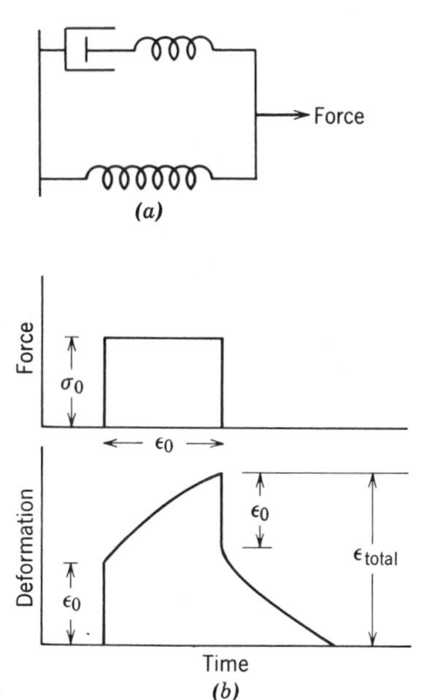

Fig. 15.3. Standard linear solid as representation of relaxation behavior. (*a*) Model; (*b*) mechanical behavior.

is insufficient time between cycles for relaxation to occur. In tests for periods of time that are long compared with the relaxation time, the lower value of the relaxed modulus is measured. At intermediate periods of time which are of the order of the relaxation time, the measured modulus is intermediate in value.

15.2 Elastic Moduli

The elastic extension of a body corresponds to uniformly increasing the separation between the atoms. As a result, elastic extension is directly related to the forces between atoms and the structure energy. Many more or less satisfactory correlations between the lattice energy and elastic moduli have been proposed and work quite well for materials with the same structure and bond type. The situation is obviously similar to that for the thermal expansion coefficient, as discussed in Chapter 12. Crystals with low thermal expansion coefficients often have high moduli of elasticity.

In two-phase systems the overall modulus is intermediate between the high- and low-modulus components. Analytical expressions for this relationship can be derived on similar bases as expansion coefficients for two-phase systems, discussed in Chapter 12. An exact treatment of the problem requires specification of the mutual interaction of many second-phase inclusions. Such a treatment has not been effected, but a number of upper and lower bounds on the elastic moduli have been evaluated. The widest possible bounds are found by assuming that the material consists of layers either parallel or perpendicular to an applied uniaxial stress. The first (Voigt) model assumes that the strain in each constituent is the same, whence the Young's modulus E_U of the composite becomes

$$E_U = V_2E_2 + (1 - V_2)E_1 \tag{15.10}$$

where V_2 is the volume fraction of the phase with modulus E_2, and E_1 is the modulus of the other phase. Similar relations may be written for the other moduli. In this case, most of the applied stress is carried by the high-modulus phase.

The second (Reuss) model assumes that the stress in each phase is the same, whence the modulus of the composite E_L is

$$\frac{1}{E_L} = \frac{V_2}{E_2} - \frac{(1 - V_2)}{E_1}. \tag{15.11}$$

Again, similar relations may be written for the other moduli. The designations E_U and E_L indicate, respectively, the upper and lower bounds on the elastic moduli.

Z. Hashin and S. Shtrikman* have determined upper and lower bounds for the moduli which are considerably narrower than the above extremes and did not include any special assumptions regarding the phase geometry. In the case of the bulk modulus, it was demonstrated that the bounds were the most restrictive that could be given in terms of only volume fraction and phase moduli. This was interpreted as meaning that the bulk modulus is indeterminate exactly, as long as only such information is available. It was considered that exact solutions could be obtained only when further factors, such as the statistical details of the phase distribution, were evaluated. It was recognized, however, that such information was rarely available, and even if available, its application would be generally uncertain. Hashin and Shtrikman's expressions were, for $K_2 > K_1$ and $G_2 > G_1$,

$$K_L = K_1 + \frac{V_2}{1/(K_2 - K_1) + [3(1 - V_2)]/(3K_1 + 4G_1)} \tag{15.12}$$

$$K_U = K_2 + \frac{1 - V_2}{1/(K_1 - K_2) + 3V_2/(3K_2 + 4G_2)} \tag{15.13}$$

J. Mech. Phys. Solids*, **11, 127 (1963).

$$G_L = G_1 + \frac{V_2}{1/(G_2 - G_1) + [6(K_1 + 2G_1)(1 - V_2)]/[5G_1(3K_1 + 4G_1)]} \tag{15.14}$$

$$G_U = G_2 + \frac{1 - V_2}{1/(G_1 - G_2) + [6(K_2 + 2G_2)V_2]/[5G_2(3K_2 + 4G_2)]}. \tag{15.15}$$

In these relations, K_U and G_U provide upper bounds, and K_L and G_L provide lower bounds on the respective moduli. These bounds, it should be noted, are not independent, since they are obtained merely by reversing the roles of matrix and inclusion in the analysis.

The relations of Eqs. 15.12 and 15.14, the lower bounds on K and G, had previously been derived in various forms by Hashin and by Kerner as exact solutions for the special case of spherical second-phase particles.

Values of Young's modulus may be obtained from the calculated bulk and shear moduli using the relation

$$E = \frac{4KG}{3K + G}. \tag{15.16}$$

Upper and lower bounds on E are obtained by inserting upper-bound or lower-bound values of K and G.

The predictions of the Voigt and Reuss expressions as well as the Hashin and Shtrikman upper and lower bounds are compared in Fig. 15.4 with the collected experimental data on the WC–Co system. Both the data and the analytical expressions are normalized as indicated. It is evident from the figure that the Hashin-Shtrikman bounds fit the data much more closely than do the Voigt and Reuss expressions.

The ultimate in adding a low-modulus material as a second phase is to add pore spaces that have approximately zero bulk modulus value. In this case the overall elasticity at porosities up to about 50% has been derived by J. K. MacKenzie.* For a typical Poisson's ratio ($\mu = 0.3$) the change in elasticity can be adequately represented for closed pores in a continuous matrix as

$$E = E_0(1 - 1.9P + 0.9P^2). \tag{15.17}$$

This relationship is compared with experimental data for aluminum oxide having uniformly distributed porosity in Fig. 15.5. For porous materials in which the pore phase is continuous and it is possible for the voids to collapse so that shifting of the solid particles relative to each other can occur, the effect of porosity on the modulus of elasticity is greater than

Proc. Phys. Soc. (London), **B63**, 2 (1950).

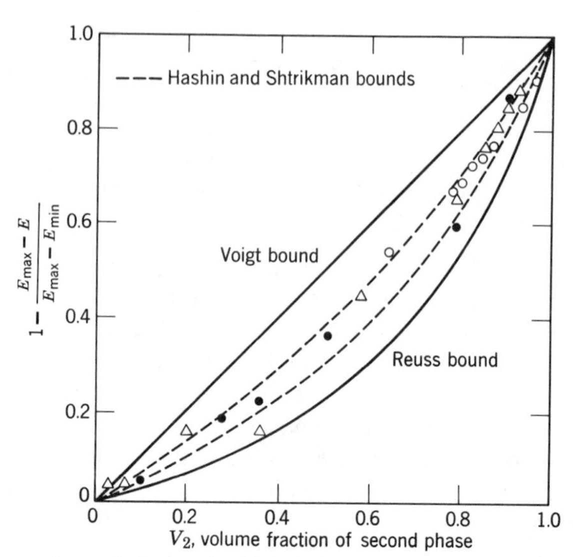

Fig. 15.4. Comparison of observed with predicted variations of Young's modulus with volume fraction of second-phase material. From R. R. Shaw and D. R. Uhlmann, *J. Non-Cryst. Solids*, **5**, 237 (1971).

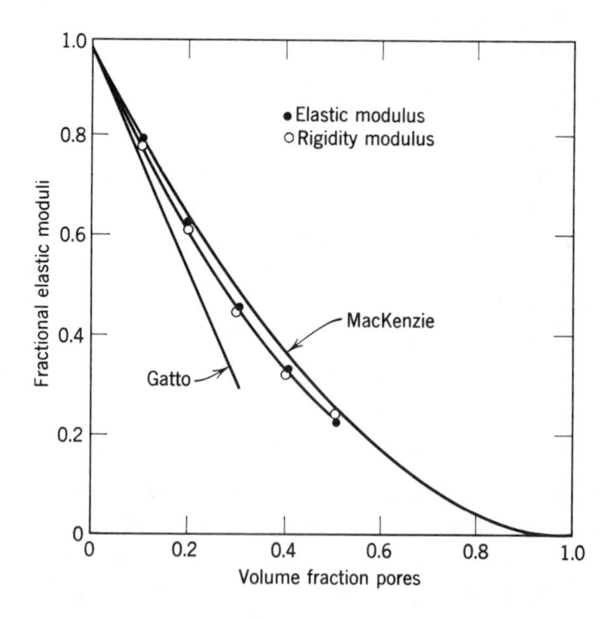

Fig. 15.5. Relative elastic moduli of alumina versus porosity. From R. L. Coble and W. D. Kingery, *J. Am. Ceram. Soc.*, **29**, 377 (1956).

these relations indicate. Elastic constants for some typical ceramic materials are collected in Table 15.1.

As the temperature level is increased, thermal expansion increases the separation between atoms and slightly decreases the force necessary for further separation. This effect is small, however, and there is but a small decrease in the modulus of elasticity as the temperature is raised, as shown in Fig. 15.7, until some temperature is reached at which anelastic relaxation phenomena become important. this temperature corresponds to a change from the unrelaxed to relaxed modulus. The anelastic relaxation strains are also frequency dependent, corresponding to new modes of accommodation to applied stress (in addition to simply increasing the separation between atoms that remain in the same relative positions).

Table 15.1. Modulus of Elasticity of Some Ceramic Materials

Material	E (psi)
Aluminum oxide crystals	55×10^6
Sintered alumina (ca. 5% porosity)	53×10^6
Alumina porcelain (90–95% Al_2O_3)	53×10^6
Sintered beryllia (ca. 5% porosity)	45×10^6
Hot pressed boron nitride (ca. 5% porosity)	12×10^6
Hot pressed boron carbide (ca. 5% porosity)	42×10^6
Graphite (ca. 20% porosity)	1.3×10^6
Sintered magnesia (ca. 5% porosity)	30.5×10^6
Sintered molybdenum silicide (ca. 5% porosity)	59×10^6
Sintered spinel (ca. 5% porosity)	34.5×10^6
Dense silicon carbide (ca. 5% porosity)	68×10^6
Sintered titanium carbide (ca. 5% porosity)	45×10^6
Sintered stabilized zirconia (ca. 5% porosity)	22×10^6
Silica glass	10.5×10^6
Vycor glass	10.5×10^6
Pyrex glass	10×10^6
Mullite porcelain	10×10^6
Steatite porcelain	10×10^6
Superduty fire-clay brick	14×10^6
Magnesite brick	25×10^6
Bonded silicon carbide (ca. 20% porosity)	50×10^6

15.3 Anelasticity

For anelastic deformation there are two limits for the elastic modulus. If the time of measurement is very short, time-dependent deformation does not have an opportunity to occur, and the initial ratio between stress and strain, called the unrelaxed modulus, is measured. However, if the load is applied and the strain is measured at very long times, the relaxed modulus will be determined. Since the long-time strain is larger than the instantaneous strain, the relaxed modulus is smaller than the unrelaxed modulus.

Experimental data on mechanical relaxation are usually not well described by the standard linear solid. Rather than a single relaxation time τ_R, distributions of relaxation times are required to represent the results. Studies of stress relaxation are, for example, generally described by a time-dependent modulus of the form

$$E(t) = E_{EQ} + \int_{-\infty}^{\infty} H(\tau_R) \exp\left(-\frac{t}{\tau_R}\right) d \ln \tau_R \qquad (15.18)$$

where E_{EQ} is the equilibrium or relaxed modulus and $H(\tau_R)$ is the distribution of relaxation times characterizing the relaxation process. Similarly, studies of deformation at constant stress are represented by a time-dependent creep compliance, $D(t) = \epsilon(t)/\sigma$, of the form

$$D(t) = D_U + \int_{-\infty}^{\infty} L(\tau_R)\left[1 - \exp\left(-\frac{t}{\tau_R}\right)\right] d \ln \tau_R \qquad (15.19)$$

where D_U is the instantaneous or unrelaxed compliance and $L(\tau_R)$ is called the distribution of retardation times ($L(\tau_R) \ln \tau_R$ is the contribution to the creep compliance from retardation times between $\ln \tau_R$ and $\ln \tau_R + d \ln \tau_R$).

The respective distribution functions can be obtained from experimental data on creep or stress relaxation, and the results from one type of study can be used to predict behavior under other experimental conditions (see discussion in reference 1). Relaxation data can also be described in terms of distributions of activation energies, and under many conditions these activation-energy spectra can be related to corresponding distributions of relaxation times.*

In addition to investigating creep and stress relaxation, studies of anelastic behavior are frequently carried out with a periodic stress or a periodic strain. A periodic experiment at frequency ω is qualitatively equivalent to a transient experiment at time $t = 1/\omega$. When a periodic

*See, for example, R. M. Kimmel and D. R. Uhlmann, *J. Appl. Phys.*, **40**, 4254 (1969).

stress is applied, the delayed extension corresponds to a lag of the strain behind the stress by some loss angle δ. That is, for a stress

$$\sigma = \sigma_0 \sin \omega t \tag{15.20}$$

the strain is

$$\epsilon = \frac{\sigma_0}{E} \sin (\omega t - \delta). \tag{15.21}$$

Since $\sin (\omega t - \delta) = \sin \omega t \cos \delta - \cos \omega t \sin \delta$, the strain is seen to consist of two components: one, of magnitude $\sigma_0/E \cos \delta$, in phase with the stress; and one, of magnitude $\sigma_0/E \sin \delta$, 90° out of phase with the stress. In terms of a complex compliance

$$\frac{\epsilon}{\sigma_o} = D^* = D_1 - iD_2 \tag{15.22}$$

where $D_1 = 1/E \cos \delta$, $D_2 = 1/E \sin \delta$, and $\tan \delta = D_2/D_1$. The real part of the compliance D_1 is proportional to the energy stored in the specimen due to the applied stress and is called the storage compliance. The imaginary part of the compliance D_2 is termed the loss compliance, since it is proportional to the energy loss. This may be seen from the energy dissipation per cycle:

$$\Delta u = \oint \sigma \, d\epsilon = -\frac{\sigma_0^2}{E} \int_0^{2\pi} \cos (\omega t - \delta) \sin \omega t \, d\omega t \tag{15.23}$$

or

$$\Delta u = \pi \frac{\sigma_0^2}{E} \sin \delta = \pi \sigma_0^2 D_2. \tag{15.24}$$

A similar treatment can be used to define a complex modulus:

$$E^* = E_1 + iE_2. \tag{15.25}$$

This is directly related to the complex compliance, since

$$E^* = \frac{1}{D^*}. \tag{15.26}$$

The moduli and compliances are functions of frequency, as shown in Fig. 15.6. In most cases, distributions of relaxation or retardation times are required to represent experimental data, just as in transient experiments. The storage and loss moduli are then written

$$E_1(\omega) = E_{EQ} + \int_{-\infty}^{\infty} \frac{H(\tau_R)\omega^2\tau_R^2}{1 + \omega^2\tau_R^2} \, d \ln \tau_R \tag{15.27}$$

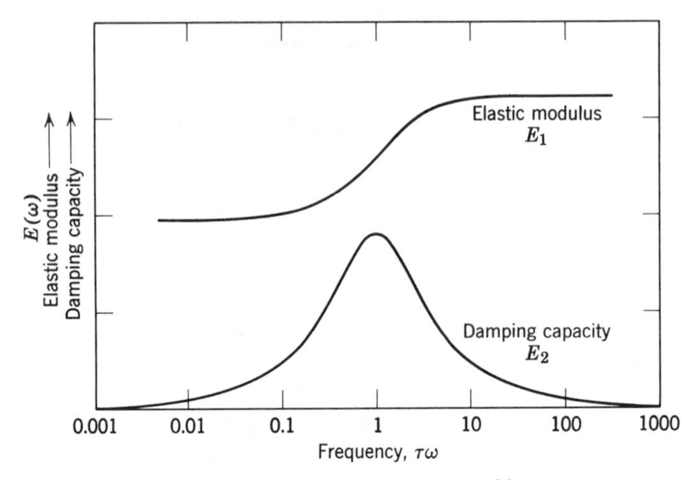

Fig. 15.6. The complex modulus E^* as a function of frequency.

and

$$E_2(\omega) = \int_{-\infty}^{\infty} \frac{H(\tau_R)\omega\tau_R}{1 + \omega^2\tau_R^2}\, d \ln \tau_R. \tag{15.28}$$

Similar expressions may be written for the storage and loss compliances (see reference 1).

In dynamic experiments, $\tan \delta$ is widely used as a measure of the energy dissipation per cycle, since

$$\frac{\Delta u}{u} = 2\pi \tan \delta \tag{15.29}$$

where u is the maximum stored energy. Another widely used measure of mechanical loss, for cases in which $\tan \delta$ is small, is the logarithmic decrement Δ. If a body is set into torsional oscillation and the free decay of the amplitude of oscillation is measured, Δ is defined by

$$\Delta = \ln \frac{A_n}{A_{n+1}} \tag{15.30}$$

where A_n and A_{n+1} are the amplitudes of successive oscillations. For small values of Δ

$$\Delta \approx \frac{1}{2} \frac{\Delta u}{u} \tag{15.31a}$$

and hence

$$\Delta \approx \pi \tan \delta. \tag{15.31b}$$

For experiments in which a body is driven into oscillation in the region of a resonant frequency ω_0 by an oscillating force of constant amplitude and variable frequency, the loss is given by

$$\frac{\Delta u}{u} = \frac{2\pi}{\sqrt{3}} \frac{\Delta\omega}{\omega_0} = 2\pi Q^{-1} \tag{15.32}$$

where $\Delta\omega$ is the half-width of the resonant peak and Q^{-1} is termed the Q of the system.

Various processes can lead to anelastic deformation. In silicate glasses containing only a single type of alkali, two loss peaks are observed at temperatures below the glass-transformation range. These are illustrated by the data in Fig. 15.7 for a $Na_2O{\cdot}3SiO_2$ glass. The peak at $-32°C$ is due to the stress-induced motion of the alkali ions and is termed the alkali peak. The smaller loss maximum at 182°C has been associated with the presence of nonbridging (singly bonded) oxygen ions and is termed the NBO peak. Each loss peak is accompanied by a relaxation of the shear modulus and is superimposed on a general decrease in modulus with

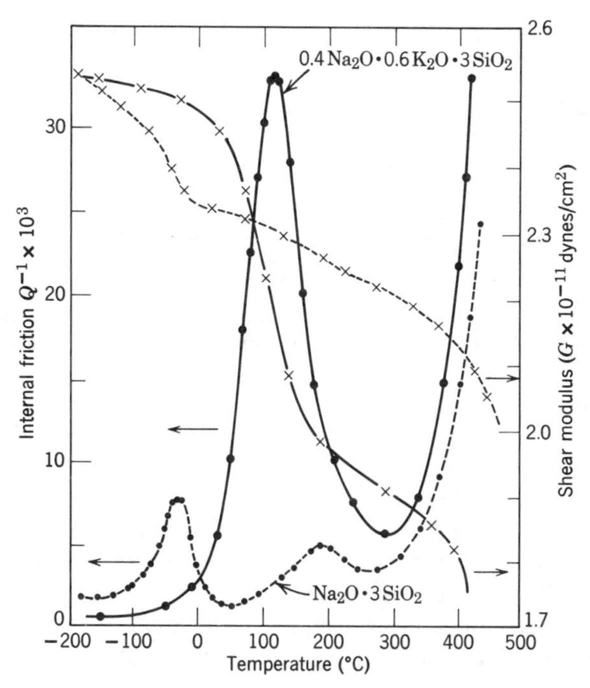

Fig. 15.7. Mechanical loss at a frequency of 0.4 Hz in single-alkali and mixed-alkali silicate glasses. From D. E. Day in *Amorphous Materials*, John Wiley & Sons, Inc., New York, 1972.

increasing temperature. The large increase in loss at temperatures above 350°C reflects viscous damping as the glass transition is approached.

In such single-alkali silicate glasses, the alkali peak becomes larger with increasing alkali content, in a manner which parallels the increase in alkali ion diffusion coefficient (Fig. 15.8). The activation energy for the alkali peak is usually close to that for electrical conductivity or alkali ion diffusion.

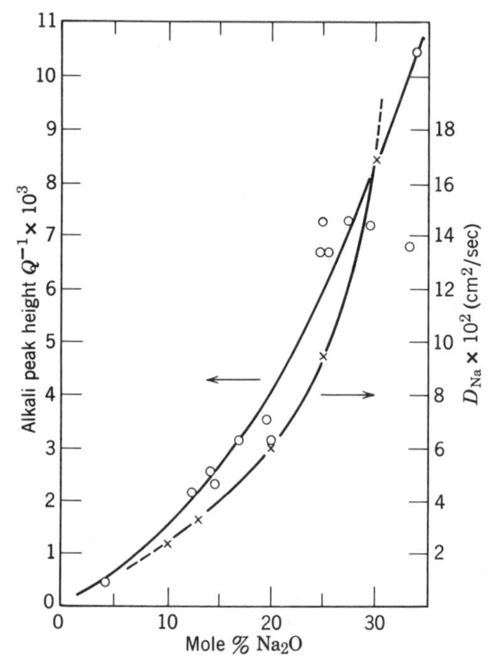

Fig. 15.8. Height of alkali loss peak and Na diffusion coefficient at 415°C in sodium silicate glasses. From D. E. Day in *Amorphous Materials*, John Wiley & Sons, Inc., New York, 1972.

On adding a second type of alkali, the alkali peak decreases in magnitude, and a new peak of greater magnitude appears in the loss curve (the peak centered close to 100°C in Fig. 15.7). The magnitude of this peak seems to be associated with the alkali diffusion coefficients, being a maximum at the composition where the diffusion coefficients of the two alkalis are equal. The increase in mechanical loss on mixing alkali ions is accompanied by a decrease in dielectric loss, and the general character of the *mixed-alkali* effect in glasses is poorly understood at the present time.* In any case, the markedly higher mechanical loss in the vicinity of

*See D. E. Day in *Amorphous Materials*, John Wiley & Sons, Inc., New York, 1972, for review and references.

ambient temperatures for mixed-alkali silicate glasses provides a rationale for the use of single-alkali glasses in thermometer applications, in which mechanical relaxation of the glass is highly undesirable.

In crystalline ceramics, the most important sources of viscoelastic relaxation are the residual glass phases, which are often located at the grain boundaries. The viscous relaxation of these boundary glasses becomes important as their glass transitions are approached, and can significantly affect the overall loss of the materials.

15.4 Brittle Fracture and Crack Propagation

The brittle fracture of a glass and the cleavage fracture of a crystal require two steps, first the production and then the propagation of a crack to final fracture. Since both processes are required, it is conceivable that either can control the overall failure process. We consider in this section the problem of crack propagation and come back to the question of the initiation of cracks in crystalline and noncrystalline materials in the next section. The fracture process is a complex one; the discussion presented here is general in nature, with the intention of providing an overall understanding of the phenomenology of the process and the important factors affecting it.

Isotropic brittle materials are observed to fracture under a critical uniaxial tensile stress. This maximum tensile stress is the criterion for fracture, in the same way that the resolved shear stress is the criterion for plastic deformation (Chapter 14). In glasses and in some crystals the fracture is noncrystallographic, following a random path through the sample. However, in many crystalline materials fracture surfaces occur along crystallographic *cleavage* planes of high atomic density. In sodium chloride and magnesium oxide, for example, the (100) plane is generally observed to be the cleavage plane. Frequently the cleavage plane is the same plane that is prominent in the normal growth habit of crystals.

Theoretical Strength. The theoretical strength σ_{th} of a body is the stress required to separate it into two parts, with the separation taking place simultaneously across the cross section. To estimate σ_{th}, consider pulling on a cylindrical bar of unit cross-sectional area. The force of cohesion between two planes of atoms varies with their separation, as shown in Fig. 15.9. Part of this curve can be approximated by the relation

$$\sigma = \sigma_{th} \sin\left(\frac{2\pi X}{\lambda}\right) \tag{15.33}$$

The work per unit area to separate the two planes of atoms is then

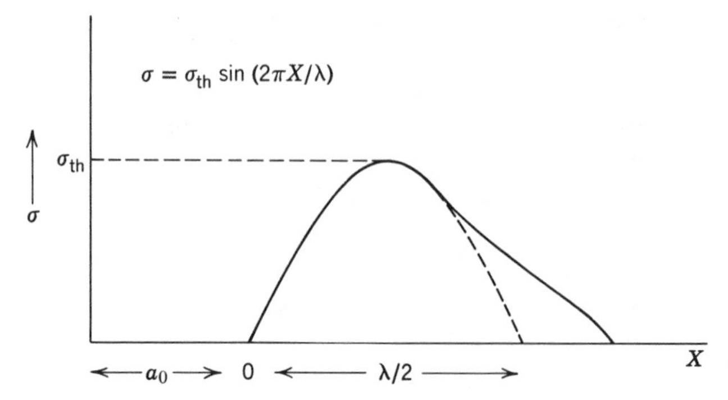

Fig. 15.9. Force versus separation relation (schematic).

$$\int_0^{\lambda/2} \sigma_{th} \sin\left(\frac{2\pi X}{\lambda}\right) dX = \frac{\lambda\sigma_{th}}{\pi} \tag{15.34}$$

This work may be equated with the surface energy 2γ of the two new surfaces, giving

$$\sigma_{th} = \frac{2\pi\gamma}{\lambda} \tag{15.35}$$

For the initial part of the curve near the equilibrium spacing, a_0, Hooke's law may be expressed

$$\sigma = E\frac{X}{a_0} \tag{15.36}$$

where E is Young's modulus. For this small X part of the curve, one obtains from Eq. 15.33

$$\frac{d\sigma}{dX} = \frac{2\pi\sigma_{th}}{\lambda} \cos\left(\frac{2\pi X}{\lambda}\right) \approx \frac{2\pi\sigma_{th}}{\lambda} \tag{15.37}$$

Equating this with $d\sigma/dX$ obtained from Eq. 15.36,

$$\frac{2\pi\sigma_{th}}{\lambda} = \frac{E}{a_0} \tag{15.38}$$

and substituting Eq. 15.38 in 15.35, one obtains

$$\sigma_{th} = \left(\frac{E\gamma}{a_0}\right)^{1/2} \tag{15.39}$$

For typical values of $E = 3 \times 10^{11}$ dynes/cm², $\gamma = 10^3$ ergs/cm², and $a_0 = 3 \times 10^{-8}$ cm, Eq. 15.39 predicts a theoretical strength of 10^{11} dynes/cm², or about 10^6 psi. Similar predictions may be obtained from Eq. 15.38 by assuming that λ is similar in magnitude to a_0, and hence

$$\sigma_{th} \approx \frac{E}{5} \text{ to } \frac{E}{10} \tag{15.40}$$

Such strengths have been obtained only with thin fibers of fused silica and whiskers of crystalline oxides such as Al_2O_3. For most commercial bodies, strengths in the range of $E/100$ to $E/1000$ or less are typically observed. For example, the strengths of window glass are generally in the range of 10^4 psi ($E/1000$); strengths of high-alumina porcelains are typically about 5×10^4 psi (again about $E/1000$).

Griffith–Orowan–Irwin Analysis. To explain this marked discrepancy between the theoretical and actual strengths of materials, use is generally made of the suggestion of Griffith* that flaws in the materials can act as stress concentrators and that the separation of surfaces in fracture takes place sequentially rather than simultaneously across the cross section.

From this concept, two approaches have been followed. The first, due initially to Griffith, is based on the suggestion that a crack propagates when the decrease in stored elastic energy associated with its extension exceeds the increase in surface energy associated with the formation of new surfaces. For elliptical cracks of major axis $2c$ in a thin plate, this condition may be expressed

$$\frac{d}{dc}\left(\frac{\pi c^2 \sigma^2}{E}\right) = \frac{d}{dc}(4\gamma c) \tag{15.41}$$

or

$$\sigma_f = \left(\frac{2E\gamma}{\pi c}\right)^{1/2} \approx \left(\frac{E\gamma}{c}\right)^{1/2} \tag{15.42}$$

The second approach considers directly the stress concentration in the vicinity of the tip of a flaw. It was shown by Inglis† that for flaws of the type considered above, the maximum stress in the vicinity of the crack tip σ_m may be expressed

$$\sigma_m = 2\sigma\left(\frac{c}{\rho}\right)^{1/2} \tag{15.43}$$

where ρ is the radius of the crack tip. On this basis, failure would be expected when the stress at the crack tip exceeds the theoretical strength of the material; that is, when $\sigma_m = \sigma_{th}$.

It was noted by Orowan‡ that the minimum radius of curvature at the tip of a crack is of the order of a magnitude of the interatomic spacing a_0. If ρ in Eq. 15.43 is replaced by a_0, the condition for failure becomes

$$\sigma_f = \left(\frac{E\gamma}{4c}\right)^{1/2} \tag{15.44}$$

Philos. Trans. R. Soc., **A221**, 163 (1920).
†*Trans. Inst. Nav. Arch.*, **55**, 219 (1913).
‡*Z. Krist.*, **A89**, 327 (1934).

It is seen by comparing Eqs. 15.42 and 15.44 that the two approaches yield similar predictions for the observed fracture strength provided the radius of curvature at the crack tip is assumed to be as small as possible (of the order of an interatomic spacing).

When the surface energy in Eq. 15.42 is evaluated from studies of fracture with intentionally introduced flaws of known size, results in the anticipated range are found for brittle oxide glasses, but for metals or glassy polymers, unreasonably large values are obtained (see discussion below). Orowan* showed that the Griffith equation could be used to describe fracture in these partially ductile materials by including a term γ_p for the plastic or viscous work required to extend a crack by unit area; that is,

$$\sigma_f = \left[\frac{E(\gamma + \gamma_p)}{c}\right]^{1/2} \tag{15.45}$$

For materials which are not completely brittle, or nearly so, γ_p is generally much larger than γ and dominates the fracture process.

Irwin† considered the crack-extension force G for the case in which a load P is applied to a plate with a crack of length $2c$:

$$\mathrm{G} = \frac{P^2}{2}\frac{d(1/m)}{dc} \tag{15.46}$$

where m is the slope of the load-extension plot. G can also be expressed as a strain-energy-release rate, that is, the rate of loss of energy from the elastic stress field during crack propagation:

$$\mathrm{G} = \frac{\pi c \sigma^2}{E} \tag{15.47}$$

Measured values of G include the energy of plastic or viscous deformation required to propagate the crack. For a given sample and test conditions G increases with crack length until a critical value G_c is reached, at which the crack becomes unstable and propagates rapidly in a brittle manner. The Griffith relation (Eq. 15.42) can be written

$$\sigma_f = \left(\frac{E\mathrm{G}_c}{\pi c}\right)^{1/2} \tag{15.48}$$

A factor related to G is the stress-intensity factor K_I:

$$K_I = \sigma(Yc)^{1/2} \tag{15.49}$$

Rep. Prog. Phys., **12**, 185 (1948).
†*Welding J.*, **31**, 450 (1952).

where Y is a parameter which depends on the specimen and crack geometry and equals unity for a central crack in a thin infinite plate. Equation 15.49 reflects the fact that the local stresses near the tip of a crack depend on the nominal stress σ and the square root of the crack depth $2c$. For a plane stress condition, $K_I^2 = \mathrm{G}E$, and in general when $\mathrm{G} = \mathrm{G}_c$, $K_I = K_{Ic}$, where K_{Ic} is called the fracture toughness.

Besides characterizing the inherent difficulty of propagating a crack in a material, K_{Ic} is used to estimate the size of the plastic zone ahead of the crack tip. Following D. S. Dugdale,* the length R of the zone is

$$R = \frac{\pi}{8}\left(\frac{K_{Ic}}{\sigma_y}\right)^2 \tag{15.50}$$

where σ_y is the yield stress of the material. This relation should provide a useful estimate for most ceramics, since their yield stresses generally exceed the stresses required for fracture.

Statistical Nature of Strength. An important consequence of the theory that brittle fracture is initiated at Griffith flaws is that the fracture strength of a brittle solid is statistical in nature, depending on the probability that a flaw capable of initiating fracture at a specific applied stress is present. This is the main explanation of the scatter of results normally found for the strength of ceramic materials. A corollary of the statistical nature of the flaws present is that the observed strength is related in some way to the volume of material under stress or the surface area of material under stress. Consequently, observed strengths vary with the manner in which tests are conducted. It is found in practice that the fracture stress of a brittle material is not so high when tested with a tensile specimen having a large surface area and volume under the maximum stress as when tested in a bend test in which the stresses decrease from a maximum at the surface to zero at the neutral axis. In the bend specimen the area under maximum stress and a volume under stress are smaller, and consequently higher strength values are observed.

There have been various attempts to develop a statistical theory for the strength of brittle solids. These all involve an assumption about the number of dangerous flaws related to the specimen volume or surface area. Direct observations of flaws that are likely to cause fracture suggest that this relationship changes from one material to another and probably depends on the fabrication method and on treatment after fabrication. Consequently, no generalized statistical strength theory can be expected to hold for all materials. The best known of these statistical theories was developed by W. Weibull.† He assumed that the risk of rupture is

J. Mech. Phys. Solids, **8**, 100 (1960).

†*Ing. Vetensk. Akad.*, Proc. 151, No. 153 (1939).

proportional to a function of the stress and the volume of the body. That is,

$$R = \int_v f(\sigma)\, dv \tag{15.51}$$

In order to obtain an explicit expression for $f(\sigma)$, he assumed that integration should be carried out over the volume under an applied tensile stress and used the form for $f(\sigma)$:

$$f(\sigma) = \left(\frac{\sigma}{\sigma_0}\right)^m \tag{15.52}$$

where σ_0 is a *characteristic strength* depending on the distribution function best fitting the data and m is a constant related to the material homogeneity. The larger the value of m, the more homogeneous the material; consequently, as m approaches zero, $f(\sigma)$ approaches unity so that the probability of failure is equal for all values of stress. As m tends toward infinity, $f(\sigma)$ is zero for all values of σ less than σ_0, and the probability of fracture becomes unity only when σ equals σ_0; that is, fracture occurs only when σ is equal to the characteristic strength. The average strength value measured is given by

$$\sigma_{R=1/2} = \left(\frac{1}{2}\right)^{1/m} \sigma_0 \tag{15.53}$$

This is the value that is normally reported in the literature as a strength measurement; the ratio of this value to σ_0 gives an indication of the dispersion of observed strength values.

This and other statistical theories of fracture* predict that larger samples should be weaker and that the dispersion in values of fracture stress should increase as the median strength increases. This is illustrated by the calculated distributions of fracture stresses shown in Fig. 15.10.

Typical distribution curves for measured strengths of ceramic materials are illustrated in Fig. 15.11. The range of observed values makes the use of average strength measurements together with some fixed safety factor an unsatisfactory design method for this class of materials. This is largely because the dispersion of strength values may change substantially for different kinds of materials; the average strength and the minimum strength may vary from a ratio near unity to one near infinity. When stresses are going to be used that are near the maximum safe value, a curve of the distribution of strength values and a ratio of the mean strength to the zero strength value such as illustrated in Fig. 15.11 are essential for design purposes.

*J. C. Fisher and J. H. Hollomon, *Trans. AIME*, **171**, 546 (1947), for example.

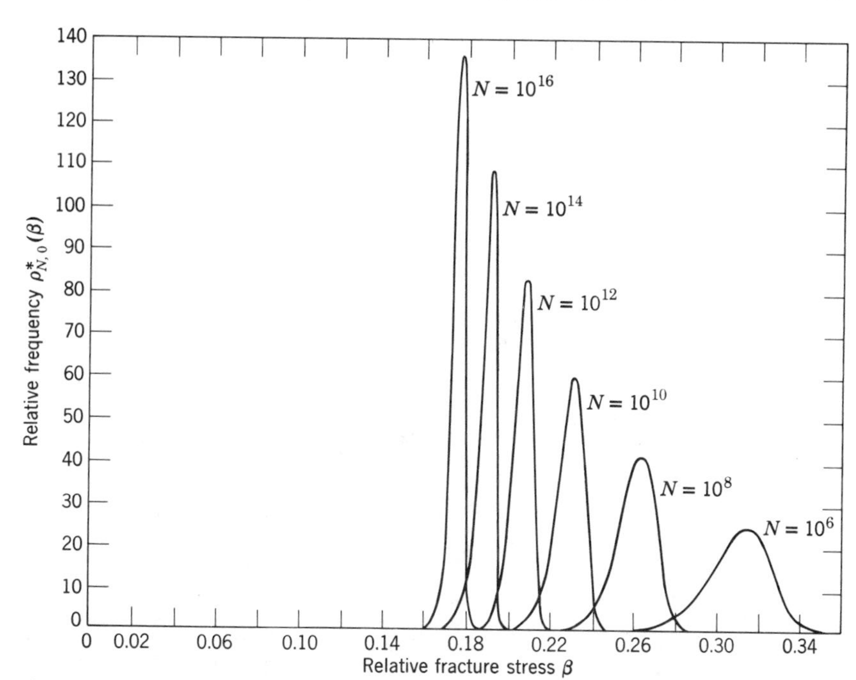

Fig. 15.10. Distribution of strengths of specimens containing N cracks. From J. C. Fisher and J. H. Hollomon, *Trans. AIME*, **171**, 546 (1947).

In the range of modest strengths in which most commercial products fail, the predictions of statistical theories are generally confirmed by experimental data. The mean strength decreases with increasing sample size, and the dispersion of breaking strengths increases with increasing mean strength. In the range of strength above 10^5 psi, however, the strength is often independent of sample size, and the dispersion may be effectively independent of mean strength. For example, both W. H. Otto* and W. F. Thomas† found that the tensile strength of glass fibers was independent of their diameter over the range of 20 to 60 microns. Mean strengths in the respective studies were 4×10^5 and 5.3×10^5 psi, and in both cases, the dispersion of strengths was small. These results very likely reflect the effect of forming conditions, but the relation between these conditions and strength remains to be elucidated.

J. Am. Ceram. Soc.*, **38, 122 (1955).
†*Phys. Chem. Glasses*, **1**, 4 (1960).

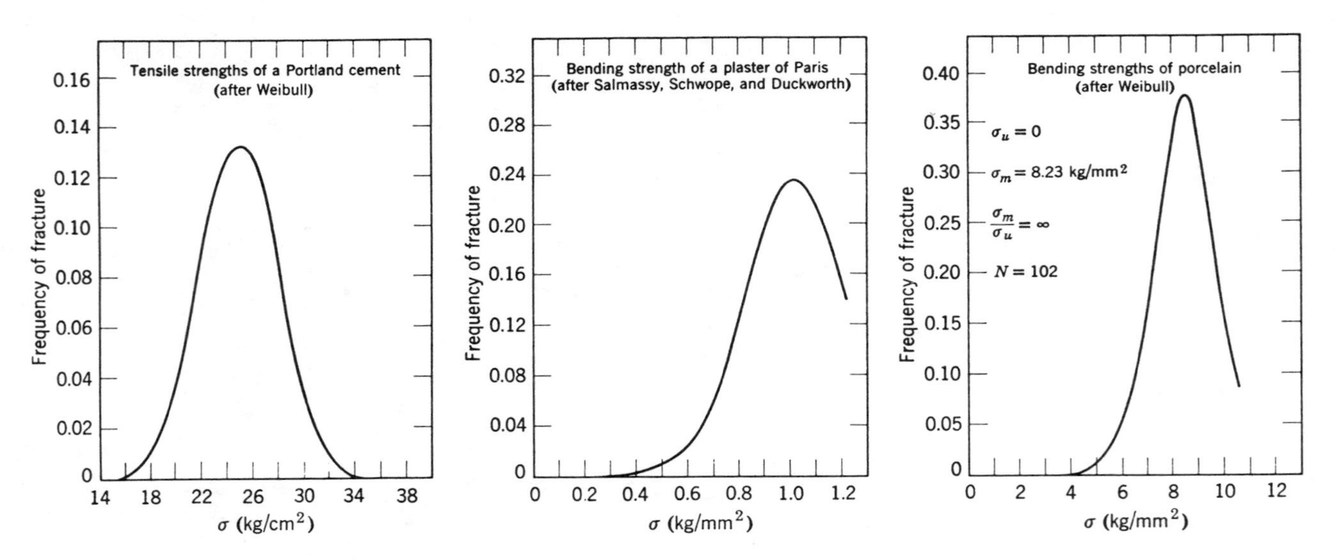

Fig. 15.11. Frequency distribution of observed strengths for some brittle materials.

15.5 Strength and Fracture Surface Work Experience

The observed strengths of ceramic materials cover a wide range, from values smaller than 100 psi for highly porous firebrick to values exceeding 10^6 psi for SiO_2 fibers and whiskers of crystalline ceramics such as Al_2O_3 prepared and tested under carefully controlled conditions. Some typical values are shown in Table 15.2.

In the case of glasses and many dense crystalline ceramics, it has been established that the flaws which are responsible for the relatively small practical strengths of bodies are associated primarily with the surfaces. To observe strengths in the 10^5 psi range, one must generally ensure that the surfaces are clean and undamaged or placed in an initial state of compression. For samples without such compression, even touching the

Table 15.2. Strength Values for Some Ceramic Materials

Material	Modulus of Rupture (psi)
Aluminum oxide crystals	50,000–150,000
Sintered alumina (ca. 5% porosity)	30,000– 50,000
Alumina porcelain (90–95% Al_2O_3)	50,000
Sintered beryllia (ca. 5% porosity)	20,000–40,000
Hot pressed boron nitride (ca. 5% porosity)	7,000–15,000
Hot pressed boron carbide (ca. 5% porosity)	50,000
Sintered magnesia (ca. 5% porosity)	15,000
Sintered molybdenum silicide (ca. 5% porosity)	100,000
Sintered spinel (ca. 5% porosity)	13,000
Dense silicon carbide (ca. 5% porosity)	25,000
Sintered titanium carbide (ca. 5% porosity)	160,000
Sintered stabilized zirconia (ca. 5% porosity)	12,000
Silica glass	15,550
Vycor glass	10,000
Pyrex glass	10,000
Mullite porcelain	10,000
Steatite porcelain	20,000
Superduty fire-clay brick	750
Magnesite brick	4,000
Bonded silicon carbide (ca. 20% porosity)	2,000
2000°F insulating firebrick (80–85% porosity)	40
2600°F insulating firebrick (ca. 75% porosity)	170
3000°F insulating firebrick (ca. 60% porosity)	290

surfaces with one's finger can reduce the strength from the 10^5 to the 10^4 psi range. Etching the surface of a damaged body, as with HF for glasses, often restores the original strength. Such a restoration is shown by the data in Table 15.3 for 0.25-in.-diameter glass rods tested for 60-min periods.

Table 15.3. Effect of Surface Condition on Strength

Surface Treatment	Strength (psi)
As received from factory	6500
Severely sandblasted	2000
Acid-etched and laquered	250,000

Source. C. J. Phillips, *Am. Sci.*, **53**, 20 (1965).

Work by D. G. Holloway* and by N. M. Cameron† has focused attention on the importance to glass strength of microscopic dirt particles bonded to the surfaces. Fractures frequently were seen to propagate from such particles; high pristine strengths were observed for regions of the samples which were free of such particles. The particles may affect strength because of a difference in moduli or thermal expansion coefficients between glass and particle or, more likely, because of local chemical attack (corrosion) at the site of the particles.

In polycrystalline ceramics, a common source of microcracks is the difference in thermal expansion coefficient between the phases present in the body, giving rise to boundary stresses, as discussed in Chapter 5. These are frequently sufficient to initiate small cracks. Stresses set up during cooling of samples from the firing temperature can also be a source of microcracks, or thermal stresses set up during use can give rise to high stresses at the surface which initiate surface cracks without leading to ultimate fracture. Surface checking of this kind on a small scale, or on a large scale in brick-sized samples, is not an uncommon phenomenon. Another source of stress concentration and crack initiation in polyphase ceramics is surface cracks resulting from thermal etching between different phases or along grain boundaries in one phase, as discussed in Chapter 5. These lead to notches in the surface and stress concentration. However, in general, mechanical abrasion of the surface and chemical attack are the major sources for crack development. These are difficult to

Phys. Chem. Glasses, **4**, 69 (1963).
†*Glass Technol.*, **9**, 14 and 121 (1968).

avoid in practical application, so that samples must usually be assumed to contain substantial numbers of flaws.

In the case of crystalline materials that are normally ductile or semibrittle, cleavage or brittle fracture can be observed under some conditions of testing. Low temperature, impact loading, and restraints on plastic deformation, such as at notches, encourage this kind of failure. In these cases it is often found that some plastic deformation does occur in all cases before the initiation of fracture. Various detailed analyses have been developed of the way dislocations generated during plastic deformation can coalesce to produce a microcrack that can lead to brittle fracture. In general, dislocations tend to pile up in large numbers at barriers such as existing slip bands, grain boundaries, and surfaces. When this occurs, high local stresses are produced that can be sufficient to force the dislocations together, forming a crack nucleus.

One particular source of cracks is the interaction of slip bands, causing a piling up of dislocations and initiation of a cleavage crack (Fig. 15.12). A

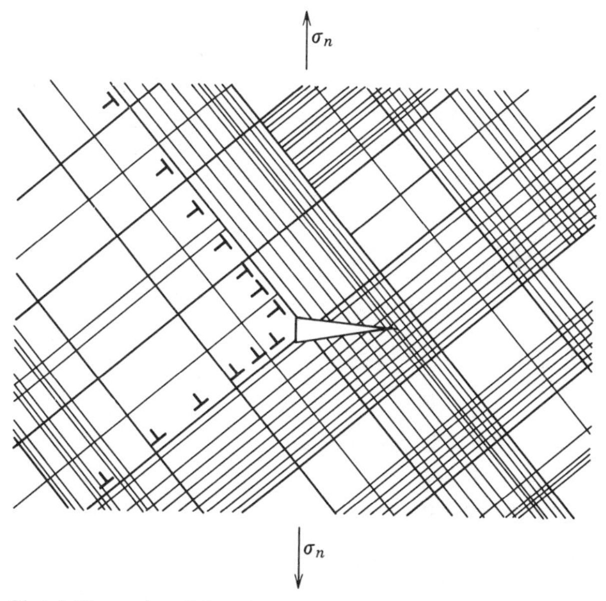

Fig. 15.12. Sketch illustrating dislocation pile-up and crack formation at intersection of two slip bands.

number of similar ways in which dislocations can coalesce to form microcracks have been observed and theoretical calculations made of the stresses required.*

*See A. N. Stroh, *Adv. Phys.*, **6**, 418 (1957).

Once microcracks are formed, the stress fields at their tips induce plastic deformation in the adjacent grain if the material is ductile; and crack growth is slow, since the stress field can be relieved by plastic flow. For semibrittle materials, or in general when flow can take place only at high stress levels, stress builds up at the grain boundary until the strength of the solid is exceeded, and fracture takes place. Assuming the slip-band length is proportional to the grain size d, the fracture stress for this case may be expressed

$$\sigma_f = \sigma_0 + k_1 d^{-1/2} \tag{15.54}$$

where

$$k_1 = \left(\frac{3\pi\gamma E}{1-\nu^2}\right)^{1/2} \tag{15.55}$$

Equation 15.54 has the form of the Petch relation and predicts that the strength should increase with decreasing grain size (as $d^{-1/2}$) of the polycrystalline array. In cases in which the sizes of the initial flaws are limited by the grains and scale with the grain size and in which brittle behavior is observed, the strength should vary with grain size as

$$\sigma_f = k_2 d^{-1/2} \tag{15.54a}$$

Equation 15.54*a* is known as the Orowan relation.

An alternative mechanism for crack initiation in polycrystalline ceramics focuses attention on the internal stresses which occur due to anisotropic thermal contraction. In the analysis by F. J. P. Clarke,* fracture is assumed to initiate at grain-boundary pores and propagate along the grain boundaries, and when the pores are much smaller than the grain size, spontaneous fracture is expected when

$$\epsilon = \left(\frac{48\gamma_b}{Ed}\right)^{1/2} \tag{15.56}$$

Here ϵ is the grain-boundary strain and γ_b is the grain-boundary surface energy.

Experimental data on the strength of polycrystalline ceramics have been summarized by S. C. Carniglia.† Of 46 sets of data examined, 6 were described by the Orowan relation (Eq. 15.54*a*), 10 were described by the Petch relation (Eq. 15.54), and 30 were described by a combination of both relations, with the large-grain-size data being of the Orowan type. This behavior is illustrated by the data on Al_2O_3 shown in Fig. 15.13.

In assessing these results, it must be recognized that failure occurs at the most severe, not at the average, flaw in a body. Hence these correlations of

***Acta Met.*, **12**, 139 (1964).

†*J. Am. Ceram. Soc.*, **55**, 243 (1972).

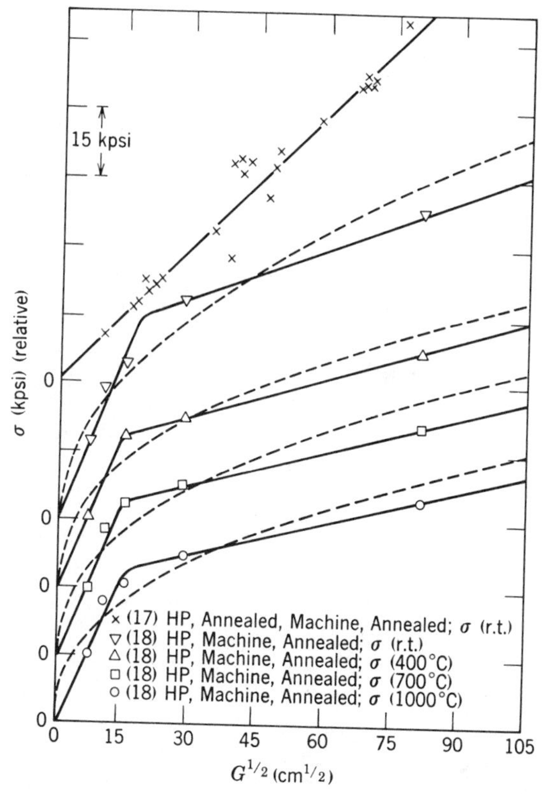

Fig. 15.13. Strength versus grain size relation for polycrystalline Al_2O_3. From S. C. Carniglia, *J. Am. Ceram. Soc.*, **55**, 243 (1972).

strength with average grain size must reflect processing techniques which produce samples in which the size of the most severe flaw scales with the average grain size. In engineering practice, the most severe flaws in a body are usually associated with porosity, surface damage, abnormal grains, or foreign inclusions, and care is often taken to minimize the size of the most severe flaw and ensure that it is close to the average flaw size.

The differences in strength behavior among the different types of ceramic materials can be viewed with a different emphasis. Once a crack has been initiated in brittle materials such as glasses and many ceramic single crystals, the stress conditions are such that the crack begins to propagate. There is no large energy-absorbing process comparable with the plastic deformation of ductile materials. Consequently, there is no mechanism by which the applied stress is limited, and the crack continues to complete failure in a uniform stress field. That is, the initiation of cracks is the critical stage of the rupture process.

In polycrystalline ceramics, obstacles such as grain boundaries can hinder crack propagation and prevent fracture of the body. The increase in stress for a crack to change direction at a boundary has been analyzed by M. Gell and E. Smith,* for typical polycrystalline arrays; increases of factors of 2 to 4 are anticipated.

For materials of high *toughness* such as metal-bonded carbide cutting tools, new cracks must be nucleated as the fracture passes each phase boundary; the ductile behavior of the metal phase provides a mechanism for energy absorption that requires increased stresses to keep a crack moving. Under these conditions the material does not fracture even when stresses exceed the crack nucleation level. A sufficiently high stress level to provide for crack propagation must be exceeded before failure occurs.

The concepts of brittleness and ductility are also reflected in the fracture surface energies ($\gamma + \gamma_p$ in Eq. 15.45) derived from studies of controlled crack propagation. Such studies of a number of silicate glasses have been carried out by S. M. Wiederhorn.† Values in the range of 3500 to 4700 ergs/cm^2 were found in tests carried out at 300°K; at 77°K the values ranged from 4100 to 5200 ergs/cm^2. For comparison, the energy represented by the formation of the new surfaces is estimated as about 1700 ergs/cm^2. These results indicate that glasses are nearly ideally brittle materials. The difference between the measured fracture energies and the estimated surface energies very likely reflects viscous deformation processes taking place in the vicinity of the crack tip. Some such deformation may be anticipated even for glasses tested well below their glass transition temperatures, since the stresses in the region ahead of the crack tip well exceed the critical stress for non-Newtonian flow (see Section 14.7). The extent of the plastic (viscous) zone is, however, quite small—in the range of 6 to 26 Å, according to Wiederhorn's estimates.

In the case of single crystals, the fracture process depends on the crystallographic orientation of the fracture plane and takes place preferentially along those planes having the lowest fracture surface energies. In many cases, this leads to a cleavage type of fracture occurring on a single crystallographic plane with much lower fracture surface energy than other planes in the crystal. Fracture surface energies of various single crystals are shown in Table 15.4. With the exception of Al_2O_3, all the data reflect cleavage fracture and indicate fracture energies which exceed the expected surface energies by factors which are similar to, or perhaps slightly larger than, those found for silicate glasses.

Acta Met., **15**, 253 (1967).

†*J. Am. Ceram. Soc.*, **52**, 99 (1969).

Table 15.4. Fracture Surface Energies of Single Crystals

Crystal	Fracture Surface Energy ($ergs/cm^2$)
Mica, vacuum, 298°K	4500
Lif, $N_2(l)$, 77°K	400
MgO, $N_2(l)$, 77°K	1500
CaF_2, $N_2(l)$, 77°K	500
BaF_2, $N_2(l)$, 77°K	300
$CaCO_3$, $N_2(l)$, 77°K	300
Si, $N_2(l)$, 77°K	1800
NaCl, $N_2(l)$, 77°K	300
Sapphire, $(10\bar{1}1)$ plane, 298°K	6000
Sapphire, $(10\bar{1}0)$ plane, 298°K	7300
Sapphire, $(11\bar{2}3)$ plane, 77°K	32,000
Sapphire, $(10\bar{1}1)$ plane, 77°K	24,000
Sapphire, $(22\bar{4}3)$ plane, 77°K	16,000
Sapphire, $(11\bar{2}3)$ plane, 293°K	24,000

In the case of polycrystalline ceramics, the measured fracture surface energies are larger by a factor of 5 to 10 than those determined for single crystals. The large $(\gamma + \gamma_p)$ of the polycrystals is associated with the irregular crack paths in the materials (see Figs. 15.21 and 15.22). These in turn reflect the obstacles which must be overcome during crack motion and which increase the work of fracture.

For small grain sizes, the fracture surface energy of polycrystalline ceramics apparently increases with increasing grain size. Since single crystals are generally stronger than polycrystals and since the strength of polycrystalline ceramics increases with decreasing grain size, it seems that the strength of polycrystals depends primarily on single-crystal or grain-boundary fracture surface energies rather than on the polycrystal $(\gamma + \gamma_p)$ values.

15.6 Static Fatigue

In addition to their brittle failure, a characteristic of the rupture of silicate glasses and crystalline ceramics is that the measured strength depends on the length of time that a load is applied or on the loading rate. This phenomenon, termed *fatigue*, is illustrated in Fig. 15.14, where the fracture stresses for different times of loading are shown for a number of glasses tested under different conditions. These data are describable by a

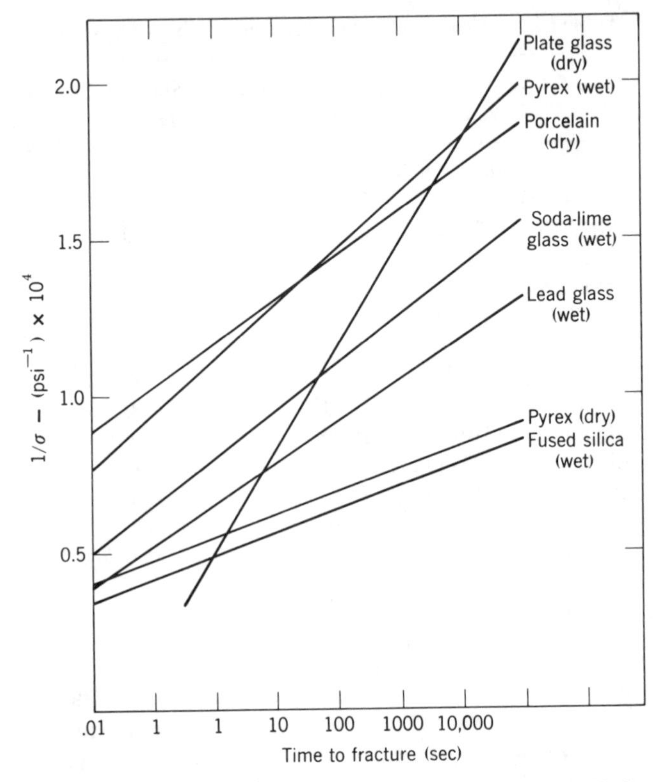

Fig. 15.14. Reciprocal of stress versus log of time to failure. From J. L. Glathart and F. W. Preston, *J. Appl. Phys.*, **17**, 189 (1946).

relation of the form

$$\log t = \frac{A}{\sigma_f} + B \tag{15.57}$$

where t is the time to failure and A and B are constants. This relation cannot be valid at very short times but provides a useful description over nearly all the range of the data.

Although ceramic materials such as the glasses cited in Fig. 15.14 can withstand a given stress for a short period of time, lower stresses ultimately lead to fracture if applied for a sufficiently long period of time. Consequently, in specifying the strength of a ceramic material, and fixing its use conditions, we must know the rate of load application or the time the stress is to be applied as well as other factors.

The nature of the process leading to delayed fracture is best seen by comparing measurements made *in vacuo*, in completely dry air, and in normal air containing a certain amount of water vapor. It is found that static fatigue is most pronounced when moisture is present in the atmosphere. The chemical nature of the process can also be seen in its temperature dependence (Fig. 15.15). The short-time strength at room temperature approaches the value found by standard testing at very low temperatures at which the rate of attack by water vapor is very low, which is also close to the value found by testing in a vacuum.

The increase in strength of glasses with falling temperature below ambient is very likely associated with decreasing atomic mobility. Below liquid nitrogen temperature, the strength changes little with changes in temperature. At temperatures above 400 to 500°K, the strength increases with increasing temperature. This increase may be due to decreasing surface adsorption of atmospheric water or to a smaller effect of occluded dirt particles or to increasing viscous work at the crack tip. In the range of temperature between +50°C and −50°C, an Arrhenius plot of the time to fracture at a constant stress (the inverse of the rate of corrosion) is a straight line with an activation energy of 18.1 kcal/mole for a soda-lime-silica glass (Fig. 15.16). That is, the process leading to fracture is a temperature-dependent activated process similar to chemical reaction or diffusion.

In tests carried out on soda-lime-silicate glasses given different abra-

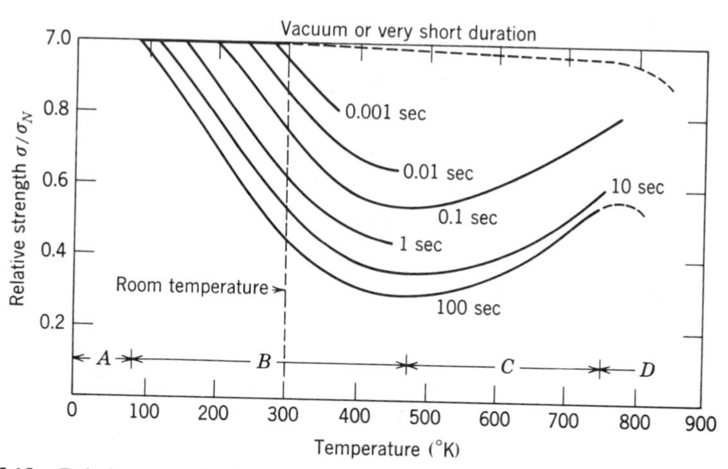

Fig. 15.15. Relative strength of glass tested in air versus temperature for loads of various durations. Semiquantitative composite curves from results of various investigators. From R. E. Mould, *Glastech. Ber.*, **32**, 18 (1959).

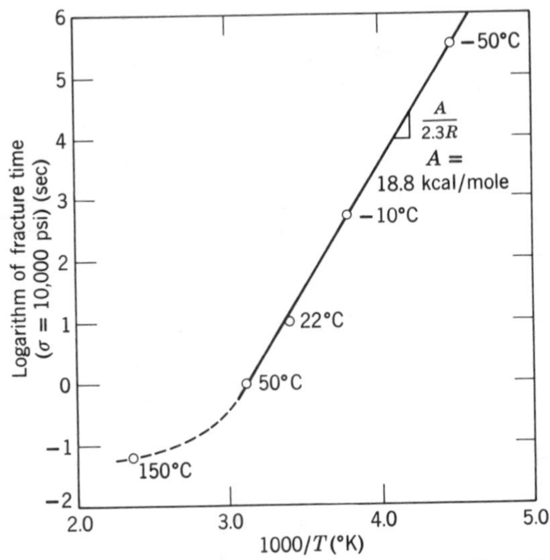

Fig. 15.16. Temperature dependence of the time to failure of sode-lime-silica rods in bending test. From R. J. Charles in *Progress in Ceramic Science*, Vol. 1, Pergamon Press, New York, 1961.

sion treatments, R. E. Mould and R. D. Southwick* found that different abrasions produced different static fatigue curves at room temperature (Fig. 15.17). If, however, the strength values for each abrasion treatment are divided by the low-temperature strength for that treatment and are plotted versus a reduced time coordinate, all the data can be fitted to a single *universal fatigue curve*. Such a curve is shown in Fig. 15.18; it indicates a long-time strength equal to approximately 20% of the short-time or low-temperature strength. Subsequent work has suggested a fatigue limit for soda-lime-silicate glasses of approximately 17% of the short-time strength. Other glasses, of different chemical compositions, are characterized by different universal fatigue curves.

The fatigue process has been associated with two phenomena: (1) a stress corrosion process, in which a sufficiently large stress enhances the rate of corrosion at the crack tip relative to that at the sides, leading to a sharpening and deepening of the crack and eventually resulting in failure, and (2) a lowering of the surface energy by the adsorption of an active species, leading to a decrease in the γ contribution to the fracture surface work. As noted above, however, even for the most brittle materials—silicate glasses—the estimated γ's represent only about 30% of the total

J. Am. Ceram. Soc., **42**, 582 (1959).

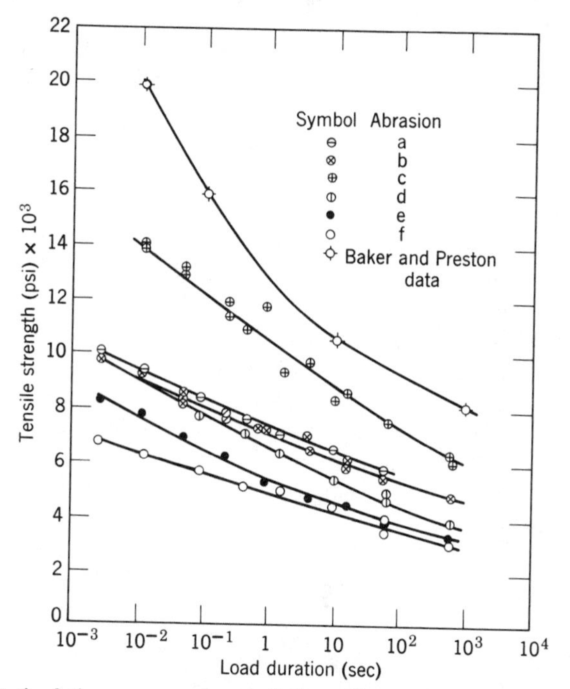

Fig. 15.17. Static fatigue curves for soda-lime-silicate glass specimens with various abrasions, tested immersed in distilled H_2O. From R. E. Mould and R. D. Southwick, *J. Am. Ceram. Soc.*, **42**, 582 (1959).

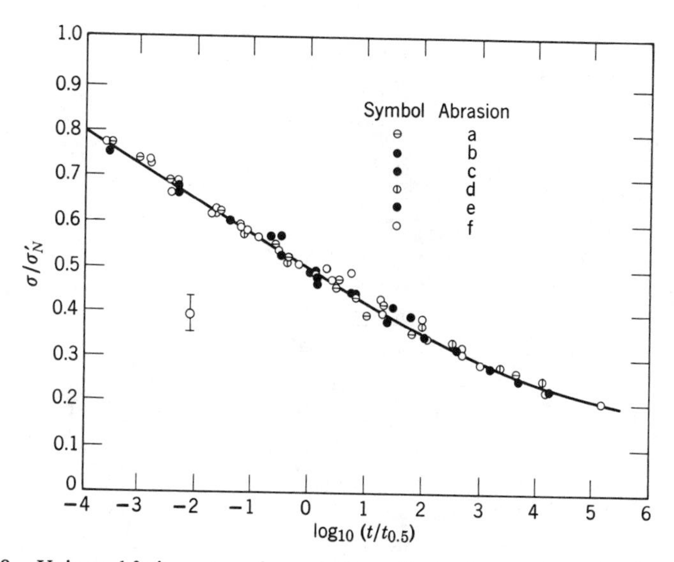

Fig. 15.18. Universal fatigue curve for soda-lime-silicate glasses, derived from the data in Fig. 15.17. σ'_N is the strength at 77°K; $t_{0.5}$ is the time with a given abrasion treatment for the strength to fall to a half of its value at 77°K.

fracture surface work, and changes in γ would be insufficient to account for the observed decrease in strength on long exposure to loads. For this reason, as well as for its generally satisfactory description of many experimental data, the stress corrosion model is preferred, although it apparently does not describe all the observed behavior (see below).

Although static fatigue seems closely related to a corrosion process, the dissolution must be of a particular type that leads to slow crack growth increasing the stress concentration at the crack tip until it grows to a size that causes fracture as required by the Griffith criterion. Etching a glass surface tends to remove surface cracks and round out crack bottoms so that the stress concentration in microcracks is reduced and the strength of the sample is increased. This result can be predicted from the fact that the small radius of curvature crack tip has a higher chemical potential and solubility than the flat side (in this connection it might be well to review the discussion of solubility and its relation to surface curvature discussed in Chapter 5).

As pointed out by R. J. Charles,* the factor causing more rapid solution at the crack tip than at the sides (this is required to give a continually increasing stress concentration) is that the high tensile stress at the tip leads to an expansion of the glass network or crystal lattice at that point and an increase in the rate of corrosion. It is known that quenched glasses which have an open structure, that is, a greater specific volume, have a higher rate of corrosion than annealed glasses with a denser structure. Similarly, the elastic extension at the tip of the crack is believed to lead to more rapid solution at this point, probably because of the increased mobility of the sodium ions in the expanded network.

The model of stress corrosion has been used by W. B. Hillig and R. J. Charles† to explain the universal fatigue curve (Fig. 15.18). It has also been used with success to describe the experiments of S. M. Wiederhorn‡ in which the crack velocity in soda-lime-silicate glasses was measured as a function of temperature, environment, and stress-intensity factor. Typical results, shown in Fig. 15.19, indicate three regions in the crack velocity versus stress-intensity factor (K_I) relation. In region I, the velocity varies exponentially with stress-intensity factor, as predicted by the stress-corrosion model, and exhibits a dependence on water-vapor pressure. In this region, the velocity is apparently limited by the kinetics of the chemical reaction at the crack tip. In region II, the crack velocity is nearly independent of K_I but depends significantly on the water-vapor pressure.

J. Appl. Phys., **29**, 1549 (1958).

†*High Strength Materials*, John Wiley & Sons, Inc., New York, 1965.

‡*J. Am. Ceram. Soc.*, **50**, 407 (1967).

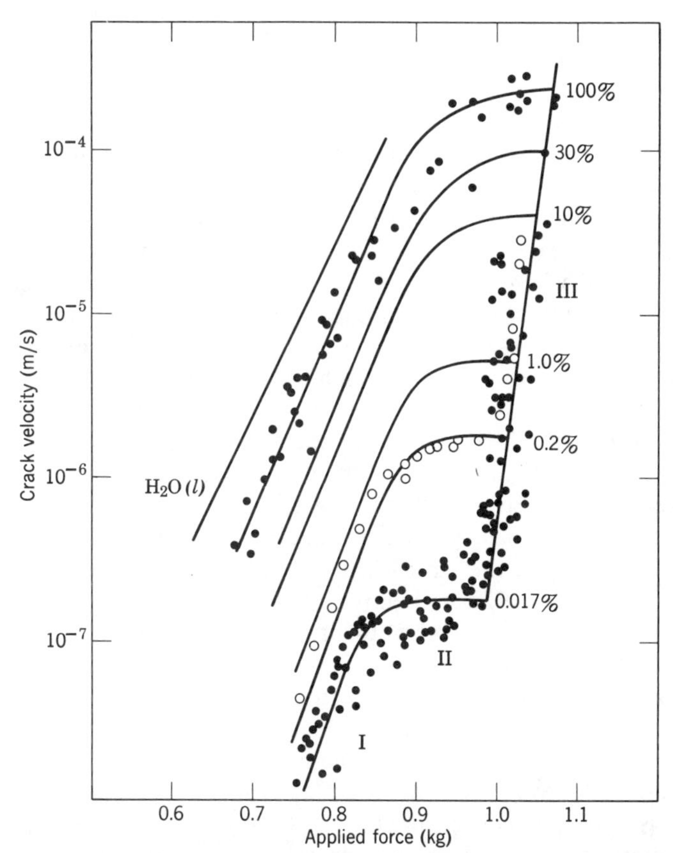

Fig. 15.19. Dependence of crack velocity on applied force for soda-lime-silicate glass. The percentage of relative humidity is given on the right-hand side of the diagram, and the Roman numerals indicate the three regions of crack propagation. From S. M. Wiederhorn, *J. Am. Ceram. Soc.*, **50**, 407 (1967).

In this region, the velocity is apparently limited by the transport of water vapor to the crack tip. In region III, the crack velocity is strongly dependent on K_I but independent of the environment. At constant K_I, the crack velocity in this region is found to depend on temperature and glass composition, as shown in Fig. 15.19.

The plot in Fig. 15.19 was made to test the predictions of the stress-corrosion model, which indicates that the crack velocity should vary exponentially with stress-intensity factor, as

$$V = V_0 \exp \frac{(-E^* + bK_I)}{RT} \tag{15.58}$$

In addition to this relation, it is often useful to describe the dependence of crack velocity on stress-intensity factor by the empirical relation

$$V = AK_I^n \tag{15.59}$$

Typical values of n determined in this way are in the range of 30 to 40, indicating a marked dependence of crack velocity on stress intensity.

Behavior similar to that shown in Fig. 15.19 has been observed in vacuum for soda-lime-silicate, aluminosilicate, borosilicate, and high-lead glasses, with rates which depend strongly on composition. For fused silica and low-alkali borosilicate glasses, on the other hand, no slow crack propagation prior to failure has been observed in vacuum. These results suggest that not all crack propagation which varies exponentially with stress-intensity factor should be interpreted in terms of the stress-corrosion model. For glasses tested in vacuum, the apparent activation energies for slow crack growth are much higher than those observed in water vapor and very likely reflect activated growth of the crack itself.

Most oxide ceramics contain an alkali silicate glassy phase so that delayed fracture and the strength dependence on loading rate are commonly observed. As indicated above, slow crack propagation, and hence delayed fracture, is also observed for systems essentially free of alkali. Results similar to those found for silicate glasses at room temperature are also found for a variety of other ceramic materials, including porcelains, glassy carbon, portland cement, high-alumina ceramics, silicon nitride, lead zirconate titanate, and barium titanate (see reference 8). In polycrystalline ceramics, slow crack growth at high temperatures seems to depend principally on impurities, particularly grain-boundary phases, rather than on the environment, and localized plasticity may also be important under these conditions. Examples of the effects of static fatigue on the strengths of ceramic bodies are shown in Table 15.5.

The dependence of strength on time of loading is closely related to its dependence on loading rate. The observed decrease in strength with decreasing loading rate can be simply explained. With slower rates of load increase, more time is allowed for slow crack growth; hence the critical stress-intensity factor for failure is reached at lower applied stresses. Quantitative expressions have been derived which relate the dependence of strength on loading rate to the dependence of the crack-propagation rate on stress-intensity factor,* and the available experimental data, admittedly limited, are in close accordance with predictions.

In addition to behavior under conditions of static loading and variable rate of loading, attention has also been directed to the behavior of ceramics

*S. M. Wiederhorn et al., *J. Am. Ceram. Soc.*, **57**, 336 (1974).

Table 15.5. Compressive (c) or Transverse (tr) Strengths under Conditions Indicated with a Machine-Loading Rate of 0.005 in./min

Material	Test Conditions (psi)			
	Liquid Nitrogen −195°C	Dry Nitrogen 240°C	Saturated Water Vapor at 240°C	Saturated Water Vapor at 25°C
Soda-lime-silica glass (tr)	22,000			11,000
Fused silica glass (c)	65,700	64,500	36,600	55,600
Granite (c)	37,400	19,700	6010	23,500
Spodumene (c)	95,200	57,300	45,700	38,500
Brazilian quartz (c)	81,600	63,800	35,800	52,200
MgO crystal (c)	30,500	26,600	8000	14,200
Al_2O_3 crystal (tr)	152,000	116,500	68,300	110,000

Source: R. J. Charles, reference 7.

under conditions of oscillating stress. The early work of Gurney and Pearson on soda-lime-silicate glasses indicated that under conditions in which the stress is varied in step functions between $-\sigma$ and $+\sigma$, the strength behavior could be described by the static-fatigue relation, with the relevant time parameter being the sum of the times which the specimens spend under the tensile stress. That is, the behavior under cyclic stress conditions could be explained in terms of crack-growth mechanisms operating under static conditions, with no evidence being found for a plasticity-associated fatigue crack growth of the type found with metals.

For a stress which varies sinusoidally with time, as $\sigma(t)=\sigma_0+\sigma_1 \sin 2\pi\nu t$, the average crack velocity per cycle can be expressed in terms of the average stress-intensity factor $K_{I_{av}}$, following Eq. 15.59:

$$V_{av}=gAK_{I_{av}}^n \tag{15.60}$$

where

$$g=\nu\int_0^{1/\nu}\left[\frac{\sigma(t)}{\sigma_0}\right]^n dt \tag{15.61}$$

Then the average crack velocity under conditions of a sinusoidal stress can be expressed:

$$V_{av}=gV_{static} \tag{15.62}$$

where V_{static} is the value obtained from Eq. 15.59 with K_I taken as $K_{I_{av}}$. Values of g for different values of n have been given by Evans and

Fuller.* Reasonable agreement has been found between predicted and experimental crack velocities under conditions of sinusoidal stress in electrical porcelain, soda-lime-silicate glass, and silicon nitride, and this agreement suggests that crack propagation under these conditions involves the same mechanisms as under static conditions. It remains possible, however, that plasticity-associated fatigue crack growth may be found for ceramics at temperatures at which plasticity is observed (see discussion in Chapter 14).

One of the most interesting developments in the application of fracture mechanics to ceramic materials has been the introduction of proof testing. Consider, for example, a proof stress σ_p, corresponding to a stress-intensity factor K_{IP}, applied to specimens. Those which survive this stress must have (recall Eq. 15.49)

$$K_{IP} = \sigma_p (Yc)^{1/2} < K_{Ic} \qquad (15.63)$$

Knowing this upper limit to the size of cracks which can be present in specimens surviving the proof test and knowing the crack-propagation velocity as a function of stress-intensity factor for the material, one can then in principle guarantee a minimum service life under given stress conditions for all samples. Use of this approach depends on factors such as ensuring (1) that additional damage does not occur to the body during subsequent handling, (2) that the conditions of the proof test approximate to the conditions of use, (3) that the in-service stress conditions for the bodies can be estimated, (4) that the crack velocity versus K_I relation is available for similar stress conditions as expected during use, (5) that account is taken of slow crack growth which occurs during unloading after the proof test, and so on. Although these factors may limit the widespread use of the technique, it does seem to offer new promise for ensuring the integrity of ceramic bodies when sufficient information is available on the materials and in-service conditions.

Considerations such as these seem likely to receive increasing attention in coming years, as structural engineers turn more to ceramic materials to withstand harsh environments. Because of their outstanding resistance to corrosion and high temperatures, ceramics will increasingly be used in structural load-bearing applications, in which the probability of failure is a critical design parameter. For such critical applications, great pressure will be placed on the ceramics engineer to provide materials whose failure characteristics are better known and understood.

*_Met. Trans._, **5**, 27 (1974).

15.7 Creep Fracture

Another kind of fracture that should be mentioned is the creep failure encountered with polycrystalline materials when they are deformed at elevated temperatures. Under these conditions a significant part of the deformation results from grain-boundary sliding. The rate of sliding of the grain boundaries is proportional to the shear stress for small deformations. For larger deformations the boundaries are uneven, and geometrical nonconformity leads to *keying* of grains to their neighbors, and the rate of boundaries sliding decreases while boundary migration occurs to accommodate the irregularity. As the stress is increased, it forces sliding to occur along the boundaries regardless of their geometrical nonconformity. As a consequence, high tensile stresses are developed in the boundary regions, and these cause the nucleation of cracks and pores. As extension continues, these small pores grow by a process which is just the reverse of the sintering phenomena discussed in Chapter 10. In crystalline materials this process is one of volume diffusion; in materials with a viscous boundary phase, most common in ceramics, the mechanism of growth is probably one of viscous flow in the boundary phase.

As the pores increase in size, the cross-sectional solid area is decreased, the unit stress increases, the fracture results. The presence of cracks and pores in a ceramic sample that has been extended at high temperatures is illustrated in Fig. 15.20. Since creep fracture is a consequence of boundary sliding, we cannot speak of a definite strength. The time required for fracture increases with lower stresses and lower temperatures. The most satisfactory way of representing experimental data is as creep-rupture curves. If time to fracture is plotted on a logarithmic scale against the applied stress, a straight line adequately represents data over a range in which this kind of process takes place.

The effects of grain size on strength have been discussed in Section 15.5. As noted there, the fracture paths in polycrystalline ceramics are often quite irregular. Examples of such fracture patterns are shown in Figs. 15.21 and 15.22. In addition to the direct effects of grain size on strength, reflected in Eqs. 15.54 and 15.54*a*, there is also the effect of boundary stresses arising from anisotropic thermal contraction, discussed in relation to Eq. 15.56. A striking example of the latter effect is provided by observations on Al_2O_3 fired at 1900°C to form grains several millimeters in diameter. Under these conditions the boundary stresses are sufficiently large so that spontaneous cracking occurs and grains can be picked out individually with a penknife. Boundary cracks in a sample of aluminum oxide can be seen in Fig. 5.15.

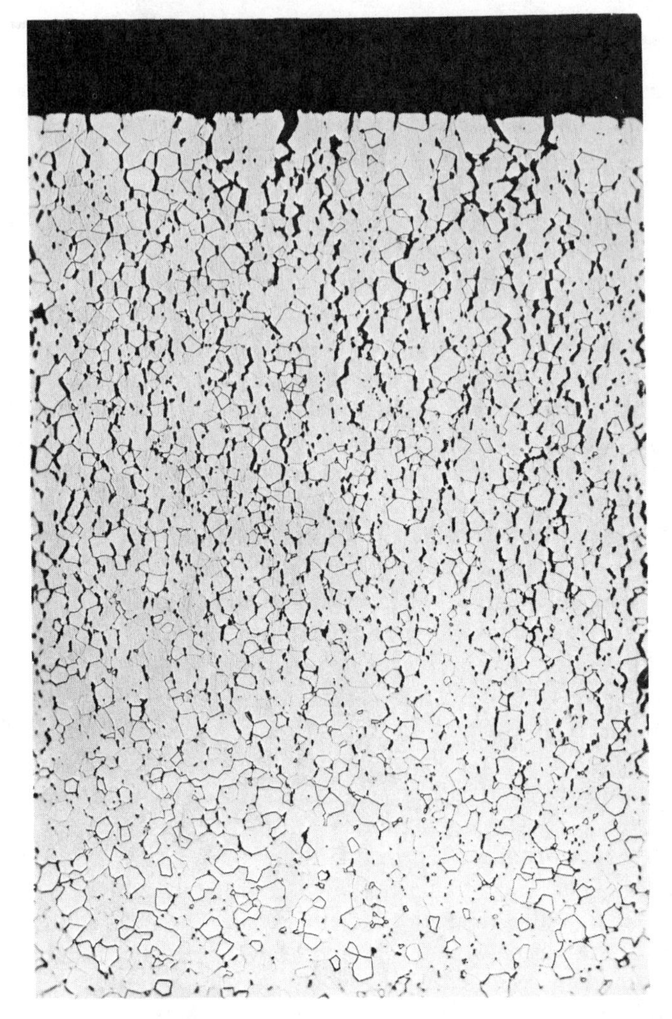

Fig. 15.20. Microstructure of initially dense Al_2O_3 which has been deformed at high temperature in a bend test. Pores appear on tensile (top) side of specimen (100×). Courtesy R. L. Coble.

15.8 Effects of Microstructure

Since fracture phenomena themselves are diverse and not clearly understood in all details, even for the simplest materials, it is apparent that no completely satisfactory and all-encompassing survey of the effects of composition and microstructure is possible.

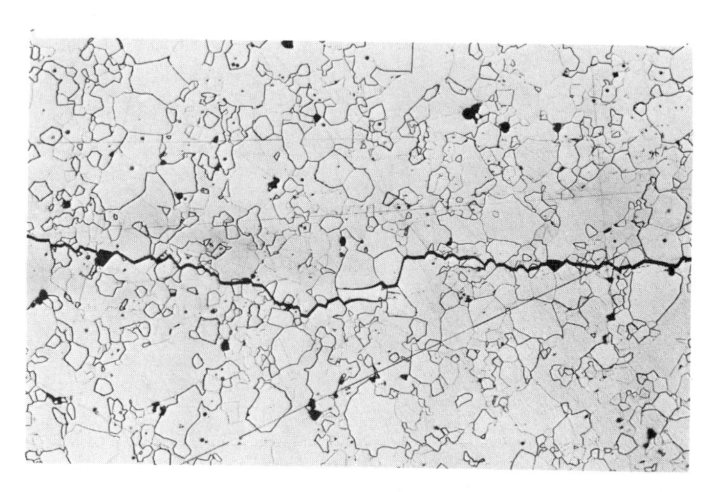

Fig. 15.21. Fracture path in a sample of thermally cracked polycrystalline aluminum oxide. Courtesy R. L. Coble.

The major effect of the structure in most ceramics is the result of porosity. Pores obviously decrease the cross-sectional area on which the load is applied but also act as stress concentrators. (For an isolated spherical pore the stress is increased by a factor of 2.) Experimentally, it is found that the strength of porous ceramics is decreased in a way that is nearly exponential with porosity. Various specific analytical relationships have been suggested for the effect of porosity. An empirical suggestion by Ryskewitsch*

$$\sigma = \sigma_0 \exp(-nP) \tag{15.64}$$

where n is in the range 4 to 7 and where P is the volume fraction porosity, approximates many data satisfactorily. That is to say, the strength is decreased to a value of half that observed for the porefree material with about 10% porosity. This amount of porosity is not uncommon. Hard porcelain has a porosity of about 3%; earthenware has a porosity of 10 to 15%. This difference in porosity is responsible for the major part of the order of magnitude difference in their strength. The relatively high strength of hard porcelain is one of its desirable features, of course. Data for several well-characterized materials illustrating the strong effect of porosity are included in Fig. 15.23.

J. Am. Ceram. Soc.*, **36, 65 (1953).

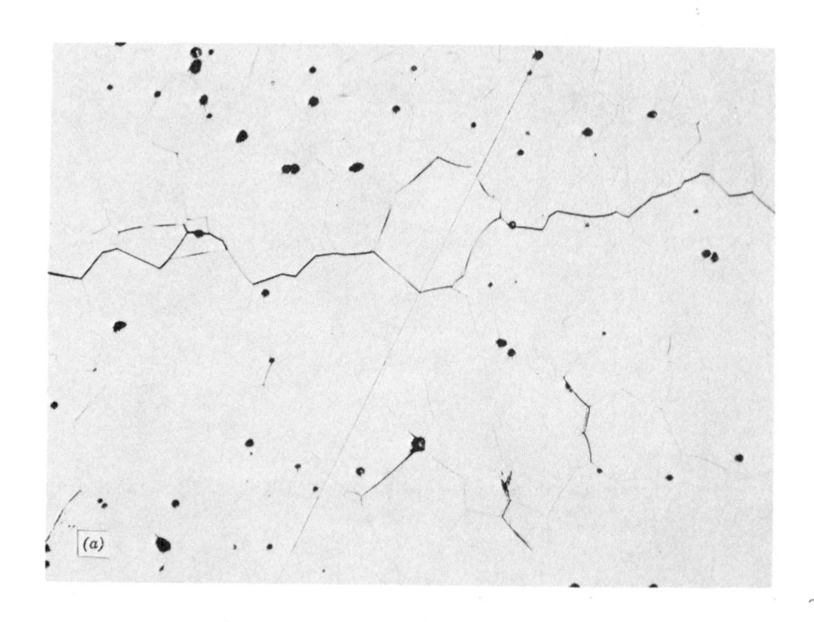

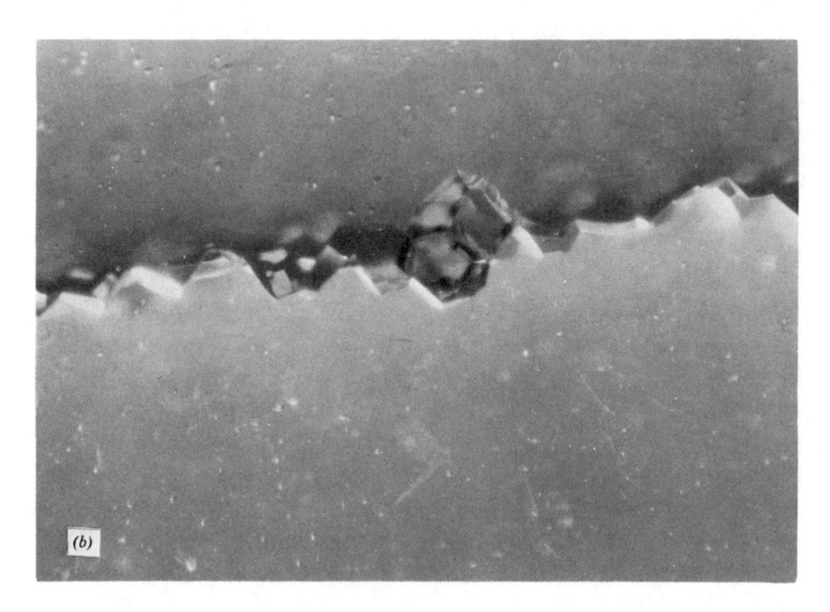

Fig. 15.22. Grain-boundary fracture path in large-grain high-density Al_2O_3. (*a*) Polished section (158×); (*b*) transmitted light (150×). Courtesy R. L. Coble.

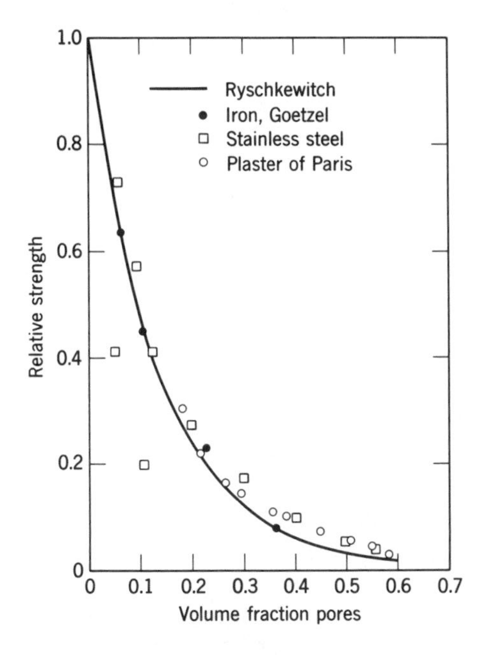

Fig. 15.23. Effect of porosity on the fracture strength of ceramics.

One place where pores can be helpful rather than harmful is in the case of a high-stress gradient. For example, under stresses such as those induced by thermal shock, discussed in Chapter 16, pores tend to hinder crack propagation so that surface checking is obtained rather than complete fracture. This occurs because the stress drops off very rapidly from a high value at the surface to a low value in the interior.

The simplest kind of two-phase system is glazed ceramics. In this case, the glaze is usually weaker than the underlying body; also, fracture is normally observed to start on the surface. Consequently, higher strengths are found with a glaze that has a lower thermal expansion coefficient than the body; this places the glaze under a compressive stress when the glaze body composite is cooled. Initial and final stress distributions in a bend specimen are as shown in Fig. 15.24. The tensile stresses in the body are small compared with the compressive stresses in the glaze so that substantial strengthening can normally be obtained in this way. In some cases fracture initiates at the glaze-body interface, particularly when an increased tensile stress is placed on the body by the glaze expansion and high glaze compression prevents fracture from starting at the surface. Consequently, excessive glaze compression is not desirable.

In multiphase systems various aspects can predominate, depending on the kind of fracture observed. The most common feature affecting

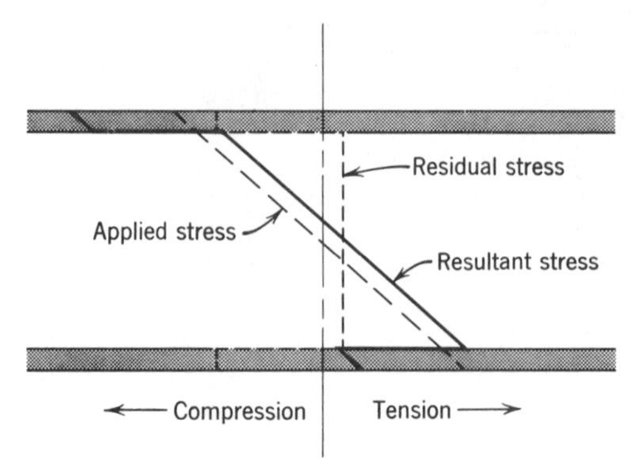

Fig. 15.24. Stresses in glaze and body after cooling and with an applied bending stress.

strength is boundary stresses between phases resulting from different thermal expansion coefficients. These stresses can lead to internal cracks such as are frequently observed for quartz grains in a triaxial porcelain (Figs. 11.11 and 11.13). In this case the effect of a second phase also depends on the particle size, as discussed in Chapter 12. Experimentally, it has been found that the strength of porcelain depends on the particle size of quartz used in their formulation. Fine-grained quartz gives higher-strength porcelains.

A ductile second phase is attractive as a mechanism stopping cracks that are initiated in a brittle phase. This is particularly true in tungsten carbide–cobalt and titanium carbide–nickel composites, which have a high strength and a high toughness, even though brittle fracture is found to occur. Observations of the fracture path indicate that fractures initiate in the carbide grains but stop at the phase boundaries, with considerable plastic deformation and strain hardening of the metal taking place before ultimate fracture occurs.

Suggested Reading

1. J. D. Ferry, *Viscoelastic Properties of Polymers, 2d ed.*, John Wiley & Sons, Inc., New York, 1970.
2. N. G. McCrum, B. E. Read, and G. Williams, *Anelastic and Dielectric Effects in Polymeric Solids*, John Wiley & Sons, Inc., New York, 1967.
3. F. A. McClintock and A. S. Argon, *Mechanical Behavior of Materials*, Addison-Wesley Publishing Company, Inc., Reading, Mass., 1966.
4. R. Houwink and H. K. de Decker, Eds., *Elasticity, Plasticity and Structure of Matter, 3d ed.*, Cambridge University Press, New York, 1971.

5. J. B. Wachtman, Ed., *Mechanical and Thermal Properties of Ceramics*, NBS Special Publication 303, U.S. Government Printing Office, 1969.
6. A. S. Tetelman and A. J. McEvily, *Fracture of Structural Materials*, John Wiley & Sons, Inc., New York, 1967.
7. B. L. Averbach et al., Eds., *Fracture*, The M.I.T. Press, Cambridge, Mass., and John Wiley & Sons, Inc., New York, 1959.
8. R. C. Bradt, D. P. H. Hasselman, and F. F. Lange, Eds., *Fracture Mechanics of Ceramics*, Vols. 1 and 2, Plenum Publishing Corporation, New York, 1974.
9. J. B. Wachtman, Jr., "Highlights of Progress in the Science of Fracture of Ceramics and Glass," *J. Am. Ceram. Soc.*, **58**, (1975).

Problems

15.1. From crack-propagation experiments at low crack velocity in fused silica, the surface energy is estimated as about 500 ergs/cm^2. In contrast, from experiments on fast-moving cracks, the surface energy is estimated as about 4000 ergs/cm^2.
  (*a*) How might these results be reconciled?
  (*b*) How might they be reconciled with the surface energy estimated from a bond-breaking model, 2000 ergs/cm^2?

15.2. When a glass is chemically etched just before testing, its strength usually increases. When it is tested submerged in an etchant solution during testing, its strength usually decreases. Explain.

15.3. Fracture in polycrystalline MgO occurs at a stress level approximately equal to the yield stress determined in single crystals. Give several mechanisms by which crack nucleation may, or has been reported to, occur in polycrystalline MgO. Are grain boundaries intrinsically weaker or stronger than the individual crystals of the polycrystalline mass? Discuss, covering all terms of an appropriate theoretical strength model.

15.4. Glass fibers, 20 microns in diameter by 100 cm long, are loaded with a weight of 100 g in lab air.
  (*a*) What elongation rate would you expect at a viscosity level of 10^{15} P?
  (*b*) If the fibers fractured during the test, Griffith flaws of what size would be indicated?
  (*c*) If the test were conducted in vacuum, what breaking stress might be expected (for the same "worst" flaws)?
  (*d*) If you were an engineer, would you design products around such strength levels? Why or why not?

15.5. Discuss the relevance of a Petch plot ($\sigma \alpha d^{-1/2}$) to room-temperature-strength data of polycrystalline (*a*) NaCl, (*b*) MgO, (*c*) Al_2O_3, and (*d*) AgBr.

15.6. Discuss the causes and effects of microstructure on strength of (*a*) a single-phase body at low temperature and (*b*) a body with 1% second-phase sintering aid at grain boundaries tested at high temperatures. Give representative strength levels and changes with typical variations with grain size and porosity.

15.7. Describe the factors which govern the strengths of commercial products of annealed glass, tempered glass, MgO bricks, and glass-ceramic materials. Also give typical strength values for these products.

15.8. Glass containers are presently fabricated from soda-lime-silicate glasses. Why are these compositions used? What is a typical range of composition? What is the structural effect of the small Al_2O_3 additions used? Why are they used? How might these compositions be strengthened? How might soda-alumina-silicate glasses be easier to strengthen? Why? Why are soda-lime-silicate glasses preferred for the container application?

15.9. Define work of fracture.

15.10. A series of glass rods of circular cross section (1/4 in. diameter) fracture at an average stress of 10,000 psi when bent. Assuming the modulus of elasticity is 10^7 psi, poisson's ratio is 0.3, surface tension is 300 ergs/cm^2, and the fictive temperature is 625°C:

(*a*) What is the average depth of the Griffith flaw?

(*b*) It is desired to coat this rod with another glass of different coefficient of thermal expansion but of essentially the same physical properties in order to double the average strength of the rods. Should the new glass have a higher or lower coefficient than the parent rod? Is there a minimum depth or thickness to the coating? If so, compute the thickness.

(*c*) For your minimum recommended thickness, compute the difference in coefficient of thermal expansion (linear) needed to double the strength of the rods.

(*d*) What will the tensile stress and compressive stress be in the various parts of the coated rod?

15.11. A set of samples of Al_2O_3 doped with Cr_2O_3 indicated that Cr_2O_3 did not affect the strength but porosity did, although there was some scatter in the data.

Wt % Cr_2O_3	% Theoretical Density	Number of Bars Broken	Average Modulus of Rupture (psi)
1.0	97.7	24	33.4
2.0	92.4	25	29.5
5.0	94.8	19	31.0
10.0	93.6	32	27.9
20.0	87.5	32	24.0
50.0	58.1	23	9.5

(*a*) Predict the strength of sintered Al_2O_3 with no porosity.

(*b*) What would be your prediction concerning modulus of elasticity for these samples; that is, how would this measurement vary from sample to sample?

(*c*) What can be said about sphericity of the pores?

15.12. In a series of samples made from powder under similar conditions, groups of samples are fired for 1 hr at successively higher temperature up to a temperature just below the melting point. Modulus-of-rupture measurements reveal an optimum temperature of firing for achieving high strengths.

(*a*) Explain these results, describing the effects which cause this behavior.

(*b*) Draw an expected curve of strength versus firing temperature for the above behavior from $T = 0.3T_m$ to $T = 0.95T_m$.

(*c*) Draw the same curve for samples formed at a higher pressure and thus have higher green density.

15.13. The modulus of elasticity of polycrystalline Al_2O_3 is 6×10^6 psi, and the surface energy is assumed to be about 1000 ergs/cm^2. The fracture stress for sintered alumina varies from 20,000 psi for alumina of 100-micron grain size to 50,000 psi for 5-micron grain size. Show whether or not the Griffith flaw could be the grain boundary.

15.14. In a typical ceramic material the observed average fracture strength and range of values found depends critically on testing conditions. Explain how and why (*a*) loading rate, (*b*) type of test (tensile versus transverse), (*c*) sample size, and (*d*) test atmosphere affect both the average value and range of values observed for a dense high-alumina porcelain. Draw on one graph the expected distribution of strength values for (*a*) large-sample-size tensile tests, (*b*) small-sample-size tensile tests, and (*c*) small-sample-size transverse tests.

16

Thermal and Compositional Stresses

The susceptibility of ceramic materials to thermal stresses and thermal shock failure is one of the main factors limiting their usefulness. For many high-temperature applications, for example, structural properties are satisfactory at the temperature at which the structure is used, but failures often occur at lower temperatures during heating and cooling. Similarly, the glass or teacup that cracks when suddenly heated or cooled is familiar to us all. In addition, there are many applications in which temperature changes result in stresses that are desirable or in which stress relief resulting from temperature changes may be detrimental. All these different situations are usually discussed as separate problems; here we first consider the problem of temperature changes and resulting stresses in a general way and only later consider specific applications.

Among these applications, two of particular interest are related to glass technology. Annealing of glass objects used for structural or optical purposes is primarily aimed at removing stresses. However, sometimes desirable patterns of residual stresses can be used to improve glass properties. Both applications involve the same fundamentals as those used in discussing thermal shock failures and are considered in the present chapter. Also to be discussed are other techniques which may be used to strengthen ceramic bodies.

16.1 Thermal Expansion and Thermal Stresses

As discussed in Chapter 12, the thermal expansion coefficients of ceramic materials vary over a wide range. Typically, the length of a 10-in. tube increases by about 0.1 in. when heated to 1000°C. If the body is homogeneous and isotropic, no stresses result from this thermal expansion. However, if the sample is restrained from expanding, by rigid cold supports, for example, considerable stresses can be developed. Under these conditions the stresses are the same as if the sample were allowed to expand freely and then compressed back to its original size by an applied restraining force. The stress required is proportional to the elasticity of the material and the elastic strain, which is equal to the product of the thermal expansion and temperature change. For a perfectly elastic rod restrained in only one direction,

$$\sigma = -Ea(T' - T_0) \tag{16.1}$$

where E is the modulus of elasticity, a the linear expansion coefficient, T_0 the initial temperature, and T' the new temperature. For the tube with $a(T' - T_0) = 0.01$ in./in. and $E = 20 \times 10^6$ psi, the resulting stress is about 200,000 psi, above the normal crushing strength for most ceramic materials.

On heating, stresses resulting from restraints are compressive, since the body tends to expand against the restraining member. On cooling, similar tensile stresses can result. Tensile stresses are developed in a glaze, for example, by firing on an underlying body of smaller thermal expansion coefficient than the glaze (discussed in Chapter 12). In the same way that stresses arise in glazes and porcelain enamels as a result of differences in expansion coefficients, the individual crystals in an anisotropic polycrystalline ceramic or the different phases in a multiphase ceramic have different expansion coefficients but are restrained in the same body. This restraint of expansion also leads to stresses (discussed in Chapter 12).

16.2 Temperature Gradients and Thermal Stresses

In addition to temperature changes under conditions that restrain free expansion, stresses also result from temperature gradients within a body when these are such that free expansion of each volume element cannot occur. Here again, the factor that leads to stresses is the restraint on free expansion. Temperature gradients along the axis of a furnace tube, for example, do not give rise to thermal stresses, since the tube is free to expand along the axis without incompatible strains. If this were not so, commonly used tube furnaces would regularly fail from thermal stresses.

Frequently, however, the temperature distribution is such that free expansion of each volume element would separate the individual volume elements so that they could not be fitted together. Since they are constrained in the same body, stresses arise. These stresses can be calculated for perfectly elastic bodies, to which many ceramics are a good approximation, from the elastic moduli, the expansion coefficients, and the temperature distribution.

Let us consider the stresses arising in a large glass plate plunged from boiling water at 100°C into an ice bath at 0°C. Under these conditions the rate of heat transfer from the surface is high; the surface reaches the new temperature instantly, but the interior remains at a uniform value, $T_0 = 100°$. The surface, if free, would contract by an amount $a(T_0 - T') = 100a$; however, it is restrained by the bulk which remains at $T_0 = 100°$, and tensile stresses arise in the surface. For stress equilibrium the surface stress must be balanced by a compressive stress in the interior.

The overall size of the sample is determined by its average temperature T_a. The stress at any point depends on the difference in temperature between that point and the average; this gives the strain at that point and fixes the stress. The restraint on free strain is similar to that in Eq. 16.1 and leads to a stress in an infinite slab given by

$$\sigma_y = \sigma_z = \frac{Ea}{1-\mu}(T_a - T) \tag{16.2}$$

For the plate plunged into ice water, the maximum stress at the surface occurs at zero time when $T_a = 100°$, $T_s = 0°$. For a typical soft glass, $E = 10^7$ psi, $a = 10 \times 10^{-6}$ in./in. °C, $\mu = 0.20$, and $\sigma_y = \sigma_z = 12{,}500$ psi. This is well above the fracture stress (about 10,000 psi), so that we would expect soft glass to fracture under these conditions. In contrast, for Pyrex glass the expansion coefficient is about 3×10^{-6}/°C, so that this treatment does not normally lead to failure but is a borderline case. For fused quartz, $a = 0.5 \times 10^{-6}$/°C, so that no dangerous stresses are developed.

In addition to sudden changes in temperature, a steady rate of temperature change can also lead to temperature gradients and thermal stresses (Fig. 16.1). When the surfaces of a plate are cooled at a constant rate, a temperature distribution results which is parabolic. The surface tempera-

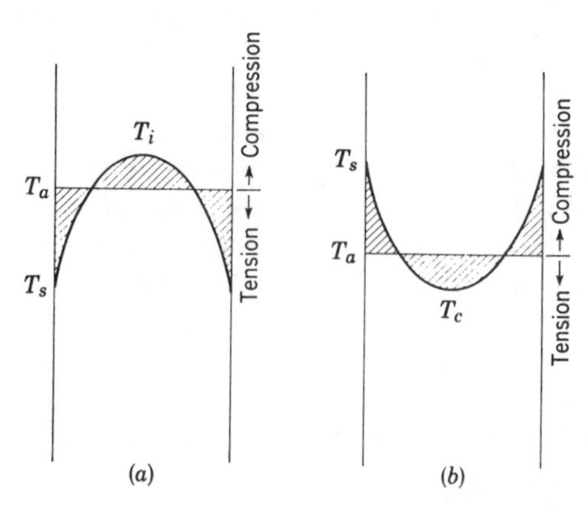

Fig. 16.1. Temperature and stress distributions for plate that is (a) cooled from the surface and (b) heated from the surface.

ture is lower than the average temperature, and surface tensile stresses result; the center temperature is higher than the average temperature, so that there are central compressive stresses. If the sample is heated, these stresses are just reversed. Since, as discussed in Chapter 15, ceramic materials are much weaker in tension than under compression, failure occurs at the surface during cooling but can occur either from the center tensile stress or from surface compressive stresses during heating. Stresses resulting from heating or cooling samples can be described for a variety of simple shapes in terms of the average temperature and the temperature extremes. These temperature extremes are usually at the surface and the center; maximum stresses for different shapes are as given in Table 16.1.

Table 16.1. Surface and Center Stresses in Various Shapes

Shape	Surface	Center
Infinite slab	$\sigma_x = 0$	$\sigma_x = 0$
	$\sigma_y = \sigma_z = \dfrac{Ea}{1-\mu}(T_a - T_s)$	$\sigma_y = \sigma_z = \dfrac{Ea}{1-\mu}(T_a - T_c)$
Thin plate	$\sigma_y = \sigma_z = 0$	$\sigma_y = \sigma_z = 0$
	$\sigma_x = aE(T_a - T_s)$	$\sigma_x = aE(T_a - T_c)$
Thin disk	$\sigma_r = 0$	$\sigma_r = \dfrac{(1-\mu)Ea}{2(1-2\mu)}(T_a - T_c)$
	$\sigma_\theta = \dfrac{(1-\mu)Ea}{1-2\mu}(T_a - T_s)$	$\sigma_\theta = \dfrac{(1-\mu)Ea}{2(1-2\mu)}(T_a - T_c)$
Long solid cylinder	$\sigma_r = 0$	$\sigma_r = \dfrac{Ea}{2(1-\mu)}(T_a - T_c)$
	$\sigma_\theta = \sigma_z = \dfrac{Ea}{1-\mu}(T_a - T_s)$	$\sigma_\theta = \sigma_z = \dfrac{Ea}{2(1-\mu)}(T_a - T_c)$
Long hollow cylinder	$\sigma_r = 0$	$\sigma_r = 0$
	$\sigma_\theta = \sigma_z = \dfrac{Ea}{1-\mu}(T_a - T_s)$	$\sigma_\theta = \sigma_z = \dfrac{Ea}{1-\mu}(T_a - T_c)$
Solid sphere	$\sigma_r = 0$	
	$\sigma_t = \dfrac{Ea}{1-\mu}(T_a - T_s)$	$\sigma_t = \sigma_r = \dfrac{2Ea}{3(1-\mu)}(T_a - T_c)$
Hollow sphere	$\sigma_r = 0$	$\sigma_r = 0$
	$\sigma_t = \dfrac{aE}{1-\mu}(T_a - T_s)$	$\sigma_t = \dfrac{aE}{1-\mu}(T_a - T_c)$

Determination of the actual stress in a ceramic material under given conditions of heat transfer requires, first, an analysis of the temperature distribution and, then, from Table 16.1 or similar relations, calculations of the resulting stresses. Perhaps the simplest stress, already discussed, occurs when the surface temperature is instantly changed without changing the average temperature from its initial value. Another stress of practical importance occurs when the surface is cooled or heated at a constant rate. Here the temperature distribution is parabolic for plate geometries; the average temperature is intermediate between the center and surface temperature so that for a half thickness r_m, a rate of cooling of ϕ°C/sec, and a thermal diffusivity $k/\rho c_p$,

$$\sigma_s = \frac{E\alpha}{1-\mu} \frac{\phi r_m^2}{3k/\rho c_p} \tag{16.3}$$

Similar expressions for other geometries are given in Table 16.2.

Table 16.2. Temperature Differences between Surface and Center of Various Shapes Cooled at a Constant Rate ($\phi = dt/d\theta$)

Shape	$T_c - T_s$
Infinite plate, half thickness = r_m	$0.50 \dfrac{\phi r_m^2}{k/\rho c_p}$
Infinite cylinder, radius = r_m	$0.25 \dfrac{\phi r_m^2}{k/\rho c_p}$
Cylinder, half length = radius = r_m	$0.201 \dfrac{\phi r_m^2}{k/\rho c_p}$
Cube, half thickness = r_m	$0.221 \dfrac{\phi r_m^2}{k/\rho c_p}$

When a surface is cooled by an air blast or by immersion in a new environment so that the average temperature changes along with the surface temperature, analytical calculations are more difficult and depend on the ratio of the product of the surface-heat-transfer coefficient h and the sample dimension r_m to the thermal conductivity k. This nondimensional ratio is called Biot's modulus, $\beta = r_m h/k$. Analysis is simplified by using a nondimensional stress equal to the fraction of the stress that would result from infinitely rapid surface cooling; that is, for a plate sample where σ is the actual stress observed,

$$\sigma^* = \frac{\sigma}{Ea(T_0 - T')/(1-\mu)} \tag{16.4}$$

This stress varies with time, as shown in Fig. 16.2. For $\beta = \infty$ it is a maximum at time equals zero, as indicated there. For other values of the

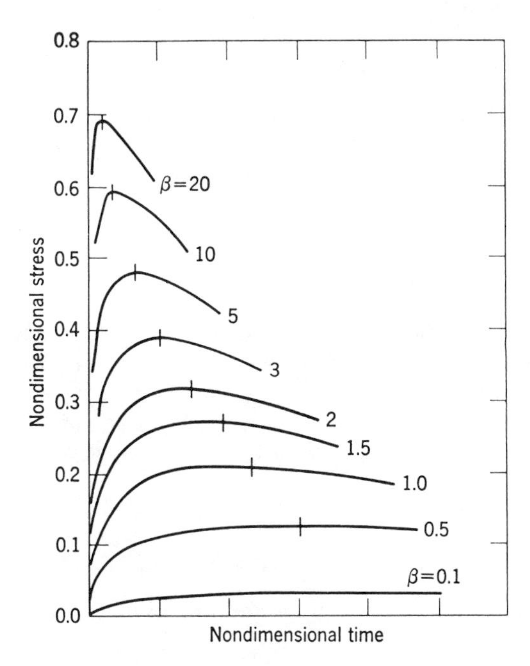

Fig. 16.2. Variation of nondimensional surface stress with dimensionless time for an infinite flat plate with different values of Biot's modulus β.

surface-heat-transfer coefficient, the stress reaches a maximum at some particular value of time (Fig. 16.2). Values of the surface-heat-transfer coefficient for various conditions are illustrated in Table 16.3.

For relatively low rates of surface-heat-transfer coefficients, which are commonly observed in convection and radiation conditions, S. S. Manson* has found that

$$\sigma^*_{\max} = 0.31\beta = 0.31\frac{r_m h}{k} \tag{16.5}$$

This relationship gives the maximum dimensionless stress in terms of the

*NACA Tech. Note 2933, July, 1937.

Table 16.3. Values of Surface Heat-Transfer Coefficient h

Conditions	h (Btu/hr/ft^2/°F)	h (cal/sec/cm^2/°C)
Air flow past cylinder:		
Flow rate 60 lb/sec/ft^2	190	0.026
Flow rate 25 lb/sec/ft^2	90	.012
Flow rate 2.5 lb/sec/ft^2	20	.0027
Flow rate 0.025 lb/sec/ft^2	2	.00027
Radiation to 0°C from 1000°C	26.0	.0035
Radiation to 0°C from 500°C	7.0	.00095
Water quenching	1000–10,000	.1–1.0
Jet turbine blades	35–150	.005–0.02

heat-transfer conditions. As shown in Fig. 16.2, the time to maximum stress increases as the dimensionless heat-transfer coefficient, Biot's modulus, decreases.

For a detailed discussion of the thermal stresses which result under a variety of heat-flow conditions for a variety of specimen geometries, the reader should consult reference 1.

16.3 Resistance to Thermal Shock and Thermal Spalling

When ceramic materials are subjected to a rapid change in temperature (thermal shock), substantial stresses develop, as discussed in the last section. Resistance to weakening or fracture under these conditions is called thermal endurance, thermal stress resistance, or thermal shock resistance. The effect of thermal stresses on different kinds of materials depends not only on the stress level, stress distribution in the body, and stress duration but also on material characteristics such as ductility, homogeneity, porosity, and pre-existing flaws. Consequently, it is impossible to define a single *thermal-stress-resistance factor* that is satisfactory for all situations.

Perhaps the simplest example, one of considerable importance, is that of an ideally elastic material which fractures when the surface stress reaches some particular level. Glasses, porcelain, whitewares, and special electronic and magnetic ceramics represent classes of materials whose resistance to thermal shock should be well represented by this criterion, and for them, the temperature conditions for fracture can be simply calculated.

For immersion in a liquid bath with rapid rates of heat transfer, we have already calculated the surface stress. Rearranging Eq. 16.2 and calling $T_0 - T'$ for fracture ΔT_f, fracture occurs when the temperature difference is such that the fracture stress is reached:

$$\Delta T_f = \frac{\sigma_f(1-\mu)}{Ea} \tag{16.6}$$

For other shapes, as indicated in Table 16.1, other geometrical constants are required, so that in general

$$\Delta T_f = \frac{\sigma_f(1-\mu)}{Ea} S = RS \tag{16.7}$$

where S is a shape factor, σ_f is the fracture stress, and $R = \sigma_f(1-\mu)/Ea$ is a material constant that can be described as a material resistance factor for thermal stresses. High fracture stress, low modulus of elasticity, and low thermal expansion coefficient indicate a good resistance to thermal stress failure on this criterion of failure.

It is worth noting that the relationship of Eq. 16.7 only applies directly when the quench is so rapid that the surface temperature reaches its final value before the average temperature changes. This occurs to a good approximation when $\beta = r_m h/k$ is equal to or greater than 20. For a glass with a water quench, for example, k equals 0.004 and h equals 0.4, so that r_m must be equal to or greater than about 0.2 cm for this relationship to apply.

Under conditions when the rate of heating is not so high, we can combine Eqs. 16.4 and 16.5:

$$(\sigma^*_{\max})_f = \frac{\sigma_f}{Ea(T_0 - T')_f/(1-\mu)} = 0.31\frac{r_m h}{k} \tag{16.8}$$

By recombining,

$$\Delta T_f = \frac{k\sigma_f(1-\mu)}{Ea}\frac{1}{0.31 r_m h} \tag{16.9}$$

In general, if we define a second thermal-stress-resistance factor $R' = k\sigma_f(1-\mu)/Ea$, again with the failure criterion of fracture occurring when the thermal stress reaches the fracture stress, then

$$\Delta T_f = R'S\frac{1}{0.31 r_m h} \tag{16.10}$$

Here the material constant includes the thermal conductivity, and the maximum-quench-temperature difference that is able to be withstood is inversely proportional to the size of the sample; that is, the situation is somewhat more complex and requires a more careful analysis of the conditions for use.

The maximum quench temperatures that can be withstood on this criterion of failure for different materials and different conditions of heat transfer are illustrated in Fig. 16.3. Data there are calculated for typical sintered compositions, normally containing about 5% porosity, and at an average temperature of about 400°. As illustrated for aluminum oxide, temperature alters the curve considerably but it does not change the shape. Similarly, changes of properties can have a considerable effect. For example, the calculation in Fig. 16.3 for aluminum oxide uses a strength value of about 20,000 psi, which is typical of a slightly porous sintered ware. However, alumina porcelains have been made having strengths of the order of 125,000 psi. This body has a thermal stress

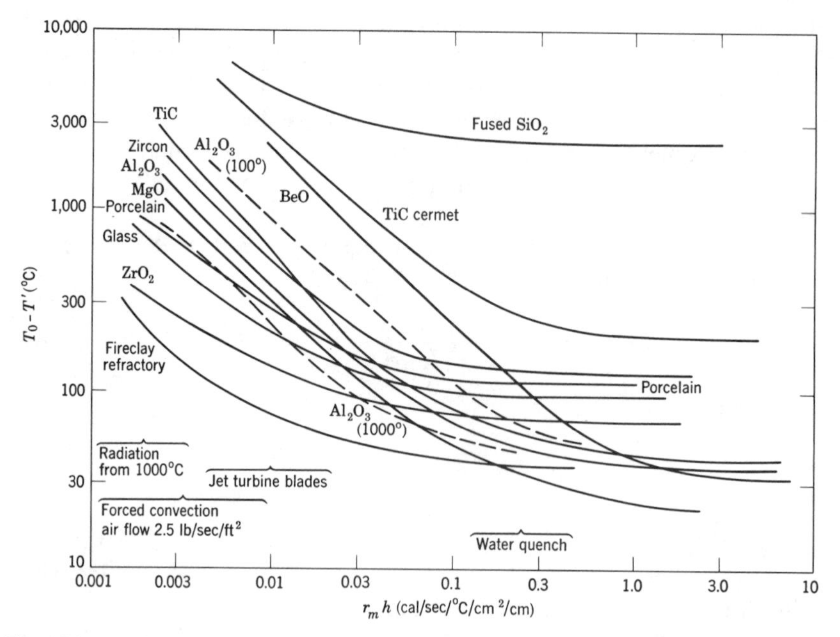

Fig. 16.3. Variation in quench temperature causing fracture for different materials under different conditions of heat transfer, assuming that failure occurs when the thermal stress reaches the fracture stress. Curves are calculated from material properties at 400°C. Dashed curves for Al_2O_3 are calculated from material properties at 100°C and 1000°C, as an indication of temperature effects.

resistance about six times as good as indicated in Fig. 16.3, so that results shown there should be regarded as illustrative rather than definitive.

An important feature of Fig. 16.3 is that the curves for a number of materials cross. This occurs for magnesium oxide and porcelain, for example, at a value of $r_m h = 0.03$. For moderate rates of surface heat transfer, magnesium oxide is better; for more rapid surface-heat-transfer coefficients or larger sample sizes porcelain has better stress resistance. Consequently, we cannot say that we are able to arrange a list of materials in a single order of thermal stress resistance, even for a given criterion of failure.

Other conditions of heat transfer are frequently encountered, such as steady-state heat flow out through the walls of a hollow cylinder, constant rate of heating from the surface, and cooling by radiation from an elevated temperature. Each of these requires a different material constant and leads to a different list of the relative advantages of different materials. For these applications it is essential to understand the conditions of heat transfer.

All the above discussion has been predicated on the assumption that failure occurs when the thermal stress reaches the fracture stress. Such an approach directs attention to the conditions which govern the nucleation of fracture. For porous materials (such as most refractories) and for nonhomogeneous materials (such as cermets), however, the large stress gradient and short stress duration mean that fracture, even though initiated at the surface, may be stopped by a pore or grain boundary or metal film before resulting in complete failure. The question "When is a crack a failure?" cannot be answered unequivocally but depends on the particular material and application. In a refractory, for example, whose main function is as a high-temperature corrosion-resistant heat container, surface cracks cause no difficulties, but it is desirable to avoid thermal spalling—the actual breaking away of pieces of brick as a result of thermal stresses. For these materials increased porosity (which decreases both R and R', since the conductivity is decreased and the strength is decreased to a greater extent than the elasticity) leads to better spalling resistance, and a porosity of 10 to 20% is often optimum.

These considerations lead to a second approach to the problem of thermal shock—treating the conditions which govern the propagation of cracks rather than their nucleation. This approach has been advanced by D. P. H. Hasselman,* who noted that the driving force for crack propagation is provided by the elastic energy stored at the moment of fracture. Considering further that crack propagation under conditions of

J. Am. Ceram. Soc., **52**, 600 (1969), and *Int. J. Fract. Mech.*, **7**, 157 (1971).

thermal stress generally occurs in the absence of external forces, a treatment of crack extension under conditions of constant deformation or strain (fixed grips) provides useful insight into the problem. Adopting a worst-case model in which the entire body is stressed to the maximum value of thermal stress, Hasselman estimated the critical temperature difference ΔT_c required for crack instability as

$$\Delta T_c = \left[\frac{\pi\gamma_{\text{eff}}(1-2\mu)^2}{2E_0\alpha^2(1-\mu^2)}\right]^{1/2}\left[1+\frac{16(1-\mu^2)Nl^3}{9(1-2\mu)}\right]l^{-1/2} \qquad (16.11)$$

where crack propagation is assumed to occur by the simultaneous propagation of N cracks per unit volume, E_0 is the Young's modulus of the crackfree material, γ_{eff} is the fracture surface work and l is the crack length. This relation is illustrated by the solid lines in Fig. 16.4. As shown there, the region of crack instability is in general bounded by two values of crack length.

For initially short cracks (lengths to the left of the minima in Fig. 16.4), the rate of energy release after initiation of crack propagation exceeds the surface energy of fracture, and the excess energy is transformed into the kinetic energy of the moving crack. When such a crack reaches the length given by Eq. 16.11, it still has kinetic energy and continues to propagate

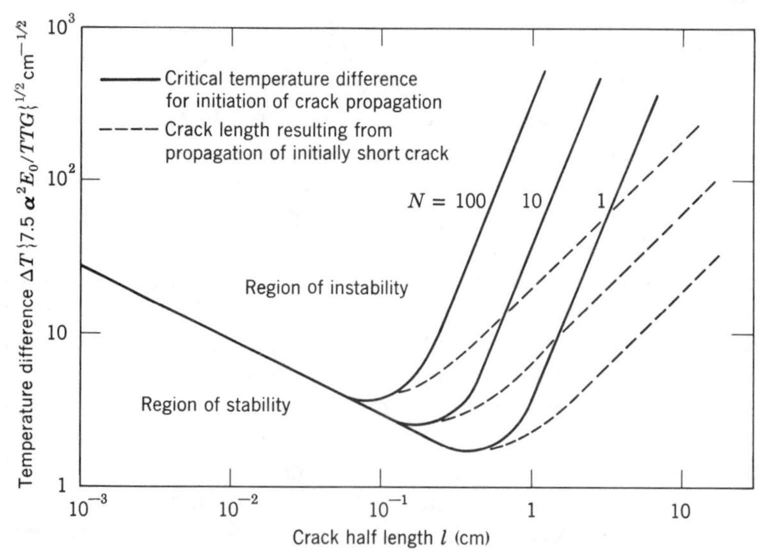

Fig. 16.4. Thermal strain required to initiate crack propagation as a function of crack length and crack density N. Poisson's ratio taken as 0.25. From D. P. H. Hasselman, *J. Am. Ceram. Soc.*, **52**, 600 (1969).

until the released strain energy equals the total surface energy of fracture. This condition is fulfilled by the final crack lengths shown by the dotted curves in Fig. 16.4. These final crack lengths are subcritical with respect to the critical temperature difference required for their initiation, and a finite increase in this temperature difference is required before the cracks again become unstable. Finally, in contrast to short cracks which propagate with significant kinetic energy, cracks with initial lengths to the right of the minima in Fig. 16.4 are expected to propagate in a quasi-static fashion.

For a material with small cracks, which propagate kinetically on initiation of fracture, the crack length is expected to change with severity of quench, as shown schematically in Fig. 16.5; the corresponding variation of strength is shown schematically in Fig. 16.6. For thermal stresses less than that required to initiate fracture, no change in strength or crack length is anticipated. At the critical stress for fracture, the cracks propagate kinetically, their length changes rapidly to a new value, and the strength shows a corresponding abrupt decrease. Since the cracks are then subcritical, the temperature difference must be increased above that required for fracture initiation, ΔT_c, before the cracks again propagate, and over the range between ΔT_c and $\Delta T_{c'}$, the range in which no further crack propagation occurs, no change in strength is expected. For more severe quenches ($\Delta T > \Delta T_{c'}$), the cracks grow quasi-statically, and the strength correspondingly decreases. Curves of the form of Fig. 16.6 have been observed in a number of studies of thermal shock damage. Figure 16.7, for example, shows results obtained on polycrystalline Al_2O_3 samples of various grain sizes.

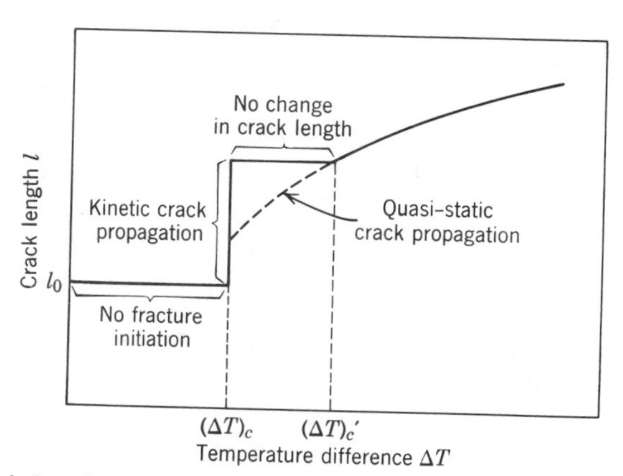

Fig. 16.5. Crack length as a function of temperature difference ΔT. From D. P. H. Hasselman, *J. Am. Ceram. Soc.*, **52**, 600 (1969).

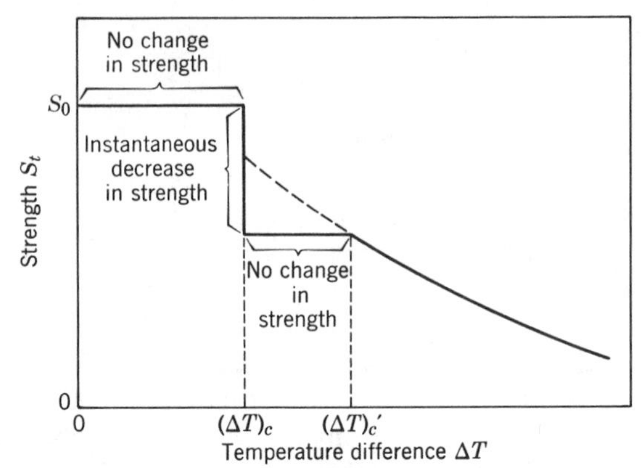

Fig. 16.6. Strength as a function of temperature difference ΔT. From D. P. H. Hasselman, *J. Am. Ceram. Soc.*, **52**, 600 (1969).

There are, then, two principal approaches to designing and selecting materials for resistance to thermal shock. The first, appropriate for glasses, porcelain, whitewares, electronic ceramics, and so on, involves the avoidance of fracture initiation. For these, the appropriate thermal-shock-resistance parameters are, depending on the heat-flow conditions,

$$R = \frac{\sigma_f(1-\mu)}{Ea} \tag{16.12a}$$

or

$$R' = \frac{k\sigma_f(1-\mu)}{Ea} \tag{16.12b}$$

For avoiding fracture initiation by thermal shock, the favorable material characteristics include high values of strength and thermal conductivity and low values of modulus and thermal expansion coefficient.

The second approach, appropriate for materials such as refractory bricks, involves the avoidance of catastrophic crack propagation. The thermal-shock-resistance parameters for this approach are the minimum elastic energy at fracture available for crack propagation:

$$R''' = \frac{E}{\sigma_f^2(1-\mu)} \tag{16.13a}$$

and the minimum distance of crack propagation on the initiation of thermal stress failure:

$$R'''' = \frac{E\gamma_{\text{eff}}}{\sigma_f^2(1-\mu)} \tag{16.13b}$$

From these parameters, the favorable material characteristics for

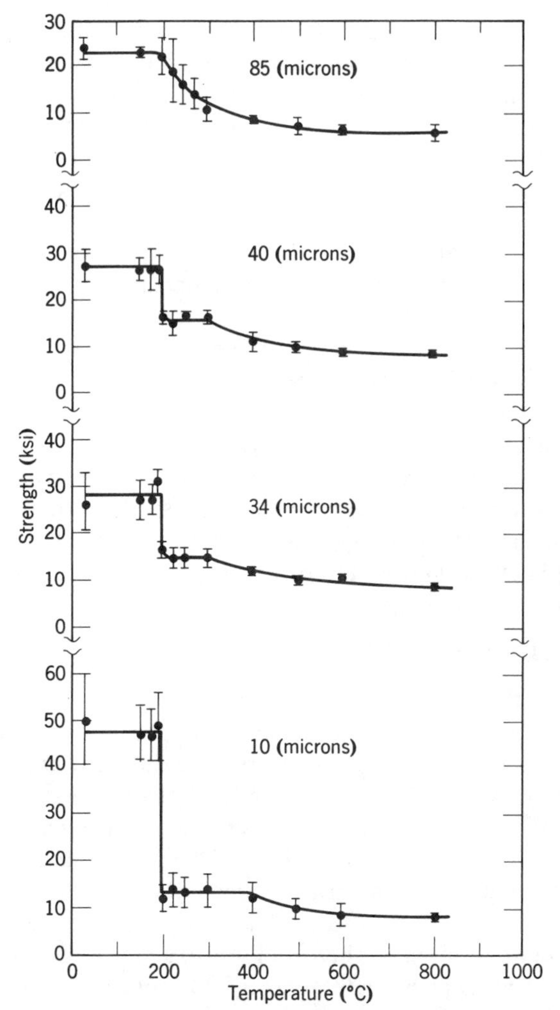

Fig. 16.7. Room-temperature moduli of rupture of Al_2O_3 specimens of various grain sizes as functions of quench temperature. From T. K. Gupta, *J. Am. Ceram. Soc.*, **55**, 249 (1972).

minimizing the extent of crack propagation are high values of modulus and fracture surface work and low values of strength. These requirements of modulus and strength stand in direct contrast to those appropriate for avoiding the initiation of fracture; hence, selection of material characteristics to avoid fracture initiation has deleterious effects on the damage resulting from fracture if it does occur.

In cases in which it can be tolerated, resistance to catastrophic crack propagation can be improved by introducing enough cracks of sufficiently large size so that crack propagation takes place quasi-statically (crack sizes near the minima in Fig. 16.4 seem ideal) or more generally by introducing microstructural heterogeneities of any form which serve as stress concentrators in the material. In this way, fracture may take place locally in the material, but catastrophic failure is avoided because of the small average stress in the material.

Recent work has confirmed the importance of microstructure in affecting the extent of thermal shock damage. In particular, blunt flaws such as those resulting from intergranular shrinkage cracking have been found to provide marked resistance to catastrophic failure; relatively sharp initial cracks from surface impacts lead to failure under less severe conditions of thermal stress. Intergranular shrinkage voids in Al_2O_3–TiO_2 ceramics serve to blunt initially sharp cracks and prevent their propagation, thus providing outstanding resistance to thermal shock damage. The intentional introduction of such flaws by using anisotropic thermal expansion behavior offers a promising route for avoiding catastrophic thermal shock failure in applications in which tensile strength is not of prime concern.

16.4 Thermally Tempered Glass

The stresses that develop on cooling a glaze or enamel, as discussed in Chapter 12, result when an initially uniform-temperature stressfree body is cooled to a new different temperature. The different expansions of the glaze and body give rise to different contractions and result in stresses. In Section 16.3 thermal stresses were described as arising when an initially stressfree uniform-temperature sample is brought into a new temperature environment, resulting in temperature gradients and thermal stresses. Just the reverse of this process is to start with a stressfree material having a nonuniform temperature and then cool this to a new uniform temperature. The temperature changes undergone by different parts of the body are different, and residual stresses result.

Residual compressive stresses are intentionally induced in the surface regions of glasses by carrying out a controlled cooling procedure. The process, known as thermal tempering, is widely used in the production of safety glass for windows and eyeglasses. The technique involves heating the glass body to temperatures above the glass transition region, but below the softening point, and rapidly cooling the surface. The cooling is most often effected by jets of cold air, although oil baths are sometimes used. The outside of the glass, which initially cools more rapidly than the interior, becomes rigid while the interior is still molten. The temperature difference between the surface and midplane of the glass generally reaches a maximum within a few seconds after quenching. With continued cooling, the interior contracts at a greater rate with changing temperature than the rigid outside (recall the change in thermal expansion coefficient on going through the glass transition) until an isothermal state is again attained at room temperature.

The initial large thermal contraction of the surface relative to the midplane tends to produce tensile stresses in the surface and compressive stresses in the midplane. In an elastic solid, such as that illustrated in Fig. 16.1, these stresses appear and then are cancelled by stresses of opposite sign during the later stages of cooling. In the case of glass, however, stresses can relax at high temperatures; the stresses induced in the later

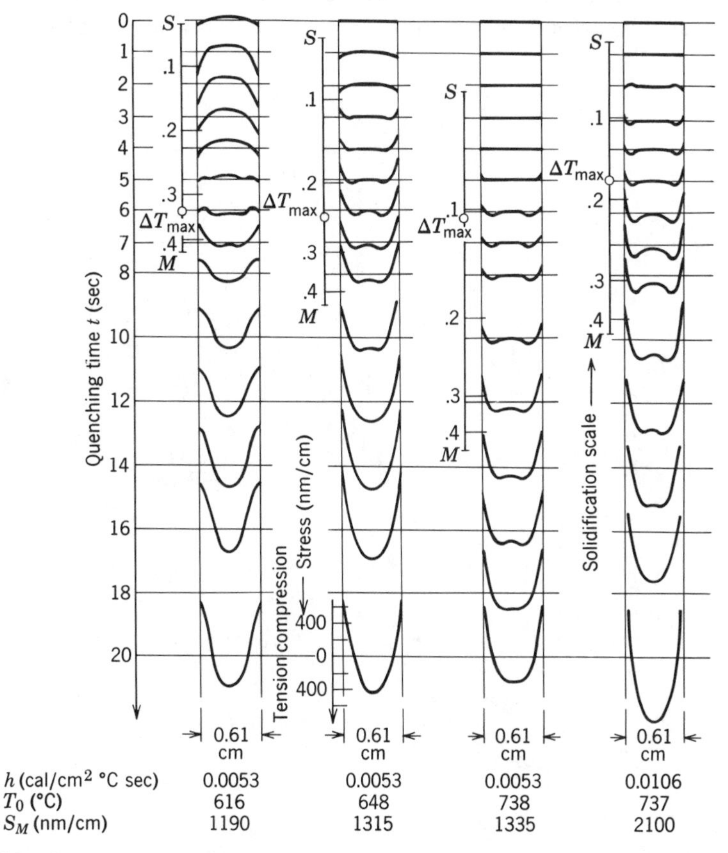

Fig. 16.8. Stress distributions through thickness of plate glass during quenching for different quench conditions. From O. S. Narayanaswamy and R. Gardon, *J. Am. Ceram. Soc.*, **52**, 554 (1969).

stages of cooling remain. The resulting stress profile is nearly parabolic in shape, with the magnitude of the compressive stress at the surface being approximately twice the maximum tension in the interior of the glass. For a flat plate cooled equally from both sides, the maximum tension occurs in the central plane of the plate. Stress profiles observed during different stages of the quenching process for different heat-transfer conditions are shown in Fig. 16.8.

The process of thermal tempering has been described quantitatively by a number of authors (see reference 2). Among the variables in thermal tempering, two of the most important are the initial temperature from which the glass is quenched, T_0, and the rate at which heat is removed from the glass surface. The latter quantity is generally related to the heat-transfer coefficient h, which increases with rate of flow of the cooling air across the surface. As illustrated in Fig. 16.9, the degree of temper S_M

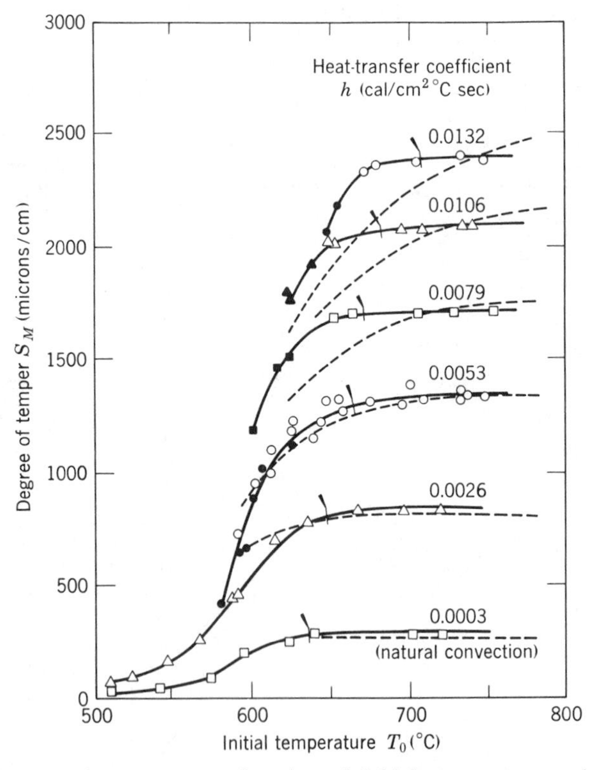

Fig. 16.9. Degree of temper as a function of initial temperature and quench rate (heat-transfer coefficient). From R. Gardon in *Proceedings of VII International Congress on Glass*, ICG, Brussels, 1966, pp. 79–83.

for a given heat-transfer coefficient increases with increasing initial temperature but eventually approaches a plateau at large values of T_0. The degree of temper also increases with increasing h, and the highest tempers are achieved with large values of h and large temperature differences over which cooling takes place.

Tempered glass is useful because, as discussed in Chapter 15, failure normally occurs under an applied tensile stress, and failure in ceramics is almost always initiated at the surface. When a residual compressive stress is placed on a surface, the applied stress must first overcome this residual compression before the surface is brought into tension at which failure can occur. The residual stress and applied stress in a bend test, and the resulting stress distribution, are as illustrated in Fig. 16.10 for a plate sample. This figure shows how residual stresses can counteract the effect of an applied load and give rise to higher strength levels.

By use of thermal tempering, the long-time mean strength of soda-lime-silicate glass can be raised to the range of about 20,000 psi. This is sufficient to permit its use in large doors and windows as well as safety lenses, but for many applications, still higher strengths are desired. In these cases, use is made of an alternative method of achieving surface compression, termed chemical strengthening, which is discussed in Section 16.6.

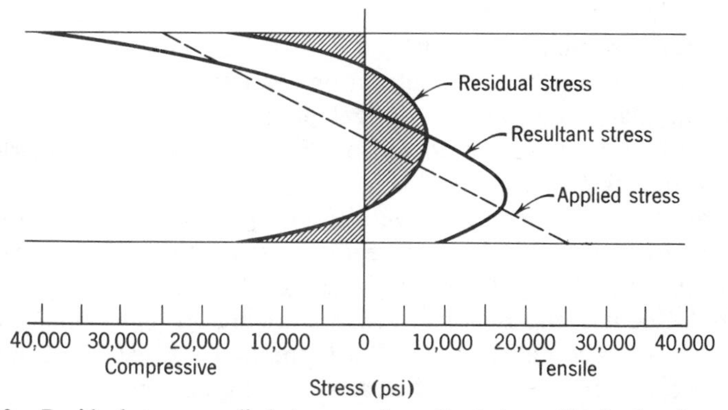

Fig. 16.10. Residual stress, applied stress, and resultant stress distribution for transverse loading of a tempered glass plate.

16.5 Annealing

Although residual stresses are sometimes desirable, as for tempered glass, they are more commonly to be avoided. Unless carefully controlled, there will be areas with high levels of concentrated residual stresses including residual tensile stresses; these lead to variation in properties and premature failure. Particularly during the forming of glasses by operations such as drawing, blowing, and pressing, the residual stress distribution will be highly variable.

Residual stresses are also a cause of optical stress birefringence. Consequently, for optical uses internal stresses must be reduced to even lower values than required for window or container glass. In addition, it is necessary that the structure of optical glass be stabilized at a low temperature to prevent changes in the refractive index over long periods of use.

Internal stresses in optical glasses are commonly measured in terms of the resulting birefringence, or difference in refractive index between light rays polarized in the plane of, and normal to, the applied stress. For a typical glass, the *stress-optical* coefficient is about 0.2 micron retardation difference per centimeter of path length for an applied stress of 1 psi. For *coarse* optical annealing a birefringence of about 75 microns/cm is satisfactory, corresponding to a residual stress of about 375 psi. For *fine* annealing a birefringence value of 10 microns/cm or less must be obtained, corresponding to residual stresses less than 50 psi. Stress-optical coefficients for a number of glasses are collected in Table 16.4.

In addition to minimizing stresses and stress birefringence, annealing is required to stabilize the glass structure to avoid property variations. Rapidly cooled glass has a greater tendency to contract on standing at room temperature than does slowly cooled glass; also, since its viscosity is abnormally low, the rate of contraction is greater. This is important for thermometer glass as well as for optical glasses. Similarly, when different parts of an optical glass are cooled at different rates, nonuniformity in physical properties results.

There are three problems involved in setting up an annealing schedule. First, the large residual stresses resulting from the large temperature gradients present during forming (pressing, blowing, drawing) must be eliminated. Second, the residual stresses arising during cooling must be kept within an acceptable level. Third, thermal stresses that are large enough to cause fracture must not arise during cooling.

One general method used for annealing is to heat the sample to a uniform temperature in the annealing range (near the transformation temperature) for a sufficient time to relieve any stresses initially present. Then the glass is cooled through the critical viscosity range at a rate sufficiently slow to keep residual stresses above some fixed limit from developing. Finally, the glass is rapidly cooled after it is below the critical temperature range. The other general method of annealing, widely used in the flat-glass industry, does not involve reheating the glass to remove

Table 16.4. Stress-Optical Coefficients for Different Glasses

Glass Type	Stress-Optical Coefficient[a] $B, \frac{\text{m}\mu/\text{cm}}{\text{psi}}$	B, brewsters
Fused silica	0.24	3.5
96% silica (Vycor)	0.26	3.65
Soda-lime-silica	0.18	2.5
Lead-alkali silicate		
40% PbO	0.19	2.7
80% PbO	−0.07	−1.0
Low expansion borosilicate	0.27	3.9

[a] Stress difference,

$$\sigma_y - \sigma_z = 10^{13} \frac{(n_z - n_y)/n}{B} \text{ dynes/cm}^2$$
$$= r/B \text{ psi}$$

where r is the retardation difference (mμ/cm) and $(n_z - n_y)/n$ is the fractional retardation difference (cm/cm).

preexisting stresses but rather involves cooling freshly formed and largely stress-free material in such a way that unduly large permanent stresses are avoided.

Analysis of the rate of decrease of stress at constant temperature during annealing is complicated by the fact that this rate is dependent on the past thermal history, which varies from spot to spot in the ware with normal glass operations. The variations with thermal history result from the fact that the glass structure and viscosity change with past history in the transition range, as discussed in Chapters 3 and 14.

If the rate of stress relief is taken as proportional to the stress

$$\frac{d\sigma}{dt} = -\frac{1}{\tau}\sigma \tag{16.14a}$$

and if the relaxation time τ is assumed independent of time, one obtains

$$\sigma = \sigma_0 \exp\left(-\frac{t}{\tau}\right) \tag{16.14b}$$

where σ_0 is the initial stress. By analogy with the Maxwell model, whose stress relaxation is also described by Eq. 16.14*a*, the relaxation time τ is often taken as proportional to the viscosity:

$$\tau = \frac{\eta}{M} \tag{16.15}$$

where M has dimensions of a modulus.

Experimentally, the annealing of stresses in glasses is not well represented by Eq. 16.14*b*. Rather than a single relaxation time, a distribution of relaxation times is required to represent the data. In these terms

$$\frac{\sigma}{\sigma_0} = \int_0^\infty H(\tau)\, e^{-t/\tau}\, d\tau \tag{16.16a}$$

where $H(\tau)$ is the distribution of relaxation times. Alternatively, the experimental data on stress relaxation can be represented by a function of the form

$$\frac{\sigma}{\sigma_0} = \exp\left[-\left(\frac{t}{\tau}\right)^{1/n}\right] \tag{16.16b}$$

where n is temperature-dependent, having a value of about 2 at temperatures near the upper part of the transformation range and increasing to about 3 at temperatures near the lower part of this range.

The fact that a distribution of relaxation times is required to describe the experimental data is associated with two phenomena: (1) a distribution of molecular processes is involved in the relaxation, and (2) the properties of the glass change during the relaxation process. The expected distribution of molecular processes is related to structural variations in the material. Foremost among the property changes which are important in annealing of glass is the variation of viscosity with time. Such a variation is shown by the data in Fig. 16.11*a*, and the corresponding variation in specific volume is shown in Fig. 16.11*b*. The variation of viscosity with time introduces an essential nonlinearity into the relaxation process, as

$$\frac{d\sigma}{dt} = -\frac{1}{\tau(\sigma)}\,\sigma \tag{16.17a}$$

where the time, and hence the stress, variation of the relaxation time is associated with the time dependence of the viscosity. The effects of such a nonlinearity have been explored by O. S. Narayanaswamy,* who introduces a reduced time scale to allow for the changing viscosity.

For many commercial glasses, the rate of stress release on annealing is

J. Am. Ceram. Soc.*, **54, 491 (1971).

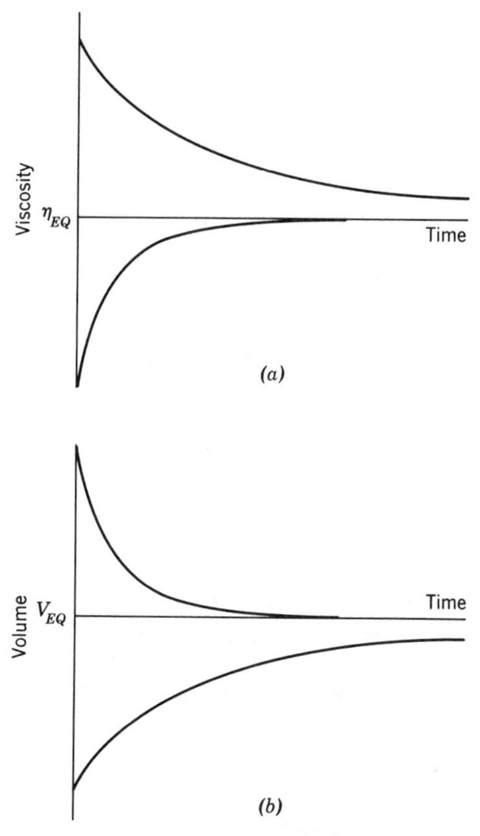

Fig. 16.11. Variation of (*a*) viscosity and (*b*) specific volume with time during annealing of a glass (schematic). Equilibrium viscosity (η_{EQ}) and equilibrium volume (V_{EQ}) indicated in figure.

found to be usefully approximated by the empirical Adams-Williamson law:*

$$\frac{1}{\sigma} - \frac{1}{\sigma_0} = At \tag{16.17b}$$

Initially large stresses decrease to a low value in about 15 min for typical glasses when the viscosity is about 10^{13} P; the temperature at which the viscosity reaches this value is called the annealing point T_a. At this temperature the annealing constant A_a in Eq. 16.17*b* is equal to about

*L. H. Adams and E. D. Williamson, *J. Franklin Inst.*, **190**, 835 (1920).

3.28×10^{-6}/psi°C for a typical soda-lime-silicate glass. The annealing constant decreases rapidly with viscosity, so that it is a function of temperature and is given for a typical soda-lime-silica glass as

$$A = A_a \exp[-C(T - T_a)] = 3.28 \times 10^{-6} \exp[-0.7(T - T_a)] \tag{16.18}$$

To obtain a rough estimate of the time required for annealing, it is convenient to assume that σ_0 is much larger than σ. Then using Eqs. 16.17*b* and 16.18, one can estimate the annealing time t for a typical soda-lime-silicate glass:

$$t \approx \frac{\exp[0.7(T - T_a)]}{3.28 \times 10^{-6}\,\sigma} \text{ sec} \tag{16.19}$$

where σ is the stress in psi. At the annealing point, $\eta = 10^{13}$, the time required for coarse annealing is about 14 min, assuming that the residual stress is reduced to 375 psi, whereas for fine annealing ($\sigma = 50$ psi) the requirement is about 204 min.

The final stress in the cooled glass will be a summation of the unreleased stress present before cooling and the residual stress resulting from the cooling process, both corrected for some stress release that occurs during cooling. If the glass is cooled at a constant rate, the temperature distribution is parabolic, as discussed earlier in this chapter. For a thick plate of half thickness r_m, $T_c - T_a$ equals $1/3(T_c - T_s)$; therefore from Table 16.2 for a cooling rate of ϕ°C/sec,

$$T_c - T_a = \frac{\phi r_m^2}{6k/\rho c_p} \tag{16.20}$$

Hence, if the cooling rate is constant, the temperature distribution is also constant; and with no changes in ΔT, no thermal stresses are generated by the cooling process. That is, if a glass is cooled at a constant rate from a high enough temperature so that all residual stresses after initially establishing the temperature gradient are annealed, the glass remains stressfree as it cools through the transformation range. As it approaches room temperature, however, ϕ and hence ΔT decrease, and the changing temperature distribution produces permanent stresses in the glass, since it is already rigid. The permanent tensile stress at the midplane is then, from Table 16.1,

$$\sigma_t = \frac{Ea'}{3(1-\nu)} \quad \Delta T = \frac{Ea' r_m^2 \phi}{18(1-\mu)k/\rho c_p} \tag{16.21}$$

where a' is the expansion coefficient in the annealing range. For a typical glass, a' is two to three times the value measured at low temperatures in the annealing range. Taking a value for $a' = 2.5a$, $E = 10^7$ psi, $k/\rho c_f =$ 0.0013 in.^{2}/sec (0.0084 cm^2/sec), and $\mu = 0.2$, which are typical values for a number of soda-lime-silica glasses, the safe rate of cooling is given as

$$\phi = 2.59 \times 10^{-10} \frac{\sigma_t}{r_m^2 a} \qquad \text{°C/sec} \tag{16.22}$$

where r_m is half the sample thickness in inches, a is the normally reported linear expansion coefficient, and σ is the maximum allowable tensile stress. For example, if $r_m = 0.25$ in., $a = 9 \times 10^{-6}$ in./in. °C, and $\sigma_t = 375$ psi as for coarse optical annealing, the maximum cooling rate must be kept below about 5°C/min, or for fine annealing ($\sigma_t = 50$ psi) the maximum allowable cooling rate is about 0.7°C/min.

The variation of stress in a soda-lime-silicate plate cooled at a nearly constant rate is shown in Fig. 16.12. The marked rise in stress observed at the end of linear cooling corresponds to that predicted by Eq. 16.21. The stresses observed at earlier times, during cooling at a constant rate, have a different origin. In particular, they are associated with the changes in physical properties, notably the thermal expansion coefficient, which characterize the transition from liquid to glass.

During cooling, the liquid-glass transition takes place earlier near the surface than in the interior. The resulting temporary differences in specific volume produce stresses whose relaxation is affected by structural relaxation. Such structural relaxation is often approximately rep-

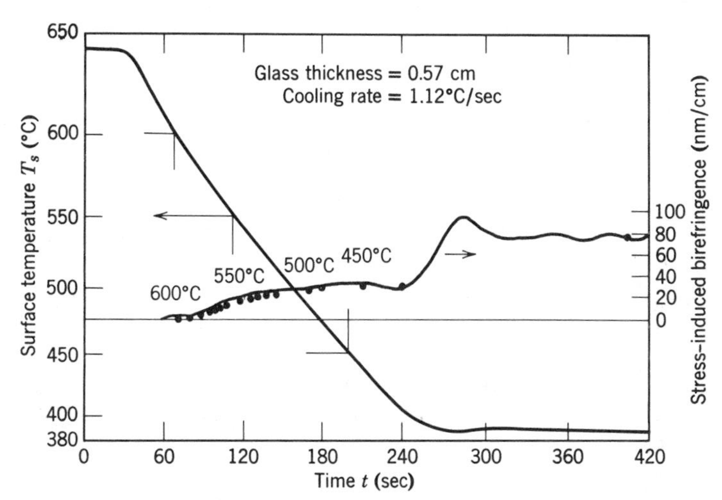

Fig. 16.12. Temperature and stress as functions of time during the annealing of a sheet glass. From R. Gardon and D. S. Narayanaswamy, *J. Am. Ceram. Soc.*, **53**, 380 (1970).

resented by changes in the fictive temperature. The fictive temperature begins to lag behind the actual temperature as the glass is cooled through the range of the glass transition and approaches a constant value near the lower end of the transition region. The approach to such a constant fictive temperature in this region should have a similar effect on stresses as the approach to a constant actual temperature in the region near ambient. Although all the stresses generated by the disappearance of actual temperature differences through the sample thickness remain as permanent stresses, only about one-third of the potential stresses generated by the disappearance of fictive temperature differences remain after viscous relaxation.*

Once the glass is somewhat below the strain point, corresponding to a viscosity of $10^{14.6}$ P, cooling can be more rapid and is mainly limited by the thermal stresses discussed in Section 16.3. A typical schedule for the annealing of ordinary commercial ware is given in Fig. 16.13. This

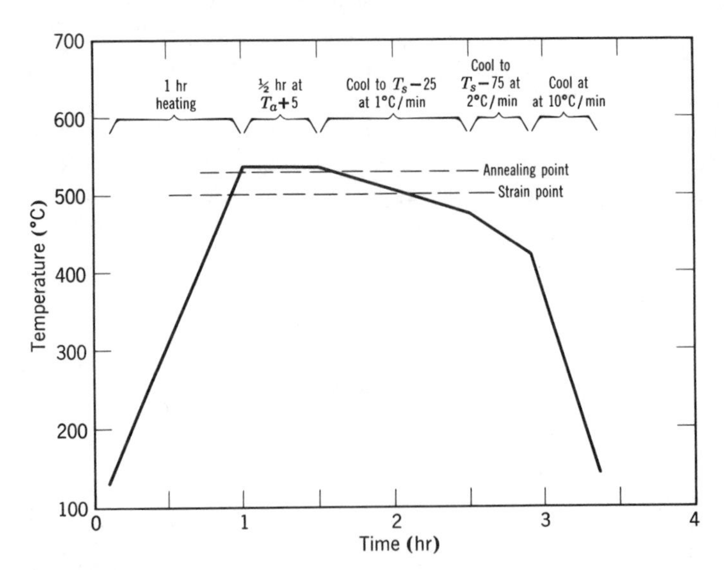

Fig. 16.13. Typical annealing schedule for $\frac{1}{2}$-in.-thick soda-lime-silica ordinary ware cooled on two sides.

schedule is purposely made conservative to permit its use for annealing ware of irregular shapes, which require more complete annealing than simple shapes.

*For further discussion of this topic, see R. Gardon and O. S. Narayanaswamy, *J. Am. Ceram. Soc.*, **53**, 380 (1970).

16.6 Chemical Strengthening

Glasses and crystalline ceramics can be strengthened by developing a state of compression in the surface regions of the materials. When strengths in the range of structural steel (yield stress, about 50,000 psi) are desired, use is made of chemical (ion exchange) techniques to achieve the desired surface compression. Such techniques produce an increase in the molar volume of the surface relative to the interior by changing the chemical composition of the surface. The increased molar volume results in a two-dimensional state of surface compression, since the expansion of the surface structure is restrained by the underlying material.

The surface stress associated with a fractional change in molar volume $\Delta V/V$ can be written approximately

$$\sigma = \frac{E}{3(1-2\mu)}\frac{\Delta V}{V} \tag{16.23}$$

Hence a 2% volumetric change in a material having a Young's modulus of 10 million psi and a Poisson's ratio of 0.25 would be expected to provide a surface compressive stress of about 90,000 psi.

As discussed in Section 16.4, the useful strength of a glass with its surface in compression is approximately equal to the sum of the surface-compression stress and the annealed strength. In applying this to chemically strengthened glass, however, consideration must be given to the effective thickness of the compressed layer relative to the depth of the surface abrasions that may be expected in manufacture or in service. The compression depth of chemically strengthened glass is controllable, in principle, but in practice it is limited to a few hundred microns. The reasons for this are associated with the fact that chemical strengthening is ordinarily accomplished by the diffusive penetration of some reagent supplied at the surface, and diffusive penetration depth is proportional to the square root of time.

The stress profile in a plate of chemically strengthened glass is ordinarily quite different from that in thermally tempered glass. In the latter, the profile is very nearly parabolic (Fig. 16.8, e.g.), and the magnitude of the maximum surface compression is approximately twice the magnitude of the maximum tension existing in the interior of the glass. In chemically strengthened glass, however, the profile is in general not parabolic. Rather, the stress tends to reproduce the error-function profile of the ion distribution, often considerably distorted by stress relaxation. It is not uncommon for a subsurface stress maximum to occur. The stress profile is also strongly dependent on the depth of penetration relative to the thickness of the plate. There is a nearly flat central region of small tension which changes abruptly to compression in the chemically altered region. The depth of compression is measured, optically, from the surface to the plane at which this crossover occurs. The depth of compression is related to the depth of ion exchange but is not necessarily identical with it. The ratio of surface compression to interior tension may be as high as several hundred. If the interior tension is small enough, chemically strengthened glass can be cut and drilled. On the other hand, if the thickness of the surface-compressed layer is about 10% of the plate thickness, the central tension is high enough to support spontaneous propagation of any crack that reaches the interior, and the material dices in the manner of thermally tempered glass.

Most chemical strengthening in commercial use involves the substitution of large ions for small. Such substitution can be effected in two ways: by diffusion and by electrically driven ion migration. From the practical standpoint, only the ions of Li, Na, K, and Ag exhibit a useful degree of mobility. In principle, the other alkali ions, polyvalent ions, and even anions should produce stuffing effects, but experimentally the results are less satisfactory.

Composition is an important variable in the practical use of this technique. Thus, alkali ions are more mobile in aluminosilicates than in silicates but less mobile in the borosilicates. The temperature at which the foreign ions are introduced is necessarily somewhat below the annealing range of the glass. The temperature used is generally a compromise between the desired increase in mobility and the undesired stress relaxation which accompany increases in temperature as the transformation region is approached. Fused salts, usually mixed nitrates, are most often used to effect the exchange.

Since the diffusing species are charged, electric fields can be used to affect the direction and rate of ion motion. In such cases, the profile of ion concentration is quite different from that which results from diffusive exchange. Since a larger, less mobile ion is generally used to displace a smaller ion to obtain the stuffing (as K^+ for Na^+), the resistivity of the volume occupied by the incoming ion is larger than that of the original glass, and the electric field is correspondingly more intense. This has the effect of sharpening the boundary between the leading and the following ions, and a square-wave concentration profile is maintained as long as the field is applied. A given depth of penetration therefore produces a higher surface stress than in the diffusive-exchange case.

The use of electric fields offers other potential advantages. Since the foreign ions can be introduced at a lower temperature than that required for diffusion, the effects of stress relaxation can be minimized. Further, thicker compression layers may be attained, since the depth of electrical penetration should tend to be proportional to the first power, rather than the square root, of time. The disadvantages of this process include the fact that glasses generally have a positive temperature coefficient of conductivity, which tends to cause channeling of current flow to any slightly overheated area of a plate-shaped conductor, resulting in perforation—an effect which is accentuated by the increase in resistivity, and corresponding increase in voltage and power for a given current, which accompanies increasing depth of penetration. The one-dimensional character of electrically induced migration can be mitigated by electrolyzing one side to twice the required depth, followed by a reversal of polarity and the passage of half the original quantity of charge in the opposite direction.

Both diffusion and electrical-migration ion-exchange techniques are limited by relaxation processes taking place at elevated temperatures. This results in the tensile strength of the body going through a maximum as the time of ion exchange is extended. The higher the temperature for a given composition, the earlier the maximum is reached and the steeper the decline thereafter. Even for ion-exchange temperatures below the nominal strain point of the glass, viscous and viscoelastic relaxation processes can still occur at appreciable rates. This is particularly true of the surface layer, where stress levels are often in the range of 100,000 psi. In addition to such stress relaxation, use of a chemically strengthened body at elevated temperatures can result in a decrease in strength associated with a decrease in the concentration gradients.

In addition to such a stuffing of the surface by substituting a large ion for a smaller ion, strengthening can also be achieved by exchanging with a smaller ion above the transformation range, as Li for Na. The surface regions containing the smaller ions are generally characterized by smaller expansion coefficients than the bulk glass and hence result in a surface compression. The magnitude of the compression achieved in this way is, however, considerably smaller than that which can be obtained by the stuffing technique. A technologically significant example of this is provided by the surface dealkalization of glasses by high-temperature treatment with a variety of acidic reagents, such as gaseous SO_2 or SO_3. The resulting surface skin has a lower thermal expansion coefficient as well as higher durability. Unfortunately, the process tends to be self-limiting, since the dealkalized surface tends to prevent further diffusive loss of alkali oxide.

Changes in the molar volume of the surface regions of a glass can also be effected by crystallization of the surface. This process, which is well covered in the patent literature but not much used in technological practice, is often coupled with an ion exchange of the surface region to provide a favorable glass phase for crystallization. In many cases, the crystallization treatment results in scattering losses associated with the presence of the crystals.

In the case of glass-ceramic materials, use of chemical strengthening techniques not only improves the strength of the small amount of residual glass but also, and more important, improves the strength of the crystalline phase or phases. The strengthening of crystalline phases may take place either by solid solution involving crowding of the crystal lattices in the surface regions or by transformations in these regions to phases of larger specific volumes. Examples of the former are provided by Na^+ for Li^+ exchange in β-spodumene solid-solution glass-ceramics, which results in increases in strength of abraded bodies from 15,000 to 50,000 psi, and by $2Li^+$ for Mg^{2+} exchange in β-quartz solid-solution glass-ceramics, which results in abraded strengths in the range of 45,000 psi. Examples of strengthening involving surface phase transformations are provided by K^+ for Na^+ exchange in nepheline glass-ceramics, which results in a transformation from nepheline to kalsilite, and by $2Li^+$ for Mg^{2+} exchange in multiphase Mg-aluminosilicate glass-ceramics, which results in phase assemblages such as those shown in Fig. 11.26. In both of these cases, abraded strengths in the range of 200,000 psi have been obtained. For further detail on these processes, see reference 3.

Suggested Reading

1. B. A. Boley and J. H. Weiner, *Theory of Thermal Stresses*, John Wiley & Sons, Inc., New York, 1960.
2. O. S. Narayanaswamy and R. Gardon, "Calculation of Residual Stresses in Glass," *J. Am. Ceram. Soc.*, **52**, 554 (1969).
3. G. H. Beall in *Advances in Nucleation and Crystallization*, L. L. Hench and S. W. Freiman, Eds., American Ceramic Society, Columbus, 1972.
4. D. P. H. Hasselman, "Unified Theory of Thermal Shock Fracture Initiation and Crack Propagation in Brittle Ceramics," *J. Am. Ceram. Soc.*, **52**, 600 (1969).
5. W. D. Kingery, "Factors Affecting Thermal Stress Resistance of Ceramics Materials," *J. Am. Ceram. Soc.*, **38**, 3 (1955).
6. J. White, "Some General Considerations in Thermal Shock," *Trans. Brit. Ceram. Soc.*, **57**, 591 (1958).
7. H. R. Lillie, "Basic Problems in Glass Annealing," *Glass Ind.*, **31**, 355 (1950).

Problems

16.1. (*a*) An uncolored base glass is to be coated with a colored glass, a process called flashing. The two glasses have the same basic composition, but the additional coloring matter in the flashing glass causes it to have a transition temperature about 10°C higher than that of the base glass. Assuming that the flashing glass is thin compared with the base glass, calculate the maximum stress to be expected in the flashing. The thermal expansion of the two glasses can be taken as equal in both the glassy and subcooled liquid regions, so that the only property difference is the transition temperature.

(*b*) It has been suggested that glassware can be strengthened by coating the base glass with a glass having a different thermal expansion or transition temperature. Is this feasible? Which should have the higher transition temperature? The thermal expansion of glass increases by a factor of about 3 at the transition temperature.

16.2. It has been proposed that high-strength structural glass be made by simultaneously drawing a three-layer sandwich of glass with the outer two layers of a composition different from the inner layer. Select a realizable combination of properties and dimensions which results in increased strength, and estimate the increase in bending strength.

16.3. In a tensile-strength test, a sample volume is subjected to a uniform tensile stress; in a transverse-strength (modulus of rupture) test, there is a stress gradient with the maximum tensile stress at the surface. In a thermal-shock test with a rapid quench rate, the stress gradient can be very steep and is transient. In terms of crack nucleation and crack growth, discuss anticipated variations in fracture behavior for these three different types of stressing for (*a*) a good-quality ceramic such as 94% Al_2O_3 porcelain and (*b*) a specialty refractory such as an 80% alumina, 20% porosity body. Make explicit assumptions about reasonable values of data (such as modulus of rupture, etc.) which may be necessary for your discussion.

16.4. A glass can be strengthened more by ion exchange than by thermal tempering.

(*a*) What functional relations underlie this fact?

(*b*) Give various "mechanisms" for ion-exchange strengthening.

(*c*) What governs the upper limit of permissible quench rate in thermal tempering?

16.5. Hot-pressed sintered alumina plates of zero porosity are 0.25 in. thick. They are glazed with a porcelain glaze of 0.025 in. thickness ($E = 10^7$ psi) and coefficient of expansion of 4×10^{-6}/°C. Assuming the fictive temperature is 825°C, what is the stress in the glaze and the body at room temperature? Assume the Poisson's ratio is 0.3 in both cases.

16.6. (*a*) A certain soda-lime-silica glass is carefully treated in HF to remove all Griffith flaws. It has a Young's modulus of 7×10^3 kg/mm^2 (10^7 psi) (assumed to be temperature-independent, which it really is not) and a Poisson's ratio of 0.35. The linear coefficient of thermal expansion is 10^{-5}/°C. The thermal conductivity is 2.5×10^3 g cal/cm^2 sec °C cm. The surface tension of the glass is estimated to be 300 dynes/cm. If the glass is quenched by dropping into ice water, what is the maximum temperature to which it can be heated before quenching without fracture due to heat shock.

(*b*) If the glass is not etched and it is known that Griffith flaws of 1 micron exist in the surface, what is the maximum temperature from which it can be quenched?

16.7. (*a*) Define by equations the coefficient of linear thermal expansion a and the coefficient of volume thermal expansion α.

(*b*) It is necessary to seal a crystalline oxide, $a = 80 \times 10^{-7}/°C$, in a butt seal to a glass, $a = 90 \times 10^{-7}/°C$. The modulus of elasticity of the glass and the crystal is 10^7 psi. The glass is carefully annealed and becomes essentially rigid at 500°C. Do you believe this seal will break? Why?

(*c*) Explain why volume changes of as little as 0.6% can cause grain fractures during phase transformations of ceramics (for example, tridymite). Assume a modulus of elasticity of 10^7 psi.

16.8. Describe what occurs in crazing of a glaze. What is the cause of crazing? How might a triaxial porcelain body or a glaze composition be altered to prevent it from occurring?

16.9. Increased strength in glass and polycrystalline ceramics has been achieved by quenching after forming. Explain this effect, giving functional relations between the important physical properties and the required quench conditions. Compare heat treatments for fused SiO_2 and a soda-lime-silica glass required for equivalent fractional strength increases.

16.10. A glass sandwich of Pyrex-soft glass-Pyrex was made and heated to form a bond. Calculate the normal stresses and the shear stress acting at the interface of the two glasses after cooling to room temperature. Pyrex has a thermal expansion coefficient of 3.6×10^{-6} and the soft glass a thermal expansion coefficient of 8.4×10^{-6}. Assume all viscous or plastic flow ceases below 500°C. The slabs are 0.01 in. thick. Young's modulus for Pyrex is 6×10^5 kg/cm^2 and for soft glass is 7×10^5 kg/cm^2.

17

Electrical Conductivity

There are a great number of applications for ceramic materials for which electrical-conduction properties are important. Semiconductor materials are used for many specialized applications such as resistance heating elements; semiconductor devices such as rectifiers, photocells, transistors, thermistors, detectors, and modulators have become an important part of modern electronics. An equally important application of ceramics is as electrical insulators; porcelains and glasses are used for low- and high-voltage insulation. Consequently, we are interested in the entire range of electrical-conduction properties.

The properties and characteristics of these materials are approached from two main points of view. Electrical engineers consider them primarily as components in electrical circuits having specified property values and characteristics with regard to electrical measurements. The physicist considers these properties in terms of a quantitative understanding of the electronic and ionic behavior. The ceramist must take an intermediate position, considering both the problems of the ultimate user and being able to understand the effects of composition, structure, and environment on properties in terms of the atomic and electronic behavior.

To prevent repetition, we restrict the discussion in this chapter to direct-current and low-frequency measurements. The equally important aspect of high-frequency measurements is discussed in Chapter 18.

17.1 Electrical-Conduction Phenomena

Mobility and Conductivity. When an electric field is applied to a ceramic, the current rapidly or slowly reaches an equilibrium *direct-current* value. We can represent equilibrium in terms of the number of charged particles present and their drift velocity in the presence of an electric field. The electric-current density j is defined as the charge transported through a unit area in a unit time. If the number of charged particles per unit volume is n and they have a drift velocity of v and a charge per particle ze, where z is the valence and e the electronic charge, then the electric-current density for the ith particle is given by

$$j_i = n_i z_i e v \tag{17.1}$$

The electrical conductivity σ is defined by the relationship

$$\sigma = \frac{j}{E} \tag{17.2}$$

where E is the electric field strength, taking any field distortion into account. (The current density, drift velocity, and field strength are vector properties, and vector notation has advantages. The nonvectorial notation is suitable for our purposes, however, and is simpler for some readers.) Consequently,

$$\sigma_i = (n_i z_i e)\frac{v}{E} \tag{17.3}$$

The drift velocity is directly proportional to the locally acting electric field strength, and this ratio is defined as the mobility:

$$\mu_i = \frac{v_i}{E_i} \tag{17.4}$$

The conductivity, then, is the product of the concentration and mobility of charge carriers:

$$\sigma_i = (n_i z_i e)\mu_i \tag{17.5}$$

Sometimes it is desirable to use an absolute mobility B (Chapter 6), defined as the drift velocity per unit of applied force. This is given by

$$B_i = \frac{v_i}{F_i} = \frac{v_i}{z_i e E} \tag{17.6}$$

In terms of the absolute mobility, then, conductivity is given by

$$\sigma_i = n_i z_i^2 e^2 B_i \tag{17.7}$$

$$\mu_i = z_i e B_i = \frac{z_i e D_i}{kT} \tag{17.8}$$

When we consider the effect of variables such as composition, structure, and temperature on the electrical conductivity, we are concerned with these two separate contributors—the concentration of charge carriers and their mobility. (Internal boundaries and blocking layers are often of great importance also, as discussed in Chapter 18.)

Most engineering data are presented in terms of volts, amperes, ohms, seconds—practical units. Other measurements or calculations are reported in terms of electrostatic units (esu), in which the fundamental units are centimeter, gram, and second, and the dielectric constant is taken as a plain number. Conversion factors are given in Table 17.1.

If we consider the charge carriers as initially having a random movement with an average drift velocity of zero, the equation of motion resulting from the application of a steady average external force, $F = zeE$, is

$$m\left(\frac{dv}{dt}+\frac{v}{\tau}\right)=F=zeE \tag{17.9}$$

where m is the mass of the particle. In the absence of external forces

$$\frac{dv}{dt}+\frac{v}{\tau}=0 \tag{17.10}$$

and by integrating,

$$v(t)=v_0 \exp\left(\frac{-t}{\tau}\right) \tag{17.11}$$

where τ is a characteristic *relaxation time* governing the time required to reach equilibrium. In Eq. 17.9 the first term describes inertial effects and must be included when v is time-dependent. The term mv/τ has the form of a frictional drag or damping force. Once the inertial effects have died out $\left(\frac{dv}{dt}=0\right)$, we have

$$v=\frac{ze\tau E}{m} \tag{17.12}$$

and by comparison with Eq. (17.3)

$$\sigma=(nze)\left(\frac{ze\tau}{m}\right)=\frac{nz^2e^2\tau}{m} \tag{17.13}$$

That is, the charge transported is proportional to the charge density (zen), the acceleration of charge in a given field (proportional to ze/m), and τ corresponding to the time that these forces act on the charge between collisions and random motion.

Kinds of Charge Carriers. The general characteristics of solids in relation to mobile charged particles have been discussed in Chapters 4 and 6. They can be described in terms of the electron energy band structure, such as those shown in Fig. 17.1. For metals there is always a finite concentration of electrons in the conduction band; for semiconduc-

Table 17.1. Conversion Factors for Practical and Esu Units

Quantity	Units		To Convert from Practical to Esu Units Multiply by:
	Esu	Practical	
Conductivity, σ	sec^{-1} [(stat ampere/statvolt)/cm]	ohm-cm^{-1} [(ampere/volt)/cm]	9×10^{11}
Current density, j	statcoulomb sec^{-1}/cm^2 [statampere/cm^2]	coulomb sec^{-1}/cm^2 [ampere/cm^2]	3×10^9
Electric field, E	statvolt/cm	volt/cm	1/300
Carrier concentration, n	carriers/cm^3	carriers/cm^3	–
Drift velocity, v	cm/sec	cm/sec	–
Mobility, μ	(cm/sec)/(statvolt/cm)	(cm/sec)/(volt/cm)	300
Electronic change, e	4.803×10^{-10} statcoulomb	1.601×10^{-19} coulomb	3×10^9

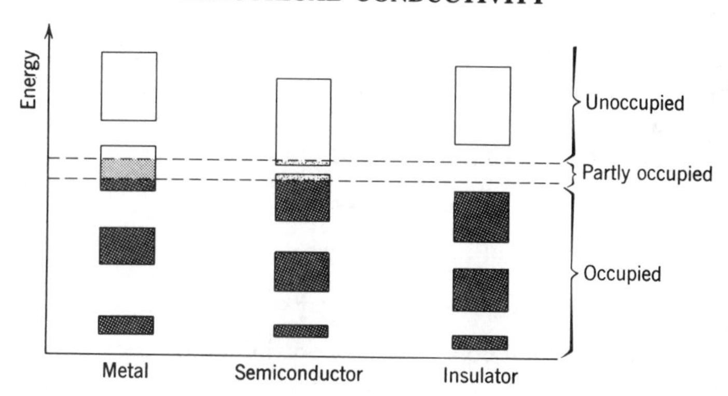

Fig. 17.1. Electron energy bands in (*a*) metals, (*b*) semiconductors, and (*c*) insulators.

tors the concentration of electrons in the conduction band depends on temperature and composition; for insulators the band gap is sufficiently large so that normally there are no electrons able to move through the crystal, and conductivity can only result from the movement of charged ions. Typical values of electrical resistivity for these classes of materials are indicated in Table 17.2.

Table 17.2. Electrical Resistivity of Some Materials at Room Temperature

Materials	Resistivity (ohm-cm)
Metals:	
Copper	1.7×10^{-6}
Iron	10×10^{-6}
Molybdenum	5.2×10^{-6}
Tungsten	5.5×10^{-6}
ReO_3	2×10^{-6}
CrO_2	3×10^{-5}
Semiconductors:	
Dense silicon carbide	10
Boron carbide	0.5
Germanium (pure)	40
Fe_3O_4	10^{-2}
Insulators:	
SiO_2 glass	$>10^{14}$
Steatite porcelain	$>10^{14}$
Fire-clay brick	10^{8}
Low-voltage porcelain	10^{12}–10^{14}

Transference Numbers. Frequently, more than one charge carrier can contribute to the electrical conduction in one material. In this case we can use the relationships given thus far to define a partial conductivity for each charged particle. That is, for particle i

$$\sigma_i = \mu_i (n_i z_i e) \tag{17.14}$$

and so on. Then the total conductivity is given by

$$\sigma = \sigma_1 + \sigma_2 + \cdots + \sigma_i + \cdots \tag{17.15}$$

The fraction of the total conductivity contributed by each charge carrier is

$$t_i = \frac{\sigma_i}{\sigma} \tag{17.16}$$

where t_i is called the transference number. The sum of the individual transference numbers must obviously be unity:

$$t_1 + t_2 + \cdots + t_i + \cdots = 1 \tag{17.17}$$

In this relationship the individual charge carriers may be charged ions, electrons, or electron holes contributing to the conduction process. Transference numbers for several materials are given in Table 17.3.

In essence, then, the problem of interpreting and controlling electrical conductivity in ceramics consists of characterizing the concentration and mobility of each possible current carrier and then summing these contributions to obtain the total conductivity.

17.2 Ionic Conduction in Crystals

One kind of charge carrier that is always present and can contribute to electrical conductivity is the ions present in crystalline materials such as the oxides and halides. As illustrated in Table 17.3, electrical conductivity resulting from ion migration is important in many ceramic materials. Its analysis requires a determination of the concentration and mobility of charge carriers, outlined in Eqs. 17.3 to 17.7 and discussed in some detail in Chapters 4 and 6. Indeed, one major result of electrical-conductivity measurements in ionic materials has been the elucidation of defect structures and determination of ion mobilities.

For an ion to move through the lattice under the driving force of an electric field it must have sufficient thermal energy to pass over an energy barrier, the intermediate position between lattice sites. For a one-dimensional case (Fig. 17.2) the current density in a forward direction due

Table 17.3. Transference Numbers of Cations t_+, Anions t_-, and Electrons or Holes $t_{e,h}$ in Several Compounds

Compound	Temperature (°C)	t_+	t_-	$t_{e,h}$
NaCl	400	1.00	0.00	
	600	0.95	0.05	
KCl	435	0.96	0.04	
	600	0.88	0.12	
KCl + 0.02% $CaCl_2$	430	0.99	0.01	
	600	0.99	0.01	
AgCl	20–350	1.00		
AgBr	20–300	1.00		
BaF_2	500	. . .	1.00	
PbF_2	200	. . .	1.00	
CuCl	20	0.00	. . .	1.00
	366	1.00	. . .	0.00
ZrO_2 + 7% CaO	>700°C	0	1.00	10^{-4}
$Na_2O \cdot 11Al_2O_3$	<800°C	1.00 (Na^+)	. . .	$<10^{-6}$
FeO	800	10^{-4}	. . .	1.00
ZrO_2 + 18% CeO_2	1500	. . .	0.52	0.48
+ 50% CeO_2	1500	. . .	0.15	0.85
$Na_2O \cdot CaO \cdot SiO_2$ glass	. . .	1.00 (Na^+)		
15% ($FeO \cdot Fe_2O_3$)$\cdot CaO \cdot SiO_2 \cdot Al_2O_3$ glass	1500	0.1 (Ca^{+2})	. . .	0.9

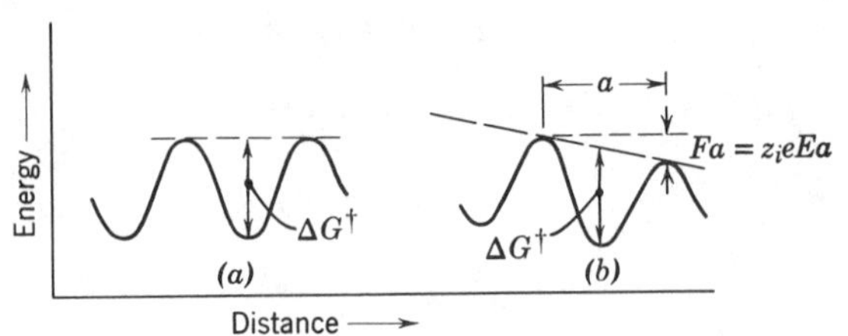

Fig. 17.2. Potential barriers (*a*) without and (*b*) with an applied field E for an ion with charge $z_i e$, and interatomic spacing a.

to biasing of the electric field is

$$j_{i_{\text{forward}}} = z_i e n_i a \nu \exp\left(\frac{-\Delta G^{\dagger} + z_i eEa/2}{kT}\right) \tag{17.18}$$

the reverse flux is

$$j_{i_{\text{backward}}} = z_i e n_i a \nu \exp\left(\frac{-\Delta G^{\dagger} - z_i eEa/2}{kT}\right) \tag{17.19}$$

where a is the jump distance, ν is the lattice vibrational frequency (approximately 10^{13}/sec), and $\Delta G^{\dagger}$ is the Gibbs free energy, the activation energy, for ion motion. The net flux is

$$j_{i_{\text{net}}} = j_{i_{\text{forward}}} - j_{i_{\text{backward}}} = 2 z_i e n_i a \nu \exp\left(\frac{-\Delta G^{\dagger}}{kT}\right) \sinh\left[\frac{z_i eEa}{2kT}\right] \tag{17.20}$$

For typical values of E, a, and T the term in brackets is small (approximately 10^{-5}), and Eq. 17.20 can be closely approximated as

$$j_{i_{\text{net}}} = 2(n_i e z_i) a \nu \left[\frac{z_i eaE}{2kT}\right] \exp\left(\frac{-\Delta G^{\dagger}}{kT}\right) = \left(\frac{n_i e^2 z_i^2 a^2 \nu}{kT} e^{-\Delta G^{\dagger}/kT}\right) E = \sigma_i E \tag{17.21}$$

By comparison with Eq. 17.5 the ion mobility is expressed as

$$\mu_i = \frac{e z_i a^2 \nu}{kT} e^{-\Delta G^{\dagger}/kT} \tag{17.22}$$

and the absolute mobility, Eq. (17.7), is

$$B_i = \frac{a^2 \nu}{kT} e^{-\Delta G^{\dagger}/kT} = \frac{D_i}{kT} \tag{17.23}$$

which is the Nernst-Einstein relationship. For a particular ion having a transference number t_i, in a crystal,

$$\sigma_i = t_i \sigma = f(n_i e z_i)(e z_i)\left(\frac{D_i}{kT}\right) = \frac{f D_i n_i z_i^2 e^2}{kT} \tag{17.24}$$

where f is a factor measuring the number of equivalent sites to which a joining may occur in a particular crystal structure ($f = 4$ for an ion vacancy in the rock salt structure).

As shown in Table 17.3 and discussed in Chapter 6, the mobility of sodium ions is much larger than the mobility of chloride ions in sodium chloride. As a first approximation we may consider that sodium ions are solely responsible for the ionic conductivity in this material and define two general regions which are illustrated in Fig. 17.3. At high temperatures in the *intrinsic* range the concentration of sodium ion vacancies (n_i in Eq. 17.24) is a thermodynamic property, and the conductivity varies with temperature as the product of the vacancy concentration and the diffusion coefficient, each of which is an exponential function of temperature. At lower temperatures the concentration of sodium ions is not in thermal equilibrium but is determined by minor solutes and previous

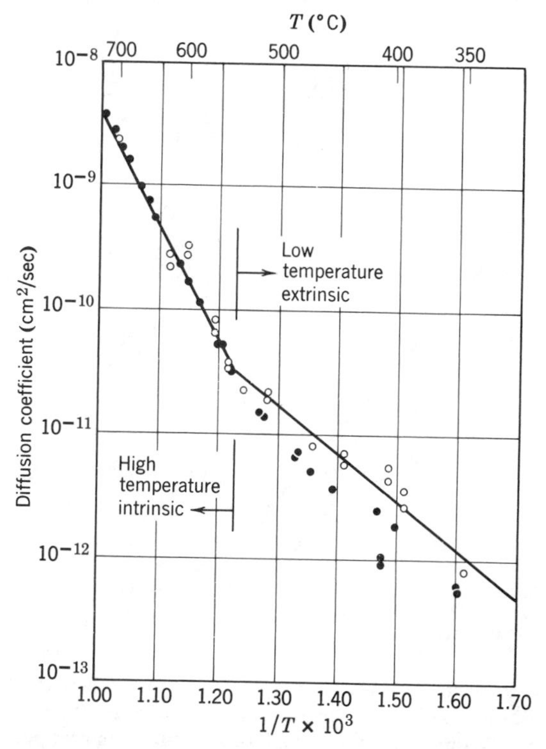

Fig. 17.3. Diffusion coefficients measured directly (open circles) and calculated from the electrical conductivity data (closed circles) for Na^+ in sodium chloride. From D. Mapother, H. N. Crooks, and R. Maurer, *J. Chem. Phys.*, **18**, 1231 (1950).

history, as discussed in Chapter 4. As a result, in the *extrinsic* range the temperature dependence of conductivity depends only on the diffusion coefficient. From data such as illustrated in Fig. 17.3 the activation energy for mobility can be determined from data in the extrinsic region; the sum of activation energies for mobility and lattice-defect formation can be determined from the temperature dependence of the intrinsic region.

However, measurements with carefully prepared samples over a wide temperature region show that the behavior is a little more complicated than this simple picture. As illustrated in Fig. 17.4 for high-purity and heavily doped samples of sodium chloride, several regions of temperature dependence of the conductivity can be identified. At a sufficiently high temperature (stage I′) an added contribution from chloride ion migration is seen. And at lower temperatures than the simple extrinsic region (stage II), there are additional regions corresponding to the association of dopants and cation vacancies (stage III) which decrease the concentration of charge carriers. For samples with substantial solute content, precipitation occurs

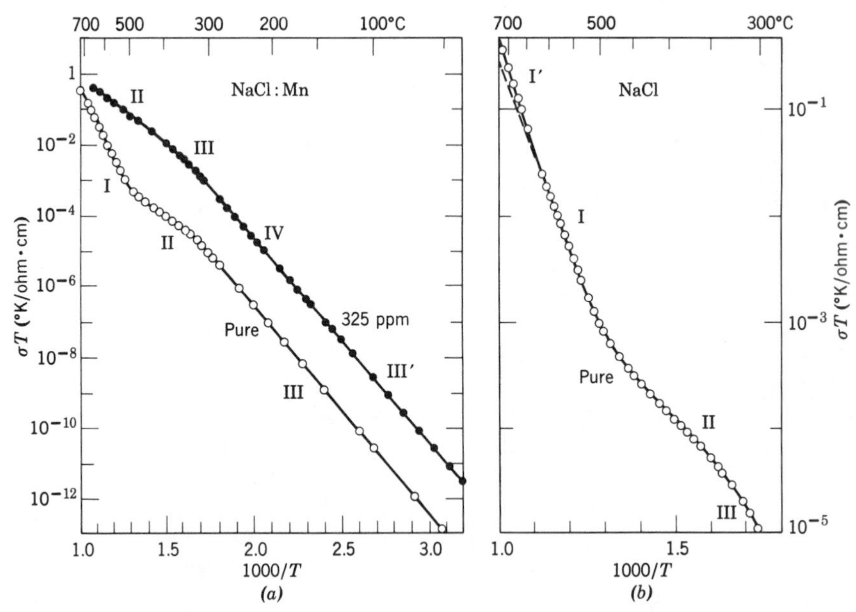

Fig. 17.4. Cation conductivity in NaCl. In (*a*) the curve for the pure crystal shows stages I (intrinsic), II (extrinsic), and III (association), and the curve for the heavily doped crystal shows stages II, III, IV (precipitation), and III′ (association with precipitation frozen). Part (*b*) for the pure crystal shows the additional rise in stage I′ from the anion contribution to the conductivity. The conductivity is expressed in ohm^{-1} cm^{-1}. From Kirk and Pratt, *Proc. Brit. Ceram. Soc.*, **9**, 215 (1967).

which further decreases the concentration of ionic carriers (stage IV), and at still lower temperatures a combination of association and precipitation further modifies the conductivity behavior (stage III′). Reference to Chapter 4, particularly Section 4.7, should clarify the nature of these different temperature regimes.

In discussing materials in which more than one defect contributes to the conduction process, it is convenient to introduce the ratio of the mobilities $\theta = B_1/B_2$ as a variable. For materials of variable composition it is convenient to use the concept of relative conductivity σ/σ_0, where σ_0 refers to the conductivity of the pure material.

In Fig. 17.5 the conductivity of AgBr containing $CdBr_2$ and Ag_2S as solutes is illustrated. In this material the predominant thermal defects are Frenkel defects, and the interstitial silver ions are more mobile than the vacant silver sites, that is, $\theta = B_{Ag_i^{\cdot}}/B_{V_{Ag}'} > 1$. Addition of $CdBr_2$ in solid solution increases the concentration of silver ion vacancies and decreases the concentration of interstitial silver ions (as a result of the Frenkel equilibrium, Chapter 4). As the concentration of $CdBr_2$ is increased, the conductivity decreases so long as the majority conduction process is interstitial migration. However, there comes a composition at which a minimum in the conductivity isotherm occurs. With further additions of

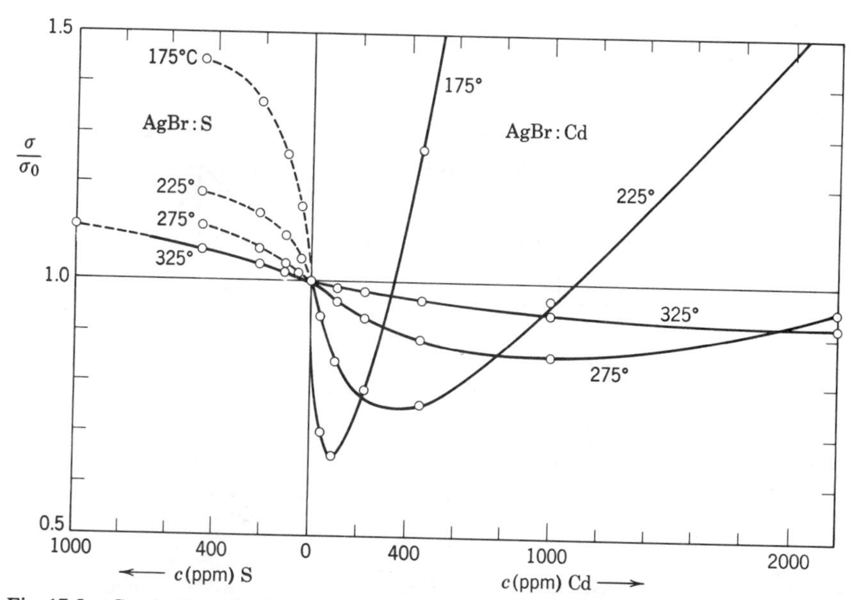

Fig. 17.5. Conductivity isotherms for AgBr doped with Cd^{2+} and S^{2-}. The dashed lines for AgBr:S are outside the equilibrium solubility limit. From J. Telfow, *Ann. Phys.*, **5**, 63 (1950), and *Z. Phys. Chem.*, **195**, 213 (1950).

$CdBr_2$, increased conductivity results from the increased vacancy concentration, which now contributes the major part of the total conductivity. Using the concepts discussed for the concentration of the individual defects and their relative mobilities, the minimum of the relative conductivity curve can be used to measure the ratio of mobilities. As Ag_2S is added to AgBr, more silver interstitials are formed, and by similar reasoning we can understand why the conductivity increases, with no minimum.

The precise application of these relationships is somewhat disturbed by chemical effects; different solutes affect the mobility, as shown in Fig. 17.6. Calcium is significantly more effective than strontium or barium in increasing the conductivity of potassium chloride at the same temperature. This may reflect the interaction of moving vacancies with a strain field around solutes, by which the larger barium ion may hold additional vacancies in its strain field to help accommodate elastic distortion. Another way of looking at this is to describe the elastic strain field as contributing to the stability of vacancy-solute associates giving rise to a greater association between vacancies and solute ions in the case of barium.

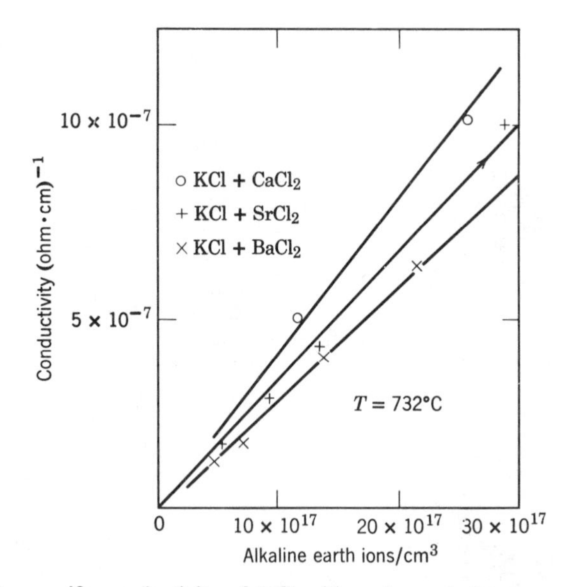

Fig. 17.6. The specific conductivity of KCl with various admixtures as indicated. The concentration of foreign cations is given as the number per cubic centimeter; this may be converted to a molar fraction by dividing by the number of K^+ ions per cubic centimeter, that is, 1.62×10^{23}. From reference 2.

The discrepancy between tracer diffusion measurements and diffusion coefficients calculated from electrical-conductivity measurements at low temperatures shown in Fig. 17.3 reflects the association of charge-carrying vacancies with solute ions to form electrically neutral vacancy-solute pairs which do not contribute to the conductivity process, that is, are not influenced by an applied electric field. The formation of such associates has been discussed in Chapter 4 and their influence on diffusion phenomena described in Chapters 6 and 9.

The uniform distribution of defects is disturbed by the tendency for each defect to be surrounded by a diffuse cloud of opposite charges. The Debye-Huckel screening constant is defined as

$$\ell^2 = \frac{8\pi n_d (z_i e)^2}{\kappa k T} \tag{17.25}$$

where n_d is the concentration of charged defects in units of cm^{-3}, and the dimensions of ℓ are cm^{-1}; the length $1/\ell$ represents a screening distance over which the excess charge of a defect is effectively neutralized. The change in energy per pair of defects because of the nonuniform charge distribution is

$$H_{DH} = -\frac{(z_i e)^2 \ell}{\kappa(1+\ell R)} \tag{17.26}$$

where R is the distance of closest approach of the charged defects (since $\ell R \ll 1$, some arbitrariness in selecting R is not too upsetting) and κ is the dielectric constant. This corresponds to a decrease in the formation enthalpy of a pair of defects, such that more defects are formed at a given temperature than would otherwise occur. In other words the defect activity a is reduced, with an activity coefficient γ given by

$$\gamma = \frac{a}{[c]} = \exp\left[-\frac{z^2 e^2 \ell}{kT(1+\ell R)}\right] = \exp\left[\frac{H_{DH}}{2kT}\right] \tag{17.27}$$

As shown in Fig. 17.7, for defect concentrations below about 1000 ppm, the effect is small, and the simple association relations discussed in Chapter 4 are completely adequate. At higher solute concentrations the simple association theory overestimates the degree of associate formation, and defect activities rather than concentrations should be used with the activity coefficient defined in Eq. 17.27.

Fast Ion Transport. Several types of compounds show exceptionally high ionic conductivity and have recently become of technological interest. Such phases fall into three broad groups: (1) halides and chalcogenides of silver and copper, in which the metal atom is disordered over several alternative sites; (2) oxides with the β-alumina structure, in which a monovalent cation is mobile; (3) oxides of the fluorite structure type, with large concentrations of defects caused either by a variable-valence cation or solid solution with a second cation of lower valence (for example, $CaO–ZrO_2$ or $Y_2O_3–ZrO_2$). Figure 17.8 shows electrical-conductivity data for some representative materials with high conductivity. The values are many orders of magnitude larger than normal ionic compounds (compare, for example, with Fig. 17.4) and are comparable with the conductivity of such liquid electrolytes as dilute solutions of suphuric acid.

The high dopant levels in the fluorite type of solid solutions leads to large defect concentrations and vacancy ordering. In such materials rapid oxygen migration occurs. This is believed to be due to the high concentration of vacancies (of the order of 15%) and also to correlated ion jumps

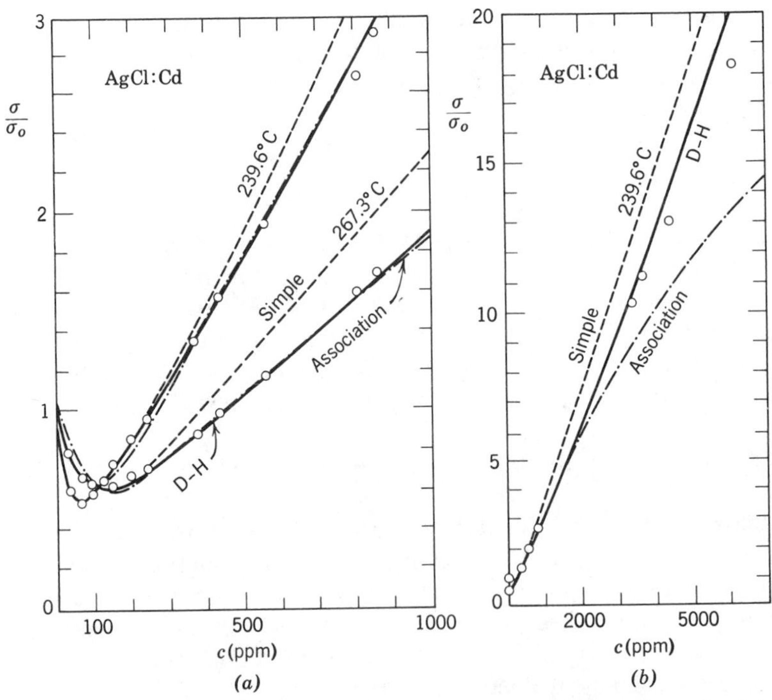

Fig. 17.7. (*a*) and (*b*) Conductivity isotherms for AgCl:Cd. The experimental points and the calculated curves with association and with Debye-Hückel interactions, including association, are taken directly from the reference. The results of the simple theory without association have been calculated for comparison with $K_F^{1/2} = 21.7$ ppm and $\theta = B_{Ag_i}/B_{V_{Ag}} = 11.13$ at 239.6°C and $K_F^{1/2} = 45.9$ ppm and $\theta = 8.63$ at 267.3°C. From H. C. Abbink and D. S. Martin, Jr., *J. Phys. Chem. Solids*, **27**, 205 (1966).

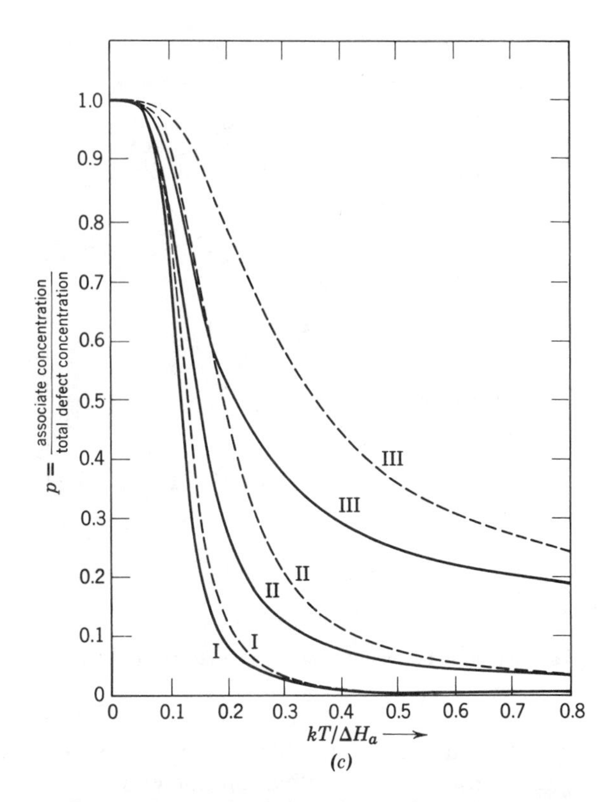

Fig. 17.7 (*Continued*). (*c*) Curves showing the degree of association p as a function of the reduced temperature $kT/\Delta H_a$ at three different concentrations: I, $c = 10^{-4}$; II, $c = 10^{-3}$; III, $c = 10^{-2}$. The dashed lines have been calculated from simple association. The full lines refer to a more elaborate calculation using Debye-Hückel theory. From reference 2.

over distances greater than an interionic separation as a result of the defect ordering. Figure 17.9 shows the measured oxygen-tracer diffusion coefficients to agree quite well with values obtained from conductivity through use of the Nernst–Einstein equation.

The silver and copper halides and chalcogenides often have simple arrays of anions. The cations occur in disorder in the interstices among the anions. The number of available sites is larger than the number of cations. In the highly conductive phases the energy barrier between neighboring sites is very small. The connectivity of such sites thus provides *channels* along which the cations are free to move. The potential energy seen by the mobile Ag ions among the body-centered cubic array of I^- ions in α-AgI has been calculated theoretically as a function of

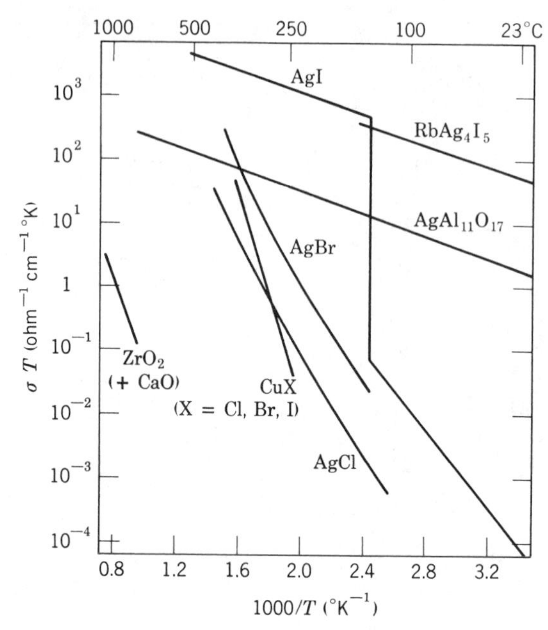

Fig. 17.8. Conductivity of some highly conducting solid electrolytes.

position within the unit cell. The calculation not only confirmed the existence of paths with very low activation barriers to Ag migration but showed that the height of the barrier increased rapidly with either a small increase or small decrease in the size of the migrating ion.

The β-aluminas are hexagonal structures with approximate composition $AM_{11}O_{17}$. The mobile ion A is a monovalent specie such as Na, K, Rb, Ag, Te, or Li, and M is a trivalent ion, Al, Fe, or Ga. Related phases also occur with approximate formulas AM_7O_{11} (β') and AM_5O_8 (β''), the latter having extremely high conductivities. The conductivities for several β-aluminas are plotted as a function of temperature in Fig. 17.10.

The crystal structure consists of planes of atoms parallel to the basal plane. Four planes of oxygens in a cubic close-packed sequence comprise a slab within which aluminum atoms occupy octahedral and tetrahedral sites as in spinel. The spinel blocks are bound together by a rather open layer of the monovalent ion and oxygen. This loosely bound layer is thought to be disordered and provide a two-dimensional path for atom motion with greater than single jump distances.

As the monovalent conduction ions become larger, their mobility becomes impeded, $\sigma(\text{Na } \beta\text{-}Al_2O_3) > \sigma(\text{K } \beta\text{-}Al_2O_3)$. As the ion becomes too small, for example, Li β-Al_2O_3, the ion "rattles around" in the

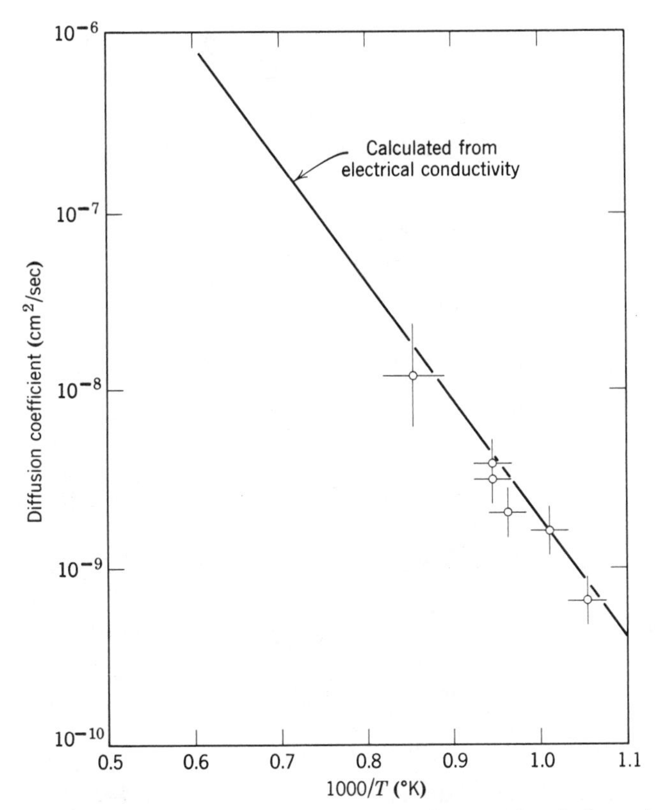

Fig. 17.9. Oxygen ion diffusion coefficient in relation to electrical conductivity in $Zr_{0.85}Ca_{0.15}O_{1.85}$. Solid line is calculated from conductivity data. Experimental points are direct measurements.

conduction channels and also has its motion impeded. As would be expected, the conductivity is extremely anisotropic, $\sigma_{\perp c} \gg \sigma_{\parallel c}$. However, polycrystalline materials show less than an order of magnitude decrease in conductivity over single crystals measured parallel to the highly conducting basal planes. This may be an indication of high-conductivity paths through grain boundaries.

Some Applications for Completely Ionic Conductors. Stable ceramics that have completely ionic conductivity ($t_i = 1$) can be used as solid-state electrolytes. Because of the precise relationship (Eq. 17.28) between the voltage and the chemical-potential gradient across the electrolyte, it can be used for batteries and fuel cells and as an ion pump or ion activity probe.

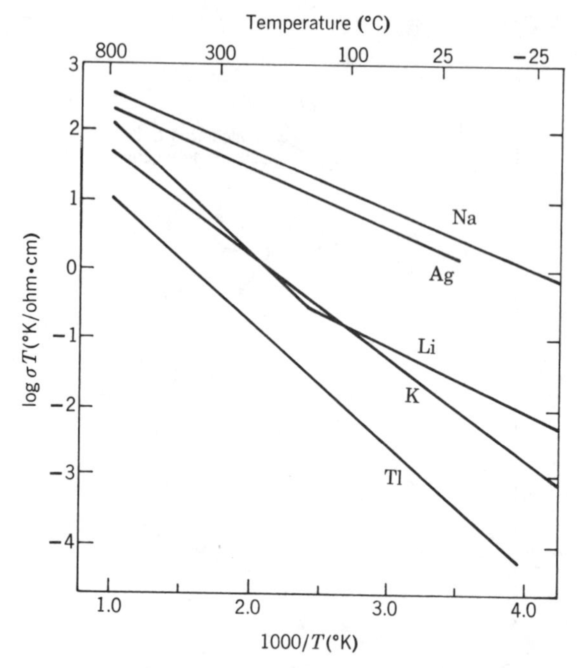

Fig. 17.10. Electrical conductivity for various β-aluminas. From R. A. Huggins.

We can derive these relationships by beginning with Eq. 9.11 for the ion flux due to a chemical-potential gradient (concentration gradient of the ion) and a voltage gradient $E = d\phi/dx$:

$$j_i = -c_i B_i \left[\frac{\partial \mu_i}{\partial x} + z_i F \frac{d\phi}{dx}\right]$$

If we set up a chemical-potential gradient across a sample which has $t_i = 1$ but allow no ions to flow because we impose an opposing voltage gradient, then Eq. 9.11 relates the voltage ϕ across the material to the chemical potential:

$$\mu_i \text{ (side I)} - \mu_i \text{ (side II)} = -\int z_i F\left(\frac{d\phi}{dx}\right) dx = -z_i F \phi \qquad (17.28)$$

For example, let us consider calcia-stabilized zirconia, in which $t_{O^{2-}} = 1.0$. If we impose an oxygen pressure $P_{O_2}{}^{I}$ on one side and $P_{O_2}{}^{II}$ on the other, then Eq. 17.25 becomes

$$\phi = \frac{RT}{4F} \ln\left(\frac{P_{O_2}{}^{I}}{P_{O_2}{}^{II}}\right) \qquad (17.29)$$

where the substitution $\mu = \mu_0 + RT \ln P_{O_2}$ has been made for the chemical potential of oxygen and $z_{O^{2-}} = -2$. If we allow current to flow (ions to migrate), this voltage is slightly less because we no longer have thermodynamic equilibrium.

Equation 17.28 is a form of the Nernst equation, which at equilibrium relates the standard free-energy change $\Delta G°$ of the virtual process (no current flow) to the voltage, $\Delta G° = -z_i F\phi$. For example, for the reactions

$$\begin{array}{ll} CO + O^{2-} = 2e' + CO_2 & \text{at side I of } ZrO_2 \\ 1/2O_2 \text{ (in air)} + 2e' \rightarrow O^{2-} & \text{at side II of } ZrO_2 \\ \hline CO + 1/2O_2 \rightarrow CO_2 & \Delta G° \end{array} \tag{17.30}$$

the voltage is

$$\phi = \frac{RT}{2F} \ln \left[\frac{P_{O_2}{}^{I}(CO_2/CO)}{P_{O_2}{}^{II}(\text{air})} \right]$$

Thus if carbon monoxide is flowed over one side of the zirconia cell and is oxidized to carbon dioxide by the flow of oxygen ions, the electron flow through an external circuit can do useful work. Operated at about 700°C electrical conversion efficiencies of ~80% are realizable.

It is clear that for power generation both the voltage and current flux are critical. The voltage is determined by the overall chemical reactions, for example, those in Eq. 17.30, the current delivery depends on the rate of diffusion of the current-carrying ions. Thus, a high conductivity is important in addition to completely ionic conductivity ($t_i = 1$). If there is electronic current conduction in the electrolyte, the measured voltage ϕ_m is less than that, ϕ, given in Eq. 17.28:

$$\phi_m = \phi - \frac{1}{z_i F} \int_{u_i{}^{II}}^{\mu_i{}^{I}} t_e \, d\mu_i \tag{17.31}$$

where t_e is the electronic transference number which may vary across the electrolyte.

The β-alumina fast ion conductors have been proposed as electrolytes for a sodium-sulfur storage battery. In this case the sodium ion is the conducting ion. Above 300°C the overall reaction is

$$Na + \tfrac{1}{2}S = \tfrac{1}{2}Na_2S \tag{17.32}$$

and the voltage is

$$\phi = -\frac{RT}{F} \ln \frac{a_{N_a}\,(Na_2S/S)}{a_{N_a}\,(Na)}$$

By application of an excess voltage, ions can be pumped from the low-concentration (activity) side of the electrolyte to the high-activity side, during which the storage battery is charged. In another application, the activity of the ion on one side can be fixed at a known value, and the activity on the other side determined for various unknown conditions.

17.3 Electronic Conduction in Crystals

When mobile electrons or electron holes are present, even in small concentrations, their relatively high mobility, several orders of magnitude greater than ionic mobilities, gives an appreciable contribution to electrical conductivity. In some cases, as shown in Fig. 17.11, *metallic* levels of conductivity result; in other cases the electronic contribution becomes vanishingly small. In all cases the electrical conductivity can be interpreted in terms of carrier concentrations and carrier mobilities, as outlined in Eqs. 17.1 to 17.7.

Electron and Electron-Hole Concentrations. In a few cases of transition metal oxides such as ReO_3, CrO_2, VO, TiO, and ReO_2 there is an overlap of electron orbitals which results in wide unfilled d or f bands such as illustrated in Fig. 17.1. This results in a concentration of 10^{22} to 10^{23} quasi-free electrons per cubic centimeter and essentially metallic conduction.

In the more usual case there is an energy gap E_g between filled and empty bands which is appreciably greater than kT. The concentration of conduction electrons in the pure stoichiometric material is equal to the concentration of electron holes and is given by

$$n = p = 2\left(\frac{2\pi kT}{h^2}\right)^{3/2}(m_e^* m_h^*)^{3/4} \exp\left(-\frac{E_g}{2kT}\right) \tag{17.33}$$

where h is Planck's constant and m_e^* and m_h^* are the *effective* masses of the electron and hole which depend on the strength of interactions between the electron and holes and the lattice and may be larger or smaller than the rest mass of the electron.

The general form of this relation can be directly obtained by considering the equilibrium achieved in the reaction

$$\text{Ground state} \rightleftharpoons \text{free electron} + \text{free hole} - E_g \tag{17.34}$$

Applying the law of mass action to this equation, we obtain

$$n = p = K_i = C_i \exp\left(-\frac{E_g}{2kT}\right) \tag{17.35}$$

from Eq. (17.33)

$$C_i = 2\left(\frac{2\pi m_e^{*1/2} m_h^{*1/2} kT}{h^2}\right)^{3/2} \tag{17.36}$$

Since

$$\sigma = |e|(n\mu_e + p\mu_h) \tag{17.37}$$

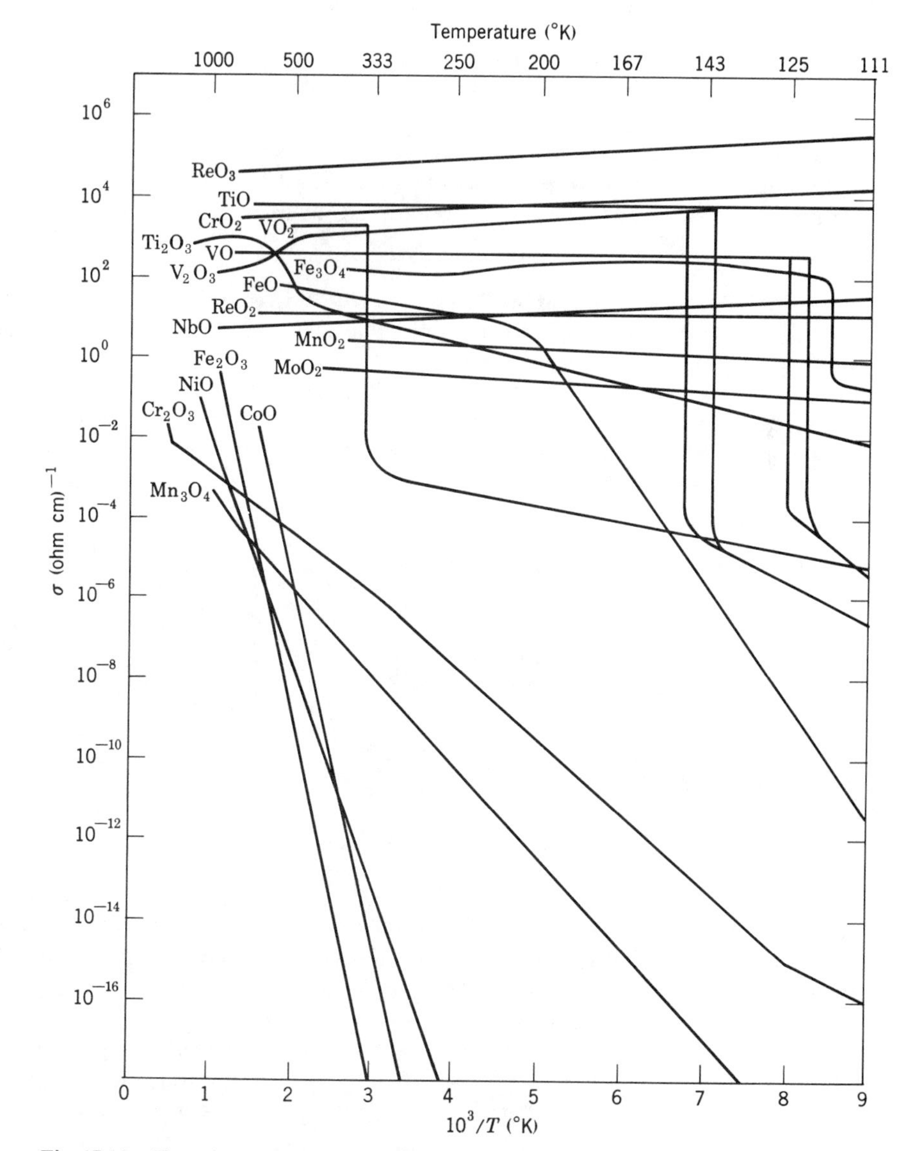

Fig. 17.11. Temperature dependence of the electrical conductivity of several electronically conducting oxides. Courtesy D. Adler.

$$\sigma_i = \left[2|e|\left(\frac{2\pi kT}{h^2}\right)^{3/2}(m_e^* m_h^*)^{3/4} \exp\left(-\frac{E_g}{2kT}\right)\right](\mu_e + \mu_h) \qquad (17.38)$$

The value of the band-gap energy at room temperature for several materials is given in Table 17.4 (the electron and hole concentrations were calculated from Eq. 17.33 in Table 4.4).

Added concentrations of electrons and electron holes may be formed in nonstoichiometric compounds and by introduction of solutes. This has been discussed in Chapter 4 and is considered again in Section 17.6.

Table 17.4. Value of the Energy Gap at Room Temperature for Intrinsic Semiconduction

Crystal	E_g (eV)	Crystal	E_g (eV)
$BaTiO_3$	2.5–3.2	TiO_2	3.05–3.8
C (diamond)	5.2–5.6	CaF_2	12
Si	1.1	BN	4.8
α-SiC	2.8–3	CdO	2.1
PbS	0.35	LiF	12
PbSe	0.27–0.5	Ga_2O_3	4.6
PbTe	0.25–0.30	CoO	4
Cu_2O	2.1	GaP	2.25
Fe_2O_3	3.1	Cu_2O	2.1
AgI	2.8	CdS	2.42
KCl	7	GaAs	1.4
MgO	>7.8	ZnSe	2.6
Al_2O_3	>8	CdTe	1.45

Electron and Hole Mobility. In an ideal covalent semiconductor, electrons in the conduction band and holes in the valence band may be considered as quasi-free particles. The environment of a periodic lattice and its periodic potential may be accounted for by the effective mass of the electron m_e^* and hole m_h^*. In this case the carriers have high drift mobilities in the range of 10 to 10^4 cm²/V sec at room temperature (Table 17.5). This is the case for both the metallic oxides and for covalent semiconductors at room temperature (e.g., Ge, Se, GaP, GaAs, CdS, CdTe, etc.)

Two types of scattering affect the motion of electrons and holes. Lattice scattering results from thermal vibrations of the lattice and increases with the increasing amplitude of vibrations at higher tempera-

Table 17.5. Approximate Carrier Mobilities at Room Temperature

	Mobility (cm^2/V sec)			Mobility (cm^2/V sec)	
Crystal	Electrons	Holes	Crystal	Electrons	Holes
Diamond	1800	1200	PbS	600	200
Si	1600	400	PbSe	900	700
Ge	3800	1800	PbTe	1700	930
InSb	10^5	1700	AgCl	50	
InAs	23,000	200	KBr (100°K)	100	
InP	3400	650	CdTe	600	
GaP	150	120	GaAs	8000	3000
AlN	...	10	SnO_2	160	
FeO, MnO, CoO, NiO	...	~0.1	$SrTiO_3$	6	
			Fe_2O_3	0.1	
			TiO_2	0.2	
			Fe_3O_4	...	0.1
GaSb	2500–4000	650	$CoFe_2O_4$	10^{-4}	10^{-8}

tures. The drift mobility in Eq. 17.13 is given by

$$\mu_e = \frac{e\tau_e}{m_e^*} \qquad \mu_h = \frac{e\tau_h}{m_h^*} \tag{17.39}$$

where τ is the characteristic relaxation time for collisions between the carrier and phonons (lattice vibrations). A second source of scattering is the presence of impurities which distort the periodicity of the lattice. The temperature dependence of τ determines the temperature dependence of the mobility. Since the mobility is proportional to the mean free path between scattering events, the total mobility is given by

$$\mu = \left[\frac{1}{\mu_T} + \frac{1}{\mu_I}\right]^{-1} \tag{17.40}$$

$$\mu_T = \mu_T^\circ T^{-3/2} \qquad \mu_I = \mu_I^\circ T^{+3/5} \tag{17.41}$$

and μ_T° and μ_I° are constants. Thus, the temperature dependence of the mobility term for quasi-free electrons and holes is much smaller than that for their concentration. As a result, the conductivity (Eq. 17.37) has a temperature dependence which is mostly determined by the concentration term.

In ionic host lattices, where there is interaction between orbitals of neighboring ions, there is a polarization of the lattice associated with the presence of electronic carriers; the associate consisting of the electronic carrier plus its polarization field is referred to as a *polaron.* When the association is weak (large polarons), conductivity similar to quasi-free electrons result with a small effective mass. When the electronic carrier plus the lattice distortion has a linear dimension smaller than the lattice parameter, it is referred to as a small polaron, and the mobility is strongly affected by the lattice distortion which must move along with the electronic carrier: this process is often referred to as a *hopping mechanism.* In this case the mobility is strongly decreased and becomes highly temperature-dependent, since the binding energy E_p of the electronic carrier to the polarized lattice must be overcome. The mobility varys exponentially with temperature:

$$\mu_{\text{polaron}} \propto \exp\left[-\frac{E_p}{2kT}\right] \tag{17.42}$$

where $E_p = e^2/\kappa_{\text{eff}} r_p$ ($1/\kappa_{\text{eff}}$ is the difference between the reciprocals of the optical and static dielectric constants and r_p is the dimension of a lattice distortion).

For conduction which primarily results from the mobility of small polarons, the temperature dependence of the conductivity is (from Eq. 17.42)

$$\sigma \propto n\mu \propto \exp\left[\left(-\frac{E_g}{2kT}\right)-\left(\frac{E_p}{2kT}\right)\right] \tag{17.43}$$

This temperature dependence is illustrated for a number of oxide semiconductors in Fig. 17.11. For intermediate degrees of coulombic attraction and for very low temperatures other regimes may result (reference 3). The mobility of small polarons is usually less than 1 $\text{cm}^2/\text{V}\cdot\text{sec}$ and may be much lower.

In order to interpret electrical-conductivity measurements in terms of atomistic processes, the mobility and concentration of electrons and holes must be determined separately, since the conductivity gives only the sum of the concentration-mobility products. The electronic-current-carrier concentrations and mobilities can be determined by combining Hall effect or Seebeck effect measurements with conductivity measurements.

Consider the current flow in the rectangular sample shown in Fig. 17.12. We apply a voltage so that the right-hand side is positive, and by convention current then flows from right to left; but if electrons are current carriers, they flow from left to right, that is, in the positive z direction. Now let us apply a magnetic field in the positive y direction which results in a force on an electron (Lorentz force) of e ($\mathbf{v} \times \mathbf{B}$). Thus the electrons flowing at a velocity $\mathbf{v}$ are deflected upward, which causes a

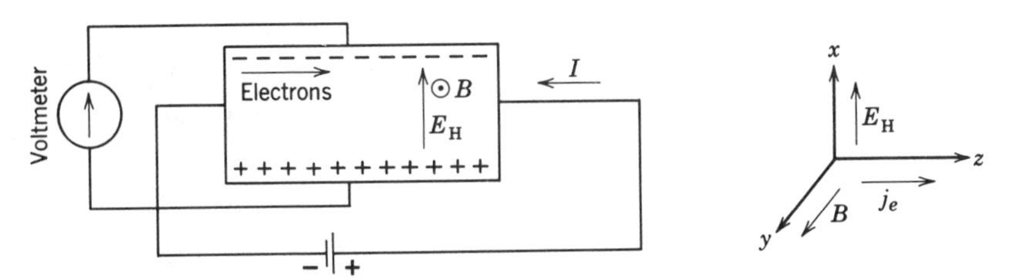

Fig. 17.12. Schematic representation of measurement of the Hall effect.

net accumulation of electrons at the top of the slab and an effective net positive charge accumulation at the bottom. At steady state the voltage gradient E_H, in the x direction opposes any further accumulation due to the motion electrons in the magnetic field:

$$E_H = vB \tag{17.44}$$

or in terms of current density j_e

$$E_H = R_H j_e B \tag{17.45}$$

where R_H is called the Hall coefficient. For electrons, $R_H = -1/ne$; for holes we have a voltage of opposite sign and $R_H = 1/pe$. The Hall mobility can be determined because we now have the concentrations n or p:

$$\mu_H = R_H \sigma \tag{17.46}$$

If both electrons and holes conduct, the Hall constant is

$$R_H = \frac{1}{|e|} \frac{[p\mu_{H,h}^2 - n\mu_{H,e}^2]}{(p\mu_{H,h} + n\mu_{H,e})^2} \tag{17.47}$$

For materials with quasi-free electrons the Hall mobility (Eq. 17.46) is the same as the drift mobility (Eq. 17.11). For polar compounds in which the electronic defects are trapped or localized at specific sites (small polarons), the drift and Hall mobilities are not identical.

Another technique for independently measuring the concentration of carriers is from the thermoelectric or Seebeck effect. When a temperature gradient is imposed on a semiconductor (Fig. 17.13), more electrons are excited into the conduction band at the higher temperature, but the *hot electrons* tend to diffuse to the colder region. When the chemical-potential gradient due to these two effects is equal but opposite to the electric-field gradient, we have a steady state. In addition to the flow of heat by phonons (lattice vibrations), a heat of transport H^* is associated with particle migration in a thermal gradient. Thus if we write an electron-current-flux equation similar to Eq. 9.11 but also include the thermal-

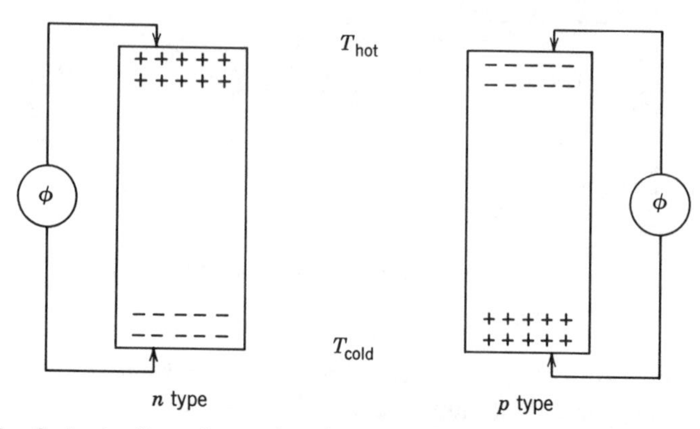

Fig. 17.13. Seebeck effect of a semiconductor. The majority carrier diffuses to the cold end, giving a $\Delta\phi/\Delta T$.

gradient effects, we have

$$j_e = |e|D_e\left[\frac{\partial n}{\partial x} + \frac{en}{kT}\frac{\partial \phi}{\partial x} - \frac{nH^*}{kT^2}\frac{\partial T}{\partial x}\right] \tag{17.48}$$

At steady state $j_e = 0$, and we can rearrange this to give us the Seebeck coefficient Q, V/deg,

$$\begin{aligned} Q = \frac{\partial \phi}{\partial T} = \frac{\partial \phi/\partial x}{\partial T/\partial x} &= \left[-\frac{k\,\partial \ln n}{\partial(1/T)} + H^*\right]\frac{1}{eT} \\ &= \left[-\frac{k\,\partial \ln n}{\partial(1/T)} + H^*\right]\frac{1}{eT} \end{aligned} \tag{17.49}$$

The first term in the brackets can be obtained from Eq. 4.50 (we would use Eq. 4.51 for electron holes). Thus the Seebeck effect is related to concentration of carriers:

$$\begin{aligned} Q_e &= \frac{1}{eT}[(E_F - E_c) + H_e^*] \\ &= \frac{k}{e}\left[\ln\frac{N_v}{n} + \frac{H_e^*}{kT}\right] \end{aligned} \tag{17.50}$$

where N_v is the density of states and $k/e = 86 \times 10^{-6}$ V/deg. The majority carrier thus accumulates at the cold end. When electrons are the majority carrier, the cold end is negative with respect to the hot end. If holes are the majority carriers, the sign of the voltage is opposite, but the magnitude is given by a similar expression:

$$Q_h = \frac{k}{e}\left[\ln\frac{N_v}{p} + \frac{H_h^*}{kT}\right] \tag{17.51}$$

When electrons and holes both contribute to conduction, the Seebeck voltage is

$$Q = \frac{n\mu_e Q_e + p\mu_h Q_h}{n\mu_e + p\mu_h} \tag{17.52}$$

Thus, Seebeck measurements give us independent information about the concentration of carriers which, combined with conductivity measurements, allows us to delineate the separate components of mobility and concentration.

17.4 Ionic Conduction in Glasses

In glasses containing significant concentrations of alkali oxides, particularly sodium, the current is carried almost entirely by alkali ions. The mobility of these ions is much larger than that of the network-forming ions at all temperatures, and at temperatures below the glass transition they are more mobile by several orders of magnitude.

When the current is carried completely by the alkali ions, their transference number is unity, and the conduction characteristics are determined by the concentration and mobility of the alkali ions. A main difference between glasses and crystals is the fact that there is no single value for the energy barrier between sodium ion positions in glasses. Rather, the energy configuration along a coordinate corresponding to sodium migration through the glass is similar to that illustrated in Fig. 17.14. There are often adjacent low-energy positions with but a small energy barrier between them; large-energy barriers occur between occasional adjacent positions in accordance with the random nature of the glass structure.

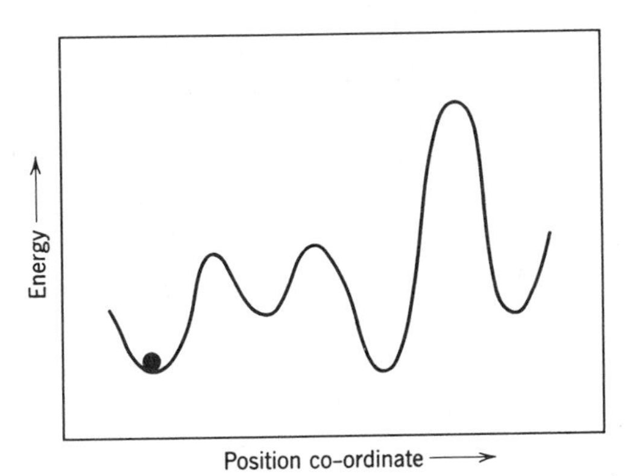

Fig. 17.14. Potential energy barriers along the path of sodium ion migration in a glassy network.

Absorption Current. For a potential ϕ applied to a simple condenser with capacitance C in a circuit with external resistance R, the current I for an ideal dielectric is

$$I = \frac{\phi}{R} \exp\left(-\frac{t}{RC}\right) \tag{17.53}$$

where t is the time. This is called a displacement or polarization current.

Many ceramic insulators, including glasses, have in addition to a large, rapid charging current, given by Eq. 17.53, and a small steady conduction current, associated with their finite resistance, a current of intermediate magnitude which decays over periods of seconds to minutes or longer at room temperature. This intermediate current is termed the absorption current and is shown for a soda-lime-silicate glass in Fig. 17.15. As indicated there, such currents are also observed when the capacitor is discharged by a short circuit; the discharge curve is closely similar in form and magnitude to the curve observed during charging.

The variation of the absorption current with time is not well represented by a single exponential function but requires a series of exponentials for its description:

$$I = A_1 \exp\left(-\frac{t}{\tau_1}\right) + A_2 \exp\left(-\frac{t}{\tau_2}\right) + A_3 \exp\left(-\frac{t}{\tau_3}\right) + \cdots \tag{17.54}$$

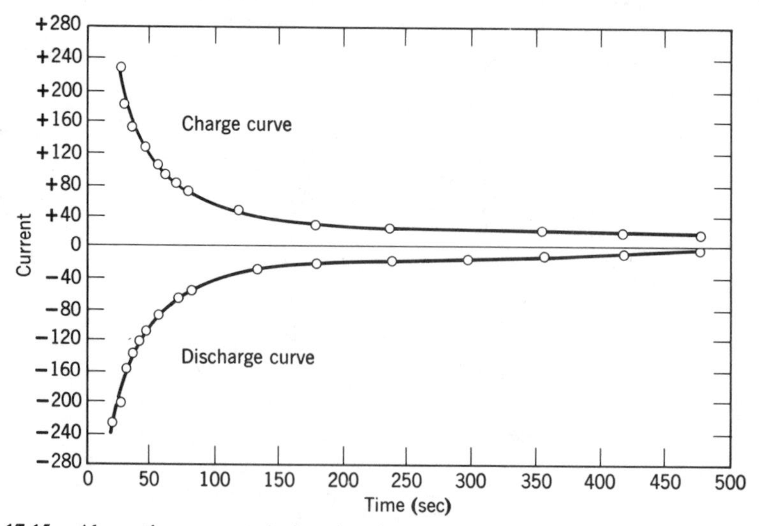

Fig. 17.15. Absorption current during charging and discharge of soda-lime-silicate glass. From E. M. Guyer, *J. Am. Ceram. Soc.*, **16**, 607 (1933).

The number of relaxation times τ required to describe the data is associated with local structural variations, and corresponding variations of local potential barriers, in the glass. As the temperature is raised and ion mobility increases, the time dependence of the absorption current shortens, and it is not commonly observed in dc measurements at temperatures above about 300°C. The absorption current in a rapidly cooled glass is about four times as large as in a well-annealed specimen of the same glass.

Electrode Polarization. In electrolytic conduction ions move from one electrode toward the other. Unless there is a source of replenishment for these ions at the electrode, we soon run out of ions, and the measured conductivity decreases. This electrode polarization requires that suitable conditions and electrodes be used if material properties are to be measured rather than electrode effects. Even then, sufficiently large current–electrode area ratios make the rate of electrode reaction a limiting factor on the amount of current flow. In order to avoid polarization effects in dc measurements of glass conductivity, it is necessary to use an anode material capable of replenishing sodium ions that migrate through the glass. Sodium amalgams and molten sodium nitrate are suitable, among others. For higher temperatures at which the conductivity is increased and the absorption current is not observed, ac measurements are suitable for preventing electrode polarization; alternatively, guard-ring electrodes may be used in dc measurements.

Examples of the potential distribution in glasses subject to a dc field are shown in Fig. 17.16. The results shown there for the alkali-containing glass indicate appreciable space-charge polarization. The greatest part of the potential drop occurs near the cathode, as expected for positively charged mobile ions. In contrast, there is almost no development of space-charge regions in the alkali-free glass, and the potential drop is nearly linear over the entire thickness of the sample.

Temperature Dependence. As the temperature is raised, the conductivity of a glass rapidly increases and over a considerable temperature range can be expressed as

$$\sigma = \sigma_o \exp\left(-\frac{E}{RT}\right) \tag{17.55}$$

where E is an experimental activation energy for conductivity. This activation energy and the temperature dependence of electrical conductivity show a discontinuity at the transformation range corresponding to the freezing of the glass structure at this temperature. In this connection it is of interest to note that the electrical conductivity of a quenched glass (an open-network structure) is larger than that of an annealed glass

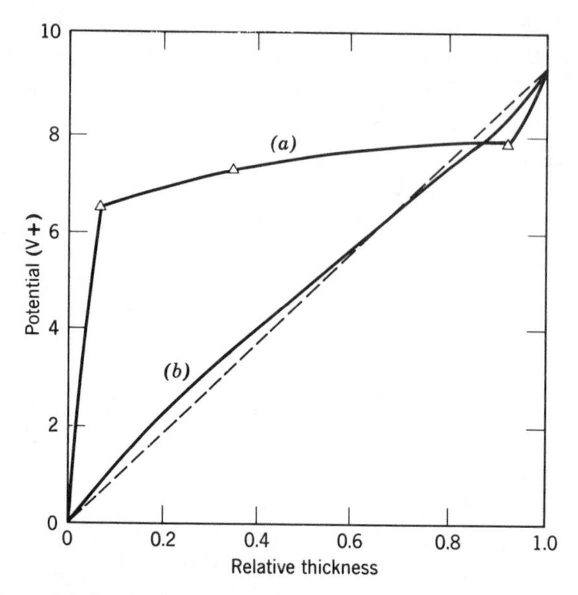

Fig. 17.16. Potential distributions in glasses. (*a*) Alkali-lead-silicate glass after 10 hr at 383°C; (*b*) alkali-free glass after 18 hr at 460°C. Field strength, 160 to 180 V/cm. From T. M. Proctor and P. M. Sutton, *J. Am. Ceram. Soc.*, **43**, 173 (1960).

(dense-network structure). In the molten range, the conductivities of glasses are sometimes shown to vary with temperature as

$$\sigma = \sigma_0 \exp\left[-AT + BT^2 + \cdots\right] \tag{17.56}$$

Effect of Composition. The overriding effect of composition on the conductivity of glasses is related to the type and amount of modifier ion present, particularly alkali ion. In a sodium silicate glass the conductivity increases in direct proportion to the sodium ion concentration (Fig. 17.17). However, for the same sodium ion concentration the conductivity is decreased when CaO, MgO, BaO, or PbO replace a part of the silica to form ternary systems. This results from the fact that the larger modifier ions fit in to plug up the migration paths through the lattice. By virtue of their larger size and higher charge, they are not themselves so easily mobile. The results of a systematic investigation of this effect in Na_2O–RO–SiO_2 glasses containing 20 mole% Na_2O and 20 mole% RO (Fig. 17.18) indicate that the effectiveness of an oxide in increasing the resistivity increases smoothly as the radius of the metal ion increases. These results are, however, in disagreement with the earlier work of M. Fulda, shown in Fig. 17.19, which indicates CaO to have the most

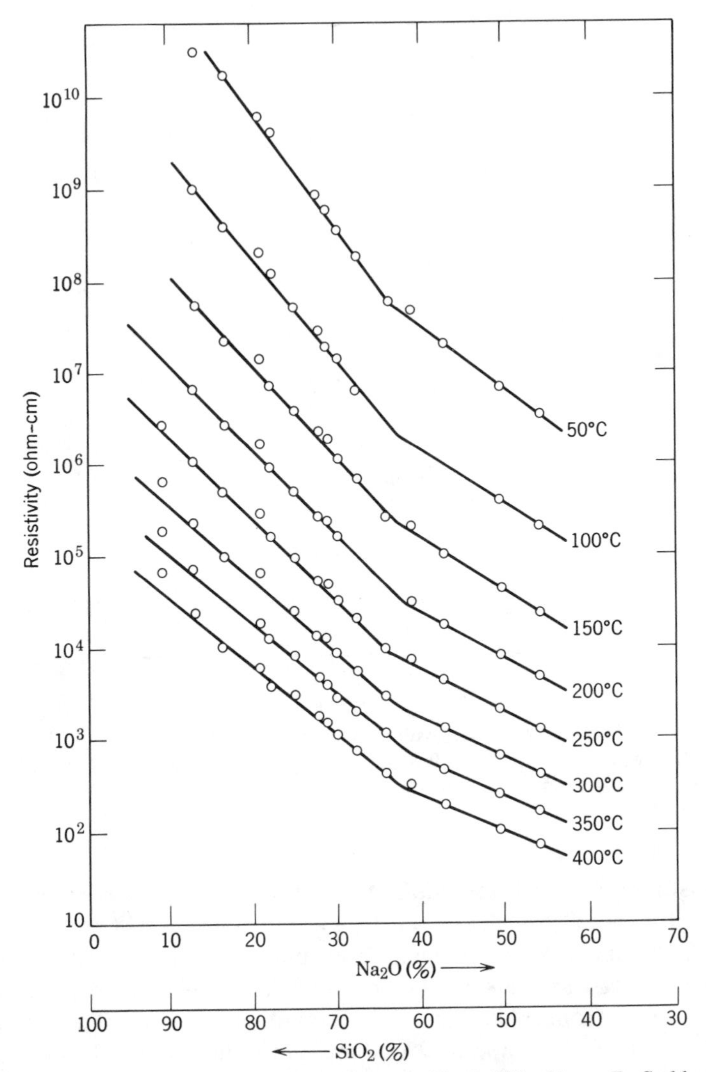

Fig. 17.17. Conductivity of glasses in the system Na_2O–SiO_2. From E. Seddon, E. J. Tippett, and W. E. S. Turner, *J. Soc. Glass Technol.*, **16**, 950 (1932).

pronounced effect in increasing the resistivity of a $0.18Na_2O$–$0.82SiO_2$ glass.

The substitution of Al_2O_3 for SiO_2 in Na_2O–SiO_2 glass has the interesting effect of increasing the conductivity, accompanied by a pronounced drop in the activation energy, which passes through a minimum for

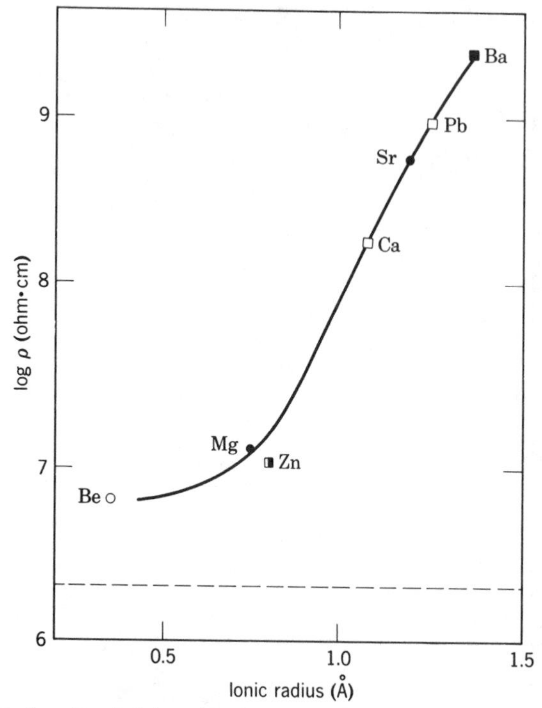

Fig. 17.18. Variation of resistivity with divalent ion radius for $0.20Na_2O–0.20RO–0.60SiO_2$ glasses. Dashed line = resistivity of $0.20Na_2O–0.80SiO_2$ glass. From O. V. Mazurin and R. V. Brailovskii, *Soviet Phys. Solid State*, **2**, 243 (1960).

compositions in which the Na_2O/Al_2O_3 ratio is unity. This shown by the data in Fig. 17.20 for glasses in the system $Na_2O \cdot X Al_2O_3 \cdot 2(4-X)SiO_2$.

A. E. Owen and R. W. Douglas* have compared the activation energies in fused silicas and alkali-silicate and alkali-lime-silicate glasses. The values range from 34 kcal/mole in a synthetic fused silica containing about 4×10^{-7}% Na to about 12 kcal/mole in a $Na_2O–SiO_2$ glass containing about 50% Na_2O. The resistivities at 350°C correspondingly vary from about 10^{12} ohm-cm for the fused silica to about 10^2 ohm-cm for the sodium silicate glass.

In alkali silicate or alkali-lime-silicate glasses, the conductivity at a given temperature is generally observed to decrease in the order Li > Na > K. The activation energies decrease with increasing alkali oxide content, as shown in Fig. 17.21 for $Na_2O–SiO_2$, $K_2O–SiO_2$, and $Cs_2O–SiO_2$

J. Soc. Glass Technol., **43**, 159 (1959).

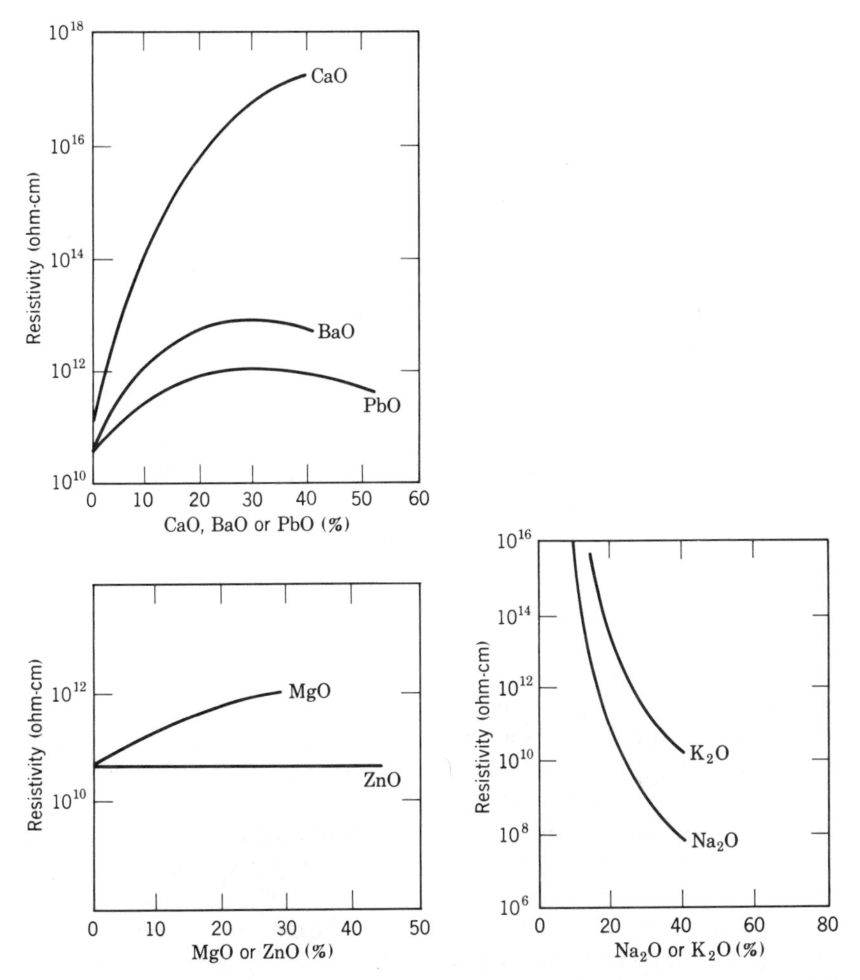

Fig. 17.19. Effect of replacement of SiO_2 by other oxides on a weight percent basis in an $0.18Na_2O$–$0.82SiO_2$ glass. From M. Fulda, *Sprechsaal*, **60**, 769, 789, 810 (1927).

glasses. The differences shown there between the results obtained by different investigators are probably associated with differences in the chemical or structural states of the glass specimens investigated. These differences are particularly pronounced in the case of glasses in the Na_2O–SiO_2 system, in which a metastable miscibility gap extends from SiO_2 out to about 20% Na_2O (see Fig. 3.15).

The effects of thermal history have been investigated in a number of studies. In typical cases, the conductivity of a well-annealed glass is

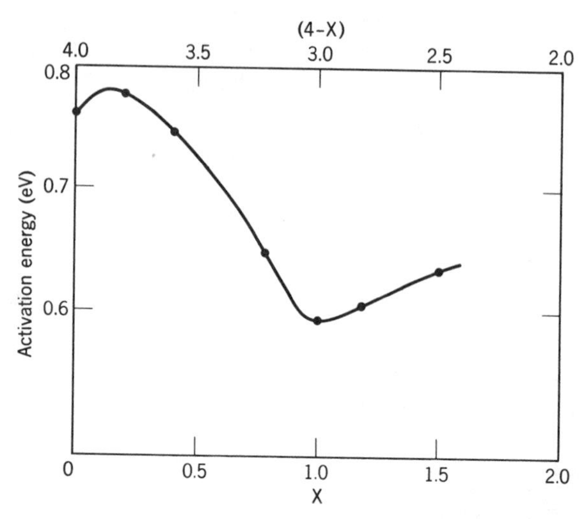

Fig. 17.20. Variation of activation energy for dc conduction with composition in the system $Na_2O{\cdot}XAl_2O_3.\ 2(4-X)SiO_2$. From J. O. Isard, *J. Soc. Glass Technol.*, **43**, 113 (1959).

found to be smaller by about an order of magnitude than an unannealed glass of the same composition.

Mixed-Alkali Effect. When one alkali oxide is progressively substituted for another, it is observed that the resistivity does not vary linearly with the fraction substituted. Rather, it goes through a pronounced maximum, often but not always in the range of composition in which the two alkalis are present in nearly equimolar amounts. This behavior is shown in Fig. 17.22 for Li_2O–Na_2O–SiO_2 glasses. The activation energies for conduction show a similar variation with the concentration ratio of the two alkalis, as shown by the data in Fig. 17.23. Of the glasses described in this figure, those in the Cs–Li and Cs–Na systems exhibit fine-scale phase separation; those in the Cs–K and Cs–Rb are homogeneous on the scale of electron-microscope observations.

In a number of cases, the diffusion coefficients of the alkali ions have been measured for glasses in the same systems in which conductivity data, and mechanical relaxation data as well, have been obtained. The results indicate that the mixed-alkali-peak maximum occurs in the range of composition in which the diffusion coefficients of the two alkali ions are equal.

The mixed-alkali effect, which appears in mechanical and dielectric relaxation as well as dc conductivity, is associated with an interaction between ions of different types in the glass. The magnitude of the effect,

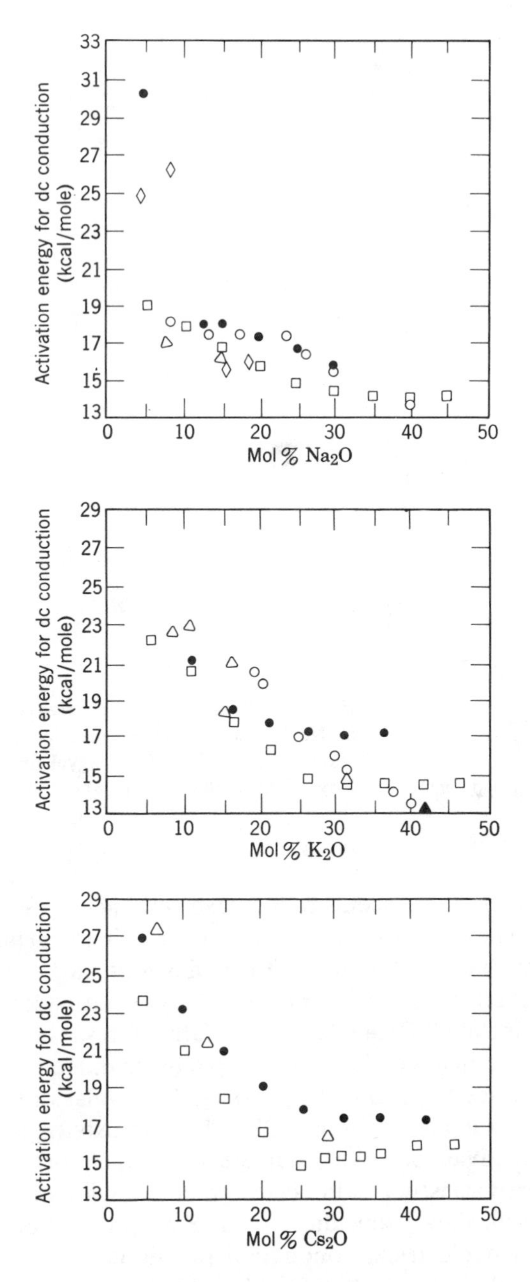

Fig. 17.21. Variation of activation energy with composition for Na_2O–SiO_2, K_2O–SiO_2, and Cs_2O–SiO_2 glasses. Different symbols in each plot indicate results of different investigators. From R. M. Hakim and D. R. Uhlmann, *Phys. Chem. Glasses*, **12**, 132 (1971).

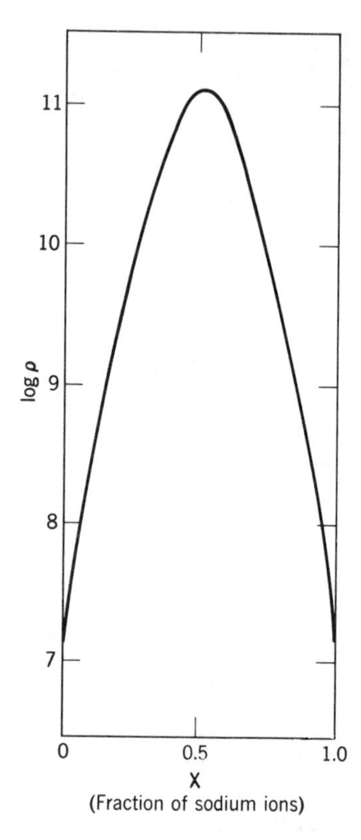

Fig. 17.22. Variation of resistivity with substitution of Na for Li in a silicate glass containing 26% total alkali oxides. From B. Longyel and Z. Boksay, *Z. Phys. Chem.*, **204**, 157 (1955).

and hence the interaction, seems to increase with increasing difference in size between the ions and to decrease with decreasing total alkali content. For sufficiently dilute solutions, with sufficiently large interionic distances, the interaction is small, and the effect is not observed.

Conductivity in Alkali-Free Glasses. Relatively few studies have been carried out of electrical conductivity in alkali-free oxide glasses, and the results obtained are far from unequivocal. For a glass of composition $PbO \cdot SiO_2$, for example, data on Pb^{2+} ion diffusivity and electrical conductivity are available over a range of temperature. The results* indicate measured resistivities in close agreement with values calculated from the diffusion data using the Nernst-Einstein relation (Eq. 17.23). This suggests that electrical conduction in this glass, and presumably other $PbO–SiO_2$ glasses as well, takes place by the motion of Pb^{2+} ions. In

*See G. C. Milnes and J. O. Isard, *Phys. Chem. Glasses*, **3**, 157 (1962).

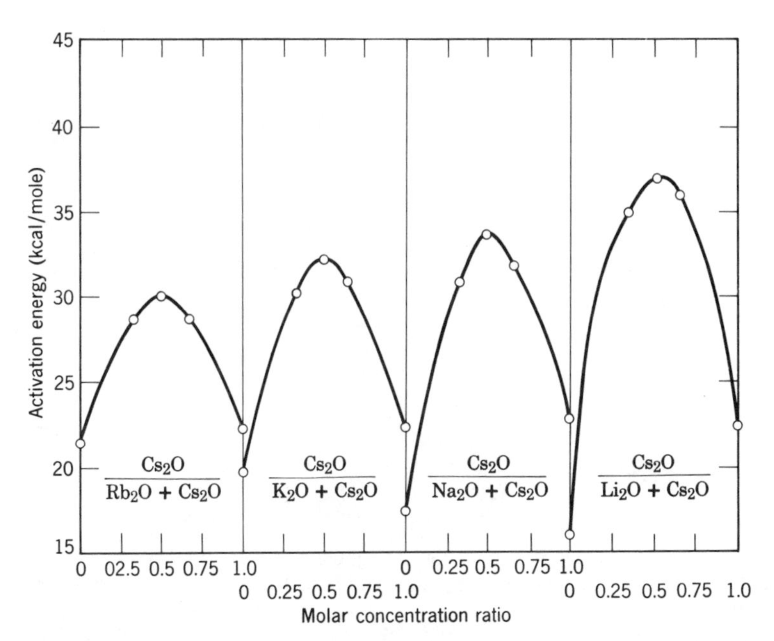

Fig. 17.23. Variation of activation energy for dc conduction with substitution of other alkali ions for Cs in binary silicate glasses containing 15% alkali oxide. From R. M. Hakim and D. R. Uhlmann, *Phys. Chem. Glasses*, **8**, 174 (1967).

contrast, subsequent electrolysis measurements* suggested that the current is carried by H^+ ions or electrons and not by Pb^{2+} ions, and it was suggested that the agreement between measured resistivities and those calculated from diffusion data was fortuitous. Still later work† has indicated, however, that variations in the oxidation-reduction state of the $PbO \cdot SiO_2$ glass have no significant effect on its electrical behavior and tends to support the original suggestion. Studies of several $MO–B_2O_3–Al_2O_3$ glasses have been carried out. The results for glasses of approximately the composition $2MO \cdot xB_2O_3 \cdot Al_2O_3$ are shown in Fig. 17.24 and indicate that the conductivity depends little, if at all, on the nature of the alkaline earth metal. Some of these glasses are noted for electrical resistivities which are higher than the purest available grade of fused silica.

When glasses of high resistivity are required, the alkali content should be kept to a minimum and divalent ions such as lead and barium used as

*K. Hughes, J. O. Isard, and G. C. Milnes, *Phys. Chem. Glasses*, **9**, 43 (1968).
†B. M. Cohen, D. R. Uhlmann, and R. S. Shaw, *J. Non-Cryst. Solids*, **12**, 177 (1973).

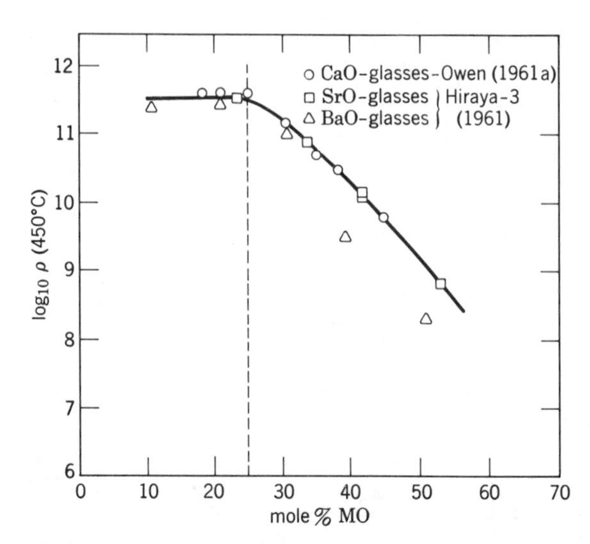

Fig. 17.24. Variation of resistivity at 450°C with composition in $MO–B_2O_3–Al_2O_3$ glasses. From O. V. Mazurin, G. A. Pavlova, E. I. Lev, and E. K. Leko, *Soviety Phys. Tech. Phys.*, **2**, 2511 (1957).

modifiers. These glasses combine a high resistivity with good working properties (viscosity-temperature dependence) and a reasonable working temperature and are free from devitrification difficulties. Lime-alumina-silica glasses of even higher resistivity can be made, but they are difficult to work and have a tendency toward devitrification. The resistivity of some commercial glasses is illustrated in Fig. 17.25.

17.5 Electronic Conduction in Glasses

Certain oxide glasses which contain multivalent transition metal ions display electronic conductivity. The best known are the vanadium phosphate and iron phosphate glasses, but electronically conducting glasses can be prepared with vanadium, iron, cobalt, or manganese in a matrix of phosphate, borate, or silicate. More recently, semiconducting chalcogenide glasses have been studied; these are based on sulfur, selenium, and tellurium either alone or in combination with phosphorus, arsenic, antimony, or bismuth. Electronic conductivity in amorphous germanium, silicon, and silicon carbide has also been well documented.

The temperature dependence of the electronic conductivity of amorphous materials is similar to that of crystalline materials. In vanadium phosphate and iron phosphate glasses, as the concentration of the transition metal oxide is increased, the conductivity increases (Fig. 17.26). For low concentrations of transition metal ions, the conductivity is very

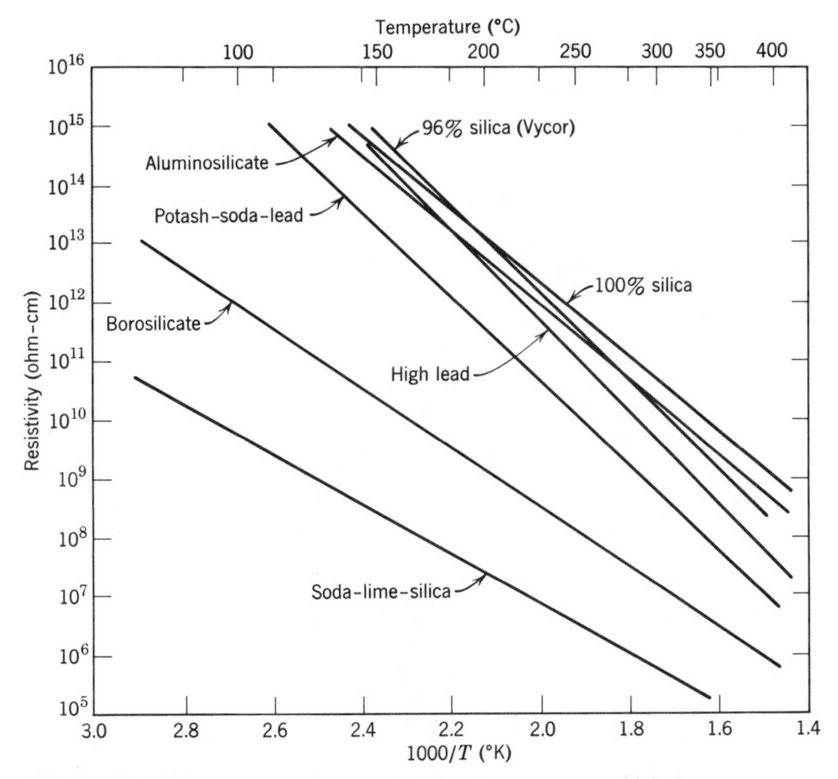

Fig. 17.25. Direct-current volume resistivity of some commercial glasses.

sensitive to the number of adjacent ions of different valence, but above a concentration of about 10%, each ion has, on the average, a neighboring transition metal ion. For higher concentrations, the change in the conductivity is more a function of the valence states of the transition metal ions. For example, in vanadium phosphate glasses the conductivity is related to the relative amounts of the pentavalent and tetravalent (V^{5+} and V^{4+}) ions. A conductivity maximum occurs when the molar ratio V^{4+}/V_{total} is about 0.1 to 0.2. Irrespective of the third component present in iron phosphate glasses, the conductivity is a maximum when a ratio of Fe^{3+}/Fe_{total} is about 0.4 to 0.6. Figure 17.27 indicates the conductivity maximum (resistivity minimum) but also a change from p type to n type conduction as the trivalent iron content is increased. The number of electrons or holes available for conduction is high; however, their mobility is low ($\ll 0.1\ cm^2/V\ sec$). The activation energy for conduction of a phosphate glass (80% V_2O_5, 20% P_2O_5) increases with increasing temperature.*

*A. P. Schmid, *J. Appl. Phys.*, **39**, 3140 (1968).

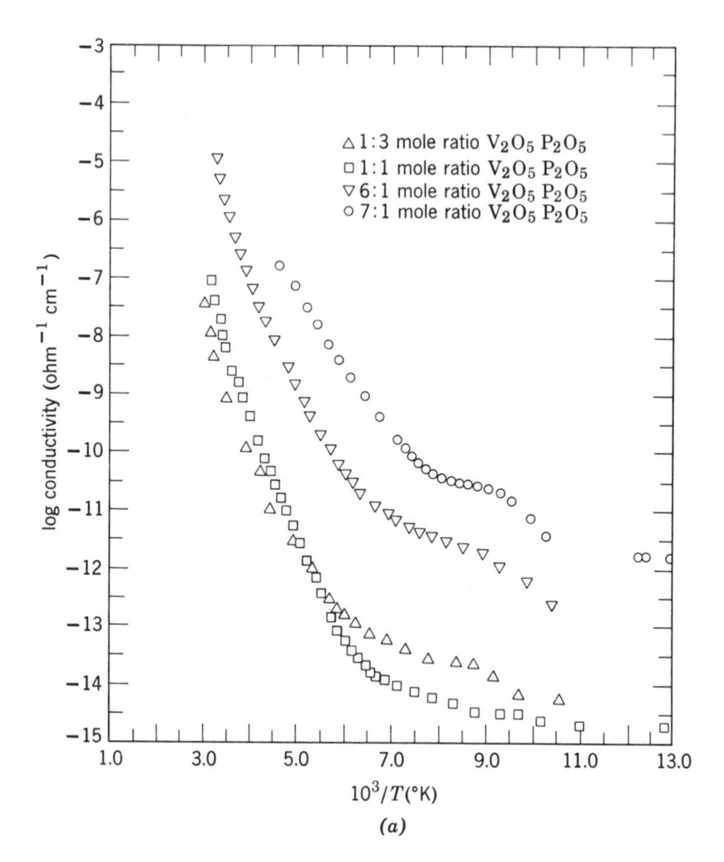

Fig. 17.26. (*a*) Log conductivity as a function of $1/T$ for four typical V_2O_5–P_2O_5 glasses. From A. B. Schmid, *J. Appl. Phys.*, **39**, 3140 (1968).

Electron-spin-resonance measurements in the glass indicate that the V^{4+}/V^{5+} ratio does not change appreciably with increasing temperature, indicating that the number of charge carriers is not a function of temperature. Thus the increase in the activation energy with temperature is attributed to the mobility term rather than the concentration term.†

The electrical conductivity of most chalcogenide glasses is also thermally activated and resembles that of intrinsic semiconductors (Fig. 17.28). The most striking characteristic of chalcogenides quenched from the melt or thin films deposited from the vapor (silicon, germanium) is the insensitivity of their conductivity to impurities. In crystalline semiconductors, tiny concentrations of foreign atoms cause large changes in the

†See N. F. Mott, *J. Non-Cryst. Solids*, **1**, 1 (1968), for discussion.

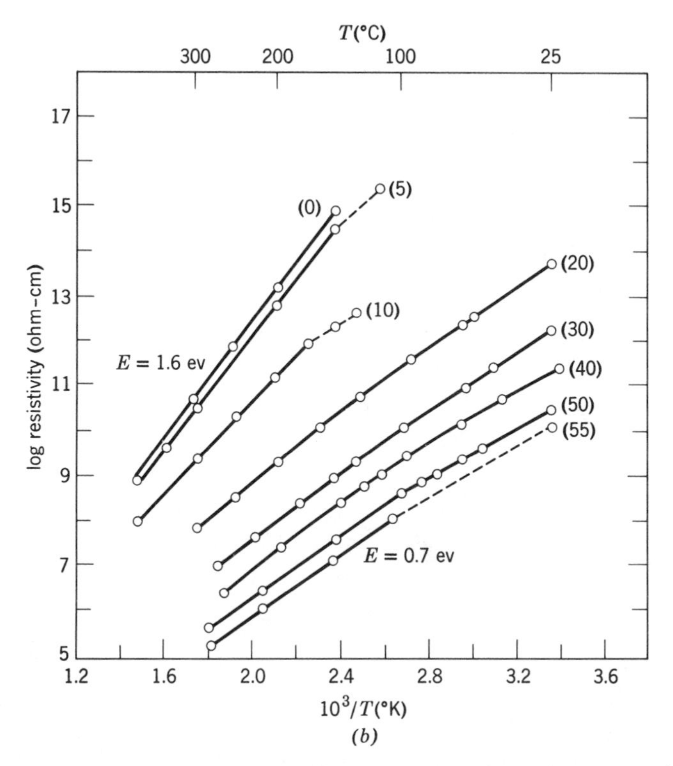

Fig. 17.26 (*Continued*). (*b*) Logarithm of resistivity versus $1/T$ for glasses containing 55% (FeO + MgO)–45% P_2O_5. The numbers in parentheses are the percentages of FeO. From K. W. Hansen, *J. Electrochem. Soc.*, **112**, 994 (1965).

conductivity and in the type of electrical charge carriers, but the amorphous nature of glassy semiconductors yields enough dangling bonds and localized charge sites so that they are insensitive to impurities. The band gap for the amorphous semiconductors is somewhat less than that of their crystalline counterparts (Fig. 17.29), but in most cases the intrinsic conductivity is higher for crystalline than for amorphous materials. The conductivity behavior, in general, is more similar to elemental semiconductors than to the small-polaron oxides discussed previously. However, the mobility of the charge carriers is usually low (< 0.1 cm²/V sec). A more detailed discussion of the theories and mechanisms of the chalcogenide type glasses is given by N. F. Mott and E. A. Davis.*

**Electronic Processes in Non-Crystalline Materials*, Oxford University Press, London, 1971.

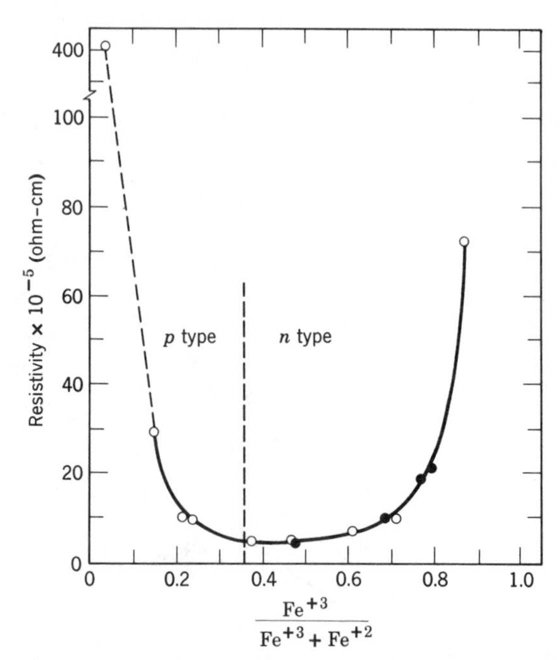

Fig. 17.27. The effect of Fe^{+3}/Fe_{total} on resistivity at 200°C for glasses containing 55% FeO–45% P_2O_5. From K. W. Hansen, *J. Electrochem. Soc.*, **112**, 994 (1965).

17.6 Nonstoichiometric and Solute-Controlled Electronic Conduction

Most oxide semiconductors are either doped to create extrinsic defects or are annealed under conditions in which they become nonstoichiometric. These effects have been carefully studied in many oxides, but the precise nature of the low mobility value is often difficult to measure. Reported conductivities are often at variance because the variable impurity effects and past thermal history overwhelm other effects. In this section we consider several electronically conducting ceramics to point out important features in their behavior. A partial list of impurity semiconductors is given in Table 17.6. The strong effect of impurities on the properties of semiconductors results from the fact that the impurity atoms introduce new localized energy levels for electrons intermediate between the valence band and the conduction band. If the new energy levels are unoccupied and lie close to the energy of the top of the valence band, it is easy to excite electrons out of the filled band into these new *acceptor* levels. This leaves an electron hole in the valence band that can contribute to electrical conductivity. Positive carrier (p-type) oxide conductors most commonly arise as a result of nonstoichiometric com-

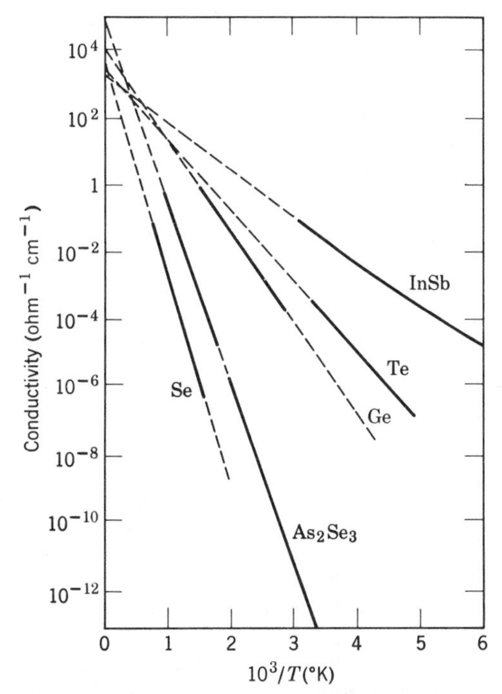

Fig. 17.28. Intrinsic electrical conductivity of various amorphous semiconductors as a function of $1/T$. Dashed lines are extrapolated. From C. C. Sartain et al., *J. Non-Crystalline Solids*, **5**, 55 (1970).

position with a decreased metal content ($Cu_{2-x}O$ is an example) and are sometimes called *deficit* semiconductors. If the impurity additions have filled electron levels close to the energy level of the conduction band, electrons may be excited from impurity atoms into the conduction band; these are called *donor* levels. The electron excited into the conduction band is able to contribute to the conductivity. Negative carrier (*n*-type) oxide conductors most commonly result from a nonstoichiometric composition with an excess metal content ($Zn_{1+x}O$ is an example) and are sometimes called *excess* semiconductors.

Strontium Titanate. For $SrTiO_{3-x}$, density, lattice parameter, and conductivity data have demonstrated that the mass action relationships which we have discussed in Chapter 4 are valid.* Single crystals were reduced at 1200 to 1400°C in gas mixtures ($P_{O_2} = 10^{-7}$ to 10^{-12} atm) in

*H. Yamada and G. R. Miller, *J. Solid State Chem.*, **6**, 169 (1973).

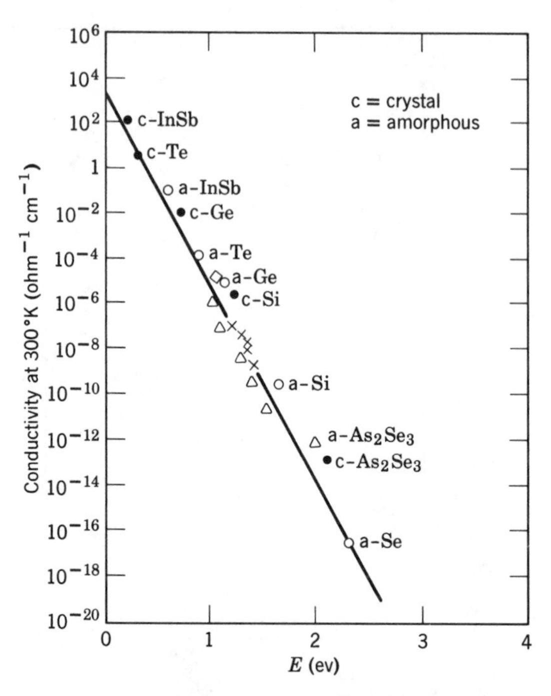

Fig. 17.29. Relation between electrical conductivity and activation energy E for conduction for various crystalline and amorphous semiconductors. From J. Stuke, *J. Non-Cryst. Solids*, **4**, 1 (1970).

Table 17.6. Partial List of Impurity Semiconductors

		n-Type			
TiO_2	Nb_2O_5	CdS	Cs_2Se	$BaTiO_3$	Hg_2S
V_2O_5	MoO_2	$CdSe$	BaO	$PbCrO_4$	ZnF_2
U_3O_8	CdO	SnO_2	Ta_2O_5	Fe_3O_4	
ZnO	Ag_2S	Cs_2S	WO_3		
		p-Type			
Ag_2O	CoO	Cu_2O	SnS	Bi_2Te_3	MoO_2
Cr_2O_3	SnO	Cu_2S	Sb_2S_3	Te	Hg_2O
MnO	NiO	Pr_2O_3	CuI	Se	
		Amphoteric			
Al_2O_3	SiC	PbTe	Si	Ti_2S	
Mn_3O_4	PbS	UO_2	Ge		
Co_3O_4	PbSe	IrO_2	Sn		

which the following reduction reaction is assumed:

$$O_O = 1/2O_2(g) + V_O^{\cdot\cdot} + 2e' \tag{17.57}$$

The equilibrium constant for the reaction is,

$$\frac{[e']^2[V_O^{\cdot\cdot}]P_{O_2}{}^{1/2}}{[O_O]} = K(T) \tag{17.58}$$

and from this the concentration of free electrons for conduction is given as

$$[e'] = (2K(T))^{1/3}P_{O_2}{}^{-1/6} = 2^{1/3}P_{O_2}{}^{-1/6}\exp\left(-\frac{\Delta G}{3RT}\right) \tag{17.59}$$

The experimental corroboration of Eq. 17.59 is shown in Fig. 17.30*a*, where electron concentration data were obtained from Hall measurements (Eq. 17.45). The temperature dependence of Eq. 17.59 (Fig. 17.30*b*) indicates that the enthalpy to create an oxygen vacancy and two conduction-band electrons is 5.76 eV (133 kcal/mole). The defect model given in Eq. 17.57 is confirmed by comparison of measured crystalline densities on quenched specimens and calculated densities (Fig. 17.30*c*). In this case the oxygen vacancies become doubly ionized donors of electrons to the conduction band. The mobility of the electrons at room temperature is 6 cm^2/V sec.

Zinc Oxide. When ZnO is heated in a reducing atmosphere in which zinc gas is present, the zinc content of the oxide increases to form an

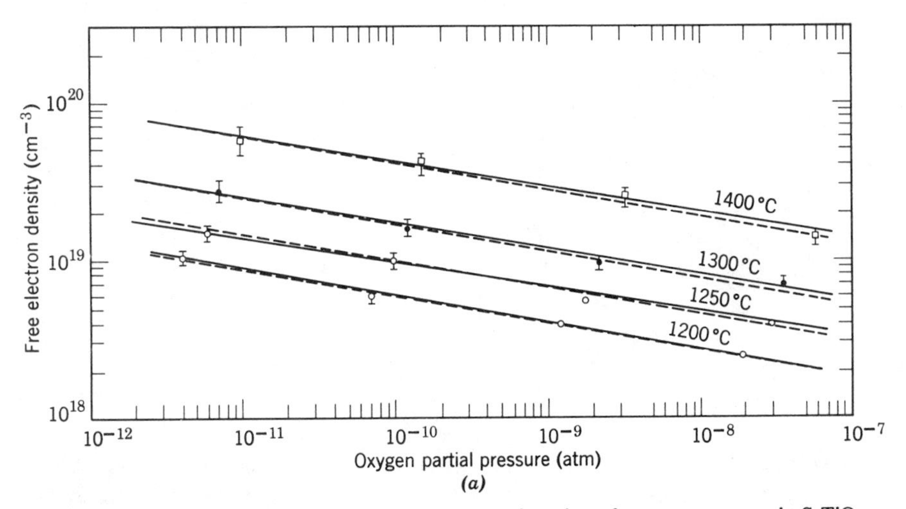

Fig. 17.30. (*a*) Concentration of free electrons as a function of oxygen pressure in $SrTiO_3$. Dashed lines indicate $P_{O_2}^{1/6}$ dependence.

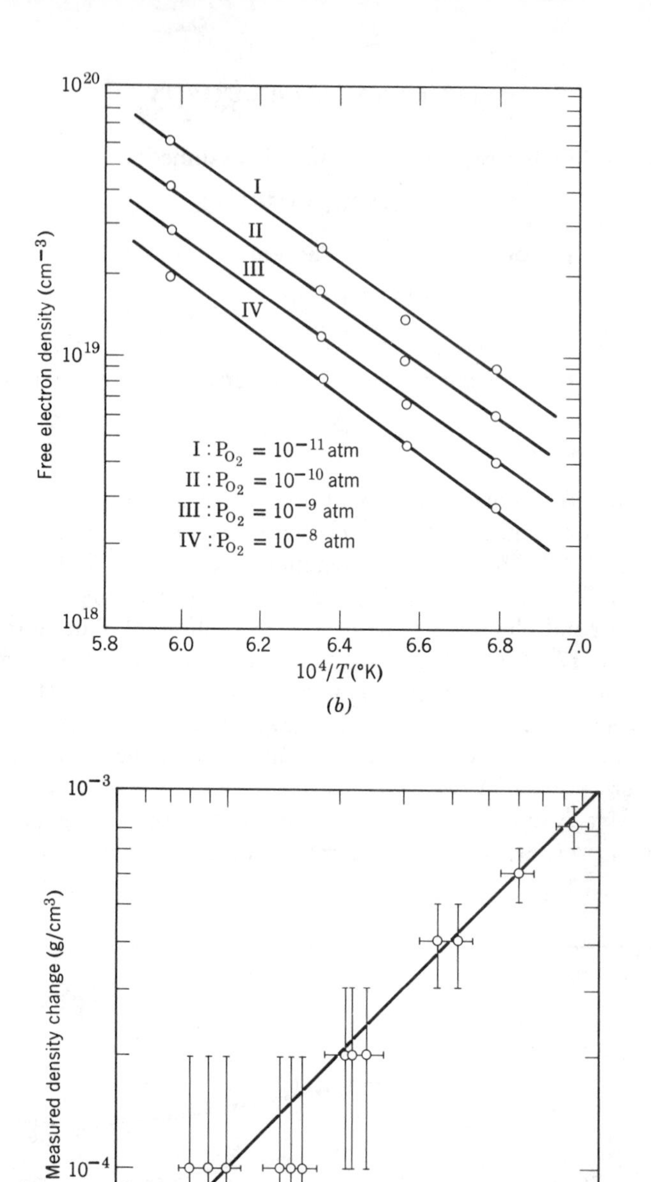

Fig. 17.30 (*Continued*). (*b*) Temperature dependence of the free electron concentration (Eq. 17.59) at fixed oxygen pressures. (*c*) Measured densities compared to those calculated from the defect reaction involving oxygen vacancy formation. From H. Yamada and G. R. Miller, *J. Solid State Chem.*, **6**, 169 (1973).

excess semiconductor. Maintaining electrical neutrality for the reaction with the gas phase, we can write

$$x\,\mathrm{Zn}(g) + \mathrm{ZnO}(c) = \mathrm{Zn}_{1+x}\mathrm{O}(c) \qquad (17.60)$$

Since we know that zinc atoms fit into interstitial position in the structure (see Chapter 4), this same relationship can be written as

$$\mathrm{Zn}(g) = \mathrm{Zn}_i \text{ (in ZnO)} \qquad (17.61)$$

These interstitial zinc atoms correspond to impurity levels in the structure. They may be expected to become ionized by a reaction such as

$$\mathrm{Zn}_i = \mathrm{Zn}_i^{\cdot} + e' \qquad K(T) \propto e^{-E_1/kT} \qquad (17.62)$$

and may become doubly ionized by a second reaction such as

$$\mathrm{Zn}_i^{\cdot} = \mathrm{Zn}_i^{\cdot\cdot} + e' \qquad K'(T) \propto e^{-E_2/kT} \qquad (17.63)$$

where E_1 and E_2 are the ionization energies. In order to maintain electrical neutrality, it is necessary that the electron concentration and impurity concentration be the same in a crystal. However, since the dielectric constant of the crystal is usually high, the electron that is dissociated in accordance with Eq. 17.62 is situated at a rather large average distance from the positive core $(\mathrm{Zn}_i^+)^{\cdot}$.

We can write mass-action expressions giving the ratios of concentrations for each of the reactions corresponding to the formation and ionization of interstitial atoms. Which species predominates depends on the energy changes involved. The overall concentration of interstitial atoms is determined by the vapor pressure, as indicated in Eq. 17.61. The extent of dissociation into electrons and positive interstitial ions is determined by their relative energies. These can be represented in terms of an energy-band scheme such as that shown in Fig. 17.31. The extent of interstitial ionization also depends on the temperature. At sufficiently low temperatures the electrons are in the lowest energy levels corresponding to filled impurity bands. As the temperature level is raised, the fraction of these atoms that are excited into the conduction bands increases. At a sufficiently high temperature this ratio is so high that substantially all the impurity atoms are ionized.

Consequently, impurity semiconductor properties depend on the ambient gas pressure (which is at equilibrium at elevated temperatures but may be quenched to low temperatures corresponding to nonequilibrium conditions) and also on the temperature (which determines how many of these are dissociated into mobile electrons). As for zinc oxide, it turns out that the ionization energy is such that at room temperature and above all the interstitial atoms have a +1 valence; therefore the overall reaction of

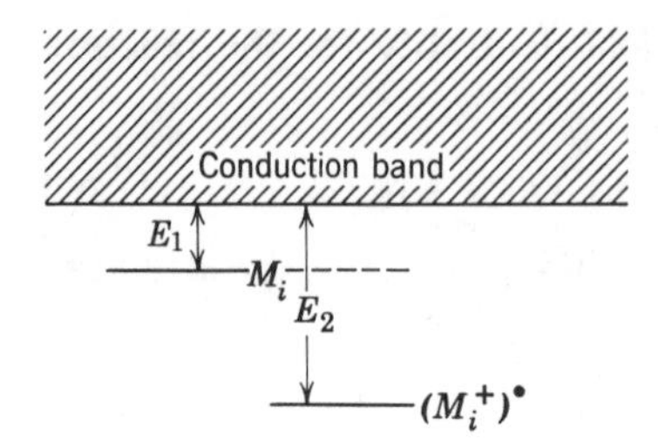

Fig. 17.31. Schematic representation of energy levels in excess semiconductors such as $Zn_{1+x}O$.

equilibrium with the gas phase can be written as

$$Zn(g) = Zn_i^{\cdot} + e' \tag{17.64}$$

which is a combination of Eqs. 17.61 and 17.62. The mass-action expression for this relationship is

$$K = \frac{[Zn_i^{\cdot}][e']}{P_{Zn(g)}} \tag{17.65}$$

If the material is in the extrinsic range in which reaction 17.64 is the main source of conduction electrons, these are equal in concentration to the interstitial zinc ions, $[Zn_i^{\cdot}] = n$, so that the concentration of electrons is proportional to the square root of zinc pressure or to $P_{O_2}^{-1/4}$:

$$[e'] = K^{1/2}P_{Zn(g)}^{1/2} = K^{1/2}K_{O_2}^{1/2}P_{O_2}^{-1/4} \tag{17.66}$$

where K_{O_2} is the equilibrium constant for $Zn(s) + \frac{1}{2}O_2(g) = ZnO(s)$.

As shown in Fig. 17.32, this is found to be true by experiment. Under these conditions the ionization energy E_1 is small compared with the energy change for Eq. 17.61. Consequently, the temperature dependence at constant zinc pressure is largely determined by the heat of reaction and follows an exponential relationship, $\log \sigma \sim 1/T$, which is determined by the impurity concentration, since the impurity dissociation is nearly complete and electron mobility does not much change with temperature.

At low temperatures, at which the equilibrium is frozen in so that the impurity concentration is fixed, the temperature dependence for conductivity is determined by electron mobility, and the conductivity is nearly independent of temperature. At even lower temperatures the ionization of

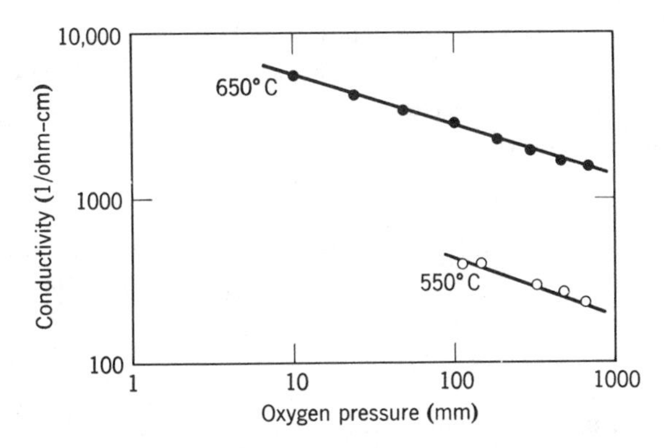

Fig. 17.32. Effect of oxygen pressure on electrical conductivity of zinc oxide. From H. H. V. Baumbach and C. Wagner, *Z. Phys. Chem.*, **B22**, 199 (1933).

impurities or the dissociation of impurities is incomplete, so that very low temperatures may again be a region in which the conductivity decreases rapidly with temperature. The details of these relationships over a wide temperature range have but seldom been completely worked out for semiconducting materials, since minor concentrations which have been difficult to control have a large effect. The main exceptions to this statement are silicon and germanium, for which extremely pure single crystals have been available as a solvent, allowing detailed studies of controlled impurity effects.

Copper Oxide. Results are essentially similar if we consider *deficit* or p-type oxide semiconductors. For example, we can write for copper oxide

$$Cu_2O = Cu_{2-x}O + xCu(g) \tag{17.67}$$

corresponding to a reduction in the copper content. This can be interpreted in terms of oxygen content by the equilibrium to form cuprous oxide:

$$xCu(g) + \frac{x}{4}O_2(g) = \frac{x}{2}Cu_2O(s) \tag{17.68}$$

Adding these equations gives

$$\left(1 - \frac{x}{2}\right)Cu_2O = Cu_{2-x}O - \frac{x}{4}O_2(g) \tag{17.69}$$

which is equivalent to Eq. 17.67, except that it is given in terms of a different variable. If we start with a stoichiometric Cu_2O and remove a copper atom, the charge balance requires that an electron from one of the adjacent oxygen ions be taken along with the copper ion when it is removed from the lattice. Formation of the impurity center (equivalent to Eq. 17.67) corresponds to

$$\mathrm{Cu_{Cu}} = \mathrm{Cu}(g) + V_{\mathrm{Cu}} \tag{17.70}$$

Replacement of the missing electron at the impurity center from the valence-band electrons corresponds to an effective negative charge at the vacancy:

$$V_{\mathrm{Cu}} = V'_{\mathrm{Cu}} + h^{\cdot} \qquad K(T) \propto e^{-E_1/kT} \tag{17.71}$$

If we allow for a possible second ionization state, we have

$$V'_{\mathrm{Cu}} = V''_{\mathrm{Cu}} + h^{\cdot} \qquad K'(T) \propto e^{-E_2/kT} \tag{17.72}$$

The schematic acceptor states are shown in Fig. 17.33.

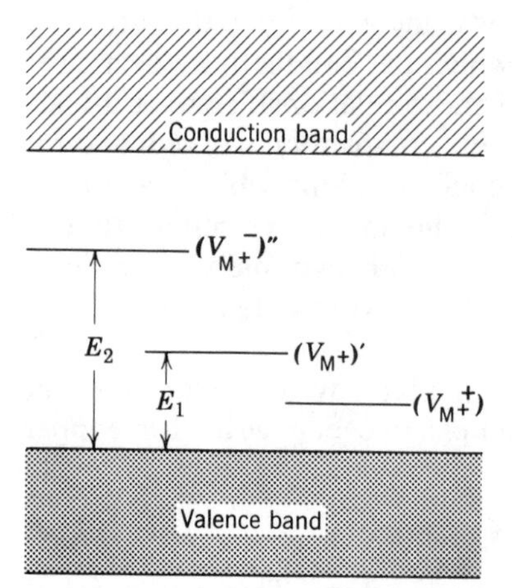

Fig. 17.33. Schematic representation of energy levels in a deficit semiconductor such as $Cu_{2-x}O$.

Relationships similar to Eqs. 17.60 to 17.66 can be derived for deficit semiconductors in which vacancies are formed in the cation lattice when the oxygen pressure is increased. The overall reaction for the formation of vacancies and electron holes corresponds to

$$\frac{1}{2}\mathrm{O_2}(g) = \mathrm{O_O} + 2V'_{\mathrm{Cu}} + 2h^{\cdot} \tag{17.73}$$

We can write a mass-action constant:

$$K(T) = \frac{[V'_{\mathrm{Cu}}]^2[h^{\cdot}]^2}{P_{\mathrm{O_2}}^{1/2}} \tag{17.73a}$$

If the concentration of vacancies is largely determined by reaction with the atmosphere, we obtain a relationship between oxygen pressure and conductivity:

$$\sigma \propto [h^{\cdot}] = K(T)^{1/4} P_{O_2}{}^{1/8} \qquad (17.74)$$

Experimentally, as illustrated in Fig. 17.34, the electrical conductivity is found proportional to $P_{O_2}^{1/7}$, which is in reasonable agreement with the prediction of Eq. 17.74.

Additional Energy Levels. Sometimes an electron freed by an impurity center may not be released to the conduction band but instead is trapped by another ion present having an electron affinity which gives it a lower energy level than the free electron. This happens, for example, in partially reduced TiO_2, in which vacancies are formed in the oxygen ion lattice and ionized to form free electrons according to a reaction such as

$$O_o = V_O^{\cdot\cdot} + 2e' + \frac{1}{2} O_2(g) \qquad (17.75)$$

However, the Ti^{4+} ions present have an electron affinity such that the conduction electrons react with the lattice ions to give

$$2e' + 2Ti_{Ti} = 2Ti'_{Ti} \qquad (17.76)$$

The relative concentration of free conduction electrons and Ti^{3+} ions, Ti'_{Ti}, depends on the relative energy levels of the conduction band and the Ti^{4+} acceptor levels. Mass-action equations can be written for reactions corresponding to Eqs. 17.75 and 17.76. In this particular case it turns out that most of the electrons are associated with specific Ti^{4+} ions and consequently have a much lower mobility than they would if they were in the conduction band.

In general, then, behavior of semiconducting ceramic materials in the impurity-controlled region depends to a large extent on atmosphere,

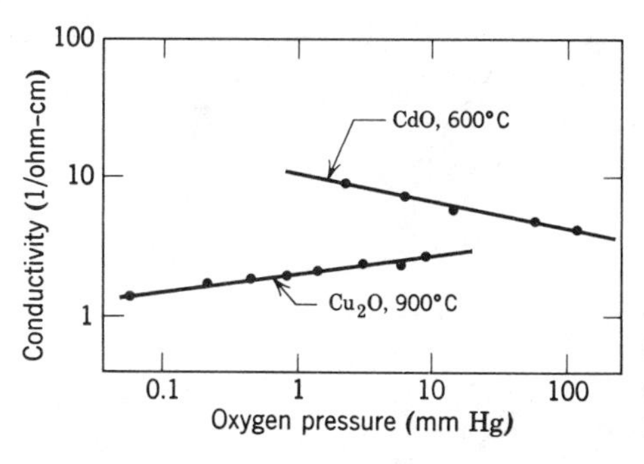

Fig. 17.34. Electrical conductivity of cuprous oxide and cadmium oxide as a function of oxygen pressure.

chemistry, temperature, and the energy levels in the individual materials under consideration. The range of effects that can be observed over a wide temperature range is illustrated for silicon carbide in Fig. 17.35. In this material the intrinsic region is not reached in the range of temperature tested. The conductivity initially increases and reaches a maximum corresponding to increased electron mobility as the temperature is lowered with nearly constant carrier concentration. At lower temperatures there are two straight-line portions in both the n-type and the p-type SiC. The conductivity varies with temperature as $T^{3/4}\exp(-E/2kT)$ with two separate activation energies. Typical values for the activation energy are 0.2 eV and 0.1 eV for the green crystals and 0.31 eV and 0.054 eV for the black crystals. The higher activation energies predominate at high temperatures, and the lower activation energies predominate at lower temperatures. These are assumed to correspond to two groups of carriers for each type of crystal. The ones of importance at low temperatures are

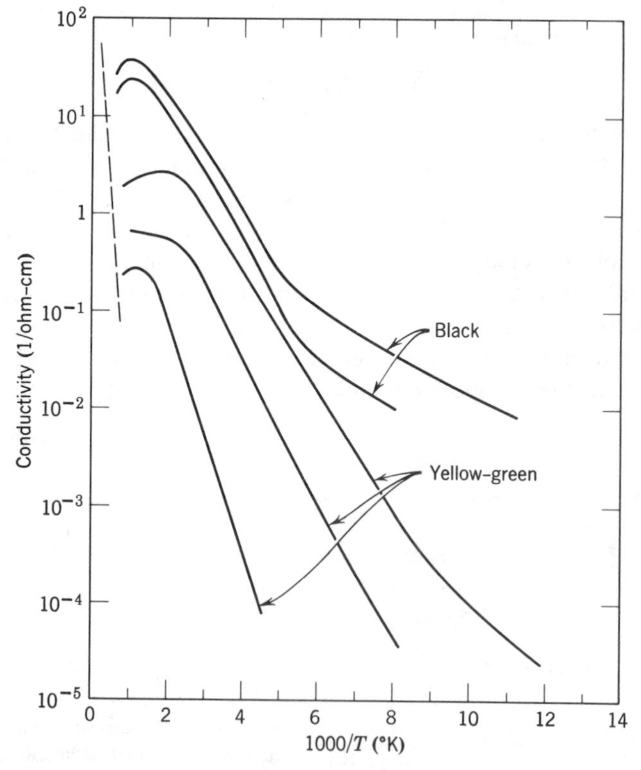

Fig. 17.35. Semiconductor properties of yellow-green n type of silicon carbide and black p type of silicon carbide. From G. Busch, *Helv. Phys. Act.*, **19**, 189 (1946).

overwhelmed at somewhat higher temperatures as a result of a larger energy of carrier formation or a higher energy of migration or both. At even higher temperatures than those indicated in Fig. 17.35, the intrinsic mobility is expected to become important corresponding to another linear portion of the curve of conductivity versus inverse temperature.

In general, then, for ceramic semiconductors in the impurity range a variety of factors is important, but for most materials they have not been studied in detail. In order to analyze particular effects precisely, it is essential to have an understanding of the chemistry of the system and an understanding of the electron energy levels in the system. For many oxide systems such as zinc oxide and copper oxide, the major effect is a change in chemical composition with temperature and oxygen pressure. For these materials the temperature dependence and composition dependence can be analyzed satisfactorily. For other materials such as TiO_2 the situation is complicated by the presence of additional energy levels intermediate between the normal valence band and the conduction band. For materials such as silicon carbide the impurity concentration and effects have not been satisfactorily analyzed chemically, so that detailed interpretation is difficult.

17.7 Valency-Controlled Semiconductors

The strong dependence of nonstoichiometric oxide semiconductors on the atmosphere and on minor concentrations of impurities makes it difficult to prepare them with reproducible properties. The rate of cooling, the atmosphere used, and minor changes in fabrication techniques all have a considerable effect on electrical properties. As a result it is found preferable to use compositions in which an appreciable amount of second component is added, preferably one with a high ionization potential and consequently a fixed valency, in order to obtain more uniform properties. The basic requirements for forming solid solutions of mixed-valency ions have been discussed in Chapter 4.

Nickel Oxide. As an example, nickel oxide forms a deficit semiconductor in which vacancies occur in cation sites similar to those discussed for cuprous oxide. For each cation vacancy there must be two electron holes formed, but these are normally associated with a lattice cation so that the reaction can best be described as

$$\frac{1}{2}O_2(g) = O_O + V''_{Ni} + 2Ni^{\cdot}_{Ni} \qquad (17.77)$$

Semiconduction results from the transfer of positive charge from cation to cation through the lattice. This charge-transfer process corresponds to a low mobility.

If a small amount of lithium oxide is added to the nickel oxide and the mixture is fired in air at the same temperature, a product with much lower resistivity is obtained. For pure nickel oxide the resistivity after firing in air is approximately 10^8 ohm-cm; the addition of 10 atomic% lithium gives a product with a resistivity of about 1 ohm-cm. Appreciable quantities of lithium are dissolved into nickel oxide lattice and result in the formation of Ni^{3+} content equivalent to the amount of lithium added to the solution. X-ray studies indicate a homogeneous crystal of the same structure as the initial nickel oxide but with a somewhat smaller unit cell. The reaction resulting in this product can be represented as

$$\frac{x}{2}Li_2O + (1-x)NiO + \frac{x}{4}O_2 = (Li_x^+Ni_{1-2x}^{2+}Ni_x^{3+})O \qquad (17.78)$$

The formation of ions of increased valency (Ni^{3+}) is promoted by the introduction of lower-valence ions at normal cation sites.

In order to obtain this result, it is necessary that the ion have nearly the same size as the ion being substituted, and furthermore the ion must have a fixed valency. The second ionization potential of lithium is more than twice as large as the third ionization potential of nickel, so that this condition is satisfactorily fulfilled in the Li_2O–NiO system.

In the same way the insulating characteristics of nickel oxide can be improved by the addition of a stable trivalent ion such as Cr^{3+} in solid solution. The effect of adding trivalent ions to the lattice is to decrease the fraction of Ni^{3+} ions which forms. Since electron transfer between Ni^{2+} and Cr^{3+} does not occur, the overall conductivity is substantially decreased.

Hematite. The same result can be obtained in a number of other systems of both n-type and p-type semiconductors. For example, Fe_2O_3 is an n-type semiconductor in which oxygen ion vacancies are formed along with electrons that tend to be associated with specific cations. This is equivalent to forming a certain fraction of Fe^{2+} ions. If Ti^{4+} is added to Fe_2O_3 in solid solution, an increased fraction of the Fe^{3+} is forced into the Fe^{2+} state—a number equal to the Ti^{4+} additions. As a result, the conductivity of the product is substantially increased; it is determined primarily by the concentration of titanium oxide added and is much less dependent on oxygen pressure and firing conditions than is the pure material. Variations of conductivity with additions of Li_2O in NiO and TiO_2 in Fe_2O_3 are illustrated in Fig. 17.36.

Spinels. Another method of obtaining semiconductors with a controlled resistivity and avoiding difficulties due to stoichiometry deviations is by forming solid solutions of two or more compounds of widely different conductivity. In particular, it is found that magnetite, Fe_3O_4, is an

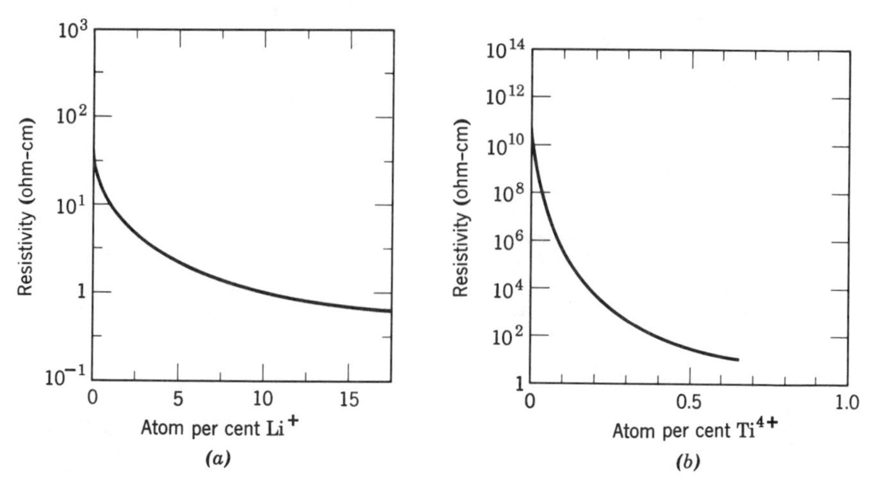

Fig. 17.36. Effect of (*a*) added Li_2O on the conductivity of NiO and (*b*) added TiO_2 on the conductivity of Fe_2O_3. From K. Lark-Horowitz, *Elec. Eng.*, **68**, 1087 (1949).

excellent semiconductor having a specific resistance of about 10^{-2} ohm-cm compared with values of the order of 10^{10} ohm-cm for most stoichiometric transition element oxides. Fe_3O_4 has a spinel type of crystal structure, as described in Chapter 2. In this structure the oxygen ions are nearly on a close-packed cubic lattice; cations are partly in octahedral sites and partly in tetrahedral sites. The structure of magnetite can be described by the notation

$$Fe^{3+}(Fe^{2+}Fe^{3+})O_4 \tag{17.79}$$

which denotes that one-third of the iron ions (all plus three valence) are on tetrahedral sites and two-thirds (both divalent and trivalent ions) are on octahedral positions. The good electrical conductivity of magnetite is related to the random location of Fe^{2+} and Fe^{3+} ions on these octahedral sites so that electron transfer from cation to cation can take place. This is best illustrated by the order-disorder transformation occurring at about 120°K. Below this temperature the Fe^{2+} and Fe^{3+} ions are distributed in an ordered pattern on the octahedral sites; above this temperature the Fe^{2+} and Fe^{3+} positions are randomly distributed (order-disorder transitions are discussed in Chapter 6). The effect of this is to substantially increase the electrical conductivity (Fig. 17.37).

In general, a condition for appreciable conductivity in the spinel structure is the presence of ions having multiple valence states at equivalent crystallographic sites. The number of these ions in Fe_3O_4 can be controlled by controlling the solid solutions in which Fe^{2+} or Fe^{3+} are

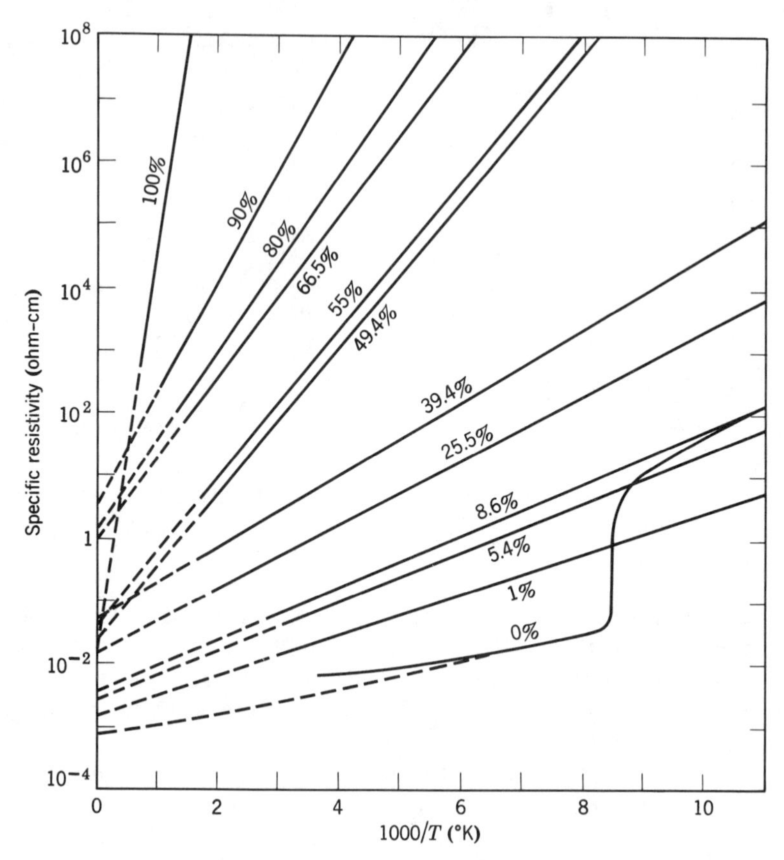

Fig. 17.37. Specific resistivity of solid solutions of Fe_3O_4 and $MgCr_2O_4$. Mole % $MgCr_2O_4$ is indicated on the curves. From E. J. Verwey, P. W. Haagman, and F. C. Romeijn, *J. Chem. Phys.*, **15**, 18 (1947).

diluted by other ions which do not participate in the electronic exchange. This is illustrated in Fig. 17.37 for solid solutions of Fe_3O_4 and $MgCr_2O_4$. With increasing resistivity the temperature coefficient or activation energy also increases. Semiconductor materials of this type with materials like $MgAl_2O_4$, $MgCr_2O_4$, and Zn_2TiO_4 as the nonconducting component can be prepared to have a controlled temperature coefficient of resistivity. Semiconductors made in this way are used as *thermistors.*

17.8 Mixed Conduction in Poor Conductors

For ceramics having electrical conductivities below about 10^{-5} (ohm · cm)$^{-1}$ it is clear that the concentration-mobility product of the charge carrier is small. As a result, minor variations in composition, impurity-content, heat treatment, stoichiometry, and other variables can have a significant effect on measured results. In addition, experimental measurement techniques become more difficult. In consequence, it is characteristic that the reported data for a single material may cover several orders of magnitude, as illustrated in Fig. 17.38. Clearly we must have the greatest reluctance in interpreting these results unless complete information is available with regard to purity and solute concentration, nonstoichiometry, experimental techniques, possible presence of high-mobility paths such as grain boundaries and dislocations, prior heat treatment, and similar data which are seldom fully described. At low temperatures, at which conductivities are in the range below 10^{-12} ohm/cm, difficulties in measurement resulting from surface conductivity, grain boundaries, and other high-conductivity paths make precise measurements formidable.

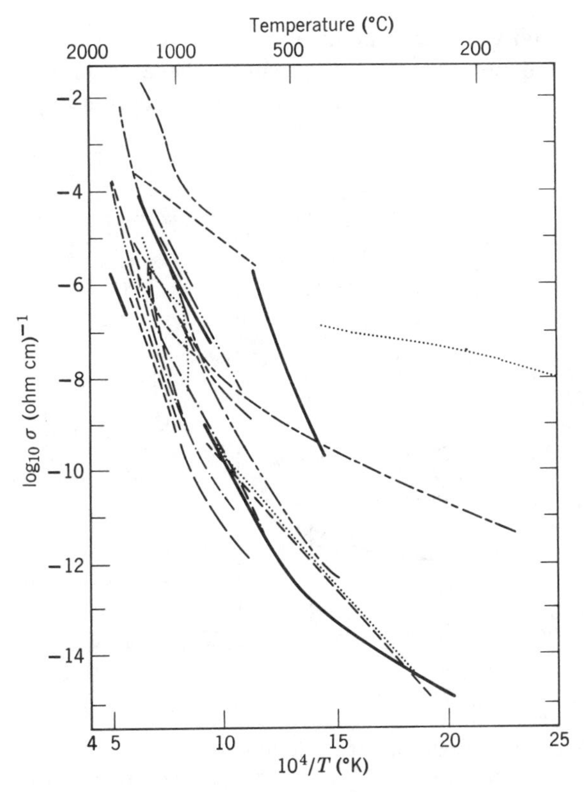

Fig. 17.38. Reported literature data for the electrical conductivity of aluminum oxide. From A. A. Bauer and J. L. Bates, *Battelle Mem. Inst. Rept.*, 1930, July 31, 1974.

At higher temperatures, at which the conductivity is above about 10^{-8} ohm/cm, the electrical conductivity is characteristically found to be a strong function of the oxygen pressure, even in compounds such as magnesium oxide and aluminum oxide in which the range of stoichiometry is known to be exceedingly small. Experimental results are usually interpreted in terms of impurity controlled and nonstoichiometric conductivity processes, using the measured slopes of the log σ–log P_{O_2} curves as a basis for interpretation. It is often found, however, that these slopes depart from simple relationships and that a relatively small range of oxygen pressure is available for their determination. These difficulties, along with the variety of impurities that is present as solutes or precipitates, have led to contradictions in the literature and leave a final interpretation uncertain. Typical results for the best data available are illustrated in Fig. 17.39.

17.9 Polycrystalline Ceramics

The conductivity characteristics of ceramic systems usually result from contributions of several phases present. These include porosity (low conductivity), semiconductors (appreciable conductivity), glasses (having

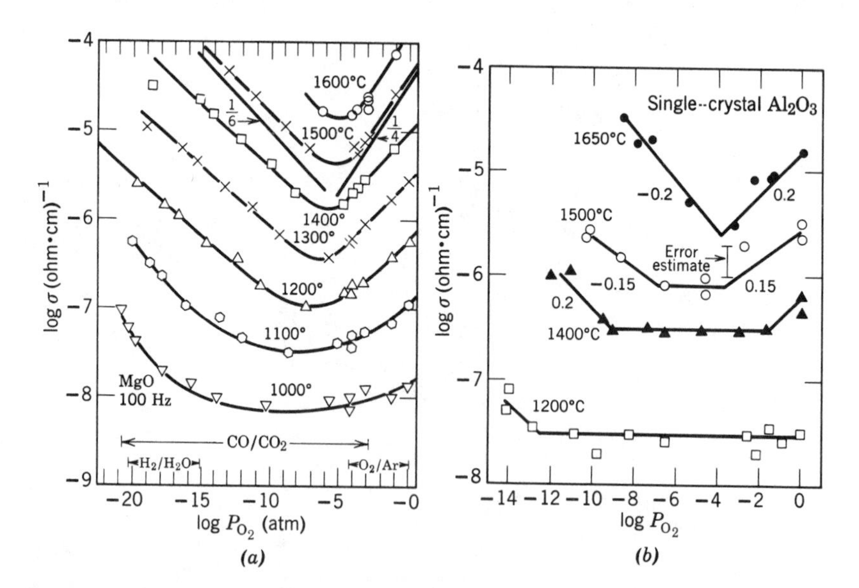

Fig. 17.39. Conductivity isotherms for single crystals of (*a*) MgO, from C. M. Osburn and R. W. Vest, *J. Am. Ceram. Soc.*, **54**, 428 (1971), and (*b*) Al_2O_3, from K. Kitazawa and R. L. Coble, *J. Am. Ceram. Soc.*, **57**, 250 (1974).

an appreciable conductivity at high temperatures), and insulating crystals (low conductivity). The range of conductivity with temperature for a number of good insulators is illustrated in Fig. 17.40; it should be recalled (Fig. 17.38) that the reported data for each of these compositions may cover a wide range of values. The range of conductivity with temperature for a number of good oxide conductors is shown in Fig. 17.41; resistivities of a wide range of commercial refractories are shown in Fig. 17.42. In these latter materials the phase composition and arrangement are most important. The effects of phase arrangement are similar to those discussed for thermal conductivity in Chapter 12, the main difference being that the range of variation for individual phases is much larger for electrical properties than for thermal properties. Effects related to high-frequency measurements are discussed in Chapter 18.

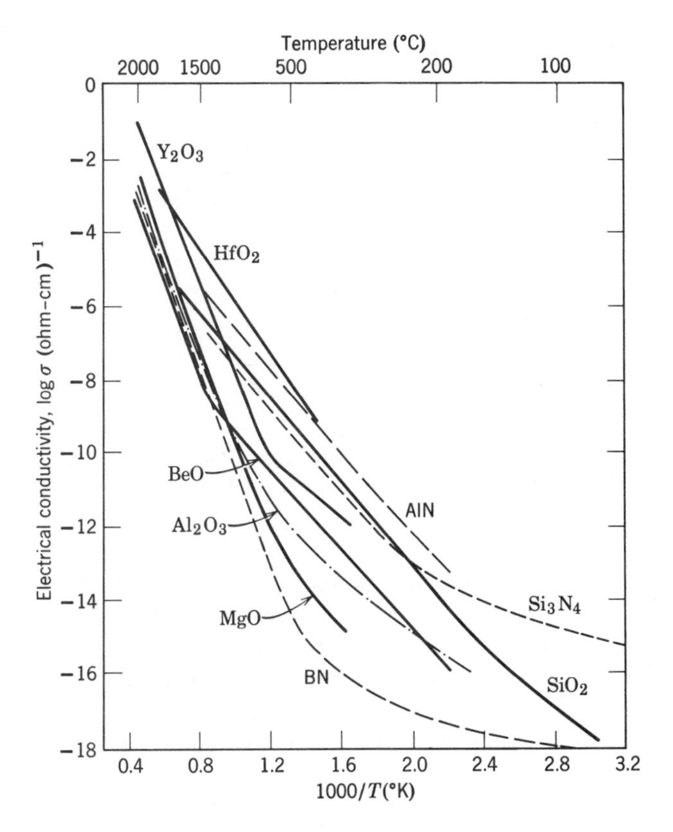

Fig. 17.40. Electrical conductivity of our best insulating materials. From A. A. Bauer and J. L. Bates, *Battelle Mem. Inst. Rept.*, 1930, July 31, 1974).

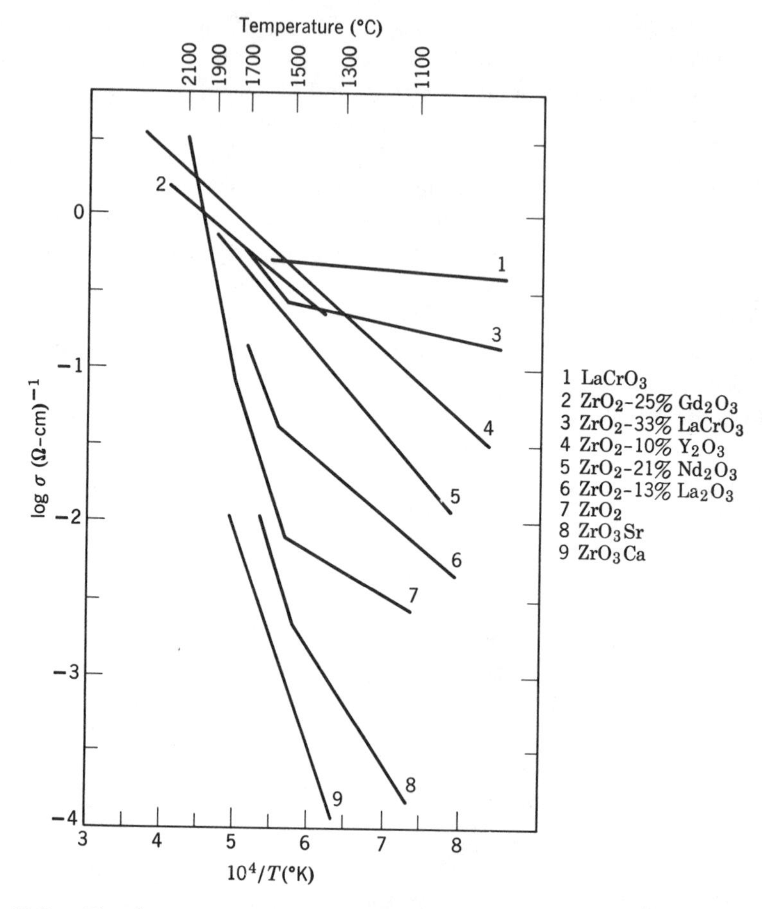

Fig. 17.41. Electrical conductivity of good high-temperature oxide conductors. From A. M. Anthony and D. Yerouchalmi, *Phil. Trans. R. Soc. (London)*, **261**, 504 (1966).

The effects of porosity are similar to those discussed for thermal conductivity in Chapter 12. As the porosity increases, the electrical conductivity decreases almost in proportion for small values of porosity corresponding to isometric uniformly distributed pores such as are normally present. For large amounts of porosity the effect of pores is more substantial, giving results similar to those already discussed in Chapter 12.

The effect of grain boundaries in polycrystalline materials is related to the mean free path of the ions or electrons between collisions. This is the order of interatomic distances for ionic conduction and is usually less than 100 to 150 Å for electronic conductivity. This means that except for very thin films or extremely fine-grained samples (less than 0.1 micron) the effects of boundary scattering are small compared with the lattice scattering. Consequently, the grain size in uniform-composition materials has but little effect.

However, substantial effects can result from impurity concentration and changes in composition in grain boundaries. Particularly for oxide materials there is a tendency to form a glassy silicate phase at the boundary between particles. A number of the microstructures given in Chapter 10 illustrate this clearly. The structure corresponds to a two-phase system, with the minor phase being the continuous one. The resulting conductivity depends on relative values of the conductivities of the individual phases. At room temperature the conductivity of both

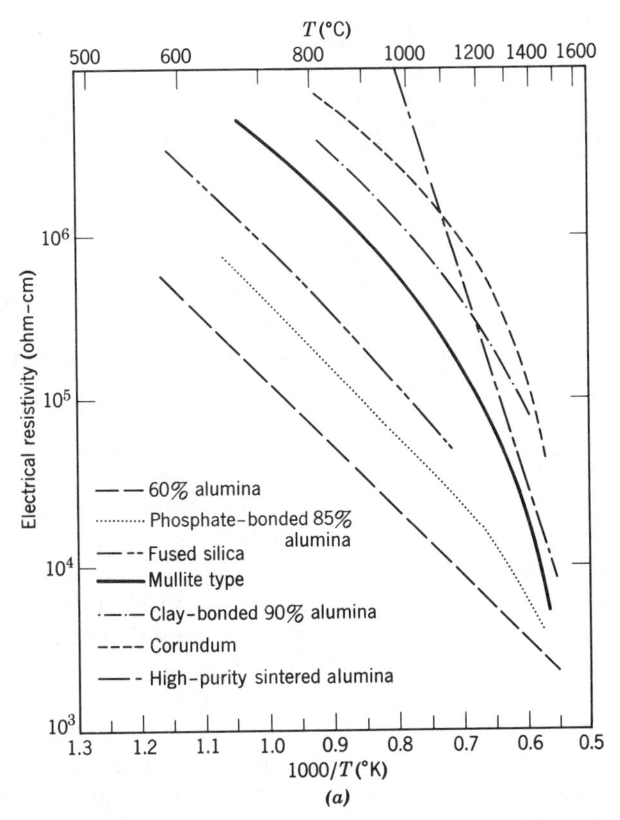

Fig. 17.42. Electrical resistivity of several refractories. From R. W. Wallace and E. Ruh, *J. Am. Ceram. Soc.*, **50**, 358 (1967).

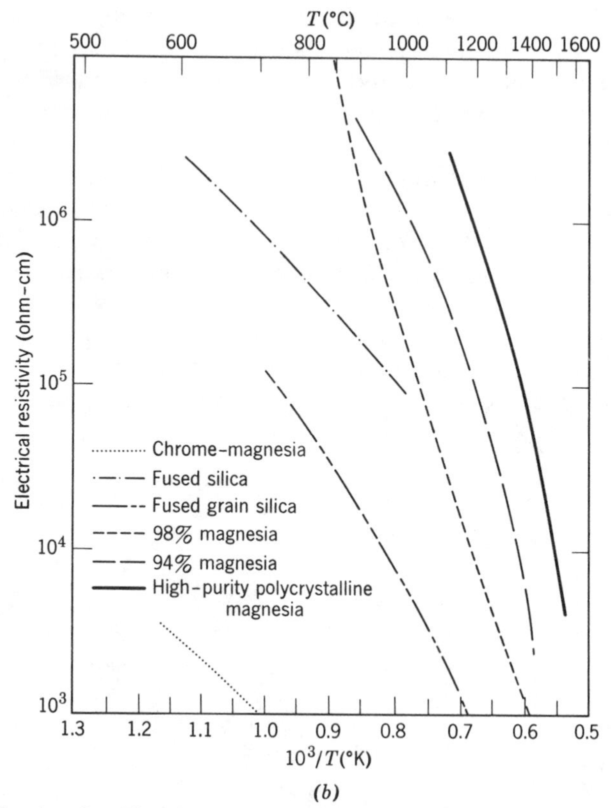

(*b*)

Fig. 17.42 (*Continued*). Electrical resistivity of several refractories. From R. W. Wallace and E. Ruh, *J. Am. Ceram. Soc.*, **50**, 358 (1967).

phases is low. As the temperature level is increased, the conductivity of the glassy phase increases and becomes substantial at a much lower temperature than the crystalline oxide phase; as a result the conductivity increases more rapidly with temperature. This is illustrated for a number of different insulator compositions in Fig. 17.43. Compositions that increase most in conductivity at moderate temperatures correspond to compositions having an appreciable glassy phase present.

Exactly the reverse effect is found for semiconductor materials in which lower-temperature equilibrium is achieved at the boundaries during cooling. As a result there is often a tendency for the boundaries to have a higher resistance than the interior of the grains. Since each grain is enveloped in a high-resistance boundary, the overall dc resistance of the material can be quite high. This same sort of structure occurs with

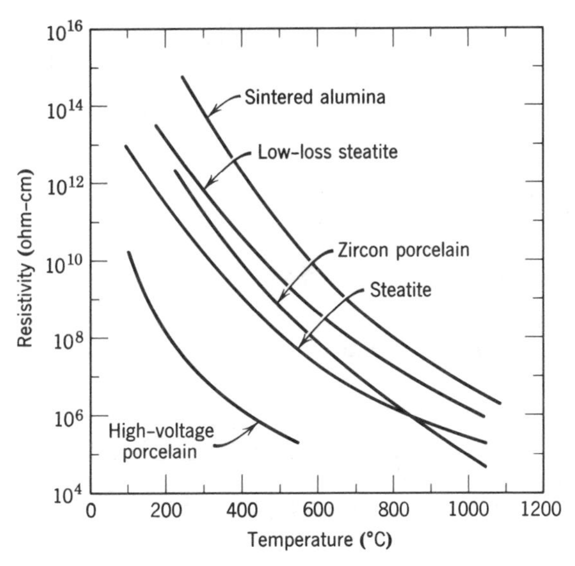

Fig. 17.43. Decrease in resistivity with temperature for some typical oxide ceramics (mostly bonded with a vitreous phase).

powdered metals that have an oxide coating and are compressed at low temperatures to a high density. The continuous oxide phase means that the overall dc conductivity remains low, even though a substantial part of the material may be present as a metallic conductor. This same sort of process occurs in silicon carbide heating elements, in which a silica layer is formed between semiconductor grains. As an increasing amount of oxide phase develops, the resistivity gradually increases. Data for some systems in which the microstructure and phase distribution have been measured are illustrated in Fig. 17.44.

Particularly for materials such as oxide semiconductors in which the equilibrium of a nonstoichiometric composition with the atmosphere determines the conductivity to a large extent, substantial changes in properties can result from different firing conditions and the extent to which the high-temperature composition is retained on cooling. Rapid cooling rates tend to retain the high-temperature, high-conductivity structures. Similarly, dense samples tend to react less on cooling than do porous samples. Consequently, electrical properties are frequently correlated with porosity, although the effect of porosity is in controlling the kinetics of compositional changes rather than any direct contribution.

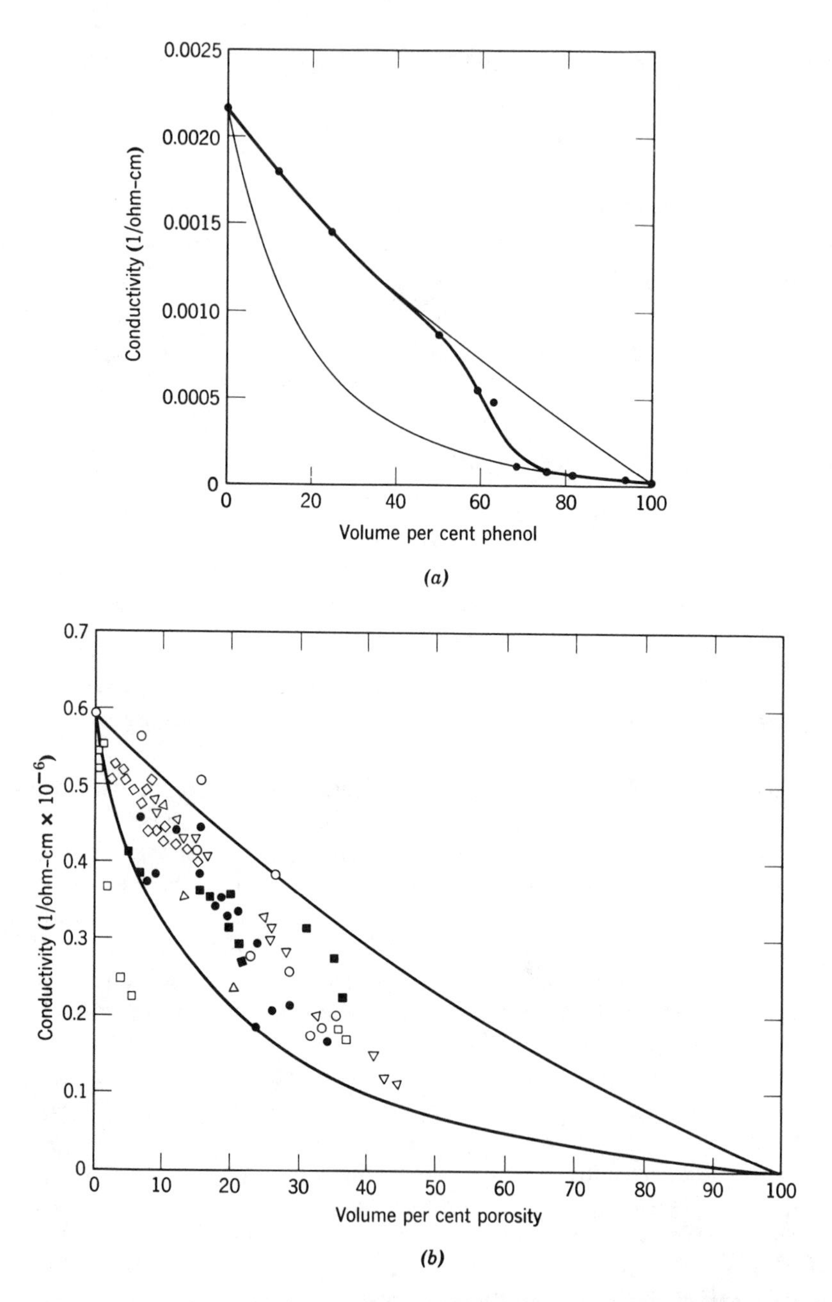

Fig. 17.44. Change in conductivity with composition for (*a*) phenol-KI emulsions and (*b*) pressed copper powder. From P. Grootenhuis et al., *Proc. Phys. Soc.* (*London*), **B65**, 502 (1952).

Suggested Reading

1. N. F. Mott and R. W. Gurney, *Electronic Processes in Ionic Crystals*, Dover Publications, Inc., New York, 1964.
2. A. B. Lidiard, "Ionic Conductivity," in *Encyclopedia of Physics* (Handbuch der Physik), S. Flugge, Ed., Vol. 20, Springer-Verlag, Berlin, 1957.
3. R. J. Friauf, "Basic Theory of Ionic Transport Processes," in *Physics of Electrolytes, Vol. 1*, J. Hladik, Ed., Academic Press, Inc., New York, 1972.
4. K. Hughes and J. O. Isard, "Ionic Transport in Glasses," same as reference 3.
5. I. G. Austin and N. F. Mott, "Polarons in Crystalline and Noncrystalline Materials," *Adv. Phys.*, **18**, 41 (1969).
6. N. F. Mott and E. A. Davis, *Electronic Processes in Non-crystalline Materials*, Clarendon Press, Oxford, 1971.
7. N. M. Tallen, Ed., *Electrical Conductivity in Ceramics and Glasses*, Marcel Dekker, Inc., New York, 1974.
8. P. Kofstad, *Nonstoichiometry, Electrical Conductivity, and Diffusion in Binary Metal Oxides*, John Wiley & Sons, Inc., New York, 1972.
9. L. L. Hench and D. B. Dove, *Physics of Electronic Ceramics*, Marcel Dekker, Inc., New York, 1971.

Problems

17.1. Rigdon and Grace* measured the electrical conductivity of single crystals of $CaWO_4$. It can be assumed that $t_e \approx 1$. Given their data:
(*a*) Give a plausible defect model for the conduction mechanism.
(*b*) Show how the transient dc conductivity for step changes in P_{H_2O}/P_{H_2} gives the chemical-diffusion coefficient $\tilde{D}$.
(*c*) How do you reconcile the ΔH (conductivity) = 136 kcal/mole and ΔH (chemical diffusion) = 6.3 kcal/mole with the fact that $t_e \approx 1$?

17.2. Although calcia-stabilized zirconia (CSZ) is an ionic conductor ($t_0 = 1$; $\sigma = 0.4$/ohm-cm at 1000°C), care must be taken when using it as an oxygen detector because of permeation of oxygen from the high-partial-pressure side to the low side. It has been suggested that the rate-limiting defect for this permeation is electron holes which have a mobility of 10^{-4} cm²/V sec and transference number of 10^{-3} at 1000°C. What is the oxygen permeability (moles/cm² sec) at 1000°C for a 0.1-cm-thick specimen? $D_0 \approx 3 \times 10^{-8}$ cm²/sec, $D_{Zr} \approx 10^{-17}$ cm²/sec at 1000°C.

17.3. (*a*) Derive a relation for the oxygen pressure dependence of electrical conductivity in NiO.
(*b*) Discuss the effects of Cr_2O_3 additions on the electrical conductivity of NiO, giving equations describing the temperature at which the conductivity changes from intrinsic to extrinsic in relation to the impurity concentration.

***J. Am. Ceram. Soc.*, **56**, 475 (1973).

17.4. If a sodium chloride crystal is annealed in Na vapor at a sufficiently high temperature, additional sodium can be incorporated in the crystal, causing a deviation from stoichiometry. The excess sodium is compensated by formation of chlorine ion vacancies which have trapped an electron (F centers or V_{Cl}^{x}). The F centers give rise to coloration in the crystal. In one particular experiment only the end portion of a high-purity crystal was annealed in Na vapor so that coloration was limited to that portion of the crystal. When a dc electric field was applied to the crystal at a sufficiently high temperature, the colored region migrated toward the anode. The migration velocity of the colored portion of the crystal was found, empirically, to be

$$v = \epsilon v_0 \exp\left(\frac{-Q}{kT}\right)$$

where ϵ was the applied field, Q was an activation energy, and v_0 was found to be essentially independent of temperature over the range of experimentation.

(*a*) Explain briefly why stoichiometric NaCl is transparent, whereas crystals annealed in Na vapor exhibit coloration.

(*b*) Explain the motion of the colored region of the crystal under the influence of an external field.

(*c*) Based on your explanation in (*b*), formulate an expression for the migration velocity v, and identify v_0 and Q in terms of physically meaningful parameters.

17.5. Plastic deformation of sodium chloride crystals results in crystals with much lower electrical resistance than undeformed crystals. In addition, the deformed crystals can be irradiated with X rays, which turn them yellow. Is this due to dislocations or some other defect generated through movement of dislocations?

17.6. The electrical conductivity of UO_2 is 0.1 $(\text{ohm-cm})^{-1}$ at 723°K. The diffusion coefficient of O^{2-} measured at this temperature is 1.0×10^{-13} cm^2/sec. Calculate the mobility of $O^{=}$ and its transference number (density of $UO_2 = 10.5$). How do you interpret this result?

17.7. (*a*) Given that the electrical conductivity of a 40 mole% Na_2O–60 mole% SiO_2 glass is 5×10^{-3} $(\text{ohm-cm})^{-1}$ at 673°C, calculate the sodium ion diffusion coefficient.

(*b*) What would be the effect of changing the Na_2O–SiO_2 ratio on the electrical conductivity in this system? Why?

17.8. Compute the temperature for a solid at which there is a 1% probability that an electron in the solid will have an energy 0.5 eV above the Fermi energy of 5 eV.

17.9. Indium antimonide has $E_g = 0.18$ eV, dielectric constant = 17, $m_e = 0.014$ m. Calculate (*a*) the donor ionization energy; (*b*) the radius of the ground-state orbit. (*c*) At what minimum donor concentration will appreciable overlap effects between the orbits or adjacent impurity atoms occur? (*d*) If $N_d = 1 \times 10^{14}/\text{cm}^3$ for a particular specimen, calculate the concentration of conduction electrons at 4°K, using the expression

$$n = (2N_d)^{1/2} \left(\frac{2\pi kT}{h^2} m_e\right)^{3/4} e^{-E_d/2kT}$$

18

Dielectric Properties

Dielectric (essentially nonconducting) and magnetic characteristics of ceramic materials are of increasing importance as the field of solid-state electronics continues to expand rapidly. This field is one in which limitations of available materials are frequently the bottleneck preventing improved designs. Also, the reliability of components is of great importance for many applications; the failure of one component can cause failure of an entire missile costing several millions of dollars, for example. At the same time there is an extensive effort to reduce the size of all communications devices. All these factors have led to an increasing interest in ceramic insulators and semiconductors (Chapter 17), dielectrics, and magnetic materials (Chapter 19).

The principal applications for ceramic dielectrics are as capacitive elements in electronic circuits and as electrical insulation. For these applications the properties of most concern are the dielectric constant, dielectric loss factor, and dielectric strength. Our discussion is restricted to these properties for the most part. New devices and new applications are continually increasing the frequency range and the range of environmental conditions, particularly temperature, that are of practical interest. Consequently, we approach these variables with a somewhat broader point of view than might be required for today's technology alone. We have also included in this chapter the frequency dependence of resistivity of both insulators and semiconductors. Most of the high-frequency applications related to electrical properties are concerned with dielectrics, so that this aspect of conductivity fits in naturally here.

For applications as capacitors and as electrical insulation, organic plastics are available; they are usually cheaper and can be fabricated with better dimensional accuracy than ceramics. The advantages of ceramics, which frequently indicate their use, are superior electrical properties, absence of creep or deformation under stresses at room temperature, greater resistance to environmental changes (particularly at high temperatures at which plastics frequently oxidize, gasify, or decompose), and the ability to form gastight seals with metals and become an integral part of an electronic device. It should be noted that the selection of a particular dielectric frequently depends on its ability to be formed as a gastight part to operate under unusual environmental conditions, on its suitable thermal-expansion characteristics, on its satisfactory thermal stress resistance and impact resistance, on its ability to be formed into complex shapes with good dimensional characteristics, and on other characteristics which are completely independent of the electrical behavior but are essential for the building of practical devices.

18.1 Electrical Phenomena

In this section we wish to define the dielectric properties that are of interest for ceramic applications. These include the *dielectric constant, magnetic permeability, dielectric loss factor, dielectric resistivity*, and *dielectric strength.* It is essential that ceramists understand why electrical engineers, their customers, find these particular sets of properties and formulations useful. This aspect is discussed here, but without going into the mathematical formulations required for actual application to circuit design. Because of space limitations, mathematical derivations are kept to those essential for understanding the phenomenological aspects. Interpretation and evaluation of material properties in terms of glass structure, crystal structure, and the microstructure are deferred until later sections.

Capacitance. The electrical engineer is most concerned with dielectric materials in relation to a capacitor in an electrical circuit. The principal characteristic of a capacitor is that an electrical charge Q can be stored. The charge on a capacitor is

$$Q = CV \tag{18.1}$$

where V is the applied voltage and C is the capacitance. The voltage is directly proportional to the amount of charge stored, and the current passing through the capacitor is given by

$$V = \frac{Q}{C} = \frac{\int I\,dt}{C} \qquad I = C\frac{dV}{dt} \tag{18.2}$$

With a sinusoidal voltage

$$V = V_0 \exp i\omega t \tag{18.3}$$

as used in ac circuits, a charging current results

$$I_c = i\omega CV \tag{18.4}$$

which is exactly 90° advanced in phase in relation to the applied voltage. In Eqs. 18.3 and 18.4, i equals $\sqrt{-1}$, and ω equals $2\pi f$, where f is the frequency in cycles per second.

The capacitance C contains both a geometrical and a material factor. For a large plate capacitor of area A and thickness d the geometrical capacitance *in vacuo* is given by

$$C_0 = \frac{A}{d}\epsilon_0 \tag{18.5}$$

where ϵ_0 is the permittivity (dielectric constant) of a vacuum. If a ceramic material of permittivity ϵ' is inserted between the capacitor plates,

$$C = C_0\frac{\epsilon'}{\epsilon_0} = C_0\kappa' \tag{18.6}$$

where κ' is the relative permittivity or relative dielectric constant. This is the material property determining the capacitance of a circuit element and is of principal concern to the ceramist.

Inductance. Parallel to the capacitance in the circuit elements available to an electrical engineer is the inductance L, which may be considered a current-storage device. The current-storage capacity of a coil results from the fact that a moving charge in the coil sets up a magnetic field parallel to the coil axis. Current changes in an inductance create opposing voltages and hence must be sustained by countervoltages:

$$V = L\frac{dI}{dt} \tag{18.7}$$

When a sinusoidal voltage is applied to an inductance, magnetization currents retarded by 90° in relation to the phase of the voltage result:

$$I_m = \frac{V}{i\omega L} \tag{18.8}$$

This is in direct contrast to the capacitance, in which the charging current is 90° advanced from the applied voltage. The inductance L, like the capacitance, contains both a geometrical and material factor. It can be written as

$$L = L_0\frac{\mu'}{\mu_0} = L_0\kappa'_m \tag{18.9}$$

where L_0 is the geometrical inductance of a long coil of N turns,

Table 18.1. Conversion Factors for Electrical and Magnetic Units

Quantity	Symbol	mks Units		Conversion Ratios	
		Primary	Derived	To Change from Rationalized mks Units to esu Multiply by:	To Change from esu to emu Multiply by:
Capacitance	$C = \frac{Q}{V}$	$sec^2\ coul^2/kg\ m^2$	farad	9×10^{11}	$c^2 (c = 3 \times 10^{10}\ cm/sec)$
Dielectric permittivity	$\epsilon' = \frac{D}{E}$	$sec^2\ coul^2/kg\ m^3$	farad/m	$4\pi \times 9 \times 10^9$	c^2
Conductivity	$\sigma = \frac{j}{E}$	$sec\ coul^2/kg\ m^3$	mho/m	9×10^9	c^2
Current density	j	$coul/sec\ m^2$	amp/m^2	3×10^5	c
Electric charge	Q	coul	coul	3×10^9	c
Electric dipole moment	$\mu = Qd$	coul m	amp sec m	3×10^{11}	c
Electric field strength	E	$kg\ m/sec^2\ coul$	volt/m	$\frac{1}{3} \times 10^{-4}$	$1/c$
Electric flux density	D	$coul/m^2$	$farad\ volt/m^2$	$12\pi \times 10^5$	c
Electric loss factor	ϵ''	$sec^2\ coul^2/kg\ m^3$	farad/m	$36\pi \times 10^9$	c^2
Electric polarization	P	$coul/m^2$	$farad\ volt/m^2$	$12\pi \times 10^5$	c
Electric potential	ϕ	$kg\ m^2/sec^2\ coul$	volt	$\frac{1}{300}$	$1/c$
Inductance	$L = \frac{V}{dI_m/dt}$	$kg\ m^2/coul^2$	henry	$\frac{1}{9} \times 10^{-11}$	$1/c^2$
Electric susceptibility	$\chi = \kappa' - 1$	ratio	ratio	–	–

Loss tangent	$\tan\delta = \frac{\epsilon''}{\epsilon'}$	ratio	ratio	–	–
Magnetic dipole moment	m	coul m^2/sec	amp m^2	3×10^{13}	–
Magnetic field strength	H	coul/m sec	amp-turn/m	$12\pi \times 10^7$	c
Magnetic flux density (magnetic induction)	B	kg/sec coul	volt sec/m^2= weber/m^2	$\frac{1}{3} \times 10^{-6}$	$1/c$
Magnetic permeability	$\mu' = \frac{B}{H}$	kg m/coul^2	henry/m	$\frac{1}{36\pi} \times 10^{-13}$	$1/c^2$
Magnetic susceptibility	$\chi_m = \kappa_m' - 1$	ratio	ratio	–	–
Magnetization	$M = \frac{B}{\mu_0} - H$	coul/m sec	amp-turn/m	3×10^7	c
Q (quality) factor	$Q = \frac{1}{\tan\delta}$	ratio	ratio	–	–
Relative dielectric constant	$\kappa' = \frac{\epsilon'}{\epsilon_0}$	ratio	ratio	–	–
Resistance	$R = \frac{V}{I}$	kg m^2/sec coul^2	ohm	$\frac{1}{9} \times 10^{-11}$	$1/c^2$
Resistivity	$\rho = \frac{1}{\sigma}$	kg m^3/sec coul^2	ohm-m	$\frac{1}{9} \times 10^{-9}$	$1/c^2$
Relative magnetic permeability	$\kappa_m' = \frac{\mu'}{\mu_0}$	ratio	ratio	–	–

$L_0 = N^2 A/d\mu_0$. Combining this with Eq. 18.9, we obtain for an inductance for a long coil with real material

$$L = N^2 \frac{A}{d} \mu' \tag{18.10}$$

In these equations μ' is the permeability of the medium, μ_0 the permeability of a vacuum, and κ'_m the relative permeability.

Units. A variety of units is used for dielectric and magnetic properties, and these are sometimes confusing. Three basic sets of units are used. In the electrostatic-units system (esu), electrical energy is related to mechanical and thermal energy in the cge system by Coulomb's law for the force exerted between electric charges. The dielectric constant is arbitrarily taken as a plain number. In electromagnetic units (emu) the permeability is taken as a plain number. These formulations lead to some inconsistencies and fractional expressions in dimensional equations; these can be eliminated by taking the electrical charge as a fourth-dimensional unit. This is done in a rationalized system in which for practical units the meter (m), kilogram (kg), second (sec), and coulomb (c) are used as primary units. This is called the mks or Georgi system. Several derived units such as amperes and ohms are used for convenience. In this system the dielectric constant of free space becomes $\epsilon_0 = (36\pi)^{-1} \times 10^{-9} = 8.854 \times 10^{-12}$ F/m; the permeability of free space is $\mu_0 = 4\pi \times 10^{-7} = 1.257 \times 10^{-6}$ H/m. Units for a number of quantities of interest are collected and compared in Table 18.1.

Polarization. A dielectric material reacts to an electric field differently from a free space because it contains charge carriers that can be displaced, and charge displacements within the dielectric can neutralize a part of the applied field. Since $V = Q/C$ and $C = \kappa' C_0$, we can write for a capacitor containing a dielectric

$$V = \frac{Q/\kappa'}{C_0} \tag{18.11}$$

That is, only a fraction of the total charge, the *free* charge Q/κ', sets up an electric field and voltage toward the outside; the remainder, the *bound* charge, is neutralized by polarization of the dielectric. We can schematically represent (Fig. 18.1) the total electric flux density D as the sum of the electric field E and dipole charge P:

$$D = \epsilon_0 E + P = \epsilon' E \tag{18.12}$$

where the polarization is the surface charge density of the bound charge, equal to the dipole moment per unit volume of material:

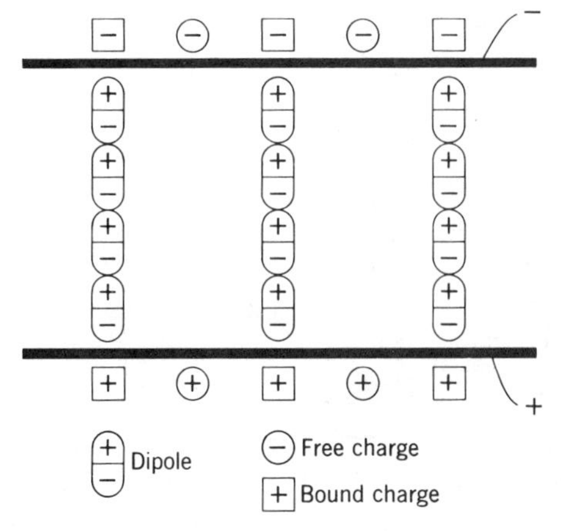

Fig. 18.1. Schematic representation of polarization by dipole chains and bound charges. From A. R. von Hippel, reference 1.

$$P = N\mu \tag{18.13}$$

where N is the number of dipoles per unit volume and μ is the average dipole moment. The electric dipole moment corresponds to two electric charges of opposite polarity $\pm Q$ separated by the distance d:

$$\mu = Qd \tag{18.14}$$

Thus polarization can equivalently designate either the bound-charge density or the dipole moment per unit volume (Fig. 18.2). From Eq. 18.12,

$$P = \epsilon' E - \epsilon_0 E = \epsilon_0(\kappa' - 1)E \tag{18.15}$$

Another measure of the ratio of polarization to applied field is the electric susceptibility:

$$\chi = \kappa' - 1 = \frac{P}{\epsilon_0 E} \tag{18.16}$$

The susceptibility is the ratio of the bound-charge density to the free-charge density.

The average dipole moment $\bar{\mu}$ of the elementary particle is proportional to the *local* electric field E' which acts on the particle:

$$\bar{\mu} = \alpha E' \tag{18.17}$$

The proportionality factor α, the polarizability, is a measure of the average dipole moment per unit of local field strength. Its dimensions are cubic meters in the mks system or cubic centimeters in the esu system. An alternate expression for the polarization then is

$$P = N\alpha E' \tag{18.18}$$

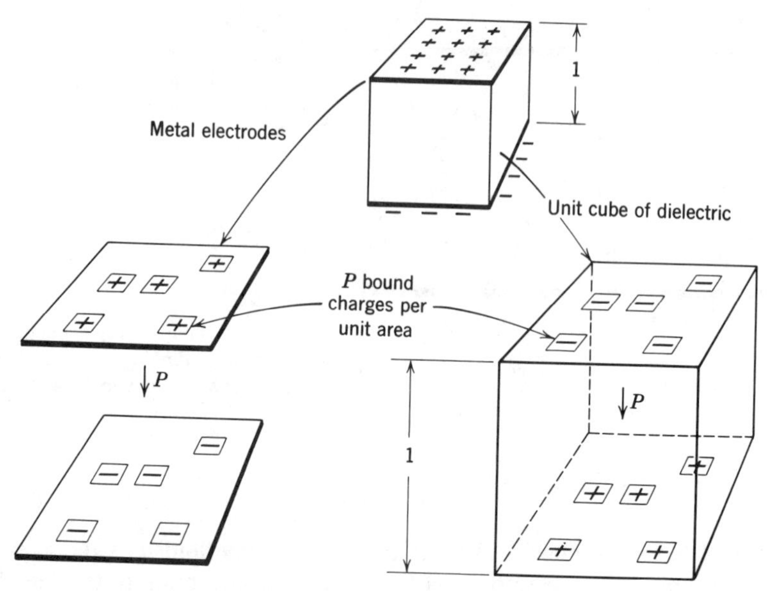

Fig. 18.2. Polarization P designates both the bound-charge density and the average dipole moment per unit volume. From A. R. von Hippel, reference 1.

For gases at low pressures, at which interactions between molecules can be neglected, the locally acting field E' is the same as the externally applied field E. However, for solids the polarization of the surrounding medium can have a marked effect on the local field acting on a particular molecule. This local field can be calculated (Fig. 18.3) when a particular molecule is surrounded by an imaginary sphere sufficiently large for the

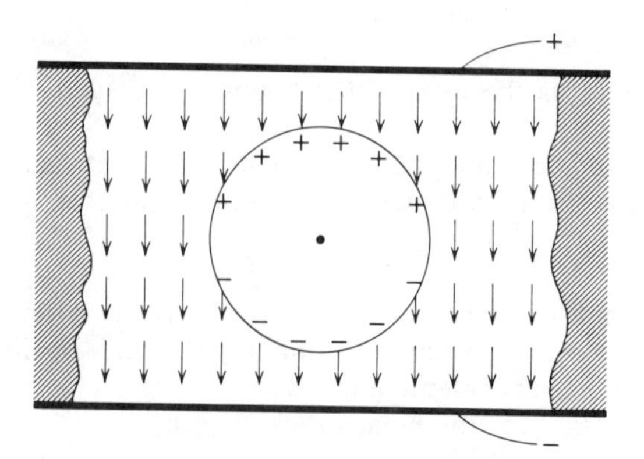

Fig. 18.3. Model for calculation of the internal field.

dielectric beyond it to be treated as a continuum. If this sphere is cut out of the solid, the polarization outside remains unchanged. The field acting on a particular molecule in the center arises from three sources—free charges at the electrodes, E, free ends of the dipole chains that lie on the cavity walls, E_1, and contributions from molecules inside the sphere so close that their individual positions have to be considered, E_2. Thus we can write

$$E' = E + E_1 + E_2 \tag{18.19}$$

where E is by definition the applied field intensity. The contribution of the polarization E_1 is determined by the normal component of the polarization vector integrated over the surface of the spherical cavity. This is

$$E_1 = \frac{P}{3\epsilon_0} = \frac{(\kappa' - 1)E}{3} \tag{18.20}$$

The contribution to the field arising from individual molecular interactions can be assumed equal to zero if they are arranged either in complete disorder or in a symmetrical array. Using this assumption, first made by Mosotti, we obtain for the local field

$$E' = E + \frac{P}{3\epsilon_0} = \frac{E}{3}(\kappa' + 2) \tag{18.21}$$

By inserting this local Mosotti field into Eqs. 18.16 and 18.18, we obtain a relation between the polarizability per unit volume $N\alpha$ and the relative dielectric constant κ'

$$\frac{N\alpha}{3\epsilon_0} = \frac{\kappa' - 1}{\kappa' + 2} \tag{18.22}$$

where N is the number of molecules per unit volume which is related to the number of molecules per mole by Avogadro's number, $N = N_0\rho/M$, where M is the molecular weight and ρ the density. Substituting in Eq. 18.22, we obtain

$$\alpha = \frac{3\epsilon_0}{N_0}\frac{\kappa' - 1}{\kappa' + 2}\frac{M}{\rho} = \frac{3\epsilon_0}{N_0}P_m \tag{18.23}$$

where P_m is the molar polarizability, directly proportional to the atomic polarizability. This relationship can be compared with Eq. 13.23 for molar refractivity.

There are various possible mechanisms for polarization in a dielectric material (Fig. 18.4). One process common to all materials is electron polarization, or the shift of the center of gravity of the negative electron cloud in relation to the positive atom nucleus in an electric field. This has

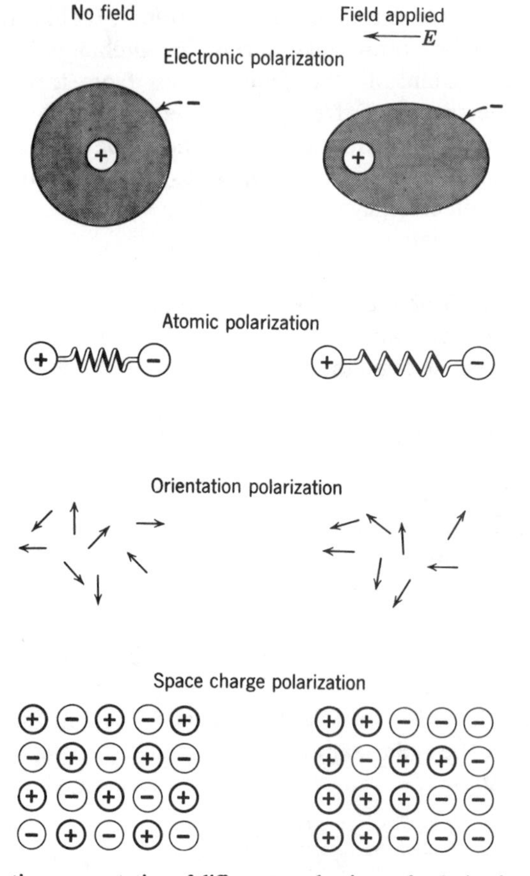

Fig. 18.4. Schematic representation of different mechanisms of polarization. From A. R. von Hippel, reference 1.

already been discussed in Chapter 13 in connection with optical properties. A second mechanism is the displacement of positive and negative ions in relation to one another, called ionic or atomic polarization. A third kind of polarization, uncommon in ceramics, is associated with the presence of permanent electric dipoles which exist even in the absence of an electric field. An unequal charge distribution between partners in a molecule or complex ion is not uncommon; when a field is applied, these tend to line up with the electric dipoles in the direction of the field, giving rise to an orientation polarization. A final source of polarization is mobile charges which are present because they are impeded by interfaces, because they are not supplied at an electrode or discharged at an electrode, or because they are trapped in the material. Space charges resulting from these phenomena appear as an increase in capacitance as far as the exterior circuit is concerned. The total polarizability of the dielectric can be represented as the sum of these:

$$\alpha = \alpha_e + \alpha_i + \alpha_0 + \alpha_s \tag{18.24}$$

where α_e is the electronic, α_i the ionic, α_0 the orientation, and α_s the space charge polarizability.

A special kind of behavior that we might anticipate is the development of spontaneous polarization or the spontaneous lining up of electric or magnetic dipoles without the application of any external field. This can occur if the polarization resulting from neighboring dipoles exerts a sufficiently large force. It is the process observed in ferromagnetism and in ferroelectricity. Materials which behave in this way are discussed in Sections 18.6 and 18.7.

In an ideal capacitor the electric charge adjusts itself instantaneously to any change in voltage. In practice, however, there is an inertia-to-charge movement that shows up as a *relaxation time* for charge transport. This is exactly analogous to the time required for elastic strain to follow an applied stress, discussed in Chapter 15. Just as we have a *relaxed* and an *unrelaxed* elastic modulus, we have a dependence of the dielectric constant on frequency (Fig. 18.5). The electronic polarization is the only process sufficiently rapid to follow alternative fields in the visible part of the spectrum; as a result the index of refraction, for example, depends only on this process, as discussed in Chapter 13. Ionic polarization

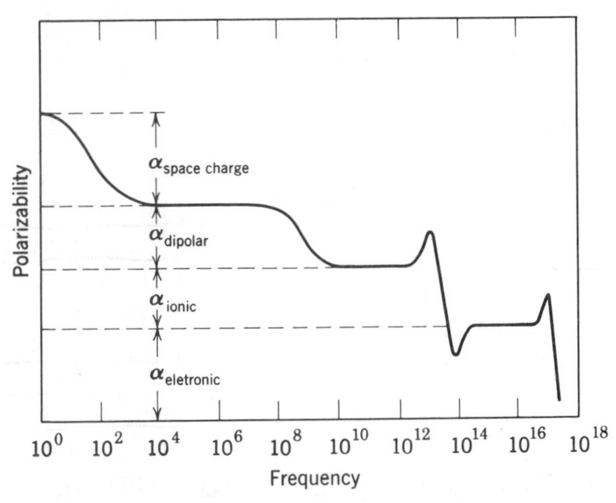

Fig. 18.5. Frequency dependence of several contributions to the polarizability (schematic).

processes are able to follow an applied high-frequency field and contribute to the dielectric constant at frequencies up to the infrared region of the spectrum. Orientation and space charge polarization have relaxation times corresponding to the particular system and process but, in general, participate only at lower frequencies.

The rate of heat generation in a dielectric is proportional to the product of voltage and current and is zero for an ideal dielectric but has some appreciable value for any dielectric with a finite response time.

For the situation depicted in Fig. 18.6, the variation of polarization with time parallels the variation of charge with time. These variations can be represented by simple physical models. During charging, for example, if the rate of change of the polarization with time is assumed proportional to the difference between its final value and its actual value

$$\frac{d(P_t - P_\infty)}{dt} = \frac{1}{\tau}[(P_s - P_\infty) - (P_t - P_\infty)] \tag{18.25}$$

then

$$P_t - P_\infty = (P_s - P_\infty)(1 - e^{-t/\tau}) \tag{18.26}$$

Here P_t is the polarization at time t, P_∞ is the instantaneous polarization on applying the field, P_s is the final value of the polarization, and τ is a

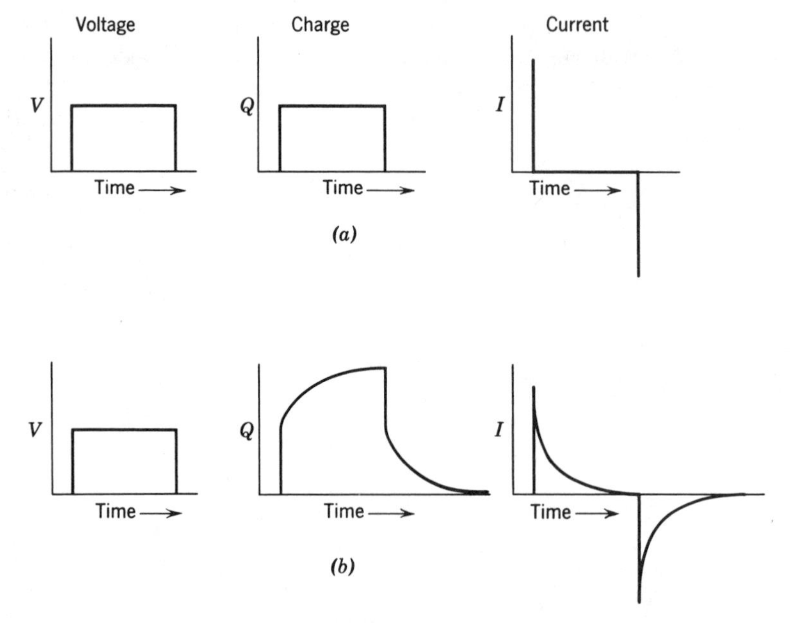

Fig. 18.6. Schematic behavior of charge buildup and current flow in (*a*) an ideal dielectric and (*b*) a real dielectric.

constant having the dimensions of time. This constant is called a relaxation time and is a measure of the time lag of the system. The response of most real dielectric materials to an applied field is not well represented by a single relaxation time. Rather, distributions of relaxation times are required to describe the experimental data. This subject is discussed further in the following sections.

Loss Factor. For an alternating field the time required for polarization shows up as a phase retardation of the charging current. Instead of being 90° advanced, as indicated in Eq. 18.4, it is advanced by some angle $90 - \delta$, where the loss angle δ is exactly comparable with the loss angle for mechanical strain discussed in Chapter 15. Representing the electric field and displacement (flux density) in complex notation:

$$E = E_0 e^{i\omega t} \tag{18.27}$$

$$D = D_0 e^{i(\omega t - \delta)} \tag{18.28}$$

and making use of the relation

$$D = \kappa^* E \tag{18.29}$$

one obtains

$$\kappa^* = \kappa_s e^{-i\delta} = \kappa_s(\cos\delta - i\sin\delta) \tag{18.30}$$

where $\kappa_s = D_0/E_0$ is the static dielectric constant. In terms of a complex dielectric constant

$$\kappa^* = \kappa' - i\kappa'' = \frac{\epsilon^*}{\epsilon_0} = \frac{1}{\epsilon_0}(\epsilon' - i\epsilon'') \tag{18.31}$$

one has from Eq. 18.30

$$\kappa' = \kappa_s \cos\delta \tag{18.32}$$

$$\kappa'' = \kappa_s \sin\delta \tag{18.33}$$

and from Eqs. 18.32 and 18.33 the loss tangent is given by

$$\tan\delta = \kappa''/\kappa' = \epsilon''/\epsilon' \tag{18.34}$$

This phase shift corresponding to a time lag between an applied voltage and induced current causes loss currents and energy dissipation in ac circuits which do not require charge-carrier migration as discussed in Chapter 17. For a simple plate capacitor with a sinusoidal applied voltage, the charging current is given by $I_c = i\omega\epsilon' E$ and the loss current by $I_l = \omega\epsilon'' E = \sigma E$, where σ is the dielectric conductivity. These components are shown in Fig. 18.7 and are combined as

$$I = (i\omega\epsilon' + \omega\epsilon'')\frac{C_0}{\epsilon_0} V = i\omega C_0 \kappa^* V \tag{18.35}$$

The corresponding energy loss per cycle is

$$W = 2\pi\epsilon' \frac{V_0^2}{2} \tan\delta \qquad \text{per cycle} \tag{18.36}$$

and the energy loss per second is

$$P = \sigma\frac{V_0^2}{2} = \omega\epsilon''\frac{V_0^2}{2} = \omega\epsilon'\frac{V_0^2}{2}\tan\delta = 2\pi f\epsilon'\frac{V_0^2}{2}\tan\delta \tag{18.37}$$

where V_0 is the maximum voltage.

As indicated in Eqs. (18.36) and (18.37), the material factors determining the energy loss in a dielectric are the product of the dielectric constant and the tangent of the loss angle. This is the *loss factor* or *relative loss factor* indicated in Eq. 18.34.

The loss factor is the primary criterion for the usefulness of a dielectric as an insulator material. For this purpose it is desirable to have a low dielectric constant and particularly a very small loss angle. For applications in which it is desirable to obtain a high capacitance in the smallest

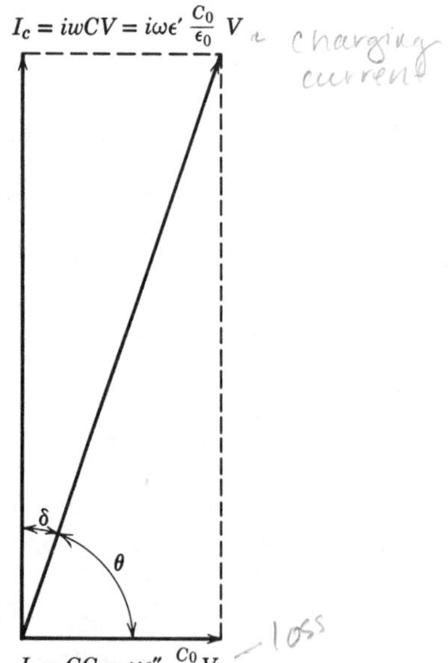

Fig. 18.7. Charging and loss current for a capacitor.

physical space, the high-dielectric-constant material must be used. For these applications it is equally important to have a low value for the dissipation factor, tan δ.

As an alternative to these expressions, the situation illustrated in Fig. 18.7 is equivalent to a parallel capacitance and resistance circuit and can be represented in terms of a dielectric permittivity ϵ' and a dielectric conductivity σ. Here, as for mechanical anelasticity, we have several equivalent ways of representing the fact that the inertia of electrical charge movement causes a time lag in the current following the applied field which results in an energy dissipation.

Which particular description of this phenomenon is convenient depends on the application. Electrical engineers concerned with power generation frequently are concerned with the dielectric constant ϵ' and the dielectric loss factor, ϵ' tan δ. Engineers concerned with radio, television, and high-frequency circuits frequently work with the dielectric constant ϵ' and the loss tangent, tan δ. Very often the inverse of the loss tangent, $Q = 1/\tan\delta$, is used as a figure of merit, the Q factor, in high-frequency problems. Engineers concerned with dielectric heating might use the dielectric constant ϵ' and the dielectric conductivity $\sigma = \omega\epsilon''$ for their purposes. These various choices are matters of convenience for engineers using these materials and refer to the same phenomena as far as material development and material properties are concerned.

As indicated in Fig. 18.5, the frequency at which a dielectric is used has a critical effect on the importance of different relaxation phenomena. As for mechanical-energy losses, the maximum dielectric loss occurs when the period of the relaxation process, whatever it may be, is the same as the period of the applied field. When the relaxation time is large compared with the period of the applied field, losses are small. Similarly, when the relaxation process is rapid compared with the frequency of the applied field, losses are small. The relative variation of the dielectric loss factor and also the dielectric constant and dielectric conductivity are illustrated in Fig. 18.8. Although the dielectric constant decreases from its relaxed value at low frequencies to its unrelaxed value at high frequencies, the dielectric conductivity is its mirror image, increasing from zero at low frequencies to its unrelaxed value at high frequencies.

In terms of the dielectric constant ϵ_∞ corresponding to the instantaneous polarization P_∞ (the dielectric constant at frequencies much higher than $1/\tau$), the complex dielectric constant may be expressed as

$$\epsilon^* = \epsilon_\infty + \frac{\epsilon_s - \epsilon_\infty}{1 + i\omega\tau} \tag{18.38}$$

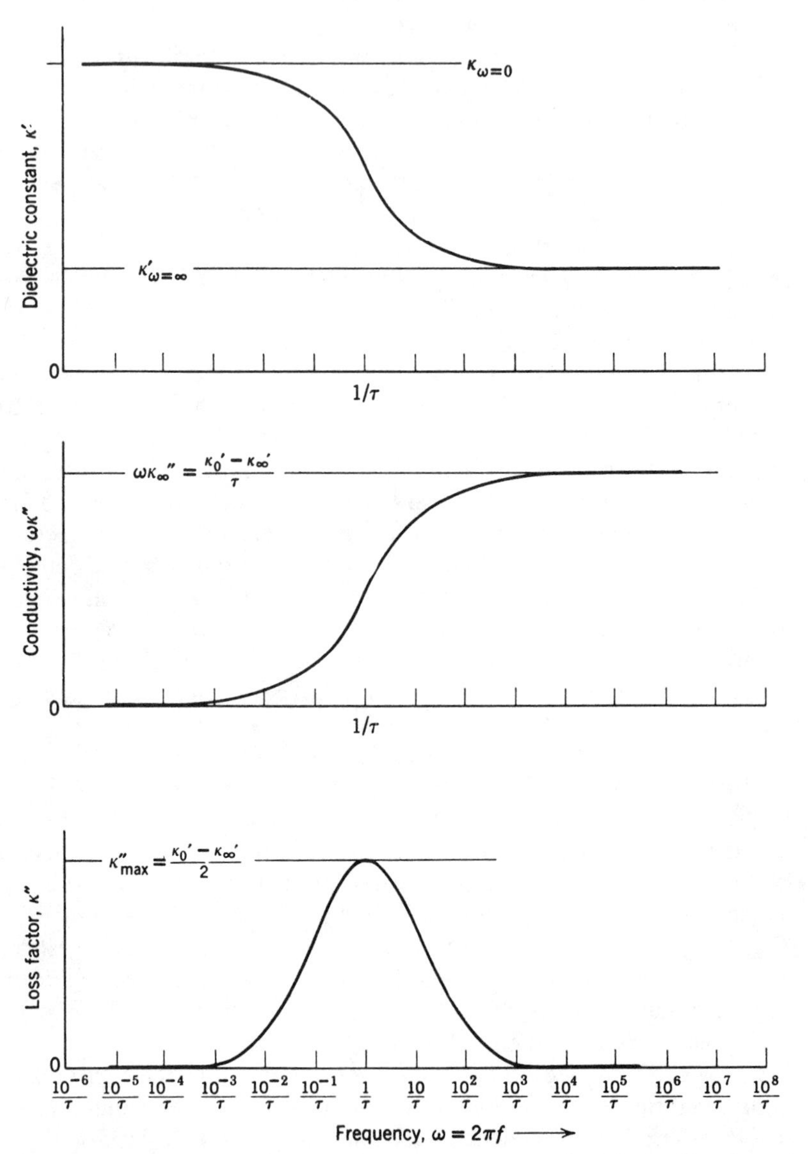

Fig. 18.8. Relaxation spectra of relative dielectric constant, conductivity, and loss factor for a simple relaxation process with a single relaxation time.

Dividing this into real and imaginary parts, as in Eq. 18.31,

$$\epsilon' = \epsilon_\infty + \frac{\epsilon_s - \epsilon_\infty}{1 + \omega^2\tau^2} \tag{18.39a}$$

$$\epsilon'' = \frac{(\epsilon_s - \epsilon_\infty)\omega\tau}{1 + \omega^2\tau^2} \tag{18.39b}$$

and

$$\tan\delta = \frac{\epsilon''}{\epsilon'} = \frac{(\epsilon_s - \epsilon_\infty)\omega\tau}{\epsilon_s + \epsilon_\infty\omega^2\tau^2} \tag{18.40}$$

Equations 18.38 to 18.40 are often termed the Debye equations. The frequency dependence of ϵ' and ϵ'' have been illustrated by the curves for κ' and κ'' in Fig. 18.8. The Debye curves for dielectric dispersion ϵ' and absorption ϵ'' are symmetric about $\omega\tau = 1$. The maximum of the absorption curve and the midpoint of the dispersion curve occur at a frequency given by $\omega_{\max} = 1/\tau$, and the half width $\Delta \log \omega\tau$ of the absorption curve is approximately 1.14 decades in frequency.

The Debye equations are based on the assumption that the transient polarization can be represented by a simple exponential with a single relaxation time. Hence any model which predicts a simple exponential rise of polarization on applying a field will in an ac field lead to dielectric dispersion and absorption curves of the form given by Eqs. 18.39 and 18.40. For most materials, however, the experimental data are not well described by the Debye equations. Rather, the dispersion of the dielectric constant occurs over a wider range of frequency, and the absorption curves are much broader and flatter than shown in Fig. 18.8.

These differences from single-relaxation-time behavior are generally associated with the fact that in condensed phases the environments of different ions are not all identical. Even in crystals, the magnitudes of the interactions between ions and of the thermal fluctuations are not identical for all places and all times. Hence it seems reasonable that a spread of relaxation times, distributed about the most probable relaxation time, is observed in experiments. The greater variation in the environments of different ions in glasses relative to crystals can be used to explain the broader distributions of relaxation times usually required to describe relaxation phenomena in amorphous materials. A sharp dichotomy between crystals and glasses should not, however, be suggested. Even in crystals a distribution of finite width should be expected, and in fluid liquids, behavior closely similar to that predicted by a single-relaxation-time model is observed.

For a distribution of relaxation times, Eq. 18.38 becomes

$$\epsilon^* = \epsilon_\infty + (\epsilon_s - \epsilon_\infty) \int_0^\infty \frac{G(\tau)\, d\tau}{1 + i\omega\tau} \tag{18.41}$$

where $G(\tau)$ is the distribution of relaxation times, i.e., where $G(\tau)\,d\tau$ is the fraction of relaxing species at a given time associated with relaxation times between τ and $\tau + d\tau$. A symmetric gaussian function is often assumed for $G(\tau)$, as

$$G(\tau)\,d\tau = \frac{b}{\sqrt{\pi}} e^{-b^2 z^2}\,dz \qquad (18.42)$$

Here b is a constant and $z = \ln(\tau/\tau_0)$, where τ_0 is the most probable relaxation time. The effects of changes in the width of the distribution (changes in b) on the dielectric loss curve have been shown.*

Although the dielectric loss curves for most materials cover a wider range of frequency than predicted by the Debye equations, they usually do not exhibit the symmetry expected for a gaussian distribution of relaxation times. Rather, experimental data indicate dielectric losses which are asymmetric to the high-frequency side of the loss maximum (the loss at a frequency of $100\omega_{max}$ is larger than that at a frequency of $\omega_{max}/100$). The implications of this result are discussed in Section 18.3.

Dielectric Strength. Another important property of dielectric materials is the ability to withstand large field strengths without electrical breakdown. At low field strengths there is a certain dc conductivity corresponding to the mobility of a limited number of charge carriers related to electronic or ionic imperfections. As the field strength is increased, this dc conduction increases, but also when some sufficiently large value of potential is reached, a field emission from the electrodes makes available sufficient electrons for a burst of current which produces breakdown channels, jagged holes, or metal dendrites bridging the dielectric and rendering it unusable. Various processes may contribute to these dielectric breakdown phenomena; however, different measurement techniques give a considerable scatter in results, and detailed interpretations are still in some doubt. For single crystals tested with carefully designed electrodes and suitable precautions, values up to 10,000 V/mil (approximately 4×10^6 V/cm) are observed. For polycrystalline polyphase ceramics tested by usual techniques without extensive precautions, values as low as 100 V/mil are observed at room temperature and considerably lower values at high temperatures. For many applications, particularly in higher temperature ranges, low values of dielectric strength are a major restriction on widespread use of insulators.

Specifications. A wide variety of ceramic dielectric materials is required for different applications. As a guide to the range of values used, the minimum requirements for the joint Army and Navy specifications for ceramics to be used as insulators in electronic devices are given in Table 18.2. It should be noted that the mechanical strength and dielectric

*W. A. Yager, *Physics*, **7**, 434 (1936).

Table 18.2. Jan-I-10 Minimum Requirements for Insulating Ceramics, Radio, Class L

Porosity:	
No liquid penetration at 10,000 psi pressure	
Thermal stress resistance:	
Type A—withstand 20 cycles from 100°C into 0° water	
Type B—withstand 5 cycles from 100°C into 0° water	
Transverse strength:	
Greater than 3000 psi	
Dielectric strength:	
Greater than 180 volts/mil	
Dielectric constant:	
Less than 12 after 48-hr water immersion	
Loss factor ($\kappa' \tan \delta$):	
Grade L-1	<0.150
L-2	<0.070
L-3	<0.035
L-4	<0.016
L-5	<0.008
L-6	<0.004

strength characteristics of most ceramics far exceed the minimum requirements.

18.2 Dielectric Constants of Crystals and Glasses

The dielectric constant of a single crystal or glass sample results from electronic, ionic, and dipole orientation contributions to the polarizability. The electronic contributions are always present and, as indicated in Fig. 18.5, are the main contributors in the optical range of frequencies. By comparing the Lorentz equation, Eq. 13.23 and the Mosotti equation, Eq. 18.23, we see that in this range the relative dielectric constant is equal to the square of the index of refraction:

$$\kappa'_e = n^2 \tag{18.43}$$

The electronic polarization and factors affecting it have already been discussed in Chapter 13. For a typical silicate crystal or glass, $n = 1.5$, so that κ'_e equals 2.25. In comparison, a typical measured value of κ' at radio and lower frequencies is in the range from 5 to 10, indicating that about one-third of the polarizability usually corresponds to electronic processes for silicate structures. For high-index-of-refraction materials the electronic polarizability is increased. For example, for barium oxide, κ'_e equals n^2, which equals 4, whereas the measured dielectric constant is about 34; that is, for these high-dielectric-constant materials ionic contributions overshadow the electronic part. In contrast to ionic crystals, completely covalent structures have no mechanism for ionic polarizability. This is illustrated in germanium, for example, in which $\kappa'_e = n^2 = 16$. Similarly, dielectric measurements at low frequencies give $\kappa' = 16$.

Ionic Polarizability. The ionic polarizability arises from the displacement of ions of opposite sign from their regular lattice sites under the influence of an applied field and also from the deformation of the electronic shells resulting from the relative displacement of the ions.

For ions that can be treated as hard spheres, the dipole moment per molecule is $ze(\delta_+ - \delta_-)$, where ze is the ionic charge and δ_+ and δ_- are the displacements from equilibrium positions. For the sodium chloride structure the volume per molecule is $2a^3$, where a is the nearest-neighbor distance. It can be shown that for the sodium chloride lattice in a uniform external field E, the ion displacement is given by

$$\delta'_+ - \delta'_- = \frac{zeE}{\nu_0^2}\left(\frac{1}{m} - \frac{1}{M}\right) \tag{18.44}$$

where ν_0 is the lattice infrared vibration absorption frequency and m and M are the mass of the sodium and chloride ions respectively. This relationship allows a calculation of the ion displacement, the resulting dipole moment per molecule, and from this and Eq. 18.13, the polarizability, and from Eq. 18.16, the susceptibility. The calculated result for sodium chloride is an ionic dielectric constant, $\kappa'_i \simeq 3$, in reasonably good agreement with the experimental value, $\kappa'_i = \kappa' - \kappa'_e = 5.62 - 2.25 = 3.37$. This agreement indicates that the main contribution to the dielectric constant is satisfactorily predicted on the basis of a hard-sphere model for sodium chloride. This is true of most silicates and aluminates of interest in ceramics for which the dielectric constant is in the range of 5 to 15 and results from this kind of ion displacement.

For ions that are highly polarizable there is an additional contribution to the dielectric constant resulting from the deformation of electronic shells following the displacements of the ions. This contribution is difficult to calculate but is qualitatively seen to be of increasing importance for ions that are polarizable and for structures in which considerable ion deformations are possible. Thus, it is found that the ratio between the low-frequency dielectric constant and the optical dielectric constant increases

Table 18.3. Dielectric Constants of Some Crystals and Glasses at 25°C and 10^6 Hz

Material	κ'	$\text{Tan } \delta = \dfrac{\kappa''}{\kappa'}$
LiF	9.00	0.0002
MgO	9.65	.0003
KBr	4.90	.0002
NaCl	5.90	.0002
TiO_2 (‖ c-axis)	170	.0016
TiO_2 (⊥ c-axis)	85.8	.0002
Al_2O_3 (‖ c-axis)	10.55	.0010
Al_2O_3 (⊥ c-axis)	8.6	.0010
BaO	34	.001
KCl	4.75	.0001
Diamond	5.68	.0002
Mullite	6.60	–
Mg_2SiO_4 (forsterite)	6.22	.0003
Fused silica glass	3.78	.0001
Vycor (96 SiO_4 4 B_2O_3) glass	3.85	.0008
Soda-lime-silica glass	6.90	.01
High-lead glass	19.0	.0057

as the magnitude of these values increases. The ratio κ'/κ'_e is about 2.5 for sodium chloride and about 8.5 for barium oxide. Room-temperature dielectric constants for a number of single crystals and glasses of interest for ceramics are collected in Table 18.3.

Ion Jump Polarization. Orientation polarization effects are of decisive importance in liquids such as water and in molecular solids and gases, but they are not of major concern for ceramic systems. They occur in glasses and crystals when two or more equivalent positions for an ion are present. This is perhaps most clearly illustrated for crystalline solids by the association between a lattice vacancy and an ion of greater valence than those normally present in the lattice. As discussed in Chapter 4, for example, additions of small amounts of calcium chloride to a crystal of KCl lead to increased concentration of cation vacancies. There is a tendency for the Ca^{2+} ions in these crystals to be associated with a positive ion vacancy, and the associated pair has a dipole moment. When an electric field is applied, the Ca^{2+} and the vacancy can exchange positions by a simple jump of the cation to a neighboring position. This

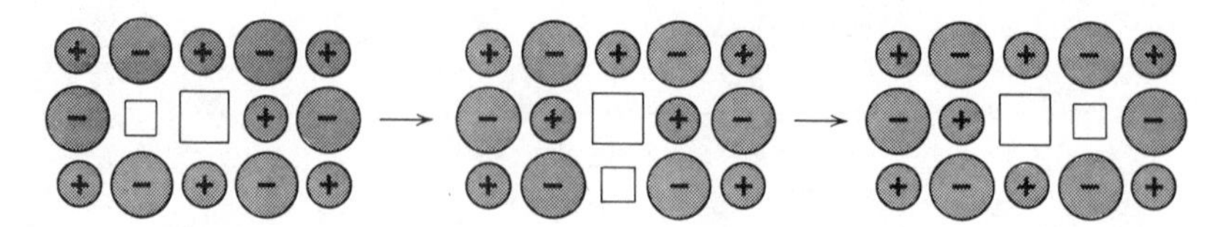

Fig. 18.9. Reorientation of a lattice vacancy pair. Other crystal defect pairs give similar results.

kind of process is shown schematically in Fig. 18.9. The average polarizability per particle resulting from this process is given by

$$\alpha_j = \frac{1}{1 + i\omega\tau} \frac{(zed)^2}{kT} \tag{18.45}$$

where d is the distance between the two atom positions, ze is the ion charge, and τ is the relaxation time for the jump process. Then for the complex dielectric constant, $\kappa^* = \kappa' - i\kappa''$, we have

$$\frac{\kappa^* - 1}{\kappa^* + 2} = \frac{N}{3\epsilon_0}\left(\alpha_e + \alpha_i + \frac{n}{N}\alpha_j\right) \tag{18.46}$$

where n is the number of ion pairs, N is the number of ions per unit volume, and ϵ_0 equals 8.85×10^{-12} F/m.

A similar effect can arise in glasses when there are multiple sites available to a modifier ion which cannot contribute to the observed dc conductivity. This is schematically illustrated in Fig. 18.10, which shows two equivalent positions in which the modifier ion can reside; charge displacement occurs by an atom jump rather than movement about an equilibrium position.

These effects in crystals are not normally observed at room temperature but can be used as a sensitive tool for investigating relaxation

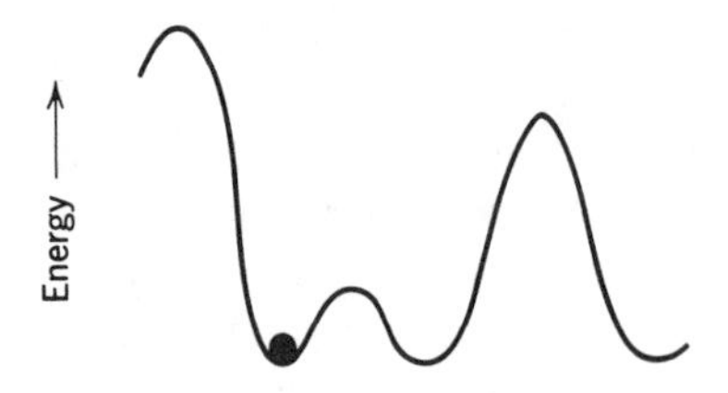

Fig. 18.10. Potential wells in a glass structure.

phenomena as shown by R. G. Breckenridge.* In glasses this process causes a modest increase in the dielectric constant at low frequencies for soda-lime-silica glasses, an effect which is absent for pure fused silica. For a typical soda-lime-silica glass, for example, the dielectric constant increases from a value of 6.90 at 1 MHz to 8.30 at 100 Hz, whereas for fused-silica glass a constant value of 3.78 is found independent of frequency over the range of 10^2 to 10^{10} Hz. The effect is more pronounced in sodium silicate glasses in which for a $30Na_2O$–$70SiO_2$ composition the dielectric constant increases from 8.5 at 10^6 Hz to 18 at 10^2 Hz. As would be expected from the ion size, for a constant mole ratio the effect decreases in the order $Li^+ > Na^+ > K^+ > Rb^+$. In all these glasses, however, the process leads to much larger changes in the loss factor and is considered from this point of view in the next section.

In this connection we may mention that the jump frequencies for ion motion in glasses at room temperature are slow, even when compared with low-frequency dielectric property measurements. As a result, the static dielectric constant may be considerably larger than that measured for frequencies as low as 100 Hz. This has been discussed in Chapter 17 in connection with the anomalous charging current which for glasses at room temperature may have a time constant of several minutes or hours. These long-charging-time characteristics are of less use to engineers than the normal higher-frequency measurements.

Effects of Frequency and Temperature. Effects of frequency and temperature are not independent. For electronic and ionic polarization the frequency effect is negligible at frequencies up to about 10^{10} Hz, which is the limit of normal uses. Similarly, the effect of temperature on electronic and ionic polarization is small. At higher temperatures, however, there is an increasing contribution resulting from ion mobility and crystal imperfection mobility (Figs. 18.9 and 18.10). Also, at higher temperature dc conductivity effects, which increase exponentially with temperature, become important. The combined effect is to give a sharp rise in the apparent dielectric constant at low frequencies with increasing temperature, corresponding to both ion jump orientation effects and space charge effects resulting from the increased concentration of charge carriers. The effectiveness of charge carriers in giving an increased dielectric constant for single crystals and glasses depends critically on the electrode materials, polarization effects at the electrodes, and the resulting space charges. For ionic conductivity electrode reactions are required at the electrode

*"Relaxation Effects in Ionic Crystals," *Imperfections in Nearly Perfect Crystals*, W. Shockley, J. H. Holloman, R. Maurer, and F. Seitz, Eds., John Wiley & Sons, Inc., New York, 1952, pp. 219–245.

surfaces to provide a source for and dissipation of the charge carriers. When this reaction cannot keep up with the large number of charge carriers arriving at or departing from an electrode during any half-cycle, polarization and an increased apparent dielectric constant result. In general, it is found in aqueous solutions that dielectric-constant measurements must be carried out at frequencies above about 1000 Hz in order to avoid this kind of electrode polarization effect.

The combined effects of temperature and frequency are illustrated for aluminum oxide in Fig. 18.11 and for a soda-lime-silica glass in Fig. 18.12.

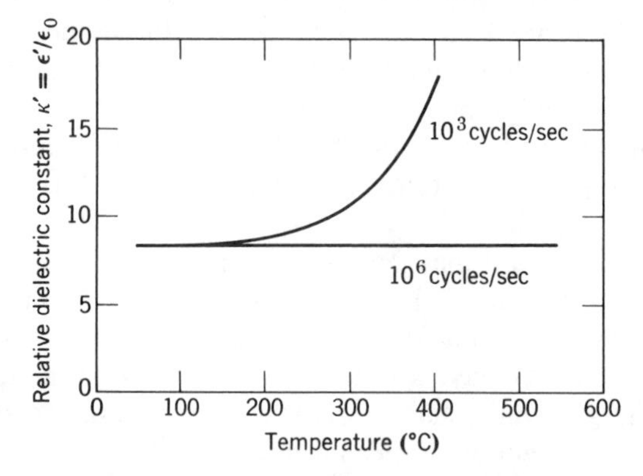

Fig. 18.11. Effect of frequency and temperature on the dielectric constant of an Al_2O_3 crystal with field normal to the *c* axis.

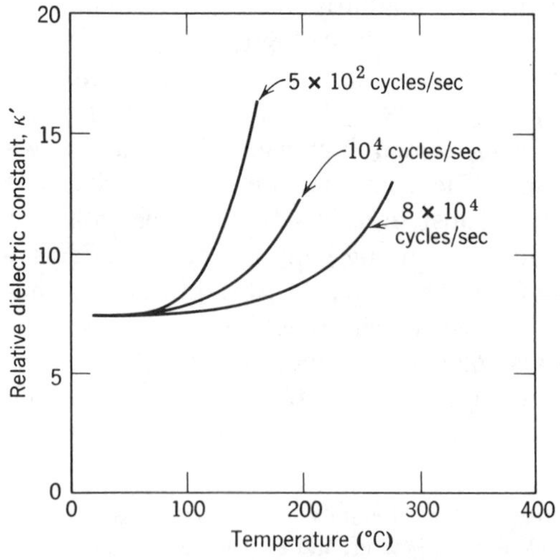

Fig. 18.12. Effect of frequency and temperature on the dielectric constant of a soda-lime-silica glass. From M. J. O. Strutt, *Arch. Elektrotech.*, **25**, 715 (1931).

When the high-temperature data are plotted against frequency, they give an apparent dielectric constant increasing at low frequencies as a result of electrode polarization. Consequently, they are not solely a material property but also depend on the electrodes used for measurements and in applications.

18.3 Dielectric Loss Factor for Crystals and Glasses

As discussed in Section 18.1, the power dissipation in an insulator or capacitor is directly proportional to the dielectric loss factor, $\epsilon' \tan \delta$. Consequently, this factor is of great concern for many applications of ceramic materials. Indeed, one of the main advantages of ceramics as dielectrics is that this loss factor is small compared with that of other available materials such as plastics. Energy losses in dielectrics result from three primary processes:

1. Ion migration losses
   a. Dc conductivity losses
   b. Ion jump and dipole relaxation losses
2. Ion vibration and deformation losses
3. Electron polarization losses

Of these, the electron polarization losses giving rise to absorption and color in the visible spectrum have already been discussed in Chapter 13. The ion vibration and deformation losses become important in the infrared but are not a major concern for frequencies below about 10^{10} Hz. By far the major factor affecting the use of ceramic materials is the ion migration losses which tend to increase at low frequencies and as the temperature is raised.

The loss factor can be written in terms of the electrical conductivity

$$\omega\epsilon'' = \sigma = \omega\epsilon' \tan \delta \tag{18.47}$$

or

$$\tan \delta = \frac{\sigma}{2\pi f \kappa' \epsilon_0} = \frac{\sigma}{(8.85 \times 10^{-14})(2\pi f)\kappa'} \tag{18.48}$$

where the conductivity is given in ohm-cm^{-1}. These conduction migration losses are normally small. For a commercial soda-lime-silica glass with $\sigma = 10^{-12}$/ohm-cm and $\kappa' = 9$ for a frequency of 1000 Hz, $\tan \delta$ is calculated as 20×10^{-4} corresponding to an experimental value of about 250×10^{-4}. In general, then, conduction losses are small at frequencies greater than 100 Hz at room temperature even for soda-lime glasses, but they may be important at (1) low frequencies and (2) high temperatures.

As indicated in Eq. 18.48, the power factor increases at low frequencies and is inversely proportional to the frequency. In general, ion jump relaxation between two equivalent ion positions is responsible for the largest part of the dielectric loss factor for crystals and glasses at moderate frequencies. If the relaxation time for an atom jump is τ, the maximum energy loss occurs for a frequency equal to the jump frequency, $1/\tau$. When the applied alternating field frequency is much smaller than the jump frequency, atoms follow the field, and the energy loss is small. Similarly, if the applied frequency is much larger than the jump frequency, the atoms do not have an opportunity to jump at all, and losses are small. The resulting expressions for ϵ', ϵ'', and $\tan\delta$ are given in Eqs. 18.38 to 18.40.

For an ion jump (Fig. 18.10) the jump frequency depends on the energy barrier separating the two ion positions, as discussed in Chapter 6. If we assume for simplicity that there is one relaxation time, this is given by

$$\tau = \tau_0 \exp\left(\frac{U}{kT}\right) \tag{18.49}$$

where τ_0 is the period of atomic vibrations, of the order of 10^{-13} sec. The value of the activation energy can vary considerably, but if it is similar to that for ionic migration processes, it is in the order of 0.7 eV, giving a loss maximum over a spectrum range corresponding to 10^3 to 10^6 Hz, a region of particular concern for many dielectric applications.

Ion vibration and deformation losses become important at room temperature only at frequencies in the infrared corresponding to 10^{12} to 10^{14} Hz. This is beyond the range of frequencies that are normally of concern for electronic applications. These processes begin to become apparent in the higher-frequency measurements, 10^{10} Hz.

The total value for $\tan\delta$ is the sum of individual contributions already discussed. At room temperature the resulting curve for a glass or a crystal having considerable impurities or defects is as illustrated in Fig. 18.13. At lower frequencies conduction losses become important, at moderate frequencies ion jump and dipole losses are most important, at intermediate frequencies dielectric losses are small, and at sufficiently high frequencies ion polarization effects give energy absorption.

For reasonably good insulators (Chapter 17) the conductivity increases exponentially with temperature. Consequently, for this process we would expect $\tan\delta$ to increase exponentially with temperature, as indicated in Eq. 18.48.

A significant part of the increase in $\tan\delta$ of the soda-lime-silicate glasses indicated in Fig. 18.14 corresponds to conduction losses. These losses are directly related to the dc conductivity of the glass. Such conductivity gives rise to currents which in an ac field are in phase with the applied voltage and hence cause dielectric losses which are indepen-

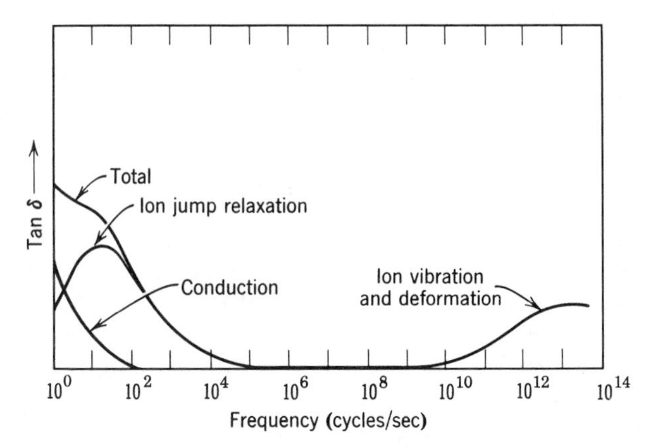

Fig. 18.13. Effect of different dielectric loss mechanisms on tan δ at room temperature.

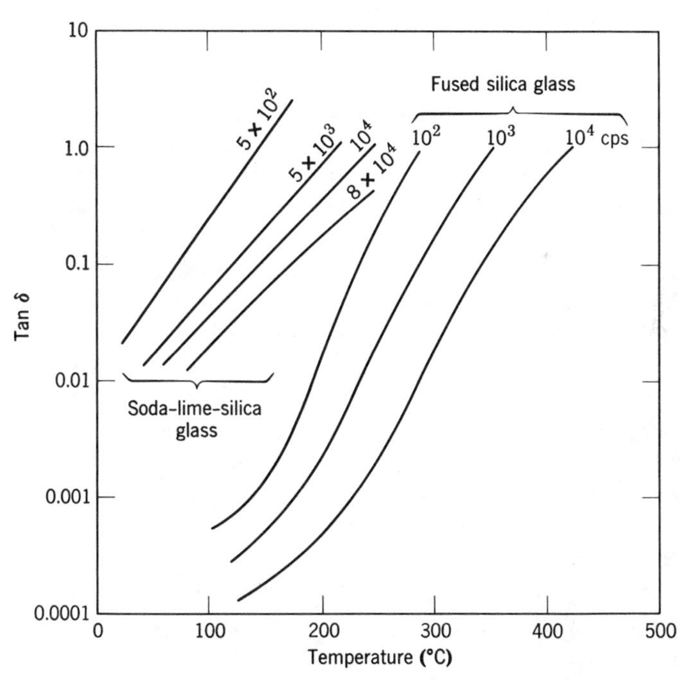

Fig. 18.14. Increase in tan δ with increasing temperature for a soda-lime-silica glass and for fused-silica glass.

dent of any other absorption mechanism. For a condenser with a glass dielectric, represented by a resistance R (equivalent to the dc resistance) in parallel with a capacitance C, tan δ in an ac field is inversely proportional to the frequency as

$$\tan \delta = \frac{1}{\omega RC} \tag{18.50}$$

At room temperature, such conduction losses are small relative to other losses, at least for frequencies of 50 to 100 Hz or higher. As an example, a soda-lime-silicate glass with a resistivity of 10^{12} ohm-cm at room temperature and a dielectric constant of 9 at a frequency of 10^3 Hz would have a tan δ of about 2×10^{-5}, which is much smaller than the measured tan δ. In contrast, at a temperature of 200°C, where the resistivity is about 2×10^7 ohm-cm, the tan δ from dc conduction losses would be about unity, which is comparable with or larger than other losses in the material. In general, conduction losses in glasses become increasingly important at elevated temperatures, as the dc resistivity decreases, and at low frequencies.

For an ion jump process with a single activation energy, the activation energy can be derived from a coincidental determination of the dependence of tan δ on temperature and on frequency, using Eqs. 18.40 and 18.49. This has been done for the dielectric losses in kaolin by J. van Keymeullen and W. D. Dekeyser,* who determined the value of tan δ over a range of frequencies at different temperatures. These gave maxima at different frequencies corresponding to values of $\omega\tau = 1$ from which the activation energy in Eq. 18.49 can be determined by plotting the log ω_{max} versus $1/T$. The results are illustrated in Fig. 18.15. These authors found an activation energy of approximately 0.69 eV which they attribute to an ion pair formed by the replacement of an aluminum ion with a divalent cation and the corresponding formation of a vacancy at a normally occupied hydroxyl ion site. Similar experiments have been carried out on sodium chloride doped with various divalent ions to form M^{2+}–vacancy pairs.

The dielectric losses of single crystals are small and are mainly determined by minor impurities which have not been extensively or satisfactorily investigated except for special cases. In contrast, the dielectric losses in glasses vary over a wide range and show an intimate relationship with the glass structure and composition. The nature of the modifying cations has a significant effect on the dielectric loss of oxide glasses. For example, the order of mobility of alkali ions is $Li^+ > Na^+ >$

J. Chem. Phys., **27**, 172 (1957).

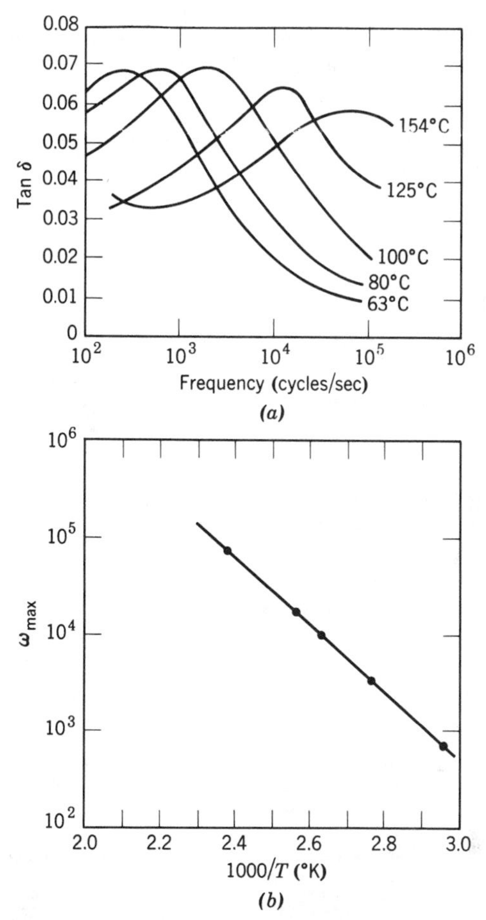

Fig. 18.15. Tan δ versus frequency at different temperatures and log ω_{max} versus $1/T$ for Georgia kaolin. From J. Van Keymeulan and W. Dekeyser, *J. Chem. Phys.*, **27**, 172 (1957).

$K^+ > Rb^+$, and the dielectric losses follow in the same order. For example, a glass composition consisting of 53.3% SiO_2, 32% M_2O, 10.2% PbO, and 4.5% CaF_2 gives values for tan δ (at 1.5×10^6 Hz) of 0.0132 for lithium, 0.0106 for sodium, and 0.0052 for potassium. In general, a substitution of large divalent ions such as barium oxide or lead oxide for the alkalis allows the formulation of reasonably low-melting glasses which have low values for tan δ. Glasses with 20 to 30% BaO or 30 to 50% PbO have values of tan δ as low as 0.0005.

In addition to the use of divalent ions alone as modifiers, they are effective in small amounts as *blocking* ions; they prevent the easy mobility of alkali ions by filling up critical sites through which the alkali ions would normally pass. This is illustrated in Fig. 18.16 for the change in

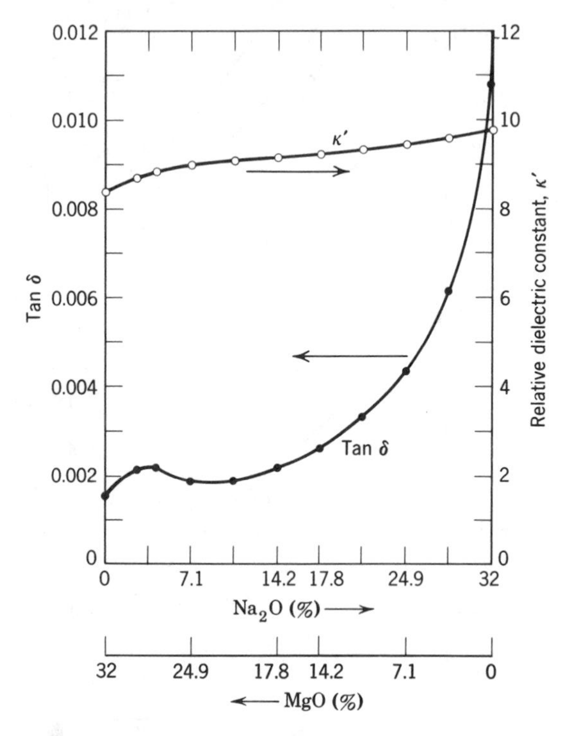

Fig. 18.16. Change in tan δ and κ' on substituting MgO and Na_2O in a silicate glass. From J. M. Stevels, reference 5.

tan δ for an alkali silicate glass when sodium oxide is replaced by magnesium oxide. Small additions of sodium oxide have little effect of increasing the loss factor, since sodium migration is blocked by magnesium ions.

Some typical values of dielectric loss factors observed in single crystals and glasses are given in Table 18.3. As indicated in the discussion thus far, these are generally inversely proportional to the conductivity. The frequency dependence of tan δ for some typical glasses is shown in Fig. 18.17.

Detailed studies of dielectric loss have been carried out in a number of oxide glass systems. H. E. Taylor* studied a number of sodium silicate

J. Soc. Glass Technol., **43**, 124 (1959).

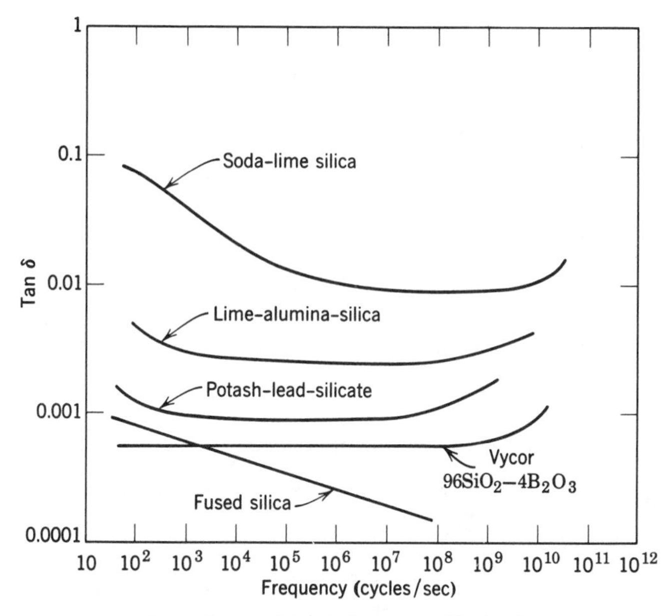

Fig. 18.17. Frequency dependence of tan δ for some silicate glasses.

glasses as well as a typical soda-lime-silicate glass over a range of frequency and temperature. After subtracting the conduction losses, he obtained dispersion and absorption curves (Fig. 18.18) which are significantly broader than predicted by a single relaxation time model (Eqs. 18.39 and 18.40). It was also found that the data obtained at different temperatures could be superimposed by shifting the curves along the frequency axis by amounts calculated from the activation energy and temperature. This result indicates that relaxation times differing by several orders of magnitude are characterized by the same activation energy. This activation energy is closely similar to that for dc conduction in the same glasses. It was also found that the dielectric-loss curves for all the glasses could be superimposed on a reduced scale, in which the frequency is scaled with ω_{max}, the frequency of the maximum loss, and the loss is scaled with $\epsilon_s - \epsilon_\infty$. Such a master curve is shown in Fig. 18.19 and indicates that the distribution of relaxation times in all the glasses is the same. The data in Fig. 18.19 also indicate a broad high-frequency tail which extends over some eight decades in frequency beyond the peak. Even allowing for the contribution of other loss mechanisms at high frequencies, the results indicate ionic relaxation losses which cover an extended range of frequency.

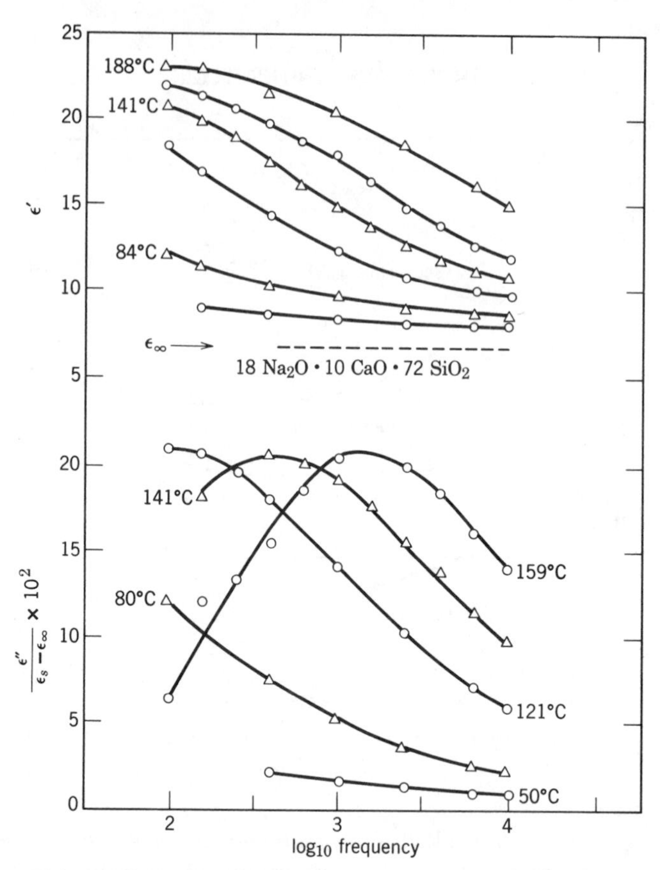

Fig. 18.18. Dielectric dispersion and absorption curves, corrected for dc conductivity, for a typical soda-lime-silicate glass. From H. E. Taylor, *J. Soc. Glass Technol.*, **43**, 124 (1959).

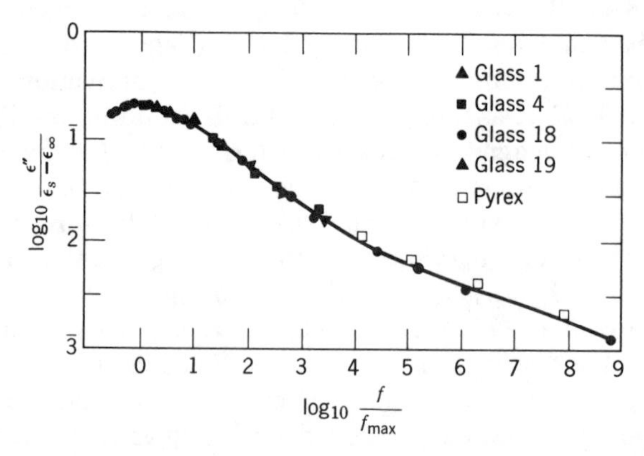

Fig. 18.19. Reduced dielectric loss curves for several glasses. Glass 1, $0.12Na_2O{\cdot}88SiO_2$; glass 4, $0.24Na_2O{\cdot}0.76SiO_2$; glass 18, $0.10Na_2O{\cdot}0.20CaO{\cdot}0.70SiO_2$; glass 19, $0.18Na_2O{\cdot}0.10CaO{\cdot}0.72SiO_2$. From H. E. Taylor, *J. Soc. Glass Technol.*, **43**, 124 (1959).

Several of the compositions investigated by Taylor have subsequently been shown to exhibit pronounced phase separation, but similar results have been obtained on a variety of other glasses, both homogeneous and phase-separated. The effects of phase separation on dielectric loss were investigated by R. J. Charles* on glasses in the Li_2O–SiO_2 system. For a glass containing 6.7 mole% Li_2O, which was characterized by a discrete, isolated second-phase submicrostructure, two peaks were seen in the dielectric loss versus frequency relations. For other compositions investigated, some of which were characterized by interconnected second-phase morphologies and some by homogeneous, featureless submicrostructures, only a single loss peak was observed.

The most satisfactory description of the derived distributions of relaxation times is provided by empirically modifying Eq. 18.38 to obtain a *skewed* dielectric constant:

$$\epsilon^* = \epsilon_\infty + \frac{\epsilon_s - \epsilon_\infty}{(1 + i\omega\tau)^\beta} \qquad (18.51)$$

For the alkali-silicate glasses investigated, the values of β were in the range of 0.21 to 0.29. A relation of this form had previously been suggested empirically by K. S. Cole and R. H. Cole† to represent dielectric relaxation data on a variety of materials, and a similar relation was obtained theoretically by S. H. Glarum,‡ who considered relaxation processes involving diffusion of defects to the relaxing species. Although this model may not be directly applicable to loss processes in oxide glasses which seem to involve local motions of the modifying cations, it seems reasonable that the relaxation of a given ion depends on the availability of other nearby vacant sites, and the probability of the ion relaxing increases when a nearby site has become vacated by another diffusive or relaxational process, thus introducing a cooperative character of an appropriate form to the relaxation process. In any case, it should be emphasized that the observed distributions of relaxation times may not reflect corresponding distributions of molecular processes.

18.4 Dielectric Conductivity

The dielectric loss phenomena discussed in the last section can also be described as an ac conductivity of solids. This is in accordance with Eq. 18.47, which defines the dielectric conductivity as $\sigma = \omega\epsilon''$. As illustrated in Fig. 18.8, this part of the conductivity increases as the frequency is

J. Am. Ceram. Soc., **46**, 235 (1963).
†*J. Chem. Phys.*, **9**, 341 (1941).
‡*J. Chem. Phys.*, **33**, 639 (1960).

raised. In fact, a whole alternative mathematical expression of material behavior can be formulated on the basis of a real (giving power losses) and an imaginary (charging current) conductivity.

In ceramic applications this charging current is particularly important when determining the conductivity of glasses at low temperatures. Near room temperature the anomalous charging current resulting from an ion jump or several ion jumps which do not contribute to the dc conductivity leads to relaxation times measured as several seconds or several minutes. As a result, measured conductivity depends on the frequency used, ac measurements giving higher values than dc measurements.

At higher temperatures the fraction of ions contributing to the dc conductivity becomes nearly equal to the fraction contributing to the ac conductivity if no blocking layer exists at the electrodes, as a result, at temperatures above about 250° the ac measurements and dc measurements are generally comparable. These effects for a typical soda-lime-silica glass are illustrated in Fig. 18.20.

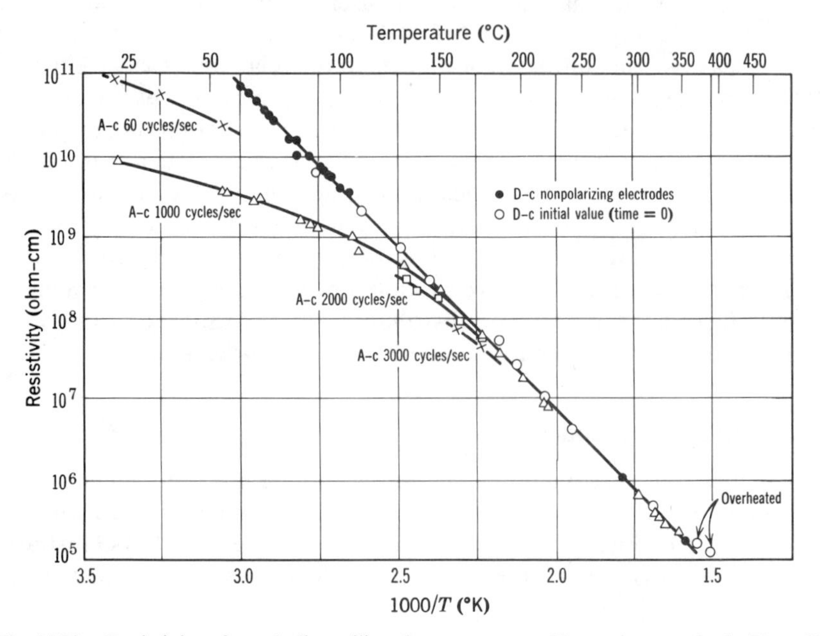

Fig. 18.20. Resistivity of a soda-lime-silica glass as measured by various methods. From D. M. Robinson, *Physics*, **2**, 52 (1932).

18.5 Polycrystalline and Polyphase Ceramics

To understand the behavior of the great majority of ceramic materials, which are polycrystalline and polyphase, we must extend our consideration of single crystals and glasses to include the effects of grain boundaries, porosity, and phase mixtures. This requires, first, consideration of dielectric properties of mixtures and, second, consideration of the space charge polarization which can result in mixtures of components having different resistivity characteristics.

Mixture Rules. Mixtures of ideal dielectrics can be most simply considered on the basis of layer materials with the layers either parallel or normal to the applied field (Fig. 18.21). When layers are parallel to the capacitor plates, the structure corresponds to capacitive elements in series, and the inverse capacitances are additive, as is true of the inverse conductivities ($\sigma = \omega\kappa''$). Then

$$\frac{1}{\kappa'} = \frac{v_1}{\kappa'_1} + \frac{v_2}{\kappa'_2}; \qquad \frac{1}{\kappa''} = \frac{v_1}{\kappa''_1} + \frac{v_2}{\kappa''_2} \tag{18.52}$$

where v_1 and v_2 are the volume fraction of each phase, equal to the relative plate thicknesses. In contrast, when the plate elements are arranged normal to the capacitor plates, the applied field is similar for each of the elements so that the capacitances are additive:

$$\kappa' = v_1\kappa'_1 + v_2\kappa'_2 \qquad \kappa'' = v_1\kappa''_1 + v_2\kappa''_2 \tag{18.53}$$

Equations 18.52 and 18.53 are special cases of a general empirical relationship:

$$\kappa^n = \sum_i v_i\kappa_i^n \tag{18.54}$$

where n is constant (-1 in Eq. 18.52 and $+1$ in Eq. 18.53) and v_i is the volume fraction of phase i; as n approaches zero, κ^n equals $1 + n \log \kappa$,

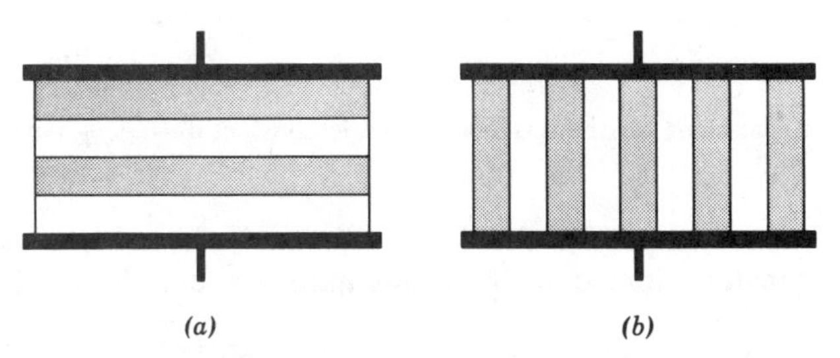

Fig. 18.21. Possible arrangement of layers having different characteristics in a dielectric.

and we have

$$\log \kappa = \sum_i v_i \log \kappa_i \tag{18.55}$$

the so-called logarithmic mixture rule. This gives a value intermediate between the extremes illustrated in Eqs. 18.52 and 18.53.

If we consider a dispersion of spherical particles of dielectric constant κ'_d in a matrix of dielectric constant κ'_m, Maxwell has derived a relationship for the mixture:

$$\kappa' = \frac{v_m \kappa'_m \left(\frac{2}{3} + \frac{\kappa'_d}{3\kappa'_m}\right) + v_d \kappa'_d}{v_m \left(\frac{2}{3} + \frac{\kappa'_d}{3\kappa'_m}\right) + v_d} \tag{18.56}$$

As shown in Fig. 18.22, this relationship comes very close to the logarithmic expression, Eq. 18.55, when the dispersed phase has a higher dielectric constant than the matrix material.

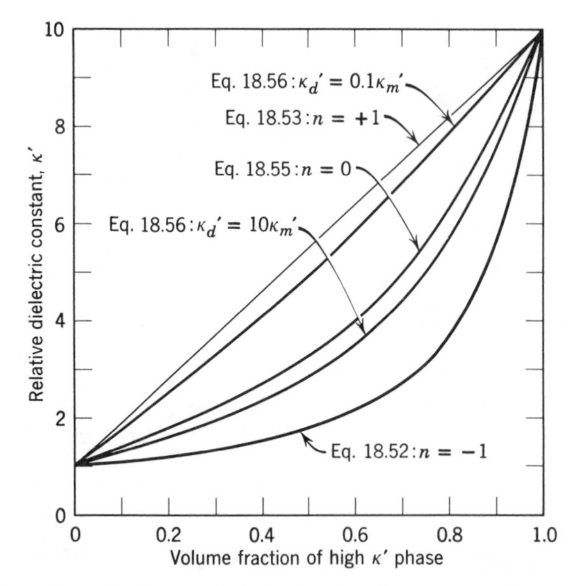

Fig. 18.22. Expressions for the resultant dielectric constant of various mixtures of two dielectrics.

Some data reported by A. Buchner* for TiO_2, which has a high dielectric constant, mixed with various matrix materials provide a good

* *Wiss. Veröff. Siemens-Werken*, **18**, 84 (1939).

test for these relationships. As shown in Fig. 18.23, the geometrical configuration of TiO_2 particles in a continuous matrix is apparently controlling. For these mixtures, and even more so when the range between dielectric constants is smaller, the logarithmic relationship and Eq. 18.56 give very similar results. For two of the mixtures either of these relationships is satisfactory. For the TiO_2–kaolin mixture, apparently as a result of the microstructure and porosity, which were not described, Eq. 18.52 most adequately describes the experimental results. If microstructure data are available, it is preferable to interpret the dielectric constant in terms of the structure observed, Eq. 18.56. In practice, the high-dielectric-constant material is almost always a crystalline phase dispersed in a lower-dielectric-constant matrix. Under these conditions the logarithmic mixture rule, Eq. 18.55, is simple and adequately represents experimental results. Its adequacy is merely fortuitous, however, and consequently it must be used with care.

Porosity is one example of practical importance in which the low-dielectric-constant phase is dispersed in a high-dielectric-constant matrix. Equation 18.53 or 18.56 is satisfactory for the small amounts of porosity

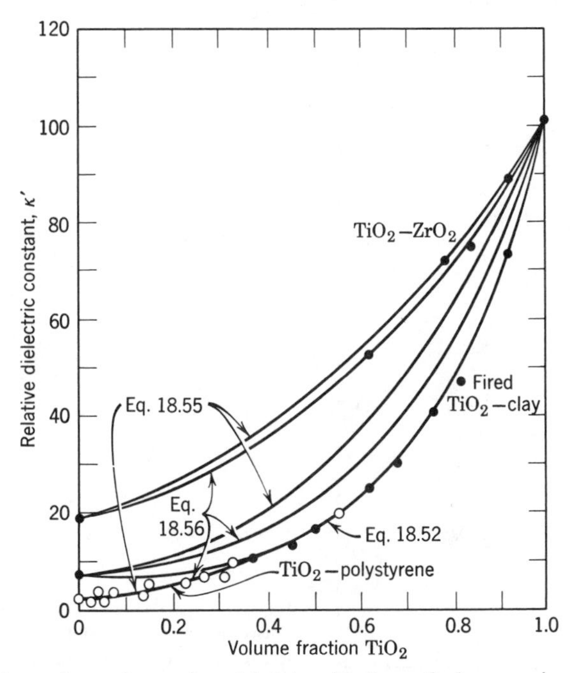

Fig. 18.23. Comparison of experimental data with theoretical expressions for dielectric constants of two-phase mixtures.

normally encountered. Using the simpler Eq. 18.53 has been found satisfactory for most purposes. If the solid phases and their distribution and individual dielectric constants are known, these can be combined in an appropriate way and then corrected for the volume fraction porosity present. In Fig. 18.24 the dielectric constant of a polycrystalline TiO_2 is shown to decrease with increasing porosity. In some of these samples a decrease in porosity with increased firing time was observed, indicating the opening of flat grain-boundary cracks resulting from the boundary stresses discussed in Chapters 5 and 12. The variation in microstructure among samples is clearly illustrated in the range of values found for any particular porosity.

Insufficient experimental data are given, but it is probable that the TiO_2–clay data in Fig. 18.23 depart from the line corresponding to Eq. 18.56 because of the presence of a certain fraction porosity. A pore fraction of 15 to 20%, which is entirely possible, would give reasonable agreement. This uncertainty is a good example of the need for complete sample description and microstructure evaluation in order to interpret resulting properties of ceramic materials properly.

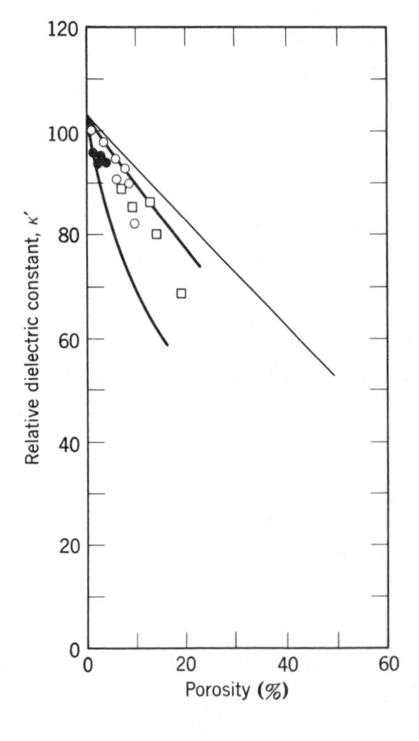

Fig. 18.24. Effect of porosity on the dielectric constant of polycrystalline TiO_2. From G. Economos, in *Ceramic Fabrication Processes*, W. D. Kingery, Ed., John Wiley & Sons, N.Y., 1958, p. 201.

As discussed in Chapter 11, most ceramic compositions consist of crystalline phases dispersed in a glass matrix or separated by a vitreous boundary layer. As a result, their properties are intermediate between those of the single crystals and those of glasses. In general, both the measured dielectric constant and the loss factor increase with temperature, particularly at the lower frequencies. Typical data for a steatite composition and for an alumina porcelain are illustrated in Figs. 18.25 and 18.26. The glassy part is the main contributor to dielectric losses, and the composition of the glass phase must be carefully controlled to obtain low-loss ceramics. This is customarily done by avoiding the use of feldspar and other alkali fluxes and substituting mixtures of clay, talc, and alkaline earth oxides.

Classes of Dielectrics. Most ceramic dielectrics can be classified as (1) insulating materials with a dielectric constant below 12, (2) capacitor

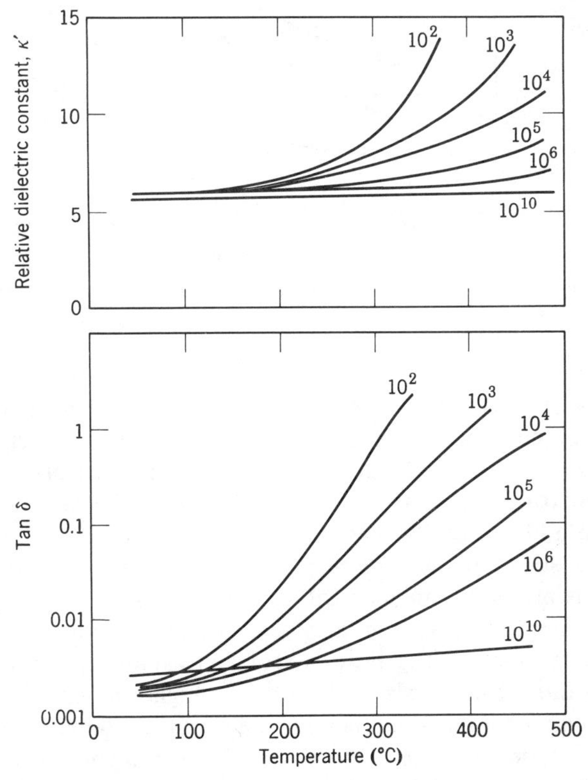

Fig. 18.25. Dielectric constant and tan δ for a steatite ceramic over a range of temperatures and frequencies.

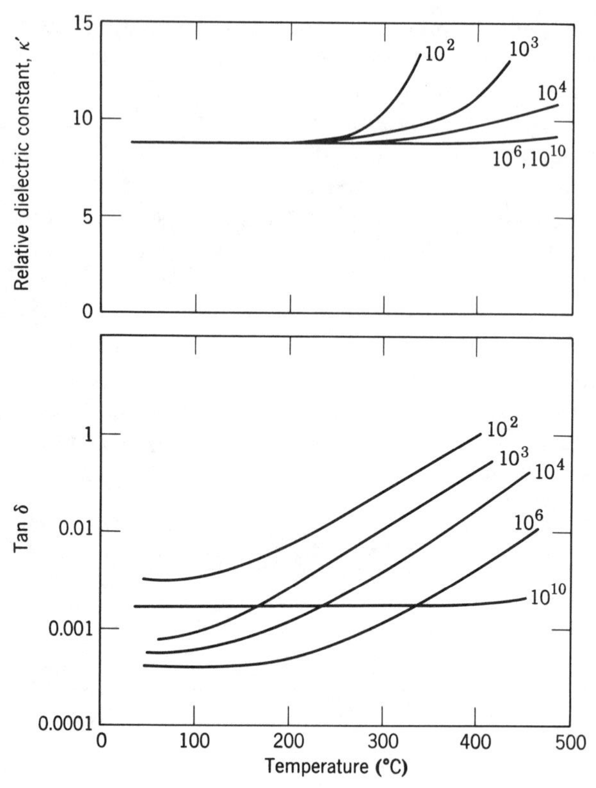

Fig. 18.26. Dielectric constant and tan δ for an alumina porcelain over a range of temperatures and frequencies.

dielectrics with a dielectric constant above 12, and (3) ferroelectric and ferromagnetic ceramics. A large part of the first group consists of low- and high-tension electrical insulators composed primarily of triaxial porcelain mixtures of clay, flint, and feldspar. These insulators are inexpensive and are easily manufactured; for high-voltage applications the water absorption must be zero in order to ensure high dielectric strength and also high mechanical strength. This is less important for low-tension applications, and although increased vitrification usually indicates better quality, it is more expensive to achieve in manufacturing. A second group of materials in this class is high-temperature porcelains used as supports for electrical heating elements; these must be free from deformation under mechanical loads and have a high resistivity at high operating temperatures. For these, the glassy phase must be kept to a minimum, and more or less porous products are generally used. A special and important application is as spark plugs for high-compression automobile and aircraft engines. These have high thermal and electrical stress resistance and operate at temperatures near 1100°C. High-alumina porcelains have been found most satisfactory.

For applications as high-frequency insulation, vitrified products are required which are dimensionally stable, have good mechanical strength, and have a low loss factor. Requirements and classifications for these materials have already been indicated in Table 18.2. Many different compositions are used for different applications. Low-loss steatite is the most widely used; it is unsurpassed for economy in manufacture, since a large fraction of talc, which is soft, readily formed, and does not cause excessive die wear, is used as a raw material. The resulting product consists of clinoenstatite, or one of its polymorphic forms, in a glassy matrix. Steatite bodies usually have a loss factor corresponding to class L-3 to L-5. For lower dielectric losses, forsterite ceramics having Mg_2SiO_4 as a main crystalline phase are frequently used. Alkaline earth oxide fluxes are used to give excellent dielectric properties. The high expansion coefficient is detrimental for thermal shock but is an advantage for forming metal ceramic seals, since it provides a good match for some of the nickel-iron alloys. Other crystalline phases used for low-loss applications are zircon, $ZrSiO_4$, and cordierite. Both of these materials have low expansion coefficients and consequently exceptionally good thermal shock resistance.

Some typical applications and property descriptions of a number of ceramic dielectrics are listed in Table 18.4. In Table 18.5 the batch compositions to form some of these are illustrated by means of typical examples.

Space Charge Polarization. Polycrystalline and polyphase aggregates exhibit an interfacial or space charge polarization arising from differences between the conductivity of various phases present. This polarization resulting from heterogeneity is particularly important for compositions such as ferrites and semiconductors in which the electrical conductivity is appreciable; also for polycrystalline and polyphase materials at higher temperatures. This manifests itself as a high dielectric constant and causes a peak in the loss factor.

If we have two layers of material of different conductivity present, as schematically illustrated in Fig. 18.21*a*, motion of the charge carriers occurs readily through one phase but is interrupted when it reaches a phase boundary. This causes a buildup of charges at the interface which, to an outside observer, corresponds to a large polarization and high dielectric constant. Electrically the actual system resembles an equivalent circuit for series layers illustrated in Fig. 18.27*a*, but the macroscopic

Table 18.4. Some Typical Physical Properties of Ceramic Dielectrics[a]

	Vitrified Products						
Material→	1 High-Voltage Porcelain	2 Alumina Porcelain	3 Steatite	4 Forsterite	5 Zircon Porcelain	6 Lithia Porcelain	7 Titania, Titanate Ceramics
Typical Applications→ Properties ↓	Power Line Insulation	Spark-Plug Cores, Thermocouple Insulation, Protection Tubes	High-Frequency Insulation, Electrical Appliance Insulation	High-Frequency Insulation, Ceramic-to-Metal Seals	Spark-Plug Cores, High-Voltage–High-Temperature Insulation	Temperature-Stable Inductances, Heat-Resistant Insulation	Ceramic Capacitors, Piezoelectric Ceramics
Specific gravity (g/cm^3)	2.3–2.5	3.1–3.9	2.5–2.7	2.7–2.9	3.5–3.8	2.34	3.5–5.5
Water absorption (%)	0.0	0.0	0.0	0.0	0.0	0.0	0.0
Coefficient of linear thermal expansion/°C (20–700)	$5.0–6.8 \times 10^{-6}$	$5.5–8.1 \times 10^{-6}$	$8.6–10.5 \times 10^{-6}$	11×10^{-6}	$3.5–5.5 \times 10^{-6}$	1×10^{-6}	$7.0–10.0 \times 10^{-6}$
Safe operating temperature (°C)	1000	1350–1500	1000–1100	1000–1100	1000–1200	1000	-
Thermal conductivity ($cal/cm^2/cm/sec/°C$)	0.002–0.005	0.007–0.05	0.005–0.006	0.005–0.010	0.010–0.015	-	0.008–0.01
Tensile strength (psi)	3000–8000	8000–30,000	8000–10,000	8000–10,000	10,000–15,000	-	4000–10,000
Compressive strength (psi)	25,000–50,000	80,000–250,000	65,000–130,000	60,000–100,000	80,000–150,000	60,000	40,000–120,000
Flexural strength (psi)	9000–15,000	20,000–45,000	16,000–24,000	18,000–20,000	20,000–35,000	8000	10,000–22,000
Impact strength (ft-lb; ½-in. rod)	0.2–0.3	0.5–0.7	0.3–0.4	0.03–0.04	0.4–0.5	0.3	0.3–0.5
Modulus of elasticity (psi)	$7–14 \times 10^6$	$15–52 \times 10^6$	$13–15 \times 10^6$	$13–15 \times 10^6$	$20–30 \times 10^6$	-	$10–15 \times 10^6$
Thermal shock resistance	Moderately good	Excellent	Moderate	Poor	Good	Excellent	Poor
Dielectric strength (volts/mil; ¼-in.-thick specimen)	250–400	250–400	200–350	200–300	250–350	200–300	50–300
Resistivity ($ohms/cm^3$) at room temperature	$10^{12}–10^{14}$	$10^{14}–10^{15}$	$10^{13}–10^{15}$	$10^{13}–10^{15}$	$10^{13}–10^{15}$	-	$10^8–10^{15}$
Te-value (°C)	200–500	500–800	450–1000	above 1000	700–900	-	200–400
Power factor at 1 Mc	0.006–0.010	0.001–0.002	0.0008–0.0035	0.0003	0.0006–0.0020	0.05	0.0002–0.050
Dielectric constant	6.0–7.0	8–9	5.5–7.5	6.2	8.0–9.0	5.6	15–10,000
L-grade (JAN Spec. T-10)	L-2	L-2–L-5	L-3–L-5	L-6	L-4	L-3	-

[a] After H. Thurnaur, reference 2.

Semivitreous and Refractory Products

Material→	8 Low-Voltage Porcelain	9 Cordierite Refractories	10 Alumina, Aluminum Silicate Refractories	11 Massive Fired Talc, Pyrophyllite
Typical Applications→ Properties ↓	Switch Bases, Low-Voltage Wire Holders, Light Receptacles	Resistor Supports, Burner Tips, Heat Insulation, Arc Chambers	Vacuum Spacers, High-Temperature Insulation	High-Frequency Insulation, Vacuum Tube Spacers, Ceramic Models
Specific gravity (g/cm^3)	2.2–2.4	1.6–2.1	2.2–2.4	2.3–2.8
Water absorption (%)	0.5–2.0	5.0–15.0	10.0–20.0	1.0–3.0
Coefficient of linear thermal expansion/°C (20–700)	$5.0\text{–}6.5 \times 10^{-6}$	$2.5\text{–}3.0 \times 10^{-6}$	$5.0\text{–}7.0 \times 10^{-6}$	11.5×10^{-6}
Safe operating temperature (°C)	900	1250	1300–1700	1200
Thermal conductivity ($cal/cm^2/cm/sec/°C$)	0.004–0.005	0.003–0.004	0.004–0.005	0.003–0.005
Tensile strength (psi)	1500–2500	1000–3500	700–3000	2500
Compressive strength (psi)	25,000–50,000	20,000–45,000	15,000–60,000	20,000–30,000
Flexural strength (psi)	3500–6000	1500–7000	1500–6000	7000–9000
Impact strength (ft-lb; ½-in. rod)	0.2–0.3	0.2–0.25	0.17–0.25	0.2–0.3
Modulus of elasticity (psi)	$7\text{–}10 \times 10^6$	$2\text{–}5 \times 10^6$	$2\text{–}5 \times 10^6$	$4\text{–}5 \times 10^6$
Thermal shock resistance	Moderate	Excellent	Excellent	Good
Dielectric strength (volts/mil; ¼-in. thick specimen)	40–100	40–100	40–100	80–100
Resistivity ($ohms/cm^3$) at room temperature	$10^{12}\text{–}10^{14}$	$10^{12}\text{–}10^{14}$	$10^{12}\text{–}10^{14}$	$10^{12}\text{–}10^{15}$
Te-value (°C)	300–400	400–700	400–700	600–900
Power factor at 1 Mc	0.010–0.020	0.004–0.010	0.0002–0.010	0.0008–0.010
Dielectric constant	6.0–7.0	4.5–5.5	4.5–6.5	5.0–6.0
L-grade (JAN Spec. T-10)	–	–	–	–

Table 18.5. Batch Compositions of Some Ceramic Dielectrics

	Ste-atite	For-sterite	Zircon-Talc	Alumina Porcelain
Talc	84	71.3	32	4
BaF_2	10	–	5	1
Ball clay	3	2.6	7	–
Bentorite	3	–	3	–
$BaCO_3$	–	6.8	–	–
$Mg(OH)_2$	–	19.5	–	–
Zircon	–	–	53	–
Al_2O_3	–	–	–	95
Firing temperature (°C)	1330	1250	1325	1600
Relative dielectric constant, κ'	6.5	6.6	7.51	8.5

dielectric properties are correlated with the different equivalent circuit illustrated in Fig. 18.27*b*.

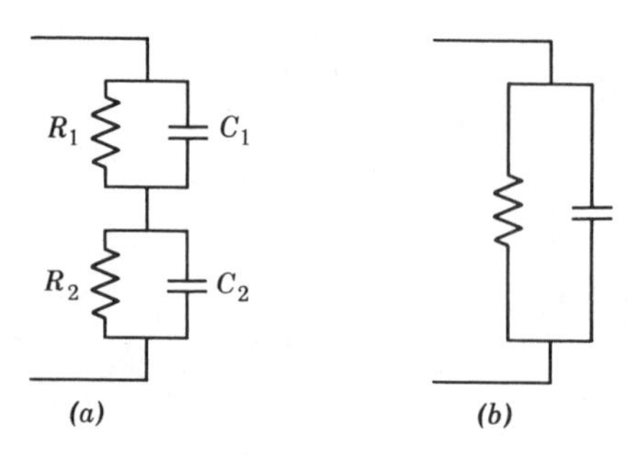

Fig. 18.27. Equivalent circuits for (*a*) two-layer capacitor and (*b*) macroscopic observations.

The geometrical shape and distribution of the two phases are important, as has been discussed by R. W. Sillars.* Perhaps the most important occurrence of this phenomenon is for semiconducting polycrystalline materials in which the grain boundaries, or a grain-boundary phase, have a high resistance. Then the structure corresponds to thin layers of boundary material. If the individual capacitance elements in Fig. 18.27 are considered, the time constant for current flow in each can be represented as $\tau_1 = C_1R_1$ and $\tau_2 = C_2R_2$. These can be related by the relative values of

J. Inst. Elec. Eng. (Lond), **80**, 378 (1937).

the resistance and capacitance, using constants a and b so that $\tau_2 = aR_1bC_1 = ab\tau_1$. The resulting resistivity and dielectric constant for the combined circuit vary with frequency (Fig. 18.28); the conductivity is the mirror image of the dielectric constant. As shown in Fig. 18.28, the low-frequency dielectric constant or resistivity resulting from interfacial polarization can be several orders of magnitude larger than the value observed at high frequency. When present, this is by far the largest source of capacitance in this kind of dielectric material.

If ρ_1 and ρ_2 are the resistivity and κ_1' and κ_2' the dielectric constants of layers 1 and 2, and the ratio of their thickness is x, such that $x \ll 1$ (corresponding to the fraction of material in a boundary layer), we can

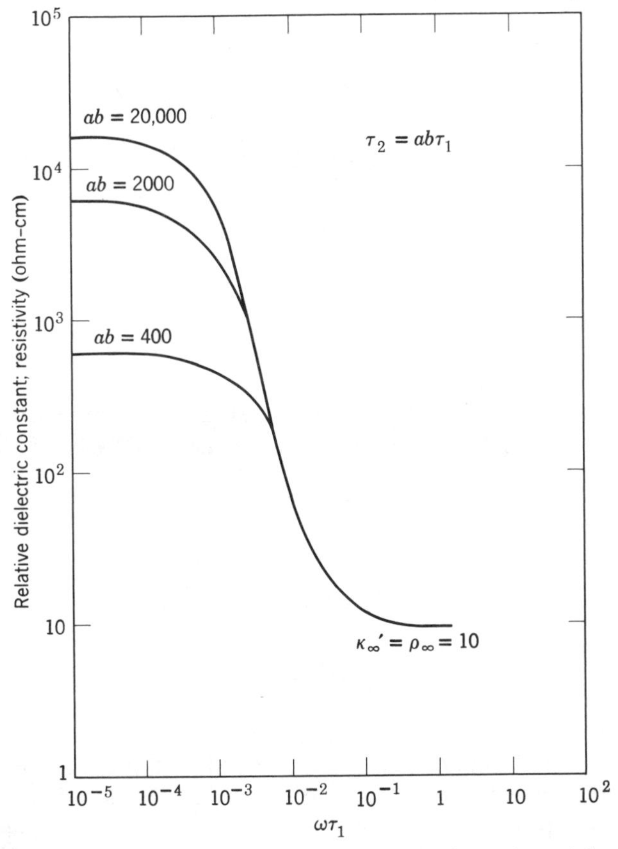

Fig. 18.28. Change in interfacial polarization with relative value of boundary-layer characteristics.

write for the resistivity

$$
\begin{aligned}
\rho &= \rho_0 + \frac{\rho_0 - \rho_\infty}{1 + \tau_\rho^2 \omega^2} \\
\rho_\infty &= \frac{\rho_1 \rho_2 (\kappa_1' + x_1 \kappa_2')^2}{\rho_1 \kappa_1'^2 + x_1 \rho_2 \kappa_2'^2} \\
\rho_0 &= x_1 \rho_1 + \rho_2 \\
\tau_\rho &= \epsilon_0 \left[\frac{\rho_1 \rho_2 (\rho_1 \kappa_1'^2 + x_1 \rho_2 \kappa_2'^2)}{x_1 \rho_1 + \rho_2} \right]^{1/2} \\
& \qquad x \ll 1
\end{aligned}
\tag{18.57}
$$

and for the dielectric constant

$$
\begin{aligned}
\kappa' &= \kappa_\infty' + \frac{\kappa_0' - \kappa'}{1 + \tau_{\kappa'}^2 \omega^2} \\
\kappa_\infty' &= \frac{\kappa_1' \kappa_2'}{\kappa_1' + x_1 \kappa_2'} \\
\kappa_0' &= \frac{x_1 \rho_1^2 \kappa_1' + \rho_2^2 \kappa_2'}{(x_1 \rho_1 + \rho_2)^2} \\
\tau_{\kappa'} &= \epsilon_0 \frac{\rho_1 \rho_2 (\kappa_1' + x_1 \kappa_2')}{x_1 \rho_1 + \rho_2} \\
\tau_{\kappa'} &= \left(\frac{\rho_\infty}{\rho_0} \right)^{1/2} \\
& \qquad x \ll 1
\end{aligned}
\tag{18.58}
$$

For the structure of most common practical interest, the resistivity of the boundary layer is substantially larger than the resistivity of the grain:

$$
\begin{aligned}
&\kappa_1' = \kappa_2' \qquad x_1 \ll 1 \qquad \rho_1 \gg \rho_2 \\
&\rho_\infty = \rho_2 \qquad \rho_0 = x_1 \rho_1 + \rho_2 \\
&\kappa_\infty' = \kappa_2' \qquad \kappa_0' = \kappa_2' \frac{x_1 \rho_1^2 + \rho_2^2}{(x_1 \rho_1 + \rho_2)^2} \\
&\tau_{\kappa'} = \epsilon_0 \kappa_2' \frac{\rho_1 \rho_2}{x \rho_1 + \rho_2}
\end{aligned}
\tag{18.59}
$$

For a typical material, nickel zinc ferrite, $Ni_{0.4}Zn_{0.6}Fe_2O_4$, C. G. Koops* found that the dielectric constant and resistivity characteristics could be satisfactorily represented by Eq. 18.59, taking $\kappa_1' = \kappa_2' = 17$, $\rho_1 = 3.3 \times 10^6$ ohm-m, $\rho_2 = 5.0 \times 10^3$ ohm-m, and $x = 0.45 \times 10^{-2}$. The ex-

Phys. Rev., **83**, 121 (1951).

perimental results together with calculations from Eq. 18.59 are illustrated in Fig. 18.29.

The time constant for this interfacial polarization, and consequently the frequency at which it becomes important, is proportional to the product of resistivities, as indicated in Eqs. 18.57 to 18.59. For most dielectrics at room temperature this product is so large that the interfacial polarization is negligible even at low frequencies. For semiconductors and semiconducting ferrites this resistivity product is not so large, and the effect is important at room temperature. For other dielectrics this factor increases as the temperature is raised. At the same time, the conduction losses and ion jump losses also increase. (These can in one sense be considered *local* space charge polarization resulting from mobility barriers, as illustrated in

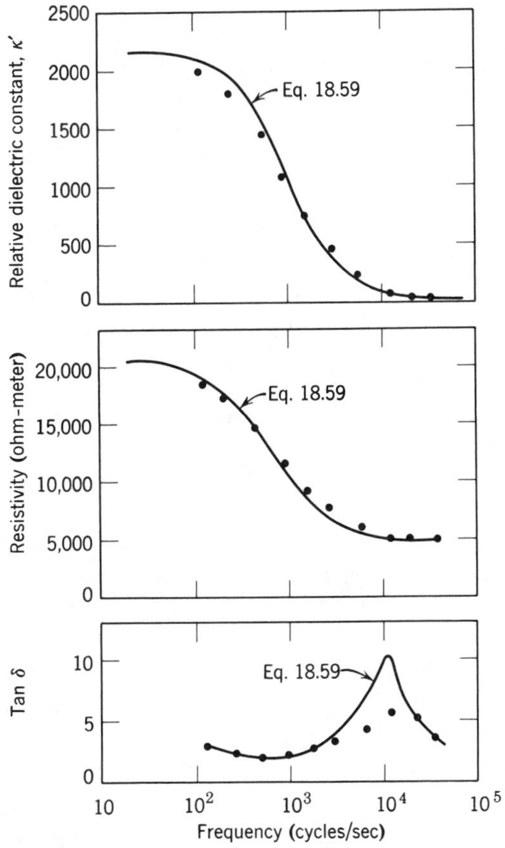

Fig. 18.29. Dielectric constant, resistivity, and tan δ for $Ni_{0.4}Zn_{0.6}Fe_2O_4$. From C. G. Koops, *Phys. Rev.*, **83**, 121 (1951).

Fig. 18.10.) The combined result is that, as conductivity increases at higher temperatures, the apparent dielectric constant also increases, the dielectric losses increase even more rapidly, and the effectiveness of insulation is decreased. To minimize this increase, compositions having the lowest possible conductivity should be used.

18.6 Dielectric Strength

The dielectric breakdown of insulating materials under an applied field takes place two different ways. The first is electronic in origin and is sometimes referred to as the intrinsic dielectric strength. The second process is caused by local overheating, arising from electrical conduction; the local conductivity increases to a point at which instability occurs and permits a rush of current, melting, and puncture; this is called thermal breakdown. The tendency toward thermal breakdown increases at higher temperatures and when voltages are applied for a longer time.

The theory of breakdown strength as a material property is complicated by the wide variations resulting from testing procedures. It is difficult to avoid nonhomogeneous fields during testing, and breakdown tends to be initiated at positions corresponding to points on electrodes, even though the average measured field is still low. Edge effects are also important, as is the sample thickness. In general, as samples decrease in thickness, the measured dielectric strength increases. Similarly, testing with spherical electrodes under oil or under a semiconducting liquid gives higher measured dielectric strengths. (Dielectric strength is calculated in terms of volts per centimeter or volts per mil.)

Intrinsic breakdown strengths, measured under conditions of careful electrode design and with thin samples, approach values of 10×10^6 V/cm. In contrast, values measured on 1/4-in.-thick samples with standard electrodes are only about 1/150 of this value, and values found for semivitreous ceramics are sometimes as low as 1/1000 of this value. This lack of control over all variables during testing makes the theoretical situation uncertain, even though a number of detailed theories have been proposed.

In electronic breakdown, failure occurs when a localized voltage gradient reaches some value corresponding to intrinsic electrical breakdown. Electrons within the structure are accelerated by the field to a velocity that allows them to liberate additional electrons by collision. This process continues at an accelerating rate and finally results in an electron avalanche which corresponds to breakdown and sample rupture. At low temperatures (below room temperature) the intrinsic breakdown strength of crystalline materials increases with rising temperature, corresponding to increased lattice vibrations and a resulting increase in electron scattering by the lattice; a greater field strength is required to accelerate electrons to a point at which an electron avalanche is initiated. In contrast, the intrinsic dielectric strength of glasses is independent of temperature at low temperatures, corresponding to a random lattice structure in which electron scattering remains independent of temperature. This characteristic of glass structure is similar to the thermal conductivity behavior (related to the phonon scattering) as discussed in Chapter 12. At low temperatures solid solutions in crystalline materials increase the electron scattering and consequently increase the dielectric strength. As the temperature increases, the intrinsic breakdown strength passes through a maximum at about room temperature, at which for most insulators sufficient electrons are available for the characteristics of avalanche formation to change. The dielectric strength for electronic breakdown at low temperatures, the intrinsic electron breakdown strength for glasses and crystals, is of the order of 1 to 10×10^6 V/cm.

Thermal breakdown behavior can be differentiated from intrinsic breakdown in that it is associated with electrical stresses of appreciable time duration for local heating to occur, and it takes place at high enough temperatures for the electrical conductivity to increase. Electrical energy losses raise the temperature even further and increase the local conductivity. This causes channeling of the current; local instability and breakdown result in the passage of high currents with resulting fusion and vaporization that constitute a puncture of the insulation.

The effect of time duration and temperature on the breakdown behavior of glasses depends critically on the conduction characteristics, and consequently the composition. In general, intrinsic behavior is usually found below about −50°C, whereas thermal breakdown usually occurs at temperatures above 150°C. At intermediate temperatures the breakdown characteristic depends on the applied voltage, the time duration of the applied voltage, and the test conditions as shown in Fig. 18.30. Results for some glasses with different soda content are compared in Table 18.6. It is seen that the intrinsic low-temperature behavior is nearly independent of composition, whereas the temperature at which thermal breakdown begins is strongly dependent on the resistivity and the alkali content of the glass. J. J. Chapman and L. J. Frisco* have found that the dielectric breakdown strength of glasses decreases as the frequency, and consequently the dielectric loss and dielectric heating, is increased. Although for one glass studied, a dielectric strength of about 1000 V/mil was found

Elec. Manuf., **53**, 136 (May, 1954).

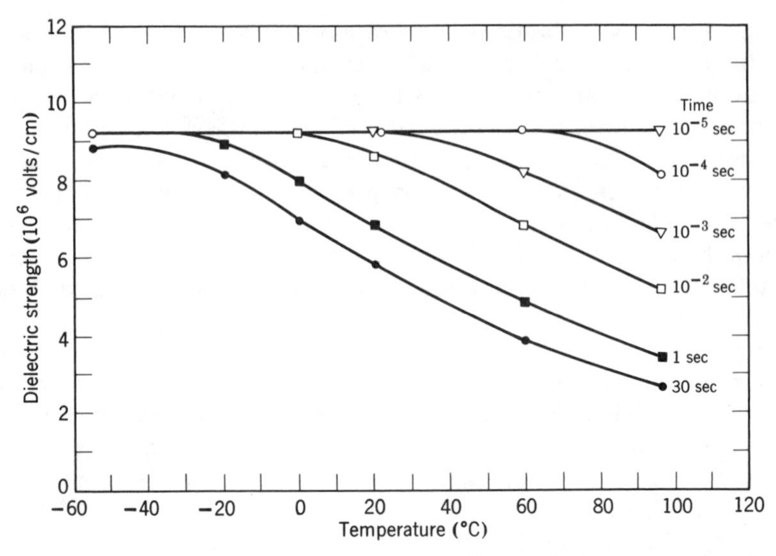

Fig. 18.30. Effect of test temperature and test duration on the breakdown strength of a Pyrex glass. From J. Vermeer, *Physica*, **20**, 313 (1954).

Table 18.6. Relationship between Thermal Breakdown and Electrical Resistivity[a]

	Glass No.			
	1	2	3	4
Intrinsic breakdown strength, E_b (volts/cm)	9×10^6	11.2×10^6	9.2×10^6	9.9×10^6
Critical temperature (°C)	−150°	−125°	−60°	+150°
Per cent Na_2O in glass	12.8	5.1	3.5	0.9
Resistivity at 200°C (ohm-cm)	2×10^8	6×10^8	1.25×10^9	2.5×10^{14}

[a] After J. Vermeer, *Physica*, **22**, 1247 (1956).

at 60 Hz, this had decreased to about a tenth of that in the megacycle range.

In the same way the testing conditions and electrode design are important for fixing the measured dielectric strength, minor heterogeneities in a normal ceramic composition can considerably decrease the dielectric strength. The most common of these is porosity, which tends to give variation in the local electrical field and leads to low measured values. This is illustrated with data for titania ceramics in Fig. 18.31. A

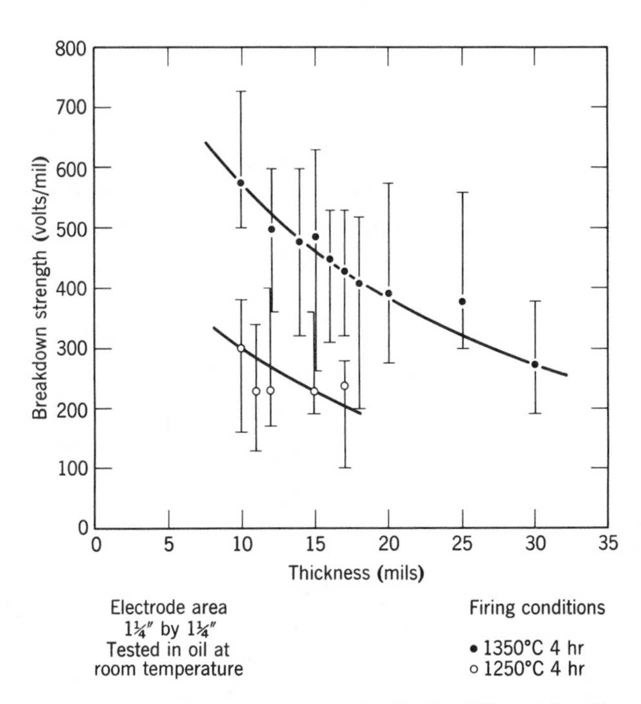

Fig. 18.31. Dielectric strength of two titania bodies fired to different densities and tested in oil with $1\frac{1}{4}$-in. × $1\frac{1}{4}$-in. electrodes. From G. Economos, in *Ceramic Fabrication Processes*, W. D. Kingery, Ed., John Wiley and Sons, N.Y., 1950, p. 205.

sample containing about 14% porosity was found to have a dielectric strength about half that of a sample with 5% porosity. Similarly, as indicated in Table 18.4, semivitreous porcelain compositions normally have a dielectric strength a third that of dense vitreous ware. Aside from the effects of porosity, the major factor affecting the dielectric strength of most ceramic compositions is the onset of thermal breakdown caused by dielectric losses or conductivity. Factors giving high values of conductivity or dielectric loss have already been discussed. High-density low-loss ceramics normally have breakdown strengths which are suitable for dielectric applications and are well in excess of minimum standards.

18.7 Ferroelectric Ceramics

Ferroelectricity is defined as the spontaneous alignment of electric dipoles by their mutual interaction. This is a process parallel to the spontaneous alignment of magnetic dipoles observed in ferromagnetism and derives its name from its similarity and features analogous to that process. The source of ferroelectricity arises from the fact that the local field E' increases in proportion to the polarization. For a material containing electric dipoles increased polarization increases the local field; spontaneous polarization is to be expected at some low temperature at which the randomizing effect of thermal energy is overcome, and all the electric dipoles line up in parallel arrays. This is a cooperative phenomenon of the same sort as discussed for order-disorder transitions in Chapter 4 and has many similar features.

The general source of spontaneous polarization is apparent from the defining equation for polarization, Eq. 18.15,

$$P = (\kappa' - 1)\epsilon_0 E = N\alpha E' \tag{18.60}$$

By introducing the Mosotti field (Eq. 18.21) for the local field:

$$E' = E + \frac{P}{3\epsilon_0} \tag{18.61}$$

we obtain for the polarization and the electric susceptibility

$$P = \frac{N\alpha E}{1 - N\alpha/3\epsilon_0} \tag{18.62}$$

$$\chi = \kappa' - 1 = \frac{P}{\epsilon_0 E} = \frac{N\alpha/\epsilon_0}{1 - N\alpha/3\epsilon_0} \tag{18.63}$$

When the polarizability term in the denominator, $N\alpha/3\epsilon_0$, approaches unity, the polarizability and susceptibility must approach infinity.

The orientation polarizability of a dipole is inversely proportional to temperature according to the relation

$$\alpha_0 = \frac{C}{kT} \tag{18.64}$$

If we consider systems in which the orientation polarizability is much larger than that for the electronic and ionic portions, a critical temperature is reached at which

$$\frac{N\alpha}{3\epsilon_0} = \frac{N}{3\epsilon_0}\left(\frac{C}{kT_c}\right) = 1 \tag{18.65}$$

This fixes the critical temperature as

$$T_c = \frac{NC}{3\epsilon_0} = \frac{N\alpha_0 T}{3\epsilon_0} \tag{18.66}$$

Below this temperature, the Curie temperature, spontaneous polarization sets in and all the elementary dipoles have the same orientation.

Combining Eqs. 18.64 and 18.65 with Eq. 18.66 gives for the susceptibility (and dielectric constant and polarization)

$$\chi = \kappa' - 1 = \frac{P}{\epsilon_0 E} = \frac{3T_c}{T - T_c} \tag{18.67}$$

This linear dependence of the inverse susceptibility on $T - T_c$ is known as the Curie-Weiss law, and T_c is the Curie temperature. Although this represents experimental measurements satisfactorily at temperatures well above the Curie temperature, at which dipole orientation is more or less random, it breaks down near and below the Curie temperature, at which dipoles are oriented. Under these conditions to assume that the near field (E_2 of Eq. 18.19) is zero is no longer satisfactory, and Eq. 18.21 and subsequent calculations do not apply. A somewhat better model designed by L. Onsager* is more complex mathematically but is still unsuitable for crystalline solids.

The result of spontaneous polarization at some critical temperature is the appearance of very high dielectric constants together with a hysteresis loop for polarization in an alternating field (Fig. 18.32). This is similar to

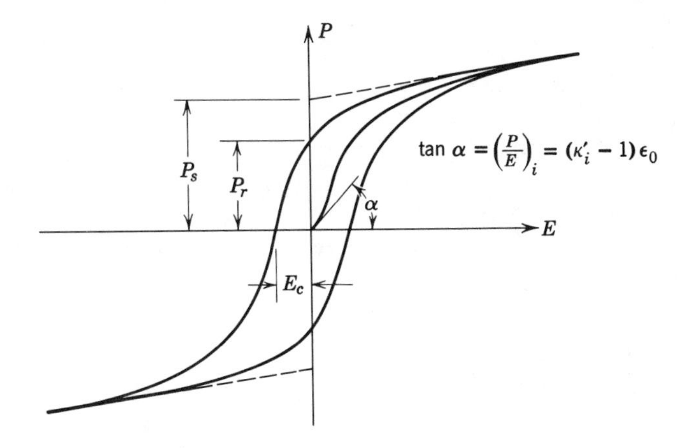

Fig. 18.32. A typical ferroelectric hysteresis loop.

J. Am. Chem. Soc., **58**, 1486 (1936).

the hysteresis loop observed for ferromagnetic materials and results from the presence of different *domains* in which there is complete alignment of electric dipoles. The boundaries between ferroelectric domains are crystallographic domain walls which have a 90° or 180° orientation in barium titanate. The domains in a polycrystalline barium titanate ceramic are illustrated in Fig. 11.30, in which differences in orientation are brought out by the etching.

At low field strengths in unpolarized material, the polarization is initially reversible and nearly linear with an applied field; the slope of the curve gives the initial dielectric constant κ_i'. At higher field strengths the polarization increases more rapidly as a result of switching of ferroelectric domains, that is, the changing polarization direction in domains by 90° or 180°; this occurs by domain boundaries moving through the crystal. At the highest field strengths the increase of polarization for a given increase in field strength is again less, corresponding to polarization saturation, having all the domains of like orientation aligned in the direction of the electric field. Extrapolation of this curve back to the $E = 0$ ordinate gives P_s, the saturation polarization corresponding to the spontaneous polarization with all the dipoles aligned in parallel.

When the electric field is cut off, the polarization does not go to zero but remains at a finite value called the remanent polarization P_r. This results from the oriented domains being unable to return to their original random state without an additional energy input by an oppositely directed field. That is, energy is required for a change in domain orientation. The strength of the electric field required to return the polarization to zero is known as the coercive field E_c.

At low temperatures the hysteresis loops become fatter, and the coercive field becomes greater, corresponding to a larger energy required to reorient the domain walls; that is, the domain configuration is frozen in. At higher temperatures the coercive force decreases until at the Curie temperature no hysteresis remains, and there is only a single value for the dielectric constant. Hysteresis loops for barium titanite, which has a Curie temperature of about 125°C, are illustrated in Fig. 18.33. In Fig.

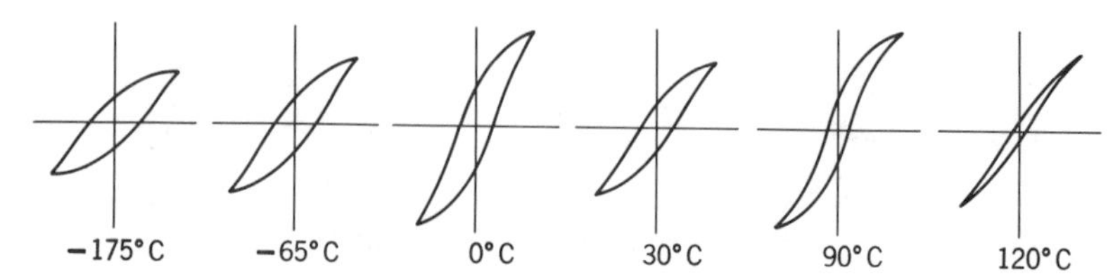

Fig. 18.33. The change in barium titanate ferroelectric hysteresis loop shape with temperature.

18.34 the temperature dependence of the dielectric susceptibility of a barium–strontium titanate ceramic is illustrated, corresponding to the Curie-Weiss law, Eq. 18.67.

An increasing number of materials are being found to demonstrate spontaneous polarization. Barium titanate is the one that has been most widely investigated. Lead titanate, which has the same perovskite structure as barium titanate, is also ferroelectric. Other ferroelectrics include Rochelle salt (potassium–sodium tartrate tetrahydrate); potassium dihydrogen phosphate, KH_2PO_4; potassium dihydrogen arsenate, KH_2AsO_4; other perovskites, $NaCbO_3$, $KCbO_3$, $NaTaO_3$, and $KTaO_3$; ilmenite structures, $LiTaO_3$ and $LiCbO_3$; and tungsten oxide, WO_3.

The crystal structure of barium titanate, the perovskite structure (named for the mineral perovskite, $CaTiO_3$), is illustrated in Fig. 18.35. Each large barium ion is surrounded by twelve nearest-neighbor oxygen ions; each titanium ion has six oxygen ions in octahedral coordination. As discussed in Chapter 2, the barium and oxygen ions together form a face-centered cubic lattice, with titanium ions fitting into octahedral interstices. The characteristic structural feature shared by barium and lead titanate, as contrasted to other perovskites, seems to be that the large size of the barium and lead ions increases the size of the cell of the face-centered cubic BaO_3 structure so that the titanium atom is at the lower edge of stability in the octahedral interstices. The *rattling titanium*

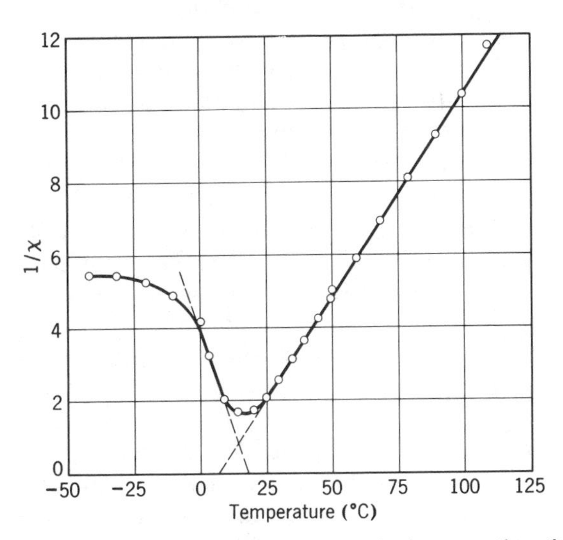

Fig. 18.34. Confirmation of the Curie-Weiss law for a barium-strontium titanate composition. From S. Roberts, *Phys. Rev.*, **71**, 890 (1947).

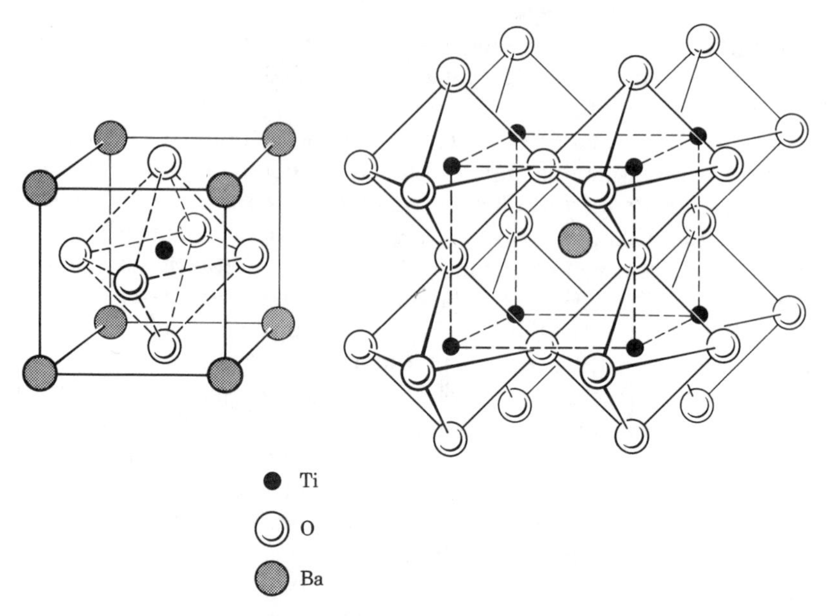

Fig. 18.35. Ion positions in ideal perovskite structure.

hypothesis suggests that there are minimum-energy positions for the titanium ion which are off center and consequently give rise to an electric dipole. At high temperatures the thermal energy is such that the titanium ion has no fixed unsymmetrical position, but the open octahedral site allows the titanium ion to develop a large dipole moment in an applied field. On cooling below the Curie temperature, the position of the titanium ion and the octahedral structure change from cubic to tetragonal symmetry with the titanium ion in an off-center position corresponding to a permanent electrical dipole. These dipoles are ordered, giving a domain structure, as already discussed. The change from the paraelectric to the ferroelectric state can be illustrated in terms of crystallographic changes and corresponding dielectric constant changes (Fig. 18.36). The relative displacements of ions in the barium titanate tetragonal phase as determined by X-ray analysis are illustrated in Fig. 18.37.

Although the very high dielectric constant of barium titanate and other ferroelectric materials offers the possibility of obtaining high capacitance with very small-sized capacitors, there are a number of problems associated with their use. The change in polarization with applied field requires domain orientation and, as illustrated in Fig. 18.33, the measured dielectric constant depends on the applied field. For high field strengths,

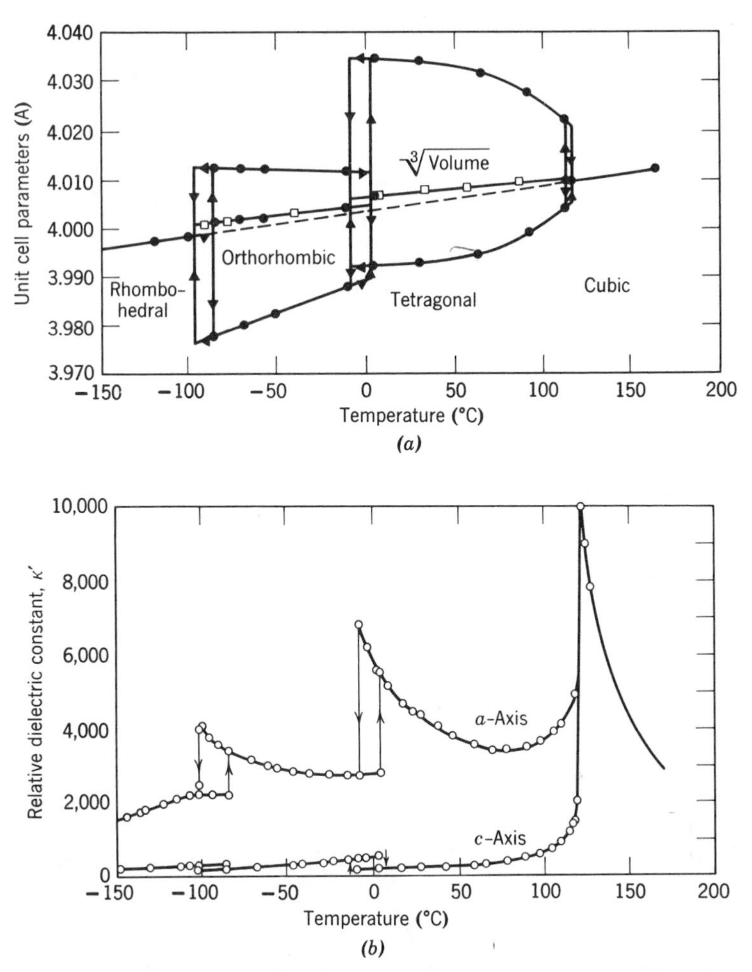

Fig. 18.36. (*a*) Dimensions of pseudocubic unit cell of $BaTiO_3$. From H. F. Kay and P. Vousdan, *Phil. Mag.*, 7, 40, 1019 (1949). (*b*) Temperature dependence of dielectric constant. From W. J. Merz, *Phys. Rev.*, **76**, 1221 (1949).

the domains are more effectively oriented, and a higher dielectric constant results. At the same time, there is a strong temperature dependence of the dielectric constant, and circuit characteristics change even over a moderate-temperature range. Fortunately, the temperature dependence and also other properties such as the dielectric constant can be modified by forming solid solutions over a wide range of compositions. In the perovskite lattice substitutions of Pb^{2+}, Sr^{2+}, Ca^{2+}, and Cd^{2+} can be made

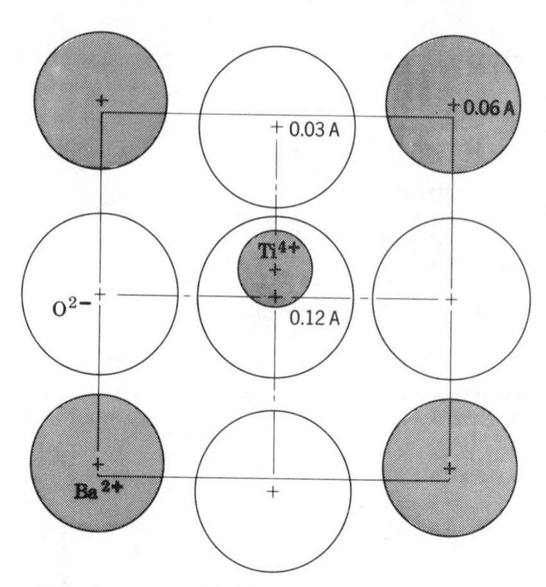

Fig. 18.37. Ion positions in tetragonal $BaTiO_3$. From G. Shirane, F. Jona, and R. Pepinsky, *Proc. I.R.E.*, **42**, 1738 (1955).

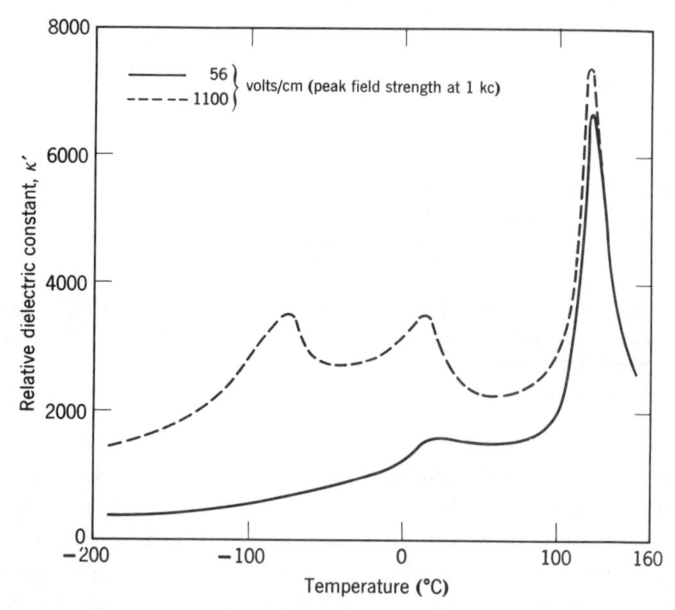

Fig. 18.38. Dielectric constant of barium titanate ceramic as a function of temperature. From A. R. von Hippel, reference 1.

for part of the Ba^{2+} ions, maintaining the ferroelectric characteristics. Similarly, the Ti^{4+} ion can be partially replaced with Sn^{4+}, Hf^{4+}, Zr^{4+}, Ce^{4+}, and Th^{4+}. In addition, the columbates and tantalates are available as ferroelectrics. The possibilities for forming solution alloys in all these structures offer a tremendous range of compositions, dielectric constants, temperature dependence, and other characteristics for development; a considerable amount of effort has gone into developing particular compositions for various uses.

The domain shifts in ferroelectrics lead to difficulties in the production of uniform and consistent materials. The strain energy introduced on cooling through and below the Curie temperature, when some domains change their orientation in relation to others, causes a time dependence for the dielectric constant known as aging. The dielectric constant as a function of the time follows a relationship, $\kappa' = \kappa_0' - m \log t$, where κ' is the dielectric constant at t, κ_0' is the initial dielectric constant, and m is a rate of decay. The rate of change increases as the initial dielectric constant increases. Details of the aging mechanism are not well understood, but variables such as composition and heat treatment strongly affect the results obtained, and these same variations affect the measured dielectric constant. The ease of domain motion depends on the strains imposed by surrounding crystals as well as on the crystal size itself. In a polycrystalline ceramic, domain reorientation is affected by the size of individual grains, by the presence of impurities and pores which prevent domain-wall movement, by stresses imposed by surrounding grains, by the nature of the grain boundaries, and by the presence of second-phase particles. As a result, the microstructure details, the purity of initial starting materials, and the manufacturing and firing procedures all have a strong effect on resulting properties. These have not been analyzed in detail; as a result, products made by similar processes may have variations of as much as 15 to 20% in the dielectric constant measured at some fixed applied voltage.

For electronic devices the relative loss factor, $\kappa' \tan \delta$, is intensified by the high dielectric constant. Consequently, the presence of second phases such as glasses which would contribute to losses is very undesirable. Barium titanate and other ferroelectric ceramics are manufactured as single-phase compositions by solid-state sintering techniques in order to avoid this difficulty.

Barium titanate is most widely used for its strong piezoelectric characteristics. Barium titanate, like other asymmetrical crystals, develops a potential difference when compressed in specific crystallographic directions. Similarly, the application of an electric voltage creates a mechanical distortion. This arises from the anisotropic nature of the dielectric characteristics (Fig. 18.36*b*). In practice, polycrystalline barium titanate piezoelectrics are made by cooling through the Curie temperature in the presence of a strong electric field which gives permanent orientation of the dipoles in the ceramic rather than the random arrangement which would otherwise occur. The piezoelectric efficiency is measured in terms of a coupling coefficient, indicating the fraction of applied mechanical force converted into an electric voltage. Barium titanate prepolarized ceramics have the advantage of a high coupling coefficient (approximately 0.5 as compared with a value of 0.1 for quartz and other useful piezoelectrics), and at the same time they are mechanically and thermally stable. Other ferroelectrics with high coupling coefficients, such as rochelle salt, have relatively poor mechanical and thermal properties. Barium titanate can be manufactured in a variety of shapes and subsequently polarized in order to obtain optimum efficiency as piezoelectric elements. Since the Curie temperature of barium titanate is relatively high, piezoelectric properties are maintained, and barium titanate can be used for these purposes, at temperatures as high as 70°C. Lead titanate with a higher Curie temperature increases the available temperature range to well above 100°.

Transducers of barium titanate are widely used for ultrasonic technical applications such as the emulsification of liquids, mixing of powders and paints, and homogenization of milk; they are also used in microphones, phonograph pickups, accelerometers, strain gauges, and sonar devices.

Suggested Reading

1. A. R. von Hippel, *Dielectrics and Waves*, John Wiley & Sons, Inc., New York, 1954.
2. A. R. von Hippel, Ed., *Dielectric Materials and Applications*, Technology Press, Cambridge, Mass., and John Wiley & Sons, Inc., New York, 1954.
3. *Encyclopedia of Physics* (*Handbuch der Physik*), S. Flügge, Ed., Vol. 17, Dielectrics, Springer, Berlin, 1956: (*a*) W. F. Brown, "Dielectrics," pp. 1–154; (*b*) W. Franz, "Dielectrischer Durchschlag" ("Dielectric Breakdown"), pp. 155–263; (*c*) P. W. Forsbergh, Jr., "Piezoelectricity, Electrostriction and Ferroelectricity," pp. 264–392.
4. E. J. Murphy and S. O. Morgan, "The Dielectric Properties of Insulating Materials," *Bell Sys. Tech. J.*, **16**, 493 (1937); **17**, 640 (1938).
5. J. M. Stevels, "The Electrical Properties of Glass," *Handbuch der Physik*, S. Flügge, Ed., Vol. 20, Electrical Conductivity II, Springer-Verlag, Berlin, 1957, pp. 350–391.
6. W. Kanzig, "Ferroelectrics and Anti-ferroelectrics," *Solid State Physics*, F. Seitz and D. Turnbull, Eds., Vol. 4, Academic Press, Inc., New York, 1957, pp. 5–199.

7. *Progress in Dielectrics*, Academic Press, Inc., New York, 1957–1967.
8. J. C. Burfoot, *Ferroelectrics: An Introduction to Physical Principles*, D. Van Nostrand Company, Inc., Princeton, N.J., 1967.
9. F. Jona and G. Shirane, *Ferroelectric Crystals*, The MacMillan Company, New York, 1962.
10. W. G. Cady, *Piezoelectricity*, Vols. 1 and 2, Dover Publications, Inc., New York, 1964.
11. L. L. Hench and D. B. Dove, *Physics of Electronic Ceramics*, Marcel Dekker, Inc., New York, 1971.

Problems

18.1. In general, glasses used as dielectric insulation are low in alkali content and high in alkaline earth or lead content.
 (*a*) What are the properties desirable for dielectric insulators?
 (*b*) Why are low-alkali glasses preferred for insulation?
 (*c*) Explain in terms of the structure why alkaline earth–containing glasses are desirable for dielectric insulation.
 (*d*) What are the disadvantages of fused quartz?

18.2. Sketch the dielectric constant and loss factor as a function of frequency from 1 to 10^{18} Hz in:
 (*a*) Germanium (high purity).
 (*b*) CaF_2.
 (*c*) Label the dispersion regions with appropriate polarization mechanisms.
 (*d*) For CaF_2 (only) give the temperature dependence for the change in frequency at which dispersion occurs for each mechanism.

18.3. If the atomic radius of atom A is twice that of atom B, the electronic polarizability of atom A is roughly ______ times that of B, other things being equal.

18.4. List the following crystalline materials in the order of their expected index of refraction $\kappa \approx n^2$ (from lowest to highest):

LiF	MgO
NaI	KF
PbO	NaBr
CaS	RbF
NaF	NaCl
CaO	CsF

18.5. A fine-grained ceramic specimen of rutile (TiO_2) has a κ' of 100 at 20°C and 100 Hz. To what do you attribute this high value? How would you experimentally distinguish the various contributing mechanisms?

18.6. It is desirable to have capacitors with high storage capacity and small size. Discuss how you would go about making a ceramic material to use as a dielectric to fit this need. What properties are important, and what factors must be controlled?

18.7. A 2-cm disk 0.025 cm thick of steatite was measured and found to have a capacitance of 7.2 μF and a dissipation factor of 72. Determine the following: (*a*) permittivity, (*b*) electric loss factor, and (*c*) electric susceptibility.

18.8. If one observes a loss-peak maxima for dipole relaxation at a temperature of 300°C when his oscillator is set at 1 Hz, at what temperature should he observe the peak when he changes to 100 Hz, assuming an activation energy of 1.0 eV?

18.9. Describe four polarization mechanisms which exist in a typical dielectric such as $BaTiO_3$ at a temperature below the Curie point. Draw a curve of dielectric constant and loss factor as a function of frequency from 10 to 10^{20} Hz.

18.10. (*a*) Explain why the dielectric constant for silicon carbide is the same as the square of the index of refraction n^2.

(*b*) For KBr would you predict $\kappa = n^2$? Why?

(*c*) For all substances, the index of refraction is equal to unity at high enough frequencies. Explain this.

18.11. Sketch a hysteresis curve for a typical virgin ferroelectric, and explain the cause for this nonlinear relationship in terms of the mechanisms involved. How would this curve change if the material in question were antiferroelectric?

19

Magnetic Properties

Many of the magnetic characteristics of ceramics are analogous to their dielectric characteristics. Magnetic polarization and dielectric polarization, permanent electric dipoles and permanent magnetic dipoles, sponteneous magnetization and a spontaneous electric moment, and other similarities between magnetic and dielectric phenomena are striking. However, individual positive and negative electric charges (monopoles) exist; there is no corresponding magnetic monopole; magnetic fields arise from *spinning* electrons or current loops.

The magnetic characteristics of ceramic materials are of increasing importance as the field of solid-state electronics continues to expand. Although the lodestone (Fe_3O_4) was known to have useful magnetic properties and used in compasses beginning in the thirteenth century, it was in 1946 that studies of J. L. Snoeck at the Philips Laboratories in Holland led to oxide ceramics with strong magnetic properties, high electrical resistivity, and low relaxation losses. Ceramic magnets are often used in high-frequency devices in which the greater resistivity of the ferrimagnetic oxide gives them a decisive advantage over metals; their use as circuit elements for radio, television, and electronic devices has increased as this technology has developed over the last 30 years. The use of ceramic magnets as memory units with rapid switching times in digital computers has been essential to the explosion of computer technology. Magnetic ceramics are important as special circuit elements in microwave devices and in devices which rely on their permanent magnet behavior.

Extensive and thorough treatments of magnetic materials and properties are given in references 1 to 3. The reader is encouraged to go to these texts for a more detailed coverage than is possible here.

19.1 Magnetic Phenomena

When a magnetic field intersects a real material, the magnetizing field H creates or aligns magnetic dipoles similar to the electric dipole chains illustrated in Fig. 18.1. As a result the total magnetic flux density B, which is analogous to the total electric flux density D, is the summation of the magnetizing field and the total effect of the magnetic dipoles:

$$B = \mu_0 H + \mu_0 M = \mu' H \tag{19.1}$$

where M is the magnetization of the material, μ_0 the permeability of free space (a vacuum), and μ' the effective permeability of the material.

In the rationalized mks system of units, the field strength H is measured in units of amperes per meter. For a vacuum

$$E = \sqrt{\frac{\mu_0}{\epsilon_0}}\, H = 120\pi H \qquad \text{V/m} \tag{19.2}$$

and the magnetic flux density in a vacuum is given by

$$B = \mu_0 H = 4\pi \times 10^{-7}\, H \text{ V/sec m}^2 = \text{Wb/m}^2 \tag{19.3}$$

Then

$$\frac{E}{B} = \frac{1}{\sqrt{\epsilon_0 \mu_0}} = \text{light velocity} = 3 \times 10^8 \text{ m/sec} \tag{19.4}$$

Frequently the magnetic field strength and flux density are measured in units of oersted and gauss. The conversion factors are

$$\begin{aligned} H\ (\text{A/m}) &= \frac{4\pi}{10^3} H\ (\text{Oe}) \\ 1\ \text{Oe} &= 79.7\ \text{A/m} \\ B\ (\text{Wb/m}^2) &= 10^4 B\ (\text{G}) \\ \frac{B}{H} = \mu_0 &= 4\pi \times 10^{-7}\ \text{H/m} = 1\ \text{G/Oe} \end{aligned} \tag{19.5}$$

Consistent sets of units are given in Table 18.1.

The magnetic dipole moment per unit volume is the product of the number of elementary magnetic dipoles per unit volume n and their magnetic moment P_m:

$$M = nP_m = n\alpha_m H \tag{19.6}$$

where α_m is the magnetizability of the elementary constituents. The magnetic moment is proportional to the magnetizing field strength. Magnetic properties, in parallel with dielectrics, can also be measured as the ratio of the magnetization to the applied field, called the magnetic susceptibility:

$$\chi_m = \frac{M}{H} = n\alpha_m = \frac{\mu'}{\mu_0} - 1 \tag{19.7}$$

which is an alternate way of expressing the relative magnetic permeability.

When a current i circles an area a, a magnetic dipole moment is created, $m = ia$, where m is a vector normal to the plane of the enclosed area. This is the situation when an electron circles a proton ν times per second in an orbit of radius r, producing a magnetic moment $-e\nu\pi r^2$; simultaneously, it has a quantized angular momentum. From Bohr's theory of the atom, when the azimuthal quantum number l is equal to 1, the combination of magnetic moment and angular momentum leads to an elementary magnetic moment of

$$P_m = \mu_B = \frac{eh}{4\pi m} \cong 9.27 \times 10^{-24} \text{ A m}^2\text{/electron}$$
$$= 9.27 \times 10^{-21} \text{ erg/G} \tag{19.8}$$

where h is Planck's constant and μ_B is defined as one *Bohr magneton*, the orbital contribution to the magnetic moment of an atom by one electron when $l = 1$. This is not the only mechanism which contributes to the magnetic moment of an atom. The electron itself has an intrinsic angular momentum which gives rise to an electron-spin contribution of approximately $2s$ Bohr magnetons, where s is the spin quantum number ($\pm 1/2$). In ceramic materials the orbitals are essentially fixed by the lattice and tied down by bonding in such a way that their moments cancel (orbital quenching); the main contribution to the magnetic moment of these materials results from electron spins which are free to orient with the magnetic field. The coupling between orbital and electron-spin moments does affect the observed results, however, so that the constant relating spin quantum number to the magnetic moment of an atom or ion is only approximately equal to 2.

According to the Pauli exclusion principle, only two electrons can fill any energy level; these have opposite spin directions ($s = +1/2$, $s = -1/2$), and their magnetic moments cancel. Permanent magnetic moments arise in systems in which unpaired electrons are present. These include metals with conduction electrons, atoms and molecules containing an odd number of electrons, and atoms and ions with partially filled inner electron shells, for the most part the transition elements, rare earth elements, and actinide elements. The electronic structures of these elements have been discussed in Chapter 2.

Diamagnetic Materials. The magnetic effect which corresponds to the induced dielectric effect discussed in Chapter 18 is called diamagnetism. The induced magnetization M is a linear function of the magnetic field strength H; χ_m is a constant and independent of the field. The direction of magnetization is opposite to the direction of the field, so that χ_m is negative. The effect is weak, and the relative permeability μ'/μ_0 is only slightly less than unity.

If we consider the classical picture of the atom as a nucleus with electronic charges circulating in definite orbits, we can gain a picture of the physical origin of diamagnetism. In a manner similar to the way that we derived Eq. 19.8 we consider how the angular velocity of the electron is altered if a magnetic field is slowly applied, assuming that the radius of the electron orbit is unchanged. The change in the angular velocity gives rise to a net magnetic moment μ_D of magnitude

$$\mu_D = -\frac{e^2}{4m}\mu_0 r^2 H \tag{19.9}$$

where e is the charge on the electron, m is the mass, r is the radius of the orbit, and the magnetic field H is applied normal to the plane of the orbit. For electronic orbits of radius of about 1 Å, the factor $e^2\mu_0 r^2/4m$ is about 10^{-28} cm^3. There are about 10^{22} to 10^{23} atoms/cm^3; thus the volume diamagnetic susceptibility χ_m should be in the order of 10^{-5} to 10^{-6}, which is about the actual measured value of solids. As for the induced electronic polarizability, the diamagnetic susceptibility is independent of temperature.

Diamagnetism is associated with all ceramic materials in which the ions have closed electronic shells, that is no unpaired electrons. This generally means that ceramics not containing transition metal ions or rare earth ions are diamagnetic.

Paramagnetic Materials. Ions from the transition series and rare earth series possess a net magnetic moment because the ion contains an odd number of electrons (Table 19.1). In the absence of a magnetic field these moments usually point in random directions, producing no macroscopic magnetization. However, in the presence of a magnetic field, the moments tend to line up preferentially in the field direction and produce a net magnetization. When the unpaired electrons are acted on individually with no mutual interaction between them, the effect is called *paramagnetism*. The paramagnetic susceptibility χ_m is positive because the moments line up in the same direction as the field and thus enhance the magnetic flux density.

For example, the Mn^{2+} ion has a half filled $3d$ shell containing five electrons; according to *Hund's rule* the five electrons are unpaired, that

Table 19.1. Outer-Shell Electron Configuration and Number of Unpaired Electrons for Several Spinel-Forming Ions

Ion	Electron Configuration	Number of Unpaired Electrons
Mg^{2+}	$2p^6$	0
Al^{3+}	$2p^6$	0
O^{2-}	$2p^6$	0
Sc^{3+}	$3p^6$	0
$Ti^{4+}(Ti^{3+})$	$3p^6(3d^1)$	0(1)
$V^{3+}(V^{5+})$	$3d^2(3p^6)$	2(0)
$Cr^{3+}(Cr^{2+})$	$3d^3(3d^4)$	3(4)
$Mn^{2+}(Mn^{3+})(Mn^{4+})$	$3d^5(3d^4)(3d^3)$	5(4)(3)
Fe^{2+}	$3d^6$	4
Fe^{3+}	$3d^5$	5
$Co^{2+}(Co^{3+})$	$3d^7(3d^6)$	3(4)
Ni^{2+}	$3d^8$	2
$Cu^{2+}(Cu^{+})$	$3d^9(3d^{10})$	1(0)
Zn^{2+}	$3d^{10}$	0
Cd^{2+}	$4d^{10}$	0

is, have the same spins, which results in a net magnetic moment per ion of five Bohr magnetons $5\mu_B$. An ionic solid containing n noninteracting Mn^{2+} ions has a paramagnetic susceptibility proportional to n and to $5\mu_B$. The attractive force for most paramagnetic materials is small, with $\chi_m \simeq 10^{-5}$ to 10^{-6}. Although the individual magnetic moments tend to align with a magnetic field, there is also an opposing tendency of thermal motion to randomize the spins.

Ferromagnetic and Ferrimagnetic Materials. In some materials the magnetic moments of the individual ions are strongly coupled, and thus there are regions in the solid in which the spins are aligned parallel even in the absence of a magnetic field. This results in a large microscopic magnetic moment for the small regions, called *Weiss domains*, even in the macroscopically demagnitized state. In the Weiss domains in a ferromagnetic material, the system energy is lowered by a parallel alignment of all the electron spins.

The exchange interactions between electron spins in a ferromagnetic material is positive; that is, all spins align in the same direction. However, in some solids the exchange between the unpaired electrons causes antiparallel alignment of spins. Several of the transition metal monoxides

(MnO, FeO, NiO, and CoO) exhibit this behavior. Thus the spins of the *d* electrons of adjacent iron ions in FeO are aligned in opposite directions. We call this behavior *antiferromagnetism.* In an FeO crystal having the rock salt structure, ions on any (111) plane have parallel spins, but ions on adjacent (111) planes have antiparallel spins. The aligned moments of the ions in the two directions cancel, and the FeO crystal as a whole has no magnetic moment. Antiferromagnetism with no net magnetic moment is a special case in which the number of spins aligned in opposite directions is just equal. In the more general case in which ions with unpaired electrons are arranged on two sublattices with antiparallel spin alignment, we must sum the net moment for each sublattice. A ferrimagnetic material is one in which these net moments for the two sublattices are unequal, which results in a net macroscopic magnetic moment. That is, we have incomplete cancelation of the antiferromagnetically arranged spins. This class of materials is the most important group of magnetic oxides.

Magnetic Domains. A ferromagnetic or ferrimagnetic material is divided up into many small regions or domains each of which is fully magnetized; that is, all the moments within each domain are aligned in the same direction. When the bulk material is unmagnetized, the net magnetization of these domains is zero. The way that the magnetization vectors, that is, these net magnetic moments, sum to zero is important in understanding magnetic oxides. The two opposing magnetic domains in Fig. 19.1*a* sum to zero; however, the energy of the material is lowered by the successive breakup of the domains shown in Fig. 19.1*b* and *c*. In each of these latter two cases the sum of the magnetization is also zero. The pie-shaped domains on the ends of material are called closure domains and complete the magnetic flux path within the solid; when the magnetic flux is kept mostly in the solid, Fig. 19.1*b* and *c*, the energy of the system is lower.

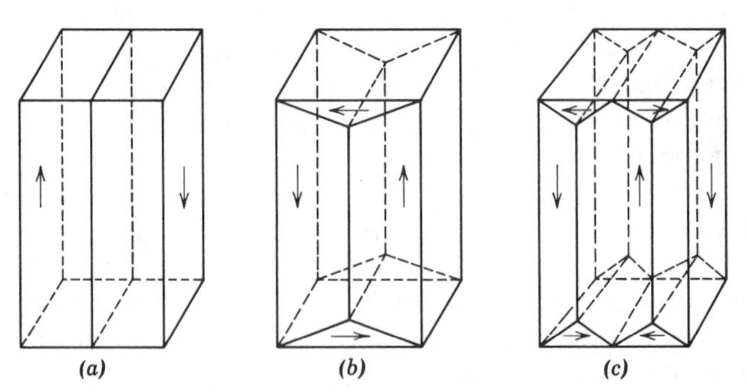

Fig. 19.1. Several domain structures of a solid, each having zero net magnetization.

Because of anisotropy in the ligand field, discussed in Chapter 13, there are preferred low-energy crystallographic directions of spontaneous magnetization. The boundary regions between these domains consist of a gradual transition in spin orientation, as shown in Fig. 19.2. The thickness of this transition region, the *domain wall*, is a balance between a tendency to have a small angle between adjacent spins, requiring a thick domain wall, and a tendency for the spins to have a particular crystallographic orientation, requiring a thin domain wall. Typically the domain wall has a thickness of about 1000 Å and an energy about 2×10^{-8} cal/cm^2 (~ 1 erg/cm^2).

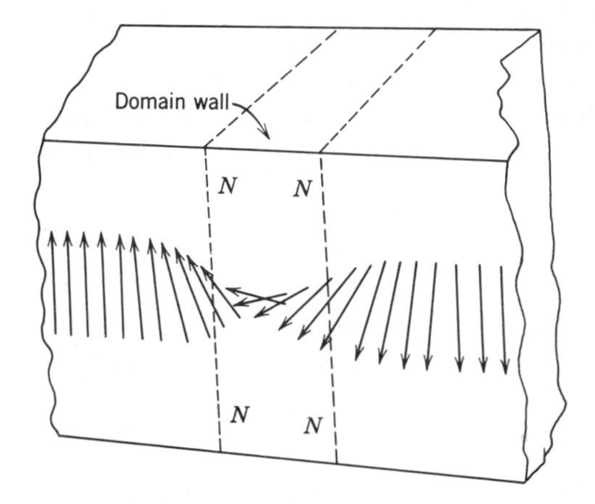

Fig. 19.2. Change in atomic-dipole orientation through a domain wall. All moments lie in the plane of the wall. The *N*'s represent poles which are formed on the surface of the material.

Just as dielectrics change their length when polarized, magnetic materials change their length when magnetized. The fractional change in length (dl/l) associated with a change in magnetization from zero to saturation (all spins aligned) is the *magnetostriction*. For example, $NiFe_2O_4$ contracts in the direction of the magnetization by about 45 parts per million at the saturation magnetization. The amount of dimensional change observed is a function of both the magnetic field strength and the crystallographic orientation.

Hysteresis. The state of magnetization of a solid is a function of the strength and direction of the magnetizing field. If we consider a ferrimagnetic material which contains many small magnetic domains but no net magnetization, we can examine what happens to the domains as the field strength is increased (Fig. 19.3). As the field is increased from zero, the effect on the solid is to displace domain boundaries in a reversible fashion. If the magnetic field is switched off, the domain boundaries return to their starting positions. Thus the initial part of the $B - H$ curve results from reversible domain boundary displacement, and the slope is called the initial permeability μ_i. As the magnetic field strength is increased, there is an irreversible boundary displacement, and at first the induced magnetization increases more rapidly than the field strength and gives a maximum slope μ_{max}. Finally in the upper part of the magnetization curve all domain boundaries have been displaced, and further increases in the magnetic field cause rotation of the domains in the direction of the applied field. At this point the material is saturated; higher fields cannot induce more magnetization.

As the magnetic field is decreased to zero, the induced magnetization does not decrease to zero, but the alignment of most of the domains during magnetization results in a remanent magnetization or remanence

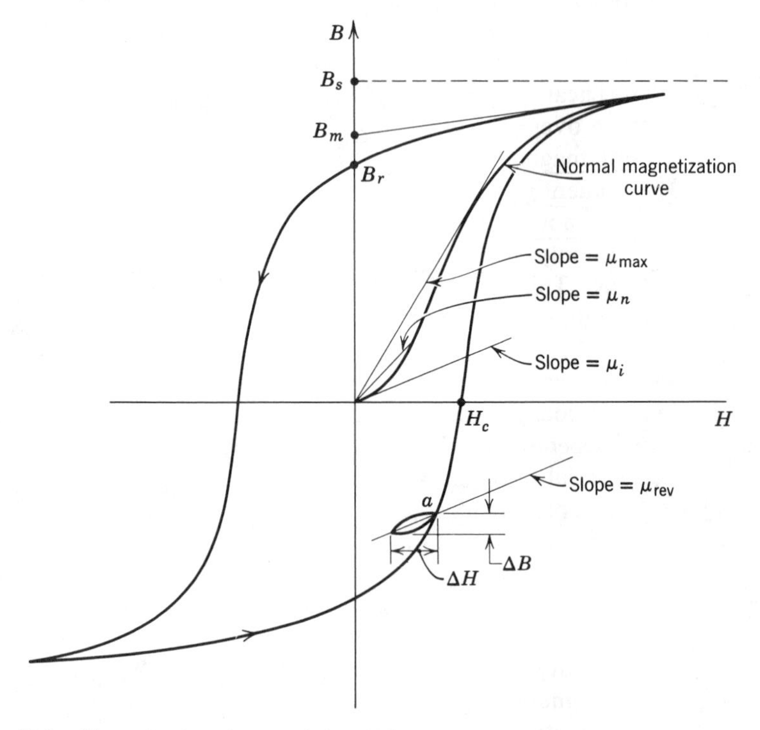

Fig. 19.3. Magnetization characteristic and hysteresis loop caused by domain action.

B_r. When the direction of the magnetic field is reversed, the induced magnetization decreases and finally becomes zero at a value of the magnetic field strength called the coercive force H_c. Further increased magnetic field strength in this opposite direction eventually causes magnetic saturation in the reverse direction and produces a saturation B_s and remanence B_r values of the same magnitude as in the first quadrant. As an applied field is cycled from one direction to the other direction, the *hysteresis loop* is followed. Since the area of the hysteresis loop represents the energy or work to bring about changes in the magnetic domain structure, the product $B \cdot H$, called the energy product, represents a net loss in the system, usually in the form of heat. In applications in which the magnetic material is cycled around the magnetization curve many times per second, hysteresis losses are critical, and *soft* magnetic materials (low B_r) are required.

In addition to the energy loss due to the hysteresis curve there are energy losses resulting from electrical currents, *eddy currents*, induced in the material. The changing magnetic flux produces a power loss in the system proportional to ϕ^2/R, where ϕ is the locally induced voltage (proportional to the time rate of change of the magnetic flux) and R is the resistance of the material. For this application, magnetic oxides which have a high electrical resistivity have small eddy-current losses and a distinct advantage over metals.

For permanent magnets, *hard* magnetic materials (high B_r) are required. For permanent magnets a high coercive force is also required, so that the material is not easily demagnetized. A single quality, the energy product, is commonly used to describe the quality of a permanent magnetic material. This is usually the maximum value of the $B \cdot H$ product. High-quality permanent magnetic materials have an energy product of about 1 cal/cm^3. From the hysteresis curve, Fig. 19.3, one notes that high values of $(BH)_{\max}$ require high values of both remnant magnetization and coercive field.

Temperature Dependence of the Paramagnetic Susceptibility. In a paramagnetic material in which there are n magnetic moments of value P_m (Eq. 19.6) we apply the classical theory of magnetism developed by Langevin to determine how these magnetic moments align as function of temperature. Consider an ideal gas in which each molecule has a net magnetic moment P_m. Since in the ideal gas there are no interactions among the molecules, except for molecular collisions, the magnetic dipole vectors are oriented at random in the absence of a magnetic field. With the application of a magnetic field they rotate to align in the field direction. This is shown schematically in Fig. 19.4, in which the torque to rotate the

$\mathbf{P}_m$

$\mathbf{B}$ θ

$\tau = \mathbf{P}_m \times \mathbf{B} = |\mathbf{P}_m||\mathbf{B}| \sin\theta$

(a)

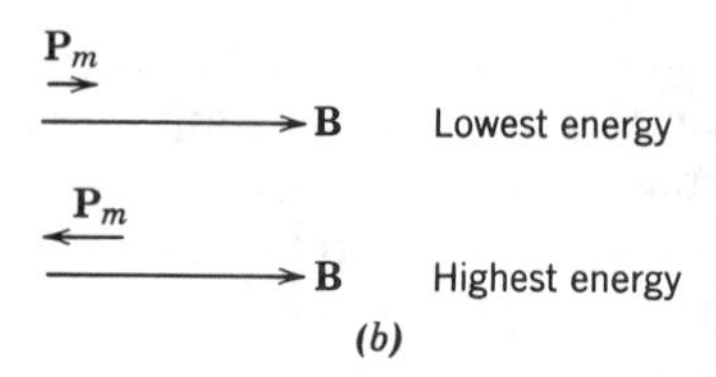

Fig. 19.4. (*a*) The torque on the magnetic moment of anion to align with the applied field *B*. (*b*) The alignments for lowest and highest energy configurations.

magnetic moments parallel to the applied field is given as $P_mB \sin\theta$. The energy distribution as a function of θ is given by

$$E(\theta) = \int_0^\theta \tau \, d\theta = \int_0^\theta P_mB \sin\theta \, d\theta = -P_mB \cos\theta \qquad (19.10)$$

The number of molecules with a net magnetic moment at a particular energy $E(\theta)$ is given from kinetic theory by the Boltzman distribution:

$$N(\theta)\, d\theta = A\, 2\pi \exp\left[\frac{P_mB\cos\theta}{kT}\right] \sin\theta \, d\theta \qquad (19.11)$$

where $N(\theta)$ is the number of magnetic dipoles per unit volume at an angle θ to the applied field and $A = n/(2\pi \int_0^\pi \sin\theta \exp[P_mB\cos\theta/kT]\, d\theta)$, where n is the total number of dipoles per unit volume. Finally the magnetization is given by

$$M = \int N(\theta) P_m \cos\theta \, d\theta = nP_m\left(\coth L - \frac{1}{L}\right) \qquad (19.12)$$

where $L = mB/kT$. The term in the parentheses is called the Langevin function. At temperatures above $\sim 10°K$ for most magnetic field strengths the magnetization is given as

$$M = nP_m\left[\frac{P_mB}{3kT}\right] = nP_m\left[\frac{P_m\mu_0H}{3kT}\right] = \chi H \qquad (19.13)$$

where the magnetic susceptibility of the paramagnetic substance is given by the Curie law:

$$\chi = \frac{nP_m^{\,2}\mu_0}{3kT} = \frac{C}{T} \qquad (19.14)$$

By plotting the inverse of the measured susceptibility as a function of temperature, the slope yields the value for the magnetic dipole moment P_m. This same expression applies to condensed-phase paramagnetic materials and is often given in reciprocal form:

$$\frac{1}{\chi} = \frac{3k}{nP_m^2\mu_0} T = \frac{T}{C} \tag{19.15}$$

Above a temperature called the Curie temperature T_c, ferromagnetic, ferrimagnetic, and antiferromagnetic materials become paramagnetic because the thermal motion is sufficient to randomize the orientation of magnetic dipoles. Thus, above T_c there is no net magnetic moment in the absence of a magnetic field. The effective field $H_{\text{eff}} = H_{\text{applied}} + \lambda M$ is acting to align the spins; thus Eq. 19.15 must be adjusted. In particular, for a ferromagnetic material we have the Curie–Weiss law for the variation of the susceptibility with temperature in the paramagnetic region above T_c. The susceptibility is given as

$$\frac{M}{H_{\text{applied}}} = \chi = \frac{C}{T - C\lambda} = \frac{C}{T - T_c} \tag{19.16}$$

where C is defined by Eq. 19.14, and $T_c = C\lambda$, which includes the effect of the exchange field.

For an antiferromagnetic material we assume opposing electron spins, and therefore the magnetic dipoles for the two sublattices are equal but opposite. However, above a transition temperature called the Neel temperature $T_N = \frac{\lambda C}{2}$, the spins become random. In this case the susceptibility is given by

$$\frac{M}{H_{\text{applied}}} = \chi = \frac{C}{T + \lambda C/2} = \frac{C}{T + T_N} \tag{19.17}$$

The important magnetic ceramic materials are of the ferrimagnetic class, and in this case we must consider the distribution of the antiparallel spins on two sublattices, for example, the tetrahedral and octahedral sites in the spinel structure. In the simple model according to Néel we consider the distribution of a particular magnetic ion, for example, Fe^{3+}, on the two sublattices. Let x be the fraction of iron ions on the a sites (tetrahedral) and y the fraction on the b sites (octahedral), where $x + y = 1$. Since the spins on the a and b sites are antiparallel, the maximum net magnetic moment is given as,

$$\mu_m = 2(x - y)P_m \tag{19.18}$$

which for Fe^{3+} ions is $2(x - y)5\mu_B$. Again by assuming a Curie law type of behavior in the paramagnetic region of a ferrimagnetic material, the temperature dependence of the susceptibility is

$$\frac{1}{\chi} = \frac{T}{C} + \frac{1}{\chi_0} - \frac{\xi}{T - \theta} \tag{19.19}$$

where $\frac{1}{\chi_0} = \lambda_{AB}(2xy - x^2\alpha - y^2\beta)$

$\theta = \lambda_{AB}xyC(2 + \alpha + \beta)$

$\xi = \lambda_{AB}^2 xyC[x(1 + \alpha) - y(1 + \beta)]^2$

and where λ_{AB} is the exchange constant for interactions of the ions on the a and b sites and is a negative term; α is defined as $\lambda_{AA}/\lambda_{AB}$ and β is defined as $\lambda_{BB}/\lambda_{AB}$. Both of these latter terms are also negative. The origin of these exchange constants (Weiss or molecular field constants) in ferrimagnetic materials is discussed in the next section. The temperature dependence of the susceptibility for these four cases is illustrated in Fig. 19.5.

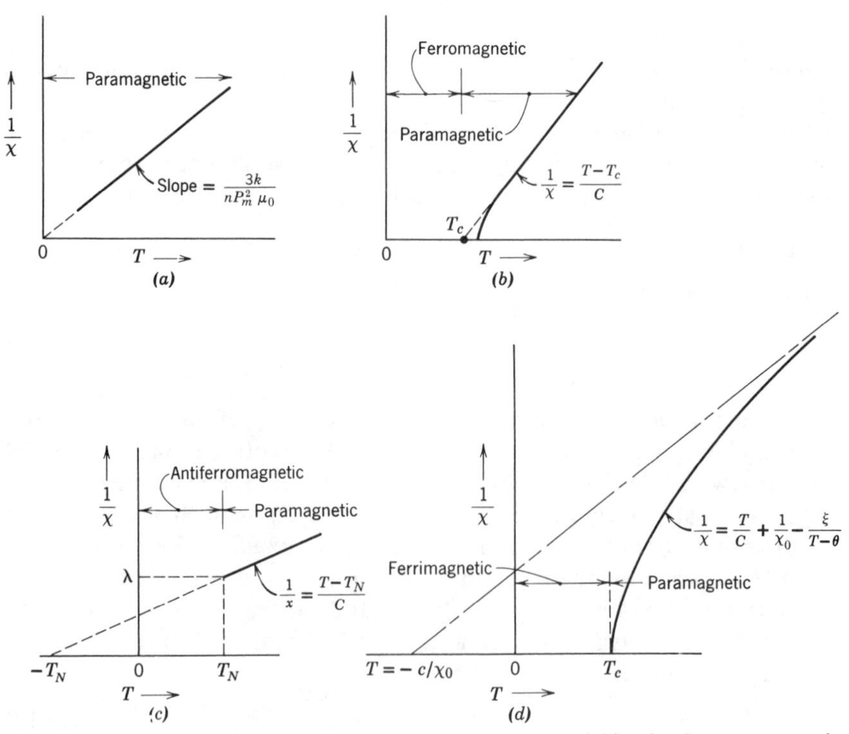

Fig. 19.5. Temperature dependence of the inverse susceptibility in the paramagnetic regime of (a) paramagnetic, (b) ferromagnetic, (c) antiferromagnetic, and (d) ferrimagnetic materials.

Temperature Dependence of the Saturation Magnetization. As the Curie temperature is approached, the magnetization M varies in a way similar to the order-disorder phenomena described in Chapter 4. The aligned spins gradually begin to randomize, and the rate of randomization increases significantly as the critical temperature is approached. For a ferromagnetic material we have the Bloch $T^{3/2}$ law:

$$M(T) = M(T = 0)[1 - FT^{3/2} + \cdots] \tag{19.20}$$

where F is a constant related to interaction at the electron spins. This behavior is shown schematically in Fig. 19.6.

For an antiferromagnetic material the magnetization depends on the direction of the applied field, that is, whether the field is applied parallel to the direction of the magnetic dipoles or perpendicular to the dipoles or applied to a random array of dipoles as in a polycrystalline solid. This behavior is shown schematically in Fig. 19.7.

Based on the assumptions used to derive Eq. 19.18 for a ferrimagnetic material, Neel predicted six types of behavior for the temperature dependence of the magnetization. Three of these predicted forms are not thermodynamically possible because of the requirements that

$$\left[\frac{\partial M}{\partial T}\right]_{T\to 0} = 0.$$

The variation with temperature of the magnetic dipoles on the two sublattices gives a net magnetization-temperature curve for the three possible situations and is shown schematically in Fig. 19.8. Each of these magnetization temperature curves has been observed in ferrimagnetic materials since Neel's predictions. In the discussion of specific magnetic materials which follows, data for each of these classes are given.

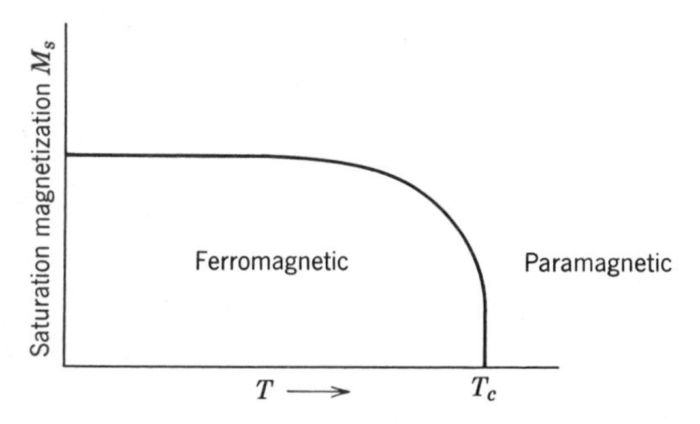

Fig. 19.6. Temperature dependence of the saturation magnetization of a ferromagnetic material.

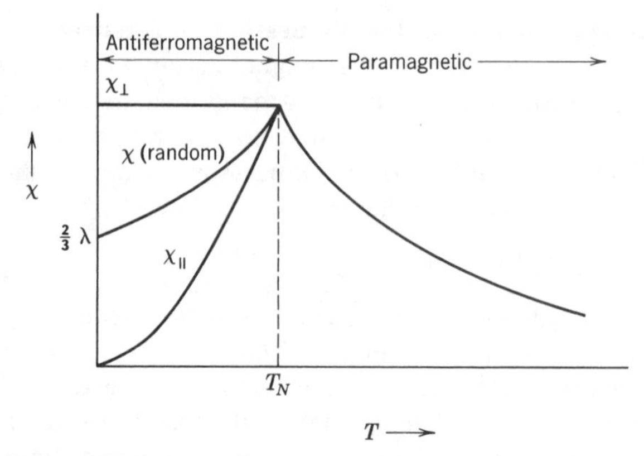

Fig. 19.7. Susceptibility of an antiferromagnetic material for cases when the field is applied perpendicular ($\chi_\perp$) and parallel ($\chi_\parallel$) to the magnetic dipoles and applied to a polycrystalline, that is, random (χ random) set of dipoles.

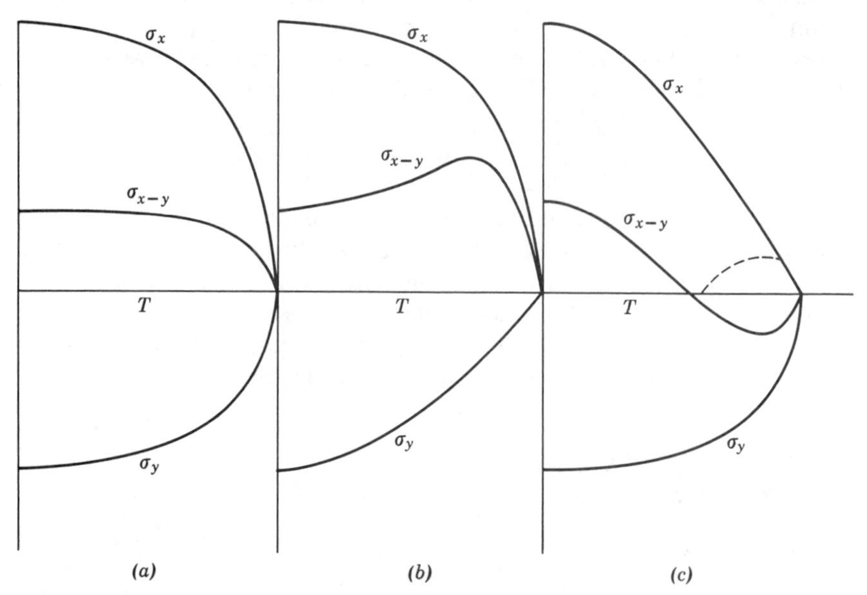

Fig. 19.8. Néel's prediction for magnetization σ_{x-y} versus temperature in a ferrimagnetic material where x and y refer to the sublattices.

19.2 The Origin of Interactions in Ferrimagnetic Materials

In Section 17.3, we discussed the importance of the interaction and overlap of the outer electrons in determining the electrical characteristics of oxides. The magnetic properties of solids also result from the outer electrons, especially transition metal ions, and the way in which these outer electrons interact with the electrons of neighboring ions.

Direct Exchange Interaction. The magnetic dipoles in a ferromagnetic material (e.g., Fe, Ni) are orientated spontaneously so that a net macroscopic magnetic moment is observed. Even in a polycrystalline material the dipoles are parallel to one another except as one approaches the Curie temperature. The coupling between the electron spins which results in this parallelism is not of magnetic origin. Such a magnetic dipole-dipole interaction would be too small by a factor of about 10^3 to explain the observed Curie temperatures (350°C for Ni, 770°C for Fe). The only adequate explanation of this interaction is based on quantum mechanics.* This direct exchange interaction may either be positive or negative (Fig. 19.9) and is given by the exchange integral. The magnitude and the sign of the exchange integral depend on the ratio D/d, where D is the atomic (or ionic) separation of the interacting atoms (or ions) and d is the diameter of the electron orbit under consideration (the $3d$ or $4f$ orbits of the transition metals). From Fig. 19.9 it is seen that $D/d < 1.5$ gives a negative

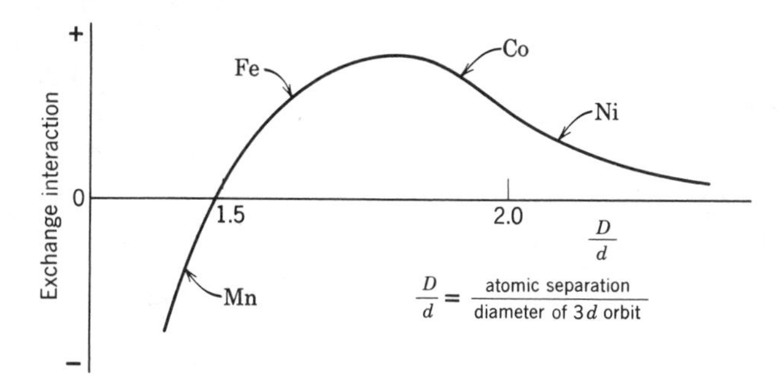

Fig. 19.9. Slater-Bethe curve showing the magnitude and sign of the exchange integral as a function of D/d. From reference 2.

exchange interaction; above this value the interaction becomes positive, and it reaches a maximum at $D/d = 1.8$ and then diminishes to very small but still positive values. In the ferrimagnetic spinels, the values of D/d are usually of the order of 2.5; that is, a moderate to weak positive interaction would be predicted from the direct exchange interaction. Yet the experimental evidence favors a strong negative interaction between the a (tetrahedral) and b (octahedral) sites.

We must conclude that a distinctly different situation occurs in the oxides from that in the ferromagnetic metals because of the oxygen ions

*R. M. Bozorth, *Ferromagnetism*, D. Van Nostrand Company, Inc., Princeton, N.J., 1951.

which separate the metal ions with partially filled outer orbitals. That is, the direct exchange between metal ions is partially or wholly obscured by the intervening oxygen ions. Two mechanisms by which a negative interaction may be obtained and in which the oxygen ion plays an important role have been suggested and are called the superexchange interaction and the double exchange interaction.

Superexchange Interaction. Since the bonding in the oxides is mainly ionic, the oxygen ion with a full $2p$ shell has an inert gas electron configuration, and its interaction in this ground state with metallic ions is small. The superexchange interaction has been proposed for the case in which there is a mechanism of excitation from this ground state.

The possible excitation mechanism involves the temporary transfer of one oxygen $2p$ electron to a neighboring metal ion. Qualitatively we can describe the superexchange interaction by considering the following example of ferric ions in an oxide. We go from a ground state of these ferric ions in which the five $3d$ electrons according to Hund's rule are all aligned. The six $2p$ electrons of the oxygen ions form three pairs in which each of the pairs have canceling spins. Recalling that the p orbit is shaped like the dumbbell (Fig. 2.5), let us consider the interaction when an excited state exists and one of the p electrons temporarily becomes one of the d electrons of the iron ion. The transfer process in which we have one ferric ion on one side of the oxygen and another ferric ion on the other side is given as follows

$$\begin{array}{cccccc} Fe^{3+}(3d^5) & O^{2-}(2p^6) & Fe^{3+}(3d^5) \rightarrow & Fe^{3+}(3d^5) & O^{1-}(2p^5) & Fe^{2+}(3d^6) \\ \downarrow\downarrow\downarrow\downarrow\downarrow & \uparrow\downarrow & \downarrow\downarrow\downarrow\downarrow\downarrow & \downarrow\downarrow\downarrow\downarrow\downarrow & \uparrow\downarrow & \downarrow\downarrow\downarrow\downarrow\downarrow \\ & \uparrow\downarrow & & & \uparrow\downarrow & \uparrow \\ & \uparrow\downarrow & & & \downarrow & \end{array}$$

The one ferric ion now becomes a ferrous (Fe^{2+}) ion. The unpaired electron of the oxygen p orbital which was directed toward the ferric ions now can interact in a negative way with the unpaired ferric ion on the other side. If the $3d$ shells of the metal ions are less than half full, the superexchange should favor a positive interaction; for $3d$ shells which are half filled or more than half filled, like our example of the ferric ion, a negative interaction with antiparallel spin is probable. It is generally assumed that this superexchange interaction diminishes rapidly as the distance between the ions increases. The dumbbell shape of the $2p$ orbital makes it reasonable to assume that the interaction for a given ionic separation is greatest when the metal oxygen-metal angle is 180° and is least when this angle is 90°. Thus in a spinel lattice we conclude that the a-b interaction is relatively strong, the a-a interaction is relatively weak, and the b-b interaction is probably intermediate.

Double Exchange Interaction. Another mechanism has also been proposed to account for the interaction between adjacent ions of parallel spins through a neighboring oxygen ion. This model is more restrictive than the superexchange interaction and requires the presence of ions of the same element but in different valence states; that is, in magnetite we have ferrous and ferric ions. The double exchange interaction involves the transfer of one of the *d* electrons on the ferrous ion to the neighboring oxygen ion with the simultaneous transfer of the *s* electron with the same spin from the oxygen ion to a neighboring ferric ion. This process is similar to the hopping conduction model discussed in Chapter 17 for the electrical conductivity in transition metal oxides. The double exchange mechanism favors only positive interaction (i.e., parallel spins on adjacent ions). It cannot account for the negative *a*-*b* interactions in ferrites but may be a contributing factor to the observed ferromagnetic (positive) interactions in certain manganites and cobaltites.

19.3 Spinel Ferrites

Crystallography and Magnetic Structures. The ferrimagnetic oxides known as ferrites have the general formula $M^{2+}O \cdot Fe_2^{3+}O_3$, where M^{2+} is a divalent metallic ion such as Fe^{2+}, Ni^{2+}, Cu^{2+}, Mg^{2+}. Mixed ferrites can also be fabricated in which the divalent cation may be a mixture of ions (e.g., $Mg_{1-x}Mn_xFe_2O_4$), so that a wide range of composition and magnetic properties is possible. The crystal structure is that of the spinels (Chapter 2) in which the oxygen ions are in a nearly close-packed cubic array. In a unit cell, which contains 32 oxygen ions, there are 32 octahedral sites and 64 tetrahedral sites; of these, 16 of the octahedral sites are filled (*b* sites), and 8 of the tetrahedral sites are filled (*a* sites), as illustrated in Fig. 19.10. The distribution of the cations in the available sites must be determined experimentally and is sensitive to the specific cations as well as to the temperature.

There are two idealized structures. In the *normal* spinel all the divalent ions are on the tetrahedral *a* sites, $(Zn^{2+})(Fe^{3+})_2O_4$. In the *inverse* spinel the 8 tetrahedral sites are filled with trivalent ions and the 16 octahedral sites are equally divided between di- and trivalent ions, $Fe(Fe^{3+}Fe^{2+})O_4$. In some systems, and particularly at high temperatures, the cation distribution may be disordered and the cations nearly randomly distributed on *b* sites and between *a* and *b* sites, but generally there is a tendency for individual ions to fit into particular sites so that either the normal or inverse arrangement is preferred.

All the ferrimagnetic spinels are more or less inverse; that is, some of the trivalent ions occupy octahedral *b* sites, and an equal fraction of the trivalent ions occupy tetrahedral *a* sites presumably because of the

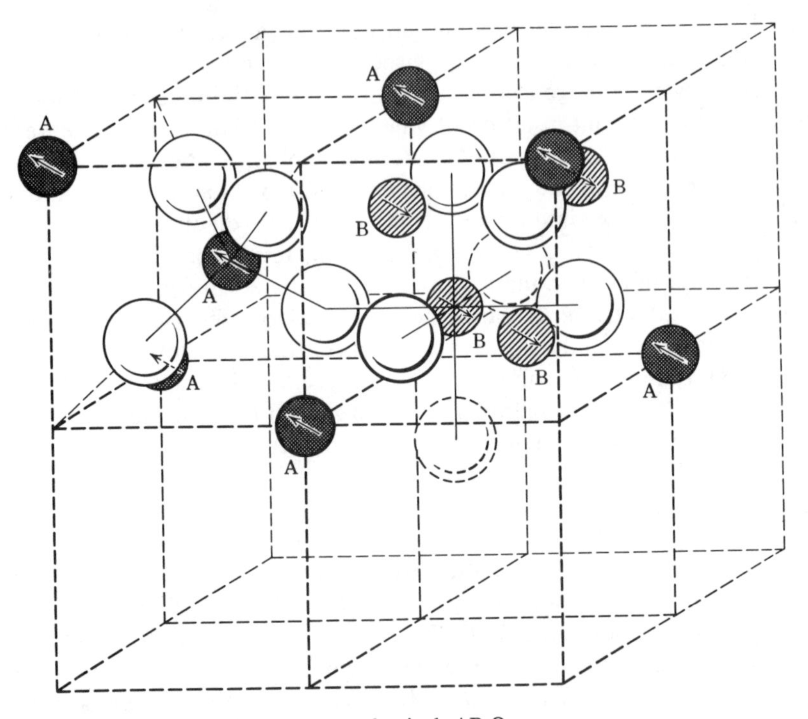

Fig. 19.10. The crystal structure of spinel, AB_2O_4.

tendency of the larger divalent ions to occupy the larger octahedral positions. The ions on the a site are coupled through their electron spin in an antiparallel fashion to those on the b sites so that the net magnetic moment is that on the divalent ion; that is,

$$Fe_a^{+3} \uparrow 5\mu_B \qquad Fe_b^{+3} \downarrow 5\mu_B \qquad M_b^{+2} \downarrow \tag{19.21}$$

For instance, if the divalent M ion is from the transition series with n electrons in the d shell, the moment is $n\mu_B$ or $(10-n)\mu_B$, depending on whether the d shell is less than or more than half filled. If the inversion in the spinel is not complete, that is, only a fraction x of the divalent M ions are on the b sites, the arrangement of the moments is

$$(1-x)M_a \uparrow x\,Fe_a \uparrow (1-x)Fe_b \downarrow Fe_b \downarrow x\,M_b \downarrow \tag{19.22}$$

The net moment is

$$\mu_m = \{n[(1-x)-x]-5[1+(1-x)-x]\}\mu_B = \{n[1-2x]-10[1-x]\}\mu_B \tag{19.23}$$

Table 19.2. Summary of Properties of Ferrites

	Molecular Weight	Density	a_0 (Å)	ρ (ohm-cm) (room temp)	μ_m (calc)	μ_m	σ_0	M_0 (0°K)	B_0	σ_s (room temp)	M_s	B_s	T_c (°C)	K_1 ($\times 10^{-3}$)	λ_s ($\times 10^6$)
$ZnFe_2O_4$	241.1	5.33	8.44	10^2	Antiferromagnetic			$T_N = 9.5°K$			$M = 5\ \mu_B$				
$MnFe_2O_4$	230.6	5.00	8.51	10^4	5	4.55	112	560	7000	80	400	5000		−40	−5
$FeFe_2O_4$ (Fe_3O_4)	231.6	5.24	8.39	4×10^{-3}	4	4.1	98	510	6400	92	480	6000	585	−130	+40
$CoFe_2O_4$	234.6	5.29	8.38	10^7	3	3.94	93.9	496	6230	80	425	5300	520	+2000	−110
$NiFe_2O_4$	234.4	5.38	8.337	10^3–10^4	2	2.3	56	300	3800	50	270	3400	585	−69	−17
$CuFe_2O_4$:															
quenched	239.2	5.42	8.37	10^5	1	2.3							455	−63	−10
slow-cooled		5.35	8.70 8.22			1.3	30	160	2000	25	135	1700		−60	
$MgFe_2O_4$	200.0	4.52	8.36	10^7	0	1.1	31	140	1800	27	110	1500	440	−40	−6
$Li_{0.5}Fe_{2.5}O_4$	207.1	4.75	8.33	10^2	2.5	2.6	69	330	4200	65	310	3900	670	−83	−8
γFe_2O_3	159.7		8.34		2.5	2.3	81			73.5	417		575		
$MnMn_2O_4$ (Mn_3O_4)	228.8	4.84	b.c.t. $a_0 = 5.75$ $c_0 = 9.42$			1.85		218			185		42°K	-10^7	
$MgMn_2O_4$	198.2							25			85				

σ_s = saturization magnetization per gram, emu/g.
M_s = saturization magnetization per cm^3, emu/cm^3.
σ_0, M_0, B_0 refer to values at $T = 0°K$.
K_1 = magnetocrystalline anisotropy constant, $ergs/cm^3$.
λ_s = the magnetostrictive constant at saturation for polycrystalline (random) material, $\lambda_s = \delta l/l$.
Source. Reference 1.

In the fully normal spinel ($x = 0$), the net moment would be $\mu_m = (n - 10)\mu_B$.

The extent to which inversion of the cations occurs depends on the heat treatment, but in general increasing the temperature of a normal spinel causes an excitation of the ions to the inverted position. Thus in the preparation of a ferrite such as $CuFe_2O_4$, quenching of the high-temperature inverse structure is necessary to obtain it at low temperature. Table 19.2 contains a summary of the properties of several ferrites, including the number of Bohr magnetons per molecule μ_B from the measured characteristics and the number of Bohr magnetons which would be calculated from Eq. 19.23. Also included are the Curie temperature, the room-temperature resistivity, the saturization magnetization per gram σ_s and per cm^3 M_s, the magnetostrictive constant λ, and the magnetocrystalline anisotropy constant K_1. The last is essentially the energy required to rotate the magnetization out of the preferred (easy magnetization) direction. The effect of temperature on the magnetization is shown in Fig. 19.11.

Manganese ferrite is about 80% normal spinel, and this arrangement does not change greatly with heat treatment. Since the Mn^{2+} ion has a moment of $5\mu_B$, Eq. 19.23 shows that inversion should not affect the net moment of $MnFe_2O_4$. This is concluded from the data in Table 19.3 for polycrystalline and single-crystalline manganese ferrites which have been

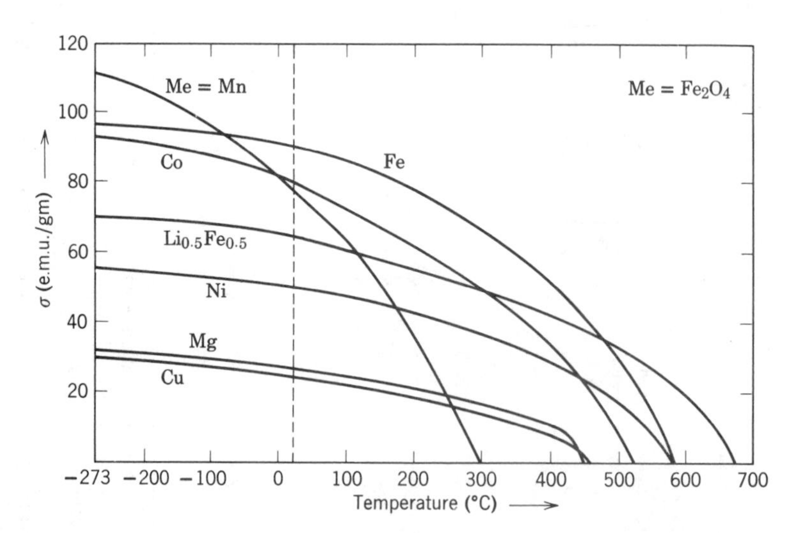

Fig. 19.11. Saturation magnetization per gram, σ, for simple ferrites with spinel structure as a function of temperature T. From reference 3.

Table 19.3. $MnFe_2O_4$ Curie Temperature T_c, Saturation Magnetization σ_0 per Gram at 4.2°K, and Corresponding Moment μ_m per Formula Unit Measured on Two Polycrystalline Samples and Single Crystal
(ϵ = amount of Mn at B sites)

Sample	Heat Treatment	T_c (°K)	σ_0 at 4.2°K (G cm^3/g)	n_B at 4.1°K (μ_B)	ϵ
Polycrystalline	1150°C quenched	610	108.5 ±0.2	4.49 ±0.01	0.11–0.12
	1150°C slowly cooled and annealed at 1000, 900, and 700°C	555	111.8 ±0.2	4.61 ±0.01	0.05–0.08
Single crystal	Rapidly cooled from melting point	585	110.4 ±0.2	4.56 ±0.01	
	1100–1200°C under pressure of 60,000 bars rapidly cooled	620	106.8 ±0.2	4.42 ±0.01	

Source. Reference 1.

quenched and slowly cooled. Nickel ferrite has similar properties, and it is 80 to 90% normal spinel.

Magnesium ferrite transforms at high temperatures to the normal spinel structure as the divalent magnesium ions are thermally excited onto the tetrahedral a sites. In this case the magnetization is strongly influenced by the cooling rate. Quenching retains the normal spinel structure; slow cooling allows the inverse spinel structure to occur because enough thermal energy and time are available to allow the magnesium ions to migrate to the preferred octahedral b sites. The saturation moment for a rapidly quenched sample is $2.23\mu_B$; a slow furnace cool results in $1.28\mu_B$. The Mg^{2+} ion has no net moment so that the inverse spinel should have zero magnetization and the normal spinel $10\mu_B$. Magnesium ferrite has high resistivity and low magnetic and dielectric losses and with its derivatives has wide applications in microwave technology.

Mixed Ferrites. The pure ferrites listed in Table 19.2 have a wide range of properties; magnetic constants differ in magnitude and often have different signs. Therefore, a solid solution of two ferrites allows even greater variation of the magnetic parameters. For instance, the gradual replacement of the divalent manganese ion in manganese ferrite with a ferrous ion reduces the magnitude of the magnetic anisotropy (λ_s and K_1) to zero and yields a material of high permeability. With a high enough concentration of Fe^{2+} ions the sign of the anisotropy and magnetostriction is reversed; ferrous ions also have the effect of decreasing the resistivity, which is usually an undesirable effect.

The effect of mixing a small proportion of cobalt ferrite (approximately 1%) with most other ferrites is to reduce the magnitude of the usually negative anisotropy or to reverse its sign. These combinations can be predicted from the data in Table 19.2.

Commercial ferrites usually require a high permeability. The initial permeability due to domain rotation is proportional to B_s^2/K_1, so that a high permeability may arise from either a high saturation magnetization or a low anisotropy. The anisotropy constant K_1 decreases rapidly with increasing temperature and in fact the permeability is a maximum near the Neel temperature. Thus high-permeability ferrites result from either a high value of B_s or a low Curie temperature, one which is just above the operational temperature (usually room temperature).

The addition of cobalt ferrite affects the magnetocrystalline anisotropy of most ferrites, but it also produces mixed ferrites which are sensitive to magnetic annealing, and this is of particular importance in applications where there is need of magnetic materials with a hysteresis loop of particular shape, for example, a square loop ferrite for switching purposes. Magnetic annealing (annealing in a magnetic field) introduces uniaxial anisotropy superimposed on any crystal anisotropy. The mechanism for this may be electron transport (hopping), diffusion of vacancies, or diffusion (interchange) of ions. It appears to be a thermally activated process; thus if anisotropy is due to electron hopping (the change of an electron between a ferrous and a ferric ion), it may be induced at room temperature, since the activation energy is low. Ion diffusion or vacancy diffusion requires higher activation energies; thus the anisotropy could only be introduced at elevated temperature and frozen in by cooling to room temperature.

In contrast to most ferrimagnetic spinels, which are largely inverse, zinc ferrite is a normal spinel (less than 5% inverse). The mixed ferrite $M_{1-x}Zn_xFe_2O_4$, where M is a divalent ion, is a solid solution in which the magnetic moment varies as a function of zinc content, as shown in Fig. 19.12. The incorporation of up to about 40% zinc increases the magnetization, and then it decreases toward zero for pure zinc ferrite. The effect cannot be explained in terms of a mere substitution of the M ion by the Zn^{2+} ion, as this would just reduce the overall moment. The zinc ions go on the a site (normal spinel), displacing an iron ion onto a b site to take up the vacated M ion position. The magnetization for the ferrite containing M which has n d electrons is given as $[n(1-x)+5(1+x)-5(1-x)]\mu_B = [n(1-x)+10x]\mu_B$. With continuing substitution a normal spinel would be produced with a net moment of $10\mu_B$, but above about 40 to 50% the antiparallelism between the diminishing number of Fe_a ions and the Fe_b ions cannot be maintained against the increasing antiparallel interaction

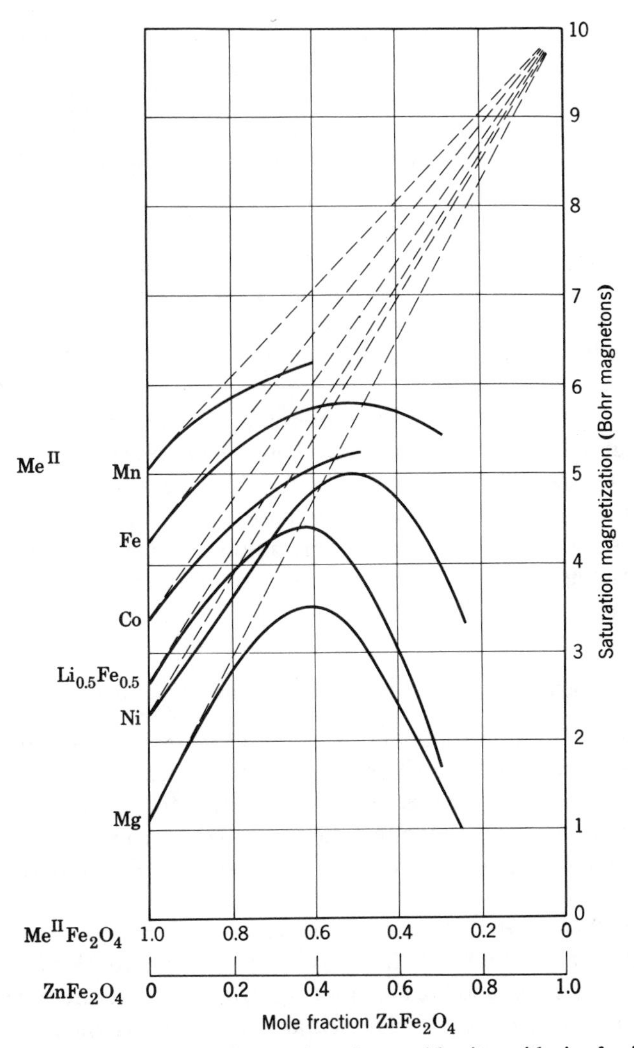

Fig. 19.12. Change in saturation magnetization of several ferrites with zinc ferrite added in solid solution. From E. W. Gorter, *Nature* (*London*), **165**, 798 (1950).

on the *b* ion sublattices; therefore the magnetization begins to fall off. However, the slope of the initial magnetization curves given an intercept of value $10\mu_B$ (Fig. 19.12).

We have emphasized that many of the properties of ceramics are sensitive to the effects of heat treatment and composition. A surplus or deficiency of Fe ions of a few percent can change the resistivity of a magnetic ceramic by several orders on magnitude. Addition of 2% cobalt to nickel ferrite increases the resistivity from 10^6 to 10^{11} ohm-cm; addition of 2% manganese to magnesium ferrite increases the resistivity from 10^4 to 10^{11} ohm-cm. The mixed zinc ferrites can be prepared with high-saturation magnetization and a low Curie point. In particular, manganese zinc ferrites have a high saturation and are good soft magnets but have a relatively low resistivity (approximately 10^2 ohm-cm for $x = 0.5$) which limits their applications to those in which the field is cycled at low frequency. However, nickel zinc ferrites have a permeability maximum at 70% zinc ferrite (maximum initial permeability ~ 4000) and a high resistivity (10^5 to 10^9 ohm-cm), allowing applications at high frequencies.

We have discussed many of the important ferrites and have indicated the range of magnetic properties possible. There are also a number of compounds with the spinel structure which have semiconducting properties, such as the calcogenides ($CdCr_2S_4$, $CdCr_2Se_4$), and a wide range of transition metal oxides, such as chromates, vanadates, and manganates. These exist with varying degrees of inversion and may be ferrimagnetic or antiferromagnetic. They are less important than the ferrites in current technology.

19.4 Rare Earth Garnets, Orthoferrites, and Ilmenites

In addition to the spinel structure there are several other crystal structures containing transition or rare earth metal ions which have interesting magnetic properties, particularly the garnet, perovskite, pseudoperovskite, and ilmenite structures. These compounds have only recently been studied extensively and are just beginning to find use in new applications.

Rare Earth Garnets. The rare earth garnets have the general formula $M_3{}^c Fe_2{}^a Fe_3{}^d O_{12}$ or $(3M_2O_3)^c(2Fe_2O_3)^a(3Fe_2O_3)^d$, where M is a rare earth ion or an yttrium ion and the superscripts *c*, *a*, *d* refer to the type of lattice sites which the ions occupy. The metal ions are all trivalent. The crystal structure is cubic, with 160 atoms per unit cell containing 8 formula units. The *a* ions are arranged on a *bcc* type of lattice with the *c* ion and *d* ion lying on cube faces (Fig. 19.13). The unit cell consists of 8 of these subunits. Each *a* ion occupies an octahedral site, each *c* ion is surrounded by 8 oxygen ions forming a dodecahedron site, and each *d* ion is on a tetrahedral site. None of these polyhedra is regular, and the oxygen lattice is severely distorted. The physical properties of a series of garnets are given in Table 19.4.

Like the spinels, a net magnetic moment arises from an uneven contribution from antiparallel spins; the *a* ions and *d* ions are aligned

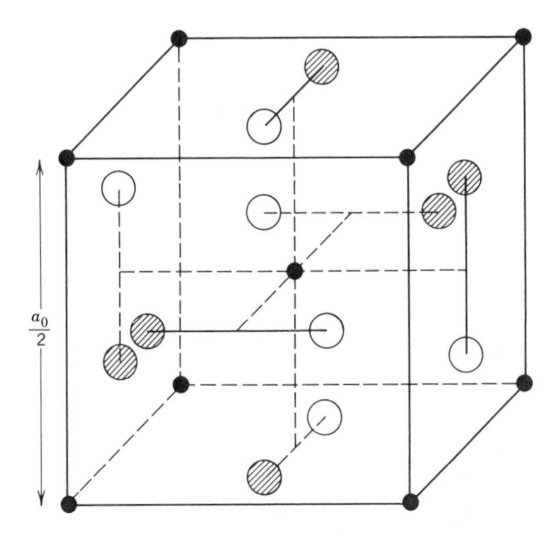

Fig. 19.13. Simplified diagram of garnet structure unit $(Fe_2)_A(Fe_3)_B(M_3)_CO_{12}$ without the oxygen ions; the unit cell side a_0 consists of eight of these units ●, a ions; ○, c ions; ◍, d ions. The a ions are arranged on a b.c.c. lattice and with c and d ions lying on the cube faces.

antiparallel, as are the c ion and d ion moments. For $M_3{}^cFe_2{}^aFe_3{}^dO_{12}$

$$\mu_{net} = 3\mu_c - (3\mu_d - 2\mu_a) = 3\mu_c - 5\mu_B \qquad (19.24)$$

if we assume a moment of $5\mu_B$ per Fe^{3+} ion. The predicted moments per formula unit are more difficult to estimate for the garnet structure than for spinel because there is a contribution due to spin-orbit coupling in addition to the contribution of the electron spin.

Magnetization-temperature curves for several garnets are shown in Fig. 19.14. First we see the behavior predicted by Neel (Fig. 19.8) for uncompensated antiferromagnetism, in which the magnetization decreases to zero at the compensation temperature and then has a moment in the opposite direction which increases and then decreases as the Curie temperature is reached. This is caused by a more rapid randomization of the moments on one sublattice relative to the other. Thus, a magnetization of opposite sign on passing through the compensation temperature refers to a change in the direction of magnetization in the crystalline lattice, but there is still a positive interaction with an external magnetic field. In Fig. 19.14 the spontaneous magnetization in terms of μ_B per formula unit is plotted only as positive values.

The open garnet structure with several kinds of cation sites allows for a wide range of substitution, mixing between rare earth garnets as well as

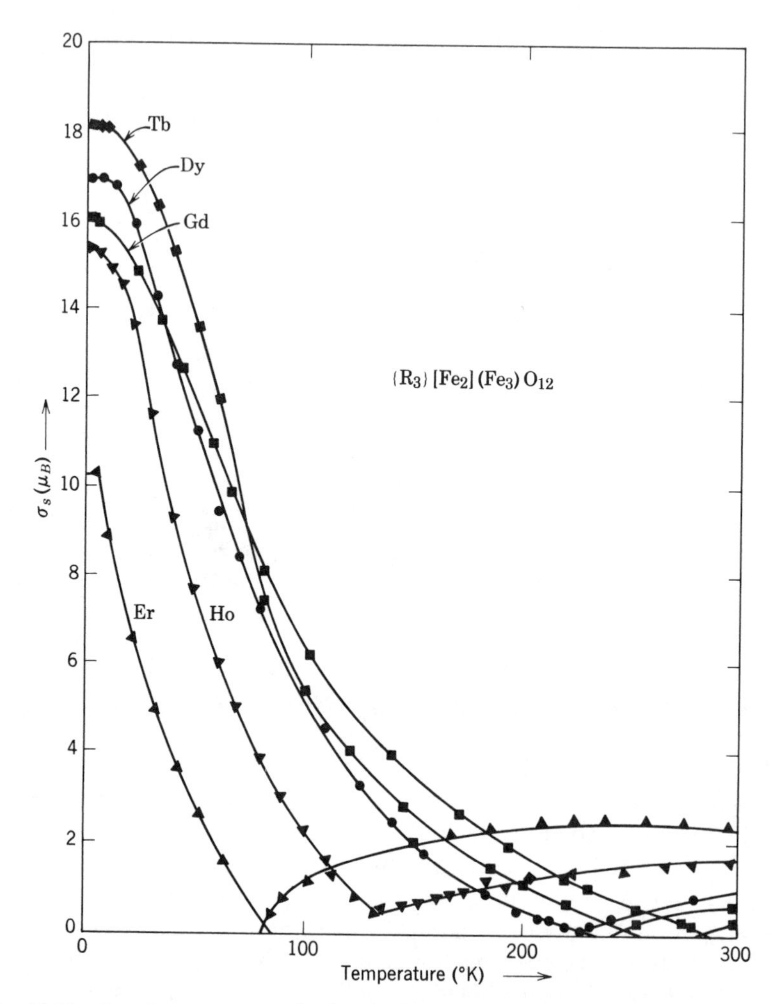

Fig. 19.14. Spontaneous magnetization in Bohr magnetons per formula units versus temperature of Gd, Tb, Dy, Ho, and Er iron garnets. From reference 1.

substitution of various transition metal ions and others (Al, Ga, Ca, Si). This has also led to a continuous variation in the lattice constants between end members.

One application of magnetic garnets involves both the magnetic characteristics and the precise control of the lattice constant by preselecting a suitable composition. Thin films (~ 5 microns) are epitaxially deposited on nonmagnetic substrates. The slight differences in thermal contraction on cooling to room temperature results in a preferred magnetization perpendicular to the plane of the film. Small magnetic domains with spin up, separated by regions with spin down, appear as *bubbles* in a polarizing microscope. The spin up–spin down array provides the binary input for a digital computer, and thus we have the term bubble memory materials. Note that the induced anisotropy due to thermal stresses is the source of this behavior rather than intrinsic anisotropy.

Orthoferrites. The crystal structure of the orthoferrites is that of perovskite (Fig. 2.28) with the general formula $MFeO_3$, $MCoO_3$, and $MMnO_3$, where M is $La^{3+}Ca^{2+}$, $Sr^{2+}Ba^{2+}$, Y^{3+}, or a rare earth ion; the Fe, Co or Mn ion is trivalent or tetravalent, depending on the valency of M. The compounds $LaMnO_3$, $LaFeO_3$, and $LaCrO_3$ are antiferromagnetic with Neel temperatures of 100, 750, and 320°K respectively. Some of the mixed oxides such as $LaMnO_3$ doped with calcium, strontium, or barium show ferromagnetism over a limited range of composition. The single oxides $RFeO_3$ in which R is yttrium or a rare earth metal also show weak ferromagnetism. There is a slight misalignment (canting) of the two antiferromagnetically coupled lattices. This canting, which is of the order of 10^{-2} radians, is sufficient to introduce a small net magnetic moment.

Ilmenites. A number of oxides with the structure of ilmenite ($FeTiO_3$) and corundum are ferromagnetic or show parasitic ferromagnetism. Some examples are α-Fe_2O_3, $MnTiO_3$, $CoTiO_3$, and $NiTiO_3$; the oxides $CoMnO_3$ and $NiMnO_3$ have the same structure but are ferrimagnetic at room temperature.

Although the uses of these magnetic ceramics have not found so wide an application as the ferrites, their wide range of magnetic characteristics will most certainly lead to greater future use.

19.5 The Hexagonal Ferrites

The hexagonal ferrites have a structure related to the spinel structure but with hexagonal close-packed oxygen ions and a unit cell made up of two formulae of $AB_{12}O_{19}$, where A is divalent (Ba, Sr, or Pb) and B is trivalent (Al, Ga, Cr, or Fe), corresponding to a molecular formula $A^{2+}O \cdot B_2^{3+}O_3$. The best known examples are magnetoplumbite, which has a formula $PbFe_{12}O_{19}$, and barium ferrite, $BaFe_{12}O_{19}$.

The barium ferrite structure consists of sections of the cubic spinel lattice (designated S) separated by a hexagonal close-packed section (designated R structure) containing the Ba-ion. Each S section consists of two layers of four oxygen ions, parallel to the hexagonal basal plane or (111) spinel planes, with three cations between each layer. The R section contains three layers of the hexagonal lattice, with one of the four oxygen ions in the center layer replaced by Ba. The unit cell consists of successive sections R, S, R*, S*, R, S, and so on, in which * denotes a rotation of 180° about the hexagonal c axis or the ⟨111⟩ direction of the corresponding spinel lattice. The unit cell thus contains ten oxygen layers, with Ba replacing the oxygen ion every five layers. In a unit cell each S section has the formula Fe_6O_8 and each R section $BaFe_6O_{11}$. The total formula is $BaFe_{12}O_{19}$ or for a unit cell RSR*S*, $2(BaFe_{12}O_{19})$. The Fe sites are tetrahedral and octahedral, and one site is surrounded by five oxygen ions forming a trigonal bipyramid.

The magnetization arises from the moment of the Fe ions, each with a spin of $5\mu_B\uparrow$ arranged per unit formula $BaFe_{12}O_{19}$ as follows: in the S section, $2\times 5\mu_B\uparrow$ due to the two Fe ions on tetrahedral sites in the spinel; the nine Fe ions on octahedral sites give $2\times 5\mu_B\uparrow$ and $7\times 5\mu_B\downarrow$; and the one Fe ion in fivefold symmetry in the R section gives a moment $1\times 5\mu_B\uparrow$. The net magnetic moment is thus $4\times 5\mu_B = 20\mu_B$ for an oxide containing the ferric ion. The spin orientation among the spinel S sections and the R sections is

S sections:	$2\uparrow$ tetrahedral	$4\downarrow$ octahedral
R section:	$1\downarrow$ fivefold	$2\uparrow 3\downarrow$ octahedral

The hexagonal ferrites are of interest because of their high magnetocrystalline anisotropy, making them suitable for permanent magnets. They have high coercivity, approaching the value $2K/B_s$ for pure rotation. The best known of these compounds is barium ferrite, which is also known as ferroxdure or hexagonal M compound. There are other more complex derivative compounds known in the literature as hexagonal X, W, Y, and Z compounds. The magnetic properties and compositions for a series of these hexagonal compounds is given in Table 19.4.

These compounds are prepared by sintering the appropriate mixture of oxides at temperatures near 1300°C. Anisotropic specimens can be prepared by the application of a magnetic field during pressing and sintering. This results in the rotation of the particles so that easy directions tend to become aligned parallel to the magnetic field. This maximizes the effects of the anisotropy field ($2K/B_s$), which is an indication of the degree of preference for a particular magnetization direction (in Table 19.5 barium ferrite has an anisotropy field of 17,000 Oe). This treatment reduces the coercivity (typically from 3000 to 1500 Oe) but increases their remanence B_r (from 2000 to 4000 G). The coercive force also depends on the particle size; if the grain size is below single-domain size, the coercivity is expected to be high because the grain boundary interacts with each domain. Typically a change in the average grain diameter from 10 to 1 micron increases the coercivity from about 100 to 2000 Oe.

Table 19.4. Magnetic and Crystallographic Properties of Garnets

μ_m is the magnetic moment in Bohr magnetons per unit formula $M_3Fe_5O_{12}$; results are sometimes given per unit formula $(3M_2O_3)(2Fe_2O_3)(3Fe_2O_3)$, i.e., $M_6Fe_{10}O_{24}$, giving twice μ_m.

Ion M	Y	La	Pr	Nd	Sm	Eu	Gd	Tb	Dy	Ho	Er	Tm	Yb	Lu
α_0(Å)	12.376	[12.767]	[12.646]	[12.600]	12.529	12.498	12.471	12.436	12.405	12.375	12.347	12.323	12.302	12.283
Density	5.17	[5.67]	[5.87]	6.00	6.23	6.31	6.46	6.55	6.61	6.77	6.87	6.94	7.06	7.14
Number of 4*f* electrons	0	0	2	3	5	6	7	8	9	10	11	12	13	14
μ_m (at 0°K)	5.01	[5.0]	[9.8]	[8.7]	5.43	2.78	16.0	18.2	16.9	15.2	10.2	1.2	0.0	5.07
B_s emu cm^3 (20°C)	139	...	...	...	135	93	135	4	43	78	103	110	130	140
Compensation temperature (°K)	...	...	...	...	...	...	286	246	226	137	83	None	0–6	
Curie temperature (°K)	553	...	...	...	578	566	564	568	563	567	556	549	548	549

Brackets [ ] indicate extrapolated data.

Source. Reference 1.

Table 19.5. Magnetic Properties of Hexagonal M, X, W, Y, and Z Compounds

						20°C			0°K	
	C_0 (Å)	α_0	Density d	Molecular Weight	B_s	K_1 or (K_1+2K_2) (10^6 ergs/cm^3)	H_K^* (kOe)	σ_0	μ_m (μ_B)	T_c (°C)
BaM	23.18	5.889	5.28	1,112	380	3.3	17.0	100	20	450
PbM	23.02	5.877	5.65	1,181	320	2.2	13.7	80	18.6	452
SrM	23.03	5.864	5.11	1,062	370	3.5	20	108	20.6	460
Mg_2W	...	...	5.10	1,512						
Mn_2W	...	...	5.31	1,573	310	...	...	97	27.4	415
Fe_2W	32.84	5.88	5.31	1,575	320	3.0	19.0	98	27.4	455
Co_2W	...	...	5.31	1,581	340	−5				
Ni_2W	...	...	5.32	1,580	330	2.1	12.7			
Cu_2W	...	...	5.36	1,590						
Zn_2W	...	...	5.37	1,594	340					
NiFeW	...	...	...	1,577	273	...	...	79	22.3	520
ZnFeW	...	...	...	1,584	380	2.4	12.5	108	30.7	430
MnZnW	...	...	...	1,583	370	1.9	10.2			
$Fe_{0.5}Zn_{1.5}W$	...	...	...	1,589	380	2.1	11.1			
$Fe_{0.5}Ni_{0.5}ZnW$	...	...	...	1,586	350	1.6	9.1			
$Fe_{0.5}CO_{0.75}Zn_{0.75}W$	...	...	...	1,584	360	(−0.4)	2.2			
$FeNi_{0.5}Zn_{0.5}W$	...	...	...	1,581	360	...	...	104	29.5	45

Mg_2Y	. . .	. . .	5.14	1,346	119	(−0.6)	10	29	6.9	280
Mn_2Y	. . .	. . .	5.38	1,406	167	. . .	. . .	42	10.6	290
Fe_2Y	43.6	5.9	5.39	1,408						
CO_2Y	. . .	. . .	5.40	1,414	185	(−2.6)	28	39	9.8	340
Ni_2Y	. . .	. . .	5.40	1,414	127	(−0.9)	14	25	6.3	390
Cu_2Y	. . .	. . .	5.45	1,424	. . .	. . .	. . .	28	7.1	
Zn_2Y	43.56	5.88	5.46	1,428	227	(−1.0)	9.0	72	18.4	130
$Fe_{0.5}Zn_{1.5}Y$	. . .	. . .	. . .	1,423	191	(−0.9)	9.5			
Mg_2Z	. . .	. . .	5.20	2,457	. . .	. . .	. . .	55	24	
Mn_2Z	. . .	. . .	5.33	2,518						
Fe_2Z	. . .	. . .	5.33	2,520						
CO_2Z	52.30	5.88	5.35	2,526	267	(−1.8)	13	69	31.2	410
Ni_2Z	. . .	. . .	5.35	2,526	. . .	. . .	. . .	54	24.6	
Cu_2Z	. . .	. . .	5.37	2,536	247	. . .	. . .	60	27.2	440
Zn_2Z	. . .	. . .	5.37	2,539	310	0.6	. . .	. . .	. . .	360
Fe_2X	84.11	5.88	5.29	2,386	. . .	1.4	9.6	92.5	60.5	400
Zn_2U	113.2	5.88	5.36	3,651	295					

BaM = $BaFe_{12}O_{19}$
$Me_2W = Me_2BaFe_{16}O_{27}$
$Me_2Y = Me_2Ba_2Fe_{12}O_{22}$
[a] H_K is the anisotropy field $\approx 2K/Bs$.

$Me_2Z = Me_2Ba_3Fe_{24}O_{41}$
$Me_2X = Me_2Ba_2Fe_{28}O_{46}$
$Me_2U = Me_2Ba_4Fe_{36}O_{60}$
Source. Reference 1.

19.6 Polycrystalline Ferrites

In commercial and polycrystalline ferrites, processing variables and resultant microstructures have important consequences on measured properties.

Applications of spinel ferrites can be divided into three main groupings according to the frequency of the magnetic field of the device in which it is used: (1) low-frequency high-permeability applications, (2) high-frequency low-loss applications, and (3) microwave applications. The low-frequency magnetic properties of cubic spinel ferrites or garnets are considerably inferior to those of magnetic metals or alloys. Generally their permeabilities are lower by a factor of 10 to 100, but their coercivities are usually about 10 times higher. Apart from the limited effect of a lower saturation magnetization, this low-frequency inferiority is a result of the difficulty in preparing ferrites in a condition as chemically homogeneous and structurally perfect as metals. However, for high-frequency applications the high electrical resistivity of ferrites compensates for these shortcomings, and oxide ceramics replace metals.

A typical material for high-permeability low-frequency applications is manganese-zinc ferrite. The initial permeability of commercial materials at low frequencies is about 1000, and the maximum permeability is about 4000. Common values of the coercivities are about 0.1 Oe. Losses are comparatively high, and the useful frequencies are limited to about 5×10^5 Hz because of relaxation effects.

Low loss high-frequency ferrites are often compositions containing nickel and zinc. These have poorer low-frequency properties than Mn–Zn ferrites; that is, they have a higher hysteresis loss, but the electrical resistivities are higher and frequency-dependent losses are lower. In particular the losses remain low, and the permeability retains its low frequency value up to 10×10^6 Hz.

Frequently used microwave materials are nickel-zinc ferrites, garnets, and some hexagonal compounds. As in the case of low-loss high-frequency ferrites the losses are due to resonance as the moments try to follow the applied field, but in addition to the oscillating field, there is a loss associated with an applied dc field. When characterizing materials for these applications, a resonance peak is measured. The important parameters are the width and position (frequency) of the peak. The width represents the extent to which the losses are concentrated around a specific resonance field and gives a qualitative indication of perfection. The smallest line widths are obtained for garnets.

In the following discussion we wish to consider the effects of processing variables on magnetic characteristics. The composition, grain size, effects of impurities and sintering atmosphere, and the size and distribution of porosity have significant consequences for the magnetic properties.

Compositional Effects. Variability may result from batching problems, but microscopic compositional variations are also important, and careful attention to powder preparation (ball milling, calcination, coprecipitation, etc.) is extremely important. Assuming that the cations are present in the correct proportions, the sintering operation affects the microstructure (grain size, porosity) and also the distribution of the cations within the crystalline lattice, at the grain boundaries, and in second phases. The sintering time and temperature, the partial pressure of oxygen in the sintering atmosphere, and the cooling rate must be controlled. The oxygen pressure should be chosen with regard to the phase equilibria. It was shown in Chapters 4, 6, and 17 that the oxygen stoichiometry influences the valence of transition metal ions and that the ratio of ferrous and ferric ion content determines the electrical conductivity. To maintain a high resistivity, a slightly oxidizing atmosphere is desirable to keep the iron in the Fe^{3+} state.

In manganese ferrites, firing in the air causes oxidation of some of the manganese to form trivalent ions and causes a sharp deterioration of the magnetic properties. Table 19.6 shows the effect of various sintering atmospheres on the resistivity, permeability, and coercivity of manganese ferrite. Another important effect of the oxygen environment arises because of the formation of cation vacancies; location of the cations on nonequivalent lattice sites leads to an induced anisotropy which can stabilize the positions of the domain walls and decrease permeability.

The cooling rate is chosen either to maintain a random distribution of ions between octahedral and tetrahedral sites (air quenching) or to allow ordering to take place (slow cooling). In manganese ferrite the magnetiza-

Table 19.6. The Effect of Sintering Atmosphere on Some Properties of Manganese Ferrite

Sintering atmosphere	Air	CO_2	Helium
Resistivity (ohm cm)	10^5	6×10^3	10^3
Initial permeability	50	228	232
Maximum permeability	138	3200	3220
Coercivity (Oe)	1.67	0.50	0.89

Source. Reference 5.

tion per formula unit is $0.76\mu_B$ after quenching from 1400°C; it increases to $2.68\mu_B$ when quenched from 1400°C.

The importance of composition on the bulk magnetic properties has been discussed for the mixed ferrites. Table 19.7 further indicates the compositional variation of properties for manganese-zinc and nickel-zinc ferrites. The low-frequency permeabilities of the manganese-zinc ferrites are higher; the nickel-zinc ferrites have much higher resistivities. However, as the frequency of measurement is raised, the difference in the permeabilities becomes less significant. Substitution of zinc for manganese affects both the saturation magnetization and the anisotropy, but it is the anisotropy which has the greatest effect on the permeability. At low temperatures the substitution of the Zn^{2+} for Mn^{2+} on *a* sites increases the magnetization of the two sublattices. The presence of zinc also reduces the strength of the *a*-*b* interactions and decreases the Curie temperature.

Grain Size and Porosity. The permeability of polycrystalline ferrites increases with increasing grain size, assuming that other factors remain constant. Evidence for this is shown in Fig. 19.15 for the initial permeability of manganese-zinc ferrite. All samples have the same composition and the same crystal anisotropy and magnetostriction. The distribution of pores does change; pores are within the grain (intragranular porosity) as

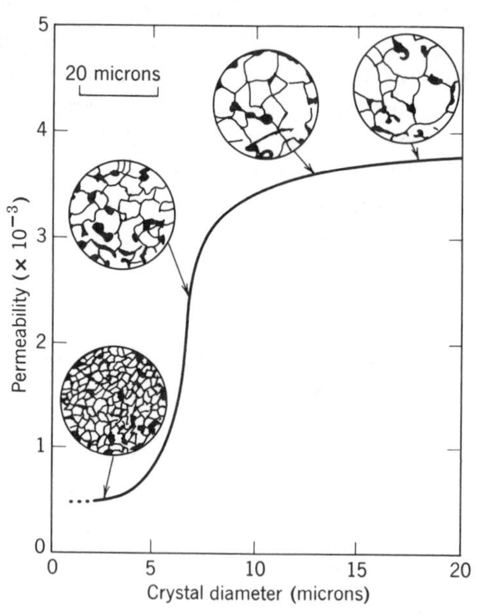

Fig. 19.15. The variation of permeability with average grain diameter of a manganese-zinc ferrite with uncontrolled porosity. From Guillaud and Paulus, reference 1.

Table 19.7. Some Comparative Properties of Manganese–Zinc and Nickel–Zinc Ferrites

Composition (mole%)		Sat. ind. (G)	Curie Temp. (°C)	Initial Perm. (g/Oe)	Density (g/cm^3)	Coercive Force (Oe)	Resistivity (ohm-cm)
$MnFe_2O_4$	$ZnFe_2O_4$						
48	52	3300	100	1400	4.9	0.2	20
58	42	4500	150	900	4.9	0.3	50
62	38	4700	150	1100	4.9	0.4	80
79	21	5100	210	700	4.8	0.5	80
$NiFe_2O_4$	$ZnFe_2O_4$						
36	64	3600	125	650	4.9	0.4	10^5
50	50	4200	250	230	4.5	0.7	10^5
64	36	4100	350	90	4.2	2.1	10^5
80	20	3600	400	45	4.1	4.2	10^5
100	. . .	2300	500	17	4.0	11.0	10^5

Source. Reference 1.

opposed to pores at grain boundaries for grain sizes greater than 20 microns.

The effect of grain size on μ_i is more complex in nickel-zinc ferrite, as shown in Fig. 19.16. When the average grain size becomes greater than about 15 microns and the percentage of grains with included pores becomes 50%, the initial permeability reaches a maximum and then begins to decrease. The technical importance of controlling the microstructure of ferrites is clear from these results. Both the grain size and the porosity and the distribution of porosity determine the initial permeability. In more recent experiments on a more nickel-rich nickel-zinc ferrite the effects of porosity were shown to be less significant to the initial permeability than the grain size. Figure 19.17 shows the temperature dependence of the initial permeability in these controlled studies. The density ρ and the grain size D_m are shown for each curve.

Although the empirical effect of grain size is well established, an understanding of the physical phenomena is not complete. If we assume very small grains, then there should be no domain walls within the grains, and thus the permeability is given solely by the rotational processes. In this case, the permeability is

$$\mu - 1 = \frac{2\pi B_s^2}{K_1} \qquad K_1 < 0$$
$$\mu - 1 = \frac{4\pi B_s^2}{3K_1} \qquad K_1 > 0 \qquad (19.25)$$

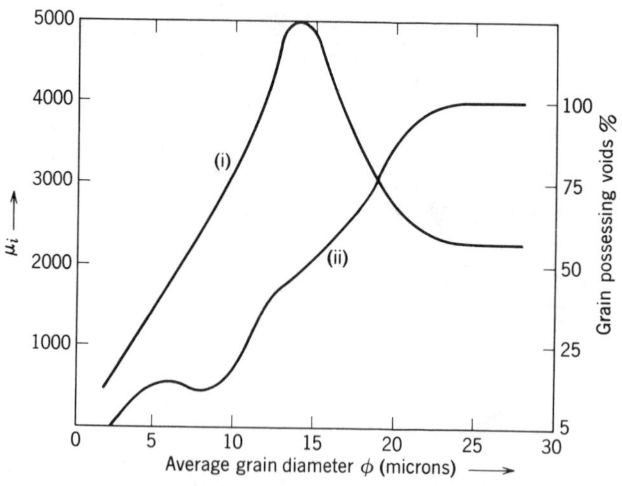

Fig. 19.16. The initial permeability of a nickel-zinc ferrite as a function of average grain diameter (i), together with an indication of the percentage of crystallites found to contain visible pores (ii). From Guillaud and Paulus, reference 1.

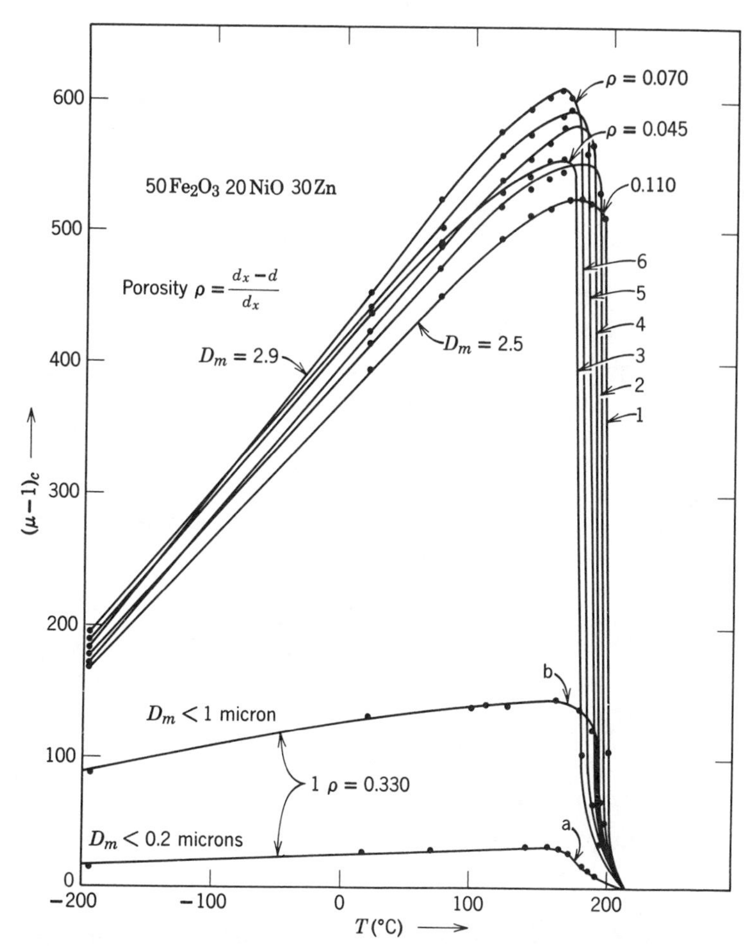

Fig. 19.17. The temperature dependence of the initial permeability of nickel-zinc ferrite ($50Fe_2O_3$, 20NiO, 30ZnO) showing the insensitivity to porosity by the upper set of curves for similar grain size and the sensitivity to grain size in specimens of the same porosity by the two lower curves. The permeability is corrected to correspond to single-crystal density. From reference 1.

As the grain size is increased, domain walls are included within the grains, and they may become pinned at grain boundaries or at intergranular pores. Thus their motion would be sensitive to the grain size and the distribution of porosity.

In addition to the strong effects on μ_i of pores within the grains and at the grain boundaries, the maximum permeability, coercivity, and hysterisis are also affected. Figure 19.18 shows the effect of porosity on the magnetization loops of nickel-zinc ferrites. This is demonstrated more

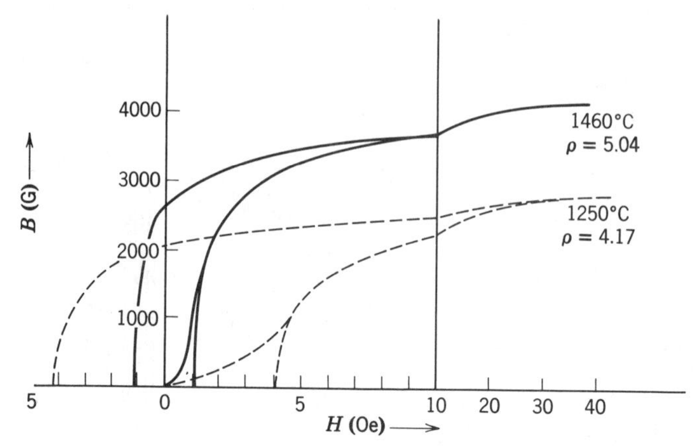

Fig. 19.18. The effect of porosity on the B/H loops of nickel-zinc ferrites (33NiO, 17ZnO, balance Fe_2O_3 and FeO). The densities indicated were obtained by firing at the two different temperatures. From reference 3.

clearly for magnesium ferrites, in which there is a regular increase in the initial permeability and a decrease in coercivity with increasing density (Fig. 19.19). In contrast to Fig. 19.18 the data for magnesium ferrite indicate that the maximum induction increases with increasing density but that the remanence decreases. If the remanence magnetization is understood as the return of the magnetization vectors to the nearest easy direction while each grain remains saturated, then remanence and saturation should retain the same ratio. The pores give rise to demagnetizing fields which can either cause rotation of the magnetization away from easy directions or nucleation of reverse domains, in which case it is surprising that the ratio of the remanence to the saturation should be greater for the more porous material. The intergranular pores affect the permeabilities and coercivities because of the impedence to the motion of domain walls.

Different structural factors are of major importance in connection with the different types of losses; for example, the hysteresis loss can be controlled by the same factors which control the low-frequency permeability and coercivity (porosity, grain size, and impurities). The lowest hysterises losses are associated with minimum anisotropy and magnetostriction, large grain sizes, and low intragranular porosity. Eddy-current losses are controlled by controlling the resistivity of the ferrite. Generally it is important to avoid the presence of ferrous ions if the resistivity is to be high and reduce eddy-current losses for high-frequency applications. This is accomplished by including manganese or cobalt in the ferrite.

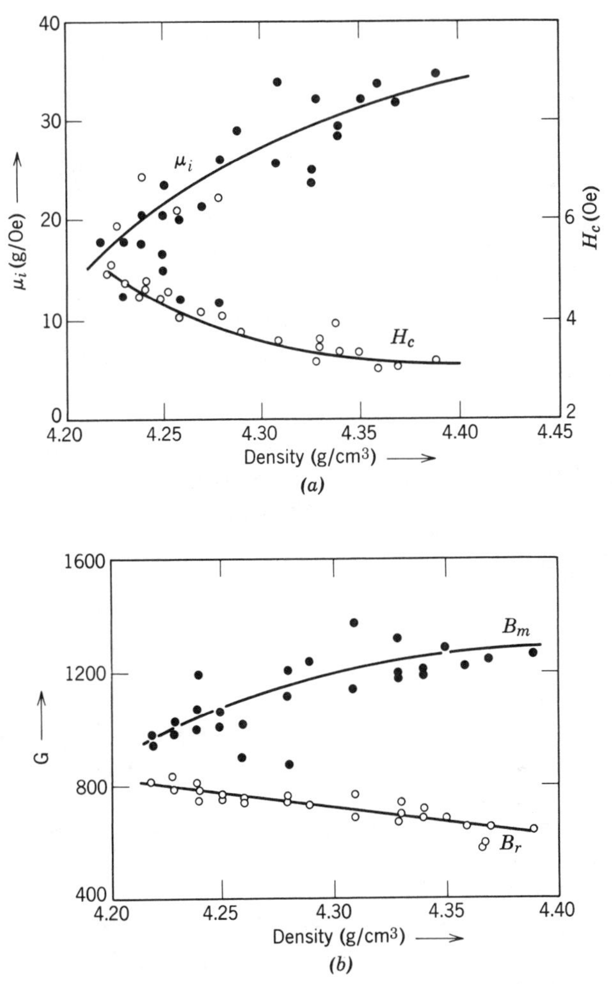

Fig. 19.19. (*a*) The variation of initial permeability, measured at 10 kHz, and coercivity, with density for specimens of Mg Fe_2O_4. (*b*) The variation of maximum induction and remanent induction with density for Mg Fe_2O_4. From reference 5.

As a final example of the effect of microstructure on magnetic properties of cubic spinels, let us consider nickel ferrite thin films prepared by reaction sputtering of a ferrite composition onto a cold substrate. If the films are deposited at temperatures below 0°C, the material is noncrystalline, and the magnetic susceptibility shows the material to behave as a paramagnet. If the nickel ferrite is deposited at higher temperatures (less than 400°C), a micropolycrystalline film results with grain sizes less than 150 Å. In this case the magnetic behavior is *superparamagnetic*. In a superparamagnetic material the Langevin equation 19.12 is calculated for the net magnetic moment of the material due to the micrograins rather than the individual dipoles. Thus they have larger moments per particle, and thus the term superparamagnetic is used. For larger-grain films (150 Å) deposited at temperatures above 400°C the films show ferrimagnetism similar to bulk nickel ferrites.

Hexagonal Ferrites. The microstructural features important for cubic ferrites also apply to hexagonal ferrites. However, because hexagonal ferrites are generally used as hard magnetic materials, we must also evaluate the influence of ceramic processing on the coercivity. A reduction of the grain size from 10 to 1 micron in barium ferrite increases the coercivity from 100 to 2000 Oe. The effect of nucleation and growth of domains is to reduce the coercivity below the value which would be obtained if only rotation of the magnetization were possible. This assumes that there is considerable crystalline and shape anisotropy. High coercivities may arise in three ways: (1) The nucleation of domains may require high fields. If the nucleation field is greater than the field required to move the domain walls, which are throughout the specimen, then we require a switching or reversal field to reverse the magnetization which results in a rectangular loop. (2) There may be barriers which inhibit wall motions and which can only be overcome by high reverse fields. (3) The material may exist in such a condition that domains cannot form (i.e., as very fine particles). In this case only rotational processes are possible. A finely divided structure (small grain size) has been demonstrated to increase the coercivity in almost all permanent magnet materials. This is an important feature.

Further improvements in properties are obtained by aligning the crystallites during pressing or slip casting by carrying out these processes under a high magnetic field. Thus in barium ferrite the hexagonal axes are preferentially aligned, giving the polycrystalline mass an anisotropic magnetic behavior. Improvements in the initial permeability by a factor of about 3 are possible by particle alignment.

Suggested Reading

1. R. S. Tebble and D. J. Craik, *Magnetic Materials*, Wiley-Interscience, New York, 1969.
2. K. J. Standley, *Oxide Magnetic Materials*, Clarendon Press, Oxford, 1962.
3. J. Smit and H. P. J. Wijn, *Ferrites*, John Wiley & Sons, Inc., New York, 1959.

4. C. A. Wert and R. M. Thomson, *Physics of Solids*, McGraw-Hill Book Company, New York, 1964.
5. G. Economos in *Ceramic Fabrication Processes*, W. D. Kingery, Ed., John Wiley & Sons, Inc., New York, 1958.
6. B. D. Cullity, *Introduction to Magnetic Materials*, Addison-Wesley, Reading, Mass., 1972.
7. C. Heck, *Magnetic Materials and Their Applications*, Crane-Russak Co., New York, New York, 1974.

Problems

19.1. List the important factors that determine the following:
(*a*) Magnitude of the Bloch wall energy.
(*b*) Magnetic domain size.
(*c*) Curie temperature for a ferrimagnetic material.
(*d*) Curie temperature for a ferroelectric material.
(*e*) Maximum *BH* product in a hysteresis loop.

19.2. Magnetobarite (often called barium ferrite), a hard magnetic material, has a hexagonal structure and is magnetically highly anisotropic, with the magnetic axis perpendicular to the basal plane. Describe two different methods by which a high-energy product ($B \times H$) can be achieved in sintered barium ferrite. Explain what factors are controlled in the processes involved.

19.3. When the normal, spinel, $CdFe_2O_4$, is added to an inverse spinel such as magnetite, Fe_3O_4, the Cd ions retain their normal configuration. Calculate the magnetic moment for the following compositions:
$Cd_xFe_{3-x}O_4$, with
(*a*) $x = 0$.
(*b*) $x = 0.1$.
(*c*) $x = 0.5$.

19.4. (*a*) Describe the effect of porosity and grain size on the properties of soft ferrites such as $MgFe_2O_4$ compared with hard ferrites such as $BaFe_{12}O_{19}$. Grain size and porosity result from the sintering procedure. What factors become important parameters in the fabrication of hard ferrites compared with soft ferrites?

19.5. Predict the saturation magnetic moment per unit volume in Bohr magnetons for the following inverse spinel structures:
(*a*) $MgFe_2O_4$.
(*b*) $CoFe_2O_4$.
(*c*) $Zn_{0.2}Mn_{0.8}Fe_2O_4$.
(*d*) $LiFe_5O_8$.
(*e*) γ-Fe_2O_3.
What would be the effect on μ_B if each composition were quenched from 1200°C?

19.6. A study of magnetic domains has been very rewarding to those interested in understanding magnetism.
(*a*) What is the nature of Bloch (domain) walls that allows them to be observed even during movements?
(*b*) How does an inclusion, especially a pore, alter the movement of a Block wall?

19.7. The magnetic moment of a ferrite, $Li_{0.5}Fe_{2.5}O_4$, has been measured and observed to be 2.6 Bohr magnetons per unit of spinel formula. How do you justify this result from the known net spins associated with the ions involved? What position(s) in the crystal lattice does Li^{+} occupy? Fe^{+3}?

19.8. (*a*) The domain wall energy (180°) for iron metal is about 10^{-3} J/m^2 (1 erg/cm^2). How much of this energy would you predict is magnetostatic ________, magnetostrictive ________, exchange energy ________, crystalline anisotropy ________, nucleus-electron interaction ________, other ________ (specify).

19.9. Ferromagnetic and ferrimagnetic behavior is observed to occur only in compounds which include ions of the transition and rare earth series. What is unique about the structure of these ions that leads to this type of magnetic behavior? Why are only some of the compounds of these ions ferromagnetic or ferrimagnetic and others are not?

19.10. (*a*) Crystals of MnF_2 are antiferromagnetic and have a Curie point of 92°K. At 150°K the magnetic susceptibility χ is 1.8×10^{-2} per mole.

(i) If temperature is decreased below 150°K, how will the susceptibility change?

(ii) At what temperature will the susceptibility have maximum value?

(iii) At what temperature will the susceptibility have minimum value?

(*b*) A magnesium ferrite has composition

$$(Mg_{0.8}Fe^{3+}_{0.2})(Mg_{0.1}Fe^{3+}_{0.9})_2O_4$$

and has a lattice constant of 8.40 Å. The Fe^{3+} ion has a magnetic moment of 5 Bohr magnetons. Compute the saturation magnetization which this material would have.

(*c*) It is desired to use the above ferrite in a solid-state device which requires high initial permeability and low coercive force. What sort of microstructural features would you aim for in processing the material?

Index